Second Edition Volume 5

Fundamental Concepts of

Inorganic Chemistry

As per the National Education Policy (NEP) 2023

Asim K Das
MSc (Gold Medalist, CU), PhD (CU), DSc (Visva Bharati)
Professor, Department of Chemistry
Visva Bharati University, Santiniketan 731235
West Bengal, India

Mahua Das
Former Research Associate
Department of Chemistry
Visva Bharati University, Santiniketan 731235
West Bengal, India

Ankita Das
MSc (Visva Bharati)
DST INSPIRE Research Fellow
IACS, Kolkata

Udita Das
MSc (Visva Bharati)
DST INSPIRE Scholar, SHE

CBS Publishers & Distributors Pvt Ltd

New Delhi • Bengaluru • Chennai • Kochi • Kolkata • Lucknow • Mumbai
Hyderabad • Jharkhand • Nagpur • Patna • Pune • Uttarakhand

ISBN: 978-93-5466-279-9

Second Edition: 2025

First Edition: 2014
Reprint: 2015, 2016, 2018, 2019, 2021

Published by Satish Kumar Jain and produced by Varun Jain for
CBS Publishers & Distributors Pvt Ltd
4819/XI Prahlad Street, 24 Ansari Road, Daryaganj, New Delhi 110 002, India.
Ph: 011-23289259, 23266838 Website: www.cbspd.com
e-mail: delhi@cbspd.com

Corporate Office: 204 FIE, Industrial Area, Patparganj, Delhi-110092
Ph: 011-4934 4934 Fax: 011-4934 4935 e-mail: publishing@cbspd.com; publicity@cbspd.com

Branches

- **Bengaluru:** Seema House 2975, 17th Cross, K.R. Road, Banasankari 2nd Stage, Bengaluru 560 070, Karnataka, India
  Ph: +91-80-26771678/79 Fax: +91-80-26771680 e-mail: bangalore@cbspd.com
- **Chennai:** 18/8B, Subbarayan Street, Shenoy Nagar, Chennai 600 030, Tamil Nadu
  Ph: +91-044-42032115, 26681266 e-mail: chennai@cbspd.com
- **Kochi:** 42/1325, 1326, Power House Road, Opp KSEB Power House, Ernakulam 682 018, Kochi, Kerala, India
  Ph: +91-484-4059061-65 Fax: +91-484-4059065 e-mail: kochi@cbspd.com
- **Kolkata:** 147, Hind Ceramics Compound, 1st Floor, Nilgunj Road, Belghoria, Kolkata 700 056, West Bengal, India
  Ph: +91-033-25633055, 033-25633056 e-mail: kolkata@cbspd.com
- **Lucknow:** Basement, Khushnuma Complex, 7-Meerabai Marg (behind Jawahar Bhawan), Lucknow 226 001, UP, India
  Ph: +91-522-400043, 9919002738 e-mail: tiwari.lucknow@cbspd.com
- **Mumbai:** PWD Shed, Gala no. 25/26, Ramchandra Bhatt Marg, Next to JJ Hospital Gate no. 2
  Opp. Union Bank of India, Noorbaug, Mumbai-400009, Maharashtra, India
  Ph: 022-60061880/89 e-mail: mumbai@cbspd.com

Representatives

• **Hyderabad**	0-9885175004	• **Jharkhand**	0-9811541605	• **Nagpur**	0-8692091830
• **Patna**	0-9334159340	• **Pune**	0-9664372571	• **Uttarakhand**	0-9716462459

Printed at: Mudrak, Noida, UP, India

Second Edition **Volume 5**

Fundamental Concepts of Inorganic Chemistry

As per the National Education Policy (NEP) 2023

Asim K Das
MSc (Gold Medalist, CU), PhD (CU), DSc (Visva Bharati)
Professor, Department of Chemistry
Visva Bharati University, Santiniketan 731235
West Bengal, India

Mahua Das
Former Research Associate
Department of Chemistry
Visva Bharati University, Santiniketan 731235
West Bengal, India

Ankita Das
MSc (Visva Bharati)
DST INSPIRE Research Fellow
IACS, Kolkata

Udita Das
MSc (Visva Bharati)
DST INSPIRE Scholar, SHE

CBS Publishers & Distributors Pvt Ltd

New Delhi • Bengaluru • Chennai • Kochi • Kolkata • Lucknow • Mumbai
Hyderabad • Jharkhand • Nagpur • Patna • Pune • Uttarakhand

Disclaimer
Science and technology are constantly changing fields. New research and experience broaden the scope of information and knowledge. The authors have tried their best in giving information available to them while preparing the material for this book. Although all efforts have been made to ensure optimum accuracy of the material, yet it is quite possible some errors might have been left uncorrected. The publisher, the printer and the authors will not be held responsible for any inadvertent errors or inaccuracies.

ISBN: 978-93-5466-279-9

Second Edition: 2025
First Edition: 2014
Reprint: 2015, 2016, 2018, 2019, 2021

Published by Satish Kumar Jain and produced by Varun Jain for
CBS Publishers & Distributors Pvt Ltd
4819/XI Prahlad Street, 24 Ansari Road, Daryaganj, New Delhi 110 002, India.
Ph: 011-23289259, 23266838 Website: www.cbspd.com
e-mail: delhi@cbspd.com

Corporate Office: 204 FIE, Industrial Area, Patparganj, Delhi-110092
Ph: 011-4934 4934 Fax: 011-4934 4935 e-mail: publishing@cbspd.com; publicity@cbspd.com

Branches

- **Bengaluru:** Seema House 2975, 17th Cross, K.R. Road, Banasankari 2nd Stage, Bengaluru 560 070, Karnataka, India
  Ph: +91-80-26771678/79 Fax: +91-80-26771680 e-mail: bangalore@cbspd.com
- **Chennai:** 18/8B, Subbarayan Street, Shenoy Nagar, Chennai 600 030, Tamil Nadu
  Ph: +91-044-42032115, 26681266 e-mail: chennai@cbspd.com
- **Kochi:** 42/1325, 1326, Power House Road, Opp KSEB Power House, Ernakulam 682 018, Kochi, Kerala, India
  Ph: +91-484-4059061-65 Fax: +91-484-4059065 e-mail: kochi@cbspd.com
- **Kolkata:** 147, Hind Ceramics Compound, 1st Floor, Nilgunj Road, Belghoria, Kolkata 700 056, West Bengal, India
  Ph: +91-033-25633055, 033-25633056 e-mail: kolkata@cbspd.com
- **Lucknow:** Basement, Khushnuma Complex, 7-Meerabai Marg (behind Jawahar Bhawan), Lucknow 226 001, UP, India
  Ph: +91-522-400043, 9919002738 e-mail: tiwari.lucknow@cbspd.com
- **Mumbai:** PWD Shed, Gala no. 25/26, Ramchandra Bhatt Marg, Next to JJ Hospital Gate no. 2
  Opp. Union Bank of India, Noorbaug, Mumbai-400009, Maharashtra, India
  Ph: 022-60061880/89 e-mail: mumbai@cbspd.com

Representatives

- **Hyderabad** 0-9885175004
- **Jharkhand** 0-9811541605
- **Nagpur** 0-8692091830
- **Patna** 0-9334159340
- **Pune** 0-9664372571
- **Uttarakhand** 0-9716462459

Printed at: Mudrak, Noida, UP, India

A tribute to

Priyadaranjan Ray (1888–1982)

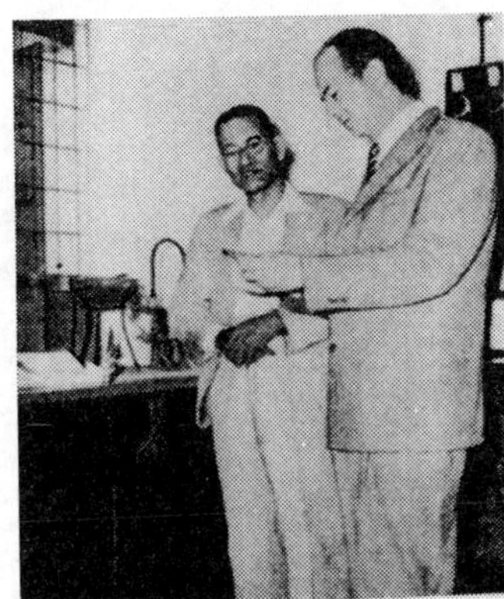

Prof Linus Pauling is discussing with Prof P Ray (left) in the laboratory of Prof P Ray (1955) at IACS, Calcutta

Prof P Ray made a significant contribution in understanding the structure and bonding in coordination compounds mainly through the magnetic moment measurements (**magnetochemistry**). Some of his landmark contributions are highlighted here.

- **Ray–Dutt twist mechanism** of racemisation of coordination compounds.
- **Stabilisation of unusual high oxidation states of different metals** like Ag(II), Ag(III) (as the ethylenedibiguanide complex), Ni(IV) (as nickel periodate $MNiIO_6$, M = alkali metal ion), etc.
- Isolation and characterization **of linkage isomers of the thiosulfate ligand** as in $H_4[Co(CN)_5(S_2O_3)]$ (η^1-O, yellow; η^1-S, red).
- Discovery of **inner metallic complexes of higher than second-order.**
- **Master of biguanide complexes.**
- **Development of different microanalytical reagents** like rubeanic acid for the detection Cu^{2+} even at the concentration level of 3 ppb (1 ppb = 1 part per 10^9 parts; most sensitive test known so far for the detection of Cu^{2+}).

श्रद्धावान् लभते ज्ञानम्

One who has shraddha acquires knowledge

"All power is within you, you can do anything and everything. Believe in that, do not believe that you are weak."

"The secret of religion lies not in theories but in practice. To be good and to do good – that is the whole of religion."

"Have faith in yourselves, and stand up on that faith and be strong; that is what we need."

Swami Vivekananda

Preface to the Second Edition

We feel delighted to record the warm response which the first edition of the book has received from the students and teachers throughout the country. We have incorporated all the aspects of New Education Policy (**NEP 2023**, **four-year undergraduate programme**) and choice based credit system (CBCS) syllabus recently introduced in all universities and institutes and the suggestions received from the readers. In this edition, each chapter has been thoroughly revised, updated and rewritten to accommodate the recent views. A new chapter on **photochemistry and photochemical reactions** has been included. In this enlarged version, scope of the book has been broadened by adding new topics and revising the earlier sections. In revising the book, we have taken all the measures to retain the basic features for which the earlier edition has been so popular.

In preparing the manuscript, we have again freely consulted the original research papers, books and reviews of the earlier authors and have borrowed their ideas whenever required. We are grateful and indebted to them.

All the facilities provided by our institute, Visva Bharati, are thankfully acknowledged. We are extremely grateful to Mr SK Jain, CMD, CBS Publishers & Distrubutors, New Delhi, for his hearty support and encouragement. We are also thankful to Mr YN Arjuna, Senior Vice President—Publishing, Editorial and Publicity, for his kind cooperation and suggestions. We are delighted to express our gratefulness to Ms Sanjubala Tripathi, Editor, for her meticulous effort to give the present shape of the production. We are extremely thankful to the concerned DTP section for taking the troubles in processing the manuscript.

In spite of our best efforts, some mistakes and misconceptions might have again crept in and for these inconveniences, we beg to be pardoned in advance. Suggestions and constructive criticisms are always welcome to better the presentation.

Asim K Das
Mahua Das
Ankita Das
Udita Das

श्रद्धावान् लभते ज्ञानम्

One who has shraddha acquires knowledge

Preface to the First Edition

We do endeavour. He gives the strength and patience

The present volumes 4–7 are in continuation of the existing ***Fundamental Concepts of Inorganic Chemistry***. The classnotes and valuable suggestions received from the esteemed readers have been shaped in these volumes.

These volumes cover the structure and bonding (VBT, CFT, and MOT), stability and reactivity, spectral and magnetic properties of metal complexes in depth. Kinetics and reaction mechanisms of ligand substitution, electron transfer and photochemical reactions have been included. Magnetochemistry and organometallic chemistry have been covered. Applications of different spectroscopic techniques (Raman, IR, NMR, ESR, Mossbauer, UV-VIS, UV-PES, etc.) have been discussed to widen the utility of the series. In developing the present extension, we have taken all the measures to retain the basic features of the existing title.

In preparing the manuscript, we have freely consulted the books and review of the earlier authors and have borrowed their ideas whenever it has been required. These sources are listed and acknowledged at the end of the text. We are grateful and indebted to these authors. In reality, we have picked up flowers from these gardens to prepare the garland to worship the goddess of learning.

We are extremely thankful and grateful to Mr SK Jain, Managing Director, CBS Publishers & Distributors, New Delhi, for his continued support. We are thankful to Mr YN Arjuna, Senior Vice-President, Publishing, Editorial and Publicity, and the DTP staff for taking the trouble in processing the manuscript.

In spite of our best efforts, some mistakes and misconceptions might have crept in for which we beg to be excused. Constructive criticism and suggestions are always welcome to better the presentation.

Asim K Das
Mahua Das

Contents

CHAPTER 5

Reaction Mechanism: Ligand Substitution, Isomerisation and Racemisation Reactions of Metal Complexes

5.1 ENERGETICS OF REACTIONS (IN GENERAL) AND DIFFERENT TYPES OF REACTIONS OF METAL COMPLEXES

5.1.1 Energetics of Reactions (in General)

For any reaction to occur, ΔG (Gibbs free energy charge under the given conditions) must be negative. **It is the thermodynamic condition**, however, the reaction may be kinetically disfavoured. At this point, we shall pay attention to only the thermodynamic requirements of a reaction to occur.

$$\Delta G = \Delta H - T\Delta S < 0$$

i. If $\Delta H < 0$ (exothermic reaction) and $\Delta S > 0$, the reaction will occur at all temperatures ($\Delta G < 0$) as **both the enthalpy and entropy drive the reaction**.

ii. If $\Delta H > 0$ (endothermic reaction) and $\Delta S < 0$, the reaction is **nonspontaneous at all temperatures** ($\Delta G > 0$).

iii. If $\Delta H > 0$ (endothermic reaction) and $\Delta S > 0$, the reaction is spontaneous only at **higher temperature** under the condition, $|T\Delta S| > |\Delta H|$, *i.e.* the reaction will be **entropy driven** under the condition.

iv. If $\Delta H < 0$ and $\Delta S < 0$, the reaction is only spontaneous at low temperature ($|T\Delta S| < |\Delta H|$), and the reaction is **enthalpy driven** under the condition.

v. If $\Delta H = 0$ (*e.g.* racemisation reaction), the reaction is **entropy driven under the condition $\Delta S > 0$.**

Note: $\Delta G < 0$ (spontaneous process), > 0 (nonspontaneous process) $= 0$ (process at equilibrium). $\Delta G = \Delta G° + RT \ln K^{o}_{eq}$, where $\Delta G°$ = standard free energy change when the reactants and products at their equilibrium states (P = 1 bar at T of interest), K^{o}_{eq} = thermodynamic equilibrium constant, a dimensionless quantity. At equilibrium, $\Delta G = 0 = \Delta G° + RT \ln K^{o}_{eq}$ and $\Delta G° = -RT \ln K^{o}_{eq}$.

The **thermodynamic equilibrium constant** (K^{o}_{eq}) is related to the standard free energy change ($\Delta G°$) as $\Delta G° = -RT \ln K^{o}_{eq}$. To make K^{o}_{eq} **dimensionless**, in the expression of equilibrium constant, each concentration term is divided by the unit quantity (*i.e.* 1.0 mol dm^{-3} denoted by C_o, standard concentration). It is illustrated for $A + B \rightleftharpoons C$ or $K_{eq} = \frac{[C]}{[A][B]}$, $K^{o}_{eq} = \frac{([C]/C_o)}{([A]/C_o)([B]/C_o)}$. K_{eq} (**stoichiometric equilibrium constant**, *i.e.* **concentration quotient**) is expressed by the concentration terms expressed in mol dm^{-3}. The unit of K_{eq} in the given reaction is mol^{-1} dm^{3}. To make it dimensionless, each concentration term is divided by the **unit quantity** C_o (= 1.0 mol dm^{-3}, **standard state concentration**). Thus the numerical values of both K_{eq} and K^o_{eq} are the same but K^o_{eq} is dimensionless. Thus to express K^o_{eq}, concentration of each **reactant** and **product is to be measured with respect to the standard state of concentration** C_o (= 1.0 mol dm^{-3}).

A. Let us illustrate some reactions to understand the effects of enthalpy change (ΔH) (*i.e.* **enthalpy driven reactions)**

a. **Lewis acid-base adduct formation:** A + :B → AB (*e.g.* BF_3 + $:NH_3$ → $F_3B \leftarrow :NH_3$), ΔS < 0, the reaction to occur must be enthalpy driven (*i.e.* ΔH < 0).

b. **Acid dissociation:** A–H + :B → BH^+ + A^- (*e.g.* H_3CO_2H + H_2O → $CH_3CO_2^-$ + H_3O^+), ΔS < 0 (**effect of electrorestriction of the ionic products on solvent**, ΔS refers to $\Delta S_{universe}$), the reaction to be spontaneous, ΔH must be negative to satisfy the condition $|\Delta H| > |T\Delta S|$. However, if proton transfer neutralises the charge, then ΔS becomes positive (*i.e.* electrorestriction over the solvent is reduced).

$$NH_4^+(aq) + PO_4^{3-}(aq) \rightarrow NH_3(aq) + HPO_4^{2-}(aq), \Delta S > 0$$

ΔH change in a reaction can be calculated from the involved **Born-Haber cycle** (see Vol. 3A). For the Lewis acid-base adduct formation, ΔH can be calculated by using the **Drago-Wayland equation** (see Vol. 3A).

c. **Solubilisation of ionic compounds in polar solvents like water** is basically an enthalpy driven process (see Sec. 11.7, Vol. 2).

d. Let us take the following electron transfer reaction occurring through the inner-sphere mechanism.

$$SO_3^{2-}(aq) + ClO_3^-(aq) \rightarrow SO_4^{2-}(aq) + ClO_2^-(aq)$$

$$\Delta H^\circ = -251 \text{ kJ}, \Delta S^\circ = -88 \text{ J K}^{-1}, \Delta G^\circ \approx -224 \text{ kJ at } 25°C$$

The reaction is **entropically disfavoured** but the reaction is **enthalpy driven,** *i.e.* $|\Delta H^\circ| > |T\Delta S^\circ|$.

e. **Formation of ionic compounds** (*e.g.* Na(s) + $0.5Cl_2$(g) → NaCl(s), lattice energy driven process), **electrode potentials** of different redox couples (*e.g.* M/M^{n+}, etc.) are the classic examples of enthalpy driven reactions (see Sec. 12.6, Vol. 3A for more details).

B. Let us illustrate the **entropy driven reactions**. In the cases where ΔH ≈ 0 or ΔH > 0, the reaction to be spontaneous, it must satisfy the following condition, $|T\Delta S| > |\Delta H|$, (of course ΔS refers to $\Delta S_{universe}$).

a. **Solubilisation of nonpolar solutes in nonpolar solvents** (see Sec. 11.7, Vol. 2 for more details). In the ideal mixture (assuming no interaction between the solvent and solute), the entropy change is given by:

$$\Delta S_{mix} = -\sum n_i R \ln x_i = -R(n_1 \ln x_1 + n_2 \ln x_2), (\Delta H_{mix} = 0)$$

x_1 and x_2 denote the mole fractions of solute and solvent respectively. For $n_1 = n_2 = 1$ and $x_1 = x_2 = 0.5$, ΔS = +1.7 kJ mol^{-1} (*cf.* **spreading of a gaseous substance** in air is an example of entropy driven process).

b. **Clathrate or cage compound formation** is basically an entropy driven process (see Sec. 11.3.2, Vol. 2).

c. **Crystal defect formation** is also an entropy driven process (see Sec. 12.15, Vol. 3A for more details).

d. **Chelate effect** and **macrocyclic effect** to increase the stability of the complex are basically the entropy driven processes (see Vol. 4).

e. Thermal decomposition of a compound producing specially the gaseous substances (see Sec. 12.8.2, Vol. 3A for details) is entropy driven.

$$MCO_3(s) \xrightarrow{\Delta} MO(s) + CO_2(g), \Delta G = \Delta H - T\Delta S \ (\Delta H = +ve, \Delta S = +ve)$$

Decomposition temperature (T_d) is obtained as follows:

$$\Delta G = 0 = \Delta H_d - T_a \Delta S_d, i.e. T_d = \frac{\Delta H_d}{\Delta S_d}$$

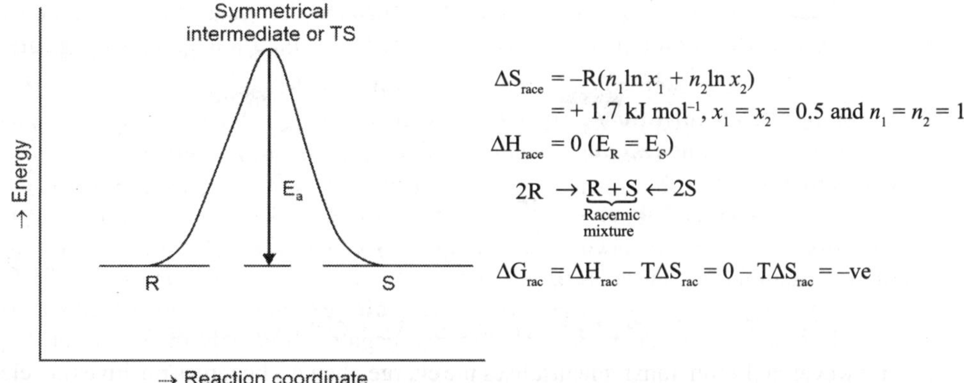

Fig. 5.1.1.1: Energy profile diagram of a racemisation reaction

Note: An **isomerisation reaction** producing a mixture of *cis-* and *trans*-isomers (*i.e.* geometric isomers) makes $\Delta S > 0$ but $\Delta H \neq 0$ as the stabilities of the geometric isomers are different (*i.e.* one form is generally more stable)

f. **Explosives** produce a large volume of gaseous substances to favour the explosion process entropically (see Appendix 16A, Vol. 3B).

g. Importance of entropy effect has been illustrated in **Ellingham diagram** (see Vol. 3B).

h. **Racemisation reaction** ($\Delta H = 0$) is purely an **entropy driven process** (see Fig. 5.1.1.1). It propduces a mixture of R- and S-conformations to increase the overall entropy of the process. If the **activation energy barrier** (E_a) is very small, then the optical isomers (*i.e.* enantiomers) cannot be separated at ordinary condition (*i.e.* room temperature). The inversion process of chiral amines and phosphines through the planar intermediate/TS (transition state) has been discussed in Sec. 10.9.1, Vol. 2 (*cf.* phosphine derivatives having the higher E_a-values are optically more stable).

5.1.2 Different Types of Reactions of Metal Complexes

Different types of reactions of metal complexes are as follows:

- ligand substitution reactions
- metal substitution reactions
- isomerisation and recemisation reactions
- redox (*i.e.* electron transfer) reactions
- reactions of coordinated ligands (*i.e.* activation of ligands).

5.2 CLASSIFICATION OF SUBSTITUTION REACTION

These are classified as follows:

- nucleophilic substitution (S_N)
- electrophilic substitution (S_E)

In **nucleophilic substitution reactions, a nucleophile** (*i.e.* Lewis base) donates its electron cloud to a positively charged centre, *i.e.* nucleus (*i.e.* Lewis acid) through the replacement of another nucleophile.

$$M—X + Y \rightarrow M—Y + X \ (S_N)$$

(Charges not shown)

Here both *X* and *Y* are the nucleophiles (*i.e.* ligands). *Y* is the *entering* or *incoming nucleophile* while *X* is the *leaving group* or *leaving nucleophile.* The metal centre (*i.e. M*) is the Lewis acid. Thus in a nucleophilic substitution reaction, one Lewis base (*i.e.* ligand) displaces another Lewis base from a Lewis acid (*i.e.* metal centre).

Thus the ligand substitution reaction is a case of nucleophilic substitution reaction. In ligand substitution reactions, other ligands (*i.e. nonleaving groups)* are called the **spectator ligands**.

$$L_5M—X + Y \rightarrow L_5M—Y + X$$

Here the ligands denoted by L_5 are called the *spectator ligands that may influence the rate and process of the substitution reaction through the steric and electronic effects.*

In an electrophilic substitution reaction, the metal centres (*i.e.* electron seeking, Lewis acids) are replaced, *i.e.* one metal centre replaces another one. In metal complexes, the *ligand substitution* (*i.e.* nucleophilic substitution) reactions are abundant while the *metal substitution* (*i.e.* electrophilic substitution) reactions are relatively rare.

5.3 DIFFERENT TYPES OF LIGAND SUBSTITUTION REACTIONS

These are illustrated with some specific examples.

Acid hydrolysis: The leaving group is replaced by H_2O (*i.e.* solvent in aqueous media) in acidic condition.

$$[Co(NH_3)_5X]^{2+} + H_2O \rightarrow [Co(NH_3)_5(OH_2)]^{3+} + X^-$$

This is also called the *aquation* or simply the *dissociation* reaction.

Base hydrolysis: The leaving group is replaced by OH^- group in aqueous media. This is also a case of dissociation reaction.

$$[Co(NH_3)_5X]^{2+} + OH^- \rightarrow [Co(NH_3)_5(OH)]^{2+} + X^-$$

Anation: The coordinated solvent molecule (*i.e.* H_2O in aqueous media) is replaced by an anion (X^-).

$$[Co(NH_3)_5(OH_2)]^{3+} + X^- \rightarrow [Co(NH_3)_5X]^{2+} + H_2O$$

This is also described as the *formation reaction.*

Solvent exchange: This is illustrated below.

$$[M(OH_2)_n] + H_2O^* \rightarrow [M(OH_2)_{n-1}(^*OH_2)] + H_2O, \text{ (in aqueous solvent)}$$

Ligand exchange: This is illustrated below.

$$[PtCl_2(py)_2] + 2^*Cl^- \rightarrow [Pt(^*Cl)_2(py)_2] + 2Cl^-$$

Formation reaction: Replacement of a coordinated solvent molecule by a ligand other that the solvent itself.

$$[M(OH_2)_n] + L \rightarrow [M(OH_2)_{n-1}(L)] + H_2O$$

Solvolysis reaction (*cf.* Hydrolysis reaction): Replacement of a ligand by the solvent itself.

$$M—L + S \text{ (solvent)} \rightarrow M—S + L$$

5.4 THERMODYNAMIC AND KINETIC STABILITY: LABILITY AND INERTNESS (*cf.* Sec. 4.1)

Thermodynamic stability of a complex is determined by β_n called **thermodynamic stability constant** (a dimensionless quantity).

$$M + nL \rightleftharpoons ML_n,\ \beta_n = \frac{[ML_n]/(C_o)}{([M]/C_o)([L]/C_o)^n} \text{ and } \beta_{n(C)} = \frac{[ML_n]}{[M][L]^n}$$

$C_o = 1.0$ mol dm^{-3} (**standard state concentration**), a unit quantity. $\beta_{n(C)}$ called **stoichiometric stability constant** (concentration quotient), where the concentrations are expressed in the unit mol dm^{-3}. To obtain β_n (dimensionless) from $\beta_{n(C)}$, each concentration term is divided by the unit quantity C_o (= 1.0 mol dm^{-3}) (see Sec., 5.1.1).

The higher value of β_n indicates its higher thermodynamic stability. Thus it gives the measure of the extent to which the equilibrium can proceed but it cannot say anything regarding the speed with which the equilibrium is attained. Thus the thermodynamic and kinetic stability are not necessarily correlated. A thermodynamically stable complex may react slowly or fast depending on the condition. This can be illustrated by considering the different cyanido complexes which are extremely stable (*i.e.* very high value of β_n) but they behave quite differently in terms of the rate of exchange of the radiocarbon labelled cyanide.

$$\left[Hg(CN)_4\right]^{2-} + 4\,^{14}CN^- \rightleftharpoons \left[Hg(^{14}CN)_4\right]^{2-} + 4CN^-, \textbf{ very fast}$$

$$\left[Ni(CN)_4\right]^{2-} + 4\,^{14}CN^- \rightleftharpoons \left[Ni(^{14}CN)_4\right]^{2-} + 4CN^-, t_{1/2} \approx 30\,s, \textbf{ very fast}$$

$$\left[Pt(CN)_4\right]^{2-} + 4\,^{14}CN^- \rightleftharpoons \left[Pt(^{14}CN)_4\right]^{2-} + 4CN^-, t_{1/2} \approx 1\text{ min}$$

$$\left[Mn(CN)_6\right]^{3-} + 6\,^{14}CN^- \rightleftharpoons \left[Mn(^{14}CN)_6\right]^{3-} + 6CN^-, t_{1/2} \approx 1\text{ h}$$

$$\left[Cr(CN)_6\right]^{3-} + 6\,^{14}CN^- \rightleftharpoons \left[Cr(^{14}CN)_6\right]^{3-} + 6CN^-, t_{1/2} \approx 24\text{ d},$$

$$\left[Fe(CN)_6\right]^{3-/4-} + 6\,^{14}CN^- \rightleftharpoons \left[Fe(^{14}CN)_6\right]^{3-/4-} + 6CN^-, \textbf{ (very slow)}$$

All the above complexes are thermodynamically very stable (*cf.* $[Ni(CN)_4]^{2-}$, log $\beta_4 \approx 29$; $[Pt(CN)_4]^{2-}$, log $\beta_4 \approx 40$; $[Hg(CN)_4]^{2-}$, log $\beta_4 \approx 42$; $[Fe(CN)_6]^{3-}$, log $\beta_6 \approx 44$; $[Fe(CN)_6]^{4-}$, log $\beta_6 \approx 37$; etc.) but their kinetic stabilities differ widely. *The thermodynamically stable complexes like* $[Ni(CN)_4]^{2-}$, $[Hg(CN)_4]^{2-}$, etc. *react fast in the ligand exchange reactions but* $[Cr(CN)_6]^{3-}$ *which is also thermodynamically stable reacts slowly* and $[Fe(CN)_6]^{4-}$ also reacts very slowly.

Between $[Fe(CN)_6]^{4-}$ (t_{2g}^6) and $[Fe(CN)_6]^{3-}$ (t_{2g}^5), $[Fe(CN)_6]^{3-}$ is relatively more labile. In fact, slow aquation of $[Fe(CN)_6]^{3-}$ produces $[Fe(CN)_5(OH_2)]^{2-}$ to release CN^- **that can induce toxicity**. In fact, **ferricyanide solution is more poisonous than ferrocyanide solution in terms of CN^- poisoning.** The lability or inertness of the complexes and the pathway of ligand substitution can be interpreted in terms of crystal field activation energy **(CFAE)** discussed later.

- For some **square planar Pt(II) – complexes**, experiencing the **associative pathway of ligand substitution,** the lability and thermodynamic stability follow the same sequence, *i.e.* **the most stable complex is the most labile one** in the corresponding ligand exchange reaction.

Complex:	$[PtCl_4]^{2-}$	$[PtBr_4]^{2-}$	$[PtI_4]^{2-}$	$[Pt(CN)_4]^{2-}$
~ logβ_4:	17	20	30	40
~ $t_{1/2}$ (min):	850	6	4	1

$\longrightarrow$ **Increasing lability** (*cf.* Secs. 5.30.5,6) **and stability**

- **Dissociation of $[Co(NH_3)_6]^{3+}$** in aqueous acidic media is thermodynamically highly favoured but the process is kinetically highly disfavoured.

$$\left[Co(NH_3)_6\right]^{3+} + 6H_3O^+ \rightleftharpoons \left[Co(OH_2)_6\right]^{3+} + 6NH_4^+, \log K_{eq} \approx 25.0$$

The tremendous thermodynamic driving force for the forward reaction arises from the protonation of the basic NH_3 ligands, but the complex remains unchanged in a fairly strong acidic medium for weeks at room temperature. In fact, the salt $[Co(NH_3)_6]Cl_3$ can be crystallised from a hot aqueous solution of HCl without any noticeable decomposition. Thus, *stability of the complex arises not from the thermodynamic stability but from the kinetic stability.*

- **$[M(en)_3]^{3+}$ and $[M(bigH)_3]^{3+}$** (M = Cr, Co): The biguanide complex is more stable than the ethylenediamine complex but with respect to the acid catalysed dissociation reaction, $[M(bigH)_3]^{3+}$ reacts much faster than the $[M(en)_3]^{3+}$ complex.

$$\left[M(en)_3\right]^{3+} \xrightarrow{H_3O^+} \left[M(OH_2)_6\right]^{3+} + 3enH^+$$

$$\left[M(bigH)_3\right]^{3+} \xrightarrow{H_3O^+} \left[M(OH_2)_6\right]^{3+} + 3bigH_2^+$$

In fact, $[Co(bigH)_3]^{3+}$ undergoes dissociation in an acidic medium at a measureable rate at 30°C (*cf.* D. Banerjea *et al*, *J. Inorg. Nucl. Chem.*, **26,** 1233, 1964) and $[Co(en)_3]^{3+}$ remains completely unchanged in 1 M $HClO_4$ at room temperature even for months.

Note: The kinetic favour for dechelation of the biguanide ring arises from the *its possibility of protonation before ring opening* but such a protonation cannot occur prior to opening of the chelate ring for the ethylenediamine complex.

- To differentiate between the thermodynamic stability and kinetic stability, H. Taube has coined the kinetic terms: **lability** and **inertness.** The kinetically stable complexes are called the **inert complexes** while the kinetically unstable complexes are called the **labile complexes.** The simple terms — stable and unstable refer to the thermodynamic parameters.
- In ligand substitution reactions, the rate can span a very wide range of time scale — **nanosecond (ns) to years.**

$[Cu(aq)]^{2+} + 4NH_3 \rightarrow [Cu(NH_3)_4]^{2+}$, **(instantaneous)**
(blue purple)

$[Co(NH_3)_6]^{3+}$ or $[Co(en)_3]^{3+} \rightarrow$ Hydrolysis or aquation **(years)**

- *There is no sharp border line to distinct between the labile and inert complexes.* However, H. Taube has described the complexes as the **labile complexes** having $t_{1/2}$ (substitution half life) $\leq$ 30 s in a particular reaction while the complexes with $t_{1/2} >$ 30 s as the **inert complexes.** Rate process for the inert complexes can be followed by the conventional techniques while for the labile complexes, it requires some special techniques (*e.g. stopped flow, P–jump, T–jump,* etc.). This is why, more information is available for the inert complexes.
- The lability or inertness depends on the **activation energy,** *i.e.* high activation energy imparts the inertness while low activation energy imparts the lability (*cf.* Fig. 4.1.1). The inertness or lability is determined by $\Delta G^{\neq}$ **(free energy of activation).**

$\Delta G^{\neq} = \Delta H^{\neq} - T\Delta S^{\neq}$, $\Delta H^{\neq}$ = enthalpy of activation, $\Delta S^{\neq}$ = entropy of activation.

The stability of a complex is determined by the free energy change (ΔG^o) (*cf.* Fig 4.1.1) in a reaction.

$$\Delta G^o = \Delta H^o - T\Delta S^o = -RT \ln K_{eq}$$

$\Delta G^{\neq}$ depends on the reaction pathway (*i.e.* reaction mechanism) while ΔG^o depends on the difference in standard free energy of the reactant and product.

Labile and inert complexes of 3*d*–series

It depends on the d^n–configuration for a particular oxidation state.

Labile centres: d^0, d^1, d^2, d^4 (h.s.), d^5 (h.s.), d^6 (h.s.), d^7, d^8, d^9

Inert centres: d^3, d^4 (l.s.), d^5 (l.s.), d^6 (l.s.)

These can be explained by CFT. These aspects will be discussed later.

5.5 NUCLEOPHILICITY *vs.* BASICITY; ELECTROPHILICITY

- It has been already mentioned that the nucleophiles are basically the Lewis bases but the term **basicity** (measured with respect to H^+ in terms of the pK_a value of the corresponding conjugate acid, YH^+) refers to the thermodynamic concept (*i.e.* equilibrium concept). The analogous kinetic term is called **nucleophilicity** (nucleophile = nucleus loving). In an associative pathway, the entering nucleophile (*i.e.* entering ligand) makes a bond with the metal centre in producing the **activated complex or transition state**. The *kinetic ease* with which a nucleophile can attack the metal centre to produce the activated complex gives the measure of its nucleophilicity. Relative nucleophilicity of a particular ligand L is measured with reference to that of a standard ligand L^0 by comparing their rates of substitution reaction (occurring through the **associative process**) at a chosen metal centre, *i.e.*

$$\text{Nucleophilicity of } L = \frac{\text{rate of substitution by } L}{\text{rate of substitution by } L^0}$$

The details of nucleophilicity scale will be discussed later (*cf.* Sec. 5.30.5)
The Lewis acids act as *electrophiles.*

- **Acidity** is a thermodynamic term while the analogous kinetic term is **electrophilicity** (electrophile = electron loving). The rate of reaction of a Lewis acid (*i.e.* metal centre in the complexes) with the Lewis bases (*i.e.* entering ligands or entering nucleophiles) gives the measure of the electrophilicity of the Lewis acid. Just like the relative nucleophilicity scale, a relative electrophilicity scale can be generated.

5.6 METHODS OF FOLLOWING KINETICS (*i.e.* RATE MEASUREMENT)

(1) For the inert systems, conventional techniques like: UV-visible spectroscopic method, potentiometry, conductometry, polarography, polarimetry, titrimetry, etc. can be used to follow the rate process. In fact, reactions of the complexes of Cr(III), Co(III), Rh(III), Ir(III), Pt(II), Pt(IV), some complexes of Ni(II), etc. can be studied by the conventional techniques. Some examples are given below.

- **Direct chemical analysis:** Direct chemical estimation of either the reactants or products at regular intervals during the progress of reaction is possible in many cases.

(a) $[PtCl(NH_3)_3]^+ + H_2O \rightarrow [Pt(NH_3)_3(OH_2)]^{2+} + Cl^-$

(b) $[CoCl_2(en)_2]^+ + H_2O \rightarrow [CoCl(en)_2(OH_2)]^{2+} + Cl^-$

(c) $[Co(NCS)(NH_3)_5]^{2+} + OH^- \rightarrow [Co(NH)_5(OH)]^{2+} + SCN^-$

(d) $[Co(NH_3)_5(S_2O_3)]^+ + L \rightarrow [Co(L)(NH_3)_5]^{n+} + S_2O_3^{2-}$

(L = different nucleophiles)

Here, we shall illustrate the direct chemical estimation of one of the products in the above reactions.

In the reactions (a) and (b), the released Cl^- can be estimated by *argentometry* (*i.e.* titration by $AgNO_3$) or by *potentiometry* (using an electrode reversible with respect to Cl^-, *e.g.* Ag/AgCl electrode). However, there is a possibility by the *direct attack* of Ag^+ *on the bound* Cl^- (*i.e.* overestimation of Cl^-) in the *argentometry method.* This can be minimised, if it is carried out at low temperature and in a mixed solvent like acetone-water mixture. In the reaction (c), released SCN^- can be also estimated similarly by titration with $AgNO_3$. In the reaction (d), the released $S_2O_3^{2-}$ can be estimated iodimetrically as usual.

- **Spectrophotometric method:** Generally, the coordination compounds are coloured and concentration of a coloured coordination compound can be determined from the optical density measurement at a suitable wavelength (A = optical density = εCl, ε = molar extinction coefficient, C = molar concentration, l = optical path length). Generally, in a particular reaction, the absorption spectra of the product and reactant are different and the most suitable wavelength for the optical density measurement is the wavelength at which absorption of the reactant and product differs most.

If the measured optical density (*OD*) is due to a particular compound whose formation or decay experiences a first order kinetics then the plot of $\log\frac{A_0 - A_\infty}{A_t - A_\infty}$ *vs.* t (time) will give the first order rate constant.

$A_t = OD$ at time t; $A_0 = OD$ at zero time; $A_\infty = OD$ at infinity (*i.e.* at the end of the reaction).

Note: For a first-order reaction, we can write:

$$C_t = C_0 \exp(-kt),\ i.e.\ \ln\frac{C_t}{C_0} = -kt \text{ or, } 2.303\log\frac{C_t}{C_0} = -kt \text{ and } k = \frac{2.303}{t}\log\frac{a}{a-x}$$

$$C_0 = a = \text{initial concentration } (i.e.\ t = 0),\ C_t = (a - x) = \text{concentration at } t.$$

If a physical property P (like *OD*), directly proportional to its concentration is available, then we can write:

$$k = \frac{2.303}{t}\log\frac{P_0 - P_\infty}{P_t - P_\infty};\ \text{slope of the plot, } \log\frac{P_0 - P_\infty}{P_t - P_\infty}\ vs.\ t \text{ is } \frac{k}{2.303}$$

● **Electrometric method** (*i.e.* measurement of conductance, emf, pH, etc.): In the following aquation reactions, conductance increases remarkably during the progress of reaction.

$$[CoCl_2(en)_2]^+ + H_2O \rightarrow [CoCl(en)_2(OH_2)]^{2+} + Cl^-$$
$$[PtCl(NH_3)_3]^+ + H_2O \rightarrow [Pt(NH_3)_3(OH_2)]^{2+} + Cl^-$$

Thus conductance measurement with time can be practised to study the above reactions. The rate of Cl^- release can also be measured potentiometrically by using an electrode sensitive to Cl^-.

In the following reaction, during the progress of reaction, pH decreases due to the release of H^+ ion.

$$M^{2+} + LH \rightarrow ML^+ + H^+$$

Thus the pH measurement with time can evaluate the rate constant of the process.

● **Polarimetric method:** The measurement of optical rotation with time can be used to determine the rate constant of a racemisation processes.

$$\Delta\text{-}\left[Co(\text{big H})_3\right]^{3+} \rightleftharpoons \Lambda\text{-}\left[Co(\text{big H})_3\right]^{3+}$$

It should be taken into consideration that the light used in the optical rotation measurements may accelerate the isomerisation process in some cases. In such cases, the method becomes erratic.

● **Isotope tracer technique:** The electron exchange reactions in the couples like $[Ru(NH_3)_6]^{3+}$/$[Ru(NH_3)_6]^{2+}$, $[Fe(CN)_6]^{3-}$/$[Fe(CN)_6]^{4-}$, $[Fe(OH_2)_6]^{3+}$/$[Fe(OH_2)_6]^{2+}$, etc. may be followed by using the labelled isotope (see Secs. 6.1, 6.3.2). Isotopic tracer techniques may also be used to follow the metal exchange and ligand exchange reactions (see Sec. 5.12).

(2) For following the fast reactions of labile centres, it needs some special techniques like stopped-flow spectrophotometry, T–jump, P–jump, NMR, etc. depending on the $t_{1/2}$ of the reactions.

● **Stopped-flow spectrophotometry:** In the stopped-flow technique, the mixing time between the reactants is about 1 ms (*i.e.* 10^{-3} s) and the reactions faster than the mixing time cannot be studied by the stopped-flow spectrophotometry. In the stopped-flow spectrophotometry, after mixing of the reactants within the reaction cell, progress of the reaction can be followed spectrophotometrically with the help of a fast recorder device.

● **Perturbation technique (*i.e.* T–jump, P–jump):** This technique is applicable for the reactions of $t_{1/2}$ in the time scale of μs (= 10^{-6} s). In this technique, the equilibrium is perturbed suddenly (within a fraction of a μs or less) by a temperature change, *ca.* 5 – 10°C (in T–jump) or a pressure change, *ca.* a few hundred atmospheres (in P–jump). This change of temperature or pressure directs the equilibrium in a direction as demanded by *Le Chatelier's principle.* **This relaxation of the system towards the**

new equilibrium position can be followed spectrophotometrically by using a fast recorder device. This is why, this perturbation technique is referred to as the **relaxation technique.**

It measures the **relaxation time** (τ) which is the time required for a reaction to cover a certain fraction of the path towards the new equilibrium state. For the following type of reaction, we have:

$$A \underset{k_{-1}}{\overset{k_1}{\rightleftharpoons}} B, \Delta x_t = \Delta x_0 \exp(-t/\tau), \text{ where } \tau = \frac{1}{k_1 + k_{-1}}$$

Here Δx_o is the perturbation or departure from the equilibrium immediately after the T-jump and Δx_t measures the departure from the new equilibrium state at the time t after the perturbation of the old equilibrium state (see Fig. 5.6.1). **Whatever may be the order of the reaction, relaxation towards the new equilibrium state always occurs exponentially like a first order reaction.**

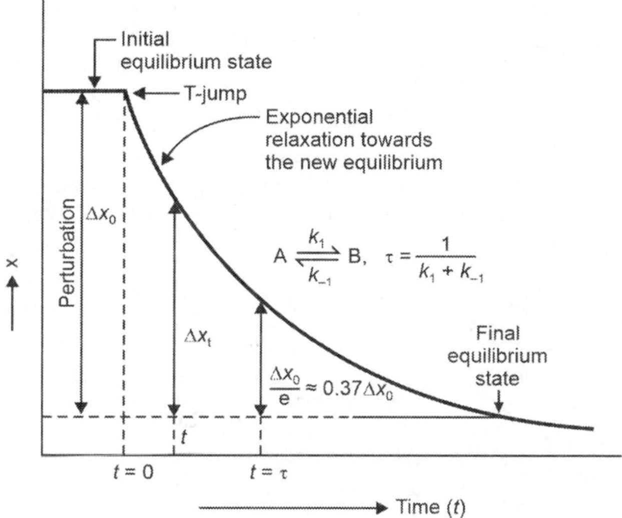

Fig. 5.6.1: Illustration of the exponential relaxation towards the new equilibrium after a perturbation as in a T-jump technique. Here x is a measurable quantity like absorbance that varies linearly with the composition of the system

Relaxation time (τ) corresponds to $\Delta x_0/\Delta x_t = e$, *i.e.* $\ln(\Delta x_0/\Delta x_t) = 1$. Thus relaxation time is the time taken by the system to drop the deviation (Δx_t) to $1/e$ (≈ 0.368) times the initial deviation (Δx_o). The expression of τ depends on the nature of reaction. By using the T-jump technique, the following reaction has been studied.

$$2H_2O \rightleftharpoons H_3O^+ + OH^-$$

After T-jump, **conductivity** increases with time and the relaxation process can be followed by measuring the conductivity ($\tau \approx 3.7 \times 10^{-5}$ s at 25°C). From this τ value, the second order rate constant for the combination of H_3O^+ and OH^- is found to be 1.4×10^{11} dm^3 mol^{-1} s^{-1} which is a very high value (the **fastest reaction in solution phase**). This is, in fact, a diffusion controlled process.

It may be mentioned that the T–jump technique can be applied to all systems (assuming $\Delta H° \neq 0$) but the P-jump technique can only be applied to the systems where there is a change in the number of species in the reaction (more correctly, **where there is a molar volume change**). P-jump technique has been utilised for the rate constant measurement of the reaction, $H^+ + OH^- \rightarrow H_2O$ ($k = 1.5 \times 10^{11}$ M^{-1} s^{-1}, 25°C). **This is the fastest reaction known in solution** and it is controlled by the rate of diffusion of the involved ions in solution. This is why, reaction between the strong acid and strong base occurs in no time.

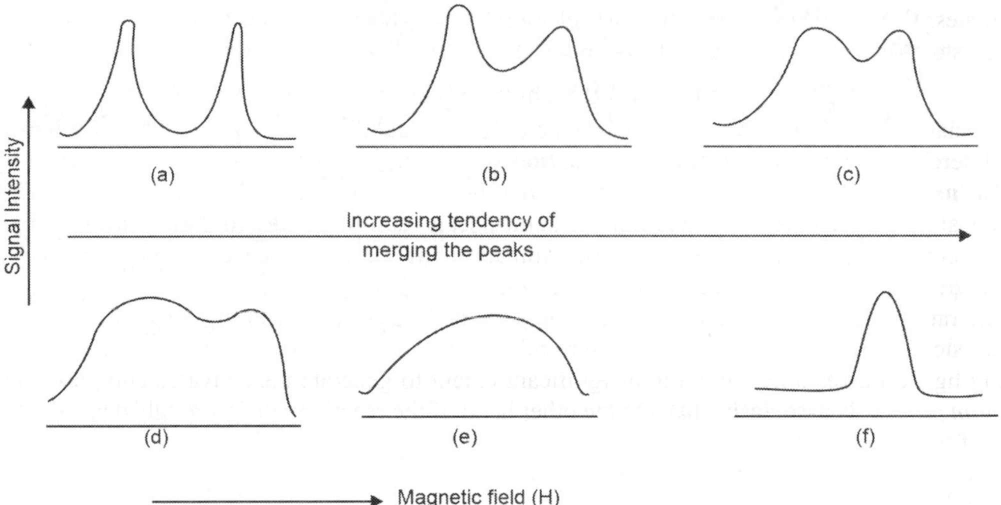

Fig. 5.6.2: Change of the NMR signals with the gradual increase of exchange rates (*e.g.* PMR signals in the proton exchange process between the two sites X—H and Y—H) from (a) → (f). (a) Exceedingly slow exchange rate (*i.e.* practically no exchange); (b) to (e) Gradually increasing rate; (f) Very rapid exchange rate (*see* Sec. 3.15, Fig. 3.15.1 in Vol. 7).

Temperature dependence of the equilibrium constant can be expressed by **van't Hoff relation** (see Sec. 5.19.3)

$$\frac{d\ln K_{eq}}{dT} = \frac{\Delta H^\circ}{RT^2}, \textit{ i.e. } \ln K_{eq} = -\frac{\Delta H^\circ}{RT} + \text{constant}$$

$$\text{or} \quad \ln\left(\frac{K_2}{K_1}\right) = \frac{\Delta H^\circ}{R}\left(\frac{T_2 - T_1}{T_1 T_2}\right)$$

$$\Delta G^\circ = -RT\ln K_{eq} = \Delta H^\circ - T\Delta S^\circ$$

$$\text{or} \quad \ln K_{eq} = -\frac{\Delta H^\circ}{RT} + \frac{\Delta S^\circ}{R} = -\frac{\Delta H^\circ}{RT} + \text{constant}$$

● **NMR technique** (*cf.* Sec. 3.15, Vol. 7): For the studies of *fast exchange reactions* (both ligand exchange and metal exchange), the NMR method is very much important. For the studies of water exchange reactions, $^{17}OH_2$ is used. The NMR spectra of a particular NMR-active centre depends on its chemical environment. Thus the NMR spectra of ^{17}O are different for metal bound $^{17}OH_2$ and free $^{17}OH_2$ present in bulk solvent. If the water exchange rate is slow, then the NMR signals of ^{17}O for two different chemical environments are quite distinct. *But with the increase of exchange rate, the signals will move to merge or overlap* (*i.e.* the peak will be broadening). If the exchange rate is extremely fast, then a single peak is noticed. Thus from the nature of merging the peaks, it is possible to determine the exchange rate constant. Merging of peaks due to an exchange reaction is qualitatively shown in Fig. 5.6.2.

5.7 MECHANISM OF LIGAND SUBSTITUTION REACTION: INTIMATE AND STOICHIOMETRIC MECHANISM

● **Stoichiometric mechanism:** It deals with the *sequence of elementary steps* leading to a chemical reaction. It looks at the *reactants, products* and *intermediates* but not at the transition states. Thus the stoichiometric mechanism looks at the species residing at the **potential minima** along the reaction coordinates. Consideration of the involved elementary steps can lead to the rate law. There are three types of stoichiometric mechanism of ligand substitution reaction. These are:

Dissociative (*D*), Interchange (*I*) and Associative (*A*)

Dissociative and associative reactions are the **two-step reactions** passing through an **intermediate,** while interchange reaction is a **one-step reaction** without the formation of a **true intermediate.**

● **Intimate mechanism:** It deals with the *activation process* leading to the formation of an activated complex at the rate determining step. It looks at the species residing at **the highest point** (*i.e.* activated complex or transition state) on the reaction coordinate. Thus it considers the *energetics of the formation of the activated complex* and *consequently the rate.*

If the rate of formation of the activated complex at the **rate determining step** (*rds*) (*i.e.* slowest reaction step) depends strongly on the nature of the entering ligand (say *L*) then it indicates that the entering ligand makes a new bond to a significant extent to generate the activated complex. Thus the *activation process* is **associative (*a*)**. On the other hand, if the reaction rate is strongly dependent on the nature of the leaving group (say *X*) and almost independent of the nature of enterging group (*L*), then it indicates that bond breaking by the leaving group is important (*i.e. rds*) to generate the activated complex. Thus the *activation process* is ***dissociative* (*d*)**.

Here it is worth mentioning that the **associative activation** and **dissociative activation** are denoted by ***a*** and ***d*** respectively (not by ***A*** and ***D*** which denote the stoichiometric associative and dissociative reaction mechanisms respectively).

● **Relationship between the stoichiometric and intimate mechanism: *D* mechanism** must involve the **dissociative (*d*) activation** while ***A* mechanism** must involve the **associative (*a*) activation. *I* mechanism** involves **both the dissociative and associative activation,** *i.e.* both bond breaking by the leaving group and bond formation by the entering group contribute to the rate determining step. If in this process of activation, bond breaking (*i.e.* dissociative activation) is the predominant factor then it is referred to as I_d. On the other hand, if in the interchange process, bond formation (*i.e.* associative activation) by the entering group is the predominant factor then it is referred to as I_a.

Thus the **combined notations** describing both the stoichiometric and intimate mechanisms are:

A, I_a, I_d, D (Langford-Gray notation)

● **Relationship between Langford-Gray notation and Hughes-Ingold notation used in organic chemistry:** According to Basolo and Pearson, *A mechanism* is equivalent to **S_N2 lim**. (*i.e.* limiting situation of S_N2) while *D mechanism* is equivalent to **S_N1 lim**. I_a and I_d correspond to S_N2 (not limiting) and S_N1 (not limiting) respectively.

Note: $S_N1 \Rightarrow$ substitution nucleophilic unimolecular; $S_N2 \Rightarrow$ substitution nucleophilic bimolecular.

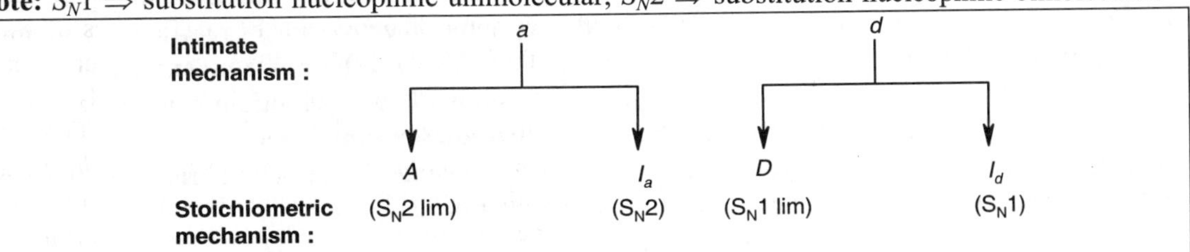

- For both the ***D*** and ***I*$_d$** processes, the intimate mechanism is ***d*** (*i.e.* ***d***-activation) but for the *D*–process, the intermediate (*i.e.* **long life time**) of lower coordination number is detectable while the intermediate is not detectable (*i.e.* **very small life time**) for the I_d process. Thus their stoichiometric mechanisms (*i.e.* sequence of elementary steps) differ, *i.e.* single step (I_d) *vs.* two steps (*D*) (cf. Fig. 5.8.1)
- Similarly for both the ***A*** and ***I*$_a$** processes, the intimate mechanism is ***a*** (*i.e.* ***a***–activation). The intermediate of higher coordination number is detectable for the ***A***–pathway but not for the I_a–pathway. Thus their stoichiometric mechanisms (*i.e.* sequence of elementary steps) differ, *i.e.* single step (I_a) *vs.* two steps (*A*) (*cf.* Fig. 5.8.1).

● **Cases involving a fast *pre-equilibrium step* followed by the rate determing step (*rds*):** The fast pre-equilibrium step may lead to the formation of reactive intermediates like **ion pair (IP), conjugate base (CB), conjugate acid (CA),** etc. and then these reactive intermediates participate at the rate determining steps. In such cases, notation of the reaction mechanism depends on both the nature of the reactive intermediate (formed in a rapid pre-equilibrium step) and the nature of rate determining step (*rds*) involving the reactive intermediate. This aspect is illustrated below.

$$\text{Reactants} \rightleftharpoons \text{Intermediate (fast pre-equilibrium)} \xrightarrow{rds} \text{Product}$$

Notation of the mechanism: Nature of the rds (*i.e. A* or *D* or *I*) – nature of the intermediate (*i.e.* IP or CA or CB).

If the intermediate is *IP*, and the rds is an *A*–pathway, then the mechanism is designated by ***A–IP*** (*i.e.* S_N2***–IP*** in old nomenclalure). Similarly, we can have:

***D–IP* (*i.e.* S_N1 lim–*IP*, in short S_N1–*IP*)** meaning *IP* as the reactive intermediate, and rds as the *D*-pathway;

***D–CA* (*i.e.* S_N1 lim–*CA*, in short S_N1–*CA*)** meaning *CA* as the reactive intermediate and rds as the *D*-pathway;

***D–CB* (*i.e.* S_N1 lim–*CB*, in short S_N1–*CB*)** meaning *CB* as the reactive intermediate, and rds as the *D*-pathway;

Similarly, we can write: I_d***–CB***, I_d***–IP***, I_a***–IP***, etc.

● **Pre-equilibrium step *vs.* steady-state (*i.e.* stationary-state) approximation:** In a consecutive reaction, steady-state approximation is applied under the condition: **rate of formation of the reactive intermediate is much slower than the rate of disappearance of the reactive intermediate**. Under this condition, the intermediate disappears as soon as it is formed and consequently, **there is no build-up of the concentration of the intermediate**. In other words, it finds no time to get accumulated during the reaction. Consequently, **very small concentration of the intermediate is assumed to remain constant** so long the formation of the intermediate continues.

On the other hand, if **the rate of formation of the intermediate is faster than the rate of its disappearance**, then there is a gradual build-up of its concentration in the system and very soon, it attains an equilibrium condition with the reactants. The intermediate remaining in an equilibrium transforms into the product slowly. Under this condition, during the progress of reaction at any time, concentration of the intermediate can be considered as the **equilibrium concentration of the intermediate** and its concentration can be expressed in terms of the equilibrium constant (K_{eq}) and concentration of the reactants. This is described as the **pre-equilibrium concept**.

Let us consider the following consecutive reaction to illustrate the pre-equilibrium concept and steady-state approximation for the concentration of the intermediate (*C*).

$$A + B \underset{k_{-1}}{\overset{k_1}{\rightleftharpoons}} C \xrightarrow{k_2} \text{product}$$

Pre-equilibrium step: If k_1 and k_{-1} are much higher than k_2, then the first step is called the **pre-equilibrium step** and concentration of the intermediate (*C*) remaining in an equilibrium with the reactants *A* and *B* is expressed as follows:

$$K_{eq} = \frac{[C]}{[A][B]}$$ (assuming the forward and backward reactions to be the **elementary reactions** to express $K_{eq} = k_1/k_{-1}$

Rate of the reaction $= k_2[C] = k_2K_{eq}[A]\,[B]$

Here $[A]$, $[B]$ and $[C]$ denote their respective equilibrium concentrations that can be expressed in terms of the initial concentrations of A and B, *i.e.* $[A]_0$ and $[B]_0$ respectively.

From the **mass balance relation**, we can write from the stoichiometry of the reaction:

$$[A]_0 = [A] + [C] \text{ and } [B]_0 = [B] + [C]$$

i.e.

$$[A] = [A]_0 - [C] \text{ and } [B] = [B]_0 - [C]$$

Under the condition, $[A]_0 \gg [B]_0$, we can reasonably write:

$$[A]_0 \approx [A] \text{ and } [B] = [B]_0 - [C]$$

$$K_{eq} = \frac{[C]}{[A][B]} = \frac{[C]}{[A]_0\{[B]_0 - [C]\}}$$

i.e. $K_{eq}[A]_0[B]_0 - K_{eq}[A]_0[C] = [C]$

or

$$[C] = \frac{K_{eq}[A]_0[B]_0}{1 + K_{eq}[A]_0}$$

$$\text{Rate} = k_2[C] = \frac{k_2 K_{eq}[A]_0[B]_0}{1 + K_{eq}[A]_0}$$

Steady-state or stationary-state approximation: If k_2 is comparable to k_1 and k_{-1}, then steady-state approximation can be applied to express the concentration of the intermediate (C) whose **very small concentration remains constant during the reaction**.

$$\frac{d[C]}{dt} = 0 = k_1[A][B] - k_{-1}[C] - k_2[C], \text{ i.e. } [C] = \frac{k_1[A][B]}{k_{-1} + k_2}$$

$$\text{Rate} = k_2[C] = \frac{k_1 k_2[A][B]}{k_{-1} + k_2}$$

Under the condition, $k_{-1} \gg k_2$, we can write:

$$\text{Rate} = \frac{k_1 k_2[A][B]}{k_{-1} + k_2} \approx \left(\frac{k_1}{k_{-1}}\right) k_2[A][B] = K_{eq} k_2[A][B]$$

Thus, the **rate equation of steady-state approximation transforms into the rate equation for the pre-equilibrium concept under the condition,** $k_{-1} \gg k_2$.

The rate equation for the steady-state approximation can be expressed in terms of the initial concentrations of the reactants, *i.e.* $[A]_0$ and $[B]_0$.

Under the condition, $[A]_0 \gg [B]_0$ we can write:

$$[A]_0 = [A] + [C] \approx [A] \text{ and } [B]_0 = [B] + [C], \text{ i.e. } [B] = [B_0] - [C]$$

$$\frac{d[C]}{dt} = 0 = k_1[A][B] - k_{-1}[C] - k_2[C], \text{ or } k_1[A][B] = [C](k_{-1} + k_2)$$

or

$$k_1[A]_0\{[B]_0 - [C]\} = [C](k_{-1} + k_2)$$

or

$$k_1[A]_0[B]_0 = [C](k_{-1} + k_2) + k_1[C][A]_0$$

or

$$[C] = \frac{k_1[A]_0[B]_0}{k_1[A]_0 + (k_{-1} + k_2)}$$

$$\text{Rate} = k_2[C] = \frac{k_1 k_2[A]_0[B]_0}{k_1[A]_0 + (k_{-1} + k_2)}$$ (*cf.* **Michaelis-Menten equation** in enzyme kinetics)

● **Derivation of rate equation for a reversible reaction:** Let us first consider the following 1st order reversible reaction.

$$A \underset{k_{-1}}{\overset{k_1}{\rightleftharpoons}} P,\ K_{eq} = \frac{k_1}{k_{-1}} = \frac{[P]_e}{[A]_e}$$

$$\frac{d[P]}{dt} = -\frac{d[A]}{dt} = k_1[A] - k_{-1}[P],$$

$$[A]_0 + [P]_0 = [A]_t + [P]_t = [A]_e + [P]_e \text{ (mass balance)}$$

Combining the above relations, we can write:

$$[A]_t + [P]_t = [A]_e + [P]_e$$

or
$$[A]_t = [A]_e + [P]_e - [P]_t$$
$$= [A]_e + \frac{k_1}{k_{-1}}[A]_e - [P]_t$$

i.e.
$$[P]_t = [A]_e\left(1 + \frac{k_1}{k_{-1}}\right) - [A]_t$$

$$-\frac{d[A]_t}{dt} = k_1[A]_t - k_{-1}[P]_t$$
$$= k_1[A]_t - k_{-1}\left\{[A]_e\left(1 + \frac{k_1}{k_{-1}}\right) - [A]_t\right\}$$
$$= k_1[A]_t - k_{-1}[A]_e - k_1[A]_e + k_{-1}[A]_t$$
$$= [A]_t(k_1 + k_{-1}) - [A]_e(k_1 + k_{-1})$$
$$= ([A]_t - [A]_e)(k_1 + k_{-1})$$

or
$$-\frac{d[A]_t}{([A]_t - [A]_e)} = (k_1 + k_{-1})dt$$

Integration of the above equation gives:

$$-\ln([A]_t - [A]_e) = (k_1 + k_{-1})t + I \text{ (constant of integration)}$$

Under the boundary condition: at $t = 0$, $[A]_t = [A]_0$

i.e.
$$I = -\ln([A]_0 - [A]_e)$$

It gives: $\ln\left(\frac{[A]_t - [A]_e}{[A]_0 - [A]_e}\right) = -(k_1 + k_{-1})t$

Negative slope of the linear plot of $\ln([A]_t - [A]_e)$ *vs.* t) (time) gives the observed rate constant (k_{obs}) as:

$$-\text{slope} = k_{obs} = k_1 + k_{-1}$$

Now let us consider the following complex formation reaction.

$$M + L \underset{k_{-1}}{\overset{k_1}{\rightleftharpoons}} ML,\ K_{eq} = \frac{k_1}{k_{-1}} = \frac{[ML]_e}{[M]_e[L]_e}$$

Under the **pseudo-first order condition,** $[L]_0 \gg [M]_0$, we can write:

$$[M]_0 + [L]_0 + [ML]_0 = [M]_t + [L]_t + [ML]_t = [M]_e + [L]_e + [ML]_e$$
$$\text{and } [L]_0 \approx [L]_t \approx [L]_e \text{ (mass balance)}$$

$$[\mathrm{M}]_t = [\mathrm{M}]_e + [\mathrm{L}]_0 + [\mathrm{ML}]_0 - [\mathrm{L}]_0 + [\mathrm{ML}]_t$$
$$= [\mathrm{M}]_e + \frac{k_1}{k_{-1}}[\mathrm{M}]_e[\mathrm{L}]_e - [\mathrm{ML}]_t$$
$$\text{or } [\mathrm{ML}]_t = [\mathrm{M}]_e + \frac{k_1}{k_{-1}}[\mathrm{M}]_e[\mathrm{L}]_e - [\mathrm{M}]_t$$
$$\frac{d[\mathrm{ML}]_t}{dt} = -\frac{d[\mathrm{M}]_t}{dt} = k_1[\mathrm{M}]_t[\mathrm{L}]_t - k_{-1}[\mathrm{ML}]_t$$
$$i.e. -\frac{d[\mathrm{M}]_t}{dt} = k_1[\mathrm{M}]_t[\mathrm{L}]_t - k_{-1}\left\{[\mathrm{M}]_e + \frac{k_1}{k_{-1}}[\mathrm{M}]_e[\mathrm{L}]_e - [\mathrm{M}]_t\right\}$$
$$= k_1[\mathrm{M}]_t[\mathrm{L}]_t - k_{-1}[\mathrm{M}]_e - k_1[\mathrm{M}]_e[\mathrm{L}]_e + k_{-1}[\mathrm{M}]_t$$
$$= [\mathrm{M}]_t(k_1[\mathrm{L}]_t + k_{-1}) - [\mathrm{M}]_e(k_1[\mathrm{L}]_e + k_{-1})$$
$$= ([\mathrm{M}]_t - [\mathrm{M}]_e)(k_1[\mathrm{L}]_0 + k_{-1}), [\mathrm{L}]_t \approx [\mathrm{L}]_0 \approx [\mathrm{L}]_e$$
$$\text{or } -\frac{d[\mathrm{M}]_t}{([\mathrm{M}]_t - [\mathrm{M}]_e)} = (k_1[\mathrm{L}]_0 + k_{-1})dt$$

Integration of the above equation leads to:

$$\ln\frac{[\mathrm{M}]_t - [\mathrm{M}]_e}{[\mathrm{M}]_0 - [\mathrm{M}]_e} = k_1[\mathrm{L}]_0 + k_{-1}$$

Negative slope of the linear plot of ln ($[\mathrm{M}]_t - [\mathrm{M}]_e$) *vs.* t (time) gives the k_{obs} under the given pseudo-first order condition.

$$\text{slope} = k_{obs} = k_1[\mathrm{L}]_0 + k_{-1}$$

From the plot of k_{obs} *vs.* $[\mathrm{L}]_0$, the forward and backward rate constants can be calculated (Fig. 5.7.1). For this purpose, using different concentration of $[\mathrm{L}]_0$ under the condition, $[\mathrm{L}]_0 \gg [\mathrm{M}]_0$, k_{obs} values are to be determined.

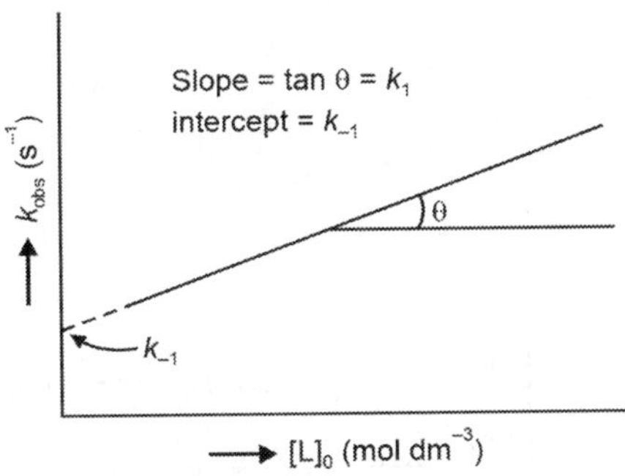

Fig. 5.7.1: Evaluation of forward and backward rate constants in a typical complexation reaction

5.8 ENERGY PROFILE DIAGRAMS FOR DIFFERENT STOICHIOMETRIC MECHANISMS

Energy profile (also known as **reaction profile)** diagram represents the plot, energy of the system *vs..* **reaction coordinate** that indicates the extent to which the reaction process has occurred (simply, the progress of the reaction). The energy profile diagram indicates the positions of reactants, transition states, intermediates and products.

A. Associative (A) process: Two step reaction with two transition states and one intermediate

$$L_5M—X \xrightarrow[(rds)]{+Y} L_5M\begin{matrix}X\\ \\Y\end{matrix} \xrightarrow[(\text{fast})]{-X} L_5M—Y$$

The entering ligand (*Y*) makes a new bond with the metal centre at the rate determining step (*rds*) to **increase the coordination number by unity.** In attaining this intermediate of higher coordination number, there is no bond breaking by the leaving group (*X*). After the rate determining step (rds), the leaving group (*X*) is lost from the intermediate at a faster step.

The reaction pathway involves the formation of a **single intermediate** ($Y—ML_5—X$) and **two transition states (T.S.)** — one for the formation of the intermediate and another for the decomposition of the intermediate to the product. The activation energy barrier for the decomposition of the intermediate is relatively lower.

B. Dissociative (*D*) process: Two step reaction with two transition states and one intermediate

$$L_5M—X \xrightarrow[(rds)]{-X} L_5M \xrightarrow[(fast)]{+Y} L_5M—Y$$

The leaving group (*X*) is lost at the rate determining step (*rds*) through the rupture of the *M—X* bond to generate an intermediate (L_5M) with **a lower coordination number** (*i.e.* coordination number decreases by unity for the unidentate leaving group). At this stage of generation of the intermediate, there is no interaction with the entering ligand (*Y*). After the rate determining step, the entering group enters into the coordination sphere at a faster step to give the final product.

As in the *A* process, in the *D*–process also, there is **a single intermediate** but of lower coordination number, and **two transition states** – one leading to the formation of the intermediate and the other leading to the product from the intermediate. The activation energy barrier leading to the formation of the intermediate is higher.

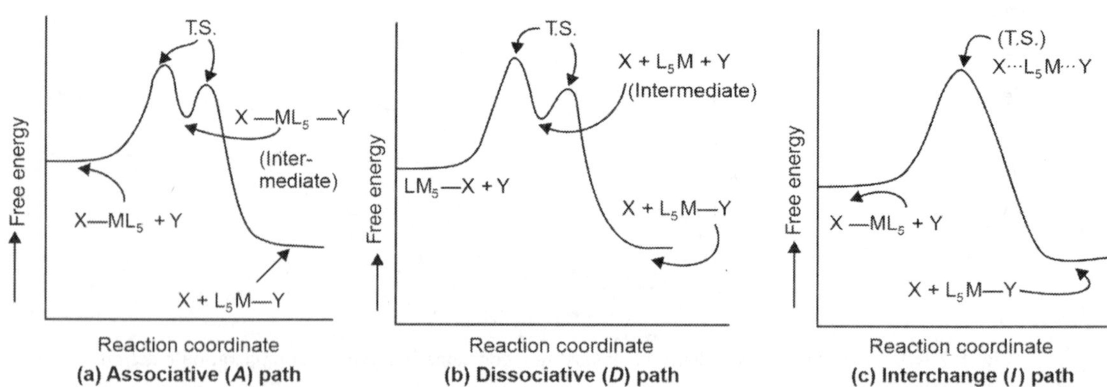

Fig. 5.8.1: Energy-profile (reaction profile) diagrams for the different mechanisms of ligand substitution reactions (see **Fig. 5.30.3.1** for different types of reaction profiles of square planar complexes)

Principle of microscopic reversibility: For a reversible process (R $\rightleftharpoons$ P, say), it states that the mechanisms of a reversible reaction in microscopic detail are the same in both directions under the same conditions. According to this principle, the energy profile diagram for the forward reaction (R→P) will be traversed from the right to left for the reverse or backward reaction (P→R). Thus the same energy profile diagram can describe the pathways of both forward and backward reactions.

C. Interchange (*I*) process: Single step reaction with a single transition state

$$L_5M—X \xrightarrow[(rds)]{+Y} L_5M \begin{smallmatrix} \cdot\cdot X \\ \\ \cdot\cdot Y \end{smallmatrix} \xrightarrow[\text{(fast)}]{-X} L_5M—Y \quad (\cdots \text{ denotes breaking or forming bond})$$

Both the leaving and entering groups are involved in **a single step** to generate the **activated complex or transition state (T.S.) (not the true intermediate because it is too short-lived to be detected).** In attaining this activated complex, both the bond breaking by the leaving group (*i.e.* **slight rate dependance on the nature of the leaving group**) and bond formation by the entering group (*i.e.* **slight rate dependance on the nature of the entering group**) are important. In other words, bond formation by the entering group and bond rupture by the leaving group go on simultaneously in a *concerted manner* and ultimately, the incoming group gets completely bound with the metal centre and the leaving group is completely dislodged. The reaction profile is characterised by a **single transition state** (T.S.).

Depending on the relative importance of the bond formation by the entering group and bond dissociation by the leaving group in the interchange process, two different situations may arise giving rise to two different types of transition states.

$$L_5M—X \xrightarrow{+Y} L_5M \begin{smallmatrix} \cdot\cdot X \\ \\ \diagdown Y \end{smallmatrix} \quad (I_a)$$

$$L_5M—X \xrightarrow{+Y} L_5M \begin{smallmatrix} / X \\ \\ \cdot\cdot Y \end{smallmatrix} \quad (I_d)$$

Bond denoted by ······· makes a less contribution in the activation process.

Thus in the $\boldsymbol{I_a}$ path, the new bond formation by the entering group is more important (*i.e.* rate is relatively more sensitive on the nature of the entering group) while in the $\boldsymbol{I_d}$ path, bond rupture by the leaving group is more important in the activation process.

5.9 RATE LAWS FOR THE DIFFERENT REACTION PATHWAYS (*i.e.* STOICHIOMETRIC MECHANISMS)

5.9.1 Dissociative Mechanism (*cf.* Sec. 5.16)

In a pure ***D*** process (*i.e.* limiting S_N1), an intermediate of lower coordination number is produced. *Life time of the intermediate is sufficiently high* so that it can be consumed in a way that is independent of its mode of formation. The elementary steps for a *D*-process (stoichiometric mechanism) are outlined below.

$$L_5M—X \underset{k_{-1}}{\overset{k_1}{\rightleftharpoons}} L_5M + X \qquad \text{...(5.9.1.1)}$$

$$L_5M + Y \xrightarrow{k_2} L_5M—Y \qquad \text{...(5.9.1.2)}$$

The rate law for this mechanism is obtained by considering the **steady-state concentration** of the intermediate L_5M (*i.e. its concentration is small at any moment and it remains virtually constant during the reaction*), *i.e.*

$$\frac{d}{dt}[L_5M] = 0 = k_1[L_5MX] - k_{-1}[L_5M][X] - k_2[L_5M][Y]$$

or

$$[L_5M] = \frac{k_1[L_5MX]}{k_{-1}[X] + k_2[Y]}$$

The rate of the overall reaction is given by:

$$-\frac{d}{dt}[L_5MX] = \frac{d}{dt}[L_5MY] = k_2[L_5M][Y]$$

$$= \frac{k_1k_2[L_5MX][Y]}{k_{-1}[X]+k_2[Y]} \qquad \ldots(5.9.1.3)$$

$$= \frac{k_1k_2[L_5MX][Y]/(k_{-1}[X])}{1+k_2[Y]/(k_{-1}[X])}$$

$$= k_{obs}[L_5MX], \text{ (when } [Y] \gg [L_5MX])$$

$$k_{\text{obs}} = \frac{k_1k_2[Y]/(k_{-1}[X])}{1+k_2[Y]/(k_{-1}[X])} \qquad \ldots(5.9.1.4)$$

- The rate equation indicates that k_{obs} (observed rate constant) shows a **complicated dependence pattern** on both **[*X*] and [*Y*]**. If *X* = solvent (*i.e.* **formation reaction**) then [*X*] = constant and the situation becomes simplified for analysis. Then we can write:

$$k_{\text{obs}} = \frac{k_1k_2[Y]/(k_{-1}[H_2O])}{1+k_2[Y]/(k_{-1}[H_2O])}; \; i.e. \; k_{obs} = \frac{m[Y]}{1+n[Y]} \qquad \ldots(5.9.1.5)$$

$$m = \frac{k_1k_2}{k_{-1}[\text{solvent}]} = \frac{k_1k_2}{k'_{-1}} = k = \text{constant}$$

$$n = \frac{k_2}{k_{-1}[\text{solvent}]} = \frac{k_2}{k'_{-1}} = \text{constant};$$ (*see* Sec. 5.16 to understand the **significance of this constant**)

i.e.

$$\frac{1}{k_{obs}} = \frac{1}{m[Y]} + \frac{n}{m} \qquad \ldots(5.9.1.6)$$

The ***Burk-Lineweaver type*** **double reciprocal plot** $\left(i.e. \; \frac{1}{k_{obs}} \; vs. \; \frac{1}{[Y]}\right)$, gives a straight line and from the slope and intercept of the plot, the constants *m* and *n* may be evaluated.

Thus for the ***D*-process of formation reaction** (*i.e. X* = solvent), in general, we can write:

$$k_{\text{obs}} = \frac{m[Y]}{1+n[Y]} \qquad (cf. \text{ Eqn. } 5.9.1.5)$$

or

$$\frac{1}{k_{obs}} = \frac{1}{m[Y]} + \frac{n}{m} \qquad (cf. \text{ Eqn. } 5.9.1.6)$$

The plots are shown in Fig. 5.9.1.1.

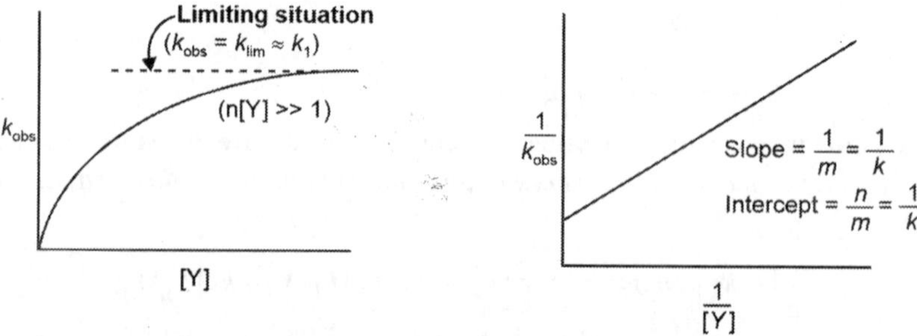

Fig. 5.9.1.1: Rate dependence on the concentration of the incoming nucleophile (*Y*) in a formation reaction occurring through the dissociative process under the condition, $[ML_5X] \ll [Y]$

- For the **formation reaction,** *i.e.* X = solvent,

$$\text{rate} = \frac{k_1 k_2 [ML_5\text{—}OH_2][Y]}{k_{-1}[H_2O] + k_2[Y]}, \text{ (say, solvent = water)}$$

At **high concentration of Y**, *i.e.* under the condition, $k_{-1}[H_2O] \ll k_2[Y]$, we have:

$$\text{rate} = k_1[ML_5\text{—}OH_2] = k_{obs}[ML_5\text{—}OH_2],$$

i.e. $k_{obs} = k_{lim} \approx k_1$ and the **rate is independent of [*Y*])**

But at **low concentration of [*Y*]**, *i.e.* under the condition, $k_{-1}[H_2O] \gg k_2[Y]$, we have:

$$\text{rate} = \frac{k_1 k_2 [ML_5\text{—}OH_2][Y]}{k_{-1}[H_2O]} = k_{obs}[ML_5\text{—}OH_2]$$

where $k_{obs} = k[Y], k = \dfrac{k_1 k_2}{k_{-1}[H_2O]}$ = constant

i.e. **first-order dependence on [*Y*]** (*cf.* Sec. 5.16).

Under the **limiting condition,** $k_{-1}[H_2O] \ll k_2[Y]$, *i.e.* at high concentration of *Y*,

$$\text{rate} = k_1[ML_5\text{—}OH_2]$$

i.e. no dependence on [*Y*].

5.9.2 Associative Mechanism

In the pure ***A***-process (*i.e.* limiting S_N2), an intermediate of higher coordination number is produced. *The life time of the intermediate is sufficiently high and it can be detected.* The elementary steps of this stoichiometric mechanism are outlined below.

$$L_5M\text{—}X + Y \underset{k_{-a}}{\overset{k_a}{\rightleftharpoons}} L_5M\begin{matrix} X \\ Y \end{matrix} \quad (i.e.\ L_5MXY) \quad \ldots(5.9.2.1)$$

$$L_5M\begin{matrix} X \\ Y \end{matrix} \xrightarrow{k_b} L_5M\text{—}Y + X \quad \ldots(5.9.2.2)$$

Applying the **steady state condition** to the intermediate, L_5MXY, we can write:

$$\frac{d[L_5MXY]}{dt} = 0 = k_a[L_5MX][Y] - k_{-a}[L_5MXY] - k_b[L_5MXY]$$

i.e.

$$[L_5MXY] = \frac{k_a[L_5MX][Y]}{k_{-a} + k_b}$$

Rate of the reaction is given by:

$$-\frac{d[ML_5X]}{dt} = \frac{d[ML_5Y]}{dt}$$

$$= k_b[L_5MXY]$$

$$= \frac{k_a k_b [ML_5X][Y]}{k_{-a} + k_b}, \quad \boxed{k_{obs} = \frac{k_a k_b [Y]}{k_{-a} + k_b}} \quad \ldots(5.9.2.3)$$

$$= k_{obs}[ML_5X], \text{ (when } [Y] \gg [ML_5X])$$

Note: This second order kinetics of A–reaction is observed provided the second step is irreversible, *i.e.* $k_{-b} = 0$. If this step is reversible, *i.e.* $k_b \neq 0$, then there will be a deviation from the second order kinetics.

5.9.3 Interchange Mechanism (Consistent with the Eigen-Wilkins Mechanism, Sec. 5.14)

In this pathway, *the intermediate is too short-lived to be detected.* Depending on the conditions, it may be I_a or I_d. This interchange mechanism operates within a *pre-formed outer-sphere (OS) complex.* The **outer-sphere (OS) complex** or **encounter complex** is produced in a rapid pre-equilibrium step (**Eigen-Wilkins mechanism** *cf.* Sec. 5.14). The process can be outlined as follows:

$$L_5M{-}X + Y \xrightleftharpoons{K_{OS}} \underset{\text{(outer-sphere complex)}}{(L_5MX, Y)} \quad \ldots(5.9.3.1)$$

$$\underset{\text{O.S. Complex}}{(L_5MX, Y)} \xrightarrow{k_i} L_5M{-}Y + X \quad \ldots(5.9.3.2)$$

The rate law can be expressed in terms of the initial concentration of the starting complex, *i.e.* $[L_5MX]_0$ and the initial concentration of Y, *i.e.* $[Y]_0$. In presence of Y, the reacting complex is also partly distributed in the outer sphere complex. Thus we can write:

$$[L_5MX]_0 = [L_5MX] + [L_5MX, Y]$$

$$= [L_5MX] + [O.S.]$$

and

$$[Y]_0 = [Y] + [O.S.] \approx [Y], \text{ (when } [Y]_0 \gg [L_5MX]_0)$$

$$K_{OS} = \frac{[O.S.]}{[L_5MX][Y]} = \frac{[O.S.]}{\{[L_5MX]_0 - [O.S.]\}[Y]_0}$$

or, $K_{OS}[L_5MX]_0[Y]_0 - K_{OS}[O.S.][Y]_0 = [O.S.]$

or,

$$[O.S.] = \frac{K_{OS}[L_5MX]_0[Y]_0}{1 + K_{OS}[Y]_0} \quad \ldots(5.9.3.3)$$

The rate of the reaction is given by:

$$\frac{d[L_5MY]}{dt} = k_i[O.S.] = \frac{k_i K_{OS}[L_5MX]_0[Y]_0}{1 + K_{OS}[Y]_0} \quad \ldots(5.9.3.4)$$

$$= k_{obs}[L_5MX]_0, \text{ (when } [Y]_0 \gg [L_5MX]_0)$$

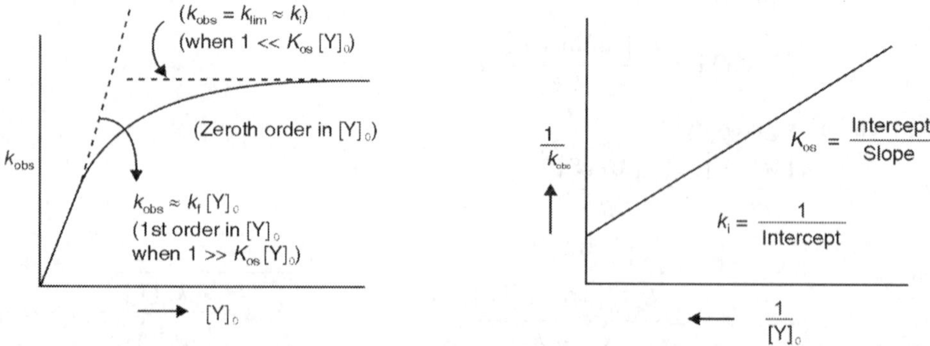

Fig. 5.9.3.1: Rate dependence on $[Y]_0$ in an interchange (I) process and evaluation of k_i and K_{OS} (**double reciprocal plot,** general procedure)

Distinction between the I_a and I_d mechanisms (*cf.* Secs. 5.14, 5.19.9)

- Rate law: $k_{obs} \approx k_i K_{OS}[Y]_0$ when $1 \gg K_{OS}[Y]_0$, *i.e.* $k_{obs} = k_f[Y]_0$; $k_i = \dfrac{k_f}{K_{OS}}$

  Here it may be noted that k_f is not a pure rate constant and it is a **composite rate constant** ($k_f = k_i K_{os}$) which is a product of an equilibrium constant and a rate constant. K_{OS} can be theoretically calculated (*cf.* Sec. 5.14). **Same form** of the rate law for both the I_d and I_a mechanisms arises.
- The **distinctions between the I_a and I_d mechanisms** are determined by the extent to which the interchange rate constant (k_i) depends upon the nature of the entering group (Y). The greater dependence on the nature of Y indicates the $\boldsymbol{I_a}$ mechanism while less sensitivity on the nature of Y indicates the $\boldsymbol{I_d}$ mechanism. In the I_d mechanism, $k_{lim} \approx k_i \approx k_{ex}$ **(water exchange rate constant)**

  $\left[Co(NH_3)_5(OH_2)\right]^{3+} \xrightarrow{\text{anation}} k_i$, almost independent of the nature of Y, *i.e.* $\boldsymbol{I_d}$ mechanism

  $\left[Rh(NH_3)_5(OH_2)\right]^{3+} \xrightarrow{\text{anation}} k_i$, sensitive to the nature of Y, *i.e.* $\boldsymbol{I_a}$ mechanism.

Distinction among the D, I_d and I_a mechanisms

This aspect has been discussed in Sec. 5.16.

$$k_{obs} = \frac{k_i K_{OS}[Y]_0}{1 + K_{OS}[Y]_0}$$

or

$$\frac{1}{k_{obs}} = \left(\frac{1}{k_i K_{OS}}\right)\frac{1}{[Y]_0} + \frac{1}{k_i}$$

Plot of $\dfrac{1}{k_{obs}}$ *vs.* $\dfrac{1}{[Y]_0}$ gives a straight line with slope $= \dfrac{1}{k_i K_{OS}}$ and intercept $= \dfrac{1}{k_i}$, *i.e.* $K_{OS} = \dfrac{\text{Intercept}}{\text{Slope}}$

$$k_{obs} = \frac{k_i K_{OS}[Y]_0}{1 + K_{OS}[Y]_0}, \text{ i.e. } k_{obs} = k_{lim} \approx k_i \text{ when } K_{OS}[Y]_0 \gg 1$$

Very often, $1 \gg K_{OS}[Y]_0$, it leads to: $\boldsymbol{k_{obs} \approx k_i K_{OS}[Y]_0 = k_f[Y]_0}$

Limiting rate laws for the *I*-process and rate dependence on *Y*: Depending on the relative values of 1 and $K_{OS}[Y]_0$, dfferent situations may arise.

- **Case-I ($1 \approx K_{OS}[Y]_0$):** $k_{obs} = \dfrac{k_i K_{OS}[Y]_0}{1 + K_{OS}[Y]_0}$, *i.e.* **nonlinear dependence** or **fractional order** in $[Y]_0$.
- **Case-II ($1 \gg K_{OS}[Y]_0$):** $k_{obs} \approx k_i K_{OS}[Y]_0$ *i.e.* **first order** in $[Y]_0$. k_{obs} depends on both the **concentration** of Y and its **nature** which determines both k_i and K_{OS}; but dependence of k_i on the nature of Y is relatively less sensitive for the I_d-path compared to that of the I_a-path.
- **Case-III ($K_{OS}[Y]_0 \gg 1$):** $k_{obs} = k_{lim} \approx k_i$, *i.e.* **zeroth order** in $[Y]_0$ but dependence of k_{obs} on the **nature** of Y arises for the interchange process; k_i is more sensitive on the nature of Y for the I_a-path.

5.9.4 General Rate Law for the *A*, *D* and *I* Processes in the Formation Reaction (*i.e.* *X* = solvent = *S*, say)

Under the condition, $[Y] \gg [L_5MX]$ (*i.e.* $[Y] \approx [Y]_0$), we can write the general rate law for the D and I processes as follows:

$$k_{obs} = \frac{pq[Y]}{1 + q[Y]}, \quad (p, q \text{ are constants}), \qquad \ldots(5.9.4.1)$$

$$\text{Rate} = k_{obs}[Y]$$

$$= \frac{pq[ML_5X][Y]}{1+q[Y]} \quad \text{(cf. } \textbf{D and I processes}\text{)} \quad \ldots(5.9.4.2)$$

A–process: $k_{obs} = \frac{k_a k_b [Y]}{k_{-a} + k_b}$, **D–process:** $k_{obs} = \frac{k_1 k_2 [Y]/(k_{-1}[S])}{1 + k_2 [Y]/(k_{-1}[S])}$,

I–process: $k_{obs} = \frac{k_i K_{OS}[Y]}{1 + K_{OS}[Y]}$, (*cf.* Sec. 5.14, **Eigen-Wilkins Mechanism**)

Thus for D and I (I_a and I_d) processes, the **rate expression is of similar type** and these processes are mainly observed for the octahedral complexes. **Pure *A*–process is rarely found in the octahedral complexes.**

- Under the condition, $q[Y] \gg 1$, Rate = $p[ML_5X]$
  i.e. first-order reaction for all types of mechanisms
- Under the condition, $1 \gg q[Y]$, Rate = $pq[ML_5X][Y]$
  i.e. second-order reaction for all types of mechanisms
- **It indicates that molecularity of the reaction is not necessarily indicated from the rate law.** For example, ***d***–intimate mechanism does not necessarily lead to the first order kinetics. Similarly, ***a***–intimate mechanism does not necessarily lead to the second-order kinetics.

Table 5.9.4.1: General rate law (*i.e.* $k_{obs} = pq[Y]/(1 + q[Y])$ for the D, I_a and I_d processes of the formation reactions through the displacement of solvent (S) (*cf.* Sec 5.16)

Parameters (p, q)	I_a	I_d	D
• $p =$	k_i	k_i	k_1
• $q =$	K_{OS}	K_{OS}	$k_2/(k_{-1}[S])$
• Dependence of p on Y	Strongly dependent	Very slightly dependent	Independent
• Dependence of p on k_{ex} (solvent exchange rate constant)	No relationship	$k_{ex} \approx p$	$k_{ex} = p$

Note: For **A–process:** $p = \left(\frac{k_a k_b}{k_{-a} + k_b}\right)\frac{1}{K_{OS}}$, $q = K_{OS}$; $1 + K_{OS}[Y] \approx 1$; $(1 >> K_{OS}[Y])$, p depends very strongly on the nature of Y; but p does not bear any relationship with k_{ex} and p may be greater than k_{ex} and even slower than k_{ex}).

5.9.5 Implications from the Rate Laws for the *D* and I_d Processes for the Solvent (S) Replacement Reactions

- For the **I_d–process,** $k_{obs} \approx k_i K_{OS}[Y] = k_f[Y]$, (if $1 \gg K_{OS}[Y]$)

It is evident that k_f is not a pure rate constant and it is a **composite rate constant** (a product of an equilibrium constant and a rate constant, $k_f = k_i K_{OS}$). For the different entering ligands, the formation rate constant (k_f) varies for the different values of K_{OS} (*outer sphere association constant*). But, k_i ($= \frac{k_f}{K_{OS}}$, K_{OS} can be calculated theoretically) remains the same for the different types of entering ligands. Thus the calculated k_i values for the different types of ligands reacting with a particular complex remain constant and become very close to k_{ex} (solvent exchange rate constant). **This is the test for the I_d mechanism and this type of mechanism is described very often as the Eigen mechanism.**

$[Ni(OH_2)_6]^{2+}$ reacting with the different types of ligands maintains a constant value of k_i (= k_f/K_{OS}) which is close to 10^4 s^{-1} at 25°C and this is comparable to k_{ex} (water exchange rate constant) of $[Ni(OH_2)_6]^{2+}$. Here it may be mentioned that though k_i is more or less constant, k_f may vary in a wide range ($10^2 - 10^5$ dm^3 mol^{-1} s^{-1}; *cf.* Table 5.14.1).

● For the ***D*–process** (*cf.* Sec. 5.9.1) of formation reaction (*i.e.* leaving group X = solvent), we have:

$$k_{\text{obs}} = \frac{k_1 k_2 [Y]}{k_{-1}[S] + k_2 [Y]} \approx \frac{k_1 k_2 [Y]}{k_{-1}[S]}; \text{ (under the condition, } k_{-1}[S] \gg k_2[Y])$$

i.e.

$$= k_f[Y] \text{ and } k_f = \frac{k_1 k_2}{k_{-1}[S]}$$

Here $k_1/(k_{-1}[S])$ is constant. **Thus the dependence of k_f on $[Y]$ arises from the ability of the lower coordinate intermediate (*i.e.* ML_5) to discriminate the competing nucleophiles, *i.e.* solvent and the entering nucleophile (*i.e.* Y).** Thus in a perfect *D*–process, the dependence of k_f on $[Y]$ arises from the sensitivity of the lower coordinate intermedite towards Y in the k_2 – step. The dislodged leaving group (X = solvent for the formation reaction) and the entering group (Y) compete for the lower coordinate intermediate and this is why, k_{obs} depends on the ratio k_2/k_{-1}.

Limiting forms of the rate laws in *d*–intimate mechanism (*cf.* Sec. 5.16)

Stoichiometric	Condition	Rate law	k_{obs}, for $[Y] \gg [ML_5X]$	Inference
D	$k_2[Y] \gg k_{-1}[X]$	$k_1[ML_5X]$	k_1	k_{obs} depends on the leaving group, but is independent of Y.
D	$k_2[Y] \ll k_{-1}[X]$ (*e.g.* X = solvent)	$\frac{k_1 k_2 [ML_5X][Y]}{k_{-1}[X]}$	$\frac{k_1 k_2 [Y]}{k_{-1}[X]}$	k_{obs} depends on the leaving group; dependence on Y due to the sensitivity of the k_2–step.
I_d	$K_{OS}[Y] \gg 1$	$k_i[ML_5X]$	k_i	k_{obs} depends on the interchange rate constant (k_i)
I_d	$K_{OS}[Y] \ll 1$	$k_iK_{OS}[ML_5X][Y]$	$k_iK_{OS}[Y]$	k_{obs} depends on both k_i and Y

5.9.6 Implications from the Rate Laws for the A and I_a Processes for the Solvent (S) Replacement Reactions (*cf.* Sec. 5.16)

In the ***A*-path (2 step process),** the **intermediate** of higher coordination number is **relatively long-lived** to be characterised but in the I_a–path (**single step process**), the intermediate is **too short-lived** to be characterised.

For the ***I_a*–path,** the rate law gives, $k_{obs} = k_iK_{OS}[Y]$ under the condition $1 \gg K_{OS}[Y]$. Thus it follows the usual second-order kinetics as in the ***A*–process** for which the k_{obs} is given by $k_ak_b[Y]/(k_{-a} + k_b)$. Thus under these conditions, it is a difficult task to discriminate between the A and I_a paths. In fact, in the anation and formation reactions of $[Pd(OH_2)_4]^{2+}$ and $[Pt(OH_2)_4]^{2+}$, the second order kinetics is maintained to complicate the distinction between the A and I_a paths. Under the conditions, $K_{OS}[Y] \gg 1$, for the I_a–mechanism, k_{obs} is given by k_i which is the *ligand interchange rate constant.* In this limiting condition, the I_a path shows the **first order kinetics** (*i.e.* rate = k_i[complex]) but the A-path **always shows the second order kinetics** (*i.e.* always first order with respect to the entering ligand).

Limiting forms of rate laws for the *a*–intimate mechanisms

Stoichiometric mechanism	*Conditions*	*Rate law*	k_{obs} *for [Y] ≫ [ML₅X]*	*Inference*
A	[*Y*] is very large or small	$\frac{k_a k_b [Y][ML_5X]}{(k_{-a}+k_b)}$	$\frac{k_a k_b [Y]}{(k_{-a}+k_b)}$	Always second-order kinetics
I_a	$K_{OS}[Y] \gg 1$	$k_i\,[ML_5X]$	k_i	k_{obs} is the ligand interchange rate constant; 1st order kinetics
I_a	$K_{OS}[Y] \ll 1$	$k_iK_{OS}[ML_5X][Y]$	$k_iK_{OS}[Y]$	k_{obs} is a composite one; 2nd order kinetics

In **both the I_a– and *A*–paths,** the rate is very much sensitive to the nature of the entering ligand. *The rate is faster than the solvent exchange rate if the entering ligand is a better nucleophile than the solvent. On the other hand, the rate is slower than the solvent exchange rate, if the entering ligand is a poorer nucleophile than the solvent.*

The ***A*– and I_a–paths** are well established in the ligand substitution processes of square planar complexes of Ni(II), Pd(II), Pt(II), Au(III), etc. *The pure A–path in the octahedral complexes is still called in question but the I_a–path is known to operate in many octahedral complexes.*

A vs. I_a Path: Stereochemistry of the Products (*cf.* Sec. 5.30.4)

- **I_a–path (single step process,** *cf.* Fig. 5.8.1) produces the higher coordinate intermediate whose **life time is too short** (*i.e.* transient species) while **A–path (two step process,** *cf.* Fig. 5.8.1) produces an intermediate (higher coordination number) which is **relatively long-lived** and it may experience a rearrangement.
- ***sp*³–C** can adopt the **I_a–path** (but not the *A*-path; *cf.* Ch. 10, Vol. 2) and the **TBP intermediate** (a 3*c*–4*e* bond along the axial directions involving the entering and leaving groups) **is too short-lived to experience any rearrangement** and it gives the product with **Walden inversion for** the optically active compound. On the other hand, ***sp*³–Si** can adopt the **A-path** and it can stabilise the TBP intermediate by using its *d*-orbital (*cf.* sp^3d hybridisation for TBP). **The TBP intermediate is sufficiently long-lived to experience the Berry pseudorotation** to produce two different TBPs remaining in an equilibrium and from these two TBPs, a racemic product is obtained.

$$\left.\begin{array}{l}(P)(Q)(R)C{-}X + Y^- \xrightarrow{I_a\text{-path}} (P)(Q)(R)C{-}Y + X^- \\ \qquad\qquad\qquad\qquad\qquad\qquad\quad \textbf{(Inversion)} \\ (P)(Q)(R)Si{-}X + Y^- \xrightarrow{A\text{-path}} (P)(Q)(R)Si{-}Y + X^- \\ \quad\textbf{(Chiral compound)} \qquad\qquad\qquad \textbf{(Racemic product)}\end{array}\right\} (cf.\ \text{Scheme } 5.30.4.3)$$

- **Retention of stereochemistry** during the ligand substitution in **Pt(II) square planar complexes** can be explained in terms of a special type I_a-mechanism not by the *A*-mechanism (*cf.* Scheme 5.30.4.2).

5.10 REPRESENTATIVE EXAMPLES OF SUBSTITUTION REACTIONS PASSING THROUGH DIFFERENT TYPES OF MECHANISMS

(A) *d*–reactions, *i.e.* *D* and I_d mechanisms: True *D*-reactions are rarely known. In fact, if any convincing evidence lacks in favour of the existence of an intermediate of lower coordination number, then it is better to describe the *d*–reactions as the I_d-reactions instead of the *D*–reactions.

- ***D* mechanism** (supported by high $\Delta V^{\neq}$ value which is not complicated by the charge effect) has been argued for the **anation** or **aquation** of some octahedral complexes having the large **negative**

charges and *trans*-labilising ligands. Anionic nature of the complex disfavours the approach of the entering **nucleophile in *a*-activation** and also disfavours the formation of an outer-sphere complex — a prerequistic condition for the I_d–path. Such examples are:

anation of $[Co(CN)_5(OH_2)]^{2-}$; aquation of $[Fe(CN)_5L]^{3-}$ ($\Delta V^{\neq} \approx +22$ cm^3 mol^{-1}); aquation of $[Cr(NCS)_6]^{3-}$ ($\Delta V^{\neq} = +16$ cm^3 mol^{-1}); aquation of $[Co(CN)_5X]^{3-}$. Anation of $[Co^{III}(corrin)(OH_2)]$ (relevant to vitamin B_{12}, **flexible corrin ring** sterically favours to dissociate the leaving group, see Sec. 5.16) also satisfies the requirements (*cf.* Sec. 5.16) of a *D*–process.

● $\boldsymbol{I_d}$ **mechanism** operates for the substitution reactions at the aqua metal ions $[M(OH_2)_6]^{2+}$ (M = Fe, Co, Ni; *i.e.* bivalent metal ions of the late members of 1st transition series) and at $[M(OH)(OH_2)_5]^{2+}$ (M = Fe, Cr). Anation of $[Co(NH_3)_5(OH_2)]^{3+}$ and hydrolysis of $[Co^{III}(NH_3)_5(X)]^{2+}$ also follow the I_d–path (see Sec. 5.19.6).

(B) *a*-reactions, *i.e.* *A* and $\boldsymbol{I_a}$ mechanisms: It has been already mentioned that when $K_{OS}[Y] \ll 1$, it becomes difficult to distinguish between the *A* and I_a mechanisms because in both the situations, the usual second-order kinetics is maintained. In absence of any strong evidence in favour of formation of an intermediate of higher coordination number, it is better to describe the *a*-reaction as the I_a–reaction rather than the *A*-reaction.

● In the $\boldsymbol{I_a}$**–process**, the life time of the higher coordinate intermediate is **too short** while in the ***A*-process**, the life time of the higher coordinate intermediate is **relatively higher**. The TBP intermediate generated from the ***sp*^{3}–C** is **too short-lived to experience any rearrangment** and it causes the **Walden inversion** (*i.e.* I_a path) in the product while in the case of ***sp*^{3}–Si compound**, the TBP generated is stable (*cf.* role of the *d*-orbital of Si) and it can experience the **Berry pseudorotation** (*i.e.* A-path) and it gives the **racemised product** (*cf.* Sec. 5.30.4).

● Complexation (including anation) at $\mathbf{[Pd(OH_2)_4]^{2+}}$ **and** $\mathbf{Pt(OH_2)_4]^{2+}}$ has been argued to pass through the ***A*–mechanism**. In fact, in such cases, the second order kinetics is maintained for an appreciable range of concentration of *Y* and even when the entering ligand, *i.e.* *Y*, is anionic. Anionic nature of *Y* makes the high value of K_{OS} and at high concentration of *Y*, the condition $K_{OS}[Y] \ll 1$ is disfavoured. Thus if the process is argued to pass through the I_a–path then for the high concentration of the anionic entering ligand, it is reasonable to expect the condition $K_{OS}[Y] \gg 1$ leading to a first-order kinetics. In reality, it does not happen so. Thus it is reasonable to conclude that complexation at $[Pd(OH_2)_4]^{2+}$ and $[Pt(OH_2)_4]^{2+}$ occurs through the ***A*–mechanism**. It is also supported by negative values of $\Delta V^{\neq}$ for complexation with neutral ligands (where no complications occurs in the interpretation of $\Delta V^{\neq}$).

● Strong dependence on the nature of entering ligand for substitution at $\mathbf{[Ti(OH_2)_6]^{3+}}$ indicates the ***A*–mechanism**. The variation of rate constant covers a wide range (*ca.* $10^2 - 10^6$ dm^3 mol^{-1} s^{-1}) for the different types of entering ligands.

● Highly negative value of $\Delta V^{\neq}$ for aquation of $\mathbf{[RuCl(NH_3)_5]^{2+}}$ and anation of $\mathbf{[Ru(NH_3)_5(OH_2)]^{3+}}$ indicate the ***A*–mechanism**.

Intimate *a*- and *d*-mechanisms for the octahedral complexes at different metal centres

***a*:** Ti(III) (d^1), V(II) (d^3), V(III) (d^2), Mo(III) (d^3), Cd(II) (d^{10}), In(III) (d^{10}, post-transition element), Ir(III) (d^6), etc.

Both ***a*** and ***d*:** Cr(III) (d^3, predominantly ***a***), Mn(II) (d^5), Ru(II) (d^6), Ru(III) (d^5), Rh(III) (d^6).

***d*:** Mg(II) (d^0), Al(III) (d^0), Fe(II) (d^6), Co(III) (d^6), Ni(II) (d^8), Zn(II) (d^{10}), Ga(III) (d^{10}).

● Trivalent metal centres tend to adopt the ***a***–process:

● Heavier congeners in a group tend to adopt the ***a***–process;

$$\xrightarrow[\text{increasing associative character}]{\text{Al(III)} < \text{Ga(III)} < \text{In(III)}} \; ; \; \xrightarrow[\text{increasing associative character}]{\text{Co(III)} < \text{Rh(III)} < \text{Ir(III)}}$$

(C) I_a and I_d mechanisms: Depending on the sensitivity of k_i on the nature of the incoming nucleophile, the I_a and I_d mechanisms can be distinguished.

● Anation of $[Co(NH_3)_5(OH_2)]^{3+}$ occurs through the I_d path (*i.e.* k_i is almost independent of the nature of the entering group) while anation of $[Rh(NH_3)_5(OH_2)]^{3+}$ follows the I_a path (*i.e.* k_i *is sensitive on the nature of the incoming group).*

● Anation of $[Cr(OH_2)_6]^{3+}$ occurs at different rates for the different entering anions and this difference cannot be explained in terms of the difference in K_{OS}. It indicates the **associative character** in the reaction pathway. This associative character is also supported by the **LFER plot** (linear free energy relationship plot, *i.e.* ΔG^0 *vs.* $\Delta G^{\neq}$ or log K_{eq} *vs.* log k) for which the slope is less than unity. This aspect will be discussed later. Interestingly, $[Cr(OH)(OH_2)_5]^{2+}$ reacts through the I_d–path. Here the π–donor OH group strongly labilises the ligand H_2O bound at the *cis*–**position** (*i.e.* ***cis–effect***, Sec. 5.21) to impart the dissociative character. The **good σ-donor property** of OH also labilises the coordinated water mole-cules to impart the dissociative activation. Besides these, the **overall less positive charge** (+2 *vs* +3) in $[Cr(OH)(OH_2)_5]^{2+}$ favours the dissociative path. ***In fact, both the σ- and π-donor properties of the nonleaving ligands reduce the positive charge on the metal centre to favour the bond breaking and to stabilise the 5-coordinate TS/intermediate to favour the d-activation*** (see Secs 5.19.6, 5.21).

● In the same way, $[Fe(OH_2)_6]^{3+}$ reacts through the I_a path while its conjugate base, $[Fe(OH(OH_2)_5]^{2+}$ reacts through the I_d path.

● $[Cr(OH_2)_6]^{3+}$ reacts through the I_a path while $[Cr(NH_3)_5(OH_2)]^{3+}$ reacts through the I_d path. It happens so due to the better σ–donor property of NH_3 than that of H_2O (see Sec. 5.19.6).

● $[M(OH_2)_6]^{2+}$ (M = late members of 1st transition series) like $[Fe(OH_2)_6]^{2+}$, $[Ni(OH_2)_6]^{2+}$, etc. follow the I_d path in the ligand substitution process while the early members of the series, *e.g.* $[V(OH_2)_6]^{2+}$, $[Mn(OH_2)_6]^{2+}$, etc. follow the I_a path. This aspect will be discussed later.

5.11 RATE LAWS: LIMITATIONS OF THE RATE LAWS AND MECHANISMS OF LIGAND SUBSTITUTION REACTION

The rate law derived from the proposed reaction mechanism does not predict anything conclusively regarding the molecularity of the reaction. **Different reaction mechanisms may lead to the same rate law.** The rate law gives only the order of a reaction. The rate determining step may be unimolecular but the overall rate expression may be a second order one. These are the limitations of the rate law. Some of these aspects have been already discussed. This aspect is further illustrated in the following cases.

(i) **1st order kinetics and participation of solvent:** It can be shown that for two different reaction mechanisms, we obtain the same rate law.

Scheme I

$$\left.\begin{aligned} L_5M{-}X &\xrightarrow[(rds)]{\text{slow }(k_1)} L_5M + X \\ L_5M + Y &\xrightarrow{\text{fast}} L_5M{-}Y \end{aligned}\right\} \; \boldsymbol{D \equiv S_N 1\ \text{lim.}}$$

Here, the rate determining step (*rds*) is a **unimolecular** one and the rate law is given by:

$$\text{rate} = \frac{[L_5MY]}{dt} = k_1[L_5M{-}X] \quad \ldots(5.11.1)$$

Scheme II

$$L_5M{-}X + H_2O \xrightarrow[(rds)]{\text{slow }(k_1)} L_5M{-}OH_2 + X$$

$$L_5M{-}OH_2 + Y \xrightarrow{\text{fast}} L_5M{-}Y + H_2O$$

Here the rate determining step is a **bimolecular** one and the rate law is:

$$\text{rate} = k_1[L_5MX][H_2O] = k_1'[L_5MX] \qquad \ldots(5.11.2)$$

where $k_1' = k_1[H_2O]$ = constant, $[H_2O] \approx 55.5$ mol dm^{-3}

Thus it becomes a **pseudo first-order reaction.**

It is evident that the two different reaction mechanisms (*i.e.* **unimolecular *rds*** in Scheme I, **bimolecular *rds*** in Scheme II) are giving the same rate law. To distinguish between these possibilities, we need other observations.

(ii) **Second order kinetics for different reaction schemes:** Several reaction schemes may be proposed to explain the second order kinetics in the ligand substitution process.

Scheme I

$$\left.\begin{array}{l} L_5M\text{—}X + Y \xrightarrow[(rds)]{\text{slow }(k_1)} L_5MXY \\ L_5MXY \xrightarrow{\text{fast}} L_5M\text{—}Y + X \end{array}\right\} \mathbf{A \equiv S_N2\ lim.}$$

The above scheme leads to the following rate law:

$$\text{rate} = k_1[L_5M\text{—}X][Y] \qquad \ldots(5.11.3)$$

This *A*–mechanism leads to the second order kinetics.

Scheme II

It involves the formation of an *outer sphere* (*O.S.*) *complex* **(remaining at steady state)** that participates at the rds.

$$\left.\begin{array}{l} L_5M\text{—}X + Y \underset{k_{-1}}{\overset{k_1}{\rightleftharpoons}} (L_5M\text{—}X)\cdot Y \ (\text{O.S. complex}) \\ (L_5M\text{—}X)\cdot Y \xrightarrow[(rds)]{k_2} L_5M\text{—}Y + X \end{array}\right\} \textbf{Formation of an O.S. complex in a steady state followed by } \textbf{\textit{rds}}.$$

Under the steady state condition of $(L_5M\text{—}X)\cdot Y$ (O.S. complex), we have the rate law (*cf.* Eqn. 5.9.2.3):

$$\text{rate} = \frac{k_1k_2[L_5MX][Y]}{k_{-1} + k_2} = k[L_5MX][Y] \qquad \ldots(5.11.4)$$

where $k = k_1k_2/(k_{-1} + k_2)$

Even if the **rds is a unimolecular one,** but the rate process follows *a second order kinetics.* A similar situation arises when **there is a pre-equilibrium step followed by a *rds*** (*cf.* Fig. 5.11.1).

Scheme III

$$\left.\begin{array}{l} L_5M\text{—}X + Y \underset{k_{-1}}{\overset{k_1}{\rightleftharpoons}} L_5M\begin{smallmatrix}X\\Y\end{smallmatrix} \\ L_5M\begin{smallmatrix}X\\Y\end{smallmatrix} \xrightarrow[(rds)]{k_2} L_5M\text{—}Y + X \end{array}\right\} \textbf{A-process}$$

Under the steady state condition of L_5MXY, an intermediate of higher coordination number, the following rate law is obtained (*cf.* Sec. 5.9.2)

$$\text{rate} = \frac{k_1k_2[L_5MX][Y]}{k_{-1} + k_2} = k[L_5MX][Y] \qquad \ldots(5.11.5)$$

where $k = k_1k_2/(k_{-1} + k_2)$

This second order kinetics is valid if the second step is irreversible (*i.e.* $k_{-2} = 0$). If it is not irreversible (*i.e.* $k_{-2} \neq 0$), then a deviation from the second order kinetics occurs.

Scheme IV

Let us consider the following formation reaction.

$$L_5M(OH_2) + Y \rightarrow L_5M—Y + H_2O$$

The above reaction may be considered to pass through the following mechanism where the rds is a unimolecular one.

$$\left.\begin{array}{l} L_5M—OH_2 \underset{k_{-1}}{\overset{k_1}{\rightleftharpoons}} L_5M + H_2O \\ L_5M + Y \xrightarrow{k_2} L_5M—Y \end{array}\right\} \textbf{\textit{D}-process}$$

Under the steady state condition of the intermediate L_5M, we have the rate law (*cf.* Sec. 5.9.1)

$$\text{rate} = \frac{k_1k_2[L_5M—OH_2][Y]}{k_{-1}[H_2O]+k_2[Y]} = \frac{k_1k_2[L_5M—OH_2][Y]}{k'_{-1}+k_2[Y]}, \quad ...(5.11.6)$$

$(k'_{-1} = k_{-1}[H_2O] = \text{constant})$

If the concentration of Y is small, *i.e.* $\boldsymbol{k_{-1}[H_2O] \gg k_2[Y]}$, we have:

$$\text{rate} \approx k[L_5M—OH_2][Y]$$

where

$$k = \frac{k_1k_2}{k'_{-1}}$$

It gives the second-order kinetics.

- At high concentration of Y, *i.e.* $\boldsymbol{k_2[Y] \gg k_{-1}[H_2O]}$, we have:

$$\text{rate} \approx k_1[L_5M—OH_2]$$

It gives the first-order kinetics.

From the above discussed Schemes, it is evident that the ***observed second-order kinetics does not necessarily suggest the bimolecularity of the process.***

(iii) **Involvement of a rapid pre-equilibrium step:** The pre-eqilibrium step may involve the formation of an **outer-sphere complex** or an **acid-base reaction.** The outer-sphere complex or the conjugate acid/base formed in the pre-equilibrium step passes through the rate determining step (*rds*) to give the product. The rds may involve a dissociative (**D**), an associative (**A**) or an interchange (**I**) process, but ***in all cases, the rate expression will incorporate the concentration term of the entering ligand (Y).*** These are illustrated below.

(a) **Formation of an outer-sphere (O.S.) complex at a rapid pre-equilibrium step:** The reactants may interact through electrostatically (*e.g.* ion-ion, ion-dipole) or hydrogen bonding to form the outer-sphere complex (**Eigen Mechanism,** Secs. 5.9.3 and 5.14).

$$L_5MX + Y \overset{K_{OS}}{\rightleftharpoons} \underset{\text{(O.S. Complex)}}{L_5MX \bullet Y} \xrightarrow[\text{(slow)}]{k} L_5M—Y + X$$

It leads to:

$$K_{OS} = \frac{[O.S.]}{[L_5MX][Y]} \text{ and } [O.S.] = \frac{K_{OS}T_C[Y]}{1+K_{OS}[Y]},$$

(T_C = total concentration of the starting complex

$= [L_5MX] + [L_5MX{\cdot}Y] = [L_5MX] + [O.S.]$)

$$\text{rate} = k[O.S.] = \frac{kK_{OS}T_C[Y]}{1+K_{OS}[Y]} \quad ...(5.11.7), (cf. \text{ Eqn. } 5.9.3.4)$$

It leads to: $$k_{obs} = \frac{kK_{OS}[Y]}{1+K_{OS}[Y]}, \text{ when } [Y] \gg T_C$$

- Under the condition, $1 \gg K_{OS}[Y]$, rate $\approx kK_{OS}T_C[Y]$
  i.e. second order kinetics (*i.e.* first order dependence on both the complex and entering ligand) though the *rds* is a unimolecular one.
  This is attained at very low concentration of *Y* or for very low values of K_{OS}.
- Under the condition, $K_{OS}[Y] \gg 1$, rate $\approx kT_C$
  i.e. first-order in the starting complex and zeroth order in the entering ligand.
  This condition is attained at very high concentration of *Y* or for very high values of K_{OS}.

Thus at very low concentration of *Y*, it shows a first-order dependence on [*Y*] while at very high concentration of *Y*, it shows a zeroth-order dependence on [*Y*]. This is illustrated in Fig. 5.11.1.

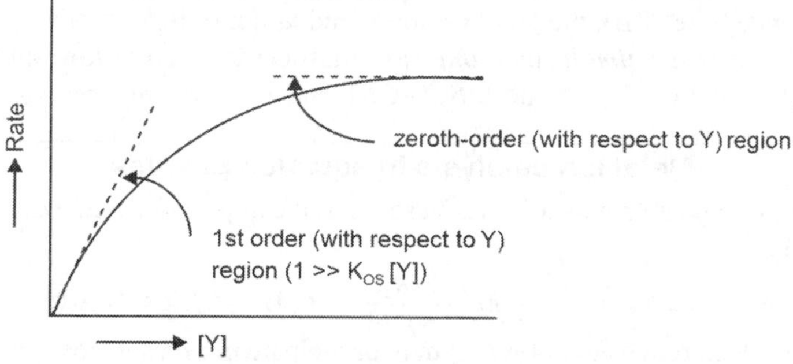

Fig. 5.11.1: Rate dependence pattern on [*Y*]

Outer-sphere complex in a steady state

If the outer-sphere complex remains in a *steady state* then the above reaction scheme is represented as follows:

$$L_5M{-}X + Y \underset{k_{-1}}{\overset{k_1}{\rightleftharpoons}} \underset{\text{complex}}{\text{O.S.}} \xrightarrow[(rds)]{k_2} L_5M{-}Y + X$$

$$\text{rate} = \frac{k_1k_2[L_5MX][Y]}{k_{-1}+k_2} = k[L_5MX][Y],\ k = \frac{k_1k_2}{k_{-1}+k_2} \quad \ldots(5.11.8)\ (cf.\ \text{Eqns. 5.11.4; 5.9.2.3})$$

The rate determining step is a unimolecular one, but it shows the second order kinetics.

If the outer sphere (*O.S.*) complex is an ion pair, then depending on the molecularity or nature of the rds, the reaction may be described as ***D–IP* (≡ S_N1 lim–*IP*, in short S_N1–*IP*)** (*i.e.* dissociative *rds*), ***A–IP* (≡ S_N2 lim–*IP*; in short, S_N2–*IP*)** (*i.e.* associative *rds*, *Y* attacks as a nucleophile) at the *rds*.

(b) Formation of a conjugate acid or base at the pre-equilibrium step: If the leaving group (*X*) possesses suitable lone pairs to act as a base, then it may lead to the formation of a **conjugate acid** (*CA*) at the pre-equilibrium step. This happens so for the **acid catalysed aquation** of the complexes like $[Cr(OH_2)_5X]^{2+}$, $[Co(OH_2)_5X]^{2+}$, $[Co(NH_3)_5X]^{2+}$ ($X^- = N_3^-$, halide, etc.), etc.

$$[(H_3N)_5Co{-}X{:}]^{2+} + H^+ \rightleftharpoons [(H_3N)_5Co{-}XH]^{3+}\ \textbf{(CA)}$$

$$\downarrow \text{(rds)}\ +H_2O$$

$$[Co(NH_3)_5(OH_2)]^{3+} + XH$$

If the CA reacts through an associative pathway (*i.e.* H_2O attacks as a nucleophile at the *rds*) then the process is described as ***A–CA*** **(≡ S_N2 lim–*CA*; in short, S_N2–*CA*).** On the other hand, if the CA participates in a dissociative process (*i.e.* XH is dislodged without the attack of H_2O at the *rds*), then the process is called ***D–CA*** **(≡ S_N1 lim–*CA*, in short S_N1–*CA*).**

● In the **base catalysed hydrolysis** of the complexes like $[Co(NH_3)_5X]^{2+}$, the conjugate base (CB) may be formed at the pre-equilibrium step and then the CB participates at the *rds* (*cf.* Sec. 5.25).

$$\left[Co(NH_3)_5X\right]^{2+} + OH^- \rightleftharpoons \underset{\textbf{(CB)}}{\left[Co(\ddot{N}H_2)(NH_3)_4(X)\right]^{+}} + H_2O$$

$$CB \xrightarrow[\text{(slow)}]{rds} \left[Co(NH_2)(NH_3)_4\right]^{2+} + X^-$$

$$\left[Co(NH_2)(NH_3)_4\right]^{2+} + H_2O \xrightarrow{\text{fast}} \left[Co(NH_3)_5(OH)\right]^{2+}.$$

Here, in the conjugate base (CB), the good σ–donor and π–donor ligand (NH_2^-) labilises the leaving group (X). *Thus, the rds is a unimolecular one* (*i.e.* **dissociative activation**) and the mechanism is described as ***D–CB*** **(≡ S_N1 lim–*CB*; in short, S_N1–*CB*).** *But it shows the second-order kinetics.*

Metal ion catalysis in aquation process

If the bound leaving group can act as a Lewis base, then it can produce a *Lewis-acid base adduct* at the pre-equilibrium step.

$$L_5M—X: + M' \rightleftharpoons L_5M—X—M' \xrightarrow[(rds)]{+H_2O} L_5M—OH_2 + M'X\,(\text{or } M' + X).$$

The adduct (*i.e.* *X*–bridged binuclear complex) may participate in a dissociative or an associative (*i.e.* nucleophilic attack by H_2O at the *rds*) path to give the product.

(iv) **Proton ambiguity and rate law:** This aspect has been separately discussed in Sec. 5.17.

5.12 KINETICS AND MECHANISM OF SOLVENT EXCHANGE REACTIONS: DIAGNOSIS OF OTHER SUBSTITUTION REACTION MECHANISM

5.12.1 Characteristics of Solvent Exchange Reactions

The solvent exchange reaction involves the exchange between the coordinated and bulk solvent molecules.

$$[M(S)_x]^{n+} + xS^* \rightarrow [M(^*S)_x]^{n+} + xS$$

These reactions are generally followed by *isotopic exchange technique, NMR line broadening technique,* and other *relaxation* techniques. For the water exchange (*i.e.* water as the solvent) reaction, $H_2{}^{17}O$ is used and O-17 NMR line broadening technique is commonly used for measuring the exchange rate constant (k_{ex} in the range $10^3 – 10^9\ s^{-1}$) of paramagnetic ions. The water exchange process is:

$$[M(OH_2)_x]^{n+} + xH_2{}^{17}O \rightarrow [M(^{17}OH_2)_x]^{n+} + xH_2O$$

The chemical and electronic environment of free and coordinated $H_2{}^{17}O$ *molecules are different and consequently, chemical shifts for free* and *coordinated* $H_2{}^{17}O$ *molecules are different.* The extent of merging the NMR peaks depends on the exchange rate (*cf.* Fig. 5.6.1).

Rate laws: The kinetic behaviour of metal ions towards the solvent exchange process is of an important consideration to understand the mechanistic aspects of other substitution reactions. The water exchange process can be simply represented as:

$$M—{}^*OH_2 + H_2O \rightarrow M—OH_2 + {}^*OH_2 \qquad \text{...(5.12.1)}$$

$$(cf.\ M—X + Y \rightarrow M—Y + X,\ i.e.\ X = {}^*OH_2,\ Y = H_2O)$$

(i) If the process occurs by a ***D* mechanism** then the rate law is given by:

$$k_{obs} = \frac{k_1 k_2 [Y]}{k_{-1}[H_2O^*] + k_2[Y]} = \frac{k_1 k_2 [H_2O]}{k_{-1}[H_2O^*] + k_2[H_2O]} \qquad (cf.\ \text{Sec. 5.9.1})$$

$$\approx k_1 \text{ (under the condition, } k_2[H_2O] \gg k_{-1}[H_2O^*])$$

(ii) If the process occurs through **the *I* process** (both I_a and I_d), the rate law is given by:

$$k_{obs} = \frac{k_i K_{OS}[Y]}{1 + K_{OS}[Y]} = \frac{k_i K_{OS}[H_2O]}{1 + K_{OS}[H_2O]}, \qquad (cf.\ \text{Sec. 5.9.3})$$

$$\approx k_i, \text{ (under the condition, } K_{OS}[H_2O] \gg 1)$$

Table 5.12.1.1: Water exchange rate constants (k_{ex}) (at 25° C) for $[M(OH_2)_6]^{n+}$ (high spin) and their activation parameters

(a) **Bivalent metal ions**

Metal ion	V^{2+}	Cr^{2+}	Mn^{2+}	Fe^{2+}	Co^{2+}	Ni^{2+}	Cu^{2+}	Zn^{2+}
d^n	$3d^3$	$3d^4$	$3d^5$	$3d^6$	$3d^7$	$3d^8$	$3d^9$	$3d^{10}$
(~) log k_{ex}	2.0	> 8.5	7.4	6.6	6.4	4.5	9.9	7.5
$\Delta H^{\neq}$ (kJ mol^{-1})	62	–	33	41	47	57	23	–
$\Delta S^{\neq}$ (J K^{-1} mol^{-1})	0	–	6.0	21.0	37.0	32.0	25.0	–
$\Delta V^{\neq}$ (cm^3 mol^{-1})	–4.1	–	–5.4	+3.8	+6.1	+7.2	–	–

(b) **Trivalent metal ions**

Metal ion	Ti^{3+}	V^{3+}	Cr^{3+}	Fe^{3+}
d^n	$3d^1$	$3d^2$	$3d^3$	$3d^5$
(~) log k_{ex}	5.3	2.7	–5.5 (–3.7)	2.2 (5.0)
$\Delta H^{\neq}$ (kJ mol^{-1})	43	49	110 (110)	64 (42)
$\Delta S^{\neq}$ (J K^{-1} mol^{-1})	1	–27.5	16.0 (55)	12.0 (5)
$\Delta V^{\neq}$ (cm^3 mol^{-1})	–12.1	–8.9	–9.3 (+2.7)	–5.4 (7.0)

Values within the parentheses for $[M(OH)(OH_2)_5]^{2+}$

(c) **Dependence of $\Delta V^{\neq}$ (cm^3 mol^{-1}) of k_{ex} on the bulkiness of solvent**

Solvent	Co(II)	Ni(II)		Solvent	Fe(III)	Cr(III)	Ga(III)	
H_2O	+6.1	+7.2	↓ increasing d-activation	H_2O	–5.4	–9.3	–	↓ increasing d-activation
MeOH	+9.0	+11.3		$OCHNMe_2$ (DMF)	–0.9	–6.3	+7.9	
				$OSMe_2$ (DMSO)	–3.1	–11.0	+13.1	

Refs. Inorg. Chem. **21**, 774, 1982; *Adv. Chem. Sev.* **49**, 55, 1965; *Coord. Chem. Rev.*, **5**, 45, 1970; **36**, 89, 1981

The important conclusions on the water exchange rate constants (k_{ex}) are given below:

(i) **Effect of LFAE (Ligand Field Activation Energy):** The **relative inertness** for the $3d^3$ (V^{2+}, Cr^{3+}) and $3d^8$ (Ni^{2+}) systems can be explained in terms of the relatively high **ligand field activation energy (LFAE).** This aspect of LFAE will be discussed later.

(ii) **Effect of J.T. distortion:** The **high lability** for the $3d^4$ (high spin) (Cr^{2+}) and $3d^9$ (Cu^{2+}) systems is due to the **Jahn-Teller distortion.** Due to this distortion, the weakly bound solvent molecules at the tetragonally elongated sites are rapidly exchanged which is almost close to the *diffusion controlled rate*. Thus, the solvent exchange preferably occurs at the tetragonally elongated sites and then a *facile vibration* leads to the rapid exchange among the six coordination sites of the octahedral geometry.

(iii) **Sign of the $\Delta V^{\neq}$ value depending on the mechanistic path:** *For the solvent exchange processes, $\Delta V^{\neq}$ (volume of activation) is a very reliable parameter to identify the mechanism. In such cases, there is no possibility of charge annihilation or charge creation in attaining the transition state in all possible routes of a- and d-processes.* Thus, there is no complication in $\Delta V^{\neq}$ from the electrorestriction over the solvent molecules because **in attaining the transition state** through the addition **(*a*-activation)** or removal **(*d*-activation)** of a neutral solvent H_2O molecule, there is no change in charge, *i.e.* no change in electrorestriction over the solvent molecules. ***It makes the $\Delta V^{\neq}$ parameter, a reliable parameter to diagnose the mechanistic path.***

For the solvent exchange process, the **straight-forward conclusions** are:

positive value of $\Delta V^{\neq}$ indicates the ***d*-activation** and *negative value of $\Delta V^{\neq}$ indicates the* ***a*-activation**.

It may be noted that molar volume of free H_2O (*i.e.* bulk solvent) is larger than that of coordinated H_2O (Sec. 5.19.5). In the *d*-activation, the **positive value of $\Delta V^{\neq}$** arises from the release of a coordinated H_2O molecule (*i.e.* one molecule splits into 2 molecular species). In the *a*-activatiion, the **negative value of $\Delta V^{\neq}$** arises because a free H_2O molecule gets fixed into the coordination sphere of the starting complex (*i.e.* 2 molecules combine into a single one).

From the $\Delta V^{\neq}$ values of k_{ex}, it is evident that for $[M(OH_2)_6]^{2+}$, the **early members** (*e.g.* V^{2+}, Mn^{2+}) of the 1st transition series adopt the **associative path** while for the **late members** (*e.g.* Fe^{2+}, Co^{2+} and Ni^{2+}), the **dissociative mechanism** is operated. The increasingly dissociative character from Mn^{2+} to Ni^{2+} is reflected in the increasingly positive $\Delta S^{\neq}$ values.

(iv) **Effect of the d^n–configuration:** For the octahedral complexes $[M(OH_2)_6]^{n+}$, the **associative path** involves the approach of the entering solvent molecule along the three-fold axes (C_3) of the octahedral complexes (Sec. 5.18.2). The two t_{2g} orbitals (*i.e.* d_{xz} and d_{yz}) are projected along the C_3-axes, *i.e.* the lobes of these two t_{2g} orbitals are projected towards the centres of trigonal faces of the octahedral complexes. The **edge attack** at the d_{xy}-t_{2g} orbital is energetically less favoured in the associative path (see Sec. 5.18.2). In terms of MOT, the t_{2g} orbitals (NBMO) or e_g^* (ABMO) orbital will act as the LUMO to receive the nucleophilic attack. Obviously, participation of the e_g^* (ABMO) orbital is energetically highly disfavoured. But in some cases, e_g^*-LUMO receives the nucleophilic attack (see Sec. 5.19.6). Thus, greater electron density in the t_{2g} level will disfavour the approach of the entering nucleophile electrostatically. *This is why, for the early members of M^{2+} (of the 1st transition series) having the less electron density in the t_{2g} level, the associative path is favoured while for the late members of the series where the t_{2g}–level is filled in, the associative path is highly disfavoured (i.e. dissociative path is favoured).* Here it may be noted that, pairing of electrons in the t_{2g} level may produce the required vacancy to accommodate the nucleophile in the associative path. In fact, it happens so for the t_{2g}^3 system of Cr(III) (t_{2g}^3) and Fe(III) (high spin, $t_{2g}^3e_g^2$).

Thus for the water exchange process at the M^{2+} ions of 1st transition series, there is a change-over of the mechanism from a dissociative path to an associative path at Mn^{2+} in moving from right to left along the 3d-period.

(v) **Effect of the bulkiness of solvent:** With the increase in bulkiness of the solvent, the associative path leading to an increase in coordination number in attaining the transition state is disfavoured more. This is why, for a particular metal centre, the $\Delta V^{\neq}$ value increases in the order: DMSO (dimethyl sulfoxide)/DMF (dimethyl formamide) > H_2O. In fact, for the solvent exchange process, $\Delta V^{\neq}$ is positive (*i.e.* dissociative path) for $[Mn(DMF)_6]^{2+}$ while $\Delta V^{\neq}$ is negative for $[Mn(OH_2)_6]^{2+}$ (*i.e.* associative path). *Probably, a perfect D-mechanism operates in the solvent exchange process of* $[Ni(DMF)_6]^{2+}$.

(vi) **Effect of charge on the metal ion:** For the M^{3+} ions, *e.g.* $[Ti(OH_2)_6]^{3+}$, $[V(OH_2)_6]^{3+}$, $[Cr(OH_2)_6]^{3+}$, $[Fe(OH_2)_6]^{3+}$, the associative path is operated (evidenced by the negative values of $\Delta V^{\neq}$). Higher positive charge favours the binding of the entering nucleophile.

(vii) **Hydrolysis of trivalent metal ions and difference in the reactivity of the unhydrolysed and hydrolysed species:** It is interesting to note that the associative path is operated for $[M(OH_2)_6]^{3+}$ (M = Fe, Cr, etc.), but for their hydrolysed species $[M(OH)(OH_2)_5]^{2+}$, the dissociative path operates, *i.e.* $[M(OH)(OH_2)_5]^{2+}$ behaves like the bivalent metal ions of the late members of the 1st transition series. This difference in their mechanisms is supported by the $\Delta V^{\neq}$ values. The dissociative character (*i.e.* positive $\Delta V^{\neq}$ value) is introduced into the hydrolysed species $[M(OH)(OH_2)_5]^{2+}$ because of two grounds:

lower positive charge and *labilising effect of the good σ–donor and π–donor OH ligand* (see Secs. 5.19.6, 5.21).

The labilising effect of the OH ligand leads to about 10^2-10^3 fold rate enhancement (*cf.* k_{ex} values).

(viii) **Effect of electron density in the t_{2g}–level:** In the case of $[M(OH_2)_6]^{2+}$ species, the associative character for the solvent exchange process decreases with the increase of electron density in the t_{2g} level. It is supported by the fact that there is a gradual increase of the $\Delta V^{\neq}$ value from Ti(III) (t_{2g}^1) to Fe(III) ($t_{2g}^3e_g^2$) for their water exchange processes.

(ix) **Unsually high k_{ex} value for $[Co(aq)]^{3+}$:** The unusually high k_{ex} value for $[Co(OH_2)_6]^{3+}$ needs an explanation. This aspect has been specially discussed in Sec. 5.13.

5.12.2 Identification of Mechanism of Ligand Substitution Reaction from the Knowledge of Solvent Exchange Rate Constant (k_{ex}) (*cf.* Secs. 5.14, 5.19.9)

When the entering ligand (*Y*) is other than the solvent (*i.e.* H_2O in aqueous media), the **upper limiting rate constant** (k_{lim}) can be obtained by making [*Y*] very large (*i.e.* $k_i \approx k_{lim} \approx k_{obs}$ at high [*Y*]) (*cf.* Sec. 5.9.5).

k_{lim}/k_{ex} for *d*- and *a*-activation (*cf.* Sec. 5.19.9)

Generally, the upper limiting rate constant (k_{lim}) is taken as the k_{obs} obtained at high concentration of the entering ligand (*Y*) and ***the ratio k_{lim}/k_{ex} can never exceed 1 for the d-activation but the ratio may exceed 1 for the a-activation, if the entering ligand is a better nucleophile than the solvent.***

$$[Co(NH_3)_5(OH_2)]^{3+} \xrightarrow[(Y=N_3^-,\ SO_4^{2-},\ Cl^-,\ NCS^-)]{\text{anation}} \text{Product};\ k_{\text{lim}}/k_{ex} \approx 1-0.16\ (i.e. = \textbf{d-activation})$$

(*cf.* Table 5.19.6.2.)

$$[Rh(NH_3)_5(OH_2)]^{3+} \xrightarrow[(Y=Cl^-,Br^-,SO_4^{2-})]{\text{anation}} \text{Product};\ k_{\text{lim}}/k_{ex} \approx 1-5.0\ (i.e.\ \textbf{a-activation})$$

$$[Ir(NH_3)_5(OH_2)]^{3+} \xrightarrow{\text{anation}} \text{Product};\ k_{\text{lim}}/k_{ex} > 1\ (i.e.\ \textbf{a-activation})$$

Thus it is the common observation that the associative character in the ligand substitution process increases for the heavier congeners in a particular group of d-block elements.

● In the process of ***d*-activation**, *M—OH*$_2$ bond breaking becomes the *rate controlling factor* for anation and other formation reactions. **Thus for such reactions, the rate cannot be greater than the water exchange rate, *i.e.*** $k_{lim}/k_{ex} \approx 1$. Obviously, the ratio k_{lim}/k_{ex} can never exceed unity. It has been observed in many cases. For the anation of $[Co(NH_3)_5(OH_2)]^{3+}$ by different types of anions (*i.e.* $Y = N_3^-$, SO_4^{2-}, Cl^-, NCS^-, etc.), the condition, $k_{lim}/k_{ex} \approx 1$, has been found true ($k_{ex} \approx 10^{-4}$ s^{-1}).

● For the **I_d-type Eigen Mechanism**, the k_f/K_{OS} ratio is constant and close to k_{ex} for different types of entering nucleophiles. This aspect will be discussed later.

● For the ***a*-activation process,** *M—Y* bond formation (*i.e.* bond formation by the entering ligand) is primarily important. Thus if *Y* is a better nucleophile than the solvent, *i.e.* water, then the situation will give $k_{lim} > k_{ex}$. On the other hand, if *Y* is a poorer nucleophile than the solvent, the situation will give $k_{lim} < k_{ex}$.

5.13 CLASSIFICATION OF THE METAL IONS IN TERMS OF WATER EXCHANGE RATE CONSTANTS

A. Classification of the metal ions: Based on the values of **water exchange rate constant** (k_{ex}), Langford and Gray have divided the metal ions in four classes.

● **Class-I** ($k_{ex} \geq 10^8$ s^{-1}): The water exchange process is exceedingly fast and the *first-order exchange rate constant* (k_{ex}) is of the order of 10^8 s^{-1} (*i.e. almost diffusion controlled rate*). In such cases, the water molecules are mainly bound to the metal centre through electrostatically and the metal ions are generally of low charge and large ionic radius (*i.e.* low value of the ratio, $Z^{*2}/r \leq 10 \times 10^{-28}$ C^2 m^{-1}).

Examples: Alkali metal ions, alkaline earth metal ions (except Be^{2+} and Mg^{2+}), Cd^{2+}, Hg^{2+}, Cr^{2+}, Cu^{2+}, and several trivalent lanthanides (*e.g.* Gd^{3+}, Ho^{3+} etc.)

● **Class-II** ($k_{ex} = 10^4$ to 10^8 s^{-1}): The water exchange process is fast and k_{ex} lies in the range 10^4 to 10^8 s^{-1}. In such cases, the *M—OH*$_2$ bond is somewhat stronger than that in the case of Class-I metal ions. To the metal-ligand bonding, cfse contribution is relatively small. The ratio Z^{*2}/r for such metal ions ranges approximately in the span 10–30 × 10^{-28} C^2 m^{-1}.

Examples: Bivalent metal ions of 1st transition series including Zn^{2+} (except for V^{2+}, Cr^{2+} and Cu^{2+}), Ti^{3+}, In^{3+}, Mg^{2+}, some trivalent lanthanides (*e.g.* Yb^{3+}), etc.

● **Class-III** ($k_{ex} = 10^0$ to 10^4 s^{-1}): The water exchange process is relatively slower compared to the cases of Class I and Class II but the rate is definitely fast on an absolute scale that spans a very wide range, *ca.*, 10^{-8} to 10^{10} s^{-1} (see Fig. 5.13.1). In such cases, in the metal-ligand interaction, besides the electrostatic factor, cfse contributes to a significant extent. The Z^{*2}/r ratio for such metal ions is greater than about 30 × 10^{-28} C^2 m^{-1}.

Examples: Tripositive metal ions of first transition series (*e.g.* Fe^{3+}, V^{3+}), Al^{3+}, Ga^{3+}, Be^{2+}, Pd^{2+}, etc.

● **Class-IV** (inert centres, $k_{ex} \approx 10^{-2}$ to 10^{-8} s^{-1}): The water exchange process is very slow. In terms of size, such metal ions are comparable to those of Class-III, but contribution of cfse to the metal-ligand interaction is more important for the metal ions of Class-IV.

Examples: Cr^{3+}, Rh^{3+}, Ru^{2+}, Ru^{3+} and Pt^{2+}.

B. Different factors controlling the k_{ex} values: The important observations from the classification of metal ions in terms of their k_{ex} values are discussed below.

● **Electrostatic interaction in the metal-ligand bond:** The parameter, **ionic potential** (= charge/radius) or Z^{*2}/r ($Z^* = Z -$ shielding constant) gives the measure of electrostatic contribution to the metal-ligand interaction. This electrostatic interaction is the main contributing factor for the metal ions of

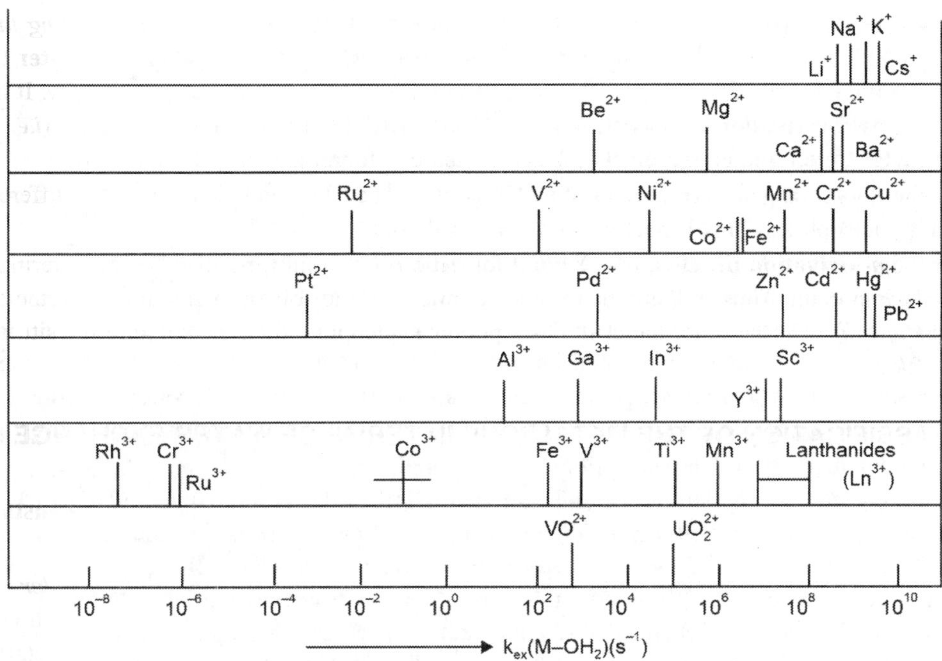

Fig. 5.13.1: Water exchange rate constants (at 25°C) for various aqua metal ions

noble gas and **pseudo-noble gas electronic configuration.** In fact, with the increase of ionic potential, the lability measured by k_{ex} decreases. It explains the following **lability sequences:**

(i) $Cs^+ > K^+ > Na^+ > Li^+$

(ii) $Ba^{2+} > Sr^{2+} > Ca^{2+} > Mg^{2+} > Be^{2+}$
(*cf.* Ba^{2+}, Sr^{2+}, Ca^{2+} in Class I; Mg^{2+} in Class-II; Be^{2+} in Class III)

(iii) $In^{3+} > Ga^{3+} > Al^{3+}$
(*cf.* In^{3+} in Class-II; Ga^{3+} in Class-III; Al^{3+} in Class III)

(iv) $Hg^{2+} > Cd^{2+} > Zn^{2+}$
(*cf.* Hg^{2+} and Cd^{2+} in Class-I; Zn^{2+} in Class-II)

(v) $Na^+ > Mg^{2+} > Al^{3+}$; $Ru^{2+} > Ru^{3+}$; $Fe^{2+} > Fe^{3+}$

(For the representative elements (*i.e.* *s*- and *p*-block elements), charge/radius ratio decreases in moving from top to bottom).

The parameter, charge/radius ratio, indicates that the stronger **Bronsted acids,** *i.e.* the $[M(OH_2)_x]^{n+}$, having the larger charge/radius ratio are possessing the lower k_{ex} values. In fact, the *nonacidic* or *feebly acidic* $[M(OH_2)_x]^{n+}$ ions are highly labile ($k_{ex} \geq 10^8\ s^{-1}$) while the acidic cations are relatively more inert.

The decrease of lability with the increase of the ratio (charge/radius) indicates the importance of **dissociative activation.**

For the transition metal ions, in addition to the contribution of the parameter, ionic potential (charge/radius ratio), **crystal field activation energy** (CFAE) is of an important consideration. However, **importance of the ionic potential parameter** is reflected in the fact that the trivalent metal ion is less labile than the bivalent metal ions. The lability sequences: $Fe^{2+} > Fe^{3+}$; $Ru^{2+} > Ru^{3+}$, etc. support the fact (exception: $V^{2+} < V^{3+}$). Here it should be mentioned that for the trivalent cations, in general, CFAE is also higher (*cf.* crystal field splitting power increases with the increase of oxidation state of the metal centre) (*cf.* exception, V(II) *vs.* V(III)). Higher oxidation state may favour the low spin complexes for which CFAE is higher.

Note: To estimate the parameter Z^{*2}/r, the effective nuclear charge may be obtained from the consideration of shielding effect (see **Slater's rule,** Vol. 1), *i.e.* effective nuclear charge (Z^*) = atomic number – shielding constant.

● **Crystal field activation energy** (CFAE) (*cf.* Sec. 5.18.2): For the *d*-block elements, CFAE is an important factor to determine the lability. Thus, increase of charge may not necessarily increase the inertness. For example, V(III) (d^2) is more labile than V(II) (d^3). In fact, for the *d*-block metal ions, charge effect is straight-forward only for the isoelectronic species like Cr^{3+} and V^{2+}; Fe^{3+} and Mn^{2+}.

For the **trivalent lanthanides** (Ln^{3+}), the crystal field effect (*i.e.* ligand field effect) on the deeply seated 4*f*-orbitals is not much important. Consequently, **CFAE or LFAE is not important** to control the lability of the trivalent lanthanides (Ln^{3+}) and in their complexes, **ionic interactions** are predominant as in the cases of nontransition metal ions (*cf.* for the transition metal ions, covalent interaction contributes significantly, and CFAE plays an important role). In general, **trivalent lanthanides are labile** and they belong to the Class-I metal ions (in terms of water exchange rate constants).

●**Jahn-Teller distortion:** The exceptionally high lability of Cr^{2+} ($t_{2g}^3e_g^1$) and Cu^{2+} ($t_{2g}^6e_g^3$) can be explained by considering the strong **Jahn-Teller distortion** in the ground state and zero CFAE (see Table 5.18.2.1). Thus in terms of *dissociative activation, the ground state geometry of such distorted octahedral complexes is very close to that of the transition state for the water exchange process.* In other words, the elongated $M—OH_2$ bond ruptures easily in the dissociative process to favour the water exchange process. Then the exchanged water molecule experiences a rapid exchange among the six positions through a **facile vibration**.

Note: Here it is worth mentioning that in the octahedral complexes, **J.T. distortion is more pronounced, if it originates from the e_g level** (see Vol. 4) as in Cr^{2+} (high spin, $t_{2g}^3e_g^1$) and Cu^{2+} ($t_{2g}^6e_g^3$). In fact, it is less pronounced when it originates from the t_{2g} level as in the cases of Fe^{2+} (high spin, $t_{2g}^4e_g^2$) and Co^{2+} (high spin, $t_{2g}^5e_g^2$).

● **Lability sequence for the M^{2+} (high spin) ions of 1st transition series:** In terms of the k_{ex} values, the lability of bivalent metal ions of 1st transition series runs as:

$V^{2+}(d^3) \ll Cr^{2+}(d^4) > Mn^{2+}(d^5) > Fe^{2+}(d^6) > Co^{2+}(d^7) \gg Ni^{2+}(d^8) \ll Cu^{2+}(d^9) > Zn^{2+}(d^{10})$

The actual lability sequence is:

	Cr^{2+} ≈	Cu^{2+} ≫	Mn^{2+} ≈	Zn^{2+} >	Fe^{2+} >	Co^{2+} ≫	Ni^{2+} ≫	V^{2+}
log k_{ex} :	*ca.* 9	*ca.* 9	7.4	7.5	6.6	6.4	4.5	2.0

The above sequence can be explained by considering **both CFAE and Z^* (effective nuclear charge)** (more correctly the parameter, Z^{*2}/r). The effective nuclear charge gradually increases from left to right along the period due to the gradual addition of **low shielding** 3*d* electrons along the series towards the late members. Here it may be mentioned that for the late members of the series (approximately starting from Cr^{2+}/Mn^{2+}), the water exchange process passes through the **dissociative path** (evidenced by +ve $\Delta V^{\neq}$ values). This dissociative path is disfavoured more for increasing the Z^* values.

The abnormally high lability at Cr^{2+} and Cu^{2+} arises due to the **J.T. distortion** and **zero CFAE** (actually highly –ve value, see Table 5.18.2.1). The CFAE for Mn^{2+}, Fe^{2+}, Co^{2+} and Zn^{2+} are considered to be zero (after ignoring the –ve values,, see Table 5.18.2.1) for their high spin $[M(OH_2)_6]^{2+}$ complexes.

● **Ni(II) (d^8) *vs.* V(II) (d^3) having the same CFAE:** In terms of the high and same CFAE (Table 5.18.2.1), both are expected to be equally inert. **But V(II) ($k_{ex} = 10^2\ s^{-1}$) is more inert than Ni(II) ($k_{ex} = 3 \times 10^4\ s^{-1}$).** For V(II) having the t_{2g}^3 electronic configuration, through pairing an electron, **a vacancy in the t_{2g} level can be generated to accommodate the nucleophile in the associative path.** On the other hand, for Ni(II) having the electronic configuration $t_{2g}^6e_g^2$, the associative path is not possible as there is no vacancy in the t_{2g} level to accommodate the nucleophile. In fact, V(II) experiences the **associative activation** ($\Delta V^{\neq}$ = negative) for which the CFAE is 4.26 Dq_o (assuming the **pentagonal bipyramid intermediate**) while Ni(II) experiences the **dissociative activation** ($\Delta V^{\neq}$ = **positive**) for which the CFAE is 2 Dq_o (assuming the **square pyramid intermediate**).

This CFAE order (**4.26 Dq_o *vs.* 2 Dq_o**) can explain their water exchange rate constant (k_{ex}) order. But, if the **associative path** involving the **trapezoidal octahedron** as the intermediate or T.S. is considered for $V(aq)^{2+}$ then CFAE becomes 1.8 Dq_o (see Table 5.18.2.1). It suggests that then the CFAE values for the water exchange process at $V(aq)^{2+}$ (associative path) and $Ni(aq)^{2+}$ (dissociative path) becomes comparable (1.8 Dq_o *vs.* 2.0 Dq_o). But the actual value of CFAE is still higher for $V(aq)^{2+}$ because the Dq_o value of $V(aq)^{2+}$ is higher (≈ 1200 cm^{-1}) than that of $Ni(aq)^{2+}$ (≈ 850 cm^{-1}). This explains qualitatively the order of water exchange rate constants of $V(aq)^{2+}$ and $Ni(aq)^{2+}$. Here it is worth mentioning that the CFAE value (in terms of Dq_o) is the same for a particular *D*- or *A*-path (Table 5.18.2.1) for the d^3 and d^8 systems of the octahedral complexes.

In reality, the water exchange process at $Ni(aq)^{2+}$ is kinetically more favoured compared to that of $V(aq)^{2+}$ by both the $\Delta H^{\neq}$ and $\Delta S^{\neq}$ values (Table 5.12.1.1). It may be pointed out that **CFAE contributing to $\Delta H^{\neq}$ only** (see Secs 5.18.2, 3) **cannot alone explain the order of k_{ex} values of $V(aq)^{2+}$ and $Ni(aq)^{2+}$**. For $Ni(aq)^{2+}$, the benefit from ΔH^{*} (**less positive value**, 57.0 *vs.* 62.0 in kJ mol^{-1}) may arise partly from CFAE due to its lower Dq_o value (*cf.* **850 cm^{-1} *vs.* 1200 cm^{-1}**) but the benefit from $\Delta S^{\neq}$ (**more positive value**, 32.0 *vs.* 0.0 in J K^{-1} mol^{-1}) arises from the dissociative activation.

● **Lability sequence among the congeners of a particular group of *d*-block metals** (see Sec. 5.18.2 for details)**:** For the transition metal ions, in a particular group, the lability decreases for the heavier congeners (*i.e.* heavier congeners are more inert). This can be explained by considering the *crystal field activation energy.* **The heavier congeners tend to form the low spin complexes for which CFAE is higher** (*cf.* $[Fe(OH)_2)_6]^{2+/3+}$ high spin while $[Ru(OH_2)_6]^{2+/3+}$ low spin). The lability sequences are:

(i) **Ni^{2+} > Pd^{2+} > Pt^{2+}** (*cf.* Ni^{2+} in Class-II; Pd^{2+} in Class-III; Pt^{2+} in class IV; see Sec. 5.30.10).

$[Ni(OH_2)_6]^{2+}$ **(octahedral)** experiences the **dissociative activation** in the water exchange process while $[Pd(OH_2)_4]^{2+}$ and $[Pt(OH_2)_4]^{2+}$ **(square planar)** adopt the **associative activation** in the water exchange process. CFAE for Pd(II) and Pt(II) is higher than that of Ni(II) (*cf.* Dq value: Pt(II) > Pd(II) > Ni(II)).

For Ni(II), Pd(II) and Pt(II), if the complexes are considered to be **square planar,** then their ligand substitution process passes through the **associative path** in which the incoming nucleophile approaches along the *z*-direction assuming the starting square planar complex to remain in the *xy*-plane (*cf.* Sec. 5.30.3)**.** In terms of **MOT**, it requires the participation of the **vacant antibonding d_{z^2} orbital** (*i.e.* a_{1g}^{*}, see Fig. 3.17.6.3, Vol. 4, assuming the energy order, $a_{1g}^{*} < a_{2u}(p_z)$; however, uncertainty in this energy order) to accommodate the entering nucleophile. The energy of the vacant antibonding d_{z^2} orbital increases from Ni(II) to Pt(II).

Energy order of the corresponding vacant antibonding d_{z^2} orbital (in terms of MOT) follows the order: Pt(II) > Pd(II) > NI(II). In other words, in terms of MOT, the antibonding character of the d_{z^2} orbital (LUMO) increases gradually from Ni(II) to Pt(II). Thus participation of the **vacant antibonding d_{z^2} orbital** (more correctly, a_{1g} in terms of MOT, see Vol. 4) to accommodate the entering nucleophile causes the loss of *lfse* (*i.e. cfse*) and this loss increases from Ni(II) to Pt(II). ***Thus the CFAE or LFAE increases from Ni(II) to Pt(II) and the lability gradually decreases from Ni(II) to Pt(II)*** (see Table 5.30.10.1, Dq-value: Pt(II) > Pd(II) > Ni(II)). Here it may be mentioned that in an alternative view, it can be argued that the incoming ligand occupies the

vacant nonbonding np_z orbital (= a_{2u} in terms of MOT assuming the energy order, $a_{2u} < a_{1g}^*$ in Fig. 3.17.6.3 of Vol. 4; *cf.* uncertainty in the energy order of a_{2u} and a_{1g}^*). In terms of VBT, participation of the vacant np_z orbital to acommodate the nucleophile produces the **square pyramid** T.S./intermediate having the hydrbdisation dsp^3 (*i.e.* $d_{x^2-y^2} + s + p_x + p_y + p_z$, see Vol. 2). The energy order is: $6p$ (Pt^{II}) > $5p$ (Pd^{II}) > $4p$ (Ni^{II}) and it explains their lability order. Participation of the **vacant *pd*-hybrid orbital** ($p_z^0 + d_{z^2}^2$ combination) to receive the nucleophilic attack can also explain the lability order as the energy order of vacant *pd*-hybrid orbital follows the sequence: *Pt*(II) > *Pd*(II) > *Ni*(II) (see Sec. 5.30.3).

> **Associative path of the square planar complexes of Ni(II), Pd(II), Pt(II), Au(III), etc. d^8-system:** Available orbital to acommodate the incoming nucleophile.
> - Vacant a_{1g}^* (*i.e.* $d_{z^2}^*$) or a_{2u} (*nb*–p_z) in terms of **MOT**.
> - Vacant p_z or *pd* ($p_z - d_{z^2}$) hybrid orbital in terms of **VBT**.

Here it is worth mentioning that Ni(II) ($3d^8$) very often adopts the octahedral geometry and its CFAE is also high (*cf.* other M^{2+} ions of $3d$-series, Table 5.18.2.1).

(ii) $Fe^{2+} > Ru^{2+}$ (*cf.* Fe^{2+} in Class-II; Ru^{2+} in Class-IV)

(iii) $Co^{3+} > Rh^{3+} > Ir^{3+}$. $(Co(aq)^{3+}$, high spin $\rightleftharpoons$ low spin; while $Rh(aq)^{3+}$ and $Ir(aq)^{3+}$, low spin).

Mean residence time of a water molecule in the coordination spheres of $[M(OH_2)_6]^{3+}$ at 25°C follows the sequence: $[Cr(OH_2)_6]^{3+}$($3d^3$, t_{2g}^3)(~100 h) < $[Rh(OH_2)_6]^{3+}$($4d^6$, t_{2g}^6) (~1.5 y) ≪ $[Ir(OH_2)_6]^{3+}$($5d^6$, t_{2g}^6)(~300 y).

Heavier congeners *vs.* lighter congeners in *d*-block: For a particular d^n configuration, the CFAE increases for the heavier congeners having the higher Dq values. Besides this, for a particular oxidation state, the heavier congeners tend to adopt the **low spin complexes** (see Vol. 4, effect of crystal field splitting power) (*e.g.* $Fe(OH_2)_6]^{2+/3+}$ **high spin** while $[Ru(OH_2)_6]^{2+\backslash 3+}$ **low spin**) and **CFAE is higher for low spin complexes**. It explains the lability order. Another important aspect is that the heavier congeners tend to adopt the **associative path** (effect of the increased Z* value).

- **Unusually high lability of $[Co(OH_2)_6]^{3+}$ – Why?** The Co(III) (d^6) complexes are generally more inert than the corresponding Cr(III) (d^3) complexes (*cf.* CFAE values). But astonishingly, the rate of water exchange in $[Cr(OH_2)_6]^{3+}$ (*i.e.* exchange with $^{17}OH_2$) is much slower (*ca.* $k_{ex} \approx 10^{-6}$ s^{-1} at 25°C) than that for $[Co(OH_2)_6]^{3+}$. In fact, k_{ex} for $[Co(OH_2)_6]^{3+}$ is fairly high (≈ 10^{-1} s^{-1}). $[Co(OH_2)_6]^{3+}$ is highly oxidising ($E^0 \approx 1.8$ V) and it can oxidise water to release oxygen slowly. Thus, even the purest form of $Co(aq)^{3+}$ sample in aqueous solution gets invariably contaminated by **$Co(aq)^{2+}$ that may catalyse the processs.** $Co(aq)^{2+}$ is labile and its water exchange rate is fast as expected.

$$4[Co(OH_2)_6]^{3+} + 2H_2O \longrightarrow 4[Co(OH_2)_6]^{2+} + 4H^+ + O_2(\uparrow) \text{ (slow reaction)}$$

Overall rate benefit in the water exchange process

$$\left[Co(OH_2)_6\right]^{3+} + 6H_2O^* \rightleftharpoons \left[Co(^*OH_2)_6\right]^{3+} + 6H_2O, \textbf{ (Apparently rapid exchange)}$$

$$\left[Co(OH_2)_6\right]^{2+} + 6H_2O^* \rightleftharpoons \left[Co(^*OH_2)_6\right]^{2+} + 6H_2O; \textbf{ (Fast, labile centre)}$$

$$\left[Co(OH_2)_6\right]^{3+} + \left[Co(^*OH_2)_6\right]^{2+} \rightleftharpoons \left[Co(OH_2)_6\right]^{2+} + \left[Co(^*OH_2)_6\right]^{3+},$$

(Rapid electron exchange through the outer sphere process)

The reason behind the *rapid electron exchange* in the couple $Co(aq)^{3+}/Co(aq)^{2+}$ will be discussed in the next chapter (*cf.* Secs. 6.3.6; 6.6).

Lability enchanced by coupling the favoured electron transfer reactions (see Sec. 6.6)

Here it may be mentioned that it is the common observation:

If the higher oxidation state of a particular metal centre is inert while its lower oxidation state is more labile, then the lower oxidation state will catalyse the ligand substitution process at the relatively inert higher oxidation state. Thus, Cr(II) can catalyse the ligand substitution process at Cr(III); Pt(II) can catalyse the ligand substitution process at Pt(IV) through the electron transfer process. This aspect will be discussed in the next chapter.

There is **another reason** for the enhanced k_{ex} value for $Co(aq)^{3+}$. In $[Co(OH_2)_6]^{3+}$, the $10Dq$ value provided by the H_2O ligands is a **critical 10Dq value** (*i.e.* $10Dq \approx P$). It allows the **spin isomerism** between $[Co(OH_2)_6]^{3+}$ (high spin, $t_{2g}^4e_g^2$) and $[Co(OH_2)_6]^{3+}$ (low spin, $t_{2g}^6e_g^0$) and the equilibrium is probably shifted towards the low spin state (see Vol. 4 and H. Taube *et al*, *JACS*, **73**, 4028, 1951). The low spin $[Co(OH_2)_6]^{3+}$ is inert because of the very high crystal field activation energy, but the high spin $[Co(OH_2)_6]^{3+}$ is more labile because of its lower crystal field activation energy (CFAE). It leads to an overall rate enhancement in the water exchange process

Overall rate benefit in the water exchange process

$$\left[Co(OH_2)_6\right]^{3+}(t_{2g}^4e_g^2) \rightleftharpoons \left[Co(OH_2)_6\right]^{3+}(t_{2g}^6e_g^0)$$

(labile) (Rapid water exchange process) — **(inert)** (Slow water exchange process)

Rapid ligand exchange in $[Co(LL)_3]^{3+}$ (LL = bpy, phen, en) (see Sec. 6.6)

Catalysis by $Co(aq)^{2+}$ in the water exchange process of $Co(aq)^{3+}$ has been explained by coupling the electron exchange process in the $Co(aq)^{3+}/Co(aq)^{2+}$ couple. A similar mechanism catalysed by labile Co(II) can operate in the rapid exchange of the free ligand LL with those in the $[Co(LL)_3]^{3+}$ complex. Here it must be mentioned that the $[Co(LL)_3]^{3+}$ complexes (low spin, t_{2g}^6) are undoubledly inert as expected from CFAE. In fact, $[Co(phen)_3]^{3+}$ can survive from its dissociation even in 5N H_2SO_4 under refluxing for a period of three hours; $[Co(en)_3]^{3+}$ remains unchanged in 1M $HClO_4$ at 30°C even for 2 months.

Sometimes, **pseudo-substitution** reactions at Co(III) without the rupture of Co(III)—ligand bond can occur and these reactions are rapid as the process does not involve the Co(III)—ligand bond rupture. In fact, such reactions occur through the **eletrophilic attack on the bound ligand.** These have been discussed separately in Sec. 5.24.

- **Effect of electron density in the t_{2g} level:** It is interesting to compare the kinetic parameters for the water exchange process at Sc(III), Ti(III), V(III) and Fe(III) – all of which follow the **associative path.** Higher positive charge on the metal centre favours the associative path.

	$[Sc(OH_2)_6]^{3+}$	$[Ti(OH_2)_6]^{3+}$	$[V(OH_2)_6]^{3+}$	$[Fe(OH_2)_6]^{3+}$
Electronic Configuration	t_{2g}^0	$t_{2g}^1e_g^0$	$t_{2g}^2e_g^0$	$t_{2g}^3e_g^2$
~ log k_{ex} (25°C)	7.2	5.3	2.7	2.2
r (M^{3+}) (pm)	89	81	78	79
CFAE	0	~0	~0	0

The ionic radii of the trivalent metal ions are more or less comparable but their water exchange rate constants differ widely. It can be explained in terms of the electron population density at the t_{2g}– level to which the entering ligand approaches in an associative path. Definitely it is most favoured for $Sc^{3+}(t_{2g}^0)$ and most disfavoured for $Fe^{3+}(t_{2g}^3)$. Here it may be pointed out that CFAE for all these $[M(OH_2)_6]^{3+}$ (high spin) complexes is zero.

Note: In terms of MOT, the t_{2g} level (*i.e.* NBMO) or e_g^* (ABMO) may act as the LUMO (in the octahedral complexes) to receive the nucleophilic attack in the associative path (see Sec. 5.18.2). If there is a vacancy in the t_{2g} level to receive the nucleophilic attack, it will act as the LUMO. Otherwise, e_g^* will act as the LUMO. Obviously, participation of the vacant ABMO–e_g^* to accommodate the nucleophile is energeticlly highly disfavoured. Hence, vacancy in the t_{2g} orbital is required to favour the associative path (see Sec. 5.18.2). This vacancy may be created through electron pairing in the t_{2g} level as in the t_{2g}^3 system of Cr(III) and Fe(III) (h.s.).

However, in some cases, participation of e_g^*–LUMO to receive the nucleophilic attack has been argued (Sec. 5.19.6).

5.14 EIGEN-WILKINS MECHANISM (1960–65)

The stoichiometric *I* (interchange) mechanism (two step reactions involving two TSs, see Fig. 5.8.1) for ligand substitution can be represented, in general, as follows:

$$L_5M—X + Y \rightarrow (Y\cdots L_5M\cdots X)^{\neq} \rightarrow L_5M—Y + X$$

The transient nature of the activated complex precludes its observation or characterisation. It may be noted that the perfect *D* and *A* mechanisms (single step reaction involving one TS, see Fig. 5.8.1) operate only in some limited cases but the ***I*–stoichiometric mechanism is the most common one**. It may be I_a or I_d depending on the situation.

The interchange mechanism for the ligand substitution reactions of octahedral complexes is very often consistent with the **Eigen-Wilkins mechanism.** This well documented reaction mechanism of ligand substitution reaction is named after M. Eigen (who shared the **1967 Nobel Prize in Chemistry**, for his contribution in understanding the fast reactions of labile metal complexes with R.G. Norrish and G. Porter) and R.G. Wilkins. This mechanism is also described as **Eigen-Tamm** or **Eigen-Wilkins-Tamm** or simply **Eigen mechanism**.

The **first step** of this mechanism considers the formation of an **outer-sphere (O.S.) complex** or **encounter complex** in a rapid (almost diffusion controlled rate) pre-equilibrium step. In the encounter complex, the leaving group or ligand remains in the ***primary coordination sphere*** while the entering ligand (say *Y*) resides in the first solvation shell, *i.e.* in the ***secondary coordination sphere*** of the reacting complex (say $L_5M—X$, where X is the leaving group). This pre-equilibrium step is rapid as the solvent molecules in the secondary coordination sphere are loosely held and they exchange very rapidly.

Formation of the *outer-sphere* or *encounter complex* may involve the different types of binding forces, *e.g.* Coulombic attraction (for the oppositely charged reactants, *i.e.* ion-pair formation), ion-dipole interaction (when one reactant is charged and the other reactant is neutral but polar), hydrogen-bonding interaction, etc. Then *at the **second step** which is the rate determining step (rds),* the encounter complex transforms into the product complex (*i.e.* $L_5M—Y$) and *X* which resides in the secondary coordination sphere of the product complex. This rate determining step leads to the ***interchange of the X and Y ligands between the first and second coordination spheres***. Finally, the leaving group *X* diffuses away from the second coordination sphere rapidly. Here it may be mentioned that theoretically transformation of the O.S. complex into the product complex at the rate determining step may occur by a *D* or an *A* or an *I*-process but **most commonly by an *I* (I_d or I_a)** process (*cf.* Sec. 5.9.3). It is reasonable because in the O.S. complex, the entering ligand (*Y*) is favourably disposed or positioned to participate in a bonding interaction with the metal centre to some extent in the transition state. Thus the whole process can be represented as follows:

$$L_5M—X + Y \underset{}{\overset{K_{OS}}{\rightleftharpoons}} (L_5MX, Y) \text{ (Outer-sphere complex)} \quad \textbf{(Step 1)}$$

$$(L_5MX, Y) \xrightarrow[(rds)]{k_i} (L_5MY, X) \text{ (O.S. Complex)} \quad \textbf{(Step 2)}$$

$$(L_5MY, X) \xrightarrow{\text{fast}} L_5M—Y + X \quad \textbf{(Step 3)}$$

Scheme 5.14.1: Eigen mechanism

Steps 1 and 3 are the **diffusion controlled processes** and they operate in the time scale of about ns (10^{-9}s). It indicates that formation and break up of the complex in step 1 occur very rapidly and the same thing happens for the step 3. Very often, for the sake of simplicity, step 2 and step 3 are not shown separately. These two steps are simply shown as follows:

$$(L_5MY, X) \xrightarrow[(rds)]{k_i} L_5M—Y + X \qquad \textbf{(Step 3)}$$

Scheme 5.14.1 leads to the following rate law (*cf.* Sec. 5.9.3).

$$\text{Rate} = \frac{k_iK_{OS}[L_5MX]_0[Y]_0}{1+K_{OS}[Y]_0}; \ ([Y]_0 \gg [L_5MX]_0)$$

and

$$k_{obs} = \frac{k_ik_{OS}[Y]}{1+K_{OS}[Y]}, \ (i.e.\ [Y]_0 = [Y]+[O.S.] \approx [Y])$$

It leads to:

$$\frac{1}{k_{obs}} = \left(\frac{1}{k_iK_{OS}}\right)\frac{1}{[Y]} + \frac{1}{k_i}$$

Linear plot of $\frac{1}{k_{obs}}$ *vs.* $\frac{1}{[Y]}$ can evaluate k_i and K_{OS} (*cf.* Fig. 5.9.3.1).

$$k_i = \frac{1}{\text{intercept}}; \ K_{OS} = \frac{\text{intercept}}{\text{slope}}.$$

Case-I: $k_{obs} \approx k_{\lim} \approx k_i$ when $K_{OS}[Y] \gg 1$. This limiting value is obtained for the high concentration of *Y*.

Case II: $k_{obs} = k_iK_{OS}[Y] = k_f[Y]$, (when $1 \gg K_{OS}[Y]$) *i.e.*, k_f (formation rate constant, a **composite rate constant**) = k_iK_{OS}.

This is the **most common form of the rate law** for the *I* (I_d and I_a) process.
An example of the reaction is:

$$[Ni(OH_2)_6]^{2+} + NH_3 \xrightleftharpoons{K_{OS}} \{[Ni(OH_2)_6]^{2+}, NH_3\} \ \text{(O.S. Complex)}$$

$$\{[Ni(OH_2)_6]^{2+}, NH_3\} \xrightarrow{k_i} [Ni(NH_3)(OH_2)_5]^{2+} + H_2O$$

and

$$k_{obs} = \frac{k_iK_{OS}[NH_3]}{1+K_{OS}[NH_3]}, \ ([NH_3]_0 \gg [[Ni(OH_2)_6]^{2+}]_0)$$

$$= k_iK_{OS}[NH_3] = k_f[NH_3], \ (\text{when } 1 \gg K_{OS}[NH_3])$$

Special Case: Y = solvent; X = Solvent, *i.e.* Solvent Exchange Reactions

If the entering ligand is the solvent then a special case arises because the encounter equilibrium remains always saturated. It happens so because the encounter complex $\{L_5MX, Y\}$ is the solvated starting complex $[L_5M(OH_2), OH_2]$ (in aqueous solvent). ***Thus the reacting complex always exists as the encounter complex,*** because the reacting complex is always surrounded by the solvent molecules in the second coordination sphere. It leads to: $k_{obs} = k_i = k_{ex}$.

The relation, $k_i = k_f/K_{OS}$, is very important to identify a reaction mechanism. Here k_f($dm^3\ mol^{-1}\ s^{-1}$) is called the **formation rate constant (a composite rate constant)** and k_i denotes the rate constant for the transformation of an outer-sphere complex into an inner-sphere complex through the interchange process. The k_f value may be different for different ligands but if k_i remains almost constant, it occurs through the I_d pathway. ***In fact, outer-sphere complexation of an aqua-complex does not remarkably alter its water exchange rate constant (*** k_{ex} ***).*** Thus the estimated k_i values ($= k_f/K_{OS}$, k_f is an experimental quantity; K_{OS} can be computed from the theoretical models, *e.g.* Fuoss-Eigen equation for an ion-pair formation) should be very close to k_{ex} for the **dissociative transformation** (*i.e.* I_d) of the outer-sphere complex.

Distinction between I_d and I_a paths in terms of Eigen-Wilkins mechanism (*cf.* Sec. 5.19.9)

For an I_d path, the k_i ($= k_f/K_{OS}$) values estimated for the different types of ligands differing in basicity, structure and binding sites should cover a narrow range close to k_{ex} while for an I_a path (*i.e.* associative transformation of the outer-sphere complex), the extimated k_i values should cover a huge span without having any relationship with k_{ex}. Formation reaction at the **bivalent metal ions** like $[M(OH_2)_6]^{2+}$ (M = Ni, Fe, Cr, etc.) passes through the I_d path (*i.e.* constancy in the estimated k_i values) while, the formation reaction at the **trivalent metal ions** like $[M(OH_2)_6]^{3+}$ (M = Ti, Fe, Cr, etc.) adopts the I_a path (*i.e.* the estimated k_i-values cover a wide range depending on the nature of the entering ligands). But their **hydrolysed products**, *i.e.* $[M(OH)(OH_2)_5]^{2+}$, very often adopt the I_d-path because of the lower charge (*cf.* +3 *vs.* +2) and labilising effect of the π-donor OH group **(*cis*-effect).**

That $[Ni(OH_2)_6]^{2+}$ reacts with the different types of ligands through the I_d path is evidenced by the constancy of k_i values (*cf.* Table 5.14.1) (*i.e.* $k_f = k_i K_{OS} \approx k_{ex} K_{OS}$).

Table 5.14.1: Kinetic parameters (25°C) for the formation reactions of $[Ni(OH_2)_6]^{2+}$ (*cf.* R.G. Wilkins, *Acc. Chem. Res.*, **3**, 408, 1970; D.B. Rorabacher, *Inorg. Chem.*, **5**, 1891, 1966)

Entering ligand (Y)	k_f ($dm^3\ mol^{-1}\ s^{-1}$)	K_{OS} ($dm^3\ mol^{-1}$)	$k_i = \left(\frac{k_f}{K_{OS}}\right)$ (s^{-1})
NH_3	5×10^3	0.15	3×10^4
F^-	8×10^3	1.0	0.8×10^4
$CH_3CO_2^-$	1×10^5	3.0	3×10^4
py	4×10^3	0.13	3×10^4
SCN^-	6×10^3	1.0	0.6×10^4
Im	5×10^3	0.31	1.6×10^4
H_2O	–	–	3×10^4 ($= k_{ex}$)

Note: The composite rate constant k_f varies in the range, $10^3 - 10^5$ $dm^3\ mol^{-1}\ s^{-1}$ but k_i covers a narrow range, 0.6×10^4 to 3×10^4 s^{-1} and this is close to the k_{ex} value (*i.e.* $k_f = k_i K_{OS} \approx k_{ex} K_{OS}$).

Statistical relationship between k_i and k_{ex}

$$[M(^*OH_2)_6] + H_2O \longrightarrow k_{ex};\ [M(OH_2)_6] + Y \longrightarrow k_i$$

It can be shown: $k_i = \frac{6}{8} k_{ex} = 0.75 k_{ex}$. This aspect has been discussed in Sec. 5.19.9.

Eigen-Fuoss equation to estimate K_{OS}: Oppositely charged reactant particles are supposed to generate the ion-pair (IP) as the encounter or outer-sphere complex. The equilibrium constant (K_{OS}) for the formation of an ion-pair outer-sphere complex or encounter complex can be very often calculated by Eigen-Fuoss equation. In fact, R.M. Fuoss developed the equation by using the concept of statistical thermodynamics while it was developed by M. Eigen based on the principle of kinetics. Their result is called Eigen-Fuoss equation as given below.

$$K_{OS} = \frac{4\pi N r^3}{3000} \exp\{-U(r)/k_B T\}$$

where $U(r)$ is the **Debye-Huckel interionic potential.**

$$U(r) = \frac{Z_1Z_2e^2}{rD} - \frac{Z_1Z_2e^2\kappa}{D(1+\kappa r)}; \quad \kappa^2 = \frac{8\pi Ne^2\mu}{1000Dk_BT}$$

Thus $U(r)$ represents the sum of the attractive and repulsive potentials between the two ions bearing the charge Z_1e and Z_2e.

(N = Avogadro's number; r = distance of closest approach of the ions of charges Z_1e and Z_2e; e = charge carried by an electron in esu units; D = bulk dielectric constant, μ = ionic strength; k_B = Boltzmann's constant).

It is evident that K_{OS} being proportional to r^3 is highly favoured for the **larger interacting ions/ species** (*i.e.* r becomes larger) and K_{OS} is also favoured for the **oppositely charged ions** (*i.e.* Z_1Z_2 becomes negative).

Thus the Eigen-Fuoss equation can be used to calculate the K_{OS} values for the **ion-dipole association** and **ion-pair association.** Taking, $\mu = 0.1$ mol dm^{-3} and $r = 500$ pm, the computed values of K_{OS}/(dm^3 mol^{-1}) are: ~0.3, ~2, ~5, ~15 and ~100 for the charge products $(Z_1Z_2) = 0, -2, -3, -4$ and -6 respectively. For the ion-dipole interaction in the O.S. complex, a different equation may be used and the estimated K_{OS} value is 0.2 dm^3 mol^{-1} (Ref. J.E. Prue, *J. Chem. Soc.*, 1965, 7534).

- **Eigen-Fuoss equation (simplified form in SI unit):**

$$K_{OS} = \frac{4}{3}\pi r^3 N \exp\left[-\frac{U(r)}{k_BT}\right]$$

$$U(r) \text{ (Coulombic potential energy)} \approx \frac{Z_1Z_2e^2}{4\pi\varepsilon r}$$

where ε = permitivity of the medium = $\varepsilon_r\varepsilon_o$ where ε_r (relative permitivity or dielectric constant) = $\varepsilon/\varepsilon_o$, ε_o = vacuum permitivity = 8.854×10^{-12} C^2 N^{-1} m^{-2}.

The above simplified equation also concludes that K_{OS} increases for the **larger interacting species** ($K_{OS} \propto r^3$) and the **oppositely charged interacting ions** ($U(r)$ becomes negative). Obviously, in the above simple form, for the ion-dipole interaction, *i.e.* one of the reactants is uncharged (*i.e.* $Z_1Z_2 = 0$) as in the case of H_2O substitution by NH_3 in $[Ni(OH_2)_6]^{2+}$, $U(r) = 0$. Taking the encounter distance (r) = 200 pm = 200×10^{-12} m, we find:

$$K_{OS} = \frac{4}{3}\pi r^3 N = \frac{4\pi}{3}(2\times10^{-10}\ \text{m})^3 \times 6.02\times10^{23}\ \text{mol}^{-1}$$

$$= 2.0\times10^{-5}\ \text{m}^3\ \text{mol}^{-1} = 2.0\times10^{-2}\text{dm}^3\ \text{mol}^{-1}$$

For the two singly charged opposite ions (*i.e.* $Z_1Z_2 = -1$) in water media (relative permitivity, $\varepsilon_r = 78$) at 298 K, K_{OS} becomes 36 times higher than that for $Z_1Z_2 = 0$ for the same interacting distance.

$$\exp\left[-\frac{U(r)}{k_BT}\right] = \exp\left[\frac{e^2}{4\pi\varepsilon rk_BT}\right] \approx 36$$

Here it may be recalled that K_{OS} for the ion-dipole association can be calculated by a different equation rather than Eigen-Fuoss equation.

Estimation of K_{OS} from kinetic data: Nonlinear dependence of k_{obs} on [Y] can lead to the evaluation of K_{OS}. The **double reciprocal plot** (*i.e.* Burk-Lineweaver plot), *i.e.* $\frac{1}{k_{obs}}$ *vs* $\frac{1}{[Y]}$ can evaluate K_{OS} (= intercept/slope) which may be compared with the values calculated theoretically.

5.15 MECHANISTIC STEPS FOR CHELATION BY A DIDENTATE LIGAND: STEADY-STATE MECHANISM *vs.* EQUILIBRIUM MECHANISM

If the entering ligand is a didentate one, then the general mechanism may be given as in Scheme 5.15.1.

$$\left[(H_2O)_5M—OH_2\right] + L—L \xrightleftharpoons{K_{OS}} \underset{(I)}{\left\{\left[(H_2O)_5M(OH_2)\right]\cdot L—L\right\}} \quad ...(5.15.1)$$

$$\left\{\left[(H_2O)_5M\underset{(I)}{(OH_2)}\right]\cdot L—L\right\} \underset{k_{-1}}{\overset{k_1}{\rightleftharpoons}} \left[\underset{(II)}{(H_2O)_5M}—L—L\right] + H_2O \quad ...(5.15.2)$$

$$\left[\underset{(II)}{(H_2O)_5M}—L—L\right] \underset{k_{-2}}{\overset{k_2}{\rightleftharpoons}} \left[\underset{(III)}{(H_2O)_4M}\langle L—L\rangle\right] + H_2O, \text{ i.e. } \left[M(OH_2)_4(L_2)\right] \quad ...(5.15.3)$$

Scheme 5.15.1: Chelation process by a didentate ligand.

Species **I** is an outer-sphere complex; species **II** is an inner-sphere complex where the didentate ligand binds as a unidentate one; species **III** is the final product.

Here k_2 and k_{-2} are the rate constants for the ring closure and ring opening step respectively. The formation (k_f) and dissociation (k_d) rate constants may be given as follows if the **steady-state condition** is maintained for the species $[(H_2O)_5M—L—L]$ (**II**).

$$\text{Rate} = k_f[M(OH_2)_6][L—L] - k_d[M(OH_2)_4(L_2)]$$

In presence of excess ligand concentration with respect to that of the metal ion, the above rate equation integrates to the following equation (see Sec. 5.7):

$$k_{obs} = k_f[L-L] + k_d = -\frac{2.303}{t}\log\frac{[M(OH_2)_4(L_2)]_e - [M(OH_2)_4(L_2)]_t}{[M(OH_2)_6]_e - [M(OH_2)_6]_0}$$

where the subscripts e and o denote the equilibrium and initial concentration respectlvely.

$$k_f = K_{OS}k_1k_2/(k_{-1}+k_2) \quad ...(5.15.4)$$

$$k_d = k_{-1}k_{-2}/(k_{-1}+k_2) \quad ...(5.15.5)$$

From the plot of k_{obs} *vs.* $[L–L]_o$ (under the condition, $[L–L]_o \gg [M]_o$, k_f and k_d can be estimated (see Fig. 5.7.1).

- **If the ring closure step is faster** than the bond fission of the unidentatedly bonded ligand in the intermediate **II** (*i.e.* $k_2 \gg k_{-1}$), then the formation rate constant is given by:

$$k_f \approx K_{OS}k_1 \quad ...(5.15.6)$$

$$\approx K_{OS}k_{ex}, \ (\textbf{for the } I_d \textbf{ path}), \quad ...(5.15.7)$$

The condition, $k_2 \gg k_{-1}$, is attained when the ring closure step (*i.e.* k_2-step) is not sterically hindered and the first bond formed by L—L is relatively strong so that tendency of the bond to break (measured by k_{-1}) is less. The above equation is equivalent to the equation found true for complexation by the unidentate ligand through the Eigen mechanism (*cf.* Sec. 5.14). This type of observation has been noted in many cases where the ring closure step is a faster one. This situation (*i.e.* faster ring closure step) leading to the Eigen mechanism for the chelation process may be referred to as **normal substitution** which indicates that the didentate or polydentate ligand resembles approximately the unidentate ligand (for which only the first step is meaningful) in terms of rate constant (k_f). The chelation by *bpy*, *phen*, *terpy*, etc. with $[Ni(OH_2)_6]^{2+}$ (*cf.* R.G. Wilkins, *Acc. Chem. Res.*, **3**, 408, 1970) follows the Eigen mechanism, *i.e.* ***normal*** I_d ***substitution,*** *i.e.* k_f (formation rate constant, a composite rate constant) ≈ $K_{OS}k_{ex}$ (*cf.* Table 5.15.1).

Table 5.15.1: Formation rate constants (k_f) for the formation of some representative Ni(II)-mono-chelates from $[Ni(OH_2)_6]^{2+}$ (data at 25°C) (*cf.* R.G. Wilkins, *Acc. Chem. Res.* **3**, 408, 1970)

Entering ligand:	*bpy*	*phen*	*terpy*	*gly*$^-$	*α–ala*$^-$
k_f (dm^3 mol^{-1}s^{-1}):	1.6×10^3	3.2×10^3	1.4×10^3	4×10^4	2×10^4

Note: The difference in the k_f values arises from the difference in the K_{OS} values but, k_f/K_{OS} remains approximately constant which is close to the water exchange rate constant (k_{ex}). For an approximate calculation, K_{OS} may be taken as 0.2 dm^3 mol^{-1} for bpy and phen; and as 2 dm^3 mol^{-1} for the anionic ligand gly$^-$ and α-ala$^-$ (see Sec. 5.14 for the K_{OS} values)

If the ring closure step contributes to k_f, then k_f calculated by Eqn. 5.15.7 differs significantly from the observed k_f value. When the ring closure step is sterically hindered or the first metal-ligand bond formed is weak (*e.g.* by the O-donor sites) to favour the k_{-1} step, this situation arises. If k_1, k_{-1} are sufficiently fast and $k_{-1} \gg k_2$, then this situation arises, *i.e.* **ring closure step contributes to k_f. Under this condition, the steady-state concept will not be applicable.** Thus the **steady state mechanism** will be shifted to an **equilibrium mechanism.**

Table 5.15.1 indicates that ***phen reacts faster than bpy.*** It is believed that this difference arises due to some contrbution of the ring closure step to k_f for complexation by *bpy*. In *phen*, the two binding sites are held at close proximity to the metal centre and consequently, the ring closure step is easier compared to the case of *bpy* where due to the rotation around the C—C single bond, the second binding site may lie away from the coordination sphere. **Thus, for bipy, the ring closure step is more hindered.** In other words, for *bpy*, contribution of the ring closure step is more probable than the case of *phen*. Similarly, contribution of the ring closure step for *β–ala*$^-$ that forms a **six membered chelate ring** (which is sterically more hindered) is more probable than for *α–ala*$^-$ that forms a **five membered chelate ring** which is more stable (see Vol. 4). **In fact, α–ala$^-$ reacts faster than β–ala$^-$** (*cf.* $k_f = 1 \times 10^4$ dm^3 mol^{-1} s^{-1} for *β–ala*$^-$, $= 2 \times 10^4$ dm^3 mol^{-1} s^{-1} for *α–ala*$^-$, in the reaction with $[Ni(OH_2)_6]^{2+}$.).

● **Steady state mechanism *vs.* equilibrium mechanism (see Sec. 5.7):** Under the condition: k_1 and k_{-1} are very fast and the ring closure step (k_2) is very slow, the **steady state mechanism is no longer valid**. The *most favourable situation for the contribution of the ring closure step appearing as the rate determining step* (*i.e.* $k_{-1} >> k_2$) arises during the formation of **ternary complexes** where the primary ligand (say *A*) is tetradentate in character, bulky in size and the entering ligand binds first through the *O*-site (*i.e.* weak bond forming site) followed by ring closure through the *N*–site. (*cf.* A.K. Das, *Int. J. Chem. Kinet,* **28**, 275-282, 1996 and the references cited therein; D. Sc. Thesis by A.K. Das, Visva Bharati). This leads to the **equilibrium mechanism** illustrated in Scheme 5.15.2 that involves a rapid pre-equilibrium between the reactants followed by the rate determining ring closure step. Here, the species II remains **in an equilibrium** with the reactants.

$$cis\text{-}[(A)Ni(OH_2)_2]^- + \overset{O^-}{\underset{N}{|}} \;\underset{}{\overset{K_{OS}}{\rightleftharpoons}}\; \left\{[(A)Ni(OH_2)_2]^- \cdot \overset{O^-}{\underset{N}{|}}\right\} \;\overset{K_1}{\rightleftharpoons}\; \left[(A)Ni{<}^{O\cdots N}_{OH_2}\right]^{2-} + H_2O$$

(*e.g.* glycinate) I (O.S. Complex) II (ICB) effect

$$\text{II} \xrightarrow[\text{(rds)}]{k_2} \left[(A)Ni{<}^{O}_{N}\right]^{2-} + H_2O$$

Scheme 5.15.2: Formation of ternary complex where A (primary ligand) is a tetradentate ligand (*e.g.* nta^{3-})

Scheme 5.15.2 leads to:

$$k_f = \frac{K_{OS}K_1k_2[\text{O—N}]}{1+K_{OS}K_1[\text{O—N}]} \quad \text{...(5.15.8)}$$

$$= \frac{K_{eq}k_2[\text{O—N}]}{1+K_{eq}[\text{O—N}]}$$

where K_{eq} (overall equilibrium constant) $= K_{OS}K_1 = \frac{k_1'}{k_{-1}'}$ (say).

In Scheme 5.15.2, the species **II** remains in an equilibrium with the reactants. The condition $k_{-1} \gg k_2$ leads to $K_1 = \frac{k_1}{k_{-1}}$ $\left(\text{or, } k_{-1}' \gg k_2 \text{ leading to } K_{eq} = \frac{k_1'}{k_{-1}'}\right)$ and *it does not allow the steady-state condition of the species* **II.** Rather, a rapid equilibrium leading to a **considerable buildup of the species II** (in which the didentate ligand binds unidentatedly through the weaker bond forming site) is established before the sluggish ring closure step (k_2). In such cases, the high value of the pre-equilibrium constant (K_{eq}) arises from the **ICB (internal conjugate base) effect** in which the free end of the amine-N (*cf.* pK_a of amine-N of glycine = 9.70) forms a H-bond with the coordinated H_2O molecule.

5.16 DISTINCTION AMONG THE *D*, I_d and I_a PATHWAYS (*cf.* Sec. 5.19.9)

It has been already mentioned that in the ligand substitution process of the octahedral metal complexes, the most commonly observed mechanisms are *D*, I_d and I_a. *They can be distinguished from the nature of rate dependence on the entering nucleophile (Y).*

A. Distinction between the *D* and I_d processes (*cf.* Secs. 5.9.1, 5.9.3., 5.14, 5.19.9)

Here it is worth mentioning that the **D- and I_d-mechanisms give the similar rate law** (*cf.* Sec. 5.9.4) and it becomes a very difficult task to distinguish between the *D*- and I_d-mechanisms. If the **lower coordinate intermediate is sufficiently long-lived to discriminate the nucleophiles**, then the *D*-mechanism can be argued. If no convincing evidence in favour of a such long-lived intermediate is available, then it is better to propose the I_d-mechanism. For an I_d-process, we have: $k_f = k_iK_{os}$ and $k_{obs} \approx k_f[Y]$ when $1 \gg K_{os}[Y]$ (*cf.* Sec. 5.14).

Let us try to illustrate by taking the complex $[L_5M(OH_2)]$ reacting with the different types of nucleophiles in a *D*-process.

$$\left.\begin{aligned} L_5M\text{—}OH_2 &\underset{k_{-1}}{\overset{k_1}{\rightleftharpoons}} L_5M + H_2O \\ L_5M + Y &\xrightarrow{k_2} L_5M\text{—}Y \end{aligned}\right\} \textbf{\textit{D}-process}$$

The rate law (*cf.* Sec. 5.9.1) is given by:

$$\text{rate} = \frac{k_1k_2[L_5M(OH_2)][Y]}{k_{-1}[H_2O]+k_2[Y]}$$

$$= \frac{k_1k_2[L_5M(OH_2)][Y]}{k_{-1}'+k_2[Y]}$$

$$= k_f[L_5M(OH_2)][Y]$$

where $$k_f = \frac{k_1k_2}{k_{-1}'+k_2[Y]}$$

Here k_1 denotes the *first-order forward rate constant* and k'_{-1} denotes the *pseudo first-order backward rate constant* for the first step ($k_{-1}[H_2O] = k'_{-1}$ = constant). It leads to (under the condition, $[Y] \gg [L_5M(OH)_2]$):

$$\text{rate} = k_{obs}\,[L_5M(OH_2)]$$

where
$$k_{obs} = k_f\,[Y] = \frac{k_1k_2\,[Y]}{k'_{-1} + k_2\,[Y]}$$

or
$$\frac{1}{k_{obs}} = \left(\frac{k'_{-1}}{k_1k_2}\right)\frac{1}{[Y]} + \frac{1}{k_1}$$

Pure dissociative process (*D*): Does it depend on the nature of the entering ligand?

- Yes, the rate depends nonlinearly on [*Y*], $k_{obs} = \dfrac{k_1k_2\,[Y]}{k'_{-1} + k_2\,[Y]}$
- Yes, the rate will depend linearly on [*Y*], under the condition $k'_{-1} \gg k_2\,[Y]$
- No, under the condition, $k_2[Y] \gg k'_{-1}$, $k_{obs} = k_{lim} = k_1$ **(limiting rate law)**

Conditions favouring the *D*-process over the I_d process

- Anation or aquation of the **anionic complexes** disfavours the outer-sphere association due to an electrostatic repulsion between the starting anionic complex and the electron rich approaching nucleophile. Outer-sphere association is a pre-requisite condition for the I_d–path (*cf.* Eigen mechanism). Such a situaion favours the *D*–process; dissociation of the leaving group from the anionic complex is favoured. It may be noted that the anionic complexes strongly disfavour the approach of the entering nucleophile in the **associative path**.
- Strong *trans* or *cis*–labilising groups favour the *D*–path. Pi-donor nonleaving groups can show the ***cis-labilising effect; σ-donor ligand can labilise at the trans-position.***
  Examples (see Sec. 5.10): Anation of $[Co(CN)_5(OH_2)]^{2-}$; anation of $[RhCl_5(OH_2)]^{2-}$; aquation of $[Fe(CN)_5L]^{3-}$; aquation of $[Cr(NCS)_6]^{3-}$; aquation of $[Co^{III}(CN)_5X]^{3-}$; etc.

Lifetime of the intermediate in the *D*- and I_d-processes

- In a *D*-process (a **two step reaction** involving two TSs), the lower coordinate intermediate is sufficiently long lived (*i.e.* finite life time) while in the I_d-process (a **single step reaction** involving one TS), the lower coordinate intermediate is too short lived to discriminate the nucleophiles.

If the concentration of the entering ligand is sufficiently low to maintain the situation, $k'_{-1} \gg k_2[Y]$, then we have:

$$k_{obs} = \frac{k_1k_2}{k'_{-1}}[Y] = k_f\,[Y]$$

Thus k_f (formation constant) depends on the ratio k_2/k'_{-1}.

What is the significance of the ratio k_2/k'_{-1} (more correctly, k_2/k_{-1})?

It means that the lower coordinate intermediate, *i.e.* L_5M, is **sufficiently long lived** to react competitively with the solvent (*i.e.* k_{-1} path) and *Y* (*i.e.* k_2 path). The kinetic competition between H_2O and *Y* for the lower coordinate intermediate generates the term k_2/k'_{-1}. ***The value of this ratio depends on the relative nucleophilicities of Y (i.e. k_2–path) and the solvent (k_{-1}-path).*** For an **I_d-process**, the intermediate is **too short lived** to discriminate the nucleophiles. Thus in a *D*-process, k_f depends on the nature of *Y* but in the I_d-process, k_f is relatively more insensitive of the nature of *Y* provided K_{OS} does not vary significantly (*cf.* $k_f = k_iK_{OS}$).

For an I_d **path** (Sec. 5.9.3), formation of the outer-sphere complex is the required condition and if this step can be made disfavoured then the *D*–process may operate. **Anation of a highly anionic complex** will disfavour the formation of outer sphere complex due to an electrostatic repulsion between the reactants. This will disfavour the I_d path but **favour the *D* path**. The following anation reactions meet this condition.

$$\left[Co(CN)_5(OH_2)\right]^{2-} + Y^- \rightarrow \left[Co(CN)_5(Y)\right]^{3-} + H_2O$$

$$[RhCl_5(OH_2)_5]^{2-} + Y^- \rightarrow [RhCl_5(Y)]^{3-} + H_2O, \ (Y^- = Cl^-, Br^-, I^-, SCN^-, N_3^-, \text{etc.}).$$

In fact, for these reactions, the $\frac{k'_{-1}}{k_2}$ $\left(= \frac{\text{slope}}{\text{intercept}}\right.$, slope and intercept obtained from the linear plot, $1/k_{obs}$ *vs.* $1/[Y^-]$; *cf.* Fig. 5.9.1.1) values determined depend on the nature of the entering anion. $k_{-1}/k_2 = 1$ for the water exchange process, *i.e.* $Y = H_2O$ and $k_2 = k_{-1}$. It supports the operation of the *D*–mechanism.

B. Distinction between the *D* and I_a processes

Now the question arises to distinguish between the *D* and I_a processes. Both of which depend on the nature of the entering ligand.

Under the condition, $k_2[Y] \gg k'_{-1}$, the **limiting rate law for the *D*–process** is given by:

rate $\approx k_1[L_5MX]$, *i.e.* $k_{obs} \approx k_{lim} = k_1$; k_1 describes the water loss step.

Thus k_{obs} *gives the rate constant for the loss of the bound* H_2O *molecule to produce the lower coordinate intermediate. Under this limiting condition,* k_{obs} *for the D–process is independent of the nature of the entering ligand.* **This situation is never attained for the I_a–process** [*cf.* $k_{obs} = \frac{k_i K_{OS}[Y]}{1+K_{OS}[Y]} \approx k_i K_{OS}[Y]$ (when $1 \gg K_{OS}[Y]$); and $k_{lim} \approx k_{obs} \approx k_i$ (when $K_{OS}[Y] \gg 1$), k_i depends on the nature of *Y* in the I_a path] **and it makes the difference between the *D* and I_a pathways.** ***Thus in a D-process,*** k_{lim} ($\approx k_1$) ***is independent of the nature of Y but in an*** I_a***-process,*** k_{lim} ($\approx k_i$) ***is dependent on the nature of Y.***

C. Example of *D*, I_d and I_a reactions (*cf.* Sec. 5.19.9)

- $\left[Co^{III}(corrin)(OH_2)\right] \xrightarrow{\text{anation}}$ (Mimicking the vitamin B_{12})
- $\left[Co(CN)_5(OH_2)\right]^{2-} \xrightarrow{\text{anation}}$

} **D-process**, *i.e.* k_{obs} depends on $[Y]$ under the condition $k'_{-1} \gg k_2[Y]$ and k_{obs} is independent of $[Y]$ under the condition, $k'_{-1} \ll k_2[Y]$; the lower coordinate intermediate is sufficiently **long lived.**

Note: The **flexible corrin ring** (see **Vit. B_{12}**, authors' book, Bioinorganic Chemistry) tilts towards the leaving group to **favour the dissociative activation** through the steric interaction between the corrin ring and leaving group.

- ***D*-process:** The ratio k'_{-1}/k_2 obtained from the double reciprocal plot (*i.e.* $1/k_{obs}$ *vs.* $1/[Y^-]$, *cf.* Fig. 5.9.1.1) has been found to vary for the different entering anions. From the estimated $\frac{k'_{-1}}{k_2}$ ratios, the $\frac{k_{-1}}{k_2}$ ratios can be calculated (*cf.* $k'_{-1} = k_{-1}[H_2O]$, $[H_2O] \approx 55.5$ mol dm^{-3}). For the water exchange process, the ratio k_2/k_{-1} becomes unity. ***The ratio*** k_2/k_{-1} ***gives the measure of the relative nucleophilicities of Y or*** Y^- ***and*** H_2O. The values of k_2/k_{-1} are as follows for the **anation of** $[Co(CN)_5(OH_2)]^{2-}$.

$$k_2/k_{-1} = 1\ (H_2O),\ 6\ (Cl^-),\ 11.7\ (1^-),\ 20.4\ (NCS^-),\ 23.9\ (py)$$

- **D-process:** Aquation of $[Cr(NCS)_6]^{3-}$ ($\Delta V^{\neq}$ = +16 cm^3 mol^{-1}), $[Co^{III}(CN)_5X]^{3-}$, $[Fe^{III}(CN)_5X]^{3-}$, ($\Delta V^{\neq} \approx$ + 22 cm^3 mol^{-1}).

- **I_d-process :** $[Co(NH_3)_5(OH_2)]^{3+} \xrightarrow{\text{anation}} \}I_d$; $k_{obs} \approx k_i K_{OS}[Y] = k_f[Y]$, $k_i \left(= \dfrac{k_f}{K_{OS}}\right) \approx$ constant (independent of Y).

In fact, the lower coordinate **intermediate $[Co(NH_3)_5]^{3+}$ cannot** live longer than a few picoseconds.

- **I_a-process :** $\left[Ti(OH_2)_6\right]^{3+}, \left[Cr(OH_2)_6\right]^{3+}, \left[Fe(OH_2)_6\right]^{3+} \xrightarrow{\text{anation}} \}I_a$; $k_{obs} \approx k_i K_{OS}[Y] = k_f[Y]$, $k_i \left(= \dfrac{k_f}{K_{OS}}\right)$ not constant and varies in a wide range depending on the nature of Y.

5.17 PROTON AMBIGUITY AND LIMITATION OF THE RATE LAW (*cf.* D. Seewald, *Inorg. Chem.*, 2, 643, 1963)

If both the metal centre and ligand participate in protonation and deprotonation equilibria, then different kinetic paths may have the same type of proton dependence giving rise to **proton ambiguity** and the paths are kinetically indistinguishable. This aspect is illustrated in Scheme 5.17.1.

$$\underset{(CA)}{\left[M(OH_2)_6\right]^{m+}} + \underset{CB}{L^{n-}} \xrightarrow[\textbf{(path-I)}]{k_1} \left[M(OH_2)_5L\right]^{(m-n)+}$$

$$-H^+, K_h \Updownarrow \qquad\qquad -H^+, K_a \Updownarrow$$

$$\underset{CB}{\left[M(OH)(OH_2)_5\right]^{(m-1)+}} + \underset{(CA)}{LH^{(n-1)-}} \xrightarrow[\textbf{(path-II)}]{k_2} \left[M(OH_2)_5L\right]^{(m-n)+}$$

Scheme 5.17.1: Illustration of proton ambiguity and isomeric reaction paths.

The rate law for the Scheme 5.17.1 can be obtained as follows:

$$[M(OH_2)_6] \overset{K_h}{\rightleftharpoons} \left[M(OH)(OH_2)_5\right] + H^+; \text{ (charges not shown).}$$

$$K_h = \frac{[H^+]\left[M(OH)(OH_2)_5\right]}{\left[M(OH_2)_6\right]}; \; T_M = \left[M(OH_2)_6\right] + \left[M(OH)(OH_2)_5\right]$$

or

$$K_h = \frac{[H^+]\left[M(OH)(OH_2)_5\right]}{T_M - \left[M(OH)(OH_2)_5\right]}$$

i.e.

$$[M(OH)(OH_2)_5] = \frac{K_h T_M}{K_h + [H^+]}$$

and

$$[M(OH_2)_6] = T_M - \left[M(OH)(OH_2)_5\right] = \frac{T_M[H^+]}{K_h + [H^+]}$$

$$LH \overset{K_a}{\rightleftharpoons} L + H^+, \text{ (charge of the ligand not shown)}$$

i.e.

$$K_a = \frac{[L][H^+]}{[LH]}, \; T_L = [L] + [LH]$$

i.e. $$K_a = \frac{[\mathrm{L}][\mathrm{H}^+]}{T_\mathrm{L} - [\mathrm{L}]}$$

i.e. $$[\mathrm{L}] = \frac{T_\mathrm{L} K_a}{K_a + [\mathrm{H}^+]}; \text{ and } [\mathrm{LH}] = T_\mathrm{L} - [\mathrm{L}] = \frac{T_\mathrm{L}[\mathrm{H}^+]}{K_a + [\mathrm{H}^+]}$$

$$r_1 \textbf{ (rate in path-I)} = k_1[\mathrm{L}][\mathrm{M(OH_2)_6}]$$

$$= \frac{k_1 T_\mathrm{M}[\mathrm{H}^+]}{(K_h + [\mathrm{H}^+])} \times \frac{K_a T_\mathrm{L}}{(K_a + [\mathrm{H}^+])}$$

$$r_2 \textbf{ (rate in path-II)} = k_2[\mathrm{LH}][\mathrm{M(OH)(OH_2)_5}]$$

$$= \frac{k_2 T_\mathrm{L}[\mathrm{H}^+]}{(K_a + [\mathrm{H}^+])} \times \frac{K_h T_\mathrm{M}}{(K_h + [\mathrm{H}^+])}$$

$$\text{Total rate} = r = r_1 + r_2$$

$$= \frac{T_\mathrm{M} T_\mathrm{L}[\mathrm{H}^+]}{x}\{k_1 K_a + k_2 K_h\}, \text{ where } x = (K_h + [\mathrm{H}^+])(K_a + [\mathrm{H}^+]),$$

$$= \frac{T_\mathrm{M} T_\mathrm{L}}{x}\{k_1 K_a[\mathrm{H}^+] + k_2 K_h[\mathrm{H}^+]\}.$$

From the above equation, it is evident that *both the k_1– and k_2–path show the same type of hydrogen ion dependence* (*i.e.* first-order) and they remain **kinetically indistinguishable in terms of [H⁺] dependence**. In fact, the plot,

$$\frac{x(\text{rate})}{T_\mathrm{M} T_\mathrm{L}} \textit{ vs. } [\mathrm{H}^+], \text{ slope } = k(say) = k_1 K_a + k_2 K_h$$

gives a straight line with the slope = $k_1K_a + k_2K_h$. From this slope, it is not possible to find the contribution of the k_1–path and k_2–path separately.

k_1 path ⇒ $(CA)_M + (CB)_L$: called ***k(CACB)* path;**
k_2 path ⇒ $(CB)_M + (CA)_L$: called ***k(CBCA)* path;**

and these are called the **isomeric reaction pathways.**

- For the metal centres like Fe(III) (having log $K_h \approx -3.0$) which can provide $[Fe(OH_2)_6]^{3+}$ (*CA*) and $[Fe(OH)(OH_2)_5]^{2+}$ (*CB*) even in a fairly strong acidic condition, such a proton ambiguity prevails. From other evidences, it can be shown that though the unhydrolysed species is the predominant one in an acidic condition, *the reaction mainly passes through the step involving* $[Fe(OH)(OH_2)_5]^{2+}$ (*cf.* A.K. Das, *Bull. Chem. Soc. Jpn.*, **65**, 2205–2210, 1992; *Transition Met. Chem.*, **17**, 484-488, 1992; and the references cited there in).

Detailed analysis indicates the following results:

$k = k_1K_a + k_2K_h \approx k_2K_h$, $k_2 \approx k_iK_{os}$ (I – path) and $k_i \approx k_{ex}$ of $[M(OH)(OH_2)_5]^{2+}$, *i.e.* $[Fe(OH)(OH_2)_5]^{2+}$ reacts in the I_d-path. If ***the k_1K_a term is argued to contribute to k, it leads to an unusually high value of k_1.***

● To solve the problem with the proton ambiguity, we are, to consider other factors because from the experimental data, we cannot determine k_1 and k_2 separately. This is illustrated in the following reaction between Fe(III) and $HCrO_4^-$ at 25°C (*cf.* J.H. Espenson *et al*, *Inorg. Chem.*, **8**, 1051, 1969).

$$k(\mathbf{CACB})\mathbf{path}\begin{cases} HCrO_4^- \rightleftharpoons CrO_4^{2-} + H^+, \ K_a \approx 3\times10^{-7} \text{ mol dm}^{-3} \\ \left[Fe(OH_2)_6\right]^{3+} + CrO_4^{2-} \xrightarrow{k_1} \left[Fe(OCrO_3)(OH_2)_5\right]^{+} + H_2O \end{cases}$$

$$k(\mathbf{CBCA})\mathbf{path}\begin{cases} \left[Fe(OH_2)_6\right]^{3+} \rightleftharpoons \left[Fe(OH)(OH_2)_5\right]^{2+} + H^+, \ K_h \approx 1.7\times10^{-3} \text{ mol dm}^{-3} \\ \left[Fe(OH)(OH_2)_5\right]^{2+} + HCrO_4^- \xrightarrow{k_2} \left[Fe(OCrO_3)(OH_2)_5\right]^{+} + H_2O \end{cases}$$

$$k = k_1K_a + k_2K_h = 15 \text{ s}^{-1} \text{ (experimentally found).}$$

If we consider that the term 'k_1K_a' contributes to k (= 15 s^{-1}), it should be at least in the order of 10^0 to 10^1. Taking $k_1K_a = 10$, we get $k_1 = 10/K_a \approx 3.3 \times 10^7$ mol^{-1} dm^3 s^{-1} **which is unusually high** (*cf.* $k_{ex} \approx 10^2$ s^{-1} for $[Fe(OH_2)_6]^{3+}$). **In fact, this calculated value is not acceptable.** On the other hand, let us consider k_2K_h to contribute to k. Taking $k_2K_h \approx 10$, *i.e.* $k_2 = 10/K_h \approx 10^4$ mol^{-1} dm^3 s^{-1}. Taking $k_2 \approx k_iK_{OS}$ (*I*-process), the estimated k_i value is in the order of 10^4 s^{-1}.This value is quite reasonable for $[Fe(OH)(OH_2)_5]^{2+}$. In fact, the water exchange rate constant (k_{ex}) is in the order of 10^4 s^{-1} for $[Fe(OH)(OH_2)_5]^{2+}$. Thus in the present case, we may reasonably conclude that the ***k(CACB) path does not pratically contribute to k, i.e. k*** $\approx \boldsymbol{k_2K_h}$.

It may be mentioned that besides the k(CACB), k(CBCA) paths, other possible reaction paths in Scheme 5.17.1 are as follows:

$$[M(OH_2)]_6 + LH \Rightarrow k(\text{CACA})$$
$$[M(OH)(OH_2)_5] + L \Rightarrow k(\text{CBCB})$$

It may be mentioned that these *k(CACA) and k(CBCB) paths are not isomeric* and *they do not introduce any proton ambiguity.*

5.18 INTERPRETATION OF LABILITY AND INERTNESS IN TERMS OF THE VALENCE BOND THEORY (VBT) AND CRYSTAL FIELD THEORY (CFT) OF OCTAHEDRAL METAL COMPLEXES

5.18.1 Application of VBT in Explaining the Kinetic Behaviour of Octahedral Complexes

(a) Outer and inner orbital octahedral complexes: Valence bond theory (VBT) classifies the octahedral complexes as the **outer orbital complexes** (sp^3d^2) and **inner orbital complexes** (d^2sp^3). The involved *d*–orbitals in hybridisation are $d_{x^2-y^2}$ and d_{z^2}. Thus in the outer orbital complexes, the *nd* orbitals participate along with the *ns* and *np* orbitals for hybridisation while in the inner orbital complexes, the $(n-1)d$ orbitals participate along with the *ns* and *np* orbitals for hybridisation.

The inner orbital complexes are formed only when at least two $(n-1)d$ orbitals are available for d^2sp^3 hybridisation. Thus the situation is straight-forward for the d^0, d^1, d^2 and d^3 configurations. For d^n ($n > 3$) configuration, it needs rearrangement of electrons through pairing and/or excitation of *d*–electrons for the inner orbital complex formation.

(b) Kinetic behaviour of the outer orbital octahedral complexes: In terms of VBT, the outer orbital complexes are considered to be labile. The outer orbital complexes formed by $Mn^{2+}(3d^5)$, $Fe^{2+}(3d^6)$, $Co^{2+}(3d^7)$, $Ni^{2+}(3d^8)$, $Cu^{2+}(3d^9)$, $Zn^{2+}(3d^{10})$, etc. are labile. In the same way, the complexes of Cd^{2+}, Hg^{2+}, Al^{3+}, Ga^{3+}, In^{3+} are expected to be labile because they form the outer orbital complexes.

The nd orbitals are of higher energy than the (n − 1)d orbitals. This is why, the outer orbital complexes (sp^3d^2) are **considered to be relatively less stable** than the inner orbital complexes. Such complexes follow the **dissociative path** (*i.e.* bond breaking mainly contributes to the activation energy) and the process gets kinetically favoured due to the lower bond dissociation energy.

However, if the **charge on the metal centre** increases, the metal-ligand bond becomes stronger and consequently, the **dissociative path gets disfavoured**. *Thus the lability of the outer orbital complexes decreases with the increase of the positive charge on the central atom.* It explains the following lability sequence:

Ligand exchange rate: $[AlF_6]^{3-} > [SiF_6]^{2-} > [PF_6]^{-} > [SF_6]$; $[Na(OH_2)_x]^{+} > [Mg(OH_2)_x]^{2+} > [Al(OH_2)_6]^{3+}$

In fact, the outer orbital complexes (using the 3*d*, 3*s* and 3*p* orbitals in sp^3d^2 hybridisation) like SF_6, $[PF_6]^-$ are very much inert because of the very high positive charge on the central atom.

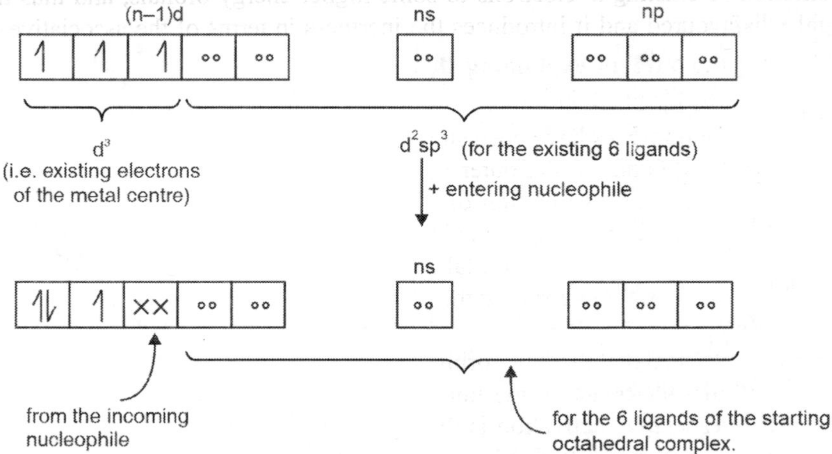

Scheme 5.18.1.1: Associative path for the d^3 system through the rearrangement of electrons

(c) Kinetic behaviour of the inner orbital octahedral complexes: The *d*-orbitals d_{xy}, d_{yz} and d_{xz} (*i.e.* t_{2g} set in terms of CFT) of the (*n* – 1) shell are kept out of hybridisation. *Participation of the inner (n – 1)d orbitals in hybridisation is considered to make* **the relatively stronger bonds** (compared to those of the outer orbital complexes). Because of this stronger metal-ligand bond in the inner orbital complexes, *the dissociative pathway is expected to be disfavoured but the* ***associative pathway will be favoured*** *through the formation of a new metal-ligand bond by the entering nucleophile to produce an intermediate of higher coordination number.*

If the substitution process is argued to pass through an **associative path** to produce a 7-coordinate intermediate from the starting octahedral complex, then the ease of formation of the intermediate depends on the **availability of a vacant *d*–orbital.** ***It indicates that at least one of the d–orbitals (i.e. d_{xy}, d_{yz} and d_{xz}) which do not participate in bonding in the starting octahedral complex, must remain vacant to accommodate the nucleophilic attack by the entering nucleophile.*** This situation is prevailing for d^0, d^1 and d^2 systems.

Sometimes, the rearrangement of the electrons may be required to make one such *d*-orbital vacant. For example, in the inner orbital complexes of the d^3 and d^4 electronic configurations, as such there is no vacant *d*–orbital to accommodate the incoming nucleophile in the associative process. However, redistribution of these electrons through pairing can make one *d*–orbital vacant for the incoming nucleophile. It is illustrated for the d^3–system in Scheme 5.18.1.1.

Thus we can conclude as follows.

- **Labile inner-orbital complexes**

d^0, d^1, d^2: vacant *d*–orbital is already available to accommodate the entering nucleophile in the associative path; and these are labile in terms of the associative activation.

- **Inert inner-orbital complexes**

d^3, d^4: redistribution of electrons through pairing can provide a vacant *d*–orbital for the entering nucleophile in the associative path; the process is relatively disfavoured compared to the cases of d^0, d^1, d^2 systems and consequently, the d^3, d^4 centres are less labile in terms of the associative activation.

d^5, d^6: to provide a vacant *d*-orbital for the entering nucleophile in the associative path, it needs the promotion of existing *d*–electrons to some higher energy orbitals, and thus the process is highly disfavoured and it introduces the inertness in terms of the associative activation.

(d) Limitations of the VBT in explaining the kinetic behaviour of the octahedral complexes: The important drawbacks are given below.

(i) The **outer orbital complexes** have been suggested to participate in a **dissociative path** based on the argument that the bonds in the outer orbital complexes are weaker than those in the inner orbital complexes. But the assumption of weaker bonds in the outer orbital complexes is not experimentally justified. For example, the hydration energies of the metal ions having the d^7, d^8, d^9 and d^{10} configurations (*i.e.* outer orbital hexaaqua complexes) are much higher than those of the d^0, d^1 and d^2 configurations forming the inner-orbital hexaaqua complexes. This fact weakens the argument of VBT.

(ii) In VBT, it was suggested that the inner orbital complexes participate in an **associative path** based on the argument that the bonds in the inner orbital complexs are stronger. Consequently, bond breaking in the dissociative activation is disfavoured for the inner orbital complexes. However, the argument is not really justified.
Thus, the ***basic assumptions: dissociative process for the outer orbital complexes and associative process for the inner orbital complexes, are not justified.***

(iii) The serious objection against the VBT treatment is that it cannot predict anything regarding the order of lability. Thus it is purely a qualitative treatment.
For example, the difference in lability among the inner orbital complexes of the d^0, d^1 and d^2 configurations cannot be predicted and these are lumped into a group provided they do not differ significantly in charge. Similarly, all the inner orbital complexes of d^3, d^4, d^5 and d^6 are lumped into a group described as the inert complexes.

5.18.2 Application of Crystal Field Theory (CFT) in Explaining the Kinetic Behaviour of the Octahedral Metal Complexes (*cf.* Sec. 5.30.10 for CFAE of the square planar complexes)

Activation energy controls the reaction rate. To the activation energy, loss of cfse (crystal field stabilisation energy) in attaining the activated complex from the starting complex contributes. This contribution is described as the **crystal field activation energy** (CFAE), more correctly **ligand field activation energy** (LFAE). This CFAE or LFAE mainly contributes to $\Delta H^{\neq}$.

CFAE = loss of cfse in forming the activated complex
= cfse of the starting complex – cfse of the activated complex.

Calculation of cfse of the starting complex whose geometry is known is straight-forward. But calculation of cfse of the activated complex is a difficult task without the exact knowledge of the structure of the activated complex. Without this knowledge, some approximations may be considered as suggested by Basolo and Pearson. For the **associative path,** the activated complex may be consi-

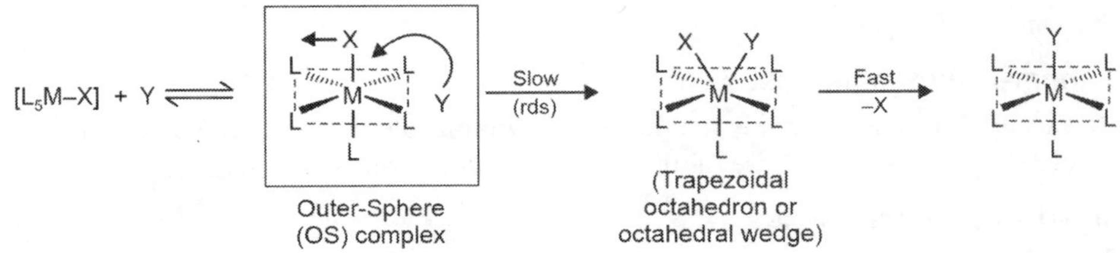

$[L_5M–X] + Y \rightleftharpoons$ Outer-Sphere (OS) complex $\xrightarrow[(rds)]{Slow}$ (Trapezoidal octahedron or octahedral wedge) $\xrightarrow[-X]{Fast}$

(i) Nucleophilic attack at the trigonal octahedral face with the **movement away of the leaving group**

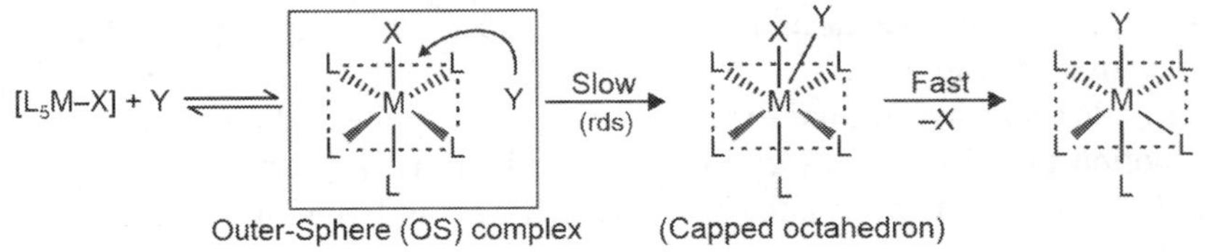

$[L_5M–X] + Y \rightleftharpoons$ Outer-Sphere (OS) complex $\xrightarrow[(rds)]{Slow}$ (Capped octahedron) $\xrightarrow[-X]{Fast}$

(ii) Nucleophilic attack at the trigonal octahedral face **without movement of the leaving group**

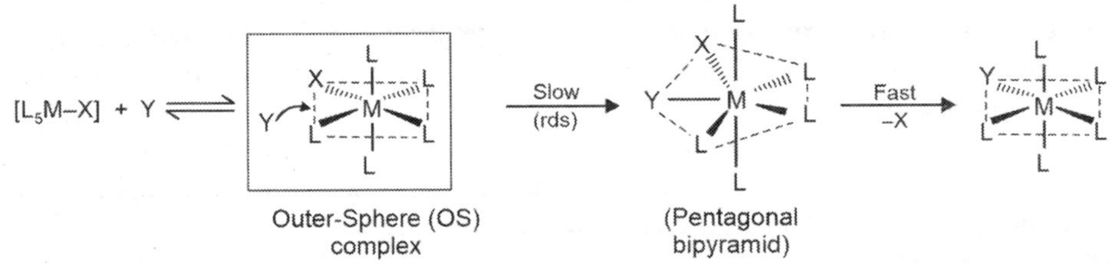

$[L_5M–X] + Y \rightleftharpoons$ Outer-Sphere (OS) complex $\xrightarrow[(rds)]{Slow}$ (Pentagonal bipyramid) $\xrightarrow[-X]{Fast}$

(iii) Nucleophilic attack through octahedral edge

Fig. 5.18.2.1a: Schematic representation of the associative path (*A*-path) of ligand substitution in the octahedral complex (ML_5—X): $ML_5—X + Y \rightarrow ML_5—Y$

dered to be **seven coordinate** of three probable geometries (see Fig. 5.18.2.1a, b): ***pentagonal bipyramid*** (*i.e.* the entering ligand approaches towards the midpoint of an edge of the octahedron) and ***trapezoidal octahedron* or *capped octahedron*** (*i.e.* the entering ligand approaches towards the centre of a trigonal face of the octahedron). The cfse of such geometries like pentagonal bipyramid or capped octahedron or trapezoidal octahedron may be calculated easily. Similarly, for the **dissociative path**, the activated complex may be approximately considered to be **five coordinate** of two probable geometries: *square pyramid* and *trigonal bipyramid.*

***D*–path:** $\underset{(O_h)}{X—ML_5} \xrightarrow{(rds)} \underset{(TBP \text{ or } SP)}{ML_5} + X$

CFAE = cfse of ML_5X – cfse of ML_5

The 5-coordinate intermediate formed from the octahedral geometry through the rupture of $M—X$ bond is a square pyramid (SP) which may change into a trigonal bipyramid (TBP). In presenting the CFAE in Table 5.18.2.1, the TBP geometry has not been considered.

***A*–path:** $X—ML_5 \xrightarrow{+Y,\ (rds)} ML_5XY$ (pentagonal bipyramid or trapezoidal octahedron or capped octahedron)

CFAE = cfse of ML_5X – cfse of ML_5XY

In Table 5.18.2.1, capped octahedron which is energetically less favoured compared to the trapezoidal octahedron in the pathway of face attack by the incoming nucleophile is not considered

Based on the above arguments, Basolo and Pearson have presented the CFAE values (Table 5.18.2.1) for both the dissociative and associative paths for both the strong and weak field ligands.

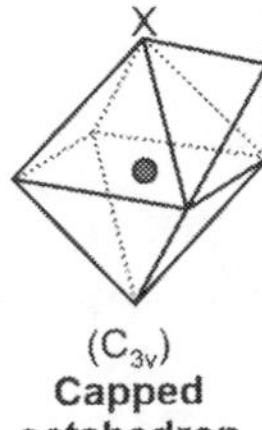

(C3v)
Capped octahedron

Resulting from the nucleophilic attack by Y at the trigonal face of the octahaderal complex without the movement of the leaving ligand (X)

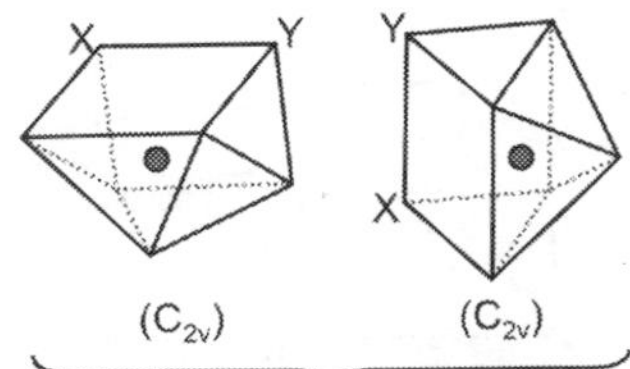

(C2v) (C2v)

Trapezoidal octahedron *i.e.* Capped trigonal prism

Resulting from the nucleophilic attack by Y at the trigonal face of the octahedral complex with the movement away of the leaving ligand (X)

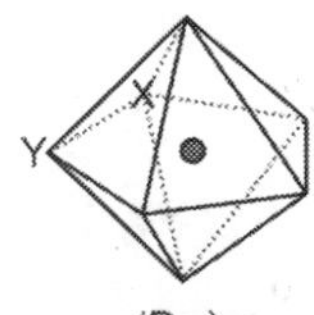

(D5h)
Pentagonal bipyramid

Resulting from the nucleophilic attack by Y at the edge (bearing the leaving ligand, X) of the octahedral complex

Fig. 5.18.2.1b: The possible 7-coordinate geometries of the intermediates/transition states (T.S.) in the associative path of nucleophilic ligand substitution of the octahedral complexes, ● denotes the position of metal centre; vertices are occupied by the ligands; only the positions of the leaving ligand (X) are entering ligand (Y) are shown (with reference to the Fig. 5.18.2.1a) here.

Note 1: For the **associative path**, the approach of a nucleophile through the mid point of an **octahedral edge** along with a minor motion of the coplanar four vertices to accommodate the nucleophile produces a **pentagonal bipyramid** (D_{5h}) T.S. or intermediate. **Face attack** (along the C_3-axis) by the nucleophile may produce either the **capped octahedron** (C_{3v}) or **capped trigonal prism** (C_{2v}) (where one ligand is capped at the square face) which is also described as the **trapezoidal octahedron** or **octahedral wedge**. In fact, capped trigonal prism is energetically more favourable than capped octahedron. **Face attack (to the d_{xz} or d_{yz} t_{2g}-orbital) is more probable than the edge attack to the d_{xy} t_{2g}-orbital producing the pentagonal bipyramidal geometry**.

Note 2: In the **associative path**, the t_{2g}-orbitals are to accommodate the incoming nucleophile because the e_g-orbitals are projected towards the ligands. Approach of the nucleophile along the edge of a basal plane is accommodated by the d_{xy} orbital while approach of the nucleophile along the face is accommodated by d_{xz}/d_{yz} orbital. In terms of σ-bonding framework of MOT, the t_{2g} orbitals are the **NBMOs** while LUMOs are the e_g^* orbitals (**ABMOs**) which are of high energy (see Sec. 5.19.4 for details). The **edge attack in the basal plane is sterically more hindered** (*cf.* 4 ligands are already in the basal plane) **than the face attack**. Obviously, **the increased electron density in the t_{2g}-level disfavours the nucleophilic attack in the associative path** (discussed and illustrated later).

Table 5.18.2.1: Crystal field activation energy (CFAE) for reactions of octahedral complexes

System	*Calculated values of CFAE (in Dq_o)**					
	Strong field ligands			*Weak field ligands*		
	D	*A*	*A*	*D*	*A*	*A*
	Square pyramid	*Pentagonal bipyramid*	*Trapezoidal octahedral*	*Square pyramid*	*Pentagonal bipyramid*	*Trapezoidal octahedral*
d^0	0	0	0	0	0	0
d^1	−0.57	−1.28	−2.08	−0.57	−1.28	−2.08
d^2	−1.14	−2.56	−0.68	−1.14	−2.56	−0.68
d^3	2.00	4.26	1.80	2.00	4.26	1.80
d^4	1.43	2.98	−0.26	−3.14	1.07	−2.79
d^5	0.86	1.70	1.14	0	0	0
d^6	4.00	8.52	3.63	−0.57	−1.28	−2.08
d^7	−1.14	5.34	−0.98	−1.14	−2.56	−0.68
d^8	2.00	4.26	1.80	2.00	4.26	1.80
d^9	−3.14	1.07	−2.79	−3.14	1.07	−2.79
d^{10}	0	0	0	0	0	0

See Sec. 5.30.10 for CFAE of the square planar complex. * The negative values of CFAE are essentially zero.

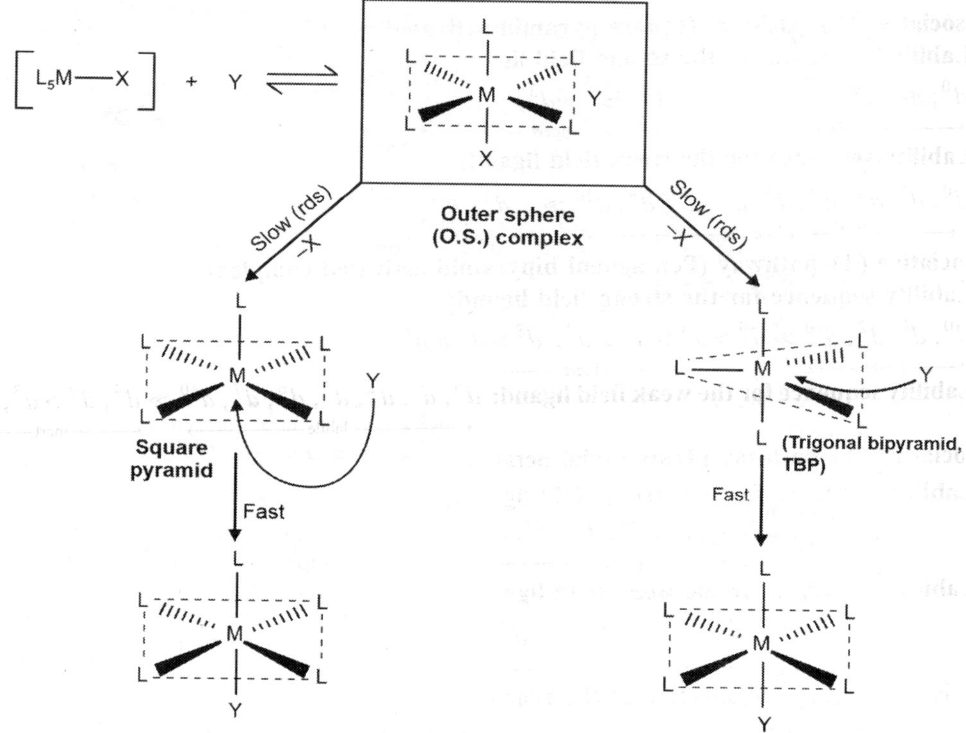

Fig. 5.18.2.1c: Schematic representation of the dissociative path (D-path) of ligand substitution in the octahedral complex (ML_5—X): $ML_5—X + Y \rightarrow ML_5—Y + X$

Note 1: For the **dissociative path, in terms of CFAE**, the square pyramid (SP) T.S. or intermediate is always more favourable than the alternative possible geometry, **trigonal bipyramid (TBP)**. In fact, $[Co(NH_3)_5X]^{2+}$ ($X^- = Cl^-, Br^-$, NCS^-, etc.) aquates in a dissociative path involving the **square pyramidal transition state**. The experimental activation energy (E_a) for their aquation nicely matches with the calculated value assuming the square pyramidal T.S./intermediate. For the TBP T.S./intermediate, the calculated E_a values are much higher than the experimental values (*JACS*, **90**, 6628, 1968). However, the TBP geometry is sterically more favoured in terms of **ligand-ligand repulsion** (*cf.* repulsion, at 120° *vs.* 90°). Besides this, the TBP geometry allows the better **π-bonding** interaction at the equatorial plane. In fact, in many cases, the square pyramid geometry initially formed rearranges to the TBP geometry to earn the better π-bonding interaction to stabilise the T.S./intermediate. For the TBP intermediate/T.S., the nucleophile (Y) at the first step enters through one of the edges of equatorial trigonal plane.

- Thus the **positive value of CFAE** indicates the loss of *cfse* in forming the activated complex or transition state and the rate becomes slow. The more positive value of CFAE indicates the slower reaction.
- The **negative value of CFAE** indicates that the *cfse* of the activated complex is more than that of the starting complex. It implies that the transition state is more stable (in terms of *cfse*) than the reactant but it is improbable. In fact, the negative value of CFAE arises in the cases where the additional *cfse* due to the J.T. distortion in the starting complex is ignored in estimating the CFAE. It may happen so for Cr(II) (d^4), Cu(II) (d^9), etc. For the sake of simplicity, the negative values of CFAE are considered as zero and such complexes react very fast.
- Prediction of lability/inertness in terms of CFAE is only meaningful for the systems **having the comparable Dq values** that depend on the **oxidation state and position of the metal centre** in the Periodic Table and nature of the ligands.

Based on the CFAE, the **major conclusions** regarding the lability orders for the reactions of octahedral complexes are given here.

(a) **Dissociative (*D*)–pathway (square pyramid activated complex)**

(i) **Lability sequence for the strong field ligand:**

$$\underbrace{d^0, d^1, d^2, d^7, d^9, d^{10}}_{\text{labile}} \gg \underbrace{d^5 > d^4 > d^3,\ d^8 > d^6}_{\text{inert}}$$

(ii) **Lability sequence for the weak field ligand:**

$$\underbrace{d^0, d^1, d^2, d^4, d^5, d^6, d^7, d^9, d^{10}}_{\text{labile}} \gg \underbrace{d^3,\ d^8}_{\text{inert}}$$

(b) **Associative (*A*)–pathway (Pentagonal bipyramid activated complex)**

(i) **Lability sequence for the strong field ligand:**

$$\underbrace{d^0, d^1, d^2, d^{10}}_{\text{labile}} \gg \underbrace{d^9 > d^5 > d^4 > d^3,\ d^8 > d^7 > d^6}_{\text{inert}}$$

(ii) **Lability sequence for the weak field ligand:** $\underbrace{d^0, d^1, d^2, d^5, d^6, d^7, d^{10}}_{\text{labile}} \gg \underbrace{d^4, d^9 > d^3,\ d^8}_{\text{inert}}$

(c) **Associative (*A*)–pathway (Trapezoidal octahedron/octahedral wedge activated complex)**

(i) **Lability sequence for the strong field ligand:**

$$\underbrace{d^0, d^1, d^2, d^4, d^7, d^9, d^{10}}_{\text{labile}} \gg \underbrace{d^5 > d^3,\ d^8 > d^6}_{\text{inert}}$$

(ii) **Lability sequence for the weak field ligand:**

$$\underbrace{d^0, d^1, d^2, d^4, d^5, d^6, d^7, d^9, d^{10}}_{\text{labile}} \gg \underbrace{d^3,\ d^8}_{\text{inert}}$$

Thus it is evident that **irrespective of the reaction pathway,** the octahedral complexes react as follows:

- d^3, d^8 and d^6 (low spin) centres are **inert**;
- d^3, d^8 and low spin complexes of d^4, d^5 and d^6 are **inert** and the reactivity order is: $d^5 > d^4 > d^3, d^8 > d^6$
- d^0, d^1, d^2, d^7 (except strong field ligand, A-path through the pentagonal bipyramidal activated complex), d^9 and d^{10} centres are **labile**;
- high spin complexes of d^4, d^5, d^6 and d^7 are **labile**.
- d^4 (high spin) and d^9 centres are **highly labile because of the J.T. distortion.**
- the lability decreases for the **heavier congeners** because of the increase of CFAE.

***d*⁵ *vs.* *d*⁴:** The above conclusions have been verified experimentally. Generally, the d^5 system reacts faster than the d^4 system **for the strong field ligands**. However in the case of **exchange of CN⁻ ligand in low spin [M(CN)₆]ⁿ⁻** (n = 3, 4), the lability order runs as: $[V(CN)_6]^{3-}$ (d^2) > $[Mn(CN)_6]^{3-}$ (d^4) ≫ $[Cr(CN)_6]^{3}$ (d^3) > $[Fe(CN)_6]^{3-}$ (d^5); $[Cr(CN)_6]^{4-}$ (d^4) > $[V(CN)_6]^{4-}$ (d^3) ~ $[Mn(CN)_6]^{4-}$ (d^5) > $[Fe(CN)_6]^{4-}$ (d^6); $[V(CN)_6]^{3-}$ (d^2) > $[V(CN)_6]^{4-}$ (d^3); and $[Fe(CN)_6]^{3-}$ > (d^5) > $[Fe(CN)_6]^{4-}$ (d^6).

The observed lability order (*cf.* CFAE order) is: $d^2(t_{2g}^2) > d^4(t_{2g}^4) > d^5(t_{2g}^5) > d^3(t_{2g}^3) > d^6(t_{2g}^6)$. It indicates that the process goes on through the *trapazoidal octahedral activated complex since only this associative pathway predicts this order where d^4 reacts faster than d^5*. It may be noted that the lability

Factors favouring the associative path

- **High effective nuclear charge** and **high overall positive charge of the complex** favouring a new metal-ligand bond formation (*cf.* higher valent metal ions *vs.* lower valent metal ions; heavier congeners *vs.* lighter congeners)
- **Large size of the central metal** to increase the coordination number (*cf.* heavier congeners *vs.* lighter congeners)
- **Low *d*–electron density** (*cf.* early members *vs.* late members of the transition series) specially **in the t_{2g} level** for the octahedral complexes; more correctly, LUMO of low energy)

order for the **square pyramid intermediate** (*i.e.* **dissociative path**) for the low spin complexes is: $d^2(t_{2g}^2) > d^5(t_{2g}^5) > d^4(t_{2g}^4) > d^3(t_{2g}^3) > d^6(t_{2g}^6)$. Here it is worth mentioning that because of the higher lability of $[Fe(CN)_6]^{3-}$ than that of $[Fe(CN)_6]^{4-}$, slow aquation of $[Fe(CN)_6]^{3-}$ releases CN^- that may cause CN^- poisoning.

● **Effect of charge (*i.e.* oxidation state) and periodic position of the metal centre:** Increase of the positive charge is expected to reduce the lability in the pathway of **dissociative activation**, but in terms of CFT, to predict the lability, we should consider the CFAE. With the increase of positive charge, the crystal field splitting power increases sufficiently (*i.e.* **$10Dq_o$ value increases)** and sometimes, it may favour the formation of low spin complexes **for which CFAE is higher**. For example, Co(II) complexes are very often high spin complexes (more labile) but the Co(III) complexes are generally low spin complexes which are more inert because of higher CFAE. **This is why, generally the metal ions of higher charge react more slowly in terms of both dissociative and associative activation. Because of the same ground, the heavier congeners having the higher crystal field splitting power react more slowly.** For example, the high spin $[Fe(OH_2)_6]^{2+/3+}$ complexes are labile while the low spin $[Ru(OH_2)_6]^{2+/3+}$ complexes are more inert. Besides this, the magnitude of CFAE (say xDq_o) increases for the heavier congeners even for the same spin state of a particular d^n configuration. For example, CFAE is $4Dq_o$ for the low spin d^6 configuration (t_{2g}^6) in the dissociative path (see Table 5.18.2.1) but the actual magnitude of the CFAE goes to increase in the order Ir(III) > Rh(III) > Co(III) (each forming the low spin complexes, t_{2g}^6) because the magnitude of Dq_o increases in the same order. It is supported by the lability orders: Co(III) > Rh(III) > Ir(III); Pt(II) < Pd(II) < Ni(II), etc. This aspect has been discussed in connection with the classification of metal ions in terms of water exchange rate constant (*cf.* Sec. 5.13).

V(III) is more labile than V(II) — why?

V^{3+} is expected to be more inert than V^{2+} in terms of the charge/radius ratio. But in reality, V(III) (d^2) is more labile than V(II) (d^3). It is expected in terms of CFAE which is much higher for V(II).

● **Aquation rate of $[Fe(CN)_6]^{3-}$ is faster than that of $[Fe(CN)_6]^{4-}$—Why?**

Macrocyclic effect

The so called labile centres may show the inertness in their macrocyclic complexes (see Sec. 5.3.4.2).

Increased inertness for the heavier congeners

● Increased tendency to form the low spin complexes for which the CFAE is higher.

● The actual magnitude of CFAE (say xDq_o) for the same spin state of a particular d^n configuration is higher for the heavier congeners.

5.18.3 Some Applications of the Knowledge of CFAE and Limitations of CFAE

● **Ignorance of the contribution of $\Delta S^{\neq}$:** $\Delta G^{\neq}$ (= $\Delta H^{\neq} - T\Delta S^{\neq}$) determines the rate constant but CFAE mainly contributes to determine $\Delta H^{\neq}$. $\Delta S^{\neq}$ is an important parameter to determine the rate constant. This is why, prediction of lability/inertness from CFAE only is not always correct.

● **Lability of the d^6, d^7 and d^8 systems:** CFAE of the low spin $d^6(t_{2g}^6)$, high spin $d^7(t_{2g}^5e_g^2)$ and $d^8(t_{2g}^6e_g^2)$ complexes runs in the sequence $d^7(\sim 0\ Dq) < d^8(2\ Dq) < d^6(4\ Dq)$ and lability also follows the reverse sequence. Thus the complexes of Fe(II), Co(II) and Ni(II) having the same oxidation state follow the lability sequence.

Co(II) (h.s.) > Ni(II) > Fe(II) (l.s.)

If both the d^7 and d^6 systems form the high-spin complexes, then the CFAE runs in the sequence: $d^6(t_{2g}^4e_g^2, 0\ Dq) \approx d^7(t_{2g}^5e_g^2 \sim 0\ Dq) > d^8(t_{2g}^6\ e_g^2, 2\ Dq)$.

It explains why the **high-spin complexes** like $[Fe(OH_2)_4(phen)]^{2+}$, $[Fe(OH_2)_3(terpy)]^{2+}$ dissociate faster than the corresponding Ni(II)–complexes.

• **Dissociation of $[M(phen)_3]^{2+}$ to its *bis*-complex (M = Fe, Ni) – failure of the prediction from CFAE:** Low spin $[Fe(phen)_3]^{2+}$ dissociates faster than $[Ni(phen)_3]^{2+}$

	Electronic Config.	*CFAE*	**k (min^{-1}) (for *tris*- to *bis*-complex)**	**$\Delta S^{\neq}$ (J mol^{-1} K^{-1})**	**E_a (kJ mol^{-1}) ($E_a \approx \Delta H^{\neq} + RT$)**
$[Fe(phen)_3]^{2+}$:	t_{2g}^6	4 *Dq*	4.8×10^{-3}	117.5	135
$[Ni(phen)_3]^{2+}$:	$t_{2g}^6 e_g^2$	2 *Dq*	1.3×10^{-4}	4.2	110

Though the CFAE is higher for the Fe(II)–complex, it dissociates faster than the corresponding Ni(II)-complex. It happens so due to the **entropic benefit,** *i.e.* higher positive value of $\Delta S^{\neq}$ for the Fe(II)–complex. It is suggested that the Fe(II) complex experiences **a change over in spin state,** *i.e.* low spin to high spin (*cf.* Sec. 4.4.4, Vol. 4) in the dissociation process (*cf. tris*-complex low spin, *bis*-complex high spin). **Thus the corresponding transition state is expanded causing elongation of the metal-ligand bond length**. Here it is worth mentioning that in the Fe(II)-bpy/phen system, the *tris*-complex is low-spin (diamagnetic) while the *bis*- and *mono*- complexes are high spin (Sec. 4.4.4, Vol. 4). There is no change over in the spin state for the Ni(II) complex and consequently, there is no such entropic benefit. Thus the higher positive value of $\Delta S^{\neq}$ explains the higher dissociation rate of the Fe(II)–complexes in spite of its higher CFAE. The similar situation occurs for their bpy–complexes but the effect is less pronouned as the entropic benefit is relatively less for $[Fe(bpy)_3]^{2+}$ ($k = 9.7 \times 10^{-3}$ min^{-1}; E_a = 119.3 kJ mol^{-1}, $\Delta S^{\neq}$ = 71.5 J K^{-1} mol^{-1}).

It indicates that to determine the relative rates, we need to consider both the activation parameters (i.e. both $\Delta H^{\neq}$ and $\Delta S^{\neq}$*) as in some cases, the prediction from only CFAE (which mainly contributes to* $\Delta H^{\neq}$*) may not be true.*

• **Dissociation of $[M(bpy)_3]^{2+}$ to its *bis*-complex (M = Cr, Ni) – failure of the prediction from CFAE:** The rates of dissociation and their activation parameters of $[Cr(bpy)_3]^{2+}$ and $[Ni(bpy)_3]^{2+}$ are quite interesting.

	CFAE	**k (min^{-1})**	**E_a (kJ mol^{-1})**	**$\Delta S^{\neq}$ (J mol^{-1} K^{-1})**
$[Cr(bpy)_3]^{2+}$ ($t_{2g}^4 e_g^0$):	1.43 *Dq*	22.8	94.9	54.5
$[Ni(bpy)_3]^{2+}$ ($t_{2g}^6 e_g^2$):	2.0 *Dq*	0.18	93.2	8.5

$E_a \approx \Delta H^{\neq} + RT$

Though the CFAE values for the said complexes are comparable, the low-spin Cr(II)-complex dissociates faster because of **an entropic benefit** that probably arises from the **change over in spin state in attaining the transition state** as in the case of $[Fe(phen)_3]^{2+}$ complex. It again indicates that to compare the rates, we need the knowledge of both the activation parameters (both $\Delta H^{\neq}$ and $\Delta S^{\neq}$) besides the CFAE.

• **Dissociation of $[V(phen)_3]^{2+}$ *vs.* $[Ni(phen)_3]^{2+}$ – failure of the prediction of CFAE:** $[V(phen)_3]^{2+}$ (d^3) dissociates faster than $[Ni(phen)_3]^{2+}$ ($t_{2g}^6 e_g^2$) though CFAE (= 2 *Dq*) is the same in both cases. It is probably due to the less effective nuclear charge and less important metal→ligand π–bonding in the V(II)–complex because **V(II) being less electron rich (t_{2g}^3) is a poorer π-door than Ni(II) ($t_{2g}^6 e_g^2$).** It makes the **metal-ligand bond weaker in the V(II)-complex.**

• **Water exchange rate of $[V(aq)]^{2+}$ *vs.* $[Ni(aq)]^{2+}$:** Interestingly, the water exchange rate for Ni(II) ($k_{ex} \approx 10^4$ s^{-1}) is higher than that for V(II) ($k_{ex} \approx 10^2$ s^{-1}). It occurs through the **associative path** (CFAE = 1.8Dq, trapezopidal octahedral intermediate/T.S.) for $[V(aq)^{2+}$ (t_{2g}^3) while it occurs through the

dissociative path (CFAE = 2Dq, the square pyramidal intermediate/T.S.) for $[Ni(aq)]^{2+}(t_{2g}^6e_g^2)$. Apparently, these CFAE values are comparable (*i.e.* 1.8Dq ≈ 2.0Dq) but the higher value of Dq (= 1200 cm^{-1}) for $[V(aq)]^{2+}$ compared to that of $[Ni(aq)]^{2+}$ (Dq = 850 cm^{-1}) makes the CFAE value of $[V(aq)]^{2+}$ higher than that of $[Ni(aq)]^{2+}$ (see Sec. 5.13 for details).

In fact, the water exchange process is more favoured for Ni(II) compared to that of V(II) by both $\Delta H^{\neq}$ and $\Delta S^{\neq}$ values (see Table 5.12.1). Thus CFAE contributing to $\Delta H^{\neq}$ only cannot alone explain the fact.

- **Lability sequence of $[M(OH_2)_6]^{3+}$ of 1st transition series:** The water exchange rates for the following $[M(OH_2)_6]^{3+}$ follow the sequence:

$$[Sc(OH_2)_6]^{3+} > [Ti(OH_2)_6]^{3+} > [V(OH_2)_6]^{3+} > [Fe(OH_2)_6]^{3+}$$

The water exchange process for these systems goes through the ***associative path*** and CFAE for all these system is zero, but the k_{ex} values differ widely (*cf.* ~10^7 s^{-1} for Sc^{3+}, ~10^2 s^{-1} for Fe^{3+}). *Thus, CFAE cannot explain this wide variation of k_{ex}.* It can be explained in terms of the **electrostatic hindrance experienced by the entering ligand approaching towards the t_{2g} orbitals** (acting as the LUMO in terms of MOT) in an associative path. The higher electron density at the t_{2g} level will repell the entering ligand more at the *rds* to increase the activation energy. The electron density in the t_{2g} level is minimum at Sc^{3+} while it is maximum at Fe^{3+}. It explains the sequence of k_{ex} (see Sec. 5.13).

$$Sc^{3+}(t_{2g}^0) > Ti^{3+}(t_{2g}^1) > V^{3+}(t_{2g}^2) > Fe^{3+}(t_{2g}^3e_g^2)$$

- **Inertness of $[Cr(OH_2)_6]^{3+}$:** The very low value of k_{ex} for $[Cr(OH_2)_6]^{3+}$ (t_{2g}^3) arises from two factors: high electron density at the t_{2g}–level to disfavour the associative path of water exchange and high CFAE. It may be noted that the actual value of CFAE (1.8Dq involving the trapezoidal octahedron as the intermediate/T.S. in the *A*-path) is high because of the +3 oxidation state of chromium (*cf.* magnitude of the Dq value increases with the increase of positive oxidation state).

5.18.4 Comparison of the Predictions on Lability from VBT and CFT

(i) **d^n (*n* = 0, 1, 2):** The lability of the systems like d^0, d^1 and d^2 having at least one $(n-1)d$ orbital unoccupied has been explained in terms of VBT by considering the **associative mechanism**. CFT also predicts the lability of these systems in terms of CFAE but it does not mean that these systems will have to react in the associative pathway. Thus in terms CFT, to explain the lability of such systems having at least one empty $(n-1)d$ orbital, it is not necessarly to impose the associative path. However, it is found that such systems very often adopt the associative path (*cf.* $\Delta V^{\neq}$ values for the water exchange processes, Table 5.12.1.1) because of the availability of a vacant t_{2g} orbital to accommodate the nucleophile.

(ii) **d^0, d^9, d^{10} and high spin d^n (*n* = 4, 5, 6):** Both the theories predict the lability for the d^0 systems (*e.g.* nontransition elements, lanthanides), d^{10} systems (*e.g.* posttransition elements, Cu(I), Zn(II), Cd(II), Hg(II), Ga(III). etc.), d^9 system, and high-spin (*i.e.* outer orbital) complexes of d^4, d^5 and d^6 but the arguments are different.

(iii) **d^3 and d^8:** There is a **disagreement between the theories** in connection with the lability of d^8–system. VBT predicts that the d^8 system forms a weaker outer orbital complex and consequently, it is labile but CFT predicts it to be inert in terms of CFAE. In terms of VBT, the d^3 system forming a stronger inner orbital complex is inert and CFT also predicts it to be inert. But the arguments are different. In fact, CFAE predicts the comparable inertness of the d^3 and d^8 systems of comparable Dq values (*cf.* same CFAE for a particular, *D*- or *A*-path for the d^3 and d^8 systems, Table 5.18.2.1).

Here it may be mentioned that the octahedral complexes of Ni(II) (d^8) react fast generally. However, the Ni(II)–centre is less labile than the many other bivalent metal centres, like Mn(II), Fe(II), Co(II) and Cu(II). But the octahedral Cr(III) (d^3) complexes react slowly. Thus apparently, the predictions for the d^3 and d^8 systems in terms of CFAE are incorrect. But, as such, this comparison is not meaningful where one centre is trivalent and the other centre is divalent. *Comparison in terms of CFAE is meaningful only when the 10Dq values (which depend on the charge of the metal, periodic position of the metal and nature of the ligand) are comparable.* Definitely, for the octahedral complexes, 10*Dq* value for Cr^{3+} is higher than that for Ni^{2+}. Thus the comparison becomes more appropriate for Cr^{3+} (d^3) and Cu^{3+} (d^8) bearing the same charge and having the position in the same period.

Note: For the octahedral d^3 and d^8 systems, the CFAE is the same for a particular *D*- or *A*-path, but the d^3 system generally adopts the **associative activation** (CFAE = 1.8Dq for the trapezoidal octahedron as the intermediate/T.S.) while the d^8-system adopts the **dissociative activation** (CFAE = 2Dq, square pyramid as the intermediate/T.S.). These CFAE values are comparable when the magnitude of Dq is comparable for the d^3 and d^8 systems.

5.19 DIFFERENT FACTORS GOVERNING THE REACTION RATE AND REACTION MECHANISMS: DIAGNOSIS OF THE REACTION MECHANISM OF LIGAND SUBSTITUTION PROCESSES

5.19.1 Effect of Charge and Size of the Metal Centre

(a) **Dissociative (*d*) and associative (*a*) activation**

(i) Higher positive charge and lower ionic radius (*i.e.* **higher ionic potential**) will increase the metal-ligand bond strength. This will favour the **associative path** and disfavour the **dissociative path.** In fact, the metal ions having the high ionic potentials tend to adopt the associative path provided there must be a suitable vacant orbital to accommodate the nucleophile in the associative path. For the metal ions bearing the low ionic potential values, the reverse is true (*i.e.* dissociative path is favoured and associative path is disfavoured).

● **The trivalent metal centres** show the more associative character than the divalent metal centres because the higher positive charge favours the bond formation by the nucleophile. Comparison of the water exchange rate parameters (Table 5.12.1.1) of $[M(aq)]^{3+}$ and $[M(aq)]^{2+}$ of the first transition series indicates that in general, the **associative path prevails for $[M(aq)]^{3+}$** while the **dissociative path prevails for $[M(aq)]^{2+}$ specially for the late members having the electronic configuration $d^n (n \geq 5)$**. The relatively weaker $M^{II}—OH_2$ bond compared to the $M^{III}—OH_2$ bond favours the dissodciative path for $[M(aq)]^{2+}$. The higher charge on $[M(aq)]^{3+}$ favours the binding of the nucleophile at the *rds* provided there is a suitable vacancy in the t_{2g}-level for the nucleophilic attack.

If we compare the substitution of H_2O in $[Ru^{III}(edta)(OH_2)]^-$ and $[Ru^{II}(edta)(OH_2)]^{2-}$ (edta acting as a **pentadentate ligand** with the sixth position occupied by a H_2O molecule to be replaced by the incoming nucleophile), then the I_a-path operates for the Ru(III)-complex and the I_d-path operates for the Ru(II)-complex. The associative path (I_a) for the Ru(III)-complex is supported by the **negative entropy of activation** ($\Delta S^{\neq}$ = –ve for the incoming neutral nucleophile like pyridine, imidazole, pyrazine, CH_3CN, etc.) and a **wide range of rate dependence on the nature of the entering ligand**. On the other hand, for the Ru(II)-complex, the rate constants do not differ significantly for the different ligands (Table 5.19.1.1). The dissociative activation (*i.e.* I_d path) for the Ru^{II}-complex is favoured due to the relatively weaker $Ru^{II}—OH_2$ bond compared to the $Ru^{III}—OH_2$ bond.

● **The heavier congeners** in a group possess the more Z^* value to favour the associative path more, *e.g.* Ir(III) adopts the more ***a***–activation while Co(III) adopts basically the ***d***–activation. For the low spin complexes of d^6-configuration (t_{2g}^6) of the metal centre M(III) (M = Co, Rh and Ir), in the **associative path**, the LUMO (e_g^*) accommodates the nucleophile. **The LUMO being an ABMO, disfavours the associative path**. It makes the low spin Ir(III) complexes so inert.

Table 5.19.1.1: Rate parameters for the substitution reactions of $[Ru^{III}(edta)(OH_2)]^-$ and $[Ru(edta)(OH)_2]^{2-}$

Ligand	k_f for $(Ru^{III}(edta)(OH_2)]^-$ ($dm^3\ mol^{-1}\ s^{-1}$)	k_f for $(Ru^{II}(edta)(OH_2)]^{2-}$ ($dm^3\ mol^{-1}\ s^{-1}$)
Isonicotinamide	$(8.3 \pm 0.6) \times 10^3$ ($\Delta S^{\neq} = -19 \pm 3$ J K^{-1} mol^{-1})	30 ± 15
CH_3CN	30 ± 7 ($\Delta S^{\neq} = -24 \pm 4$ J K^{-1} mol^{-1})	13 ± 1
Pyridine	$(6.3 \pm 0.5) \times 10^3$	
Pyrazine	$(20.0 \pm 1.0) \times 10^3$ ($\Delta S^{\neq} = -20 \pm 3$ J K^{-1} mol^{-1})	
SCN^-	$(2.7 \pm 0.2) \times 10^2$ ($\Delta S^{\neq} = -18 \pm 3$ J K^{-1} mol^{-1})	2.7 ± 0.2

Ref.: T. Matsubara and C. Cerutz, *Inorg. Chem.*, **18**, 1956, 1979.

- Effect of the **charge/radius ratio** on the **water exchange process** has been illustrated (*cf.* Secs. 5.12.1 and 5.13). $[M(OH_2)_6]^{3+}$(M = Cr, Fe) participates in the I_a process while $[M(OH)(OH_2)_5]^{2+}$ participates in the I_d process (*cf. cis*-effect of the pi-donor OH group).
- **Larger size of the metal ion** will favour the accommodation of the approaching nucleophile to favour the associative path (*cf.* heavier congeners *vs.* lighter congeners).

(ii) For the complexes, the **effect of overall charge** on the rate process is very much important. The ligands already present can neutralise the charge to different extents. The σ– and π–bonding properties of these ligands can significantly monitor the electronic environment of the metal centre.

A significant increase in rate with the decrease of the overall positive charge of the complex supports the **dissociative path.** It is observed in the **aquation** of Cl^- ligand in the chlorido-ammine complexes of Co(III). On the other hand, for the corresponding Cr(III)–complexes, the rate is less sensitive with the decrease of overall positive charge of the complex. It supports the **different mechanisms** for the said Co(III) and Cr(III) complexes.

Complex:	$[CoCl(NH_3)_5]^{2+}$	$[CoCl_2(NH_3)_4]^+$	$[CoCl(en)_2(NH_3)]^{2+}$	$[CoCl_2(en)(NH_3)_2]^+$
k_{aq} (s^{-1}, 25°C):	$\sim 10^{-6}$	$\sim 10^{-2}$ to 10^{-3}	$\sim 10^{-7}$	$\sim 10^{-5}$
Complex:	$[CrCl(NH_3)_5]^{2+}$	$[CrCl_2(NH_3)_4]^+$	$[CrCl_2(en)_2]^+$	
k_{aq} (s^{-1}, 25°C):	9.7×10^{-6}	4×10^{-5}	$\sim 10^{-4} - 10^{-5}$	

(For Co(III), the *cis*-complex reacts faster (Sec. 5.21). Here for the sake of simplicity, the *cis– trans* isomers are not specified).

It is evident that for the Co(III)–complexes, the rate significantly increases with the decrease of the overall positive charge of the complex. It supports the **dissociative activation.** For the corresponding Cr(III) complexes, the rate increases less significantly with the decrease of overall positive charge. It indicates that for the Cr(III)–complexes, bond breaking leading to the transition state is not so important compared to that for the Co(III)–complexes. ***In fact, the Cr(III) complexes experience the more associative activation*** (see Sec. 5.19.6).

(b) Interchange process

It is quite expected that for an associative (*A*) mechanism, bond formation by the entering ligand leading to the activated complex will be favoured with the increase of positive charge on the complex. It will enhance the rate. But for an **interchange (*I*) process,** the expectation is not so straight-forward. In the interchange process, both the bond formation by the entering nucleophile and bond dissociation by the leaving group are important at the rate determining step (*rds*). With the increase of positive charge of the complex, bond formation by the entering nucleophile will be favoured but bond dissociation by the leaving group will be disfavoured. These **two opposing factors** will operate and consequently, the effect of charge on the interchange process cannot be concluded so easily. Even in the I_a

process, effect of charge does not produce any straight-forward conclusion. To identify the I_a process, we should rely on the effect of other parameters like sensitivity on the nature of the entering ligand rather than the effect of charge.

(c) Effect of charge — complication in interpretation due to the change in electronic environment by the introduced ligand

To change the overall charge of a complex, it is essentially required to introduce the differently charged ligands which may vary significantly in their σ– and π– bonding properties that may influence the rate and even the reaction mechanism. For example, $[M(OH)(OH_2)_5]^{2+}$ is more reactive than $[M(OH_2)_6]^{3+}$ (M = Fe, Cr) and $[M(OH)(OH_2)_5]^{2+}$ adopts the I_d path due to the presence of the better π–donor ligand (OH^-) that shows the ***cis*–labilising effect** (*i.e.* ***cis*–effect**). On the other hand, $[M(OH_2)_6]^{3+}$ adopts the I_a path. Obviously, the **less overall positive charge** on $[M(OH)(OH_2)_5]^{2+}$ (compared to that of $[M(OH_2)_6]^{3+}$) favours the dissociative activation of the hydrolysed species which may behave kinetically like the bivalent metal ions, *i.e.*, $[M(OH_2)_6]^{2+}$.

5.19.2 Effect of *d*–Electronic Configuration

This aspect has been discussed in Secs. 5.18.1–2 in terms of CFAE to estimate the relative lability.

The trivalent metal centres adopt the associative character more (effect of more positive charge) than the divalent metal centres. In a period, the **heavier congeners** having the larger size and more Z^* (effective nuclear charge) adopt the more associative character.

5.19.3 Effect of Temperature on Rate Constant: Arrhenius Equation and Eyring Equation: Transition State Theory and Activation Parameters

The temperature dependence of rate constants (k) in terms of **Arrhenius relation** is given by:

$$k = A \exp(-E_a/RT) \text{ or, } \ln k = \ln A - E_a/RT = \text{constant} - E_a/RT \qquad \ldots(5.19.3.1)$$

The pre-exponential term A is called the **Arrhenius frequency factor** which is related to the collision frequency (Z) in collision theory and related to $\Delta S^{\neq}$ in transition state theory; E_a is called the **Activation energy** which can be determined from the linear plot of ln k *vs.* $1/T$ (Fig. 5.19.3.1).

> **Note:** Arrhenius equation is applicable for the rate constant (k) of all types of reaction (*e.g.* zero order, 1st order, 2nd order, 3rd order, etc.). The ratio (E_a/RT) is dimensionless and unit of A is the unit of the rate constant (k).

Arrhenius emperical equation, $k = A \exp(-E_a/RT)$ finds justification from **van't Hoff equation** of temperature dependence of K_{eq}.

For an **elementary reaction**, *i.e.* single step reaction (say $A \rightleftharpoons B$), we can write:

$$K_{eq} = \frac{k_1}{k_{-1}}$$

and

$$\frac{d \ln K_{eq}}{dT} = \frac{\Delta H^\circ}{RT^2} \quad \text{(van't Hoff equation, Gibbs-Helmholtz equation)}$$

ΔH° = heat of reaction (see Fig. 5.19.3.1).

Expressing K_{eq} in terms of the forward and backward rate constants, we can write:

$$\frac{d \ln k_1}{dT} - \frac{d \ln k_{-1}}{dT} = \frac{\Delta H^\circ}{RT^2} = \frac{E_1 - E_2}{RT^2} = \frac{E_1}{RT^2} - \frac{E_2}{RT^2}$$

See Fig. 5.19.3.1

$$E_1 = H^* - H_R^o; \Delta H^\circ = H_P^o - H_R^o$$

$$E_2 = H^* - H_P^o = E_1 + H_R^o - H_P^o = E_1 - \Delta H^o$$

$$\Delta H^o = E_1 - E_2$$

Temperature dependence of thermodynamic equilibrium constant (K^o)

We have: $\Delta G^\circ = -RT \ln K_P^o$, *i.e.* $\dfrac{d \ln K_P^o}{dT} = \dfrac{\Delta H^\circ}{RT^2}$ (van't Hoff equation)

For the gaseous reaction, $K_P^o = K_C^o(C_o RT/p_o)^{\Delta n}$

where Δn gives the difference in the stoichiometric number of gaseous molecules between the products and reactants. $C_o = 1.0$ mol dm^{-3} **(standard state concentration)**, $p_o = 1$ bar ≈ 1 atmospheric pressure = 101.32 kN m^{-2} = 101.32 kPa **(standard state pressure)**. It leads to:

$$\frac{d \ln[K_c^o(C_o RT/p_o)]^{\Delta n}}{dT} = \frac{\Delta U^o + \Delta nRT}{RT^2} = \frac{\Delta U^o}{RT^2} + \frac{\Delta n}{T}$$

$$(cf.\ H = U + PV, \Delta H = \Delta U + P\Delta V + V\Delta P,$$

$$i.e.\ \Delta H_P = \Delta U + P\Delta V = \Delta U + \Delta nRT, \text{ assuming } \Delta U_V = \Delta U_P)$$

$$\text{or } \frac{d \ln K_C^o}{dT} + \Delta n \frac{d \ln(C_o RT/p_o)}{dT} = \frac{\Delta U^o}{RT^2} + \frac{\Delta n}{T}$$

$$\text{or } \frac{d \ln K_C^o}{dT} + \Delta n \frac{(C_o R/p_o)}{(C_o RT/p_o)} = \frac{\Delta U^o}{RT^2} + \frac{\Delta n}{T} \quad \text{or } \frac{d \ln K_C^o}{dT} + \frac{\Delta n}{T} = \frac{\Delta U^o}{RT^2} + \frac{\Delta n}{T}$$

$$\text{or } \frac{d \ln K_C^o}{dT} = \frac{\Delta U^o}{RT^2}$$

For the reactions in **solution**, ΔU° ≈ ΔH°.

Transition state (T.S.) theory: Rack mechanism and entatic mechanism in enzyme activity

Rack mechanism: Binding of a substrate to the enzyme causes a distortion towards the geometry/structure of the T.S.

Entatic mechanism: The structure of active site of the enzyme at the ground state is distorted as required in the T.S. to reduce the activation energy. In **metalloenzymes**, the coordination geometry of the metal ion at the active site is very often distorted (*i.e.* entatic state). The distorted tetrahedral geometry of Zn(II) in carboxypeptidase-A and carbonic anhydrase represents the entatic state of the active site of the enzymes.

E_1 and E_2 denote the activation energy for the forward and backward reaction respectively. Comparing both sides of the above equation, we can write:

$$\frac{d \ln k_1}{dT} = \frac{E_1}{RT^2} + I \text{ and } \frac{d \ln k_{-1}}{dT} = \frac{E_2}{RT^2} + I$$

In the above relations, I represents a constant which is experimentally found to be zero, *i.e.*

$$\frac{d \ln k_1}{dT} = \frac{E_1}{RT^2} \text{ and } \frac{d \ln k_{-1}}{dT} = \frac{E_2}{RT^2}$$

Integrating the above expressions, we can write

$$\ln k_1 = -E_1/RT + \text{constant}$$

and

$$\ln k_{-1} = -E_2/RT + \text{constant}$$

$$k_1 = A_1 \exp\left(-\frac{E_1}{RT}\right) \text{ and } k_{-1} = k_2(\text{say}) = A_2 \exp\left(-\frac{E_2}{RT}\right)$$

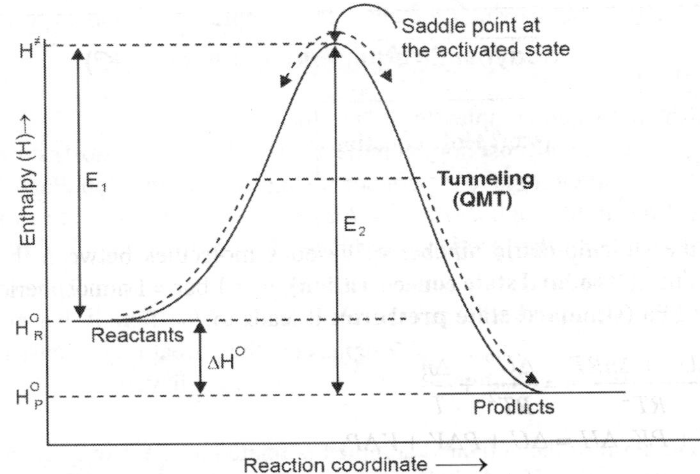

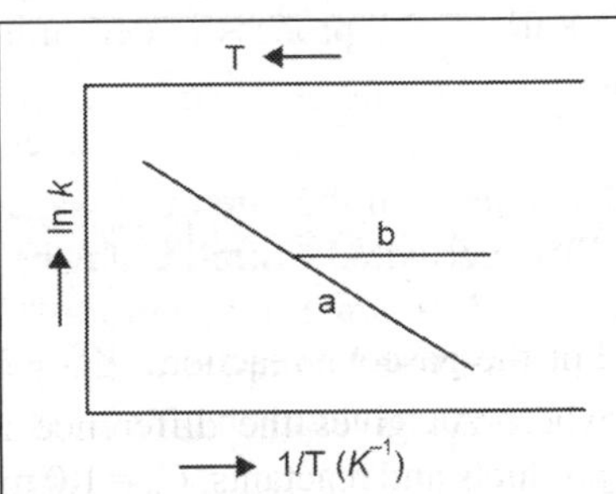

a. Classical reaction pathway over the top of the barrier predominating over the QMT **at higher temperature** (T)
b. Reaction though QMT predominating over the classical pathwazy **at very low temperature**

Fig. 5.19.3.1: Variation of enthalpy (energy) in a reaction and tunneling mechanism in some particular reactions

In general, $$k = A\exp\left(-\frac{E_a}{RT}\right) \text{ or, } \ln k = \ln A - \frac{E_a}{RT} = \text{constant} - \frac{E_a}{RT}$$

For the differential and integrated forms of Arrhenius equation and van't Hoff equation, see Sec. 5.19.4.

Nonlinear plot of ln *k vs.* 1/*T*: According to the Arrhenius equation, the ln k *vs.* $1/T$ plot should be linear with the slope = $-E_a/R$. In some cases, at very low temperature, the available thermal energy is insufficient to cross the activation energy barrier and then the reaction may occur only through the **quantum mechanical tunneling (QMT)** (Fig. 5.19.3.1) and the rate constant shows the **temperature independence** (*i.e.* deviation from the linear ln k vs. $1/T$ plot). The probability of QMT is high for the reactions **involving the lighter particles with the smaller activation barrier width**. QMT has been discussed in Vol. 1. The **temperature independence of rate constant and very high kinetic isotope effect (KIE)** (*cf.* probability of QMT increases abruptly for the lighter particles, see Vol. 1) are the convincing footprints of QMT controlled reactions.

In terms of **collision theory of reaction,** $A = PZ$ where P is called **the probability factor** or **steric factor** and Z is the **collision number**. The factor P cannot be related with any physical quantity, specially when the reactions are studied in solution. This is why, **transition state theory** is preferred. In terms of transition state (T.S.) theory, the reactants are in equilibrium with an **activated complex** which leads to the product formation. This activated complex lies in the region of highest energy along the reaction coordinate in the energy-profile diagram (*cf.* Fig. 5.8.1). ***This region of highest energy state along the reaction coordinate in the process of conversion of the reactant(s) to the product(s) represents the transition state (T.S.)*** T.S. theory of a **bimolecular reaction** may be represented as:

$$A + B \underset{}{\overset{K^{\neq}}{\rightleftharpoons}} \underset{\text{(Activated Complex)}}{X^{\neq}} \xrightarrow{k^{\neq}} \text{product} \quad \ldots(5.19.3.2)$$

$$K^{\neq} = \frac{[X^{\neq}]/C_o}{([A]/C_o)([B]/C_o)}, \text{ or, } [X^{\neq}] = K^{\neq}[A][B]/C_o \quad \ldots(5.19.3.3)$$

Note: Concentrations are expressed in the unit, mol dm^{-3}. To make the **thermodynamic equilibrium constant** ($K^{\neq}$) **dimensionless**, each concentration term is divided by the unit quantity, C_o = 1 mol dm^{-3} (standard state). In the general expression of [$X^{\neq}$], the C_o term appears as C_o^n where $n = 0$ for the unimolecular reaction, $n = 1$ for the bimolecular reaction, and so on, *i.e.* the term C_o^n appears for the reaction of ($n + 1$) molecularity. For the **single step reactions** (*i.e.* elementary reactions), molecularity = order of the reaction.

Rate of the process is determined by the rate at which the activated complex decomposes to the product, *i.e.*

$$\text{rate} = k^{\neq}[X^{\neq}]$$

$k^{\neq}$ is related to the rate of passage of the activated complex to the product.

The **activated complex** may be regarded as an **almost** normal molecule (*cf.* the word **almost** is used because **it is short in one mode of vibrational energy**) to experience the molecular vibration. Passage of the activated complex to the product occurs through a *crucial vibration of frequency* ($\nu^{\neq}$) that will carry the activated complex to the product along the reaction coordinate. It leads to:

$$\text{rate} = k^{\neq}[X^{\neq}] = \kappa\nu^{\neq}[X^{\neq}] \quad \text{...(5.19.3.4)}$$

Here κ is the proportionality constant which is described as the **transmission coefficient** that may vary in the range 0.0 to 1.0. The transmission coefficient actually represents the probability of passing of the activated complex into the product. **If κ is considered to be unity** then we have:

$$\text{rate} = \nu^{\neq}[X^{\neq}]$$

The frequency ($\nu^{\neq}$) of this crucial vibration of the activated complex is related to the energy quanta $h\nu^{\neq}$. The mean energy ($\bar{\varepsilon}$) of an oscillator is:

$$\bar{\varepsilon} = \frac{h\nu}{\exp(h\nu/k_BT) - 1} \approx \frac{h\nu}{(1 + h\nu/k_BT) - 1} \approx k_BT,$$

(when $h\nu/k_BT$ is small, at high temperature, *i.e.* $\frac{h\nu}{k_BT} \to 0$)

Under the condition we have:

$$\bar{\varepsilon} = h\nu^{\neq} = k_BT, \text{ or, } \nu^{\neq} = k_BT/h \quad \text{...(5.19.3.6)}$$

Approximation in Conventional Eyring Equation

- **Transmission coefficient κ:** It has been considered to be unity. But the activated state may experience two possibilities: it may move to the product and it may also back to the reactant. If κ is taken to be unity then it ignores the probability of returning the activated state to the reactant. In reality, it may not happen.
- **Quantum mechanical tunneling (QMT) effect:** Tunneling is not possible in classical mechanics, but it is possible quantum mechanically. In the conventional Eyring equation, tunneling is ignored. The tunneling is favoured when the potential energy barrier is sufficiently thin and the particle to tunnel must be light **Thus tunneling effect is important for the electron transfer; H–atom transfer, H^+ transfer and H^- transfer processes.** Tunneling is not favoured for the heavier isotope deuterium. *In fact, the high kinetic isotope effect ratio (H vs. D) is an indication of tunneling effect.*

  At very low temperature, the rate process over the energy barrier is less important than the rate process through the tunneling. **The rate constant for QMT is temperature independent** (*cf.* Fig. 5.19.3.1). This is a foot-print of QMT.

It leads to:

$$\text{rate} = \left(\frac{k_BT}{h}\right)[X^{\neq}] = \left(\frac{k_BT}{h}\right)K^{\neq}[A][B]/C_o \quad \text{...(5.19.3.7)}$$

Thus assuming the transmission coefficient to be unity, we can write:

$$\text{rate} = \left(\frac{k_BT}{h}\right)[X^{\neq}] = \left(\frac{k_BT}{h}\right)K^{\neq}[A][B]/C_o = k_2[A][B] \quad \text{...(5.19.3.8)}$$

k_B = Boltzmann's constant; h = Planck's constant;
k_2 = second-order rate constant (in $dm^3\ mol^{-1}\ s^{-1}$) in the present case
$K^{\neq}$ can be expressed with the thermodynamic function $\Delta G^{\neq}$.

$$\Delta G^{\neq} = -RT\ln K^{\neq}\ ;\ -\frac{\Delta G^{\neq}}{RT} = \ln K^{\neq}\ ;\ \text{or } K^{\neq} = \exp(-\Delta G^{\neq}/RT)$$

It leads to:

$$\text{rate} = \left(\frac{k_B T}{h}\right)\exp\left(-\frac{\Delta G^{\neq}}{RT}\right)[A][B]/C_o;$$

$$= \left(\frac{k_B T}{hC_o}\right)\exp\left(-\frac{\Delta H^{\neq}}{RT}\right)\exp\left(\frac{\Delta S^{\neq}}{R}\right)[A]\,[B];$$

$$\Delta G^{\neq} = \Delta H^{\neq} - T\Delta S^{\neq} \qquad \text{...(5.19.3.9)}$$

$\Delta G^{\neq}$ = Standard free energy of activation
$\Delta H^{\neq}$ = Standard enthalpy of activation
$\Delta S^{\neq}$ = Standard entropy of activation
} The superscript zero to indicate the standard state is dropped for the sake of simplicity.

It gives:

$$k_2 = \left(\frac{k_B T}{C_o h}\right)\exp\left(-\frac{\Delta H^{\neq}}{RT}\right)\exp\left(\frac{\Delta S^{\neq}}{R}\right),$$

(Called **Eyring Equation**, for the 2nd order reaction)...(5.19.3.10)

or, $$\frac{k_2 h}{k_B T} = \left(\frac{1}{C_o}\right)\exp\left(-\frac{\Delta H^{\neq}}{RT}\right)\exp\left(\frac{\Delta S^{\neq}}{R}\right)$$

or, $$\ln\left(\frac{k_2 h C_o}{k_B T}\right) = -\frac{\Delta H^{\neq}}{RT} + \frac{\Delta S^{\neq}}{R}$$

or, $$-\ln\left(\frac{k_2 h C_o}{k_B T}\right) = \frac{\Delta H^{\neq}}{RT} - \frac{\Delta S^{\neq}}{R},$$ (another form of **Eyring Equation** for the 2nd order reaction)

...(5.19.3.11)

In general, the Eyring equation for a reaction of $(n + 1)$ order is:

$$\frac{kh}{k_B T} = \frac{1}{C_o^n}\exp\left(-\frac{\Delta G^{\neq}}{RT}\right)$$

or $$\frac{C_o^n kh}{k_B T} = \exp\left(-\frac{\Delta H^{\neq}}{RT}\right)\exp\left(\frac{\Delta S^{\neq}}{R}\right)$$

or $$-\ln\left(\frac{C_o^n kh}{k_B T}\right) = \frac{\Delta H^{\neq}}{RT} - \frac{\Delta S^{\neq}}{R}$$

where k = rate constant, C_o = 1.0 mol dm^{-3} (standard state), $n = 0$ for the first order reaction, $n = 1$ for the 2nd order reaction, etc. It makes the ratio $\left(\frac{C_n^o kh}{k_B T}\right)$ **dimensionless**. The unit of $k_B T$ = J, unit of h = J s, unit of k = s^{-1} (1st order reaction), = dm^3 mol^{-1} s^{-1} (2nd order reaction), etc.

(A) Evaluation of $\Delta H^{\neq}$ and $\Delta S^{\neq}$ from Eyring equation:

Thus from the linear plot of $-\ln\left(\frac{khC_o^n}{k_B T}\right)$ *vs.* $\frac{1}{T}$, $\Delta H^{\neq}$ can be obtained from the slope, *i.e.* slope $= \frac{\Delta H^{\neq}}{R}$

The temperature is generally varied within a span of 30°C. **The intercept** $\left(= -\frac{\Delta S^{\neq}}{R}\right)$ at $\frac{1}{T} = 0$ (*i.e.* $T = \infty$) **cannot be experimentally attained by simple extrapolation method** (*i.e.* extrapolation of the experimental linear plot upto $1/T = 0$ where T = infinity). To determine $\Delta S^{\neq}$, the estimated $\Delta H^{\neq}$ value is used in the above Eyring equation for a known value of k at a particular temperature.

It may be shown that uncertainty in the evaluation of $\Delta S^{\neq}$ is more than that in $\Delta H^{\neq}$. In fact, uncertainty in $\Delta S^{\neq}$ is usually at least $\pm$ 2.5 J K^{-1} mol^{-1}.

(**Note:** It may be noted that the linear plot, ln k vs. $\frac{1}{T}$ may deviate under the conditions when tunneling effect (QMT) contributes to the rate process significantly, see Fig. 5.19.3.1).

$\Delta S^{\neq}$ can be correlated with the pre-exponential factor A of Arrhenius equation while $\Delta H^{\neq}$ is correlated with E_a as follows:

From the T.S. theory, we can write (in general):

$$k = \frac{k_B T}{hC_o^n} K^{\neq} \text{ (molecularity of the reaction} = n + 1)$$

i.e.
$$\ln k = \ln\left(\frac{k_B}{hC_o^n}\right) + \ln T + \ln K^{\neq}$$

or
$$\frac{d\ln k}{dT} = \frac{1}{T} + \frac{d\ln K^{\neq}}{dT} = \frac{1}{T} + \frac{\Delta U^{\neq}}{RT^2} \quad \left(cf.\ \textbf{van't Hoff isochore,}\ \frac{d\ln K^{\neq}}{dT} = \frac{\Delta U^{\neq}}{RT^2}\right)$$

or
$$\frac{d\ln k}{dT} = \frac{RT + \Delta U^{\neq}}{RT^2}$$

From Arrhenius equation, $k = A$ exp $(-E_a/RT)$, we have:

$$\ln k = \ln A - \frac{E_a}{RT} \text{ or } \frac{d\ln k}{dT} = \frac{E_a}{RT^2}$$

Comparing the above two relations, we can write:

$$E_a = RT + \Delta U^{\neq} = RT + \Delta H^{\neq} - P\Delta V^{\neq} = RT + \Delta H^{\neq} - \Delta x RT$$

where Δx = change in the number of molecules in the formation of the activated complex in an ideal gas phase reaction.

It gives: $\Delta H^{*} = E_a + (\Delta x - 1)RT$.

Δx = change in the number of molecules in forming the activated complex, *i.e.* $\Delta x = 0$ (unimolecular process), $\Delta x = -1$ (bimolecular process). Thus, $E_a = \Delta H^{\neq} + RT$ for the **unimolecular gas phase reaction** and for the **reactions in solutions where $\Delta V^{\neq}$ is small.** In general, we have:

$$k = \left(\frac{k_B T}{h}\right)\exp\left(-\frac{\Delta H^{\neq}}{RT}\right)\exp\left(\frac{\Delta S^{\neq}}{R}\right) \text{(for the } \textbf{unimolecular reaction}, C_o^n = C_o^0 = 1, \text{ and } k \text{ in s}^{-1})$$

$$= \left(\frac{k_B T}{h}\right)\exp\left(\frac{\Delta S^{\neq}}{R}\right)\exp\left(-\frac{E_a - RT}{RT}\right)$$

$$i.e.\ k = e\left(\frac{k_B T}{h}\right)\exp\left(\frac{\Delta S^{\neq}}{R}\right)\exp\left(-\frac{E_a}{RT}\right)$$

$$cf.\ k = A\exp\left(-\frac{E_a}{RT}\right) \text{(Arrhenius equation)}$$

For the rate constant of **a single step 1st order or unimolecular reaction**, comparison between the Eyring equation and Arrhenius equation leads to:

$$A \text{ (pre-exponential factor)} = e\left(\frac{k_B T}{h}\right)\exp\left(\frac{\Delta S^{\neq}}{R}\right)$$

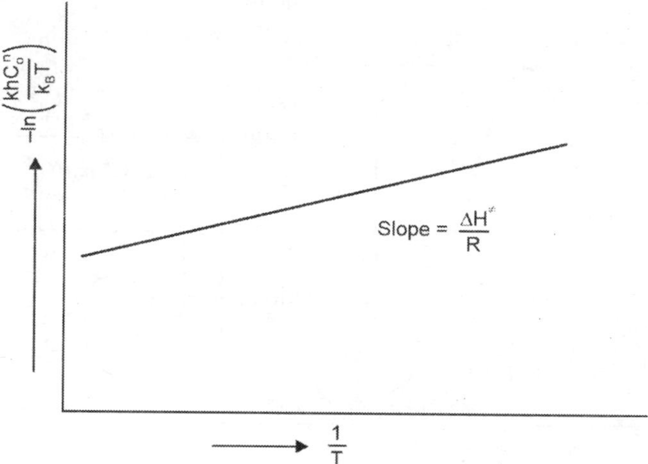

Fig. 5.19.3.2: Schematic representation of Eyring plot and evaluation of $\Delta H^{\neq}$

Note: $\Delta S^{\neq}$ cannot be obtained directly from the plot by extrapolation of the experimental line upto $1/T = 0$ where $T =$ infinity. Intercept $= \frac{-\Delta S^{\neq}}{R}$ when $\frac{1}{T} = 0$, *i.e.* $T = \infty$. This condition cannot be realised experimentally. To find the value of $\Delta S^{\neq}$, the value of $\Delta H^{\neq}$ (obtained from the slope of the plot) is used in the Eyring equation for a known value of rate constant at a particular temperature.

$$\Delta S^{\neq} = R\left[\ln A + \ln\left(\frac{h}{k_B T}\right) - 1\right]$$

For the **bimolecular (or 2nd order single step reaction)**, $E_a = \Delta H^* + 2RT$ and it leads to:

$$A = e^2\left(\frac{k_B T}{C_o h}\right)\exp\left(\frac{\Delta S^{\neq}}{R}\right)$$

In general, $\Delta H^{\neq} = E_a - yRT$, ($y = 1$ for unimolecular reaction; $= 2$ for bimolecular reaction); etc. It may be pointed out that $\Delta H^{\neq}$ is usually of the order of 10–30 kcal mol^{-1} for **reactions in solution** while RT is 0.6 kcal mol^{-1} (at 25°C). **Thus it is reasonable to assume $\Delta H^{\neq} \approx E_a$** for the solution phase reactions.

(B) Significance of $\Delta H^{\neq}$ and $\Delta S^{\neq}$ in understanding the reaction mechanism

(a) Factors controlling $\Delta H^{\neq}$: $\Delta H^{\neq}$ may include **different energy factors** like electrostatic interaction, bond breaking, bond formation, internal reorganisation within the reactant molecules, solvation and desolvation, etc. involved in producing the transition state from the starting reactants. If a pre-equilibrium is involved in the process then the ΔH change in the pre-equilibrium process is also included in $\Delta H^{\neq}$ of the **composite rate constant** (say $k_f = kK_{eq}$) (see Sec. E).

(b) Enthalpy of transition (ΔH_T): The parameter called **enthalpy of transition** (ΔH_T) is given by the change of enthalpy in the formation of the product from the transition state. Thus ΔH_T is given by the energy difference between the transition state and product, *i.e.* $\Delta H^{\neq} - \Delta H$ where ΔH gives the enthalpy change in the formation of the product from the reactants (see Fig. 5.19.3.3).

● **Identification of the mechanistic path of aquation of X^- of $[Co(NH_3)_5X]^{2+}$ by considering the $\Delta H^{\neq}$ and ΔH_T parameters:** The knowledge of ΔH_T and $\Delta H^{\neq}$ may be helpful in understanding the reaction mechanism. **In aquation of $[Co(NH_3)_5X]^{2+}$** (X^- = different univalent anions like Cl^-, Br^-, NO_3^-, NO_2^-, NCS^-, etc.) leading to the same product $[Co(NH_3)_5(OH_2)]^{3+}$, **ΔH_T is more or less the same** (*ca.* 25 kcal mol^{-1}). It indicates the **common transition state** involving $[Co(NH_3)_5]^{3+}$ attained in a ***dissociative path*** and the leaving group (X^-) remains as a free solvated species in both transition state

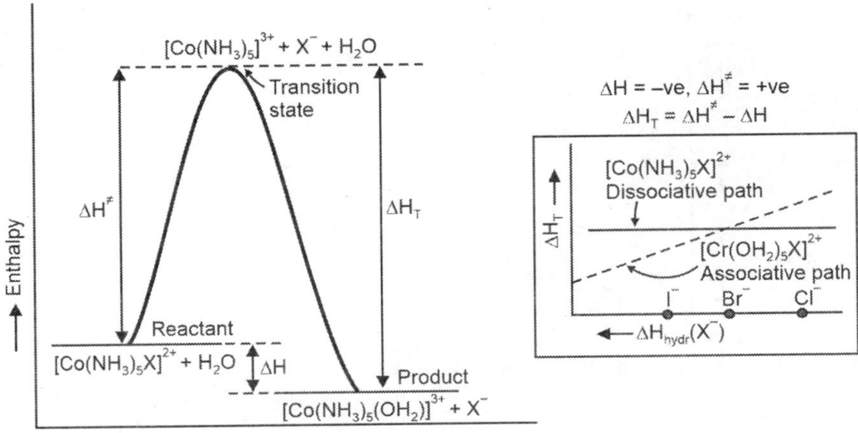

Fig. 5.19.3.3: Qualitative illustration of relations among the parameters $\Delta H^{\neq}$, ΔH_T and ΔH in the aquation of $[Co(NH_3)_5X]^{2+}$ ($X^- = Cl^-, Br^-, NO_3^-$, etc.) through a **dissociative path (ΔH_T independent of X^-)** (*Ref. J. Inorg. Nucl. Lett.*, **8**, 891, 1972).

Note: A constant value of ΔH_T (*ca.* 25 kcal mol^{-1}) for the aquation of $[Co(NH_3)_5X]^{2+}$ and water exchange of $[Co(NH_3)_5(OH_2)]^{3+}$ suggests that the transition state involves $[Co(NH_3)_5]^{3+}$ as the common Co(III)-species in the transition state and the common product $[Co(NH_3)_5(OH_2)]^{3+}$. It supports the **dissociative path** and the leaving group (X^-) remains almost as a free solvated species in the transition state as in the product. This is why, nature of X^- does not affect ΔH_T as X^- remains in the same solvated state in both the transition state and product. This is why, the plot of ΔH_T *vs.* enthalpy of hydration (ΔH_{hydr}) of X^- is linear with a slope = 0 (zero) (inset).

- For the **dissociative path of aquation** of $[Co(NH_3)_5X]^{2+}$, $\Delta H^{\neq}$ and ΔH are different for different X^- but ΔH_T (= $\Delta H^{\neq}$ – ΔH) remains constant.
- For the **associative path of aquation** as in the case of $[Cr(OH_2)_5X]^{2+}$, ΔH_T is not constant and the plot ΔH_T *vs.* ΔH_{hydr} (X^-) is linear with a nonzero slope. ΔH_T increases with the increase of $\Delta H_{hydr}(X^-)$ (*i.e.* more negative value). In the associative path, the leaving group remains bound with Cr(III) in the transition state and it is released as the solvated X^- ion in the product. It explains: **more the hydration enthalpy of X^-, more the ΔH_T in the associative path.** It explains the order of ΔH_T in the aquation of $[Cr(OH_2)_5X]^{2+}$: $Cl^- > Br^- > I^- > NO_3^-$.

The degree of hydration depends on the charge density (charge/radius ratio) of X^-. It explains the order of hydration enthalpy of X^- (= Cl^-, Br^-, I^- and NO_3^-).

In the plot, ΔH_T *vs.* $\Delta H_{hydr}(X^-)$, the slope deviates more from zero (for the ideal dissociative process) for the more associative path. It indicates the order of associative character in the aquation of the complexes: $[Cr(OH_2)_5X]^{2+}$ > $[Cr(NH_3)_5X]^{2+}$ (effect of σ-donor ligand; **σ-basicity:** $NH_3 > OH_2$, see Sec. 5.19.6).

- $[Co(NH_3)_5X]^{2+}$ (**dissociative path**) *vs.* $[Rh(NH_3)_5X]^{2+}$ (relatively **more associative character**): heavier congeners generally favour the more associative path (see Sec. 5.18.2).
- For the **base hydrolysis** of $[Co(NH_3)_5X]^{2+}$ (**D-CB mechanism**, discussed later), the independence of ΔH_T (≈ 32 kcal mol^{-1}) of the leaving group (X^-) and the zero slope of the plot, ΔH_T *vs.* $\Delta H_{hydr}(X^-)$ support the **dissociative activation.**

and product. For aquation of $[Co(NH_3)_5X]^{2+}$, the $\Delta H^{\neq}$ value increases with the increase of the Co(III)—X bond strength because it experiences a dissociative (*d*) activation. In *fact, $\Delta H^{\neq}$ bears a linear relationship with* $\bar{\nu}_1$ (wave number in cm^{-1} for the first absorption band, $^1A_{1g} \rightarrow {}^1T_{1g}$, $h\nu \approx 10Dq_o - 4B'$, see Sec. 7.15.1)) *which is related with* $10Dq_o$ (average value). All these support the *dissociative activation* in the aquation of X^- of $[Co(NH_3)_5X]^{2+}$.

(c) Factors controlling $\Delta S^{\neq}$: The magnitude of $\Delta S^{\neq}$ depends mainly on two factors:

(i) change in the number of particles in forming the transition state from the reactants;

(ii) relative **electrorestriction** over the solvents by the reactants and transition state.

- In a **dissociative path,** in attaining the transition state, the number of particles increases (*i.e.* **translational entropy** increases) to make **$\Delta S^{\neq}$ positive**. On the other hand, for an **associative path,** the number of particles decreases in producing the transition state to **make $\Delta S^{\neq}$ negative.** These straightforward conclusions operate in the reactions where in attaining the transition state, **there is no creation** or **annihilation of charge, so that electrorestriction over the solvent does not change significantly, in producing the transition state.** This may happen for the solvent exchange reactions, formation reactions where the metal bound solvent molecule is replaced by the *neutral ligands* (*e.g.* $[Ni(OH_2)_6]^{2+} + NH_3 \rightarrow$ product) and other ligand substitution reactions where both the leaving and entering nucleophiles are neutral.

Note: In an intramolecular path, cyclisation (*i.e.* formation of a cyclic T.S.) causes a **conformational entropy loss** (see Vol. 4) to make $\Delta S^{\neq}$ negative.

● When two reactants of same charge combine to generate the transition state, the T.S. will be more solvated than the reactants. Similarly, if a reactant dissociates (*e.g.* aquation of X^- of $[Co(NH_3)_5X]^{2+}$) to attain the T.S. which is nothing but an ion pair, the T.S. will be more solvated. *Thus it leads to separation of charge in producing the T.S.* These are the examples of the processes **leading to creation of charges in attaining the T.S. Such processes tend to make $\Delta S^{\neq}$ negative because the T.S. is more solvated than the reactants.**

● When two reactants of opposite charges combine to generate the T.S., the **charge neutralisation** (*i.e.* **charge annihilation**) will slacken the eletrorestriction over the solvent molecules. **This will tend to make the $\Delta S^{\neq}$ positive.**

If there is any pre-equilibrium, the ΔS value of the pre-equilibrium step will also contribute to $\Delta S^{\neq}$ of the composite rate constant (*e.g.* $k_f = kK_{eq}$).

(C) $\Delta S^{\neq}$ and diagnosis of reaction mechanism:

● **$\Delta S^{\neq}$ for the dissociative and associative process:** We conclude that ***after correction for the electrorestriction effect,*** a large negative value of $\Delta S^{\neq}$ is an indication of the associative mechanism, and a large positive value of $\Delta S^{\neq}$ is an indication of the dissociative mechanism.

● **Mechanism of aquation of X^- of $[M(NH_3)_5X]^{2+}$ and $\Delta S^{\neq}$** (see Sec. 5.21)**:** In the aquation of $[M(NH_3)_5X]^{2+}$ (X = uninegative anion), if the **dissociative process** goes on, then the departed X^- in the T.S. undergoes solvation. In such cases, $\Delta S^{\neq}$ bears a linear relationship with the ΔS values of solvation of X^-. For $X^- = Cl^-$, Br^- and I^-, the extent of hydration (*i.e.* solvation) varies as: $Cl^- > Br^- > I^-$, *i.e.* Cl^- is most extensively solvated, *i.e.* electrorestriction effect causes the **highest loss of entropy in the solvation of Cl^-**. Consequently, ΔS for solvation of these halides follows the sequence: $I^- > Br^- > Cl^-$. In fact, $\Delta S^{\neq}$ values also follow the same sequence to support the dissociative mechanism and such mechanisms may be described as the **solvent assisted dissociative** (SAD) mechanism (see Fig. 5.21.2).

Aquation of $[Cr(NH_3)_5X]^{2+}$ (Sec. 5.22):

$X^- = Cl^-$	Br^-	I^-
$\Delta S^{\neq}$ (e.u.) = −9.0	−6.0	−2

1 e.u. = 1 cal mol^{-1} K^{-1} = 4.184 JK^{-1} mol^{-1}.

The SAD mechanism also prevails in the aquation of the corresponding Co(III)-complex, $[Co(NH_3)_5X]^{2+}$ (see Sec. 5.21 and Fig. 5.21.2).

● **Dissociation of $[M(LL)_3]^{2+}$(M = Fe, Ni; Cr; LL = bpy, phen):** In these cases, importance of $\Delta S^{\neq}$ has been illustrated in Sec. 5.18.3.

● **Solvent exchange reaction and $\Delta S^{\neq}$:** For the **solvent exchange reactions** where there is no complication from the standpoint of charge creation or annihilation on attaining the transition state, the $\Delta S^{\neq}$ value is a reliable parameter to diagnose the reaction mechanism (*cf.* Table 5.12.1.1).

● **Ring closure rate determining step and $\Delta S^{\neq}$:** Ring closure causes the loss of entropy (*i.e.* loss of **conformational entropy**). In the chelation process, if the **ring closure** step contributes significantly to the rate determining step, *the $\Delta S^{\neq}$ value becomes negative.*

● **Reliability of the $\Delta S^{\neq}$ values to diagnose a reaction mechanism:** It has been already mentioned that in the experimental procedure of evaluation of $\Delta S^{\neq}$, the uncertainty is relatively large (at least

±2.5 J K^{-1} mol^{-1}). **This is why, too much reliance should not be put on the small values of $\Delta S^{\neq}$ to diagnose a reaction mechanism.**

(D) Effect of $\Delta H^{\neq}$ and $\Delta S^{\neq}$ on the rate constant:

More the positive value of $\Delta H^{\neq}$, more the rate retardation; more the positive value of $\Delta S^{\neq}$, more the rate acceleration.

(E) Isokinetic trend and reaction mechanism:

In comparable systems, different reactions passing through a **similar reaction mechanism,** a linear relationship (called **isokinetic trend**) between $\Delta H^{\neq}$ and $\Delta S^{\neq}$ is maintained (see Fig. 5.19.3.4). It may be mentioned that, ***$\Delta H^{\neq}$ and $\Delta S^{\neq}$ can balance each other's effect*** on the rate. Thus similar rates may be obtained for different $\Delta H^{\neq}$ and $\Delta S^{\neq}$ values. The isokinetic trend is expressed as:

$$\Delta H^{\neq} = \alpha + \beta \Delta S^{\neq}$$

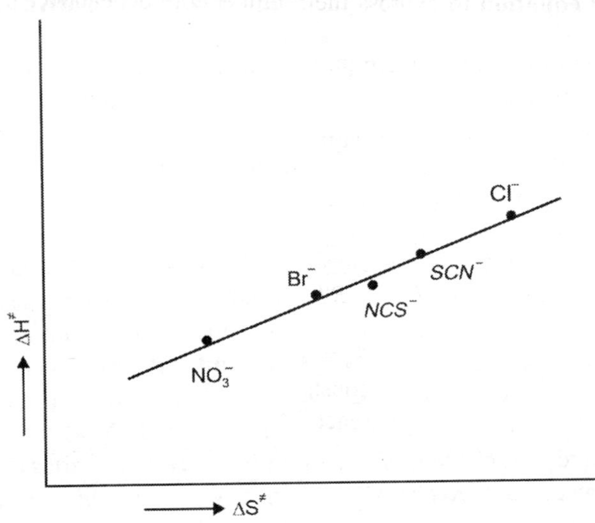

Fig. 5.19.3.4: Qualitative representation of the isokinetic plot for the anation of $[Cr(OH)(OH_2)_5]^{2+}$ by different ligands like Cl^-, Br^-, SCN^-, NO_3^-, etc. (*cf.* D. Thusius, *Inorg. Chem.*, **10**, 1106, 1971).

Here, α is actually $\Delta G^{\neq}$ (= $\Delta H^{\neq} - T\Delta S^{\neq}$) and β is called the **isokinetic temperature (= slope of the isokinetic plot)** at which all the related reactions occur at the same rate. Above this temperature, the rates are controlled by the $\Delta S^{\neq}$ values and below this temperature, the rates are controlled by the $\Delta H^{\neq}$ values. The isokinetic trend is also described as **compensation effect** where the effect of $\Delta H^{\neq}$ is balanced by $\Delta S^{\neq}$ (*i.e.* higher $\Delta H^{\neq}$ is accompanied with the higher positive $\Delta S^{\neq}$). It may be mentioned that the conclusion regarding the reaction mechanism from the isokinetic relationship has been criticised by may workers.

(F) Temperature Dependence of Composite Rate Constant of a Complex Reaction (*i.e.* Multistep Reaction)

For the complex reactions, the **composite rate constant (*k*) may decrease with the increase of temperature.** It may happen in the reaction involving a rapid pre-equilibrium step followed by the rate determining step (see Sec. 5.7).

$$A + B \underset{k_{-1}}{\overset{k_1}{\rightleftharpoons}} C \text{ (in a rapid equilibrium)}$$

$$C \xrightarrow{k_2} \text{product (rds)}$$

The observed **composite rate constant** is given by (see Sec. 5.7):

$$k = \left(\frac{k_1}{k_{-1}}\right)k_2 = K_{eq}k_2$$

If the pre-equilibrium step is **exothermic**, K_{eq} decreases with temperature (*cf.* **van't Hoff's equation,** $d\ln K_{eq}/dT = \Delta H°/RT^2$, $\Delta H°$ = heat of reaction). If this decrease can outweigh the increase in k_2 (with the increase of temperature), **then the observed composite rate constant (*k*) decreases with temperature**. Thus in such cases, temperature dependence of the **composite rate constant (*k*) decreases with temperature.** Thus in such cases, temperature dependence of the **composite rate constant** (*k*) depends on the temperature dependence of both K_{eq} and k_2.

We can express the rate constant (*k*) in terms of **Arrhenius equation** and equilibrium constant (K_{eq}) in terms of **van't Hoff equation** to express their temperature dependence.

$$\frac{d\ln k}{dt} = \frac{E_a}{RT^2}$$ (Arrhenius Eqn. differential form; $k = A\exp(-E_a/RT)$ integrated form)

$$d\ln K_{eq} = \frac{\Delta H°}{RT^2}$$ **(van't Hoff equation),** $\Delta H°$ = heat of reaction)

For $k = k_2K_{eq}$, or $\ln k = \ln k_2 + \ln K_{eq}$

i.e. $$\frac{d\ln k}{dT} = \frac{d\ln k_2}{dT} + \frac{d\ln K_{eq}}{dT} = \frac{E_{a2}}{RT^2} + \frac{\Delta H°}{RT^2}$$

i.e. $$\frac{E_a}{RT^2} = \frac{E_{a2}}{RT^2} + \frac{\Delta H°}{RT^2}$$ E_a = activation energy for the observed composite rate constant (*k*), E_{a2} = activation energy for k_2, $\Delta H°$ = heat of reaction for K_{eq} (*i.e.* pre-equilibrium step)

or $E_a = E_2 + \Delta H°$

If $\Delta H° = -\text{ve}$ and $|\Delta H°| > E_2$

$E_a = -\text{ve}$

Sometimes for a multistep complex reaction, the rate expression derived from the steady-state approximation may appear as follows:

$$k = (k_1k_2k_3/k_4)$$

where k_1, k_2, k_3 and k_4 are the rate constants of **various elementary steps**. If the temperature dependence of the rate constants of the elementary steps is expressed in terms of Arrhenius equation, then we can write:

$$k = \left(\frac{A_1A_2A_3}{A_4}\right)\exp\left(-\frac{E_{a1}+E_{a2}+E_{a3}-E_{a4}}{RT}\right)$$

$$= A\exp\left(-\frac{E_a}{RT}\right)$$

It gives $$A = \left(\frac{A_1A_2A_3}{A_4}\right),\ E_a = E_{a1}+E_{a2}+E_{a3}-E_{a4}$$

The sign of E_a depends on the relative values of E_{a1}, E_{a2}, E_{a3} and E_{a4}. In fact, $E_a = 0$ when $E_{a4} = E_{a1} + E_{a2} + E_{a3}$; $E_a = -\text{ve}$, when $E_{a4} > E_{a1} + E_{a2} + E_{a3}$; $E_a = +\text{ve}$, when $E_{a4} < E_{a1} + E_{a2} + E_{a3}$.

5.19.4 The Linear Free-Energy Relationship (LFER) and Leaving Group Effect and Reaction Mechanism

● **Quantitative interpretation of LFER in terms of $\Delta G^{\neq}$ (transition state theory) and ΔG°:** The kinetic and thermodynamic aspects of a reaction are distinctly different. Rate constant (k) is the kinetic parameter while equilibrium constant (K_{eq}) is the thermodynamic parameter (see Sec. 3.19.3).

$$\Delta G^{\neq} = -RT\ln\left(\frac{khC_o^n}{k_BT}\right),\ i.e.\ \ln k = -\frac{\Delta G^{\neq}}{RT} - \ln\left(\frac{h}{k_BT}\right) - n\ln C_o \quad \textbf{(Transition state theory)}$$

Here $C_o = 1.0$ mol dm^{-3}, $n = 0$ for 1st order rate constant, $= 1$ for 2nd order rate constant etc. (see Sec. 5.19.3). It leads to:

$$\log k = -\frac{\Delta G^{\neq}}{2.303\ RT} - \log\left(\frac{h}{k_BT}\right)$$

and,
$$\Delta G^{\circ} = -RT\ln K_{eq},\ i.e.\ \log K_{eq} = -\frac{\Delta G^{\circ}}{2.303RT}$$

However, in some cases, $\Delta G^{\neq}$ and ΔG° (*i.e.* log k and log K_{eq}) bear a linear relationship which is called the **Linear free energy relationship** (LFER) having the form.

$$\Delta G^{\neq} = a\Delta G^{\circ} + b,\ \text{or}\ \ \log k = A\log K_{eq} + B \qquad \ldots(5.19.4.1)$$

To understand the origin of *LFER*, let us consider the aquation of $L_5M—X$, *i.e.*

$$L_5M—X + H_2O \underset{k_{-1}}{\overset{k_1}{\rightleftharpoons}} L_5M—OH_2 + X^-, \qquad \ldots(5.19.4.2)$$

Taking $K_{eq} = \frac{k_1}{k_{-1}}$, we can write:

$$\log k_1 = \log K_{eq} + \log k_{-1} \qquad \ldots(5.19.4.3)$$

If the aquation process (*i.e.* forward reaction of the above equilibrium Eqn. 5.19.4.2) goes on through a dissociative path, **then the activated complex and products resemble,** *i.e.* X^- is substantially dissociated in the transition state. Thus in the transition state, fate of X^- is almost comparable to that in the product. In fact, free X^- remains highly solvated in both the transition state and product.

If the anation reaction (*i.e.* backward reaction of the above equilibrium) passes through the *d*–activation (*i.e.* $M—OH_2$ bond breaking is the main contributing factor to the rate determining step), then the nature of X^- is less important to describe the nature of the transition state (see Table 5.19.6.2, for **insensitivity of the anation rate constants** of $[Co(NH_3)_5(OH_2)]^{3+}$ towards the nature of the entering ligand). Thus it is reasonable to conclude that the ***value of log k_{-1} is approximately constant for the d-mechanism of anation at a particular metal centre***. Thus for the *d*–activation of both the forward (*i.e.* aquation) and backward (*i.e.* anation) processes of the above equilibrium (*cf.* **principle of microscopic reversibility**, Sec. 5.8), we can write (*cf.* Eqns. 5.19.4.1 and 5.19.4.3).

$$A = 1.0 \text{ and } B = \log k_{-1} = \text{constant.}$$

Then it gives a linear plot of log k *vs.* log K_{eq} with the slope = 1 and intercept = log k_{-1} (constant). Thus we can conclude that for the said aquation reactions, the variation of ΔG° values with the variation of X^- arises mainly from the variation of the Co(III)—X bond strength and solvation of X^-. These are quantitatively reflected in the same way in $\Delta G^{\neq}$. Thus LFER demonstrates the **leaving group effect (*i.e.* nucleofugality).**

The unit slope of the plot supports the ***d***-activation. Some workers have proposed that the slope less than unity indicates the ***a***-activation. However, this prediction of *a*-activation has been criticised by many workers.

● **Qualitative interpretation of the LFER in terms of Arrhenius equation and van't Hoff equation** or **Gibbs–Helmholtz equation** (*cf.* Sec. 5.19.3):

Temperature dependence of rate constant:

$$\ln k = \ln A - \frac{E_a}{RT}, \text{ i.e. } k = A\exp\left(-\frac{E_a}{RT}\right) \text{ (Arrhenius equation for rate constant)}$$

Temperature dependence of equilibrium constant: It can be obtained by using the following thermodynamic relationships:

$$\Delta G° = -RT \ln K_{eq}, \Delta G° = \Delta H° - T\Delta S°$$

i.e. $RT \ln K_{eq} = -\Delta H° + T\Delta S°$ (a similar relationship is obtained from van't Hoff equation, see Box)

or $$\ln K_{eq} = -\frac{\Delta H°}{RT} + \frac{\Delta S°}{R}, \text{ i.e. } K_{eq} = A'\exp\left(-\frac{\Delta H°}{RT}\right), \ln A' = \frac{\Delta S°}{R} \text{ (a constant)}$$

By comparing the above two equations for ln k and ln K_{eq}, we can explain the origin of LFER.

If the **pre-exponential factor** (A) and **entropy change** ($\Delta S°$) are almost constant and activation energy (E_a) depends on the enthalpy change ($\Delta H°$) of the reaction, then there will be linear relationship between ln k and ln K_{eq}. In other words, such a linear relationship in the log-log plot indirectly indicates that the activation energy (E_a) depends on the thermodynamic parameter ($\Delta H°$). In terms of the molecular bonding, a stronger bond between the metal and leaving group requires a higher activation energy for a dissociative path. In other words, the transition state represents almost the product like state. In acid hydrolysis of $[Co(NH_3)_5X]^{2+}$, in the transition state, X^- is almost completely dissociated to exist as the solvated anion and water molecule (entering group) remains weakly bound to the metal centre.

LFER: Arrhenius equation and van't Hoff equation (Sec. 5.19.3)

$$k = A\exp\left(-\frac{E_a}{RT}\right) \text{ (integrated form of **Arrhenius equation**)}$$

i.e. $$\ln k = \ln A - \frac{E_a}{RT}$$

$$\frac{d \ln k}{dT} = \frac{E_a}{RT^2} \text{ (differential form of **Arrhenius equation**)}$$

$$K_{eq} = A'\exp\left(-\frac{\Delta H°}{RT}\right) \text{ (integrated form of **van't Hoff equation**)}$$

i.e. $$\ln K_{eq} = \ln A' - \frac{\Delta H°}{RT} \text{ and } \ln A' = \frac{\Delta S°}{R} \text{ (a constant)}$$

$$\frac{d \ln K_{eq}}{dT} = \frac{\Delta H°}{RT^2}$$ (differential form of **van't Hoff equation known also as Gibbs-Helmholtz equation**)

It gives $$\frac{d \ln K_{eq}}{d(1/T)} = -\frac{\Delta H°}{R}\left[cf.\, d\left(\frac{1}{T}\right) = -\frac{dT}{T^2}\right] \Rightarrow \int d \ln K_{eq} = -\int \frac{\Delta H°}{R} d\left(\frac{1}{T}\right)$$

(assuming $\Delta H°$ independent of T)

i.e. $$\ln K_{eq} = -\frac{\Delta H°}{RT} + \text{integration constant } (= \ln A', \text{ say})$$

Linear relationship between ln k and ln K_{eq} indicates that $\Delta H°$ (heat of reaction) controls the E_a (activation energy) in the aquation/acid hydrolysis of $[Co(NH_3)_5X]^{2+}$.

● **LFER for aquation of $[Co(NH_3)_5X]^{2+}$ and $[Rh(NH_3)_5X]^{2+}$:** For aquation of $[Co(NH_3)X]^{2+}$ (X^- = monovalent anions like NCS^-, Cl^-, Br^-, I^-, NO_3^-, etc.), the linear plot of log k_{aq} *vs.* log K_{aq} with slope = 1 supports the **dissociative process.** For the bivalent anions (X^{2-}) like SO_4^{2-}, $C_2O_4^{2-}$, etc. the LFER line is parallel to the LFER line for the monovalent anions.

For the aquation of $[Rh(NH_3)_5X]^{2+}$ complexes, the LFER line gives a slope slightly less than unity. It indicates some associative character in the rate process. Interestingly, the reactivity for the halides as the leaving groups is different for the Co(III) and Rh(III) centres.

$[Co(NH_3)_5X]^{2+}$: k_{aq}: $I^- > Br^- > Cl^-$; Bond strength: Co—Cl > Co—Br > Co—I

$[Rh(NH_3)_5X]^{2+}$: k_{aq}: $Cl^- > Br^- > I^-$; Bond strength: Rh—I > Rh—Br > Rh—Cl

Rh(III) being softer than Co(III), prefers the halides in the sequence: $I^- > Br^- > Cl^-$. It indicates that for the aquation of $[Rh(NH_3)_5X]^{2+}$, ***the bond breaking still contributes significantly.***

● **Aquation of $[Cr(OH_2)_5X]^{2+}$ *vs.* $[Cr(NH_3)_5X]^{2+}$ and LFER plots:** For the aquation of $[Cr(OH_2)_5X]^{2+}$, the slope of the LFER plot is only 0.56 while for the aquation of $[Cr(NH_3)_5X]^{2+}$, the corresponding slope is *ca.* 0.91. It indicates the more associative character in the aquation process of $[Cr(OH_2)_5X]^{2+}$. The more dissociative character for $[Cr(NH_3)_5X]^{2+}$ arises due to the presence of the better σ–donor ligands (*cf.* NH_3 is a better σ–donor ligand than H_2O) (*see* Sec. 5.19.6 for explanation in terms of energy of the LUMO).

● **Co(III) centre *vs.* Cr(III) centre** (see Sec. 5.19.6): From the slope of the LFER plots, we can have some idea regarding the extent of *d*– and *a*–activation. In the above mentioned aquation processes, the entering ligand, *i.e.* H_2O makes a stronger bond (*i.e.* I_a path) with the Cr(III) centre in the transition state (slope less than unity) while H_2O makes a very weak interaction with the Co(III) centre (*i.e.* I_d path) in the transition state.

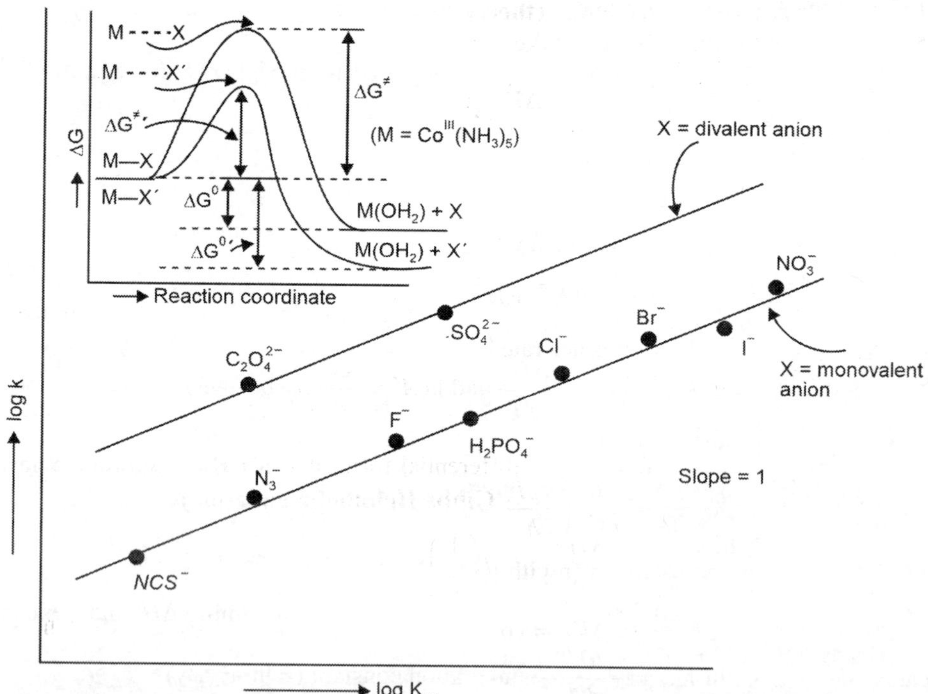

Fig. 5.19.4.1: Qualitative representation of LFER plot for aquation of $[Co(NH_3)_5X]^{n+}$, *i.e.* $[Co(NH_3)X]^{n+} + H_2O \rightarrow [Co(NH_3)_5(OH_2)]^{3+} + X^{-/2-}$ (*cf.* C.H. Langford, *Inorg. Chem.*, **4**, 265. 1965; T.W. Swaddle *et.al*, *Inorg. Chem.*, **8**, 1604, 2504, 1969; A. Haim, *Inorg. Chem.*, **9**, 426, 1970)

In the associative process, the approaching nucleophile towards the t_{2g}-orbital (Sec. 5.18.2) experiences relatively less repulsion for the Cr(III)-complexes (t_{2g}^3) than for the Co(III)-complexes (t_{2g}^6). **Thus the associative path is relatively less favoured for the Co(III)-complexes in terms of the t_{2g}-electron density (t_{2g}^6 *vs.* t_{2g}^3).** In terms of MOT, the LUMO to receive the nucleophilic attack is either the partially filled NBMO, *i.e.* SOMOs (*i.e.* t_{2g}^3) or the vacant ABMO (*i.e.* e_g^*) for the Cr(III) complexes while for the Co(III)-complexes, the **high energy vacant ABMO** (*i.e.* e_g^*) **can only act as the LUMO to receive the nucleophilic attack** because the low energy NBMO (t_{2g}^6) is completely filled in. It indicates that the LUMO (*i.e.* e_g^*) of the Co(III)-complexes is of much higher energy than the LUMO (which may be the partially filled t_{2g}^3-orbitals, *i.e.* SOMOs or the vacant e_g^*) of the Cr(III)-complexes (*cf.* Sec. 5.19.6 for $[Cr(NH_3)_5X]^{2+}$ *vs.* $[Cr(OH_2)_5X]^{2+}$, energy of the e_g^*-LUMO is higher in the case of $[Cr(NH_3)_5X]^{2+}$). Thus, in this line of argument (*i.e.* **energy of the LUMO**) also, the associative process is relatively less favoured for the low-spin Co(III) complexes (t_{2g}^6).

5.19.5 Effect of Pressure and Volume of Activation ($\Delta V^{\neq}$)

The pressure dependence of rate constant (k) can be described as follows:

$$\ln k = A' - \frac{P\Delta V^{\neq}}{RT}, \text{ i.e. } \log k = \frac{A'}{2.303} - \frac{P\Delta V^{\neq}}{2.303RT}, \ A' = \text{some constant.}$$

It leads to: $$\ln\left(k_2/k_1\right)_T = \frac{-\Delta V^{\neq}\left(P_2 - P_1\right)}{RT}$$

or $k = A\exp\left(-\frac{P\Delta V^{\neq}}{RT}\right)$ where A = proportionality constant and $A' = \ln A$, *i.e.* $k \propto \exp\left(-\frac{P\Delta V^{\neq}}{RT}\right)$

Rate dependence on pressure and $\Delta V^{\neq}$

We have the relationship: $\Delta G = -RT\ln K_{eq}$, **(thermodynamic parameters at standard condition, the zero superscript is dropped for the sake of simplicity)**

$$V = \left[\frac{\partial G}{\partial P}\right]_T \text{ i.e. } \left(\frac{\partial \Delta G}{\partial P}\right)_T = \Delta V, \ (P = \text{hydrostatic pressure})$$

It leads to: $$\left(\frac{\partial \ln K_{eq}}{\partial P}\right)_T = -\frac{\Delta V}{RT}$$

From the conventional **transition state theory** (see Sec. 5.19.3), we have:

i.e. $$k = \frac{k_B T}{C_o^n h} K^{\neq}, \text{ i.e. } \ln k = \ln K^{\neq} + \ln\left(\frac{k_B T}{h}\right) - n\ln C_o$$

$C_o = 1.0$ mol dm^{-3}, $n = 0$ for the first order rate constant, = 1 for the 2nd order rate constant and so on (see Sec. 5.19.3).

It leas to $$\ln k = \ln K^{\neq} + \ln\left(\frac{k_B T}{h}\right)$$

i.e. $$\left(\frac{\partial \ln k}{\partial P}\right)_T = \left(\frac{\partial \ln K^{\neq}}{\partial P}\right)_T, \ \left(\frac{k_B T}{h} \text{ is a constant}\right)$$

The variation of equilibrium constant ($K^{\neq}$) with pressure is given by:

$$\left(\frac{\partial \ln K^{\neq}}{\partial P}\right)_T = -\frac{\Delta V^{\neq}}{RT}, \ \Delta V^{\neq} = \text{volume of activation}$$

Thus we can write:

$$\left(\frac{\partial \ln k}{\partial P}\right)_T = \left(\frac{\partial \ln K^{\neq}}{\partial P}\right)_T = -\frac{\Delta V^{\neq}}{RT} \quad \text{i.e.} \quad \ln\left[\frac{(k_2)_{P_2}}{(k_1)_{P_1}}\right] = -\frac{\Delta V^{\neq}}{RT}(P_2 - P_1)$$

where k_1 and k_2 denote the rate constants at pressure P_1 and P_2 respectively at constant temperature T and $\Delta V^{\neq}$ is the **volume of activation** measuring the molar volume change that occurs on going to the transition state from the reactants. The linear plot of ln k *vs.* P gives a straight line of slope $= -\Delta V^{\neq}/RT$.

(a) Factors controlling $\Delta V^{\neq}$: As in the **case of $\Delta S^{\neq}$**, the following two factors play the important roles to determine the $\Delta V^{\neq}$ values.

(i) **Change in the number of free particles in attaining the transition state:** In a dissociative path, the leaving group is at least partially dislodged and released to the bulk solvent in attaining the transition state. It should make $\Delta V^{\neq}$ positive. On the other hand, in an associative path, the entering ligand formerly present in the bulk solvent gets coordinated to the metal centre. It should make $\Delta V^{\neq}$ negative. *These predictions are made without considering any other factor like change of electrorestriction over the solvent molecules.* Based on the above argument, *i.e.* consideration of only ***intrinsic bond angles and bond strength,*** we may conclude as follows:

	***D* – path**	***I_d* – path**	***I_a* – path**	***A* – path**
$\Delta V^{\neq}$:	more positive	less positive	less negative	more negative

These are illustrated in Fig. 5.19.5.1.

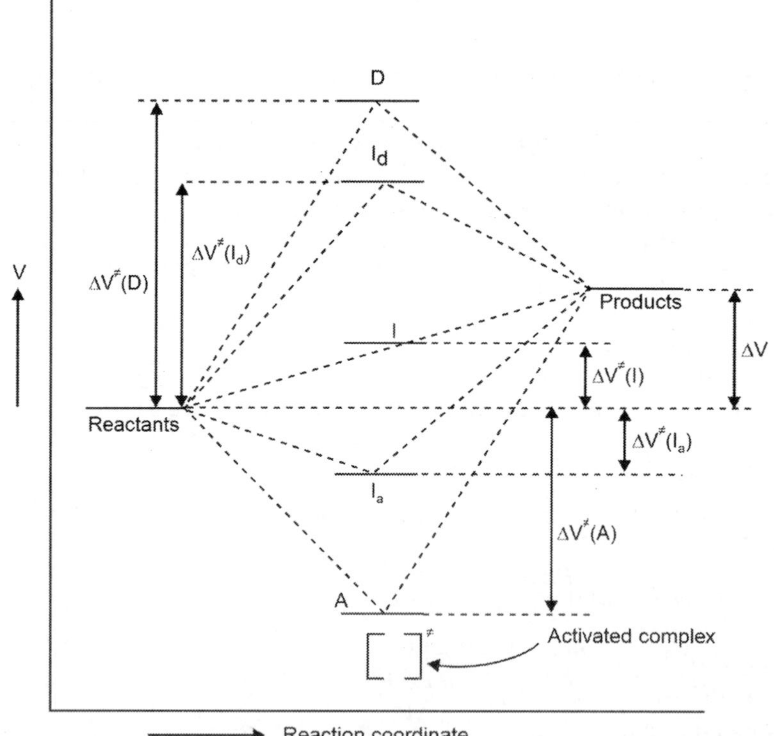

Fig. 5.19.5.1: Volume profile for different possible mechanisms of a substitution reaction (*e.g.* $[M(OH_2)_6]^{n+} + L \rightleftharpoons [M(OH_2)_5(L)]^{n+} + H_2O$) where the molar volume (V) of the products is larger than that of reactants, *i.e.* ΔV = molar volume change of the reaction = +ve)

Note: Here the volume change due to the outer-sphere complex formation is not shown. In fact, if the entering ligand is neutral, then the $\Delta V_{OS,}$ *i.e.* molar volume change due to the outer-sphere complex formation is negligible. In the present diagram, along the reaction coordinate, position of the outer-sphere complex is not separately shown.'

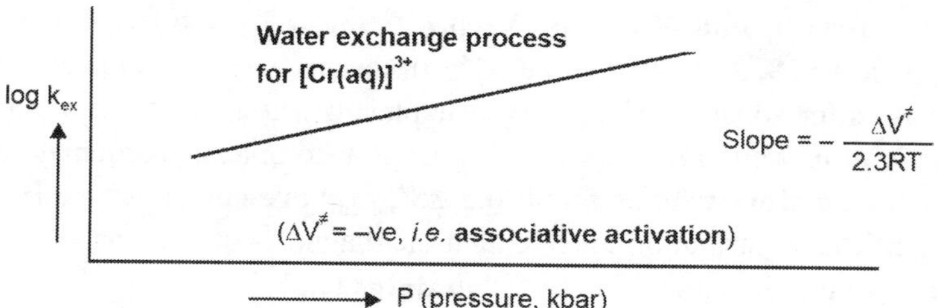

Fig. 5.19.5.2: Qualitative representation of the effect of pressure on k_{ex} for $[Cr(H_2O)_6]^{3+} + H_2O^* \rightarrow [Cr(OH_2)_5(^*OH_2)]^{3+} + H_2O$

(ii) **Change of electrorestriction over the solvent in attaining the transition state:** In attaining the transition state, if charge separation or charge creation occurs (*as in the case of dissociative aquation of* $[M(NH_3)_5X]^{2+}$) then the T.S. exercises a more electrorestriction over the solvent than the starting reactants, *i.e.* the T.S. is more solvated than the reactants. In other words, in attaining the T.S., some relatively freer water molecules (*i.e.* bulk water molecules) are **immobilised in the hydration sphere of the T.S.** ***It may be noted that molar volume of bulk water is more than the molar volume of water present in the hydration sphere. It makes*** $\Delta V^{\neq}$ ***negative.*** In attaining the T.S., if charge neutralisation or charge annihilation occurs (*as in the case of associative anation of* $[M(OH_2)_6]^{3+}$), the reactants exercise a more electrorestriction than the T.S. over the solvent, *i.e.* the reactants are more solvated than the T.S. It makes $\Delta V^{\neq}$ positive.

Thus $\Delta V^{\neq}$ is arising from two factors:

$$\Delta V^{\neq} = \Delta V^{\neq}_{\text{intrinsic}} + \Delta V^{\neq}_{\text{solvent}}$$

Table 5.19.5.1: Rate constants (at 25°C) and activation parameters for aquation of *L* of $[M(NH_3)_5L]^{3+}$

L		$10^5 k_{aq}$ (s^{-1})	$\Delta S^{\neq}$ ($J\ K^{-1}\ mol^{-1}$)	$\Delta V^{\neq}$ ($cm^3\ mol^{-1}$)	
M = Cr	H_2O^*	5.2	0	−5.8	associative path
	$OSMe_2$	1.95	−15.1	−3.2	
	$OCHNH_2$	5.1	−12.2	−4.8	
	$OC(NH_2)_2$	2.0	−22.3	−8.2	
M = Co	H_2O^*	0.59	+28.1	+1.2	dissociative path
	$OSMe_2$	1.8	+10.1	+2.0	
	$OCHNH_2$	0.58	+12.2	+1.1	
	$OC(NH_2)_2$	5.5	+10.1	+1.3	

(*cf.* N.J. Curtis, *et al, Inorg. Chem.*, **28,** 329, 1989) * water exchange process

(b) **Cases where the electrorestriction factor** (*i.e.* $\Delta V^{\neq}_{\text{solvent}}$) **absent**: If $\Delta V^{\neq}_{\text{solvent}}$ (*i.e.* electrorestriction factor) remains absent as in the case of ***solvent exchange reaction, aquation of neutral ligand, formation of complexes by the neutral ligands***, then $\Delta V^{\neq}$ ($\approx \Delta V^{\neq}_{\text{intrinsic}}$) parameter appears as a good parameter to diagnose the reaction mechanism (*i.e.* ***+ve value of*** $\Delta V^{\neq} \Rightarrow$ ***d–activation; –ve value of*** $\Delta V^{\neq} \Rightarrow$ ***a–activation***). These are illustrated in the following cases:

(i) **Solvent exchange reaction:** Diagnosis of the mechanism of solvent exchange reaction in terms of $\Delta V^{\neq}$ (Table 5.12.1.1) has been discussed in detail in Sec. 5.12.1.

(ii) **Aquation of neutral ligands (*L*) in $[M(NH_3)_5L]^{3+}$** (*cf.* N.J. Curtis, G.A. Lawrance, and R. van Eldik, *Inorg. Chem.,* **28**, 329, 1989): The $\Delta V^{\neq}$ values are given in Table 5.19.5.1.

From the $\Delta V^{\neq}$ values for aquation of uncharged ligands (*L*), it appears that the Co(III)–complexes participate in ***d*–activation** while the corresponding Cr(III)–complexes participate in ***a*–activation.**

(c) **Cases where the electrorestriction factor (*i.e.* $\Delta V^{\neq}_{solvent}$) present:** If $\Delta V^{\neq}_{solvent}$ (*i.e.* electrorestriction factor) makes a significant contribution to $\Delta V^{\neq}$, then the diagnosis of reaction mechanism from $\Delta V^{\neq}$ is not so straight-forward. It is illustrated in the following examples.

(i) **For the aquation of anionic ligand (*X*) in $[Co(NH_3)_5X]^{x+}$:** In $[Co(NH_3)_5(SO_4)]^+$ (*i.e.* $X^{2-} = SO_4^{2-}$) and $[Co(NH_3)_5Cl]^{2+}$, the $\Delta V^{\neq}$ values are –18.3 and –10.6 $cm^3\ mol^{-1}$ respectively. *But there are other evidences that the processes pass through the dissociative activation.* It may be mentioned that the $\Delta V^{\neq}$ value for the water exchange process in $[Co(NH_3)_5(OH_2)]^{3+}$ is +1.2 $cm^3\ mol^{-1}$ and it is an indication of the *dissociative character* in the water exchange process (*cf.* ΔV^{*} values for the water exchange process at $[Cr(NH_3)_5(OH_2)]^{3+}$ and $[Rh(NH_3)_5(OH_2)]^{3+}$ are –5.8 and –4.1 $cm^3\ mol^{-1}$ respectively to support the **associative activation**). In fact, in the dissociative pathway of aquation of the above Co(III)–ammine complexes, the charge creation in the T.S. makes a powerful electrorestriction over the solvent. ***Thus, in such cases,*** $\boldsymbol{\Delta V^{\neq}_{solvent}}$ ***dominates over*** $\boldsymbol{\Delta V^{\neq}_{intrinsic}}$ ***to make*** $\boldsymbol{\Delta V^{\neq}}$ ***negative*** (Fig. 5.19.5.3).

(ii) **Aquation of X^- in $[M(NH_3)_5X]^{2+}$ (M = Co, Cr):** Aquation of $[M(NH_3)_5X]^{2+}$ (M = Co, Cr) experiences –ve value of $\Delta V^{\neq}$ in spite of the dissociative mechanism due to the more solvated nature of the T.S., a consequence of charge separation.

	$[Co(NH_3)_5X]^{2+}$			$[Cr(NH_3)_5X]^{2+}$			
X^- :	Cl^-	Br^-	NCS^-	Cl^-	Br^-	I^-	NCS^-
$\Delta V^{\neq}$: ($cm^3\ mol^{-1}$)	–10.6	–9.2	–4.0	–10.8	–10.2	–9.4	–8.6
ΔV ($cm^3\ mol^{-1}$)	–11.5	–10.5		–8.4	–7.2	–6.0	

It may be noted that ΔV (molar volume change of the reaction) of the aquation reaction is also negative.

$$[M(NH_3)_5X]^{2+} + H_2O \rightarrow [M(NH_3)_5(OH_2)]^{3+} + X^-,\ \Delta V = -ve.$$

The slope of the linear plot, $\Delta V^{\neq}$ *vs.* ΔV (Fig. 5.19.5.3) for the aquation of $[Co(NH_3)_5X]^{2+}$ is close to 1.0 (*i.e.* dissociative path) while slope of the corresponding plot for the aquation of $[Cr(NH_3)_5X]^{2+}$

Relation between $\Delta V^{\neq}$ and $\Delta S^{\neq}$

From our discussions, it has been pointed out that the same factors contribute to the both parameters. ***A negative*** $\boldsymbol{\Delta V^{\neq}}$ ***value corresponds to a negative*** $\boldsymbol{\Delta S^{\neq}}$ ***value and a positive*** $\boldsymbol{\Delta V^{\neq}}$ ***value corresponds to a positive*** $\boldsymbol{\Delta S^{\neq}}$ ***value. In fact, a good correlation between the*** $\boldsymbol{\Delta V^{\neq}}$ ***and*** $\boldsymbol{\Delta S^{\neq}}$ ***values prevails for the reactions where the electrorestriction effects are important.*** But the uncertainty in the estimated $\Delta S^{\neq}$ value is relatively more. Thus $\Delta V^{\neq}$ (obtained from the slope of the plot, ln *k vs. P*, *cf.* Fig. 5.19.5.2) appears a better parameter, specially when there is no complication to diagnose the reaction mechanism. But determination of $\Delta S^{\neq}$ is very easy (*i.e.* temperature dependence of the rate constant) while determination of $\Delta V^{\neq}$ needs a very costly equipment which can allow us to follow the pressure dependence of rate constant. A quantitative relationship between $\Delta V^{\neq}$ and $\Delta S^{\neq}$ will help us to estimate $\Delta V^{\neq}$ from the $\Delta S^{\neq}$ values. A relationship has been found (*cf.* M.V. Twigg, *Inorg. Chim. Acta.*, **24**, L84, 1977) for the aquation and water exchange processes of the complexes of d^3 and d^6 (low spin) systems of the type $[M(NH_3)_5X]^{n+}$ (X = H_2O, halides and other anions).

$$\Delta V^{\neq} = (1.04 \pm 0.10)\Delta S^{\neq} - (4.4 \pm 0.3);\ (\Delta V^{\neq} \text{ in } cm^3\ mol^{-1},\ \Delta S^{\neq} \text{ in cal } K^{-1}\ mol^{-1})$$

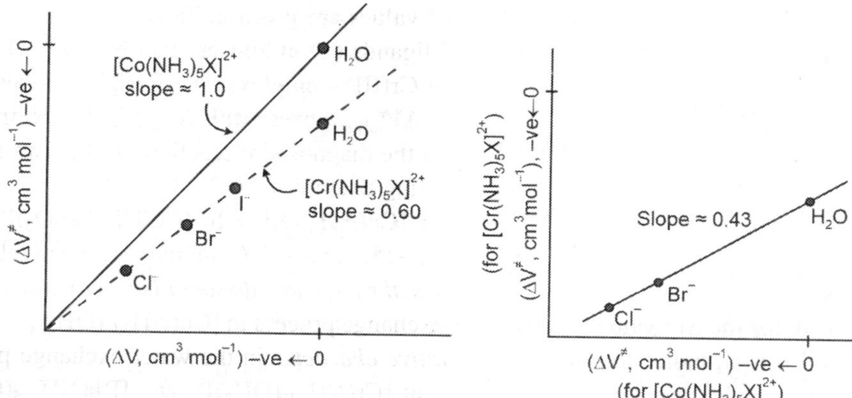

Fig. 5.19.5.3 (*Ref.* T.W. Swaddle *et al. Can. J. Chem.*, **51**, 821, 1973)): (a) Qualitative illustration of the plot of ΔV^{*} (more correctly, $\Delta V_o^{\neq}$ at zero pressure) *vs.* ΔV for the aquation of $[M(NH_3)_5X]^{2+}$, *i.e.* $[M(NH_3)_5X]^{2+} + H_2O \rightarrow [M(NH_3)_5(OH_2)]^{3+} + X_{hydr}^-$ (M = Co, Cr and X^- = Cl^-, Br^-, I^-, NCS^-, etc.). Slope ≈ 1.0 for the Co^{III}-complexes (*i.e.* $\Delta V^{\neq} \approx \Delta V$) indicates the **dissociative process** (*i.e.* dissociation and solvation of the leaving group X^- are complete to the 100% extent in the T.S.). Slope ≈ 0.60 for the corresponding Cr^{III}-complexes indicates about 60% dissociation and solvation of the leaving group (X^-) in the activated complex (*i.e.* T.S.).
(b) Plot of $\Delta V^{\neq}$ for the aquation of $[Cr(NH_3)_5X]^{2+}$ *vs.* $\Delta V^{\neq}$ for the aquation of $[Co(NH_3)_5X]^{2+}$. Slope ≈ 0.43 suggests that if, in the case of Co^{III}-complexes, dissociation and solvation of the leaving group are 100% complete in the T.S. then for the corresponding Cr^{III}-complexes, the dissociative process occurs about 50% to attain the T.S.
Note: ΔV values for the water exchange process: +1.2 cm³ mol⁻¹ (**dissociative path**) for $[Co(NH_3)_5(OH_2)]^{3+}$, –5.8 cm³ mol⁻¹ (**associative path**) for $[Cr(NH_3)_5(OH_2)]^{3+}$; $\Delta V = 0$ for both the Co^{III}- and Cr^{III}-complexes. The given plots can accommodate the $\Delta V^{\neq}$ and ΔV values of the water exchange process.
Water exchange: $[M(NH_3)_5(OH_2)]^{3+} + H_2O^* \rightarrow [M(NH_3)_5(^*OH_2)] + H_2O$, $\Delta V = 0$
$\Delta V^{\neq}$ = +ve for M = Co (***d*-activation**), = –ve for M = Cr, Rh (***a*-activation**)
Aquation: $[M(NH_3)_5X]^{2+} + H_2O \rightarrow [M(NH_3)_5(OH_2)]^{3+} + X_{hydr}^-$, ΔV = –ve
ΔV (molar volume change in the reaction) is **negative** because the reaction produces charge separation and the **products are more solvated than the reactants**. For this type of aquation reactions (*i.e.* aquation of anions), ΔV is always negative whether the reaction occurs through the *d*-activation or *a*-activation.

is about 0.6 suggesting separation of X^- and its solvation are only 60% complete in the activated complex (*i.e.* less dissociative character compared to that of Co(III)-complex).

For M = Co, ΔV^{*} values are close to the ΔV values. It supports the dissociative path. due to more electrorestriction imposed by the products on the solvent.

(iii) **Aquation of $[RuCl(NH_3)_5]^{2+}$ and anation (*i.e.* reverse reaction) of $[Ru(NH_3)_5(OH_2)]^{3+}$:**

$$\left[RuCl(NH_3)_5\right]^{2+} \xrightarrow{\text{(aquation)}} \left[Ru(NH_3)_5(OH_2)\right]^{3+}, \Delta V^{\neq} = -30 \text{ cm}^3 \text{ mol}^{-1}$$

$$\left[Ru(NH_3)_5(OH_2)\right]^{3+} + Cl^- \xrightarrow{\text{(anation)}} \left[RuCl(NH_3)_5\right]^{2+}, \Delta V^{\neq} = -20 \text{ cm}^3 \text{ mol}^{-1}$$

The $\Delta V^{\neq}$ values for the above mentioned aquation and anation reactions are in **conformity with the associative path.** Aquation through the associative path (*i.e.* bond formation by a neutral H_2O molecule at the *rds*) does not lead to any charge creation or charge annihilation in attaining the T.S. Thus the highly –ve value of $\Delta V^{\neq}$ for aquation is a strong evidence in favour of the associative path. Aquation through the dissociative path (*i.e.* separation of Cl^- through bond breaking at the *rds*) will generate two opposing contributions of $\Delta V_{intrinsic}$ (+ve) and $\Delta V^{\neq}_{solvent}$ (–ve due to the charge creation to attain the T.S.). Thus the resultant $\Delta V^{\neq}$ can never be so highly negative in a dissociative path.

$\Delta V^{\neq}$ for anation through the dissociative path (*i.e.* dissociation of the neutral H_2O molecule at the *rds*) will not get any contribution from $\Delta V^{\neq}_{solvent}$ (*i.e.* electrorestriction factor) and $\Delta V^{\neq}$ will be positive as usual. Anation through the associative path (*i.e.* bond formation by Cl^- at the *rds* to cause charge neutralisation) will tend to make $\Delta V^{\neq}$ positive due to charge neutralisation at the transition state. This will tend to make $\Delta V^{\neq}$ positive (electrorestriction factor, $\Delta V^{\neq}_{solvent}$) but the associative activation will tend to make $\Delta V^{\neq}$ negative ($\Delta V^{\neq}_{intrinsic}$). These are the two opposing factors. ***Thus the –ve value of $\Delta V^{\neq}$ is a convincing evidence in favour of the associative path.*** The *relatively less negative value for the anation* reaction compared to that for the aquation reaction is reasonable in terms of relaxation of electrorestriction in generating the associative transition state, *i.e.* positive contribution of $\Delta V^{\neq}_{solvent}$ (electrorestriction factor) to $\Delta V^{\neq}$. It is opposed by $\Delta V^{\neq}_{intrinsic}$ which makes a negative contribution in an associative path. Here $\Delta V^{\neq}_{intrinsic}$ predominates over $\Delta V^{\neq}_{solvent}$ to make $\Delta V^{\neq}$ negative.

5.19.6 Effects of the Nature of Leaving and Entering Group

(a) **Dissociative activation:** For the **dissociative activation mechanism,** the nature of the **nucleofugal,** *i.e.* leaving group (X), is very much important because it needs the rupture of the M—X bond to attain the transition state. Then the entering group (Y) enters into the coordination sphere in a fast step. Thus the dissociative process is more or less independent (**under the limiting condition**; *cf.* Secs. 5.9.5, 5.16) of the nature of entering ligand. This aspect is illustrated in some cases.

● **Aquation of L in $[Co(NH_3)_5L]^{n+}$:** The aquation rates for $[Co(NH_3)_5(L)]^{n+}$ (L = NO_3^-, I^-, Cl^-, SO_4^{2-}, F^-, N_3^-, NCS^-, etc.) cover a huge range *ca.* 10^{-5} to 10^{-10} s^{-1} (*cf.* Table 5.19.6.1), *i.e.* the aquation rate is highly sensitive to the nature of the leaving group (L). This suggests the **dissociative activation.**

Table 5.19.6.1: Aquation rates of $[Co(NH_3)_5(L)]^{n+}$ complexes (at 25°C) (*cf.* M.L. Tobe, *Adv. Inorg. Bionorg. Mech.*, **2**, 1, 1984)

L	k_{aq} (s^{-1})	*L*	k_{aq} (s^{-1})
NO_3^-	2.4×10^{-5}	SO_4^{2-}	8.9×10^{-7}
I^-	8.3×10^{-6}	F^-	8.6×10^{-8}
Cl^-	1.7×10^{-6}	N_3^-	2.1×10^{-9}
Br^-	6.3×10^{-6}	NCS^-	3.7×10^{-10}
		$H_2PO_4^-$	2.6×10^{-7}

$k_{ex}(H_2O)$ for $[Co(NH_3)_5(OH_2)]^{3+} = 5.8 \times 10^{-6}$ s^{-1}

Rate dependence on the nature of entering ligand (*Y*) in *d*-activation

Here it should be cautioned that the rate is independent of the entering ligand in the *d*-activation process under the *limiting conditions only* [*i.e.* $k_2[Y] \gg k_{-1}[X]$ for the *D*-process, $K_{OS}[Y] \gg 1$ for the I_d -process, *cf.* Secs. 5.9.5, 5.16).

The dissociative activation for the aquation of X^- in $[Co(NH_3)_5X]^{2+}$ (X^- = NO_3^-, halides, etc.) has also been supported by an almost constant value (~105 kJ mol^{-1}; Sec. 5.19.3b) of ΔH_T (***enthalpy of transition***). ΔH_T gives the energy difference between the transition state and product (see Fig. 5.19.3.3 for details).

● **Anation of $[Co(NH_3)_5(OH_2)]^{3+}$:** Anation rates for $[Co(NH_3)_5(OH_2)]^{3+}$ for Y^{n-} = N_3^-, SO_4^{2-}, Cl^-, NCS^- cover a relatively a narrower range (1.6 – 10.0) × 10^{-5} s^{-1} (*cf.* Table 5.19.6.2)

Table 5.19.6.2 Limiting rate constants (k_a) for anation of $[Co(NH_3)_5(OH_2)]^{3+}$ by Y^{n-} (at 45°C).

Y^{n-}	k_a (s^{-1})	k_{ex}/k_a	Y^{n-}	k_a (s^{-1})	k_{ex}/k_a
N_3^-	10^{-4}	1.0	Cl^-	2.1×10^{-5}	4.76
SO_4^{2-}	2.4×10^{-5}	4.2	NCS^-	1.6×10^{-5}	6.25

k_{ex} (45°) ≈ 10^{-4} s^{-1}

The almost *insensitivity of the anation rate constants* (*cf.* Table 5.19.6.2) towards the nature of entering ligand indicates the dissociative activation. It is again further supported by the fact that the *anation rate is almost close to the water exchange rate* (*cf.* k_{ex}/k_a varies in the range 1–6).

● **Substitution at $[M(OH_2)_6]^{2+}$ (M = late members of the 1st transition series, *e.g.* Co, Ni):** In the formation of $[M(H_2O)_5Y]^{n+}$ from $[M(OH_2)_6]^{2+}$ (M = Co, Ni), the second order formation rate constant (a **composite rate constant,** $k_f = k_iK_{os}$) is also fairly insensitive for the different types of entering ligand *Y* (both charged and anionic) provided the K_{OS} values do not differ significantly (see Table 5.14.1). This supports the dissociative activation.

Dependence on the nature of the entering ligand in *D*, I_d and I_a process (*cf.* Sec. 5.16): In an I_d path, the formation rate constant (a **composite rate constant,** $k_f = k_iK_{os}$) does not show any strong dependence on the nature of the entering ligand provided the K_{OS} values for the different entering ligands do not vary significantly (*cf.* Secs. 5.9.3, 5.14, 5.16 and Table 5.14.1). ***In a perfect D process, there is a slight dependence on the nature of the entering ligand because for the long-lived 5-coordinate intermediate, there will be a competition between the solvent and the entering ligand.*** However, it differs from the I_a process. This aspect has been discussed in detail in Sec. 5.16.

(b) **Associative activation:** For the **associative activation mechanism,** the rate constant is highly sensitive to the nature of the entering nucleophile. This is illustrated in the following cases.

● **Substitution at $[Ti(OH_2)_6]^{3+}$:** The rate of ligand substitution process at $[Ti(OH_2)_6]^{3+}$ is very much sensitive to the nature of the entering ligand (*cf.* Table 5.19.6.3). This is an indication of the associative character.

Table 5.19.6.3 Rate constants for anation of $[Ti(OH_2)_6]^{3+}$ (at 13°C).

Y^{n-}	k (dm^3 mol^{-1} s^{-1})
NCS^-	8×10^3 (at *ca.* 9°C)
$CH_3CO_2^-$	1.8×10^6
$ClCH_2CO_2^-$	2.1×10^5

k_{ex} (H_2O) = 4.8×10^5 s^{-1}, *i.e.* k_{ex} (in dm^3 mol^{-1} s^{-1}) = $4.8 \times 10^5/55.5 = 8.7 \times 10^3$; *cf.* H. Diebler, *Z. Phys. Chem.*, **68**, 64, 1969.

● **Substitution at $[V(OH_2)_6]^{2+}$:** $[V(OH_2)_6]^{2+}$ (d^3) participates in **an associative path** in the anation process. It is evidenced from the rate dependence on the nature of the entering ligand. The second order rate constant (k_f in $mol^{-1}dm^3$ s^{-1}) for the different anions follows the following sequence:

$$k_{f(Cl^-)} : k_{f(NCS)^-} : k_{f(N_3^-)} = 1 : 2 : 10$$

In terms of the interchange associative path (*i.e.* I_a-path), under the condition, $1 \gg K_{OS}[Y]$, we have k_f (second order formation rate constant) = k_iK_{OS} (*cf.* Sec. 5.9.3). Here the entering (Y^-) ligands are the monovalent anions and they are of comparable size. Thus it is reasonable to conclude that K_{OS} (*i.e.* ion pair association constant) remains more or less the same for Cl^-, NCS^- and N_3^-. Thus the difference in k_f arises from the difference in k_i which depends on the nature of the entering nucleophile.

● **Anation of $[Cr(OH_2)_6]^{3+}$ *vs.* $[Cr(NH_3)_5(OH_2)]^{3+}$:** Anation of $[Cr(OH_2)_6]^{3+}$ passes through the associative process and the rate constant is highly sensitive to the nature of the entering ligand (*Y*). But anation of $[Cr(NH_3)_5(OH_2)]^{3+}$ passes through the dissociative process. For the anation of $[Cr(OH_2)_6]^{3+}$ by the anions like Cl^-, Br^-, I^-, NO_3^-, NCS^- and SCN^-, the formation rate constants (k_f) cover a wide range, 7.3×10^{-7} to 8×10^{-10} dm^3 mol^{-1} s^{-1}. On the other hand, for the anation of $[Cr(NH_3)_5(OH_2)]^{3+}$, the rate constants (k_f) cover relatively a narrower range $(0.7 - 4.2) \times 10^{-4}$ dm^3 mol^{-1} s^{-1} for the anions like Br^-, Cl^-, NCS^- (see Table 5.19.6.4).

Table 5.19.6.4 Rate constants for the anation of $[Cr(OH_2)_6]^{3+}$ and $[Cr(NH_3)_5(OH_2)]^{3+}$

Y^{n-}	$[Cr(OH_2)_6]^{3+}$ *k* ($dm^3\ mol^{-1}\ s^{-1}$)		$[Cr(NH_3)_5(OH_2)]^{3+}$ *k* ($dm^3\ mol^{-1}\ s^{-1}$)	
Cl^-	3×10^{-8}	Associative character	0.7×10^{-4}	Dissociative character
Br^-	1×10^{-8}		3.7×10^{-4}	
NCS^-	1.8×10^{-6}		4.2×10^{-4}	

It is evident that not only the mechanisms of ligand substitution process at $[Cr(OH_2)_6]^{3+}$ and $[Cr(NH_3)_5(OH_2)]^{3+}$ differ but their reactivities also differ significantly. The reactivity of the ammine complex is increased by a factor about 10^4. The NH_3 ligands are much better σ–donor ligands than the H_2O ligands due to the **lower electronegativity** of sp^3–N (in NH_3) compared to that of sp^3–O (in H_2O). This **better σ–donor properties of the NH_3 ligands** labilise the sixth aqua ligand through the dissociative process. These good σ-donor ligand reduces the positive charge on the metal centre to favour the bond dissociation at the *rds*. The better σ–donor NH_3 ligands can also stabilise the activated complex of lower coordination number while the aqua complex requires the **nucleophilic assistance** to stabilise the activated complex. In the aqua complex, the relatively lower energy of the e_g^*–LUMO also favours the nucleophilic attack in the associative path as explained below.

● **Possibility of the nucleophilic attack on Cr(III) depending on the energy of the vacant orbital (*i.e.* LUMO):** In the associative path, the incoming nucleophile will occupy the LUMO which may be the **nonbonding t_{2g} orbitals** or the e_g^* orbitals depending on the d^n-configuration of the metal centre (see Vol. 4 for the MO diagram of octahedral complexes). For the $d^{0,1,2}$ systems, the NBMO (*i.e.* t_{2g} orbitals) may act as the LUMO to receive the nucleophilic attack. For the d^3 system (t_{2g}^3), the NBMOs can act as the **SOMOs** (singly occupied MOs) and they may receive the nucleophilic attack through pairing of the electron in the t_{2g}-level. This is an energy requiring process. To avoid this, the incoming nucleophile (in an associative path) may occupy the empty e_g^* orbital (in terms of MOT) of Cr(III). In the **aqua complex**, because of the weak σ–donor property of the aqua ligands, the empty e_g^* orbital is not raised much (*cf.* Fig. 3.17.4.1, Vol. 4). ***Thus, this low lying e_g^* orbital may be occupied by the incoming nucleophile in the associative path***. In the corresponding ammine complex, energy of the vacant e_g^* orbital is raised higher because of the better σ–donor property of the NH_3 ligands. **Thus in the ammine complex, the associative path becomes relatively more disfavoured.**

● **Substitution of the coordinated H_2O molecule in $[Ru^{III}(edta)(OH_2)]^-$ and $[Ru^{II}(edta)(OH_2)]^{2-}$ ($edta^{4-}$ acting as a pentaentate ligand):** For the Ru^{III}-complex (t_{2g}^5), the process passes through the **associative path** (I_a) characterised by **negative entropy of activation** ($\Delta S^{\neq}$) and a wide range of formation rate constants depending on the nature of the entering ligand. On the other hand, the Ru^{II}-complex (t_{2g}^6) experiences the **dissociative path** (I_d) characterised by the insensitivity of the formation rate constants towards the nature of the entering ligand (see Table 5.19.1.1). The **dissociative activation is favoured for the Ru^{II}-complex** due to the weaker Ru^{II}—OH_2 bond compared to the Ru^{III}—OH_2 bond.

● **Formation reactions at $[M(OH_2)_6]^{3+}$ and its hydrolysed product, *i.e.* $[M(OH)(OH_2)_5]^{2+}$ (M = Fe, Cr):** The unhydrolysed species having a higher positive charge adopt the I_a-path while the hydrolysed species adopt the I_d-path because of the factors: ***lower positive charge of the complex (+3 vs. +2) favouring the d-activation and disfavouring the a-activation; good σ-donor and π-donor property of the OH^- group labilising the coordinated water molecules through the d-activation to stabilise the 5-coordinate T.S. (see Sec. 5.21).***

$[Cr(NH_3)_5(OH_2)]^{3+}$ *vs.* $[Cr(OH_2)_6]^{3+}$: Relative strength of the $Cr^{III}—OH_2$ bonds and relative energy of the e_g^*-LUMO to accommodate the nucleophile.
$[M(OH_2)_6]^{3+}$ *vs.* $[M(OH)(OH_2)_5]^{2+}$ (M = Fe, Cr): Effect of overall positive charge (+3 *vs.* +2), stabilisation of the 5-coordinate T.S. in the dissociative path (Sec. 5.21).

Difference in the pathways of reactions of the Co(III) (low spin) and Cr(III) Complexes (*cf.* Secs. 5.19.4, 5.21-22, 24, 5.25.3)

- Generally, the Co(III) complexes (t_{2g}^6) adopt the dissociative path like $\boldsymbol{I_d}$ **path** for their ligand substitution reactions.
- The Cr(III) complexes (t_{2g}^3) generally adopt the associative path like the $\boldsymbol{I_a}$ **path** for their ligand substitution reactions like aquation of $[Cr(OH_2)_5X]^{2+}$, formation reactions at $[Cr(OH_2)_6]^{3+}$, solvent exchange reactions at Cr(III), etc. However, $[Cr(OH)(OH_2)_5]^{2+}$ adopts the $\boldsymbol{I_d}$**–path** (due to the ***cis*-effect**, *i.e.* labilising effect of the good σ- and π-donor property of the OH^- group to **stabilise the 5-coordinate T.S.** (see Sec. 5.21) and reduction of overall positive charge (+3 to +2) to favour the M–OH_2 bond breaking at the *rds*); anation of $[Cr(NH_3)_5(OH_2)]^{3+}$ and aquation of $[Cr(NH_3)_5(X)]^{2+}$ adopt also the $\boldsymbol{I_d}$**-path.** The reason behind the changeover from the I_a path to the I_d path has been explained already (*cf.* Sec. 5.19.6).
- The six electrons (*i.e.* t_{2g}^6) in the Co(III)–complexes offer a more repulsion to the approaching nucleophile in the associative path compared to the Cr(III) complexes bearing three electrons (*i.e.* t_{2g}^3).
- The **vacant antibonding orbital (*i.e.* LUMO) to accommodate the incoming nucleophile** in the associative path is energetically raised more in the Co(III) (low-spin) complexes compared to that in the Cr(III)–complexes. For the Cr(III) complexes, even the partially filled NBMO (*i.e.* t_{2g}^3) can act as the LUMO to receive the nucleophilic attack through the rearrangement of electrons in the t_{2g}-level. Thus the associative path is less favoured for the Co(III) complexes compared to that for the Cr(III) complexes (*cf.* Sec. 5.19.6).
- In terms of **CFAE**, the Co(III) (low spin) complexes are more inert than the Cr(III) complexes.

5.19.7 Steric Effects of the Spectator (Nonleaving) Ligands

Steric acceleration and steric retardation: If the starting complex is sterically crowded then in the dissociative process, the rate experience a **steric acceleration** while in the associative process, the rate experiences a **steric retardation.** In a dissociative process, the sterically crowded complex experiences **a steric relaxation** in attaining the activated complex through the dissociation of the leaving group. On the other hand, in an associative process, in the activated complex, the **steric crowding is enhanced** to disfavour the process. These aspects are illustrated in the following cases.

(i) **Amino acid exchange reaction at the square planar complexes of Cu(II):** The amino acid exchange rate at the square planar complexes of Cu(II), decreases with the increase of steric crowding (*i.e.* **steric retardation**). The rate sequence for the amino acids is:

$H_2N–CH_2–C(=O)OH$ > $MeNH–CH_2–C(=O)OH$ > $Me_2N–CH_2–C(=O)OH$

(Glycine) (Sarcosine) (Dimethylglycine)

The above rate sequence supports the **associative activation** in the ligand exchange process of the amino acids at the square planar complexes of Cu(II).

(ii) **Amino acid exchange reaction at the octahedral complexes of Co(II) and Ni(II):** On the other hand, for the octahedral complexes of Co(II) and Ni(II), the opposite rate sequence, *i.e.* dimethyl glycine ⟩ sarcosine (*i.e.* N-methyl glycine) > glycine supports the **dissociative activation** experiencing the **steric acceleration.**

(iii) **Steric acceleration in the aquation of Co(III)-amine complexes:** The aquation reactions of *trans*–$[CoCl_2(N—N)_2]^+$ (where N—N denotes the didentate ethylene diamine ligand bearing different substituents) experience a **steric acceleration** (*cf.* Sec. 5.21, Table 5.21.1).

$$trans\text{–}[CoCl_2(N—N)_2]^+ + H_2O \rightarrow [CoCl(N—N)_2(OH_2)]^{2+} + Cl^-$$

Substitution on the ethylenediamine skeleton may occur on both the 'N' and 'C' site but *the rate is more sensitive to the substitution on N.* The steric acceleration is in conformity with the dissociative activation.

(iv) **Aquation of *trans*-$[CoCl_2(N_4)]^+$ and *trans*-$[CoCl(N_4)X]^+$:** The aquation rates (for the loss of Cl^-) of *trans*–$[CoCl(N_4)(X)]^+$ and *trans*–$[CoCl_2(N_4)]^+$ increase from *cyclam* (N_4 = cyclam, *i.e.* tetraazacyclotetradecane) complex to *tet–b* (N_4 = hexamethyl derivative of cyclam) complex. This *steric acceleration (cf.* Sec. 5.21, Table 5.21.2) is in conformity with the dissociative activation.

(v) **Aquation reactions of Co(III)- *vs.* Cr(III)-ammine complexes:** Steric effects on the aquation rates of Co(III)–ammine and Cr(III)–aminine complexes are of different types.

	$[CoCl(NH_3)_5]^{2+}$	$[CoCl(MeNH_2)_5]^{2+}$	$[CrCl(NH_3)_5]^{2+}$	$[CoCl(MeNH_2)_5]^{2+}$
$k_{aq}(s^{-1})$ (50°C)	3.6×10^{-5}	6.5×10^{-4}	1.8×10^{-4}	1.4×10^{-5}
	(Steric acceleration)		(Steric retardation)	

It indicates that the Co(III)–complexes experiences the **dissociative activation** while the corresponding Cr(III)–complexes experience the **associative activation.**

(vi) **Aquation of $[MCl(dien)(en)]^{2+}$ and $[MCl(dien)(tn)]^{2+}$ (M = Co *vs.* Cr)** (*Inorg. Chim. Acta.*, **19**, 23, 1976)): $[MCl(dien)(tn)]^{2+}$ and $[MCl(dien)(en)]^{2+}$ complexes can produce three geometrical isomers (Fig. 5.19.7.1).

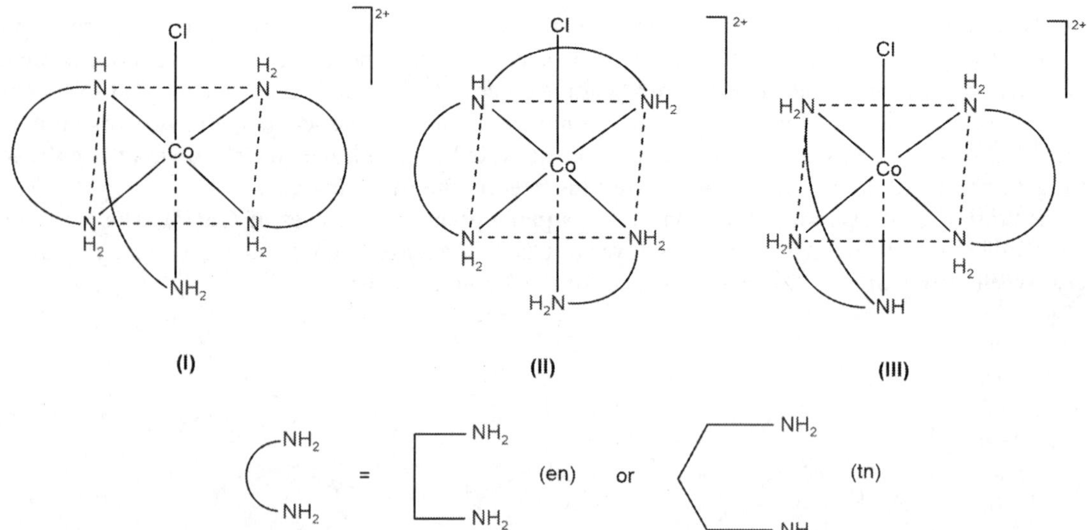

Fig. 5.19.7.1: Three geometrical isomers of $[CoCl(dien)(N—N)]^{2+}$ where N—N = en or tn.

For Co(III), *tn–complex aquates much faster than the corresponding en–complex* (*e.g.* for the **II** isomer, $k_{tn}/k_{en} \approx 32$ at 25°C). **Interestingly, for the corresponding Cr(III)–complexes, both the *tn*– and *en*–complexes react at comparable rates.**

In the dissociative path of aquation, the five coordinate activated complex adopts the trigonal bipyramidal (TBP) geometry where the diamine (N—N) chelate ring is to be stretched considerably from 90° to 120° bite angle (for the isomer **II**) in going from the octahedral to the trigonal bipyramidal geometry (see Fig. 5.19.7.2). *This bite angle expansion is more constrained for the 5–membered diamine chelate ring (en) than for the 6–membered diamine (tn) chelate ring.*

Here it is worth mentioning that if the aquation occurs through the dissociative path involving the **square pyramid T.S.**, then no distortion of the **chelate ring is required to attain the T.S.** In such cases, both the *en*- and *tn*-complexes would react at the identical/comparable rates. The higher aquation rate for the *tn*-complex supports the **trigonal bipyramidal T.S. in the dissociative path**. The different rate ratios for the three isomers: k_{tn}/k_{en} = 17.0 (I), 32.0 (II) and 2.4 (III) can be explained by considering the relative difficulties to attain the TBP T.S. It is most difficult for the isomer II of *en*-complex. It explains the highest rate ratio for the isomer II.

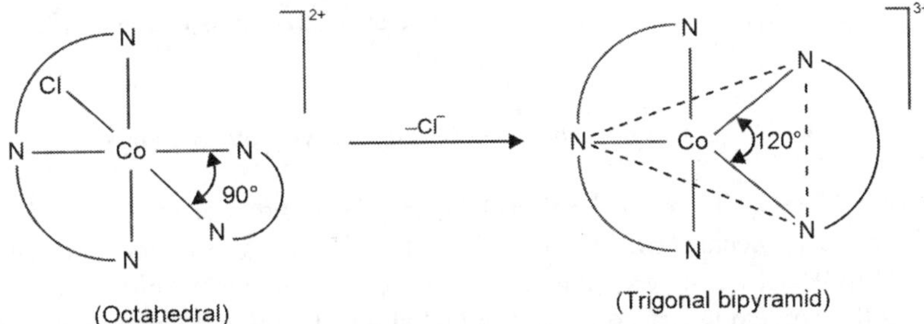

Fig. 5.19.7.2: Structural change (illustrated for the isomer II) in attaining the TBP activated complex through the dissociative path for $[CoCl(dien)(N—N)]^{2+}$ where N—N = *en* or *tn* (*Inorg. Chim. Acta.*, **19**, 23, 1976)

The higher aquation rate for the *tn*–complex of Co(III) is in comformity with the **dissociative path involving the trigonal bipyramidal T.S./intermediate.** Absence of such effect for the Cr(III)–complexes indicates the **associative path.** In the **associative path**, for attaining the 7–coordinate activated complex (which may be pentagonal bipyramidal or trapezoidal octahedral, see Fig. 5.18.2.1), **no such stretching of bond angle of the diamine chelate ring is required.**

5.19.8 Electronic Effects of the Spectator or Nonleaving Group

Electronic effects (*i.e.* electron pushing and withdrawing) exerted by the nonleaving groups can control the reaction rate and mechanism of the ligand substitution process. This aspect is illustrated in the following cases.

- **Base hydrolysis of the Co(III)-ammine complexes:** It passes through the **D-CB or I_d-CB mechanism** in which the CB (conjugate base generated at the pre-equilibrium step) experiences the **dissociative activation by the labilising effect of** the NH_2 or NH group of CB (which is a good σ- and π-donor group). The different aspects of this D–CB mechanism have been discussed in Sec. 5.25.
- **Aquation of $[CoCl(en)_2(L)]^{n+}$** (*cf.* Sec. 5.21): If the nonleaving π–donor L–group (*e.g.* OH^-, Cl^-, etc.) occupies the *cis*–position then the aquation rate is faster than the corresponding *trans*–complex. The higher reactivity of the *cis*–isomer arises through the ***cis*–effect** by the nonleaving L-group. It indicates the **dissociative pathway.** In fact, to stabilise the 5–coordinate intermediate by the nonleaving π–donor ligand (L), the corresponding square pyramid for the *cis*–complex is in the right configuration but the square pyramid for the corresponding *trans*–complex is not in the right configuration. In fact,

the square pyramid obtained from the *trans*–complex is to be changed into the trigonal bipyramidal geometry *so that the vacant orbital (available due to the departure of the leaving group) of the metal centre can participate in π-bonding with the nonleaving π-donor group (L) (cf. Fig. 5.21.1)*. It raises the activation energy.

The *cis*–complexes having the π–donor nonleaving ligand (L) react faster than the complexes where L is a π–acceptor ligand (*e.g.* CN^-) or a non π–bonding ligand like NH_3. It can be explained by the dissociative mechanism.

If L is a *π–withdrawing group (e.g. NO_2^-, CN^-, etc.) then the trans–isomer* is more reactive than the *cis*–isomer. It can be explained by considering the **associative mechanism.**

Effect of L (nonleaving group) on the aquation of $[CoCl(en)_2L]^{n+}$

Thus in the aquation of $[CoCl(en)_2L]^{n+}$, if the nonleaving group **L is a π donor** ligand, then the process passes through the **dissociative path** and the *cis*–complex reacts faster than the *trans*–isomer. On the other hand, if **L is a π acceptor ligand,** then it adopts the **associative path** where the *trans*–isomer reacts faster. These aspects have been discussed in detail in Sec. 5.21.

π-donor L : $\ddot{L}$ — $Co(en)_2$ — Cl **(dissociative path, presumably I_d)**

:Nu

π-acceptor L : L — $\ddot{Co}(en)_2$ — Cl **(associative path, presumably I_a)**

● **Reactivity of $[M(OH_2)_6]^{3+}$ *vs.* $[M(OH)(OH_2)_5]^{2+}$ (M = Fe, Cr, *cf.* Sec. 5.19.6):** The higher reactivity and preferred I_d mechanism for $[M(OH)(OH_2)_5]^{2+}$ while the lower reactivity and preferred I_a mechanism for $[M(OH_2)_6]^{3+}$ can be explained by considering the electronic effect of the nonleaving OH group. In fact, the nonleaving π–donor group OH shows the ***cis*–effect** to impart the dissociative path by stabilising the 5-coordinate T.S. Here it may be mentioned that H_2O is a poor π–donor ligand. It may be mentioned that the better σ-donor property of OH group (compared to that of OH_2) also labilises the coordinated H_2O molecule and imparts the dissociative activation. The decrease of overall positive charge (+3 *vs.* +2) of the complex, favours also the dissociative path in $[M(OH)(OH_2)_5]^{2+}$.

● **Reactivity of $[Cr(OH_2)_6]^{3+}$ *vs.* $[Cr(NH_3)_5(OH_2)]^{3+}$ (*cf.* Sec. 5.19.6):** $[Cr(OH_2)_6]^{3+}$ adopts the I_a path while $[Cr(NH_3)_5(OH_2)]^{3+}$ adopts the I_d path. It is due to the better σ–donor property of NH_3 compared to that of H_2O.

5.19.9 Comparison between the Rate of Formation Reaction (*i.e.* Replacement of Metal Bound Water Molecule) and Water Exchange Reaction (*cf.* Sec. 5.12.2)

Comparison of the formation rate constants with the water exchange rate constants (k_{ex}) is helpful in many cases to diagonse the reaction pathway. This aspect has been illustrated in Sec. 5.12.2 and some aspects are discussed below.

(a) **Interchange (*I*) process (Secs. 5.9.3; 5.14) :** It has been already pointed out that compared to the perfect *D* and *A* processes, the interchange processes (*i.e.* I_d and I_a) are more well documented in the ligand substitution reactions of octahedral complexes.

Under the limiting condition, $K_{OS}[Y] \ll 1$, we have $k_f = K_{OS}\, k_i$ (*cf.* Secs. 5.9.3, 5.14) and the ratio k_f/K_{OS} is more or less insensitive to the nature of the entering ligand for the I_d process. K_{OS} (outer sphere association constant) depends on the nature of the entering group (Y) for a particular metal complex and consequently, k_f (in dm^3 mot^{-1} s^{-1}) depends on the nature of Y but k_i ($= k_f/K_{OS}$) remains more or less constant and close to k_{ex}. This aspect has been illustrated (*cf.* Table 5.14.1) for the following reactions where Y varies in structure and basicity widely.

$$[Ni(OH_2)_6]^{2+} + Y^{x-} \longrightarrow [Ni(OH_2)_5(Y)]^{n+} + H_2O$$

Statistical Relationship between k_i and k_{ex}

k_i = rate constant for the ligand penetration leading to the inner-sphere complex from the outer-sphere complex in the interchange process (*cf.* Secs. 5.9.3, 5.14).

k_{ex} = first order rate constant for the water exchange process.

$$[M(OH_2)_6] + L : k_i \,;\; [M(^*OH_2)_6] + H_2O : k_{ex}$$

k_i and k_{ex} are statictically related (*cf.* B. Permutter – Hayman *et. al*, *J. Coord. Chem.*, **6**, 31, 1976) as: the ligand L residing in the outer coordination sphere can exchange with any one of the six water molecules residing in the inner coordination sphere but the ligand penetration competes with the water molecules present in the outer coordination sphere. Consequently, the chance of ligand penetration is lowered by a factor of eight (8 faces of octahedral structure, where the entering ligand can approach; ligands approach along the C_3-axes passing through the centres of opposite trigonal octahedral faces; the d_{xz} and d_{yz} t_{2g}-orbitals are projected along the C_3-axes, see Sec. 5.18.2). Thus:

$$k_i = \frac{6}{8}k_{ex} = 0.75k_{ex}$$

(b) **Identification of the I_d and I_a paths in terms of the limiting ratio k_{lim}/k_{ex} (*cf.* Secs. 5.9.5-6, 5.12.2):** Under the upper limiting condition, $K_{OS}[Y] \gg 1$, we have: $k_{obs} = k_{lim} \approx k_i$; and under the limiting condition, $1 \gg K_{OS}[Y]$, we have: $k_{obs} \approx k_i K_{OS}[Y]$, $k_f \approx k_i K_{OS}$.

In fact, in the I_d pathway, the rate determining step is the M—OH_2 bond rupture for both the formation reaction and water exchange reaction. ***Thus in the I_d path, the ratio k_f/K_{OS} (= k_i) is constant and close to the k_{ex} value. This is the diagnostic test for the I_d path.***

Relationship between k_i or k_{lim} and k_{ex} for the Interchange Processes

- Under the condition, $K_{OS}[Y] \ll 1$, $k_{obs} = k_i K_{OS}[Y]$ and $k_f = k_i K_{OS}$; k_f (in $dm^3\ mol^{-1}\ s^{-1}$):

  $k_i = \dfrac{k_f}{K_{OS}}$ = constant and close to k_{ex} for the different entering ligands (**I_d path).**

  $k_i = \dfrac{k_f}{K_{OS}}$ = dependent on the nature of the entering ligand and no relationship with k_{ex} **(I_a path)**
- Under the condition, $K_{OS}[Y] \gg 1$, $k_{obs} = k_{lim} \approx k_i$ (in s^{-1}); generally, the upper limiting rate constant (k_{lim}) is taken as the k_{obs} at high [*Y*].
- k_{lim}/k_{ex} can be less than or greater than unity depending on the relative nucleophilicities of the entering ligand and water (**I_a path**).
- For the **I_d-path,** $k_i = k_{lim} \approx k_{ex}$.
- k_{lim}/k_{ex} can never exceed unity and the ratio is close to unity (**I_d path**).

For the **associative activation** (*i.e.* I_a path), under the condition $K_{OS}[Y] \ll 1$, we can apply the condition, $k_f = K_{OS}k_i$ but in such cases, the ratio k_f/K_{OS} (= k_i) is not constant and it depends on the nature of the entering ligand. There is no relationship between $k_i = k_f/K_{OS}$ or = k_{lim} (under the condition $K_{OS}[Y] \gg 1$) and k_{ex}. In the I_a pathway, the rate determining step is mainly controlled by the bond formation by the entering ligand during its interchange with the metal bound solvent molecules. If the entering ligand is a better nucleophile than the solvent, *i.e.* H_2O, then k_i or k_{lim} is greater than k_{ex}. On the other hand, if the entering ligand is a poorer nucleophile than the solvent then k_i or k_{lim} is less than k_{ex}.

Table 5.19.9.1: Identification of reaction mechanism from the knowledge of k_{lim}/k_{ex} ratio

(a) **$[Co(NH_3)_5(OH_2)]^{3+} + Y^{n-}$**

Y^{n-}	k_{lim} (s^{-1}), 45°C	k_{lim}/k_{ex}	
Cl^-	2.1×10^{-5}	0.21	
NCS^-	1.6×10^{-5}	0.16	**Dissociative activation, *i.e.* I_d path**
N_3^-	1.0×10^{-4}	1.0	(The ratio does not exceed the value 1)
SO_4^{2-}	2.4×10^{-5}	0.24	

$k_{ex} = 1 \times 10^{-4}\ s^{-1}$, $k_{lim} = k_{obs}$ (obtained at high [*Y*])

(b) **$[Rh(NH_3)_5(OH_2)]^{3+} + Y^{n-}$** (*cf.* F. Monacelli, *Inorg. Chim. Acta.*, **2**, 263, 1968)

Y^{n-}	k_{lim} (s^{-1}), 65°C	k_{lim}/k_{ex}	
Br^-	7.9×10^{-3}	4.9	
Cl^-	4.2×10^{-3}	2.6	**Associative activation, *i.e.* I_a path**
SO_4^{2-}	1.7×10^{-3}	~1.0	(The ratio can exceed the value 1)

$k_{ex} = 1.6 \times 10^{-3}\ s^{-1}$, $k_{lim} = k_{obs}$ (obtained at high [*Y*])

(c) **$[Ir(NH_3)_5(OH_2)]^{3+} + Y^{n-}$** (*cf.* E. Borghi, *et al*, *Inorg. Nucl. Chem. Lett.*, **6**, 667, 1970)

Y^{n-}	k_{lim} (s^{-1}), 85°C	k_{lim}/k_{ex}	
Cl^-	9.2×10^{-4}	4.2	**Associative activation, *i.e.* I_a path** (The ratio can exceed the value 1)

$k_{ex} = 2.2 \times 10^{-4}\ s^{-1}$, $k_{lim} \approx k_{obs}$ (obtained at high concentration of [*Y*])

Table 5.19.9.2: Kinetic parameters for the formation reactions at $[Fe(OH_2)_6]^{3+}$ and $[Fe(OH)(OH_2)_5]^{2+}$ (Refs. M. Grant and R.B. Jordan, *Inorg. Chem.*, **20,** 55, 1981; S. Gouger and J. Stuehr, *Inorg. Chem.*, **13**, 379, 1974)

Entering ligand	**$[Fe(OH_2)_6]^{3+}$ (I_a path)**		**$[Fe(OH)(OH_2)_5]^{2+}$ (I_d path)**	
(Y^-)	k_f	k_f/K_{OS}	k_f	k_f/K_{OS}
Cl^-	9.4	1.88	1.1×10^4	0.55×10^4
Br^-	50	10.0	4.1×10^4	2×10^4
SCN^-	127	25.4	1×10^4	0.5×10^4
$CH_2ClCO_2^-$	4.9×10^3	10^3	2.8×10^4	1.4×10^4

(k_f in $dm^3\ mol^{-1}\ s^{-1}$, $k_i = k_f/K_{OS}$, k_i in s^{-1}), $k_{ex} \approx 10^2\ s^{-1}$ for $\left[Fe(OH_2)_6\right]^{3+}$, $\approx 10^4\ s^{-1}$ for $\left[Fe(OH)(OH_2)_5\right]^{2+}$
$K_{OS} = 2\ dm^3\ mol^{-1}$ (for $Z_AZ_B = -2$), $= 5\ dm^3\ mol^{-1}$ (for $Z_AZ_B = -3$)

Table 5.19.9.3 Kinetic parameters for the formation reactions at $[Cr(OH_2)_6]^{3+}$ and $[Cr(OH)(OH_2)_5]^{2+}$ centres

Entering ligand	**$[Cr(OH_2)_6]^{3+}$ (I_a path)**		**$[Cr(OH)(OH_2)_5]^{2+}$ (I_d path)**	
(Y^-)	k_f	k_f/K_{OS}	k_f	k_f/K_{OS}
Cl^-	2.9×10^{-8}	5.8×10^{-9}	2.8×10^{-5}	1.4×10^{-5}
Br^-	9.0×10^{-9}	4.5×10^{-9}	1.7×10^{-5}	0.85×10^{-5}
NCS^-	1.8×10^{-6}	3.6×10^{-7}	4.9×10^{-5}	2.5×10^{-5}
k_{ex} (s^{-1})	2.4×10^{-6}		1.8×10^{-4}	

k_f in $dm^3\ mol^{-1}\ s^-$; $k_i = k_f/K_{OS}$; k_i in s^{-1}; $K_{OS} = 2$ (for $Z_AZ_B = -2$), $= 5$ (for $Z_AZ_B = -3$); *cf.* J.H. Espenson, *Inorg. Chem.*, **8**, 1534, 1969.

Thus in the I_a path, k_{lim}/k_{ex} may be greater or less than unity depending on the relative nucleophilicities of the entering ligand and water. But in the I_d pathway, $k_i = k_{lim} \approx k_{ex}$ and the ratio k_{lim}/k_{ex} can never exceed unity and it is close to unity.

The said criteria for the I_a path are well documented for the substitution reactions at $[Ti(OH_2)_6]^{3+}$, $[Cr(OH_2)_6]^{3+}$, $[Rh(NH_3)_5(OH_2)]^{3+}$, $[Ir(NH_3)_5(OH)_2]^{3+}$.

(c) **Reactivity of $[M(OH_2)_6]^{3+}$ *vs.* $[M(OH)(OH_2)_5]^{2+}$:** For $[M(OH_2)_6]^{3+}$, (M = Fe, Cr), variation of the ratio k_f/K_{OS} in a wide range for the different ligands indicates the **associative mechanism** (*i.e.* I_a). On the other hand, the ratio k_f/K_{OS} is more or less constant for the different ligands entering at the $[M(OH)(OH_2)_5]^{2+}$ centre. This suggests the **dissociative path** (*i.e.* I_d) at $[M(OH)(OH_2)_5]^{2+}$. In $[M(OH)(OH_2)_5]^{2+}$, the *cis-effect of the π–donor OH group introduces the higher reactivity and dissociative character* (see Secs. 5.19.6, 5.21). The better σ-donor property of OH^- (compared to that of H_2O) also labilises the remaining coordinated water molecules and it imparts a dissociative character towards the substitution reaction. The overall lower charge (+2 *vs.* +3) also favours the dissociative path at $[M(OH)(OH_2)_5]^{2+}$ (see Sec. 5.19.6 for details).

(d) **Rate dependance on the nature of the entering ligand for a pure D-process (*cf.* Secs. 5.9.1, 5.16):** For the perfect *D*-process (where the I_d path is disfavoured because of the disfavour in the formation **of the outer-sphere complex**, *e.g.* anation or aquation of highly negatively charged complex) as in the anation of the complexes like $[Co(CN)_5(OH_2)]^{2-}$, $[RhCl_5(OH_2)]^{2-}$, etc. the ratio k_2/k_{-1} obtained from the double reciprocal plot, $1/k_{obs}$ *vs.* $1/[Y]$ depends on the nature of the entering ligand *Y*.

$$\left.\begin{array}{l} L_5M{-}OH_2 \underset{k_{-1}}{\overset{k_1}{\rightleftharpoons}} L_5M + H_2O \\ L_5M + Y \xrightarrow{k_2} L_5M{-}Y \end{array}\right\} \begin{array}{l}\textbf{D-process, } \boldsymbol{Y} = \text{labelled water} \\ \text{or any other ligand } (cf.\ \text{Secs. 5.9.1 and 5.16}).\end{array}$$

$$\textbf{Note: } (k_2/k_{-1})_{\text{exchange}} = 1 \ \textbf{(Water exchange reaction)}$$

The variation of the ratio (k_2/k_{-1}) with the nature of *Y* arises from the ***competition between the solvent and Y for the long lived 5-coordinate intermediate.***

I_a vs. *D* (*cf.* Sec. 5.16): What is the difference?

Rate depends on the nature of the entering ligand (*Y*) for both the *D* and I_a process. But under the **limiting condition** (*i.e.* high [*Y*]):

$k_{obs} \approx k_{\lim} = k_1$ (**D-process**; $k_2[Y] \gg k_{-1}[H_2O]$ and k_1, *i.e.* water loss process is independent of *Y*); while $k_{obs} \approx k_{\lim} = k_i$ (**I_a-process**; $K_{OS}[Y] \gg 1$) which depends on the nature of *Y*.

- $k_{lim} \approx k_1$ (independent of Y) ⇒ **D-process:** $k_{lim} \approx k_i$ (dependent of *Y*) ⇒ **I_a process**

Table 5.19.9.4 (k_2/k_{-1}) ratio for the formation reaction at $[RhCl_5(OH_2)]^{2-}$ through the *D*–process.

Y	Cl^-	Br^-	SCN^-	N_3^-	H_2O (exchange)
k_2/k_{-1}:	2.1×10^{-2}	1.6×10^{-2}	7.8×10^{-2}	0.14	1.0

(*cf.* D. Robb *et al*, *Inorg. Chim. Acta.*, **3**, 383, 1969)

Table 5.19.9.5 (k_2/k_{-1}) ratio for anation of $[Co(CN)_5(OH_2)]^{2-}$ through the *D*-process.

Y	Cl^-	I^-	NCS^-	*py*	H_2O (exchange)
k_2/k_{-1}	6.0	11.7	20.4	23.9	1

(*cf.* A. Haim, *et al.*, *Adv. Chem. Ser.*, **49**, 31, 1965)

Table 5.19.9.6 (k_2/k_{-1}) ratio for the anation of $[Co^{III}(corrin)(OH_2)]$ ($k_{ex} = 14\ s^{-1}$) through the *D*-process.

Y	SCN^-	I^-	Br^-	N_3^-	HSO_3^-	H_2O (exchange)
k_2/k_{-1}	1.64×10^{-2}	1×10^{-2}	7.1×10^{-3}	8.6×10^{-3}	1.2×10^{-3}	1

(*cf.* D. Thusius, *J. Am. Chem. Soc.*, **93**, 2629, 1971)

5.19.10 Effect of Ionic Strength on Rate Constant

From the **Debye-Huckel theory,** the rate dependence of a bimolecular reaction on ionic strength (I) may be expressed as follows:

$$\log k_I \approx \log k_0 + 2QZ_AZ_B\sqrt{I}\ ,\ \log\left(\frac{k_I}{k_0}\right) = 2QZ_AZ_B\sqrt{I} \qquad \ldots(5.19.10.1)$$

k_I = rate constant at ionic strength I; k_0 = rate constant at zero ionic strength;

$Q \propto \varepsilon^{-3/2}$ where ε = dielectric constant of the solvent;

Z_A and Z_B are the ionic charges of the reactant A and B.

I = ionic strength $= \frac{1}{2}\Sigma C_i Z_i^2$ where C_i is the molar concentration of the ion i having the charge Z_i.

It is evident that the plot of log k_I *vs.* $\sqrt{I}$ gives a straight line with the slope = $2QZ_AZ_B$.

In aqueous solvent, $Q = 0.509$ (at 25°C) and thus the slope becomes $1.02Z_AZ_B$.

● Thus it is evident that if the product Z_AZ_B (*i.e.* **reacting species are of opposite charges**) is negative, the rate will linearly decrease with $\sqrt{I}$.

Examples: $[CoCl(NH_3)_5]^{2+} + OH^- \rightarrow$; $[Co(C_2O_4)_3]^{3-} + Fe^{2+} \rightarrow$; $[Cr(OH_2)_6]^{3+} + SCN^- \rightarrow$ etc.

● If the **reacting species are of the same type of charge** (*i.e.* Z_AZ_B = +ve), then the rate increases with $\sqrt{I}$.

Example: $[CoBr(NH_3)_5]^{2+} + Hg^{2+} \rightarrow$; etc.

Origin of the Effect of Ionic Strength on Rate Constant

$$A + B \underset{}{\overset{K^{\neq}}{\rightleftharpoons}} X^{\neq} \xrightarrow{k^{\neq}} \text{Product, } (cf.\ \text{Transition State Theory, Sec. 5.19.3})$$

The rate is expressed in terms of the concentration term not in terms of the activity term. It gives:

$$\text{rate} \propto [X^{\neq}],\ i.e.\ \text{rate} = k^{\neq}[X^{\neq}];\ K^{\neq} = \frac{a_{X^{\neq}}}{a_A a_B} = \frac{[X^{\neq}]}{[A][B]}\frac{f_{\neq}}{f_A f_B}$$

It leads to:

$$\text{rate} = k^{\neq}[X^{\neq}] = k^{\neq}K^{\neq}[A][B]f_A f_B / f_{\neq}$$

or,

$$= k_0[A][B]\frac{f_A f_B}{f_{\neq}} = k[A][B],\ k_0 = k^{\neq}K^{\neq}$$

Here,

$$k = k_0\frac{f_A f_B}{f_{\neq}};\ \text{or}\ \log k = \log k_0 + \log\frac{f_A f_B}{f_{\neq}}$$

According to **Debye-Huckel theory:**

$$\log f = -QZ^2\sqrt{I}$$

Using this relationship, we get:

$$\begin{aligned}\log k &= \log k_0 + \log f_A + \log f_B - \log f_{\neq}\\ &= \log k_0 - \sqrt{I}Q\left[(Z_A)^2 + (Z_B)^2 - (Z_A + Z_B)^2\right]\\ &= \log k_0 + 2QZ_AZ_B\sqrt{I}\end{aligned}$$

This is why, the kinetic studies are carried out at a *fixed* and *fairly high ionic strength* (*ca.* 0.1–1 mol dm^{-3}) maintained by the *inert electrolytes*. The inert electrolytes (*e.g.* $NaClO_4$, KNO_3, etc.) used for maintaining the ionic strength are expected not to interact with the reactants. Here it may be mentioned, that the Eqn. 5.19.10.1 is valid at low I-values.

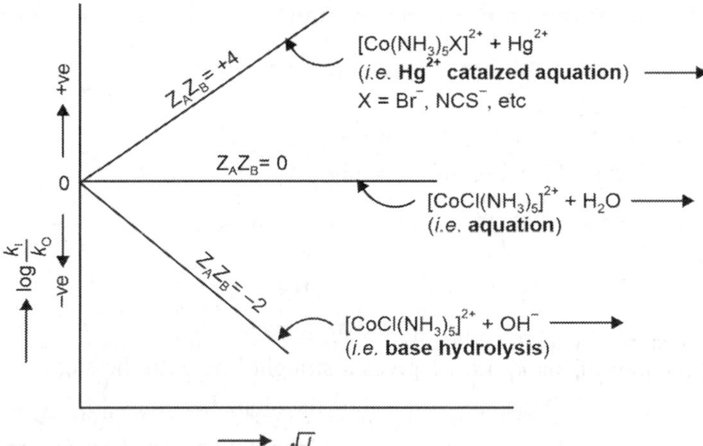

Fig. 5.19.10.1: Qualitative representation of the effect of ionic strength on the rate constant (*i.e.* linear plot of $\log(k_I/k_0)$ *vs.* $\sqrt{I}$) ignoring the **secondary salt effect**

Primary and secondary salt effect: The effect of inert electrolytes on the rate process has been expressed in Eqn. 5.19.10.1 in terms of ionic strength (I). This is called the *primary salt effect.*

Sometimes, the electrolyte (used for maintaining the ionic strength) may participate in some specific interactions with the reactants. **Such electrolytes show the secondary salt effect.** In fact, when the reactants are highly charged, they may participate in **ion-pair formations** with the ions obtained from the inert electrolytes used for maintaining the ionic strength. Reactivity of such ion pairs is different compared to that of the free reacting ions. For example, in the formation reactions of $[Cr(OH_2)_6]^{3+}$, it forms ion pairs with the anions (X^{n-}) of the electrolyte used for maintaining the ionic strength. Reactivities of $[Cr(OH_2)_6]^{3+}$ and the ion pair like $\{[Cr(OH_2)_6]^{3+}\cdot X^{n-}\}$ are different. **It introduces the secondary salt effect.** Similarly, in the Hg^{2+} catalysed aquation of $[CoBr(NH_3)_5]^{2+}$, *i.e.*

$$\left[CoBr\left(NH_3\right)_5\right]^{2+} + Hg^{2+} \xrightarrow{H_2O} \left[Co\left(NH_3\right)_5\left(OH_2\right)\right]^{3+} + HgBr^+,$$

the ion associations like $\{[CoBr(NH_3)_5]^{2+}\cdot X^{n-}, Hg^{2+}\cdot X^{n-}\}$ (X^{n-} represents the anion coming from the electrolyte) may also participate in the reaction with different reactivities. *Existence of the secondary salt-effect is indicated by the deviation from the linear plot of log k_I vs.* $\sqrt{I}$.

To avoid the secondary salt effect (*e.g.* ion-pair formation), generally **perchlorate and nitrate salts** are used. Because of the large size of the monovalent anions like ClO_4^-, NO_3^- (*i.e.* ***low charge density***), possibility of ion-pair formation is reduced. Moreover, NO_3^- and ClO_4^- are of very **poor complexing anions.** Consequently, untoward complexation by NO_3^- or ClO_4^- may be avoided.

5.19.11 Hammett Relationship: Linear Gibbs Energy Relationship

It expresses the relation of reactivities of a series of *meta*– and *para*– substituted aromatic compounds with a common reactant.

$$\log\left(\frac{k}{k_0}\right) = \log\left(\frac{K}{K^0}\right) = \rho\log\left(\frac{K_a}{K_a^0}\right) = \rho\sigma,\ i.e.\ \log k = \log k_0 + \rho\sigma \qquad \ldots(5.19.11.1)$$

k and k_0 are the rate constants for the reactions of the *meta*– or *para*– R–substituted and *unsubstituted aromatic compounds* respectively; K and K^0 are the equilibrium constants of the corresponding reactions; K_a and K_a^0 are the acid dissociation constants of the substituted and unsubstituted benzoic acid respectively; σ called *substituent constant* depends only on the substituent but not on the reaction series; ρ called *reaction constant* depends on the reaction under consideration.

It is evident that for the **ionisation reactions,** by definition, ρ **is taken as unity**, *i.e.*

$$\log\left(\frac{K_a}{K_a^0}\right) = \rho_a\sigma = \sigma,\ (\rho_a = 1)$$

The above equation may be expressed as follows:

$$\mathbf{\log k = \log k_0 + \rho\sigma \text{ and } \log K = \log K^0 + \rho\sigma} \qquad \text{...(5.19.11.1)}$$

where $\sigma = \log K_a - \log K_a^0 = pK_a^0 - pK_a$ ...(5.19.11.3)

The linear plot of log k *vs.* σ or log K vs. σ gives the slope = *reaction constant* (ρ). The *substituent constants* (σ) of different groups for *meta*– and *para*– positions are available in literature.

The above relations indicate that **for two homologus series of reactions,** the rate constants (say k_1 and k_2) and the equilibrium constants (K_1 and K_2) are related as follows:

$\log k_1 = \text{constant} + \rho \log k_2$ and $\log K_1 = \text{constant} + \rho \log K_2$ where $\rho = \rho_1/\rho_2$

The rate constant (k) is related with the standard free energy of activation ($\Delta G^{\neq}$; the zero subscript to indicate the standard state is dropped for the sake of simplicity) and the equilibrium constant (K) is related with the standard free energy (ΔG^0) of reaction.

$$\Delta G^0 = -RT \ln K, \textit{i.e.} \log K = -\frac{\Delta G^0}{2.303\,RT}$$

$$k = \frac{(k_B T)}{C_o^n h}\exp\left(-\Delta G^{\neq}/RT\right) \text{ (see Sec. 5.19.3)}$$

$$\textit{i.e.} \quad \ln k = \ln\left(\frac{k_B T}{h}\right) - \frac{\Delta G^{\neq}}{RT} - n \ln C_o$$

In terms of **transition state theory** (see Sec. 5.19.3), $C_o = 1.0$ mol dm^{-3} (standard state concentration), $n = 0$ for the 1st order rate constant, = 1 for the second order rate constant, and so on. Using the value of C_o, from the above equation, we get:

$$\log k = \log\left(\frac{k_B T}{h}\right) - \frac{\Delta G^{\neq}}{2.303RT}$$

Thus the Hammett relationships indicate the linear relationship between the Gibbs free energy, of activation or reaction, for different reaction series. Thus for two series of homologus reactions, a linear relationship between $\Delta G^{\neq}_{(1)}$ and $\Delta G^{\neq}_{(2)}$ or $\Delta G_{0(1)}$ and $\Delta G_{0(2)}$ exists.

i.e. $\Delta G_1^{\neq} = \text{constant} + \rho\Delta G_2^{\neq}$; or $\Delta G_{0(1)} = \text{constant} + \rho\Delta G_{0(2)}$, $\rho = \rho_1/\rho_2$

● ***Positive σ-value*** *indicates the electron withdrawing property of the group* (*e.g.* $-NO_2$, –Cl, etc.) *compared to that of hydrogon.* It means that the substituted benzoic acid is stronger than the unsubstituted benzoic acid, *i.e.* $pK_a^0 > pK_a$,

● ***Negative σ-value*** *indicates the electron pushing effect of the group* (*e.g.* $-CH_3$, $-C_2H_5$). It means that the substituted benzoic acid is weaker than the unsubstituted one, *i.e.* $pK_a^0 < pK_a$.

It may be mentioned that for a particular substituent R, the σ–value may be different for the *meta*– and *para*– positions because of the different electronic effect arising from the *meta*– and *para*–positions. The group R can show only the **inductive effect** at the *meta*–position but at the para-position, it can show both the **inductive effect** and **mesomeric effect** (if possible, as in the case of OH group, NO_2 group, etc).

The linear plot indicates that the series of reaction pass through a common mechanism. Positive value of the *reaction constant* (which is constant for a particular series of similar reactions), indicates that the k-value increases as σ becomes more positive.

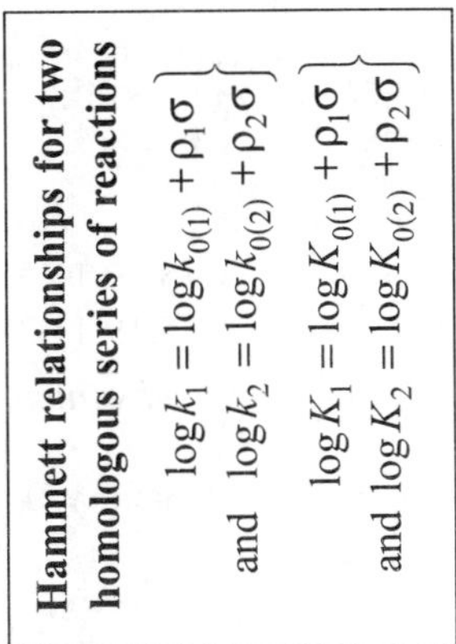

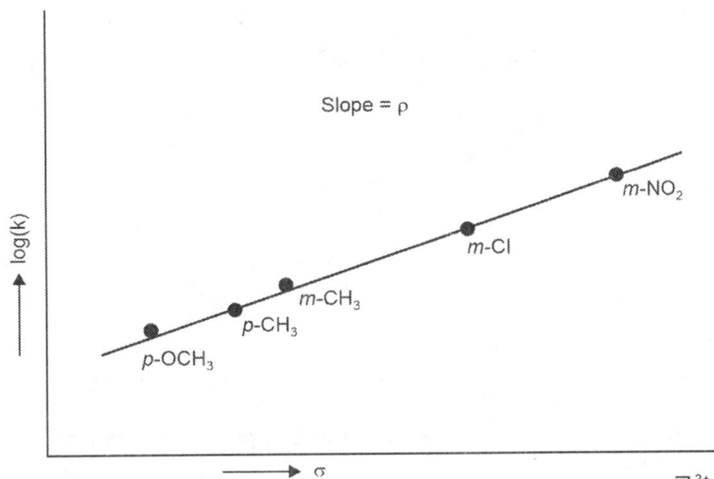

Fig. 5.19.11.1 Qualitative representation of Hammett plot for the base hydrolysis of $[(H_3N)_5Co—O—C(=O)—C_6H_4R]^{2+}$ in aqueous methanol

Determination reaction constant (ρ) helps us establish the nature of reaction. It is illustrated below.

$$[(H_3N)_5Co^{III}—O—C(=O)—C_6H_4R]^{2+} \xrightarrow[\text{(Base hydrolysis)}]{OH^-} [(H_3N)_5Co^{III}—OH]^{2+} + R—C_6H_4—CO_2^-$$

(R at the *m*– or *p*–position)

The linear plot of log *k vs.* σ for different substituents (R) indicates a **common mechanism** of the above base hydrolysis reaction (Fig. 5.19.11.1). The reaction can occur either through the 'Co—O' or 'C—O' bond fission. The ρ-value for 'C—O' bond fission (*i.e.* alkaline hydrolysis of ester) is in the range 1.8–2.5. But, *the* ρ-*value* (*ca.* 0.75–0.9) *experimentally determined for the* above mentioned *base hydrolysis reactions is much smaller than the above range* (*cf.* F. Aprile *et al*, *J. Inorg. Nucl. Chem.*, **21,** 325, 1961). This smaller value of ρ has been argued for the Co—O bond cleavage. The ρ-value indicates the **sensitivity of the reaction series to the ring substitution.** In alkaline hydrolysis of ester, the C—O bond which ruptures is closer to the aromatic ring and it makes the ρ-value larger. On the other hand, in the said base hydrolysis, the Co—O bond which ruptures lies relatively further away (compared to the C—O bond) from the ring and it makes the ρ value relatively less (*cf.* 0.75 – 0.9 *vs.* 1.8 – 2.5). Thus in the base hydrolysis of complex, the reaction is relatively less sensitive to the ring substitution.

5.19.12 Taft Relationship

Hammett equation was found inadequate for the reactions of aliphatic compounds and *ortho*–substituted compounds. To overcome these limitations, Hammett equation was modified by Taft as follows:

$$\log(k/k_0) = \sigma^*\rho^* \qquad \ldots(5.19.12.1)$$

σ^* called *polar substituent constant;* ρ^* called *reaction constant* bearing the same significance as in Hammett equation. σ^* is defined as:

$$\sigma^* = 0.4[\log(k/k_0)_{\text{basic}} - \log(k/k_0)_{\text{acidic}}]$$

The factor 0.4 is arbitrarily chosen to bring the σ^* values to nearly the same scale of Hammet σ-values.

k denotes the rate constant for acid or base hydrolysis of an ester having a substituent (R) alpha to the carbonyl group and k_0 denotes the corresponding value for the Me-substituted ester. Thus, $\sigma^* = 0$ for Me-substitution by definition.

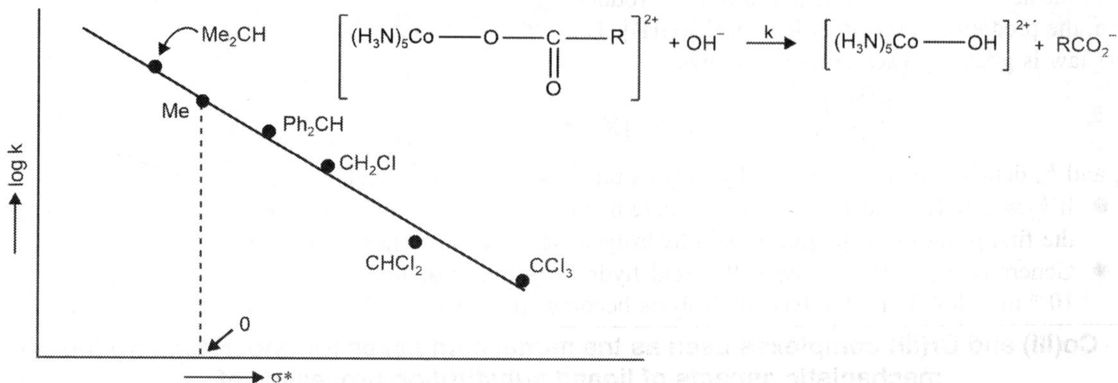

Fig. 5.19.12.1 Qualitative representation of Taft plot for the given reaction (*cf.* W.E. Jones, *et al*, *J. Chem. Soc., A*, 1481, 1966).

Determination of ρ* **(Taft reaction constant)** for the following reaction can decide whether the 'Co—O' or 'C—O' bond ruptures. This aspect has been discussed in Sec. 5.19.11.

$$\left[(H_3N)_5 Co—O—\overset{\overset{\displaystyle O}{\|}}{C}—R\right]^{2+} \xrightarrow[\text{(Base hydrolysis)}]{\text{(k), OH}^-} \left[(H_3N)_5 Co—OH\right]^{2+} + RCO_2^-$$

R = CH_3, C_2H_5, $(CH_3)_2CH$, etc.

5.19.13 Effect of Solvent on the Rate Process

If the two reacting species bear the charges Z_A and Z_B, then effect of the **dielectric constant** (ε) of the medium on the rate constant is expressed as follows:

$$\log k_\varepsilon = \log k_\infty - \frac{Z_A Z_B e^2}{2.303\varepsilon r_{AB} k_B T} \quad \ldots(5.19.13.1)$$

e = electronic charge; r_{AB} = distance between the species in the transition state; k_ε = rate constant in a medium of dielectric constant ε; k_∞ = rate constant in a medium of dielectric constant ∞; k_B = Boltzmann constant; T = temperature in Kelvin scale.

For infinite dielectric constant, the last term of the equation becomes zero, *i.e.* when there is no electrostatic force between the reacting species.

The above relation indicates that the plot of log k_ε *vs.* 1/ε is linear and sign of the slope depends on the sign of the product Z_AZ_B. The calculated value of r_{AB} from the slope of the plot is quite reasonable.

It is evident that if the product Z_AZ_B is positive (*i.e.* reacting species are of the same charge) then the rate decreases with the increase of 1/ε (*i.e.* reciprocal of the dielectric constant). If the product Z_AZ_B is negative, then the rate constant increases with the increase of 1/ε. If the product Z_AZ_B is zero (*i.e.* one of the reacting species is neutral), then the rate constant does not depend on ε.

5.20 HYDROLYSIS OF THE OCTAHEDRAL COBALT(III)-AMMINE COMPLEXES

Hydrolysis of the cobalt(III)-ammine complexes (say, $[Co(N_5)X]$) may be represented simply as:

$$[Co(N_5)X] \xrightarrow{\text{water}} [Co(N_5)(OH_2)] + X^-$$

$$\Big\downarrow\!\!\Big\uparrow \; +H^+ \,/\, -H^+$$

$$[Co(N_5)(OH)]$$

(N_5 stands for the non-replaceable amine/ammine ligands occupying the five coordination sites)

In the acidic condition, the predominant product is $[Co(N_5)(OH_2)]$ while in a relatively basic condition, the predominant product is $[Co(N_5)(OH)]$. The reaction pathways also depend on the pH and the rate law is given by two terms as follows:

$$-\frac{d[Co(N_5)X]}{dt} = k_A[Co(N_5)X] + k_B[Co(N_5)X][OH^-]$$

k_A and k_B denote the acid and base hydrolysis rate constants respectively.

- If $k_A \gg k_B[OH^-]$, the second term, *i.e.* base hydrolysis path, is less important but when $k_A \ll k_B[OH^-]$, the first term, *i.e.* aquation or acid hydrolysis path, becomes less important.
- Generally, $k_B \approx 10^6 k_A$, hence the acid hydrolysis becomes important at pH < 8 (*i.e.* $[OH^-] = 10^{-6}$ mol dm^{-3}) and the base hydrolysis becomes important at pH greater than 8.

Co(III) and Cr(III) complexes used as the model complexes for understanding the mechanistic aspects of ligand substitution processes of octahedral complexes—why?

In fact, coordination chemistry started to develop with mainly the Co(III)–complexes. The important factors behind this choice are:
● they can be easily prepared; ● they are both kinetically and thermodynamically stable; ● the compounds are not costly; ● because of their inertness, they can be used for kinetic studies by using the conventional techniques; ● for the kinetic studies of labile centres like Co(II), Ni(II), Cu(II), etc. it needs special techniques which are costly.

5.21 AQUATION OR HYDROLYSIS UNDER THE ACIDIC CONDITION OF COBALT(III) AMMINE/AMINE COMPLEXES

In terms of the **principle of microscopic reversibility** (see Sec. 5.8), aquation or acid hydrolysis reactions are the reverse of anation reactions. The acid hydrolysis (pH < 5) of acidopentaamminecobalt(III) salt is shown below.

$$[Co(NH_3)_5X]^{2+} + H_2O \underset{\text{Anation}}{\overset{\text{Hydrolysis}}{\rightleftharpoons}} [Co(NH_3)_5(OH_2)]^{3+} + X^-$$

The rate law for aquation of $[Co(N_5)X]$ shows a first order dependence on the starting complex. There are **different possibilities** to explain the observed rate law. The simplified schemes are given below. For details of the *D*, *A* and *I* mechanisms, Secs. 5.9.1–3 are to be considered.

(i) $$\left.\begin{array}{l}[Co(N_5)X]^{2+} \xrightarrow[(rds)]{k} [Co(N_5)]^{3+} + X^- \\ [Co(N_5)]^{3+} + H_2O \xrightarrow{\text{fast}} [Co(N_5)(OH_2)]^{3+}\end{array}\right\} \textbf{Dissociative path}$$

i.e. rate $= k[Co(N_5)X^{2+}] = k$[Complex] ...(5.21.1)

(ii) $$\left.\begin{array}{l}[Co(N_5)X]^{2+} + H_2O \xrightarrow[(rds)]{k} [Co(N_5)(OH_2)(X)]^{2+} \\ [Co(N_5)(OH_2)X]^{3+} \xrightarrow{\text{fast}} [Co(N_5)(OH_2)]^{3+} + X^-\end{array}\right\} \textbf{Associative path}$$

i.e. rate $= k[Co(N_5)X^{2+}][OH_2]$
$= k \times 55.5 \times [Co(N_5)X^{2+}]$
$= k'[Co(N_5)X^{2+}] = k'$[Complex], Pseudo-first-order reaction ...(5.21.2)

(iii) $$\left.[Co(N_5)X]^{2+} + H_2O \xrightarrow[(rds)]{k} [Co(N_5)(OH_2)]^{3+} + X^-\right\} \textbf{Interchange path}$$

rate $= k[Co(N_5)X^{2+}][H_2O]$
$= k'[Co(N_5)X^{2+}] = k'$[Complex] ...(5.21.3)

Note: The above rate laws are the simplified forms (under some conditions). The actual schemes and rate laws for the *D*, *A* and *I* processes have been discussed in detail earlier.

$$Y = H_2O \text{ (in the present case)} \begin{cases} \textbf{\textit{D}-process} \text{ (Sec. 5.9.1)}: \text{rate} = \dfrac{k_1 k_2 [\text{Complex}][Y]}{k_{-1}[X] + k_2[Y]} \approx k_1[\text{Complex}], \\ \qquad \text{when } k_2[Y] \gg k_{-1}[X] \\ \textbf{\textit{A}-process} \text{ (Sec. 5.9.2)}: \text{rate} = \dfrac{k_a k_b [\text{Complex}][Y]}{k_{-a} + k_b} = k[\text{Complex}][Y],\ k = \dfrac{k_a k_b}{k_{-a} + k_b} \\ \textbf{\textit{I}-process} \text{ (Sec. 5.9.3)}: \text{rate} = \dfrac{k_i K_{OS}[\text{Complex}][Y]}{1 + K_{OS}[Y]} \approx k_f[\text{Complex}][Y], \\ \qquad \text{when } 1 \gg K_{OS}[Y],\ k_f = k_i K_{OS} \end{cases}$$

The above associative and interchange paths will lead to the *pseudo-first order conditions for the aquation process (i.e. $Y = H_2O$). **Thus the three possible pathways lead to the same rate law.*** However, these possibilities can be distinguished by considering the effects of different factors on the rate process. These are discussed below. *The evidences support the dissociative path (more correctly, I_d path) for the said aquation reactions.*

(A) Application of CFT and CFAE: Experimental and calculated activation energy in the aquation of X^- in $[Co(NH_3)_5X]^{2+}$ ($X^- = Cl^-$, Br^- and NCS^-) (*JACS*, **90**, 6626, 1968): Experimental activation energy for the given aquation reaction has been compared with the theoretically calculated values (by considering the CFSE effect and interelectronic repulsion energy in the formation of activated complex) for the different possibilities like the **dissociative path** involving the square pyramidal (C_{4v}) and trigonal bipyramidal (D_{3h}) activated complexes, **associative path** involving the trapezoidal octahedral (C_{2v}) and pentagonal bipyramidal (D_{5h}) activated complexes (see Fig. 5.18.2.1). The comparison indicates that the experimental activation energy values are close to the calculated values for the dissociative path involving the **square pyramidal activated complex**. It supports the dissociative path through the square pyramidal activated complex.

Note: ● For the aquation of the corresponding Cr(III) complexes, *i.e.* $[Cr(NH_3)_5X]^{2+}$, the experimental activation energies are close to the calculated values of the **associative path** involving the pentagonal bipyramidal activated complex.

● In the dissociative path, the square pyramidal activated complex leads to **no change in the stereochemistry** of the product but the trigonal bipyramidal activated complex causes a change in the stereochemistry of the product (discussed later, *cf.* Fig. 5.21.3–5).

(B) Group replacement factor ($GRF = k_{R-X}/k_{R-Cl}$) in the aquation of R—X and R—Cl, $R = Co(NH_3)_5^{3+}$ and X = F, Cl, Br and I: For the more electron withdrawing X (leaving group), the positive charge on the metal centre of $[Co(NH_3)_5X]^{2+}$ increases to favour the associative activation and to disfavour the dissociative activation. The observed sequence of GRF values (I > Br > Cl > F) rules out the associative activation and it supports the dissociative activation in terms of accumulation of positive charge on the Co(III)-centre and bond dissociation energy (BDE) of the Co^{III}—X bond (smaller values of BDE lowers the activation energy in the *d*-activation process, bond strength order: Co^{III}—F > Co^{III}—Cl > Co^{III}—Br > Co^{III}—I). It may be pointed out that for the **associative path of aquation**, the GRF values are expected to run in the sequence: F > Cl > Br > I.

(C) Effect of Charge (*cf.* Sec. 5.19.1): The aquation rate of $[CoCl(NH_3)_5]^{2+}$ (Cl^- leaving group) is about *ca.* 300 times slower than that of *trans*–$[CoCl_2(NH_3)_4]^+$; and *cis*–$[CoCl_2(NH_3)_4]^+$ aquates *ca.* 3 times faster than the corresponding *trans*–isomer.

Aquation rate: *cis*-$[CoCl_2(NH_3)_4]^+$ > *trans*-$[CoCl_2(NH_3)_4]^+$ ≫ $[CoCl(NH_3)_5]^{2+}$.

Table 5.21.1: Steric effect of the nonleaving ligands (N—N) on the aquation rate of *trans*–$[CoCl_2(N—N)_2]^+$ (at 25°C) (*cf.* R.G. Pearson, *et al, J. Am. Chem. Soc.*, **75**, 3089, 1953)

	N—N	10^5k (s^{-1})		*N—N*	10^5k (s^{-1})
(i)	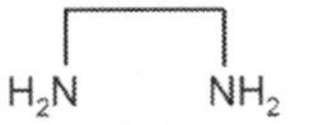 H_2N–CH₂–CH₂–NH_2 (en)	3.2	(iv)	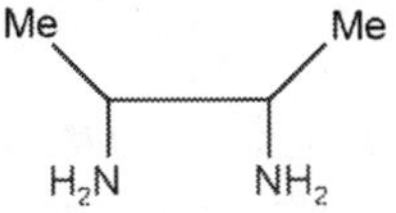 H_2N–CH(Me)–CH(Me)–NH_2	42.0
(ii)	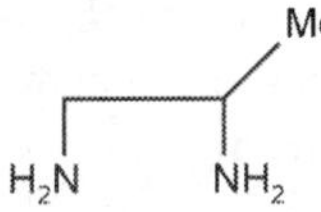 H_2N–CH₂–CH(Me)–NH_2 (pn)	6.2	(v)	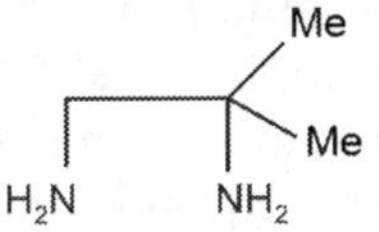 H_2N–CH₂–C(Me)₂–NH_2	22.0
(iii)	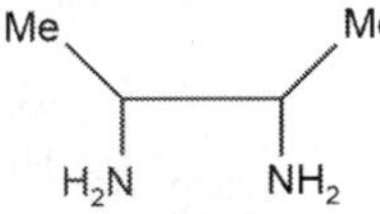 H_2N–CH(Me)–CH(Me)–NH_2 (*dl*-bn)	15.0	(vi)	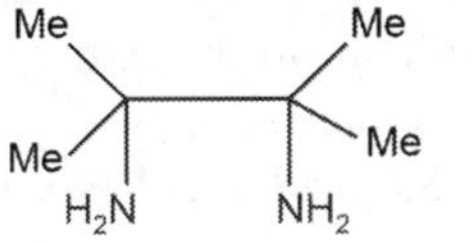 H_2N–C(Me)₂–C(Me)₂–NH_2	very fast (instantaneous)

	N—N	10^5k (s^{-1})
(i)	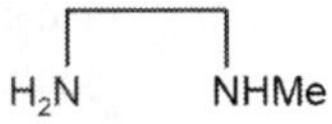 H_2N–CH₂–CH₂–NHMe	1.7
(ii)	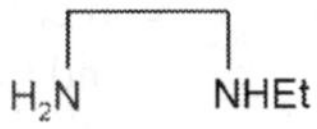 H_2N–CH₂–CH₂–NHEt	6.0
(iii)	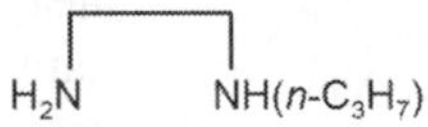 H_2N–CH₂–CH₂–NH(n-C_3H_7)	11.8

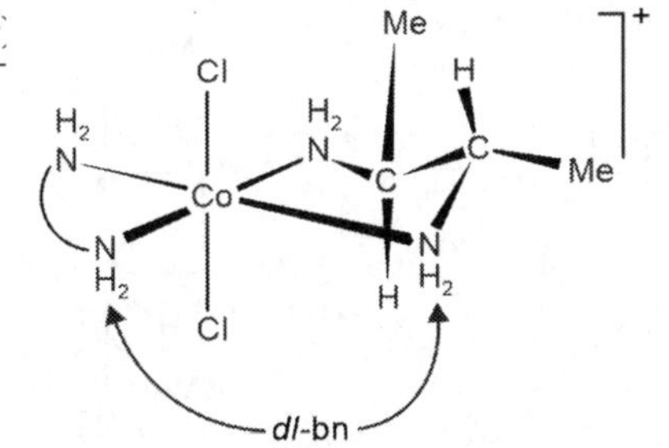

Chiral-bn sterically less crowded; Me-groups on the opposite sites of the chelate ring

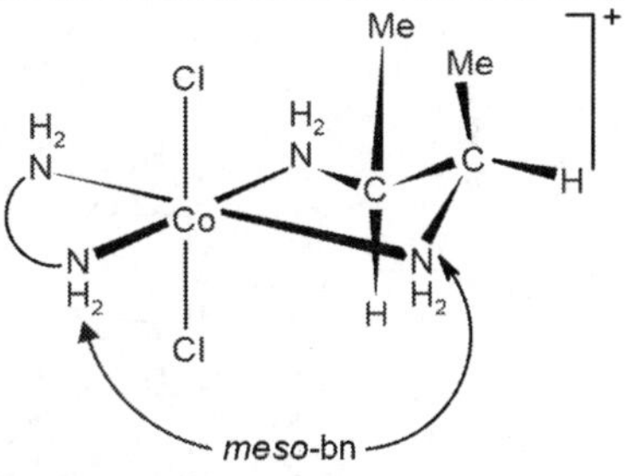

Achiral-bn sterically more crowded (Aquation rate faster)

Table 5.21.2: Aquation rates of $[CoCl(N_4)X]^+$ to illustrate steric acceleration

Complex	*k* (s^{-1}) *25°C (for loss of Cl^-)*
trans–$[CoCl_2(cyclam)]^+$	1.1×10^{-6}
trans–$[CoCl_2(tet–b)]^+$	9.3×10^{-4}
trans–$[CoCl(cyclam)(NCS)]^+$	1.1×10^{-9}
trans–$[CoCl(NCS)(tet–b)]^+$	7.0×10^{-7}
trans–$[Co(CN)Cl(cyclam)]^+$	4.8×10^{-7}
trans–$[Co(CN)Cl(tet–b)]^+$	3.4×10^{-4}

Aquation rates of other complexes follow the similar sequences:

$$[CoCl_2(NH_3)_4]^+ \gg [CoCl(NH_3)_4(OH_2)]^{2+};$$

$$cis\text{–}[CoCl_2(en)_2]^+ \gg cis\text{–}[CoCl(en)_2(NH_3)]^{2+}$$

Thus, it is evident that the aquation rate remarkably increases with the decrease of the overall positive charge of the complex. *This nature of dependence supports the dissociative activation (i.e. bond breaking is primarily important in attaining the transition state).* It is justified because removal of an anionic ligand like Cl^- from a more positively charged complex is more difficult.

If the **associative activation** is argued, then the more positively charged complex should react faster.

(D) Effect of steric crowding (causing steric accleration) (*cf.* Sec. 5.19.7):

● **Aquation of *trans*-$[CoCl_2(N—N)_2]^+$ (N—N = en or substituted en) (*cf.* Sec. 5.19.7):** Aquation of the complexes of the type *trans*-$[CoCl_2(N—N)_2]^+$ where N—N denotes ethylenediamine or substituted ethylenediamine experiences a *steric acceleration* (*i.e.* with the increase of steric crowding, the rate of aquation increases).

$$trans\text{-}\left[CoCl_2(N—N)_2\right]^+ + H_2O \longrightarrow trans\text{-}\left[CoCl(N—N)_2(OH_2)\right]^{2+} + Cl^-$$

This **steric acceleration** is an indication of the **dissociative activation**.

● **Aquation of *trans*-$[CoCl_2(N_4)]^+$ and $[CoCl(N_4)X]^+$ (N_4 = cyclam or its derivative):** Similarly, Co(III)–complexes of the type $[CoCl_2(N_4)]^+$ and $[CoCl(N_4)(X)]^+$ where N_4 denotes the macrocyclic ligand *cyclam* or its hexamethyl derivative, *tet–b* (which provides more steric crowding than cyclam) experience a **steric acceleration** in the aquation process.

$$trans\text{–}[CoCl(N_4)X]^+ + H_2O \rightarrow trans\text{–}[Co(N_4)(OH_2)X]^{2+} + Cl^-$$

This **steric acceleration** is in conformity with the **dissociative activation.**

Cyclam = 1,4,8,11-tetraazacyclotetradecane

tet–b = *dl*–1,4,8,11-tetraaza-5,5,7,12,12,14-hexamethylcyclotetradecane (the corresponding *meso*-isomer is called **tet-a**)

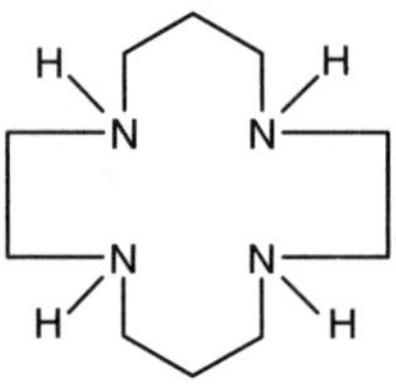

(cyclam)

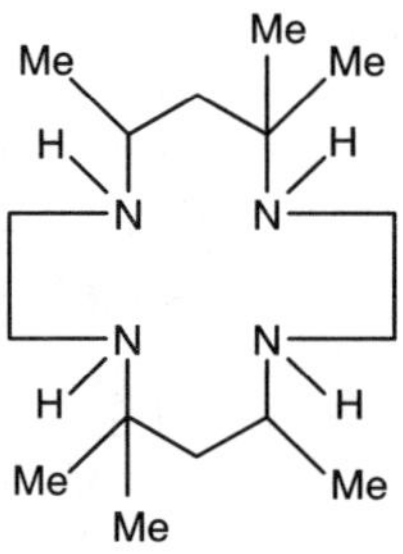

(*tet-b*)

● **Aquation of $[CoCl(MeNH_2)_5]^{2+}$ *vs.* $[CoCl(NH_3)_5)]^{2+}$:** $[CoCl(MeNH_2)_5]^{2+}$ aquates faster than $[CoCl(NH_3)_5]^{2+}$ by a factor *ca.* 20 at 50°C suggesting the **dissociative activation (steric acceleration).** *It may be noted that the corresponding Cr(III)–complex experiences steric retardation in aquation supporting the associative activation* (*cf.* Sec. 5.19.7) .

● **Aquation of $[CoCl(dien)(tn)]^{2+}$ *vs.* $[CoCl(dien)(en)]^{2+}$:** $[CoCl(dien)(tn)]^{2+}$ aquates much faster than $[CoCl(dien)(en)]^{2+}$. This rate enhancement (*i.e.* **steric acceleration**) can be explained by considering the dissociative activation involving the **trigonal bipyramidal transition state** (not square pyramidal transition state). The corresponding 5-coordinate transition state of trigonal bipyramidal geometry is sterically more stable for $[CoCl(dien)(tn)]^{2+}$. This aspect has been discussed in Sec. 5.19.7 (*cf.* Figs. 5.19.7.1–2).

(E) Electronic effect (*e.g.* *cis*-effect and *trans*-effect) of the nonleaving group (L) (*cf.* Sec. 5.19.8): The observations for aquation of $[CoCl(en)_2(L)]^{n+}$ (*i.e.* replacement of Cl^- by H_2O) are:

(i) **Pi–donor nonleaving groups** like Cl^-, OH^-, NCS^- at the *cis*–position can accelerate the aquation process. The *cis*-complex aquates faster than the corresponding *trans*–isomer, *i.e.* $k_{cis} > k_{trans}$.

(ii) If the **pi–donor nonleaving group** resides at the ***cis*-position**, there is a **100% retention of configuration,** *i.e.* the product is of *cis*–configuration (*cf.* Figs. 5.21.3-4).

(iii) If the **pi-donor nonleaving group** resides at the *trans–position,* the aquated product is a mixture of *cis*– and *trans*–configuration (*i.e.* **it leads to isomerisation**; *cf.* Fig. 5.21.3-4).

(iv) If the nonleaving group is a **weak π–donor** (*e.g.* H_2O) or having no π-bonding property (*i.e.* neither π–acceptance nor π–donation as for the ligand NH_3), then both the *cis*- and *trans*-isomers react at comparable rates, *i.e.* $k_{cis} \approx k_{trans}$.

(v) If the nonleaving group is a **strong π–acceptor ligand** (*e.g.* CN^-, NO_2^-) then the corresponding *trans*-isomer aquates faster, *i.e.* $k_{trans} > k_{cis}$. However, this effect (*i.e.* ***trans*-effect**) is less pronounced than the *cis*– effect (shown by the π–donor ligand residing at the *cis*-position).

Table 5.21.3: Aquation rate constants of $[CoCl(en)_2(L)]^{n+}$ at 25°C showing the electronic effect of the nonleaving group (L). $[CoCl(en)_2(L)]^{n+} + H_2O \rightarrow [Co(en)_2(L)(OH_2)]^{(n+1)+} + Cl^-$

L	*Electronic effect of L*	*cis*–$[CoCl(en)_2(L)]^{n+}$		*trans*-$[CoCl(en)_2(L)]^{n+}$		k_{cis}/k_{trans}
		k (s^{-1})	*% of cis-product*	*k* (s^{-1})	*% of trans-product*	
OH^-	**good π-donor**	1.2×10^{-2}	100	1.6×10^{-3}	25	7.5
Cl^-		2.4×10^{-4}	100	3.5×10^{-5}	65	6.85
NCS^-		1.1×10^{-5}	100	5.0×10^{-8}	30–50	220
H_2O	**weak π-donor**	1.6×10^{-6}		2.5×10^{-6}		0.64
$*N_3^-$		2.5×10^{-4}	100	2.4×10^{-4}	80	1.04
NH_3	**non π-bonding**	5×10^{-7}	100	3.4×10^{-7}	100	1.47
NO_2^-	**strong π-acceptor**	1.1×10^{-4}	100	1.0×10^{-3}	100	0.11
CN^-		6.2×20^{-7}	100	8.2×10^{-5}	100	7.5×10^{-3}

* N_3^- (azide ion) can act either as a pi-donor or as a pi–acceptor. However, in the present case, it does not act strongly in either capacity. (*cf.* M.L. Tobe, *Inorg. Chem.*, **7**, 1260, 1968; W.G. Jackson *et al*, *Inorg. Chim. Acta.*, **60**, 115, 1982; C.K. Poon *et al*, *J. Chem. Soc. A*, 1549, 1968).

Note: For the *cis*-complex with the π-donor nonleaving group (L), the square pyramidal (SP) intermediate or T.S. is stabilised by the π-donor ligand (***cis*-effect**) and this SP intermediate explains the 100% retention of configuration. But, the SP intermediate may be partly converted to a TBP intermediate, keeping L at one equatorial position (*cf.* Figs. 5.21.4, 5.28.2.1) and from the TBP intermediate, both the *cis*- and *trans*-products are produced in the 2:1 ratio.

● **Mechanistic path for the π–donor ligand (L):** When the leaving group is dissociated, it produces a square pyramidal geometry (CN = 5). The starting octahedral complex uses six sp^3d^2 or d^2sp^3 hybrid orbitals to bind the 6 ligands. Thus dissociation of a Cl^- ligand from the octahedral complex gives a vacant d^2sp^3 or sp^3d^2 orbital. If the π–donor ligand L resides at the *cis*-position of the leaving group, then the lone pair of L can be donated to the vacant orbital to stabilise the 5-coordinate square pyramidal intermediate or activated complex. In other words, if the reaction experiences a **dissociative activation,** the 5-coordinate activated complex or the transition state (square pyramidal) is stabilised. Thus the pi-donor ligand (L) favours the dissociation of the leaving group at the *cis*–position. This phenomenon is described as the ***cis*-effect in dissociative activation.**

cis-effect: *cis*-$(en)_2Co(L)(Cl)]^{n+}$ **(octahedron)** $\xrightarrow{rds}$ *cis*-$Co(en)_2L]^{(n+1)+}$ **(square pyramid)** + Cl^-

The π-donor properties of some representative ligands can be represented as follows:

$Co{-}N{\equiv}C{-}\ddot{S}: \longleftrightarrow Co{-}N{=}C{=}S: \longleftrightarrow Co{=}N{=}C{=}S:$

$Co{-}N{=}N{=}N: \longleftrightarrow Co{=}N{=}N{=}N: \longleftrightarrow$; $Co{-}OH \longleftrightarrow Co{=}OH$

$Co{-}Cl: \longleftrightarrow Co{=}Cl:$

Charges not shown for the sake of simplicity, OH group possesses **three lone pairs**

If the π-donor nonleaving group L resides at the *trans*-position, then the 5–coordinate square pyramidal geometry (obtained due to the dissociation of Cl^-) cannot be stabilised by L as such. But transformation of the **square pyramidal (SP)** geometry into a **trigonal bipyramidal (TBP) geometry** allows the stabilisation of the intermediate or T.S. by L occupying an equatorial position of the TBP. In other words, if the reaction is supposed to pass through the dissociative activation, the activated complex or transition state is to be stabilised by the π-donor ligand L through a geometrical transformation. ***This required transformation requires energy and it raises the activation energy to slow down the reaction, i.e.*** $k_{trans} < k_{cis}$.

$$trans\text{-}\left[CoCl(en)_2(L)\right]^{n+} \xrightarrow{-Cl^-} \underset{(SP)}{\left[Co(en)_2L\right]^{(n+1)^+}} \rightleftharpoons \underset{(TBP)}{\left[Co(en)_2L\right]^{(n+1)^+}} \xrightarrow{+H_2O} \underset{(cis+trans)}{\text{Product}}$$

This transformation (SP to TBP) leads to **isomerisation.**

Thus if the nonleaving group (L) is a π–donor ligand, then the aquation process goes on through the dissociative path. This pathway explains the observations mentioned above. For the **non-π–bonding ligand or weak π–donor ligands**, there is no question of such stabilisation of the 5–coordinate transition state. Consequently, in such cases, the *cis*- and *trans*-isomers aquate at comparable rates (*cf.* Table 5.21.3 for L = H_2O, N_3^-, NH_3).

The said ***cis*–effect** also explains why ***cis*–$[CoCl_2(NH_3)_4]^+$** aquates faster than ***trans*–$[CoCl_2(NH_3)_4]^+$**.

● **Mechanistic path for the π–acceptor nonleaving group (L):** The π–acceptor nonleaving ligands (L) withdraw the t_{2g}-electron cloud from the metal centre and *thus it favours the nucleophilic attack at the metal centre at the trans position where the electron cloud is depleted.* In fact, the metal → ligand π-bonding effectively increases positive charge on the metal centre and *it favours the associative path.*

L— $\ddot{C}o(en)_2$ —Cl **(associative path)**

:Nu

The π-acceptor properties of some representative ligands can be represented as follows (charges not shown for the sake of simplicity):

Co—C≡N: ⟷ Co=C=$\ddot{N}$:, Co—N(=O:)(—$\ddot{O}$:) ⟷ Co=N(—$\ddot{O}$:)(—$\ddot{O}$:)

For the nonleaving π-acceptor ligands like NO_2^-, CN^-, etc., the electrons from the nonbonding *d*–orbitals (*i.e.* filled t_{2g}^6-orbitals) of the metal centre are pushed into the **vacant orbitals** of NO_2^- or CN^- (*cf.* Fig. 5.21.1c, d). For the *trans*-isomer, the **electron density is lowered in both the octahedral trigonal faces** adjacent to the leaving group (*i.e.* Cl^-). On the other hand, for the *cis*–isomers, the electron withdrawal causes the **electron depletion in one of the octahedral trigonal faces** adjacent to the leaving group. In other words, electron depletion surrounding the leaving group is more for the *trans*-isomer. ***Thus, the trans-isomer favours better than the cis–isomer for the approach of the incoming nucleophile*** (*i.e.* H_2O) to displace the leaving group in an associative path as argued for the Pt(II)-square planar complexes with the strong *trans*-directing π-acceptor ligands (*cf.* Fig. 5.30.15.2). In other words, the *trans*–isomer stabilises the transition state (attained through the associative activation) better than the *cis*–isomer.

For the π–acceptor nonleaving ligand (L), the nucleophile (*i.e.* H_2O in the aquation process) attacks at the *trans*–positions of L, *i.e.* at the *cis*-positions to X (leaving group) for the *trans*-isomer and at the *cis*-position of X for the corresponding *cis*-isomer (*cf.* Fig. 5.21.1c, d). **Thus this associative activation** (*i.e.* bond making by the incoming nucleophile at the *rds*) **is totally stereoretentive (***i.e. cis* → *cis, trans*

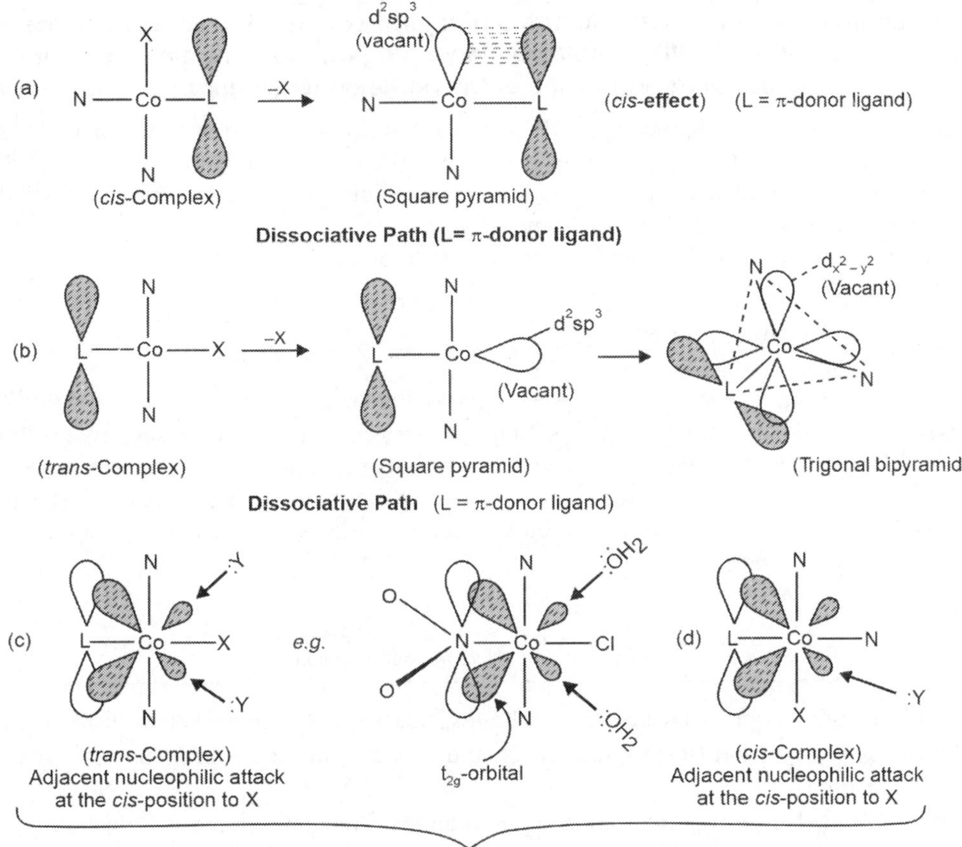

Fig. 5.21.1: Effect of the π–donor and π–acceptor nonleaving group (L) on the nucleophilic substitution process of *cis–/trans*–$[CoL(N)_4X]$ (X = leaving group). (Here, only the ligands in the *xy*–plane are shown for the sake of simplicity)

→ *trans*) **and it becomes important for the π–acceptor nonleaving ligand.** This associative activation is well understood for the substitution reactions of square planar complexes of Pt(II) in the presence of strongly *trans*–directing π–acceptor ligands (see Sec. 5.30.15).

● **Mechanistic path for the non-π-bonding nonleaving group (L):** The *non-π-bonding nonleaving ligands* like NH_3 may also lead to a such associative path but they cannot show any special effect to stabilise the transition state or activated complex in the aquation process for both the *cis*- and *trans*-isomers. In fact, such non-π-bonding nonleaving ligands fail to favour both the *d*- and *a*-mechanisms and consequently, both these isomers react at the comparable rates. It is quite reasonable to argue that for the non-π-bonding nonleaving group, the Co(III)-complex aquates through the dissociative path (I_d) as in the case of $[Co(NH_3)_5X]^{2+}$.

(F) Electronic effect (inductive effect) of the nonleaving group: The bond strength between the Co(III) centre and leaving group is partially controlled by the electronic effect such as **inductive effect** of the nonleaving group. This is illustrated in the following aquation process.

$$[CoCl(en)_2(R\text{–}py)]^{2+} + H_2O \rightarrow [Co(en)_2(OH_2)(R\text{–}py)]^{3+} + Cl^-$$

With the increase of basicity of the pyridine ligand (R–py), the electron pushing inductive effect of the ligand increases. This enhanced electron density on the metal centre weakens the 'Co—Cl' bond

Difference in the mechanistic paths of ligand substitution process in the octahedral Co(III) complexes depending on the nature of the π–bonding properties of the nonleaving group (L)

(i) If L is a strong **pi–donor ligand** (*e.g.* OH^-, Cl^-, etc.), it favours the **dissociative activation** (*i.e.* bond breaking by the leaving group is primarily important in producing the transition state). The *cis*-isomer (*i.e.* L and leaving group at the *cis*-positions) react faster (***cis*-effect**) than the corresponding *trans*-isomer. The *cis*-isomer tends to retain the stereochemical configuration in the product while the *trans*-isomer leads to isomerisation (*i.e.* mixed products of *cis*- and *trans*-isomers).

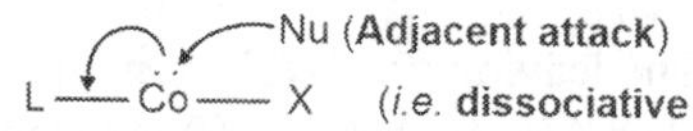

L—Co—X, Nu (Adjacent attack) (*i.e.* dissociative activation) } $k_{cis} < k_{trans}$; 100% retention of sterochemical configuration in both the *cis*- and *trans*-isomers

(ii) If L is a strong **pi–acceptor ligand** (*e.g.* NO_2^-, CN^-, etc.), it favours the **associative activation** (*i.e.* bond formation by the incoming nucleophile is primarily important in attaining the transition state). The *trans*-isomer (*i.e.* L and leaving group at the *trans*-positions) reacts faster (*trans*-effect) than the *cis*-isomer. There is a 100% retention of stereochemical configuration for both the *cis*- and *trans*-isomers.

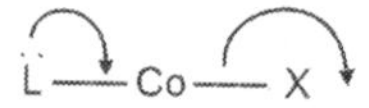

L—Co—X (*i.e.* dissociative activation) } $k_{cis} > k_{trans}$; *cis*-: 100% retention of sterochemical *trans*-: isomerisation

'strength'. If the aquation process experiences the **dissociative activation,** then with the increase of basicity of the pyridine ligand (R–py), the rate should increase. In fact, it happens so to support the dissociative activation.

Basicity order of R–py: pyridine < 3–methylpyridine < 4–methylpyridine < 4–methoxypyridine

	pyridine	3–methylpyridine	4–methylpyridine	4–methoxypyridine
10^5k (s^{-1}) at 50°C:	1.1	1.3	1.4	1.5

Note: It is the general observation that the **good σ-donor nonleaving ligands** (*i.e.* good basic ligands) impart the favourable **dissociative character** at the metal centre, *e.g.* $[Cr(OH_2)_6]^{3+}$ adopts the associative activation while $[Cr(NH_3)_5(OH_2)]^{3+}$ adopts the dissociative activation (*cf.* σ-basicity: $NH_3 > OH_2$) (see Sec. 5.19.6). It may be noted that the π-donor ligands like OH^-, Cl^- are also the good σ-donor ligands and they favour the dissociative activation.

(G) Effect of the leaving group (X) in $[Co(NH_3)_5X]^{n+}$: The aquation rate decreases with the increase of the thermodynamic bond strength of the 'Co—X' bond. The aquation rate follows the sequence: X^-: $NO_3^- > I^- > Br^- > H_2O > Cl^- > SO_4^{2-} > F^- > CH_3COO^- > NH_3 > NCS^-$ (N–bonded) $> N_3^- > NO_2^-$ (*N*–bonded)

Thus with the increase of thermodynamic affinity of Co(III) for X, the aquation rate decreases. It has been noted that the activation energy (E_a measuring the rate constant) increases linearly with the average ligand field strength measured by $\bar{\nu}$ (wave number in cm^{-1}) of the first $d \rightarrow d$ absorption band of the complex (*cf.* ν_1: $^1A_{1g} \rightarrow {}^1T_{1g}$, $\nu_1 \approx 10Dq_o - 4B'$, see Sec. 7.15.1).

$$E_a : \xrightarrow[\text{Increasing ligand field strength}]{Br^- < NCS^- \text{(N-bonded)} < NO_2^- \text{(N-bonded)}}$$

All these indicate that rupture of the 'Co—X' bond is playing an important role in attaining the transition state and it supports the *d*-activation. However, it cannot tell us the role of H_2O (*i.e.* entering nucleophile) in the transition state.

(H) Linear free energy relationship (LFER) for aquation of $[Co(NH_3)_5X]^{n+}$ (*cf.* Sec. 5.19.4): For this aquation process, the LFER plot, *i.e.* plot of log k_{aq} *vs.* log K_{eq}, gives a good straight line having the slope ≈ 1. This supports the *dissociative activation.* This aspect has been discussed in Sec. 5.19.4.

(I) Knowledge of enthapy of transition ($\Delta H_T = \Delta H - \Delta H^{\neq}$) (*cf.* Sec. 5.19.3B and Fig. 5.19.3.3 for details): ΔH_T (difference in enthalpy between the transition state and product) has been found to be more or less constant (*ca.* 25 kcal mol^{-1}) for aquation of $[Co(NH_3)_5X]^{2+}$ (X^-= Cl^-, Br^- and NO_3^-) and it is also very much comparable to ΔH_T for the water exchange process at $[Co(NH_3)_5(OH_2)]^{3+}$. *This supports the existence of a* ***common transition state*** *involving $[Co(NH_3)_5]^{3+}$ as the Co(III) species for both the aquation process and water exchange process.* In the said aquation and water exchange reactions, Co(III) exists as the common species $[Co(NH_3)_5]^{3+}$ **(in the transition state)** and $[Co(NH_3)_5(OH_2)]^{3+}$ **(in the product)** in the dissociative path. The leaving group (X^-) remains as a free solvated species in both the transition state and product. It makes ΔH_T (= $\Delta H^{\neq} - \Delta H$) constant for the dissociative path of aquation for different X^-. For the associative path of aquation of $[Cr(OH_2)_5X]^{2+}$, ΔH_T is not constant. Thus independence of ΔH_T on the nature of the leaving group supports the **dissociative activation.**

(J) Electronic effect of the leaving group (X = acetate group) indicating some associative activation: In the aquation of $[Co(NH_3)_5X]^{2+}$ (X = acetate or substituted acetate), the basicity of the acetates was varied by a factor *ca.* 10^4 in moving from $CF_3CO_2^-$ to $Me_3CCO_2^-$. Thus it is reasonable to assume that the 'Co—X' bond strength should vary by a such large factor. If the 'Co—X' bond rupture is supposed to determine the activation energy solely then the rate constants for the aquation process should cover a large span. Interestingly, the aquation rate constants cover only a narrow span of 10. *This wide span of basicity of X^- against the narrow span of aquation rate constants indicates that the process is not a pure dissociative process.* Thus nucleophilic attack by H_2O is also partly important to produce the transition state.

If R—CO_2^- (*i.e.* X^-) is a weaker base, it will make a weaker Co—X bond (*favouring the dissociative activation*). But at the same time, it will create a lesser amount of electron density on the Co(III)–centre and it will facilitate the nucleophilic attack by H_2O (*i.e. associative activation*).

R — CO — O — Co ← : Nu

If the associative activation is considered to operate then the aquation will be favoured most for X^- = CF_3COO^- causing the maximum electron depletion at the Co(III) centre and will be disfavoured most for X^- = Me_3CCOO^-. It happens so for the aquation process. *But the polar nonionic H_2O is not a very good nucleophile to make a huge difference in rate constants for the aquation process.*

Table 5.21.4: Aquation rate constants of some $[Co(NH_3)_5X]^{2+}$ (X^- = acetate and substituted acetates) complexes

X^-		pK_a for HX		k_{aq} (min^{-1}) (70°C)	
$CF_3CO_2^-$	↑ Increasing electron withdrawing inductive effect	0.3	↓ Increasing basicity of X^-	3.3×10^{-3}	↑ Increasing aquation rate
$CCl_3CO_2^-$		0.7		3.2×10^{-3}	
$CHCl_2CO_2^-$		1.3		9.6×10^{-4}	
$CH_2ClCO_2^-$		2.85		3.5×10^{-4}	
$CH_3CO_2^-$		4.74		4.9×10^{-4}	
$(CH_3)_2CHCO_2^-$		4.82		1.6×10^{-4}	
$(CH_3)_3CCO_2^-$		5.0		2.6×10^{-4}	

(*cf.* F. Basolo, *et al.*, *J. Phys. Chem.*, **56**, 22, 1952)

It may be noted that in presence of a better nucleophile than H_2O, the span of variation of rate constants becomes larger.

The narrow span of aquation rate constants (k_{aq}) for the said series of complexes may be explained by considering the nucleophilic attack by the poor nucleophile H_2O to produce the transition state. ***It only rules out the perfect D–process but not the interchange process where both the bond rupture and bond formation by H_2O (as a nucleophile) contribute to determine the activation energy.***

Note: For the dissociative (*d*) activation, the reactivity order will also be the same, *i.e.* highest k_{aq} for $X^- = CF_3CO_2^-$ and lowest k_{aq} for $X^- = Me_3CCO_2^-$, but the rate constants (k_{aq}) would cover a much wider range because the Co^{III}—X bond rupture is mainly important in the *d*-path to produce the transition state.

(K) Effect of chelation by $(N)_4$ or $(N)_5$ ligand on the aquation rate of *cis-trans*-$[Co(N)_4Cl_2]^+$ and $[Co(N)_5Cl]^{2+}$ (R.G. Pearson, *et al.*, *JACS*, **59**, 304, 1955): It has been noted that with the increase of the extent of chelation by $(N)_4$ or $(N)_5$ ligand, the acid catalysed aquation rate (k_{aq}) decreases.

Aquation rate order:

(i) *cis*-$[Co(NH_3)_4Cl_2]^+ \gg$ *cis*-$[Co(en)_2Cl_2]^+ >$ *cis*-$[Co(trien)Cl_2]^+$

(ii) *trans*-$[Co(NH_3)_4Cl_2]^+ >$ *trans*-$[Co(en)(NH_3)_2Cl_2]^+ >$ *trans*-$[Co(en)_2Cl_2]^+$

For the dichlorido-complexes, aquation rate considers the **aquation of first chloride:**

$$[Co(N)_4Cl_2]^+ \xrightarrow[-Cl^-]{\text{aquation}} [Co(N)_4Cl(OH_2)]^{2+}$$

(iii) $[Co(NH_3)_5Cl]^{2+} >$ *cis*-$[Co(en)_2(NH_3)Cl]^{2+} >$ *cis*-$[Co(trien)(NH_3)Cl]^{2+} > [Co(en)(dien)Cl]^{2+} > [Co(tetraen)Cl]^{2+}$.

- **Possibility of the associative path in terms of the steric effect:** With the increase of extent of chelation, the nucleophilic attack along the C_3-axes (projected through the octahedral trigonal faces, *cf.* projection of the d_{xz} and d_{yz} t_{2g}-orbital lobes) in the **associative path** becomes more difficult (see Sec. 5.18.2). For $[Co(en)(dien)Cl]^{2+}$ and $[Co(tetraen)Cl]^{2+}$, such an **associative path is completely ruled out**. Hence, the **associative path apparently can explain the rate order but it fails to explain the aquation rates** for the complexes $[Co(tetraen)Cl]^{2+}$ and $[Co(en)(dien)Cl]^{2+}$ that aquate **only slightly** more slowly than others in the series of $[Co(N)_5Cl]^{2+}$. In fact, in the associative path, the rate retardation for $[Co(tetraen)Cl]^{2+}$ and $[Co(en)(dien)Cl]^{2+}$ is expected to be more than the experimental result.
- **Inductive effect of the amines:** The pK_a values of the onium ions are: 9.28 (NH_4^+), 9.93 (enH^+), 9.98 ($dienH^+$), 9.92 ($trienH^+$). It indicates that NH_3 is the **weakest base** among these and consequently, it gives the least amount of electron through σ-donation to Co(III). Then it is expected that in $[Co(NH_3)_4Cl_2]^+$ and $[Co(NH_3)_5Cl]^+$, the Co—Cl bond is the **strongest one** in the respective series. However, these complexes are the most reactive ones. **Apparently, it contradicts the dissociative path of aquation.**
- **Solvation effect of the transition state in the dissociative path:** In the dissociative path, the transition state (T.S.) is considered to be a five coordinate one (square pyramidal) with the increase of cationic charge by one unit, *i.e.* for $[Co(N)_4Cl_2]^+$, the T.S. is $[Co(N)_4Cl]^{2+}$ while for $[Co(N)_5Cl]^{2+}$, the T.S. is $[Co(N)_5]^{3+}$. **Hence, the solvation effect is more important in the T.S. which is more positively charged than the starting precursor complex.**

As the charge density decreases, the solvation (*i.e.* hydration) of the cation decreases through the **ion-dipole interaction**. Replacement of the NH_3 ligands by the chelating polyamines leads to the **increase of the overall size of the cation** and it reduces the charge density of the cation to undergo solvation/hydration. It concludes that the **T.S. is less solvated for the starting complexes which are more chelated by the $(N)_4$ or $(N)_5$ ligands** (Fig. 5.21.2a).

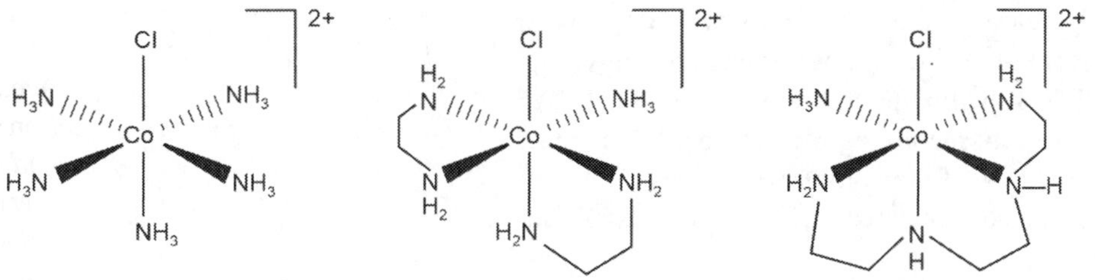

A
No. of N–H bonds: 15
No. of 5 membered chelate rings: 0

B
No. of N–H bonds: 11
No. of 5 membered chelate rings: 02

C
No. of N–H bonds: 09
No. of 5 membered chelate rings: 03

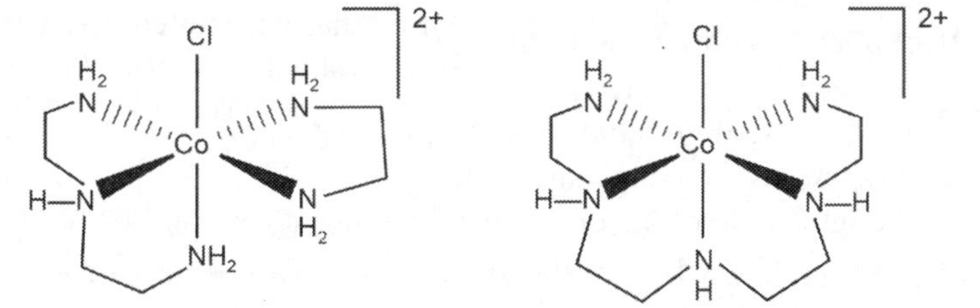

D
No. of N–H bonds: 09
No. of 5 membered chelate rings: 03

E
No. of N–H bonds: 07
No. of 5 membered chelate rings: 04

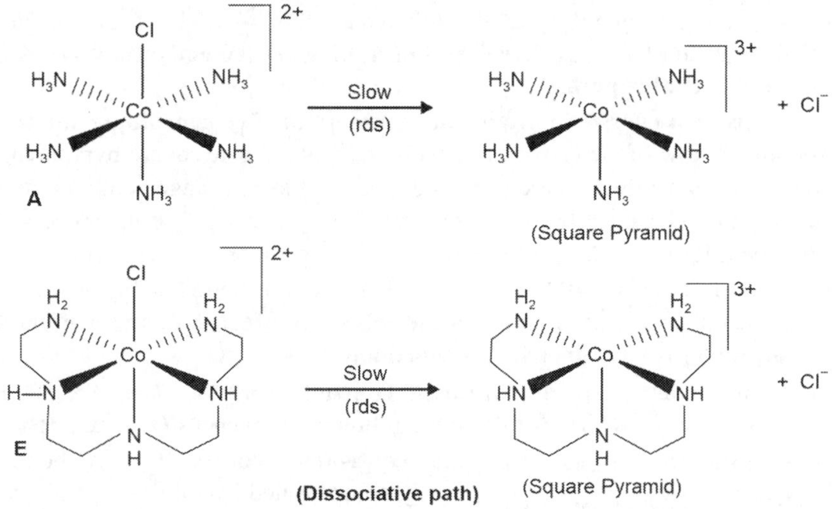

A —Slow (rds)→ (Square Pyramid) + Cl^-

E —Slow (rds)→ (Square Pyramid) + Cl^-

(Dissociative path)

$\mathbf{A} = [Co(NH_3)_5Cl]^{2+}$, $\mathbf{B} = cis\text{-}[Co(en)_2(NH_3)Cl]^{2+}$, $\mathbf{C} = [Co(trien)(NH_3)Cl]^{2+}$

$\mathbf{D} = [Co(dien)(en)Cl]^{2+}$, $\mathbf{E} = [Co(tetraen)Cl]^{2+}$

Relative aquation rate : 100 (**A**) : 21 (**B**) : 10 (**C**) : 7.8 (**D**) : 3.8 (**E**)

Fig. 5.21.2a: Effect of chelation on the aquation rate of $[Co(N)_5Cl]^{2+}$

Besides the above **charge density effect**, the extent of hydration of the cation depends on the H-bonding interaction between water (solvent) and the N—H bonds (bearing the protonic hydrogen) of the complex. In fact, it has been noted that the log (aquation rate) *vs.* number of N—H bonds or the number of chelate rings plot is linear (Fig. 5.21.2b). The plot indicates that the aquation rate increases with the increase of the number of N—H bonds or with the decrease of the number of chelate rings (Fig. 5.21.2b). This solvation/hydration effect on the T.S. (5 coordinate, in a **dissociative path**) can explain the rate order (*i.e.* the T.S. of the complex becomes more solvated when it is less chelated) in terms of the **dissociative path**.

Note: In the I_d-path, in the T.S., the leaving Cl^--group is partly lost to increase the charge separation in the T.S. However, the solvation effect is of the same kind for both the *D*- and I_d-paths.

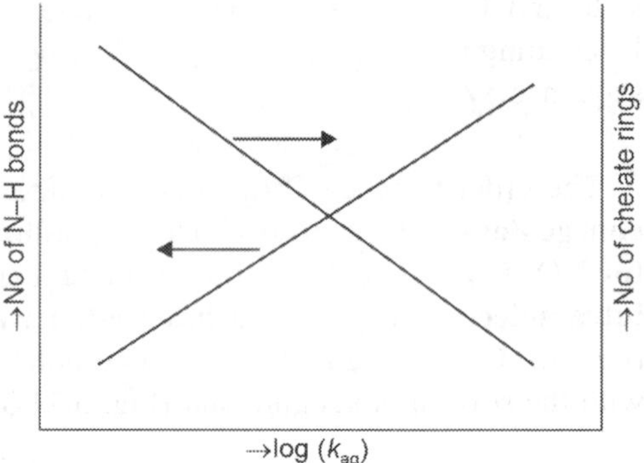

Fig. 5.21.2b: Qualitative representation of the log(k_{aq}) vs. number of N—H bonds and number of chelate rings for the aquation of $[Co(N)_4Cl_2]^+$ and $[Co(N)_5Cl]^{2+}$

● **Bite angle effect of the chelate ring on the dissociative path:** It has been already mentioned that with the increase of the extent of chelation, the steric factor disfavours the associative path more. Now let us consider the bite angle effect on the dissociation of the **Co—Cl bond at the dissociative path**. In $[Co(N)_4Cl_2]^+$ and $[Co(N)_5Cl]^{2+}$, the bond angles $N—\widehat{Co}—N$, $N—\widehat{Co}—Cl$ and $Cl—\widehat{Co}—Cl$, are expected to be close to 90° (*i.e.* ideal bond angle in the O_h complex). **For a chelate ring, the bite angle** $N—\widehat{Co}—N$ **is less than the ideal bite angle 90°**. Thus in terms of the bite angle effect, with the increase of the extent of chelation, the $N—\widehat{Co}—Cl$ bond angle increases to decrease the steric effect on the Co—Cl bond. Thus in terms of bond angle, the **Co—Cl bond is most sterically crowded in $[Co(NH_3)_4Cl_2]^+$ and $[Co(NH_3)_5Cl]^{2+}$** to favour the Co—Cl bond dissociation at the *rds* of the dissociative activation (*cf.* **steric acceleration in dissociative activation**). Thus the bite angle effect supports **the dissociative path**.

● **Reorganisation energy to attain the symmetrical five coordinate square pyramidal T.S. in the dissociative path:** In the dissociative path, the 5 coordinate square pyramidal T.S. tends to have a **symmetrical configuration**. The energy required for this rearrangement/reorganisation in the T.S. is more for the more chelated complex. Obviously, this energy requiring process disfavours more the dissociative path for the complexes having the larger number of chelate rings.

Conclusion: The chelate effect by the nonleaving $(N)_4$ and $(N)_5$ ligands in the given complexes supports the dissociative path in terms of the **solvation effect, bite angle effect and reorganisation energy in attaining the symmetrical 5 coordinate T.S.**

(L) Role of the solvent (in favour of the SAD mechanism) (*J. Chem. Soc.*, 3776, 1959)**:** Aquation rates for *trans*-$[Co(en)_2(NH_3)X]^{2+}$ follow the following sequences (X^- = Cl^-, Br^-, NO_3^-).

Aquation rate:	$NO_3^- > Br^- > Cl^-$	**$\Delta S^{\neq}$ controlled process.** 'Co—X' bond breaking energy is partly compensated by solvation of X^- in attaining the T.S. It makes $\Delta H^{\neq}$ relatively less important (*ca.* 23.5–25.5 kcal mol^{-1}, almost constant) for controlling the rate sequence.
$E_a \approx \Delta H^{\neq}$:	$NO_3^- > Br^- > Cl^-$	
$\Delta S^{\neq}$:	$NO_3^- > Br^- > Cl^-$	

Note: Arrhenius equation, $k = A \exp(-E_a/RT)$. Here the orders of E_a and A (related to $\Delta S^{\neq}$) are opposite, *i.e.* log *A*: $NO_3^- > Br^- > Cl^-$. In the present series, in terms of Arrhenius equation, the aquation rate order is controlled by the **pre-exponential term *A***. In terms of the transition state theory, it is explained in terms of ΔS^* because $\Delta H^{\neq}$ ($\approx E_a$) remains almost constant for the given leaving groups (X^-).

The above observation can be explained by considering the **solvent assisted dissociation** (SAD) **mechanism**. H–bonding interaction with the leaving group by the solvent will favour the rupture of the Co—X bond. Because of this H–bonding interaction, the 'Co—X' bond is elongated in the transition state and finally breaks down completely and H_2O enters into the vacated coordination site. The H–bonding interaction between the leaving group and water (solvent) in the transition state, follows the sequence:

$$Cl^- > Br^- > NO_3^-$$

The order between Cl^- and Br^- is quite reasonable in terms of charge density on X^- (*i.e.* **higher charge denstiy measured by charge/radius ratio causes higher solvation). The very low solvation for NO_3^- is believed to be due to the fact that the O–atoms of the ligated nitrate participate in an intramolecular H–bonding interaction with the —NH_2 groups of the coordinated amine ligand** (*i.e.* en). Consequently, before the complete dislodgement of the NO_3^- ligand, H-bonding interaction with the solvent is insignificant (Fig. 5.21.3).

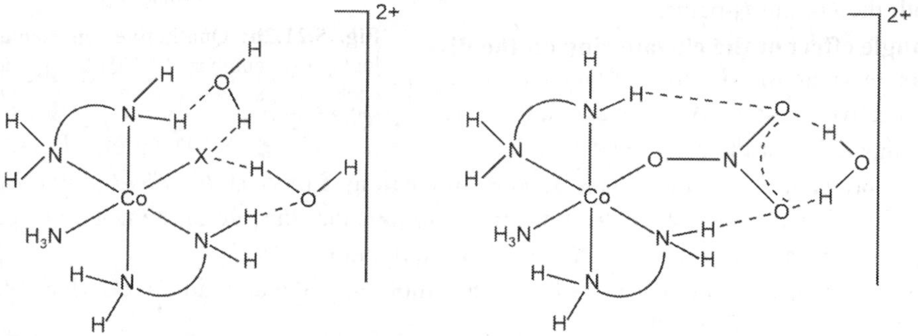

(a) (b)

Fig. 5.21.3: Solvent assisted dissociation (SAD) of X^- in aquation of *trans*–$[Co(en)_2(NH_3)X]^{2+}$(X^- = halide, NO_3^-). (a) Hydration of leaving group for X^- = halide; (b) Intramolecular H–bonding of the leaving group, X^- = NO_3^- leading to a less hydration of the leaving group NO_3^- at the stage of bond rupture. (*cf.* M.L. Tobe, *J. Chem. Soc.*, 3776, 1959; F. Basolo et al., *J. Am., Chem. Soc.*, **82,** 1077, 1960)

The aquation rate (k_{aq}) sequence is governed by the $\Delta S^{\neq}$ sequence. More solvation of the departing ligand in the transition state will lower both the $\Delta H^{\neq}$ value (*i.e.* **enthalpic favour**) and $\Delta S^{\neq}$ value (*i.e.* **entropic disfavour;** lowering of $\Delta S^{\neq}$ arises from the electrorestriction over the solvent by H–bonding with the departing ligand). **Thus the loss of entropy in attaining the transition state is minimum for X^- = NO_3^- which is least solvated in the transition state.**

(**Note:** SAD mechanism for the aquation of $[Cr(NH_3)_5X]^{2+}$ (X^- = Cl^-, Br^-, I^-) has also been argued. The aquation rate constants follow the sequence:

$$\begin{cases} k_{aq} : \ I^- > Br^- > Cl^- \\ \Delta S^{\neq} \text{ (e.u.)}: \ I^-(-1.9) > Br^-(-6.1) > Cl^-(-9.2) \end{cases}$$

(*i.e.* least solvation of the departing ligand in the transition state gives the maximum entropic favour).

$(1 \text{ e.u.} = 1 \text{ cal K}^{-1} \text{ mol}^{-1} = 4.184 \text{ J K}^{-1} \text{ mol}^{-1})$

Interestingly, **$\Delta H^{\neq}$ values remain more or less the same** (*ca.* 21–22 kcal mol^{-1}) for X^- = Cl^-, Br^- and I^- (*i.e.* the rate difference is controlled by $\Delta S^{\neq}$). The energy required to rupture the Co—X bond varies as: Co—Cl > Co—Br > Co—I but the energy released through the solvation of departing halide follows the opposite order. It makes $\Delta H^{\neq}$ more or less the same. Thus in the SAD mechanism, the bond breaking energy is partially compensated by the solvation energy of the leaving group.

(M) Effect of pressure on the aquation process (*i.e.* interpretation of $\Delta V^{\neq}$) (*cf.* Sec. 5.19.5): In the aquation process for the *anionic leaving group*:

$$[Co(NH_3)_5X]^{n+} + H_2O \rightarrow [Co(NH_3)_5(OH_2)]^{(n+1)+} + X^-,$$

if the **dissociative path** operates, **separation of charge (*i.e.* creation of charge) in the transition state** will tend to make $\Delta V^{\neq}$ negative because of the enhanced electrorestriction over the solvent imposed by the transition state. Thus, here, the $\Delta V^{\neq}$ value will be less positive compared to that of the other dissociative processes (*e.g.* water exchange process, aquation of neutral leaving group) where there is no change of charge in attaining the **dissociative transition state.**

If the aquation process passes through the **associative path** (*i.e.* nucleophilic attack by H_2O at the *rds*), there will be ***no change of overall charge in attaining the transition state. Consequently, $\Delta V^{\neq}$*** **will be negative as usual.**

Thus, in the aquation process of an anionic leaving group, it is complicated to conclude firmly the mechanistic path in terms of the $\Delta V^{\neq}$ value. These aspects are discussed below.

$$[Co(NH_3)_5X]^{2+} + H_2O \rightarrow [Co(NH_3)_5(OH_2)]^{3+} + X^-$$

$$\left.\begin{array}{llll} X^- = & Cl^- & Br^- & NCS^- \\ \Delta V^{\neq}\left(cm^3\,mol^{-1}\right): & -10.6 & -9.2 & -8.6 \end{array}\right\} \begin{array}{l} \Delta V^{\neq} \text{ for water exchange process in} \\ [Co(NH_3)_5(OH_2)]^{3+} = +1.2\ cm^3mol^{-1} \end{array}$$

The positive value of $\Delta V^{\neq}$ for the water exchange process conclusively supports the dissociative path. The negative value of $\Delta V^{\neq}$ for the said aquation processes is roughly the **same as the molar volume change (ΔV)** for the completed aquation reaction (*i.e.* $\Delta V^{\neq} \approx \Delta V$). The plot, $\Delta V^{\neq}$ *vs.* ΔV, for the aquation of $[Co(NH_3)_5X]^{2+}$ is linear with slope $\approx$ 1.0 (see Fig. 5.19.5.3). *It indicates that the transition state resembles the final products.* Thus it can be concluded that in spite of the negative value of $\Delta V^{\neq}$, the said aquation processes pass through the dissociative path.

Similarly the negative value of $\Delta V^{\neq}$ ($= -18.3\ cm^3\ mol^{-1}$) for the aquation of $[Co(NH_3)_5(SO_4)]^+$ can be argued for the dissociative path.

● Let us consider the **aquation of the neutral leaving group** in the following reaction.

$$[Co(NH_3)_5(L)]^{3+} + H_2O \rightarrow [Co(NH_3)_5(OH_2)]^{3+} + L,\ \Delta V^{\neq} = \text{positive value,}$$

(leaving group is neutral).

$$L = OSMe_2,\ OCHNH_2,\ OC(NH_2)_2, \text{ etc.}$$

In the above cases, there is no complication in interpreting the $\Delta V^{\neq}$ value as there is no possibility of creation or destruction of charge in attaining the transition state through the both *d*– and *a*–activation. ***Thus the positive value of $\Delta V^{\neq}$ conclusively supports the dissociative path.***

(N) Stereochemical change in the aquation process (*cf.* Secs. 5.25.3, 5.28.2): The important observations of aquation of *cis–/trans–* $[CoCl(en)_2(L)]^{n+}$ are given below (L = non-leaving group).

● For the ***cis*–complex**, there is a **100% retention of configuration irrespective of the nature of L** (*i.e.* there is no isomerisation). There are some exceptions.

● For the ***trans*–complex**, the degree of **isomerisation** depends on the π–bonding properties of L. With the increase of the π–donor property of L, the tendency to isomerise increases.

It follows the order.

$$L = OH^- > NCS^- > Cl^-, \ (\textit{cf.}\ \text{Tables 5.21.3, 5–6})$$

● If the non-leaving group L is a *π–acceptor* ligand (*e.g.* NO_2^-, CN^-) or a non–π–bonding ligand (*e.g.* NH_3), there is a **100% retention of configuration for both the *cis*– and *trans*–complexes** (*cf.* Tables 21.5.3, 5–6).

● For the rigid tetradentate macrocyclic ligand (N_4) like cyclam, there is a **100% retention of configuration for the complex which can exist only as the *trans*–complex,** *i.e.*

$$\left[CoCl(cyclam)(L)\right]^{n+} \xrightarrow{+H_2O} trans\text{-}\left[Co(cyclam)(OH_2)\right]^{(n+1)^+}$$

(100% retention of configuration irrespective of the nature of L).

(a) Prediction of Stereochemical Results in Terms of the Dissociative Path: The stereochemistry of the products is outlined in Figs. 5.21.4–6.

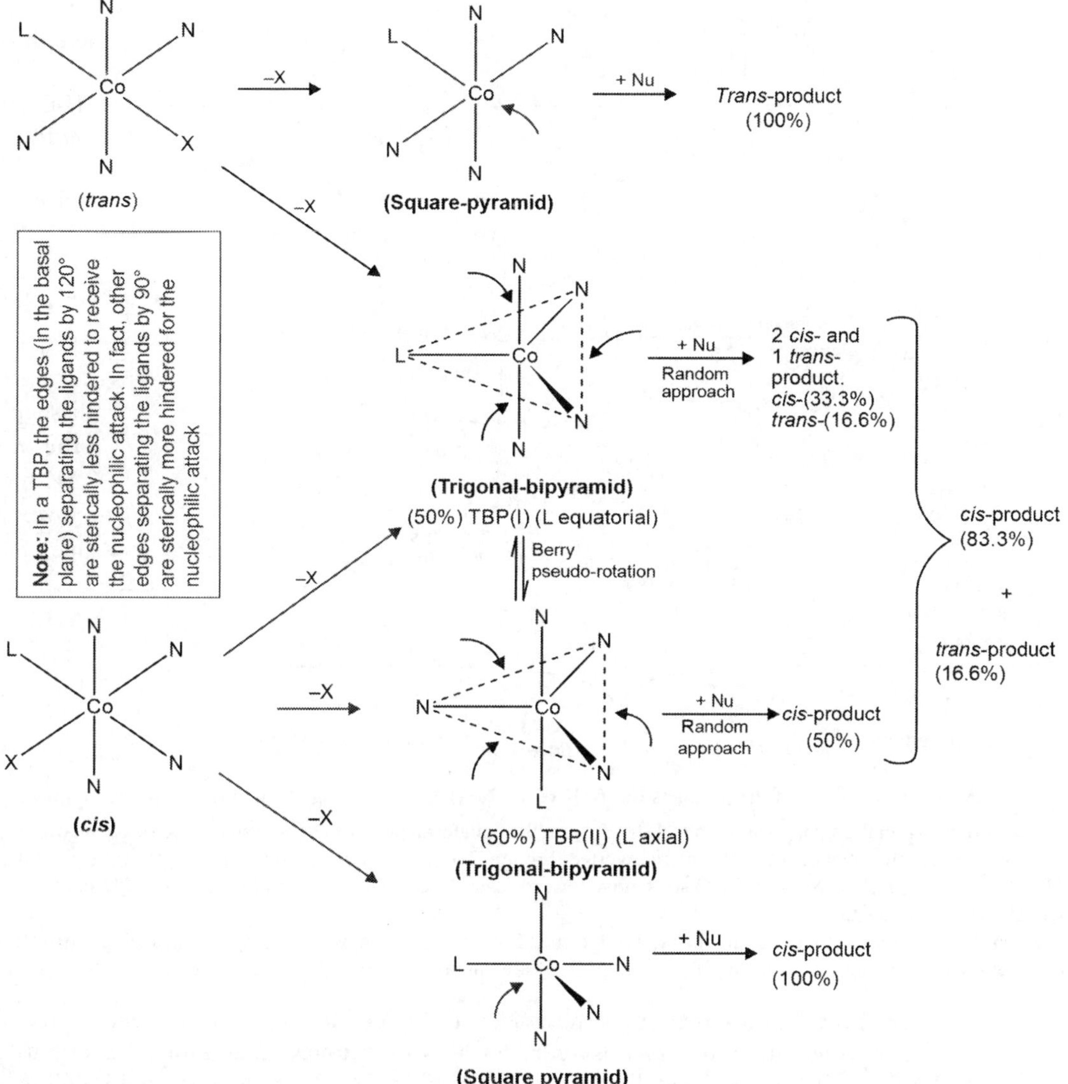

Fig. 5.21.4: Stereochemistry of the products of *cis–/trans–* $[Co(N)_4LX]^{n+}$ reacting through the dissociative (D) path

Note: ● N = nonleaving groups like NH_3; X = leaving group like Cl^-; L = Nonleaving group. **TBP(I)** (L at the equatorial position) and **TBP(II)** (L at the axial position) are supposed to be equally probable (*i.e.* 50% each) and they are mutually interconvertible through Berry's pseudorotation (*cf.* Fig. 5.28.2.2)

The important features of the dissociative path (*i.e.* formation of the 5–coordinate intermediate or activated complex) are given below:

(i) If the 5–coordinate intermediate is a **square pyramid** (*i.e.* removal of one octahedral position) **then there will be a 100% retention of configuration.**

(ii) If the 5–coordinate intermediate adopts the **trigonal bipyramidal geometry**, **then there is a stereochemical change.**

(iii) If the nonleaving group (L) is a **π–donor ligand** (*e.g.* OH^-, NCS^-, Cl^-, etc.) then the square pyramidal intermediate obtained from the ***cis*-complex** is stabilised as the vacant orbital can

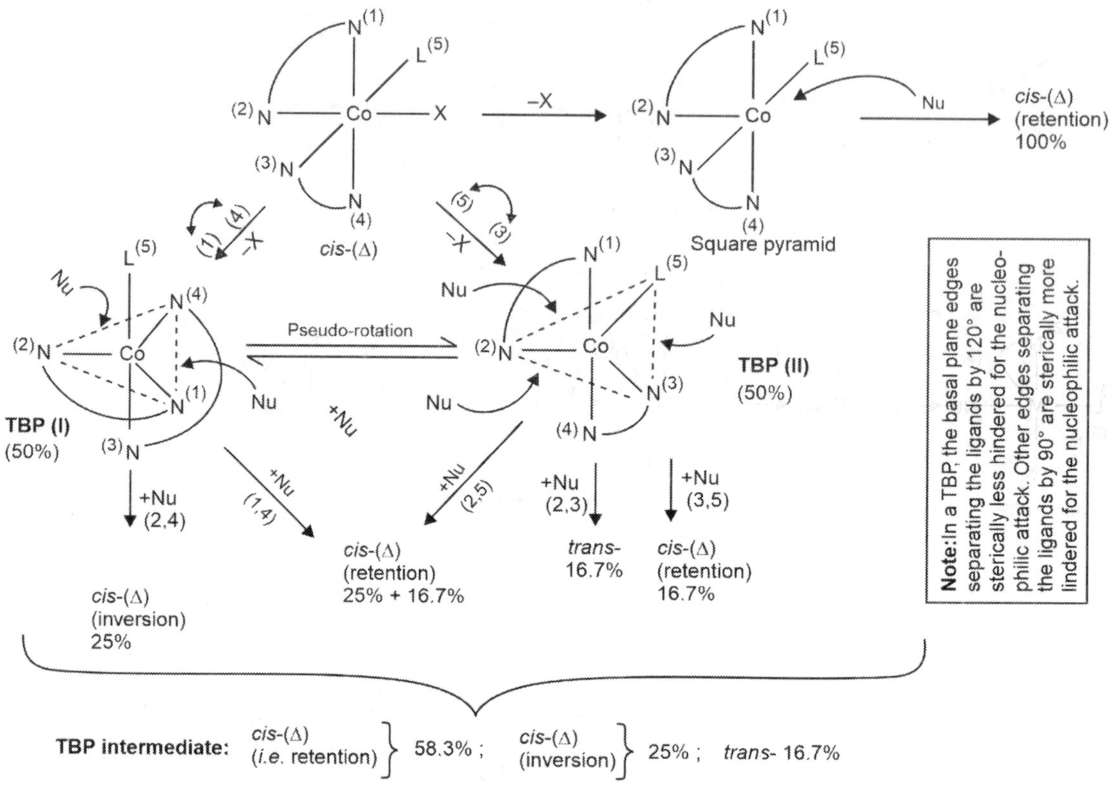

Fig. 5.21.5: Stereochemistry of the products of *cis*–$[Co(N—N)_2(L)X]^{n+}$ reacting through the dissociative process.

Note: ● **TBP(I)** (L in the axial position) and TBP(II) (L in the equatorial position) are considered to be equally probable (*i.e.* 50% each) and they can be mutually interconverted through Berry's pseudo-rotation taking N(2) as the pivotal site.
● TBP(I) is generated, if the N(1) and N(4) sites move towards each other. TBP(II) is generated, if the N(3) and L(5) sites move towards each other.
● For TBP(I), nucleophilic attacks along the (1, 4) edge and (2, 4) edge are supposed to be equally probable. For TBP(II), nucleophilic attacks along the three edges (3, 5; 2, 3; 2, 5) are supposed to be equally probable.

receive back the electron cloud from the adjacent π–donor ligand L. The square pyramidal intermediate obtained from the ***trans*–isomer in which L occupies an equatorial position** (see Fig. 5.21.1 for the L→metal π-bonding) cannot be stabilised by the π–donor ligand L unless it is converted into a trigonal bipyramidal geometry. ***This is the driving force for the conversion of the square pyramidal geometry into the trigonal bipyramidal geometry*** in which L occupies an equatorial position (see Fig. 5.21.1 for the L → metal pi-bonding). It explains: better the π–donor property of L, more the chance for conversion into the trigonal bipyramidal geometry; *i.e.* more the isomerisation for the *trans*-complex.

Note: For **isomerisation** of *cis*-$[CoCl(N)_4L]^{n+}$ even for the π-bonding L (nonleaving group), it passes through the TBP intermediate (see Sec. 5.28.2 for details).

(iv) In the **trigonal bipyramidal intermediate**, there are **two types of edges** — edges separating the ligands by 90° and edges separating the ligands by 120°. ***Obviously, the three edges at the basal or equatorial plane are the better positions to receive the nucleophilic attack because other edges separating the ligands by 90° are sterically more hindered.***

Thus in a trigonal bipyramidal intermediate obtained from *trans*-$[CoCl(N)_4(L)]^{n+}$, *e.g.* *trans*-$[CoCl(NH_3)_4(L)]^{n+}$, there are three possible ways to receive the nucleophilic attack. **There are two possible trigonal bipyramidal geometries remaining in an equilibrium through**

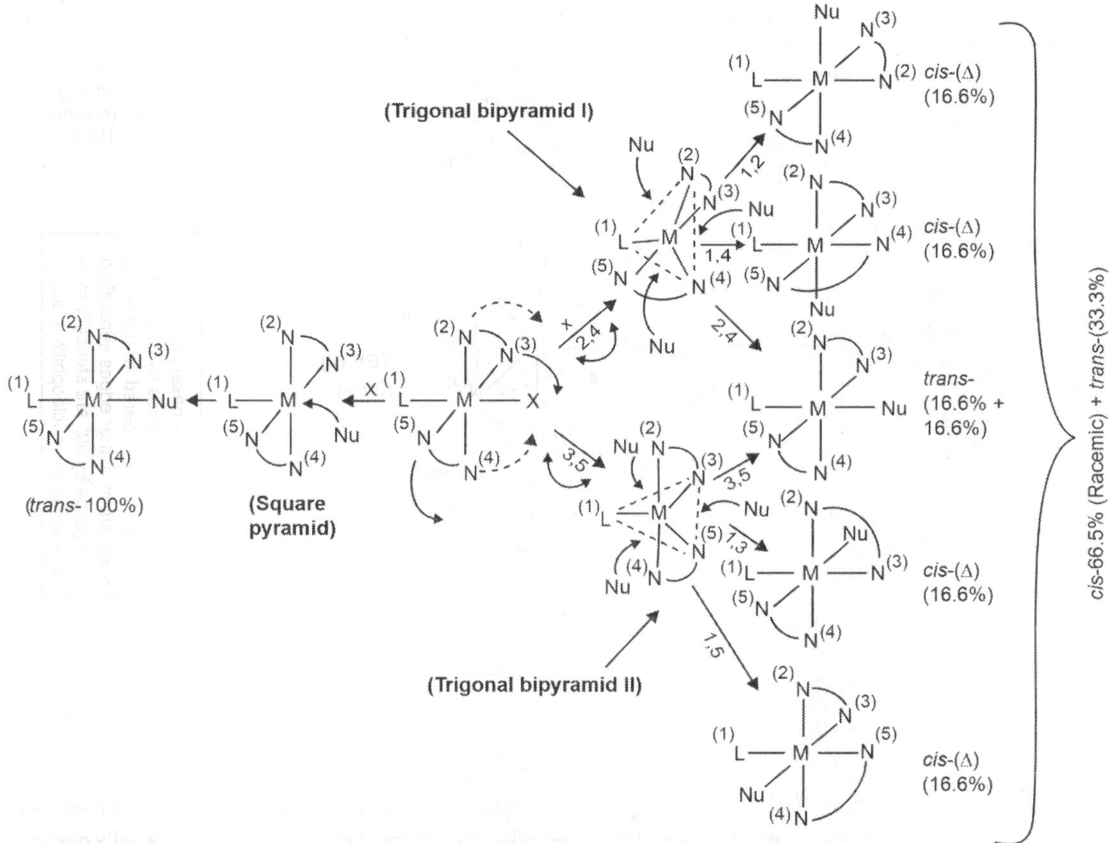

Fig. 5.21.6: Stereochemistry of products of *trans*–$[Co(N—N)_2(L)(X)]^{n+}$ reacting through the dissociative path.
Note: ● Dissociation of X can lead two TBP–geometries *i.e.* TBP(I) and TBP(II). The TBP(I) is generated, if the N(2) and N(4) sites move towards each other. The TBP(II) is generated, if the N(3) and N(5) sites move towards each other. In TBP(I), N(3) and N(5) are kept at the axial positions while in TBP(II), N(2) and N(4) are kept at the axial positions.
● In calculating the product distribution, TBP(I) and TBP(II) are considered to be equally probable and each nucleophilic attack has been considered to be equally probable (*e.g.* for TBP (II), all the possible three nucleophilic attacks are equally probable).

Berry's pseudorotation (*cf.* Figs. 5.21.4-5). There are also two possible trigonal bipyramidal geometries for *trans*-$[CoCl(N—N)_2(L)]^{n+}$, *e.g.* *trans*-$[CoCl(en)_2(L)]^{n+}$.

Note: Obviously, for the good π-donor nonleaving group (L), the trigonal bipyramidal (TBP) activated complex bearing **L at the equatorial position is more favoured** over the other TBP keeping L at an axial position (Sec. 5.21).

(v) If the nonleaving group (L) is a non–π–bonding ligand (*e.g.* NH_3) or a π–acceptor ligand (*e.g.* CN^-, NO_2^-, etc.), the square pyramidal geometry **(if formed)** does not have any driving force for its conversion into a trigonal bipyramidal geometry. Consequently there will be no isomerisation even if the aquation process goes on through the dissociative path. In fact, other evidences indicate the **associative path,** specially for the π–accepter nonleaving ligand. However, this associative path also explains the 100% retention of configuration (*cf.* Fig. 5.21.1c, d).

(vi) **Isomerisation (*cf.* Tables 5.21.3, 5-6) arises due to the dissociative path when the 5–coordinate intermediate (activated complex) adopts the trigonal bipyramidal structure.** Thus isomerisation, a consequence of the conversion of square pyramidal intermediate into the trigonal

Table 5.21.5 Stereochemical change in aquation of $[CoCl(N)_4(L)]^{n+}$, $[CoCl(N-N)_2(L)]^{n+}$ and $[CoCl(N_4)(L)]^{n+}$. (leaving group = Cl^-, *i.e.* aquation of Cl^-) (*cf.* M.L. Tobe, *Inorg. Chem.*, **7**, 1260, 1968; C.K. Poon, *et al.*, *J. Chem., Soc. A*, 1549, 1968).

***cis*–complex**	***Retention of Configuration (%)***	**$\Delta S^{\neq}$ (J K^{-1} mol^{-1})**
cis–$[CoCl(en)_2(CN)]^+$	100	–20
cis–$[CoCl(en)_2(OH)]^+$	100	–42
cis–$[CoCl_2(en)_2]^+$	100	–20
cis–$[CoCl(en)_2(NCS)]^+$	100	–59
cis–$[CoCl(en)_2(NO_2)]^+$	100	–13
cis–$[CoCl(en)_2(NH_3)]^{2+}$	100	–46
cis–$[CoCl(en)_2(N_3)]^+$	100	–17
cis–$[CoCl_2(cyclam)]^+$	100	–25
***trans*–complex**		
trans–$[CoCl_2(NH_3)_4]^+$	50 ± 10	+36
trans–$[CoCl(en)_2(CN)]^+$	100	–8
trans–$[CoCl(en)_2(NO_2)]^+$	100	–8.5
trans–$[CoCl(en)_2(OH)]^+$	25	+84
trans–$[CoCl_2(en)_2]^+$	65	+60
trans–$[CoCl(en)_2(N_3)]^+$	80	0
trans–$[CoCl(en)_2(NH_3)]^{2+}$	100	–46
trans–$[CoCl_2(cyclam)]^+$	100	–13
trans–$[CoCl(cyclam)(OH)]^+$	100	–29

* Aquation of Cl^- through a dissociative path leads to the separation (*i.e.* creation) of charges in attaining the T.S. Hence, T.S. imposes more electrorestriction than the starting complex on the surrounding solvent (*i.e.* **T.S. is more solvated**, see Sec. 5.19.5) and it makes $\Delta S^{\neq}$ negative. But conversion of the square pyramidal T.S. to the trigonal bipyramidal T.S. leading to the isomerisation makes $\Delta S^{\neq}$ more positive.

Note: ● It has been mentioned that for the *cis*–$[CoCl(N—N)_2(L)]^{n+}$ complex, there is a complete retention of configuration, but in some cases, 15 – 30% conversion of the *cis*– into the *trans*–product has been noted. **This finding contradicts the earlier observation.** *It indicates that the 5–coordinate square pyramid intermediate is sufficiently long-lived in such cases to convert it into a trigonal bipyramidal geometry* (see Fiog. 5.21.5).

● If the *cis*–product retains the **same enantiomeric configuration** of the starting *cis*–complex, then it can be assumed that the *cis*–product originates from the square pyramid intermediate and trigonal bipyramidal intermediate where **L-occupies the equatorial position** (*cf.* Fig. 5.21.5). The trigonal bipyramidal intermediate bearing **L at the equatorial position** partly also produces the optically inactive *trans*-product. Thus the loss of optical activity is due to only isomerisation to the *trans*–product and not due to the recemisation.

● It may be noted that the trigonal bipyramidal intermediate bearing **L at the axial position** will lead to a racemic mixture (*cf.* Fig. 5.21.5).

● **No change in stereochemistry** in the product indicates the **square pyramidal activated complex** (*i.e.* T.S.) but change in stereochemistry in the product indicates the **trigonal bipyramidal T.S.** in terms of the dissociative path of aquation (see Table 5.21.6)

bipyramidal intermediate, leads to the more **positive $\Delta S^{\neq}$ value** (M.L. Tobe, 1968). On the other hand, for the square pyramidal intermediate (*i.e.* retention of configuration), **$\Delta S^{\neq}$ value is less and negative** (*cf.* aquation of anionic group X^- through the dissociative path leads to the –ve value of ΔS^* because of the more electrorestriction effect on solvent by T.S., Sec. 5.19.5).

Table 5.21.6 Prediction of stereochemical results in terms of the dissociative path for the aquation of Cl^- in the given complexes (*cf.* Figs. 5.21.3–5).

Complex	*Square pyramid intermediate*		*Trigonal bipyramidal intermediate**	
	cis–product	*trans–product*	*cis–product*	*trans–product*
cis–$[CoCl(N)_4(L)]^{n+}$	100%	0%	83.3%	16.7%
trans–$[CoCl(N)_4(L)]^{n+}$	0%	100%	83.3%	16.7%
Λ-*cis*–$[CoCl(N—N)_2(L)]^{n+}$	100%	0%	Λ = 58.3% Δ = 25%	16.7%
trans–$[CoCl(N—N)_2(L)]^{n+}$	0	100%	66.6% (Racemic)	33.3%

* Considering the equal stabilities of both the trigonal bipyramidal intermediates (*cf.* **Figs. 5.21.4–6**).

(b) Prediction of Stereochemical Results in Terms of the Associative Path: The observation is: *100% retention of configuration for both the cis- and trans-complexes* (*cf.* Fig. 5.21.1 c, d). In fact, this has been experimentally verified for Cr(III), Rh(III), Ir(III) and Ru(III) complexes experiencing the associative activation. Even for the Co(III)–complexes like $[Co(en)_2(L)X]^{2+}$ where **L is a strong π–acceptor ligand,** the associative activation is quite reasonable.

$$\left.\begin{array}{l} cis\text{-}\left[RuCl_2(NH_3)_4\right]^+ \xrightarrow{\text{aquation}} 100\%\ \text{retention} \\ trans\text{-}\left[RuCl_2(NH_3)_4\right]^+ \xrightarrow{\text{aquation}} 100\%\ \text{retention} \end{array}\right\} \text{(aquation of } Cl^-\text{)}$$

The 100% retention of configuration can be explained by considering the adjacent attack (i.e. adjacent to the leaving group) by the incoming nucleophile (*cf.* Fig. 5.21.1c, d).

(L) Conclusion: The evidences given above (**A** to **N**) support the fact that the Co(III)–complexes aquate mainly through the dissociative path (most probably I_d path). However an associative activation has been argued in some special cases when the nonleaving group is a good π-acceptor ligand (*cf.* **N(b)** of this section) and the leaving group is a powerful electron withdrawing group (*cf.* **J,** Table 5.21.4 of this section).

5.22 AQUATION OR ACID HYDROLYSIS OF THE OCTAHEDRAL CHROMIUM(III) COMPLEXES (*cf.* Secs. 5.19.5–6)

(A) Solvent assisted dissociation (SAD) mechanism for $[Cr(NH_3)_5X]^{2+}$ (X^- = Cl^-, Br^-, I^-): The characteristic features of SAD mechanism for aquation of $[Co(NH_3)_5X]^{2+}$ have been discussed in Sec. 5.21. The same features are prevailing here also.

k_{aq} : $I^- > Br^- > Cl^-$

$\Delta S^{\neq}$: $I^- > Br^- > Cl^-$

In the SAD path, the departing ligand is solvated through H-bonding and *the entering nucleophile* H_2O *simultaneously slips into the position vacated by* X^-. The rate sequence can be explained in terms of $\Delta S^{\neq}$. This aspect of SAD mechanism has been discussed in Sec. 5.21.

(B) Group replacement factor (GRF) (see Sec. 5.21)**:** In the aquation of $[Cr(NH_3)_5X]^{2+}$ (X^- = halide), the GRF (= k_{R-X}/k_{R-Cl}) (R = $Cr(NH_3)_5^{3+}$) values run in the sequence: $I^- > Br^- > Cl^- \gg$ F. It rules out the pure associative path for which the GRF values should run in the sequence: $F^- \gg Cl^- > Br^- > I^-$. F^- (most electron withdrawing) should make the metal centre most suitable for the nucleophilic attack. The order of GRF values indicates the importance of bond breaking in the activation process (*cf.* the conclusion of **SAD mechanism**). It may be noted that for the aquation of $[Co(NH_3)_5X]^{2+}$, a similar trend of GRF values has been observed to support the dissociative path (see Sec. 5.21).

(C) LFER plot: The slopes of LFER plots (*i.e.* log k_{aq} *vs.* log K_{eq}) for the aquation of $[Cr(NH_3)_5X]^{2+}$ and $[Cr(OH_2)_5X]^{2+}$ are 0.91 and 0.56 respectively. It indicates that separation of X^- is about 90% complete in the transition state for the aquation of $[Cr(NH_3)_5X]^{2+}$ while the separation is only 50–60% complete for the aquation of $[Cr(OH_2)_5X]^{2+}$. It indicates the more dissociative character in the aquation of $[Cr(NH_3)_5X]^{2+}$ (*cf.* Secs. 5.19.4, 6 for explanation in terms of energy of the LUMO).

(D) Steric retardation and activation parameter: $[CrCl(NH_2Me)_5]^{2+}$ aquates slower than $[CrCl(NH_3)_5]^{2+}$ because of the **steric retardation** as expected in an associative activation. Higher $\Delta H^{\neq}$ and $\Delta S^{\neq}$ values for the sterically more crowded complex may be due to the **weaker bond formation by H_2O at the associative T.S.** The enthalpic disfavour (in spite of the entropic favour) makes the aquation rate of $[CrCl(NH_2Me)_5]^{2+}$ slower.

	$10^6 k_{aq}$ (s^{-1})	$\Delta H^{\neq}$ (kJ mol^{-1})	$\Delta S^{\neq}$ (J K^{-1} mol^{-1})
$[CrCl(NH_3)_5]^{2+}$:	8.7	93	–29
$[CrCl(NH_2Me)_5]^{2+}$:	0.25	110 ↓	–2 ↓
		Enthalpic disfavour	**Entropic favour**

Here it is worth mentioning that $[CoCl(NH_2Me)_5]^{2+}$ aquates faster than $[CoCl(NH_3)_5]^{2+}$ because of the **dissociative activation experiencing the steric acceleration** (*cf.* Sec. 5.21D).

(E) Volume of activation (see Sec. 5.19.5): Aquation of $[Cr(NH_3)_5L]^{3+}$ (L = neutral ligands like DMSO, DMF, urea, etc.) experiences the negative $\Delta V^{\neq}$ values (*cf.* analogous Co-compounds experience the positive $\Delta V^{\neq}$ values; *cf.* Sec. 5.21M). It indicates the associative activation for such Cr(III) complexes. For the aquation of $[Cr(NH_3)_5X]^{2+}$ (X = Cl, Br, I), the ΔV^{*} values are negative (*ca.* –10 cm^3 mol^{-1}). **This can be apparently explained by both the associative path and dissociative path leading to the charge separation in the T.S.** (*cf.* $\Delta V^{\neq}$ = –ve for the aquation of $[Co(NH_3)_5X]^{2+}$ through the dissociative path, see Sec. 5.21M). It may be noted that for the aquation of $[Cr(NH_3)_5X]^{2+}$, the slope of the plot, ΔV^{*} *vs.* ΔV, is about 0.6 while for the aquation of $[Co(NH_3)_5X]^{2+}$, slope of the corresponding plot is about 1.0 (see Fig. 5.19.5.3). It suggests the less dissociative character in the aquation of $[Cr(NH_3)X]^{2+}$.

A plot of $\Delta V^{\neq}$ values of aquation of $[Co(NH_3)_5X]^{2+}$ versus the $\Delta V^{\neq}$ values of aquation of $[Cr(NH_3)_5X]^{2+}$ gives the slope far away from unity (see Fig. 5.19.5.3). It indicates that $[Cr(NH_3)_5X]^{2+}$ complexes aquate in a different mechanism compared to theanalogous $[Co(NH_3)_5X]^{2+}$ complexes which adopt the dissociative path.

(F) Effect of charge: The aquation rates follow the following sequences.

$[MCl(NH_3)_5]^{2+} < [MCl_2(NH_3)_4]^{+}$; rate enhancement by a factor about 10^3–10^4 (for M = Co), and by a factor about 10^1 only (for M = Cr).

Thus for the Co(III)-complex, the rate is very much sensitive to the overall charge of the complex while for the Cr(III)-complex, the effect is less significant. ***Thus for the Co(III)-complexes, dissociative activation is primarily important but for the Cr(III) complex, it is not so important.***

(G) Comparison between calculated and experimentally found activation energy (E_a) (see Sec. 5.21A)**:** For the aquation of $[M(NH_3)_5X]^{2+}$ (M = Co, Cr), activation energy has been calculated by considering the possible intermediates (*i.e.* activated complex) like *square pyramid* (*i.e.* dissociative path), *pentagonal bipyramid* (*i.e.* associative path), *trapezodial octahedron* (*i.e.* associative path). For this calculation of activation energy, change of cfse, and interelectronic repulsion energy in the formation of the transition state were mainly considered. For the Co(III)-complexes, the observed E_a values were found in good agreement with those calculated for the square pyramidal intermediate. On the other hand, for the Cr(III) complexes, observed E_a values were in good agreement with those calculated for the pentagonal bipyramid intermediate. These support the **dissociative path for the Co(III)-complexes** and the **associative path for the Cr(III) complexes** (*Ref. JCAS*, **90**, 6626, 1968).

(H) Enthalpy of transition ($\Delta H_T = \Delta H^{\neq} - \Delta H$, see Fig. 5.19.3): For the dissociative path of aquation of $[M(NH_3)_5X]^{2+}$, ΔH_T remains constant (*i.e.* independent of X^-) but it depends on the nature of X^- for the associative path. By considering this fact, it has been suggested that $[Co(NH_3)_5X]^{2+}$ aquates through a dissociative path but $[Cr(OH_2)_5X]^{2+}$ aquates through an associative path (see Fig. 5.19.3).

Conclusion: Aquation of the Cr(III)-complexes generally experiences the associative activation (most probably the I_a-path) while the analogous Co(III) complexes generally experience the dissociative activation. The reason behind this difference has been discussed in Sec. 5.19.6.

5.23 ELECTROPHILIC CATALYZED AQUATION OF THE OCTAHEDRAL COMPLEXES

Very often, the leaving groups (X) are the conjugate bases of weak acids. They can bind with the **electrophiles** (E) (*e.g.* H^+, M^{n+}) even when coordinated in a metal complex. Sometimes, the polydentate leaving groups can bind with such electrophiles either in partially dechelated condition (*e.g. en, bpy*) or in fully coordinated condition (*e.g.* ox^{2-}, $acac^-$, etc.). This binding interaction with the electrophile favours the removal of the group (X). The process can be illustrated as follows in a general way.

(i) $M—X + E \underset{}{\overset{K_E}{\rightleftharpoons}} M—X—E \xrightarrow[(+H_2O)]{k_E} M—OH_2 + E—X,$

$$[M—X—E] = \frac{K_E[E][M—X]}{1+K_E[E]} \text{ when } [E] \gg [M—X], \text{ (cf. Sec. 5.9.3)}$$

(ii) $M—X \xrightarrow[(+H_2O)]{k_0} M—OH_2 + X^-$

$$\text{Rate of aquation} = k_0[M—X] + k_E[M—X—E] = k_0[M—X] + \frac{k_E K_E[E][M—X]}{1+K_E[E]}$$

$$= \left\{k_0 + \frac{k_E K_E[E]}{1+K_E[E]}\right\}[M—X].$$

The second term involving the k_E term represents the contribution of the catalyzed path and the first term involving the k_0 term represents the contribution of the uncatalyzed path to the aquation process. Depending on the condition, the above rate equation may be reduced as follows:

$$\text{rate} \approx \{k_0 + k_E K_E[E]\}[\text{complex}];\quad 1 \gg K_E[E]$$

$$\approx \{k_0 + k_E\}[\text{complex}];\quad K_E[E] \gg 1$$

Very often k_E increases with K_E (measuring the stability of interaction between the electrophile and leaving group). This stability interaction measured by K_E can be easily understood by HSAB principle. If the leaving group provides the hard donor sites (*e.g.* O, F, N, etc.) then the hard electrophiles, *i.e.* hard acids (*e.g.* H^+, Be^{2+}, Al^{3+}, etc.), are suitable. On the other hand, if the leaving group is a softer one (*e.g.* Cl, Br, I, S, etc.), then the soft acids (*e.g.* Hg^{2+}, Ag^+, Tl^+, etc.) are more suitable.

5.23.1 Brönsted Acid Catalyzed Aquation Reactions

The catalyzed path is:

$$M—X + H^+ \rightleftharpoons \underset{\text{CA}}{\left[M—X—H\right]^+} \xrightarrow{+H_2O} M—OH_2 + XH$$

If the protonated species $[M—XH]^+$ aquates through the *dissociative path,* then the process is referred to as the **D–CA mechanism** (≡ $\mathbf{S_N1}$**–CA** in old nomenclature); if the protonated species aquates through the associative path, then the process is referred to as the **A–CA** (≡ $\mathbf{S_N2}$**–CA**) mechanism.

Some examples of the acid catalyzed aquation reactions are discussed below.

- **Aquation of $[M(NH_3)_5X]^{n+}$, $[M(NH_3)_4X_2]^{n+}$ (X having the free basic sites):** Aquation of the metal complexes $[M(NH_3)_5X]^{n+}$ (M = Co, Cr; X = F^-, NO_2^-, N_3^-, CN^-, CO_3^{2-}, SO_4^{2-}, oxalate, acetylacetonate, etc.) is acid catalyzed. Here the metal bound leaving group X undergoes protonation because of the presence of free basic sites. Obviously, no acid catalysis is observed for NH_3 as the leaving group because coordinated NH_3 does not have any free basic site or lone pair for protonation.
- **Aquation of $[M(NH_3)_5X]^{2+}$ (X = Cl, Br, I) *vs.* $[M(NH_3)_5F]^{2+}$ (M = Co, Cr):** $[M(NH_3)_5X]^{2+}$ (X = Cl, Br, I) does not experience the acid catalyzed aquation but $[M(NH_3)_5F]^{2+}$ experiences the

H^+ catalyzed aquation. Protonation of the bound Cl, Br or I is less likely because of two reasons: *these softer sites do not like* H^+ *for interaction;* HCl, HBr and HI *are much stronger acids than* HF (*i.e.* basicity order: $F^- \gg Cl^- > Br^- > I^-$) *and consequently, protonation of the heavier halides is less likely.*

- **Aquation of $[M(OH_2)_5X]^{2+}$ (M = Cr, Co, Rh, etc., X = F, Cl, Br, etc.):** It is illustrated for $[Cr(OH_2)_5F]^{2+}$ (*cf.* acid catalysis is favoured in the sequence, $F^- \gg Cl^- > Br^- > I^-$ as explained above).

$$[Cr(OH_2)_5{-}F]^{2+} + H^+ \rightleftharpoons [Cr(OH_2)_5\cdots F\cdots H]^{3+}$$
$$[Cr(OH_2)_5\cdots F\cdots H]^{3+} + H_2O \rightarrow [Cr(OH_2)_6]^{3+} + HF$$

Rate constant of the above aquation reaction is $1.4 \times 10^{-8}\ s^{-1}$ in an acidic solution while it is $6.2 \times 10^{-10}\ s^{-1}$ in a neutral solution.

- **Aquation of $[M(L{-}L)_3]^{n+}$:** For some ligands (L—L) like en, bpy, phen, etc. there is **no free basic** site for protonation when they remain coordinated. In such cases, acid catalysis is possible, if the ligand undergoes **partial dechelation** to provide the protonation sites. On the other hand, the ligands like oxalate, acetylacetonate, biguanide, etc. may undergo protonation even when they remain coordinated because of the presence of **free basic sites.** Acid catalyzed aquation of such ligands becomes relatively easier. In fact, the biguanide complexes of Co(III) or Cr(III) are thermodynamically more stable than their corresponding ethylenediamine complexes. But in terms of the acid catalyzed aquation reaction, ***the biguanide complexes are more labile than the ethylenediamine complexes***.

(**biguanide chelate** possesses the free basic sites for protonation)

(**ethylenediamine chelate** possesses no tree basic site for protonation)

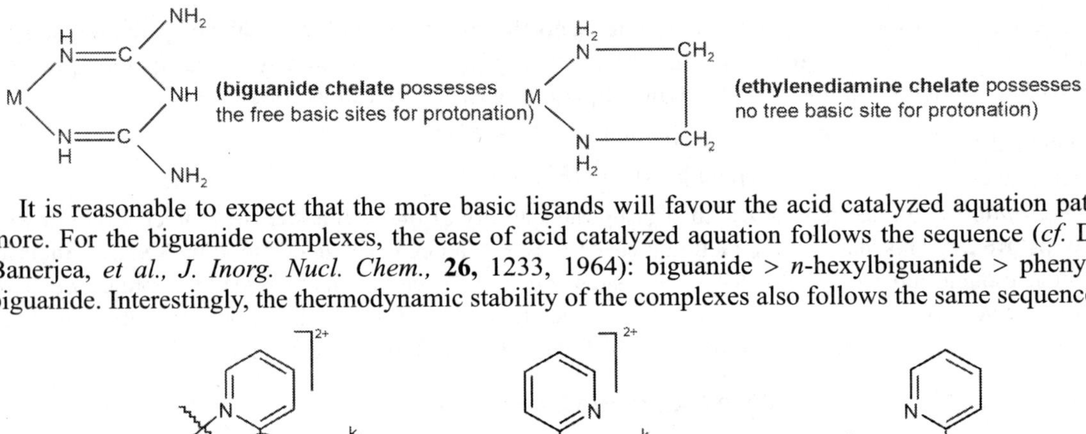

It is reasonable to expect that the more basic ligands will favour the acid catalyzed aquation path more. For the biguanide complexes, the ease of acid catalyzed aquation follows the sequence (*cf.* D. Banerjea, *et al., J. Inorg. Nucl. Chem.*, **26,** 1233, 1964): biguanide > *n*-hexylbiguanide > phenylbiguanide. Interestingly, the thermodynamic stability of the complexes also follows the same sequence.

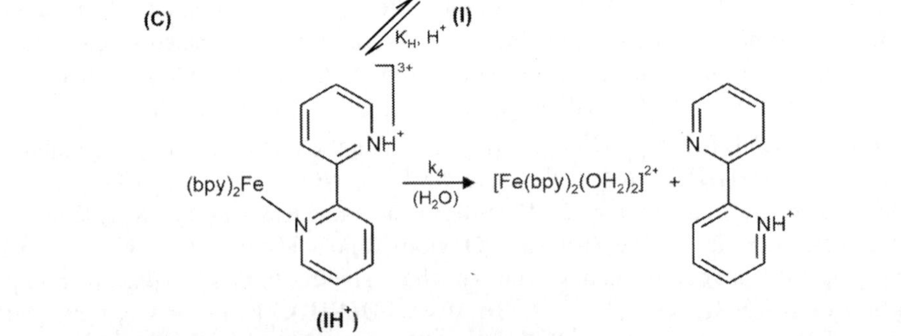

Scheme 5.23.1.1: Acid catalyzed dissociation of $[Fe(bpy)_3]^{2+}$ to $[Fe(bpy)_2(OH_2)_2]^{2+}$ by involving the **one ended dissociation** of the *bpy* chelate ring

- **Acid catalysed aquation of $[M(bpy)_3]^{2+}$ *vs.* $[M(phen)_3]^{2+}$ (M = Fe, Ni):** Acid catalyzed aquations of $[M(bpy)_3]^{2+}$, $[M(phen)_3]^{2+}$ (M = Fe, Ni) are quite intersecting in terms of acid dependence. ***The bpy-complex shows the acid dependence while the phen-complex does not show any acid dependent path*** (*cf.* J.H. Baxendale *et al.*, *Trans. Faraday Soc.*, **46,** 736, 1950; J.E. Dickens *et al.*, *J. Am. Chem. Soc.*, **75,** 1286, 1957).

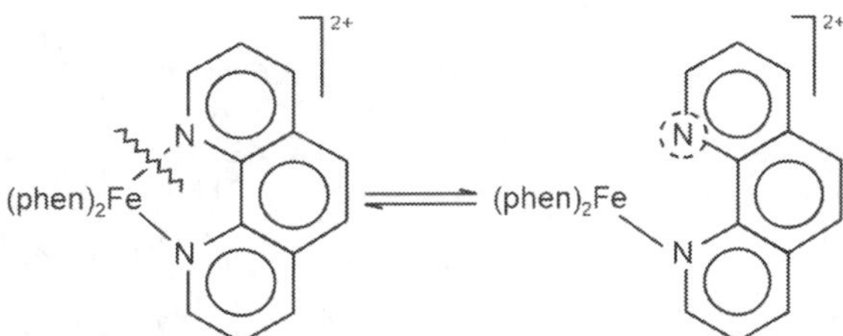

Scheme 5.23.1.2 Possibility of one-ended dissociation of the phen-chelate ring in $[Fe(phen)_3]^{2+}$. The dislodged N-site denoted by :N: is still within the coordination sphere as the rigidity of the structure prevents this dislodged site to go away from the coordination sphere. ***Thus, there is no real existence of the one-ended opening of the phen-chelate ring.***

In the *bpy* and *phen* complexes, for protonation of their complexes, partial dechelation (*i.e. one ended dissociation*) of the coordinated ligand is required. The ***bpy* complex** may experience such a **one ended dissociation** accompanied by protonation of the released end of the ligand. Then it experiences a fast rupture of the remaining metal-ligand bond leading to the complete loss of the ligand. This pathway is illustrated in scheme 5.23.1.1. ***It may be noted that the intermediate having a one ended dissociated ligand experiences a rotation around the C—C single bond to take away the dislodged N-site from the coordination sphere.*** This intermediate may also experience the uncatalyzed aquation.

In the case of *phen*-complexes, one ended dissociation pathway is not possible. *In phen, the rotation around the C—C bond is prevented due to its rigid structure where the third ring remains fused to the bpy-skeleton.* Thus, even if the *phen*-chelate ring experiences the said one ended dissociation, the dislodged N-site still remains wthin the coordination sphere. Thus, practically there is no meaning of such one ended dissociation of the *phen* ring. *For the loss of the phen ligand, both the 'M—N' bonds must rupture simultaneously at the rate determining step.* Then the completely dislodged *phen* ligand undergoes protonation rapidly and this prevents its reentry into the coordination sphere.

Thus presence of acid is required to prevent the overall backward reaction but not to catalyze the forward dechelation reaction kinetically.

$$[M(N—N)_3]^{2+} \rightleftharpoons [M(N—N)_2]^{2+} + N—N, \textbf{(slow)}\ (N—N = phen)$$

$$N—N + H^+ \rightleftharpoons N—NH^+, \textbf{(fast)}, (pK_a \ of\ N—NH^+ \approx 4)$$

By considering the **steady-state condition** to the species **(I)** (*cf.* Scheme 5.24.1.1) having a **half-bonded *bpy* ligand** (*i.e.* partially dissociated *bpy* ring), the observed rate constant of aquation is given by:

$$\frac{d(I)}{dt} = 0 = k_1[C] - k_2[I] - k_3[I] - k_4[IH^+]$$

$$K_H = \frac{[IH^+]}{[I][H^+]} \text{ or, } [IH^+] = K_H[I][H^+]$$

$$= k_1[C] - k_2[I] - k_3[I] - k_4K_H[I][H^+]$$

or

$$[I] = \frac{k_1[C]}{(k_2+k_3)+k_4K_H[H^+]}$$

$$\text{Rate} = k_3[I] + k_4[IH^+]$$

$$= \frac{k_1k_3[C]}{(k_2+k_3)+k_4K_H[H^+]} + \frac{k_1k_4K_H[H^+][C]}{(k_2+k_3)+k_4K_H[H^+]}$$

$$= \frac{k_1\left(k_3+k_4K_H[H^+]\right)[C]}{(k_2+k_3)+k_4K_H[H^+]}$$

i.e. $$k_{obs} = \left\{\frac{k_3+K_Hk_4[\mathrm{H}^+]}{(k_2+k_3)+K_Hk_4[\mathrm{H}^+]}\right\}k_1 \text{ when, } [\mathrm{H}^+] \gg [C]$$

- At high acidity, $K_Hk_4[H^+] \gg k_3$ and $K_Hk_4[H^+] \gg (k_2 + k_3)$ and it leads to:

  $k_{obs} = k_1$, (*i.e.* **acid independent**)

- At very low acidity, $k_3 \gg K_Hk_4[H^+]$ and $(k_2 + k_3) \gg K_Hk_4[H^+]$ and it leads to:

  $$k_{obs} = \frac{k_1k_3}{k_2+k_3} \text{ (\textit{i.e.} \textbf{acid independent}).}$$

- At the intermediate range of $[H^+]$, k_{obs} shows a **complicated pattern of acid dependence** (*cf.* Fig. 5.24.1.1).

The **limiting values** of k_{obs} are:

$$\mathbf{k_{obs}}\ (\textbf{lower limit}) = \frac{k_1k_3}{k_2+k_3} \text{ (at very low acidity)}$$

$$\mathbf{k_{obs}}\ (\textbf{upper limit}) = k_1 \text{ (at very high acidity)}$$

$$\frac{k_{obs}\ (\text{lower limit})}{k_{obs}\ (\text{upper limit})} = \frac{k_3}{k_2+k_3} \approx \frac{k_3}{k_2} \approx 0.16$$

(experimentally found)

It signifies that at very low acidity, from the 5-coordinate intermediate (I), approximately 84% chances are for reforming the bond to give back the *tris*-chelate and approximately 16% chances are for the complete dissociation to give the *bis*-complex.

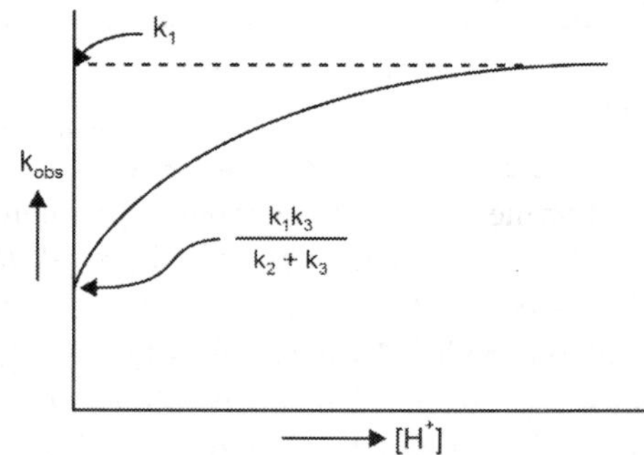

Fig. 5.23.1.1 Acid dependence of the aquation rate of $[Fe(bpy)_3]^{2+}$.

- **Aquation of $[Ni(N—N)_3]^{2+}$ *vs.* $[Fe(N—N)_3]^{2+}$:** (i) For dechelation of the *tris*-chelate of Ni(II) to the *bis*-chelate, it follows the same behaviour as found for the corresponding Fe(II) complexes.

(ii) The Ni(II) complexes differ from the Fe(II) complexes in the following aspect.

$[Fe(N—N)_3]^{2+}$ (low spin) —Step-I (slow), H^+, $-$ N—NH$^+$→ $[Fe(N—N)_2(OH)_2]^{2+}$ (high spin) —Step-II (fast), H^+, $-$ N—NH$^+$→ $[Fe(N—N)(OH_2)_4]^{2+}$ (high spin) —Step-III (fast), H^+, $-$ N—NH$^+$→ $[Fe(OH_2)_6]^{2+}$

(N——N = bpy, phen)

Here the lability sequence runs as:

Step I ≪ Step-II, Step-III

i.e. ***the bis– and mono-chelates are more labile than the tris-chelate.*** *It happens so because the tris-chelate is a low-spin complex having a very high crystal field activation energy while the lower complexes* *are the high-spin complexes having a relatively smaller crystal field activation energy.* In the case of corresponding Ni(II)-complexes, it does not occur so because there is no such spin-state change in moving from the higher complexes to the lower complexes.

5.23.2 Metal Ion Catalyzed Aquation Reactions (*cf.* D. Banerjea, *Transition Met. Chem.*, 12, 97, 1987)

The didentate and the multidentate ligands like oxalate, malonate, acetylacetonate, aminopolycarboxylate may bind with the catalyzing labile metal ions (like Mn^{2+}, $Fe^{2+/3+}$, Co^{2+}, Ni^{2+}, Cu^{2+}, Zn^{2+}, etc. which are **labile**) to produce the **binuclear bridged intermediates** which subsequently undergo aquation. It is illustrated below for the **oxalato complex** like the Cr^{III}- or Co^{III}-oxalato complex (**inert complex**).

M(oxalate) + M_{cat} ⇌ M(μ-oxalate)M_{cat} —(slow) (H_2O)→ M(aq) + M_{cat}(oxalate) → M(aq) + M_{cat}(aq) + oxalate (**for the labile M_{cat}**)

(Charges of the metal ions not shown; M_{cat} denotes the catalysing metal ion)

For the **biguanide complex,** the intermediate binuclear complex may be represented as follows:

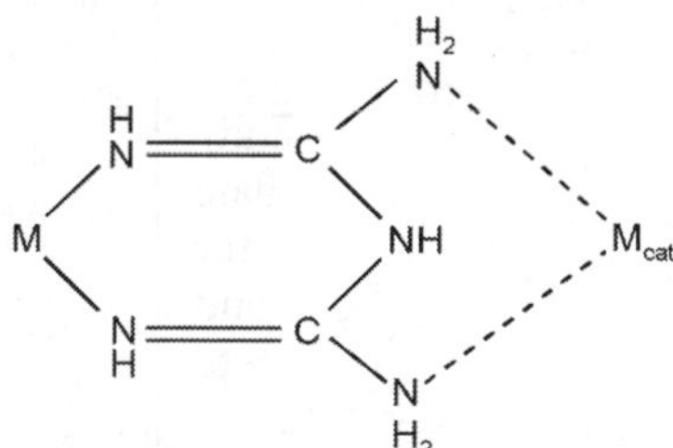

The catalyzing power of M_{cat} depends on HSAB matching and its complexing power that can be measured by Z^{*2}/r where $Z^* = Z -$ shielding constant and **Irving-Williams stability order** (see Vol. 4).

The catalysing power of metal ions parallels the thermodynamic stabilities of their complexes as illustrated for the metal ion catalysed dissociation of the oxalato complexes of Co(III) or Cr(III): Mn(II) < Co(II) < Ni(II) < Cu(II) > Zn(II) (**Irving-Williams stability order. The catalysing power follows the order: Fe(III) > Al(III) > In(III)** (*cf.* Z^{*2}/r ratio).

Effect of the nature of the catalyzing metal ion on the reaction pathway: The metal ion catalyzed aquation of $[Co(NH_3)_5X]^{2+}$ (X = Cl, Br, I) furnishes some information regarding the transition state. In presence of Hg^{2+} as a catalyst, the aqua product $[Co(NH_3)_5(OH_2)]^{3+}$ bears the same ratio of $^{16}O/^{18}O$ for different halides when the aquation reaction is carried out in water enriched in ^{18}O content. This observation supports the **perfect D-process,** *i.e.* in attaining the transition state, the leaving group X^- is completely dislodged so that, it cannot influence the aquation process.

Interestingly, in presence of the catalyzing metal ions like Ag^+ or Tl^+, the ratio of $^{16}O/^{18}O$ content in the aqua product was found to depend on the nature of the leaving group X^-. ***It indicates that though a common intermediate is produced for the Hg^{2+} catalyzed aquation of $[Co(NH_3)_5X]^{2+}$, no such***

common intermediate is produced for the Ag^+ or Tl^{3+} catalyzed aquation process. In fact, for the Ag^+ or Tl^+ catalyzed aquation reaction, in the transition state, both the leaving group and entering nucleophile (H_2O) remain bound as in the **interchange process** where the nucleophilic attack by H_2O is relatively less important than the bond breaking by the leaving group. Thus the suggested mechanism is I_d in which the leaving group is not completely dislodged at the transition state. ***This is why, the partially dislodged leaving group at the transition state influences the entry of the nucleophile (i.e. H_2O).*** This extent of dislodgement (catalyzed by the metal ion) of the leaving group depends on the nature of the leaving group and catalyzing metal ion. For this reason, for the reaction passing through the I_d-path, the ratio of ^{16}O and ^{18}O in the product depends on the nature of the leaving group for a particular metal ion catalysing the process.

5.24 PSEUDO-SUBSTITUTION REACTION WITHOUT THE METAL-LIGAND BOND BREAKING THROUGH THE ELECTROPHILIC ATTACK ON THE BOUND LIGAND

The ligand substitution process has been found in many cases to proceed fast apparently even for the inert metal centres like Co(III), Rh(III), Ir(III). These are explained by considering the **pseudo-substitution** reaction where the actual metal-ligand bond does not rupture. Some representative examples are shown below:

(i) **Acid catalyzed aquation of $[M(CO_3)(NH_3)_5]^+$ (M = Co, Rh, Ir)** (*cf.* A.B. Lamb *et al.*, *J. Am. Chem. Soc.*, **67**, 468, 1945): The said carbonato complexes experience a rapid aquation (*i.e.* acid hydrolysis) to form the aqua-complex, $[M(NH_3)_5(OH_2)]^{3+}$.

$$[M(CO_3)(NH_3)_5]^+ \xrightarrow{H_3O^+} [M(NH_3)_5(OH_2)]^{3+} + CO_2\,(aq).$$

$$\left[(H_3N)_5M{-}O{-}C(=O)O^-\right]^+ + H^+ \underset{}{\overset{\text{(fast)}}{\rightleftharpoons}} \left[(H_3N)_5M{-}O{-}C(=O)OH\right]^{2+}$$

$$\left[(H_3N)_5M{-}O{-}C(=O)OH\right]^{2+} \overset{\text{(fast)}}{\rightleftharpoons} \left[(H_3N)_5M{-}\overset{+}{O}(H){-}C(=O)O^-\right]^{2+}$$

$$\left[(H_3N)_5M{-}OH\right]^{2+} \underset{+CO_2\,\text{(rds)}}{\overset{-CO_2\,\text{(rds)}}{\rightleftharpoons}} \left[(H_3N)_5M{-}\overset{+}{O}(H){-}C(=O)O^-\right]^{2+}$$

$$\left[(H_3N)_5M{-}OH\right]^{2+} \underset{-H^+}{\overset{+H^+\,\text{(fast)}}{\rightleftharpoons}} \left[(H_3N)_5M{-}OH_2\right]^{3+}$$

When the said aquation reactions are carried out in presence of H_2O^*, no O* is found in the aqua product. It supports that the M—O bond remains intact and cleavage of the 'C—O' bond gives off CO_2. The probable mechanism involves the *electrophilic attack* by the proton (*i.e.* H_3O^+) on the O-atom of the bound carbonate ligand.

In fact, formation and aquation of the bicarbonato/carbonato complexes can be shown as follows:

$$\left[(H_3N)_5M{-}{}^*OH_2\right]^{3+} + HCO_3^- \underset{k_d}{\overset{k_f}{\rightleftharpoons}} \left[(H_3N)_5M{-}O^*{-}CO_2^-\right]^{+} + H^+ + H_2O$$

The k_f (in s^{-1}) and k_d (in $dm^3\,mol^{-1}\,s^{-1}$) values at 25°C for the said complexes are as follows:

M(III) =	**Co(III)**	**Rh(III)**	**Ir(III)**
k_f (s^{-1}):	1.10	1.13	1.45
k_d ($dm^3\,mol^{-1}\,s^{-1}$):	220	470	590

It is evident that the k_f and k_d values are very much comparable for Co(III), Rh(III) and Ir(III). The lability order is: Co(III) ≫ Rh(III) ≫ Ir(III). *Thus, from the comparable values of k_f and k_d, it is reasonable to conclude that for the said processes, the metal-ligand bond* (*i.e.* M—O bond) *replacement does not occur.* For the aquation of $[MBr(NH_3)_5]^{2+}$, the aquation rates follow the sequence.

$$k_{aq}\ (s^{-1}),\ 25°C: 6.3 \times 10^{-6}\ (Co),\ 1 \times 10^{-8}\ (Rh)\ and\ 2 \times 10^{-10}\ (Ir).$$

In these bromido-complexes, aquation goes through the M—Br bond replacement. Similarly, aquations of $[Co(NH_3)_5(SO_4)]^+$, $[Co(NH_3)_5(NO_3)]^{2+}$ and $[Co(CH_3COO)(NH_3)_5]^{2+}$ proceed through the 'Co—O' bond replacement and the processes are slow as usual.

(ii) **Formation of $[M(NH_3)_5(—ONO)]^{2+}$ from $[M(NH_3)_5(OH_2)]^{3+}$ by HNO_2–NO_2^- buffer:** The nitrito-isomer, $[M(NH_3)_5(ONO)]^{2+}$ is thermodynamically less stable than the corresponding nitro or nitrito(κN)-complex (*i.e.* linkage isomer). In the said reaction, the nitrito(κO) isomer (*kinetically controlled product*) is produced at a fairly fast reaction through the **pseudo-substitution process:**

$$\text{Rate} \propto [\text{complex}][HNO_2]^2 \quad (i.e.\ \text{second order dependence on } [HNO_2])$$

It may be noted that **nitrosation of amines** also shows the second order dependence on $[HNO_2]$ (see Chapter 10, Vol. 2).

The pseudo-substitution reaction **maintaining the second order dependence** on $[HNO_2]$ can occur through the following mechanistic paths.

Dehydration of HNO_2 produces N_2O_3 that readily undergoes acid catalysed decomposition to produce NO^+ and HNO_2. NO^+ and/or N_2O_3 can participate to carry out an O-nitrosation type reaction desdcribed as the **pseudo-substitution reaction**.

Dehydration of HNO_2 to produce N_2O_3 (nitrogen trioxide) can occur in the following pathway (see Vol. 2, Sec. 10.10.5).

$$HNO_2 + H^+ \longrightarrow H_2O^+\text{—}\ddot{N}{=}O \quad (i.e.\ H_2NO_2^+)$$

$$O{=}\ddot{N}\text{—}OH + {}^+H_2O\text{—}\ddot{N}{=}O \xrightarrow{\text{I-process}} O_2N\text{—}\ddot{N}{=}O + H_2O + H^+ \quad (i.e.\ N_2O_3)$$

$$H^+ + N_2O_3 \longrightarrow {}^+H\cdots O_2N\text{—}NO \longrightarrow HNO_2 + NO^+$$

The overall reaction involving NO^+ is:

$$2HNO_2 \rightleftharpoons H_2O + NO_2^- + NO^+$$

$$[(H_3N)_5M\text{—}OH_2]^{3+} \rightleftharpoons [(H_3N)_5M\text{—}\ddot{O}H]^{2+} + H^+$$

$$\downarrow \text{O-Nitrosation} \quad NO^+$$

$$[(H_3N)_5M\text{—}ONO]^{2+} \xleftarrow{-H^+} [(H_3N)_5M\text{—}O(H)(NO)]^{3+}$$

(M = Cr, Co, Rh, Ir, etc.)

Formation of the said nitrito(κO) complex may be also explained by taking the reactive species N_2O_3 as follows (illustrated for the Co(III)-complex):

O-Nitrosation:

$$\left[(H_3N)_5Co^{III}—{}^*O(H)\cdots O{=}N—N(O)O\right]^{2+} \rightleftharpoons \left[(H_3N)_5Co^{III}—{}^*O(H)\cdots O{=}N\cdots N(O)O\right]^{2+} \xrightarrow{rds} [(H_3N)_5Co—{}^*ONO]^{2+} + HNO_2$$

N_2O_3 (blue liquid or solid): It can exist in two isomeric forms: **asymmetric** and **symmetric**. The asymmetric form is the predominant one. This is why, it readily dissociates to form NO and NO_2 through the **homolytic fission** of the N—N bond. In presence of H^+, the **heterolytic cleavage** of the N—N bond produces NO^+ and NO_2^-. The **asymmetric $ONNO_2$ is converted into the symmetric isomer $O(NO)_2$** when irradiated at 720 nm and is reconverted to the asymmetric isomer when irradiated at 380 nm.

$$\underset{\text{(asymmetric)}}{O_2N—NO} \underset{380\ nm}{\overset{720\ nm}{\rightleftharpoons}} \underset{\text{(symmetric)}}{O{=}N—O—N{=}O}$$

The reaction involving NO^+ occurs through the nucleophilic attack by the coordinated-OH group on NO^+ followed by the release of a proton.

$$[(H_3N)_5Co^{III}—{}^*O(H)]^{2+} + NO^+ \xrightarrow{-H^+} [(H_3N)_5Co—{}^*ONO]^{2+}$$

Thus the real substitution reaction outlined below is ruled out.

$$[(H_3N)_5Co—{}^*OH]^{2+} + NO_2^- \rightarrow [(H_3N)_5Co—NO_2]^{2+} \text{ or } [(H_3N)_5Co—ONO]^{2+} + {}^*OH^-$$

For Co(III), the nitrito(κO) isomer is **red coloured** and the nitro isomer is **yellow coloured.** In this pseudo-substitution process, the M—O bond does not break down. The proposed mechanism has been supported by the fact that when the reaction is carried out in water enriched with H_2O^*, there is no incorporation of O* in the product.

Here it may be mentioned that formation of the nitrito(κO) isomer occurs fairly fast even at very low temperature but on heating, the nitrito(κO) isomer **slowly** (as expected from the inertness of low-spin d^6 metal centres) isomerizes to the nitro or nitrito(κN) isomer through the M—O bond rupture.

$$\underset{\textbf{(Red)}}{[(H_3N)_5M—OH_2]^{3+}} \xrightarrow[\text{fast, (low temp.)}]{HNO_2—NO_2^-} [(H_3N)_5M—ONO]^{2+}, \textbf{ (Kinetically controlled product)}$$

$$\downarrow \text{(slow) Heating}$$

$$\underset{\textbf{(Yellow)}}{[(H_3N)_5M—NO_2]^{2+}}, \textbf{ (Thermodynamically controlled product)}$$

(iii) Catalyzed aquation of $[(H_3N)_5M(-ONO)]^{2+}$: The aquation process can be catalyzed by Cl^-, Br^- in acidic condition. It goes on through the pseudo-substitution process as outlined below.

$$\left[(H_3N)_5M—\overset{*}{O}NO\right]^{2+} \xrightarrow{H^+,\ X^-} \left[(H_3N)_5M—\overset{*}{O}(H)—N(X)(=O)\right]^{2+}$$

(M = Co, Cr)

$$\xrightarrow{-\ NOX} \left[(H_3N)_5M—\overset{*}{O}H\right]^{2+} \xrightarrow{H_3O^+} \left[(H_3N)_5M—\overset{*}{O}H_2\right]^{3+}$$

NOX is rapidly hydrolyzed to HNO_2 and HX. In this aquation process also, there is no metal-ligand bond rupture.

5.25 BASE HYDROLYSIS OF THE OCTAHEDRAL COBALT(III) AMMINE COMPLEXES

We have already mentioned (*cf.* Sec. 5.20) that in the rate law for the hydrolysis of the complex like $[Co(NH_3)_5X]^{2+}$, the base hydrolysis term is: k_B**[complex][OH⁻]**. The base hydrolysis rate constant is about 10^5–10^6 times higher than the acid hydrolysis rate constant (*i.e.* aquation process).

For $[CoCl(NH_3)_5]^{2+}$, $k_{aq} \approx 10^{-6}\ s^{-1}$, $k_B \approx 0.25\ dm^3\ mol^{-1}\ s^{-1}$ at 25°C.

The term $k_B[\text{complex}][OH^-]$ may arise either due to a **genuine S_N2 (*i.e.* A) rate determining step or due to the existence of a pre-equilibrium involving OH^- followed by a S_N1 (*i.e.* D) type rate determining step.** In the pre-equilibrium, proton abstraction from the coordinated NH_3 ligand leads to the formation of ***a conjugate base (CB)*** $[Co(NH_2)(NH_3)_4X]^+$ which subsequently participates in the rate determining dissociative step leading to the expulsion of the leaving group X^-. Then the 5-coordinate intermediate experiences a nucleophilic attack at a faster step. This idea was first proposed by *Garrick* and then developed and popularized by *Basolo and Pearson*. This process of mechanism of base hydrolysis is called **S_N1–CB** (*i.e.* **D–CB** in modern nomenclature). The mechanistic steps **(Garrick mechanism)** are outlined below:

X^- = halide, N_3^-, $OSO_2CF_3^-$ etc.

$$\underset{\text{(Complex)}}{\left[Co(NH_3)_5X\right]^{2+}} + OH^- \underset{}{\overset{K_{CB}\ \text{(fast)}}{\rightleftharpoons}} \underset{\textbf{(CB)}}{\left[Co(NH_2)(NH_3)_4X\right]^{+}} + H_2O$$

$$\left[Co(NH_2)(NH_3)_4X\right]^{+} \xrightarrow[(rds)]{k} \underset{\textbf{(5 coordinate intermediate)}}{\left[Co(NH_2)(NH_3)_4\right]^{2+}} + X^-$$

$$\left[Co(NH_2)(NH_3)_4\right]^{2+} \xrightarrow[\text{(fast)}]{+H_2O} \left[Co(NH_2)(NH_3)_4(OH_2)\right]^{2+} \xrightarrow[\text{fast}]{-H^+} \left[Co(NH_3)_5(OH)\right]^{2+}$$

$K_{CB} = \dfrac{K_a}{K_w}$, K_a = Corresponding acid dissociation constant of the coordinated NH_3 ligand

The rate law of the above process is given below.

$$\text{Rate} = k[CB] \text{ and } K_{CB} = \frac{[CB]}{[\text{complex}][OH^-]}$$

$$[\text{complex}]_T = [\text{complex}] + [\text{CB}]$$

$$= \frac{[\text{CB}]}{K_{CB}[\text{OH}^-]} + [\text{CB}] = [\text{CB}]\left\{\frac{1 + K_{CB}[\text{OH}^-]}{K_{CB}[\text{OH}^-]}\right\}$$

i.e. $$[\text{CB}] = \frac{[\text{complex}]_T K_{CB}[\text{OH}^-]}{1 + K_{CB}[\text{OH}^-]}, [\text{complex}]_T \ll [\text{OH}^-], \textit{ i.e. } [\text{OH}^-]_T \approx [\text{OH}^-]$$

i.e. $$\text{Rate} = \frac{kK_{CB}[\text{complex}]_T[\text{OH}^-]}{1 + K_{CB}[\text{OH}^-]} = k_B[\text{complex}]_T[\text{OH}^-].$$

Free NH_3 is a very weak Brönsted acid, *i.e.* its proton releasing power is negligibly small. But the metal coordinated NH_3 is a relatively a stronger acid. It happens so for the metal coordinated H_2O. Obviously, this acid strength increases for the higher valent metal centre (*cf.* pK_a (free water) = 15.74, pK_a of $[Fe(OH_2)_6]^{3+} = 2.2$).

pK_a of the ammine proton in the complex is *ca.* 15–16 (*cf.* pK_a of free $NH_3 \ggg 15$). Consequently, $\boldsymbol{K_{CB}}$ **will have a very low value** (*ca.* $10^{-1} - 10^{-2}$ mol^{-1} dm^3) and the condition, $1 \gg K_{CB}[OH^-]$, will be maintained even for the reasonably high concentration of $[OH^-]$. Thus we can write:

$$\text{Rate} = -\frac{d[\text{complex}]}{dt} = kK_{CB}[\text{complex}]_T[\text{OH}^-]$$

$$= k_B[\text{complex}][\text{OH}^-];$$

(when K_{CB} is very small, $[\text{complex}]_T \approx [\text{complex}]$ and $[\text{OH}^-] \gg [\text{complex}]_T$)

(**Note:** If the conjugate base (*CB*) is supposed to remain at a steady state $\left(\textit{i.e. } \frac{d[CB]}{dt} = 0\right)$ having $K_{CB} = \frac{k_1}{k_{-1}}$, we get the following rate expression as follows:

$$\frac{d[\text{CB}]}{dt} = 0 = k_1[\text{complex}][\text{OH}^-] - k_{-1}[\text{CB}][\text{H}_2\text{O}] - k[\text{CB}]$$

i.e. $$[\text{CB}] = \frac{k_1[\text{complex}][\text{OH}^-]}{k_{-1}[\text{H}_2\text{O}] + k}; \text{Rate} = k[\text{CB}].$$

It gives: $$\text{Rate} = \frac{k_1k[\text{complex}][\text{OH}^-]}{k_{-1}[\text{H}_2\text{O}] + k}$$

$$= k_B[\text{complex}][\text{OH}^-], \text{ where } k_B = \frac{kk_1}{k + k_{-1}[\text{H}_2\text{O}]}$$

- **Acid hydrolysis *vs.* base hydrolysis** of *trans*-$[Co(en)_2Cl_2]^+$:

trans- $[Co(en)_2Cl_2]^+$ (green) —[H_2O, pH « 7, (acid hydrolysis)]→ $[Co(en)_2Cl(OH_2)]^{2+}$ (**pink**), $t_{½}$ = seven hours (Simple **dissociative path**)

trans- $[Co(en)_2Cl_2]^+$ (green) —[H_2O, pH » 7, (**base hydrolysis**)]→ $[Co(en)_2Cl(OH)]^+$ (**purple**), (Instantaneous, **D-CB path**)

Base hydrolysis in terms of the D–CB path is outlined as follows (see Fig. 5.25.3.1 for details of the **stereochemistry of the products):**

$$[Co(en)_2Cl_2]^+ + OH^- \underset{}{\overset{\text{fast}}{\rightleftharpoons}} \underset{\text{(CB)}}{[Co(en)(en\text{-}H)Cl_2]} + H_2O; \text{ (en–H denotes } H_2N—CH_2CH_2—NH^-)$$

$$[Co(en)(en\text{–}H)Cl_2] \xrightarrow[\text{(rds)}]{\text{slow}} [Co(en)(en\text{–}H)Cl]^+ + Cl^-$$

$$[Co(en)(en\text{–}H)Cl]^+ \xrightarrow[\text{(fast)}]{H_2O/OH^-} [Co(en)_2Cl(OH)]^+$$

5.25.1 Characteristics of the D–CB Process

Proton abstraction can occur from the coordinated NH_3, RNH_2 or R_2NH ligand depending on the condition. Here it is illustrated for proton abstraction from the coordinated NH_3 ligand in $[Co(NH_3)_5X]^{2+}$.

- **Enhanced reactivity:** k_{OH} (for base hydrolysis) $\gg$ k_{aq} (acid hydrolysis) and $k_{anation}$ (acidic condition). The proposed D–CB mechanism is to explain the above reactivity order:

$$\left.\begin{array}{l}[Co(NH_3)_5Cl]^{2+} + OH^- \longrightarrow [Co(NH_3)_5(OH)]^{2+} + Cl^-, k_{OH} \\ [Co(NH_3)_5Cl]^{2+} + F^- \longrightarrow [Co(NH_3)_5F]^{2+} + Cl^-, k_F\end{array}\right\} k_{OH} \gg k_F$$

  F^- and OH^- are comparable **in terms of size and nucleophilicity** but replacement of Cl^- by OH^- is much faster than the replacement by F^- (*i.e.* $k_{OH} \gg k_F$). It happens so because of their different mechanistic paths of substitution.

- **Strict first-order in $[OH^-]$:** Here it should be mentioned that in the said D–CB mechanism, K_{CB} value is exceedingly small (*ca.* 10^{-1}–10^{-2} mol^{-1} dm^3). Consequently, **saturation kinetics** will never be attained, *i.e.* it will maintain strictly a first-order dependence on $[OH^-]$ for an appreciable range of OH^-.

- **Deprotonation site to generate the CB:** The experimental evidences indicate that deprotonation occurs generally from the NH_3 group present in the *trans*-position to the leaving group *X*. But there are some evidences for proton abstraction from the *cis*-NH_3 group (with respect to the leaving group).

- **Role of the $—\ddot{N}H_2$ group (amido group) in CB for rate enhancement:** Formation of the CB ***reduces the overall cationic charge*** (*cf.* $[CoCl(NH_3)_5]^{2+}$ *vs.* $[CoCl(NH_2)(NH_3)_4]^+$) and it favours the **dissociative expulsion** of the leaving group (*i.e.* **charge effect**). Besides this charge effect, the NH_2^- group (amido group) shows the **electronic effects** to favour the rate process. The NH_2^- group is a strong **σ-donor** ligand in the CB and also a strong **pi-donor ligand** (*cf.* NH_3 is only a σ-donor ligand). **This strong σ-donor property labilizes the *trans*-position (*trans*-effect)** and the π-donor property labilizes the *cis*-position (*i.e.* ***cis*-effect**). This labilizing effect of the NH_2^- group present in the CB favours the **dissociative activation** and explains the unusually high rate of the base hydrolysis process. **The *trans*-effect (*i.e.* σ-donor property) is probably more important** than the *cis*-effect (*i.e.* π-donor property) in this labilisation process in the CB. ***This explains why deprotonation preferably occurs at the trans-prosition*** (see Scheme 5.25.1.1). However, deprotonation from the *cis*-position **(with respect to *X*) is not completely ruled out** (discussed later).

- **Geometry of the five coordinate intermediate:** Through the expulsion of the leaving group from the CB at the rate determining step, it adopts the *trigonal bipyramidal geometry* where the NH_2^- group occupies one equatorial position (unless the other nonleaving ligands are too rigid to rearrange the square pyramid initially formed to the trigonal bipyramid). The trigonal bipyramidal (TBP) geometry is **sufficiently long-lived** to allow the Berry's pseudorotation. ***The trigonal***

bipyramidal intermediate is stabilized by the π-donor NH_2^- group present at the equatorial site (see Fig. 5.21.1b). In a TBP geometry (*cf.* $sp^2 + pd$; sp^2-hybrid orbitals towards the equatorial ligands; ***pd***-hybrid orbitals towards the axial ligands); ***the metal centre can receive electrons from the equatorial ligands better than from the axial ligands through the σ-bond (cf. Bent's rule). Thus the σ-donor property of the NH_2^- group residing at the equatorial position of the TBP also stabilises the TBP intermediate.*** Formation of this long-lived five-coordinate intermediate (trigonal bipyramidal geometry) explains the **isomerization that occurs for both the *cis*- and *trans*-complex.** [*cf.* In the aquation process (see Sec. 5.21), isomerization is noted generally in the *trans*-complexes like *trans*-$[Co(NH_3)_4LX]^{n+}$ (L = π-donor nonleaving group). For the corresponding *cis*– complex, 100% retention of the configuration is noted in the aquation process].

Intermediate TBP *vs.* SP: The TBP intermediate is stabilised by both the σ- and π-donor properties of the NH_2^- group placed at the equatorial position while the SP intermediate is mainly stabilised by the π-donor property of the NH_2^- group (*cis*-effect). The initially produced SP through the expulsion of the leaving group from the *trans*-position of the NH_2^- group of CB (σ-*trans*-effect) is more readily converted into the TBP geometry than the SP initially produced through the expulsion of the leaving group from the *cis*-position of the NH_2^- group of CB because this SP is already stabilised to some extent by the π-donor property of the NH_2^- group (*cis*-effect).

- **High $\Delta S^{\neq}$:** The initially generated square pyramidal (SP) intermediate (through the expulsion of the leaving group from the CB) rearranges to the **long-lived and more stable trigonal bipyramidal (TBP) intermediate** (stabilised by the NH_2^- group occupying one equatorial position by its both σ- and π-donor property) which may experience the Berry pseudorotation. Formation of such rearranged TBPs is indicated by high $\Delta S^{\neq}$.
- **Comparable reactivity and isomerisation of both *cis*- and *trans*-$[Co(en)_2L–X]^{n+}$** (*cf.* Table 5.25.3.1)**:** If L is a nonleaving π-donor ligand, then the *cis*-complex reacts much faster than the *trans*-isomer in the aquation process. *But in contrast to the aquation process* (*i.e.* acid hydrolysis), *both the cis- and trans-isomers react at comparable rates in base hydrolysis.* A good π-donor ligand at the *cis*-position to the leaving group can stabilize the 5-coordinate square pyramid (*cis*-effect) without rearrangement (*cf.* acid hydrolysis) to a trigonal bipyramid.

  In the base hydrolysis, one of the amine groups (RNH_2) of the coordinated en-ligands loses a proton to produce an **amido group (RNH^-) at the *trans*-position of the leaving group and this RNH^- group of the CB can act as a strong σ-donor site as well as a strong π-donor ligand to stabilise the trigonal bipyramidal (TBP) intermediate (produced from the CB by the expulsion of the leaving group X) by occupying an equatorial position** (See Figs. 5.21.1b, 5.21.4, 5, 5.25.3.1 and 5.28.2.1).

  The stabilized TBP intermediate is sufficiently long-lived to experience the **Berry's pseudorotation** keeping the amido group (RNH^-) (having the strong **σ-donor and π-donor property**) at the equatorial position and the two TBP intermediates remain in an equilibrium (Fig. 5.25.3.1). This rearrangement is supported by **high entropy of activation ($\Delta S^{\neq}$)**. Both the *cis*- and *trans*-isomers can produce the same TBP intermediates through the Berry's pseudorotation. This rearrangement of the TBP intermediate explains the **comparable reactivity** and the ***cis-trans* isomerism** for both the *cis*- and *trans*-isomers (*cf.* Table 5.25.3.1 and different observations in aquation or acid hydrolysis).
- **Steric acceleration:** Base hydrolysis of $[CoCl(RNH_2)_5]^{2+}$ is faster than that of $[CoCl(NH_3)_5]^{2+}$. This steric acceleration supports the dissociative activation of the CB. The steric crowding for

the R–NH_2 ligands favours (*i.e.* **steric acceleration**) the formation of 5 coordinate intermediate in the dissociative path.

- **LFER plot:** The linear plot **log k_{OH} *vs.* log K_{OH}** with the slope close to unity supports the dissociative activation. The linear plot, **log k_{OH} *vs* log k_{aq}** indicates the common mechanism (*i.e.* *d*-activation) for both the base hydrolysis and aquation.
- **ΔH_T (enthalpy of transition = $\Delta H^{\neq} - \Delta H$):** In the base hydrolysis of $[Co(NH_3)_5X]^{2+}$, ΔH_T is found constant for the different leaving groups (X^-) (see Fig. 5.19.3). This supports the dissociative activation as in acid hydrolysis.
- **Entering ligand is H_2O not OH^-:** In the base hydrolysis product, OH^- group is present but OH^- is not the entering group. In fact, H_2O as a nucleophile attacks at the 5-coordinated intermediate at a faster step.

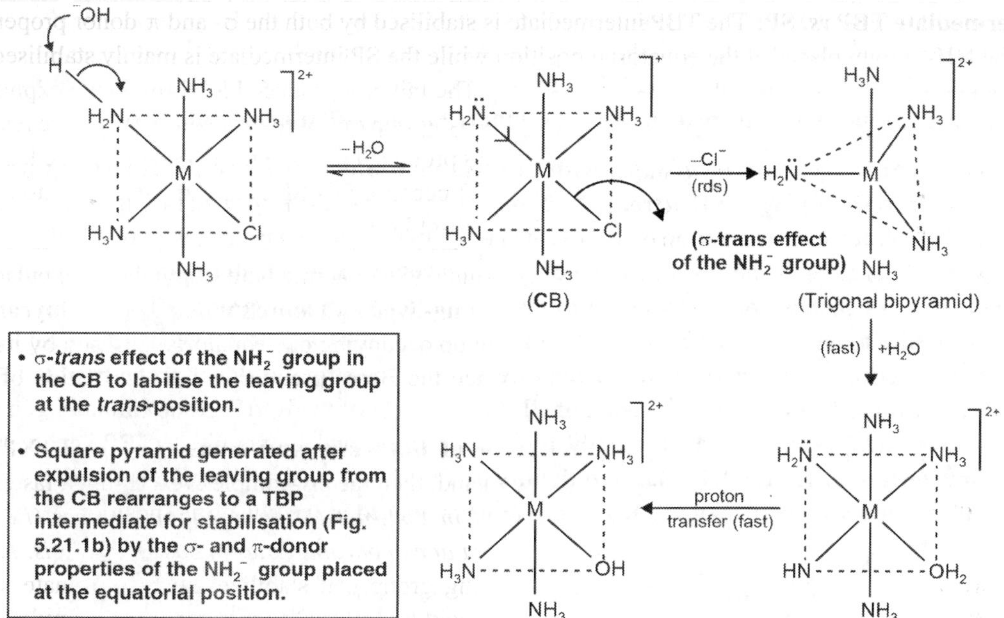

Scheme 5.25.1.1: Schematic representation of the D–CB mechanism **(Garrick mechanism)** for the base hydrolysis of $[CoCl(NH_3)_5]^{2+}$

That the OH group incorporated into the product comes from the nucleophilic attack (at a faster step) by H_2O not by ^-OH is evidenced from the use of ^{18}O labelled water. The isotope exchange equilibrium is:

$$H_2{}^{16}O + {}^{18}OH^- \rightleftharpoons H_2{}^{18}O + {}^{16}OH^-, \ K_{ex} = 1.04$$

The equilibrium constant ($K_{ex} > 1$) indicates that $H_2{}^{18}O$ is less acidic than $H_2{}^{16}O$**,** so hydroxide contains the less ^{18}O content, *i.e.*

$$\left(\frac{{}^{18}O}{{}^{16}O}\right)_{hydroxide} < \left(\frac{{}^{18}O}{{}^{16}O}\right)_{water}$$

The $^{18}O/^{16}O$ ratio incorporated into the product complex is in conformity with the ratio found in water not with the ratio found in hydroxide. It supports the fact that water (not hydroxide) enters as the nucleophile.

- **Base catalysed proton exchange in $[Co(NH_3)_5X]^{2+}$:** It has been observed that the base catalysed proton exchange between the solvent (H_2O) and coordinated NH_3 ligands **occurs faster at the *trans*-**

position of X^- than at the *cis*-position due to the higher acidity of N—H group at the *trans*-position (an electron withdrawing inductive effect of *trans*-ligand X^-). This suggests that formation of the CB is more favoured through the proton abstraction from the *trans*-position to X^- than from the *cis*-position to X^-.

Driving forces behind the preferred deprotonation at the *trans*-position of the leaving group to generate the reactive intermediate CB: (i) Strong σ-*trans*-effect of the deprotonated site (NH_2^- or RNH^-) in the CB favours the dissociative expulsion of the leaving group to produce a square pyramidal (SP) intermediate **which can be more easily converted into the more stable TBP intermediate** keeping the NH_2^- or RNH^- group at an equatorial position. (ii) **Higher acidity** of the N—H group at the *trans*-position (an electron withdrawing inductive effect of the leaving group at the *trans*-position) (*cf.* base catalysed proton exchange rates). (iii) Deprotonated site of the CB at the *cis*-position (to the leaving group) produces a SP intermediate after the dissociative expulsion of the leaving group and this SP intermediate is stabilised to some extent by the π-donor property of the deprotonated site (*i.e.* NH_2^- or RNH^-) (*cis*-effect, see Sec. 5.21). Consequently, **this SP faces a higher energy barrier to transform into the more stable TBP intermediate.**

● **Requirement of the ammine groups (with dissociable protons) *trans* to the leaving group and the base hydrolysis of $[CoCl(NH_3)(tren)]^{2+}$:** In most of the cases, CB is formed through the abstration of 'N—H' proton at the *trans*-position of the leaving group. In fact, the **red isomer** of $[CoCl(NH_3)(tren)]^{2+}$ having the —NH_2 group at the *trans*-position of 'Cl', **hydrolyses much faster** than the **purple isomer** lacking in such dissociable proton(s) at the *trans*-position of the leaving group (Cl^-).

For the **red isomer,** the CB generated through the removal of a proton at the *trans*-position (to Cl^-) can experience the **strong *trans*-labilising effect** due to the strong σ-donor property of the RNH^- group. This favoured dissociative expulsion of Cl^- (*trans*-effect of the RNH^- group) initially generates a square pyramidal geometry which readily produces the trigonal bipyramidal (TBP) geometry (**a favourable geometry for the tripodal *tren* ligand,** *cf.* Fig. 2.11.7.8, Vol. 4) which is stabilised by the both σ- and π-donor properties of the RNH^- group when placed at the equatorial position of the TBP (*cf.* Fig. 5.21.1b).

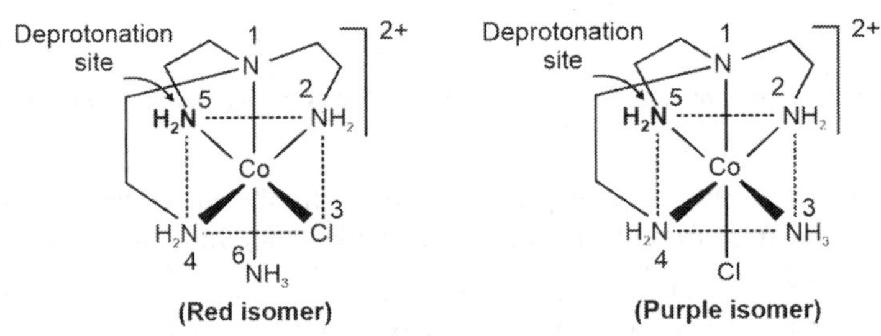

Reactivity order (in base hydrolysis): $\frac{k_{red}}{k_{purple}} \approx 10^4$

Acidity order (N—H): *tren* > NH_3 (*cf.* electron withdrawing inductive effect of tertiary-N in *tren*)

Fig. 5.25.1.1: Red and purple isomers of $[CoCl(NH_3)(tren)]^{2+}$ and their reactivity order in base hydrolysis (*cf.* Fig. 2.5.3.11 for structures, Vol. 4) (*cf.* D.A. Buckingham *et al.*, *Inorg. Chem.*, **14**, 1485, 1975)

In the **purple isomer**, the dissociable N—H proton is available only at the *cis*-position (to Cl^-) from the R—NH_2 group (*tren*) and NH_3 ligand. The N—H proton of the R—NH_2 group of *tren* is more acidic due to the electron withdrawing inductive effect of tertiary-N) than that of the NH_3 ligand. Thus for the **purple isomer,** the CB generated through the removal of a proton at the *cis*-position to the leaving

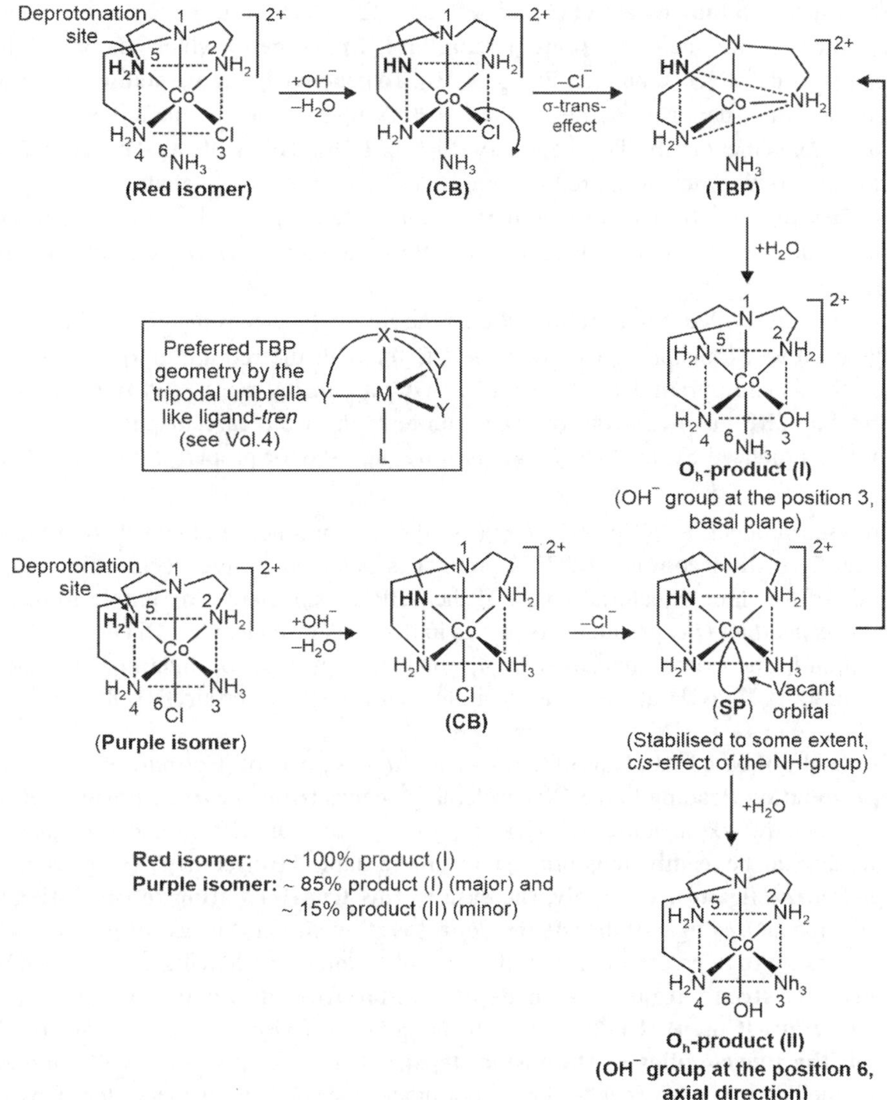

Fig. 5.25.1: Base hydrolysis products of the red and purple isomers of $[CoCl(NH_3)(tren)]^{2+}$ (*cf.* Fig. 5.25.1.1 for numbering of the positions in the product)

Note: The square pyramidal (SP) geometry initially produced from the CB of **red isomer** is suitably oriented to generate the **tripodal form of *tren*** in the TBP geometry. This is why, for the red isomer, the initially produced SP geometry readily converts into the **more stable TBP intermediate**. This SP is favourably generated by the **strong σ-*trans*-effect** of the deprotonated–NH site but this SP cannot be stabilised by the π-donor property of the NH-group located at the *trans*-position to the vacant coordination site.

On the other hand, for the **purple isomer**, the initially produced SP geometry from the CB is not suitably oriented to give the tripodal form of the *tren* ligand in the desired TBP geometry. Before its conversion to the TBP geometry, **it requires the rearrangement of the NH_3 ligand** so that ligand *tren* can adopt its favourable tripodal form in the TBP geometry. **This causes an energy barrier for the conversion of the SP (purple isomer) to the TBP intermediate**. Moreover, the SP (purple isomer) is itself stabilised to some extent by the π-donor property of the NH-group located at the *cis*-position to the vacant coordination site (*cis*-effect). This further raises the energy barrier for its conversion into the TBP.

Thus the **tripodal nature of the nonleaving *tren* ligand** plays an important role to direct the deprotonation site (to produce the CB) and reactivity order of the red- and purple-isomers.

group can show the *cis*-**labilising effect** (*i.e.* ***cis*-effect** of the RNH^- group) to cause an expulsion of the leaving group (*i.e.* Cl^-) producing a **square pyramidal (SP) geometry** which can be stabilised by the π-donor property of the RNH^- group (*cf.* Fig. 5.21.1b). To generate the **more stable TBP intermediate** where both the σ- and π-donor properties of the RNH^- group can operate, the initially formed SP is to experience a rearrangement to the TBP geometry (Fig. 5.21.1b). **This will raise the activation energy.** Thus from the reactivity order of the red and purple isomers, we can conclude:

(i) ***trans*-effect** (due to the σ-donor property) of the RNH^- group is **more important** than its *cis*-effect (due to the π-donor property) in the CB to cause a **dissociative expulsion of the leaving group**.

(ii) the 5-coordinate TBP is **more stabilised** than the SP by the electronic effect of the RNH^- group.

(iii) The dissociative expulsion of Cl^- from the CB for both the red and purple isomers, produces initially the square pyramidal (SP) geometry **which can be more readily converted into the more stable TBP geometry for the red isomer** than in the case of purple isomer for which the initially produced SP is already stabilised by the π-donor property of the RNH^- group (*cis*-effect).

The hydrolysed product, *i.e.* $[Co(NH_3)(OH)(tren)]^{2+}$ (*i.e.* product **I**) obtained from the red isomer **retains the configuration completely.** This product is obtained by the nucleophilic attack of OH_2 along any side of the three equatorial edges of the TBP intermediate. The purple isomer gives **two products: major product** (*i.e.* product **I**) (same as in the case of red isomer; OH at the basal plane and at the *cis*-position to the axial tertiary-N of *tren*) obtained from the TBP intermediate, and the **minor product** (*i.e.* product **II**) (OH at the axial position-6, *trans* to the tertiary-N) obtained from the SP intermediate.

Note: The results of base hydrolysis of the *cis*- and *trans*-isomers of $[Co(tren)(NH_3)Cl]^{2+}$ support the fact that **deprotonation** (leading to the CB) preferably **occurs from the *trans*-position of the leaving group**. The dissociative expulsion of Cl^- (leaving group) frrom the CB produces a square pyramidal (SP) geometry that can be readily transformed into a more stable TBP geometry **(a preferred geometry for the tripodal *tren* ligand). Probably, the ease of this transformation to the TBP geometry** (a demand of the tripodal *tren* ligand) **directs the deprotonation site in the case of $[Co(tren)(NH_3)Cl]^{2+}$.** However, the concept of deprotonation from the *trans*-position of the leaving group may not be true in all cases. **There are strong arguments for deprotonation from the *cis*-position in $[Co(NH_3)_5Cl]^{2+}$ and in *cis*- and *trans*-$[Co(en)_2(L)X]^{n+}$** (*Ref.* F.R. Nordmeyer, *Inorg. Chem.*, **8,** 2780, 1969).

- **Entry of the nucleophiles at the faster steps:** Other nucleophiles (say Y^-) besides H_2O (in aqueous media) may also invade the 5-coordinate intermediate at the faster steps to give the product containing Y. However, their concentration terms are not involved in the rate expression. Thus these nucleophiles are the *spectators* of the five coordinate intermediate complex and they are not involved in the generation of this intermediate. *However, OH^- or any other base (B) must be present to generate the CB.*

$$[Co(NH_3)_5X]^{2+} + B \rightleftharpoons [Co(NH_2)(NH_3)_4X]^{+}\,\textbf{(CB)} + BH^+$$

$$\downarrow -X^-,\ \text{rds}$$

$$[Co(NH_2)(NH_3)_4Y]^{+} \xleftarrow[\text{(fast)}]{Y} [Co(NH_2)(NH_3)_4]^{2+} \xrightarrow[\text{(fast)}]{+H_2O} [Co(NH_3)_5(OH)]^{2+}$$

$$[Co(NH_2)(NH_3)_4Y]^{+} \xrightarrow[\text{(fast)}]{+H_2O} [Co(NH_3)_5(Y)]^{2+}$$

- **Presence of an acidic proton in the starting complex:** To produce the CB, the parent complex must possess a dissociable proton to **generate a good σ-donor as well as a π-donor amido group** (*e.g.* NH_2^- or RNH^- amido group). Otherwise, the proposed base hydrolysis mechanism will not operate. This is why, the complexes like **$[CoCl_2(py)_4]^+$, $[CoCl(CN)_5]^{3-}$**, etc. **lacking in such dissociable protons** do not respond to the said D–CB mechanism. However, their hydrolysis simply occurs through an interchange process (most probably I_d path) slowly without any special rate benefit.

In fact, such complexes experience aquation slowly through the **I_d-path followed by the proton abstraction by OH^- from the aqua product in a fast step.** Consequently, the rate does not depend on $[OH^-]$.

$$\left[CoCl(py)_5\right]^{2+} \xrightarrow[(I_d\text{-path})]{\text{aquation}} \left[Co(OH_2)(py)_5\right]^{3+} \xrightarrow[-H_2O,\ (\text{fast})]{OH^-} \left[Co(OH)(py)_5\right]^{2+}$$

$$\text{Rate} = k_A[\text{complex}][OH^-]^0 = k_A[\text{complex}]$$

- **Base hydrolysis, Co(III)- *vs.* Cr(III)-ammine complexes:** In terms of CFAE, the low spin Co(III) (t_{2g}^6) complexes are more inert than the corresponding Cr(III) (t_{2g}^3) complexes but in the base hydrolysis of their ammine complexes, **their reactivity order is reversed.** In the base hydrolysis process, compared to the Co(III)-ammine complexes, the corresponding Cr(III)-ammine complexes react *ca.* 3×10^3 times slower. *Probably, the lower electron affinity of Cr(III) (compared to that of Co(III)) makes the electron donating NH_2^- or RNH^- amido group (of the CB) less efficient to labilise the leaving group in the CB of the Cr(III)-complexes at the rate determining step.* The higher electron affinity of Co(III) compared to that of Cr(III) is evidenced by the fact that under the comparable condition, it is easier to produce Co(II) from Co(III) while it is more difficult to produce Cr(II) from Cr(III). The E^0 values also support this fact. The higher electronegativity/electron affinity of Co(III) compared to that of Cr(III) can be rationalised in terms of their positions in the periodic table. **This is why, the electron pushing effect (both by σ- and π-bond) of the NH_2^- or RNH^- amido group towards M(III) in the corresponding CB is expressed better in the Co(III) complex** and this effect plays the most important role in the **favoured dissociative activation of the CB**. Besides this, the protonic character of the N—H bond of the coordinated ammine/amine ligand is more for the Co(III)-complexes compared to that of the corresponding Cr(III)-complexes because of the more electron withdrawing inductive effect of Co(III). **Thus formation of the CB is thermodynamically more favoured for the Co(III)-complexes.** This factor favours the rate process of the D–CB mechanism for the Co(III)-complexes in base hydrolysis. Here it is worth mentioning that for the heavier members like Ru(III), the A–CB mechanism rather than the D–CB mechanism operates (*i.e.* base hydrolysis: $[CoCl(NH_3)_5]^{2+}$ **(D–CB)**; $[RuCl(NH_3)_5]^{2+}$ **(A–CB)**).

- **Base hydrolysis of the complexes having no acidic proton**
The complexes like $[CoCl_2(py)_4]^+$, $[Co(bpy)_2(NO_2)_2]^+$, $[CoBr(CN)_5]^{3-}$, etc. experience the base hydrolysis very slowly without showing any rate dependence on $[OH^-]$ in the rate process. In fact, such complexes simply experience **the aquation through the I_d path followed by the rapid transformation into the hydroxo species through the acid-base reaction.** It is illustrated below:

$$\left[CoBr(CN)_5\right]^{3-} \xrightarrow[(I_d\ \text{path})]{\text{aquation (slow)}} \left[Co(CN)_5(OH_2)\right]^{2-} \xrightarrow[\substack{\text{(deprotonation of}\\ \text{the aqua ligand)}}]{OH^-\,(\text{fast})} \left[Co(CN)_5(OH)\right]^{3-}$$

- **Reactivity order (in base hydrolysis): Co(III)- *vs.* Cr(III)-ammine complexes**
In terms of CFAE, the Co(III)-complexes (low-spin) should be more inert than the Cr(III)-complexes. **The reversal order of reactivity in base hydrolysis of their ammine complexes can be explained by considering** the fact that the higher electron affinity of Co(III) makes the NH_2 (*i.e.* amido) group **quite active** in terms of electron donation to Co(III) (by both

σ- and π-bond). In fact, because of the much lower electron affinity of Cr(III), the NH_2^- group in the CB of Cr(III)-complexes becomes **quite ineffective** to **labilise the leaving group**. Formation of the CB is also relatively easier for the Co(III)-complexes due to the stronger electron withdrawing inductive effect of Co(III). The relative electron affinities of Co(III) and Cr(III) may be understood from their relative ease to form Co(II) and Cr(II) respectively in solution. It is also supported by their E^0 values of the aqua-complexes, *i.e.* Co(III) is more easily reduced to Co(II).

$$Co^{3+}/Co^{2+}\ (E^0 = 1.8\ V),\ Cr^{3+}/Cr^{2+}\ (E^0 = -0.41\ V)$$

- **D–CB *vs.* I_d–CB:** Stereochemical results indicate that the leaving group shows some sort of influence on the stereochemical change (*cf.* Table 5.25.3.1). It indicates that the leaving group is not completely dislodged (as expected in a pure *D*-process) in the transition state, *i.e.* the leaving group is still partly bonded with the metal centre to influence the entry of the nucleophile. Thus the process may be considered as a concerted process which may be better described as the *I_d-CB* rather than the *D–CB*. Other evidences in favour of I_d-CB have been discussed in Sec. 5.25.2.
- **Application of Hammett relationship and Taft relationship in base hydrolysis of $[Co(NH_3)_5(RCO_2)]^{2+}$:** These are discussed in Secs. 5.19.11 and 5.19.12.

5.25.2 Relative Importance of the σ-Donor and π-Donor Property of the NH_2^- Group (generated from coordinated NH_3 ligand) to Stabilise the 5-Coordinate T.S. in the D-CB Mechanism: Deprotonation Site — *Cis-* or *Trans-*Position?

- The **π-donor property** of the NH_2^- group can stabilise both the 5-coordinate geometries—square pyramid (SP) and trigonal bipyramid (TBP) in the T.S or intermediate. To stabilise the SP, the **NH_2^- group should reside at the *cis*-position of the leaving group** (*cf.* Sec. 5.21 and Fig. 5.21.1a) but to stabilise the TBP, **the NH_2^- group must reside at the equatorial position** (*cf.* Fig. 5.21.1b) of the TBP.
- If the **π-donor property** of the NH_2^- group were the main factor to stabilise the 5-coordinate T.S. or intermediate, **then the NH_2^- group would be preferably generated** (through proton abstraction) **at the *cis*-position of the leaving group** (*cf.* *cis*-effect in aquation process) and the T.S. or intermediate would be preferably SP. But the experimental findings do not support this prediction.
- If the **π-donor property** of the NH_2^- group were the main factor to stabilise the 5-coordinate T.S. or intermediate, then the $M\text{—}\ddot{N}R_2$ (M—N̈ bonded to two R groups) **planar moiety** (*i.e.* *sp^2*–N) **would be the most efficient one** but the experimental findings do not support this expectation.
- All these findings indicate the necessity of **an alternative explanation** which suggests that **the σ-donor (*i.e.* σ-basicity) property** of the amino group of the CB plays the main role for the dissociative expulsion of the leaving group from the CB and to stabilise the TBP T.S. or intermediate. **In fact, a very good σ-donor ligand like the NH_2^- group can stabilise the TBP structure** and it happens also so for the substitution reactions of **Pt(II)-square planar complexes** where the T.S. or intermediate adopts the TBP structure.
- In the octahedral CB, the NH_2^- (*i.e.* amido) group can strongly labilise the *trans*-group (*i.e.* ***trans* effect**) by dint of its σ-donor property while it can labilise the *cis*-position (*i.e.* **cis-effect**) by its π-donor property. In most of the cases, the NH_2^- group is generated through the proton removal at the *trans*-position to the leaving group. **It indicates that the σ-*trans*-effect of the NH_2^- group (*i.e.* σ-donor property) is more important than its *cis*-effect (*i.e.* π-donor property) in the D-CB mechanism.**

- The TBP T.S. or intermediate can be stabilised by both the σ- and π-donor property of the NH_2^- group (when placed at the equatorial position of the TBP) while the SP T.S. or intermediate is stabilised by the π-donor property only of the NH_2^- group (when placed at the *cis*-position of the leaving group). **Thus the NH_2^- group stabilises the TBP T.S. or intermediate more compared to the SP T.S. or intermediate.**
- If the SP structure is produced from the CB where the NH_2^- group is generated at the *cis*-position to the leaving group, then the SP structure (stabilised to some extend by the π-donor NH_2^- group, ***cis*-effect**) will have **to rearrange to attain the more stable TBP structure** (*cf.* electronic property of the NH_2 group stabilises the TBP structure more than the SP structure). **This energy requiring structural rearrangement will raise the activation energy.** Alternatively, if the NH_2^- group is generated at the *trans*-position to the leaving group, then **the CB can more readily attain the TBP structure** through the dissociative expulsion of the leaving group (*cf. trans*-labilising effect of the NH_2^- group). Here it is important to mention that the dissociative expulsion of the leaving group from the *cis*- or *trans*-position of the NH_2^- group of the CB always first produces the 5 coordinate square pyramid (SP) geometry. But, the SP generated through the expulsion of the leaving group from the *trans*-position of the NH_2^- group, can more readily rearrange to the desired TBP geometry (which is better stabilised by the NH_2^- group) compared to the alternative possibility where the SP initially produced through the expulsion of the leaving group from the *cis*-position of the NH_2^- groupm is itself stabilised to some extent by the π-donor property of the NH_2^- group (*cis*-effect). Besides this, the strong **σ-*trans* effect** of the NH_2^- group preferably labilises the *trans*-position to favour the dissociative expulsion of the leaving group at the *rds*. This is why, proton abstraction preferably occurs at the *trans*-position of the leaving group to produce the CB.
- **More acidity of the N—H proton** at the *trans*-position to the leaving group compared to that of the *cis*-position (*cf.* proton exchange rate, Sec. 5.21.2) favours more the deprotonation from the *trans*-position (see Sec. 5.21.2 for details).
- All these propositions are experimentally verified by comparing the base hydrolysis of the **red and purple isomers of $[CoCl(NH_3)(tren)]^{2+}$** (Figs. 5.25.1.1–2).

5.25.3 Evidences in Favour of the D–CB or I_d–CB Pathway Rather than the A-Pathway in the Base Hydrolysis of Cobalt(III)-Ammine Complexes

The observed rate law (= k_B[complex][OH^-]) can be attained either from the *associative (A) path* or from the *D–CB* or *I_d–CB* path discussed above. There are evidences to support the *D–CB* or *I_d–CB* path and to reject the associative path. These are discussed below.

(i) **Rate enhancement – charge effect and electronic effect of the NH_2^- group (generated from the coordinated NH_3 ligand through deprotonation) of CB:** Just by considering the better nucleophilic character of OH^-, compared to that of H_2O, the tremendous rate acceleration by a

σ-*trans* effect: $H_2\ddot{N} \rightarrow Co - X$; Stabilisation of the TBP intermediate:

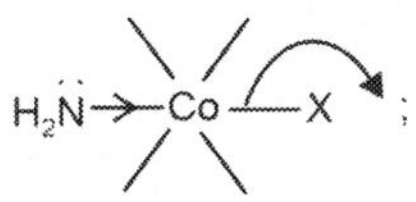

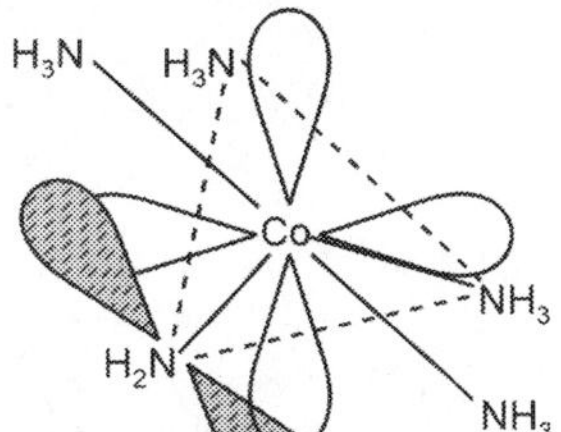

(*cf.* Fig. 5.21.1.1b)

factor *ca.* 10^6 (compared to that of the acid hydrolysis) cannot be explained. This rate benefit arises from the **charge effect** (+2 charge of the starting complex becomes +1 in the CB) favouring the dissociative activation and **the electronic activation effect of the NH_2^--group** (generated in the *CB*) *that shows the trans-labilizing effect* in the CB and stabilising effect on the TBP T.S. (by both the good σ- and π-donor property).

(ii) **No rate enhancement in the base hydrolysis of $[Co(CN)_5X]^{3-}$ or *trans*–$[CoCl_2(py)_4]^+$:** If the simple associative path involving the nucleophilic attack by OH^- were operative, then these complexes would also react at comparable rates with those of $[Co(NH_3)_5X]^{2+}$. *In fact, the complexes lacking in dissociable protons cannot participate in the D–CB mechanism and consequently, they do not gain any rate benefit.*

In fact, such complexes lacking in dissociable protons experience the base hydrolysis very slowly and the processes are also independent of $[OH^-]$ (*i.e.* **simple dissociative path**). It suggests that the complexes like $[Co(CN)_5X]^{3-}$, *trans*–$[CoCl_2(py)_4]^+$ experience the simple dissociative path (without any additional rate enhancement) while the complexes like $[Co(NH_3)_5X]^{2+}$ experience the *D–CB* path (with a tremendous rate benefit) showing a rate dependence on $[OH^-]$ in the base hydrolysis process.

(iii) **Entry of other nucleophiles:** In the D-CB path, OH^- acts in the proton abstraction process to produce the CB in a rapid pre-equilibrium step. The CB **generates the 5-coordinate intermediate through the dissociative expulsion of the leaving at the rate determining step.** Then the nucleophiles present in the media will attack the 5-coordinate intermediate. Thus it is reasonable to expect that if the base hydrolysis of $[Co(NH_3)_5X]^{2+}$ is carried out in presence of other nucleophiles Y^- (*e.g.* SCN^-, N_3^-, NO_2^-, etc.), the product will contain the Y-group but the rate will not depend on such nucleophiles. It has been experimentally observed. *It may be pointed out that without any base, entry of such nucleophiles under the conditions is exceedingly slow.*

In nonaqueous solvents like dimethylsulfoxide (DMSO), the replacement of X^- by Y^- (*e.g.* N_3^-, SCN^-, etc.) in $[Co(NH_3)_5X]^{2+}$ occurs very slowly in absence of any base capable of abstracting the N–H protons. But in presence of a trace amount (*i.e.* catalytic amount) of base *B* (*e.g.* OH^- or pipyridine), the process is tremendously catalyzed.

$$[Co(NH_3)_5X]^{2+} + Y^- \xrightarrow{\text{(in } DMSO)} [Co(NH_3)_5Y]^{2+} + X^- \ (t_{1/2} \text{ in hr.})$$

$$[Co(NH_3)_5X]^{2+} + Y^- \xrightarrow[\left(\substack{\textbf{catalytic amount} \\ \textbf{of base B}}\right)]{\text{(in DMSO)}} [Co(NH_3)_5Y]^{2+} + X^- \ (t_{1/2} \text{ in min.})$$

The rate of the catalyzed reaction depends on [B] but not on $[Y^-]$. These findings can be explained in terms of the *D-CB* mechanism.

$$[Co(NH_3)_5X]^{2+} + B \xrightleftharpoons{\text{(fast)}} [Co(NH_2)(NH_3)_4X]^+ \ \textbf{(CB)} + BH^+$$

$$CB \xrightarrow{rds} [Co(NH_2)(NH_3)_4]^{2+} + X^-$$

$$\downarrow \text{(fast)} \ +Y^-$$

$$[Co(NH_2)(NH_3)_4Y]^+ \xrightarrow[\text{(fast)}]{+BH^+} [Co(NH_3)_5Y]^{2+} + B$$

The proportion of the 5 coordinate intermediate captured by Y^- depends on its nucleophilicity and concentration. By using this procedure, $[Co(en)_2(NO_2)_2]^+$ has been prepared from *trans-*

$[Co(en)_2Cl(NO_2)]^+$ dissolved in DMSO in presence of added NO_2^- and a catalytic amount of OH^- (*i.e.* B).

$$[Co(en)_2(NO_2)Cl]^+ + OH^- \xrightleftharpoons{\text{fast}} [Co(en)(en\text{–}H)(NO_2)Cl](\mathbf{CB}) + H_2O$$

$$[Co(en)(en\text{–}H)(NO_2)Cl] \xrightarrow{\text{slow}} [Co(en)(en\text{–}H)(NO_2)]^+ + Cl^-$$

$$[Co(en)(en\text{–}H)(NO_2)]^+ + NO_2^- \xrightarrow{\text{fast}} [Co(en)(en\text{–}H)(NO_2)_2]$$

$$[Co(en)(en\text{–}H)(NO_2)_2] + H_2O \xrightarrow{\text{fast}} [Co(en)_2(NO_2)_2]^+ + OH^-$$

(en–H denotes deprotonated en, *i.e.* $H_2NCH_2CH_2NH^-$, *cf*. Fig. 5.25.3.1)

(iv) **Competition results in the products in the base hydrolysis of $[Co(NH_3)_5X]^{n+}$ in presence of other nucleophiles and D–CB *vs.* I_d–CB mechanism:** When the base hydrolysis of $[Co(NH_3)_5X]^{n+}$ (X = neutral group like $OSMe_2$; uninegative anion like I^-, $OSO_2CF_3^-$; dinegative anion like OSO_3^{2-}) is carried out in presence of SCN^-, then the mixed product, *i.e.* $[Co(NH_3)_5(OH)]^{2+}$ and $[Co(NH_3)_5Y]^{2+}$ (Y = NCS^- which may bind through the *N*– or *S*– end) is obtained. The proportion of NCS^- entry (as NCS^- and SCN^-) depends on the charge of the leaving group X^{m-}. **With the increase of the anionic charge of the leaving group, the proportion of $[Co(NH_3)_5(OH)]^{2+}$ increases and the proportion of NCS-content in the product decreases.** For the neutral leaving group, the NCS^- content product is *ca.* 18%, for the monoanionic leaving group, *i.e.* X^-, it is *ca.* 13% and for X^{2-} it is decreased to 7%.

These observations can be explained by the D-CB or I_d – *CB mechanism not by the associative mechanism.* The leaving group (X^{m-}) remains in the secondary coordination sphere of the long-lived 5-coordinate intermediate. Consequently, more negative X^{m-} will discourage the anionic nucleophiles like NCS^- to attack the 5-coordinate intermediate. In other words, it may be said that the leaving group X^{m-} remains partly bonded in the transition state to prevent/discourage the approaching anionic nucleophile Y^- more compared to the approaching neutral nucleophile H_2O. **This second view supports the *I_d-CB* path rather than the *D-CB* path.** However, the competition results in the product cannot be explained by the associative path.

Anation in acidic and in basic solution

$$\left[Co(NH_3)_5X\right]^{2+} + Y^- \rightarrow \left[Co(NH_3)_5Y\right]^{2+} + X^-$$ **(anation rate is much faster in a basic condition)**

- **Acidic condition:** $\left[Co(NH_3)_5X\right]^{2+} \xrightarrow[\text{(dissociative process)}]{\text{aquation}} \left[Co(NH_3)_5(OH_2)\right]^{3+} \xrightarrow[\text{(anation)}]{+Y^-\text{ (dissociative process)}} \left[Co(NH_3)_5Y\right]^{2+}$

  (*i.e.* aquation followed by anation; **both steps dissociative**)

- **Basic condition:** $\left[Co(NH_3)_5X\right]^{2+} \xrightleftharpoons{+OH^-} CB \xrightarrow[rds]{-X^-} \left[Co(NH_2)(NH_3)_4\right]^{2+} \xrightarrow[\text{(fast)}]{+Y^-,\, H_2O} \left[Co(NH_3)_5Y\right]^{2+}$

  (*i.e.* through the CB formation)

  i.e. **D-CB** mechanism.

(v) **Comparable reactivities and isomerisation of both *cis*- and *trans*-$[Co(en)_2(L)X]^{n+}$ (X = leaving group):** In the D-CB mechanism, it has been suggested that after expulsion of the leaving group, the 5-coordinate square pyramidal intermediate rearranges to the ***long-lived trigonal bipyramidal geometry*** *where the RNH^- amido group (produced through the deprotonation of one of the RNH_2 groups of coordinated en-ligand) occupies one equatorial position to show its both σ- and π-donor property. The trigonal bipyramid is sufficiently long-lived to allow the* **Berry's pseudorotation** and the two TBP intermediates remain in an equilibrium (see Fig. 5.25.3.1). Thus the ligands both *cis*- and *trans*- to the leaving group (present in the starting complex) occupy the equivalent positions in the trigonal bipyramidal geometry. This is why, both the *cis*- and *trans-isomers react at the comparable rates accompanied with the cis-trans isomerisation in base hydrolysis* (*cf.* aquation rates and stereochemistry of the products very often differ for the *cis*- and *trans*-isomers). It is illustrated for the base hydrolysis of $[Co(en)_2(L)X]^{m+}$ in Fig. 5.25.3.1. Here it may be pointed out that if the nonleaving group (L) is a π-donor ligand, then the trigonal bipyramid will be stabilized more (by keeping L at an equatorial position of the TBP, see

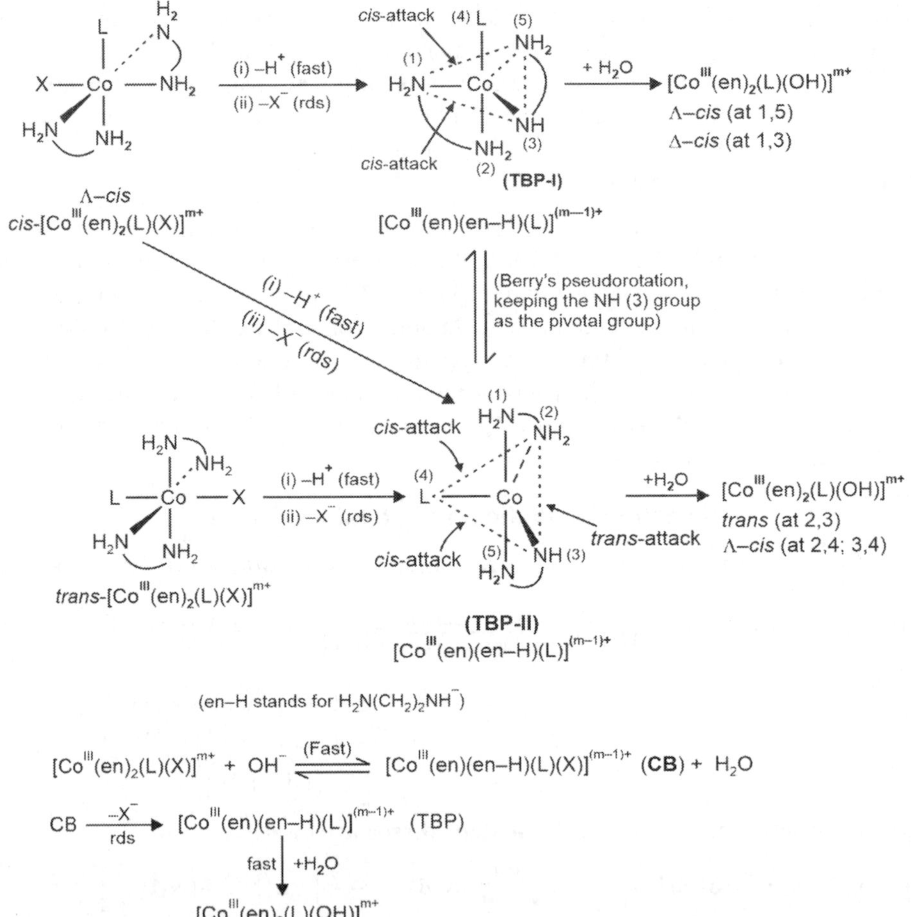

Fig. 5.25.3.1: Reaction mechanism (*cf.* Figs 5.21.4.5, 5.28.2.1, 2) to explain the comparable reactivity and isomerisation in the base hydrolysis of *cis*/*trans*-$[Co(en)_2(L)(X)]^{m+}$ (where X^- is the leaving group) (*Ref.* R.G. Pearson and F. Basolo, *Inorg. Chem.*, **4**, 1522, 1965).

Note: For L = NH_3, deprotonation may also occur from the coordinated NH_3 ligand (*cf.* NH_3: $pK_b \approx 4.7$; en: $pK_{b(1)} \approx 4.0$, $pK_{b(2)} \approx 6.8$)

Sec. 5.21) compared to the system having the π-acceptor nonleaving group (*e.g.* L = CN^-). **This is why, the complexes bearing the π-acceptor nonleaving groups show the low reactivity.**

(vi) **Steric acceleration:** In the base hydrolysis, $[M(NH_2Me)_5(OSO_2CF_3)]^{2+}$ (M = Co, Rh, Ir, Cr) reacts much faster than $[M(NH_3)_5(OSO_2CF_3)]^{2+}$. This **steric acceleration** supports the **dissociative activation.** It may be pointed out that in aquation (*i.e.* acid hydrolysis), the Co(III)-complexes show also the **steric acceleration** but the Cr(III), Ir(III)-complexes show the **steric retardation** suggesting the **associative activation.**

(vii) **Nucleophilicity *vs.* basicity—importance of the N–H proton abstraction:** That the N–H proton abstraction is an important step in the base hydrolysis reaction is established from the observation in the reaction between $[Co(NH_3)_5X]^{2+}$ and H_2O_2/OH^- mixture. In the presence of H_2O_2, the following *acid-base interaction* (*i.e.* proton transfer) in alkaline condition occurs:

$$H_2O_2(A_1) + OH^-(B_2) \rightleftharpoons HO_2^-(B_1) + H_2O(A_2)$$

HO_2^- *(hydroperoxide ion) is a better nucleophile than OH^- but it is a weaker Brönsted base than* OH^-. Thus with the addition of H_2O_2, the better nucleophile HO_2^- is generated at the cost of the better base, *i.e.* OH^- (which can abstract the N–H proton better).

If the reaction proceeds through the **associative path**, the rate should increase with the generation of the better nucleophile. On the other hand, if the reaction passes through the **D-CB or I_d-CB mechanism,** then the rate should decrease with the decrease of the amount of better base (*i.e.* proton abstracting agent). Thus the observation supports the D-CB or I_d-CB mechanism not the associative mechanism.

(viii) **Activation parameters:** The proposed mechanism involving the **dissociative transformation** of the CB (produced at a rapid pre-equilibrium step) **into the 5-coordinate trigonal bipyramid** at the *rds* has been supported by the *large positive values of* $\Delta S^{\neq}$ *and* $\Delta V^{\neq}$. The pre-equilibrium step leading to the CB through the **charge destruction** makes a positive contribution to both $\Delta S^{\neq}$ and $\Delta V^{\neq}$ of the **composite rate constant** k_B (= kK_{CB}, see rate equation); the **dissociative transformation** of the CB to a square pyramid geometry followed by its transformation into a trigonal bipyramid makes also a positive contribution to $\Delta V^{\neq}$ of the composite rate constant (k_B). But this dissociative transformation step leading to the **separation of an anionic ligand (X^{n-})** creates charge which makes a negative contribution to $\Delta V^{\neq}$ of the composite rate constant k_B from the electrorestriction imposed on the solvent. Thus the resultant $\Delta V^{\neq}$ value is less positive.

The factors to contribute to $\Delta V^{\neq}$ of the composite rate constant k_B (= kK_{CB}) in the D-CB mechanism are: (i) charge destruction in the formation of CB (pre-equilibrium step) causing a positive contribution; (ii) dissociative transformation of the CB to TBP causing a positive contribution; (iii) charge separation in the transformation of the CB into a TBP causing a negative contribution.

Complex:	$[CoCl(NH_2Me)_5]^{2+}$	$[CoCl(NH_2Et)_5]^{2+}$	*cis*–$[CoCl_2(en)_2]^+$
$\Delta V^{\neq}$ ($cm^3\ mol^{-1}$) (for base hydrolysis):	+32.7	+31.0	+28

Relationship between the $\Delta G^{\neq}$ values for acid hydrolysis and base hydrolysis

- For the **Co(III)-ammine complexes**, there is a linear relationship between ln k_H and ln k_{OH} (k_H and k_{OH} denote the rate constants for acid hydrolysis and base hydrolysis respectively). It indicates the linear relationship between $\Delta G_H^{\neq}$ and $\Delta G_{OH}^{\neq}$. It suggests a **common mechanistic path** (which is the dissociative activation) for both the acid hydrolysis (k_H) and base hydrolysis (k_{OH}) path.
- There is no such linear relationship between k_H and k_{OH} for the Cr(III)-ammine complexes. It indicates that the acid hydrolysis and base hydrolysis for the **Cr(III)-complexes** occur through the **different mechanisms**. In fact, their acid hydrolysis mainly passes through the I_a path while the base hydrolysis occurs through the dissociative path.
- **Reactivity order:** $k_H(Co^{III}) < k_H(Cr^{III})$ but $k_{OH}(Co^{III}) > k_{OH}(Cr^{III})$. This aspect has been explaied in Sec. 5.25.1.

Stereochemical Change for $[Co^{III}(en)_2(L)X]$: Acid Hydrolysis *vs.* Base Hydrolysis

(A) Acid Hydrolysis (*cf.* Sec. 5.21): 100% retention of configuration for the *cis*-compounds irrespective of the electronic nature of the **non-leaving group** L. 100% retention of configuration for the *trans*-compound when L is a non-π-bonding ligand (*e.g.* NH_3) or a π-acceptor ligand (*e.g.* CN^-). The *trans*-compounds experience a geometrical isomerisation for L = π-donor ligands (*e.g.* OH^-).

- For L = π-acceptor ligand, an **associative path** leads to stereoretentivity for both the *cis*- and *trans*- compounds.
- For L = π-donor ligand, stereoretentivity is attained through the formation of a **square pyramidal transition state** (dissociative path) as in the case of *cis*-compounds. Geometrical isomerisation occurs for the *trans*-compounds producing a **trigonal bipyramidal transition state/intermediate from the square pyramidal structure.**
- For L = non-π-bonding ligand, stereoretentivity can be attained in a dissociative path passing through the **square pyramidal transition state;** and associative path also leads to stereoretentivety.

(B) Base Hydrolysis (*cf.* Sec. 5.25): Both the *cis*- and *trans*-compounds experience the geometrical isomerisation due to the formation of a **long-lived trigonal bipyramidal intermediate** which is stabilised by the good σ-donor and π-donor properties of the RNH^- group (generated from the deprotonation of one of the RNH_2 groups of coordinated en-ligand) when placed at the equatorial position.

Reactivity of $[Co^{III}(en)_2(L)X]$: Acid Hydrolysis *vs.* Base Hydrolysis

(A) Acid Hydrolysis: L = π-donor ligand, $k_{cis} > k_{trans}$ **(dissociative path)**; L = weakly π-bonding ligand or non-π-bonding ligand, $k_{cis} \approx k_{trans}$; L = π-acceptor ligand, $k_{trans} \rangle k_{cis}$ **(associative path)**.

(B) Base Hydrolysis: Irrespective of the nature of L, $k_{cis} \approx k_{trans}$.

D–CB vs. I_d–CB in Base Hydrolysis of the Co(III) Complexes

- Distribution of product in presence of different nucleophiles depends on the nature of the leaving group. It suggests the *I_d–CB* path rather than the *D–CB* path.
- Distribution of stereochemical configuration in the product depends on the nature of the leaving group. It suggests the *I_d–CB* path rather than the *D–CB* path.

Comparison of the Reactivity of Co(III) (low-spin) and Cr(III) (*cf.* Secs. 5.19.5–6, 5.22)

In terms of CFAE, the Co(III) (low-spin) complexes are more inert than the Cr(III) complexes. It is experimentally verified in many reactions such as:

Ligand exchange: $[M(CN)_6]^{3-} + 6{*}CN^- \rightleftharpoons [M({*}CN)_6]^{3-} + 6CN^-$

Aquation: $[M(NH_3)_3X]^{2+} + H_2O \rightarrow [M(NH_3)_5(OH_2)]^{3+} + X^-(X^- = Cl^-, Br^-, NCS^-)$

Dissociation: $[M(AA)_3]^{3+} + H_3O^+ + H_2O \rightarrow [M(AA)_2(OH_2)_2]^{3+} + AAH^+$
(acid catalyzed) (AA = en, bigH, etc.)

However, there are **some exceptions** where the reactivity of Co(III) is found more than that of Cr(III). These are:

- **Water exchange reaction:** It occurs through a special mechanism discussed earlier.
- **Base hydrolysis:** It does not occur through the simple associative or dissociative path. It has been already discussed.
- **Pseudo-substitution:** Sometimes, due to the **pseudo-substitution process**, apparently the Co(III) complexes may appear more labile. These aspects have been discussed already (*cf.* Sec. 5.24).

(ix) **Stereochemical change in base hydrolysis and *D–CB vs. I_d–CB* mechanisms:** Stereochemical change in the base hydrolysis of the complexes like $[Co(en)_2(NH_3)X]^{n+}$, $[Co(en)_2(L)X]^{m+}$ (L and X are anionic, X^- is replaced by OH^-) has been extensively studied. *It has been found that isomerization occurs for both the cis– and trans– isomers.* This can be explained by considering the formation of a **long-lived 5-coordinate** trigonal bipyramid intermediate (Fig. 5.25.3.1) that experiences the pseudo-rotation to produce two possible trigonal bipyramids remaining in an equilibrium (*cf.* Figs. 5.21.3–5).

However, the extent of retention of configuration depends partly on the nature of the leaving group X. *It indicates that the leaving group probably remains partly bonded to the transition state to influence the stereochemistry of the product* (*i.e.* nucleophilic attack by H_2O is influenced by the leaving group). It indicates some sort of interchange process. **Thus the *D-CB* mechanism should be better described as the *I_d-CB* mechanism.**

Table 5.25.3.1: Stereochemical change in the base hydrolysis of $[Co(en)_2(NH_3)X]^{n+}$ at 25° C (*cf.* Figs. 5.21.3–5) (*cf.* D.A. Buckingham, *et al. Inorg. Chem.*, **18**, 1985, 1979; *J. Am. Chem. Soc.*, **90,** 6654, 1968)

	(a) cis-$[Co(en)_2(NH_3)X]^{n+}$		***cis-product***	
X	***%-cis product***	***%-trans product***	***% (Retention)***	***% (Inversion)***
$OSMe_2$	77	23	84.5	15.5
Cl^-	78	22	81	19
Br^-	78	22	77	23
NO_3^-	77	23	80	20

	(b) trans-$[Co(en)_2(NH_3)X]^{n+}$	
X	***%-cis product***	***%-trans product***
Cl^-	76	24
Br^-	81	19

Base hydrolysis of *cis*-$[Co(en)_2(amine)Cl]^{2+}$ and deprotonation site to produce the CB (*Trans. Met. Chem.*, **12**, 97, 1987, *J. Inorg. Nuch. Chem.* **38**, 1959, 1976): From the distribution of stereochemistry of the products, it appears that **deprotonation can occur from all the possible sites rather that only at the amino group** *trans* to the leaving group Cl^-. Thus deprotonation from both the coordinated en and R—NH_2 (amine) ligands is possible.

(x) **Competition results in the products of base hydrolysis of $[Co(NH_3)_5X]^{n+}$ dependent on the nature of the leaving group:** This aspect has been discussed in Sec. 5.25.3 and the results support the I_d–CB mechanism not the *D–CB* mechanism.

5.25.4 *D–CB vs. D–IP* Pathway in Base Hydrolysis

The rate law for base hydrolysis (*D-CB* path) of $[M(NH_3)_5X]^{n+}$ is:

$$\text{Rate} = \frac{kK_{CB}[\text{complex}][OH^-]}{1+K_{CB}[OH^-]}$$

It was argued that the value of K_{CB} is very small and consequently, the condition $1 \gg K_{CB}[OH^-]$ is always maintained to show the strict first-order dependence on $[OH^-]$ even for a fairly high concentration range of OH^-. *In other words, for the D-CB path, there would be never saturation kinetics or mixed order behaviour.* But, for base hydrolysis of $[Cr(NCS)(NH_3)_5]^{2+}$, the rate *vs.* $[OH^-]$ plot deviates from the linearity and the plot levels off to a constant value at the relatively higher value of $[OH^-]$. *This mixed order or saturation kinetics cannot be explained in terms of the D-CB mechanism.*

To explain the mixed order kinetics of base hydrolysis of $[Cr(NCS)(NH_3)_5]^{2+}$, the following *D-IP* mechanism has been suggested (D. Banerjea *et al, Z. Anorg. Allgm. Chem.,* **99**, 361, 1968).

$$[Cr(NCS)(NH_3)_5]^{2+} + OH^- \xrightleftharpoons{K_{OS}} \underset{\text{(Ion pair, } IP)}{[Cr(NCS)(NH_3)_5]^{2+}\cdot OH^-}$$

$$[Cr(NCS)(NH_3)_5]^{2+}\cdot OH^- \xrightarrow{k} [Cr(NH_3)_5(OH)]^{2+} + NCS^-$$

$$\text{rate} = \frac{kK_{OS}[\text{complex}][OH^-]}{1+K_{OS}[OH^-]}$$

$$= \frac{kK_{IP}[\text{complex}][OH^-]}{1+K_{IP}[OH^-]} \text{(when } [OH^-] \gg [\text{complex}]) \text{ (taking } K_{OS} = K_{IP})$$

where $$k_{obs} = \frac{kK_{IP}[OH^-]}{1+K_{IP}[OH^-]}$$

or $$\frac{1}{k_{obs}} = \frac{1}{k} + \frac{1}{(kK_{IP}[OH^-])}, \text{(when } [OH^-] \gg [\text{complex}])$$

The plot of $\frac{1}{k_{obs}}$ *vs.* $\frac{1}{[OH^-]}$ can estimate K_{IP} (≈ 10 mol^{-1} dm^3). The estimated value is in good agreement with the theoretically calculated value. *The relatively high value of K_{OS} (i.e. K_{IP}) is responsible for the observed* ***mixed order kinetics.***

Here it is interesting to note that though $[Cr(NCS)(NH_3)_5]^{2+}$ adopts the *D-IP* path, $[Cr(NH_3)_5(N_3)]^{2+}$ adopts the *D-CB* path. **It is suggested the conjugate base (CB) and ion pair (IP) are the two limiting situations of the following system.**

$$M—NH_2^-\cdots\cdots H^+\cdots\cdots OH^-$$

If the proton is primarily under the control of the —NH_2^- group, it leads to the IP. On the other hand, if the proton is primarily under the control of OH^-, it leads to the CB.

$$M — NH_2^- - - - H^+ - - - OH^-$$

$$\underset{\textbf{(Ion pair)}}{M — NH_3 - - - - OH^-} \rightleftharpoons \underset{\textbf{(Conjugate base)}}{M — NH_2^- - - - HOH}$$

- If the protonic character of the 'N—H' proton is sufficiently high then the CB is favoured over the IP.
- If the protonic character of the N—H proton is relatively less then the IP formation is favoured over the CB.

In $[M(NH_3)_5X]^{2+}$, the protonic character of the 'N—H' proton depends on the electron withdrawing/pushing character of 'X'. *More electron pushing character of X will reduce the protonic character.* ***NCS^- is more electron pushing than N_3^-.*** Consequently, in $[M(NH_3)_5(N_3)]^{2+}$, the acidic or protonic character of the N—H proton is more compared to that in $[M(NCS)(NH_3)_5]^{2+}$ to favour the *D–CB* path for the azido complex.

The difference in electron pushing character of NCS^- and N_3^- can be understood by considering their resonating structures.

$$^-N{=}C{=}S \longleftrightarrow :N{\equiv}C{-}S^-$$

$$:N{\equiv}\overset{+}{N}{-}N^{2-} \longleftrightarrow {}^-N{=}\overset{+}{N}{=}N^- \longleftrightarrow {}^{2-}N{-}\overset{+}{N}{\equiv}N:$$

In N_3^-, because of the positively charged central N-atom, the terminal N-atoms are the weaker electron donors. This is supported by the fact that N_3^- is more electronegative than NCS^-; ***and N_3^- is a poorer nucleophile than NCS^-***. It makes the following difference.

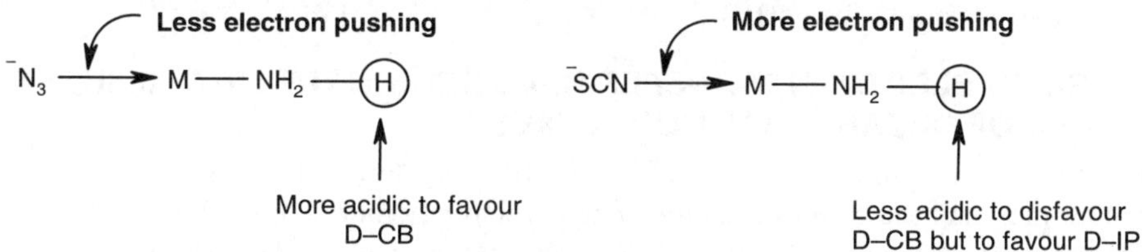

Thus *D-CB vs. D-IP* mechanistic profile depends on the electronegativity or nucleophilicity of X^-. It has been experimentally observed in the base hydrolysis of *trans*-$[Rh(en)_2X_2]^+$ where the electronegativity sequence of X^- runs as: $I^- < SCN^- < N_3^- < Br^- < Cl^-$. The change-over from *D-IP* to *D-CB* is noted between SCN^- (*N*-bonded) and N_3^-, *i.e.*

$$X^-: \quad \xleftarrow{\text{favouring } D\text{–}IP} \quad I^- \; SCN^- \; N_3^- \; Br^- \; Cl^- \quad \xrightarrow[\text{increasing electronegativity and favouring } D\text{–}CB]{}$$

Note: For the Co(III)-ammine complexes, because of the stronger electron withdrawing inductive effect of Co(III) (*cf.* **electronegativity order** in terms of positions of Co and Cr in periodic table), the N—H proton of coordinated NH_3 ligand is more acidic compared to that of the Cr(III)-ammine complexes (see Sec. 5.25.1). **Consequently, for Co(III), the *D–IP* mechanism is quite unlikely**.

5.26 KINETIC ASPECTS OF FORMATION REACTIONS AT IRON(III)

$[Fe(OH_2)_6]^{3+}$ shows a very strong tendency to undergo hydrolysis even in a fairly strong acidic condition.

$$[Fe(OH_2)_6]^{3+} + H_2O \rightleftharpoons [Fe(OH)(OH_2)_5]^{2+} + H_3O^+, \; pK_a \approx 2.0$$

Thus the observed reactivity is the sum of reactivity of $[Fe(OH_2)_6]^{3+}$ (*i.e.* unhydrolyzed species) and $[Fe(OH)(OH_2)_5]^{2+}$ (*i.e.* hydrolyzed species). In acidic condition, $[Fe(OH_2)_6]^{3+}$ is definitely more abundant than $[Fe(OH)(OH_2)_5]^{2+}$, *but reactivity of the hydrolyzed species is about 10^3 times higher than that of the unhydrolyzed species* (*cf.* $k_{ex} \sim 10^2\ s^{-1}$ for $[Fe(OH_2)_6]^{3+}$; $k_{ex} \sim 10^5\ s^{-1}$ for $[Fe(OH)(OH_2)_5]^{2+}$). *Thus very often, contribution to the observed total reactivity is mainly coming from the less abundant species*, $[Fe(OH)(OH_2)_5]^{2+}$. Here it may be mentioned that though **$[Fe(OH_2)_6]^{3+}$ reacts through the I_a-path, $[Fe(OH)(OH_2)_5]^{2+}$ reacts through the I_d-path.** (*cf.* A.K. Das, *Bull. Chem. Soc. Jpn.*, **65**, 2205, 1992; *Transition Met. Chem.*, **17**, 484, 1992 and the references cited therein).

$$\begin{array}{ccc} [Fe(OH_2)_6]^{3+} & \xrightarrow[(I_a\text{-path})]{+\text{ligand}} & \text{Product} \\ {\scriptstyle -H^+}\big\downarrow\big\uparrow & & \\ [Fe(OH)(OH_2)_5]^{2+} & \xrightarrow[(I_d\text{-path})]{+\text{ligand}} & \text{Product} \end{array}$$

The higher cationic charge of $[Fe(OH_2)_6]^{3+}$ imparts the **associative activation** to it while the overall **less cationic charge** of $[Fe(OH)(OH_2)_5]^{2+}$ (+3 *vs.* +2) favours it to adopt the **dissociative activation** (*cf.* effect of charge, Sec. 5.19.1). Besides this, the **better σ-donor and π-donor property of the $^-$OH group** (compared to those of OH_2) introduce the dissociative character in $[Fe(OH)(OH_2)_5]^{2+}$. In fact, the ***cis*-effect** (*cf.* Sec. 5.21) (π-donor property of the $^-$OH group) labilises the H_2O molecule at the *cis*-position.

If the ligand (say LH) remains in the following equilibrium under the experimental conditions,

$$LH \rightleftharpoons L^- + H^+$$

then complication may arise in analysing the rate constants due to **proton ambiguity** that has been discussed separately (*cf.* Sec. 5.17).

Other metal ions like $[Cr(OH_2)_6]^{3+}$, $[Ce(OH_2)_6]^{4+}$, etc. having a strong tendency for hydrolysis show the similar behaviour observed for Fe(III).

5.27 CATALYSIS BY ELECTRON TRANSFER REACTIONS IN THE LIGAND SUBSTITUTION REACTIONS OF OCTAHEDRAL COMPLEXES

If a particular metal centre can exist in two different oxidation states where the higher oxidation state is relatively more inert and the lower oxidation state is relatively more labile, then the lower oxidation state of the metal can catalyze the ligand substitution process at the relatively more inert centre of higher oxidation state. This catalysis originates from the electron transfer reactions (through the inner-sphere mechanism) between the two different oxidation states of the metal centre. Thus, Co(II) can catalyze the substitution reactions at Co(III); Cr(II) can catalyze the substitution reactions at Cr(III); Pt(II) can catalyze the substitution reactions of Pt(IV); etc.

Catalysis by Co(II) in the water exchange process of Co(III) has been already discussed (Sec. 5.13). The other examples will be discussed in the next chapter.

5.28 ISOMERIZATION REACTIONS IN THE OCTAHEDRAL COMPLEXES

5.28.1 Linkage Isomerization (see Sec. 2.2.5, Vol. 4)

- **Nitrito $\rightleftharpoons$ Nitro (intramolecular path):** The nitrito complex, $[Co(NH_3)_5(ONO)]^{2+}$ (**Red**) undergoes isomerization even in solid state to the thermodynamically more stable (*cf.* **symbiotic effect**, Chapter 2, Vol. 4)) nitro complex, $[Co(NH_3)_5(NO_2)]^{2+}$ (**Yellow**)

$$\left[Co(NH_3)_5({}^*ONO)\right]^{2+} (red) \xrightarrow[\text{(slow)}]{\text{heating}} \left[Co(NH_3)_5(NO_2)\right]^{2+} (yellow), \textit{ i.e. } [Co(NH_3)_5(NO{}^*O)]^{2+}$$

The rearrangement occurs through an **intramolecular path**. It is evidenced by the fact that there is no uptake of the labelled nitrite added to the solution during the isomerisation process. The intramolecular path of isomerisation may be outlined as follows:

$$\left[(H_3N)_5Co—{}^*ONO\right]^{2+} \rightleftharpoons \left[(H_3N)_5Co\begin{smallmatrix}\cdot\cdot {}^*O\\ \vdots \\ \cdot\cdot N{=}O\end{smallmatrix}\right]^{2+} \rightleftharpoons \left[(H_3N)_5Co—NO{}^*O\right]^{2+}$$

nitrito, *i.e.* nitrito(κO) isomer — nitro, *i.e.* nitrito(κN) isomer

The above mechanism has been described as S_Ni (*substitution, nucleophilic, internal displacement*). The similar mechanism has been suggested for the isomerisation of nitrito to nitro complexes of Rh(III) and Ir(III).

For the isomerisation of nitrito to nitro complex of Co(III), the $\Delta V^{\neq}$ value is negative (*ca.* –5 to –7 $cm^3\ mol^{-1}$ for such isomerisation reactions of Co(III), Rh(III) and Ir(III) complexes). In fact, all the complexes experiencing the nitrito to nitro isomerisation through the S_Ni path shows the **negative value of $\Delta V^{\neq}$**. In the **cyclic transition state** both the N– and O– sites remain bonded to the metal centre. The cyclisation in the transition state causes a ***conformational entropy loss*** to make $\Delta S^{\neq}$ also negative and it also reduces the molar volume of the group experiencing the cyclisation **to make $\Delta V^{\neq}$ also negative.** In fact, these are the common features of the cyclic transition state. It may be mentioned that the molar volume of the nitro complex is less than that of the nitrito isomer.

$$\left[(H_3N)_5Co—N\begin{smallmatrix}O^*\\ \\ O\end{smallmatrix}\right]^{2+} \qquad \left[(H_3N)_5Co—\overset{*}{O}—N{=}O\right]^{2+}$$

Molar volume: 69.4 $cm^3\ mol^{-1}$ — 82.3 $cm^3\ mol^{-1}$

- **Thiocyanato ⇌ Isothiocyanato (intramolecular path):** Isomerisation of $[Co(NH_3)_5(SCN)]^{2+}$ to the more stable (*cf.* symbiotic effect) isomer $[Co(NH_3)_5(NCS)]^{2+}$ occurs through the **intramolecular path.**

$$\left[(H_3N)_5Co—S—C\equiv N\right]^{2+} \longrightarrow \left[(H_3N)_5Co\cdots S\equiv C\cdots N\right]^{2+} \longrightarrow \left[(H_3N)_5Co—N=C=S\right]^{2+}$$

thiocyanato, *i.e.* thiocyanato(κ*S*) isomer   isothiocyanto, *i.e.* thiocyanato(κ*N*) isomer

This intramolecular path is again supported by the **negative value of** $\Delta V^{\neq}$ (= -5.3 cm^3 mol^{-1}).

- **Thiocyanato ⇌ Isothiocyanato (intermolecular path):** It is illustrated for the **square planar complex,** $[(Et_4dien)Pd—XCN]^+ \rightarrow [(Et_4dien)Pd—NCX]^+$ (X = S or Se). Its rate of isomerisation is identical with that of its solvolysis or substitution of XCN^- by Br^-. It can be explained by considering the **intermolecular** (dissociation mechanism).

$$[(Et_4dien)Pd—XCN]^+ + S\,(solvent) \xrightarrow[slow]{k_d} [(Et_4dien)Pd(S)]^{2+} + XCN^-$$

$$[(Et_4dien)Pd(S)]^{2+} + XCN^- \xrightarrow[fast]{k_{isomer}} [(Et_4dien)Pd—NCX]^+ + S$$

$$[(Et_4(dien)Pd(S)]^+ + Br^- \xrightarrow[fast]{k_{Br}} [(Et_4dien)Pd—Br]^+ + S$$

$$k_d\ (solvolysis\ rate\ constant) = k_{isomer} = k_{Br}$$

See Sec. 2.2.6, Vol. 4 for the relative stabilities of these two isomers.

Intermolecular *vs.* intramolecular path for the isomerisation M—SCN ⇌ M—NCS and M—ONO ⇌ M—NO₂

$$M—S—C\equiv N \rightleftharpoons M—N=C=S \longleftrightarrow M—N\equiv C\rightarrow S$$

$$M—O—N=O \rightleftharpoons M—N(=O)\rightarrow O$$

It is obvious that for the **intramolecular path** (*i.e.* S_Ni), the **cyclic transition state** experiences a less strain for the interconversion between the nitro and nitrito isomers compared to that for the interconversion between the thiocyanto and isothiocyanto isomers. In fact, cyclisation by the **linear inflexible** SCN^- moiety is energetically unfavourable. This is why, the S_Ni process is more likely in M—ONO ⇌ M—NO_2 isomerisation than in M—SCN ⇌ M—NCS isomerisation

It is assumed that the anation process (that leads to two different linkage isomers) is relatively faster than the aquation process. *It makes the isomerisation rate equal to the aquation rate.* Aquation releases SCN^- whose reentry through anation leads to the linkage isomers.

5.28.2 Geometrical Isomerisation and Racemisation for the $[M(A—A)_2X_2]$ Type Complexes

(a) ***Cis–trans*** **isomerisation of the octahedral complexes of the type** $[M(A—A)_2X_2]^+$**:** It is illustrated for $[CoCl_2(en)_2]^+$ (see Sec. 1.16.9, Vol. 4).

$[Co(CO_3)(en)_2]^+ + HCl(aq) \rightarrow$ mixture of *cis*- and *trans*-$[CoCl_2(en)_2]Cl$.

A particular isomer that will separate out depends on its solubility. The *less soluble isomer*, *cis*-$[CoCl_2(en)_2]Cl$ separates out from an aqueous solution. On the other hand, the *trans*–isomer, *trans*-$[CoCl_2(en)_2]Cl{\cdot}HCl$ separates out from an aqueous solution of HCl. It may be mentioned that from an aqueous solution of the *cis*- and *trans*-isomers of $[CoCl_2(en)_2]NO_3$, the *trans*-isomer being less soluble separates out first. Thus, the solubility of an isomer depends on the ion residing in the outer-sphere.

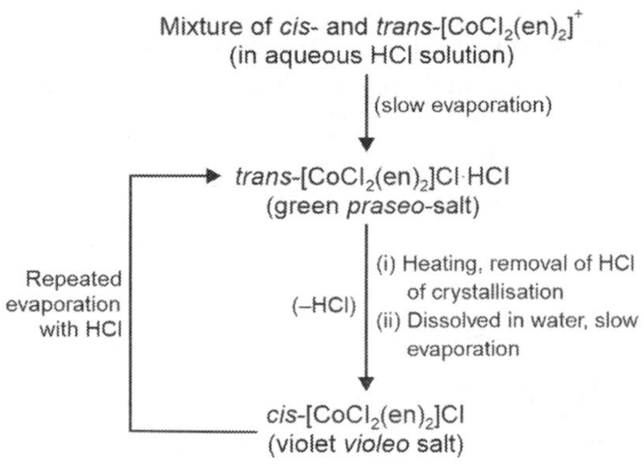

The *cis–trans* isomerisation of $[CoCl_2(en)_2]^+$ is believed to pass through the aquation process.

$$cis\text{-}\left[CoCl_2(en)_2\right]^+ + H_2O \xrightleftharpoons{\text{SP intermediate}} cis\text{-}\left[CoCl(en)_2(OH_2)\right]^{2+} + Cl^-$$

$$\upharpoonleft\downharpoonright \text{ TBP intermediate}$$

$$trans\text{-}\left[CoCl_2(en)_2\right]^+ + H_2O \xrightleftharpoons{\text{SP intermediate}} trans\text{-}\left[CoCl(en)_2(OH_2)\right]^{2+} + Cl^-$$

The above **intermolecular path** is supported by the fact that when the said isomerisation is carried out in presence of labelled chloride, *i.e.* $*Cl^-$, there is a random distribution of $*Cl^-$ in the complex. *It may be mentioned that replacement of one anion by another anion occurs through the aquation (i.e. direct replacement does not occur).* When the isomerisation is carried out at a relatively higher temperature, further aquation of $[CoCl(en)_2(OH_2)]^{2+}$ to $[Co(en)_2(OH_2)_2]^{3+}$ may occur. Then the successive anation of $[Co(en)_2(H_2O)_2]^{3+}$ may lead to an isomerisation. It may be noted that aquation of *cis*-$[CoCl_2(en)_2]^+$ produces 100% *cis*–$[CoCl(en)_2(OH_2)]^{2+}$ as the initial product while aquation of *trans*-$[CoCl_2(en)_2]^+$ gives a mixture of *cis*- (*ca.* 35%) and *trans*-product (*ca.* 65%) of $[CoCl(en)_2(OH_2)]^{2+}$ (*cf.* Sec. 5.21).

Retention of stereochemistry can be explained by considering the **square pyramidal** (SP) intermediate in the dissociative path. But for the *cis* $\rightleftharpoons$ *trans* **isomerisation**, it requires the formation of a **trigonal bipyramidal** (TBP) intermediate in the dissociative path.

(b) **Mechanism of geometrical isomerisation of the $[M(A—A)_2X_2]$ type complexes through the TBP intermediate (*i.e.* activated complex):** In such cases, a **dissociative mechanism** *leading to a trigonal bipyramid intermediate* causes the isomerisation. This trigonal bipyramidal intermediate having X at an equatorial site provides the path of isomerisation (Fig. 5.28.2.1).

In the trigonal bipyramid intermediate (*cf.* Fig. 5.28.2.1), the nucleophile enters along the edges of the basal triangle where the ligands are separated by 120° (*cf.* other edges separate the ligands by 90°; *i.e.* **edges of the basal triangle are sterically less hindered towards the nucleophilic attack)**. Entry along the edge (2–3) produces the *trans*-isomer while entry along the edges (2–4, 3–4) produces the *cis*– isomers. If all these sites of nucleophilic attack are equally accessible then the entry of X will give the *cis*- and *trans*-isomers in 2:1 ratio. The *cis*-product possesses the same optical activity of the starting *cis*-complex. Thus the loss of optical activity in the mechanism given in Fig. 5.28.2.1 is due to the formation of optically inactive *trans*-product.

It is expected that if optically pure (*i.e.* Δ or Λ) *cis*–$[CoCl_2(en)_2]^+$ is allowed to isomerise as in Fig. 5.28.2.1, then it will be accompanied with the **loss of optical activity** due to the isomerisation to the optically inactive *trans*-isomer (*cf.* Figs. 5.21.4, 5.28.2.2). In fact, the rate of *cis*- to *trans*-isomerisation is almost comparable to the rate of loss of optical activity and also to the rate of chloride exchange (using $*Cl^-$).

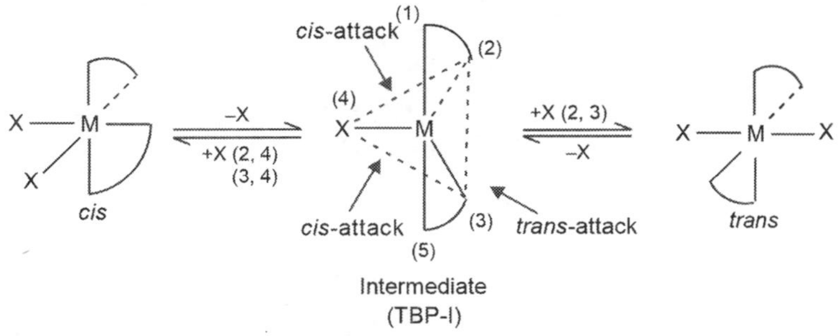

Fig. 5.28.2.1: Dissociative mechanism (intermolecular path for *cis– trans* isomerisation of the complexes like $[Co(en)_2X_2]^+$ (*cf.* Figs 5.21.4, 5.21.5 and 5.28.2.2)

Note: The TBP given in Fig. 5.28.2.1 gives the *cis*-product with the same optical activity (*i.e.* Δ gives the Δ isomer). There is another possible trigonal bypyramid (having X at one axial site) that will produce a *cis*–**racemic mixture** without geometrical isomerisation. In fact, these two TBPs may be interconverted through Berry pseudorotation (*cf.* Figs. 5.21.4 and 5.28.2.2)

Thus the conclusions are:

(i) isomerisation (*i.e. cis-trans*) is accompanied with the loss of optical activity.

(ii) k (*cis-* to *trans*-isomerisation) ≈ k (rate of loss of optical activity) ≈ k (Cl^- exchange).

These indicate that there is a **common rate determining step** for the geometrical isomerisation, and Cl^- exchange processes. This common rate determining step is the dissociative transformation (through the expulsion of one X) of the octahedral complex to a trigonal bipyramidal intermediate **keeping preferably one X (which is a π–donor ligand) at an equatorial position**. These are in agreement with the mechanism outlined in Fig. 5.28.2.1.

Here it is worth mentioning that if X is a good π-donor ligand, then for the *cis*-complex, the dissociative expulsion of one X preferably produces the **square pyramidal** (SP) **intermediate** which is stabilised by the π-donor remaining X (*i.e. cis*-effect). In such cases, **there will be no isomerisation** (see Sec. 5.21). To explain the **isomerisation**, the SP intermediate initially produced must be converted into the TBP geometry where X occupies an equatorial position.

(c) **Mechanism of isomerisation accompanied with the racemisation for the *cis*-$[M(A—A)_2X_2]^+$ type complexes:** The dissociative expulsion of one X initially produces a square pyramidal (SP) intermediate that may be converted into a trigonal bipyramidal (TBP) intermediate to **cause the isomerisation** (*cf.* no isomerisation for the SP intermediate). Here it may be mentioned that if the trigonal bipyramid intermediate is supposed to experience the **Berry's pseudorotation** (*i.e.* the unidentate ligand X may occupy both the equatorial and axial positions), then two different trigonal bipyramids (**TBP(I)** and **TBP(II)**) may be formed; *cf.* Fig. 5.28.2.2). The Berry's pseudorotation leading to two different trigonal bipyramidal geometries is favoured if the said 5-coordinate TBP intermediate is **long-lived and X (unidentate ligand) does not have any strong preference to occupy an equatorial or axial position.** In fact, if X is a good π–donor ligand, it will have a preference to occupy the equatorial position (*i.e.* TBP (I)).

Regeneration of $[M(A—A)_2X_2]$ from the five coordinate TBP intermediate **structure II** having X in the axial position (Fig. 5.28.2.2) will give only the *cis*–product which is a **racemic mixture.** On the other hand, the TBP intermediate, **structure I** keeping X at one equatorial position can lead to *cis*–(which is optically pure and **retention of the same chirality**) and *trans*-product in 2:1 ratio.

The important conclusions are as follows:

- If X is a **good π-donor ligand** then TBP (I) (**X at the equatorial position**, see Fig. 5.21.1) will be preferred over the TBP(II) **(X at the axial position).**

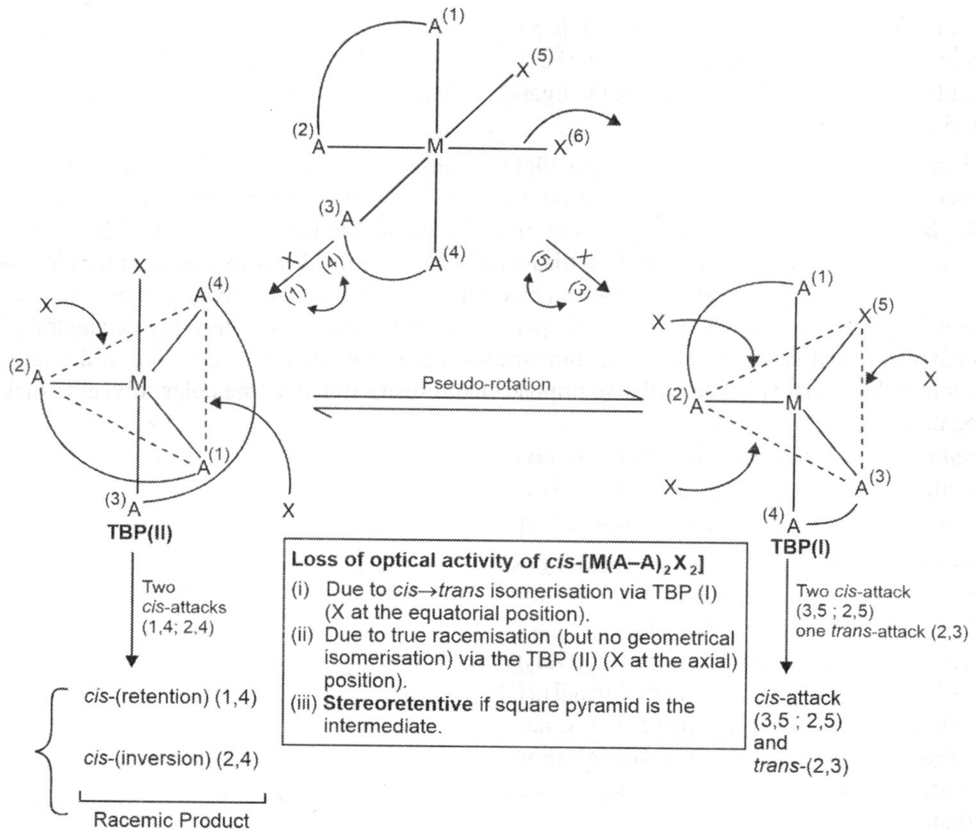

Fig. 5.28.2.2 Trigonal bipyramids as the intermediates for *cis–trans* isomerisation of the $[M(A—A)_2X_2]$ type complexes along with the racemisation of the optically active *cis*–complex, (*cf.* Fig. 5.21.4).

- TBP(I) leads to the **geometrical isomerisation** (*cis* → *trans*) and the *cis*-product with the retention of the same optical activity. The loss of optical activity along this route is due to the isomerisation to the *trans*-complex **which is optically inactive.**
- TBP(II) causes the **true racemisation without any geometrical isomerisation.**
- Measured **loss of optical activity** (some times called **apparent** or **measured racemisation**) can occur due to both the geometrical isomerisation (via TBP(I)) and true racemisation (via TBP (II)).
- Rate (loss of optical activity), *i.e.* Rate (measured racemisation) ≈ Rate (*cis* → *trans*) indicates TBP(I) to be the predominant species (when X is a good π-donor ligand).
- Rate (loss of optical activity), *i.e.* Rate (measured racemisation) > Rate (*cis* → *trans*) (≠ 0) indicates both TBP(I) and TBP(II) to contribute.
- Rate (loss of optical activity), *i.e.* Rate (measured racemisation) ≫ Rate (*cis* → *trans*) (≈ 0, *i.e.* practically no geometrical isomerisation) indicates TBP(II) to be the predominant species.

(d) **Isomerisation and racemisation of *cis*-$[Cr(C_2O_4)_2(OH_2)_2]^-$ in neutral solution:** Here it may be pointed out that both the intermediate TBP structures I and II are not equally stable. Thus the product distribution depends on the relative proportions of these two structures. In fact, for *cis*– $[Cr(C_2O_4)_2(OH_2)_2]^-$ (optically active), the rate of racemisation is much faster than the rate of *cis*– to *trans*– isomerisation. It suggests that **the loss of optical activity of the starting *cis*-complex is not only due to its isomerisation to the *trans*-complex via TBP(I).** In other words, both the TBP(I) and TBP(II) (causing the true racemisation but no geometrical isomerisation) are involved and the TBP(II) may predominate over the TBP(I). In fact, oxalate-O ($C_2O_4^{2-}$) is a much better π-donor ligand than H_2O. Thus oxalate as a bidentate chelating ligand occupying the two equatorial positions stabilises TBP(II) better (*cf.* π-bonding can efficiently occur when the ligand occupies the equatorial positions of a TBP geometry, See Fig. 5.21.1b).

Here it is worth mentioning that in **neutral or slightly acidic condition** (pH ≈ 3–7), *cis*-$[Cr(C_2O_4)_2(OH_2)_2]^-$ experiences isomerisation and recemisation through the TBP intermediate generated from the **Cr—OH$_2$ bond cleavage** ($\Delta V^{\neq}$ = positive) but in **acidic condition** (pH ≈ 0 – 2), the TBP intermediate is generated by **acid catalysed one-ended dissociation (*cf.* Fig. 5.29.2) of one oxalate chelate ring** making a free —CO_2^- group which imposes an electrorestriction over the solven to make $\Delta V^{\neq}$ negative. Subsequently, ring closing by the protonated open end of the oxalate gives back the octahedral product with racemisation and isomerisation. The unneutralised positive charge on Cr^{III} due to the one ended dissociation of the oxalate chelate will also impose electrorestriction over the polar solvent to make $\Delta V^{\neq}$ more negative.

(e) **Isomerisation and racemisation for *cis*-$[Co(en)_2(OH_2)X]^{2+}$ (X = OH^-, Cl^-, Br^-, N_3^-):** The experimental observations are (recemisation is measured by the **loss of optical activity):**

k(*cis* → *trans*) ≈ *k*(measured recemisation); $\Delta V^{\neq} \approx +7$ cm^3 mol^{-1} (*i.e.* dissociative path)

It indicates the dissociative process (cleavage of the Co—OH$_2$ bond) predominantly passes through the TBP(I) where X occupies an equatorial position of the TBP structure. It is reasonable because of the π-donor property of the given X-ligands.

(f) ***Trans* → *cis* isomerisation of $[Cr(OH_2)_2(ox)_2]^-$ *vs.* $[Cr(mal)_2(OH_2)_2]^-$:** The *trans*– → *cis*–isomerisation of $[Cr(OH_2)_2(ox)_2]^-$ and $[Cr(mal)_2(OH_2)_2]^-$ (ox = oxalate, mal = malonate) occurs through the dissociative transformation into a 5–coordinate intermediate. *But in the case of oxalato complex, the chelate ring opens* (*i.e.* **one ended dissociation of the chelate ring)** to give a five coordinate TBP intermediate while for the malonato complex, the aqua ligand dissociates to give the 5–coordinate TBP intermediate.

$$[Cr(C_2O_4)_2(OH_2)_2]^- \xrightarrow{rds} \left[Cr(C_2O_4)(OH_2)_2 - O - C(=O) - CO_2^-\right]^- \text{ (TBP)}$$

(one ended dissociation of the oxalate chelate ring)

$$[Cr(C_3H_2O_4)_2(OH_2)_2]^- \xrightarrow{rds} \underset{\text{(TBP)}}{[Cr(C_3H_2O_4)_2(H_2O)]^-} + H_2O$$

It is suggested that the six membered chelate ring formed by the malonate is more stable than the five membered chelate ring formed by the oxalate in the TBP intermediate where the chelate ring spans the equatorial and axial position. *This is why, the oxalato chelate ring experiences the one–ended dissociation while the malonato complex experiences the dissociation of one aqua ligand.*

The above propositions are experimentally supported by the $\Delta V^{\neq}$ values. For the oxalato complex, $\Delta V^{\neq}$ is **highly negative** (= –16.6 cm^3 mol^{-1}) while for the malonato complex, it is **positive**

- **Racemisation** producing a mixture of Δ and Λ forms is a purely **an entropy driven process** ($\Delta H_{race} = 0$ as $E_\Delta = E_\Lambda$), ΔS_{race} = +ve) (see Fig. 5.1.1.1).
- **Isomerisation** producing a mixture of *cis*- and *trans*-forms (*i.e.* geometric isomers) increases also entropy (*i.e.* $\Delta S > 0$) but the energy of the geometric isomers may be different (*i.e.* one form may be more stable). Thus sign of ΔH for an isomerisation process depends on the relative stability of the isomers (*i.e.* $\Delta H \neq 0$).

(= + 8.9 cm^3 mol^{-1}) as expected for the dissociative process. For the oxalato complex, one ended dissociation of the chelate ring generates a free —CO_2^- group which can impose an electrorestriction over the solvent molecules to make $\Delta V^{\neq}$ negative. On the other hand, dissociation of one neutral ligand (*i.e.* H_2O) neither creates nor destructs any charge. Thus $\Delta V^{\neq}$ is positive as usual.

The isomerisation of *trans* → *cis* for $[Cr(C_2O_4)_2(OH_2)_2]^-$ can be catalysed by the bivalent metal ions (*e.g.* Cu^{2+}, Ni^{2+}, Co^{2+}, Ca^{2+}, etc.) which favour **the one-ended dissociation of the oxalate ring** through complexation with the partially dissociated oxalate moiety. Thus the TBP generated gives back the octahedral complex through ring closure and simultaneous release of the metal ion.

(g) ***Trans* → *cis* isomerisation of $[Co(CH_3COO)(en)_2(OH_2)]^{2+}$:** The *trans*→*cis* isomerisation of the complex is associated with the positive $\Delta V^{\neq}$ value (= +7.9 cm^3 mol^{-1}). It suggests the generation of a 5–coordinate TBP intermediate through the dissociation of the neutral aqua ligand.

5.29 RACEMISATION OF THE OCTAHEDRAL COMPLEXES (*cf.* Sec. 5.33 for the tetrahedral complexes)

(1) ***cis*–$[M(L—L)_2X_2]$:** Racemisation of such optically active complexes can be explained in terms of the generation of a trigonal bipyramid intermediate produced through the dissociation of one X ligand. These aspects have been discussed in Sec. 5.28.2 (*cf.* Figs. 5.21.4; 5.28.2.2).

(2) **$[M(L—L)_3]$:** Racemisation of such *tris*–chelates can occur through two possible routes:

(A) intermolecular mechanism and **(B) intramolecular mechanism**

(A) **Intermolecular dissociative mechanism:** This process involves the reversible dissociation of a ligand giving rise to a symmetrical *bis*–chelate.

Reentry of L—L to the **symmetrical intermediate** $[M(L—L)_2]$ can give both the Δ and Λ isomers with an equal probability. This leads to racemisation. It is evident that this process will also lead to the ligand exchange with the free ligand L—L present in solution.

In this process, we can reasonably expect that the rate of racemisation should be comparable (within the experimental error limit) to the rate of dissociation of the *tris*– to *bis*– complex and also the rate of ligand exchange of the $[M(L—L)_3]$ complex with the free ligand L—L (*cf.* ligand exchange process carried out by using the labelled ligand)

$$k_{rac} \approx k_{diss}\,(tris \rightarrow bis) \approx k_{ex}$$

$$\underset{\Delta}{[M(L—L)_3]} \rightleftharpoons [M(L—L)_2] + L—L \rightleftharpoons \underset{\Lambda}{[M(L—L)_3]}$$

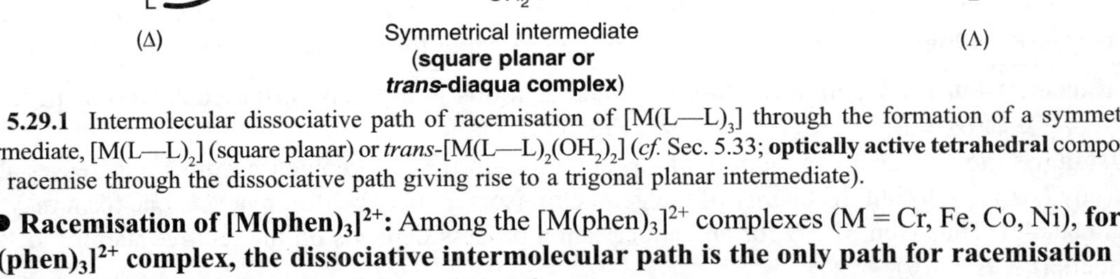

Fig. 5.29.1 Intermolecular dissociative path of racemisation of $[M(L—L)_3]$ through the formation of a symmetrical intermediate, $[M(L—L)_2]$ (square planar) or *trans*-$[M(L—L)_2(OH_2)_2]$ (*cf.* Sec. 5.33; **optically active tetrahedral** compounds may racemise through the dissociative path giving rise to a trigonal planar intermediate).

● **Racemisation of $[M(phen)_3]^{2+}$:** Among the $[M(phen)_3]^{2+}$ complexes (M = Cr, Fe, Co, Ni), **for the $[Ni(phen)_3]^{2+}$ complex, the dissociative intermolecular path is the only path for racemisation and it is supported by the fact:** $k_{diss}\,(tris \rightarrow bis) \approx k_{rac}$.

For all other $[M(phen)_3]^{2+}$ complexes (M = Cr, Fe, Co), $k_{rac} > k_{diss}$ *indicates that the racemisation process probably passes through the different path, i.e.* **intramolecular path.** Here it may be mentioned that in such cases, *both the inter– and intramolecular paths may also go on simultaneously* to make the racemisation rate faster than the dissociation rate; or the racemisation through the intramolecular path is faster than the intermolecular dissociative path. In general, we may write:

$$k_{rac} = k_{intra} + k_{inter} \ (= k_{diss})$$
$$= k_{diss}, \ (\text{when } k_{intra} = 0)$$

● **Racemisation of $[Ni(bpy)_3]^{2+}$ *vs.* $[Ni(phen)_3]^{2+}$:** Both these complexes undergo racemisation through the **dissociative intermolecular pathway.** In fact, for both the complexes, the kinetic parameters (*i.e.* k, $\Delta H^{\neq}$ and $\Delta S^{\neq}$) are identical (within the experimental error limit) for dissociation (*tris*– to *bis*– complex) and racemisation in acid media under the comparable conditions, *i.e.* $k_{rac} \approx k_{diss}$.

Table 5.29.1: Comparison of rate constants for dissociation (k_{diss}) of *tris*- to *bis*-complex and racemisation (k_{rac}), (at 25°C) of $[M(phen)_3]^{2+}$ complexes (*cf.* E.L. Blinn *et al., Inorg. Chem.*, **15,** 2952, 1976; R.G. Wilkins *et al., J. Chem. Soc. (London),* 1763, 1957)

M	k_{diss} (s^{-1})	k_{rac} (s^{-1})
Cr	1.7×10^{-2}	0.12
Fe	7.0×10^{-5}	6.5×10^{-4}
Co	0.16	20.0
Ni	7.5×10^{-6}	9.4×10^{-6}

However, acid dependence of rates is of different types for $[Ni(bpy)_3]^{2+}$ and $[Ni(phen)_3]^{2+}$. The rates are independent of acid concentration for $[Ni(phen)_3]^{2+}$ while the rates are dependent on acid concentration for $[Ni(bpy)_3]^{2+}$ which can experience the **one ended dissociation of the chelate ring** accompanied by a rotation around the C—C bond and protonation of the released end of the ligand followed by the complete loss of the ligand. In the phen-complex, because of the rigid structure of the ligand, no such one-ended dissociation process is possible. This makes the difference in acid dependence pattern of their rates. This aspect has been discussed in detail in Sec. 5.23.1.

In acidic media, the dissociated chelating ligand (*i.e.* **phen**) remains protonated and it cannot recombine with the *bis*–chelate to regenerate the *tris*–chelate, *i.e.* **racemisation is not noticeable in a strongly acidic condition.** In neutral water, the dissociated ligand does not remain protonated, and recombination of the dissociated ligand to produce the *tris*–chelate is quite feasible. ***In fact, in neutral media, dissociation (i.e. tris–complex to bis–complex) is negligible and racemisation is predominant.*** *In acidic media where the dissociated ligand remains in a protonated condition, the reverse is true.* As, the rate determining step is the same for both the raceimisation and dissociation processes, **the rate of recemisation in neutral media is the same as the rate of dissociation in acidic media for $[Ni(phen)_3]^{2+}$.**

In general, in acidic condition, racemisation of $[M(N—N)_3]^{2+}$ (N—N = bpy, phen) through the chelate ring opening (both one end and two end opening) stops due to protonation of the dislodged ligand. This aspect has been discussed later.

(B) **Intramolecular mechanism:** This can occur in three possible ways.

(i) Formation of a **symmetrical or stereochemically nonrigid intermediate of lower coordination number** (*e.g.* 5, 4) through the one ended dissociation of the chelate ring(s);

(ii) Formation of a **symmetrical intermediate of higher coordination number** (*e.g.* 7, 8) through the ligation by the monodentate auxiliary ligand(s);

(iii) Formation of a **symmetrical intermediate without any change of coordination number** through the **distortion or twisting of the octahedral structure**.

(a) Formation of an intermediate of lower coordination number through the one–ended opening of a chelate ring: One-ended dissociation of a chelate ring can generate a **stereomobile** (*i.e.* fluxional or stereochemically nonrigid) **5–coordinate** (*e.g.* trigonal bipyramid, square pyramid) intermediate. These 5-coordinate intermediates (*i.e.* SP and TBP) can undergo Berry pseudorotation with the scrambling of ligand sites. Religation by this free end of the partially dissociated ligand can lead to both Δ and Λ

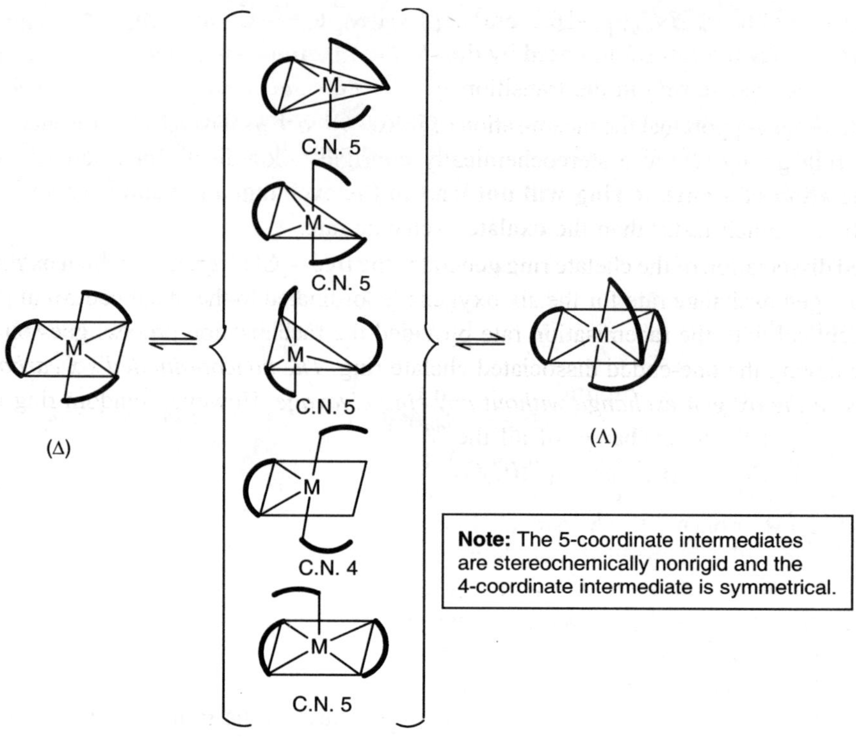

Fig. 5.29.2: Intramolecular path of racemisation of the [M(L—L)$_3$] type chelate through the one-ended dissociation of the chelate ring.

isomers. One-ended dissociation in two chelate rings produces a **4-coordinate symmetrical intermediate** from which both Δ- and Λ isomers can be generated through the religation by the free ends. These are illustrated in Fig. 5.29.2.

• **Racemisation of $[M(C_2O_4)_3]^{3-}$ (M = Co, Cr, Rh) in solution:** $[M(C_2O_4)_3]^{3-}$ (M = Co, Cr, Rh) experiences racemisation in this pathway. In fact, for such complexes, the racemisation rate is much faster than the ligand exchange rate with the labelled oxalate, $^*C_2O_4^{2-}$. It indicates that the racemisation for such complexes does not occur through the **dissociative intermolecular path.** However, the acid-catalysed exchange of ^{18}O in $H_2{}^{18}O$ enriched water with all the oxygens in $[Cr(C_2O_4)_3]^{3-}$ – $^{18}OH_2$ system occurs almost in the same rate of racemisation under the comparable conditions. Thus the observation is:

$$k(\text{racemisation}) \approx k(\mathbf{O}\text{-exchange}) \gg k(\text{oxalate exchange}), \ (\text{for } [\mathbf{Cr(ox)_3}]^{3-})$$

In $[Co(C_2O_4)_3]^{3-}$ and $[Rh(C_2O_4)_3]^{3-}$, the six outer oxygens (*i.e.* noncoordinating carbonyl oxygens) exchange with $^{18}OH_2$ much faster than the remaining six oxygens (*i.e.* coordinated O–atoms) and than the racemisation process.

The outer noncoordinating carbonyl O-atoms may exchange with ^{18}O of $H_2{}^{18}O$ (through the nucleophilic attack by $H_2{}^{18}O$ on the carbonyl-C) without any M—O bond cleavage. But the said O-exchange for the metal coordinated O-centres can occur only when the M—O bond cleaves. Co(III) and Rh(III) are kinetically more inert than Cr(III) and this is why, the Co(III) and Rh(III) complexes experience the O-exchange in **both ways**: in fully coordinated oxalate and in one end open oxalate. In Cr(III), the O-exchange process mainly goes through the one-ended dissociation of the oxalate ring.

The acid catalysed or M^{2+}-catalysed recemisation of *cis*-$[Cr(C_2O_4)_2(OH_2)_2]^-$ also occurs through the TBP intermediate produced from the one-ended dissociation of the oxalate ring (*cf.* Sec. 5.28.2). The highly **negative value of $\Delta V^{\neq}$** (= −16.3 cm^3 mol^{-1}) for racemisation of $[Cr(ox)_3]^{3-}$ is due to the electrorestriction over the solvent imposed by the free $—CO_2^-$ group generated through the one-ended dissociation of the oxalate ring in the transition state.

All these findings support that the racemisation of $[M(ox)_3]^{3-}$ occurs through the one-ended dissociation of a chelate ring giving rise to a stereochemically nonrigid 5-coordinate intermediate. **Such a one-ended dissociation of a chelate ring will not lead to the exchange of oxalate.** This is why, the rate of racemisation is much faster than the oxalate exchange rate.

One-ended dissociation of the chelate ring generates the free $—CO_2^-$ group at which oxygen exchange can occur. Oxygen exchange rate for the six oxygens (coordinated to the metal centre) in $[M(C_2O_4)_3]^{3-}$ should be identical with the racemisation rate provided the racemisation process goes on through the intermediate having the one-ended dissociated chelate ring. *The noncoordinated six carbonyl oxygens may experience the oxygen exchange without any ring cleavage.* However, random ring opening and ring closure will allow the exchange of all the 12 oxygens of the three bound oxalate ligands at a comparable rate in $[M(C_2O_4)_3]^{3-}$ as in $[Cr(C_2O_4)_3]^{3-}$.

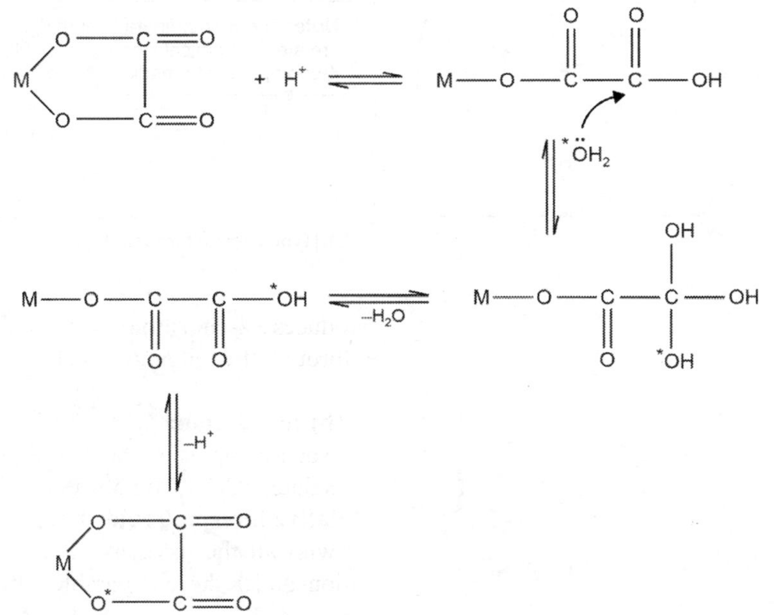

Fig. 5.29.3 O-exchange in $[M(ox)_3]^{3-}$ through the one ended dissociation of the oxalate ring. Racemisation occurs through the formation of the same 5-coordinate intermediate.

● **Racemisation and linkage isomerisation in a Co(III) complex, *tris*(acetylacetylacetonato)-cobalt(III) where one CH_3 group is substituted by CD_3:** Dissociative pathway (*i.e.* **one-ended dissociation of the chelate ring**) has been argued for racemisation of the compound.

The complex (in chlorobenzene) undergoes also linkage isomerisation but the *the rate of racemisation is faster than the rate of linkage isomerisation.* It is proposed that the 5–coordinate intermediate is common for the both processes. However, for the linkage isomerisation, it needs a rotation around the 'C—C' bond of the one-ended dissociated chelate. This additional operation makes the linkage isomerisation process slower. The process is outlined in the scheme (Fig. 5.29.4).

A → B → C: (linkage isomerisation + racemisation) where a rotation around the C—C bond is required.

A → B → D: (racemisation without linkage isomerisation) where the rotation around the C—C bond is not required.

(b) **Formation of an intermediate of higher coordination number:** Formation of a symmetrical intermediate like **cubic structure** (C.N. 8; Charonnat, 1931) (Fig. 5.29.5) by the ligation of two monodentate **auxiliary ligands** (which may be the coordinating solvents) can lead to both Δ and Λ isomers through the expulsion of these auxiliary ligands from the cubic structure. Similarly, a symmetrical intermediate like **capped octahedron** (C.N. 7; Bush *et al.*, 1962) may be also formed for racemisation (Fig. 5.29.5).

● **Racemisation of $K_3[Cr(C_2O_4)_3]{\cdot}2H_2O$ *vs.* $K_3[Cr(C_2O_4)_3]$ in solid state:** Optically active, $K_3[Cr(C_2O_4)_3] \cdot 2H_2O$ undergoes racemisation on heating (in solid state) at about 120°C in an evacuated sealed tube. It is believed that the molecules of water of crystallisation ligate as the auxiliary ligands to give the intermediate of cubic structure that leads to racemisation. From the **symmetrical cubic intermediate,** both the Δ and Λ isomers can be generated with the equal probability through the dissociation of the two auxiliary ligands (H_2O) (see Fig. 5.29.5). In fact, the anhydrous complex, $K_3[Cr(C_2O_4)_3]$ does not experience racemisation under this condition. For the anhydrous

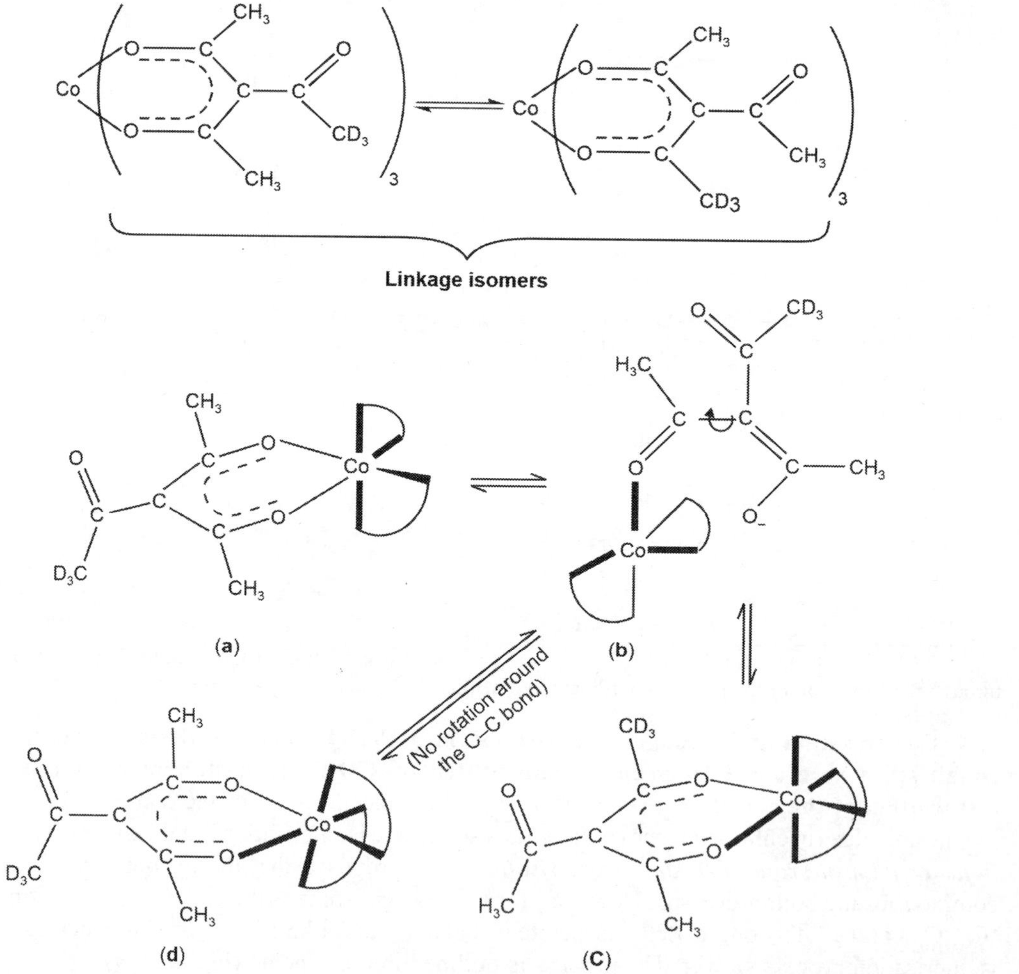

Linkage isomers

(a) (b) (C) (d)

(No rotation around the C–C bond)

Fig. 5.29.4: Linkage isomerisation and racemisation through the 5-coordinate intermediate (*cf.* A.Y. Girgis, *et al.*, *J. Am. Chem. Soc.*, **92,** 7061, 1970)

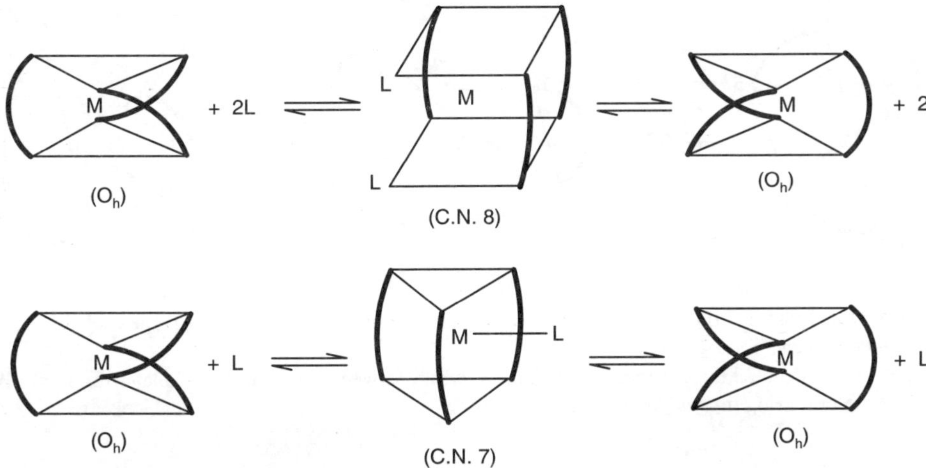

Fig. 5.29.5: Intramolecular path of racemisation of the [M(L—L)$_3$] type complexes through the formation of **symmetrical intermediate** of higher coordination number through the ligation by the auxiliary ligands (L) acting as the catalysts. (*cf.* R. Charonnat, *Ann. Chim.*, **16**, 202, 1931; C.H. Johnson, *et al.*, *Trans. Faraday Soc.*, **31**, 1621, 1935; D.H. Bush *et al.*, *Inorg. Chem.*, **1**, 13, 1962). **Note:** Here octahedron is drawn as a **trigonal antiprism** (see Fig. 5.18.2.1b for **trapezoidal octahedron**, *i.e.* capped trigonal prism).

complex, there is no such auxiliary ligand for expansion of coordination number to pave the way for racemisation.

● **Base catalysed racemisation of [Co(edta)]$^-$:** Base catalysed racemisation of [Co(edta)]$^-$ is believed to pass through the formation of a symmetrical 7-coordinate intermediate **(capped trigonal prism)** (see Fig. 5.29.5). The activation parameters ($\Delta H^{\neq}$, $\Delta S^{\neq}$) can support this proposed mechanism.

$$\Delta H^{\neq} = +134 \text{ kJ mol}^{-1};\ \Delta S^{\neq} = +230 \text{ J K}^{-1} \text{ mol}^{-1}$$

The high enthalpy of activation is quite reasonable because formation of the proposed 7–coordinate intermediate needs considerable stretching of the metal-ligand bonds. Fixing of OH^- (which is highly solvated in free condition) requires desolvation that will make a large positive value of $\Delta S^{\neq}$. Dehydration (*i.e.* desolvation) of OH^- requires a fairly large amount of energy and it also contributes to make ΔH^{*} so high.

(c) Twist mechanism: Distortion of the octahedral complex can generate a ***symmetrical intermediate*** that can give both the Δ and Λ isomers.

Ray and Dutt first proposed (1943) the twist mechanism (which is described as the **tetragonal or rhombic twist)** for the racemisation of *tris*-(biguanide)cobalt(III) ion, $[Co(bigh)_3]^{3+}$. The complex is exceedingly stable both kinetically and thermodynamically. This is why, Ray and Dutt argued that racemisation through the dissociation of the chelate ring is not possible. Ligation by the auxiliary ligands to produce a symmetrical intermediate of higher coordination number is also quite unlikely in terms of the inherent inertness of low-spin Co(III). Rather twisting is more probable for the racemisation.

In Ray-Dutt twist, keeping one chelate ring fixed, other two chelate rings rotate in opposite directions along their own planes by 45° to give a distorted octahedral geometry which may be described as a transition state (C_{2v}) for the [M(L—L)$_3$] type chelate. In fact, **Ray-Dutt twist** is described as the **rhombic twist because of this C_{2v} symmetry of the transition state.** This transition state may be also attained by rotating (through 60°) a trigonal face which is not associated with a three fold axis (C_3). Actually, this rotation is carried out about the **imaginary C_3 axis.** From this intermediate, retracing its step will give the starting optical isomer or continuation of rotation by the same magnitude will give the isomer of opposite chirality. Ray-Dutt twist is illustrated in Fig. 5.29.6–7.

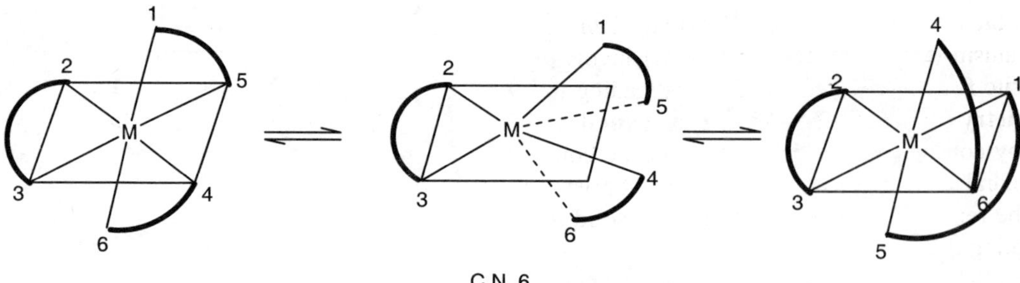

C.N. 6

Fig. 5.29.6: Ray-Dutt twist (*i.e.* rhombic twist) mechanism for racemisation of $[M(A—A)_3]$ type chelates (*cf.* P. Ray and N.K. Dutt, *J. Indian Chem. Soc.,* **20,** 81, 1943).

Note: One chelate ring denoted by 2, 3 remains unchanged while between the other two chelate rings, one chelate ring moves up while the other ring moves down simultaneously.

> **Racemisation in $[M(AB)_2]$ type tetrahedral complexes** (*cf.* Sec. 5.33). It passes through the **intramolecular rearrangement** giving rise to a symmetrical square planar intermediate.
>
> $$\underset{(T_d\Delta)}{[M(AB)_2]} \rightleftharpoons \underset{\text{planar}}{[M(AB)_2]} \rightleftharpoons \underset{(T_d\Lambda)}{[M(AB)_2]}$$
>
> **This path is similar to twisting mechanism of racemisation.**
>
> **Racemisation of A(X)(Y)(Z)** (A = N, P) (pyramidal structure) occurs through the trigonal planar intermediate without bond breaking.

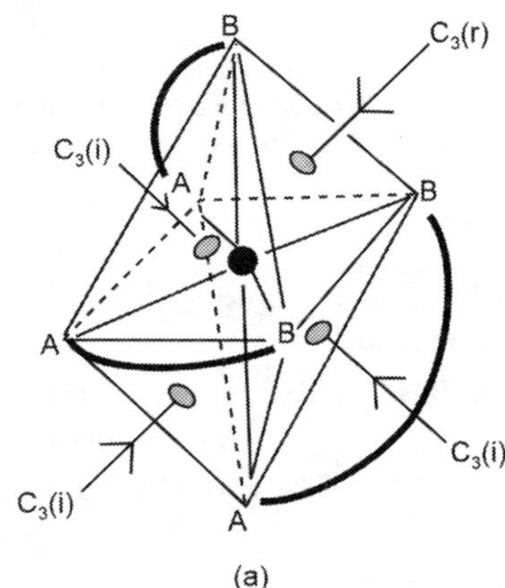

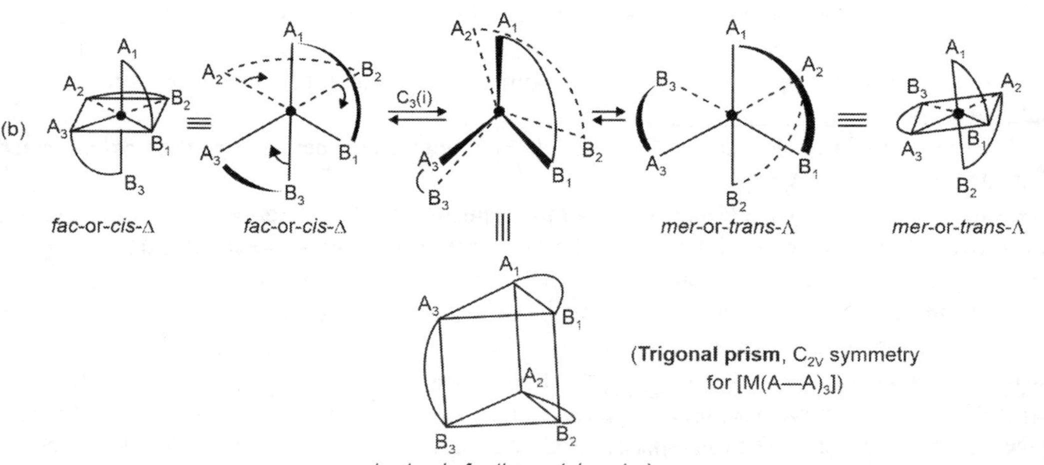

Fig. 5.29.7 (a) Position of the imaginary C_3 axis [*i.e.* C_3(i)] and real C_3 axis [*i.e.* C_3(r)] in *fac*- or *cis*-Δ isomer $[M(AB)_3]$. (b) Schematic operative stages of Ray-Dutt twist in $[M(A–B)_3]$ type chelates (*cf.* Sec. 10.9.3, Vol. 2).

Note: Here racemisation is coupled with the geometrical isomerisation. **If A = B, the intermediate possesses the C_{2v} symmetry.**

Bailar later proposed (*cf.* J.C. Bailar, *J. Inorg. Nucl. Chem.*, **8,** 165, 1958) another type of the twist mechanism (called the **trigonal twist**) involving the rotation of a trigonal face with respect to the opposite face by 60° about a C_3–axis leading to a trigonal **prismatic transition state** having the $\boldsymbol{D_{3h}}$ **symmetry (for the [M(A—A)$_3$] type complex).** The Bailar twist may be also rationalised in another way by considering the octahedral chelate complex as the **trigonal antiprism** complex. In fact, the octahedral chelate appears as a trigonal antiprism when viewed through a C_3–axis. The trigonal antiprism may be represented by two triangles — solid triangle indicating the upper plane and dashed triangle indicating the plane below the paper (*cf.* Fig. 5.29.8). Twisting of the upper triangle gives the trigonal prism (D_{3h} symmetry for the [M(A—A)$_3$] type complexes). Because of this $\boldsymbol{D_{3h}}$ **symmetry of the transitions state**, this twist mechanism is called the *trigonal twist.* From this transition state, retracing will give the starting optical isomer while continution by another 60° will give the optical isomer of opposite chirality. This is illustrated in Figs. 5.29.8–9.

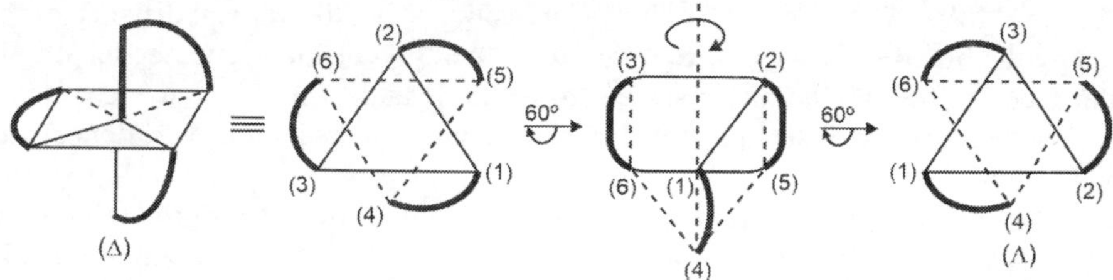

Fig. 5.29.8: Schematic representation of **Bailar twist** ('*M*' placed at the centre of the geometry, not shown)

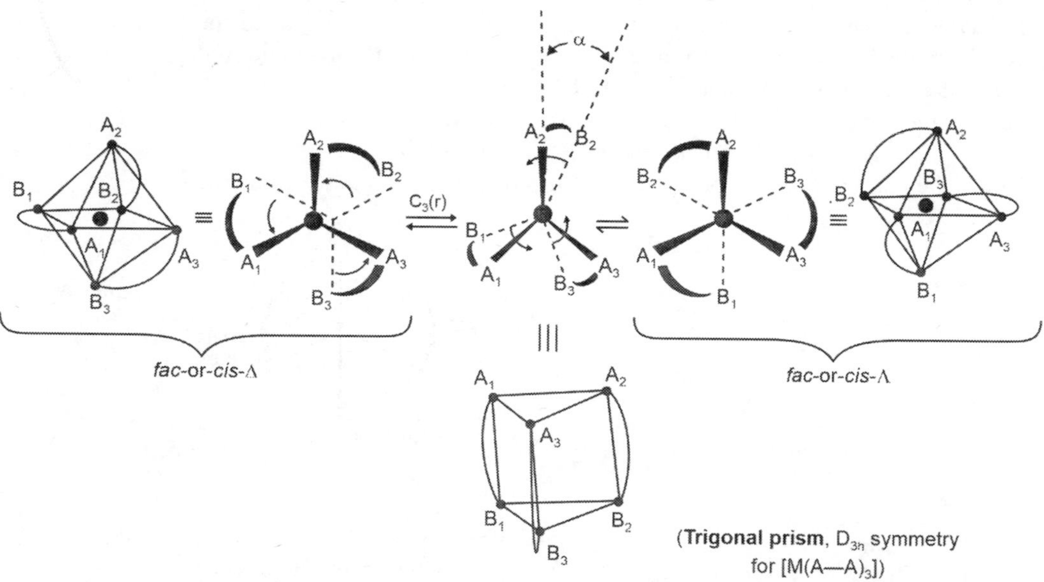

Fig. 5.29.9: Bailar twist (*i.e.* trigonal twist) and twist angle α for racemisation of a *tris*–chelate. Illustrated for the [M(A—B)$_3$] type chelates; rotation along the C_3–axis, *i.e.* C_3(r) passing through the octahedral face bearing the three ligating sites A's or three B's. (If A = B, the intermediate is a trigonal prism having the D_{3h} symmetry). (M placed at the centre of the geometry) (*cf.* Sec. 10.9.3, Vol. 2)

Note: ***For the [M(A—B)$_3$] type chelate, Bailar twist leads to racemisation without any cis–trans or fac–mer isomerisation while Ray-Dutt twist leads to both racemisation and fac–mer isomerisation simultaneously.***

(d) Distinction between the Bailar twist and Ray-Dutt twist: This distinction cannot be done for the [*M*(*A*—*A*)$_3$] type complexes having the symmetrical chelate rings but it can be done for the [*M*(*A*—*B*)$_3$] type complexes bearing the unsymmetrical chelating ligands.

The Bailar twist on $[M(A—B)_3]$ will lead to **only racemisation without any geometrical isomerisation** (*i.e. cis–trans,* or *fac–mer* isomerisation). Thus it will convert *cis*–Δ to *cis*–Λ. On the other hand, the Ray-Dutt twist on [M(A—B)$_3$] will lead to both inversion of chirality and geometrical isomerisation. The *cis*–Δ will lead to **both inversion of chirality and geometrical isomerisation.** Thus *fac*- or *cis*–Δ will lead to *mer*- or *trans*-Λ.

$$\underset{cis\text{-}\Lambda \text{ or } fac\text{-}\Lambda}{\left[M(A\text{—}B)_3\right]} \xleftarrow[\text{twist}]{\text{Bailar}} \underset{cis\text{-}\Delta \text{ or } fac\text{-}\Delta}{\left[M(A\text{—}B)_3\right]} \xrightarrow[\text{twist}]{\text{Ray-Dutt}} \underset{trans\text{-}\Lambda \text{ or } mer\text{-}\Lambda}{\left[M(A\text{—}B)_3\right]}$$

Thus we may conclude that the intramolecular twist mechanism will lead to the inversion of chirality always but not necessarily the *cis–trans, i.e. fac–mer* isomerisation.

(e) Factors favouring the Bailar twist or Ray-Dutt twist: Bailar twist leads to a **transition state (D_{3h}) of higher symmetry** while Ray-Dutt twist leads to a **transition state (C_{2v}) of relatively lower symmetry** for the [M(A—A)$_3$] complex. In the D_{3h} transition state, the vertical distance of the trigonal prism is given by the **bite distance (*b*) of the ligand**. In the C_{2v} transition state, the length of one edge of a triangular face is given by the bite distance (*b*) of the ligand, while the lengths of the other two edges of the triangular face are dictated by the distance (*l*) between the ligating (*i.e.* donor) atoms of the neighbouring ligands.

- **Chelating ligands with small bite distances (*b*) are known to stabilise the trigonal prismatic geometries** (D_{3h} symmetry) (see Chapter 2, Vol. 4). In fact, the favourable condition for the Bailar twist is attained when:

  b* is sufficiently smaller than *l (distance between the donor atoms of neighbouring ligands). In fact, when *b* is small (*i.e.* small twist angle, *cf.* Fig. 5.29.8), the ground state structure is already considerably distorted towards the transition state, *i.e.* trigonal prismatic configuration (*cf.* **entatic state** of enzymes, ground state close to the transition state).
- Favourable condition for the Ray-Dutt twist is attained when:

  ***b* is sufficiently larger than *l*.**
- Under the condition, $b \approx l$, both the twist mechanisms can go on simultaneously.

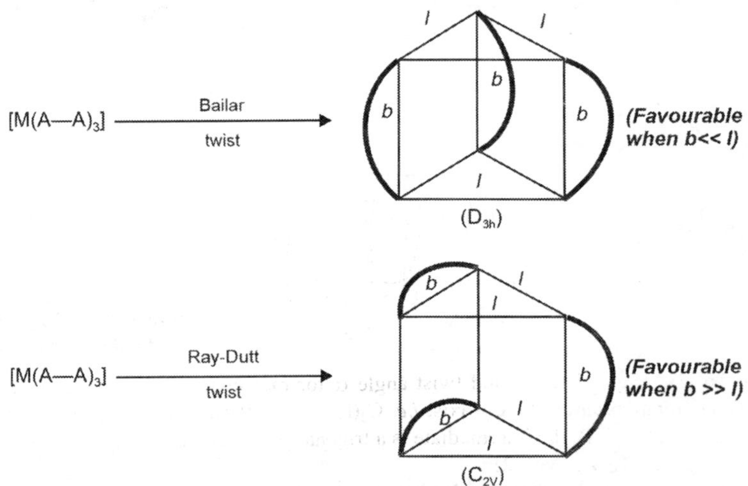

Fig. 5.29.10 Structural parameters of the transition states generated in Ray-Dutt twist and Bailar twist.

(f) Examples of complexes experiencing racemisation through the twist mechanism:

- ***cis*-[CoCl$_2$(trien)]$^+$ (in solid state):** *cis*– α–[CoCl$_2$(trien)]$^+$ undergoes slow isomerisation in solid state at 157°C to the *cis*-β-isomer (*cf.* Fig. 2.5.3.11, Vol. 4 for structures) through the trigonal twist (*i.e.*

Bailar twist) (Bailar *et al.* 1980). For the tetradentate trien ligand, the tetragonal twist (*i.e.* Ray-Dutt twist) is not possible.

• **$[Ni(N—N)_3]^{2+}$ *vs.* $[Fe(N—N)_3]^{2+}$ (N—N = bpy, phen):** The rate of raceimisation is comparable to that of dissociation (*tris* to *bis*) for the Ni(II)–complexes while the rate of racemisation is greater than the rate of dissociation for the Fe(II) complexes. In fact, for these Fe(II)–complexes, racemisation occurs through **both the intramolecular and intermolecular paths** while for the Ni(II)–complexes, the racemisation occurs exclusively through the *intermolecular path*.

$k_{rac} \approx k_d$ (*tris* $\rightarrow$ *bis* complex) $\approx k_{ex}$ (Ligand exchange) for **$[Ni(bpy)_3]^{2+}$** and **$[Ni(phen)_3]^{2+}$**;

$k_{rac} > k_d \approx k_{ex}$ for **$[Fe(bpy)_3]^{2+}$** and **$[Fe(phen)_3]^{2+}$**.

• **$[Ni(N—N)_3]^{2+}$ (Racemisation process):** Only the intermolecular path as shown in Fig. 5.29.1.

• **$[Fe(N—N)_3]^{2+}$ (Racemisation process):** Both the intermolecular and intramolecular paths.

The intramolecular path may involve the ***one-ended dissociation of the chelate ring*** or ***twisting mechanism.***

• **$[Fe(phen)_3]^{2+}$ *vs.* $[Fe(bpy)_3]^{2+}$:** For **$[Fe(phen)_3]^{2+}$**, one ended dissociation of the chelate ring is not possible (*cf.* Sec. 5.24.1). Thus its intramolecular path involves the **twisting mechanism** only.

For **$[Fe(bpy)_3]^{2+}$**, ***both the twisting mechanism and one ended dissociation of the chelate ring are possible for the intramolecular path.*** The life-time of the 5–coordinate intermediate produced through the one ended dissociation of the chelate ring decreases with the increase of acidity (see Scheme 5.23.1.1 for acid catalysed dissociation of $[Fe(bpy)_3]^{2+}$ through one ended dissociation of the chelate ring). *Thus the rate of racemisation through this intramolecular path decreases with the increase of acidity and it becomes almost zero at high acidity.* At high acidity, racemisation through the dissociative intermolecular path also stops as the dissociated ligand gets protonated and **religation by the protonated ligand is not possible.** Consequently, regeneration of the *tris*–complex becomes difficult. Thus at high acidity, both the intermolecular dissociative path and the intramolecular one-ended dissociation of chelate ring path fail to contribute to racemisation. **Thus at high acidity, the limiting residual racemisation observed for $[Fe(bpy)_3]^{2+}$ is due to the twisting mechanism only.**

• **Racemisation of $[Ni(bpy)_3]^{2+}$ or $[Ni(phen)_3]^{2+}$:** Through the formation of a symmetrical *bis*–complex (*i.e.* intermolecular path), *i.e.* $k_{rac} \approx k_d \approx k_{ex}$.

• **Racemisation of $[Fe(phen)_3]^{2+}$:** Through the formation of a symmetrical *bis*–complex (*i.e.* intermolecular path) + through the twisting mechanism (*i.e.* intramolecular path).

• **Racemisation of $[Fe(bpy)_3]^{2+}$:** Through the formation of a symmetrical *bis*–complex (*i.e.* intermolecular path) + through the one-ended dissociation of chelate ring (*i.e.* intramolecular path) + through the twisting mechanism (intramolecular path).

• **Twisting mechanism for $[Fe(N—N)_3]^{2+}$ *vs.* $[Ni(N—N)_3]^{2+}$ (N—N = bpy, phen):** The twisting mechanism operates for the racemisation of $[Fe(N—N)_3]^{2+}$ but it remains absent for $[Ni(N—N)_3]^{2+}$. The Fe(II)–complex may remain in a spin equilibrium, $l.s.\left(t_{2g}^6 e_g^0\right) \rightleftharpoons h.s.\left(t_{2g}^4 e_g^2\right)$. The **high-spin state** (where the metal-ligands bonds are relatively weaker) probably allows the twisting mechanism. The said spin-equilibrium is quite reasonable as the corresponding *bis*–complex is known to exist in high-spin state. For the Ni(II)–complex ($t_{2g}^6 e_g^2$), such a spin-isomerism is not possible.

(**Note:** The intramolecular path of twisting mechanism definitely contributes for both the $[Fe(N—N)_3]^{2+}$ complexes. This contribution is the limiting rate of racemisation found at high acidity where no dissociative path can contribute to the racemisation process. For $[Fe(phen)_3]^{2+}$, one-ended dissociation of the chelate ring (*i.e.* intramolecular path) is not possible for its racemisation. For $[Fe(bpy)_3]^{2+}$, racemisation through the intramolecular path may occur by both twisting mechanism and one-ended dissociation of the chelate ring.

- ***cis*–[MCl$_2$(en)$_2$]Cl (M = Co, Cr) (in solid state):** *cis*–[MCl$_2$(en)$_2$]Cl (M = Co, Cr) undergoes racemisation slowly at 150 – 180°C in solid state **without any *cis–trans* isomerisation** (D. Banerjea and J.C. Bailar, *Transition Met. Chem.*, **10,** 331, 1985). The *trans*–compound does not experience isomerisation under the similar condition. Interestingly, both the Co(III) and Cr(III) complexes show the comparable rates and activation parameters while in normal ligand substitution process, Co(III) reacts much more slowly than Cr(III).

These observations can be explained by considering the tetragonal twist (*i.e.* Ray-Dutt twist) mechanism. It may be mentioned that operation of Bailar twist on *cis*–[MCl$_2$(en)$_2$]$^+$ will lead to **racemisation along with the *cis-trans* isomerisation.** If racemisation occurs through the one-ended dissociation of a chelate ring or dissociation of one Cl$^-$ leading to a 5-coordinate intermediate, then it will also experience the concommitant *cis–trans* isomerisation (*cf.* Figs 5.28.2.1, 2). All these support the recemisation of *cis*-[MCl$_2$(en)$_2$]$^+$ in solid state through the **Ray-Dutt twist** mechanism.

- **[Co(bigH)$_3$]$^{3+}$:** The robust (kinetically inert) complexes like [Co(bigH)$_3$]$^{3+}$, etc. undergo racemisation through the intramolecular twist mechanism (Ray and Dutt, 1943).
- **[Cr(L—L)$_3$]$^{3+}$ (L—L = bpy, phen) *vs.* [Cr(ox)$_3$]$^{3-}$:** The very **small positive values of $\Delta V^{\neq}$** (+3.3 cm^3 mol^{-1}) for racemisation of [Cr(bpy)$_3$]$^{3+}$ and [Cr(phen)$_3$]$^{3+}$ support the **intramolecular twist mechanism** of racemisation because in this path, there is no appreciable change in molar volume in attaining the transition state (T.S.). The **highly negative values of $\Delta V^{\neq}$** (*ca.* –12 to –16 cm^3 mol^{-1}) for racemisation of the anionic complexes like [Cr(ox)$_3$]$^{3-}$, [Cr(N—N)(ox)$_2$]$^-$ can be rationalised in terms of the **one-ended dissociation of the oxalate chelate ring** in the transition state giving rise to the TBP structure. The released —CO_2^- group (in the T.S.) puts **a strong electro-restriction over the solvent** to cause this negative value of $\Delta V^{\neq}$.

5.30 LIGAND SUBSTITUTION REACTIONS IN THE SQUARE PLANAR COMPLEXES

Transition metal ions of d^8–configuration (*e.g.* Ni^{II}, Pd^{II}, Rh^{I}, Ir^{I}, Au^{III}) generally form the 16*e*-square planar complexes. Ni^{II} forms the square planar complexes with the very strong field ligands only; otherwise, it has a tendency to form the octahedral complexes. The Pd^{II} centre has a tendency to be reduced to Pd^0. Au(III) is a fairly strong oxidising agent. Relatively very few square planar complexes of Rh^{I} and Ir^{I} have been studied. In fact, understanding of ligand substitution process in the square planar complexes emerges basically from the studies of Pt(II) complexes.

5.30.1 Characteristics of the Ligand Substitution Process in Square Planar Complexes

(i) **Associative activation (see Secs. 5.13, 5.30.3):** The coordinatively unsaturated and electronically unsaturated 16*e*-square planar Pt(II) complexes generally adopt the associative path to produce the 5-coordinate 18*e*-transition state (T.S.)/intermediate. The 16*e*-square planar complexes are produced by using the dsp^2 hybrid orbitals of the central atom (*cf.* *xy*-molecular plane, dsp^2 hybridisation: $(n-1)d_{x^2-y^2} + ns + np_x + np_y$) and thus there remains **a vacant *p*–orbital,** *i.e.* np_z orbital, perpendicular to the molecular plane **(in terms of VBT).** This **vacant p_z–orbital** can accommodate the incoming nucleophile to allow the *associative activation.* The **vacant antibonding d_{z^2} orbital in terms of MOT** (*xy*-plane as the molecular plane) can also receive the nucleophilic attack. In other words, in terms of VBT, the **vacant hybrid *pd*-orbital** (developed from the $np_z^0 \pm (n-1)d_{z^2}^2$ combination) can also perform the task (see Sec. 5.30.3 for details) to accommodate the nucleophile in the associative path.

The five coordinate 18*e*-intermediates or transition states for the d^8–systems are not unlikely. Characterisation of the trigonal bipyramidal species like [Pt(SnCl$_3$)$_5$]$^{3-}$, [Ni(CN)$_5$]$^{3-}$ in the crystals of [Cr(en)$_3$]$^{3+}$–salt, [Fe(CO)$_5$], [Co(CO)$_3$L$_2$]$^+$, etc. supports the possibility of expansion of C.N. 4 to 5 during the reaction. The **strong π–acceptor ligands** like CN$^-$, SnCl$_3^-$, etc. can stabilise the trigonal bipyramidal structure. Five coordinate 18*e*-compounds of the d^8 systems like Fe(0), Ru(0), Os(0), Rh(I), Ir(I), etc. with the π–acid ligands are now well known. In fact, in the reaction of $[AuCl_4]^-$ with NCS^-, the proposed five coordinate intermediate has been detected.

(ii) **Two term rate law:** The following substitution reaction follows the two-term rate law.

$$ML_3X + Y \rightarrow ML_3Y + X$$

$$-\frac{d[ML_3X]}{dt} = (k_s + k_Y[Y])[ML_3X]$$

i.e. $k_{obs} = k_s + k_Y[Y]$, when $[Y] \gg [ML_3X]$.

Experimental evidences indicate that the k_s-path (independent of [Y]) involves the solvent (S) as the first entering nucleophile. The overall reaction may be simply represented as follows to get the two term rate law.

$$L_3M\text{—}X + S \underset{k_{-1}}{\overset{k_1}{\rightleftharpoons}} L_3M\text{—}S + X;\ L_3M\text{—}S + Y \xrightarrow{k_2} L_3M\text{—}Y + S$$

$$L_3M\text{—}X + Y \xrightarrow{k_3} L_3M\text{—}Y + X$$

Under the **steady state condition** applied to L_3MS, we have:

$$\frac{d[L_3MS]}{dt} = 0 = k_1[L_3MX][S] - k_{-1}[L_3MS][X] - k_2[L_3MS][Y]$$

or

$$[L_3MS] = \frac{k_1[L_3MX][S]}{k_{-1}[X] + k_2[Y]} = \frac{k_1'[L_3MX]}{k_{-1}[X] + k_2[Y]},\ k_1[S] = \text{constant} = k_1'(\text{say})$$

$$\text{Rate} = k_2[L_3MS][Y] + k_3[L_3MX][Y]$$

$$= \frac{k_1'k_2[L_3MX][Y]}{k_{-1}[X] + k_2[Y]} + k_3[L_3MX][Y]$$

$$\approx k_1'[L_3MX] + k_3[L_3MX][Y];$$

(when $k_{-1}[X] << k_2[Y]$ which is most likely, if not X is added externally), *i.e.*

$k_{obs} = k_s + k_Y[Y]$, (under the pseudo-first order conditions of excess Y)

where $k_1' = k_s$, $k_3 = k_Y$.

It may be noted that this rate law derivation does not tell us regarding the mechanistic paths of the substitution process.

The k_s–path does not depend on the nature of the entering ligand but **it depends on the nucleophilicity of the solvent.** On the other hand, the k_Y-path is strongly dependent on the nature of Y but not so strongly on the nature of the solvent. However, if **the nucleophilicity of Y depends on the nature of solvent, then k_Y shows its dependence on the nature of the solvent.** The two term rate law arises from two parallel pathways, both experiencing the associative activation.

(iii) **Stereospecific substitution:** In most of the cases, stereochemistry of the starting complex is retained fully in the product, *i.e.* a *trans*-reactant gives a *trans*-product and a *cis*-reactant gives a *cis*-product.

Retention of this stereochemistry can be explained by considering the formation of an intermediate of ***trigonal bipyramidal geometry*** *in which the entering group, leaving group and the group originally present to the trans-position of the leaving group occupy the equatorial positions.* Thus retention of stereochemistry again supports the associative activation.

Note: In some cases (*e.g.* square planar phosphine complexes of Pt(II)), there is a change of stereochemistry (*i.e.* case of isomerisation). In such cases, ***the proposed trigonal bipyramid intermediate is sufficiently long lived to experience the Berry pseudorotation to interchange the axial and equatorial ligands***.

Chemistry of Platinum and Russia

Platinum is a very costly metal. The abundance of platinum ores in Russia led the Russian chemists to explore the coordination chemistry of platinum extensively. In fact, many characteristic aspects (like stereospecific ligand displacement) of the reactions of square planar complexes of platinum(II) were first discovered by the Russian chemists.

(iv) **Effects of the entering group and leaving group:** The rate ***strongly*** depends on the **nucleophilicity of the entering ligand.** It follows the following sequence.

Y: CN^-, CO > PR_3 > H^- > I^- > Cl^- > H_2O **(strong dependence)**

The rate also **mildly** depends on the nature of the leaving group. It follows the following sequence:

X: H_2O > Cl^- > I^- > H^- > PR_3 > CO, CN^- **(mild dependence)**

It is interesting to note that the above mentioned sequences follow the reverse order. ***Thus, the good entering groups (i.e. good nucleophiles) are the poor leaving groups (i.e.* poor nucleofugals).**

Here it is important to compare the extent or degree of rate dependence on the nature of the entering group and leaving group. It is illustrated for the following reactions.

$$trans\text{–}[PtCl_2(py)_2] + Y \rightarrow trans\text{–}[PtCl(py)_2(Y)]^{n+} + Cl^-$$, (*cf.* Table 5.30.5.1)

k_Y ($dm^3\ mol^{-1}\ s^{-1}$):	4.7×10^{-4}	4.5×10^{-4}	3.7×10^{-3}	0.107	6.0
Y:	NH_3	Cl^-	Br^-	I^-	tu

Here the rate constant changes by a factor 10^4 in moving from NH_3 to tu as the entering ligand. For changing the entering nucleophile Cl^- to I^-, the rate constant increases by a factor of about 10^3. On the other hand, for changing the leaving group Cl^- to I^-, *the rate constant changes by a factor only 3.5*

Other possible ways of ligand replacement in square planar complexes of Pt(II) (*cf.* R.J. Cross, *Chem. Soc. Rev.*, 14, 197, 1985)

In addition to nucleophilic substitution, other processes like *electrophilic substitution and oxidative addition followed by reductive elimination* can also lead to the ligand replacement reactions. These are illustrated for the complex, $[Pt^{II}(Cl)(L)_2(Me)]$, (L = PMe_2Ph)

$$[Pt^{II}Cl_2L_2] \xleftarrow[-MeHgCl]{+HgCl_2} [Pt^{II}(Cl)L_2(Me)] \xrightarrow[-Cl]{+N_3^-} [Pt^{II}L_2(Me)N_3]$$

(Electrophilic substitution) (Nucleophilic substitution)

$\downarrow$ $+Cl_2$ (oxidative addition)

$$[Pt^{IV}Cl_3L_2(Me)] \xrightarrow[\text{(Reductive elimination)}]{-MeCl} [Pt^{II}Cl_2L_2]$$

(Octahedral)

In the reaction with $HgCl_2$, the electrophilic attack by Hg(II) on the Pt^{II}—Me bond leads to the ligand substitution at Pt(II).

$$-Pt^{II}-Me + Cl-Hg-Cl \longrightarrow -Pt^{II}-Cl + Cl-Hg-Me$$

In some special cases, **dissociative pathway** of ligand substitution of the square planar complexes of Pt(II) has been proposed. This aspect is discussed later (see Sec. 5.31).

under the comparable conditions (*cf.* Table 5.30.6.1). ***Thus the rate process is strongly sensitive to the nature of the entering group but it depends mildly on the nature of the leaving group (cf. Table 5.30.6.1).*** These observations are in favour of the *associative activation.*

(v) **Effect of the spectator ligands (*i.e.* nonleaving groups):** If the *trans–* ligand (to the leaving group) is a good nucleophile, then the rate is favoured. This phenomenon is described as the ***trans–*effect**. The nonleaving *cis–*ligand can also mildly influence the reation rate. This phenomenon is called the ***cis–*effect.** ***It may be mentioned that the trans–effect is more pronounced than the cis–effect.***

5.30.2 Characteristic Features of the Two-term Rate law

(i) **k_s and k_Y rate constants in a particular solvent:** The rate law for the reaction.

$$ML_3{-}X + Y \rightarrow ML_3{-}Y + X$$

is : $$k_{obs} = k_s + k_Y[Y], \text{ when } [Y] \gg [ML_3X]$$

From the linear plot of k_{obs} *vs.* $[Y]$, we get k_s (= intercept) and k_Y(= slope). In a particular solvent, the k_Y values are different for the different entering groups in a particular reaction while k_s remains the same for all entering nucleophiles. It may be illustrated (Fig. 5.30.2.1) for the following reaction.

$$trans\text{-}\left[PtCl_2(py)_2\right] + Y \longrightarrow trans\text{-}\left[PtCl(py)_2(Y)\right] + Cl^-$$ (charge of the product complex not shown)

(ii) **k_s and k_Y rate constants in different solvents:** If a particular reaction is carried out in different solvents then the k_s–value (= intercept) will change depending on the coordinating behaviour of the solvent and the k_Y value may also change depending on the nature of the solvent. *The change in the k_s- value arises from the change of nucleophilicity of the solvent* (S) which will first enter as a ligand followed by the substitution of the coordinated solvent by the entering ligand in a relatively faster path.

k_s-path: $$ML_3{-}X + S \xrightarrow[(-X)]{\text{slow}} ML_3{-}S \xrightarrow[(-S)]{+Y\,(\text{fast})} ML_3{-}Y$$

k_Y-path: $$ML_3{-}X + Y \xrightarrow[(-X)]{} ML_3{-}Y$$

In a particular solvent, for the different entering ligands, the k_Y–value differs depending on the nature of the nucleophilicity of the entering ligand.

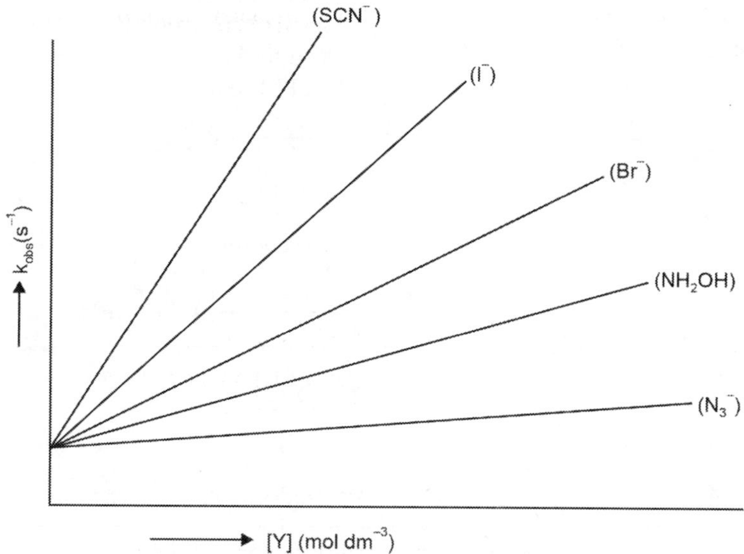

Fig. 5.30.2.1: Qualitative representation of the two term rate law for the nucleophilic substitution of Cl^- by *Y* from *trans*–$[PtCl_2(py)_2]$ in methanol solvent.

The change in the k_Y–value for a particular entering ligand in different solvents arises from the fact that the nucleophilicity of the entering nucleophile (i.e. Y) depends on the degree of its solvation in the given solvent. Better solvation makes a nucleophile a weaker nucleophile because in making the nucleophilic attack, the solvated nucleophile is to be desolvated first. **The higher desolvation energy makes it a poorer nucleophile** (*e.g.* nucleophilicity order of the halides in making the C—X bond in S_N2 reaction in polar protic solvents (*e.g.* H_2O, ROH, etc.): $I^- > Br^- > Cl^- > F^-$; while the **reverse order** in polar aprotic solvents lacking in OH groups and nonpolar solvents incapable of solvating the halides: $F^- > Cl^- > Br^- > I^-$). *Thus, the nucleophilicity of a particular ligand is different in different solvents.*

(iii) **Relative contributions of the k_s and k_Y paths:** Sometimes, the k_Y path may have a negligible contribution and sometimes, the k_s–path may have also a negligible contribution depending on the situation. These are all experimentally verified.

$$[Pt(bpy)Cl_2] + py \rightarrow [Pt(bpy)Cl(py)]^+ + Cl^-$$

i.e.

For the above reaction in methanol solvent, k_s is almost zero while k_Y is *ca.* 6×10^{-3} dm^3 mol^{-1} s^{-1}.

For the following ligand exchange reaction, k_s is almost zero in the noncoordinating solvent hexane while in the coordinating solvent methanol, k_Y is almost zero.

$$trans\text{–}[PtCl_2(N^*HEt_2)(PPr_3)] + NHEt_2 \rightarrow trans\text{–}[PtCl_2(NHEt_2)(PPr_3)] + {}^*NHEt_2$$

Hexane (a noncoordinating solvent) cannot act as a nucleophile. Hence, absence of the k_s–path is quite reasonable. Interestingly, in the coordinating solvent methanol, the reaction mainly occurs through the solvent dependent path (*i.e.* k_s–path). *It is believed that the k_Y–path which is an associative path is* ***disfavoured for the bulky entering nucleophile like diethylamine*** **(Et_2NH)** *and the reaction proceeds through the alternative path, i.e. k_s–path* (which is also an associative path) involving the relatively less bulky nucleophile like MeOH (*i.e.* solvent).

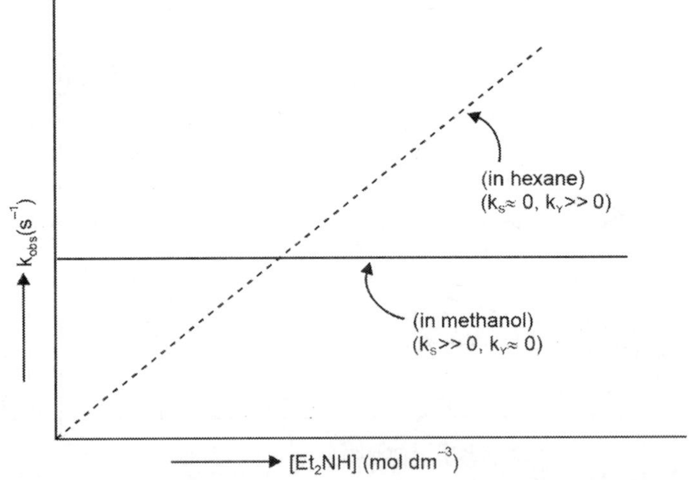

Fig. 5.30.2.2 Qualitative represention of the relative importance of k_s– and k_Y–paths in different solvents for the ligand exchange reaction: *trans*–$[PtCl_2(N^*HEt_2)(PPr_3)] + NHEt_2 \rightarrow$

(iv) **Reaction course in the k_s-path:** The k_s–path (*i.e.* solvent dependent path) involves a **double displacement process** in which each step is an associative path.

$$[ML_3X] + S \xrightarrow[(k_1)]{\text{slow}} [ML_3(S)] + X,\ S = \text{solvent}\ \textbf{(associative path)}$$

$$[ML_3(S)] + Y \xrightarrow{\text{fast}} [ML_3(Y)] + S\ \textbf{(associative path)}$$

The rate of this path is: k_1[complex][S] = k_s [complex] because [S] is a constant.

- The k_s–path involves the formation of the intermediate $ML_3(S)$ *that can be trapped in some cases to give the experimental evidence in favour of the proposed routes.* In fact, it has been possible to trap the aqua-intermediate in the following reaction by adding a base (B) that can abstract a proton from the coordinated water molecule to produce a hydroxo species. ***The hydroxido group is a poor leaving group to allow the characterisation of the hydroxido complex.***

$$[\text{Pt(dien)X}]^+ \xrightarrow[(-X^-),\ \text{slow}]{H_2O} [\text{Pt(dien)}(OH_2)]^{2+} \xrightarrow[(-H_2O),\ \text{fast}]{+Y^-} [\text{Pt(dien)(Y)}]^+$$

$$[\text{Pt(dien)}(OH_2)]^{2+} + B \rightleftharpoons [\text{Pt(dien)(OH)}]^+ + BH^+\ \text{(fast)}$$

(inert to substitution by Y^-).

Obviously, contribution of the k_s-path increases with the increase of coordinating power of the solvents having the comparable steric factors.

(v) **Relative contributions of the k_s and k_Y paths for different metal centres of d^8 configuration:** Here it should be mentioned that the relative contributions of the k_s- and k_y-paths depend on the nature of the starting complex and metal centre also in a particular solvent. In the Au(III)-system (which experiences more *associative activation* due to its higher positive charge), the k_Y-path is more predominant than the k_s-path compared to those in the analogous Pt(II)-complex. This is illustrated in the following data. Compared to the Pt(II) complex, the **higher positive charge in the Au(III) complex** favours more the addition of the nucleophile (*i.e.* lower $\Delta H^{\neq}$) at the *rds*, and more charge neutralisation (*i.e.* more desolvation) makes $\Delta S^{\neq}$ more positive. This **entropic favour** for the k_S-path involving the neutral nucleophile is less important compared to the k_Y-path involving the anionic nucleophile for the Au(III) complex of higher positive charge.

Reaction	M	$k_{H_2O}\ (s^{-1})$	$k_Y\ (dm^3\ mol^{-1}s^{-1})$	$\frac{k_Y}{k_s}\ (dm^3 mol^{-1})$
$[MCl(dien)]^{n+} + Br^- \longrightarrow [MBr(dien)]^{n+} + Cl^-$	M = Au(III)	0.50	154	308
	M = Pt(II)	8.0×10^{-5}	5.3×10^{-3}	66
$[MCl_4]^{m-} + {}^*Cl^- \longrightarrow$ (ligand exchange reaction)	M = Au(III)	6.0×10^{-3}	1.47	245
	M = Pt(II)	3.8×10^{-5}	–	–

Rate constant are at 20°C for Au(III), and at 25°C for Pt(II) (*cf.* F. Basolo, *et al.*, *Inorg. Chem.*, **3**, 1087, 1964; **2**, 921, 1963; L.I. Elding *et al.*, *Acta. Chem. Scand. Sect. A*, **32**, 867, 1978).

5.30.3 Mechanism of the Reaction (see Secs. 5.13, 5.30.1)

The coordinatively and electronically unsaturated 16*e* square planar complexes of Pt(II) (d^8 system) generally adopt the associative path to produce the 5-coordinate 18*e* T.S./intermediate. Considering the ***dsp*² hybridisation** (*i.e.* $(n-1)d_{x^2-y^2} + ns + np_x + np_y$, see Vol. 2) of the starting square planar complex (in the *xy*-plane), the **vacant np_z orbital** (*i.e.* **$6p_z$ orbital for Pt(II)**, *i.e.* nonbonding orbital (= a_{2u} in terms of MOT) or the **vacant *pd*-hybrid orbital** (developed from the combination of vacant np_z^0 and filled $(n-1)d_{z^2}^2$-orbital, *cf.* because of the **relativistic expansion** of 5*d*-orbitals of Pt(II), the energy difference between the 5*d*- and 6*p*-orbitals decreases to favour their hybridisation) may receive the incoming nucleophile from the *z*-direction. In terms of VBT, participation of the vacant np_z-orbital to accommodate the nucleophile produces the square pyramid T.S./intermediate having the hybridisation *dsp*³ (*i.e.* $d_{x^2-y^2} + s + p_x + p_y + p_z$, see Vol. 2). In terms of MOT, the **vacant antibonding** $(n-1)$ d_{z^2}

orbital (more correctly a_{1g}^{*}) or the vacant nonbonding np_z orbital (*i.e.* a_{2u}) may also receive the nucleophilic attack from the *z*-direction. It may be noted that in the MO-diagram of square planar complexes, there is an uncertainty in the energy order of a_{1g}^{*} and a_{2u} (see Fig. 3.17.6.3, Vol. 4). Thus any one of these two vacant orbitals (a_{1g}^{*} and a_{2u}) can accommodate the nucleophilic attack in the associative path in terms of MOT. The nucleophilic attack (leading to the **loss of CFSE or LFSE measuring the CFAE or LFAE)** along the *z*-direction in the **associative path** initially produces the **5-coordinate square pyramid geometry**. Now, the entering group present in the axial position of the square pyramid geometry initially formed can move in any of the four possible directions.

The possible directions (*cf.* Scheme 5.30.3.1) are: towards L_1, towards L_2, towards *X* (which is the leaving group), towards *T* (*trans*–position with respect to the leaving group *X*). ***The most favoured direction of movement of Y is in the direction of the ligand which is least strongly bound.***

Trigonal bipyramidal transition state

Formation of a trigonal bipyramidal transition state bearing the entering group (*Y*), leaving group (*X*) and its *trans*-group (*T*) in the trigonal equatorial plane and two *cis*-groups (L_1 and L_2, with respect to the leaving group in the starting square planar complex) in the axial positions can explain the *trans*-effect, *cis*-effect, stereochemistry of the product, etc.

Transition state (T.S.): Trigonal bipyramid (TBP) *vs.* Square pyramid (SP)

In terms of **CFAE**, the SP T.S. is more favoured than the TBP T.S. (see Table 5.30.10.1). However, the TBP T.S. is likely to be formed because of its **better stabilisation through the π-bonding** at the equatorial plane and **lesser steric crowding** (*cf.* ligand–ligand repulsion: 6 at 90° and 3 at 120° in TBP; 8 at 90° in SP; repulsion at 120° not important, see VSEPR Vol. 2).

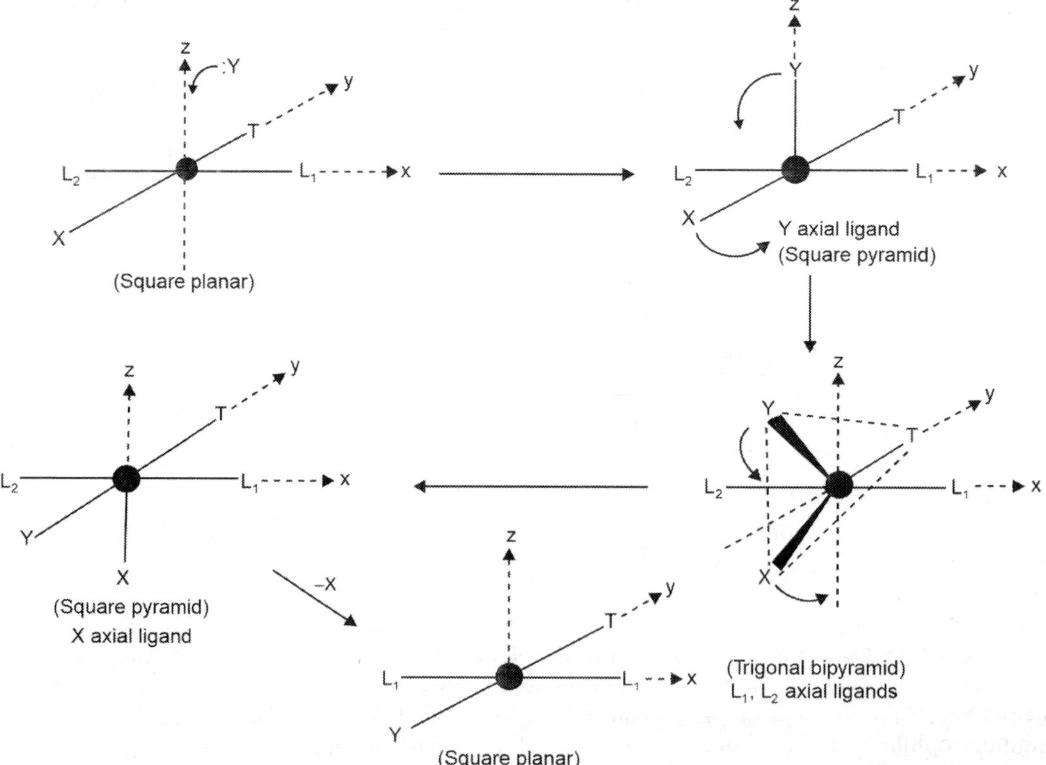

Scheme 5.30.3.1: Schemetic representation of the associative mechanism of ligand substitution at the square planar complex of Pt(II)

This approach can explain the phenomena *trans*-effect and *trans*-influence to be discussed later. Here the leaving group X gradually weakens its attachment with the metal centre and bends away. This will transform the square pyramid (SP) into a trigonal bipyramid (TBP) having the three ligands (Y, T and X), *i.e.* entering group, leaving group and the *trans*–group (with respect to the leaving group) in the equatorial plane. The other two groups (L_1 and L_2, *trans*–ligands at the starting square planar complex) will occupy the axial positions of the trigonal bipyramid (TBP). *As the leaving group X moves away from the trigonal plane of the trigonal bypyramid, the angle Y—M—T opens up to* 180° *leading to a square pyramid geometry placing the leaving group (X) at the axial position of a square pyramid (SP) geometry.* Then the expulsion of X generates from the square pyramid (SP) geometry, the square planar geometry.

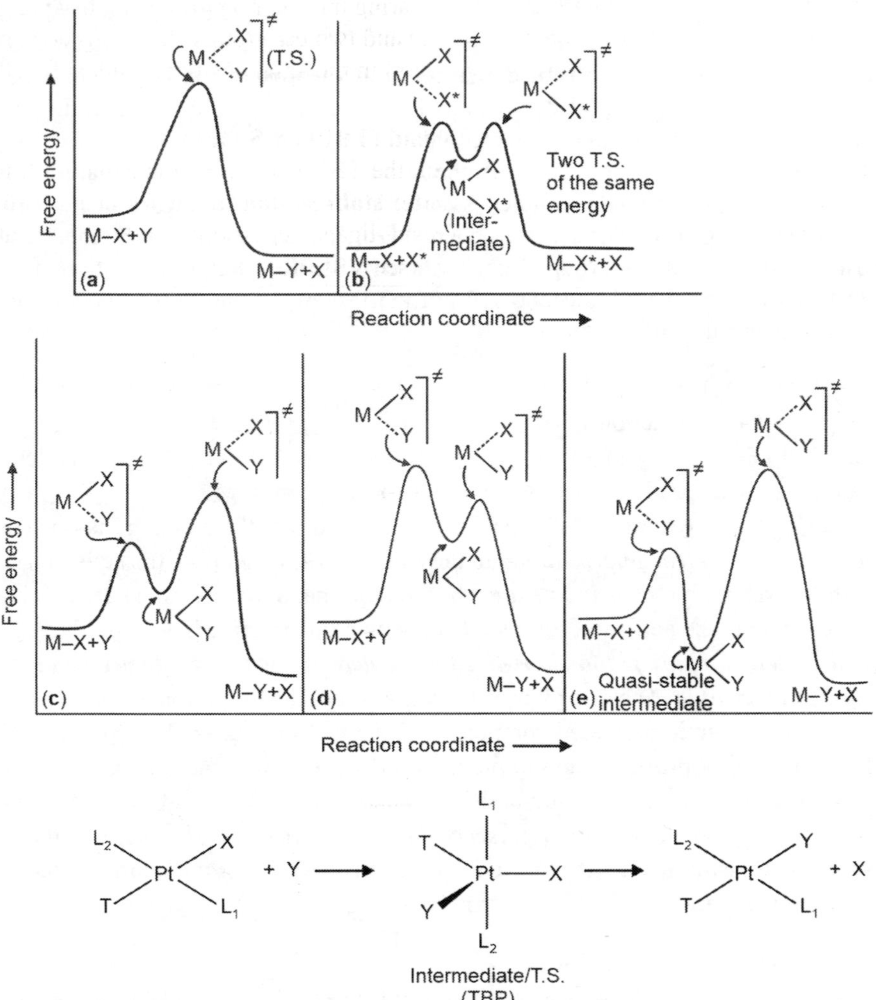

Fig. 5.30.3.1: Possible reaction energy profile diagrams for the nucleophilic substitution reactions of the square planar complexes of Pt(II).

Fig 5.30.3.1: (a) **Single step interchange** I_a path. (b) Ligand exchange reaction through a **two step associative path.** Energies of the reactants and products are the same and both the transition states are identical and are of the same energy. (c) M—X bond breaking is more important in the two step process involving a 5 coordinate **unstable intermediate** which is less stable than both the reactants and products. (d) M—Y bond formation is more important in the two step process (as in the pure associative **A path** involving a **5-coordinate unstable intermediate**). (e) A two step process through the formation of a **quasi-stable 5-coordinate intermediate** (more stable than the reactant but less stable than the final product); bond breaking by the leaving group from the stable intermediate is the rate determining step; noted in some Rh(I) and Ni(II) complexes as in the following reaction (*JACS*, **90**, 4800, 1968).

(cod) Rh(Cl)(SbR_3) + py ⟶ $[Rh(cod)Cl(SbR_3)(py)]$ ⟶ Rh(Cl)(py) + SbR_3

The 2nd step is much slower than the first step leading to the formation of a stable 5-coordinate intermediate. For this reaction, **the solvent dependent path is absent** (*cf.* in solvent dependent path, the slower first step produces the solvent substituted complex from which the coordinated solvent is substituted in a faster step by the entering nucleophile).

Note: In the pure *A*-process as shown in (d), the rate does not depend on the nature of the leaving group but for the square planar Pt(II) complexes, **rate depends on the nature of both the entering nucleophile and leaving group**. The reaction energy profile (c) can explain the rate dependence on both the entering group and leaving group. Probably, it involves the reversible addition of the incoming nucleophile (Y) followed by the *rds* (second T.S.) of the loss of leaving group accompanied with the structural rearrangement. It is illustrated as:

$$[L_3Pt—X] + Y \rightleftharpoons [Y—PtL_3—X] \text{ (TBP intermediate)} \xrightarrow{rds} L_3Pt—Y + X$$

Thus the sequence of the reaction is:

Square planar complex + Y → SP (axial-Y) → TBP (equatorial-Y) → SP (axial-X), leaving group) → Square planar complex (product) (see Scheme 5.30.3.1).

The proposed trigonal bipyramid (TBP) species produced during the reaction may be considered as an *activated complex* or as a *true intermediate.* In the energy profile diagram, the activated complex lies at a peak and thus it refers to a *transition state.* On the other hand, the intermediate is relatively more stable and long-lived than the activated complex from which this intermediate is produced. ***In terms of life time, the activated complex is too short-lived to be detected while the intermediate is relatively long-lived and it is detectable.*** The energy profile diagram for substitution in a square planar complex involving a trigonal bipyramid activated complex or a trigonal bipyramidal intermediate is shown in Fig. 5.30.3.1. In the energy profile diagram, *presence of two maxima and one minimum in the curve suggests the existence of a true trigonal bipyramid intermediate.* This represents the **A-process** involving **two steps.** In the energy profile diagram, presence of a single maximum represents the existence of trigonal bipyramidal transition state. It represents the I_a**–process** involving a **single step.**

The two term rate law arises from the mechanistic paths of Scheme 5.30.3.2.

5.30.4 Stereochemistry of the Product

Retention of stereochemistry is a characteristic property of the ligand substitution process in square planar complexes. Thus it leads to:

cis-complex → *cis*-product,

trans-complex → *trans*-product.

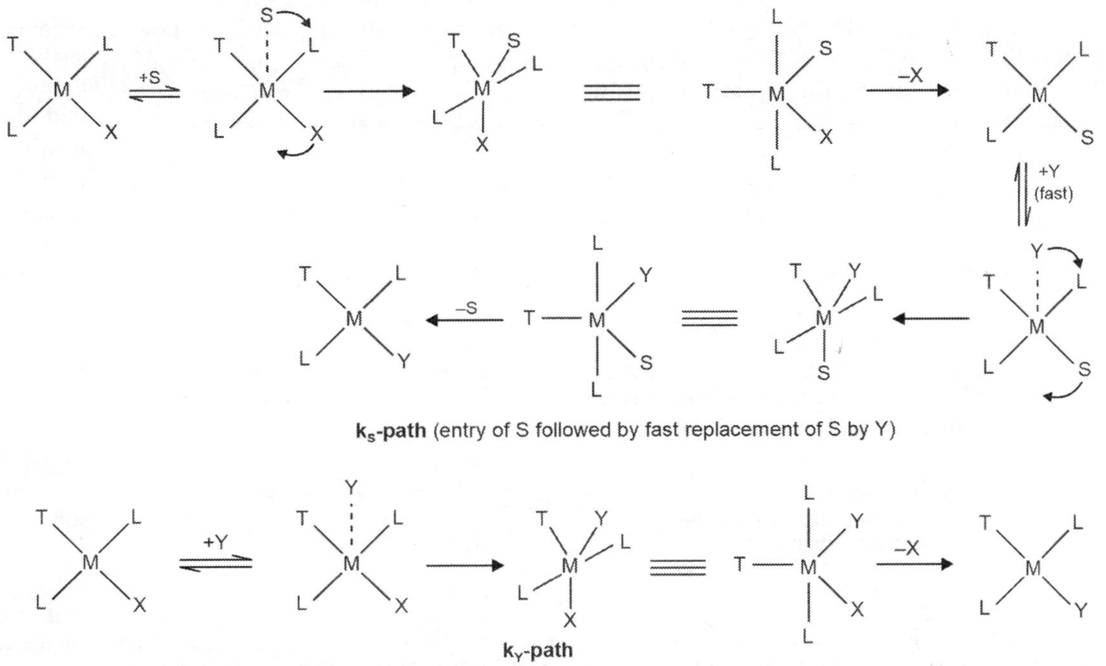

k_S-path (entry of S followed by fast replacement of S by Y)

k_Y-path

Scheme 5.30.3.2: Associative pathways for both the k_S- and k_Y-paths in the ligand substitution process in square planar complexes (For details of the mechanistic steps, *see* Scheme 5.30.3.1).

Retention of stereochemistry can be explained by considering the *trigonal bipyramidal activated complex* (*i.e.* I_a **process**) bearing the entering group (*Y*), leaving group (*X*) and *trans*–group (*T*) in the trigonal plane. If the trigonal bipyramid is a **long lived intermediate** (*i.e.* **A–process**), then it can experience an interchange between the equatorial and axial ligands. Thus for the **long-lived stereomobile** (*i.e.* fluxional) **trigonal bipyramid intermediate,** the *trans*–(*T*) and *cis*–ligand (*L*) can exchange their positions in the trigonal bipyramid. *Under the circumstances, the starting trans–complex will give a mixture of cis– and trans– products* (*i.e.* an isomeric mixture). This aspect is illustrated in Scheme 5.30.4.2. Two successive substitution reactions will occur to carry out the ligand catalysed isomerisation

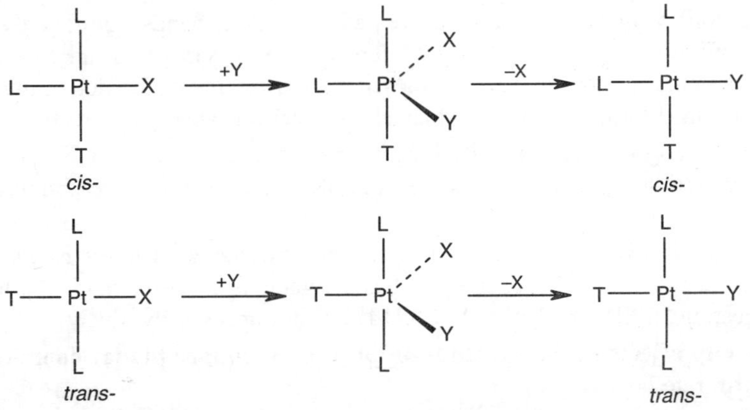

cis- *cis-*

trans- *trans-*

Scheme 5.30.4.1: Retention of stereochemical configuration in substitution reaction of square planar complexes of Pt(II) reaction (*cf.* Sec. 5.30.11) through the formation of a **stereomobile TBP intermediate.** The leaving group (*X*), its *trans*-ligand (*T*) and entering ligand (*Y*) occupy the equatorial positions of the TBP and **if they strongly favour the equatorial positions** (specially in terms of π-bonding), then the Berry pseudorotation is not favoured. **This will lead to the retention of stereochemistry**.

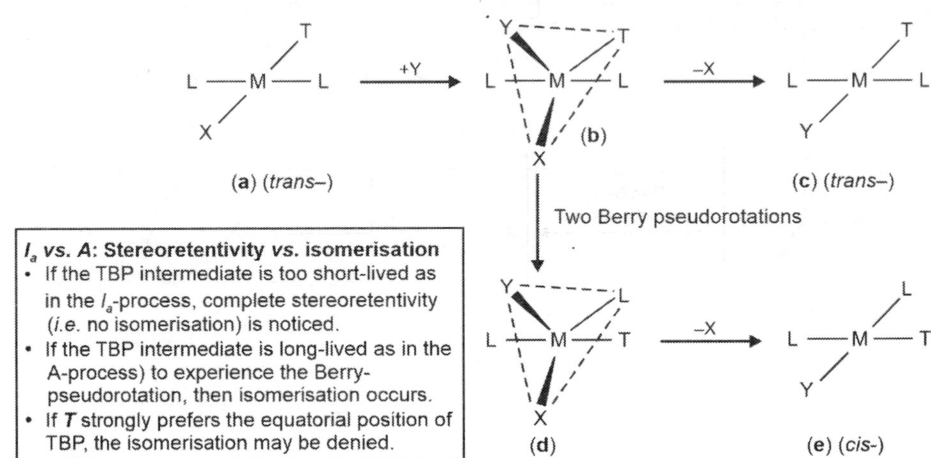

Scheme 5.30.4.2: Schematic representation of isomerisation in substitution reactions of the square planar complexes ***I*$_a$-path:** (a) → (b) → (c); ***A*-path:** (a) → (b) → (d) → (e) **Note:** For the conversion, (b) → (d), it needs **two successive Berry-pseudorotations** – keeping ***T* as the pivotal group** in the first pseudorotation and ***L* as the pivotal group** in the second pseudorotation

I_a *vs.* *A*: Tetrahedral C-compound *vs.* Tetrahedral Si-Compound

For the C-compounds, ***I*$_a$-process** is possible but ***A*-process is not possible;** for the Si-compounds, ***A*-process is possible** (*see* Vol. 2, Ch. 10). Thus for the C-compounds, in nucleophilic substitution, a short-lived TBP (sp^2 hybridisation for the equatorial positions, and 3*c*–4*e* bonds along the axial directions by using the remaining *p*-orbital of C) intermediate is produced and it leads to **Walden inversion.** By using the vacant *d*-orbital, Si can stabililise the TBP (sp^3d hybridisation) intermediate (*A*-process) which is sufficiently **long-lived to experience the Berry pseudorotation and it may lead to** racemisation.

5.30.5 Entering Group Effects and Nucleophilicity Parameter

(a) **Nucleophilicity and basicity:** In an associative path, the rate depends significantly on the nature of the entering nucleophile (*cf.* Sec. 5.3). This rate dependence is very often measured in terms of the *nucleophilicity* of the entering group. Thus it is a ***kinetic parameter*** which determines the ability of a Lewis base to act as an entering group to influence the reaction rate.

Basicity gives the thermodynamic affinity for protonation of the entering Lewis base (Y). This is a ***thermodynamic parameter*** and it is measured by the pK_a value of the corresponding conjugate acid (YH^+).

Thus nucleophilicity is a *kinetic parameter* and basicity is a ***thermodynamic parameter***. Very often, these two parameters run parallel, *i.e.* a more basic ligand is a better nucleophile (*cf.* **Brönsted relationship** suggesting LFER; *see* Vol. 3A, Ch. 14) but not necessarily always.

(b) **Entering group effects in the substitution process of square planar complexes of Pt(II):** It follows the two term rate law.

$$k_{obs} = k_s + k_Y[Y],\ [Y] = \text{concentration of the entering ligand} \gg [\text{complex}]$$

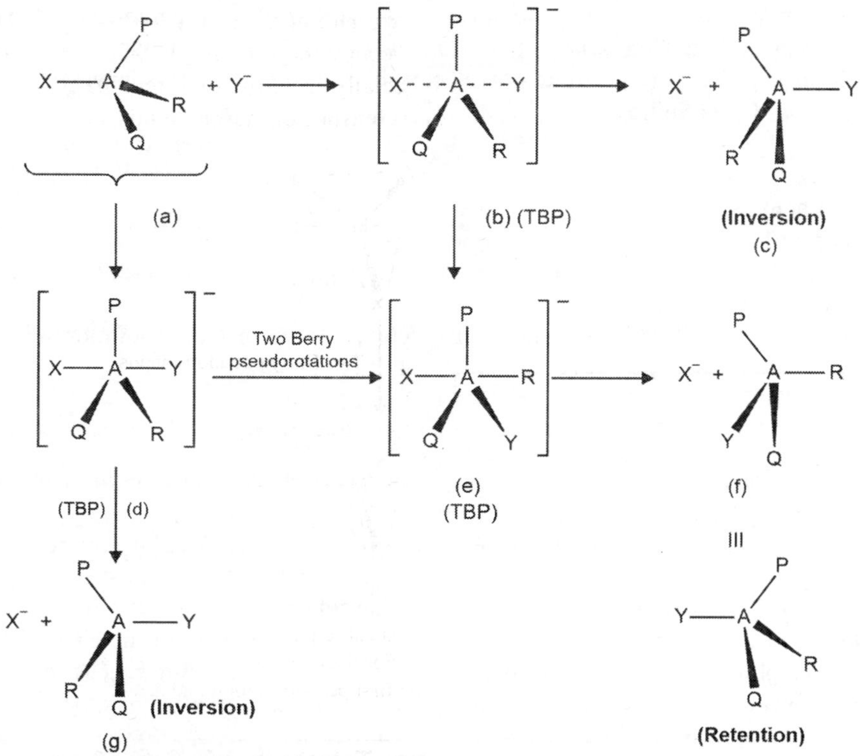

Scheme 5.30.4.3: ***I_a*-path** (a → b → c) for the teterahedral C-compounds leads to inversion; **A-path** (a → d → e → f, a → d → g) for the tetrahedral Si-compounds causes racemisation.

Note 1: For the regeneration of the tetrahedral geometry from the TBP, the leaving group (X^-) **must occupy one axial position of the TBP**. If X^- (leaving group) and Y^- (entering group) are **highly electronegative**, then they will preferably occupy the axial directions of the TBP **(Bent's rule)** and will try to deny the Berry pseudorotation. **It will lead to inversion**. It also happens so for the good leaving group (X^-) that preferably occupies the axial position of TBP.

Note 2: For the conversion, (d) → (e), it needs two successive Berry pseudorotations keeping R and Y as the pivotal groups respectively in these successive psuedorotations. In fact, in presence of a nucleophile, the tetrahedral Si-compounds undergo racemisation easily. **This is why, it is difficult to have the optically pure Si-compounds**. On the other hand, for the tetrahedral C-compounds, optical purity can be maintained.

There are two paths: first-order (which is actually a pseudo-first order) k_s–path and second order k_Y-path. This k_Y–path shows the dependence on the nucleophilicity of the entering ligand. This second order path may be considered to compare the relative nucleophilicity of the entering groups.

(c) Swain-Scott equation to the nucleophilic substitution reactions of C-compounds: To measure the relative nucleophilicities, the earliest attempt was given by Swain-Scott equation.

$$\log\left(\frac{k_Y}{k_0}\right) = sn, \; s = 1 \text{ for } CH_3Br, \quad \text{where } k_0 = \frac{k_{H_2O}}{[H_2O]}$$

where k_Y is the rate constant (for the associative process) of the following reaction.

$$CH_3\text{—}Br + Y \rightarrow CH_3\text{—}Y^+ + Br^- \; (s = 1)$$

Similarly, $k_0 (= k_{H_2O}/[H_2O])$ is obtained for the reaction, $CH_3Br + H_2O$, where H_2O enters as the nucleophile.

Thus, Swain-Scott equation gives the relative ability of a nucleophile (say *Y*) to replace Br^- from CH_3Br compared to that of the reference nucleophile H_2O. Here, it is assumed that both *Y* and H_2O participate in an associative path to replace Br^- from CH_3Br.

The constant n is described as the *nucleophilic constant,* a characteristic parameter of the nucteophile Y. On the other hand, s is a *discrimination constant* that depends on the nature of the substrate (*i.e.* electrophile). Thus it varies from substrate (*i.e.* electrophile) to substrate. This is why, s is described as a *substrate constant.* By definition:

$$n = 0 \text{ for } H_2O \text{ (nucleophile)}; \; s = 1 \text{ for } CH_3Br \text{ (electrophile)}$$

The knowledge of s and n expressing the characteristic features of electrophiles and nucleophiles respectively can help us predict the reaction rate of an associative reaction.

(d) **Extension of Swain-Scott approach to nucleophilic substitution reactions of square planar complexes of Pt(II):** Swain-Scott approach was utilised by Basolo, Pearson and their coworkers to develop the **nucleophilicity scale (n_{Pt})** by considering the nucleophilic substitution reactions (at 30°C) at *trans*–$[PtCl_2(py)_2]$ (*i.e.* **reference substrate or electrophile**) in methanol solution where MeOH is taken as the *reference nucleophile.* It is defined as follows:

$$sn_{Pt} = \log\frac{k_Y}{k_{CH_3OH}} = \log k_Y - \log k_{CH_3OH}$$

Here k_Y is the second order rate constant (having unit $dm^3\ mol^{-1}\ s^{-1}$) of the following reaction.

$$trans\text{–}[PtCl_2(py)_2] + Y \rightarrow trans\text{–}[PtCl(py)_2(Y)]^{n+} + Cl^-, \; (s = 1)$$

k_{CH_3OH} denotes the rate constant for the same reaction involving CH_3OH as the reference nucleophile where CH_3OH is the solvent. Obviously, k_{CH_3OH} is a pseudo first-order rate constant (having unit s^{-1}). This is why, k_{CH_3OH} is divided by the concentration of the solvent methanol (= 25 mol dm^{-3} at 30°C), so that the **nucleophilicity parameter becomes dimensionless.** It may be noted that to make the Swain-Scott nucleophilicity parameter (n) dimensionless, the rate constant k_{H_2O} is to be divided by 55.5 mol dm^{-3}, concentration of water as a pure solvent. This is why, it is customary to express the nucleophilicity parameter for the Pt(II)–substrate as follows:

$$sn^0_{Pt} = \log\left[\frac{k_Y}{k_{CH_3OH}/[CH_3OH]}\right], \; s = 1 \text{ for } trans\text{-}[PtCl_2(py)_2]$$

The ***substrate constant (s)*** regarded as the ***nucleophilic discrimination factor*** (NDF) is taken 1.0 for the reference substrate *trans*–$[PtCl_2(py)_2]$. The n^0_{Pt} values for various nucleophiles are given in Table 5.30.5.1.

Table 5.30.5.1: Effect of entering ligand on the rate constant (k_Y) in the reaction *trans*–$[PtCl_2(py)_2]$ + Y → *trans*–$[PtCl(py)_2(Y)]^{n+}$ + Cl^-, at 30°C (if not mentioned otherwise) in methanol and n^0_{Pt} values (*cf.* U. Belluco *et. al., J. Am. Chem. Soc.,* **87**, 241, 1965; R.G. Pearson *et al., ibid,* **90**, 319, 1968; M. Becker *et al., Inorg. Chim. Acta.,* **116**, 47, 1986)

Y	k_Y **($dm^3\ mol^{-1}\ s^{-1}$)**	n^0_{Pt}	pK_a *of YH*$^+$	$n^0_{CH_3I}$
CH_3OH*	2.7×10^{-7} (25°C)	0.0	−1.7	0.0
Cl^-	4.5×10^{-4}	3.04	−5.7	4.37
NH_3	4.7×10^{-4}	3.07	9.25	5.50
pipyridine	–	3.13	11.21	7.30
pyridine	5.5×10^{-4}	3.19	5.23	5.23
NO_2^-	6.8×10^{-4}	3.22	3.37	5.35
N_3^-	1.55×10^{-3}	3.58	4.74	5.78
Br^-	3.7×10^{-3}	4.18	−7.7	5.80
I^-	1.07×10^{-1}	5.46	−10.7	7.42
CN^-	4.0 (25°C)	7.14	9.3	6.70
Ph_3P	249.0 (25°C)	8.93	2.73	7.0
Ph_3As	–	6.90	–	4.77

* The rate constant value refers to that of k_{CH_3OH}.

In Table 5.30.5.1, $n^0_{CH_3I}$ values are also given and it is defined as for n^0_{Pt}. For $n^0_{CH_3I}$, the substrate is CH_3I where the replacement of I^- by different nucleophiles is considered. There is a co-relationship between n^0_{Pt} and $n^0_{CH_3I}$ in general but there are exceptions.

> $\boldsymbol{n^0_{Pt}}$ ***vs.*** $\boldsymbol{n^0_{CH_3I}}$ **:** It is evident that these two nucleophilicity parameters do not bear any *close correlationship.* It is due to the fact that the centres receiving the nucleophilic attack in the associative path of reaction are different: one is sp^3-C while the other is polarisable Pt(II) (*i.e.* **soft centre)**. The LUMOs to receive the nucleophilic attack are different in two cases.

It is evident from the Table 5.30.5.1, that the nucleophilicity of the entering ligands towards the soft Pt(II) centre measured by n^0_{Pt}, depends on the **softness of the ligands**. It indicates that the **soft-soft matching** is playing an important role to determine the rate constant k_Y and consequently, the nucleophilicity of Y. The rate constants cover the span about 10^9 (*i.e.* 10^2 to 10^{-7}). The rate constants and n^0_{Pt} vary in the sequence for the different donor atoms as:

$$F^- \ll Cl^- < Br^- < I^-;\ O < S;\ NH_3 < AsPh_3 < PPh_3.$$

It explains the order of **ligand exchange rate in** $\boldsymbol{[PtX_4]^{2-}}$**:** k_{ex} for $X^- = CN^- > I^- > Br^- > Cl^-$.

- $\boldsymbol{n^0_{Pt}}$ ***vs.*** $\boldsymbol{pK_a}$ **(YH⁺) and polarisability of Y:** Here it is interesting to note that the nucleophilicity measured by n^0_{Pt} does not bear any correlationship with the basicity of the ligands (measured by the pK_a values of the conjugate acids). But, interestingly, n^0_{Pt} bears a good correlation with the **polarisability** (*i.e.* **softness**) the ligands. ***With the increase of softness of Y, the rate constant*** $\boldsymbol{(k_Y)}$ ***and nucleophilicity*** $\boldsymbol{(n^0_{pt})}$ ***increase*** (*cf.* $I^- > Br^- > Cl^-$).

- $\boldsymbol{n^0_{Pt}}$ **and π-bonding property of the ligand:** Effect of the nature of the entering ligand on the rate constant can be rationalised in terms of the ***stabilisation of the trigonal bipyramidal activated complex*** in which one equatorial position is occupied by the entering ligand. The equatorial ligands can stabilise the trigonal bipyramidal activated complex through π-bonding. *Thus a good π-bonding ligand is a good entering ligand and at the same time, a good polarisable ligand* (*cf.* Pt^{II}–centre is a polarisable centre, effect of the *d*- and *f*-electrons — **specially the electrons present in the relativistically expanded 5*d*- and 4*f*-orbitals,** $\boldsymbol{5d^8}$ **and** $\boldsymbol{4f^{14}}$) *is also a good entering ligand.* The general order of the entering ligands to favour the rate is:

$$Ph_3P > tu > CN^- > SCN^- > I^- > Br^- > N_3^- > Cl^-$$

- **Nucleophilic discrimination factor (NDF, *s*):** The nucleophilic discrimination factor (*s*) is obtained from the following equation.

$$sn^0_{Pt} = \log\left[\frac{k_Y}{k_{CH_3OH}/[CH_3OH]}\right] = \log\left(\frac{k_Y}{k^0_{CH_3OH}}\right) = \log k_Y - \log k^0_{CH_3OH}$$

or,
$$\log k_Y = sn^0_{Pt} + \log k^0_{CH_3OH} = sn^0_{Pt} + \log k_s,\ k^0_{CH_3OH} = k_s\ \text{(say)}$$

Plot of log k_Y *vs.* n^0_{Pt} gives a straight line having slope = *s*. For the substrate, *trans*–$[PtCl_2(py)_2]$, *s* is taken as unity. ***The larger value of s indicates that the reaction rate is more sensitive to the change of nucleophilicity.*** It has been noted that the Pt(II) complexes having the softer ligands show the higher *s*–values (*cf.* Table 5.30.5.2). In fact, **high reactivity of the starting reference complex leads to the low sensitivity** (*i.e.* **smaller value of *s*).** In other words, the less reactive reference complexes are more sensitive (*i.e.* higher values of *s*) to the change of nucleophilicity of the entering ligand. In fact, **there is an inverse correlation** between *s* and k_s (solvent rate constant in $dm^3\ mol^{-1}\ s^{-1}$) (Table 5.30.5.2). The smaller k_s values indicate the **lower intrinsic reactivity** of the complex towards the poorest nucleophile (*i.e.* solvent.).

Table 5.30.5.2: Nucleophilic discrimination factors (*k_s value in water; otherwise in methanol)

Complex	*s(NDF)*	*log k_s*
$[PtCl_2(en)]$*	0.64	–5.26
trans-$[PtCl_2(py)_2]$	1.00	–5.80
trans-$[PtCl_2(AsEt_3)_2]$	1.25	–7.50
trans-$[PtCl_2(PEt_3)_2]$	1.43	–8.82

The complexes bearing the **phosphine** and **arsine ligands** possess the highest *s* values and lowest k_s values (Table 5.30.5.2). These **π-bonding soft ligands** can stabilise the transition state (see Fig. 5.30.15.4) and such complexes exhibit a higher sensitivity to the change of nucleophilicity of the entering ligands.

5.30.6 Effect of the Leaving Group: Relative Effects of the Entering and Leaving Group

There is also a marked effect on the rate for the variation of the nature of the leaving group. This indicates that there is a *considerable perturbation of the M—X bond* in *producing the* transition state. ***However, this bond breaking/purturbation by the leaving group is relatively less important than the bond formation by the incoming nucleophile to produce the transition state***. This is why, the rate is more sensitive to the nature of the incoming nucleophile than to the nature of the leaving group (*cf.* the relative effects of the π-bonding properties of the entering and leaving group; Sec. 5.30.15B). It supports the I_a mechanism.

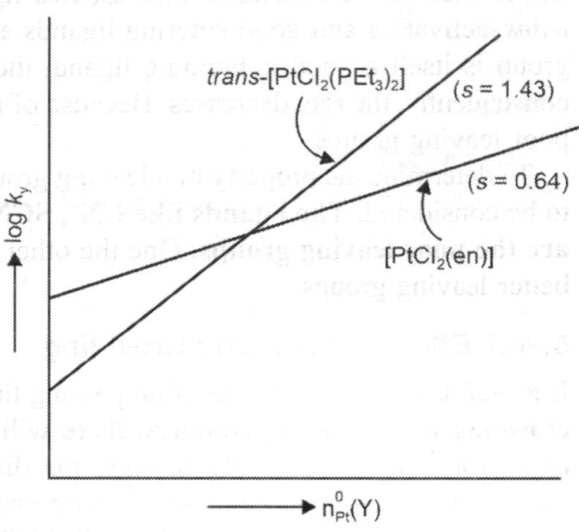

Fig. 5.30.5.1: Qualitative representation of the responsiveness of the Pt(II)-complexes towards the nucleophilicity of the entering groups

Ligand exchange reactions — relative effects of the same entering and leaving group: The above arguments can explain the order of ligand exchange rate at $[PtX_4]^{2-}$ (*cf.* more **importance as the entering group):**

$$k_{ex}: \quad [Pt(CN)_4]^{2-} > [PtI_4]^{2-} > [PtBr_4]^{2-} > [PtCl_4]^{2-}$$
$$\sim t_{1/2}\ (\text{min}): \quad 1 \qquad 4 \qquad 6 \qquad 850$$

cf. **Nucleophilicity order:** $CN^- > I^- > Br^- > Cl^-$; (Table 5.30.5.1)

cf. **Leaving group order:** $Cl^- > Br^- > I^- > CN^-$

It indicates the more importance of the entering group (*i.e.* I_a or *A* process) to determine the rate.

Table 5.30.6.1: Effect of the nature of the incoming nucleophile and leaving group on the rate constant (25°C).

(a) $[PtBr(dien)]^+ + Y \rightarrow [Pt(dien)Y]^{n+} + Br^-$		(b) $[Pt(dien)X]^+ + py \rightarrow [Pt(dien)(py)]^{2+} + X^-$ ($[py] = 5.9 \times 10^{-3}$ mol dm^{-3})	
Y	k_Y *(dm³ mol⁻¹ s⁻¹)*	X^-	k_{obs} (s^{-1})
Cl^-	8.8×10^{-4}	Cl^-	3.5×10^{-5}
I^-	2.3×10^{-1}	Br^-	2.3×10^{-5}
py	3.3×10^{-3}	I^-	1.0×10^{-5}
N_3^-	7.7×10^{-3}	N_3^-	8.3×10^{-7}
SCN^-	4.3×10^{-1}	SCN^-	3.0×10^{-7}

(*cf.* H.B. Gray *et al., Inorg. Chem.* **1**, 481, 1962)

From Table 5.30.6.1, it is evident that in changing Y (*i.e.* entering group), from Cl^- to I^-, the second order rate constant changes by a **factor about 10^3** while in changing X^- (*i.e.* leaving group) from I^- to Cl^-, the observed rate constant changes by **a factor only 3.5**.

Effect of the leaving group on the rate process follows the sequence:

$$NO_3^- > H_2O > Cl^- > Br^- > I^- > N_3^- > SCN^- > NO_2^- > CN^-$$

It may be noted that the good entering ligands are the poor leaving groups. It can be explained by considering the formation of a trigonal bipyramidal activated complex in which the equatorial plane is occupied by the leaving group, entering group and the *trans*–directing ligand. ***The trigonal bipyramidal activated complex is stabilised by the σ- and π–bonding property of these equatorial ligands***. In fact, out of these three equatorial ligands, two ligands can simultaneously participate in π–bonding. The *trans*–activating and good entering ligands are very often the good π–bonding ligands. If the leaving group is itself a good π–bonding ligand, then it cannot be easily dislodged to give the product and consequently, the rate decreases. Because of the same ground, the *trans*–activating ligands are also the poor leaving groups.

To determine the property of a leaving group, both the σ– and π–bonding properties of the ligand are to be considered. **The ligands like CN^-, SCN^-, etc. form both the strong σ- and π-bonds and these are the poor leaving groups.** One the other hand, halides which form only the weaker bonds are the better leaving groups.

5.30.7 Effect of the Steric Crowding

It is well known that for a reaction passing through the associative path, with the increase of the steric crowding in the starting complex, there will be a rate retardation (**steric retardation**). On the other hand, for a reaction passing through the dissociative path, there will be a rate acceleration (**steric acceleration**) with the increase of steric crowding in the starting complex. Let us consider the rate constant for the hydrolysis of *cis*–$[PtCl(L)(PEt_3)_2]^+$, *i.e.*

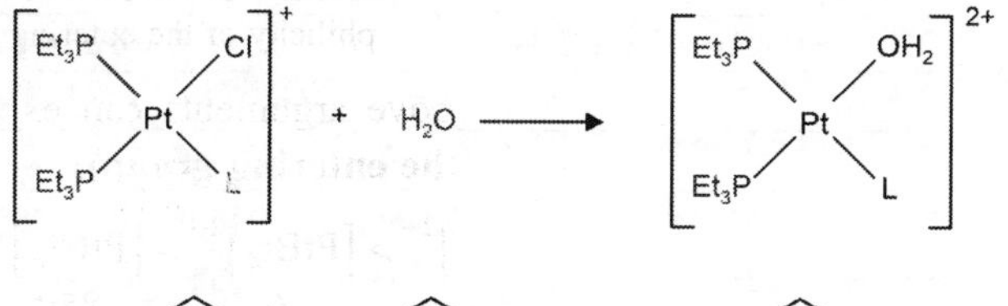

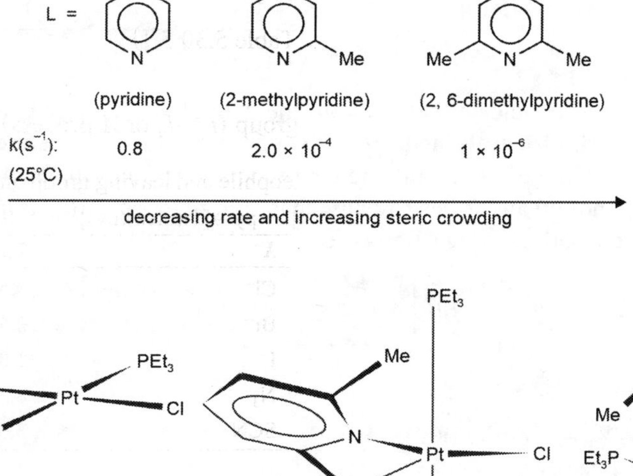

$$[PtCl(L)(PEt_3)_2]^+ + H_2O \longrightarrow [Pt(OH_2)(L)(PEt_3)_2]^{2+}$$

L = (pyridine), (2-methylpyridine), (2, 6-dimethylpyridine)

	(pyridine)	(2-methylpyridine)	(2, 6-dimethylpyridine)
$k(s^{-1})$: (25°C)	0.8	2.0×10^{-4}	1×10^{-6}

decreasing rate and increasing steric crowding →

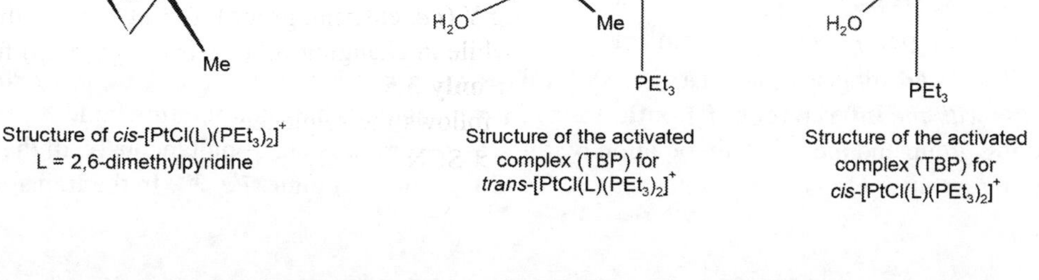

Structure of *cis*-$[PtCl(L)(PEt_3)_2]^+$ L = 2,6-dimethylpyridine

Structure of the activated complex (TBP) for *trans*-$[PtCl(L)(PEt_3)_2]^+$

Structure of the activated complex (TBP) for *cis*-$[PtCl(L)(PEt_3)_2]^+$

For L = 2, 6–dimethylpyridine, in the *cis*-complex, the *ortho*-Me groups can block the positions above and below the plane of the complex to hinder the approach of the entering ligand in the associative path. For L = 2-methylpyridine, the *ortho*-Me group can block either the above or below position to cause this hindrance. Definitely, for the unsubstituted pyridine, the hindrance is minimum. ***Thus the steric retardation of rate is in conformity with the associative path.*** *Here it may be mentioned that this steric retardation is relatively less for the corresponding trans–complex* (*i.e.* L is *trans*–to Cl^-). This can be explained by considering the formation of trigonal bipyramidal activated complex. **For the *cis*–complex**, the activated complex bears Cl^- (leaving group), PEt_3 (*trans*–to the leaving group) and H_2O (incoming nucleophile) in the trigonal plane of the trigonal bipyramid keeping L (*i.e. py* or substituted *py*) and PEt_3 in the axial directions. **Thus L causes the steric hindrance** (*i.e.* ligand–ligand repulsion) **at 90°.** For the corresponding ***trans*–complex,** the activated complex places L in one equatorial position to cause the steric hindrance at 120°C. ***Obviously, the steric crowding is more pronouned in the activated complex, for the starting cis–complex,*** *i.e.* the *trans*-complex aquates faster than the *cis*-complex.

5.30.8 Activation Parameters

The activation parameters of some representative substitution reactions of square planar complexes are given in Table 5.30.8.1.

Table 5.30.8.1: Activation parameters for the reaction (in methanol). $[Pt^{II}Br(L)(PEt_3)_2] + Y \rightarrow [Pt^{II}(L)(PEt_3)_2(Y)] + Br^-$, L = 2, 4, 6–trimethylphenyl, *i.e.* 2, 4, 6–$Me_3C_6H_2$ (*cf.* R. van Eldik *et al.*, *Inorg. Chem.*, **18**, 572, 1979)

	cis–Complex + Y →				*trans–Complex + Y →*			
Y =	I^-		$SC(NH_2)_2$		I^-		$SC(NH_2)_2$	
	k_s	k_Y	k_s	k_Y	k_s	k_Y	k_s	k_Y
$\Delta H^{\neq}$ (kJ mol^{-1}):	84	63	79	59	80	59	71	43
$\Delta S^{\neq}$ (J K^{-1} mol^{-1}):	–59	–121	–71	–121	–52	–115	–80	–130
$\Delta V^{\neq}$ (cm^3 mol^{-1}):	–67	–63	–71	–54	–16	–16	–17	–11

k_s (s^{-1}), k_Y ($dm^3 mol^{-1} s^{-1}$)

The data in Table 5.30.8.1 indicate that for both the neutral and anionic entering ligands, both the $\Delta S^{\neq}$ and **$\Delta V^{\neq}$ values are significantly negative**. It supports the **associative mechanism.** Here it may be noted that the starting complex is neutral and consequently, entry of the nucleophile (which may be neutral or anionic) does not create or destroy any charge in the associative transition state.

The $\Delta V^{\neq}$ values for the following reactions are:

HN, NH_2, Pt, H_2N, X $\xrightarrow[-X^-]{+Y^-}$ HN, NH_2, Pt, H_2N, Y

$[Pt(dien)X]^+ + Y^- \rightarrow [Pt(dien)(Y)]^+ + X^-$

$X^- = Cl^-, Br^-, I^-, N_3^-$; $Y^- = OH^-, I^-, N_3^-, NO_2^-, SCN^-$, etc.

$\Delta V^{\neq}$ (k_s–path) = –8 to –15 cm^3 mol^{-1}

$\Delta V^{\neq}$ (k_Y–path) = –6 to –12 cm^3 mol^{-1}

Here it is interesting to note that the $\Delta V^{\neq}$ values *are more negative for the k_s-path and it is in conformity with the* **interchange (I_a) path**. In the interchange mechanism, both the bond formation by the entering group and the bond dissociation by the leaving group occur simultaneously. In the k_s–path, the entering group is H_2O (*i.e.* solvent) and the leaving group is anionic, *i.e.* X^-. In the transition state, the leaving group X^- (which is dissociating) is solvated to cause the electrorestriction to make $\Delta V^{\neq}$ negative; and fixation of H_2O as the entering group also makes $\Delta V^{\neq}$ negative. Thus in the k_s-path (**interchange process**), both the leaving group and entering group make the $\Delta V^{\neq}$ value negative. In the k_Y-path, fixing of Y^- in the transition state makes $\Delta V^{\neq}$ negative definitely but it needs desolvation to make a positive contribution to $\Delta V^{\neq}$; and the dissociating X^- (in the **interchange process**) experiences solvation in the transition state to make a negative contribution to $\Delta V^{\neq}$ through electrorestriction over the solvent. Thus, $\Delta V^{\neq}$ (resultant) value becomes less negative for the k_Y-path. ***These data support the simultaneous bond rupture by the leaving group (X^-) and bond formation by the entering group (Y^- or H_2O) in the transition state.***

$$k_Y\text{-path: } Y^- \cdots\cdots Pt^{2+} \cdots\cdots X^-_{aq}\text{, and } k_s\text{-path: } H_2O\cdots Pt^{2+}\cdots X^-_{aq}$$

If the **simple associative (*A*) mechanism** is considered in the k_s-path, then fixing of H_2O as a nucleophile in the trasnition state (T.S.) makes $\Delta V^{\neq}$ negative and there is **no electrorestriction effect** (*i.e.* neither charge is created nor destroyed in attaining the T.S.). In the k_Y-path, fixing of the anionic nucleophile in the cationic complex in the associative T.S. leads to **charge neutralisation.** It makes $\Delta V^{\neq}$ more positive. Thus the simple associative path can also explain the less negative $\Delta V^{\neq}$ for the k_Y-path. But relatively **the less negative value of $\Delta V^{\neq}$** (for both k_s and k_Y) indicate the **I_a process rather than the *A*-process**. In the pure ***A***-process, fixing of the entering group leads to the T.S. while in the I_a process, both fixing of the entering nucleophile and departure of the leaving group occur simultaneously to produce the T.S. **It makes $\Delta V^{\neq}$ more negative in the *A*-process, in general.**

5.30.9 Effect of Overall Charge of the Complex

The ligand substitution rate is hardly influenced by the overall charge of the complex.

Complex:	$[PtCl(NH_3)_3]^+$	*trans*–$[PtCl_2(NH_3)_2]$	$[PtCl_4]^{2-}$
k_{ex} (s^{-1}) (for *Cl^-): (25°C)	1.5×10^{-5}	3.5×10^{-5}	3.8×10^{-5}

Complex:	$[PtCl_4]^{2-}$	$[PtCl_3(NH_3)]^-$	$[PtCl_2(NH_3)_2]$	$[PtCl(NH_3)_3]^+$
~ $t_{1/2}$ (aquation):	300 min	increasing →		700 min.

(aquation of Cl^- for 0.1 mol dm^{-3} reactant)

For the *Cl^- exchange reaction or aquation reactions of the above complexes, the overall charge changes from +1 to –2 (*i.e.* 3 unit) or –2 to +1 (*i.e.* 3 unit), but the rate marginally changes only (**by a factor only 2-3;** ***cf.*** D. Banerjea, F. Basolo and R.G. Pearson, *J. Am. Chem. Soc.*, **79,** 4055, 1957). Here it may be mentioned that for aquation or anation of Co(III)–complexes, the rate constant drastically depends on the charge, *e.g.* aquation rate changes ***by a factor about 10^3*** in going from $[CoCl(NH_3)_5]^{2+}$ to $[CoCl_2(NH_3)_4]^+$ for a change of only one unit of positive charge.

The fairly insensitiveness of the rate constants with the variation of overall charge of the square planar complexes can be explained by considering the interchange mechanism where both bond breaking by the leaving group and bond formation by the entering group contribute to the activation process. If bond breaking were the predominant factor to the activation process (as in the purely dissociative process), then the rate constant would decrease remarkably with the increase of positive charge of the complex. In fact, it happens so for the octahedral Co(III)–complexes. If the bond formation by the entering nucleophile were the predominant factor to the activation process, then the rate constant would increase significantly with the increase of the overall positive charge. If both the bond rupture and bond formation simultaneously contribute to the activation process as in the **interchange (I) mechanism** then the effect of charge is mutually cancelled due to the opposing phenomena, *i.e.* bond formation and bond rupture.

Note: In the given above examples, besides the effect of the overall charge of the complex, the effect of the *trans*-labilising power of the ligand (*i.e. trans*-effect) is to be considered.

The *trans*-labilising power (*cf. trans*-effect series) of the concerned ligands follows the sequence:

$$Cl^- > NH_3$$

In fact, *trans*-$[PtCl_2(NH_3)_2]$ is more labile than its *cis*-isomer. Thus, in the given complexes, the same *trans*-effect operates for $[PtCl_3(NH_3)]^-$, *trans*-$[PtCl_2(NH_3)_2]$ and $[PtCl_4]^{2-}$ (*i.e.* Cl^- is the *trans*-group) while for $[PtCl(NH_3)_3]^+$, NH_3 (less *trans*-labilising) is the *trans*-ligand.

5.30.10 Effect of the Central Metal and Crystal Field Activation Energy (CFAE)

(i) **Reactivity order for different d^8-systems:** Reactivity of the square planar complexes depends strongly on the nature of the central metal ion. The reactivity order for Ni(II), Pd(II) and Pt(II) follows the sequence:

$$\text{Ni(II)} \gg \text{Pd(II)} \gg \text{Pt(II)}$$

It is found in many reactions:

(a) $[M(C_6H_4CH_3)(PEt_3)_2Cl]^0$ + Y (= py) (in EtOH) → $[M(C_6H_4CH_3)(PEt_3)_2(py)]^+$ + Cl^-

k_Y ($dm^3\ mol^{-1}\ s^{-1}$): M = Pt(II): 6.7×10^{-6}; Pd(II): 0.58; Ni(II): 33.0

(b) $[M(dien)X]^+$ + py → $[M(dien)(py)]^{2+}$ + X^-

k_{py} ($dm^3\ mol^{-1}\ s^{-1}$), 25°C, in aqueous media

X^-	M = Pd(II)	Pt(II)
I^-	3.5×10^{-2}	1.0×10^{-5}
SCN^-	0.31	3.0×10^{-7}
NO_2^-	6.5	5×10^{-8}

(c) The rate constant (24°C) for the exchange of $^*CN^-$ with $[M(CN)_4]^{n-}$ follows the following sequence.

$$k_{ex}\left(dm^3\ mol^{-1}\ s^{-1}\right): \quad \frac{\text{Au(III)}}{3.9\times10^3} \quad \frac{\text{Pt(II)}}{26.0} \quad \frac{\text{Pd(II)}}{1.2\times10^2} \quad \frac{\text{Ni(II)}}{5\times10^5}$$

(ii) **Reactivity order in terms of the energy of the vacant orbital (*i.e.* LUMO) to receive the nucleophilic attack** (see Secs 5.13, 5.30.3): The observed reactivity order can be explained by considering the ease of formation of a 5-coordinate activated complex by using the vacant (non-bonding in nature) $(n)p_z$-orbital (*i.e.* a_{2a} in terms of MOT, LUMO) in the associative pathway assuming the starting square planar complex to lie in the *xy*-plane. For Ni(II), Pd(II) and Pt(II), the said orbitals are $4p_z$, $5p_z$ and $6p_z$ respectively. Obviously, for Pt(II), it is energetically most disfavoured while it is energetically most favoured for Ni(II). For Au(III), it is $6p_z$ but its energy is decreased significantly because of its higher positive charge. It causes the enhanced reactivity for Au(III) compared to that for Pt(II) which also uses the $6p_z$ orbital. Besides this, the higher positive charge (*i.e.* less negative charge) of the Au(III)-complex (*cf.* $[Au(CN)_4]^-$ *vs.* $[M(CN)_4]^{2-}$, M = Pd, Pt) also favours more the nucleophilic attack at *rds* of the associative path of ligand exchange in the Au(III) complex.

In an alternative view in terms of MOT, it may be argued that in the associative path, the incoming group will occupy the vacant antibonding $d^*_{z^2}$-orbital (*i.e.* a^*_{1g}). The energy order of this a^*_{1g} orbital follows the sequence: Pt(II) > Pd(II) > Ni(II). It explains the reactivity order: Ni(II) > Pd(II) > Pt(II) (see Sec. 5.13 for details). It may be pointed out that, in terms of CFAE (expressed in Dq), the inertness increases for the heavier congeners due to the higher Dq values.

(iii) **CFAE and reactivity order:** The reactivity of square planar complexes in ligand substitution process can be better understood in terms of *crystal field activation energy* (CFAE). In the associative path, two possible geometries for the activated complex may be considered. These are *trigonal bipyramid* (TBP) and *square pyramid* (SP). CFAE for these possible two geometries may be calculated and compared (*cf.* Table 5.30.10.1). It is evident that the TBP activated complex (T.S.) or intermediate is energetically less favoured (*i.e.* higher CFAE) compared to the SP T.S. or intermediate. But the TBP geometry is stabilised better through the π-bonding which is very much important and in terms of **steric crowding** or **ligand-ligand repulsion** (*cf.* **VSEPR**, TBP *vs.* SP, Vol. 2). Evidences support the TBP T.S. or intermediate.

Table 5.30.10.1: Crystal field activation energy (CFAE) for the ligand substitution reactions of square planar complexes (assuming the strong field ligands) (see Vol. 4 for calculation of CFSE)

d^n	*cfse (Dq)*			*CFAE (Dq)*	
	Square planar (a)	*Trigonal bipyramidal (TBP) (b)*	*Square pyramidal (SP) (c)*	*(a) – (b)*	*(a) – (c)*
d^7	26.84	13.34	19.14	13.50	7.70
d^8	24.56	14.16	18.28	10.40	6.28
d^9	12.28	7.09	9.14	5.19	3.14
d^{10}	0	0	0	0	0

In terms of CFAE, the reactivity order should be (for comparable *Dq* values):

$$d^{10} \gg d^9 > d^8 > d^7$$

It explains the **lability sequence: Zn(II) > Cu(II) > Ni(II)** (for the square planar complexes of some tetradentate Schiff base type ligand).

However, complication arises, if the starting complex is in higher coordination number due to the ligation by solvents.

The CFAE is expressed in terms of *Dq*. The *Dq*–value increases for the heavier congeners and it makes the heavier congeners more inert. It explains the **lability sequence: Ni(II) > Pd(II) > Pt(II).** The lability order for their **square planar complexes** can also be understood in terms of the energy order of their LUMOs to receive the nucleophilic attack at the *rds* **in the associative path** (see Secs 5.13 and 5.30.3 for details).

5.30.11 Nucleophile Catalysed *Cis–Trans* Isomerisation in Square Planar Complexes of Platinum(II)

The geometrical isomerisation of Pt(II) complexes is catalysed by a ligand added to the solution. It passes through an **intermolecular process.** Here it may be mentioned that the substitution process in the square planar complexes of Pt(II) is **stereospecific** and *it can be concluded that the isomerisation occurs by a* **double displacement process** *(each step involves a trigonal bipyramidal activated complex).*

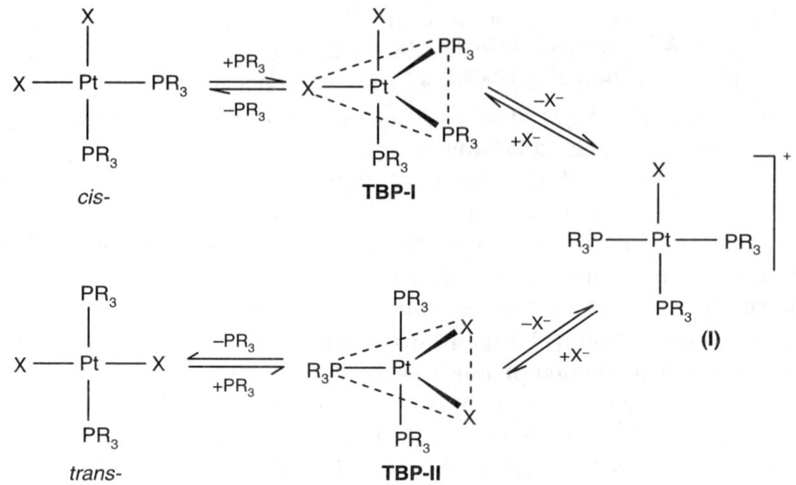

Scheme 5.30.11.1: Mechanism of *cis–trans* isomerisation in $[Pt(PR_3)_3X_2]$ through a *double displacement process* in the presence of catalytic amount of PR_3 (*see* Scheme 5.30.3.1 for the detailed mechanistic paths)

The *cis–trans* isomerisation of $[Pt(PR_3)_2X_2]$ (X^- = halide) in ***presence of the catalytic amount of*** $\boldsymbol{PR_3}$ in CH_2Cl_2 solution is well documented. This is illustrated in Scheme 5.30.11.1. The two step process is:

1st step: $cis\text{-}[Pt(PR_3)_2X_2] + PR_3 \rightleftharpoons [Pt(PR_3)_3X]^+ + X^-$ (fast)

2nd step: $[Pt(PR_3)_3X]^+ + X^- \rightleftharpoons trans\text{-}[Pt(PR_3)_2X_2] + PR_3$, (slow)

Thus in the 1st step, PR_3 acts as the entering group while X^- acts as the leaving group. In the 2nd step, the reverse event occurs. In fact, from the **intermediate** $[Pt(PR_3)_3X]^+$ (I), addition of X^- can produce **TBP-I** and **TBP-II. From the TBP-II, release of PR_3 as the leaving group gives the *trans*-isomer while from the TBP-1, release of PR_3 as the leaving group produces the *cis*-isomer.** Formation of the TBP-II (from I) leads to the *trans*-isomer, and PR_3 *trans* to X^- (in I) acts as the leaving group while to produce the *cis*-isomer, through the formation of TBP-I (from I), PR_3 *trans* to PR_3 (in I) acts as the leaving group. In the same way, $[PtCl_2(SEt_2)_2]$ undergoes isomerisation in solution in presence of SEt_2 as a catalyst.

A general mechanism of the **nucleophile (Nu) catalysed *cis–trans* isomerisation** of the square planar complexes is outlined in Scheme 5.30.11.2 where **TBP-I** and **TBP-II** remain in an equilibrium through **Berry pseudorotation** which leads to an exchange between the axial and equatorial ligands. The conversion TBP-I → TBP-II needs two successive Berry pseudorotations keeping L and X as the pivotal groups respectively in these successive pseudorotations. TBP-I generates the *cis*-isomer while TBP-II generates the *trans*-isomer.

- *Cis-trans* isomerisation of *bis*(glycinato-κ^2NO)copper(II):

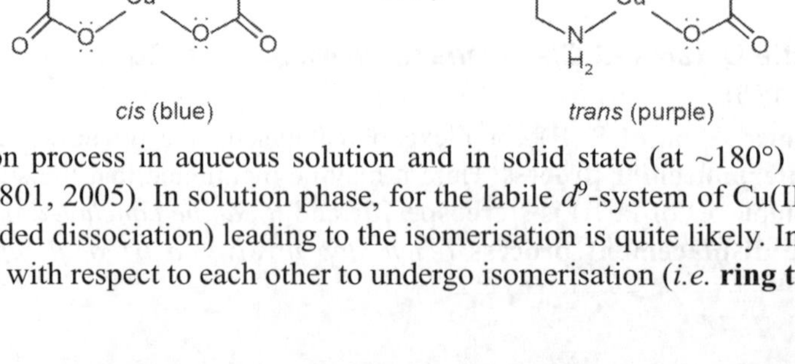

cis (blue) → *trans* (purple)

This isomerisation process in aqueous solution and in solid state (at ~180°) is well documented (*J. Chem. Edu.*, **97**, 801, 2005). In solution phase, for the labile d^9-system of Cu(II), **the ring opening mechanism** (one ended dissociation) leading to the isomerisation is quite likely. In the solid phase, the glycinate rings twist with respect to each other to undergo isomerisation (*i.e.* **ring twisting mechanism**

Mechanistic Paths of *cis-trans* Isomerisation in Square Planar Complexes

The probable routes are discussed in short. (i) **Nucleophilic attack giving rise to a stereomobile TBP intermediate:** This mechanism is outlined in Scheme 5.30.11.2 and it works well for the square planar complexes of Ni(II), Pd(II) and Pt(II). For the **more inert Pt(II)-systems,** it needs a good nucleophile as a catalyst while for the **more labile Ni(II) or Pd(II) complexes, even the solvent** can act as the required nucleophile for the isomerisation. (ii) **Ligand dissociation giving rise to a symmetrical trigonal planar intermediate:** In some rare cases, this route has been found to operate. In general, this dissociative path is most unlikely for the square planar complexes which generally experience the associative activation in the ligand substitution process. (iii) **Intra-molecular rearrangement with a tetrahedral intermediate:** This route is well documented for the **Ni(II)-complexes** where stabilities of the square planar and tetrahedral structures do not differ significantly (*cf.* Sec. 2.3 conformational isomerism).

(*cis-*) ⇌ ⇌ (*trans-*)

Trigonal bipyramid *vs.* square pyramid activated complex in the ligand substitution process of square planar complexes

Here it is interesting to note that in terms of cfse, square pyramidal geometry is favoured over the trigonal bipyramidal geometry (see Table 5.30.10.1 for CFAE). ***But the trigonal bipyramidal geometry is favoured because of the following reasons:***

- ***lower steric repulison*** among the ligands in the trigonal bipyramidal geometry (*cf.* VSEPR, Vol. 2).
- two of the equatorial ligands in the trigonal bipyramid can simultaneously participate in the π–bonding interaction; **in fact the stabilisation earned through the π-bonding can easily compensate the loss of cfse.**

Probably, because of the above mentioned two factors, the trigonal bipyramidal activated complex occurs in the ligand substitution process of the square planar complexes. However, if the equatorial ligands cannot participate in efficient π–bonding, then square pyramid geometry may also describe the activated complex.

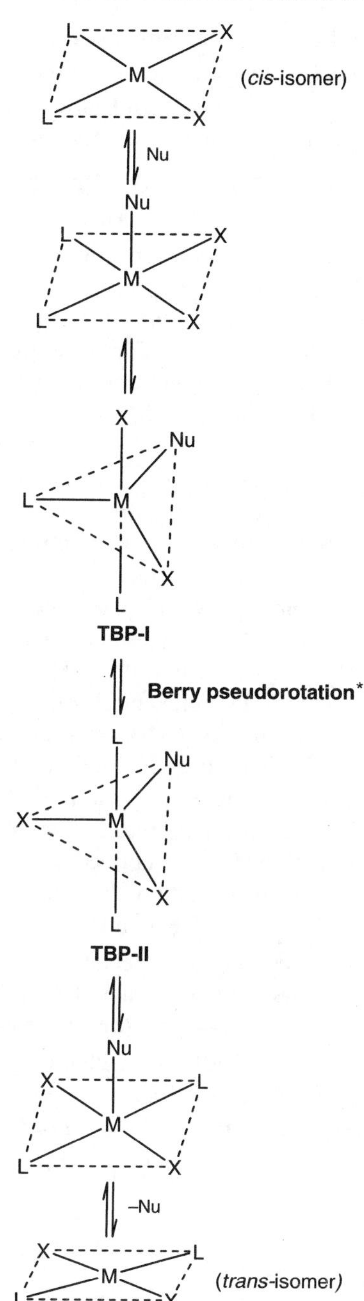

Scheme 5.30.11.2: A general mechanism for the nucleophile (Nu) catalysed *cis-trans* isomerisation of square planar complexes $[ML_2X_2]$ (**see* Note of Scheme 5.30.4.2)

rather than the ring opening mechanism, *Ref. J. Phys. Chem. B*, **108**, 2098, 2004). The ring twisting may lead to a tetrahedral intermediate (see Sec. 6.15.12). It may be noted that the ring twisting mechanisms of racemisation of octahedral complexes are well documented (see **Ray-Dutt twist** and **Bailer twist**). The mechanism of **photoisomerisation of [Pt(gly)$_2$]** has been discussed in Sec. 6.15.12.

5.30.12 Effect of the Spectator Ligand *trans-* to the Leaving Group: *Trans*-effect and *Trans*-influence

It has been already mentioned that the ligand substitution process in the square planar complexes of Pt(II) is stereospecific, *i.e. the entering ligand occupies the position strictly vacated by the leaving group.* There is another factor: *spectator ligand tends to labilise its trans position.* Thus the ligand which will act as the leaving group depends on the *trans*–labilising effect of the spectator ligands.

These two important characteristic features of the ligand substitution reactions of square planar Pt(II) complexes are:

stereospecificity, and ***trans*–labilising effect**

These two factors largely control the chemistry of square planar complexes of Pt(II). These two aspects were demonstrated in classical synthesis of *cis*– and *trans*–isomers of [PtCl$_2$(NH$_3$)$_2$].

● **Illustration of *trans*-effect in the synthesis of *cis*- and *trans*- isomers of [PtCl$_2$(NH$_3$)$_2$]:** The *cis*-isomer (called also α-isomer in old literature) was prepared by Peyrone, by treating an aqueous solution of K$_2$[PtCl$_4$] with excess NH$_3$ + NH$_4$Cl buffer at low temperature (*ca.* 0°C) for a few days. The *trans*–isomer (*i.e.* β-isomer) was prepared by Reiset by heating [Pt(NH$_3$)$_4$]Cl$_2$ at about 250°C for about 10 min in a current of dry HCl–vapour in absence of air.

In both Peyrone and Reiset reactions, the first step is a simple ligand replacement reaction where all the four groups (Cl$^-$ or NH$_3$) are the same. Thus at this first-step, there is only one possible product. At the 2nd–step, in each reaction, there are two possible products but in reality, one product is produced. It is important to note that in each case, ***at the 2nd step, the entering ligand goes to the trans–position of the chloride ligand***. In each case, the other possible product through the substitution of the ligand *trans* to NH$_3$ doest not occur. In other words, we may conclude in the language of kinetics as follows:

the *trans*–ligand with respect to the Pt—Cl bond is more labile than the *trans*–ligand with respect to the Pt—NH$_3$ bond.

This ability of a spectator ligand to direct the incoming group to its *trans*–position is described as the ***trans*–effect** (Chernyaev, 1927). By considering the Peyrone reaction and Reiset reaction, it is evident that the ***trans–effect (i.e. trans–labilising effect) of Cl$^-$ is higher than that of NH$_3$.***

● **Kinetically controlled product (*trans*-effect):** *Here it must be mentioned that the above mentioned trans–effect is a kinetic phenomenon. It gives the kinetically controlled product which may not be the thermodynamically stable product.*

cis-[PtCl$_2$(NH$_3$)$_2$]:
(Peyrone, 1845)

[PtCl$_4$]$^{2-}$ **(Red)** —(+ NH$_3$, − Cl$^-$)→ [PtCl$_3$(NH$_3$)]$^-$ **(Orange)**

+ NH$_3$, − Cl$^-$ (*trans*-effect) → [PtCl$_2$(NH$_3$)$_2$]^0 **(Exclusively *cis*-product)** **(Orange-yellow)**

+ NH$_3$, − Cl$^-$ → *trans*-[PtCl$_2$(NH$_3$)$_2$]^0 **(Not produced)**

● **Dhara synthesis of *cis*-platin** (S.C. Dhara, *Indian J. Chem.*, **8**, 193, 1970; R.A. Alderden *et al, J. Chem. Edu.*, **83**, 728, 2006): By using the principle of *trans*-effect, Dhara developed a rapid method of synthesis of *cis*-platin (an **effective anticancer drug**). In fact, most of the subsequent methods of synthesis of *cis*-platin are based on the **Dhara method** that uses $K_2[PtCl_4]$ as the starting material which is rapidly converted into $K_2[PtI_4]$ by treating with a saturated solution of KI. $K_2[PtI_4]$ on being treated with NH_3 produces *cis*-$[PtI_2(NH_3)_2]$ as in Peyrone synthesis (1845). The **yellow compound**, *cis*-$[PtI_2(NH_3)_2]$, is then heated with an aqueous solution of $AgNO_3$ leading to Ag^+-catalysed aquation to produce *cis*-$[Pt(OH_2)_2(NH_3)_2]^{2+}$ with the separation of yellow AgI precipitate. The filtrate containing *cis*-$[Pt(OH_2)_2(NH_3)_2]^{2+}$ is treated with KCl to produce the desired product.

● **Determination of *trans*-effect series:** By carrying out a large number of reactions, it is possible to compare the *trans*-directing or *trans*-labilising power of the ligands. It is illustrated below to determine the relative *trans*-directing power of *A* and *B* (assuming the higher *trans*-directing power of *A* and *B* than that of Cl^-).

The starting complex through the substitution of one Cl^- by NH_3 can lead to two possible products I and II. If the *reaction rate* (r_A) leading to the product I is faster than the *reaction rate* (r_B) leading to the product II, then we can conclude that the *trans*–effect of A is greater than that of B. Thus we have:

$$r_A > r_B \Rightarrow A > B > Cl^- \text{ (trans-effect).}$$
$$r_B > r_A \Rightarrow B > A > Cl^- \text{ (trans-effect).}$$

cis-product: $[PtCl_3(NH_3)]^-$ (H₃N, Cl / Cl, Cl) $\xrightarrow[-\,Cl^-]{+\,NH_3}$ $[PtCl_2(NH_3)_2]^0$ (H₃N, NH₃ / Cl, Cl) (k_{Cl})

trans-product: $[PtCl_3(C_2H_4)]^-$ (H₄C₂, Cl / Cl, Cl) $\xrightarrow[-\,Cl^-]{+\,NH_3}$ $[PtCl_2(C_2H_4)(NH_3)]^0$ (H₄C₂, Cl / Cl, NH₃) $(k_{C_2H_4})$

$k_{C_2H_4} >> k_{Cl}$

The above rate data indicate that the *trans*-effect of C_2H_4 is larger than that of Cl^-. **Determination of the *trans*-effect** series is illustrated in the following reaction.

(T, PEt₃ / Et₃P, Cl)Pt + Y $\xrightarrow[\text{(in MeOH)}]{-\,Cl^-}$ (T, PEt₃ / Et₃P, Y)Pt (Charges not shown)

(*trans*-$[Pt^{II}Cl(PEt_3)_2(T)]$) → *trans*-$[Pt^{II}(PEt_3)_2(T)(Y)]$

Y	$k_Y\,(dm^3\,mol^{-1}\,s^{-1})$		
	$T = Cl^-$	$T = C_6H_5^-$	$T = CH_3^-$
N_3^-	2×10^{-4}	8×10^{-3}	7×10^{-2}
Br^-	9.3×10^{-4}	1.8×10^{-2}	11.5×10^{-2}

The relative values of k_Y in the above reactions depend on the nature of the *trans*–ligand (T). The *trans*–directing or *trans*–labilising sequence of T from the above reactions is:

$$T: CH_3^- > C_6H_5^- > Cl^-$$

The same order is maintained in the above reactions for the relatively weaker nucleophile $Y = NO_2^-$, N_g^- and py.

Effect of the nucleophilicity of the entering ligand on *trans*-effect seris. Here it may be mentioned that the reaction rate is jointly determined by the *trans*–ligand (*T*), incoming nucleophile (Y) and the leaving group. For comparison, the leaving group may be kept the same but the effects of the groups *T* and *Y* may not be always compartmentalized and they may work in some sort of interdependent manner, specially *when the incoming nucleophile* (*Y*) *is a very good nucleophile*. Thus the *trans*–directing series based on the k_Y values may be different for the different entering nucleophiles. In fact, in the reaction between *trans*–$[Pt^{II}Cl(PEt_3)_2(T)]$ and I^-, the *trans*- directing series appears:

$$T: CH_3^- > Cl^- > C_6H_5^-$$

This is the limitation to determine the trans–directing series in terms of the k_Y values.

By considering the relative reaction rates, the ligands (T) can be arranged in the following **order of *trans*–directing power.**

***trans*-effect series:** CN^-, CO, NO, C_2H_4 > PR_3, AsR_3, H^- > CH_3^- > $C_6H_5^-$, $SC(NH_2)_2$ > NO_2^-, I^-, SCN^-, N_3^- > Br^- > Cl^- > NH_2OH, py > RNH_2, NH_3 > OH^- > H_2O.

● ***Trans*–influence:** This is a *thermodynamic term* measuring the bond strength at the *trans*–position of a particular ligand (T). The greater bond weakening power at the *trans*–position indicates the higher *trans*–influence of the ligand T. It is illustrated below:

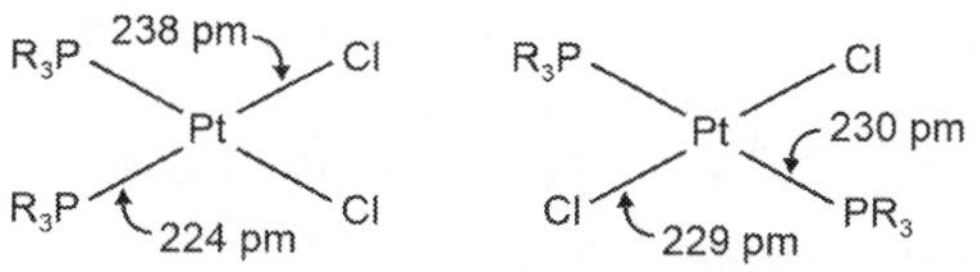

Pt—Cl bond is longer when it is *trans* to PR_3 than that *trans*- to Cl^-. Pt—P bond is also longer in the *trans*-isomer. Thus the *trans*-influence order is: $PR_3 > Cl^-$.

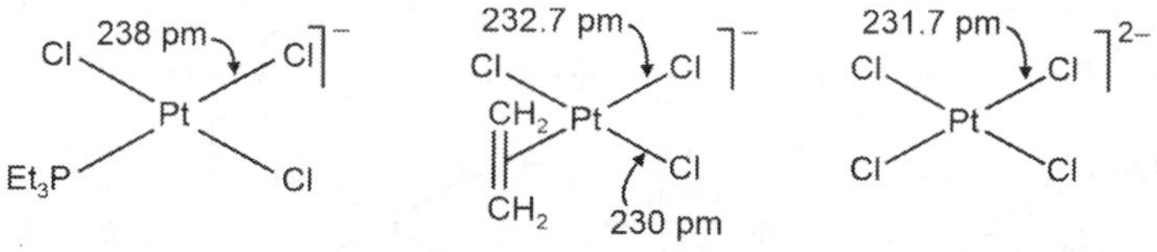

Here the Pt—Cl bond length increases for the *trans*-ligand (T) in the following sequence:

$$T = PEt_3 > C_2H_4 > Cl^-$$

Thus the *trans*–Pt—Cl bond strength changes in the sequence: $T = Cl^- > C_2H_4 > PEt_3$

In *trans*-$[PtCl_2(NH_3^-)(T)]$, the Pt–N bond strength decreases with T (*trans*-ligand) as follows:

$$T: H_2O \geq H_2S \geq Cl^- \gg PH_3 \geq H^- \geq CH_3^-$$

Trans-effect _vs._ _Trans_-influence in Square Planar Complexes

- ***Trans*– effect (a kinetic phenomenon):** It indicates the effect of a coordinated ligand upon the rate of substitution of a ligand at its *trans*–position. It is very often described as the **kinetic *trans*–effect (KTE)**.
- ***Trans*–influence (a thermodynamic phenomenon):** It describes the effect of a coordinated ligand on the bond strength at its *trans*–position in the ground state. It is also described as the ***static trans-effect* (STE).**

Trans–effect and _Trans_–influence in Octahedral Complexes

These phenomena are also known in the octahedral complexes. In the octahedral complexes, *trans*-$[Co(en)_2(OH_2)X]^{n+}$ (X = $S_2O_3^{2-}$, RSO_2^-, SO_3^{2-} where R = alkyl group; and X binds through the *S*-end), the *trans*- Co—O bond (with respect to the Co—S bond) experiences both the *trans*–effect (*i.e.* kinetic labilisation) and *trans*–influence (*i.e.* bond lengthening) (*cf.* Fig. 5.30.15.7).

The ligand which can weaken its *trans*–bond more is considered to have its more *trans*–influence. Thus in terms of the bond weakening phenomenon, the order of common ligands with regard to their *trans*–influence is:

***trans*-influence series:** $H^- > PR_3 > SCN^- > I^-$, CH_3^-, CO, C_2H_4, $CN^- > Br^- > Cl^- > NH_3 > {}^-OH$.

The strong **σ-bonding and polarizable ligands show** the strong *trans*-influence. These aspects have been explained and rationalized later (Sec. 5.30.15).

5.30.13 Application of Trans–Effect in the Synthesis of Isomers of Platinum(II) Complexes

Common Guidelines for the Routes of Synthesis

(i) **Synthesis of *cis*- and *trans*-$[PtL_2X_2]$:**

- If *trans*–effect, X > L,

cis–isomer: $[PtX_4] \xrightarrow[-2X]{+2L} cis\text{-}[PtL_2X_2]$ (*cf.* synthesis of *cis*-$[PtCl_2(NH_3)_2]$)

trans–isomer: $[PtL_4] \xrightarrow[-2L]{+2X} trans\text{-}[PtL_2X_2]$ (*cf.* synthesis of *trans*-$[PtCl_2(NH_3)_2]$)

(*trans*-effect: $Cl^- > NH_3$)

- If *trans*–effect, L > X,

cis - isomer : $[PtL_4] \xrightarrow[-2L]{+2X} cis\text{-}[PtL_2X_2]$

trans - isomer : $[PtX_4] \xrightarrow[-2X]{+2L} trans\text{-}[PtL_2X_2]$

cf. synthesis of *cis*– and *trans*–isomers of $[PtCl_2(PR_3)_2]$ where *trans*–effect: $PR_3 > Cl^-$

(ii) **Synthesis of *cis*- and *trans*-[PtX$_2$(Y)(Z)]**

If *trans*–effect: Z < X < Y

- ***cis*–isomer:** $[PtX_4] \xrightarrow[-X]{+Z} [PtX_3(Z)] \xrightarrow[-X]{+Y} cis\text{-}[PtX_2(Y)(Z)]$
- ***trans*–isomer:** $[PtX_4] \xrightarrow[-X]{+Y} [PtX_3(Y)] \xrightarrow[-X]{+Z} trans\text{-}[PtX_2(Y)(Z)]$

cf. synthesis of *cis*– and *trans*– isomers of

$[PtCl_2(NH_3)(NO_2)]^-$ (where *trans*-effect: $NO_2^- > Cl^- > NH_3$),

$[Pt(C_2H_4)Cl_2(NH_3)]$ (where *trans*-effect: $C_2H_4 > Cl^- > NH_3$)

$[Pt(CO)Cl_2(py)]$ (where *trans*-effect: $CO > Cl^- > py$)

(a) **Synthesis of the *cis*- and *trans*-isomers of $[Pt(C_2H_4)Cl_2(NH_3)]$ from $K_2[PtCl_4]$ by considering the *trans*-effect series:** $C_2H_4 > Cl^- > NH_3$

$[PtCl_4]^{2-} \xrightarrow[-Cl^-]{+C_2H_4} [PtCl_3(C_2H_4)]^- \xrightarrow[-Cl^-]{+NH_3} [PtCl_2(NH_3)(C_2H_4)]^0$ (*trans*-isomer)

$[PtCl_4]^{2-} \xrightarrow[-Cl^-]{+NH_3} [PtCl_3(NH_3)]^- \xrightarrow[-Cl^-]{+C_2H_4} [PtCl_2(NH_3)(C_2H_4)]^0$ (*cis*-isomer), **(Pt—N bond is stronger than Pt—Cl bond)**

The ligand to be replaced at the second step according to the *trans*–effect series is indicated by the dotted circle.

(b) **Synthesis of the** *cis*- **and** *trans*-**isomers of $[PtCl_2(NH_3)(NO_2)]^-$ from $K_2[PtCl_4]$:**

The *trans*–effect series is: $NO_2^- > Cl^- > NH_3$

***cis*-isomer:** NH_3 followed by NO_2^-;

***trans*-isomer:** NO_2^- followed NH_3

$$[PtCl_4]^{2-} \xrightarrow[-Cl^-]{+NH_3} [PtCl_3(NH_3)]^- \xrightarrow[-Cl^-]{+NO_2^-} cis\text{-}[PtCl_2(NH_3)(NO_2)]^-$$

$$[PtCl_4]^{2-} \xrightarrow[-Cl^-]{+NO_2^-} [PtCl_3(NO_2)]^{2-} \xrightarrow[-Cl^-]{+NH_3} trans\text{-}[PtCl_2(NH_3)(NO_2)]^-$$

(c) **Synthesis of the three isomers of** $[Pt(NH_3)(NH_2OH)(NO_2)(py)]^+$ **having the *trans*–effect series:** $NO_2^- > Cl^- > py \geq NH_3 \sim NH_2OH$

Bond strength: Pt—N > Pt—halogen

The possible three geometrical isomers are:

$[Pt(NH_3)(NH_2OH)(NO_2)(py)]^+$ isomers:

I: H_3N trans to py, O_2N trans to NH_2OH

II: O_2N trans to py, H_3N trans to NH_2OH

III: H_3N trans to NO_2, py trans to NH_2OH

- **Synthesis of the isomer I**

$$[PtCl_4]^{2-} \xrightarrow[-2Cl^-]{+2NH_3} [PtCl_2(NH_3)_2]^0 \xrightarrow[-2Cl^-]{+2py} [Pt(py)_2(NH_3)_2]^{2+}$$

$$\xrightarrow[-py,\ -NH_3]{+2Cl^-} [PtCl_2(py)(NH_3)]^0 \xrightarrow[-2Cl^-]{+2NO_2^-} [Pt(NO_2)_2(py)(NH_3)]^0 \xrightarrow[-NO_2^-]{+NH_2OH} [Pt(py)(NH_2OH)(O_2N)(NH_3)]^+$$

Here it is evident all the steps (except the 2nd step leading to replacement of two Cl^- ligands by *py*) are in conformity with the *trans*-effect series. At the 2nd step, because of the higher *trans*-labilising effect of Cl^-, it is expected that the coordinated NH_3 ligands (lying at the *trans*–positions of Cl^-) should be replaced by the entering ligands (*i.e. py*). But it does not happen so. To rationalise this step, it is required to consider the relative bond stength which runs in the sequence:

$$Pt—NH_3 > Pt—Cl$$

It makes the Pt—Cl bond more labile than the Pt—NH_3 bond. *Thus it is very often found that when other things are equal, i.e. the Pt—X* (= Br, Cl) *bond is generally more labile than the Pt—N bond.* This lability order (originating from the bond strength effect) may oppose the lability order expected from the *trans*-effect series.

The **isomer-I** can also be synthesised in the following route.

$$[PtCl_4]^{2-} \xrightarrow[-Cl^-]{+NO_2^-} [PtCl_3(NO_2)]^{2-} \xrightarrow[-Cl^-]{+NO_2^-} [PtCl_2(NO_2)_2]^{2-}$$

$$\xrightarrow[-2Cl^-]{(i)\ +NH_3;\ (ii)\ +py} [Pt(H_3N)(NO_2)(O_2N)(py)]^0 \quad \text{(cf. Pt — N > Pt — Cl in terms of bond strength)}$$

$$\xrightarrow[-NO_2^-]{+Cl^-,\ i.e.\ HCl} [Pt(H_3N)(Cl)(O_2N)(py)]^0 \xrightarrow[-Cl^-]{+NH_2OH} [Pt(H_3N)(NH_2OH)(O_2N)(py)]^+$$

- **Synthesis of the isomer II**

$[PtCl_4]^{2-}$ $\xrightarrow[-Cl^-]{+NO_2^-}$ $[PtCl_3(NO_2)]^{2-}$ $\xrightarrow[-2Cl^-]{+2NH_2OH}$ $[PtCl(NO_2)(NH_2OH)_2]^0$ (Cl, NO_2; HOH_2N, NH_2OH)

$\downarrow$ $+NH_3$, $-Cl^-$ (Bond strength effect: Pt—N > Pt—Cl)

$[Pt(NH_3)(NO_2)(NH_2OH)_2]^+$ (H_3N, NO_2; HOH_2N, NH_2OH) $\xrightarrow[-NH_2OH]{+Cl^-,\ i.e.\ HCl}$ $[Pt(NH_3)(NO_2)Cl(NH_2OH)]^0$ (H_3N, NO_2; Cl, NH_2OH) $\xrightarrow[-Cl^-]{+py}$ $[Pt(NH_3)(NO_2)(py)(NH_2OH)]^+$ (H_3N, NO_2; py, NH_2OH)

- **Synthesis of the isomer III**

$[PtCl_4]^{2-}$ $\xrightarrow[-Cl^-]{+NO_2^-}$ $[PtCl_3(NO_2)]^{2-}$ $\xrightarrow[-2Cl^-]{+2NH_2OH}$ $[PtCl(NO_2)(NH_2OH)_2]^0$ (Cl, NO_2; HOH_2N, NH_2OH)

$\downarrow$ $+py$, $-Cl^-$ (Bond strength effect: Pt—N > Pt—Cl)

$[Pt(py)(NO_2)(NH_2OH)_2]^+$ (py, NO_2; HOH_2N, NH_2OH) $\xrightarrow[-NH_2OH]{+Cl^-,\ i.e.\ HCl}$ $[Pt(py)(NO_2)Cl(NH_2OH)]^0$ (py, NO_2; Cl, NH_2OH) $\xrightarrow[-Cl^-]{+NH_3}$ $[Pt(py)(NO_2)(NH_3)(NH_2OH)]^+$ (py, NO_2; H_3N, NH_2OH)

(Isomer III)

Synthesis of the isomers-I, II and III in more direct routes

By considering the *trans*-effect order as $NO_2^- > Cl^- > NH_3$ and relative lability order Pt—Cl > Pt—N (controlled by bond strength), the said isomers can be synthesised in **more direct routes.**

$[PtCl_4]^{2-}$ $\xrightarrow[-Cl^-]{+NO_2^-}$ $[PtCl_3(NO_2)]^{2-}$ $\xrightarrow[-Cl^-]{+NH_3}$ $[PtCl_2(NO_2)(NH_3)]^-$ (Cl, NO_2; H_3N, Cl)

$\downarrow$ $+py;\ NH_2OH$, $-2Cl^-$ (Bond strength effect Pt—N > Pt—Cl)

$[Pt(py)(NO_2)(NH_3)(NH_2OH)]^+$ (py, NO_2; H_3N, NH_2OH)

(Isomer-III)

$[PtCl_4]^{2-}$ $\xrightarrow[-\,Cl^-]{+\,NO_2^-}$ $[PtCl_2(Cl)(NO_2)]^{2-}$ (Cl, Cl, NO₂ around Pt) $\xrightarrow[-Cl^-]{+py}$ $[PtCl(py)(Cl)(NO_2)]^{-}$

+ NH_3; NH_2OH (Bond strength effect Pt—N > Pt—Cl) – $2Cl^-$

$[Pt(H_3N)(NO_2)(py)(NH_2OH)]^{+}$

(Isomer-II)

$[PtCl_4]^{2-}$ $\xrightarrow[-\,Cl^-]{+\,NO_2^-}$ $[PtCl_3(NO_2)]^{2-}$ $\xrightarrow[-\,Cl^-]{+\,NH_2OH}$ $[PtCl(NO_2)(HOH_2N)(Cl)]^{-}$

(Bond strength effect: Pt—N > Pt—Cl) + NH_3, py – $2Cl^-$

$[Pt(py)(NO_2)(HOH_2N)(NH_3)]^{+}$

(Isomer-I)

(d) **Synthesis of the three isomers of [PtBrCl(NH_3)(py)]**: To design the routes of their synthesis, we are to consider both the *trans*–effect series and bond-strength series.

***trans*-effect:** $Br^- > Cl^- >$ py $> NH_3$

bond-strength: Pt—NH_3 > Pt—py > Pt—Br > Pt—Cl

In the synthesis of the **isomer I,** at the 2nd step, the *trans*-'Pt—Cl' bond is preferably replaced rather than the *trans*-Pt—NH_3 bond. In fact, it occurs so due to the higher Pt—NH_3 bond stength than that of the Pt—Cl bond. Other steps are in accordiance with the *trans*-effect series.

In the synthesis of the **isomer II,** at the 3rd step, the *trans*-effect operates (*cf.* though the Pt—*py* bond is stronger than the Pt—Cl bond), *i.e.* the bond *trans* to 'Cl^-' is replaced. But at the 2nd step, the bond strength effect operates *i.e.* the Pt—Cl bond is replaced rather than the Pt—py bond. Thus it is very difficult to rationalise the observations: ***at the 2nd step, the bond strength effect wins while at the 3rd step, the trans-effect wins***.

In the synthesis of the **isomer III**, at the 2nd step, the *trans*-Pt—Cl bond is replaced rather than the *trans*-Pt—py bond (*cf.* both Cl^- and py are *trans* to Cl^-) because of the higher Pt—py bond strength. The 3rd step is in conformity with the *trans*–effect series.

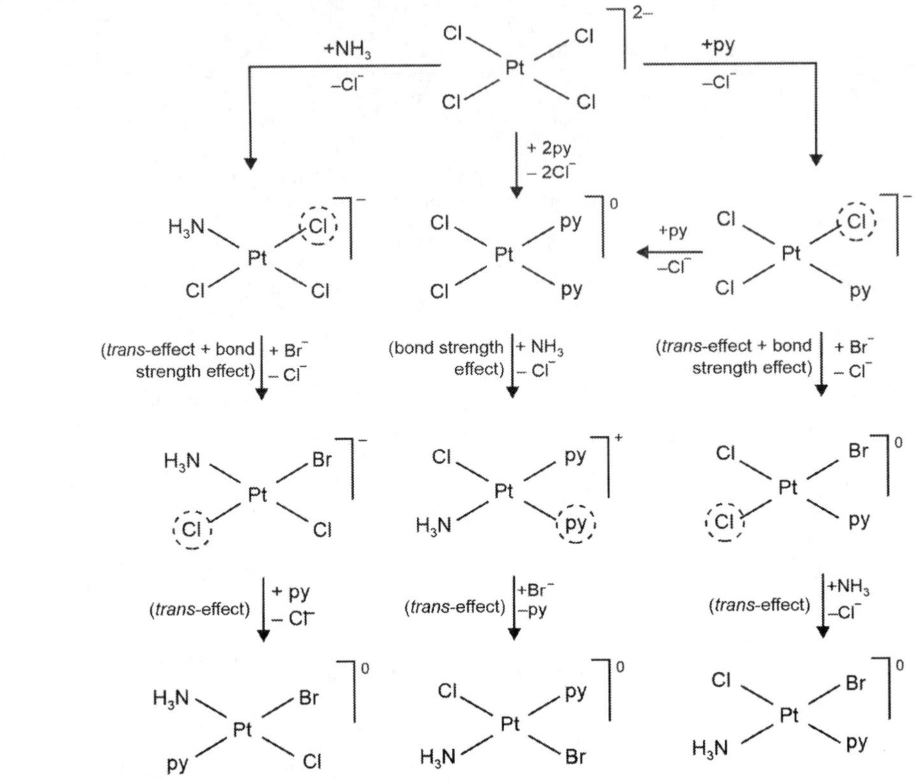

(e) **Synthesis of the three isomers of $[PtCl(NH_3)(RNH_2)(NO_2)]$**

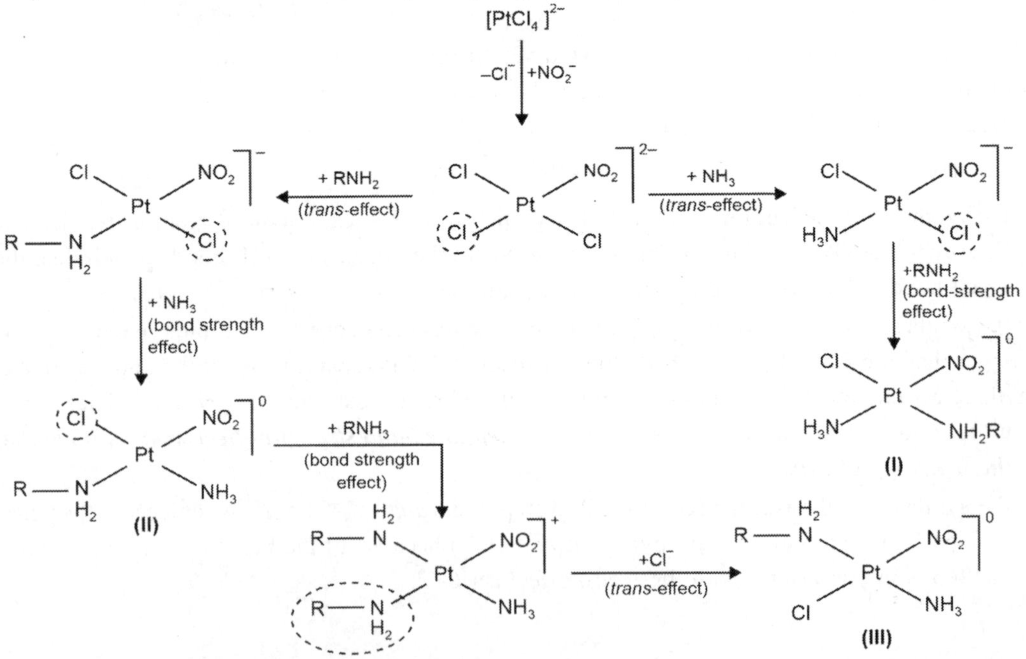

***Trans*-effect:** $NO_2^- > Cl^- > NH_3 \sim RNH_2$

Bond strength: Pt—N > Pt—halogen.

In the synthesis of the the above mentioned three isomers (I, II, III), both the *trans*-effect and bond-strength effect are kept in mind.

5.30.14 Application of *Trans*-effect in the Distinction of *cis-trans* Isomers of Pt(II) Complexes

This methodology was developed by the Russian chemists by using thiourea (tu) and $S_2O_3^{2-}$ (by considering the **very high *trans*–effect of tu and $S_2O_3^{2-}$**). This principle is illustrated below.

Using tu [*i.e.* $SC(NH_2)_2$] that coordinates through the S-site to Pt^{II} (HSAB matching, **soft-soft interaction) (N.S. Kurnakov, 1894).**

$$[Pt(NH_3)_2Cl_2]^0 \text{ (cis-isomer)} \xrightarrow[\text{(bond-strength effect)}]{+2tu,\ -2Cl^-} [Pt(NH_3)_2(tu)_2]^{2+} \xrightarrow[\text{(trans-effect)}]{\text{excess tu},\ (-2NH_3)} [Pt(tu)_4]^{2+}$$

$$[Pt(NH_3)_2Cl_2]^0 \text{ (trans-isomer)} \xrightarrow[\text{(trans-effect)}]{+2\,tu,\ -2Cl^-} [Pt(NH_3)_2(tu)_2]^{2+} \xrightarrow{\text{excess tu}} \text{No reaction}$$

The use of this thiourea reaction to determine the *cis–trans* structures of diacidodiammineplatinum(II) is known as **Kurnakov test.**

Using $S_2O_3^{2-}$ (D.I. Ryabehikov, 1941)

$$[Pt(NH_3)_2Cl_2]^0 \text{ (cis-isomer)} \xrightarrow[\text{(bond-strength effect)}]{+S_2O_3^{2-},\ (-Cl^-)} [Pt(NH_3)_2(S_2O_3)]^0 \xrightarrow[\text{(trans-effect)}]{\text{excess } S_2O_3^{2-},\ (-2NH_3)} [Pt(S_2O_3)_2]^{2-}$$

i.e. $[Pt(S_2O_3)_2]^{2-}$

The *cis*-isomer in presence of excess thiosulfate can also partly produce $[Pt(S_2O_3)_4]^{6-}$ where $S_2O_3^{2-}$ acts as a monodentate ligand (through the S-site).

$$[Pt(NH_3)Cl_2(NH_3)]^0 \text{ (trans-isomer)} \xrightarrow[\text{(trans-effect)}]{+S_2O_3^{2-},\ (-Cl^-)} [PtCl(SSO_3)(NH_3)_2]^- \text{ (unstable intermediate)} \xrightarrow[\text{(trans-effect)}]{+H_2O \text{ (solvent)},\ (-Cl^-)} [Pt(H_2O)(SSO_3)(NH_3)_2]^0 \text{ (stable)}$$

$$\text{(unstable intermediate)} \xrightarrow[-Cl^- \text{ (trans-effect)}]{+S_2O_3^{2-}} trans\text{-}[Pt(NH_3)_2(S_2O_3)_2]^{2-}$$

5.30.15 Theories of *Trans*-effect (*i.e.* Kinetic *Trans*-effect) and *Trans*-influence (*i.e.* Static *Trans*-effect)

***Trans*-influence** (*a thermodynamic phenomenon*) indicates the bond weakening phenomenon at the *trans*–position in the ground state.

***Trans*-effect** (*a kinetic phenomenon*) indicates the labilisation at the *trans*–position during the nucleophilic substitution reaction (Werner and Chernyaev)

In most of the cases, the ligands showing the high trans-influence show also the high trans-effect. Exception: olefins show the high trans-effect but the low trans-influence.

Lowering of activation energy and *trans*-effect

The enhanced rate for a good *trans*-activating ligand is due to the reduced activation energy. This lowering of activation energy can occur **in two ways**:

- the ground state may be destabilised due to the weakening of a metal-ligand bond by a *trans*-influencing ligand; it makes the activation energy (which is the energy difference between the ground state and transition state) less; **thus the *trans*-influence may also reduce the activation energy.**
- the transition state may be stabilised to lower the activation energy.

To reduce the activation energy, either of the two factors or combination of both factors may operate.

The *trans*-effect and *trans*-influence very often run parallel in the same direction. The reaction rate (kinetic process) depends on the *activation energy. **Thus a good trans-directing/activating ligand must be able to lower the activation energy for the enhanced rate***. If a good *trans*–influencing ligand shows the rate benefit then it will also contribute to lower the activation energy. Thus any theory to be proposed must take into consideration of the above fact.

A. Grinberg's polarisation theory (1935) and *trans*–influence: This polarisation theory explains the *trans*–influence (*i.e.* bond weakening at the *trans*–position). In the Pt—L bond, Pt(II) will polarise the ligand (L) to induce a dipole in the group L. This induced dipole in L will also polarise the metal centre, *i.e.* Pt(II), to induce a dipole. **This mutual polarisation works in a synergistic fashion.** In the L—Pt—X segment, if X is polarisable then the dipole induced at Pt(II) by X opposes the dipole induced by L. Thus in a symmetrical compound $[PtL_4]$, there will be no net polarisation in any direction of the Pt—L bond. In $[PtL_3X]$ (where X is more polarisable than L), there will be a net polarisation causing a repulsion between the negative charges developed in the Pt—L bond *trans*- to the Pt—X bond. This will weaken the Pt—L bond *trans* to the Pt—X bond. The degree of bond weakening in the Pt—L bond *trans* to the Pt—X bond depends on the polarisability of X, *i.e.* **greater the polarisability of X, greater the bond weakening at its *trans*–position due to the electrostatic repulsion**. Thus the *trans*-influence series of halides follows their polarisability sequence mainly controlled by their **electronegativities and sizes**.

***trans*–influence:** $I^- > Br^- > Cl^-$; $NH_3 > H_2O$

***polarisability*:** $I^- > Br^- > Cl^-$; $NH_3 > H_2O$

- ***Trans*-influence *vs.* *trans*-effect in terms of the polarisation theory:** *This bond weakening phenomenon (i.e. trans-influence), i.e. destabilisation in the ground state, helps in lowering the activation energy* (*i.e.* the energy difference between the ground state and transition state). For the **dissociative process** (*i.e.* bond breaking by the leaving group at the *rds*), Grinberg's polarisation theory directly reduces the activation energy. Thus Grinberg's polarisation theory explains the *trans*-influence and *trans*-effect (*i.e.* labilisation of the *trans*-bond in the **dissociative activation**) and lowering of the activation energy for both the dissociative and associative processes **through the destabilisation of the ground state** (see Fig. 5.30.15.2). In fact, if bond breaking by the leaving group contributes to attain the activated complex, then the *trans*-influence causes the *trans*-effect as in the **dissociative process.** In the **interchange (*I*) process,** bond breaking is only partly important to determine the activation energy. Thus the *trans*-influence can directly explain the *trans*-effect (kinetic factor) for the *D* and I_d processes as in the case of octahedral Co(III) complexes (*cf.* Fig. 5.30.15.8).

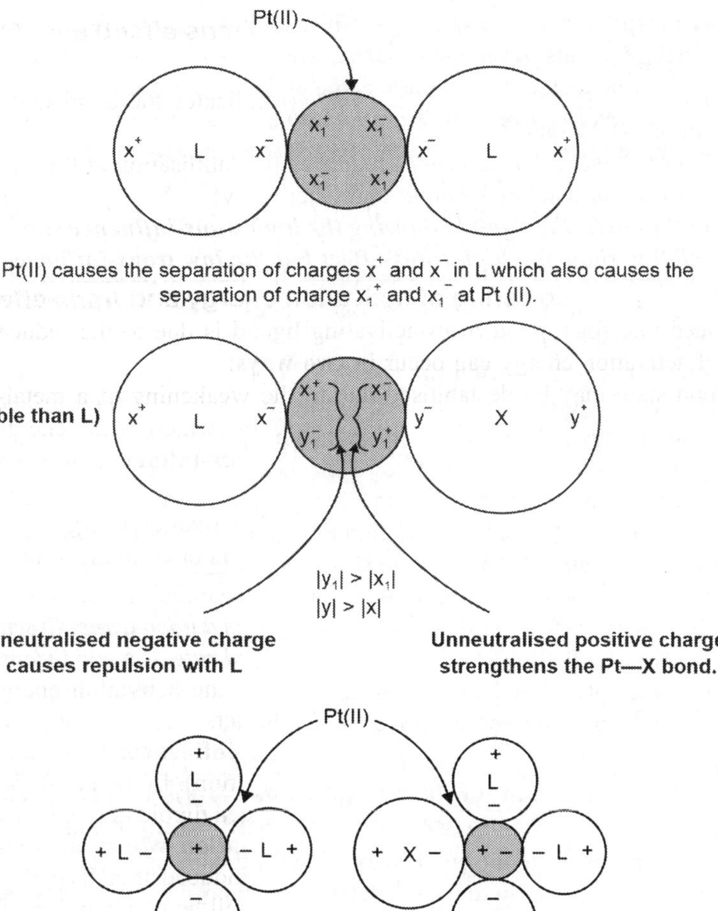

Fig. 5.30.15.1: Schematic representation of Grinberg's polarisation theory to explain the *trans-influence* (*i.e.* ground state effect) and *trans*-effect.

For the square planar Pt(II) cmplexes, the ligand substitution primarily involves the **associative activation** (mainly I_a path) where bond breaking by the leaving group is of minor importance only to determine the activation energy. Hence, to understand the origin of *trans*-effect in the square planar complexes of Pt(II), we should consider the **stabilisation of the 5-coordinate (TBP) activated complex, *i.e.* transition state.** But the *trans*-influence cannot tell anything regarding the **stabilisation of the transition state** which is more important than the **ground state destabilisation** in unserstanding the kinetic aspects of the reaction. Ground state destabilisation (*i.e.* **ground state is of higher energy**) is expected to reduce the activation energy mainly for breaking the bond in the dissociative process. Thus, *trans*-influence (*i.e.* **ground state effect**) can only partly contribute to determine the *trans*-effect in the I_a-process. Of course, the ground state destabilisation can partly reduce the activation energy in the **interchange process** like I_a-process (see Fig. 5.30.15.2), through **lowering the activation energy in terms of the ground state destabilisation** and favouring the bond breaking by the leaving group in the T.S. Definitely, these are of minor importance in the I_a process.

Note: The very high polarisability of H^- is expected to show the high *trans*-influence and *trans*-effect (due to the ground state destabilisation).

● **Relative importance of the *trans*-influence and *trans*-effect among the congeners Ni(II), Pd(II) and Pt(II) in terms of the polarisation theory:** To execute the *trans*-influence according to this polarisation theory advocating the simultaneous polarisation of both the metal and ligand (*i.e.* mutual polarisation), ***the metal centre must be polarisable***. In a group of the *d*-block elements, the heavier congeners are more polarisable due to the **relativistic expansion** of the *d*- and *f*-orbitals (Das *et al. Coord. Chem. Rev.*, 2023). Among Ni(II), Pd(II) and Pt(II), the most polarisable centre is Pt(II) and the *trans*-effect is mainly known to operate in the Pt(II)-complexes. In fact, the *trans*-effect is almost absent in the Ni(II) and Pd(II) complexes. Because of the same ground, the *trans*-effect is not known in the Pt(IV)-complexes (*cf.* Pt(IV) is much less polarisable than Pt(II) because of the higher positive charge of Pt(IV)). These indicate that the polarisation theory of *trans*-influence is, at least, partly important to understand the *trans*-effect.

Limitations of the polarisation theory: ● There is a huge difference in the *trans*-labilising powers of H^- and H_2O but in terms of their bond weakening power at their respective *trans*-positions, they differ not so dramatically. Thus, the bond weakening (*i.e.* ground state destabilisation due to the *trans*-influence) cannot be the sole factor in realising the *trans*-effect.

● It has been already mentioned that the theory expects a high *trans*–effect for the ligands showing the high *trans*–influence (*i.e.* bond weakening or ground state destabilisation). In fact, in many cases, both the effects run parallel. But olefins show the very strong *trans*-effect with a very poor *trans*-influence (*cf.* Pt—Cl bonds in Zeise's salt anion).

In fact, the theory cannot explain the origin of high *trans*–effect shown by the other π–bonding ligands like C_2H_4, CO, PR_3, CN^- etc.

● Thus the concept of bond weakening (*i.e.* ground state destabilisation) is not the total argument in understanding the *trans*-effect. It must consider the stabilisation of the transition state in understanding the *trans*–effect.

Trans-influence (_i.e._ static _trans_-effect) _vs._ _Trans_-effect (_i.e._ kinetic _trans_-effect) in square planar Pt(II)-complexes

The *trans*-influence leads to the bond weakening for the leaving group. Thus, if the process of ligand substitution experiences a **dissociative activation,** then the *trans*-influence also causes the *trans*-effect (*i.e.* lowering of activation energy) as in the octahedral Co(III) complexes (*cf.* Fig. 5.30.15.7). But the ligand substitution at the square planar complexes of Pt(II) primarily involves the **associative activation** (I_a path) where bond breaking by the leaving group is of only minor importance. Thus, the *trans*-influence fails to explain the *trans*-effect for the Pt(II)-complexes.

● *Trans*-influence ⇒ Ground state of higher energy, *i.e.* ground state destabilisation.

● *Trans*-effect ⇒ Transition state of lower energy, and lowering of activation energy that may be partly due to the ground state destabilisation (*i.e. trans*-influence, see Fig. 5.30.15.2).

Note: The strong sigma bonding ligands weaken their *trans*-bonds (*i.e. trans*-influence). The **sigma bonding** effects have been discussed later (*cf.* Fig. 5.30.15.7). In fact, the strong σ-bonding ligands show the strong *trans*-influence. It explains the high *trans*-influence of the σ-bonding ligands like H^-, PR_3, I^-, CH_3^-, etc. (see *trans*-influence sseries). **In fact, the strong σ-bonding and highly polarizable ligands show the strong *trans*-influence.**

B. **Pi-bonding theory and *trans*-effect in terms of the stabilisation of transition state:** This theory was developed by Orgel (1955) and Chatt (1956) independently in terms of stabilisation of the transition state to explain the high *trans*-effect shown by the π-bonding ligands like C_2H_4, CO, CN^-, etc. Such π-bonding ligands (*i.e.* **π-acid ligands**) participate in metal (*d*-orbital) → ligand π-back bonding. For this π-back bonding, the metal *dp* hybrid orbitals (*e.g.* 5*d* + 6*p* orbitals for Pt(II), *cf.* **relativistic effect** to favour this *p–d* hybridisation), are probably more efficient than the pure *d*-orbitals. However, for the sake of simplicity, we may consider the simple *d*-orbitals of the central metal to explain the importance of such π–bonding in explaining the *trans*-effect.

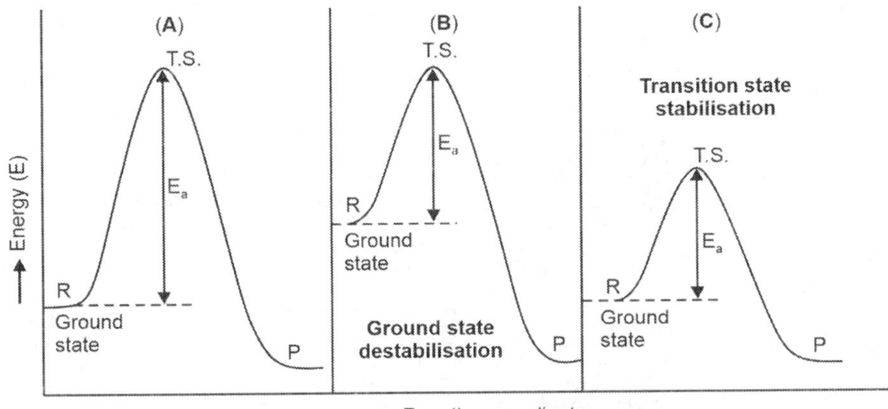

Fig. 5.30.15.2: Qualitative illustration of activation energy (E_a) and *trans*-effect in terms of a single step interchange (I_a) mechanism. (A) Weak *trans*-effect: low energy ground state and high energy transition state (T.S.), *i.e.* high activation energy. (B) *Trans*-influence effect (ground state destabilisation) reducing the activation energy, *i.e. trans*-influence causing the *trans*-effect (**as in the case of polarisable and good σ-bonding ligands**). (C) *Trans*-effect through the stabilisation of transition state, *i.e.* low E_a (**as in the case of π-bonding *trans*-ligands**).

Note: There is a debate regarding the relative contributions of ground state destabilisation and stabilisation of transition state to reduce the activation energy to explain the *trans*-effect.

Relativistic effect and *trans*-effect/influence (Das *et al. Coord. Chem. Rev.*, 2023; *CTI* (IUPAC), 2023: Here it may be noted that because of the **relativistic expansion** of the 5*d*-orbitals and **relativistic contraction** of 6*p*-orbitals of Pt(II), the energy difference between the 5*d*- and 6*p*-orbitals of Pt(II) decreases to favour their hybridisation (see Vol. 1). This relativistic effect to favour the *p–d* hybridisation occurs in the sequence: Pt(II) > Pd(II) > Ni(II). It explains the relative importance of *trans*-effect: Pt(II) > Pd(II) > Ni(II). It may be noted that the relativistic effect leads to the **polarisability order**: Pt(II) > Pd(II) > Ni(II) that explains their relative importances in *trans*-effect in terms of polarisation theory already discussed.

The metal (*d*-orbital) → ligand π-back bonding can favour the process kinetically in two ways.

(i) **Favouring the approach of the nucleophile through electron depletion:** Because of the metal (*d*-orbital) → ligand (*i.e.* M → L) π-back bonding, the electron density in the *trans*-regions of L is reduced *to favour the approach of the incoming nucleophile* (Y). This is illustrated in Fig. 5.30.15.3. Moreover, because of this π–back bonding, the metal centre becomes more electron deficient to attract the incoming nucleophile in the associative path.

In fact, the two ligands (L and X) at the *trans*-positions **compete for two metal *d*-orbitals**, if they possess two orthogonal vacant π-orbitals (as in CO and C_2H_4 having two vacant orthogonal π*-MOs). On the other hand, if L and X are located at the *cis*-positions, they **compete for one metal *d*-orbital.** If the complex lies in the *xy*-plane and the L—Pt—X segment lies along the *x*-axis (as shown in Fig. 5.30.15.3), then the competing interactions are (L = X = CO, say):

$$\left.\begin{array}{ll} & \pi^*(p_y)(\mathrm{CO})_{\mathrm{trans}} — d_{xy}(\mathrm{Pt}) — \pi^*(p_y)(\mathrm{CO})_{\mathrm{trans}} \\ \text{and} & \pi^*(p_z)(\mathrm{CO})_{\mathrm{trans}} — d_{xz}(\mathrm{Pt}) — \pi^*(p_z)(\mathrm{CO})_{\mathrm{trans}} \end{array}\right] \text{See Fig. 9.4.11.5, Vol. 6}$$

If the L and X are at the *cis*-positions (say, L—Pt along the *x*-axis, Pt—X along the *y*-axis), then these **two *cis*-ligands compete for one metal *d*-orbital**. The competing interaction is:

$$\pi^*(p_y)(\mathrm{CO})_{\mathrm{cis}}(\mathrm{Pt—CO},\ x\text{-axis}) — \mathrm{d}_{xy}(\mathrm{Pt}) — \pi^*(p_x)(\mathrm{CO})_{\mathrm{cis}}(\mathrm{Pt—CO},\ y\text{-axis}\ ]\ \text{See Fig. 9.4.11.5, Vol. 6}$$

$$\pi^*(p_z)(\mathrm{CO})_{\mathrm{cis}}(\mathrm{Pt—CO},\ y\text{-axis}) — d_{yz}(\mathrm{Pt})\ \text{no suitable orbital on}\ (\mathrm{CO})_{\mathrm{cis}}(\mathrm{Pt—CO},\ x\text{-axis})\ \text{for}\ d_{yz}(\mathrm{Pt})$$

This is why, if L is a stronger π-acid ligand than X, then L weakens better the *trans*-Pt—X bond than the *cis*-Pt—X bond.

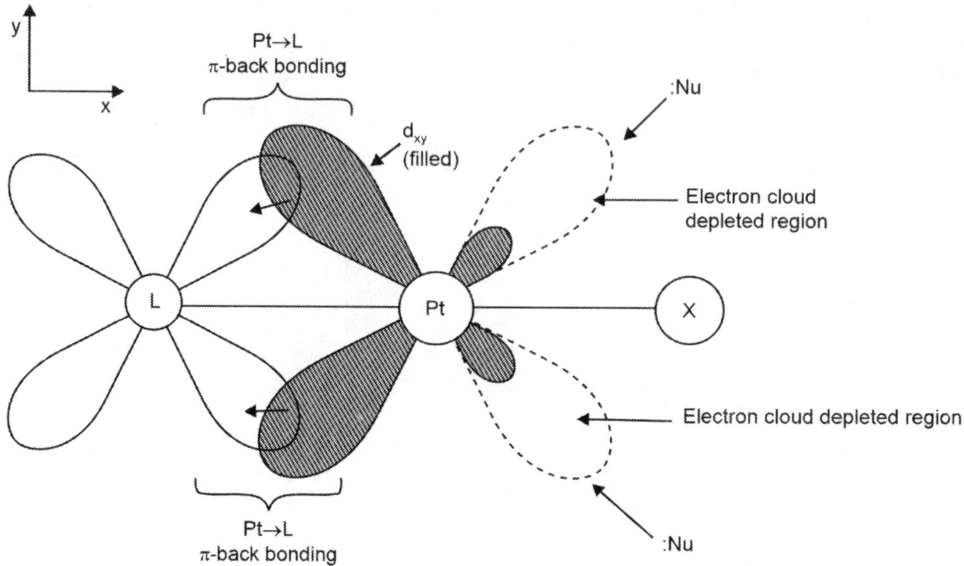

Fig. 5.30.15.3: Pi–bonding property of the *trans*–directing ligand (L), *i.e.* pi–bonding *trans*–effect theory of Orgel and Chatt (*cf.* J. Chatt, *et al., J. Chem. Soc.* (London), 4456, 1955; L.E. Orgel, *J. Inorg. Nucl. Chem.,* **1,** 137, 1956. **Note:** Pt → L π–back bonding reduces the electron density at the *trans*–regions of L to favour the approach of the nucleophile, Nu (= Y) at the *trans*-position of L.

- **Pi-bonding ligands (pi-*trans*-effect):** Mainly T.S. stabilisation along with some extent of ground state destabilisation (*i.e. trans*-influence) (Fig. 5.30.15.2).
- **Sigma-bonding ligands (sigma-*trans*-effect):** Mainly ground state destabilisation (*i.e. trans*-influence) along with some extent of T.S. stabilisation (Fig. 5.30.15.2).

Note: Strong π-bonding competition between the two *trans*-CO groups in $[M(CO)_6]$ (M = Cr, MO) labilises their *trans*-positions (see Vol. 6, and Sec. 5.32, Vol. 5). A similar situation is noted for $[Ni(CO)_4]$ (see Sec. 5.32).

(ii) **Stabilisation of the 5-coordinate TBP activated complex in the associative path** (*cf.* Fig. 5.30.15.2)**):** In the trigonal bipyramidal transition state, the incoming group (Nu = Y), the *trans*-directing ligand (L) and the leaving group (X) reside at the equatorial positions. **The ligands at the equatorial positions can stabilise the transition state through the π-bonding**. Now, if L is a good π-acceptor ligand, it can reduce the electron density along the Pt—X and Pt—Y (where Y = nucleophile Nu) bond directions (*cf.* Fig. 5.30.15.4). It will stabilise the transition state and consequently, the activation energy will be reduced. This will favour the rate process.

(iii) **Role of the π-bonding property of the entering nucleophile in the associative path:** It is evident that in the trigonal bipyramidal transition state, in the trigonal plane, L (*trans*-activating ligand), Y (incoming ligand, Nu) and X (leaving group) occupy the equatorial positions and they can stabilise the transition state. If the incoming group is a π-bonding ligand (*i.e.* π-acceptor ligand) then the transition state may be stabilised to favour the process kinetically even when L is not a good *trans*–directing ligand. ***It makes the good π-bonding ligands as the good entering ligands.***

(iv) **Role of the π-bonding property of the leaving group in the associative path:** Here it may be mentioned that if the leaving group (X) is a good π-bonding ligand then the trigonal bipyramidal geometry of the transition state will be stabilised but expulsion of the leaving group will be difficult. ***This is why, a good π-bonding ligand is a poor leaving group.*** Thus a good π-bonding leaving group (X) favours the process through the stabilisation of T.S. but disfavours the process by its poor leaving group property. This is why, π-bonding property of the entering group is more important than that of the leaving group to determine the overall rate.

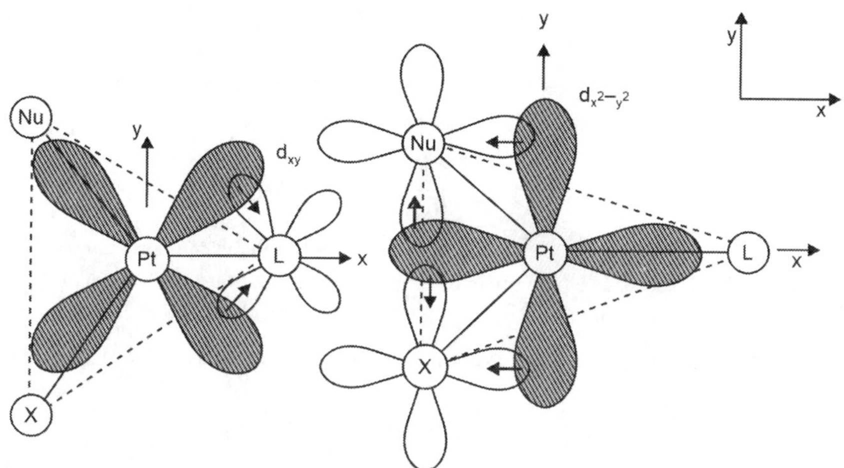

Fig. 5. 30.15.4: Positions of the d_{xy} and $d_{x^2-y^2}$ orbitals of Pt(II) in the trigonal bipyramidal transition state bearing the *trans*–directing group (L), leaving group (X) and incoming nucleophile (Nu, *i.e.* Y) at the trigonal plane. The *trans*-axial ligands of TBP are not shown.

Note:
- $Pt(d_{xy}) \xrightarrow{\pi} L$ bonding will reduce the electron density along the Pt—X and Pt—Nu bonds in the transition state
- $Pt(d_{x^2-y^2}) \xrightarrow{\pi} X$ or Nu bonding will reduce the electron density along the Pt—L bond.
- The d_{xz} and d_{yz} orbitals can also contribute to π-bonding to stabilise the trigonal bipyramidal transition state.

(v) **Relative *trans*-directing powers of the groups in terms of their π-bonding properties:** For the square planar complex, [PtLL′X$_2$], there are two possible structures of the transtion state. These are: one having X, Y and L in the equatorial plane and another having X, Y and L′ in the equatorial plane (*cf.* Fig. 5.30.15.5). The relative stabilities of these two transition states depend on the relative π-bonding properties of L and L′. Energetically, the better π-bonding ligand between L and L′ will preferably occupy the equatorial position (provided the steric factor and other factors remian the same). Thus stereochemistry of the product, [PtLL′XY] will depend on the relative π-accepting power of L and L′. If L is a better π-acceptor ligand than L′, then Y will enter at the *trans*-position of L. If L′ is a better π-acceptor ligand than L, then Y will enter at the *trans*-position of L′.

The following **order of π-acceptor abilities** of the ligands runs nearly parallel to the *trans*-effect series.

$$C_2H_2,\ C_2H_4,\ CO > CN^- > NO_2^- > SCN^- > I^- > Br^- > Cl^- > NH_3 > OH^-$$

The *trans*–effect shown by the π-acidic ligands (*e.g.* C_2H_4, CO, CN^-, etc.) as the nonleaving groups (L and L′) is very often described as the ***π-trans*–effect.**

(vi) ***Trans*-influence in terms of the π-bonding properties of the *trans*-ligands:** The π–bonding theory (called **static π–bonding theory**) can also explain the bond weakening phenomenon (*i.e. trans–influence).* If two *trans*–ligands are the π–acceptor ligands (say L and L′) then they will mutually compete for the same **two *d*–orbitals**. In fact, ***in such cases, the cis–isomer (keeping L and L′ at the cis-positions*** mutually competing for **one *d*-orbital**, i.e. competition for two *d*-orbitals *vs*. one *d*-orbital; discussed already) ***will be more stable***. If the two *trans* ligands (L and L′) are to compete for the same metal *d*-orbital, then the stronger π-acceptor ligand will predominate to weaken the bond at its *trans-*

(TBP-I)

(TBP-II)

(L > L′ in terms of π-acceptance)

(L′ > L in terms of π-acceptance)

Fig. 5.30.15.5: Illustration of pi–acceptance property to determine the relative *trans*–labilising power of the nonleaving groups (L and L′) in ligand substitution process of square planar complexes of Pt(II) (*i.e.* M = Pt)

position. However, this *trans*-influence to explain the *trans*-effect suffers from the limitations as in Grinberg's polarisation theory.

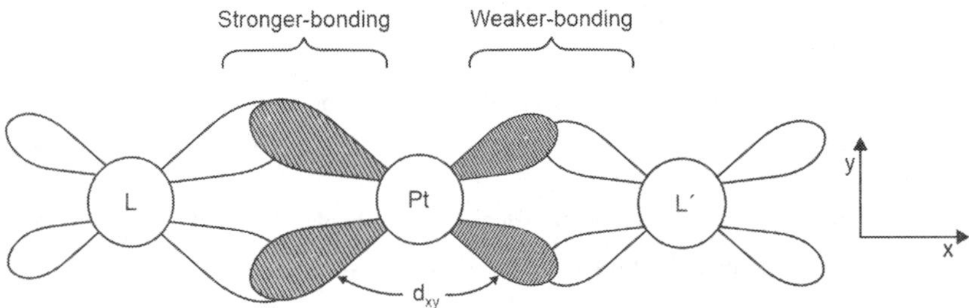

Fig. 5.30.15.6: Bond weakening at the *trans*–positions due to competition between the two *trans*-ligands (L and L′) for π-bonding with the metal d_{xy} orbital. Stronger π-bonding ligand (say L) predominates to weaken the bond at its *trans*–position (*i.e.* Pt—L′ bond) more.

Note: Here L and L′ are competing for the metal d_{xy} orbital. They can also compete for the metal d_{xz} orbital.

(vii) **Relative importance of the *trans*-effect among the congeners Ni(II), Pd(II) and Pt(II) in terms of the π-bonding theory:** Here it may be noted that the *dp*-hybrid orbitals of the metal centre actually participate in π-bonding with the ligands. Formation of these *dp*-hybrid orbitals depends on the energy difference between the $(n-1)d$ and *np* orbitals. For Pt(II), energy difference between the 5*d* and 6*p* orbitals is sufficiently small to favour their mixing. It can be explained in terms of **relativistic expansion** of $(n-1)d$ orbitals and **relativistic contraction** of *np* orbitals (see Vol. 1). **The relativistic effect is more important for the heavier congeners** and its importance runs in the sequence: Pt(II) > Pd(II) > Ni(II). In fact, this relativistic effect is less important for the lighter congeners Ni(II) and Pd(II) (*Ref.* Das *et al., Coord. Chem. Rev.* **479**, 215000, 2023) and consequently, the energy difference between their $(n-1)d$ and *np* orbitals is sufficiently large to disfavour their mixing. It explains why only Pt(II) shows the *trans*–effect in the series: Ni(II), Pd(II) and Pt(II). Relativistic effect can also explain this observation in terms of the **polarisation theory** discussed already.

(C) Sigma bonding theory and *trans*-effect in terms of the transition state effect: This theory was developed by Langford and Gray (1965) to explain the high *trans*–effect shown by the σ–bonding ligands like H^-, CH_3^-, etc.

The two *trans*-ligands (L and X) are effectively competing for the same $d_{x^2-y^2}$ and p_x-orbitals for σ-bonding along the L—M—X axis (say *x*–axis). The better sigma–bonding ligand (say L) will definitely try to monopolise these orbitals to weaken the bond at its *trans*–position (Fig. 5.30.15.7). This will lead to ***trans*-influence (*i.e.* higher energy of the ground state and ground state *trans*–effect)** (*cf.* Fig. 5.30.15.2). The electron deficiency developed in the *trans* region of the better σ-bonding ligand **will favour the approach of the entering nucleophile** in that region. Thus the incoming nucleophile will move towards the most weakly bound ligand (X) and away from the most strongly bound ligand (L). However, the σ-bonding theory can also explain the **stabilisation of the trigonal bipyramidal transition state produced in the associative path.** For the sake of simplicity, here we shall illustrate (Fig. 5.30.15.8) the effect by considering only the metal *p*-orbitals. In the L—M—X segment, if L enjoys the larger share of the same *p*–orbital for σ–bonding, then there will be a tendency in the system to change the configuration so that the same *p*–orbital can be more availed of by the better σ–bonding ligand L. This is attained by moving X out of the region of overlap in presence of the approaching ligand (Nu = Y). In the transition state, the leaving group X and entering group (Y = Nu) move to share the p_y-orbital allowing the L to have a larger share of the p_x orbital to produce the trigonal bipyramidal structure bearing L, X and Y in the equatorial plane which is the *xy*-plane (Fig. 5.28.15.8). This will stabilise the transition state, *i.e.* activation energy will be less.

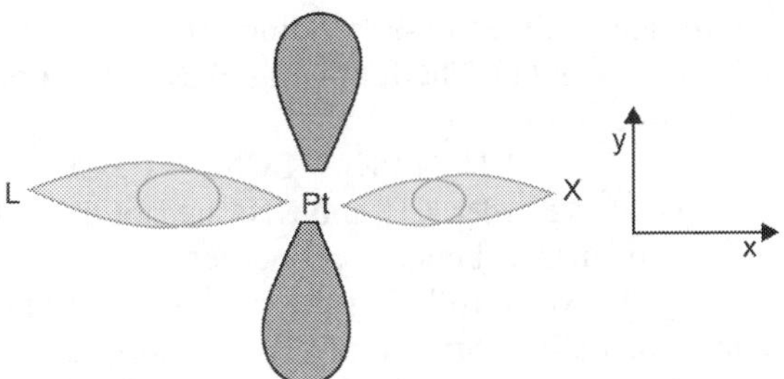

Fig. 5.30.15.7: Stronger σ-bonding in the Pt—L bond weakening the Pt—X bond, ground *trans*-effect. In the present illustration, the *trans*-ligands L and X can compete for the metal $d_{x^2-y^2}$ and p_x-orbitals for σ-bonding

Note: σ-*trans*-influence order in terms of Pt—N bond weakening by T in *trans*-$[PtCl_2(NH_3)T]$, **Pt—N bond strength order:** T = $H_2O > H_2S > Cl^- \gg PH_3 > H^- > CH_3^-$. The reverse order is the *trans*-influence order.

Stabilisation of the TBP activated complex by the better σ-bonding ligand (*cf.* Fig. 5.30.15.2): Stabilisation of a trigonal bipyramidal structure by the strong σ-bonding ligand (L) occupying an equatorial position can be rationalised in the following way. The *s*-orbital (spherical) may be considered to be equally available to the both axial and equatorial ligands. For the two axial ligands (which were initially *cis*- to the leaving group in the starting square planar complex), there is only one *p*-orbital, *i.e.* one half of the metal *p*-orbital is available to each of the axial ligands. The remaining two metal *p*-orbitals are available for the three equatorial ligands, *i.e.* each equatorial ligand can share two-third of a *p*-orbital (*cf.* sp^3d hybridisation of TBP ≡ $sp^2 + pd$, sp^2-hybrid orbitals for the equatorial bonds and *pd*-hybrid orbitals for the axial bonds). ***This is why, a better σ–bonding ligand will preferably occupy an equatorial position to stabilise the transition state to exert the trans–directing effect.*** The relative

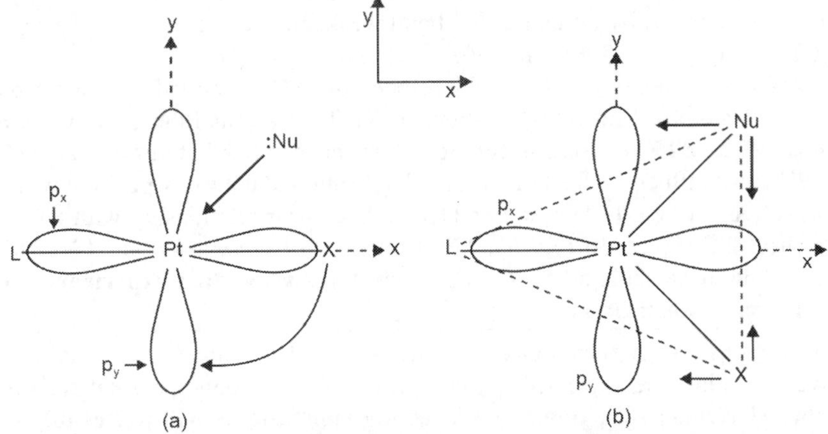

Fig. 5.30.15.8: Illustration of σ-bonding theory of *trans*-effect. (a) In the *xy*-plane (*i.e.* square plane), the ligands *L* and *X* compete for the same p_x orbital of Pt(II) for σ-bonding. (b) In the trigonal bipyramid transition state, L (*trans*-directing ligand), Nu (entering ligand) and X (leaving group) occupying the *xy*-plane (*i.e.* trigonal plane) bind with the p_x and p_y orbitals of Pt(II). The *trans*–axial ligands along the *z*-axis in the *TBP* are not shown for the sake of simplicity

trans-effects of the non-π-bonding groups (*i.e.* σ-bonding groups) are: H^- ~ PEt_3 (*ca.* 18,000) ≫ CH_3^- (*ca.* 175) > Cl^- (1). The following **order of σ-donor ability** of the ligands approximately parallels the *trans*-effect series.

$$H^- > PR_3 > SCN^- > I^-, CH_3^-, CO, CN^- > Br^- > Cl^- > NH_3 > OH^-.$$

In the above order, CO and CN^- are not the strong σ-donor ligands but show the strong *trans*-effect because of their π-bonding properties.

Note: ● By considering the dsp^2–hybridisation in the starting square planar complex, the *trans*–ligands actually share the $d_{x^2-y^2}$, *s* and one *p*–orbital rather than one *p*–orbital only. However, the strong sigma bonding ligand takes the larger shares of these orbitals. For the sake of simplicity, we have considered the *p*–orbital only for which the two *trans*–ligands are competing.

● In terms of the sp^3d hybridisation ($sp^2 + pd$; the sp^2-hybrid orbitals projected along the equatorial directions interact better with the ligands compared to the *pd*-hybrid orbitals projected towards the axial directions) of the TBP structure, it is expected that the better σ-donor ligands will occupy the equatorial positions (*cf.* ***Bent's rule***, Vol. 2; electron withdrawing electronegative groups tend to occupy the axial position of a TBP structure.).

Sigma- *vs.* pi-bonding effects in understranding the *trans*-effect

There is a strong controversy regarding the relative importance of the σ- and π-bonding properties of the *trans*-directing ligand to cause the *trans*-effect because both the strong σ-bonding ligands (*e.g.* H^-, CH_3^- etc. having no π-bonding property) and the π-bonding ligands (*e.g.* CO, CN^-, C_2H_4 etc.) are the strong *trans*-directing ligands. Some workers have argued in favour of the σ-bonding theory because the good π-bonding ligands also exercise a strong sigma bonding effect through the synergistic interaction. The σ-bonding ligands reduce the activation energy through the **ground state destabilisation** while the π-bonding ligands reduce the activation energy through the **stabilisation of transition state** (see Fig. 5.30.15.2). However, the σ-bonding ligands can also stabilise the TBP transition state and the π-bonding ligands can also cause the ground state destabilisation (*i.e. trans*-influence). These properties of the σ- and π-bonding ligands also reduce the activation energy.

Trans–influence *vs. Trans*–effect in the Octahedral Complexes

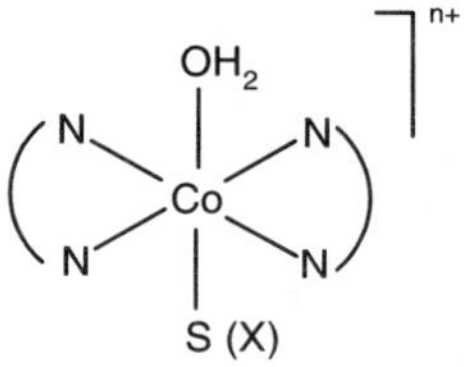

Though these phenomena are well documented in the square planar complexes of Pt(II), these are noted also in some octahedral complexes like *trans*–$[Co(en)_2(OH_2)X]^{n+}$ (X = $S_2O_3^{2-}$, RSO_2^-, SO_3^{2-}, etc., R = alkyl group; all X–ligands bind through the S-end). The S–end of the ligands (*X*) are polarisable to induce the *trans*-influence, *i.e.* bond weakening at the *trans*-position (*cf.* Grinberg's polarisation theory).

OH2 N N Co N N S (X) n+

It is noted that the *trans*–bond 'Co—O' (with respect to the Co—S bond) is elongated depending on the nature of the *trans*–axial ligand (*i.e.* S-bonded X). This is called the **ground state *trans*–effect or *static trans–effect* (STE) or *trans*-influence**. It has been noted that this bond lengthening phenomenon, *i.e.* STE also influences the kinetics, *i.e.* labilisation at the *trans*-position which is called the **kinetic *trans*–effect (KTE).** In fact, log (KTE) bears a linear relationship with STE ,*i.e.* **more the STE, more the KTE** (Fig. 5.30.15.9, *cf.* J.N. Cooper *et al., Inorg. Chem.*, **19**, 2265, 1980). This relationship is quite reasonable as the **Co(III) complexes very often experience the dissociative activation in ligand substitution.**

D. Cardwell's proposition in terms of electronegativity of the substituents: It is suggested that in the associative activation, the square planar complex of Pt(II) adopts the **TBP activated complex** in which the **more electronegative non-*trans*-directing substituents will preferably occupy the two axial positions (Bent's rule).** The **least electronegative group (L)** acting as the *trans*-directing group, leaving group (X) and entering group (Y) will occupy the equatorial positions of the TBP (activated complex). Loss of the leaving group (X) will give back the square planar complex where L and Y

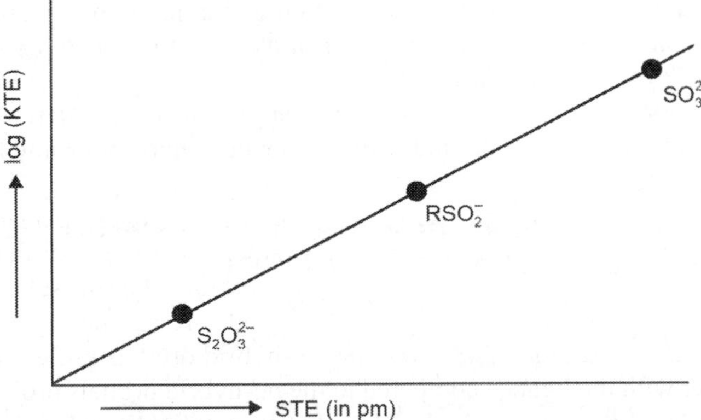

Fig. 5.30.15.9: Qualitative representation of the plot of log(KTE) *vs.* STE. KTE is measured by the water exchange rate constant; STE is measured by bond lengthening in pm in the Co—O bond; X = S-bonded ligands

remain at the *trans*-positions. Obviously, the highly electronegative groups (*e.g.* F, O, N donor sites) will preferably occupy the axial positions of the TBP activated complex **(Bent's rule)** and **such electronegative groups are the poor *trans*-directing groups**. Cl^- as a leaving group occupies an equatorial position (*cf.* neutral Cl is highly electronegative but Cl^- is not so electronegative). Here it may be mentioned that it is very difficult to estimate the **effective electronegativities** of different groups present in the complex. This limitation has reduced the popularity of this approch compared to that of other theories like the π-bonding theory (Orgel and Chatt) and σ-bonding theory (Langford and Gray).

Fig. 5.30.15.10: TBP activated complex bearing two **non-*trans*-directing groups** (A, A) of high electronegativity at the axial directions; *trans*-directing group (L) of low electronegativity at the equatorial position

Conclusions

- **Order of *trans*-influence** (mainly controlled by the σ-bonding effect and polarisability of the ligand):
  $H^- > PR_3 > SCN^- > I^-, CH_3^-, CO, CN^- > Br^- > Cl^- > OH^-$
  Note: The *trans*-effect series approximately follows the same sequence with some striking exceptions for CO and CN^- showing the high *trans*-effect by dint of their strong π-bonding property (*cf.* **stabilisation of T.S.**). *Trans*-influence (*i.e.* ground state destabilisation) reduces the activation energy to show the *trans*-effect (see Fig. 5.30.15.2). Thus the thermodynamic effect leads to the kinetic effect.
- **Ability of π-acceptance:**
  $C_2H_4, CO > CN^- > NO_2^- > SCN^- > I^- > Br^- > Cl^- > NH_3 > OH^-$
- ***Trans*-effect series (average order):** It can be explained by considering the *trans*-influence series, σ-bonding effect and π-bonding effect of the ligands.
  $CN^-, CO, NO, C_2H_4 > R_2S, AsR_3, PR_3, H^- > CH_3^-, C_6H_5^-, SC(NH_2)_2 > NO_2^-, I^-, N_3^-, SCN^- > Br^- > Cl^- > py > RNH_2, NH_3 > OH^- > H_2O$.
  - Thus the observed *trans*-effect series can be explained by considering both the **σ-bonding and π-bonding effect** of the *trans*-activating ligand. But, their relative contribution is a matter of debate (see Fig. 5.30.15.2).
  - The *trans*-activating ligand always occupies an equatorial position of the trigonal bipyramidal (TBP) activated complex and the non-*trans*-directing ligands occupy the axial positions of the TBP.

- The σ-bonding *trans*-ligand lowers the activation energy through the ground state destabilisation and by *favouring the approach of the entering nucleophile in the trans-region made electron deficient and stabilising the transition state.*
- The π-bonding *trans*-ligand lowers the activation energy by *favouring the approach of the entering nucleophile in the trans-region due to the removal of electron density from the metal centre which becomes electron deficient and stabilising the transition state through π-bonding.*

5.30.16 *Cis*-effect in the Substitution Reactions of Square Planar Complexes (*cf.* F. Basolo, *et al., J. Chem. Soc.*, 2207, 1961)

It has been established that a good *trans*-activating group can also activate the *cis*-position. However this *cis*-effect is not so strong as the *trans*-effect. The *cis*-effect is illustrated in some examples.

In the reactions of $[PtCl_3(NH_3)]^-$, $[PtCl_3(py)]^-$, the leaving group Cl^- is *trans*- to the Pt—Cl bond, but one *cis*–ligand of the leaving group is different (*cf.* NH_3 *vs.* py) in the two reactions. The rate constant k_{py} (where *py* is the *cis*–ligand) is about 40% higher than k_{NH_3} (where NH_3 is the *cis*-ligand). ***It indicates that py shows the better cis-labilising effect than NH₃.*** Here it may be mentioned that in terms of *trans*-effect, *py* is also a better *trans*-labilising group.

$$[PtCl_3(NH_3)]^- \xrightarrow[-Cl^-]{+py} [PtCl_2(NH_3)(py)]^0, \quad (k_{NH_3})$$

$$[PtCl_3(py)]^- \xrightarrow[-Cl^-]{+py} [PtCl_2(py)_2]^0, \quad (k_{py})$$

$$k_{py} > k_{NH_3}$$

Interestingly, in the above reaction, replacement of Cl^- *cis*– to NH_3 is about 20% faster in comparison to replacement of Cl^- *cis*- to NO_2^- (*i.e.* $k_{NH_3} > k_{NO_2}$). It may be noted that NO_2^- is a stronger *trans*-activating group. Thus we can write *cis*-effect: py > NH_3 > NO_2^- (*cf. trans*-effect: NO_2^- > py > NH_3). This finding of a limited study led to the conclusion at the early date as: ***a good trans-directing ligand is a poor cis-activating ligand.*** However, the extensive studies indicated that the said conclusion was not a general one. Rather, in most of the cases (with some exceptions), it was found: ***a good trans-activating ligand is also a good cis-activating ligand but the trans-effect is more pronounced than the cis-effect***. It is illustrated in the following examples (Basolo *et al., J. Chem. Soc.*, 2207, 1961).

$$trans\text{-}[PtL(Cl)(PEt_3)_2] \xrightarrow[\text{(in EtOH)}]{+py,\ -Cl^-} trans\text{-}[PtL(py)(PEt_3)_2] \quad \text{(Charges not shown)}$$

$$L = Cl^- \qquad C_6H_5^- \qquad CH_3^- \qquad H^-$$

$$\frac{k_{Y(L)}}{k_{Y(Cl)}} = 1 < 30 < 200 < 100{,}000 \quad \textbf{(\textit{i.e. trans-effect is large})}$$

(effect of the *trans*–labilising group L)

$$cis\text{-}[PtL(Cl)(PEt_3)_2] \xrightarrow[\text{(in EtOH)}]{+py,\ -Cl^-} cis\text{-}[PtL(py)(PEt_3)_2] \quad \text{(Charges not shown)}$$

$$L = Cl^- \qquad C_6H_5^- \qquad CH_3^-$$

$$\frac{k_{Y(L)}}{k_{Y(Cl)}} = 1.0 < 2.3 < 3.6 \quad \textbf{(\textit{i.e. cis}-effect is small)}$$

In the above reactions, the *trans*-labilising effect and *cis*-labilising effect of some groups have been measured against the standard ligand Cl^-. Both the *trans*- and *cis*-labilising effects very often run (but not always) in the same sequence, *i.e.* ***a better trans-labilising ligand is also a better cis-labilising ligand***. But for the ligands, Cl^-, CH_3^- and $C_6H_5^-$, the *trans*-labilising effect covers the span 1–200 while the *cis*-labilising effect covers the narrower span 1 to 3.6.

Based on the kinetic data on aquation of Cl^- in different square planar complexes of Pt(II), the relative orders of *cis*- and *trans*-effect are as follows:

***cis*-effect:** C_2H_4 (0.05) < Br^- (0.3) < Cl^- (0.4) < NH_3 (1.0) ≈ H_2O (1.0) < dmso, *i.e.* Me_2SO (5).

***trans*-effect:** H_2O (1.0) < NH_3 (200) < Cl^- (300) < Br^- (3,000) < I^- (4×10^4) < SCN^- (1×10^5) < Me_2SO (2×10^6) ≪ C_2H_4 (*ca.* 10^{11}).

It is evident that **the *cis*-effect is much poorer (in magnitude) than the *trans*-effect and the two series do not run always parallel** (*cf.* C_2H_4 is a very strong *trans*-activating group while it is a very poor *cis*-activating group).

● **Common origin of both the *cis*-effect and *trans*-effect:** It is believed that origin of both the *cis*–effect and *trans*–effect is the same but **the extent (*i.e.* degree) is different**. It is quite reasonable because in a square planar complex, in terms of the σ-bonding interaction (see Fig. 5.30.15.7), the *trans* ligands communicate via the metal $d_{x^2-y^2}$ orbital and one metal *p*-orbital (p_x or p_y) while the *cis*-ligands communicate via the only the $d_{x^2-y^2}$ orbital. The metal *s*-orbital (spherical) is equally available for both the *trans*-ligands and *cis*-ligands. This orbital involvement is understandable by considering the fact that the square planar complex is produced through the *dsp*² hybridisation $\left(d_{x^2-y^2} + s + p_x + p_y\right)$. Similarly, in terms of **π-bonding interaction** (see Figs 5.30.15.3, 4, 6), the two *trans*-ligands can communicate via the two metal *d*-orbitals, *i.e.* d_{xy} and d_{xz} (assuming the two *trans*-ligands along the *x*-axis) while the two *cis*-ligands (one along the *x*-axis and other ligand along the *y*-axis) can communicate via the one metal *d*-orbital, *i.e.* d_{xy}.

Thus roughly we may conclude that the two *cis*-ligands mutually interact through the metal orbitals in the same way as the two *trans*-ligands interact. But the degree of interaction is different in two cases (*cf.* via **2 metal orbitals** for the *trans*-ligands, and **one metal orbital** for the *cis*-ligands in both the σ- and π-interactions). Thus if a ligand can show the bond weakening and labilising effect at the *trans*-position, then the same ligand will also show the bond weakening and labilising phenomenon at the *cis*-positions also, *i.e.* ***trans*-influence, *trans*-effect** and ***cis*-influence, *cis*-effect** will appear simultaneously. They will also show the *trans*-effect and *cis*-effect in the same way.

Note: A **π-bonding ligand** like C_2H_4 as a *trans*-ligand can stabilise the TBP T.S. through π-bonding by occupying an equatorial position of the TBP T.S., but the π-bonding ligand as a *cis*-ligand cannot stabilise the TBP T.S. through π-bonding by occupying an axial position of the TBP T.S. (*cf.* **C_2H_4: strong *trans*-effect but relatively weak *cis*-effect**).

5.31 DISSOCIATIVE PATHWAY OF LIGAND SUBSTITUTION OF SQUARE PLANAR COMPLEXES OF PLATINUM(II) (*Ref.* R. Romeo *et al,. Inorg. Chim. Acta.*, **350**, 143, 2003 and Comments on Inorganic Chemistry, **11**, 21, 1990)

For the square planar complexes of Pt(II) (16*e* species), the associative pathway of ligand substitution through the formation of a 5 coordinate T.S./intermediate (18*e* species) is well documented. But, the catalytic activity of many square planar complexes of *d*⁸-system (*e.g.* **Wilkinson catalyst**, $[RhCl(PPh_3)_3]$, 16*e* species, see Vol. 6) involves the dissociation of a ligand to produce the coordinatively unsaturated and electronically unsaturated species (14*e* species), *i.e.* tricoordinate intermediate as the **catalytically active species**. Thus ligand dissociation in the **organometallic square planar complexes** is an important aspect of both theoretical and experimental research.

For the complexes like *cis*-[Pt(C, C)(S, S)] (C deonting a **strong σ-donor C-group** like alkyl or aryl group, S denoting a S-donor ligand like thioether SR_2 or sulfoxide $S(O)(R_2)$, the ligand substitution occurs through the dissociative path (*Inorg. Chem.*, **23**, 4428, 1984). The **following factors promote the dissociative path** for the square planar Pt(II) complexes.

(i) Presence of **very good σ-donor ligands** like R^- to show the **σ-*trans*-effect** (*i.e. trans*-influence) causing the bond weakening at the *trans*-position (see Fig. 5.30.15.7); **ligand dissociation at the *rds* becomes favourable.**

(ii) Presence of good σ-donor ligand makes the Pt(II) centre electron rich to disfavour the nucleophilic attack in the associative path but favours the ligand dissociation in the dissociative path.

(iii) **Stabilisation of the tricoordinate 14*e* intermediate or T.S. (T-shaped)** by the good σ-donor/π-donor nonleaving groups, by the **agostic interaction** arising from the nonleaving R-groups or from the *ortho*-hydrogens of the phenyl rings of the nonleaving group like PPh_3 (see Vol. 6 for agostic interaction).

(iv) **Sterically crowded** starting complex to disfavour the approach of the axially incoming nucleophile in the associative path (*cf.* **steric retardation** in the associative path and **steric acceleration** in the dissociative path.)

(v) **Presence of the π-acceptor ligands** (*e.g.* CO) favours the associative path through the stabilisation of TBP T.S./intermediate (see Figs. 5.30.3–6) and consequently, **disfavours the dissociative path**. For example, ligand replacement of thioether (SEt_2) in *cis*-[$PtPh_2(CO)(SEt_2)$] occurs in an associative path while it occurs in a dissociative path in *cis*-[$PtPh_2(SEt_2)_2$].

The dissociative pathway of ligand substitution (*i.e.* replacement of *dmso* by the bidentate ligands like bpy, phen, etc.) in *cis*-[$PtR_2(dmso)_2$] is illustrated in Fig. 5.31.1.

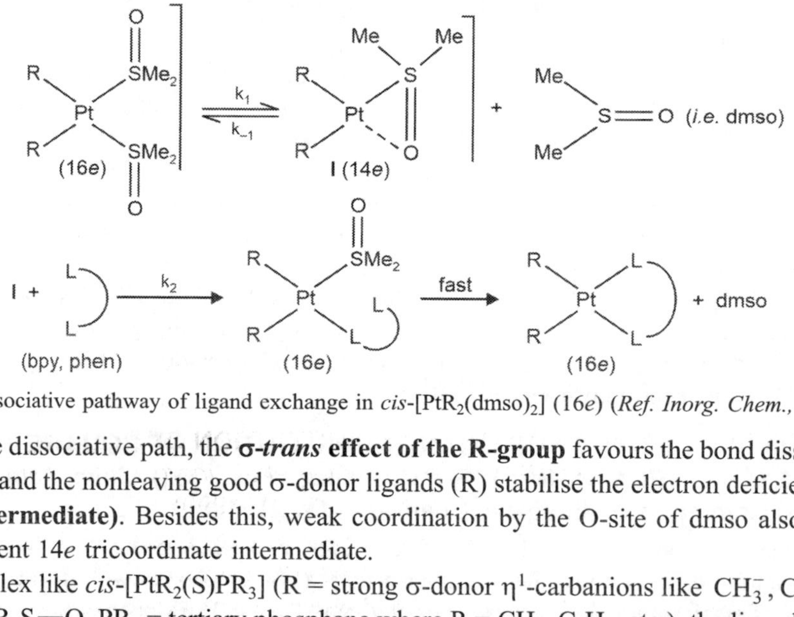

Fig. 5.31.1: Dissociative pathway of ligand exchange in *cis*-[$PtR_2(dmso)_2$] (16*e*) (*Ref. Inorg. Chem.*, **22**, 4428, 1984)

In the above dissociative path, the **σ-*trans* effect of the R-group** favours the bond dissociation at the *trans*-position and the nonleaving good σ-donor ligands (R) stabilise the electron deficient 14*e*-species **(T-shaped intermediate)**. Besides this, weak coordination by the O-site of dmso also stabilises the electron deficient 14*e* tricoordinate intermediate.

In the complex like *cis*-[$PtR_2(S)PR_3$] (R = strong σ-donor η^1-carbanions like CH_3^-, $C_6H_5^-$, etc., S = sulfoxide, *i.e.* $R_2S{=}O$, PR_3 = tertiary phosphane where R = CH_3, C_6H_5, etc.), the ligand exchange and ligand substitution reactions occur through the dissociative path. The strong σ-*trans* effect of the R^- groups favours the dissociative path. The T-shaped intermediate (14*e*) is stabilised by the strong σ-donor R^- ligands and PR_3. For the PPh_3 nonleaving group, the *ortho*-hydrogens of the phenyl ring of PPh_3 can also stabilise the 14*e* intermediate through the **agostic interaction** (see Vol. 6). The dissociative pathway of ligand substitution is shown in Fig. 5.31.2.

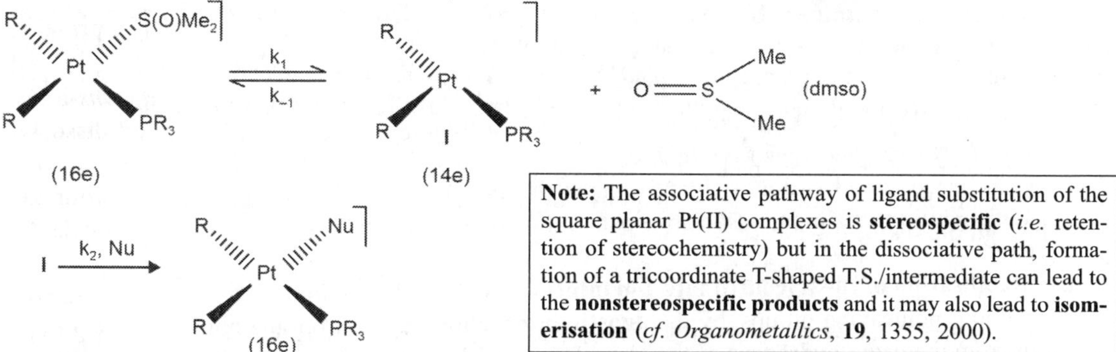

Note: The associative pathway of ligand substitution of the square planar Pt(II) complexes is **stereospecific** (*i.e.* retention of stereochemistry) but in the dissociative path, formation of a tricoordinate T-shaped T.S./intermediate can lead to the **nonstereospecific products** and it may also lead to **isomerisation** (*cf. Organometallics*, **19**, 1355, 2000).

Fig. 5.31.2: Dissociative pathway of ligand substitution in *cis*-[PtR_2(dmso)(PR_3)] (*Ref. Inorg. Chim. Acta.*, **350**, 143, 2003).

5.32 SUBSTITUTION REACTIONS OF TETRAHEDRAL COMPLEXES
(*cf.* Chapter 9 in Vol. 6 on Reaction Pathways of Organometallic Compounds)

The tetrahedral complexes possess relatively less cfse and consequently, less CFAE and very often, they react very fast. In fact, substitution reactions of the tetrahedral complexes are relatively less studied.

● **Dissociative path at Ni(0)-tetrahedral complexes:** The complexes like [$Ni(CO)_4$], [$Ni(CO)_2(PR_3)_2$], etc. maintaining a formal 18e–count experience **a dissociative path** passing through the **trigonal planar activated complex.**

$$[Ni(CO)_4] + {}^*CO \rightarrow [Ni(CO)_3({}^*CO)] + CO$$

$$[Ni(CO)_2(PR_3)_2] + {}^*PR_3 \rightarrow [Ni(CO)_2(PR_3)({}^*PR_3)] + PR_3$$

For this ligand exchange reaction, the rate law is:

rate = $k[Ni(CO)_2(PR_3)_2]$, (*i.e.* in conformity with the **dissociative path**)

Ligand replacement at [$Ni(CO)_4$] also passes through the ***dissociative path***. This is supported by the positive $\Delta S^{\neq}$ and $\Delta V^{\neq}$ values. Both the reactants are neutral and consequently, there is no change in charge in attaining the transition state (T.S.) in both the associative and dissociative paths. Thus there is no complication in rationalising the ΔV^* and ΔS^* values. **Their positive values support the dissociative path.**

$$[Ni(CO)_4] + PR_3 \rightarrow [Ni(CO)_3(PR_3)] + CO$$

$$\text{rate} = k[Ni(CO)_4]$$

The reaction pathway may be outlined as follows:

$$Ni(CO)_4 \xrightarrow[\text{(rds)}]{-CO} Ni(CO)_3 \xrightarrow[\text{(fast)}]{+PR_3} Ni(CO)_3(PR_3)$$

(trigonal planar)

The $\Delta H^{\neq}$ value (*ca.* 105 kJ mol^{-1}) is close to the average Ni—CO bond energy. This supports the dissociative path.

In [$Ni(CO)_4$]($e^4t_2^6$), the e^4-orbitals (*i.e.* filled $d^2_{x^2-y^2}$ and $d^2_{z^2}$ orbitals) participate in d(Ni)→π*(CO) back bonding. Thus among the four CO groups, there is a competition for the π-back bonding and this

competitive π-back bonding weakens the Ni—CO bonds. Replacement of the CO group by the non-π bonding or weaker π-bonding ligands strengthens the remaining Ni—CO bonds while their replacement by the stronger π-acceptor ligands weakens the remaining Ni—CO bonds. Gradual substitution of CO by PR_3 strengthens the remaining Ni—CO bonds as PR_3 is a weaker π-acceptor ligand and a better σ-donor ligand than CO. Consequently, the subsequent steps of substitution by PR_3 becomes more difficult in a dissociative path. On the other hand, gradual substitution of CO by a better π-acceptor ligand like PF_3, weakens the remaining Ni—CO bonds to favour the subsequent substitution steps through the dissociative path.

$$[Ni(CO)_4] \xrightarrow[-CO]{+PR_3(k_1)} [Ni(CO)_3(PR_3)] \xrightarrow[-CO]{+PR_3(k_2)} [Ni(CO)_2(PR_3)_2]\ [k_1 > k_2]$$

$$[Ni(CO)_4] \xrightarrow[-CO]{+PF_3(k_1)} [Ni(CO)_3(PF_3)] \xrightarrow[-CO]{+PF_3(k_2)} [Ni(CO)_2(PR_3)_2]\ [k_2 > k_1]$$

In fact, in presence of excess PF_3, all the CO groups can be readily replaced while substitution of all CO groups by PR_3 becomes a difficult task.

$$[Ni(CO)_4] \xrightarrow{+PR_3(\text{excess})} [Ni(CO)_{4-x}(PR_3)_x]\,(x = 1, 2)$$

$$[Ni(CO)_4] \xrightarrow{+PF_3(\text{excess})} [Ni(CO)_{4-x}(PF_3)_x]\,(x = 1, 2, ..., 4)$$

● **Dissociative path for the 18*e*-system:** For the **electronically saturated 18*e*-system**, the **dissociative (*D*) path** is favoured because of the following factors.

(i) **Very low oxidation state** does not encourage the nucleophilic attack, *i.e.* the incoming nucleophile is not attracted in the associative path.

(ii) The filled *d*-orbitals [*e.g.* d^{10} configuration as for Ni(0)] cannot accommodate the nucleophilic attack. *i.e.* there is **no vacant orbital to accommodate the nucleophilic attack** in the associative path.

(iii) In the dissociative path, change of the tetrahedral (sp^3) structure to the trigonal planar structure (sp^2) favours the better π-bonding, *i.e.* **metal-ligand π-bonding is better in the trigonal planar structure.** This will stabilise the transition state to favour the dissociative path.

Note 1: Some optically active tetrahedral complexes, [M(A)(B)(C)(D)], *e.g.* $[Mn(CH_3CO_2)(Cp)(NO)(PPh_3)]$ racemise through the ligand dissociation giving rise to a **trigonal planar intermediate.** In the said Mn-complex, PPh_3 dissociates and here the rates of ligand (PPh_3) exchange and racemisation are the same.

Note 2: In $[M(CO)_6]$ (M = Cr, Mo; $t_{2g}^6e_g^0$), the 6 CO groups compete for the $d(M)\rightarrow\pi^*(CO)$ π-back bonding by using the three filled t_{2g}^6 metal orbitals. This **competitive π-back bonding** mutually weakens the M—CO bonds more strongly in the *trans*-positions than in the *cis*-positions (see Fig. 5.30.15.3, Vol. 5 and Fig. 9.4.11.5, Vol. 6). This is why, the replacement of CO group by the weaker π-acceptor and better σ-donor PR_3 ligand occurs at the *trans*-positions of CO through the **dissociative path**. It produces *fac*-$[M(CO)_3(PR_3)_3]$ with the rate constants $k_1 > k_2 > k_3$ as in $[Ni(CO)_4]$.

$$[M(CO)_6] \xrightarrow[k_1]{PR_3} [M(CO)_5(PR_3)] \xrightarrow[k_2]{PR_3} [M(CO)_4(PR_3)_2] \xrightarrow[k_3]{PR_3} fac\text{-}[M(CO)_3(PR_3)_3]$$

fac-isomer

M = Cr, Mo; $k_1 > k_2 > k_3$ (dissociative path) (see vol.6 for details)

● **Associative path for the systems having the formal electron count less than 18:** Complexes showing the formal electron count less than 18 experience the **associative path** in the nucleophilic substitution process. Such complexes are coordinatively and electronically unsaturated and to gain the 18*e*–count, they have a tendency to increase the coordination number.

$[FeBr_4]^-$ (13*e* species, electronically unsaturated) undergoes the halide exchange through the **associative path.** The intermediate/activated complex is a trigonal bipyramidal one.

$$[FeBr_4]^- + X^- \xrightarrow{\text{(rds)}} [FeBr_4X]^{2-} \xrightarrow[\text{(fast)}]{-Br^-} [FeBr_3X]$$

The rate law for the said ligand exchange reaction is:

rate = $k[FeBr_4^-][X^-]$, (in conformity with the associative path).

This associative path is supported by the large negaive value of $\Delta S^{\neq}$.

Note: In the associative path, the leaving group and entering group occupy the axial directions of the TBP T.S./intermediate for the **tetrahedral starting complex** (*cf.* for the square planar starting complex, they occupy the equatorial positions of the TBP, see Scheme 5.30.4.2 *vs*. Scheme 5.30.4.3). However, for the **long-lived TBP intermediate** in both cases (tetrahedral and square planar complexes), Berry pseudorotation may lead to the exchange between the axial and equatorial ligands.

● **Associative path for the mixed nitrosyl-carbonyls:** The nitrosyl–carbonyls very often favour the *associative path* rather than the dissociative path. This is illustrated for the isoelectronic species (where NO has been formally considered to act as a 3e donor ligand).

$$\underbrace{[Ni(CO)_4]}_{\textbf{\textit{D}-path}},\ \underbrace{[Co(CO)_3(NO)],\ [Fe(CO)_2(NO)_2]}_{\textbf{\textit{A}-path}}$$

$$[Ni(CO)_4] + L \xrightarrow{-CO} [Ni(CO)_3L], \quad \text{rate} = k[Ni(CO)_4]$$

$$[Co(CO)_3(NO)] + L \xrightarrow{-CO} [Co(CO)_2(NO)L], \quad \text{rate} = k[Co(CO)_3(NO)][L]$$

$$[Fe(CO)_2(NO)_2] + L \xrightarrow{-CO} [Fe(CO)(NO)_2L], \quad \text{rate} = k[Fe(CO)_2(NO)_2][L]$$

($L = PR_3$, $P(OR)_3$, AsR_3, etc.)

The factors favouring the associative path for the said Co– and Fe– complexes are discussed below.

(i) The formal oxidation states for the given isoelectronic complexes are:

Fe(–II), Co(–I), Ni(0)

Oxidation state: ⟶ increasing trend

Metal to ligand π–bonding: ⟵ increasing trend

It is quite reasonable that in Ni(0), the metal → ligand π-bonding interaction is the weakest one while it is the strongest one in Fe(–II) having the lowest oxidation state. Thus, in terms of π–bonding, in the Ni–complex, the metal-ligand bond is the weakest one to favour the dissociative path. ***For the Co– and Fe–complexes, because of the higher metal-ligand bond strength, the dissociative path is relatively disfavoured and rather the associative path is favoured.***

(ii) NO can be present as NO^+ (*i.e.* effectively **3e-donor**) and it may be also present as NO^- (*i.e.* effectively **1e-donor**). If it exists as NO^- (*i.e.* bent M—N—O arrangement) then the metal centre possesses the 16e–count (*i.e.* the **metal centre becomes electronically unsaturated**).

(Linear linkage) M=N=Ö ⇌ M—N̈=Ö **(Bent linkage)**

Electronically saturated (18e-count, NO⁺)

Electronically unsaturated (16e-count, NO⁻)

It may be argued that in the starting tetrahedral complex, NO predominantly exists as NO^+ to maintain the 18e–rule but with the approach of the incoming nucleophile, the π–electron shifting to NO makes the metal centre electron deficient to attract the nucleophile **(*cf.* electromeric effect).**

M═N═Ö: ⟶ M—N̈(═O) with L; :L attacking M

This peculiar behaviour of the 'M—NO' linkage allows the **associative character** at the metal centre. The mechanistic steps may be outlined as follows:

$Co(CO)_3(NO)$ (18e) + L —(rds)→ [OC, CO, CO, L, Co, N═O] (18e) —(−CO, fast)→ $Co(CO)_2(NO)(L)$ (18e)

The rate law is:

$$\text{rate} = k[Co(CO)_3(NO)][L],\ L = PPh_3$$

The associative path is supported by the large negative value of $\Delta S^{\neq}$ (*ca.* –92 J K^{-1} mol^{-1})

Note: Here it may be mentioned that in some metal nitrosyl-carbonyls, both the dissociative and associative path may occur simultaneously. **It will lead to the *two term rate law*.** In fact, for the exchange of CO ligand in $[Co(CO)_2(NO)(PR_3)]$, the two term rate law has been found.

$$\text{rate} = (k_d + k_a[CO])[Co(CO)_2(NO)(PR_3)]$$

k_d stands for the dissociative path while k_a stands for the associative path.

These can be explained by considering the existence of 16e and 18e species in an equilibrium.

$[Co(CO)_2({=}N{=}O)(PR_3)] \rightleftharpoons [Co(CO)_2(—N{=}O)(PR_3)]$

Electronically saturated (18e-count, NO^+) — **Favouring the dissociative path**

Electronically unsaturated (16e-count, NO^-) — **Favouring the associative path**

5.33 RACEMISATION OF THE TETRAHEDRAL COMPLEXES
(*cf.* Sec. 5.29 for the octahedral complexes)

Very often, the optically active tetrahedral complexes rapidly undergo racemisation and they cannot be resolved. However, two types of optical isomers in the tetrahedral complexes are known (*cf.* Sec. 2.8). These are: **[M(A)(B)(C)(D)]** (having four different ligands) and **$[M(AB)_2]$** (AB = unsymmetrical didentate ligand). The different routes of racemisation for the optically active tetrahedral complexes are discussed here.

(i) **Intermolecular path—formation of a trigonal planar intermediate (*i.e.* decrease of C.N.) through dissociation of a ligand:** From the trigonal planar intermediate, both the optical isomers can be generated. $[Mn(CH_3CO_2)(Cp)(NO)(PPh_3)]$ racemises in benzence through the formation of a trigonal planar intermediate produced from the **dissociation of PPh_3**. This view is supported by the fact that the rates of racemisation and ligand (*i.e.* PPh_3) exchange are the same.

(ii) **Intermolecular path—formation of a stereomobile trigonal bipyramidal (TBP) intermediate** (*i.e.* increase of C.N.) **through a nucleophilic attack:** Nucleophilic attack can generate a **long-lived TBP intermediate** which can experience the Berry pseudorotation producing two different TBP configurations. Then expulsion of the nucleophile **(acting as a catalyst)** from the TBPs gives a racemic mixture. It happens so for the optically active **tetrahedral Si-compounds** (*cf.* Scheme 5.30.4.3) which can form the TBP intermediate by using the vacant *d*-orbital (*i.e.* sp^3d hybridisation).

This is why, for the Si-compounds, to find the optically pure compound is a difficult task while for the C-compounds (being unable to form the long-lived and stable TBP intermediate), there are so many optically pure compounds.

(iii) **Intramolecular path—formation of a square planar structure (*i.e.* no change in C.N.) through an intramolecular rearrangement (*cf.* twist mechanism, Sec. 5.29):** For the $[M(AB)_2]$ type tetrahedral complexes, if the steric hindrance and rigidity of the chelate ring do not prevent the formation of the square planar structure, then the process operates.

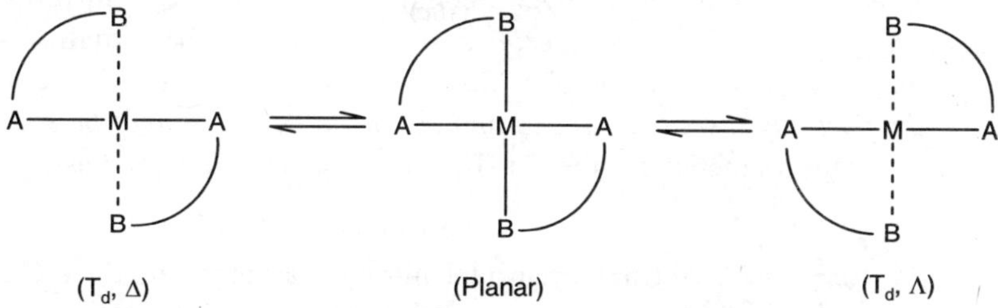

Ligand substitution Process in the Tetrahedral Compounds of C, Si, and P (*cf.* Chapter 10, Vol. 2)

Tetrahedral C-compounds can experience an **interchange path** (I_a or I_d) (*cf.* Fig. 5.30.3.2) producing a **TBP transition state (too short lived)** which cannot undergo the structural rearrangement. It leads to **Walden Inversion** for the optically active C-compounds (*cf.* Sec. 5.30.4) in the I_a-path. For the optically active tetrahedral Si-compounds, the product is a **racemic mixture.** In fact, it passes through the ***A*-path** where the TBP intermediate (sp^3d hybridisation) is sufficiently **long-lived to undergo the Berry pseudorotation** to give the racemic product (*cf.* Scheme 5.30.4.3).

5.34 ACID CATALYSED DISSOCIATION OF THE MACROCYCLIC AND OPEN CHAIN COMPLEXES: MACROCYCLIC EFFECT

5.34.1 Origin of the Macrocyclic Effect

Macrocyclic effect (*cf.* Sec. 4.5.4, Vol. 4)**:** The **macrocyclic effect** indicates the relatively enhanced stability of the metal complexes with the macrocyclic ligands compared to that of the complexes formed by the comparable open-chain ligands. This macrocyclic effect originates from both the **entropic favour** and **enthalpic favour.**

Entropic favour: The macrocyclic ligand is less flexible than the analogous open-chain ligand. During complexation by an open-chain ligand, the *cyclisation* makes ΔS more negative (loss of **conformational entropy**, see Vol. 4). The macrocyclic ligand which is **preorganised** and possesses the diminished flexibility experiences the less negative ΔS on complexation.

Enthalpic favour: The free macrocyclic ligand is more **preorganised** than the corresponding open-chain ligand. Consequently, it requires a less conformational change of the macrocyclic ligand for complexation. This **less preorganisation energy** gives an enthalpic favour. Moreover, compared to the open-chain ligand, the macrocyclic ligand in free state is less solvated because of its diminished flexibility. Thus it requires **less desolvation energy** during the complexation. This gives also another contribution to the enthalpic favour.

Macrocyclic effect in terms of dissociation kinetics: The **disfavoured dissociation process** for the macrocyclic complex also explains the macrocyclic effect in terms of kinetics (*cf.* stability constant = formation rate constant/dissociation rate constant = k_f/k_d, see Vol. 4).

5.34.2 Mechanism of Acid Catalysed Dissociations of Some Macrocyclic and Open-chain Tetraamine Complexes as the Representative Examples

In terms of the dissociation kinetics, the macrocyclic complexes are more stable compared to the open-chain counterparts. The representative examples (Fig. 5.34.2.1) of tetraamine ligands are: *tet*–a (meso), *tet*–b (*d, l*), *cyclam* and 3, 2, 3–*tet* (open-chain analogue).

For the macrocyclic complexes, during dechelation, the partially dislodged ligand still remains within the coordination sphere to **allow its religation** to regenerate the complex. To avoid this religation, the partially dislodged ligand is required to fold so that the dislodged ligating sites may move away from the metal centre. This is why, for the macrocyclic complexes, the dissociation process very often requires the *energetically unfavourable rearrangements* such as folding of the partially dislodged ligand within the coordination sphere before the complete dislodgement of the ligand. Moreover, because of the steric crowding, *the partially dislodged ligand and the metal centre cannot be stabilised through solvation.* Here the dissociation kinetics for the Cu(II)–complexes are compared as the representative examples.

$$\left[\mathrm{Cu(cyclam)}\right]^{2+} \underset{k_{-1},(-\mathrm{H}^+)}{\overset{k_1,(+\mathrm{H}^+)}{\rightleftharpoons}} \left[\mathrm{Cu(Hcylam)}\right]^{3+} \xrightarrow{k_2,(+\mathrm{H}^+)} \mathrm{H_4cyclam}^{4+} + \mathrm{Cu}_{\mathrm{aq}}^{2+}$$

$$\left[\mathrm{Cu(3,\,2,\,3\text{-}tet)}\right]^{2+} \xrightarrow{k',(+\mathrm{H}^+)} \mathrm{H_4(3,\,2,\,3\text{-}tet)}^{4+} + \mathrm{Cu}_{\mathrm{aq}}^{2+}$$

The important observations are:

- **[Cu(3, 2, 3 –tet)]$^{2+}$** (*i.e. open-chain complex*) dissociates much faster than the macrocyclic complex [Cu(cyclam)]$^{2+}$.
- In contrast to **[Cu(cyclam)]$^{2+}$** (*i.e.* macrocyclic complex), dissociation of the open chain counterpart, *i.e.* [Cu(3, 2, 3–tet)]$^{2+}$ occurs *in a single stage.*
- For the **open-chain complex**, the **acid independent solvation path** makes the *major contribution* while for the corresponding **macrocyclic complex**, the **acid dependent path** is the *main contributing factor.*

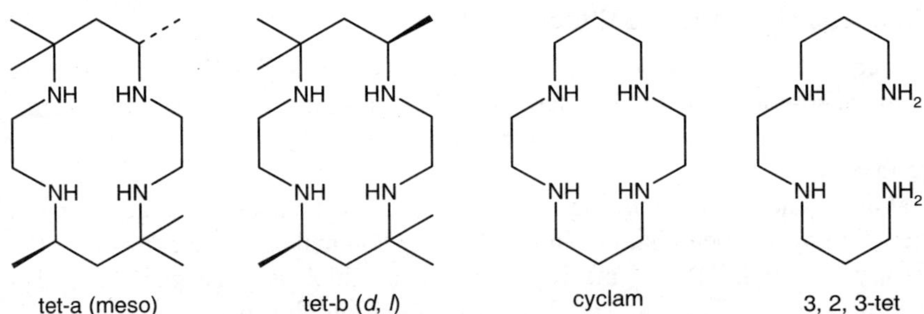

Fig. 5.34.2.1: Some representative macrocycles like 1, 4, 8, 11–tetraazacyclotetradecane (*i.e.* cyclam), C–*meso*–5, 5, 7, 12, 12, 14–hexamethyl–1, 4, 8, 11–tetraazacyclotetradecane (*i.e.* tet–*a*) (which gives both red and blue isomeric complexes); C–*rac*–5, 5, 7, 12, 12, 14–hexamethyl–1, 4, 8, 11–tetraazacyclotetradecane (*i.e.* tet–*b*) and open-chain analogue like 1, 5, 8, 12–tetraazadodecane (*i.e.* 3, 2, 3–tet)

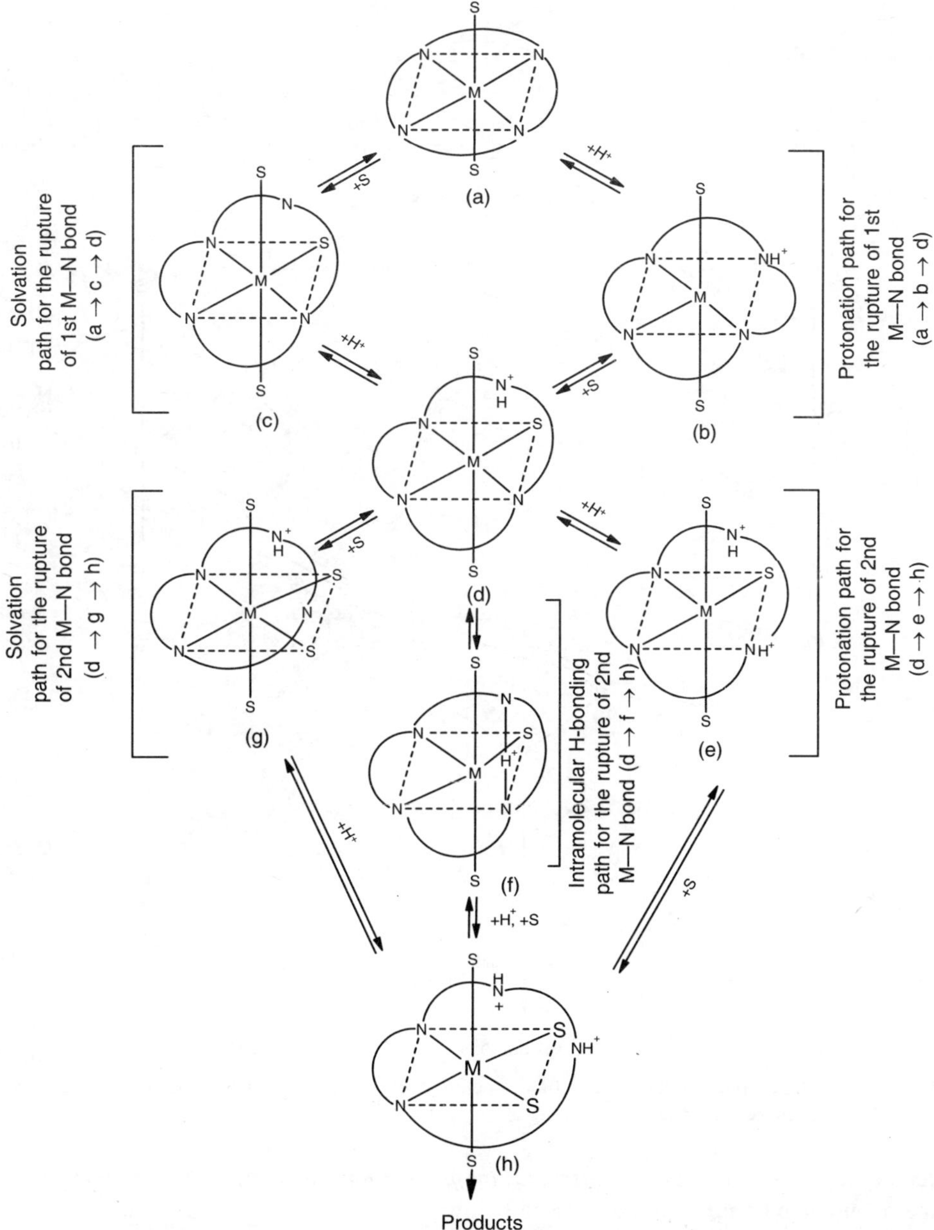

Scheme 5.34.2.1: Generalised mechanistic routes of dissociation of a macrocyclic tetraamine complex. S = protic solvent like H_2O (Overall charges not shown)

Mechanistic steps of dissociation of the macrocyclic and open-chain complex: For the dissociation of tetraamine macrocyclic and open chain complexes of Cu(II), the general mechanistic steps can be outlined as in Schemes 5.33.2.1-2 (*cf.* C.S. Chung *et al.*, *Inorg. Chem.* **20,** 2152, 1981; **22,** 1017, 1983; **25,** 1940, 1986; **27,** 1880, 1988). For the macrocyclic complexes, the first step leads to the cleavage of a Cu—N bond and it can happen through the *direct protonation path* and *solvent assisted path*. The second step cleaves the second Cu—N bond that can occur through the direct protonation, solvation and intramolecular H–bonding path. Dissociation of the corresponding open-chain complex, can be also considered to occur in two steps, *i.e.* rupture of the first Cu—N bond followed by the second Cu—N bond rupture.

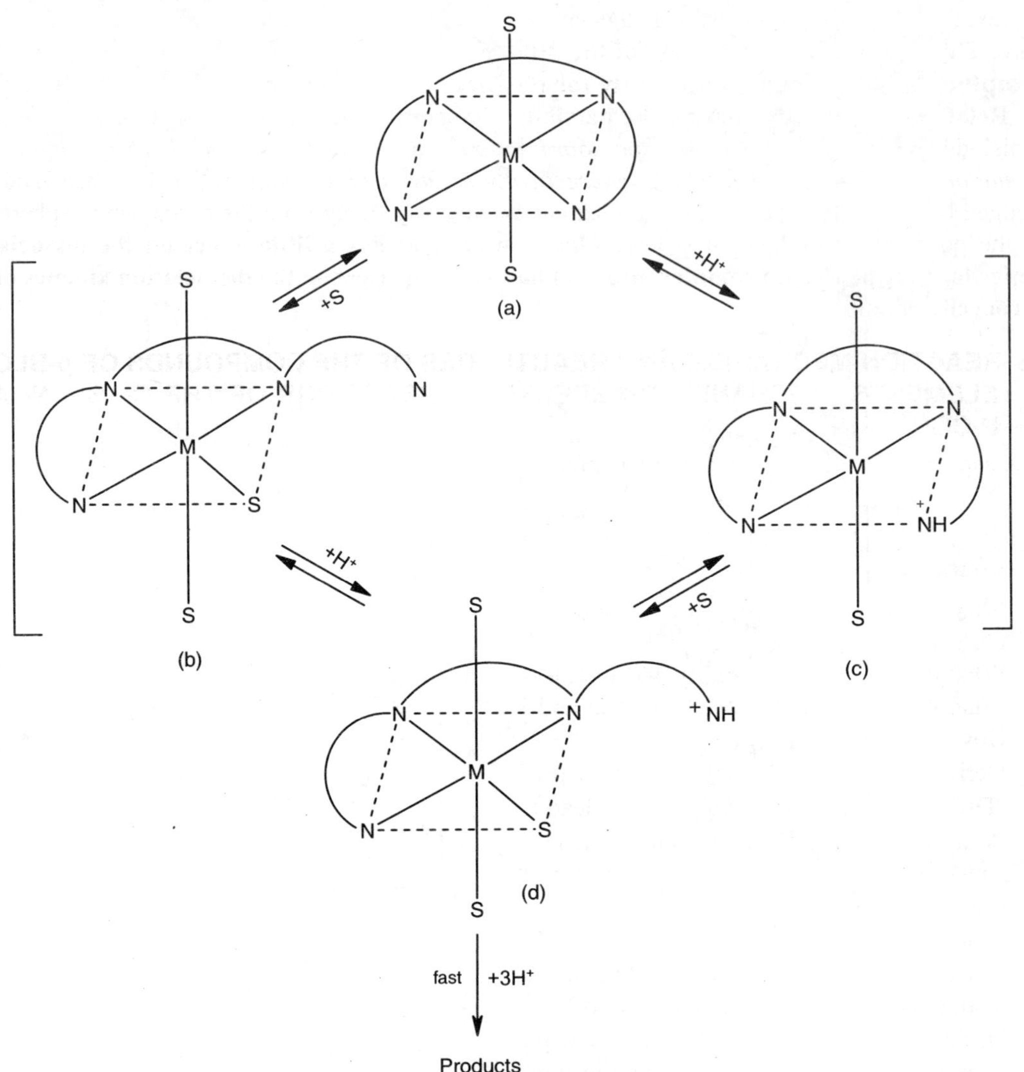

Scheme 5.34.2.2: Generalised mechanistic routes of dissociation of an open-chain tertraamine complex. S = protic solvent like H_2O (Overall charges not shown)

The *different factors disfavouring the dissociation process for the macrocyclic complex* compared to those for the open-chain complex are discussed below.

● **Relative loss of stability of the complex for the rupture of the first Cu—N bond:** Here it is important to note that rupture of the first Cu—N bond destroys only one chelate ring for the open-chain complex while it destroys two chelate rings for the corresponding macrocyclic complex. **Thus rupture of the first Cu—N bond is energetically more costly for the macrocyclic complex, and consequently, this step is much slower for the macrocyclic complex.**

● **Relative solvation effect during the dissociation process:** After the cleavage of first Cu—N bond, for the open-chain complex, the dislodged N–donor site does not experience any restriction to move away from the coordination sphere and this dislodged N–site can be solvated and it also does not produce any barrier for solvation at the vacant coordination site. On the other hand, for the macrocyclic complex, the movement of the dislodged N–site is restricted and it still remains within the coordination sphere. **This disfavours the solvation of the dislodged site and vacant coordination site. It makes the rupture of Cu—N bond more costly for the macrocyclic complex.**

- **Relative ease of religation by the partially dissociated ligand:** For the macrocyclic complex, the dislodged *N–site present within the coordination sphere can be prevented from religation by protonation* or *by rearrangement* (*e.g.* ***energetically unfavourable folding***). On the other hand, the dislodged N-site for the open-chain complex can move away freely from the coordination sphere and thus the possibility of religation is less. **This is why, acid has a little effect on the dissociation kinetics for the open-chain complex and acid has a strong effect on the dissociation kinetics of the macrocyclic complex.**

5.35 REACTION MECHANISM AND REACTIVITIES OF THE COMPOUNDS OF *p*-BLOCK ELEMENTS: MECHANISTIC ASPECTS OF REACTIONS OF THE C-, Si-, N- AND P-COMPOUNDS

These aspects have been discussed in Chapter 10 of Vol. 2.

EXERCISE 5

A. General Type Questions

1. What are the different types of ligand substitution reactions in terms of reaction mechanism? Discuss with examples.
2. What do you mean by lability and inertness?
3. What do you mean by stoichiometric and intimate mechanism of ligand substitution reaction? How are they related?
4. Derive the rate laws for the *D*, *A* and *I* processes of ligand substitution reaction.
5. "Different reaction mechanisms may lead to the same rate law". Illustrate with examples.
6. What do you mean solvent exchange reaction? Discuss the importance of solvent exchange rate constants (k_{ex}) in the diagnosis of reaction mechanisms of other ligand substitution processes.
7. Classify the metal ions in terms of water exchange rate constants (k_{ex}). Discuss the different factors to control the water exchange rate constants.
8. What do you mean by Eigen-Wilkins mechanism of ligand substitution process? How can you distinguish between the I_a and I_d paths?
9. During chelation by a didentate ligand, depending on the nature of the rate determing step, it may experience the steady-state or equilibrium mechanism (*i.e.* intermediate complex bearing the didentate ligand as a mondentate ligand may remain in steady-state or equilibrium state). Explain.
10. How can you distinguish among the *D*, I_d and I_a and *A* pathways of ligand substitution reaction?
11. What do you mean by *proton ambiguity* in kinetics of ligand substitution reaction? How can you solve the problem? Illustrate with an example.
12. How can you attempt to explain the lability or inertness of transition metal complexes in terms of VBT? What are the limitations in this approach?
13. What do you mean by crystal field activation energy (CFAE)? How can you explain the lability or inertness of metal complexes in terms of CFT?
14. What do you mean by $\Delta H^{\neq}$, $\Delta S^{\neq}$ and $\Delta V^{\neq}$? How can you apply the knowledge of these activation parameters to diagnose a reaction mechanism?
15. How can you experimentally determine $\Delta H^{\neq}$, $\Delta S^{\neq}$ and $\Delta V^{\neq}$? What are the relationships between E_a (activation energy in Arrhenius equation) and $\Delta H^{\neq}$; $\Delta V^{\neq}$ and $\Delta S^{\neq}$?
16. Discuss the factors controlling the $\Delta V^{\neq}$ and $\Delta S^{\neq}$ values.

17. What do you mean by LFER plot? How can you apply this plot to diagnose a reaction mechanism?
18. What do you mean by isokinetic trend? Why is it called **compensation effect**? What does it signify?
19. What do you mean by steric accleration and steric retardation? How can you interpret these in terms of reaction mechanism?
20. Discuss the effect of charge on a complex to diagnose a ligand substitution reaction mechanism.
21. Discuss the mechanistic aspects of aquation or acid hydrolysis of cobalt(III) complexes. Discuss the steric and electronic effects of the spectator ligands.
22. Discuss the kinetic aspects of acid catalysed and metal ion catalysed aquation of metal complexes.
23. Discuss the pseudo-substitution reaction in octahedral metal complexes.
24. Discuss the stereochemical changes in the acid hydrolysis of Co(III)–complexes.
25. Discuss the mechanism of base hydrolysis of cobalt(III)–ammine complexes. Give the experimental evidences in favour of the proposed mechanism. Critically comment on the *D*-CB, I_d-CB and *D*-IP processes in relation to such base hydrolysis reaction.
26. What are the stereochemical changes in base hydrolysis of Co(III)–ammine complexes? Compare these results with those of the acid catalysed aquation reaction.
27. Discuss the kinetic aspects of formation reactions at Fe(III).
28. Discuss the mechanism of isomerisation (both linkage and geometrical isomerisation) reaction in octahedral complexes.
29. Discuss the possible mechanistic routes of racemisation of octahedral and tetrahedral complexes.
30. What are the characteristic features of the ligand substitution reactions of square planar complexes of Pt(II)?
31. Discuss the nucleophilicity scale, *trans*–influence, *trans*–effect (both σ– and π– *trans* effect), stereospecificity, *cis–trans* isomerisation in relation to the reactivity of square planar complexes of Pt(II).
32. Discuss the application of *trans*—effect series in the preparation of isomers of Pt(II) complexes.
33. Discuss the mechanistic aspects of ligand substitution reaction in carbonyl and carbonyl-nitrosyl complexes of tetrahedral geometry.
34. Discuss the mechanism of nucleophile catalysed *cis–trans* isomerisation of square planar complexes of Pt(II). What are the other possible routes of *cis- trans-* isomerisation in square planar complexes?
35. Discuss the σ– and π–bonding theory of *trans*–effect.
36. Outline the routes to synthesise the following complexes.
    (i) isomers of $[Pt(C_2H_4)Cl_2(NH_3)]$;
    (ii) isomers of $[Pt(CO)Cl_2(py)]$.
    (iii) isomers of $[Pt(NH_3)(NH_2OH)(NO_2)(py)]^+$;
    (iv) isomers of $[PtCl(NH_3)(RNH_2)(NO_2)]$
    (v) isomers of $[PtBrCl(NH_3)py]$
37. What do you mean by macrocyclic effect? Explain its origin.
38. Compare the acid catalysed dissociation of the macrocyclic and open-chain complexes.
39. Cu(II)–complexes are, in general, labile with respect to acid catalysed dissociation. But many Cu(II)–macrocyclic complexes dissociate slowly—explain.
40. Compare the stereochemical aspects of ligand substitution process of the tetrahedral C-compounds, tetrahedral Si-compounds and square planar Pt(II)-complexes.
    - Optically pure C-compounds are well known but optically pure Si-compounds are rarely known. How can you explain it?
41. Explain the following facts in base hydrolysis of the Co(III)-ammine complexes.

(i) proton abstraction preferably occurs from the *trans*-position of the leaving group, (ii) rate enhancement by the CB, (iii) entry of H_2O not OH^- as a nucleophile at the fast step, (iv) stereochemistry of the product.

42. Give the basic principles of the techniques to follow the fast kinetics: perturbation technique (T- and P-jump), NMR technique and stopped flow technique.
43. Discuss the favourable factors for the dissociative path of ligand substitution of square planar complexes of Pt(II).

B. Justify the Statements

1. Thermodynamically stable complexes are not necessarily kinetically stable.
2. The origins of thermodynamic stability and kinetic stability are different.
3. Nucleophilicity and basicity do not always run parallel.
   - *Trans*-influence and *trans*-effect do not always run parallel (*cf.* octahedral Co(III)-complexes *vs.* square planar Pt(II)–complexes).
4. Associative (*A*) and dissociative (*D*) are the two-step reactions while interchange (*I*) reaction is a single-step reaction.
5. The Langford-Gray notation, *A*, I_a, I_d and *D* are the combinations of notations of stoichiometric and intimate mechanisms.
   - Intimate mechanism looks at the maxima while stoichiometric mechanism looks at the minima of the free energy *vs.* reaction coordinate plot.
6. Langford-Gray notation and Hughes-Ingold notation can be correlated.
7. Associative (*A*) and dissociative (*D*) pathways involve two transition states while interchange (*I*) pathway involves a single transition state.
8. In a dissociative process, the rate law can show a complicated dependence pattern on the living group and entering group.
9. A reaction occurring in a dissociative process may show the rate dependence on the entering ligand depending on the condition.
   - Pre-equilibrium step and steady state approximation for the intermediate lead to the different rate laws but under some conditions, the steady state kinetics may shift to the pre-equilibrium kinetics.
   - Sometimes, the observed rate constant may show the negative activation energy (*i.e.* rate decreases with the increase of temperature).
10. An associative reaction shows a rate law that indicates a strict first-order dependence on the entering ligand concentration.
11. The interchange pathways I_a and I_d can be distinguished by considering the rate dependence on the entering ligand concentration and in terms of k_{ex} values.
12. The rate law for the *D* and *I* processes can be expressed by a general form.

$$L_5M{-}X + Y \rightarrow L_5M{-}Y + X$$

$$\text{rate} = \frac{pq[L_5M{-}X][Y]}{1+q[Y]}$$

   - The rate law for an associative process (*A*), can also be expressed in the above form under some restrictions.
   - In the energy profile diagram (free energy *vs.* reaction coordinate), the activated complexes or transition states reside at the maxima while the intermediates reside at the minima.
   - The transition states and intermediates differ in their life times.
   - For the reaction, $M + L \underset{k_{-1}}{\overset{k_1}{\rightleftharpoons}} ML$ (under the condition, $[L]_0 \gg [M]_0$), it gives: $k_{obs} = k_1[L]_0 + k_{-1}$.
13. *D* and I_d pathways can lead to the same rate law under some conditions.

14. A formation reaction (*i.e.* replacement of the coordinated solvent) passing through a *D*–process can show a rate dependence on the nature of the entering ligand.
15. In a perfect *D*–process, the dependence of formation rate constant (k_f) on the nature of the entering ligand arises from the competition between the centering ligand and leaving group for the lower coordinate intermediate.
16. A *D*–reaction may show or may not show the dependence on the nature of the entering ligand depending on the condition.
17. An I_d-reaction may show or may not show the rate dependence on the nature of the entering ligand depending on the condition.
18. Very often, an I_d-path may be characterised by considering the solvent exchange rate constant (k_{ex}).
19. Both the I_a– and *A*–paths can develop the same rate law under some conditions.
20. In the I_d– path of a formation reaction, the interchange rate constant can never exceed the solvent exchange rate constant (k_{ex}) but for the I_a–path, it may be less than or greater than the k_{ex} value depending on the condition.
21. *A*–path always follows a second order kinetics but I_a–path may follow both the first-order and second kinetics depending on the conditions.
22. In the octahedral complex, perfect *D*–process is generally found for the anation of anionic complexes like $[Fe(CN)_5(OH_2)]^{3-}$, $[Co(CN)_5(OH_2)]^{2-}$.
23. Perfect *D*–process is also likely to operate in the octahedral complexes bearing the strong *trans*-labilising ligands.
24. To distinguish between the *A* and I_a paths, it may be required to follow the effect of the concentration of the entering ligand in a *wide range*.
25. In the formation reactions of $[Pt(OH_2)_4]^{2+}$, $[Pd(OH_2)_4]^{2+}$, $[Ti(OH_2)_6]^{3+}$, A–process is argued based on the fact: no deviation from the strict first-order dependence on the entering ligand even when concentration of the ligand is very high; the said strict first-order is maintained even when the entering ligand is highly anionic to make the K_{OS} value very high.
26. Fractional order in the entering ligand concentration is an indication of the I_a–path.
27. The same rate law can be obtained for different reaction mechanisms.
28. The second order rate law is obtained for the following reaction.
    $L_5M—X + Y \rightarrow L_5M—Y + X$; rate $\propto$ $[L_5MX][Y]$
    There are several mechanisms to generate the above rate law.
29. $L_5M—X + Y \rightarrow L_5M—Y + X$
    The fractional order in $[Y]$ or saturation kinetics at a higher value of $[Y]$ is an indication of a preequilibrium step involving the both reactants followed by the rate determining step.
30. NMR peak broadening phenomenon can be applied for the determination of water exchange rate constants of metal ions.
31. $\Delta V^{\neq}$ is a very much reliable parameter to diagnose the solvent exchange reaction mechanism.
32. V^{2+} and Cr^{3+} show the very low values of k_{ex}.
33. The water exchange rate is faster for Ni^{2+} than that for Cr^{3+} though they are having the same CFAE.
34. The k_{ex} values of Cr^{2+} and Cu^{2+} are tremendously high.
35. For the bivalent metal ions of 1st transition series, the water exchange process involves the more associative character for the early members while the late members of the series adopt the more dissociative character.
36. The mechanism of solvent exchange reaction depends also on the bulkiness of solvent molecules.
37. The trivalent metal ions of 1st transition series very often adopt the associative path in water

exchange process. However, their hydrolysed species, *i.e.* $[M(OH)(OH_2)_5]^{2+}$ (M = Fe, Cr, etc.), adopt the dissociative process like the bivalent metal ions in water exchange process.

38. $[Fe(OH_2)_6]^{3+}$ and $[Fe(OH)(OH_2)_5]^{2+}$ adopt different mechanistic paths in ligand substitution process including the water exchange process.
    - $[Fe(CN)_6]^{3-}$ solution may induce the CN^- poisoning but it is unlikely for $[Fe(CN)_6]^{4-}$ solution.
    - $[Ru^{III}(edta)(OH_2)]^-$ and $[Ru^{II}(edta)(OH_2)]^{2-}$ experience different mechanistic paths for substitution of H_2O.
39. $[Ti(OH_2)_6]^{3+}$ shows the more associative character than $[Fe(OH_2)_6]^{3+}$ in ligand substitution process.
40. For the trivalent metal ion, $[Co(OH_2)_6]^{3+}$, the water exchange process is unusually fast.
41. $\Delta S^{\neq}$ is a reliable parameter to identify the solvent exchange path.
42. Solvent exchange rate constant (k_{ex}) is an important parameter to diagnose the reaction mechanism for different types of ligand substitution processes.
43. In ***d***-activation, the ligand substitution rate constant can never exceed the k_{ex} value.
44. In ***a***-activation, ligand substitution rate constant may be greater or less than the k_{ex} value depending on the condition.
45. Based on the water exchange rate constants (k_{ex}), the metal ions have been classified into four classes.
46. The parameter *ionic potential* or Z^{*2}/r plays an important role in determining the k_{ex} values.
47. Among the alkaline earth cations, Ba^{2+}, Sr^{2+} and Ca^{2+} are in Class-I, Mg^{2+} is in Class-II while Be^{2+} is in Class-III.
48. In terms of the k_{ex} values, Hg^{2+} and Cd^{2+} are placed in Class-I while Zn^{2+} is placed in Class-II.
49. The pK_a–values of $[M(OH_2)_6]^{n+}$ (acting as the Brönsted acids) can be correlated with their k_{ex} values.
50. Cr^{2+} and Cu^{2+} are highly labile in terms of their k_{ex} values.
    - Trivalent lanthanides (Ln^{3+}) are labile.
    - V(II) and Ni(II) show the same CFAE but V(II) is more inert than Ni(II) in water exchange process.
51. The lability sequence is: Ni(II) > Pd(II) > Pt(II)
    - Ni^{2+} is a Class-II metal ion; Pd^{2+} is a Class-III metal ion and Pt^{2+} is a Class-IV metal ion.
52. The k_{ex} values run in the sequence:

$$V^{2+} \ll Cr^{2+} > Mn^{2+} > Fe^{2+} > Co^{2+} \gg Ni^{2+} \ll Cu^{2+} > Zn^{2+}$$

- $Cr^{2+} \approx Cu^{2+} \gg Mn^{2+} \approx Zn^{2+} > Fe^{2+} > Co^{2+} \gg Ni^{2+} > V^{2+}$

	Cr^{2+}	Cu^{2+}	Mn^{2+}	Zn^{2+}	Fe^{2+}	Co^{2+}	Ni^{2+}	V^{2+}
log k_{ex} : (25°C)	~9	~9.0	7.5	7.5	6.6	6.4	4.5	2.0

53. Generally, the Co(III) complexes are more inert than the Cr(III) complexes, but in terms of the k_{ex} values, the water exchange process in $Co(aq)^{3+}$ is much faster than that in $Cr(aq)^{3+}$.
54. Co(II) can catalyse the ligand substitution process in Co(III); Pt(II) can catalyse the ligand substitution process in Pt(IV); Cr(II) can catalyse the ligand substitution process in Cr(III).
55. Water exchange rate constants follows the sequece:

$$Cr(aq)^{3+} > Rh(aq)^{3+} \gg Ir(aq)^{3+}.$$

56. $[Co(phen)_3]^{3+}$ and $[Co(en)_3]^{3+}$ can survive for a long time even in a strongly acidic solution but $[Co(bigH)_3]^{3+}$ undergoes aquation slowly under the comparable conditions.
57. The ionic radii of Sc^{3+}, Ti^{3+}, V^{3+} and Fe^{3+} are comparable but their k_{ex} values differ significantly.

	$[Sc(OH_2)_6]^{3+}$	$[Ti(OH_2)_6]^{3+}$	$[V(OH_2)_6]^{3+}$	$[Fe(OH_2)_6]^{3+}$
log k_{ex} ~(25°C)	7.2	5.3	2.7	2.2

58. Eigen-Wilkins mechanism actually represents the stoichiometric interchange (I) mechanism.
59. A ligand substitution reaction passing through the interchange process shows the rate law:

$$k_{obs} \approx k_i K_{os}[Y],\ Y = \text{entering ligand.}$$

- On the other hand, a solvent exchange reaction passing through the interchange process gives the rate law:

$$k_{obs} = k_i = k_{ex}$$

(k_i, k_{ex}, K_{os} bear the usual significance).

60. $[Ni(OH_2)_6]^{2+} + L \rightarrow [Ni(L)(OH_2)_5]^{n+} + H_2O$.
In the above reaction, k_f (second order formation rate constant) varies with the nature of the entering ligand but k_f/K_{OS} remains more or less constant.
61. In the formation reaction, if the entering ligand is a didentate chelating ligand then the normal Eigen mechanism may operate or may not operate depending on the contribution of the ring closure step to the rate determining step.
62. $[Ni(OH_2)_6]^{2+}$ + L⌒L $\longrightarrow$ $[Ni(OH_2)_4(L\frown L)]^{n+}$ + $2H_2O$.

- In the above reaction, bpy reacts slower than phen.
- In the above reaction, α–ala⁻ reacts faster than β–ala⁻.

63. $[(A)Ni^{II}(OH_2)_2]^{x+} + gly^- \rightarrow [(A)Ni^{II}(gly)]^{y+} + 2H_2O$.
(**cis**–complex)
In the above ternary complex formation reaction, the normal Eigen mechanism does not operate. (A = a tetradentate and bulky ligand).
64. In the chelation reaction, sometimes the intermediate formed by the binding of the didentate ligand as a unidentate one may remain in **steady state** or **equilibrium state** depending on the conditions.
65. Both the D and I_d processes experience the **d**-activation. In the D-process, the rate constant shows some sort of dependence on the nature of the entering ligand but in the I_d process, the rate constant does not show any such dependence on the nature of the entering ligand.
66. The D and I_d processes can be distinguished by following the rate dependence on the nature of the entering ligand.
67. Both the D and I_a processes show the rate dependence on the nature of the entering ligand. They can be distinguished under some limiting conditions where the D-process becomes independent of the entering ligand but such a situation never arises for an I_a–process.
68. Proton ambiguity arises due to the presence of *isomeric reaction paths.*
69. Proton ambiguity arises generally for the trivalent metal ions prone to hydrolysis.
70. Proton ambiguity may be solved by considering the *reasonableness of the rate constant values*.
71. VBT assumes that the outer orbital complexes participate in a dissociative process for the ligand substitution reactions while the inner orbital complexes participate in an associative process.
72. The lability sequence is:
$AlF_6^{3-} > SiF_6^{2-} > PF_6^- > SF_6$.
73. The inner orbital complexes of d^3 and d^4 configurations can also participate in an associative process in ligand substitution reaction in terms of VBT.
74. The assumption, '*outer-orbital complexes are weaker than the inner orbital complexes*' is not justified always.

75. Sometimes, CFAE (crystal field activation energy) becomes negative.
    - In the associative (A) path of ligand substitution in the octahedral complexes, **face attack** leading to the 7-coordinate **trapezoidal octahedron** T.S. is energetically more feasible than the **edge attack** leading to the pentagonal bipyramidal T.S.
    - In the dissociative (D) path of ligand substitution in the octahedral complexes, the square pyramidal T.S. is energetically more probable but in some cases, it passes through the trigonal bipyramidal T.S.
76. Generally, the low spin complexes are more inert than the high spin complexes.
77. d^4 (h.s.) d^7 (l.s.) and d^9 systems are highly labile.
78. The heavier congeners are more inert than the lighter congeners for the *d*-block elements.
79. The associative character increases in moving down in a group while the associative character decreases in a period in moving from left to right for the *d*-block elements.
80. CFT also predicts that the higher valent metal ions are more inert.
81. Generally, with the increase of oxidation state, the lability decreases but V(III) is more labile than V(II).
82. The lability orders are: Co(III) > Rh(III) > Ir(III); Ni(II) > Pd(II) > Pt(II).
83. The low spin Fe(II) (d^6) complexes dissociate slower than the Ni(II) complexes but the high spin Fe(II) complexes dissociate faster than the Ni(II) complexes.
84. Generally the low spin Fe(II)–complexes dissociate slower than the Ni(II)–complexes because of the higher CFAE (*cf.* $4Dq$ for Fe^{II} *vs.* $2Dq$ for Ni^{II}), but low spin $[Fe(phen)_3]^{2+}$ and $[Fe(bpy)_3]^{2+}$ dissociate faster than the corresponding complexes of Ni(II).

$$\left.\begin{array}{l}\left[Fe(LL)_3\right]^{2+} \longrightarrow \left[Fe(LL)_2\right]^{2+} + LL;\ k_{Fe} \\ \left[Ni(LL)_3\right]^{2+} \longrightarrow \left[Ni(LL)_2\right]^{2+} + LL;\ k_{Ni}\end{array}\right\} k_{Fe} > k_{Ni}\ (LL = \text{bpy, phen})$$

    - For the square planar complexes of Pt(II) (d^8-system), the prediction from CFAE does not conform to the experimental finding.

85. $$\left.\begin{array}{l}\left[Fe(LL)\right]^{2+} \xrightarrow{k_{Fe}} Fe^{2+} + LL \\ \left[Ni(LL)\right]^{2+} \xrightarrow{k_{Ni}} Ni^{2+} + LL\end{array}\right\} k_{Fe} > k_{Ni};\ (LL = \text{phen, bpy, terpy})$$

86. $$\left.\begin{array}{l}\left[Cr(bpy)_3\right]^{2+} \xrightarrow{k_{Cr}} \left[Cr(bpy)_2\right]^{2+} + bpy \\ \left[Ni(bpy)_3\right]^{2+} \xrightarrow{k_{Ni}} \left[Ni(bpy)_2\right]^{2+} + bpy\end{array}\right\} k_{Cr} > k_{Ni}$$

    CFAEs for both the systems are more or less comparable but the Cr(II)–complex dissociates faster than the Ni(II)–complex.

87. $[V(phen)_3]^{2+}$ dissociates faster than $[Ni(phen)_3]^{2+}$ though their CFAEs are almost the same. Interestingly, the water exchange rate is much higher for Ni(II) than that for V(II).

    Lability order: $$\underbrace{\left[V(phen)_3\right]^{2+} > \left[Ni(phen)_3\right]^{2+}}_{\text{Dissociation process}};\ \underbrace{\left[V(OH_2)_6\right]^{2+} < \left[Ni(OH_2)_6\right]^{2+}}_{\text{Water exchange process}}$$

88. The water exchange rate follows the sequence:
    $[Sc(OH_2)_6]^{3+} > [Ti(OH_2)_6]^{3+} > [V(OH_2)_6]^{3+} > [Fe(OH_2)_6]^{3+} \gg [Cr(OH_2)_6]^{3+}$
89. The systems like d^0, d^1 and d^2 generally prefer to adopt the associative path. It can be explained in terms of both VBT and CFT.
90. The d^{10} systems are expected to be labile in terms of both VBT and CFT.
91. In terms of VBT, Ni(II) (d^8) is expected to be labile but in terms of CFT, it should be inert.
92. The CFAEs of Ni(II) and Cr(III) are more or less the same, but generally, the Cr(III) complexes react much slower than the Ni(II)–complexes.

93. Ni(II) centre is less labile than many other bivalent metal ions of first-transition series.
94. Generally, d^5-system reacts faster than d^4 system. But in the CN^- ligand exchange reactions of $[M(CN)_6]^{n-}$, the lability order runs as: $d^4 > d^5 > d^6$
Hints: Trapezoidal octahedral intermediate.
95. Effect of charge on the complex is straight-forward for the dissociative (*D*) and associative (*A*) paths but the effect is not so straight-forward for an interchange (*I*) process.
96. For the aquation of Cl^- in the chlorido-ammine complexes of Co(III) and Cr(III), the rate drastically increases for the Co(III) complexes with the decrease of overall positive charge on the complex, but the said change is less sensitive for the corresponding Cr(III) complexes.
97. The reactivity and choice of the mechanistic paths in the formation reactions of $[M(OH_2)_6]^{3+}$ and $[M(OH)(OH_2)_5]^{2+}$ (M = Cr^{III}, Fe^{III}) are totally different. This difference is not only due to the difference in overall charge of the complexes.
98. In the so called Eyring Equation:

$$-\ln\left(\frac{C_o^n kh}{k_B T}\right) = \frac{\Delta H^{\neq}}{RT} - \frac{\Delta S^{\neq}}{R}, \text{ (order of the reaction } n+1)$$

the assumption of transmission coefficient (κ) to be unity is not always justified.
99. Eyring equation is not applicable always correctly for the electron transfer process, $H/H^+/H^-$–transfer processes.
100. Tunneling effect (QMT) may complicate the interpretation of Eyring equation.
101. The entropy of activation ($\Delta S^{\neq}$) cannot be directly obtained from the intercept of the plot, $-\ln\left(\frac{C_o^n kh}{k_B T}\right)$ *vs.* $\frac{1}{T}$, though the enthalpy of activation ($\Delta H^{\neq}$) can be directly obtained from the slope of the plot.
102. $\Delta S^{\neq}$ (of T.S. theory) can be correlated with the preexponential factor (*A*) of Arrhenius equation.
103. For the reactions occurring in solution, $\Delta H^{\neq}$ may be roughly taken as the E_a (activation energy obtained from Arrhenius equation).
104. In the aquation of X^- in $[Co(NH_3)_5X]^{2+}$ (X^- = Cl^-, Br^-, NO_3^-, NO_2^-, NCS^-, etc.), the **enthalpy of transition** (ΔH_T) is more or less the same.
   - For the aquation of X^- in $[Co(NH_3)_5X]^{2+}$, the slope of the plot, ΔH_T *vs.* $\Delta H_{hydr}(X^-)$ is zero while for $[Cr(OH_2)_5X]^{2+}$ and $[Rh(NH_3)_5X]^{2+}$, slope of the corresponding plot is nonzero.
105. In the aquation of X^- in $[Co(NH_3)_5X]^{2+}$ (X^- = different uninegative ligands), $\Delta H^{\neq}$ bears a linear relationship with the $\bar{\nu}$ of $[Co(NH_3)_5X]^{2+}$ ($\bar{\nu}$ = wavenumber in cm^{-1} for the first absorption band in the electronic spectra).
106. In the aquation of $[M(NH_3)_5X]^{2+}$ (M = Co, Cr; X = Cl, Br, I), the $\Delta S^{\neq}$ values follow the sequence: $\Delta S^{\neq}$: $I^- > Br^- > Cl^-$.
107. In the solvent exchange reactions or formation reactions involving the replacement of a coordinated solvent molecule by a neutral ligand, $\Delta S^{\neq}$ value is a good parameter to diagnose the reaction pathway.
108. In solvent assisted dissociation (SAD) of $[M(NH_3)_5X]^{2+}$, solvation of the leaving group (X^-) is an important factor in attaining the transition state.
109. In a dissociative process, $\Delta S^{\neq}$ is expected to be positive while in an associative process, $\Delta S^{\neq}$ is expected to be negative but these straightforward conclusions may not operate always.
110. To interpret the value of $\Delta S^{\neq}$, it is very important to consider the possibility of *charge creation* (*i.e.* charge separation) or *charge destruction* (*i.e.* charge neutralisation) in attaining the transition

state.

111. To diagnose a reaction pathway, the small values of $\Delta S^{\neq}$ are not the reliable diagnositic parameters.
112. Reactions passing through a similar pathway may be characterised by their **isokinetic trends**.
113. Isokinetic trend arises from the mutual compensation effect of $\Delta H^{\neq}$ and $\Delta S^{\neq}$.
114. For aquation of $[Co(NH_3)_5X]^{2+}$, LFER holds good with unit slope.

$$[Co(NH_3)_5(-X)]^{2+} + H_2O \rightleftharpoons [Co(NH_3)_5(OH_2)]^{3+} + X^-$$

- LFER plot can be interpreted by considering the Arrhenius equation and van't Hoff equation showing the temperature dependence of rate constant and equilibrium constant respectively.

115. $[M(NH_3)_5X]^{2+} \rightarrow [M(NH_3)_5(OH_2)]^{3+} + X^-$
   - Slope of LEER plot: ~1.0 (for M = Co), < 1.0 (for M = Cr, Rh).
   - Aquation rate sequence: X = I > Br > Cl (for M = Co),
     X = Cl > Br > I (for M = Rh).
116. $[Cr(OH_2)_5X]^{2+} \rightarrow [Cr(OH_2)_6]^{3+} + X^-$; slope of LEER plot = ~0.55
   $[Cr(NH_3)_5X]^{2+} \rightarrow [Cr(NH_3)_5(OH_2)]^{3+} + X^-$; slope of LEER plot = ~ 0.90
117. Volume of activation ($\Delta V^{\neq}$) can be obtained from the pressure dependence of rate constant at constant temperature.

$$\textbf{Hints: } k \propto \exp\left(\frac{-P\Delta V^{\neq}}{RT}\right).$$

118. There are two important factors to determine the $\Delta V^{\neq}$ values: (i) change in the number of particles in attaining the T.S.; (ii) change of electrorestriction over the solvent in attaining the T.S.
   $\Delta V^{\neq} = \Delta V^{\neq}_{\text{intrinsic}} + \Delta V^{\neq}_{\text{solvent}}$
119. $\Delta V^{\neq}$ is not always a good parameter for the diagnosis of a reaction mechanism, but in some cases, it is a reliable parameter.
120. $\Delta V^{\neq}$ and $\Delta S^{\neq}$ bear a linear correlationship.
121. $[M(NH_3)_5L]^{3+} \rightarrow$ aquation of L (neutral ligand)
   $\Delta V^{\neq}$ = –ve for M = Cr; $\Delta V^{\neq}$ = +ve for M = Co.
122. Aquation of X^- in $[M(NH_3)_5X]^{2+}$ (M = Co, Cr):
   $\Delta V^{\neq}$ = –ve for both M = Co, Cr.
   - $[M(NH_3)_5X]^{2+} + H_2O \rightarrow [M(NH_3)_5(OH_2)]^{3+} + X^-$ (hydrated), ΔV = –ve.
   - For the aquation reaction $[M(NH_3)_5X]^{2+} + H_2O \rightarrow [M(MH_3)_5(OH_2)]^{3+} + X^-$, (M = Co and Cr), slope of the plot, $\Delta V^{\neq}$ *vs.* $\Delta V \approx 1.0$ (for M = Co) and ≈ 0.60 (for M = Cr).
   - Slope of the plot, $\Delta V^{\neq}$ (M = Cr) *vs.* $\Delta V^{\neq}$ (M = Co) ≈ 0.43.
123. $[Co(NH_3)_5(SO_4)]^+$ and $[Co(NH_3)_5X]^{2+}$ aquate through the dissociative path but their $\Delta V^{\neq}$ values are negative (*cf.* $\Delta V^{\neq}$ of k_{ex} for $[Co(NH_3)_5(OH_2)]^{3+}$ is positive).
124. - $\left[RuCl(NH_3)_5\right]^{2+} \xrightarrow[a\text{-activation}]{\text{(aquation)}} \left[Ru(NH_3)_5(OH)_2\right]^{3+} + Cl^-, \Delta V^{\neq} = -30 \text{ cm}^3 \text{ mol}^{-1}$
   - $\left[Ru(NH_3)_5(OH_2)\right]^{3+} + Cl^- \xrightarrow[a\text{-activation}]{\text{(anation)}} \left[RuCl(NH_3)_5\right]^{2+}, \Delta V^{\neq} = -20 \text{ cm}^3 \text{ mol}^{-1}$
   - Both the aquation and anation paths adopt the **a**–activation, but $\Delta V^{\neq}$ is less negative for the anation reaction.
125. The aquation rates of $[Co(NH_3)_5(L)]^{n+}$ differ significantly for the various types of leaving groups (L) while the formation rates of $[Co(NH_3)_5(L)]^{n+}$ from $[Co(NH_3)_5(OH_2)]^{3+}$ are more or less insensitive towards the nature of the entering group (L).
126. Anation rates of $[Ti(OH_2)_6]^{3+}$ and $[V(OH_2)_6]^{2+}$ drastically depend on the nature of the entering anion.

127. $[Cr(NH_3)_5(OH_2)]^{3+}$ and $[Cr(OH_2)_6]^{3+}$ adopt different paths for anation and their reactivities differ widely. This difference can be explained by considering the relative σ–basicity of NH_3 and H_2O; relative energies of the LUMOs.
128. Associative path is more favoured for the Cr(III)–complexes than for the low spin Co(III) complexes.
    - In terms of the energy of the LUMO, it is quite reasonable to expect a more associative character in the ligand substitution process of Cr(III) (t_{2g}^3) than that in the low spin complexes of Co(III) (t_{2g}^6).
129. Anation of $[Ti(OH_2)_6]^{3+}$ experiences a more associative character than that of $[V(OH_2)_6]^{2+}$.
130. $[V(OH_2)_6]^{2+} + X^- \rightarrow [V(OH_2)_5(X)]^+ + H_2O$; $X^- = Cl^-, N_3^-, NCS^-$
    k_f ($dm^3\ mol^{-1}\ s^{-1}$) = 1 (Cl^-): 2 (NCS^-): 10 (N_3^-).
131. Dissociative path experiences the **steric acceleration** while associative path experiences the **steric retardation**.
132. $[Cu(N—O)_2]$ (square planar complex) shows the amino acid exchange rate sequence: glycinate > sarcosinate > N, N–dimethylglycinate.
    - $[M(N—O)_3]^-$ (octahedral complex; M = Co, Ni) shows the amino acid exchange rate sequence: glycinate < sarcosinate < N, N–dimethylglycinate; (N—O = deprotonated form of the amino-acid)
133. Aquation rate follows the sequences:
    - *trans*–$[CoCl_2(cyclam)]^+$ < *trans*-$[CoCl_2(tet\text{-}b)]^+$
    - $[CoCl(NH_3)_5]^{2+} < [CoCl(MeNH_2)_5]^{2+}$
    - $[CrCl(NH_3)_5]^{2+} > [CrCl(MeNH_2)_5]^{2+}$
    - $[CoCl(dien)(tn)]^{2+} > [CoCl(dien)(en)]^{2+}$
    - $[CrCl(dien)(tn)]^{2+} \approx [CrCl(dien)(en)]^{2+}$
    - *cis*–$[Co^{III}Cl(en)_2(L)]^{n+}$ > *trans*–$[Co^{III}Cl(en)_2(L)]^{n+}$
      (L = spectator ligand with the π–donor properties; *e.g.* OH^-, Cl^-, NH_2^- etc.)
    - *cis*–$[Co^{III}Cl(en)_2(L^1)]^{n+}$ > *cis*–$[Co^{III}Cl(en)_2(L^2)]^{n+}$
      L^1 = π–donor ligand (*e.g.* Cl^-, OH^-)
      L^2 = π–acceptor ligand (*e.g.* NO_2^-, CN^-)
    - *trans*– $[Co^{III}Cl(en)_2(L)]^{n+}$ > *cis*–$[Co^{III}Cl(en)_2(L)]^{n+}$
      L = π–acceptor ligand (*e.g.* NO_2^-, CN^-)
134. $\underset{(CA)}{[M(OH_2)_6]^{3+}} + H_2O \rightleftharpoons \underset{(CB)}{[M(OH)(OH_2)_5]^{2+}} + H_3O^+$; M = Fe, Cr
    CB is more reactive than CA in ligand substitution process. CA adopts the I_a-path while CB adopts the I_d-path.
135. In an interchange process, we have the following relationships:
    - $k_i = \frac{6}{8}k_{ex}$; $k_i = \frac{k_f}{K_{OS}}$, k_f = 2nd order formation rate constant
    - $k_{lim} = k_i$, ($K_{OS}[Y] \gg 1$, Y = entering ligand).
    - k_i or k_{lim} can never exceed the k_{ex} value in the I_d process.
    - k_i or k_{lim} may be greater or less than k_{ex} depending on the conditions in the I_a process.
136. Anation of $[Co(NH_3)_5(OH_2)]^{3+}$: $k_{lim}/k_{ex} \leq 1.0$
    - Anation of $[M(NH_3)_5(OH_2)]^{3+}$: k_{lim}/k_{ex} can exceed 1.0, (M = Rh, Ir).
    - Formation reaction at $[Fe(OH_2)_6]^{3+}$: k_f/k_{OS} = no constancy
    - Formation reaction at $[Fe(OH)(OH_2)_5]^{2+}$: k_f/K_{OS} = constant and close to k_{ex}

- Anation of $[Co(CN)_5(OH_2)]^{2-}$ or $[RhCl_5(OH_2)]^{2-}$; the double reciprocal plot, $\frac{1}{k_{obs}}$ *vs.* $\frac{1}{[Y^{n-}]}$ is linear (Y^{n-} = entering ligand) with a finite intercept and slope; $\frac{\text{slope}}{\text{intercept}} = x(\text{say})$

  x depends on the nature of Y^{n-}

  (**Hints:** *D*–process.)

137. The relationship expressing the effect of ionic strength (I) on rate constant can be obtained from transition state theory and Debye-Huckel theory.

$$\log k_I \approx \log k_0 + 2QZ_A Z_B \sqrt{I}$$

138. The rate dependence pattern on ionic strength depends on the nature of charge bearing properties of the reactant species in a bimolecular reaction.
139. Deviation from the linear plot of log k vs. $\sqrt{I}$ is an indication of secondary salt effect.
140. To compare the rate constants, the reactions are to be carried out at a fixed and fairly high ionic strength to be maintained by the inert electrolytes.
141. Nitrate (NO_3^-, *e.g.* KNO_3) and perchlorate (ClO_4^-, *e.g.* $NaClO_4$) salts are generally used to maintain the ionic strength.
142. $[(H_3N)_5Co^{III}—O—C(=O)—C_6H_4—R]^{2+} \xrightarrow{\text{Base hydrolysis}} [(H_3N)_5Co^{III}—OH]^{2+} + R—C_6H_4—CO_2^-$

     Use of Hammett equation and determination of Hammett reaction constant (ρ) for different R–groups can decide whether the base hydrolysis occurs through the Co—O or O—C bond rupture.
143. $[(H_3N)_5Co^{III}—O—C(=O)—R]^{2+} \xrightarrow{\text{Base hydrolysis}} [(H_3N)_5Co^{III}—OH]^{2+} + R—CO_2^-$

     (R = alkyl groups)

     Use of Taft relationship and determination of Taft reaction constant (ρ*) can decide whether the Co—O or C—O bond rupture occurs in the above mentioned base hydrolysis reaction.
144. Solvent effect (*i.e.* dielectric constant effect) on the rate constant can tell us the charge bearing properties of the kinetically active species.
     - The plot of log k_ε *vs.* $\frac{1}{\varepsilon}$ is linear (ε = dielectric constant of the medium) and sign of the slope depends on the sign of the product of charges of the reactants in a bimolecular process.
145. $L_5M—OH_2 + Y \rightarrow L_5M—Y + H_2O$

     The above formation reaction occurring through the *D* or I_a path will slow the rate dependence on the nature of *Y*. But the *D*– and I_a paths can be distinguished by analysing the rate dependence pattern on [Y].
146. $[Co(NH_3)_5—X]^{2+} + H_2O \rightarrow [Co(NH_3)_5(OH_2)]^{3+} + X^-$

     Rate $\propto [Co(NH_3)_5(X)]^{2+}$

     Different mechanistic paths can lead to the above rate law but they can be discriminated by considering the factors controlling the rate.
147. The aquation rates follow the following sequences:
     - *cis*-$[CoCl_2(NH_3)_4]^+$ > *trans*-$[CoCl_2(NH_3)_4]^+ \gg [CoCl(NH_3)_5]^{2+}$.

- *trans*-$[CoCl_2(N—N)_2]^+$:
  N—N = *meso*-bn > *dl*-bn > pn > en.
- *trans*–$[CoCl_2(\text{tet-}b)]^+ \gg$ *trans*-$[CoCl_2(\text{cyclam})]^+$

148. *cis–trans*–$[CoCl(en)_2(L)]^{n+}$ aquates (*i.e.* replacement of Cl^- by H_2O)
- $k_{cis} > k_{trans}$, if L is a π-donor ligand, *e.g.* Cl^-, OH^-, NCS^-;
- aquation of *cis*-isomer leads to *cis*-product while aquation of *trans*-isomer leads to a mixture of *cis*- and *trans*-isomers for L = π-donor ligand.
- $k_{cis} \approx k_{trans}$, if L is a weakly π-donor ligand (*e.g.* H_2O, N_3^-) and L is a non-π-bonding ligand (*e.g.* NH_3); for L = NH_3, there is a 100% retention of configuration for both isomers.
- $k_{trans} > k_{cis}$, if L is a strong π-acceptor ligand (*e.g.* CN^-, NO_2^-) and there is a 100% retention of configuration for both the *cis*- and *trans*-isomers.

149. $[CoCl(en)_2(R—py)]^{2+}$ aquates and the aquation rate increases with the increase of basicity of the pyridine derivative.

150. $[Co(NH_3)_5X]^{2+}$ aquates and the aquation rate follows the order:

$$k_{aq}: \xrightarrow[\text{Ligand field strength}]{X = Br^- > F^- > NCS^- > NO_2^-}$$

- The LFER plot yields slope = 1.0
- ΔH_T (= $\Delta H^{\neq} - \Delta H$) remains more or less constant for different leaving groups and it is close to $\Delta H^{\neq}$ for the water exchange process of $[Co(NH_3)_5(OH_2)]^{3+}$.

151. $$\left[Co(NH_3)_5\left(O-\underset{\underset{O}{\|}}{C}-R\right)\right]^{2+} \xrightarrow{H_2O} \left[Co(NH_3)_5(OH_2)\right]^{3+} + RCO_2^-$$

Basicity of RCO_2^- covers a huge range (*ca.* by a factor 10^4) while the aquation rate covers a narrower span (*ca.* by a factor 10).

- $k_{aq}: \xrightarrow[\text{Increasing basicity}]{CF_3CO_2^- > CH_2ClCO_2^- > CH_3CO_2^- > Me_3CCO_2^-}$
- It supports an interchange pathway (*i.e.* I_a) where the nucleophilic attack by H_2O plays an important role to give the transition state.
- If a better nucleophile than H_2O becomes available, then the span of rate constants will cover a larger range.
- Relative rates of aquation: 100 (A) : 21 (B) : 10 (C) : 7.8 (D) : 3.8 (E). A = $[Co(NH_3)_5Cl]^{2+}$, B = *cis*-$[Co(en)_2(NH_3)Cl]^{2+}$, C = $[Co(trien)(NH_3)Cl]^{2+}$, D = $[Co(dien)(en)Cl]^{2+}$, E = $[Co(tetraen)Cl]^{2+}$.

152. Aquation of *trans*-$[Co(en)_2(NH_3)(X)]^{2+}$ (X^- = NO_3^-, Cl^-, Br^-)
k_{aq}: $NO_3^- > Br^- > Cl^-$; $\Delta H^{\neq} \approx E_a$; $NO_3^- > Br^- > Cl^-$; $\Delta S^{\neq}$: $NO_3^- > Br^- > Cl^-$
Hints: k_{aq} sequence is controlled by $\Delta S^{\neq}$ not by E_a; consider SAD mechanism.

153. Aquation of $[Cr(NH_3)_5X]^{2+}$: X^- = Cl^-, Br^-, I^-
k_{aq}: $I^- > Br^- > Cl^-$; $\Delta S^{\neq}$: $I^- > Br^- > Cl^-$
Hint: SAD—solvent assisted dissociation mechanism.

154. $[Co(NH_3)_5X]^{n+} + H_2O \rightarrow [Co(NH_3)_5(OH_2)]^{3+} + X^{m-}$
X^{m-} = Cl^-, Br^-, NCS^-, SO_4^{2-}
$\Delta V^{\neq}$ ($\approx \Delta V$) = –ve; $\Delta V^{\neq}$ (k_{ex} for $[Co(NH_3)_5(OH_2)]^{3+}$ = +ve).
In spite of the negative value of $\Delta V^{\neq}$, the process can be explained in terms of a dissociative path.

- The negative value of $\Delta V^{\neq}$ may be also explained in terms of an associative activation. However, this argument may be ruled out by comparing the $\Delta V^{\neq}$ values with the ΔV values.

155. $[Co(L)(NH_3)_5]^{3+} + H_2O \rightarrow [Co(NH_3)_5(OH_2)]^{3+} + L$
L = neutral leaving group like $O{=}SMe_2$, $O{=}CH(NH_2)$; $O{=}C(NH_2)_2$, etc.
$\Delta V^{\neq}$ ($\approx \Delta V$) = +ve
The observation is in conformity with the dissociative activation.

156. *cis–/trans–*$[Co(Cl)(en)_2(L)]^{n+} + H_2O \rightarrow [Co(en)_2(L)(OH_2)]^{n+1} + Cl^-$
L = nonleaving group
   - L = π-donor ligand (*e.g.* Cl^-, NH_2^-, NCS^-), there is a 100% retention of configuration for the *cis*-complex; tendency of isomerisation for the *trans*–complex: L = $OH^- > NCS^- > Cl^-$
   - L = π-acceptor ligand (*e.g.* NO_2^-, CN^-), a non-π-bonding ligand (*e.g.* NH_3), there is a 100% retention of configuration for both the *cis*- and *trans*-complexes.
   - *cis*-complex experiences 100% retention of configuration irrespective of the nature of the nonleaving group L.

157. $[CoCl(L)(N_4)]^{n+} + H_2O \rightarrow [CoL(N_4)(OH_2)]^{(n+1)+} + Cl^-$
N_4 = cyclam or tet–*b*; L = nonleaving group;
There is a 100% retention of configuration irrespective of the nature of nonleaving group (L).

158. Aquation of $[Co^{III}Cl(L)(N{—}N)_2]^{n+}$ to $[Co^{III}(L)(N{—}N)_2(OH_2)]^{(n+1)+}$
   - Isomerisation is accompanied with the positive $\Delta S^{\neq}$ value.
   - Retention of configuration is accompanied with the relatively less and negative $\Delta S^{\neq}$ value.

159. In some cases, aquation of *cis*–$[Co^{III}Cl(L)(N{—}N)_2]^{n+}$ may lead to isomerisation into the *trans*–complex to some extent.
Hint: Long lived 5-coordinate square pyramid intermediate converts into the TBP intermediate.

160. Aquation of *cis-/trans-*$[CoCl(en)_2(L)]^{n+}$:
If the dissociative path operates, then the following stereochemical results are expected.
   - Λ–*cis*–complex leads to Λ–*cis*–product (58.3%), Δ – *cis* – product (25%) and *trans*–product (16.7%), *i.e.* loss of optical activity due to both racemisation and isomerisation (considering the long-lived trigonal bipyramidal intermediate).
   - *trans*-complex gives 66.6% *cis*–product and 33.3% *trans*–product (considering the long lived trigonal bipyramidal intermediate).

161. Aquation of $[Cr(NH_3)_5X]^{2+}$ (X = Cl, Br, I, etc.)
   - k_{aq}: $I^- > Br^- > Cl^-$ and $\Delta S^{\neq}$: $I^- > Br^- > Cl^-$
   Hint: SAD path
   - Group replacement factor, $GRF\left(=\dfrac{k_{R—X}}{k_{R—Cl}}\right)$: $I^- > Br^- > Cl^- >> F^-$

   (**Hint:** Importance of bond breaking in the activation process)

162. Aquation of $[MCl(NH_3)_5]^{2+}$ (**I**) and $[MCl_2(NH_3)_4]^+$ (**II**)

$$\frac{k_{aq}(\text{II})}{k_{aq}(\text{I})} \approx 10^3\,(\text{M} = \text{Co}), \approx 10^1\,(\text{M} = \text{Cr})$$

163. Aquation of $[MCl(NH_3)_5]^{2+}$ (M = Co, Cr)
The plot of $\Delta V^{\neq}$(Co) *vs.* $\Delta V^{\neq}$(Cr) is linear, but the slope is far away from unity.

164. Hydrolysis of the Co(III)–ammine complexes show a two term rate law.
rate = k_A[complex] + k_B[complex][OH^-]
   - The first term (*i.e.* first-order path) is important at pH < 8.0 while the 2nd term (*i.e.* second order path) is important at pH > 8.0.

165. An electrophile can also catalyse the aquation of $[Co(N_5)—X]^{2+}$.
166. The complexes $[M^{III}(NH_3)_5X]^{n+}$ can experience an acid catalysed aquation (M = Co, Cr; X^{m-} = F^-, NO_2^-, N_3^-, CN^-, CO_3^{2-}, SO_4^{2-}, ox^{2-}, $acac^-$, etc.)
167. $[M^{III}(NH_3)_5X]^{2+}$ (M = Co, Cr) can experience an acid catalysed aquation for X^- = F^- but not for X^- = Cl^-, Br^-, I^-.
168. $[Co(bigH)_3]^{3+}$ is thermodynamically more stable than $[Co(en)_3]^{3+}$, but in acidic condition, $[Co(bigH)_3]^{3+}$ dissociates faster than $[Co(en)_3]^{3+}$.
169. The acid catalysed aquation rates of the biguanide complexes of Co(III) follow their thermodynamic stability orders.
    - k_{aq} and thermodynamic stability order: biguanide > phenyl-biguanide.
170. In acid catalysed aquation of $[M(bpy)_3]^{2+}$ (M = Ni, Fe), the rate process shows the acid dependence while the corresponding phen-complex, *i.e.* $[M(phen)_3]^{2+}$, shows no acid dependence of rate under the comparable conditions.
171. In acidic condition, $[M(phen)_3]^{2+}$ (M = Fe, Ni) experiences aquation. The rate of aquation process is independent of $[H^+]$ but the aquation can only occur in acidic condition.
172. Stepwise aquation (acid catalysed):

$$\left[Fe(N—N)_3\right]^{2+} \xrightarrow{k_1} \left[Fe(N—N)_2(OH_2)_2\right]^{2+} \xrightarrow{k_2} \left[Fe(N—N)(OH_2)_4\right]^{2+} \xrightarrow{k_3} \left[Fe(OH_2)_6\right]^{2+}$$

k_2, $k_3 \gg k_1$, N—N = bpy, phen

Hint: $[Fe(bpy)_3]^{2+}$ and $[Fe(phen)_3]^{2+}$-**low spin complex**; *bis*- and *mono*-complex—high spin.
    - For the corresponding Ni(II)–complex, no such step-wise rate order is noticed.
173. In the bpy–complexes, though the acid catalysed one-ended dissociation of the chelate ring is possible, it is not possible for the corresponding phen-complex.
    - For the acid catalysed dissociation of $[Fe(bpy)_3]^{2+}$ and $[Fe(phen)_3]^{2+}$, the acid dependence patterns of rates are different.
    - For the acid catalysed dissociation of $[Fe(bpy)_3]^{2+}$, the limiting values of k_{obs} at very high $[H^+]$ and low $[H^+]$ can tell us the kinetic behaviour of the 5-coordinate intermediate.
174. Different metal ions (as electrophiles) can catalyse the aquation of metal complexes bearing the chelating ligands like oxalate, malonate, acetylacetonate, aminopolycarboxylate, etc.
    - The catalysing efficiency of the metal ions depends on mainly two factors: HSAB matching and complexing power.
175. $\left[Co(NH_3)_5X\right]^{2+} \xrightarrow[(^{18}O \text{ enriched water solvent})]{Hg^{2+}} \left[M(NH_3)_5(OH_2)\right]^{3+}$

The product bears the same $^{16}O/^{18}O$ ratio for X^- = Cl^-, Br^-, I^-.
(Hint: D–process)
    - If the process is catalysed by the metal ions like Ag^+ or Tl^{3+}, the $^{16}O/^{18}O$ ratio in the product is found to depend on the nature of X^-.
    (Hint: I_d–process)
176. $\left[(H_3N)_5M—{}^*OH_2\right]^{3+} + HCO_3^- \rightleftharpoons \left[(H_3N)_5M(—{}^*O—CO_2^-)\right]^+ + H^+ + H_2O$

M = Co(III), Rh(III) and Ir(III).
    - The above mentioned reactions are basically the pseudo-substitution reactions.
    - The forward (*i.e.* formation reaction) and backward (*i.e.* aquation reaction) rate constants are more or less comparable for Co(III), Rh(III) and Ir(III).

176a. $[M(NH_3)_5X]^{(3-n)+} \xrightarrow{\text{aquation}} [M(NH_3)_5(OH)_2]^{3+} + X^{n-}$

- $X^{2-} = CO_3^{2-}$: k_{aq} values vary within a very narrow span for M = Co(III), Rh(III) and Ir(III).
- $X^- = Br^-, NO_3^-, CH_3CO_2^-$; $X^{2-} = SO_4^{2-}$: k_{aq} values cover a wide span for M = Co(III), Rh(III) and Ir(III).

177. The formation rate of $[M(CO_3)(NH_3)_5]^{2+}$ and its aquation rate are unexpectedly high for M = Co(III), Rh(III) and Ir(III).
178. When $[Co(NH_3)_5(OH_2)]^{3+}$ is treated with $HNO_2 - NO_2^-$ buffer, it rapidly produces a red coloured solution which on standing slowly changes into the yellow coloured solution.
179. $[M(NH_3)_5(OH_2)]^{3+} \xrightarrow[\text{(1st step)}]{HNO_2-NO_2^-} [M(NH_3)_5(ONO)]^{2+} \xrightarrow[\text{(2nd step)}]{} [M(NH_3)_5(NO_2)]^{2+}$

M = Co, Rh, Ir.

- The first-step leading to the nitrito isomer is unusually fast (compared to the expected inertness of the metal centre).
- The 2nd step leading to isomerisation of the nitrito isomer to the nitro isomer is a slow reaction and the corresponding rate is compatible with the expected CFAE of the metal centre.

180. Cl^-, Br^-, etc. can catalyse the aquation of $[Co(NH_3)_5(ONO)]^{2+}$ in acidic condition.

- $[(H_3N)_5Co(ONO)]^{2+} \xrightarrow{H_2O^*,\ H^+,\ X^-} [(H_3N)_5Co(OH_2)]^{3+} + HNO_2 + HX$

(No incorporation of *O in the product).

181. $[Co(NH_3]_5X]^{2+} \xrightarrow{^-OH} [Co(NH_3)_5(OH)]^{2+}$

Rate $\propto$ [complex][OH^-]

- The said rate law can be attained by considering both the A–process and D–CB process.
- For the said base hydrolysis reaction, **enthalpy of transition** ($\Delta H_T = \Delta H^{\neq} - \Delta H$) is independent of the nature of X^-.

182. In the base hydrolysis of $[Co(NH_3)_5X]^{2+}$, though it passes through a pre-equilibrium step involving OH^-, it never shows the saturation kinetics or fractional order in [OH^-].
183. Base hydrolysis of $[Co(NH_3)_5X]^{2+}$ occurs through the formation of a conjugate base (CB). The CB may be considered either to remain in a preequilibrium step or in a steady state. Both the concepts lead to the same rate law:
rate $\propto$ [complex][OH^-].
184. Base hydrolysis of $[Co(NH_3)_4(L)X]^{n+}$ leads to isomerisation for both the *cis*– and *trans*– isomers while acid hydrolysis of the complex leads to 100% retention of configuration for the *cis*–isomer and isomerisation in the *trans*–complex depending on the nature of the nonleaving group L.
185. Base hydrolysis of $[Co(NH_3)_4(L)X]^{n+}$ occurs at comparable rates for both the *cis*– and *trans*–isomers. On the other hand, in acid hydrolysis, the *cis*– and *trans*–isomer generally react at different rates and this rate difference largely depends on the nature of the nonleaving group L.
186. In acid hydrolysis of $[Co(NH_3)_5X]^{2+}$ or $[Co(L)(NH_3)_4X]^{n+}$, the 5–coordinate intermediate/activated complex may adopt the square pyramidal or trigonal bipyramidal geometry depending on the conditions but in base hydrolysis, the corresponding 5–coordinate intermediate/activated complex is always a trigonal bipyramidal.
187. If base hydrolysis of $[Co(NH_3)_5X]^{2+}$ is carried out in presence of other nucleophiles, there is a mixed product, *i.e.* $[Co(NH_3)_5(OH)]^{2+}$ and $[Co(NH_3)_5(Y)]^{2+}$, though there is no rate dependence on [Y^-].

$$[Co(NH_3)_5X]^{2+} \xrightarrow{^-OH,\ Y^-} [Co(NH_3)_5(OH)]^{2+} + [Co(NH_3)_5Y]^{2+}$$

$$\left(Y^- = N_3^-, NCS^-, \text{etc.}\right)$$

$$\text{rate} \propto [Co(NH_3)_5X^{2+}][OH^-][Y^-]^0$$

- Without OH^-, the entry of Y^- occurs very slowly.
- Deprotonation by OH^- preferably occurs from the NH_3 group *trans* to X (leaving group).
- The $-NH_2$ group in the CB favours the dissociative activation by its both σ- and π- donor properties; but the σ-donor property is more important than the π-donor property of the $-NH_2$ group.

188. The Co(III) complexes are generally more inert than the Cr(III)-complexes, but base hydrolysis of $[Co(NH_3)_5X]^{2+}$ is faster than that of $[Cr(NH_3)_5X]^{2+}$.
    - In terms of MOT and energy of the LUMO, an associative process is more disfavoured in the low spin Co(III) complexes than in the Cr(III) complexes.
189. Aquation of $[Co(NH_3)_5X]^{2+}$ is slower than that of $[Cr(NH_3)_5X]^{2+}$ but the reverse is true for base hydrolysis (X = Cl, Br^-, etc.).
190. Base hydrolysis of $[CoCl(NH_3)_5]^{2+}$ is much faster than that of $[Co(CN)_5Cl]^{3-}$ or $[CoCl(py)_5]^{2+}$.
191. Base hydrolysis of $[CoCl(NH_3)_5]^{2+}$ occurs through the D-CB path while base hydrolysis of $[CoCl(py)_5]^{2+}$ occurs through the I_d-path.
    - Base hydrolysis of $[CoCl(NH_3)_5]^{2+}$: rate $\propto$ [complex]$[OH^-]$
    - Base hydrolysis of $[CoCl(py)_5]^{2+}$: rate $\propto$ [complex]$[OH^-]^0$
    - Base hydrolysis of $[CoBr(CN)_5]^{3-}$: rate $\propto$ [complex]$[OH^-]^0$
192. For base hydrolysis of $[Co(NH_3)_5X]^{2+}$ or $[Co(NH_3)_4X_2]^+$, there are evidences to support the I_d–CB mechanism rather than the D–CB mechanism.
193. For $[Co(NH_3)_5X]^{2+}$, the base hydrolysis rate constant is much faster than the acid hydrolysis rate constant. This rate enhancement cannot be explained by simple consideration of the better nucleophilic character of the OH^- group than that of H_2O.
194. $[Co(NH_3)_5X]^{2+} + Y^- \xrightarrow{\text{(in DMSO)}} [Co(NH_3)_5(Y)]^{2+} + X^-$

    The reaction is tremendously slow but it can be potentially catalysed by a trace amount of base (B) like pipyridine or OH^-. The rate law is:
    rate $\propto$ [complex][B]$[Y^-]^0$.
195. $[Co(NH_3)_5X]^{n+} \xrightarrow{SCN^-,\ OH^-} [Co(NH_3)_5(OH)]^{2+} + [Co(NH_3)_5(Y)]^{2+}$

    X^{m-} = $OSMe_2$, Cl^-, Br^-, OSO_3^{2-} etc., Y = N– or S–bonded SCN^-
    - It is evident that with the increase of negative charge of the leaving group, % of $[Co(NH_3)_5(Y)]^{2+}$ decreases.

X^{m-} :	$OSMe_2$	I^-	OSO_3^{2-}
% of $[Co(NH_3)_5Y]^{2+}$:	~20%	~13%	~7%

    - This observation can be explained by the I_d–CB mechanism not by the A–mechanism.
    - The above observation indicates that the base hydrolysis occurs through the I_d–CB path not through the D–CB path.
    - **Competition results** in the products of base hydrolysis of $[Co(NH_3)_5X]^{2+}$ support the I_d–CB mechanism not the D–CB mechanism.
196. Anation of $[Co(NH_3)_5X]^{2+}$:

    $$[Co(NH_3)_5X]^{2+} + Y^- \rightarrow [Co(NH_3)_5(Y)]^{2+} + X^-$$

    The said anation reaction occurs in different paths in acidic and basic conditions.
197. $[Co(N)_5X]^{2+}$ experiences *steric acceleration* in both acid hydrolysis and base hydrolysis.
    - $[M(N)_5X]^{2+}$ (M = Cr, Rh, Ir) experiences steric retardation in acid hydrolysis while steric acceleration is noted in their base hydrolysis.
    - $(N)_5$ denotes the amine ligands differing in steric crowding; X^- = $^-OSO_2CF_3$.

198. Base hydrolysis of $[Co(NH_3)_5X]^{2+}$ is retarded by the addition of H_2O_2.
   - In base hydrolysis, addition of H_2O_2 produces a better nucleophile HO_2^- at the cost of OH^-, but still the process is retarded.

199. There are several factors to contribute to $\Delta V^{\neq}$ and $\Delta S^{\neq}$ in base hydrolysis of $[Co(NH_3)_5X]^{2+}$ through the D–CB mechanism.

200. Base hydrolysis of $[Co(NH_3)_5X]^{2+}$ is accompanied with the positive $\Delta V^{\neq}$ values.
   - If the base hydrolysis is carried out in presence of ^{18}O-labeled water, the ($^{18}O/^{16}O$) ratio in the product matches with that of water but not with that of hydroxide.
   - The reactivities of the two isomers (red and purple) of $[CoCl(NH_3)(tren)]^{2+}$ in base hydrolysis are different, *i.e.* the red isomer reacts faster than the purple isomer.
   - The base hydrolysis products of the red and purple isomers of $[CoCl(NH_3)(tren)]^{2+}$ are different.

201. 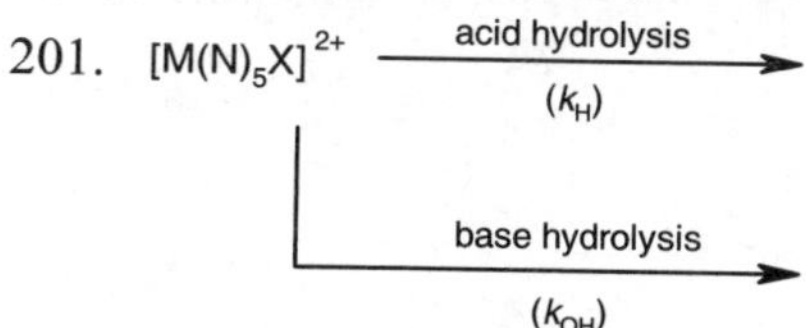

   - For M = Co, $\ln k_H$ and $\ln k_{OH}$ bear a linear relationship
   - For M = Cr, Rh and Ir; there is no linear relationship between $\ln k_H$ and $\ln k_{OH}$.
   - Stereochemical results for the aquation and base hydrolysis of $[CoCl_2(en)_2]^+$ are different. (**Hints:** Electronic effect (σ- and π-donor property) of the 'NH' group of CB; long-lived 5-coordinate intermediate in base hydrolysis).

202. *cis-/trans-*$[Co(en)_2(L)X]^{m+} \xrightarrow{\text{base hydrolysis}}$ Isomerisation for both the *cis-* and *trans*-isomers.
   - For acid hydrolysis of the complex, the stereochemical results are different.

203. $[M(NH_3)_5X]^{2+} \xrightarrow{\text{Base hydrolysis}}$
   - The base hydrolysis through the *D*–CB path can never show the saturation kinetics or mixed order kinetics with respect to $[OH^-]$.
   - The *D*–CB and *D*–IP paths are the two limiting situations of a common intermediate. The choice of the mechanism depends on the protonic character of the N—H bond and it depends largely on the electron pushing or electron withdrawing character of X^- in the case of $[Cr(NH_3)_5X]^{2+}$.

204. $[Cr(NH_3)_5X]^{2+} \xrightarrow{\text{Base hydrolysis}}$
   - X^- = NCS^-, it occurs through the D–IP mechanism, *i.e.* mixed order kinetics or saturation kinetics in $[OH^-]$.
   - X = N_3^-, it occurs through the D–CB mechanism, *i.e.* no mixed-order kinetics or saturation kinetics in $[OH^-]$.
   - D–IP path is quite unlikely in base hydrolysis of $[Co(NH_3)_5X]^{2+}$ but it may happen in some cases of $[Cr(NH_3)_5X]^{2+}$.

205. Base hydrolysis of *trans*-$[Rh(en)_2X_2]^+$

X^-: I^- SCN^- N_3^- Br^- Cl^-

$\xrightarrow{\text{favouring D–CB path}}$

$\xleftarrow{\text{favouring D–IP path}}$

206. Though $[M(OH_2)_6]^{3+}$ and $[M(OH)(OH_2)_5]^{2+}$ are correlated in terms of a protonation-deprotonation equilibrium, but their reactivities in ligand substitution reactions are totally different (M = Fe, Cr).

207. Isomerisation: (in both solid and solution phase) $\underset{\text{(Red)}}{\left[Co(NH_3)_5(ONO)\right]^{2+}} \xrightarrow{\text{heating}} \underset{\text{(yellow)}}{\left[Co(NH_3)_5(NO_2)\right]^{2+}}$

The process passess through an intramolecular path having the negative $\Delta V^{\neq}$ values (*cf.* Molar volumes for the *nitro-* and *nitrito*-somers are: 69.4 cm^3 mol^{-1} and 82.3 cm^3 mol^{-1} respectively).

208. Isomerisation: $[Co(NH_3)_5(SCN)]^{2+} \rightarrow [Co(NCS)(NH_3)_5]^{2+}$. It occurs through an intramolecular path ($\Delta V^{\neq}$ = –ve).

- $[(Et_4dien)Pd—XCN]^+ \longrightarrow [(Et_4\,dien)Pd—NCX]^+$

  $[(Et_4dien)Pd—XCN]^+ \xrightarrow[-XCN^-]{Br^-} [(Et_4dien)Pd—Br]^+$.

  $k_{isomer} \approx k_{Br} \approx k_{solvolysis}$, (X = S, Se)

  It indicates the intermolecular path of isomerisation.

209. $M—ONO \rightleftharpoons M—NO_2$; $M—SCN \rightleftharpoons M—NCS$

The S_Ni–path (*i.e.* intramolecular path) is more probable for the nitrito-nitro isomerisation, than for the thiocynato-isothiocynato isomerisation.

210. From the solution obtained by the interaction of dil. HCl with $[Co(CO_3)(en)_2]^+$, two different isomers (*i.e.* green and violet coloured) may be isolated depending on the conditions.

211. *cis*–$[CoCl_2(en)_2]^+$ slowly isomerises to its *trans*–isomer in solution through an intermolecular path.

- The rate of isomerisation is close to the rate of Cl^{*-} exchange reaction.
- The loss of optical activity of *cis*-$[M(A—A)_2X_2]$ in solution may occur in two ways.
- The rate of loss of optical activity of *cis*-$[M(A—A)_2X_2]$ may be close to or greater than the rate of geometrical isomerisation (*i.e.* *cis* → *trans* isomerisation) depending on the condition.

212. If *cis*–$[Co(N—N)_2(X)(Y)]^{n+}$ (optically pure) undergoes isomerisation in a dissociative process through a **long-lived trigonal bipyramidal intermediate**, it will yield the following results:

cis–(*ca.* 83%) and *trans*–(*ca.* 17%)

↓

$$\underbrace{50\%\ (\text{racemic}) + 33\%\ (\text{retention})}_{58\%\ (cis\text{-retention}) + 25\%\ (cis\text{-inversion}) + 17\%\ (trans)}$$

- If *trans*-$[Co(N—N)_2(X)(Y)]^{n+}$ undergoes isomerisation in a dissociative process through a **long-lived trigonal bipyramidal** intermediate, it will give the following results:

$$\underbrace{cis\text{-}(ca.\ 67\%)}_{\text{Racemic mixture}} \text{ and } trans.\ (ca.\ 33\%)$$

213. Isomerisation: *trans*-$\left[Cr(AA)_2(OH_2)_2\right]^- \rightleftharpoons$ *cis*-$\left[Cr(AA)_2(OH_2)_2\right]^-$

For A—A = malonate, $\Delta V^{\neq}$ is positive

For A—A = oxalate, $\Delta V^{\neq}$ is negative.

- The *cis-trans* isomerisation of $[Cr(C_2O_4)_2(OH_2)_2]^-$ can be catalysed by the bivalent metal ions (*e.g.* Cu^{2+}, Ni^{2+}, Co^{2+}, Ca^{2+}, etc.)
- The isomerisation and racemisation of *cis*-$[Cr(C_2O_4)_2(OH_2)_2]^-$ show $\Delta V^{\neq}$ = +ve (in neutral or slightly acidic condition), and –ve (in acidic condition).

214. Isomerisation of $[Co(CH_3COO)(en)_2(OH_2)]^{2+}$ is accompanied with positive $\Delta V^{\neq}$.
215. Racemisation of *cis*–$[M(L—L)_2X_2]$ occurs through the dissociative transformation to a trigonal bipyramidal geometry with the expulsion of one monodentate ligand X.
216. For $[Ni(N—N)_3]^{2+}$, the rate of racemisation, the rate of dissociation of *tris*– to *bis*–complex, and the rate of ligand exchange are comparable.
(N—N = bpy, phen)
217. For $[M(N—N)_3]^{2+}$, the rate of racemisation is greater than the rate of dissociation (*tris*- → *bis*-). M = Fe, Co; N—N = bpy, phen.
218. Racemisation of $[M(C_2O_4)_3]^{3-}$ (M = Co, Cr, Rh).
    - Rate of racemisation is greater than rate of oxalate exchange.
    - Rate of recemisation is more or less comparable to the O-exchange in acidic condition in ^{18}O-enriched water for $[Cr(C_2O_4)_3]^{3-}$.
    - The noncoordinating 6 oxygen atoms of $[M(C_2O_4)_3]^{3-}$ (M = Co, Rh) exchange faster than the remaining 6 oxygen atoms.
219. $K_3[Cr(C_2O_4)_3]\cdot 2H_2O$ can undergo racemisation when it is heated in a sealed tube but the racemisation is not at all efficient, if heated in an open tube.
220. $[Co(edta)]^-$ undergoes racemisation in an alkaline solution and the process is characterised by both the high positive values of $\Delta H^{\neq}$ and $\Delta S^{\neq}$.
221. For the $[M(A—A)_3]$ type chelates, Ray-Dutt twist leads to a transition state of C_{2v} symmetry while Bailar twist leads to a transition state of D_{3h} symmetry.
    - Ray-Dutt twist is described as rhombic or tetragonal twist and Bailar twist is described as a trigonal twist.
222. From the result, it is not possible to identify the Bailar twist or Ray-Dutt twist for the $[M(A—A)_3]$ type chelates but these can be distinguished for the $[M(A—B)_3]$ type chelates.
223. In the $[M(A—A)_3]$ or $[M(A—B)_3]$ type chelates, racemisation through the Bailar twist is favoured if the bite distance of the chelating ligands is small.
Hint: In such cases, the octahedral geometry is already distorted towards the required transition state.
    - Ray-Dutt twist is favoured for the chelating ligands of larger bite distance.
224. *cis*-$[MCl_2(en)_2]Cl$ (M = Co, Cr) undergoes racemisation in solid state when heated without experiencing any geometrical isomerisation. The kinetic parameters (*i.e. k*, $\Delta H^{\neq}$, $\Delta S^{\neq}$) for the racemisation are more or less comparable for the Co(III)- and Cr(III)-complexes. These findings can be explained by the Ray-Dutt twist mechanism.
    - For *tris*-(acetylacetylacetonato)cobalt(III) (where one CH_3 group is substituted by CD_3), both the linkage isomerisation and racemisation go on simultaneously but the linkage isomerisation rate is relatively slower.
225. The racemisation of the Cr(III)–complexes like $[Cr(bpy)_3]^{3+}$, $[Cr(phen)_3]^{3+}$, $[Cr(bpy)_2(ox)]^+$, $[Cr(ox)(phen)_2]^+$ etc. experience small positive $\Delta V^{\neq}$ while their anionic complexes like $[Cr(ox)_3]^{3-}$, $[Cr(ox)_2(phen)]^-$, $[Cr(bpy)(ox)_2]^-$ experience negative $\Delta V^{\neq}$ during racemisation.
226. Racemisation of $[M(N—N)_3]^{2+}$, M = Fe, Ni; N—N = bpy or phen.
    - Rate of dissociation (*tris* → *bis*) ≈ Rate of racemisation, (M = Ni)
    - Rate of dissociation (*tris* → *bis*) < Rate of racemisation, (M = Fe)
    - $[Fe(N—N)_3]^{2+}$ can partly racemise through the twist mechanism but it is not possible for $[Ni(N—N)_3]^{2+}$
    - In a strongly acidic condition, the residual rate of racemisation of $[M(N—N)_3]^{2+}$ is probably due to the twist mechanism.
    - $k_{rac} > k_{diss}$ indicates $k_{rac} = k_{rac\ (inter)} + k_{rac\ (intra)}$. The intramolecular path of racemisation for the bpy-complex can occur in two ways but for the phen-complex, it can occur only in one way.

227. There are many evidences that the d^8–systems can form the stable 5-coordinate complexes.
228. ML_3—X + Y → ML_3—Y + X
(ligand substitution process at the square planar complexes of Pt(II))
It shows a two term rate law:
rate = k_s [complex] + k_Y [complex][Y]
- k_s–path depends on the nature of the solvent not on the nature of the entering ligand Y.
- k_Y–path varies for the different entering ligands in a particular solvent.
- For a particular entering ligand Y, the k_Y value may differ for different solvents.

229. The rate of ligand substitution process in a square planar complex of Pt(II) depends on both the nature of entering ligand and leaving ligand.
- The rate process is strongly dependent on the nature of entering ligand while the process is mildly dependent on the nature of the leaving group.
- The ligand exchange rate of $[PtX_4]^{2-}$ follows the order: X^- = CN^- > I^- > Br^- > Cl^-.

230. *Trans*–effect and *trans*–influence are the two characteristic phenomena of the square planar complexes of Pt(II).
231. Ligand exchange reaction:
trans–$[PtCl_2(*NHEt_2)(PPr_3)]$ + $NHEt_2$ → *trans*–$[PtCl_2(NHEt_2)(PPr_3)]$ + $*NHEt_2$
In a noncoordinating solvent like hexane, the k_s–path makes no contribution to the overall rate process while in a coordinating solvent like methanol, the k_Y–path makes a negligible contribution to the overall rate process.
232. In the ligand substitution process of square planar complexes of Pt(II), the k_s–path (*i.e.* solvent dependent path) involves the formation of an intermediate having a solvent molecule as a ligand. In some cases, the proposed intermediate can be trapped.

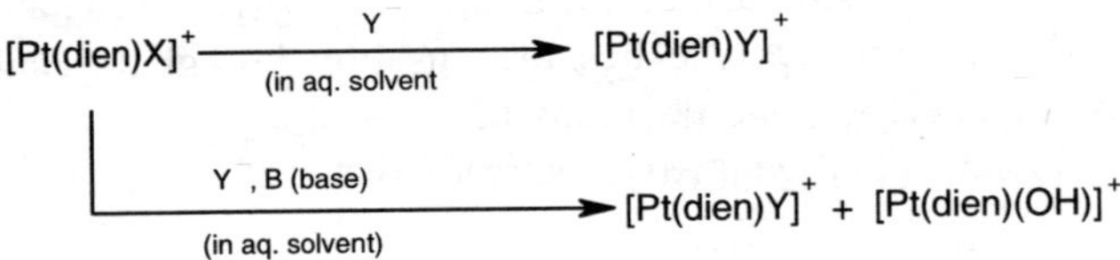

233. Compared to the Pt(II)–complexes, the Au(III)–complexes show the more associative character in the ligand substitution process.
234. A two term rate law operates for the following ligand substitution process.
$[MCl(dien)]^{n+}$ + Y^- → $[M(Y)(dien)]^{n+}$ + Cl^-
Rate = k_s [complex] + k_Y [complex]$[Y^-]$
$k_Y/k_s \gg 1$ [for M = Au(III)]; $k_Y/k_s \approx 1.0$ [for M = Pt(II)].
235. Retention of stereochemical configuration is an important characteristic feature of the ligand substitution process in the square planar complexes of Pt(II).
236. Electrophilic substitution and oxidative addition followed by reductive elimination may also lead to ligand replacement in some square planar complexes of Pt(II).
237. Racemisation of $[M(N—N)_3]^{2+}$: M = Fe, Ni; N—N = bpy, phen.
At high acidity: $k_{rac} \approx 0$ for $[Ni(N—N)_3]^{2+}$; $k_{rac} > 0$ for $[Fe(N—N)_3]^{2+}$.
- For $[Fe(N—N)_3]^{2+}$, racemisation can partly occur through the intramolecular twist mechanism but this remains absent for $[Ni(N—N)_3]^{2+}$.
- There are two possible intramolecular routes for racemisation of $[Fe(bpy)_3]^{2+}$ but for $[Fe(phen)_3]^{2+}$, there is only one possible intramolecular route of racemisation.

238. Nucleophilicity is a kinetic parameter while basicity is a thermodynamic parameter.
239. The nucleophilicity parameters n^0_{Pt} and $n^0_{CH_3I}$ do not run always parallel.

240. The nucleophilicity parameter n^0_{Pt} is defined as:

trans–$[PtCl_2(py)_2]$ + Y → *trans*–$[PtCl(py)_2(Y)]^{n+}$ + Cl^-

$$sn^0_{Pt} = \log\left[\frac{k_Y}{k_{CH_3OH}/[CH_3OH]}\right]$$

- Plot of log k_Y *vs.* n^0_{Pt} gives a straight line whose slope is given by *s* (nucleophilic discrimination constant, NDF).
- The larger value of *s* (NDF) indicates that the reaction is more sensitive to the change of nucleophilicity of the entering ligand.

s–value of the reacting complexes changes as follows:

trans-$[PtCl_2(PEt_3)_2]$ > *trans*-$[Pt(AsEt_3)_2Cl_2]$ > *trans*-$[PtCl_2(py)_2]$ > $[PtCl_2(en)]$.

241. The ligands like CN^-, SCN^- are the poor leaving groups while the halides are the better leaving groups in the substitution reactions of square planar complexes of Pt(II).

242. $[PtCl(L)(PEt_3)_2]^+$ + Y → $[Pt(L)(PEt_3)(Y)]^{n+}$ + Cl^-

L = pyridine and its substituted derivatives.

- Both the *cis*- and *trans*-complexes show the steric retardation in the substitution process, *i.e.* with the increase of bulkiness of L, the rate decreases.
- The *cis*–complex is more sensitive to the steric crowding on L than the *trans*-complex in the said substitution reaction.

243. $[Pt(dien)X]^+ + Y^- \rightarrow [Pt(dien)(Y)]^+ + X^-$

$X^- = Cl^-, Br^-, I^-, N_3^-$; $Y^- = OH^-, I^-, N_3^-, NO_2^-$,

- $\Delta V^{\neq}$ is negative for both the k_s and k_Y paths; but $\Delta V^{\neq}$ is more negative for the k_s-path.
- The observations are in conformity with the I_a mechanism rather than with the A-mechanism.

244. In the Pt(II)-square planar complexes, the ligand substitution reaction or ligand exchange reaction is almost insensitive to the change of overall charge of the complex.

- *Cl^--exchange rate in $[PtCl(NH_3)_3]^+$, *trans*-$[PtCl_2(NH_3)_2]$ and $[PtCl_4]^{2-}$ covers the range 1.5×10^{-5} s^{-1} to 3.8×10^{-5} s^{-1}, though the overall charge changes by 3 unit (+1 to –2). This observation is not in conformity with the pure D- or A-process.
- Pure **associative (*A*) path** is known to occur for the following reaction of square planar complex of Rh(I).

$$[Rh(cod)(Cl)(SbR_3)] + py \rightarrow [Rh(cod)Cl(py)] + SbR_3.$$

- The ligand exchange rate for the $[PtX_4]^{2-}$ complexes follows the order:

$$[Pt(CN)_4]^{2-} > [PtI_4]^{2-} > [PtBr_4]^{2-} > [PtCl_4]^{2-}$$

(**Hints:** Compare the relative effects of the entering group and leaving group on the rate process).

245. In terms of CFAE, formation of the square pyramidal geometry as the activated complex/intermediate in the ligand substitution process of the square planar Pt(II)–complexes is more feasible, but in reality, the process involves the trigonal bipyramidal geometry.

246. For the square planar complexes, the lability sequence runs as:

$d^{10} \gg d^9 > d^8 > d^7$, *e.g.* Zn(II) > Cu(II) > Ni(II).

247. For the square planar complexes of d^8–system, the lability sequence runs as: Ni(II) > Pd(II) > Pt(II).

248. For the *cis-trans* isomerisation in the square planar complexes like $[PtCl_2(PR_3)_2]$:

- The isomerisation process is catalysed by a trace amount of PR_3.

- Ligand replacement process in the square planar complexes of Pt(II) is very much stereospecific. Hence, to explain the isomerisation process, we must consider a double displacement process.
- For the relatively more labile complexes $[MCl_2(PR_3)_2]$ (M = Pd, Ni), the *cis-trans* isomerisation can occur in solution even without the addition of any nucleophile.
- The corresponding Ni(II) complex may isomerise through an intramolecular path.

249. *Stereospecificity, trans*-effect and *trans*-influence—these are the three important characteristic features of the reactivity of square planar complexes of Pt(II).

250. The *trans*–effect series is constructed by comparing the relative rates of ligand substitution at the *trans*–position of the ligands.
   - Thus the construction of *trans*–effect series is not absolute because the labilisation at the *trans*–position is determined by the leaving group, *trans*–directing group and incoming group jointly ***in an interdependent process*** in a particular complex.
   - *trans*–$[PtCl(PEt_3)_2(T)] + Y^- \rightarrow$ *trans*–$[Pt(PEt_3)_2(T)Y] + Cl^-$
     (i) For $Y^- = N_3^-$ or Br^-, *trans*-effect series of T: $CH_3^- > C_6H_5^- \rangle\ Cl^-$
     (ii) For $Y^- = I^-$, *trans*-effect series of T: $CH_3^- > Cl^- > C_6H_5^-$

251. In $[PtCl_3(T)]^{n-}$; the Pt—Cl bond length changes as follows with the change of its *trans*-ligand (T). (*trans*- to T): $PEt_3 > C_2H_4 > Cl^-$
   (**Hints:** It is due to the *trans*-influence).

252. *Trans*–influence (a thermodynamic effect) is referred to as the static *trans*-effect (STE).
   - *Trans*-influence can also explain the *trans*-effect (kinetic phenomenon) through the ground state destabilisation.
   - *Trans*-effect can also be explained through the stabilisation of transition state.

253. Knowledge of both the *trans*-effect series and metal-ligand bond strength are required to design the routes of synthesis of different isomers of Pt(II)-square planar complexes.

254. The *cis/trans*–isomers of $[PtCl_2(NH_3)_2]$ can be distinguished by following their substitution reactions with thiourea and thiosulfate.

255. The *cis/trans*-effect, *i.e.* lowering of activation energy can arise from two factors: ground state destabilisation, *i.e.* static kinetic effect or *trans*–influence and stabilisation of the transition state.

256. Grinberg's polarisation theory can explain the *trans*–influence better than the *trans*–effect.
   - Grinberg's polarisation theory can explain the *trans*–effect series: $I^- > Br^- > Cl^-$; very high *trans*-directing power of H^- (assuming the *trans*-influence to cause the *trans*-effect).
   - Among Ni(II), Pd(II) and Pt(II), only the Pt(II) centre shows the *trans*-influence and *trans*-effect.

257. The π-bonding ligands like C_2H_4, CN^-, CO, etc., show a weak *trans*-influence but a strong *trans*-effect.

258. In terms of *trans*-influence, the ligands like H^- and H_2O differ marginally but their *trans*–labilising powers differ drastically.

259. The π-acid ligands C_2H_4, CO, CN^-, etc. are the very powerful *trans*-directing ligands.
   - The π-acid ligands show both the *trans*–influence and *trans*-effect.
   - The π-acid ligands are the poor leaving groups.
   - The highly electronegative groups are the weak *trans*-directing ligands but the low electronegative groups are the better *trans*-directing ligands.
     (**Hint:** *See* Cardwell's theory).

260. The strong σ-bonding ligands like H^-, CH_3^- etc. are the powerful *trans*-directing ligands.
- The σ-bonding ligands can also show the ground state *trans*-effect.
- Compared to the π-bonding theory, the σ-bonding theory is more general to explain the *trans*-effect of the Pt(II) complexes.

261. A *trans*-directing group can labilise (in general) its both *cis*- and *trans*-positions in the square planar complexes of Pt(II).
- For the π-bonding ligand like C_2H_4, the *trans*-effect is more pronounced than *cis*-effect.

262. For a particular ligand, its *trans*-effect is more pronounced than its *cis*-effect in the square planar complexes of Pt(II).
- The static *trans*-effect (STE) bears a good correlationship with the kinetic *trans*-effect (KTE) for the octahedral Co(III)-complexes but such a good correlationship is not maintained for the square planar complexes of Pt(II).

263. Origin of the *cis*-effect and *trans*-effect is the same for the square planar complexes of Pt(II).

264. *cis*-$[PtMe_2(OSMe_2)_2]$ can experience the dissociative path in ligand substitution reactions.
- Sterically crowded square planar complexes of Pt(II) bearing some strong σ-bonding ligands like R^- may undergo the dissociative path for ligand substitution.
- Strong σ-*trans* effect is the key factor for the **dissociative path** of ligand substitution in the square planar complexes of Pt(II).
- *cis*-$[PtR_2(S)_2]$ and *cis*-$[PtR_2(S)(PR_3)]$ (S = R_2SO, R_2S) **adopt the dissociative path not the associative path** in the ligand substitution process.

265. • $[Ni(CO)_4]$ experiences the ligand exchange:

  rate ∝ $[Ni(CO)_4]$; $\Delta V^{\neq}$ and $\Delta S^{\neq}$ = +ve.

  $\Delta H^{\neq}$ is close to the Ni—CO average bond energy.

- $$[Ni(CO)_4] \xrightarrow{PR_3\ (excess)} [Ni(CO)_{4-x}(PR_3)_x]\ (x = 1, 2)$$

  $$[Ni(CO)_4] \xrightarrow{PF_3\ (excess)} [Ni(CO)_{4-x}(PF_3)_x]\ (x = 1, 2, 3, 4)$$

- $[FeBr_4]^-$ experiences the ligand exchange:

  rate ∝ $[FeBr_4^-][X^-]$; $\Delta V^{\neq}$ and $\Delta S^{\neq}$ = –ve

- The 18e–complexes like $[Ni(CO)_4]$, $[Ni(CO)_2(PR_3)_2]$ participate in dissociative paths in the ligand substitution process.
- Very often, the nitrosyl–carbonyl mixed ligand complexes like $[Co(NO)(CO)_3]$, $[Fe(CO)_2(NO)_2]$ adopt the associative path in the ligand substitution process.
- $[Ni(CO)_4]$ adopts the dissociative path while $[Co(NO)(CO)_3]$, $[Fe(CO)_2(NO)_2]$ adopt the associative path. This difference can be explained in terms of *effective electron count, metal-ligand bond strength and variable coordinating behaviour of NO.*
- Some metal nitrosyl-carbonyl complexes like $[Co(CO)_2(NO)(PR_3)]$ can experience both the *dissociative* and *associative* paths simultaneously in the ligand substitution process.
- Very often, it is a very difficult task to resolve the optically active pure tetrahedral complexes.
- Optically active tetrahedral complexes may undergo racemisation through both the intermolecular (both dissociative and associative) and intramolecular (*i.e.* twisting) paths.
- Optically active C-compounds are well known but optically pure Si-compounds are rarely known.
- $[M(CO)_6]$ (M = Cr, Mo) + PR_3 (excess) → *fac*-$[M(CO)_3(PR_3)_3]$ + 3CO.

266. Macrocyclic effect (*i.e.* stability of the complexes for the macrocyclic ligand *vs.* open-chain analogous ligand) rests on both the entropic and enthalpic favour.

267. Macrocyclic complexes are more resistant (kinetically) towards the dissociation compared to the corresponding open-chain complexes.
   - $[Cu(cyclam)]^{2+}$ dissociates much more slowly than $[Cu(3,2,3\text{-tet})]^{2+}$ in acidic media.
   - For the dissociation of $[Cu(cyclam)]^{2+}$, the acid dependent path is more important than in the case of $[Cu(3,2,3\text{-tet})]^{2+}$.

Inorganic Reaction Mechanisms and the World's Most Felicitated Coordination Chemists

Understanding of the reaction mechanism in organic chemistry is quite systematic but the inorganic chemists are yet to achieve that level. In fact, it is a very difficult task to systematize the versatile types of reactivities of more than one hundred elements. However to explore the reaction mechanisms of the metal complexes including the organometallics, a large number of world famous coordination chemists have contributed. Some of them are: M. Eigen (**Noble Prize, 1967**), F. Basolo, R.G. Pearson, R.G. Wilkins, H. Taube (**Nobel Prize, 1983**), H.B. Gray, C.H. Langford, R.A. Marcus (**Nobel Prize, 1992**), M.L. Tobe, J.H. Espenson, A.G. Sykes, R.D. Cannon, R. van Eldik, T.W. Saddle, A.W. Adamson, E.S. Gould, A.M. Sargeson, D.W. Maegerum, J.C. Bailar, P. Ray, etc.

Reaction Mechanism: Electron Transfer Reactions

6.1 TYPES OF ELECTRON TRANSFER REACTIONS

- **Electron exchange reactions:** There are some electron transfer reactions in which the reactants and products are identical and there is **no net chemical change**. Such reactions are called **electron exchange reactions.** In such cases, $\Delta G^0 = 0$, *i.e.* equilibrium constant = 1. Such reactions can be studied by isotope labelling experiment. This is illustrated in the following example.

$$[Fe(OH_2)_6]^{3+} + [Fe^*(OH_2)_6]^{2+} \rightleftharpoons [Fe^*(OH_2)_6]^{3+} + [Fe(OH_2)_6]^{2+}$$

Examples of Electron Exchange Reactions

$[Fe(CN)_6]^{3-} - [Fe(CN)_6]^{4-}$; $[Co(en)_3]^{3+} - [Co(en)_3]^{2+}$; $[Co(phen)_3]^{3+} - [Co(phen)_3]^{2+}$; $[Fe(bpy)_3]^{3+} - [Fe(bpy)_3]^{2+}$; $[Os(bpy)_3]^{3+} - [Os(bpy)_3]^{2+}$; $[Mo(CN)_8]^{3-} - [Mo(CN)_8]^{4-}$

- **Net redox reactions:** There are some reactions where electron transfer leads to a **net chemical change.** Such reactions are commonly described as the **redox reactions** and for such reactions, $\Delta G^0 = -ve$, *i.e.* $K_{eq} > 1$. It is illustrated in the following examples.

$$\left[Fe(phen)_3\right]^{3+} + \left[Fe(CN)_6\right]^{4-} \rightleftharpoons \left[Fe(phen)_3\right]^{2+} + \left[Fe(CN)_6\right]^{3-}$$

Examples of Redox Reactions

$[Mo(CN)_8]^{3-} - [Os(bpy)_3]^{2+}$; $[IrCl_6]^{2-} - [Fe(CN)_6]^{4-}$; $[Mo(CN)_8]^{3-} - [Fe(CN)_6]^{4-}$; $[Fe(CN)_6]^{3-} - [W(CN)_8]^{4-}$; $[Co(NH_3)_5Cl]^{2+} - [Cr(OH_2)_6]^{2+}$; $[Ru(NH_3)_6]^{3+} - [V(OH_2)_6]^{2+}$; etc.

6.2 MECHANISMS OF ELECTRON TRANSFER REACTIONS

There are two established mechanistic paths: **outer sphere mechanism** and **inner sphere mechanism.**

- **Outer sphere electron transfer (OSET) mechanism:** Electron transfer takes place from the reductant to the oxidant keeping the coordination spheres of both the reactants intact. In this path, at no stage of the electron transfer process, the two reactive species remain chemically bonded through the coordination spheres.
  Such a mechanism generally operates for the substitutionally inert complexes.
- **Inner sphere electron transfer (ISET) mechanism:** During the electron transfer process, oxidants and reductants remain chemically bonded in a binuclear complex through a bridging ligand. Generally, the substitutionally inert oxidant centre provides the bridging ligand for making a

bridge with the labile reductant centre. Thus in this route, the coordination spheres of both the reactants are modified during the course of electron transfer.

Nobel Prizes: Electron Transfer Reactions

H. Taube: In understanding the mechanism of electron transfer process, **H. Taube** and his coworkers made the most significant contribution. This contribution was recognized by the **Nobel Prize in 1983 to H. Taube.**

R. A. Marcus: He was given the **Nobel Prize** in 1992 for his contribution to the theory of electron transfer reactions.

6.3 OUTER SPHERE ELECTRON TRANSFER (OSET) PROCESS

6.3.1 Steps of Outer Sphere Electron Transfer Reaction

For the substitutionally inert metal complexes, in the OSET process, the oxidant and reductant complexes never chemically bind through their coordination spheres. The electron transfer process may be explained in terms of a **simple collision model** in which the required steps are outlined below.

- **Step-I (Formation of a precursor cage or outer sphere complex):** The reactants, *i.e.* oxidant (*Ox*) and reductant (*Red*), diffuse together to produce an **Outer Sphere** (O.S.) **complex** keeping their inner coordination spheres unchanged. This step is fast. In this O.S. complex, an electrostatic interaction as in ion-pair, ion-dipole interaction, dipole-dipole interaction, etc. can operate.

$$\mathrm{Ox} + \mathrm{Red} \rightleftharpoons [\mathrm{Ox} \cdots\cdots \mathrm{Red}], \ (\textbf{O.S. Complex})$$

- **Step II (Chemical activation of the precursor complex):** Bond distances and bond angles around the each metal centre (*i.e.* in each inner coordination sphere) within the O.S. complex change to attain the geometrical parameters as in the products. *This step basically reorganizes (through the bond length and bond angle changes) the inner coordination spheres of each metal centre towards those of the products.*

At this step, the *solvent shell around the outer sphere complex is also reorganized.* This step may be represented as follows:

$$[\mathrm{Ox} \cdots\cdots \mathrm{Red}] \rightleftharpoons [\mathrm{Ox} \cdots\cdots \mathrm{Red}]^{\neq} \ (\textbf{Activated O.S. Complex})$$

The required reorganization energy needed for chemical activation of the reactants (*i.e.* oxidant and reductant) will be consumed for:

bond length changes; bond angle changes (if the stereochemical configurations of the reactants and products differ); ***spin state change*** (if the spin states of the products and reactants differ); ***solvent reorganization in the outer sphere.***

Obviously, for the electron exchange in the couple, **$[Fe(CN)_6]^{3-}$ – $[Fe(CN)_6]^{4-}$** (where there is no spin state change; there is no change in geometrical configurations, *i.e.* both the oxidized and reduced centres are octahedral), chemical activation (*i.e.* reorganization) will **mainly involve the bond length adjustment.**

- **Step III (Electron transfer and relaxation to the successor O.S. complex):** Electron transfer occurs within the activated outer sphere complex $[\mathrm{Ox}\cdots\mathrm{Red}]^*$, *i.e.*

$$[\mathrm{Ox} \cdots\cdots \mathrm{Red}]^{*} \rightleftharpoons [\mathrm{Ox}^{-} \cdots\cdots \mathrm{Red}^{+}] \ (\textbf{Successor O.S. Complex})$$

- **Step IV (Dissociation of the successor O.S. complex to the products):** The products diffuse away from the successor O.S. complex and this step is generally fast.

$$[\mathrm{Ox}^{-} \cdots\cdots \mathrm{Red}^{+}] \rightleftharpoons \mathrm{Ox}^{-} + \mathrm{Red}^{+}$$

- **Illustration of the OSET process in the $[Fe(phen)_3]^{3+}$ – $[Fe(CN)_6]^{4-}$ system:** The above steps are illustrated in the outer-sphere oxidation of $[Fe(CN)_6]^{4-}$ by $[Fe(phen)_3]^{3+}$.

Formation of the precursor O.S. complex (rapid step): The ion-pair formation is electrostatically highly favoured here.

$$\left[Fe(phen)_3\right]^{3+} + \left[Fe(CN)_6\right]^{4-} \rightleftharpoons \underset{\textbf{Precursor O.S. complex (ion-pair)}}{\left[Fe(phen)_3\right]^{3+} \cdot \left[Fe(CN)_6\right]^{4-}}; (K_{OS}\text{ very high})$$

Chemical activation followed by the electron transfer (*rds*):

$$\left[Fe(phen)_3\right]^{3+} \cdot \left[Fe(CN)_6\right]^{4-} \rightleftharpoons \underset{\textbf{Successor O.S. Complex}}{\left[Fe(phen)_3\right]^{2+} \cdot \left[Fe(CN)_6\right]^{3-}}$$

Fission of the successor O.S. complex (rapid step):

$$\left[Fe(phen)_3\right]^{2+} \cdot \left[Fe(CN)_6\right]^{3-} \rightleftharpoons \left[Fe(phen)_3\right]^{2+} + \left[Fe(CN)_6\right]^{3-}$$

● **OSET process in the $[Co(NH_3)_5(OH_2)]^{3+}$ – $[Fe(CN)_6]^{4-}$ system:** Formation of the O.S. complex (*i.e.* precursor O.S. complex) may be sometimes electrostatically favoured as in the present case. In the reduction of $[Co(NH_3)_5(OH_2)]^{3+}$ by $[Fe(CN)_6]^{4-}$, the K_{OS} value is as high as 1500 dm³ mol⁻¹.

$$\left[Co(NH_3)_5(OH_2)\right]^{3+} + \left[Fe(CN)_6\right]^{4-} \rightleftharpoons \underset{\text{O.S. Complex}}{\left[Co(NH_3)_5(OH_2)\right]^{3+} \cdot \left[Fe(CN)_6\right]^{4-}}, \; K_{OS} \approx 1500 \text{ dm}^3\text{mol}^{-1}$$

$$\downarrow \text{e-transfer (slow, rds)}$$

$$\text{Products} \xleftarrow{\text{fast}} \left[Co(NH_3)_5(OH_2)\right]^{2+} \cdot \left[Fe(CN)_6\right]^{3-}$$

The rate constant of the said OSET reaction is 2×10^{-1} s^{-1}.

● **Energetics of electron exchange reaction:** In the *exchange reaction* (where $\Delta G^0 = 0$), the free energy of activation ($\Delta G^{\neq}$) may be expressed as follows:

$$\Delta G^{\neq} = \Delta G_r^{\neq} + \Delta G_i^{\neq} + \Delta G_s^{\neq}$$

$\boldsymbol{\Delta G_r^{\neq}}$ = energy required to bring the oxidant and reductant at a state separated by the required distance r at which the electron transfer occurs; it will take care of **electrostatic interaction** between the reactants.

$\boldsymbol{\Delta G_i^{\neq}}$ = energy required for **chemical activation** to change the bond lengths and bond angles, *i.e.* energy required to reorganise the inner coordination spheres of the reactants.

$\boldsymbol{\Delta G_s^{\neq}}$ = energy required for **solvent reorganization** in attaining the activated O.S. complex (*i.e.* transition state).

Formation of the precursor O.S. complex and specially the changes of bond length and bond angle in the precursor O.S. complex need the solvent reorganization. This solvent reorganization makes a significant contribution to $\Delta G^{\neq}$.

(**Note:** For the cross-reactions having $\Delta G^o \neq 0$, ΔG^o also contributes to $\Delta G^{\neq}$; *cf.* Marcus theory)

● **Rate law for the OSET path:** The process can be simply represented as follows:

$$\text{Ox} + \text{Red} \xrightleftharpoons{K_{OS}} [\text{Ox} \cdots\cdots \text{Red}] \xrightarrow{k} \text{Products}$$

Here, k-stands for the rate constant of the overall process involving activation of the precursor O.S. complex, electron transfer within the activated precursor O.S. complex and dissociation to the products. In fact, the successor O.S. complex produced through the electron transfer rapidly dissociates into the products. The rate law is given by:

$$\text{rate} \approx \frac{kK_{OS}[Ox][\text{Red}]}{1 + K_{OS}[\text{Red}]}, \; ([\text{Red}] \gg [\text{Ox}]), \textit{ see } \text{Sec. 5.9.3 for derivation of the rate law})$$

$$\approx kK_{OS}[\text{ox}][\text{Red}], \; (1 \gg K_{OS}[\text{Red}], \textit{ i.e. } \text{when } K_{OS} \text{ is small}).$$

6.3.2 Electron Transfer and Franck-Condon Principle: Chemical Activation to Overcome the Franck-Condon Barrier

(A) Activation energy to satisfy the demand of Franck-condon principle: Franck-Condon principle (*cf.* Fig. 7.7.2.1) indicates that the electronic motion is so rapid that during the electron transfer process, the heavier nuclei including the metals and ligands do not have the time to move because **the nuclear motion is much slower than the electronic motion**. Thus during the electron transfer process, there will be no change in atomic arrangements in the reactants (*cf. thermal reaction* vs. *photochemical reaction*; in photochemical reaction, Franck-Condon principle is not applicable; *see* Sec. 6.9.1, Figs 6.9.1.1–2). To illustrate the case, let us consider the electron exchange process in the couple $[Fe(CN)_6]^{3-}/[Fe(CN)_6]^{4-}$.

$$\left[Fe(CN)_6\right]^{3-} + \left[{}^{*}Fe(CN)_6\right]^{4-} \rightleftharpoons \left[Fe(CN)_6\right]^{4-} + \left[{}^{*}Fe(CN)_6\right]^{3-};\ \Delta G^0 = 0$$

The Fe^{II}—C bond in $[Fe(CN)_6]^{4-}$ is slightly shorter (192 pm) than that in $[Fe(CN)_6]^{3-}$ (193 pm) due to the stronger Fe→CN π-back bonding for the lower oxidation of iron. If the electron transfer occurs at the ground state configurations (without any reorganization or chemical activation), then according to the Franck-Condon principle, we shall get $[Fe(CN)_6]^{3-}$ with the shorter Fe^{III}—C bonds and $[Fe(CN)_6]^{4-}$ with the longer Fe^{II}—C bonds. ***It happens so because during the time scale of electron transfer process, there will be no change of bond length***. Both these species thus produced are unstable and they will subsequently relax to the ground state with the release of energy. **This will contradict the 1st law of thermodynamics** because the process produces energy out of nothing. *It indicates that there must be some input of energy for the electron transfer process to occur even when* $\Delta G^0 = 0$.

The above discussion indicates the requirement of input of energy for the electron transfer process to take place. This will be consumed to reorganize the nuclei of the reactants before the electron transfer. This **reorganization of the nuclei** called **chemical activation** can be attained vibronically. This reorganization process will lead to lengthening of the Fe^{II}—C bond in $[Fe(CN)_6]^{4-}$ and shortening of the Fe^{III}—C bond in $[Fe(CN)_6]^{3-}$. *This bond length shortening and elongation will continue until the 'Fe—C' bond length becomes equal in both the reactant species and under this condition, the participating orbitals become of the same energy to allow the electron transfer.* This requirement of reorganization of the nuclei of the reactants through the change of bond length and bond angle gives the height of the *Franck-Condon barrier* of the electron transfer process.

In the electron exchange process, in the couple $[Fe(CN)_6]^{3-}$– $[Fe(CN)_6]^{4-}$, an electron from the t_{2g} level of Fe(II) (t_{2g}^6 low-spin) is to be transferred to the t_{2g} level of Fe(III) (t_{2g}^5, low-spin). Because of the bond length differences between the species, energies of the participating orbitals differ. According to the Franck-Condon principle, **energy of these participating orbitals must be the same** for the efficient electron transfer process. The reorganization process prior to the electron transfer is illustrated in Fig. 6.3.2.1.

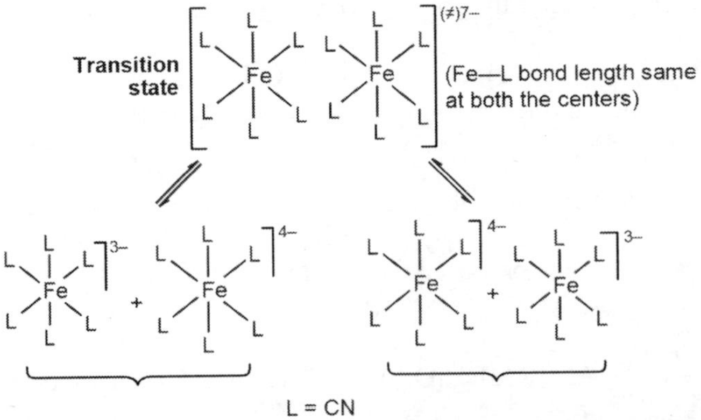

Transition state: $[L_6Fe\ \ FeL_6]^{(\neq)7-}$ (Fe—L bond length same at both the centers)

$[FeL_6]^{3-} + [FeL_6]^{4-} \rightleftharpoons$ Transition state $\rightleftharpoons [FeL_6]^{4-} + [FeL_6]^{3-}$

L = CN

Fig. 6.3.2.1: Shrinkage of the Fe^{III}—CN bond in $[Fe(CN)_6]^{3-}$ and elongation of the Fe^{II}—CN bond in $[Fe(CN)_6]^{4-}$ to produce the transition state in which the metal-ligand bond distances are equal at the both centres

These are obtained through the vibrational stretching and compression of the metal-ligand bonds.

Obviously, the amount of the required **reorganization energy will be less** if the geometric and structural parameters of the electron donor (*i.e.* reductant) and electron acceptor (*i.e.* oxidant) are more or less comparable in the electron exchange process. In such cases, the activation energy will be less and the rate will be faster. But if the structural parameters differ largely, then the reorganization energy measuring the activation energy will be higher and consequently, the electron transfer rate will be slower.

(B) Different factors to determine the magnitude of chemical activation (*i.e.* height of the Franck-Condon barrier): In the couples like $[Fe(OH_2)_6]^{3+}$–$[Fe(OH_2)_6]^{2+}$, $[Fe(CN)_6]^{3-}$–$[Fe(CN)_6]^{4-}$, etc. the **stereochemical configurations** of the Fe(III) and Fe(II) complexes are the same (*i.e.* both are octahedral), both are having the **same spin state** in each couple, but the ***bond lengths are different.*** This is why, in such cases, only the ***bond lengths (not the bond angles)*** are to be adjusted in the activation process of OSET. But when the ***stereochemical configurations*** change on addition or removal of electron, ***both the bond angles and bond lengths*** are to be adjusted in the activation process. Besides these, in OSET, the electron transfer can occur from $\pi^* \rightarrow \pi^*$ (*i.e.* $t_{2g} \rightarrow t_{2g}$). If the ground state electronic configuration does not allow this $\pi^* \rightarrow \pi^*$ transfer, then the reactants must be ***electronically activated*** to allow the $\pi^* \rightarrow \pi^*$ transfer. This **electronic activation** will also contribute to the chemical activation process. In some cases, removal or addition of electron may change the spin state. To accommodate this ***spin state change,*** in the activation process, electronic configurations of the reactants are to be changed accordingly. Thus the possible contributing factors in the ***chemical activation process*** of OSET are:

(i) *Bond length adjustment.*

(ii) *Bond angle adjustment (if there is a change in stereochemical configuration).*

(iii) *Electronic activation to allow the* $\pi^* \rightarrow \pi^*$ *(i.e.* $t_{2g} \rightarrow t_{2g}$*) transfer when the ground state configurations lead to* $\sigma^* \rightarrow \sigma^*$ *(i.e.* $e_g \rightarrow e_g$*),* $\sigma^* \rightarrow \pi^*$ *(i.e.* $e_g \rightarrow t_{2g}$*) and* $\pi^* \rightarrow \sigma^*$ *(i.e.* $t_{2g} \rightarrow e_g$*) electron transfer.*

(iv) *Electronic activation to accommodate the spin state change (if any).*

Orbital designation in the O_h-complexes: In terms of MOT, the t_{2g} orbitals (CFT) appear as the **nonbonding MOs** when only metal-ligand σ-bonding interactions operate as in $[Co(NH_3)_6]^{3+}$. When both the σ- and π-bonding interactions operate, depending on the nature of the ligands, the t_{2g} orbitals of π-symmetry may appear as the **π-BMOs (*i.e.* π–t_{2g})** or **π-ABMOs (*i.e.* π*–t_{2g})** (see Fig. 3.17.3.4, Vol. 4). But, the e_g-orbitals (CFT) always appear as the σ-ABMOs (*i.e.* σ*-e_g, simply σ*). Thus designation of the t_{2g} orbitals of **π-symmetry** in terms of MOT is different in different cases. In this book, ignoring the nature of ligands (*i.e.* σ-bonding or σ + π-bonding), the t_{2g}-orbitals have been designated as the π*-orbitals and the e_g-orbitals as the σ*-orbitals in terms of MOT.

(C) Franck-Condon Principle and the metalloproteins as the electron carriers: The metalloproteins acting as the electron carriers generally use the Fe(II)/Fe(III) couple (as in ***ferredoxins, cytochromes***) and Cu(II)/Cu(I) couple (as in ***blue proteins*** like azurin, plastocyanin, stellacyanin, etc.). The electron exchange in these couples generally occur through the *outer-sphere mechanism.*

To minimize the required ***reorganization energy***, in each couple, the following conditions should be satisfied.

- The spin state should not change during the electron transfer.
- The metal-ligand bond length in the oxidized and reduced forms of the couple should be comparable.
- The stereochemical configurations of both forms should be the same (at least comparable).

  The above requirements are maintained in the electron carrier metallo-proteins. These are illustrated below.
- High-spin distorted tetrahedral Fe(III)/Fe(II) couple as in ***rubredoxin and ferredoxins***.

- Low-spin octahedral Fe(III)/Fe(II) couple as in ***cytochromes***.
- Distorted tetrahedral Cu(II)/Cu(I) couple as in ***blue proteins***.

(i) **Rubridoxin and ferredoxin:** Tetrahedral Fe(III) ($e^2t_2^3$) does not experience any Jahn-Teller distortion but Fe(II) ($e^3t_2^3$) experiences the distortion. This is why, nature has placed the Fe-centre in such electron carriers ***in a distorted tetrahedral geometry to make a reasonable compromise between the preferences of Fe(II) and Fe(III)***. The bond lengths in this geometry for the oxidized and reduced centres are also comparable.

(ii) **Cytochromes:** *The octahedral geometry and spin state do not change during the electron exchange process*. The bond lengths in this geometry for the oxidized and reduced centres are also comparable. Thus, the reorganization energy is minimized to favour the electron exchange process.

(iii) **Blue proteins:** Cu(II) (d^9), the oxidized form of the couple, has a strong preference for the square-planar geometry while Cu(I) (d^{10}), the reduced form of the couple, has a preference to adopt the tetrahedral geometry. In a square planar geometry, Cu(II) enjoys a high *crystal field stabilization energy (cfse)* though there is a relatively more steric crowding compared to that in a tetrahedral geometry (*cf.* 90° *vs*. 109° in terms of bond angle). Cu(I) without having any cfse prefers the tetrahedral geometry to minimize the steric crowding. Here to make a reasonable compromise between the preferences of Cu(II) and Cu(I) to adopt the geometry, nature has placed the Cu-centre in a ***flattened tetrahedral geometry, an intermediate structure between the two idealized geometries, i.e. square planar and tetrahedral***. In fact, in the said two idealized geometries, the metal-ligand bond distances also differ significantly but in the distorted geometry, *i.e.* flattened tetrahedral geometry, the bond lengths of the oxidized and reduced forms do not differ significantly. In fact, if this distortion were not done, then a large amount of reorganization energy would be required to adjust the geometry and bond length parameter.

6.3.3 Potential Energy Diagram for the Electron Transfer Process

(a) Energetics of the electron exchange reaction pathway: Now let us consider the following representative electron exchange reaction.

$$[Fe(OH_2)_6]^{3+} + [^*Fe(OH_2)_6]^{2+} \rightleftharpoons [Fe(OH_2)_6]^{2+} + [^*Fe(H_2O)_6]^{3+}$$

$$k_{ex} \approx 4.0 \text{ dm}^3 \text{ mol}^{-1} \text{ s}^{-1} (25°\text{C}); \text{ activation energy} \approx 32 \text{ kJ mol}^{-1}$$

In this electron exchange process, an electron from the t_{2g} level of Fe(II) ($t_{2g}^4e_g^2$) is transferred to the t_{2g} level of Fe(III) ($t_{2g}^3e_g^2$). The bond lengths are unequal, *i.e.* Fe^{III}—O (205 pm) < Fe^{II}—O (221 pm). *Thus the participating orbitals are of different energy.* According to the demand of Franck-Condon principle, in the transition state, both the 'Fe—O' bond lengths should be the same. It will make the participating orbitals of the same energy. For this requirement, the Fe^{III}—O bond is to be elongated and the Fe^{II}—O bond is to be compressed in attaining the transition state (T.S.).

Free energy of activation ($\Delta G^{\neq}$) for the electron exchange in the couple, $[Fe(OH_2)_6]^{3+} - [Fe(OH_2)_6]^{2+}$ is the sum of three energy terms.

$$\Delta G^{\neq} \approx \Delta G_i^{\neq} + \Delta G_s^{\neq} + \Delta G_r^{\neq} \text{ (where } \Delta G^o = 0 \text{ in the electron exchange reaction)}$$

ΔG_i = Gibbs energy required for changing the bond lengths in the reactants to make them more **product-like**, *i.e.* **for reorganising the inner coordination spheres of the reactants;** $\Delta G_s^{\neq}$ = Gibbs energy required for **solvent reorganisation in the outer spheres of the reactants to attain the T.S.**; $\Delta G_r^{\neq}$ = Gibbs energy required to bring the reactants into a state where they are separated by the distance r at which the redox reaction occurs (for the charged reactants, it includes the Coulombic interactions also).

Note: More correctly, $\Delta G^{\#}$ in general, is given by:

$$\Delta G^{\#} = \Delta G_i^{\#} + \Delta G_s^{\#} + \frac{\Delta G_r^o}{2} + \frac{(\Delta G_r^o)^2}{16(\Delta G_i^{\#} + \Delta G_s^{\#})}$$

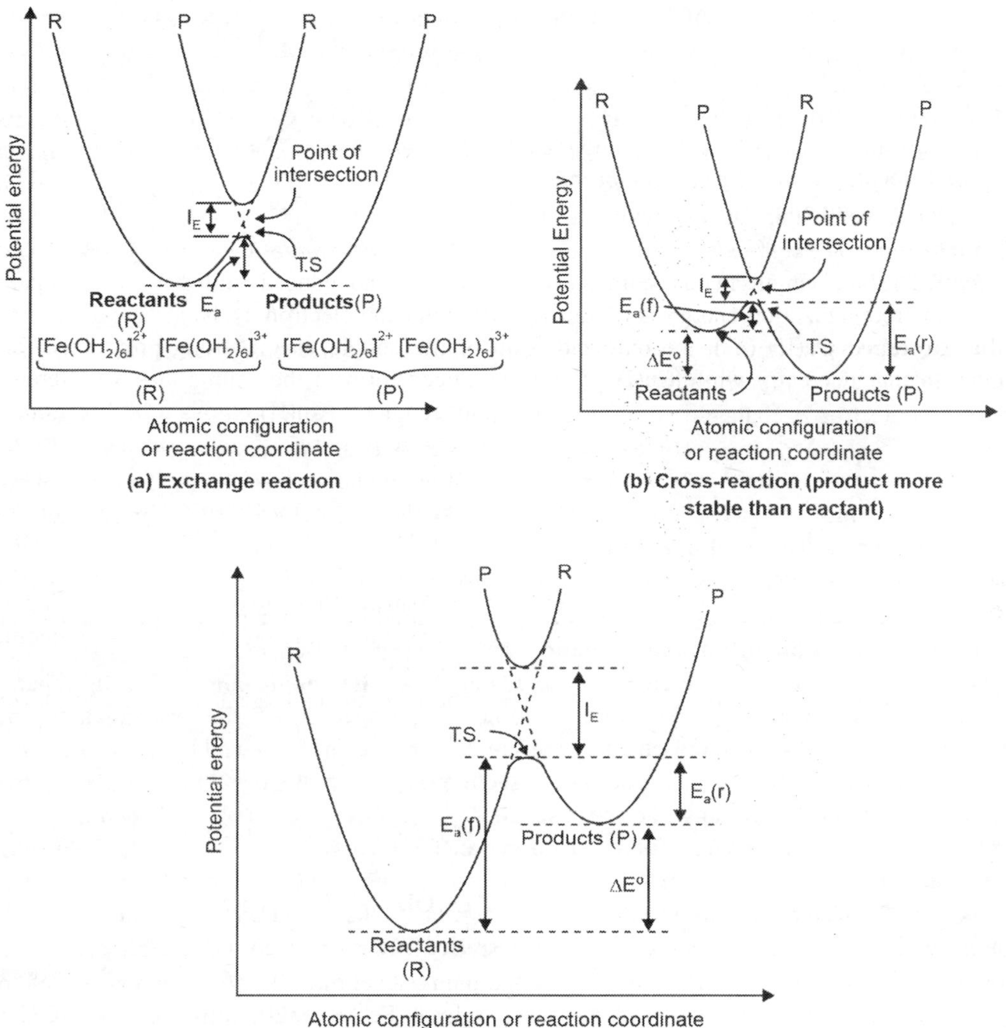

Fig. 6.3.3.1: Potential energy diagram for the outer-sphere electron transfer (OSET) process. (E_a = activation energy; T.S. transition state; P = products; R = reactants): (a) Electron exchange in a redox couple, *i.e.* homonuclear electron transfer process as in the $[Fe(OH_2)_6]^{2+}$–$[Fe(OH_2)_6]^{3+}$ couple. (b) and (c) Cross-reaction (or redox reaction), *i.e.* heteronuclear electron transfer process. $E_{a(f)}$ for the forward reaction and $E_{a(r)}$ for the reverse reaction.

Note: Electron transfer occurs at the **point of intersection** of the potential energy surfaces of two reactants because at the point of intersection, energies of the participating orbitals become equal.

$\Delta G_r^o = \Delta G^o + NU(r)$, where ΔG_r^o stands for the overall free energy change (ΔG^o) of the reaction corrected with $NU(r)$. $U(r)$ denotes the energy required to bring the reactants together in the T.S. and $U(r)$ is involved to calculate K_{os} (**Fuoss equation** already discussed, see Sec. 5.14). It is the sum of the attractive and repulsive potentials for the reactants as they approach towards the T.S. ΔG^o = overall free energy change in the reaction.

For the electron exchange reactions, ΔG^o is 0 and $\Delta G^{\neq} \approx \Delta G_i^{\neq} + \Delta G_s^{\neq}$.

The rate constant (k_{el}) for the electron transfer reaction can be approximately given by **Marcus equation**:

$$k_{el} \approx K_{OS}\kappa_{el}\nu_n \exp(-\Delta G^{\neq}/RT) = K_{OS}\kappa_{el}\nu_n \exp[-(\Delta G_i^{\#} + \Delta G_s^{\#} + 0.5\Delta G_r^o)/RT]$$

Assuming $(\Delta G_r^o)^2 \ll 16(\Delta G_i^{\neq} + \Delta G_s^{\neq})$ as in the case of electron exchange reactions (where $\Delta G^o = 0$) and for the reactions where ΔG^o is small; in such cases, the 4th term of the expression of $\Delta G^{\neq}$ can be neglected.

κ_{el} = **electron transmission coefficient** measuring the probability of electron transfer at the transition state to produce the products. It is also known as the **electronic factor**. Its value may be in the range, 0 to 1. It is taken as = 1 for the calculation of rate constant.

ν_n = **nuclear frequency factor** representing the frequency of vibrational mode of the T.S. that converts the T.S. into the products. It is often taken as 1×10^{13} s^{-1} for calculation. Assuming $\Delta G_r^o \approx \Delta G^o$, we can write:

$$k_{el} = K_{OS}\kappa_{el}\nu_n \exp[-(\Delta G_i^{\neq} + \Delta G_s^{\neq} + 0.5\Delta G^o)/RT]$$

In the rate expression, the pre-exponential factor consists of three components: K_{OS}, ν_n (**nuclear frequency** factor) and κ_{el} (electronic factor).

For the self electron exchange process as in the $[Fe(OH_2)_6]^{3+/2+}$ couple, $\Delta G^o = 0$, it is reasonable to assume $\Delta G^{\#} = \Delta G_i^{\#} + \Delta G_s^{\#}$ (neglecting the 3rd and 4th terms of the expression $\Delta G^{\neq}$).

All the energy terms (*i.e.* $\Delta G_i^{\#}$, $\Delta G_s^{\#}$, ΔG^o and ΔG_r^o) can be calculated theoretically. Thus the calculated value of k_{el} (3 dm^3 mol^{-1} s^{-1} at 25°C) for electron exchange in the $[Fe(OH_2)_6]^{3+/2+}$ couple nicely agrees with the experimental value of 4 dm^3 mol^{-1} s^{-1}.

(b) Potential energy diagrams for the outer sphere redox reactions: The metal-ligand bond deformations (required in attaining the T.S.) may be compared to the **harmonic oscillations** and the potential energy curves are of **parabolic shapes**. Thus the potential energy diagrams of the solvated reactants and products may be represented approximately by the **harmonic potential wells** of parabolic shapes. For the electron exchange reaction (*i.e.* no net chemical reaction), the reactants and products having the same energy can be represented by the same harmonic potential well but these are separated along the reaction coordinate. **At the point of intersection (shown by the dotted lines) of the potential wells** of the products and reactants, *the requirement of equal energies of the participating orbitals is attained.* At the point of intersection, for the electron transfer process to occur, there must be a **coupling of the vibrational and electronic motion**. *The point of intersection of the potential wells represents the position of* ***transition state*** *along the reaction coordinate.* The reaction coordinate represents the change of bond length and bond angle in the inner sphere of the involved metal centres.

In reality, the transition state (T.S.) lies below the point of intersection of the two potential energy curves representing the reactants and products. It can be explained by considering the **noncrossing rule** which states that the potential energy curves of two states of the same symmetry cannot cross and they mutually repell strongly at the hypothetical point of intersection giving rise to two energy curves—upper curve and lower curve. *The reactants follow the pathway of this lower curve (along the reaction coordinate) and transform into the products.*

The above phenomenon can also be realised in terms of mixing of the wave functions of the reactants and products. As the reaction coordinate approaches towards the T.S., *the wave-functions of the reactants and products mix and it leads to two separate energy states (separated by I_E energy) instead of two intersecting parabolas.* If there were no mutual interaction (*i.e.* no mixing), there would be two parabolas. The courses of such noninteracting parabolas are shown by the dashed lines. The extent of interaction or mixing is measured by I_E. At the T.S., the motion pushes the reactant well to slide down into the product well through the Franck-Condon barrier.

(c) I_E and position of T.S.: In terms of I_E, position of the T.S. with respect to the point of intersection of the parabolas of the reactants and products can be determined as follows:

- **Large I_E** (good coupling interaction, such systems are described as the **adiabatic systems**): T.S. far below the point of intersection, *i.e.* low activation energy because of low reorganisation energy (*i.e.* **low $\Delta G_i^{\neq}$ and low Franck–Condon barrier**) and in such cases κ_{el} (electron transmission coefficient) ≈ 1.

● **Small I_E** (poor coupling interaction, such systems are described as the **nonadiabatic systems**): T.S. close to the point of intersection, *i.e.* high activation energy because of high reorganisation energy (*i.e.* **high $\Delta G_i^{\neq}$ and high Franck–Condon barrier**) and in such cases, $\kappa_{el} \ll 1$.

(d) Franck-Condon barrier height and I_E: If the coupling interaction measured by I_E is ***strong, then the height of Franck-Condon barrier is small*** (*i.e.* small $\Delta G_i^{\neq}$: condition for small bond distortions and less inner sphere reorganisation) and the electron transfer process becomes favourable. If I_E is small (*i.e.* weak interaction), ***then the height of Franck-Condon barrier is large*** (*i.e.* large $\Delta G_i^{\neq}$: condition for large bond distortions and more inner sphere reorganisation) and the electron transfer process becomes less favourable.

(e) I_E and κ_{el}: There is a relationship between the ***electron transmission coefficient*** **(κ_{el})** (which indicates the probability of electron transfer at the transition state to give the product) and I_E. When I_E is large, $\kappa_{el} \approx 1.0$, *i.e.* the reactants smoothly pass into the products. If I_E is small, the transition state may roll back down to the reactant side without giving the product (*i.e.* $\kappa_{el} \ll 1$). In fact, when I_E is small (*i.e.* poor coupling interaction), the T.S. practically lies close to the point of intersection and consequently, even at the T.S., the reactant lies practically on the reactant parabola. In such cases, the reactant can easily roll back down along the reactant parabola without experiencing transformation into the product. ***Obviously, when the T.S. lies far below the point of intersection (i.e. I_E is large for good coupling interaction), possibility of such 'roll back' from the T.S. is less.***

(f) Activation energy and cross reaction: In the same way, potential energy diagram for a cross reaction (*i.e.* net reaction, *i.e.* $\Delta G^o \neq 0$) can be constructed. If the product is thermodynamically less stable than the reactant (*i.e. product curve runs above the reactant curve*), then the point of intersection of the potential energy curves of the reactants and products moves up and **consequently, the activation energy becomes higher.** On the other hand, if the product becomes thermodynamically more stable than the reactant (*i.e. product curve lies below the reactant curve*), then the point of intersection is lowered down and consequently, the **activation energy becomes less.** It leads to the conclusion: **the thermodynamically favoured outer-sphere redox reaction is also kinetically favoured** (*cf.* Marcus theory and LFER).

(g) Marcus equation for the thermodynamically favoured electron transfer reactions (ΔG^o = –ve): Marcus equation for the activation energy in terms of ΔG^o can be expressed as:

$$k_{el} \propto \exp(-\Delta G^{\neq}/RT),\ i.e.\ k_{el} = K_{OS}\nu_n\kappa_{el}\exp(-\Delta G^{\neq}/RT)$$

$$\Delta G^{\neq} = \frac{\lambda}{4}\left(1+\frac{\Delta G^o}{\lambda}\right)^2 \approx \frac{\lambda}{4}\left(1+\frac{2\Delta G^o}{\lambda}\right) = \frac{1}{4}(\lambda + 2\Delta G^o)$$

Under the condition, $|\lambda| \gg |\Delta G^o|$ (*cf.* λ is positive and ΔG^o is negative for the thermodynamically favoured reaction).

Here λ denotes the **reorganisation energy** that depends on $\Delta G_i^{\neq}$ (inner sphere reorganisation energy of the reactants) and $\Delta G_s^{\neq}$ (outer sphere solvent reorganisation energy) already described. Thus the magnitude of λ is determined jointly by $\Delta G_i^{\neq}$ and $\Delta G_s^{\neq}$. It can be shown that $\lambda = 4(\Delta G_i^{\neq} + \Delta G_s^{\neq})$.

Marcus equation: $\Delta G^{\neq} = \dfrac{(\lambda + \Delta G^o)^2}{4\lambda} = \dfrac{1}{4\lambda}[\lambda^2 + (\Delta G^o)^2 + 2\lambda\Delta G^o]$

i.e. $\Delta G^{\neq} = \dfrac{\lambda}{4} + \dfrac{\Delta G^o}{2} + \dfrac{(\Delta G^o)^2}{4\lambda} = \Delta G_i^{\neq} + \Delta G_s^{\neq} + \dfrac{\Delta G^o}{2} + \dfrac{(\Delta G^o)^2}{16(\Delta G_i^{\neq} + \Delta G_s^{\neq})}$

where $\dfrac{\lambda}{4} = \Delta G_i^{\neq} + \Delta G_s^{\neq}$ or, $\lambda = 4(\Delta G_i^{\neq} + \Delta G_s^{\neq})$, *cf.* $\Delta G_r^o = \Delta G^o + NU(r) \approx \Delta G^o$

For the electron exchange reaction, $\Delta G^\circ = 0$, and it leads to $\Delta G^{\neq} = \lambda/4$.

For the thermodynamically favoured reaction (*i.e.* $\Delta G^\circ = -$ve, but λ is positive), the **three following situations** may arise depending on the relative values of ΔG° and λ.

(i) $|\Delta G^\circ| \ll |\lambda|$: The rate constant ($k_{el}$) increases exponentially as ΔG° becomes more negative. This is the **Marcus normal region**.

(ii) $|\Delta G^\circ| \approx |\lambda|$: The reaction becomes **activationless** (*i.e.* $\Delta G^{\neq} \approx 0$) and the **rate constant becomes maximum**.

(iii) $|\Delta G^\circ| \gg |\lambda|$: The activation energy starts to increase and the rate starts to decline for the more thermodynamically favoured reaction. This represents the **Marcus inverted region**.

Thus if the rate constant (k_{el}) is plotted against ΔG° for the thermodynamically favoured reaction, the rate constant becomes maximum under the condition $|\Delta G^\circ| \approx |\lambda|$ leading to $\Delta G^{\neq} \approx 0$ (**activationless**) (see Fig. 6.3.3.2).

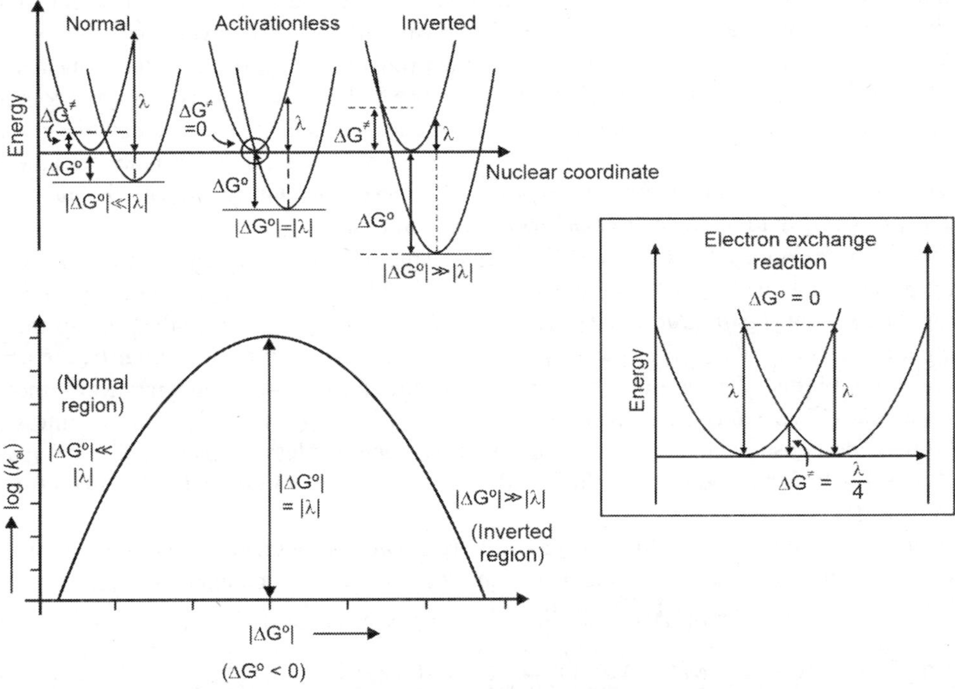

Fig. 6.3.3.2: Schematic representation of relationship between electron transfer rate constant (k_{el}) or activation energy ($\Delta G^{\neq}$) and free energy of reaction ($\Delta G^\circ = -$ve) in the normal and inverted Marcus regions. Inset (Box): Electron exchange reaction ($\Delta G^\circ = 0$) with $\Delta G^* = \lambda/4$. λ = Reorganisation energy determined by $\Delta G_i^{\neq}$ and $\Delta G_s^{\neq}$.

6.3.4 Marcus Equation for Outer Sphere Cross-Reactions

By considering the different factors to contribute to $\Delta G_{12}^{\neq}$ (free energy of activation for the cross-reaction between Ox_1 and Red_2), Marcus and Hush theoretically developed an equation to calculate the rate constant (k_{12}) of the reaction between Ox_1 and Red_2.

$$Ox_1 + Red_2 \rightleftharpoons Red_1 + Ox_2$$

$$k_{12} = \sqrt{(k_{11}k_{22}K_{12}f_{12})} \text{(for Macrus cross-reaction)}$$

where $\log f_{12} = \dfrac{(\log K_{12})^2}{4\log(k_{11}k_{22}/Z^2)} \approx 0$; *i.e.* $f_{12} \approx 1.0$

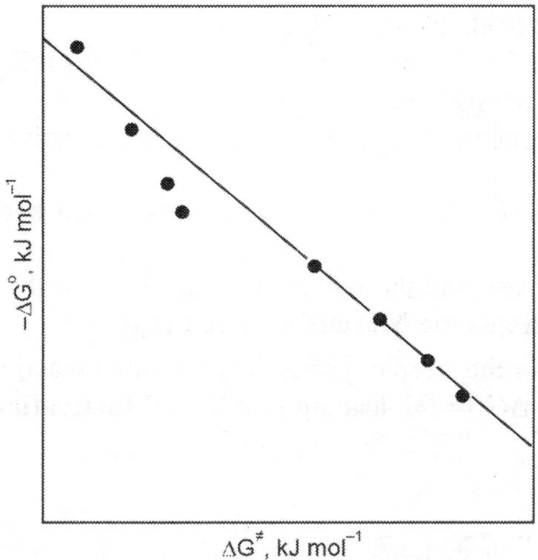

Fig. 6.3.4.1: Illustration of Marcus relationship (*i.e.* LFER plot) for the oxidation of $[Fe(L)_3]^{2+}$ (L = phen or substituted phen) complexes by Ce(IV) in aqueous sulfuric acid media (*cf.* G. Dulz *et al, Inorg. Chem. 2,* 917, 1963; J.D. Miller *et al, J. Chem. Soc. A,* 1370, 1966)

Z = number of collisions per second between the practicles (~ 10^{11} dm^3 mol^{-1} s^{-1}); k_{11} and k_{22} are the electron exchange rate constants for the redox couples Ox_1–Red_1 and Ox_2–Red_2 respectively; K_{12} is the equilibrium constant which can be calculated from the E^o values of the redox couples, Ox_1–Red_1 and Ox_2–Red_2.

$$\Delta G_{12}^{o} = -RT\ln K_{12} = -nFE_{cell}^{o};\ E_{cell}^{o} = E_1^o - E_2^o$$

where E_1^o and E_2^o are the standard reduction potentials of the couples, $Ox_1 - Red_1$ and $Ox_2 - Red_2$, *i.e.*

$$Ox_1 + ne \rightleftharpoons Red_1,\ E_1^o;\ Ox_2 + ne \rightleftharpoons Red_2,\ E_2^o$$

Taking $f_{12} \approx 1$, the **Marcus-Hush equation** (simply called Marcus equation) is given by:

$$k_{12} \approx \sqrt{(k_{11}k_{22}K_{12})} \text{ or, } k_{12}^2 = k_{11}k_{22}K_{12}$$

i.e.

$$\mathbf{2\ln k_{12} = \ln k_{11} + \ln k_{22} + \ln K_{12} \Rightarrow 2\Delta G_{12}^{\neq} = \Delta G_{11}^{\neq} + \Delta G_{22}^{\neq} + \Delta G_{12}^{o}}$$

The Marcus equation indicates that the rate constant of a redox reaction depends on the *self electron exchange rate constants* (k_{11} and k_{12}) and the *equilibrium constant* (K_{12}).

The free energy of activation ($\Delta G_{12}^{\neq}$) for the electron transfer process in the said redox reaction can be given by:

$$\Delta G_{12}^{\neq} = \frac{(\Delta G_{11}^{\neq} + \Delta G_{22}^{\neq})}{2} + \frac{\Delta G_{12}^{o}}{2},\ (\text{assuming } f_{12} \approx 1)$$

The first term involving $\Delta G_{11}^{\neq}$ and $\Delta G_{22}^{\neq}$ gives the measure of **intrinsic contribution** while the second term involving ΔG_{12}^{o} gives the measure of ***thermodynamic contribution***. $\Delta G_{11}^{\neq}$ and $\Delta G_{22}^{\neq}$ are the free energies of activation for the self electron exchange process in the couples Ox_1–Red_1 and Ox_2–Red_2 respectively. Very often, *the above equation for* $\Delta G_{12}^{\neq}$ *is dominated by the thermodynamic part* ΔG_{12}^{o} (*i.e.* free energy change in the cross-reaction) and it gives a *linear relationship* between $\Delta G_{12}^{\neq}$ and ΔG_{12}^{o}, *i.e.* linear relationship between log k_{12} and log K_{12}. Thus the Marcus equation is an example of **Linear Free Energy Relationship (LFER).**

The Marcus equation indicates: Higher the equilibrium constant (K_{12}), higher the rate constant (k_{12}), *i.e.* ***thermodynamically favoured reactions are also kinetically favoured*** (*cf.* Fig. 6.3.4.2). This can also be concluded by considering the potential wells of the reactants and products (*cf.* Fig. 6.3.3.1).

Table 6.3.4.1 Comparison of calculated and observed second order rate constants for the outer-sphere cross-reactions (*cf.* R. A. Marcus *et al*, *Biochim. Biophys. Acta.*, **811**, 265, 1985; T. J. Meyer *et al*; *Inorg. Chem.*, **7**, 2369, 1968; T. Przystas *et al*; *J. Am. Chem. Soc*; **95**, 5545, 1973).

Reaction	***logK₁₂***	k_{12} **(dm³ mol⁻¹ s⁻¹)**	
		(observed)	***(calculated)***
$[Co(phen)_3]^{3+} + [Ru(NH_3)_6]^{2+}$	5.42	1.5×10^4	1×10^5
$[Co(en)_3]^{3+} + [V(OH_2)_6]^{2+}$	0.25	5.8×10^{-4}	7×10^{-4}
$[IrCl_6]^{2-} + [Mo(CN)_8]^{4-}$	2.18	1.9×10^6	8×10^5
$[IrCl_6]^{2-} + [Fe(CN)_6]^{4-}$	4.08	3.8×10^5	1×10^6
$[MnO_4]^- + [Fe(CN)_6]^{4-}$	3.40	1.7×10^5	6×10^4
$[Co(bpy)_3]^{3+} + [Co(terpy)_2]^{2+}$	0.55	64	40
	E^0_{cell} **(V)**		
$[Co(phen)_3]^{3+} + [Cr(OH_2)_6]^{2+}$	0.83	3×10^1	$\approx 1.1 \times 10^4$
$[Co(en)_3]^{3+} + [Cr(OH_2)_6]^{2+}$	0.17	3.4×10^{-4}	$\approx 5.1 \times 10^{-4}$
$[Co(NH_3)_6]^{3+} + [Cr(OH_2)_6]^{2+}$	0.51	1×10^{-3}	$\approx 1.5 \times 10^{-3}$
$[Ru(NH_3)_6]^{3+} + [Cr(OH_2)_6]^{2+}$	0.51	2×10^2	$\approx 1.5 \times 10^3$
$[Ru(NH_3)_6]^{3+} + [V(OH_2)_6]^{2+}$	0.355	80	$\approx 4.2 \times 10^3$
$[Co(NH_3)_6]^{3+} + [V(OH_2)_6]^{2+}$	0.355	1×10^{-2}	$\approx 2.3 \times 10^{-3}$
$[Co(en)_3]^{3+} + [V(OH_2)_6]^{2+}$	0.015	7.2×10^{-4}	$\approx 6.0 \times 10^{-4}$

$E^0_{Cr^{3+}/Cr^{2+}} = -0.41V$; $E^0_{V^{3+}/V^{2+}} = -0.26V$ *i.e.* $Cr(aq)^{2+}$ is a better reducing agent thermodynamically.

Note: Thermodynamically $Eu(aq)^{2+}$ $(4f^7)$ is comparable to $Cr(aq)^{2+}$ as a reducing agent (*cf.* $E^0_{Eu^{3+}/Eu^{2+}} = -0.37V$). But it reduces $[Co(NH_3)_6]^{3+}$ faster ($k_{12} = 2\times10^{-2}\,dm^3mol^{-1}\,s^{-1}$) than $Cr(aq)^{2+}$ (*cf.* $k_{12} = 1 \times 10^{-3}$ dm³ mol⁻¹ s⁻¹).

Let us apply the Marcus equation for the following reactions:

$$\underset{(Ox_1)}{[Co(bpy)_3]^{3+}} + \underset{(Red_2)}{[Co(terpy)_2]^{2+}} \rightleftharpoons \underset{(Red_1)}{[Co(bpy)_3]^{2+}} + \underset{(Ox_2)}{[Co(terpy)_2]^{3+}}$$

Electron exchange process
$$\begin{cases} [Co(terpy)_2]^{2+} - [{}^*Co(terpy)_2]^{3+} & : k_{11} = 48\ dm^3\ mol^{-1}\ s^{-1} \\ [Co(bpy)_3]^{2+} - [{}^*Co(bpy)_3]^{3+} & : k_{22} = 9.0\ dm^3\ mol^{-1}\ s^{-1} \end{cases}$$

$$E^0[Co(bpy)_3]^{3+}/[Co(bpy)_3]^{2+} = +\ 0.34\ V$$

$$E^0[Co(terpy)_2]^{3+}/[Co(terpy)_2]^{2+} = +\ 0.30\ V$$

$$E^0_{cell} = (0.34 - 0.31) = 0.03\ V$$

$$-nFE^0_{cell} = -RT\ln K_{12}\ \text{or,}\ \frac{nFE^0_{cell}}{RT} = \ln K_{12}$$

It leads to:
$$\log K_{12} = 0.553\ \text{and}\ K_{12} = 3.6$$

$$\log f = \frac{(0.553)^2}{4\log\frac{9\times48}{10^{22}}}\text{, it leads to } f \approx 1.0$$

$$k_{12} = \sqrt{9\times48\times3.6\times1} \approx 40\ dm^3\ mol^{-1}\ s^{-1}$$

(*cf.* experimental value, 64 dm³ mol⁻¹ s⁻¹)

Conclusion: In fact, the agreement between the experimental value and calculated value is good under the conditions $K_{12} \leq 10^6$ and $f \geq 0.3$. **This agreement is regarded as an evidence in favour of the outer sphere reaction.**

The considerable disagreement indicates a mechanism other than outer-sphere mechanism. However, **for the highly exo- and endothermic cross-reactions, a modified Marcus equation is needed** (see Fig. 6.3.3.2).

By using the Marcus equation, it is also possible to calculate the self electron exchange rate constants. *This technique is specially important for the couples like metalloproteins for which it is difficult to directly measure the electron exchange rate constants.*

6.3.5 Nature of the Donor (*i.e.* Reductant) and Receptor (*i.e.* oxidant) Molecular Orbitals in Outer Sphere Electron Transfer (OSET) Process

The donor MO (generally HOMO) of the reductant and receptor MO (generally LUMO) of the oxidant must be of the same type for the electron transfer. ***It is quite reasonable that if these involved MOs are the π^*–MOs (i.e. t_{2g} in terms of CFT), then the electron transfer process will be favourable.*** The reasons behind this preference are discussed below.

- (i) **Metal-ligand bond length change and inner-sphere reorganisation energy:** Change of the electronic configuration (for the octahedral complexes) in the π^* orbital (*i.e.* t_{2g} level) does not seriously affect the metal-ligand bond length, but the metal-ligand bond length is very much sensitive to the electronic configuration in the σ^* orbital (*i.e.* e_g level in terms of CFT). It is due to the fact that the ligands are directly projected towards the lobes of e_g orbitals in the octahedral geometry.

  Thus donation of electron from the t_{2g} level ($\equiv \pi^*$) of the reductant and accommodation of the electron in the t_{2g}-level ($\equiv \pi^*$) of the oxidant do not change the bond lengths significantly at the metal centres.

Different contributing factors to determine the activation energy and rate of electrons transfer in the outer sphere process

- **Electrostatic barrier in forming the precursor complex:** Sometimes, the oxidants and reductants are bearing the same charge to cause an electrostatic repulsion in the precursor complex (*i.e.* outer sphere complex) as in the electron exchange in the couple $[Fe(CN)_6]^{3-}$–$[Fe(CN)_6]^{4-}$. To overcome this repulsive barrier, some energy will be required.
- **Reorganization energy (*i.e.* Franck-Condon barrier):** To meet the demands of Franck-Condon principle, the ***bond lengths and bond angles of the reactants*** are to be changed (*i.e.* inner coordination spheres of the reactants are to be reorganised) properly before the electron transfer. In other words, the structural parameters in the transition state become close to **product-like** to satisfy the Franck-Condon principle. Sometimes, the ground state electronic configuration may be required to change (*i.e.* **electronic activation** of the reactants) to allow the $\pi^* \rightarrow \pi^*$ electron transfer and to accommodate the spin state change, if any. This inner coordination sphere reorganization energy contributes to the activation energy.
- **Nature of the orbitals:** The $\pi^* \rightarrow \pi^*$ electron transfer is the most favourable path, *i.e.* it needs the least activation energy. It arises so because of the ***least reorganization required and favoured orbital– orbital overlapping interaction.*** Obviously, the $\sigma^* \rightarrow \sigma^*$ electron transfer makes the activation energy very high.
- **Solvent reorganization energy:** Solvent reorganization in the outer sphere at the events of precursor complex formation and reorganization process makes also an important contribution to the activation energy.
- **Thermodynamic favour:** According to the Marcus equation: Mmore the equilibrium constant, more the rate constant.

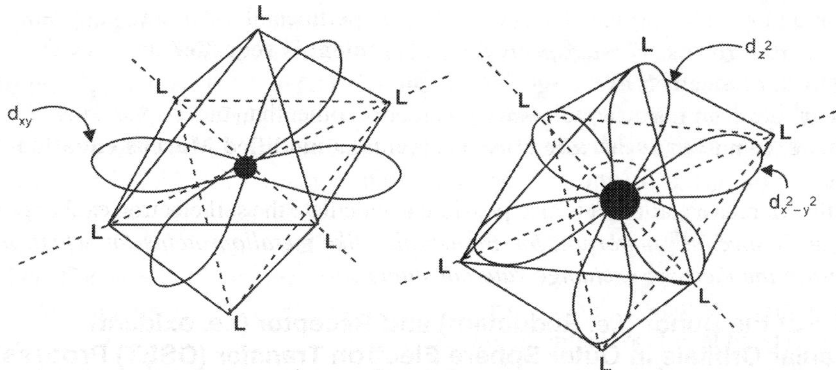

Fig. 6.3.5.1: Spatial orientation of the d_{xy} (*i.e.* t_{2g}) and $d_{x^2-y^2}$, d_{z^2} (*i.e.* e_g set) orbitals with respect to the ligands (*L*) in the octahedral geometry. The other two t_{2g}-orbitals (*i.e.* d_{xz} and d_{yz}) are projected towards the centres of octahedral trigonal faces

This is why, π^* (reductant)→ π^* (oxidant), i.e. t_{2g}→t_{2g} electron transfer needs the less reorganization energy (i.e. low Franck-Condon barrier) and it is the most favourable path. Other possible paths like $\sigma^*\rightarrow\sigma^*$, $\sigma^*\rightarrow\pi^*$ (*i.e.* $e_g \rightarrow t_{2g}$) require the higher activation energy in terms of reorganization energy. Definitely, the ***$\sigma^*\rightarrow\sigma^*$ electron transfer is the most unfavourable path.***

- (ii) **Overlap integral:** In the octahedral geometry, the π^* orbitals (*i.e.* t_{2g}) are exposed everywhere except the vertices where the ligands are positioned (*cf.* Fig. 6.3.5.1). This makes the π^*–π^* overlapping interaction better. ***Consequently, in terms of orbital overlapping interaction, $\pi^*\rightarrow\pi^*$ (i.e. $t_{2g}\rightarrow t_{2g}$) electron transfer is the most favourable pathway.*** The σ^*-orbitals (*i.e.* e_g orbitals) are, in fact, projected towards the vertices, *i.e.* ligands. Thus the σ^* orbitals are more shielded compared to the π^* orbitals. In other words, direct overlap between the σ^*-donor orbital and σ^*-receptor orbital is not possible because the σ^* (*i.e.* e_g) orbitals are directly projected towards the ligands.
- (iii) **Relative ease of electron transfer among the possible pathways ($\sigma^*\rightarrow\sigma^*$, $\sigma^*\rightarrow\pi^*$, $\pi^*\rightarrow\sigma^*$ and $\pi^*\rightarrow\pi^*$):** It is evident that in the outer sphere electron transfer process, the electron transfer can occur from π^* to π^*, *i.e.* t_{2g} to t_{2g}. If the ground state electronic configurations of the oxidant and reductant can allow this $\pi^*\rightarrow\pi^*$ electron transfer (*e.g.* Fe^{II}–Fe^{III} couple, V^{II}–Ru^{III} cross reaction) *then there is no need to activate electronically the reactants, i.e.* no need of change of the ground state electronic configurations for the electron transfer process. If the ground state electronic configuration leads to the $\sigma^*\rightarrow\sigma^*$ electron transfer (*e.g.* Co^{II}–Co^{III} couple, Cr^{II}–Co^{III} cross reaction) or $\sigma^*\rightarrow\pi^*$ electron transfer (*e.g.* Cr^{II}–Ru^{III} cross reaction) or $\pi^*\rightarrow\sigma^*$ electron transfer (*e.g.* V^{II}–Co^{III} cross reaction), then the reactants must be properly electronically excited to allow the $\pi^*\rightarrow\pi^*$ electron transfer.

  Obviously, for the net $\sigma^*\rightarrow\sigma^*$ electron transfer, both the oxidant and reductant need electronic activation; for the net $\pi^*\rightarrow\pi^*$ electron transfer, there is no need of electronic activation for the both oxidant and reductant.

6.3.6 Illustration of the Importance of Different Factors to Control the Rate Constants of Outer Sphere Electron Transfer (OSET) Reactions

(a) **$\pi^*\rightarrow\pi^*$ *vs.* $\sigma^*\rightarrow\sigma^*$ electron transfer:** Table 6.3.6.1 indicates that in general, the $\pi^*\rightarrow\pi^*$ electron transfer process is kinetically more favoured over the $\sigma^*\rightarrow\sigma^*$ electron transfer process. The $\sigma^*\rightarrow\sigma^*$ electron transfer process needs more *reorganization energy* as the bond lengths differ generally significantly for the oxidant and reductant centres. Moreover, direct overlap between the donor and receptor orbitals is prevented because of the ligands which are projected towards the σ* (*i.e.* e_g) orbitals. *These aspects have been already discussed.*

(b) **Unusually slow electron exchange rate in the $[Fe(OH_2)_6]^{3+}$ – $[Fe(OH_2)_6]^{2+}$ couple in spite of the $\pi^* \rightarrow \pi^*$ electron transfer:** Though the process leads to the $\pi^* \rightarrow \pi^*$ electron transfer, the rate is very slow compared to the other systems involving the $\pi^* \rightarrow \pi^*$ transfer (*cf.* Table 6.3.6.1). In fact, besides the nature of the participating orbitals, many other factors are also important to determine the activation energy. In the present case, the **large bond length difference** between the two species (*cf.* Fe^{III}—OH_2 = 205 pm, Fe^{II}—OH_2 = 221 pm; Ru^{III}—NH_3 = 210 pm, Ru^{II}—NH_3 = 214 pm) requires ***a large reorganization energy*** to overcome the Franck-Condon barrier. In fact, this reorganization energy is consumed to contract the Fe^{II}—O bond length and to elongate the Fe^{III}—O bond length so that in the *transition state* the 'Fe—O' bond lengths become identical for the both centres (*cf.* Fig. 6.3.3.1). It may be noted that in the $[Ru(NH_3)_6]^{3+}$–$[Ru(NH_3)_6]^{2+}$ couple, the bond lengths differ by only 4 pm (*cf.* 16 pm for the Fe(III)-Fe(II) system). Consequently, it needs less reorganization energy to attain the transition state.

(c) **Unusually high electron exchange rate in the couple $[M(phen)_3]^{3+}$–$[M(phen)_3]^{2+}$ (M = Fe, Ru):** In fact, this rate benefit is common for the complexes bearing the π-*acceptor ligands* like CN^-, phen, bpy, etc. Such unsaturated π-acid ligands bear the vacant π^*–MOs that can overlap with the π^*–MOs of the complex. This delocalization of the π^*–MO of the complex onto the

Table 6.3.6.1 2nd order rate constants for electron exchange process occurring through the outer sphere electron transfer (OSET) process (*cf.* D.R. Stranks, *Discuss Faraday soc.*, **29**, 73, 1960., T. J. Meyer *et al.*, *Inorg. Chem.*, **7**, 2369, 1968).

Reaction/System	**Rate Constant*** **(dm^3 mol^{-1} s^{-1})**	***Net electron transfer***
(a) $[Fe(OH_2)_6]^{2+}$ + $[Fe(OH_2)_6]^{3+}$ $t_{2g}^4 e_g^2$ $t_{2g}^3 e_g^2$	2.0	
(b) $[Fe(CN)_6]^{4-}$ + $[Fe(CN)_6]^{3-}$ (t_{2g}^6) (t_{2g}^5)	10^3	
(c) $[Fe(phen)_3]^{2+}$ + $[Fe(phen)_3]^{3+}$ (t_{2g}^6) (t_{2g}^5)	$> 10^7$	
(d) $[Os(bpy)_3]^{2+}$ + $[Os(bpy)_3]^{3+}$ (t_{2g}^6) (t_{2g}^5)	$> 10^5$	$t_{2g} \rightarrow t_{2g}$ *i.e.* $\pi^* \rightarrow \pi^*$
(e) $[IrCl_6]^{3-}$ + $[IrCl_6]^{2-}$ (t_{2g}^6) (t_{2g}^5)	3×10^2	
(f) $[Ru(NH_3)_6]^{2+}$ + $[Ru(NH_3)_6]^{3+}$ (t_{2g}^6) (t_{2g}^5)	10^2	
(g) $[Ru(phen)_3]^{2+}$ + $[Ru(phen)_3]^{3+}$ (t_{2g}^6) (t_{2g}^5)	$> 10^7$	
(h) $[Co(OH_2)_6]^{2+}$ + $[Co(OH_2)_6]^{3+}$ $(t_{2g}^5 e_g^2)$ (t_{2g}^6)	3	
(i) $[Co(NH_3)_6]^{2+}$ + $[Co(NH_3)_6]^{3+}$ $(t_{2g}^5 e_g^2)$ (t_{2g}^6)	$< 10^{-8}$	
(j) $[Co(en)_3]^{2+}$ + $[Co(en)_3]^{3+}$ $(t_{2g}^5 e_g^2)$ (t_{2g}^6)	1.4×10^{-4}	$\sigma^* \rightarrow \sigma^*$ *i.e.* $e_g \rightarrow e_g$
(k) $[Co(ox)_3]^{4-}$ + $[Co(ox)_3]^{3-}$ $(t_{2g}^5 e_g^2)$ (t_{2g}^6)	9×10^{-7}	
(l) $[Co(phen)_3]^{2+}$ + $[Co(phen)_3]^{3+}$ $(t_{2g}^5 e_g^2)$ (t_{2g}^6)	1.1	

* Rate constants in the temperature range 0–25°C.

phenanthroline skeleton allows the mixing between the donor π^*–MO of the reductant (M_{red}) and receptor π^*–MO of the oxidant (M_{ox}), through the π^*-MOs of phenanthroline, *i.e.*

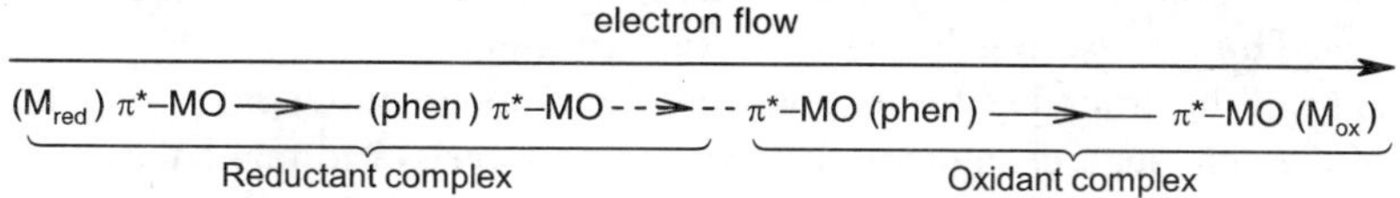

Thus it can be assumed that the electron from the donor π^*-MO of the reductant (M_{red}) is first transferred to the π^*-MO of its ligand (phen); thence to the π^*-MO of the phen ligand attached to the oxidant and thence to the receptor π^*-MO of the oxidant (*i.e.* M_{ox}). Thus participation of the vacant π^*-MOs of the phen ligands present in the coordination spheres of the oxidant (M_{ox}) and reductant (M_{red}) centres favour the electron transfer process. It happens so for the other complexes bearing the π-acid ligands having the vacant π^*–MOs.

(d) **Bond length reorganisation energy in the complexes bearing the π-acid ligands:** Besides the participation of the vacant π^*-MOs of the π-acid ligands in the electron transfer process as stated above, the π-acid ligands (*i.e.* **strong field ligands**) produce **another favourable effect,** *i.e.* geometrical parameters including the bond length for the both oxidized and reduced centres are very much comparable. For example, in the couples $[M(CN)_6]^{3-}$–$[M(CN)_6]^{4-}$ (M = Fe, Mn), $[Fe(phen)_3]^{3+}$–$[Fe(phen)_3]^{2+}$, the oxidized and reduced centres differ by one electron **in the nonbonding t_{2g} level** (*e.g.* t_{2g}^6 *vs.* t_{2g}^5 for the Fe(II)–Fe(III) couple, t_{2g}^5 *vs.* t_{2g}^4 for the Mn(II)–Mn(III) couple). *Consequently, bond lengths do not differ significantly in each couple. Moreover, the* ***metal → ligand π-back bonding*** *is more important for the lower valent complex. It will contract the bond length and the bond length becomes comparable to that in the higher valent complex, i.e. oxidized centre.* This explains the higher electron exchange rate in the $[Fe(CN)_6]^{4-}$–$[Fe(CN)_6]^{3-}$ couple (*cf.* Fe^{II}—C = 192 pm ≈ Fe^{III}—C = 193 pm) than that in the $[Fe(OH_2)_6]^{3+}$–$[Fe(OH_2)_6]^{2+}$ couple (*cf.* Fe^{II}—O = 221 pm > Fe^{III}—O = 205 pm).

Favourable factors (summary) in the electron exchange process in the couples $[Fe(CN)_6]^{4-}$–$[Fe(CN)_6]^{3-}$, $[FeL_3]^{2+}$–$[FeL_3]^{3+}$ (L = bpy, phen).

(i) ***No electronic activation*** is required for the favourable $\pi^* \rightarrow \pi^*$ (*i.e.* $t_{2g} \rightarrow t_{2g}$) electron transfer.

(ii) Bond length difference in Fe^{II}—L and Fe^{III}—L (L = π-acid ligand like CN^-, bpy, phen, etc.) is small. Generally, the Fe^{II}—L bond is lognger than the Fe^{III}—L bond, but the more favoured ligand → metal π-back bonding in the Fe(II)-complexes reduces the bond length. ***It makes the bond lengths in both the Fe(II) and Fe(III) complexes comparable.***

(iii) Electronic configurations of the low spin Fe(II) ($t_{2g}^6e_g^0$) and Fe(III) (t_{2g}^5) differ only in the t_{2g}-level. It makes also the ***bond lengths in the Fe(II) and Fe(III) complexes comparable.***

(iv) In the electron transfer process, there is ***no change in the spin state.***
All these above factors make the ***inner-sphere reorganization energy*** **less,** *i.e.* height of the Franck-Condon barrier becomes low.

(v) Besides the above factors, participation of the vacant π^*–MOs of the ligands (*i.e.* CN^-, phen, bpy) in mixing with the π^*–MO of the donor site (M_{red}) and π^*–MO of the receptor site (M_{ox}) facilitates the electron transfer process.

Note: It is worth mentioning that in the Co(III)/Co(II) couple, in the presence of strong field ligands like the π-acid ligands, ***electronic activation of Co(II) becomes easier.*** This gives an additional rate benefit (*cf.* (e) of this section).

(e) **Slow electron exchange rate in the Co^{III}–Co^{II} couple:** It experiences the net $\sigma^* \rightarrow \sigma^*$ electron transfer. In the octahedral geometries, the σ^* (*i.e.* e_g) orbitals are directly projected towards the ligands. Consequently, the σ^*– σ^* overlap between the receptor and donor orbital is not possible.

To execute the electron transfer process, the electronic configurations are to be changed so that the $\pi^* \rightarrow \pi^*$ transition can occur. It makes the activation energy high. Besides this, the oxidized centre Co^{III} is generally in low spin state (*i.e.* t_{2g}^6) while the reduced centre Co^{II} is in high spin state (*i.e.* $t_{2g}^5 e_g^2$). **Thus it also needs the spin state change.**

In fact, change of the ground state electronic configuration to allow the $\pi^* \rightarrow \pi^*$ electron transfer and requirement of the spin state change are the **two major barriers** to make the activation energy high.

Let us compare the following two redox couples:

$[Ru(NH_3)_6]^{3+}$–$[Ru(NH_3)_6]^{2+}$		$[Co(NH_3)_6]^{3+}$–$[Co(NH_3)_6]^{2+}$	
$(t_{2g}^5 e_g^0)$	(t_{2g}^6)	$(t_{2g}^6 e_g^0)$	$(t_{2g}^5 e_g^2)$
$t_{2g} \rightarrow t_{2g}$ **transfer (Net)**		$e_g \rightarrow e_g$ **transfer (Net)**	
Ru^{III}—NH_3 = 210 pm	Ru^{II}—NH_3 = 214 pm	Co^{III}—NH_3 = 194 pm	Co^{II}—NH_3 = 211 pm
i.e. bond length difference by 4 pm		*i.e.* bond length difference by 17 pm	

Bond length effect: It is evident that in terms of the bond length difference, the required *inner sphere reorganization energy* to overcome the Franck-Condon barrier is much higher for the $[Co(NH_3)_6]^{3+}$–$[Co(NH_3)_6]^{2+}$ couple.

Spin state effect: Co(III) is in low spin state while Co(II) is in high spin state ($t_{2g}^5 e_g^2$). Simple addition of one electron to Co(III) gives $t_{2g}^6 e_g^1$ which is not the ground state electronic configuration of Co(II). To attain the ground state electronic configuration of Co^{II}, it needs a rearrangement of electronic configuration.

$$\underset{\text{(Ground state, low spin)}}{Co(III)(t_{2g}^6 e_g^0)} \xrightarrow{+e} \underset{\text{(Excited state, low spin)}}{[Co(II)(t_{2g}^6 e_g^1)]} \longrightarrow \underset{\text{(Ground state, high spin)}}{Co(II)(t_{2g}^5 e_g^2)}$$

Similarly, simple removal of an electron from Co(II) does not generate the oxidized centre Co(III).

$$\underset{\text{(Ground state, high spin)}}{Co(II)(t_{2g}^5 e_g^2)} \xrightarrow{-e} \underset{\text{(Excited state)}}{[Co(III)(t_{2g}^5 e_g^1)]} \longrightarrow \underset{\text{(Ground state, low spin state)}}{Co(III)(t_{2g}^6 e_g^0)}$$

Thus the reaction between Co(III) and Co(II) at their ground states produce both the products in the excited states. ***This causes a barrier.***

Possible paths of electron transfer to the execute the $\pi^* \rightarrow \pi^*$, *i.e.* $t_{2g} \rightarrow t_{2g}$ electron transfer:

$$\textbf{Path I: } \left.\begin{matrix} Co(II) \\ + \\ Co(III) \end{matrix}\right\}: \begin{matrix} t_{2g}^5 e_g^2 \\ + \\ t_{2g}^6 e_g^0 \end{matrix} \rightleftharpoons \begin{matrix} t_{2g}^6 e_g^1 \\ + \\ t_{2g}^5 e_g^1 \end{matrix} \xrightleftharpoons{\pi^* \rightarrow \pi^*} \begin{matrix} t_{2g}^5 e_g^1 \\ + \\ t_{2g}^6 e_g^1 \end{matrix} \rightleftharpoons \begin{matrix} t_{2g}^6 e_g^0 \\ + \\ t_{2g}^5 e_g^2 \end{matrix} : \left\{\begin{matrix} Co(III) \\ + \\ Co(II) \end{matrix}\right.$$

$$\textbf{Path II: } \left.\begin{matrix} Co(II) \\ + \\ Co(III) \end{matrix}\right\}: \begin{matrix} t_{2g}^5 e_g^2 \\ + \\ t_{2g}^6 e_g^0 \end{matrix} \rightleftharpoons \begin{matrix} t_{2g}^5 e_g^2 \\ + \\ t_{2g}^4 e_g^2 \end{matrix} \xrightleftharpoons{\pi^* \rightarrow \pi^*} \begin{matrix} t_{2g}^4 e_g^2 \\ + \\ t_{2g}^5 e_g^2 \end{matrix} \rightleftharpoons \begin{matrix} t_{2g}^6 e_g^0 \\ + \\ t_{2g}^5 e_g^2 \end{matrix} : \left\{\begin{matrix} Co(III) \\ + \\ Co(II) \end{matrix}\right.$$

$$\textbf{Path III: } \left.\begin{matrix} Co(II) \\ + \\ Co(III) \end{matrix}\right\}: \begin{matrix} t_{2g}^5 e_g^2 \\ + \\ t_{2g}^6 e_g^0 \end{matrix} \rightleftharpoons \begin{matrix} t_{2g}^6 e_g^1 \\ + \\ t_{2g}^4 e_g^2 \end{matrix} \xrightleftharpoons{\pi^* \rightarrow \pi^*} \begin{matrix} t_{2g}^5 e_g^1 \\ + \\ t_{2g}^5 e_g^2 \end{matrix} \rightleftharpoons \begin{matrix} t_{2g}^6 e_g^0 \\ + \\ t_{2g}^5 e_g^2 \end{matrix} : \left\{\begin{matrix} Co(III) \\ + \\ Co(II) \end{matrix}\right.$$

$$\textbf{Path IV: } \left.\begin{matrix} Co(II) \\ + \\ Co(III) \end{matrix}\right\}: \begin{matrix} t_{2g}^5 e_g^2 \\ + \\ t_{2g}^6 e_g^0 \end{matrix} \rightleftharpoons \begin{matrix} t_{2g}^5 e_g^2 \\ + \\ t_{2g}^5 e_g^1 \end{matrix} \xrightleftharpoons{\pi^* \rightarrow \pi^*} \begin{matrix} t_{2g}^4 e_g^2 \\ + \\ t_{2g}^6 e_g^1 \end{matrix} \rightleftharpoons \begin{matrix} t_{2g}^6 e_g^0 \\ + \\ t_{2g}^5 e_g^2 \end{matrix} : \left\{\begin{matrix} Co(III) \\ + \\ Co(II) \end{matrix}\right.$$

Path-I needs activation of both centres and both the products are in the excited state.

Path-II needs activation of one centres, *i.e.* Co(III), and one product remains in the excited state.

Path-III needs activation of both centres and one product remains in the excited state.

Path-IV needs activation of one reactant, *i.e.* Co(III), but both the products remain in the excited state.

Co(III):	$t_{2g}^6e_g^0$ $\longrightarrow$	$t_{2g}^5e_g^1$ $\longrightarrow$	$t_{2g}^4e_g^2$	
cfse:	$24Dq_o - 2P$ (I)	$14Dq_o - P$ (II)	$4Dq_o$; (III)	$10Dq_o$ and P for Co(III) $10Dq_o > P$
Co(II):	$t_{2g}^5e_g^2$ $\longrightarrow$	$t_{2g}^6e_g^1$		
cfse:	$8Dq_o$ IV	$18Dq_o - P$; V		$10Dq_o$ and P for Co(II) $10Dq_o < P$

(To calculate the cfse, only the additional pairing (P) compared to that in the free state is considered.)

Path-I: $\text{IV} + \text{I} \rightleftharpoons \text{V} + \text{II} \xrightleftharpoons{\pi^* \to \pi^*} \text{II} + \text{V} \rightleftharpoons \text{I} + \text{IV}$

Path-II: $\text{IV} + \text{I} \rightleftharpoons \text{IV} + \text{III} \xrightleftharpoons{\pi^* \to \pi^*} \text{III} + \text{IV} \rightleftharpoons \text{I} + \text{IV}$

Path-III: $\text{IV} + \text{I} \rightleftharpoons \text{V} + \text{III} \xrightleftharpoons{\pi^* \to \pi^*} \text{II} + \text{IV} \rightleftharpoons \text{I} + \text{IV}$

Path-IV: $\text{IV} + \text{I} \rightleftharpoons \text{IV} + \text{II} \xrightleftharpoons{\pi^* \to \pi^*} \text{III} + \text{V} \rightleftharpoons \text{I} + \text{IV}$

Co(III) activation: It leads to ***low spin to high spin,*** *i.e.* $t_{2g}^6e_g^0 \rightarrow t_{2g}^5\, e_g^1 \rightarrow t_{2g}^4\, e_g^2$. The increase in electron density in the e_g-orbital will elongate the bond length and there is a loss of cfse.

Co(II) activation: It will lead to ***high spin to low spin,*** *i.e.* $t_{2g}^5e_g^2 \rightarrow t_{2g}^6e_g^1$. The decrease in electron density in the e_g-level will contract the bond length and there is a loss of cfse (because $P > 10Dq_o$).

Thus the bond length change in the activation of Co(III) and Co(II) centre will be reflected in the reorganization energy to increase the activation energy.

It is evident that activation energy for Co(III) (*i.e.* low spin to high spin, I→ II→ III) *increases with the increase of crystal field strength while activation energy for* Co(II) (*i.e.* high spin to low spin) *decreases with the increase of crystal field strength.* It indicates ***an opposing effect of the crystal field strength.*** In fact, it is very difficult to correlate the electron exchange rate in the Co(III)-Co(II) couple with the crystal field strength. The crystal field strength of the ligands runs in the sequence:

$$H_2O < \text{oxalate} < \text{edta} < NH_3 < \text{en} < \text{phen}.$$

The electron exchange rate in the Co(III)-Co(II) couple has been found **very fast at both ends of the series** (*i.e.* for both the weak and strong field ligands) and the rate is **very slow at the middle of the series.** In fact, the rate becomes minimum at NH_3 and it is not surprising. At this intermediate ligand field strength, activation is not favoured energetically both for Co(II) and Co(III). For the strong field ligands, activation of Co(II) (h.s. → l.s.) is favoured and for the weak field ligands, activation of Co(III) (l.s. → h.s.) is favoured.

Co(II)-Activation *vs.* Co(III)-Activation

If cfse of the low-spin Co(III) complexes does not differ drastically from that of the high spin Co(III) complexes (*i.e.* under the condition, $\Delta_o \geq P$), Co(II)-activation plays a crucial role in the electron exchange process in the Co(II)-Co(III) couple. For most of the $[CoL_6]^{2+}$ complexes, $\Delta_o \approx 10$ kK and $P \approx 20$ kK. Thus the Co(II) activation energy is given by $P - \Delta_o \approx 10$ kK ≈ 125 kJ mol^{-1} for the conversion of high spin ($t_{2g}^5\, e_g^2$) to low spin ($t_{2g}^6e_g^1$).

(f) **Unusually high electron exchange rate in the couple $[Co(OH_2)_6]^{3+}$ – $[Co(OH_2)_6]^{2+}$:** For $[Co(OH_2)_6]^{3+}$, the $10Dq_o$ value and P (pairing energy) are pretty close (*i.e.* $10Dq_o \approx$ P, critical $10Dq_o$ value) and it allows the spin isomerism (see Vol. 4, Table 3.8.1.2 and H. Taube *et al., JACS*, **73**, 4028, 1951).

$$\left[Co(OH_2)_6\right]^{3+}\left(l.s.,\ t_{2g}^6e_g^0\right) \rightleftharpoons \left[Co(OH_2)_6\right]^{3+}\left(h.s.,\ t_{2g}^4e_g^2\right)$$

It has been estimated that the low spin $[Co(OH_2)_6]^{3+}$ is more stable than the high spin $[Co(OH_2)_6]^{3+}$ by only 25 kJ mol^{-1}. ***Thus it is possible to activate $[Co(OH_2)_6]^{3+}$ for the electron transfer at the cost of a small amount of energy.*** The high spin Co(III) can react **(Path-II)** with the high spin Co(II) easily to give the observed rate benefit.

The reaction path may be outlined as follows:

$$\underset{(t_{2g}^6e_g^0)}{\left[Co(OH_2)_6\right]^{3+}} + \underset{(t_{2g}^5e_g^2)}{\left[Co(OH_2)_6\right]^{2+}} \xrightleftharpoons{\text{Activation of Co(III)}} \underset{(t_{2g}^4e_g^2)}{\left[Co(OH_2)_6\right]^{3+}} + \underset{(t_{2g}^5e_g^2)}{\left[Co(OH_2)_6\right]^{2+}}$$

$$\xrightleftharpoons{\pi^* \rightarrow \pi^*} \underset{(t_{2g}^5e_g^2)}{\left[Co(OH_2)_6\right]^{2+}} + \underset{(t_{2g}^4e_g^2)}{\left[Co(OH_2)_6\right]^{3+}}$$

$$\xrightleftharpoons{\text{Relaxation of Co(III) to ground state}} \underset{(t_{2g}^5e_g^2)}{\left[Co(OH_2)_6\right]^{2+}} + \underset{(t_{2g}^6e_g^0)}{\left[Co(OH_2)_6\right]^{3+}}$$

Path - II

Here the required activation energy for Co(III) (*i.e.* conversion of low spin state to high spin state) is very small. Relaxation of Co(III) (*i.e.* h.s. → l.s.) is energetically favourable. **It makes the rate so high.**

Here it is worth mentioning that the electron exchange in the couple $[Co(OH_2)_6]^{3+}/[Co(OH_2)_6]^{2+}$ can also partly go through the **inner sphere electron transfer path** (ISET path) by using the weak bridging property of H_2O (*cf.* Secs. 6.4.4, 6.6 for detailed discussion).

(g) **Unusually high electron exchange rate in the $[Co(phen)_3]^{3+}$–$[Co(phen)_3]^{2+}$ couple:** It has been already mentioned that for the very strong field ligands, the electron exchange process in the Co(III)/Co(II) couple is kinetically favoured. ***It is mainly due to the low cost of energy for activation of Co(II).*** In the presence of strong field ligand, high spin Co(II) can be easily converted into low spin Co(II) ($t_{2g}^6\, e_g^1$). For the π-acid ligands, the better Co(II) → L π-bonding makes the Co(II)—L bond relatively shorter, *i.e.* the ***Co(III)—L and Co(II)—L bond lengths do not differ significantly.*** It makes the inner sphere reorganisation energy less. In addition to this, the vacant π^*–MO of phen ligand can participate in the electron transfer process. The role of π-acid ligands in the rate acceleration of electron transfer in the OSET path has been discussed in detail in (c) of this section. Electron-flow occurs as follows through the vacant π^*–MOs of phen.

$$\text{Co(II)}(\pi^*\text{–MO}) \longrightarrow \text{phen}(\pi^*\text{-MO}) \dashrightarrow (\text{phen})(\pi^*\text{–MO}) \longrightarrow \text{Co(III)}(\pi^*\text{–MO})$$

(h) **$[Cr(OH_2)_6]^{2+}$ *vs.* $[V(OH_2)_6]^{2+}$ as the reducing agents:** The important kinetic and thermodynamic features of the reducing agents $[Cr(OH_2)_6]^{2+}$ and $[V(OH_2)_6]^{2+}$ are compared below.

(i) E^0 (Cr^{3+}/Cr^{2+}) = –0.41 V, E^0 (V^{3+}/V^{2+}) = –0.26 V, *i.e.* **Cr^{2+} is a better reducing agent than V^{2+} in terms of thermodynamics.** V^{3+}, V^{2+}, Cr^{3+} and Cr^{2+} simply denote their aqua complexes.

(ii) k_{ex} (Cr^{3+}–Cr^{2+}) ≈ 2 × 10^{-5} dm^3 mol^{-1} s^{-1} < k_{ex} (V^{3+}–V^{2+}) ≈ 10^{-2} dm^3 mol^{-1} s^{-1}. It indicates that **kinetically V^{2+} is a better reducing agent** in a cross-reaction because the rate constant (k_{12}) for the cross-reaction depends on the exchange rate constants k_{11} and k_{22} in terms of Marcus Equation: $k_{12} \approx \sqrt{k_{11}k_{22}K_{12}}$.

(iii) $[Cr(OH_2)_6]^{3+}$ ($t_{2g}^3 e_g^0$)–$[Cr(OH_2)_6]^{2+}$ ($t_{2g}^3 e_g^1$): σ^* (*i.e.* e_g) $\rightarrow \sigma^*$ (*i.e.* e_g) transfer.
$[V(OH_2)_6]^{3+}$ ($t_{2g}^2 e_g^0$)–$[V(OH_2)_6]^{2+}$ ($t_{2g}^3 e_g^0$): π^* (*i.e.* t_{2g}) $\rightarrow \pi^*$(*i.e.* t_{2g}) transfer.

This is why, k_{ex} for the couple V^{3+}–V^{2+} is higher than that for the Cr^{3+}–Cr^{2+} couple.

Table 6.3.6.2 Kinetic parameters of some cross-reactions occurring through the outer sphere electron transfer (OSET) process (*cf.* T. J. Meyer *et al., Inorg. Chem.,* **7,** 2369, 1968; R. A. Marcus *et al, Biochim. Biophys. Acta.,* **811,** 265, 1985).

Oxidant	*Reductant*	*Net electron transfer*	E^0_{12} **(volt)**	k_{12} **(dm³ mol⁻¹ s⁻¹)**
$[Co(NH_3)_6]^{3+}$	$[Cr(OH_2)_6]^{2+}$	$\sigma^* \rightarrow \sigma^*$	0.51	1×10^{-3}
$[Co(en)_3]^{3+}$	$[Cr(OH_2)_6]^{2+}$	$\sigma^* \rightarrow \sigma^*$	0.17	3.5×10^{-4}
$[Co(phen)_3]^{3+}$	$[Cr(OH_2)_6]^{2+}$	$\sigma^* \rightarrow \sigma^*$	0.83	3×10^{1}
$[Co(NH_3)_6]^{3+}$	$[V(OH_2)_6]^{2+}$	$\pi^* \rightarrow \sigma^*$	0.37	1×10^{-2}
$[Co(en)_3]^{3+}$	$[V(OH_2)_6]^{2+}$	$\pi^* \rightarrow \sigma^*$	0.015	7.2×10^{-4}
$[Co(phen)_3]^{3+}$	$[V(OH_2)_6]^{2+}$	$\pi^* \rightarrow \sigma^*$	0.675	4×10^{3}
$[Fe(OH_2)_6]^{3+}$ +	$[Cr(OH_2)_6]^{2+}$	$\sigma^* \rightarrow \pi^*$	1.18	2.3×10^{3}
$[Fe(OH_2)_6]^{3+}$ +	$[V(OH_2)_6]^{2+}$	$\pi^* \rightarrow \pi^*$	1.03	2×10^{4}
$[Ru(NH_3)_6]^{3+}$	$[V(OH_2)_6]^{2+}$	$\pi^* \rightarrow \pi^*$	0.355	8×10^{1}
$[Ru(NH_3)_6]^{3+}$	$[Cr(OH_2)_6]^{2+}$	$\sigma^* \rightarrow \pi^*$	0.51	2×10^{2}
$[Co(NH_3)_6]^{3+}$	$[Cr(bpy)_3]^{2+}$	$\pi^* \rightarrow \sigma^*$	–	7×10^{2}
$[Co(NH_3)_6]^{3+}$	$[Ru(NH_3)_6]^{2+}$	$\pi^* \rightarrow \sigma^*$	≈0.0	1×10^{-2}
$[Co(NH_3)_6]^{3+}$	$[Eu(OH_2)_6]^{2+}$		0.48	2×10^{-2}

(i) **Electronic activation for different paths in OSET:** OSET occurs through the $\pi^* \rightarrow \pi^*$ electron transfer. By keeping this view in mind, we can conclude: for the $\sigma^* \rightarrow \sigma^*$ net transfer, both the oxidant and reductant are to be electronically activated; for the $\pi^* \rightarrow \sigma^*$ net transfer, only the oxidant is to be electronically activated; for the $\sigma^* \rightarrow \pi^*$ net transfer, the reductant is to be electronically excited.

(j) **$[Ru(NH_3)_6]^{3+}$–$[Cr(OH_2)_6]^{2+}$ *vs.* $[Co(NH_3)_6]^{3+}$–$[Cr(OH_2)_6]^{2+}$:** Both the reactions are thermodynamically equally favourable ($E^0_{12} \approx 0.51$ V) but the rate of reduction of $[Ru(NH_3)_6]^{3+}$ is about 10^5 faster than the rate of reduction of $[Co(NH_3)_6]^{3+}$. In terms of Marcus equation, this rate benefit is mainly due to the higher electron exchange rate constant of the couple $[Ru(NH_3)_6]^{3+}$–$[Ru(NH_3)_6]^{2+}$.

k_{ex} in $[Ru(NH_3)_6]^{3+}$–$[Ru(NH_3)_6]^{2+} \approx 2 \times 10^2 \gg k_{ex}$ in $[Co(NH_3)_6]^{3+}$–$[Co(NH_3)_6]^{2+} \approx 10^{-3}$

$\pi^* \rightarrow \pi^*$ transfer	$\sigma^* \rightarrow \sigma^*$ transfer.
Ru^{III}—N = 210 pm,	Co^{III}—N = 194 pm
Ru^{II}—N = 214 pm,	Co^{II}—N = 211 pm.

The difference in the k_{ex} values (in dm³ mol⁻¹ s⁻¹) is quite reasonable in terms of the above bond length parameters (discussed earlier).

In other words, for the first reaction (*i.e.* reduction of Ru^{III}), the $\sigma^* \rightarrow \pi^*$ electron transfer process needs only the electronic activation of the reductant while for the second reaction (*i.e.* reduction of Co^{III}), the $\sigma^* \rightarrow \sigma^*$ electron transfer process needs the electronic activation of both the reductant and oxidant. It makes the rate of the second reaction slower.

(k) **$[Ru(NH_3)_6]^{3+}$–$[V(OH_2)_6]^{2+}$ *vs.* $[Co(NH_3)_6]^{3+}$–$[V(OH_2)_6]^{2+}$:** The first reaction is about 10^4 times faster than the second reaction though both the reactions are equally favoured thermodynamically. This can be explained by considering the higher electron exchange rate in the $[Ru(NH_3)_6]^{3+}$–$[Ru(NH_3)_6]^{2+}$ couple (*cf.* Marcus equation) and the favourable $\pi^* \rightarrow \pi^*$ electron transfer in the first reaction. The second reaction experiencing the $\pi^* \rightarrow \sigma^*$ electron transfer, needs the electronic activation of the oxidant.

(l) **Co(III)–Ru(II) *vs.* Ru(III)-Ru(II):** The examples are:

$\left[Ru(NH_3)_5(py)\right]^{3+} + \left[Ru(NH_3)_6\right]^{2+} \left(k_{12} = 1.4 \times 10^6\, dm^3\, mol^{-1}\, s^{-1},\ \log K_{12} \approx 4.4\right)$;

$\left[Ru(bpy)(NH_3)_4\right]^{3+} + \left[Ru(NH_3)_5(py)\right]^{2+} \left(k_{12} = 1.1 \times 10^8\, dm^3\, mol^{-1}\, s^{-1},\ \log K_{12} = 3.4\right)$;

$\left[Co(phen)_3\right]^{3+} + \left[Ru(NH_3)_6\right]^{2+} \left(k_{12} = 1.5 \times 10^4\, dm^3\, mol^{-1}\, s^{-1},\ \log K_{12} = 5.4\right)$, etc.

The electronic configurations are:

$$Co(III)\ (t_{2g}^6 e_g^0),\ Co(II)\ (t_{2g}^5 e_g^2);\ Ru(III)\ (t_{2g}^5 e_g^0),\ Ru(II)\ (t_{2g}^6 e_g^0)$$

Thus the Co(III)–Ru(II) system involves the net electron transfer $\pi^* \rightarrow \sigma^*$ (*i.e.* $t_{2g} \rightarrow e_g$) while the Ru(III)–Ru(II) reaction involves the net electron transfer $\pi^* \rightarrow \pi^*$. This is why, the Ru(III)-Ru(II) reaction is faster than the Co(III)–Ru(II) reaction, in general, in spite of the more thermodynamic favour in the Co(III)–Ru(II) reaction. *In other words,* $[Ru(aq)]^{2+}$ or $[Ru(NH_3)_6]^{2+}$ *is a good reducing agent in the OSET path as it requires no electronic activation. It may be noted that Ru(II) being inert* (t_{2g}^6) *cannot generally act as a reducing agent in the ISET process.*

(m) **Comparison of reducing activity of $[Cr(OH_2)_6]^{2+}$ and $[V(OH_2)_6]^{2+}$:** Thermodynamically, Cr^{2+} is a better reducing agent but kinetically, V^{2+} is a better reducing agent in the OSET process. In fact, V^{2+} reduces Co(III) or Ru(III) complexes much faster than Cr^{2+}. For the reduction of Co(III) complex by Cr^{2+}, it experiences the $\sigma^* \rightarrow \sigma^*$ net electron transfer and it requires the electronic activation of both the oxidant and reductant. While in the reduction of Co(III) complexes by V^{2+} through the $\pi^* \rightarrow \sigma^*$ net electron transfer, it needs the electronic activation of the oxidant only. Similarly, for the reduction of Ru(III) complexes by V^{2+} through the $\pi^* \rightarrow \pi^*$ net electron transfer, it needs no electronic activation of the both oxidant and reductant. On the other hand, for the reduction of Ru(III) complexes by Cr^{2+}, through the $\sigma^* \rightarrow \pi^*$ net electron transfer, it needs an electronic activation of the reductant. In terms of Marcus equation, the rate benefit for the reduction by V^{2+} over the reduction by Cr^{2+} arises from the higher electron exchange rate constant in the V^{3+}–V^{2+} couple ($\pi^* \rightarrow \pi^*$ electron transfer) compared to that in the Cr^{3+}–Cr^{2+} couple ($\sigma^* \rightarrow \sigma^*$ electron transfer). In other words, for the required $\pi^* \rightarrow \pi^*$ electron transfer, in cross-reactions, the reductant V^{2+} needs no electronic activation while the reductant Cr^{2+} needs an electronic activation. It makes the activation energy less for the reduction by V^{2+}.

For a common oxidant, if the rate of reduction by V^{2+} is found faster than the rate of reduction by Cr^{2+}, then it can be concluded that the redox process involves the outer sphere electron transfer mechanism.

6.3.7 Effect of External Ions on Electron Transfer Rate in the OSET Process

Sometimes, there is an **electrostatic barrier in forming the precursor outer sphere complex** as in the case of electron exchange in the couples $[Co(NH_3)_6]^{3+}$– $[Co(NH_3)_6]^{2+}$, $[Fe(CN)_6]^{3-}$– $[Fe(CN)_6]^{4-}$, $[MnO_4]^-$– $[MnO_4]^{2-}$, etc. Oppositely charged ions can bridge the oxidant and reductant bearing the same type of charge to reduce the electrostatic repulsive barrier.

- **Alkali metal ions (M^+)** can catalyze the electron exchange rate in the couple MnO_4^-– MnO_4^{2-}. M^+ probably bridges them in the outer-sphere complex as follows:
  $[MnO_4^{2-}\cdots M^+\cdots MnO_4^-]$, *i.e.* $\{[MnO_4]\cdots M\cdots[MnO_4]\}^{2-}$. The observed catalytic efficiency runs as:

$$Cs^+ > K^+, Na^+ > Li^+.$$

- **Alkali metal ions (M^+) and alkaline earth metal ions (M^{2+})** are also found to catalyze the electron exchange process in the $[Fe(CN)_6]^{4-}$–$[Fe(CN)_6]^{3-}$ couple. The transition state may involve the following bridges (with M^+).

$$\{[Fe(CN)_6]\cdots M\cdots[Fe(CN)_6]\}^{6-},\ \{M\cdots[Fe(CN)_6]\cdots M\cdots[Fe(CN)_6]\}^{5-},$$
$$\{M\cdots[Fe(CN)_6]\cdots M\cdots[Fe(CN)_6]\cdots M]\}^{4-}.$$

The catalyzing efficiency runs as:

$$Cs^+ > Rb^+ > K^+, NH_4^+ > Na^+ > Li^+ > H^+;\ Sr^{2+} > Ca^{2+} > Mg^{2+}.$$

It is suggested that to minimize the electrostatic repulsion in the outer-sphere complex of transition state, the **partially desolvated cation** bridges between the exchanging anions. The bridging cation should not be very large. *In fact, the large cation like Ph_4As^+ is not effective in catalyzing the process.* ***At the same time, the energy required to desolvate the catalyzing ion should not be large. This is why, Li^+ which requires a high desolvation energy acts as a poor catalyst.***

- For the exchanging cations, a suitable anion can bridge to catalyze the process. In fact, it has been found that OH^- can catalyze the electron exchange process in the $[Co(NH_3)_6]^{3+}$–$[Co(NH_3)_6]^{2+}$ couple. This can occur in **two ways to stabilize the outer sphere complex in the transition state.**

$$[Co(NH_3)_6]^{3+}\cdots OH^-\cdots[Co(NH_3)_6]^{2+}$$

or

$$[Co(NH_3)_6\cdots \underset{H}{O}\cdots(H_3N)_6Co]^{4+}$$

I

Simple bridging by OH^-

$$[Co^{III}(NH_3)_6\cdots \underset{H}{O}—Co^{II}(NH_3)_5]^{4+}$$

II

Bridging by the coordinated OH^- group of the Co(II)-centre with the Co(III)-centre

Co(II) centre is **labile** and it can produce $[Co(NH_3)_5(OH)]^+$ and this coordinated OH^- group can bridge with the coordination sphere of Co(III) through H-bonding (**II**). Participation of the substituted Co(II)-complex, *i.e.* $[Co(NH_3)_5(OH)]^+$, in the electron exchange process is supported by the fact that the product contains the $[Co(NH_3)_5(OH)]^{2+}$ species. Formation of the Co(III)-complex as the $[Co(NH_3)_5(OH)]^{2+}$ can be explained by considering the transition state (**II**). Cl^- is also found to catalyze the electron exchange process in the $[Co(NH_3)_6]^{2+}$–$[Co(NH_3)_6]^{3+}$ couple and the catalysis may originate in the same way.

In fact, in the outer sphere complex, aqua and hydroxo-aqua complexes can participate in H-bonding interaction.

```
       H        H                 H
       |        |                 |
  M—O····H—O—M'         M—O····H—O—M'
       |                          |
       H                          H
```

aqua-aqua (III) aqua-hydroxy (IV)

Obviously the 'OH' group is more effective than H_2O in carrying out the said H-bonding interaction as in **IV**.

6.4 INNER SPHERE ELECTRON TRANSFER (ISET) PROCESS
(*cf.* H. Taube, *J. Chem. Edu.*, **45**, 452, 1968)

The reduction of $[Co(NH_3)_6]^{3+}$ by $[Cr(OH_2)_6]^{2+}$ is slow ($k \approx 10^{-4}$ dm^3 mol^{-1} s^{-1}) while the reduction of $[CoCl(NH_3)_5]^{2+}$ by the same reducing agent is reasonably much faster ($k \approx 6 \times 10^5$ dm^3 mol^{-1} s^{-1}) under the comparable conditions. It may be noted that the thermodynamic driving force for both the reactions is more or less comparable.

This tremendous rate enhancement by a factor about 10^9 indicates that the reduction pathways are different for $[Co(NH_3)_6]^{3+}$ and $[CoCl(NH_3)_5]^{2+}$ (*cf.* Sec. 6.4.4 for the classic work of **H. Taube** to establish the ISET path).

$$\left[Co(NH_3)_6\right]^{3+} + \left[Cr(OH_2)_6\right]^{2+} \xrightarrow{H^+} \left[Co(OH_2)_6\right]^{2+} + \left[Cr(OH_2)_6\right]^{3+} + 6NH_4^+$$

$$(t_{2g}^6 e_g^0) \qquad (t_{2g}^3 e_g^1) \qquad (t_{2g}^5 e_g^2) \qquad (t_{2g}^3 e_g^0)$$

i.e. $e_g \rightarrow e_g$ electron transfer, *i.e.* **$\sigma^* \rightarrow \sigma^*$ electron transfer.**

$$\left[CoCl(NH_3)_5\right]^{2+} + \left[Cr(OH_2)_6\right]^{2+} \xrightarrow{H^+} \left[Co(OH_2)_6\right]^{2+} + \left[CrCl(OH_2)_5\right]^{2+} + 5NH_4^+$$

$$(t_{2g}^6 e_g^0) \qquad (t_{2g}^3 e_g^1) \qquad (t_{2g}^5 e_g^2) \qquad (t_{2g}^3 e_g^0)$$

i.e. $e_g \rightarrow e_g$ electron transfer $\equiv$ **$\sigma^* \rightarrow \sigma^*$ electron transfer.**

Besides the rate enhancement, in the second reaction, another important observation is that it produces the chlorido-complex of Cr(III) which is a very inert centre. It indicates that during the electron transfer process, chloride is transferred to the reducing agent. From the knowledge of anation rate of $[Cr(OH_2)_6]^{3+}$ (*i.e.* $[Cr(OH_2)_6]^{3+} + Cl^- \rightarrow [CrCl(OH_2)_5]^{2+} + H_2O$, $k \approx 10^{-8}$ dm^3 mol^{-1} s^{-1} which is much slower than the rate constant for the said redox reaction coupled with the Cl^- transfer) and isotope labelling experiment, it was proved that Cl^- does not enter into the coordination sphere of Cr(III) after its formation. To explain the experimental findings, *i.e.* ***rate enhancement and electron transfer accompanied with the chloride transfer to the reducing agent,*** it was required to consider that the second reaction does not go on through the outer-sphere mechanism and it occurs through a different mechanism which involves the electron transfer and chloride transfer in a concommitant step. This mechanistic path is described as the **inner sphere electron transfer (ISET) process.**

6.4.1 Steps of the Inner Sphere Electron Transfer Process

The inner sphere mechanism involves the formation of a *binuclear complex* (called **precursor complex**) from the reactants followed by the electron transfer across the bridging ligand within the binuclear complex. This electron transfer produces the *successor binuclear complex* that undergoes fission to give the products.

To allow the formation of the *binuclear precursor complex,* generally the oxidant (that is substitutionally inert) possesses a bridging ligand (say X) that leads to a nucleophilic substitution at the reducing centre (which is substitutionally labile). Thus, the **pre-requisite conditions** for the formation of the binuclear precursor complex are: *presence of a ligand having the bridging capacity (i.e. capable of making a bridge) in the coordination sphere of the inert oxidant and sufficient lability of the reducing centre to experience a nucleophilic substitution by the bridging ligand provided by the oxidant.* In some rare cases (*e.g.* reduction of $HCrO_4^-$ by $[Fe(CN)_6]^{4-}$) the inert reducing centre provides the bridging ligand towards the labile oxidant.

Thus between the reactants, one (generally the oxidant) is inert and the other (generally the reductant) is labile. The inert centre will provide the bridging ligand. This is the condition in general.

The *elementary steps* of an ISET process can be represented as follows in general.

Note: Presence of a bridging ligand with the inert oxidant does not always ensure its reduction by a reductant through the ISET path, *e.g.* **$[Ru(NH_3)_5X]^{2+}$ + $[V(aq)]^{2+}$: OSET path** (*cf.* V^{2+} : d^3 inert).

Step-I (Formation of the Binuclear Precursor Complex)

$$Ox—X + H_2O—Red \rightleftharpoons [Ox—X—Red]\,(\text{precursor complex}) + H_2O$$

Considering the reaction in aqueous solution, the reducing centre (Red) is to be sufficiently labile to experience a nucleophilic substitution by the ligand X (that can act as a bridging ligand) present in the coordination sphere of the starting oxidant (*i.e.* Ox—X) which is itself inert.

(**Note:** Generally, the inert oxidant centre provides the bridging ligand. In some rare cases, *e.g.* reduction of $[HCrO_4]^-$ by $[Fe(CN)_6]^{4-}$, the **inert reductant provides the bridging ligand** to the labile oxidant, see Sec. 6.4.13.).

Step-II (Chemical Activation of the Precursor Complex)

$$[\text{Ox—X—Red}] \rightleftharpoons [\text{Ox—X—Red}]^{\neq} \ (\textbf{Activated Complex})$$

This step needs the necessary inner sphere reorganizations required for the electron transfer.

This step may require the change of electronic configuration (*i.e. electronic activation*), bond length adjustment and adjustment of other geometrical parameters (*i.e.* **product-like** structural parameters are attained in the activated complex). ***All these changes are required to overcome the Franck-Condon barrier.***

Step-III (Electron transfer within the activated precursor complex)

This leads to the formation of a *binuclear successor complex.*

$$[\text{Ox} \overset{e}{\curvearrowleft} \text{X—Red}]^{\neq} \xrightleftharpoons{e\text{-transfer}} [\text{Ox}^-\text{—X—}^+\text{Red}] \ (\textbf{Successor complex})$$

In the electron transfer step within the binuclear successor complex, **reorganisation energy is required to cross the Franck-Condon barrier** as in the OSET process. It also requires solvent reorganisation energy and others as in the case of OSET process. In fact, for the electron transfer step, a **Marcus type relationship** is also applicable.

Step-IV (Fission of the Binuclear Successor Complex to the Products)

This step will lead to the mononuclear products. The position of bond cleavage depends on the relative ease (both thermodynamically and kinetically) of cleavage of the bonds Ox^-—X and X—$^+$Red.

$$\text{Ox}^- \underset{\uparrow}{—}\text{X—}^+\text{Red} + \text{OH}_2 \rightleftharpoons \text{Ox}(\text{OH}_2)^- + \text{X—}^+\text{Red}$$

Point of cleavage (**leading to the transfer of '*X*' to the reducing agent**).

OR

$$\text{Ox}^-\text{—X}\underset{\uparrow}{—}^+\text{Red} + \text{H}_2\text{O} \rightleftharpoons \text{Ox}^-\text{—X} + \text{H}_2\text{O—}^+\text{Red}$$

Point of cleavage (**leading to the retention of '*X*' with the oxidizing agent**).

Thus possibility of transfer or retention of the bridging ligand (X) depends on the relative lability of the Ox^- and Red^+ centres and the relative bond strengths of Ox^-—X and Red^+—X bonds. **Thus the ligand transfer is not an essential condition for the ISET process.**

It may be noted that in the OSET process, during the course of electron transfer, the coordination spheres of the reactants remain unchanged and *no chemical bonding interaction occurs between the coordination spheres.* On the other hand, in the ISET process, coordination spheres of the reactants are linked through a bonding interaction. *Thus besides the electron transfer, in the ISET process, bond formation and bond breaking are also required in the overall process.*

● **Outer-sphere *vs.* inner-sphere electron transfer reaction (OSET *vs.* ISET paths):** No guarantee for the inner-sphere reduction of an inert oxidant bearing a suitable bridging ligand (e.g. $[Co^{III}(NH_3)_5X]^{2+}$, $[Ru^{III}(NH_3)_5X]^{2+}$:

(a) **Reduction of Co^{III}-complexes (*e.g.* $[Co(NH_3)_5X]^{2+}$) by the inert reductants like $[Cr(bpy)_3]^{2+}$ and $[Ru(NH_3)_6]^{2+}$ through the OSET path instead of the ISET path:** The reduction of Co^{III}-complexes, *i.e.* $[Co(NH_3)_5X]^{2+}$, by the **coordinatively saturated and substitutionally inert reductants** like $[Cr(bpy)_3]^{2+}$ (t_{2g}^4, high CFAE) and $[Ru(NH_3)_6]^{2+}$ (t_{2g}^6, high CFAE) through the ISET path is kinetically highly unfavourable because in the ISET path, the first step involving the ligand substitution at the reductant by the bridging ligand (X) bound to Co^{III} to produce the binuclear precursor complex is unfavourable in such cases. In fact, in these cases, the OSET path is favoured (Table 6.4.1.1).

Table 6.4.1.1: Rate constants (in $dm^3 mol^{-1} s^{-1}$ at 25°C) for the reduction of $[Co(NH_3)_5X]^{3+/2+}$ by $[Cr(bpy)_3]^{2+}$ and $[Ru(NH_3)_6]^{2+}$ through the OSET path

Oxidant	*k* ($dm^3 mol^{-1} s^{-1}$) for the reductants	
	$[Cr(bpy)_3]^{2+}$	$[Ru(NH_3)_6]^{2+}$
$[Co(NH_3)_5(NH_3)]^{3+}$	6.9×10^2	1.1×10^{-2}
$[Co(NH_3)_5(OH_2)]^{3+}$	5×10^4	3.0
$[Co(NH_3)_5(OH)]^{2+}$	3×10^4	4×10^{-2}
$[Co(NH_3)_5(NO_3)]^{2+}$	–	3.4×10^1
$[CoF(NH_3)_5]^{2+}$	1.8×10^3	–
$[CoCl(NH_3)_5]^{2+}$	8×10^5	2.6×10^2
$[CoBr(NH_3)_5]^{2+}$	5×10^6	1.6×10^3
$[CoI(NH_3)_5]^{2+}$	–	6.7×10^3
$[Co(NH_3)_5(N_3)]^{2+}$	4.1×10^4	1.2

In the given OSET reactions, the **net electron transfer** is: $\pi^* \rightarrow \sigma^*$. It requires the electronic activation of the low-spin Co^{III}-complexes as the oxidants. This electronic activation energy is given by the **excitation energy:** $t_{2g}^6 \rightarrow t_{2g}^5 e_g'$. This varies with the ligand field strength indicated by the shift of λ_{max} (*i.e.* **blue shift** with the increase of ligand field strength). It explains the reactivity order:
$[CoI(NH_3)_5]^{2+} > [CoBr(NH_3)_5]^{2+} > [CoCl(NH_3)_5]^{2+} > [CoF(NH_3)_5]^{2+}$ (*cf.* crystal field strength: $F^- > Cl^- > Br^- > I^-$).

(b) **Reduction of Ru^{III}-complexes by $[V(aq)]^{2+}$ through the OSET path instead of the ISET path:** V(II) (t_{2g}^3) is inert to the substitution reactions because of the relatively high CFAE. For the reduction of $[Ru(NH_3)_5X]^{2+}$ complexes by V(II), the net electron transfer is $\pi^* \rightarrow \pi^*$ which is the **most favourable condition for OSET**. If the said reduction is supposed to occur through ISET, it faces two energy barriers.
(i) electronic activation of both the oxidant and reductant (*cf.* ISET passes through the $\sigma^* \rightarrow \sigma^*$ electron transfer); (ii) V(II) is substitutionally inert (*cf.* CFAE for the d^3-system is relatively high).
Because of the said energetic disfavours in the ISET path, the **reaction preferably goes through the OSET path**.

$$[Ru(NH_3)_5Br]^{2+} + [V(aq)]^{2+} \xrightarrow{\text{OSET}} [Ru(NH_3)_5Br]^{+} + [V(aq)]^{3+}$$

The rate constant ($k \approx 5 \times 10^3$ $dm^3 mol^{-1} s^{-1}$) for the above reaction, is much faster than that of the **typical anation reaction** at $[V(aq)]^{2+}$ (*cf.* $k \approx 28$ $dm^3 mol^{-1} s^{-1}$ for anation by NCS^- at $[V(aq)]^{2+}$. This experimental finding rules out the possibility of ISET path that involves the anation at $[V(aq)]^{2+}$ at the first step. This aspect will be discussed in detail later.
Note: There are evidences in some cases to support the reduction of $[Co^{III}(NH_3)_5X]^{2+}$ complexes by $[V(aq)]^{2+}$ to go through the ISET path. Here, the net electron transfer, $\pi^* \rightarrow \sigma^*$, requires only the electronic activation of the reductant for the ISET path.

(c) **Reduction of Fe(III) complexes by $[V(aq)]^{2+}$ through the OSET path:** The rate constants for the reduction of $[Fe^{III}(OH_2)_5X]^{2+}$ ($t_{2g}^3 e_g^2$) by $[V(aq)]^{2+}$ (t_{2g}^3) are much faster than the typical anation rate constants for $[V(aq)]^{2+}$. It rules out the possibility of the ISET path. The second order rate constants for the reduction of $[Fe(OH_2)_5X]^{2+}$ ($X = Cl^-$, N_3^- and NCS^-) by $[V(aq)]^{2+}$ are in the range (4.5–6.5) $\times 10^5$ $dm^3 mol^{-1}s^{-1}$ (*cf.* for the anation by NCS^- at $[V(aq)]^{2+}$, $k \approx 28$ $dm^3 mol^{-1} s^{-1}$). In these electron transfer reactions, the net electron transfer, $\pi^* \rightarrow \pi^*$ requires the electronic activation at both the oxidant and reductant for the ISET path ($\sigma^*\rightarrow\sigma^*$) while it requires no electronic activation at both the reductant and oxidant for the OSET path ($\pi^*\rightarrow\pi^*$). The electronic activation and inertness of the reductant (V^{II}, d^3 system) direct the reaction to pass through the OSET path. A similar situation arises for the reduction of Ru^{III}-complexes by $[V(aq)]^{2+}$.

6.4.2 Rate Law for the ISET Process

By considering the overall process to be irreversible, the process can be simply represented as follows:

$$\text{Ox—X} + \text{Red} \underset{k_{-1}}{\overset{k_1}{\rightleftharpoons}} \underset{\textbf{I}}{[\text{Ox—X — Red}]} \xrightarrow{k} \text{Products}$$

Here k denotes the overall rate constant for the 2nd, 3rd and 4th steps shown in Sec. 6.4.1. Thus k stands for the overall process involving activation of the binuclear precursor complex, electron transfer within the activated complex and dissociation of the successor complex into the final products. Under the **steady state condition** of the intermediate binuclear complex (**I**), we can write:

$$\frac{d[\text{I}]}{dt} = 0 = k_1[\text{Ox—X}][\text{Red}] - k_{-1}[\text{I}] - k[\text{I}]$$

i.e.

$$[\text{I}] = \frac{k_1[\text{Ox—X}][\text{Red}]}{k_{-1} + k}$$

Thus the rate is given by:

$$\text{rate} = k[\text{I}] = \frac{kk_1[\text{Ox—X}][\text{Red}]}{k_{-1} + k}$$

Under different conditions, the above rate law reduces as follows:

Case-I: When $k \gg k_{-1}$, the rate is given by:

$$\text{rate} \approx k_1[\text{Ox—X}][\text{Red}]$$

In such cases, the rate determining step is the formation of the precursor binuclear complex. *Thus the rate determining step is shifted to the step of nucleophilic substitution at the reducing centre.* If the reducing centre experiences the substitution through a **dissociative path**, then the rate becomes insensitive to the nature of the bridging ligand *X* (that acts as the entering ligand in the substitution process).

Obviously, shifting of the rate determining step to the nucleophilic substitution step at the reducing centre occurs when the reducing agent is relatively inert (e.g. $[V(aq)]^{2+}$, d^3 system).

Case-II: When $k_{-1} \gg k$, the rate is given by:

$$\text{rate} \approx \frac{kk_1}{k_{-1}}[\text{Ox—X}][\text{Red}] = kK[\text{Ox—X}][\text{Red}],$$

$$\left(K = \frac{k_1}{k_{-1}}\right).$$

In such cases, the rate determining step involves the activation of the binuclear precursor complex and electron transfer within the binuclear precursor complex leading to the binuclear successor complex *(considering fission of the successor complex into the products to be fast).*

Here it may be noted that at the electron transfer step within the binuclear precursor complex, activation energy is required to **cross the Franck-Condon barrier**. It also requires the solvent reorganisation energy and others as in the OSET process. In fact, for this electron transfer step, a **Marcus type relationship** is also applicable.

Note: In some cases, the step leading to the fission of the successor complex is very much kinetically disfavoured (compared to the other steps – formation of the binuclear precursor complex, and electron transfer within the binuclear precursor complex) and the rate determining step is shifted to this step. In such cases, the overall reaction can be represented as follows:

$$\text{Ox—X} + \text{Red} \underset{}{\overset{K_{eq}}{\rightleftharpoons}} [\text{Ox}^-\text{—X—}^+\text{Red}], \text{ (cumulative expression of Steps I, II and III)}$$

$$[\text{Ox}^-\text{—X—}^+\text{Red}] \xrightarrow[(rds)]{k} \text{product (mononuclear complexes)}.$$

The rate of the overall process can be given as follows.

$$\text{rate} = \frac{kK_{eq}[\text{Ox—X}][\text{Red}]}{1 + K_{eq}[\text{Red}]}, \text{ (when } [\text{Red}]_T \gg [\text{Ox—X}]_T\text{), (see Sec. 5.9.3 for derivation)}$$

$$\approx kK_{eq}[\text{Ox—X}][\text{Red}], \text{ when } 1 \gg K_{eq}[\text{Red}]$$

where K_{eq} is the overall equilibrium constant of the steps I, II and III leading to the formation of the binuclear successor complex, and *k actually stands for the rate constant of the step leading to the product from the successor complex through the bond fission.* Such a situation arises when both the metal centres (*i.e.* Ox^- and ^+Red) present within the successor complex are very much kinetically inert in the substitution process.

6.4.3 Effect of the Nature of Donor Orbital (*i.e.* HOMO) of the Reductant and Receptor Orbital (*i.e.* LUMO) of the Oxidant in the ISET Process

It has been already pointed out that in the OSET process, the electron transfer occurs from the π^*-*MO* of the reductant to the π^*-*MO* of the oxidant in the activated outer-sphere complex. Thus if the $\pi^* \rightarrow \pi^*$ (*i.e.* $t_{2g} \rightarrow t_{2g}$) electron transfer is possible from the ground state electronic configurations of the reactants then the process can go on without any electronic activation in the OSET process. *Such OSET reactions are very fast.* In the OSET process, the $\sigma^* \rightarrow \sigma^*$ net electron transfer (*i.e.* $e_g \rightarrow e_g$ transfer) needs electronic activation of both the oxidant and reductant. *The process is energetically costly and such OSET reactions are tremendously slow.*

In the ISET process, for the $\sigma^* \rightarrow \sigma^*$ net electron transfer (i.e. $e_g \rightarrow e_g$), the process is highly favoured and for the $\pi^* \rightarrow \pi^*$ net electron transfer (i.e. $t_{2g} \rightarrow t_{2g}$), the process is tremendously disfavoured. In fact, no ISET reaction involving the $\pi^* \rightarrow \pi^*$ net electron transfer is known and **all the known reactions experiencing the $\pi^* \rightarrow \pi^*$ net electron transfer undergo the OSET process.** Table 6.4.3.1 illustrates the effect of the nature of HOMO and LUMO of the reactants on the reaction rates of the OSET and ISET processes.

Table 6.4.3.1 Effect of the nature of the donor and receptor orbitals on the rate constants of the reactions passing through the OSET and ISET process.

Donor orbital of the reductant (i.e. HOMO of the reactant)	*Receptor orbital of the oxidant (i.e. LUMO of the oxidant)*	*Net electron transfer*	*Examples*	$\frac{k_{ISET}}{k_{OSET}}$ *ratio*
π^*	π^*	$\pi^* \rightarrow \pi^*$	V^{2+}–Ru^{3+}	Data not available
π^*	σ^*	$\pi^* \rightarrow \sigma^*$	V^{2+}–Co^{3+}	~10^4
σ^*	π^*	$\sigma^* \rightarrow \pi^*$	Cr^{2+}–Ru^{3+}	~10^2
σ^*	σ^*	$\sigma^* \rightarrow \sigma^*$	Cr^{2+}–Co^{3+}	~10^{10}

$V^{2+}(t_{2g}^3e_g^0)$; $Cr^{2+}(t_{2g}^3\,e_g^1)$; $Co^{3+}(t_{2g}^6\,e_g^0$, l.s.); $Co^{2+}(t_{2g}^5\,e_g^2$, h.s.); $Ru^{3+}(t_{2g}^5\,e_g^0$, l.s.); $Ru^{2+}\,(t_{2g}^6\,e_g^0$, l.s.).

The preference behind the $\sigma^* \rightarrow \sigma^*$ electron transfer (*i.e.* $e_g \rightarrow e_g$) in the ISET process can be explained by considering the **orbital requirement for the formation of the bridging segment M_{ox}—**

***X*—*M*$_{red}$** in the binuclear precursor complex. Considering the bridge along the *z*-axis, *the* d_{z^2} *orbitals* (*i.e.* e_g orbitals) *of* M_{ox} *and* M_{red} *can combine with the filled* p_z *orbital of the bridging ligand* X^- (say Cl^-). This will produce **three three-centre sigma *MO*s** (one bonding, one nonbonding and one antibonding). This **bonding scheme** is very much comparable to that in the XeF_2 compound. The bonding MO (σ_b) is mainly enriched with the p_z-orbital character of the bridging ligand (X^-) which is more electronegative than the metal centres while the NBMO (σ_{nb}) is mainly concentrated on the metal centres (see Secs. 9.12.12 and 9.23, Vol. 2).

If it is assumed that the d_{z^2} orbital of reductant (M_{red}) bears an unpaired electron; the d_{z^2} orbital of the oxidant (M_{ox}) is vacant and the p_z orbital of X^- (say Cl^-) is filled in, then these three electrons are distributed as σ_b^2, σ_{nb}^1 and σ_{ab}^0. (*i.e.* **3*c*–3*e* bonding**). *The placement of the unpaired electron from the* d_{z^2} *orbital of the reductant into the nonbonding MO* (*i.e.* σ_{nb}) *shifts partially the electron to the* M_{ox} *because the three centre nonbonding MO* (*i.e.* σ_{nb}) is mainly enriched with the terminal metal orbital characters of the bridging segment. The bond order in each M_{ox}—X and X—M_{red} bond in the precursor complex is half. Thus the precursor complex formation yields the following results:

> *partial transfer of the electron from* M_{red} *to* M_{ox}; *bond weakening in the original 2c–2e* M_{ox}—X *bond (i.e. bond order decreases from 1 to 1/2) and a new bond formation (of half bond order) by the bridging atom with the* M_{red}.

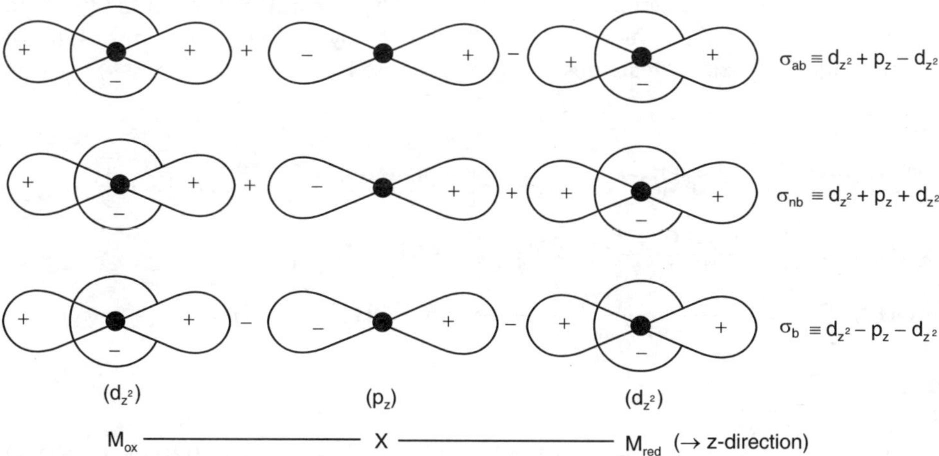

Fig. 6.4.3.1: Orbital picture for the formation of the bridging segment M_{ox}—X—M_{red} in the precursor complex (σ_b, σ_{nb} and σ^* denote the three three-centre MOs: bonding, nonbonding and antibonding respectively).

Now, if the precursor complex approaches smoothly towards the products along the reaction coordinate, the electron in the nonbonding *MO* (*i.e.* σ_{nb}^1) moves out of the M_{red}—X region into the region of M_{ox}—X. This leads to the electron transfer from M_{red} to M_{ox}.

Thus for the electron transfer from the reductant to the oxidant in the ISET process, the donor orbital (HOMO) of the reductant and the receptor orbital (LUMO) of the oxidant must be of σ^* type (*i.e.* e_g in terms of CFT). Formation of the three centre delocalized σ-*MOs* allows **the electron transfer through the delocalized MOs.** It explains that in the ISET process, electron transfer occurs from $\sigma^* \rightarrow \sigma^*$ (*i.e.* $e_g \rightarrow e_g$).

If the ground state electronic configurations of the reactants do not allow this $\sigma^* \rightarrow \sigma^*$ *electron transfer path then the reactants will be required to be electronically activated.*

Obviously, for the net $\pi^* \rightarrow \pi^*$ electron transfer process, it needs an electronic activation of both the oxidant and reductant and consequently, the activation energy in the ISET process will be very high. In fact, the systems having the $\pi^*\rightarrow\pi^*$ net electron transition adopt the OSET pathway where there is no requirement of electronic activation. In the same line of argument, for the $\sigma^*\rightarrow\pi^*$ net electron transfer process, the oxidant is to be electronically activated for the required $\sigma^*\rightarrow\sigma^*$ electron transfer path in the ISET process. Similarly, for the $\pi^*\rightarrow\sigma^*$ net electron transfer, the reductant centre needs an **electronic activation** in the ISET path.

6.4.4 Comparison of Reaction Rates for the Reduction of Co(III) Complexes by $[Cr(OH_2)_6]^{2+}$ through the OSET and ISET Processes

The most authentic evidence in favour of the ISET process was provided by **H. Taube** and coworkers (1953–54) who observed the tremendous rate enhancement for the reduction of $[CoCl(NH_3)_5]^{2+}$ by $[Cr(OH_2)_6]^{2+}$ compared to the reduction of $[Co(NH_3)_6]^{3+}$ by the same reducing agent.

$$\left[Co(NH_3)_6\right]^{3+} + \left[Cr(OH_2)_6\right]^{2+} \xrightarrow{H^+} \left[Co(OH_2)_6\right]^{2+} + \left[Cr(OH_2)_6\right]^{3+} + 6NH_4^+ \quad \left(k \approx 10^{-4}\,dm^3\,mol^{-1}\,s^{-1}\right)$$

$$\left[CoCl(NH_3)_5\right]^{2+} + \left[Cr(OH_2)_6\right]^{2+} \xrightarrow{H^+} \left[Co(OH_2)_6\right]^{2+} + \left[CrCl(OH_2)_5\right]^{2+} + 5NH_4^+ \quad \left(k \approx 6\times10^5\,dm^3\,mol^{-1}\,s^{-1}\right)$$

They argued that the difference in the rate and in the nature of the product, *i.e.* Cr(III), can be explained by considering the reduction of $[CoCl(NH_3)_5]^{2+}$ by $[Cr(aq)]^{2+}$ through the ISET path.

To explain the products in the reduction of $[CoCl(NH_3)_5]^{2+}$ by $[Cr(aq)]^{2+}$, in terms of the OSET and ISET paths, the following sets of reactions are to be proposed.

OSET path:

$$\text{(a)}\quad \left[Co^{III}Cl(NH_3)_5\right]^{2+} + \left[Cr^{II}(OH_2)_6\right]^{2+} \longrightarrow \left[Co^{II}Cl(NH_3)_5\right]^{+} + \left[Cr(OH_2)_6\right]^{3+}$$

$$\text{(b)}\quad \left[Co^{II}Cl(NH_3)_5\right]^{+} \xrightarrow[\text{(fast)}]{H^+,\,H_2O} \left[Co(OH_2)_6\right]^{2+} + Cl^- + 5NH_4^+$$

$$\text{(c)}\quad \left[Cr(OH_2)_6\right]^{3+} + Cl^- \xrightarrow[\left(k\approx 3\times10^{-8}\,dm^3\,mol^{-1}\,s^{-1}\right)]{\text{anation}} \left[CrCl(OH_2)_5\right]^{2+} + H_2O$$

ISET path:

$$\text{(d)}\quad \left[(H_3N)_5Co^{III}\text{—}Cl\right]^{2+} + \left[Cr(OH_2)_6\right]^{2+} \rightleftharpoons \left[(H_3N)_5Co^{III}\text{—}Cl\text{—}Cr^{II}(OH_2)_5\right]^{4+} + H_2O \quad \text{(Precursor complex)}$$

$$\text{(e)}\quad \left[(H_3N)_5Co^{III}\text{—}Cl\text{—}Cr^{II}(OH_2)_5\right]^{4+} \xrightarrow{rds} \left[(H_3N)_5Co^{II}\text{—}Cl\text{—}Cr^{III}(OH_2)_5\right]^{4+} \quad \text{(Successor complex)}$$

(e: electron transfer from Cr^{II} to Co^{III})

$$\xrightarrow[\text{dissociative path (fast) (f)}]{H_2O} \left[Co(NH_3)_5(OH_2)\right]^{2+} + \left[CrCl(OH_2)_5\right]^{2+}$$

$$\text{(g)}\quad [Co(NH_3)_5(OH_2)]^{2+} \xrightarrow[\text{(fast)}]{H_2O,\,H^+} [Co(OH_2)_6]^{2+} + 5NH_4^+$$

To explain the products in the OSET path, it is to be considered that after the generation of Co(II) and Cr(III) centres through the electron transfer step, they are to experience the ligand substitution process. Co(II) (high spin) being labile can rapidly respond to the reaction (b), but $[Cr(OH_2)_6]^{3+}$ (d^3 system) being kinetically inert undergoes anation (c) very slowly. In fact, the 2nd order rate constant for anation of $[Cr(OH_2)_6]^{3+}$ by Cl^- is of the order of 10^{-8} $dm^3\,mol^{-1}\,s^{-1}$ which is about 10^{13} times smaller than the overall reaction rate constant. It indicates that the independent anation of $[Cr(OH_2)_6]^{3+}$ after its generation through the electron transfer step cannot explain the result. **If the OSET path were true, then the overall reaction rate could not generate the 2nd order rate constant greater than 10^{-8} $dm^3\ mol^{-1}\ s^{-1}$ which is the 2nd order rate constant for anation of $[Cr(OH_2)_6]^{3+}$** by Cl^-.

We can conclude that to explain the experimental findings in terms of the OSET process, the overall redox reaction rate must be at least comparable to the substitution rate at the Cr(III) centre. **In fact, this is the necessary condition for the OSET process to explain the above experimental findings**.

Moreover, the **isotope labeling experimental** result is not tenable with the OSET path. If the reaction is carried out by using the labeled *Cl in the oxidant (*i.e.* $[Co(NH_3)_5{}^*Cl]^{2+}$), then the final product $[Cr(OH_2)_5{}^*Cl]^{2+}$ exclusively retains the radioactivity (*i.e.* **no free radioactive $^*Cl^-$ is noticed in the solution**). The above findings can be explained by considering the reaction to pass through the ISET path where the ligand transfer and electron transfer occur simultaneously. Actually in the ISET path, electron transfer within the activated precursor complex having the bridge Co^{III}—Cl*—Cr^{II} (when $[(H_3N)_5Co^{III}Cl^*]^{2+}$ used as the oxidant) produces a binuclear successor complex of Cr^{III} and Co^{II} having the bridge Cr^{III}—Cl*—Co^{II}. In this binuclear complex, Cr^{III} (d^3) is inert, but Co^{II} ($t_{2g}^5e_g^2$) is labile. In fact, as soon as the successor complex is produced, it undergoes fission through the dissociative cleavage of the Co^{II}—Cl bond and it leads to the transfer of Cl^- to the reducing centre giving rise to $[Cr^*Cl(OH_2)_5]^{2+}$ as the final product. ***In fact, this step reaction (i.e. reaction f) is so fast that it is reasonable to conclude that the electron transfer (i.e. reaction e) and ligand transfer (reaction f) occur simultaneously.*** This pathway does not require the separate anation at $[Cr(OH_2)_6]^{3+}$ through a different reaction.

It may be noted that if the reduction of $[Co(NH_3)_5X]^{2+}$ by $[Cr(aq)]^{2+}$ occurs through the ISET path, often the groups (say Y^{n-}) present in solution but not serving as the bridging ligand get incorporated into the final Cr(III) products. For example, if the reduction of $[CoCl(NH_3)_5]^{2+}$ by $[Cr(aq)]^{2+}$ is carried out in presence of $P_2O_7^{4-}$, both Cl^- and $P_2O_7^{4-}$ have been found in the final Cr(III) product. The rates are also changed by such nonbridging ligands. The following transition state involving the labile $[Cr(aq)]^{2+}$ as the reductant has been proposed in the ISET path to explain the findings.

$$\left[(H_3N)_5Co^{3+}\cdots\cdots Cl^-\cdots\cdots Cr(OH_2)_4^{2+}\cdots\cdots OPO_2—O—PO_3^{4-}\right]$$

In the proposed transition state, ligand substitution by both the bridging Cl^- ligand and nonbridging free ligand $P_2O_7^{4-}$ present in soluution occurs at the labile Cr(II) centre. Thus they get incorporated within the coordination sphere of Cr(III) as the final product obtained through the ISET path. In general, we can write for the **ISET process** in presence of Y^{n-} ligand:

$$\underset{\text{(Binuclear precursor complex)}}{[(H_3N)_5Co^{III}—X—Cr(OH_2)_4—Y]} \xrightarrow{H^+,\,H_2O} [Co(OH_2)_6]^{2+} + 5NH_4^+ + \underset{\text{(Final product)}}{[X—Cr^{III}(OH_2)_4—Y]}$$

(charges not shown for the sake of simplicity)

Here it is reasonably assumed that in presence of free Y^{n-} ligand, the **labile reductant** $[Cr(aq)]^{2+}$ will produce $[Cr^{II}(OH_2)_5Y]$ that may act as the reducing agent.

If the reaction passes through the OSET process in presence of the free ligand Y^{n-}, then the Cr(III) product may be $[Cr^{III}(OH_2)_5Y]$ that will not bear the ligand X originally present with the Co^{III}-oxidant. Thus, we can write the OSET process:

$$[Cr^{II}(OH_2)_6]^{2+} + Y^{n-} \xrightarrow{\text{fast}} [Cr^{II}(OH_2)_5Y]$$

$$[Co(NH_3)_5X]^{2+} + [Cr^{II}(OH_2)_5Y] \xrightarrow[H^+,\,H_2O]{\text{OSET}} [Co^{II}(NH_3)_5X]^+ + [Cr^{III}(OH_2)_5Y]$$

$$[Co^{II}(NH_3)_5X]^+ \xrightarrow[\text{(fast)}]{H^+,\,H_2O} [Co(OH_2)_6]^{2+} + 5NH_4^+ + X^-$$

Thus, we can conclude that in presence of Y^{n-}, the **labile reductant** may be $[Cr^{II}(OH_2)_x(Y)_y]$ ($x + y = 6$), (say $[Cr^{II}(OH_2)_5(Y)]$):

OSET: $[Co(NH_3)_5X]^{2+} + [Cr(aq)]^{2+} \xrightarrow{H^+, Y^{n-}} [Co(OH_2)_6]^{2+} + 5NH_4^+ + X^- + [Cr^{III}(OH_2)_5Y]$

(no incorporation of X^-)

ISET: $[Co(NH_3)_5X]^{2+} + [Cr(aq)]^{2+} \xrightarrow{H^+, Y^{n-}} [Co(OH_2)_6]^{2+} + 5NH_4^+ + [Cr^{III}(X)(Y)(OH_2)_4]$

(incorporation of both X^- and Y^{n-})

$[Co(NH_3)_6]^{3+}$ fails to react with $[Cr(OH_2)_6]^{2+}$ through the ISET path because the NH_3 ligands present with the Co(III) (*i.e.* oxidant) *fail to act as the bridging ligands.* This is why, the reaction goes through the OSET path for which the activation energy is very high because of the $\sigma^* \rightarrow \sigma^*$ net electron transfer in the reaction.

Characteristic features of reduction of $[Co^{III}(NH_3)_5X]^{n+}$ by $[Cr(OH_2)_6]^{2+}$ in the ISET path: In fact, the reduction rate of $[Co^{III}(NH_3)_5X]^{n+}$ by $[Cr(OH_2)_6]^{2+}$ depends largely on the nature of X.

(i) **For X = NH_3:** ISET is not possible because it cannot act as a bridging ligand. The reaction occurs through the OSET path (with a very high activation energy).

(ii) **For X = H_2O, OH^-:** ^{18}O-labeling experiment indicates the transfer of X, during the electron transfer process.

$$[(H_3N)_5Co{-}{}^*OH_2]^{3+} + [Cr(OH_2)_6]^{2+} \xrightarrow{H^+} [Co(OH_2)_6]^{2+} + 5NH_4^+ + [(H_2O)Cr({}^*OH_2)]^{3+}$$

'^-OH' is a much better bridging ligand than H_2O and the rate is much increased for X = OH^-. This is why, **$[Co^{III}(NH_3)_5(OH_2)]^{3+}$ reacts much faster at the relatively higher pH** where the aqua-complex produces the hydroxido-complex through deprotonation.

OSET *vs.* ISET for the reaction $[Co(NH_3)_5(OH_2)]^{3+} + [Cr(aq)]^{2+}$

Though for X = OH_2, the rate is increased by a factor about 10^4 compared to the rate for X = NH_3, there is a doubt whether the reaction switches over from outer-sphere (for X = NH_3) to inner-sphere (for X = OH_2). In fact, H_2O is not expected to act as a good bridging ligand. It may be argued that for both X = NH_3, OH_2, the reaction, *i.e.* $[Co^{III}(NH_3)_5X]^{2+} + [Cr(aq)]^{2+}$ (X = NH_3, H_2O) goes through the OSET path. *The rate benefit for X = H_2O is probably due to the better H-bonding interaction between the coordination spheres of the two reactants in the outer-sphere precursor complex.* In fact, this H-bonding interaction gives the very high K_{OS} value to favour the overall rate process in the OSET process. However, there are some evidences to support the fact that the reaction $[Co(NH_3)_5(OH_2)]^{3+} + [Cr(aq)]^{2+}$ passes through the ISET path involving the following precursor inner sphere complex.

$$\left[(H_3N)_5Co\cdots\overset{*}{O}(H)(H)\cdots Cr(OH_2)_5\right]^{5+}$$

The transfer of labeled H_2O^* to the Cr-centre, *i.e.* formation of $[Cr(OH_2)_5({}^*OH_2)]^{3+}$ when $[(H_3N)_5Co({}^*OH_2)]^{3+}$ reacting with $[Cr(aq)]^{2+}$, supports the ISET path of the reaction. There are also other evidences to support the proposed ISET path. The rate increases with the increases of pH. At higher pH, $[Co(NH_3)_5(OH_2)]^{3+}$ hydrolyses to generate $[Co(NH_3)_5(OH)]^{2+}$ that can more favourably participate in the ISET path. **$[Co(NH_3)_5({}^*OH_2)]^{3+}$ (labeled with ^{18}O) reacts slower than the unlabeled complex $[Co(NH_3)_5(OH_2)]^{3+}$** (*i.e.* predominantly $H_2{}^{16}O$, in terms of abundance of ^{18}O and ^{16}O isotopes). Heavier $H_2{}^{18}O$ moves slower than $H_2{}^{16}O$. It also supports the transfer of the bridging H_2O^* along with the electron transfer (*i.e.* supporting the ISET path). Interestingly, both the labeled and unlabeled $[Ru(NH_3)_6]^{3+}$ reacts at comparable rates with $[Cr(aq)]^{2+}$. It is reasonable because $[Ru(NH_3)_6]^{3+}$ reacts with $[Cr(aq)]^{2+}$ in the OSET path where there is no question of ligand transfer.

Table 6.4.4.1 The second order rate constant for the reduction of $[Co^{III}(NH_3)_5X]^{n+}$ by $[Cr(OH_2)_6]^{2+}$

X =	*NH_3*	*OH_2*	*^-OH*	*Cl^-*	*N_3^-*	*NCS^-*	*SCN^-*
k ($dm^3\ mol^{-1}\ s^{-1}$) = (25°C)	9×10^{-5}	0.5	1.5×10^6	6×10^5	3×10^5	20	1.8×10^5

Here it is worth mentioning that though there are evidences to support the fact that H_2O as a bridging ligand can mediate the ISET process in some cases as in the reduction of $[Co(NH_3)_5(OH_2)]^{3+}$ by $[Cr(OH_2)_6]^{2+}$, very often its poor bridging capacity leads to the OSET process. This is why, the electron exchange in the couple, $[Fe(OH_2)_6]^{3+} - [Fe(OH_2)_6]^{2+}$ occurs through the OSET process, but at the relatively higher pH, the couple $[Fe(OH)(OH_2)_5]^{2+} - [Fe(OH_2)_6]^{2+}$ experiences the ISET process for the electron exchange because the OH group attached to Fe(III) can act as a good bridging ligand (*cf.* Sec. 6.4.10).

The aspect of **H_2O *vs.* OH^- as a bridging ligand** in the ISET process has been discussed in Sec. 6.4.10.

(iii) **For X = Cl^-, N_3^-:** The rate is significantly high because they can act as the good bridging ligands.

(iv) **For X = NCS^- and SCN^-:** $[Co(NCS)(NH_3)_5]^{2+}$ reacts much slower than $[Co(NH_3)_5(SCN)]^{2+}$ by a factor about 10^4. The Cr(III) products are also different.

$$[Co(NCS)(NH_3)_5]^{2+} \xrightarrow{[Cr(aq)]^{2+},\ H^+} [Co(OH_2)_6]^{2+} + 5NH_4^+ + \underset{\textbf{(main product)}}{[Cr(OH_2)_5(SCN)]^{2+}}$$

$$[Co(NH_3)_5(SCN)]^{2+} \xrightarrow{[Cr(aq)]^{2+},\ H^+} [Co(OH_2)_6]^{2+} + 5NH_4^+ + \underset{\textbf{(main product)}}{[Cr(NCS)(OH_2)_5]^{2+}}$$

The product analysis indicates that for the first reaction, the free *S*-end of NCS^- binds with the Cr(II) centre in the precursor binuclear complex while in the second reaction, the free *N*-end binds with the Cr(II) centre. These are:

$$[(H_3N)_5Co^{III}—NCS]^{2+} + [Cr(OH_2)_6]^{2+} \rightleftharpoons [(H_3N)_5Co^{III}—NCS—Cr^{II}(OH_2)_5]^{4+} + H_2O$$

$$\downarrow \text{rds}$$

$$[Co(OH_2)_6]^{2+} + 5NH_4^+ + \underset{\substack{i.e.\ [Cr(OH_2)_5(SCN)]^{2+} \\ \textbf{(green)}}}{[(H_2O)_5Cr(SCN)]^{2+}} \xleftarrow[\text{fast}]{H^+,\ H_2O} [(H_3N)_5Co^{II}—NCS—Cr^{III}(OH_2)_5]^{4+}$$

$$[(H_3N)_5Co^{III}—SCN]^{2+} + [Cr(OH_2)_6]^{2+} \rightleftharpoons [(H_3N)_5Co^{III}—SCN—Cr^{II}(OH_2)_5]^{4+} + H_2O$$

$$\downarrow \text{rds}$$

$$[Co(OH_2)_6]^{2+} + 5NH_4^+ + \underset{\substack{i.e.\ [Cr(NCS)(OH_2)_5]^{2+} \\ \textbf{(purple)}}}{[(H_2O)_5Cr(NCS)]^{2+}} \xleftarrow{H^+,\ H_2O} [(H_3N)_5Co^{II}—SCN—Cr^{III}(OH_2)_5]^{4+}$$

$[Cr(OH_2)_6]^{2+}$ being a relatively harder centre prefers the N-binding site and the *S*-site of NCS^- is not preferred by $[Cr(OH_2)_6]^{2+}$. Thus, coordinated —*S*CN can act as a better bridging ligand than *coordinated —N*CS. **This explains the rate enhancement for Co^{III}—SCN.**

(**Note:** For Co^{III}—NCS, only ***remote attack*** on S-end is possible but for Co^{III}—SCN, both ***adjacent attack*** on S site and ***remote attack*** on N-site are possible. However, for Co^{III}—SCN, the remote attack is the predominant path. The aspect has been discussed in Sec. 6.4.9).

The **purple isomer**, $[Cr(NCS)(OH_2)_5]^{2+}$ **thermodynamically is more stable** than the **green isomer**, $[Cr(OH_2)_5(SCN)]^{2+}$ (*cf.* **symbiotic theory** of HSAB principle, see Vol. 4).

(v) **$X = N_3^-$ and NCS^- (N-bonded NCS^-):** For $X = N_3^-$, the rate is much faster than that for $X = NCS^-$ in the ISET process (*i.e.* $k_{N_3}/k_{NCS} \gg 1$, see Sec. 6.4.10).

6.4.5 Types of Inner Sphere Electron Transfer (ISET) Reactions Depending on the Position of the Rate Determining Step

(A) Type-I (electron transfer step is the rate determining step): The overall process consists of different steps (*cf.* Secs. 6.4.1, 6.4.2). If the electron transfer step within the precursor complex to generate the successor complex is the slowest step (*i.e.* formation of the precursor complex in the pre-equilibrium step is fast; fission of the successor complex into the products is also fast), then it gives the following rate law (*cf.* Sec. 6.4.2 for the rate law derivation):

$$\text{rate} \approx k\left(\frac{k_1}{k_{-1}}\right)[\text{Ox}][\text{Red}], \left(\text{under the condition, } k_{-1} \gg k\right)$$
$$= kK[\text{Ox}][\text{Red}]$$

Most of the ISET reactions belong to this type of reaction.

For the electron transfer within the precursor complex in the inner-sphere process, it may require the electronic activation of the reactants; reorganisation in the inner coordination spheres of the reactants for energy matching between the donor orbital of the reductant and the receptor orbital of the oxidant (*cf.* Sec. 6.4.14). *It also needs the reorganization* in the inner-spheree of the reactants *to overcome the Franck-Condon barrier.* To overcome this **Franck-Condon barrier**, it needs the bond length and bond angle adjustment in both the oxidant and reductant in the activated complex so that the coordination geometries of the reactants in the activated complex become **productlike** (as explained for the OSET process). This bond length and bond angle rearrangement energy depend on the ligand field strength and electronic configurations, the reactants (*cf.* Sec. 6.4.14). Thus, for this electron transfer step, the activation energy is determined by the factors like reorganisation of the inner coordination spheres of the reactants (*i.e.* to overcome the Franck-Condon barrier), solvent reorganisation, etc. as in the OSET process. In fact, for this electron transfer step, a **Marcus type relationship** also exists.

The dinuclear complex, $\left[\text{Cl(bpy)}_2\text{Ru}^{\text{II}}\text{—N}\bigcirc\text{N—Ru}^{\text{III}}\text{(bpy)}_2\text{Cl}\right]^{3+}$ involving the Ru^{II} and Ru^{III} centres has been treated as a **model complex** of the precursor binuclear complex.

(B) Type-II (formation of the binuclear precursor complex is the rate determining step): If the *reductant* (*e.g.* $[V(aq)]^{2+}$) is relatively inert, then the substitution at the reductant leading to the dinuclear precursor complex is the rate determining step (*i.e.* electron transfer leading to the dinuclear successor complex and fission of the successor complex into the product are the relatively faster steps), then the rate for this **substitution controlled ISET process** is given by:

$$\text{rate} \approx k_1[\text{Ox}][\text{Red}], \text{ (under the condition, } k \gg k_{-1}; \textit{see} \text{ Sec. 6.4.2)}$$

(a) $[V(aq)]^{2+}$ as a reducing agent in the substitution controlled ISET process: ***Substitution controlled inner sphere reduction*** by $[V(aq)]^{2+}$ (d^3) is well documented. Let us consider the lability of some common reducing agents often employed in the ISET process.

Water exchange: $\left[M(OH_2)_6\right]^{2+} + H_2O^* \longrightarrow \left[M(OH_2)_5(^*OH_2)\right]^{2+} + H_2O$; M = V, Cr, and Fe

	$[V(OH_2)_6]^{2+}$	$[Cr(OH_2)_6]^{2+}$	$[Fe(OH_2)_6]^{2+}$
k_{ex} (s^{-1}):	10^2	10^9	10^6

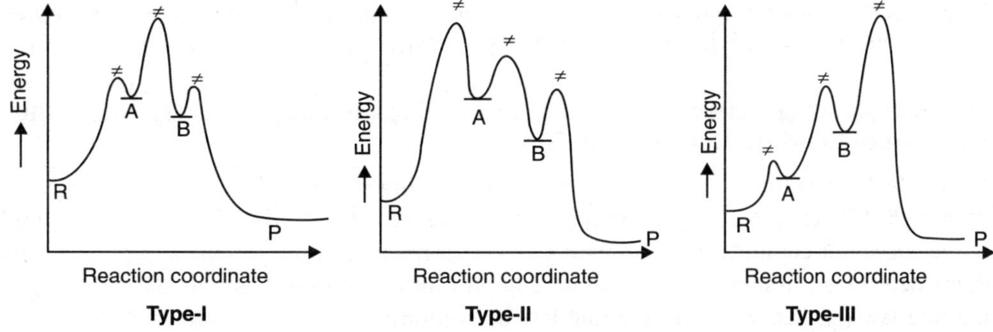

R = Reactants; A = Binuclear precursor intermediate complex, B = Binuclear successor intermediate complex, P = product. ≠ denotes the corresponding activated state or transition state.

Type-I : Electron transfer step (*i.e.* A→ B) rate determining step.
Type-II : Formation of the binuclear precursor complex (*i.e.* R→ A) rate determining step.
Type-III: Formation of products from the binuclear successor complex (*i.e.* B→ P)

Fig. 6.4.5.1: Reaction profile of inner sphere electron transfer reactions.

The reaction rates and their activation parameters for the reduction of different Co(III)–complexes by $[V(aq)]^{2+}$ are very much comparable with those of water exchange and nucleophilic substitution process of $[V(aq)]^{2+}$ (*cf.* Tables 6.4.5.1–2). These features for reduction by $[V(aq)]^{2+}$ are in direct contrast with those found for the reductions of Co(III)-complexes by $[Fe(aq)]^{2+}$ and $[Cr(aq)]^{2+}$. In fact, for reductions by $[Fe(aq)]^{2+}$ and $[Cr(aq)]^{2+}$, the rate determining step is the electron transfer step leading to the binuclear successor complex **(*i.e.* Type I ISET).**

Reduction of $[Ru^{III}(NH_3)_5X]^{2+}$ by $[V(aq)]^{2+}$: OSET or ISET

The presence of a suitable bridging ligand in the coordination sphere of an inert oxidant does not ensure its reduction by a reductant through the ISET path. The situation is illustrated in the following redox reaction.

$$[Ru^{III}(NH_3)_5X]^{2+} + [V(OH_2)_6]^{2+} \longrightarrow [Ru^{II}(NH_3)_5X]^{+} + [V(OH_2)_6]^{3+},$$
$$k \approx 5 \times 10^3 \text{ dm}^3 \text{ mol}^{-1} \text{ s}^{-1} \ (\pi^* \rightarrow \pi^* \text{ net transfer})$$

The reaction rate is much faster than the rate of nucleophilic substitution process at $[V(aq)]^{2+}$ (see Table 6.4.5.2). *It clearly indicates that the above redox reaction does not pass through the ISET path.* In fact, the **necessary condition for the OSET process is that the redox reaction rate must be faster than the substitution reaction rate on the metal centres**. This condition is attained for this reaction (reduction of Ru^{III} by V^{II}).

In spite of the presence of a suitable bridging ligand (X) with the oxidant Ru^{III}-complex, its reduction by $[V(aq)]^{2+}$ does not occur through the ISET process. Rather, the reaction occurs through the **OSET process**. In fact, the given reaction experiences the net $\pi^* \rightarrow \pi^*$ electron transfer which is the **most favourable path in the OSET process** (*cf.* reduction of the low-spin Co(III)-complexes by V(II) experiences the $\pi^* \rightarrow \sigma^*$ net electron transfer which requires only the electronic activation at the oxidant in the OSET process). The **$\pi^* \rightarrow \pi^*$ electron transfer is the most unfavourable path for the ISET process** because it needs the electronic activation of both the oxidant and reductant. Moreover, **inertness of $[V(aq)]^{2+}$ puts a mild barrier for the ISET process.**

Thus, if the reduction of Ru(III) by V(II) is supposed to pass through the ISET path, it will require the electronic activation of both the oxidant and reductant for the required $\sigma^* \rightarrow \sigma^*$ electron transfer at the kinetic step. On the other hand, if the reaction occurs through the OSET path, it needs no electronic activation. Consequently, it passes through the OSET process in which the activation energy is relatively less.

Tables 6.4.5.1–2 indicate that the kinetic parameters for the reduction of Co(III)-complexes by $[V(aq)]^{2+}$ are comparable with those of the ligand substitution reactions of $[V(aq)]^{2+}$.

Table 6.4.5.1: Rate constants and activation parameters for reduction of Co(III)–complexes by $[V(aq)]^{2+}$ (25°C) **Net electron transfer:** $\pi^* \rightarrow \sigma^*$, *i.e.* $t_{2g} \rightarrow e_g$. (*cf.* J. Espenson, *Inorg Chem.*, **4**, 121, 1965; J. A. Stritar *et al. ibid*, **8**, 2284, 1969).

Oxidant *Co(III)-complex*	***k*** **($dm^3\ mol^{-1}\ s^{-1}$)**	***ΔH***$^{\neq}$ **($kJ\ mol^{-1}$)**	***ΔS***$^{\neq}$ **($J\ K^{-1}\ mol^{-1}$)**
$[CoF(NH_3)_5]^{2+}$	2.6		
$[CoCl(NH_3)_5]^{2+}$	8.0		
$[CoBr(NH_3)_5]^{2+}$	25.0		
$[CoI(NH_3)_5]^{2+}$	120.0		
$[Co(NH_3)_5N_3]^{2+}$	13.0		
$[Co(NCS)(NH_3)_5]^{2+}$	0.3		
$[Co(NH_3)_5(SCN)]^{2+}$	30.0		
$[Co(C_2O_4H)(NH_3)_5]^{2+}$	12.5	51.0	–54.0
cis–$[Co(en)_2(NH_3)(N_3)]^{2+}$	10.3	53.0	–50.0
cis–$[Co(en)_2(N_3)(OH_2)]^{2+}$	16.5	51.0	–50.0
trans–$[Co(en)_2(N_3)_2]^{+}$	26.5	51.0	–46.0

$[V(aq)]^{2+}$ ($t_{2g}^3 e_g^0$); $[V(aq)]^{3+}$ ($t_{2g}^2 e_g^0$); Co^{3+} (t_{2g}^6, l.s.); Co^{2+} ($t_{2g}^5 e_g^2$, h.s.).

Table 6.4.5.2: Kinetic parameters (25°C) of substitution at $[V(OH_2)_6]^{2+}$

Process	*Rate Constant*	**$\Delta H^{\neq}$ ($kJ\ mol^{-1}$)**	**$\Delta S^{\neq}$ ($J\ K^{-1}\ mol^{-1}$)**
Water exchange*	$1 \times 10^2\ s^{-1}$	68.5	–23
Anation by NCS^-	$28\ dm^3\ mol^{-1}\ s^{-1}$	56.5	–29

* $k_{ex}/55.5 \approx 1.8\ dm^3\ mol^{-1}\ s^{-1}$ taking $[H_2O] = 55.5\ mol\ dm^{-3}$ in dilute solution; water exchange in $[V(aq)]^{2+}$ (t_{2g}^3) through the **associative path**.

It is interesting to note that for the reductions of $[Co^{III}(NH_3)_5X]^{2+}$ by $[V(aq)]^{2+}$, *the rates are more or less insensitive to the nature of the bridging ligands.* It is quite reasonable. The rate determining step is the nucleophilic substitution at $[V(aq)]^{2+}$ not the electron transfer step which is highly sensitive to the nature of bridging ligand in the ISET path (*cf.* $k_{SCN}/k_{NCS} \approx 10^4$, $k_{N_3}/k_{NCS} \approx 10^4$ for the **Type I** ISET reactions, Tables 6.4.10.1–2).

(b) Magnitude of the rate constants, and kinetic parameters as the diagnostic parameters for the mechanistic paths of reductions by $[V(aq)]^{2+}$: If the rate constants for the reduction of metal complexes by $[V(aq)^{2+}]$ are close to those of nucleophilic substitution process at $[V(aq)]^{2+}$, then the said redox reactions are reasonably considered to pass through the ISET path where the ***rate determining step is the nucleophilic substitution at $[V(aq)]^{2+}$ to produce the binuclear precursor complex.*** On the other hand, if the rate constants of the redox reactions are substantially higher than the rate constants of substitution reactions of $[V(aq)]^{2+}$, then the redox reactions are supposed to pass through the OSET path rather than the ISET path.

$[Co^{III}(NH_3)_5X]^{2+} + [V(aq)]^{2+}$ ($\pi^* \rightarrow \sigma^*$ net electron transfer): $k_{el} \approx k_{subs}$; **substitution controlled ISET process**

$[Ru(NH_3)_5X]^{2+} + [V(aq)]^{2+}$ ($\pi^* \rightarrow \pi^*$ net electron transfer): $k_{el} \gg k_{subs}$; **OSET process**

$[Fe(aq)]^{3+} + [V(aq)]^{2+}$ ($\pi^* \rightarrow \pi^*$ net electron transfer): $k_{el} \approx 2 \times 10^4\ dm^3\ mol^{-1}\ s^{-1} \gg k_{subs}$; **OSET process**

$k_{el} \Rightarrow$ rate constant for the given redox reaction

$k_{subs} \Rightarrow$ rate constant for the substitution process of $[V(aq)]^{2+}$.

(c) Critical comments on the mechanistic path of reduction of the Co^{III}-complexes having the suitable bridging ligands (X) by $[V(aq)]^{2+}$: There are some cases where the V^{III}—X intermediates have been detected. However, on standing, the V^{III}—X intermediate undergoes aquation (*cf.* V^{III}, d^2-system, labile, small CFAE). These reactions definitely pass through the ISET path. Such Co^{III}-oxidants are:

$[Co(NH_3)_5(SCN)]^{2+}$, $[Co(C_2O_4)(NH_3)_5]^+$, $[Co(CN)_5(N_3)]^{3-}$, $[Co(CN)_5(SCN)]^{3-}$, *cis*-$[Co(en)_2(N_3)_2]^+$ etc.

But there are many cases where the intermediate complexes have not been detected and there is a difference in opinion regarding the mechanistic path of reduction of the Co^{III} complexes by $[V(aq)]^{2+}$. In fact, for the complexes $[Co(NH_3)_5X]^{2+}$ (X = F, Cl, Br and I), there is a close resemblance between the reductants $[V(aq)]^{2+}$ and $[Cr(bpy)_3]^{2+}$. But $[Cr(bpy)_3]^{2+}$ undoubtedly reduces the Co^{III}-complexes through the OSET path (see Table 6.4.1.1). For both the reductants, the reactivity order (called **halide effect**) of the Co^{III}-complexes is: $[CoI(NH_3)_5]^{2+} > [CoBr(NH_3)_5]^{2+} > [CoCl(NH_3)_5]^{2+} > [CoF(NH_3)_5]^{2+}$. This reactivity order has been explained by considering the energy of **electronic activation** of the Co^{III}-complex (*i.e.* $t_{2g}^6 \rightarrow t_{2g}^5 e_g^1$) required for the execution of OSET path (see Sec. 6.4.1 and Table 6.4.1.1 for more details). This **resemblance between the reductants $[V(aq)]^{2+}$ and $[Cr(bpy)_3]^{2+}$**, for the reduction of the said Co^{III}-complexes supports the execution of OSET path for the reduction by $[V(aq)]^{2+}$.

Here it may be argued that the above **halide effect can also be explained by considering the substitution controlled ISET reduction** of the $[Co^{III}(NH_3)_5X]^{2+}$ complexes by $[V(aq)]^{2+}$ which experiences the **associative activation in the substitution step** as the rate determining step of Type-II ISET reactions. In the associative substitution process, the rate increases with the **nucleophilicity order of the bridging ligand** X (*cf.* nucleophilicity order: $I^- > Br^- > Cl^- > F^-$).

Reduction of $[Co^{III}(NH_3)_5X]^{2+}$ by $[V(aq)]^{2+}$: OSET or ISET?

The above discussion indicates that even in the presence of the suitable bridging ligands in the Co^{III}-complexes, **their reduction by $[V(aq)]^{2+}$ may pass through the OSET path or the substitution controlled ISET path**. The ISET path needs the electronic activation of V(II) ($t_{2g}^3 \rightarrow t_{2g}^2 e_g^1$) and CFAE for ligand substitution at V(II) (*cf.* d^3 system having the relatively high CFAE). On the other hand, the OSET path needs the electronic activation of Co(III) ($t_{2g}^6 \rightarrow t_{2g}^5 e_g^1$). Both these mechanisms can explain the halide effect also. **Thus without the concrete experimental evidence** (*e.g.* detection of intermediates), **it is very difficult to predict whether the reduction of Co^{III}-complexes occurs by V(II) through the OSET or ISET path**.

(d) $[Cr(aq)]^{2+}$ and $[Fe(aq)]^{2+}$ as the reducing agents in the ISET process: The kinetic parameters for the reduction of Co(III)-complexes by these reducing agents in the ISET process are given in Table 6.4.5.3.

Table 6.4.5.3 Rate constants (at 25° C) for the reduction of some Co(III)-complexes by $[Cr(aq)]^{2+}$ ($t_{2g}^3 e_g^1$) and $[Fe(aq)]^{2+}$ ($t_{2g}^4 e_g^2$) (*cf.* J. A. Stritar *et al, Inorg Chem.*, **8**, 2284, 1969; J. Espenson *et al, ibid,* **4**, 121, 1965).

Oxidant	**k ($dm^3\ mol^{-1}\ s^{-1}$) (*for* Cr^{II})**		**k ($dm^3\ mol^{-1}\ s^{-1}$) (*for* Fe^{II})**	
$[CoF(NH_3)_5]^{2+}$	2.5×10^5	Net $\sigma^* \rightarrow \sigma^*$ electron transfer	6.6×10^{-3}	Net $\pi^* \rightarrow \sigma^*$ electron transfer
$[CoCl(NH_3)_5]^{2+}$	6×10^5		1.5×10^{-3}	
$[CoBr(NH_3)_5]^{2+}$	1.5×10^6		7.3×10^{-4}	
$[Co(NH_3)_5N_3]^{2+}$	3×10^5		8.5×10^{-3}	
$[Co(NCS)(NH_3)_5]^{2+}$	20		3×10^{-3}	
$[Co(NH_3)_5(SCN)]^{2+}$	2×10^5		0.12	
$[Co(C_2O_4H)(NH_3)_5]^{2+}$	4×10^2		4.5×10^{-1}	

k_{ex} for water exchange process: k_{ex} for $[Fe(OH_2)_6]^{2+} = 10^6\ s^{-1}$; k_{ex} for $[Cr(OH_2)_6]^{2+} = 10^9\ s^{-1}$.

Note: In the reduction of Co(III)-complexes by Fe(II), to understand the net transfer of electron, let us consider their ground state electronic configurations.

$$\mathrm{Co(III)}\left(t_{2g}^6 e_g^0\right) \xrightarrow{+e} \mathrm{Co(II)}\left(t_{2g}^5 e_g^2\right);\ \mathrm{Fe(II)}\left(t_{2g}^4 e_g^2\right) \xrightarrow{-e} \mathrm{Fe(III)}\left(t_{2g}^3 e_g^2\right)$$

The net electron transfer involves $t_{2g} \rightarrow e_g$, *i.e.* $\pi^* \rightarrow \sigma^*$. Thus it needs the electronic activation of Fe(II) for the ISET path. But, if we consider the transfer of an electron from Fe(II) in the following way:

$$\left(\mathrm{Fe^{II}}\right)t_{2g}^4 e_g^2 \xrightarrow{-e} \left(\mathrm{Fe^{III}}\right)t_{2g}^4 e_g^1\ \textbf{(excited state)} \xrightarrow{\text{relaxation}} \mathrm{Fe(III)}\left(t_{2g}^3 e_g^2\right)$$

then it will not need the electronic activation of Fe(II) to execute the $\sigma^* \rightarrow \sigma^*$ electron transfer in the ISET path, but it will generate Fe(III) in an excited state.

For the reduction of $[Co(NH_3)_5X]^{2+}$ (X = F, Cl, Br) by Fe^{2+}, the reactivity order **(halide effect)** is: $[Co(NH_3)_5F]^{2+} > [Co(NH_3)_5Cl]^{2+} > [Co(NH_3)_5Br]^{2+}$ (*cf.* reverse reactivity order of **halide effect** in the reduction of Co^{III}-complexes through the OSET path requiring the electronic activation of Co^{III}, $t_{2g}^6 \rightarrow t_{2g}^5 e_g^1$, ligand field strength order, $F^- > Cl^- > Br^-$). This reactivity order is the thermodynamic stability order of the product halide complex FeX^{2+} (*cf.* stability constants: $FeF^{2+} > FeCl^{2+} > FeBr^{2+}$). Bond stability order in the binuclear precursor complex also follows the same sequence: Fe^{II}—F > Fe^{II}—Cl > Fe^{II}—Br. It indicates that in the transition state of the **proposed ISET path**, bond formation by the bridging ligand with the reductant is more important (*cf.* strength of bond: $F^- > Cl^- > Br^-$). **The similar trend of the reactivity order has been noted for the reduction by Eu(II)** (*cf.* stability order of EuX^+ and EuX^{2+}: $F^- > Cl^- > Br^-$). This aspect has been discussed later. It is argued that the **higher stability of the Red—X bond in the binuclear transition state gives the higher reaction rate** for the reductants like $[Fe(aq)]^{2+}$ and $[Eu(aq)]^{2+}$. Here it is worth mentioning that the products Fe^{III}—X and Eu^{III}—X undergo rapid hydrolysis due to their lability (*cf.* high spin Fe^{III}–d^5 having no CFAE and Eu^{3+}, a representative lanthanide, extremely labile because of no CFAE contribution; crystal field effect on the deeply seated 4*f*-orbitals of lanthanides is not important and practically no CFAE for the lanthanides, Ln^{3+}). Thus the products FeX^{2+} and EuX^{2+} cannot be detected.

- **Reduction of the Co(III) complexes by $[Cr(aq)]^{2+}$ is much faster than the reduction by $[Fe(aq)]^{2+}$**. Both $[Cr(aq)]^{2+}$ and $[Fe(aq)]^{2+}$ are the labile centres (*cf.* in contrast to $[V(aq)]^{2+}$) and consequently, the step leading to the formation of the binuclear precursor complex cannot be the rate determining step. In fact, for the reduction by $[Cr(aq)]^{2+}$ and $[Fe(aq)]^{2+}$, the electron transfer step leading to the binuclear successor complex is the rate determining step (**Type-I**, *cf.* Fig. 6.4.5.1).

Thermodynamically, $[Cr(aq)]^{2+}$ is a better reducing agent than $[Fe(aq)]^{2+}$. It is evident from the reduction potentials of the couples. It also favours $[Cr(aq)]^{2+}$ kinetically in terms of **Marcus equation**.

$$E^0_{Cr^{3+}/Cr^{2+}} = -0.41\ \mathrm{V};\quad E^0_{Fe^{3+}/Fe^{2+}} = +0.76\ \mathrm{V}$$

Reduction of the Co^{III}-complexes by $[Cr(aq)]^{2+}$ involves the net $\sigma^* \rightarrow \sigma^*$ electron transfer while reduction of the Co^{III}-complexes by $[Fe(aq)]^{2+}$ involves the net $\pi^* \rightarrow \sigma^*$ electron transfer. Thus for the reduction by Cr^{2+} through the ISET path, there is no need of electronic activation while for the reduction by $[Fe(aq)]^{2+}$ through the ISET path, it needs the electronic activation of the reductant. ***Thus the thermodynamic disfavour and unfavourable nature of the HOMO of the reductant are jointly responsible for the lower rates of reduction of the Co(III)-complexes by Fe(II) compared to those for the reduction by Cr(II).***

(e) Reduction of $[Co^{III}(NH_3)_5X]$ by Eu^{2+}: There are evidences to support the fact that in mamy cases, the reduction of Co^{III}-complexes having the suitable bridging ligands by $Eu(aq)^{2+}$ ($4f^7$) (which is **extremely labile**; $k_{ex} \approx 10^9\ s^{-1}$ at 25°C, $\Delta V^{\neq} = -11.3\ cm^3\ mol^{-1}$, associative path; *cf.* CFAE is of no importance for the lanthanides) occurs through the ISET path. The reactivity order for the reduction of the Co^{III}-complexes by $[Eu(aq)]^{2+}$ is: $[Co(NH_3)_5F]^{2+} > [Co(NH_3)_5Cl]^{2+} > [Co(NH_3)_5Br]^{2+}$. The same reactivity order has been noted for their reduction by $[Fe(aq)]^{2+}$ through the ISET path. The stability order of EuX^+ and EuX^{2+} ($F^- > Cl^- > Br^-$) suggests that the bond formation by the bridging ligand is important in the binuclear transition state. It is suggested that the **thermodynamic stability order** of the **Eu—X bond** in the binuclear transition state bearing the Co—X—Eu bridging segment controls the reactivity order (*i.e.* **halide effect**) as in the case of reduction by $[Fe(aq)]^{2+}$. Here it may be noted that because of the extreme lability of Eu(III), the product is immediately hydrolysed and in fact, Eu^{III}—X cannot be detected. It also happens so for the other reducing agents like Fe^{2+}, V^{2+} and Cu^+.

Step I: $[Co(NH_3)_5X]^{2+} + [M(aq)]^{2+} \xrightarrow[\text{(ISET)}]{H^+} [Co(aq)]^{2+} + [(aq)MX]^{2+} + 5NH_4^+$

Step II: $[(aq)MX]^{2+} \xrightarrow[\text{rapid}]{\text{aquation}} [M(aq)]^{3+} + X^-$; (M = Fe, V, and Eu)

The above overall ISET reactions can be considered as the **two step consecutive reactions:** electron transfer producing the $[(aq)MX]^{2+}$ intermediate followed by its rapid aquation. The kinetic conditions for the detection of $[(aq)MX]^{2+}$ have been discussed separately.

Table 6.4.5.4 Second order rate constants (25°C) for the reduction of Co(III)-complexes by $[Eu(aq)]^{2+}$ and $[Co(CN)_5]^{3-}$

Complex	***k*** **($dm^3\ mol^{-1}\ s^{-1}$)**	
	$[Eu(aq)]^{2+}$	**$[Co(CN)_5]^{3-}$**
*$[Co(en)_3]^{3+}$	~5×10^{-3}	–
*$[Co(NH_3)_6]^{3+}$	2×10^{-2}	8×10^4
*$[Co(NH_3)_5(OH_2)]^{3+}$	1.5×10^{-1}	–
$[Co(NH_3)_5(OH)]^{2+}$	–	9.3×10^4
$[CoF(NH_3)_5]^{2+}$	2.6×10^4	1.8×10^4
$[CoCl(NH_3)_5]^{2+}$	3.9×10^2	~5×10^7
$[CoBr(NH_3)_5]^{2+}$	2.5×10^2	~2×10^9
$[CoI(NH_3)_5]^{2+}$	1.2×10^2	–
$[Co(NCS)(NH_3)_5]^{2+}$	~7×10^{-1}	1.1×10^6
$[Co(NH_3)_5(N_3)]^{2+}$	1.9×10^2	1.6×10^6

* OSET path because of the absence of a suitable bridging ligand in the Co^{III}-complex.

Note: The 2nd order rate constants for the reduction of $[Co(NH_3)_5X]^{2+}$ by $[Eu(aq)]^{2+}$ cover a **smaller range** compared to the reductions by $[Cr(aq)]^{2+}$ and $[Co(CN)_5]^{3-}$. In other words, for their reduction by $[Eu(aq)]^{2+}$, **the rate is relatively less sensitive to the nature of the bridging ligand (X)**. This is why, some authors have suggested the OSET path for the reduction of the Co^{III}–complexes by $[Eu(aq)]^{2+}$. However, the reactivity of $[Co(NH_3)_6]^{3+}$ or $[Co(en)_3]^{3+}$ having no bridging ligand is much less than that of $[Co(NH_3)_5X]^{2+}$ having the bridging ligand. Thus it is reasonable to consider that $[Co(NH_3)_6]^{3+}$ or $[Co(en)_3]^{3+}$ reacts through the OSET path but $[Co(NH_3)_5X]^{2+}$ reacts through the ISET path with $[Eu(aq)]^{2+}$.

(f) Reduction of $[Co^{III}(NH_3)_5X]^{2+}$ by $[Co(CN)_5]^{3-}$: For the good bridging ligand X, the reduction of the Co^{III}-complexes by $[Co(CN)_5]^{3-}$ having a vacant coordination site occurs through the ISET path (Table 6.4.5.4). The weakly bound solvent molecule in $[Co(CN)_5]^{3-}$ ($t_{2g}^6 e_g^1$, ***z*-out J.T. distortion**) is highly labile. The product $[Co(CN)_5X]^{3-}$ is characterised and it is kinetically and thermodynamically highly stable.

Note 1: $[Co(NH_3)_6]^{3+}$ having no bridging ligand definitely reacts with $[Co(CN)_5]^{3-}$ through the OSET path. But the rate constant ($8 \times 10^4\ dm^3\ mol^{-1}\ s^{-1}$) is similar to the rate constants of many other reactions involving $[Co(NH_3)_5X]^{2+}$ to undergo the ISET path (Table 6.4.5.4).

Note 2: Co(II) is softened in $[Co(CN)_5]^{3-}$ due to the **symbiotic effect** (*cf.* C-bonded CN^- ligands soften the Co(II)-centre). In the reaction $[(H_3N)_5Co—NCS]^{2+} – [Co(CN)_5]^{3-}$ through the ISET path, the

'NCS' ligand bridges through the free S-end with the softened Co(II)-centre (**soft-soft matching**) and it gives the final product $[Co^{III}(CN)_5(SCN)]^{3-}$. But in the reaction $[(H_3N)_5Co—SCN]^{2+}$–$[Co(CN)_5]^{3-}$, through the ISET path, 'SCN' bridges through the free N-end (*i.e.* **remote attack** to avoid the steric crowding in the **adjacent attack** through the S-end, discussed later, Sec. 6.4.9) with the softened Co(II) centre (*i.e.* **poor HSAB matching**, N-end is a hard donor end). In fact, the kinetically controlled linkage isomer product $[Co^{III}(CN)_5(NCS)]^{3-}$ isomerises to the thermodynamically more stable product, $[Co^{III}(CN)_5(SCN)]^{3-}$. Obviously, for this reductant $[Co(CN)_5]^{3-}$, the rate ratios k_{SCN}/k_{NCS}, k_{N_3}/k_{NCS} are not high in contrast to other reductants (*e.g.* $[Cr(aq)]^{2+}$) reducing $[Co(NH_3)_5X]^{2+}$ through the ISET path (see Sec. 6.4.10).

Note 3: In the ISET reaction, $[Co(NH_3)_5(NO_2)]^{2+}$–$[Co(CN)_5]^{3-}$, the kinetically controlled product $[Co^{III}(CN)_5(ONO)]^{3-}$ isomerises to the thermodynamically controlled product, $[Co(CN)_5(NO_2)]^{3-}$ (*cf.* N-site is softer than O-site of NO_2^- ligand). This aspect has been illustrated later (see Sec. 6.4.9).

Note 4: In presence of **free** $\mathbf{CN^-}$, formation of $[Co(CN)_6]^{4-}$ may lead to the OSET path ($[Co(CN)_5]^{3-} + CN^- \rightleftharpoons [Co(CN)_6]^{4-}$). In the reduction of $[Co^{III}(NH_3)_5X]$, depending on the bridging property of X, both the OSET path involving $[Co(CN)_6]^{4-}$ (X poor bridging ligand) and ISET path involving $[Co(CN)_5]^{3-}$ (X good bridging ligand) can operate (see Sec. 6.6).

(g) Possibility of detection of the intermediate Red$^+$—X like Eu^{III}—X, Fe^{III}—X and V^{III}—X produced in the reduction of Co^{III}-complexes having the bridging ligand X through the ISET path: For the reducing agents like Cu^+ and Eu^{2+}, the intermediates Cu^{II}—X and Eu^{III}—X produced respectively in the ISET path **can hardly be detected because of their extreme lability**. However, the Fe^{III}—X and V^{III}—X intermediates produced respectively for the reductions by Fe^{2+} and V^{2+} in the ISET path, can be detected in some cases by **flow methods** because of their relatively less lability.

In fact, the intermediate species $[(aq)Fe(C_2O_4)]^+$ produced in the reduction of $[Co(C_2O_4)_3]^{3-}$ by $[Fe(aq)]^{2+}$ through the ISET path has been detected. The detection of such intermediates is possible only when the **electron transfer step is not significantly slower than the decomposition (*i.e.* aquation) rate of the intermediate.**

Step I: $[Fe(aq)]^{2+} + [Co(C_2O_4)_3]^{3-} \xrightarrow[ISET]{H^+} [(aq)Fe(C_2O_4)]^+ + [Co(aq)]^{2+} + 2H_2C_2O_4$

Step II: $[(aq)Fe(C_2O_4)]^+ \overset{H^+}{\rightleftharpoons} [Fe(aq)]^{3+} + H_2C_2O_4$

$[(aq)Fe(C_2O_4)]^+$ can be detected under the condition.

Step I is not appreciably slower than the Step II. It allows the **build-up of the intermediate** $\mathbf{[(aq)Fe(C_2O_4)]^+}$.

The inner-sphere reduction of $[(H_3N)_5Co(SCN)]^{2+}$ by $[Fe(aq)]^{2+}$ produces the intermediate $\mathbf{[(aq)Fe(NCS)]^{2+}}$ **which can be detected** but the intermediate $[(aq)Fe(SCN)]^{2+}$ produced in the inner-sphere reduction of $[(H_3N)_5Co(NCS)]^{2+}$ by $[Fe(aq)]^{2+}$ cannot be detected. This difference can be realised by considering the overall reaction to **proceed in two discernible steps (*i.e.* two step consecutive reaction):**

Step I: $[(H_3N)_5Co^{III}—SCN]^{2+} + [Fe(aq)]^{2+} \xrightarrow[ISET]{H^+} [(aq)Fe(NCS)]^{2+} + [Co(aq)]^{2+} + 5NH_4^+$

Step II: $[(aq)Fe(NCS)]^{2+} \rightleftharpoons [Fe(aq)]^{3+} + SCN^-$

Step I ($k_{el} \approx 0.12\ dm^3\ mol^{-1}\ s^{-1}$) is not much slower than Step II and it allows the detection of Fe^{III}—NCS intermediate (*i.e.* $[(aq)Fe(NCS)]^{2+}$. But for the reduction of $[(H_3N)_5Co(NCS)]^{2+}$ by $[Fe(aq)]^{2+}$, the electron transfer rate (*i.e.* Step I) is very slow ($k_{el} < 3 \times 10^{-6}\ dm^3\ mol^{-1}\ s^{-1}$) and the intermediate Fe^{III}—SCN (*i.e.* $[(aq)Fe(SCN)]^{2+}$) undergoes decomposition (Step II) before it can be accumulated and detected. This is the principle of any consecutive reaction.

($k_{SCN}/k_{NCS} \geq 4 \times 10^4$, an evidence for the ISET path. This aspect has been discussed in Sec. 6.4.10).

(h) Effect of halides (halide effect) as the bridging ligands in the ISET path (*Refs*: *Inorg. Chem.*, **4**, 1029, 1965; **7**, 1475, 1968)): The generalised ISET path of reduction of Co^{III}—X complexes by the reductant (Red) is:

$$Co^{III}—X + Red \longrightarrow [Co^{III}\cdots X \cdots Red]^{\neq} \longrightarrow Co^{II} + Red^{+}—X$$

(Red^+—X may undergo aquation depending on the lability of Red^+).

For the reductants like $[Cr(aq)]^{2+}$ and $[Co(CN)_5]^{3-}$, the reactivity order (**called normal order**) is:

$$[Co(NH_3)_5I]^{2+} > [Co(NH_3)_5Br]^{2+} > [Co(NH_3)_5Cl]^{2+} > [Co(NH_3)_5F]^{2+}$$

It may be noted that this **normal order** of halide effect also persists for the **OSET reduction** of $[Co(NH_3)_5X]^{2+}$ by $[Cr(bpy)_3]^{2+}$ and $[Ru(NH_3)_6]^{2+}$. This can be explained by considering the electronic activation of the Co^{III}-complex (*i.e.* $t_{2g}^6 \rightarrow t_{2g}^5 e_g^1$, ligand field strength order: $F^- > Cl^- > Br^- > I^-$) required for the $\pi^* \rightarrow \pi^*$ electron transfer in the OSET path.

But for the reductants like $[Fe(aq)]^{2+}$ and $[Eu(aq)]^{2+}$, the reverse reactivity order (**called inverse order**) is noted, *i.e.*

$$[Co(NH_3)_5F]^{2+} > [Co(NH_3)_5Cl]^{2+} > [Co(NH_3)_5Br]^{2+} > [Co(NH_3]_5I]^{2+}$$

The order of reactivity of the reductants for a particular oxidant (say, $[Co(NH_3)_5Cl]^{2+}$) is:

$$[Co(CN)_5]^{3-} > [Cr(aq)]^{2+} > [Eu(aq)]^{2+} > [Fe(aq)]^{2+}$$

In the formation of 3*c*–3*e* bridging segment, Ox—X—Red (bond order = 0.5 in each bond) in the binuclear precursor complex, the initial Ox—X 2*c*–2*e* bond (bond order = 1) partially breaks and a new Red—X bond forms. It has been suggested that for the **mild reductants** (in terms of reactivity) like $[Fe(aq)]^{2+}$ and $[Eu(aq)]^{2+}$, formation of a bond by the bridging ligand (*i.e.* Red—X bond formation) in the binuclear transition state is more important than bond breaking (*i.e.* Co^{III}—X bond breaking). The bond strength increases in the order: Red—F > Red—Cl > Red—Br > Red—I. This is the **stability order of the binuclear transition state**. It explains the reactivity order for the reductions by $[Fe(aq)]^{2+}$ and $[Eu(aq)]^{2+}$.

On the other hand, for the **more reactive reductants** like $[Cr(aq)]^{2+}$ and $[Co(CN)_5]^{3-}$, the Co^{III}—X bond breaking is more important than the Red—X bond formation in attaining the binuclear transition state. The bond strength order increases: Co^{III}—F > Co^{III}—Cl > Co^{III}—Br > Co^{III}—I. It may be noted that the ability to transmit an electron is greatest for I^- and least for F^- acting as the bridging ligand. It explains the reactivity order of the Co^{III}-complexes for their reduction by $[Co(CN)_5]^{3-}$ and $[Cr(aq)]^{2+}$.

- **Reactivity order of $[Co(NH_3)_5X]^{2+}$ (X = F, Cl, Br, I) for their reduction by $[Cr(bpy)_3]^{2+}$ and $[Ru(NH_3)_6]^{2+}$ through the OSET path** (see Table 6.4.1.1): The reactivity order, $[Co(NH_3)_5I]^{2+} > [Co(NH_3)_5Br]^{2+} > [Co(NH_3)_5Cl]^{2+} > [Co(NH_3)_5F]^{2+}$ (**normal order of halide effect**) has been explained by considering the **electronic activation** ($t_{2g}^6 \rightarrow t_{2g}^5 e_g^1$) energy of the oxidant. This electronic activation is required for the OSET path passing through the $\pi^* \rightarrow \pi^*$ electron transfer.
- **Halide effect on the OSET path:** Electronic activation at Co(III).
- **Halide effect on the ISET path:** Relative importance of the Co^{III}—X bond breaking *vs.* Red—X bond formation in attaining the binuclear transition state. **Normal order** ($I^- > Br^- > Cl^- > F^-$): Co^{III}—X bond breaking energy, nucleophilicity order of the halides and their electron transmitting efficiency order. **Inverse order** ($F^- > Cl^- > Br^- > I^-$): thermodynamic stability order of the bond Red—X in the binuclear transition state.
- **Halide effect for the reduction by $[V(aq)]^{2+}$ through the substitution (associative path) controlled ISET path:** Nucleophilicity order of the bridging halides.

Table 6.4.5.5 Second order rate constants (25°C) for the reduction of Co(III)-complexes by $[V(aq)]^{2+}$ and $[Eu(aq)]^{2+}$

Co(III)-complex (oxidant)	***k* (dm³ mol⁻¹ s⁻¹) by [V(aq)]²⁺**	***k* (dm³ mol⁻¹ s⁻¹) by [Eu(aq)]²⁺**	***k* (dm³ mol⁻¹ s⁻¹) by [Cr(bpy)₃]²⁺**
$[CoBr(NH_3)_5]^{2+}$	25	2.5×10^2	5×10^6
$[CoCl(NH_3)_5]^{2+}$	8	4×10^2	8×10^5
$[Co(NH_3)_5(N_3)]^{2+}$	13	2×10^2	4×10^4
$[Co(NH_3)_5(NO_3)]^{2+}$	–	1×10^2	–

$E^0_{V^{3+}/V^{2+}} = -0.26$ V; $E^0_{Eu^{3+}/Eu^{2+}} = -0.37$ V

(i) Electronic activation of $[V(aq)]^{2+}$ in the ISET path: Here it may be mentioned that for the reduction of the Co(III)-complexes by $[V(aq)]^{2+}$, the rate determining step is the substitution at $[V(aq)]^{2+}$ centre by the bridging ligand present with the Co(III)-centre. The net electron transfer $\pi^* \rightarrow \sigma^*$ (*i.e.* $t_{2g} \rightarrow e_g$) also requires the electronic activation of $[V(aq)]^{2+}$ from $t_{2g}^3 e_g^0$ to $t_{2g}^2 e_g^1$ to execute the $\sigma^* \rightarrow \sigma^*$ electron transfer (*i.e.* $e_g \rightarrow e_g$) at the step of electron transfer in the inner sphere path. *This electronic activation also puts a barrier compared to the reduction by* $[Cr(aq)]^{2+}$ *where the net electron transfer is* $\sigma^* \rightarrow \sigma^*$. Here it may be mentioned that $[V(aq)]^{2+}$ is a poorer reducing agent than $[Cr(aq)]^{2+}$ in a thermodynamic sense. (*cf.* Secs. 6.3.6 and Table 6.3.6.2). The **halide effect** (discussed already) for the reduction of $[Co(NH_3)_5X]^{2+}$ by $[V(aq)]^{2+}$ in the substitution controlled ISET path has been explained by considering the nucleophilicity order of the bridging halides because $[V(aq)]^{2+}$ adopts the **associative path** of ligand substitution.

(j) Reduction of $[Co^{III}(NH_3)_5X]^{2+}$ by $[V(aq)]^{2+}$, $[Eu(aq)]^{2+}$ and $[Cr(bpy)_3]^{2+}$ (Table 6.4.5.5): Reduction by $[V(aq)]^{2+}$ through the OSET/substitution controlled ISET path; reduction by $[Eu(aq)]^{2+}$ through the ISET path; reduction by $[Cr(bpy)_3]^{2+}$ (no vacant coordination site, t_{2g}^4, kinetically inert) through the OSET path.

(C) **Type-III (fission of the successor complex to the final products is the rate determining step) and detection of the binuclear successor complex:** It happens so when the electron transfer generates *both the metals centres inert in the binuclear successor complex*. It is classically illustrated in the reduction of $[Ru^{III}(NH_3)_5X]^{2+}$ by $[Cr(aq)]^{2+}$.

$$\left[(H_3N)_5Ru^{III}{-}X\right]^{2+} + \left[Cr(OH_2)_6\right]^{2+} \rightleftharpoons \left[(H_3N)_5Ru^{III}{-}X{-}Cr^{II}(OH_2)_5\right]^{4+} + H_2O$$

$$\left[(H_3N)_5Ru^{III}{-}X{-}Cr^{II}(OH_2)_5\right]^{4+} \xrightarrow[e\text{-transfer}]{\text{fast}} \left[(H_3N)_5Ru^{II}{-}X{-}Cr^{III}(OH_2)_5\right]^{4+} \xrightarrow{k\ (\text{rds})} \text{Product}$$

Thus the above reaction scheme can be simply represented as follows:

$$[(H_3N)_5Ru^{III}{-}X]^{2+} + [Cr(OH_2)_6]^{2+} \overset{K_{eq}}{\rightleftharpoons} [(H_3N)_5Ru^{II}{-}X{-}Cr^{III}(OH_2)_5]^{4+} + H_2O$$

$$[(H_3N)_5Ru^{II}{-}X{-}Cr^{III}(OH_2)_5]^{4+} + H_2O \xrightarrow[(\text{rds})]{k} [(H_3N)_5Ru^{II}{-}X]^{+} + [Cr(OH_2)_6]^{3+}$$

(Assuming the Ru^{II}—X bond thermodynamically and kinetically more stable than the Cr^{III}—X bond.
Thus the above reaction scheme can be simply represented as follows:

$$[(H_3N)_5Ru^{III}{-}X]^{2+} + [Cr(OH_2)_6]^{2+} \xrightarrow{K_{eq}} [(H_3N)_5Ru^{II}{-}X{-}Cr^{III}(OH_2)_5]^{4+} + H_2O$$

$$[(H_3N)_5Ru^{II}{-}X{-}Cr^{III}(OH_2)_5]^{4+} + H_2O \xrightarrow[(rds)]{k} [(H_3N)_5Ru^{II}{-}X]^{+} + [Cr(OH_2)_6]^{3+}$$

(Assuming, the R^{II}—X bond thermodynamically and kinetically more stable than the Cr^{III}—X bond).

The above scheme leads to the following rate law (*cf.* Secs. 5.9.3; 6.4.2):

$$\text{Rate} = \frac{kK_{eq}}{1+K_{eq}\left[\text{Cr}^{\text{II}}\right]}\left[\text{Ru}(\text{NH}_3)_5\text{X}^{2+}\right]_T\left[\text{Cr}^{\text{II}}\right];\ \text{in presence of excess reductant}\ \left(i.e.\ \left[\text{Cr}^{\text{II}}\right]_T \gg \left[\text{Ru}^{\text{III}}\right]_T\right)$$

i.e. $k_{obs} = \dfrac{kK_{eq}\left[\text{Cr}^{\text{II}}\right]}{1+K_{eq}\left[\text{Cr}^{\text{II}}\right]}$, K_{eq} = overall equilibrium constant of the process leading to the formation of the binuclear successor complex

In such cases, there is a bright possibility to *characterize the binuclear successor complex.*

Note: The lone pair on the 'N' site of the –$CONH_2$ group cannot be used for the nucleophilic attack on the reductant to make a bridge because the **lone pair is engaged in resonance with the carbonyl group.** Moreover, attack by the N-site of the –$CONH_2$ group will produce a **nonconjugated system** that will not allow the ISET process. These aspects have been discussed in Sec. 6.4.9. In the binuclear successor complex, the Ru^{II}—N bond is both kinetically and thermodynamically more stable than the Cr^{III}—O bond.

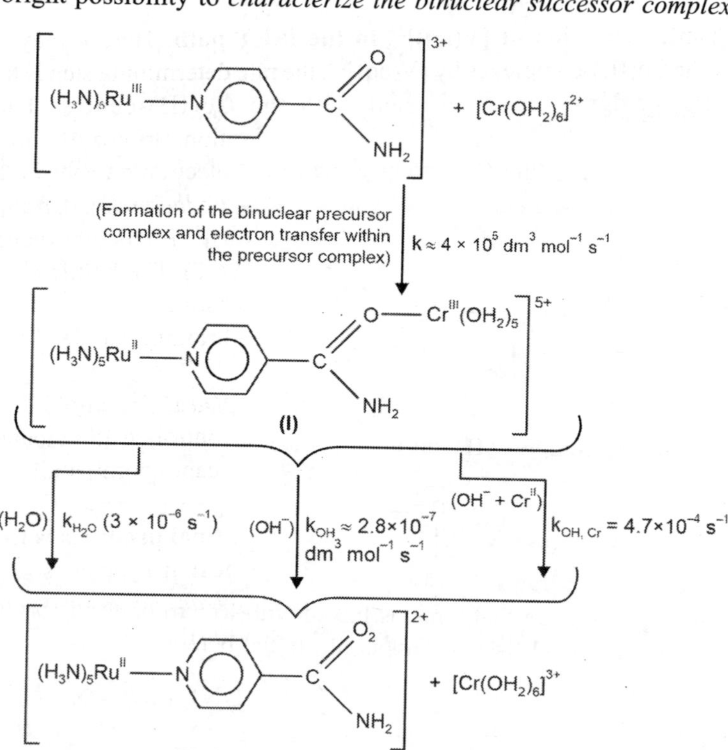

(a) Reduction of $[Ru^{III}(NH_3)_5X]$ by $[Cr(aq)]^{2+}$ through the Type-III ISET process (X = isonicotinamide) (*cf.* R. G. Gaunder *et al, Inorg. Chem.*, **9**, 2627, 1970; **14**, 1283, 1975)**:** In the reduction of pentaammineisonicotinamideruthenium(III) by $[Cr(aq)]^{2+}$, formation of the precursor complex and electron transfer within the binuclear precursor complex leading to the binuclear successor complex bearing Ru(II) and Cr(III) become complete within the time scale of mixing. *It gives the orange-yellow binuclear successor complex which is sufficiently stable to be characterized.* In fact, in this successor complex, both the metal centres are inert. These inert centres are: Ru(II) ($t_{2g}^6e_g^0$) and Cr(III) ($t_{2g}^3e_g^0$). This successor complex slowly decomposes to $[Ru(NH_3)_5(\text{isonicotinaimde})]^{2+}$ and $[Cr(aq)]^{3+}$. Fission of this successor is catalyzed by OH^- alone and jointly by Cr(II) + OH^-.

$(H_3N)_5Ru^{II}$—N ... C(=O—$Cr^{III}(OH_2)_5$)NH_2 $^{5+}$ $\xrightarrow{k_{H_2O}}$ Products

I

Between the Ru^{II} ($t_{2g}^6\,e_g^0$) and Cr(III) ($t_{2g}^3\,e_g^0$) centres, ***the Ru^{II}-centre is more inert*** and the bond fission (*i.e.* nucleophilic substitution by water) occurs at the Cr^{III}-centre. More correctly, the Ru^{II}—N bond is more inert than the Cr^{III}—O bond. In fact, the **relative inertness** of the Ru^{II}—X and Cr^{III}—X bonds depends on the crystal field strength of X (to be discussed later).

(i) In the k_{H_2O} path of decomposition of the binuclear successor complex, the Cr^{III}—O bond undergoes cleavage by the nucleophile H_2O.

(ii) In the k_{OH} path of fission of the successor complex, at higher pH, deprotonation of the coordinated H_2O molecule present with the Cr^{III}-centre generates the OH group that can show the ***cis*-labilizing effect** to favour the nucleophilic replacement of the Cr^{III}–O bond.

$$I \xrightarrow[\text{(proton abstraction)}]{+OH^-} \left[(H_3N)_5Ru^{II}\text{—N}\bigcirc\text{—C}(\text{=O—}Cr^{III}(OH_2)_4(\text{:OH}))(NH_2)\right]^{4+} \xrightarrow{H_2O} \text{Product}$$

(iii) In the $k_{OH,\,Cr}$ path, the OH group generated in the coordination sphere of Cr(III) can act as a bridging ligand to make a bridge with the catalyst $[Cr(aq)]^{2+}$ which is labile.

$$I \xrightarrow{+OH^-,\,[Cr(aq)]^{2+}} \left[(H_3N)_5Ru^{II}\text{—N}\bigcirc\text{—C}(NH_2)\text{=O—}Cr^{III}(OH_2)_4\text{—O(H)—}Cr^{II}(OH_2)_5\right]^{6+}$$

(II)

In the trinuclear complex (**II**), an inner sphere electron transfer occurs in the segment,

$$\text{—C}(NH_2)\text{=O—}Cr^{III}(OH_2)_4\text{—O(H)—}Cr^{II}(OH_2)_5$$

to generate the Cr^{II}-centre that remains linked with Ru^{II} through the bridging ligand. Cr^{II} being labile experiences a rapid bond fission. This is illustrated below.

$$\text{—C}(NH_2)\text{=O—}Cr^{III}(OH_2)_4\text{—O(H)—}{}^{*}Cr^{II}(OH_2)_5 \xrightarrow{\text{e-transfer}}$$

(e-transfer from $^{*}Cr^{II}$ to Cr^{III})

$$\text{—C}(NH_2)\text{=O} \mid Cr^{II}(OH_2)_4 \mid \text{O(H)—}{}^{*}Cr^{III}(OH_2)_5 \xrightarrow{+H_2O}$$

(labile centre) **(inert centre)**

$$\left[(H_3N)_5Ru^{II}\text{—N}\bigcirc\text{—C}(\text{=O})(NH_2)\right]^{2+} + [{}^{*}Cr(OH)(OH_2)_5]^{2+} + [Cr(OH_2)_6]^{2+}$$

It is evident that the above step regenerates the catalyst $[Cr(aq)]^{2+}$.

For the fission of the binuclear successor complex of Ru^{II} and Cr^{III} bridged by the isonicotinamide ligand, if labeled $[{}^{*}Cr(aq)]^{2+}$ is added as a catalyst, then the radioactivity gets confined in the Cr(III)-product. It supports the above mentioned step.

Characterization of the dinuclear successor complexes in the ISET process

Reactants	Species characterized*
$[RuCl_6]^{3-} + [Cr(OH_2)_6]^{2+}$	$[Cl_5Ru^{II}—Cl—Cr^{III}(OH_2)_5]^-$
$[VO(OH_2)_5]^{2+} + [Cr(OH_2)_6]^{2+}$	$[(H_2O)_4V^{III}(\mu\text{-OH})_2Cr^{III}(OH_2)_4]^{4+}$
$[VO(OH_2)_5]^{2+} + [V(OH_2)_6]^{2+}$	$[(H_2O)_4V^{III}(\mu\text{-OH})_2V^{III}(OH_2)_4]^{4+}$
$[Fe(CN)_6]^{3-} + [Co(edta)]^{2-}$	$[(NC)_5Fe^{II}—CN—Co^{III}(edta)]^{5-}$
$[IrCl_6]^{2-} + [Cr(aq)]^{2+}$	$[Cl_5Ir^{III}—Cl—Cr^{III}(OH_2)_5]$
$[IrCl_6]^{2-} + [Co(CN)_5]^{3-}$	$[Cl_5Ir^{III}—Cl—Co^{III}(CN)_5]^{5-}$
$[Fe(CN)_6]^{3-} + [Co(CN)_5]^{3-}$	$[(NC)_5Fe^{II}—CN—Co^{III}(CN)_5]^{6-}$

* Both the **metal centres in the binuclear successor complex are substitutionally inert.** $edta^{4-}$ can show the **flexidentate character** to accommodate the bridging ligand.

(b) **Reduction of $[VO(OH_2)_5]^{2+}$ by $[V(aq)]^{2+}$ through the Type-III ISET process:** Another example of **Type-III** ISET process (*i.e. formation of a stable binuclear successor complex that decomposes slowly*) is the reduction of $[VO(OH_2)_5]^{2+}$ by $[V(aq)]^{2+}$. The binuclear successor complex bearing the V(III) centres (t_{2g}^2) bridged by two OH groups is both kinetically and thermodynamically stable.

$$[VO(OH_2)_5]^{2+} + [V(OH_2)_6]^{2+} \longrightarrow \left[(H_2O)_4V^{III}(\mu\text{-OH})_2V^{III}(OH_2)_4\right]^{4+} \xrightarrow[\text{(slow)}]{\text{fission}} \text{Product}$$

(Stable successor complex)

The said successor complex can also be generated from the hydrolysis of V(III)-solution.

(c) **Reduction of *cis*-$[Ru^{III}Cl_2(NH_3)_4]^+$ by $[Cr(aq)]^{2+}$ through the Type-III ISET process:** The successor complex has been spectroscopically characterized in the following reaction.

$$cis\text{-}\left[Ru^{III}Cl_2(NH_3)_4\right]^+ + \left[Cr(OH_2)_6\right]^{2+} \xrightarrow{\text{fast}} \left[(H_3N)_4ClRu^{II}—Cl—Cr^{III}(OH_2)_5\right]^{3+} \text{ (Stable)}$$

$$\text{Stable successor} \xrightarrow{\text{slow}} \text{Products}$$

In the successor complex, both the metal centres, *i.e.* Ru^{II} ($t_{2g}^6e_g^0$) and Cr^{III} ($t_{2g}^3e_g^0$) are inert.

(d) **Reduction of $[Fe(CN)_6]^{3-}$ by $[Co(CN)_5]^{3-}$ through the Type-III ISET process:** In the reduction of $[Fe(CN)_6]^{3-}$ by $[Co(CN)_5]^{3-}$, the produced binuclear successor complex is thermodynamically and kinetically so stable (*cf.* high CFSE and inertness of the low spin t_{2g}^6 configuration with a very high CFAE) that it can be isolated as a Ba^{2+}-salt.

$$\left[Fe(CN)_6\right]^{3-} + \left[Co(CN)_5\right]^{3-} \xrightarrow{\text{fast}} \left[(NC)_5Fe^{II}—CN—Co^{III}(CN)_5\right]^{6-} \textbf{ (stable)}$$

$$\downarrow H_2O,\ \text{slow (rds)}$$

$$\left[Fe(CN)_6\right]^{4-} + \left[Co(CN)_5(OH_2)\right]^{2-}$$

In the successor complex, both the low-spin metal centres, *i.e.* Fe^{II} ($t_{2g}^6e_g^0$) and Co^{III} ($t_{2g}^6e_g^0$) are kinetically inert in the ligand substitution process. Here the **relatively weaker bond 'Co^{III}—N' undergoes cleavage** by the nucleophilic attack of H_2O (*cf.* 'N'-end of CN^- is a weak field ligand). **Note:** Detection of the binuclear successor complexes which are kinetically and thermodynamically stable is a strong evidence in favour of the ISET path.

6.4.6 Possibility of Characterization of the Binuclear Precursor Complex

There are only very few cases where the binuclear precursor complex is sufficiently stable to be characterized. The favourable situations are: *thermodynamically stable precursor complex and the electron transfer is not very fast* within the binuclear precursor complex. If formation of the complex is thermodynamically favoured then it will lead to an accumulation of the complex to a significant extent and if the electron transfer is not very fast then the life time of the precursor complex increases.

- Reduction of the nitrilotriacetic acid complex of pentaaminecobalt(III) by $[Fe(aq)]^{2+}$ produces a binuclear precursor complex (life time ≈ 10 s) that can be characterized.

$$\left[(H_3N)_5Co^{III}(ntaH_2)\right]^{2+} + \left[Fe(aq)\right]^{2+} \longrightarrow \left[Co(aq)\right]^{2+} + \left[Fe^{III}(nta)(OH_2)_2\right] + 2H^+ + 5NH_3$$

$$\left[Fe(OH_2)_6\right]^{2+} + \left[(H_3N)_5Co^{III}{-}O{-}\overset{\overset{O}{\|}}{C}{-}CH_2{-}N\begin{matrix}CH_2CO_2H\\CH_2CO_2H\end{matrix}\right]^{2+} \underset{k_{-1}}{\overset{k_1}{\rightleftharpoons}} 2H^+ +$$

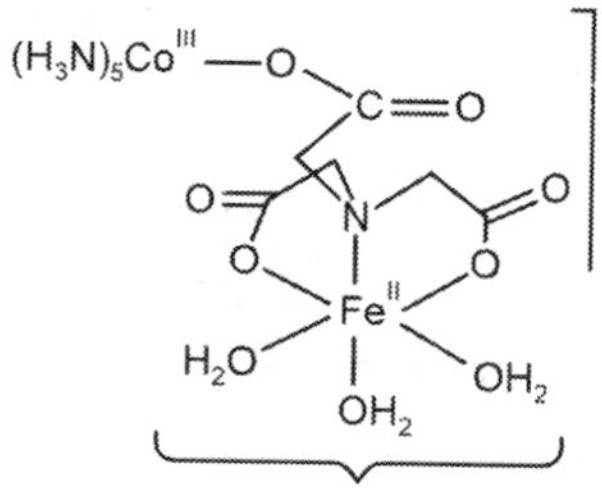

$$\left[\text{(H}_3\text{N)}_5\text{Co}^{III}\text{–O–C(=O)–CH}_2\text{–N(CH}_2\text{CO}_2)_2\text{Fe}^{II}(\text{OH}_2)_3\right]^{2+} \xrightarrow[\text{e-transfer and fission of the cpmplex}]{k} [Co(aq)]^{2+} + [Fe^{III}(nta)(OH_2)_2] + 5NH_3$$

The overall rate constant is: $k_f = K_{eq}k = 1 \times 10^5$ dm^3 mol^{-1} s^{-1} where $K_{eq} = \frac{k_1}{k_{-1}}$ (25° C) and this second order rate constant is very much higher compared to those of other reactions of reductions of Co(III)-complexes by $[Fe(aq)]^{2+}$ (*cf.* Table 6.4.5.3).

	$[CoCl(NH_3)_5]^{2+}$–$[Fe(aq)]^{2+}$	$[Co(SCN)(NH_3)_5]^{2+}$–$[Fe(aq)]^{2+}$	$[Co(C_2O_4H)(NH_3)_5]^{2+}$–$[Fe(aq)]^{2+}$
k_f: (dm^3 mol^{-1} s^{-1}) 25°C	1.5×10^{-3}	1.2×10^{-1}	4.5×10^{-1}

The apparently very high value of k_f for the reduction of $[Co(NH_3)_5(ntaH_2)]^{2+}$ by $[Fe(aq)]^{2+}$ is due to the very high value of K_{eq} (which was estimated to be 1.1×10^6 dm^3 mol^{-1}). The high value of K_{eq} arises from the **chelate effect** of the bridging *nta*-ligand that forms two 5-membered chelate rings with the Fe^{II}-centre. This yields $k = k_f/K_{eq} = 9.4 \times 10^{-2}$ s^{-1}. This value for the electron transfer step is quite reasonable compared to the rate constants for the reductions of other Co(III)-complexes by $[Fe(aq)]^{2+}$ through the ISET path (*cf.* Table 6.4.5.3).

- Fairly stable precursor complex has been reported in the following reaction. In the precursor complex, both the metal centres (t_{2g}^6, low spin) are kinetically inert.

$$\left[(H_3N)_5Co^{III}\text{—}N\langle\bigcirc\rangle\text{—}\langle\bigcirc\rangle N\right]^{3+} + \left[Fe(CN)_5(OH_2)\right]^{3-} \rightleftharpoons$$

$$\left[(H_3N)_5Co^{III}\text{—}N\langle\bigcirc\rangle\text{—}\langle\bigcirc\rangle N\text{—}Fe^{II}(CN)_5\right]^{0} + H_2O \xrightarrow[\text{and fission}]{e\text{-transfer}} \text{Products}$$

(Precursor complex)
(4,4′–bpy as the bridging ligand)

Note: Here it may be pointed out that characterization of the stable (both kinetically and thermodynamically) **binuclear precursor or successor complex** is regarded as a **strong evidence** in favour of the ISET path. Here it is worth mentioning that identification of the binuclear successor complex is a **stronger evidence in favour of the ISET path**. On the other hand, the reaction between an inert complex (bearing a bridging ligand) and a labile complex may produce a binuclear complex like a precursor complex but it does not ensure the redox reaction through the ISET path. This is illustrated below.

Here it should be mentioned that detection of the **short-lived binuclear species** like $[(H_3N)_5CoClCr(OH_2)_5]^{4+}$ (in the reaction between $[(H_3N)_5CoCl]^{2+}$ and $[Cr(OH_2)_6]^{2+}$) where the possible oxidation states of the metals are either Cr(II) (labile) and Co(III) (inert) or Cr(III) (inert) and Co(II) (labile) **does not prove that the reaction passes through the ISET path**. In such binuclear complexes, **one metal centre is inert while the other metal centre is highly labile and the complexes are kinetically short-lived**. The bridged binuclear species may remain in equilibrium with the reactants or the products which may also be produced in an OSET path. Besides these, the OSET path not the ISET path occurs within some stable binuclear precursor complexes bearing the **saturated bridging ligands** because in such cases, the saturated bridging ligands cannot mediate the *e*-transfer for the ISET path (Sec. 6.4.11). It proves that **detection of even the stable binuclear precursor complexes does not ensure the ISET path.**

6.4.7 Effect of the Nature of HOMO of the Reductant and LUMO of the Oxidant on the Rate Process of Inner-Sphere Reaction

It has already been pointed out that in the ISET path, the kinetic step of electron transfer occurs from the σ^*-HOMO ($\equiv e_g$) of the reductant to the σ^*-LUMO ($\equiv e_g$) of the oxidant.

Thus if the ground state electronic configurations allow the $\sigma^* \rightarrow \sigma^*$ electron transfer then there will be no requirement of electronic activation of the reactants. In other cases, proper electronic activation will be needed for the required $\sigma^* \rightarrow \sigma^*$ electron transfer.

The above requirement of orbital matching explains the following rate order (*cf.* Table 6.4.5.1, 3).

	Co(III)—Cr(II)	>	Co(III)—V(II)	>	Co(III)—Fe(II)
Net electron transfer:	$(\sigma^* \rightarrow \sigma^*)$		$(\pi^* \rightarrow \sigma^*)$		$(\pi^* \rightarrow \sigma^*)$

Here it may be pointed out that for the said reduction by V(II) in the ISET path, the overall process is controlled by the substitution rate at V(II). In terms of the E^0 values of the M^{3+}/M^{2+} couples, the reducing power (thermodynamic sense) runs as: $[Cr(aq)]^{2+} > [V(aq)]^{2+} \gg [Fe(aq)]^{2+}$. **Thus to explain the order of rate constants, both the nature of the HOMO-LUMO and E^0 values are to be considered simultaneously.**

6.4.8 Possibility of Ligand Transfer in the ISET Process

Possibility of the ligand transfer depends on the relative lability and bond strength of the metal centres present in the successor binuclear complex formed after the electron transfer. There may be **three possibilities** (A, B and C).

(A) Examples where the ligand transfer occurs to the reducing agent: It happens so when the reduced metal centre is more labile than the oxidized metal centre in the binuclear successor complex.

- **Reduction of the Co(III)-complexes by $[Cr(aq)]^{2+}$:** It leads to the ligand transfer to the Cr-centre.

$$\left[(H_3N)_5Co^{III}—X\right]^{n+} + \left[Cr(OH_2)_6\right]^{2+} \rightleftharpoons \left[(H_3N)_5Co^{III}—X—Cr^{II}(OH_2)_5\right]^{(n+2)+} + H_2O$$

$$\downarrow e\text{-transfer}$$

$$\left[Co(OH_2)_6\right]^{2+} + 5NH_4^+ + \left[(H_2O)_5Cr^{III}—X\right]^{n+} \xleftarrow[(+H_3O^+)]{\text{fast, fission}} \left[\underset{(t_{2g}^5e_g^2,\ \text{labile})}{(H_3N)_5Co^{II}}—X—\underset{(t_{2g}^3e_g^0,\ \text{inert})}{Cr^{III}(OH_2)_5}\right]^{(n+2)+}$$

Point of nucleophilic substitution

$X = OH^-$, halide, pseudohalide, SO_4^{2-}, PO_4^{3-}, etc.

That the entry of X^- in the coordination sphere of Cr(III) does not occur through an independent nucleophilic substitution at Cr(III) was proved by an isotope labeling experiment and comparing the substitution rate of Cr(III) with the rate of the overall redox reaction leading to the ligand transfer. These aspects have been discussed in Sec. 6.4.4.

- **Reduction of the Co(III)-complexes by $[V(aq)]^{2+}$:** Reduction of the Co^{III}-complexes by $[V(aq)]^{2+}$ also leads to the ligand transfer through the ISET path.

$$\left[L_5Co^{III}—X\right] + \left[V(OH_2)_6\right]^{2+} \rightleftharpoons \left[L_5Co^{III}—X—V^{II}(OH_2)_5\right] + H_2O$$

$$\downarrow e\text{-transfer}$$

$$\left[L_5Co^{II}—OH_2\right] + \left[X—V^{III}(OH_2)_5\right] \xleftarrow[\text{fission}]{\text{fast}} \left[\underset{\substack{\textbf{More labile}\\(t_{2g}^5e_g^2)}}{L_5Co^{II}}—X—\underset{\substack{\textbf{More inert}\\(t_{2g}^2)}}{V^{III}(OH_2)_5}\right] + H_2O$$

$$\downarrow \text{fast (further aquation)}$$

$$[Co(OH_2)_6]^{2+} + 5L$$

(*cf.* k_{ex} of the water exchange process is higher for $[Co(aq)]^{2+}$ than that for $[V(aq)]^{3+}$).

- **Reduction of the Co(III)-complexes by $[Fe(aq)]^{2+}$:** Similarly, reduction of the Co(III)-complexes by $[Fe(aq)]^{2+}$ is expected to give the ligand transfer product through the ISET path. In the successor complex, there are Co(II) and Fe(III) centres but Co(II) is more labile than Fe(III). It will lead to the ligand transfer.

  $$\left[L_5Co^{III}—X\right] + \left[Fe(OH_2)_6\right]^{2+} \xrightarrow{ISET} \left[Co(aq)\right]^{2+} + 5L + \left[Fe^{III}(OH_2)_5(X)\right]$$

  (*cf.* k_{ex} of water exchange : $[Co(OH_2)_6]^{2+} > [Fe(OH_2)_6]^{3+}$).

  $$\left[(H_3N)_5Co^{III}(ntaH_2)\right]^{2+} + \left[Fe(OH_2)_6\right]^{2+} \rightarrow \left[Co(aq)\right]^{2+} + \left[Fe^{III}(nta)(OH_2)_2\right] + 5NH_3 + 2H^+.$$

  Here the Fe^{III}–product is thermodynamically and kinetically quite stable due to the **chelate effect** of the *nta*-ligand that forms 3 five-membered chelate rings with the Fe^{III}–centre.

  Note: Because of the **lability of V(III) (t_{2g}^2) and Fe(III) ($t_{2g}^3e_g^2$)**, the products V^{III}—X and Fe^{III}—X will hydrolyse to give the free ligand (X) in solution. Here it should be pointed out that detection of the products like Fe^{III}—X, V^{III}—X, etc. is possible only when their **aquation rates are significantly slower than their formation rates** in the electron transfer step (see Sec. 6.4.5).

- **Reduction of the Cr(III)-complexes by $[Cr(aq)]^{2+}$:** Oxidation of $[Cr(aq)]^{2+}$ by Cr(III) leads to the ligand transfer in the ISET path. This can be understood by an isotope labeling experiment.

$$\left[L_5Cr^{III}—X\right] + \left[{}^*Cr(aq)\right]^{2+} \rightleftharpoons \left[L_5Cr^{III}—X—{}^*Cr^{II}(aq)\right]$$

$$\downarrow e\text{-transfer}$$

$$\left[Cr(aq)\right]^{2+} + 5L + \left[{}^*Cr^{III}(aq)(X)\right] \xleftarrow[\text{(fast)}]{\text{fission}} \left[L_5Cr^{II}—X—{}^*Cr^{III}(aq)\right]$$

In the binuclear successor complex, Cr(II) ($t_{2g}^3 e_g^1$) is more labile than Cr^{III} (t_{2g}^3).

(B) Examples where the bridging ligand is retained with the oxidizing centre (*i.e.* **no ligand transfer**)**:** It happens so when the reduced centre is more inert than the oxidized centre.

- **Reduction of $[IrCl_6]^{2-}$ by $[Cr(aq)]^{2+}$:** Reduction of hexachloridoiridate(IV) by $[Cr(aq)]^{2+}$ leads to retention of the bridging ligand (*i.e.* ***no bridging ligand transfer***) with the oxidant.

$$[Cl_5IrCl]^{2-} + \left[Cr(OH_2)_6\right]^{2+} \rightleftharpoons \left[Cl_5Ir^{IV}—Cl—Cr^{II}(OH_2)_5\right]^0 + H_2O$$

$$\downarrow \text{e-transfer}$$

$$\left[IrCl_6\right]^{3-} + \left[Cr(OH_2)_6\right]^{3+} \xleftarrow{\text{fission}} \left[\underset{\substack{\text{More inert}\\(t_{2g}^6)}}{Cl_5Ir^{III}}—Cl—\underset{\substack{\text{Relatively less inert}\\(t_{2g}^3)}}{Cr^{III}(OH_2)_5}\right]^0$$

Point of nucleophilic substitution

The Ir(III)—Cl bond is more inert and stronger than the Cr(III)—Cl bond. This is why, fission of the successor complex occurs through the nucleophilic attack on the Cr(III)—Cl bond.

- **Oxidation of $[Co(CN)_5]^{3-}$ by $[Fe(CN)_6]^{3-}$:** Oxidation of pentacyanidocobaltate(II) by hexacyanidoferrate(III) through the ISET path does not lead to transfer of the bridging ligand.

$$\left[(NC)_5Fe^{III}—CN\right]^{3-} + \left[Co(CN)_5\right]^{3-} \rightleftharpoons \left[(NC)_5Fe^{III}—C\equiv N—Co^{II}(CN)_5\right]^{6-}$$

$$\downarrow e\text{-transfer}$$

$$\left[Fe(CN)_6\right]^{4-} + \left[Co(CN)_5(OH_2)\right]^{2-} \xleftarrow{\text{fission}} \left[\underset{(t_{2g}^6 e_g^0)}{(NC)_5Fe^{II}}—C\equiv N—\underset{(t_{2g}^6 e_g^0)}{Co^{III}(CN)_5}\right]^{6-}$$

Strong field site (C) — Weak field site (N)

***C*-bound CN^- is a strong field ligand while *N*-bound CN^- is a relatively weaker field ligand** (*cf.* vacant π*–MO on CN^- remains mainly concentrated on the C-end, *i.e.* C-bound CN^- acts as a **π-acid ligand** and consequently, as a **strong field ligand**). This is why, the C-bound CN^- ligand stabilizes Fe(II) (t_{2g}^6) in $[Fe(CN)_6]^{4-}$ more than the N-bound CN^- ligand stabilizing Co(III) (t_{2g}^6) in $[Co(CN)_5(NC)]^{3-}$. This is also expected from the **principle of symbiosis and principle of HSAB theory**. This is why, the bond cleavage preferably occurs at the 'Co^{III}—N' bond.

Here it is worth mentioning that the binuclear successor complex $[(NC)_5Fe^{II}—CN—Co^{III}(CN)_5]^{6-}$ is **kinetically so stable** (*cf.* both the metal centres with the t_{2g}^6 electronic configuration are kinetically inert) with respect to hydrolysis (*i.e.* fission of the binuclear complex), that it can be isolated as the Ba-salt.

- **Oxidation of $[Co(edta)]^{2-}$ by $[Fe(CN)_6]^{3-}$:** Oxidation of $[Co(edta)]^{2-}$ by $[Fe(CN)_6]^{3-}$ through the ISET path does not lead to transfer of the bridging ligand.

$$[Co(edta)]^{2-} + [Fe(CN)_6]^{3-} \longrightarrow \underset{\text{(Successor complex)}}{[(edta)\underset{t_{2g}^6 \text{ (less inert)}}{Co^{III}}—N{\equiv}C—\underset{t_{2g}^6 \text{ (more inert)}}{Fe^{II}}(CN)_5]^{5-}} \xrightarrow{\text{slow}} [Co(edta)]^{-} + [Fe(CN)_6]^{4-}$$

Here it should be pointed out that both $[Fe(CN)_6]^{3-}$ and $[Fe(CN)_6]^{4-}$ are inert. The oxidant $[Fe(CN)_6]^{3-}$ provides the bridging ligand to the labile reductant $[Co(edta)]^{2-}$. *It may require opening of one acetate arm in $[Co(edta)]^{2-}$ to accommodate the bridging ligand. After the electron transfer, the low spin Co(III)-centre bearing the relatively weaker field ligands (3N + 3O) is more labile than the low spin Fe(II)-centre in $[Fe(CN)_6]^{4-}$ (bearing the very strong field ligands).* It leads to retention of the bridging ligand.

- **Reduction of the Ru(III)-complexes by $[Cr(aq)]^{2+}$:** Reduction of $[Ru^{III}(NH_3)_5X]^{n+}$ by $[Cr(aq)]^{2+}$ through the ISET path may or may not experience the bridging ligand transfer depending on the nature of X.

$$[(H_3N)_5Ru^{III}—X] + [Cr(aq)]^{2+} \rightleftharpoons [(H_3N)_5Ru^{III}—X—Cr^{II}(OH_2)_5] \xrightarrow{\text{e-transfer}}$$

$$\left[\underset{(t_{2g}^6,\ \text{inert})}{(H_3N)_5Ru^{II}}—X—\underset{(t_{2g}^3,\ \text{inert})}{Cr^{III}(OH_2)_5}\right] \textbf{(inert successor complex)}$$

$$\underset{\text{(Successor complex)}}{[(H_3N)_5Ru^{II}—X—Cr^{III}(OH_2)_5]} \xrightarrow[\text{slow }(rds)]{+H_2O} [(H_3N)_5Ru^{II}—X] + [Cr^{III}(OH_2)_6]^{3+}$$

$$\text{or } [(H_3N)_5Ru^{II}—OH_2]^{2+} + [(H_2O)_5Cr^{III}—X]$$

In the successor complex, both the metal centres (*i.e.* low spin Ru^{II} of t_{2g}^6 configuration and Cr^{III} of t_{2g}^3 configuration) are inert to nucleophilic substitution. Thus the successor complex is sufficiently stable to be characterized. Here, it is important to mention that $Ru^{II}(t_{2g}^6, 4d^6)$ being a member of the 4*d*-series is expected to be more inert than $Cr^{III}(t_{2g}^3, 3d^3)$ which is a 3*d*-series member. However, depending on the nature of the bridging ligand (X), the nucleophilic substitution can occur either at the Ru^{II}-centre or Cr^{III}-centre or at both the centres. In fact, *nucleophilic substitution at both the centres are known to cause fission of the successor complex* in some cases.

For *X* = Cl^-, formation of $[CrCl(OH_2)_5]^{2+}$ (*ca.* 88%) in the product indicates rupture of the Ru^{II}—Cl bond, *i.e.* **ISET reaction with the ligand transfer** (*Ref. JACS*, **91**, 7307, 1969). To explain the **minor product** $[Cr(OH_2)_6]^{3+}$ (*ca.* 12%), the slower aquation of the Cr^{III}—Cl bond in the binuclear successor complex has been suggested. This slower route leads to **no ligand transfer**. It has been noted that at the relatively higher pH, aquation of the Cr^{III}—Cl bond is favoured due to the formation of **Cr^{III}—OH bond that labilises (*cis*-effect) the Cr^{III}—Cl bond** in the binuclear successor complex (discussed in Sec. 6.4.5). This leads to more retention of Cl^- with the Ru-centre.

For *X* = isonicotinamide, formation of the product $[Ru(NH_3)_5(\text{isonicotinamide})]^{2+}$ indicates the rupture of the Cr^{III}—X, *i.e.* Cr^{III}—O bond, *i.e.* **ISET reaction without the ligand transfer**. In the binuclear successor complex, the Ru^{II}—N bond is more inert than the Cr^{III}–O bond (see Sec. 6.4.5). This is why, in the present case, the Cr^{III}—O bond ruptures (this example has been discussed already).

$$\left[(H_3N)_5Ru^{II}—N\langle\bigcirc\rangle—C(=O—Cr^{III}(OH_2)_5)(NH_2)\right]^{5+} \xrightarrow[\text{rds}]{\text{slow}} \left[(H_3N)_5Ru^{II}—N\langle\bigcirc\rangle—CONH_2\right]^{2+} + [Cr(OH_2)_6]^{3+}$$

(Successor complex)

Note: The relative inertness of the Ru^{II}—X and Cr^{III}—X bonds depends on the crystal field strength of the bridging ligand (X). **For the stronger field ligand (X), the Ru^{II}—X bond becomes more inert.**

(C) Examples where the bridging ligand is released to the solution when both the oxidized metal centre and the reduced metal centre (*e.g.* Co^{II}, $t_{2g}^5 e_g^2$) **are labile:** Reduction of $[Co(NH_3)_5X]^{2+}$ by Fe^{2+}, V^{2+}, Eu^{2+} and Cu^+ through the ISET path (with the transfer of X^-) will produce the **labile products FeX^{2+}, VX^{2+}, EuX^{2+} and CuX^+** which will hydrolyse to release the bridging ligand in solution. Among these, EuX^{2+} and CuX^+ are **extremely labile** to hydrolyse and in some cases FeX^{2+} and VX^{2+} have been characterised. Conditions for detection of such products have been discussed in Sec. 6.4.5. It may be noted that the product Co(II) is highly labile.

6.4.9 Electrophilic Remote and Adjacent Attack on the Bridging Ligand by the Reducing Agent

(A) Remote Attack

- For the **monoatomic bridging ligands** like Cl^-, Br^-, OH^- (assuming H not to bridge), etc, the same atom bridges the metal centres. There is no other possibility. This is described as the **adjacent attack** on the bridging ligand/site (generally present with the oxidising centre) by the **reducing agent as an electrophile.**
- For the N-bonded NCS^- (*i.e.* $M_{ox}—\ddot{N}=C=\ddot{S}$) and CN^- ligand (*i.e.* $M_{ox}—C≡N$), the free ligating site (*i.e.* S for –NCS; N for –CN) present at the remote area acts as the nucleophile. This is described as the **remote attack.**

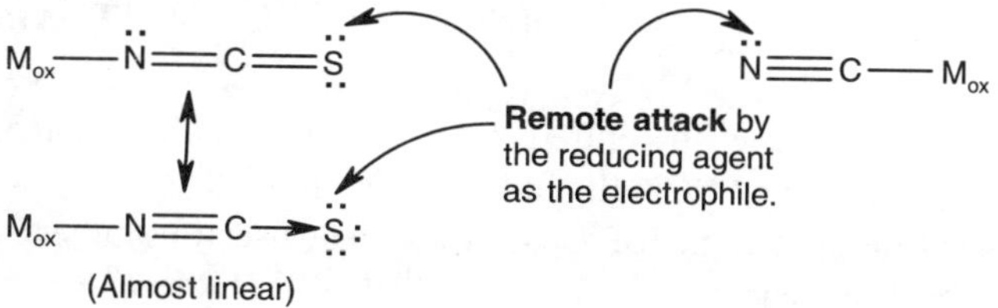

(**Note:** The lone pair on N of the –NCS linkage cannot be used for the nucleophilic attack because of the steric reason and this lone pair is engaged in resonance as shown).

In such cases, depending on the possibility of remote and adjacent attack, the ligand transfer in the ISET process can lead to the generation of linkage isomers which may not be thermodynamically stable. It is illustrated in the following examples.

(i) **$[Co^{III}(NCS)(NH_3)_5]^{2+}$ + $[Cr(OH_2)_6]^{2+}$:** Reduction of $[Co(NCS)(NH_3)_5]^{2+}$ by $[Cr(aq)]^{2+}$ produces the linkage isomer $[Cr(OH_2)_5(SCN)]^{2+}$ which on standing slowly isomerizes into the thermodynamically more stable linkage isomer, $[Cr(OH_2)_5(NCS)]^{2+}$.

$$\left[(H_3N)_5Co^{III}—NCS\right]^{2+} + \left[Cr(OH_2)_6\right]^{2+} \rightleftharpoons \left[\underset{\textbf{(Remote attack on —NCS)}}{(H_3N)_5Co^{III}—NCS—Cr^{II}(OH_2)_5}\right]^{4+}$$

$$\downarrow \textit{e-transfer}$$

$$5NH_4^+ + \left[Co(aq)\right]^{2+} + \underset{\textbf{(green isomer, kinetically controlled product)}}{\left[Cr^{III}(OH_2)_5(SCN)\right]^{2+}} \xleftarrow[\text{(fission)}]{+H_3O^+,\ \text{fast}} \left[(H_3N)_5Co^{II}—NCS—Cr^{III}(OH_2)_5\right]^{4+}$$

$$\downarrow \text{slow (isomerisation)}$$

$$\underset{\textbf{(purple isomer)}}{\left[(H_2O)_5Cr^{III}(NCS)\right]^{2+}},\ \textit{i.e.}\left[Cr(NCS)(OH_2)_5\right]^{2+}\ \textbf{(thermodynamically controlled product)}$$

Cr^{III}-being hard prefers the harder donor site *N* compared to the softer site *S*. This is why, $[Cr(OH_2)_5—SCN]^{2+}$ slowly isomerizes into the more stable linkage isomer bearing the *N*-bonded SCN^-.

(ii) $[(H_3N)_5Co^{III}—NCS]^{2+} + [Co^{II}(CN)_5]^{3-} \rightleftharpoons [(H_3N)_5Co^{III}—NCS—Co^{II}(CN)_5]^-$ **(Remote attack)**

$$\downarrow \textit{e-transfer}$$

$$5NH_4^+ + [Co(aq)]^{2+} + \underset{\text{(kinetically controlled product)}}{[Co(CN)_5(SCN)]^{3-}} \xleftarrow[\text{(fast)}]{H_3O^+} [(H_3N)_5Co^{II}—NCS—Co^{III}(CN)_5]^-$$

In the present case, the kinetically controlled product is also the thermodynamically controlled product in terms of **symbiotic theory and HSAB principle**. The C-bonded 5 CN^- ligands soften the Co-centre that prefers the S-bonded SCN^- ligand (*cf.* the reaction of $[(H_3N)_5Co—SCN]^{2+}$ with $[Co(CN)_5]^{3-}$).

(iii) $[(H_3N)_5Co^{III}—SCN]^{2+} + [Co(CN)_5]^{3-} \rightleftharpoons [(H_3N)_5Co^{III}—SCN—Co^{II}(CN)_5]^-$ **(Remote attack)**

$$\downarrow \textit{e-transfer}$$

$$5NH_4^+ + [Co(aq)]^{2+} + \underset{\text{(kinetically controlled linkage isomer)}}{[SCN—Co^{III}(CN)_5]^{3-}} \xleftarrow[\text{(fast)}]{H_3O^+} [(H_3N)_5Co^{II}—SCN—Co^{III}(CN)_5]^-$$

$$[SCN—Co^{III}(CN)_5]^{3-} \xrightarrow[\text{isomerisation}]{\text{slow}} [NCS—Co^{III}(CN)_5]^{3-}$$ **(Thermodynamically controlled product)**

Due to the **symbiotic effect**, the Co-centre is softened by the 5 CN^- ligands and consequently, the Co-centre prefers the S-bonded SCN^- ligand.

Here it is worth mentioning that the ISET pathway can be used to synthesise the S-bonded linkage isomer, $[Co(CN)_5(SCN)]^{3-}$ easily.

$$[Co(NH_3)_5(NCS)]^{2+} + [Co(CN)_5]^{3-} \xrightarrow{H_3O^+} [Co(CN)_5(SCN)]^{3-} + [Co(aq)]^{2+} + 5NH_4^+$$

$$k \approx 10^6 \text{ dm}^3 \text{ mol}^{-1} \text{ s}^{-1} \text{ (25°C)}$$

On the other hand, to synthesise the above linkage isomer through the direct substitution is a **very difficult task**.

$$[Co^{III}(CN)_5(OH_2)]^{2+} + NCS^- \longrightarrow [Co(CN)_5(NCS)]^{3-} + H_2O$$

$$\downarrow \text{Heating at} \sim 150°C \text{ for 6 hrs. in solid state}$$

$$[Co(CN)_5(SCN)]^{3-}$$

(iv) $[(H_3N)_5Co^{III}—CN]^{2+} + [Co(CN)_5]^{3-} \rightleftharpoons [(H_3N)_5Co^{III}—CN—Co^{II}(CN)_5]^-$

$$\downarrow \textit{e-transfer}$$

$$5NH_4^+ + [Co(aq)]^{2+} + [(CN)Co^{III}(CN)_5]^{3-} \xleftarrow{H_3O^+, \text{ fast}} [(H_3N)_5Co^{II}—CN—Co^{III}(CN)_5]^-$$

(*i.e.* $[Co(CN)_5(NC)]^{3-}$ **Kinetically controlled linkage isomer**)

$$\downarrow \text{slow (isomerisation)}$$

$$[Co(CN)_6]^{3-}$$

(Thermodynamically controlled product)

(v) **Remote attack on the $-NO_2$ (*N*-nitro) linkage** is illustrated below.

$$\left[(H_3N)_5Co-NO_2\right]^{2+} + [Co(CN)_5]^{3-} \rightleftharpoons \left[(H_3N)_5Co^{III}-N(=O)-O-Co^{II}(CN)_5\right]^-$$

↓ e-transfer

$$5NH_4^+ + [Co(aq)]^{2+} + \left[O{=}N-O-Co(CN)_5\right]^{3-} \xleftarrow[\text{(fast)}]{H_3O^+,\ \text{fission}} \left[(H_3N)_5Co^{II}-N(=O)-O-Co^{III}(CN)_5\right]^-$$

(*i.e.* $[Co(CN)_5(ONO)]^{3-}$)
(Nitrito isomer, kinetically controlled product)

↓ **(slow isomerisation)**

$$\left[O_2N-Co(CN)_5\right]^{3-}$$

(*i.e.* $[Co(CN)_5(NO_2)]^{3-}$)
(Thermodynamically controlled product, Nitro isomer)

The softened Co-centre (**symbiotic effect** of 5 CN^- ligands) prefers the softer N-end of the NO_2^- ligand.

(vi) **Azide (N_3^-) can experience the remote attack.** The corresponding binuclear complex may be represented as follows:

$$\left[L_5M-\ddot{N}{=}\overset{+}{N}{=}\ddot{N}-M'L_5\right]$$

- **Failure of the product analysis technique to distinguish between the remote and adjacent attack:** *For the symmetrical bridges as in the case of azide and carboxylate, it is not possible to distinguish between the remote attack and adjacent attack from the product analysis.* Because, both the possibilities will lead to the same product.

$M_{ox}-N{=}N^+{=}\ddot{N}:$ — **Remote attack** by the reductant (as an electrophile) at the terminal N; **Adjacent attack** by the reductant (as an electrophile) at the N bound to M_{ox}.

$M_{ox}-\ddot{O}-C(R){=}\ddot{O}:$ — **Adjacent attack** at the O bound to M_{ox}; **Remote attack** at the C=O oxygen.

In such cases, other arguments are needed to distinguish between the possibilities: adjacent and remote attack. **Isotope labeling experiment** may be helpful to solve the problem.

$$M^{III}—O—C(=O^*)R + [Cr(aq)]^{2+} \longrightarrow M(II)\ (labile) + Cr^{III}—O^*—C(=O)R \quad (i.e.\ \textbf{remote attack})$$

(vii) **Isonicotinamide (*i.e.* pyridine carboxamide) as a bridging ligand experiences a remote attack on the carbonyl–O.**

$(H_3N)_5Co^{III}—N\langle C_5H_4\rangle—C(=\ddot{O})\ddot{N}H_2$ — **Remote attack** by the labile reductant (electrophile)

(**Note:** In such cases, the remote attack on 'N' fails for the ISET process because of **two reasons: poor basicity** (*i.e.* poor nucleophilicity) of 'N' and **failure to resonance electron transfer mechanism** requiring an extended conjugation (see Sec. 6.4.11). The lone pair on the $—\ddot{N}H_2$ group is engaged in resonance with the 'CO' group. Thus, the $—\ddot{N}H_2$ group is not sufficiently basic to act as the nucleophile $—C(=O)\ddot{N}H_2$. These aspects are illustrated in the following example where the bridging property of the **formamide group** is discussed).

(vii) **Remote attack on the coordinated formamide group:**

$$[(H_3N)_5Co^{III}—NH_2—C(=\ddot{O})H]^{3+} + [Cr(aq)]^{2+} \rightleftharpoons [(H_3N)_5Co^{III}—NH_2—C(H)=\ddot{O}—Cr^{II}(OH_2)_5]^{5+}$$

$$\xrightarrow[(H_3O^+)]{\text{e-transfer and ligand transfer}} 5NH_4^+ + [Co(aq)]^{2+} + [(H_2O)_5Cr^{III}—O=C(H)NH_2]^{3+}$$

i.e. **ISET involving the remote attack.**

$$[(H_3N)_5Co^{III}—O=C(\ddot{N}H_2)H]^{3+} + [Cr(aq)]^{2+} \Rightarrow$$

Remote attack is not possible to allow the ISET process because the NH_2 group is not sufficiently basic and the remote attack on N fails to give a conjugated system required for the resonance transfer of electron.

Reaction passing through the OSET path

(ix) **Formation of a conjugated system through the remote attack—a prerequisite condition for ISET (*i.e.* resonance mechanism of ISET):** It has been suggested that the remote attack in the

ISET process must produce a **conjugated system** involving the oxidant, bridging ligand and reductant. This is called the **resonance electron transfer mechanism** of ISET reaction. In fact, the remote attack by the S-end of —NCS, N-end of —CN, N-end of —N_3, O-end of —NO_2, O-end of $—C(=O)NH_2$ can constitute the conjugated system required for the resonance mechanism. In such cases, participation of the π-electron cloud and π-orbitals of the bridging ligand mediate the electron transfer.

(The bridging atoms are having the **intermediate hybridisation state** because of the resonance. The bond angle M—S—C, 107° close to the tetrahedral angle, 109° indicates that S predominantly exists in sp^3-state).

(4,4′–bpy as the bridging ligand) (pyrazine as the bridging ligand)

Conjugated system and participation of π-electron system through the remote attack
(Resonance mechanism of ISET)

(Nonconjugated system and no π-electron system in the attack, failure of the Resonance electron transfer mechanism

(B) Adjacent Attack

S-bonded thiocyanate (*i.e.* –SCN) can lead to both the possibilities– adjacent attack and remote attack by the reducing agent.

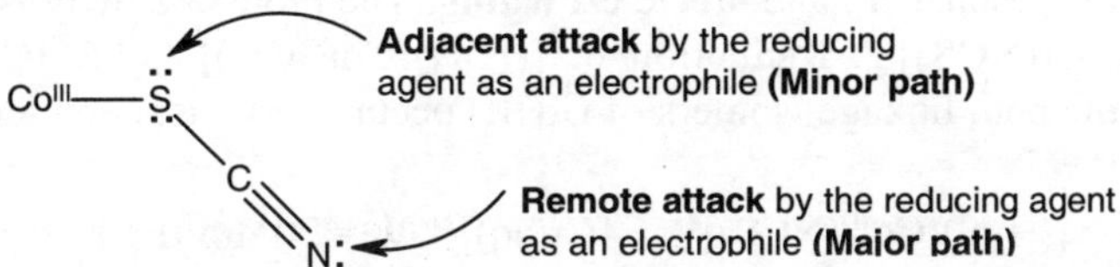

• **$[Co(NH_3)_5(SCN)]^{2+}$ + $[Cr(OH_2)_6]^{2+}$:** Here the ISET process can proceed through the both remote and adjacent attack. **Attack on the S-end** is not thermodynamically favoured by the **HSAB principle** for the hard reducing agent like $[Cr(aq)]^{2+}$. Attack on the S-end (*i.e.* adjacent attack) is also **sterically disfavoured.** But the S-site is kinetically good to mediate the electron transfer.

$$\left[(H_3N)_5Co(SCN)\right]^{2+}+\left[Cr(aq)\right]^{2+}\xrightarrow{\text{(Both remote and adjacent attack)}}\underset{ca.\ 70\%}{\left[(H_2O)_5Cr^{III}(NCS)\right]^{2+}}+\underset{ca.\ 30\%}{\left[(H_2O)_5Cr^{III}(SCN)\right]^{2+}}$$

Both Cr^{III} and Cr^{II} being hard prefer the *N*-end of SCN^-. Thus $[(H_2O)_5Cr^{III}(NCS)]^{2+}$, *i.e.* $[Cr(NCS)(OH_2)_5]^{2+}$, is the **major product.**

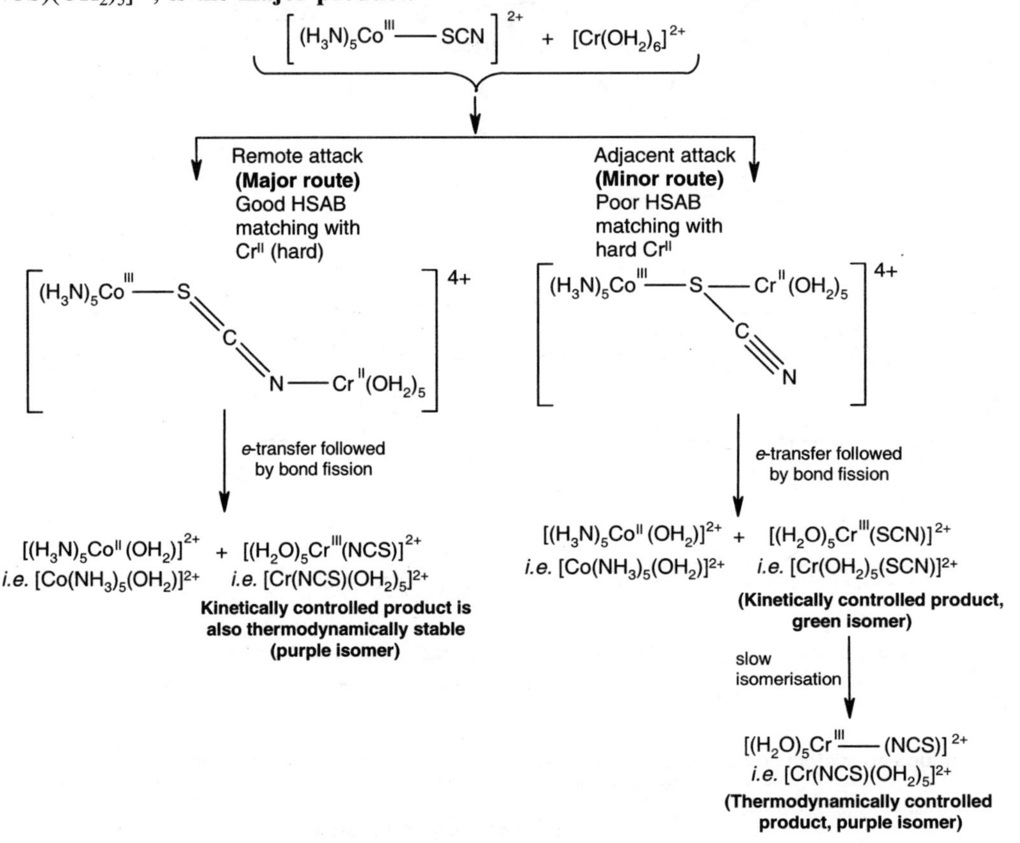

$$[Co(NH_3)_5(OH_2)]^{2+}\xrightarrow[\text{(fast)}]{\text{aquation }(H_3O^+)}[Co(OH_2)_6]^{2+}+5NH_4^+$$

• **Reduction of $[Co(NH_3)_5(SCN)]^{2+}$ *vs.* $[Co(NCS)(NH_3)_5]^{2+}$ by $[Cr(OH_2)_6]^{2+}$:** Here it may be mentioned that the reduction of $[(H_3N)_5Co(NCS)]^{2+}$, *i.e.* $[Co(NCS)(NH_3)_5]^{2+}$ by $[Cr(aq)]^{2+}$ exclusively produces $[(H_2O)_5Cr(SCN)]^{2+}$, *i.e.* $[Cr(OH_2)_5(SCN)]^{2+}$. *Here only the remote attack on the S-site is possible.* Here the **adjacent attack** on the N-site is not possible because of the **nonavailability of the N-lone pair** (engaged in resonance) and **steric crowding**. The product slowly isomerizes to the more stable isomer $[(H_2O)_5Cr(NCS)]^{2+}$. Reduction of $[(H_3N)_5Co(SCN)]^{2+}$, *i.e.* $[Co(SCN)(NH_3)_5]^{2+}$ by $[Cr(aq)]^{2+}$ producing the both linkage isomers of Cr(III) occurs very fast compared to the reduction of $[(H_3N)_5Co(NCS)]^{2+}$.

$$(cf.\ \text{Table 6.4.10.2})\left\{\begin{array}{l}[(H_3N)_5Co^{III}(NCS)]^{2+}+[Cr(aq)]^{2+}:\ k=20\ dm^3\ mol^{-1}\ s^{-1}\ \text{at } 25°C.\\ [(H_3N)_5Co^{III}(SCN)]^{2+}+[Cr(aq)]^{2+}:\ k=2\times10^5\ dm^3\ mol^{-1}\ s^{-1}\ \text{at } 25°C.\end{array}\right.$$

The higher rate for the second reaction is due to the favoured remote attack involving the harder N-site. The slowness of the first reaction is due to the remote attack on the softer *S*-site. The first reaction can only lead to this remote attack while the second reaction can lead to both the remote and adjacent attack. The rate benefit in the second reaction is due to the **better HSAB matching (both Cr^{II} and Cr^{III} hard) and less steric crowding in the remote attack.**

- **$[Co(NH_3)_5(SCN)]^{2+}$–$[Co(CN)_5]^{3-}$ *vs.* $[Co(NH_3)_5(NCS)]^{2+}$–$[Co(CN)_5]^{3-}$:** 5 soft CN^- ligands soften (symbiotic effect) the Co-centre in $[Co(CN)_5]^{3-}$ and it prefers the S-end of the bridging SCN^- ligand. This is why, $[Co(NH_3)_5(NCS)]^{2+}$ reacts faster than $[Co(NH_3)_5(SCN)]^{2+}$ with $[Co(CN)_5]^{3-}$ through the sterically favoured remote attack in the ISET path (see example iii). The linkage isomer $[Co(CN)_5(NCS)]^{3-}$ produced in the reaction with $[Co(NH_3)_5(SCN)]^{2+}$ isomerises slowly to the thermodynamically controlled product $[Co(CN)_5(NCS)]^{3-}$. Thus, the reactivity order of $[Co(NH_3)_5(NCS)]^{2+}$ and $[Co(NH_3)_5(SCN)]^{2+}$ depends on the nature of the reducing agent.

6.4.10 Role of the Bridging Ligand in the ISET Reaction (*cf.* H. Taube *et al.*, *Acc. Chem. Res.*, **2**, 321, 1969; A. Haim, *ibid*, **8**, 265, 1965)

The important functions of the bridging ligands in the ISET process are:

(i) **Kinetic Contribution**– favouring the **electron tunneling** from the reducing agent to the oxidant; this occurs for the monoatomic bridges and conjugated systems (**resonance transfer**) involving the oxidant, bridging ligand and oxidant.

(ii) **Kinetic Contribution**– reducible bridging ligand can participate in the redox process to favour the ISET process through the **chemical mechanism.**

(iii) **Kinetic contribution**– reducing the electrostatic repulsion between the positively charged oxidant (*e.g.* M^{III}) and reductant (*e.g.* M^{II}) metal centres.

(iv) **Thermodynamic Contribution**– through chelation, stability of the binuclear precursor complex may be enhanced to favour the overall rate process (*cf.* $k_f = kK_{eq}$, rate law).

Role of the bridging ligands in **chemical and tunneling mechanism** of the electron transfer will be discussed separately in Secs. 6.4.11, 12. Here we shall mainly discuss the effect of stabilization of the binuclear precursor complex on the overall rate process (*cf.* Sec. 6.4.6).

- **(a) Effect of the bridging ligand depending on the position of *rds* in the ISET process:** All the bridging ligands are not equally efficient in stabilizing the binuclear complexes and mediating the electron transfer step. ***However, effect of the bridging ligand is noticeable only when formation of the precursor complex is rapid*** (*i.e.* reducing centre is substitutionally labile) and the overall rate is given by (under the condition, $k_{-1} \gg k$):

$$\text{rate} \approx \frac{k_1}{k_{-1}} k[\text{Ox—X}][\text{Red}] = kK[\text{Ox—X}][\text{Red}] \quad (cf.\ \text{Sec. 6.4.2})$$

If the reducing centre (*e.g.* $[V(aq)]^{2+}$) is substitutionally inert then the overall rate process is controlled by the nucleophilic substitution rate at the reducing centre and the overall rate is given by:

$$\text{rate} \approx k_1[\text{Ox—X}][\text{Red}] \quad (cf.\ \text{Sec. 6.4.2})$$

If the nucleophilic substitution experiences the **dissociative activation**, then the nucleophilic substitution rate is insensitive to the nature of the bridging ligand that acts as the entering ligand. ***In such a substitution controlled ISET process, effect of the bridging ligand nature is insignificant.*** If the nucleophilic substitution at the reducing centre like $[V(aq)]^{2+}$ occurs through the **associative path** like the I_a path, then, the rate shows a mild sensitiveness to the nature of the bridging ligand (see **halide effect**, Sec. 6.4.5).

- **(b) H_2O *vs.* ^-OH as the bridging ligand:** ^-OH is a much better bridging ligand than water.

$$M—O(H)(H)\text{-----}M' \quad vs. \quad M—O(H)\text{-----}M'$$

This is why, reduction of $[Co(NH_3)_5(OH)]^{2+}$ by $[Cr(aq)]^{2+}$ is much faster than the reduction of $[Co(NH_3)_5(OH_2)]^{3+}$ by $[Cr(aq)]^{2+}$ (*See* Sec. 6.4.4 for discussion of the ISET path in the reduction of $[Co(NH_3)_5(OH_2)]^{3+}$ by $[Cr(aq)]^{2+}$).

The electron exchange rate in the couple $[Fe(OH_2)_6]^{3+}$–$[Fe(OH_2)_6]^{2+}$ is highly pH dependent. The rate increases with the pH. With the increase of pH, $[Fe(aq)]^{3+}$ hydrolyses to $[Fe(OH)(OH_2)_5]^{2+}$ which can provide the 'OH' group for the bridge formation with the labile centre $[Fe(aq)]^{2+}$. *In fact, the electron exchange in the aqua couple, i.e.* $[Fe(H_2O)_6]^{3+}$–$[Fe(H_2O)_6]^{2+}$ *occurs through the OSET path while the electron transfer in the couple,* $[Fe(OH)(OH_2)_5]^{2+}$–$[Fe(OH_2)_6]^{2+}$ *occurs more rapidly through the ISET path.*

$[Fe(OH_2)_6]^{3+}$–$[Fe(OH_2)_6]^{2+}$: OSET path (H_2O cannot act as a good bridging ligand)

$[Fe(OH)(OH_2)_5]^{2+}$–$[Fe(OH_2)_6]^{2+}$: ISET path (OH group can act as a much better bridging group; k_{ex} (ISET) >> k_{ex} (OSET)).

Effect of the externally added anions on the ISET path

Sometimes, the rate is altered in presence of the externally added ions and the product is also altered. Reduction of $[CoCl(NH_3)_5]^{2+}$ by $[Cr(aq)]^{2+}$ in presence of the foreign anions like $P_2O_7^{4-}$, gives the Cr(III) product bearing the $P_2O_7^{4-}$ ligand. This happens so due to the substitution at the labile Cr(II)-centre before the electron transfer step. Thus the precursor complex may be proposed as follows:

$$\left[(H_3N)_5Co^{III}—Cl—Cr^{II}(OH_2)_4(P_2O_7^{4-})\right]$$

$$\downarrow \text{e-transfer and fission of the successor complex}$$

$$\left[Co(aq)\right]^{2+} + 5NH_3 + \left[ClCr^{III}(OH_2)_4(P_2O_7)\right]^{2-}, \text{ i.e. } \left[CrCl(OH_2)_4O—PO_2—O—PO_3\right]^{2-}$$

- **(c) Reduction of $[Co(edta)Cl]^{2-}$ and $[Co(edta)(OH_2)]^-$ ($edta^{4-}$ acting as a pentadentate ligand) by different types of reducing agents** (H. Ogino *et al*, *J. Chem. Soc. Dalton Trans*, 894, 1983)**:** The rate ratio (k_{Cl}/k_{aq}) for the reduction of the chloro- and aqua-complexes by different reducing agents can predict the pathway of electron transfer process.

Reductant:	$[Ti(aq)]^{3+}$	$[Fe(CN)_6]^{4-}$	$[Fe(aq)]^{2+}$	$[Cr(aq)]^{2+}$
k_{Cl}/k_{aq}:	31	33	$> 3 \times 10^2$	2×10^3
Net electron transfer:	$\pi^* \rightarrow \sigma^*$	$\pi^* \rightarrow \sigma^*$	$\pi^* \rightarrow \sigma^*$	$\sigma^* \rightarrow \sigma^*$

The oxidant Co(III) (t_{2g}^6) is inert and the chloro-ligand can act as a good bridging ligand for the substitutionally labile reductants like $[Cr(aq)]^{2+}$ and $[Fe(aq)]^{2+}$ in the ISET path, but H_2O cannot act as a good bridging ligand in the ISET path. $[Fe(CN)_6]^{4-}$ (t_{2g}^6) being substitutionally inert cannot participate in the ISET path.

Thus for both the chloro- and aqua-complexes, the reduction by $[Fe(CN)_6]^{4-}$ is expected to pass through the OSET path and the small rate difference arises for their slightly different K_{OS} (outer-sphere association constant in the OSET path) values and reduction potentials of the Co^{III}/Co^{II}

couples of the Co^{III}-complexes. This is why, the reduction rates of the given Co(III)-complexes are quite similar and the rate ratio is quite small. The rate ratio for the reduction by $[Ti(aq)]^{3+}$ is also similar to that of reduction by $[Fe(CN)_6]^{4-}$. Thus it is reasonable to assume that the OSET path operates for the reduction of the given Co(III)-complexes by both $[Ti(aq)]^{3+}$ and $[Fe(CN)_6]^{4-}$. Here it is surprising to note that $[Ti(aq)]^{3+}$ is labile (*cf.* k_{ex} for water $\approx 10^5\ s^{-1}$) but still, it does not adopt the ISET path. It is probably due to the $\pi^* \rightarrow \sigma^*$ net electron transfer. Thus it requires the electronic activation at Ti(III) for the ISET path.

For the reduction of the chloro-complex of Co(III), by substitunally labile $[Cr(aq)]^{2+}$ and $[Fe(aq)]^{2+}$, the favourable ISET path operates. For the $\sigma^* \rightarrow \sigma^*$ electron transfer in the ISET path, no electronic activation is required for $[Cr(aq)]^{2+}$ ($t_{2g}^3 e_g^1$) but it requires the electronic activation for $[Fe(aq)]^{2+}$ ($t_{2g}^4 e_g^2$). For the reduction of the corresponding aqua complex, the ISET path is not favoured because of the poor bridging ligand property of H_2O and probably, the reduction of the aqua complex by $[Cr(aq)]^{2+}$ and $[Fe(aq)]^{2+}$ passes through the OSET path. However, instead of the OSET path, the disfavoured ISET path may go on slowly. Both will lead to the high k_{Cl}/k_{aq} ratio.

Note: For the ISET path, the required electronic activation energy for $[Ti(aq)]^{3+}$ (*i.e.* $t_{2g} \rightarrow e_g$) is higher than that for $[Fe(aq)]^{2+}$ because of the higher 10Dq value of $[Ti(aq)]^{3+}$ (*cf.* effect of +3 *vs.* +2 oxidation states). This is why, for $[Ti(aq)]^{3+}$, the ISET path is relatively more disfavoured compared to that for $[Fe(aq)]^{2+}$.

- **(d) N_3^- *vs.* NCS^- as the bridging ligands:** Remote attack occurs for both the azide and N-bonded thiocyanate but azide is a better bridging ligand than the N-bonded thiocyanate.

M_{ox}—$\ddot{N}$=$\overset{+}{N}$=$\ddot{N}$:$^-$ vs. M_{ox}—$\ddot{N}$=C=$\ddot{S}$: ⟷ M_{ox}—$\overset{+}{N}$≡C—$\ddot{S}$:$^-$

M_{ox}—$\ddot{N}$=$\overset{+}{N}$=$\ddot{N}$—M_{red} vs. M_{ox}—$\overset{+}{N}$≡C—$\ddot{S}$—M_{red} ⟷ M_{ox}—$\ddot{N}$=C=$\ddot{S}^+$—M_{red}

Attack on the S-end by the reducing agents like $[Cr(aq)]^{2+}$, $[V(aq)]^{2+}$, $[Fe(aq)]^{2+}$, etc. does not give a good **hard-soft matching** in terms of the HSAB principle. The S-end is soft while the said reducing agents and their oxidized centres are hard. ***Thus azide gives a better hard-soft matching.***

Besides the **HSAB matching**, azide makes a **symmetrical bridge** but NCS^- cannot make a such symmetrical bridge. Moreover, in terms of the **resonance mechanism of ISET** (involving a conjugated system), the azide bridged complex can act better than the thiocyanate bridged complex. ***This is why, for X = N_3^-, the rate is higher than that for X = NCS^-.*** The conclusions are:

$$k_{N_3}/k_{NCS} >> 1 \Rightarrow \textbf{ISET process}$$

$$k_{N_3}/k_{NCS} \approx 1 \Rightarrow \textbf{OSET process or substitution controlled ISET process}$$

Note: Compared to the N-bonded thiocyanate, the S-bonded thiocyanate can give the better bridge for the remote attack on N. This gives also a better hard-soft matching. In fact, Co^{III}—SCN is reduced faster than Co^{III}—NCS in the ISET process. This is true for the **hard reducing centres** like $[Cr(aq)]^{2+}$, $[V(aq)]^{2+}$, $[Fe(aq)]^{2+}$, etc. but not true for the soft reducing centre like $[Co(CN)_5]^{3-}$ **(symbiotic effect)**.

$$[Co(NH_3)_5(NCS)]^{2+} + [Co(CN)_5]^{3-} \longrightarrow [Co(CN)_5(SCN)]^{3-} + [Co(aq)]^{2+} + 5NH_3$$

$$k \approx 10^6\ dm^3\ mol^{-1}\ s^{-1}$$

This aspect has been already discussed (Sec. 6.4.5).

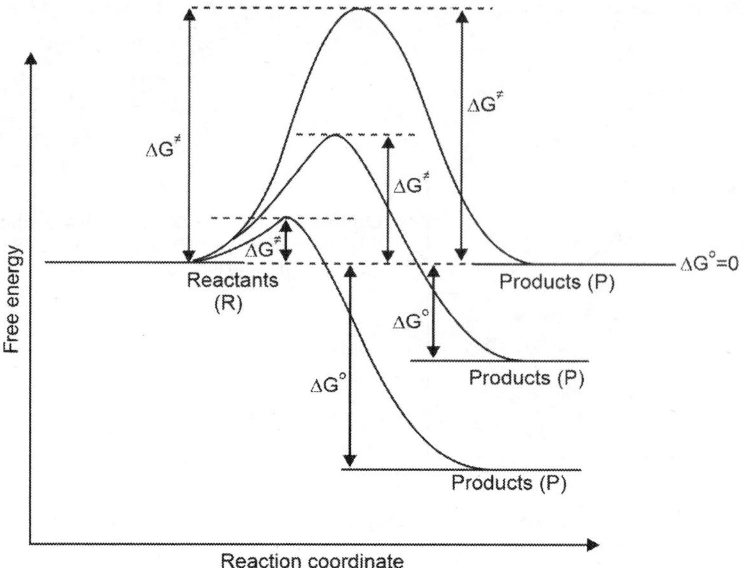

Fig. 6.3.4.2: Illustration of Marcus relationship (*i.e.* LEER plot) for the outer sphere cross-reaction. ΔG^{o} = free energy change in a reaction; $\Delta G^{\neq}$ = free energy of activation; higher $-\Delta G^{o}$ indicates lower $\Delta G^{\neq}$.
Note: Same conclusion from Fig. 6.3.3.1

The Marcus equation can theoretically calculate the rate constant of a redox reaction from the exchange rate constants and equilibrium constant that may be obtained from the E^0 values. This is of much importance in understanding the electron transfer process and Marcus's work was recognized by **Nobel Prize in 1992.**

Let us illustrate the application of Marcus equation to calculate the rate constant of the following reaction, *i.e.* oxidation of hexacyanidoferrate(II) by octacyanidomolybdate(V).

$$\underset{(Ox_1)}{\left[Mo(CN)_8\right]^{3-}} + \underset{(Red_2)}{\left[Fe(CN)_6\right]^{4-}} \rightleftharpoons \underset{(Red_1)}{\left[Mo(CN)_8\right]^{4-}} + \underset{(Ox_2)}{\left[Fe(CN)_6\right]^{3-}}; K_{12} = 1\times10^2$$

The corresponding electron exchange rate constants are:

$$\left[Fe(CN)_6\right]^{4-} + \left[{}^{*}Fe(CN)_6\right]^{3-} \longrightarrow \left[Fe(CN)_6\right]^{3-} + \left[{}^{*}Fe(CN)_6\right]^{4-};\ k_{11} = 7.4\times10^2\ dm^3\ mol^{-1}\ s^{-1}$$

$$\left[Mo(CN)_8\right]^{3-} + \left[{}^{*}Mo(CN)_8\right]^{4-} \longrightarrow \left[Mo(CN)_8\right]^{4-} + \left[{}^{*}Mo(CN)_8\right]^{3-};\ k_{22} = 3\times10^4\ dm^3 mol^{-1}\ s^{-1}$$

$$\log f = \frac{\left[\log\left(10^2\right)\right]^2}{4\log\left(\dfrac{7.4\times10^2\times3\times10^4}{10^{22}}\right)} \text{ leads to } f \approx 0.9$$

Taking, $k_{12} = \sqrt{k_{11}k_{22}K_{12}}$ (assuming $f = 1$)

$$= \sqrt{7.4\times10^2\times3\times10^4\times1\times10^2} \approx 4.7\times10^4\ dm^3\ mol^{-1}\ s^{-1}$$

The experimental value of k_{12} under the same condition is 3×10^4 $dm^3\ mol^{-1}\ s^{-1}$ which is very much comparable to the value obtained from Marcus equation. The Marcus equation has been tested in many cases (*cf.* Table 6.3.4.1).

Table 6.4.10.1 Second order rate constants (25° C) for the reduction of azido and N-bonded thiocyanato complexes by different reducing agents. (*cf.* J. Espenson, *Inorg. Chem.*, **4**, 121, 1965).

Oxidant	*Reductant*	*k* (dm³ mol⁻¹ s⁻¹)		k_{N_3}/k_{NCS}	*Mechanism*
		k_{N_3}	k_{NCS}		
$[Co(NH_3)_5X]^{2+}$	$[Cr(aq)]^{2+}$	3×10^5	20.0	1.5×10^4	ISET
$[Co(NH_3)_5X]^{2+}$	$[V(aq)]^{2+}$	13.0	0.30	43.5	substitution controlled*
$[Co(NH_3)_5X]^{2+}$	$[Eu(aq)]^{2+}$	2×10^2	0.70	2.8×10^2	ISET** or OSET**
$[Co(NH_3)_5X]^{2+}$	$[Fe(aq)]^{2+}$	8.5×10^{-3}	$< 3 \times 10^{-6}$	$>10^3$	ISET
$[Co(NH_3)_5X]^{2+}$	$[Cr(bpy)_3]^{2+}$	4×10^4	1×10^4	4.0	OSET***
$[Cr(OH_2)_5X]^{2+}$	$[Cr(aq)]^{2+}$	6.1	1.4×10^{-4}	4.3×10^4	ISET
$[Co(NH_3)_5X]^{2+}$	$[Co(CN)_5]^{3-}$	1.5×10^6	1×10^6	1.5	ISET****
$[Co(NH_3)_5X]^{2+}$	$[Ru(NH_3)_6]^{2+}$	1.2	–	–	OSET

* Reducing centre V^{2+} (t_{2g}^3) is relatively inert; ISET process is concluded from the evidences like comparison with the kinetic parameters of the **substitution reaction at this reducing centre** and product analysis (transfer of the bridging ligand to the reducing centre).

** For Eu^{2+}, it may be simple OSET or ISET (*cf.* **halide effect**, Sec. 6.4.5).

*** Low spin (t_{2g}^4) **$[Cr(bpy)_3]^{2+}$ being substitutionally inert** participates in OSET.

**** For the reducing agent $[Co(CN)_5]^{3-}$ where the Co-centre is **softened** by the 5 soft CN^- ligands (**symbiotic effect**), the bridging by the soft-S end of NCS^- is relatively more favoured in the ISET path (see Sec. 6.4.5). It makes the k_{N_3}/k_{NCS} ratio **smaller even for the ISET path**.

Table 6.4.10.2 Second order rate constants (25°C) for the reduction of *N*-bonded and *S*-bonded thiocyanato complexes (–X = –*N*CS or –*S*CN).

Oxidant	*Reductant*	*k* (dm³ mol⁻¹ s⁻¹)		k_{SCN}/k_{NCS}	*Mechanism***
		k_{SCN}	k_{NCS}		
$[Co(NH_3)_5X]^{2+}$	$[V(aq)]^{2+}$	30	0.3	10^2	ISET*
$[Co(NH_3)_5X]^{2+}$	$[Cr(aq)]^{2+}$	2×10^5	20	10^4	ISET
$[Co(NH_3)_5X]^{2+}$	$[Fe(aq)]^{2+}$	0.12	$< 3 \times 10^{-6}$	$> 4 \times 10^4$	ISET

* Substitution controlled ISET and this is why, the ratio k_{SCN}/k_{NCS} is less.

** It may be noted that this high k_{SCN}/k_{NCS} ratio for the ISET process is true for the **hard reducing metal centres** like $[M(aq)]^{2+}$ (M = V, Cr, Fe, etc.) but not true for the **softened reducing agent** like $[Co(CN)_5]^{3-}$ that prefers to bind through the S-end of the NCS^- bridging ligand (see Table 6.4.5.4).

- **(e) —*N*CS⁻ *vs.* —*S*CN⁻ as the bridging ligands:** For M_{ox}—NCS, an adjacent attack on 'N' is not possible because of the reasons: the lone pair on 'N' is engaged in **resonance** and **steric hindrance.** Thus, for M_{ox}—NCS, ***only the remote attacks on S*** by the reducing agent is possible while for M_{ox}—SCN, ***both the remote attack on 'N' and adjacent attack on 'S'*** are possible giving rise to a mixture of isomeric products (if ligand transfer in the ISET path is possible). If the reducing agents are **hard** (*e.g.* $[V(aq)]^{2+}$, $[Cr(aq)]^{2+}$, $[Fe(aq)]^{2+}$, etc.) then they prefer the *N*-site over the *S*-site. Consequently, the rate of reduction of M_{ox}—SCN is much faster than the rate of reduction of M_{ox}—NCS which can only provide the *S*-site for bridging (*cf.* Sec. 6.4.9 for reduction of $[Co(NH_3)_5(SCN)]^{2+}$ *vs.* $[Co(NCS)(NH_3)_5]^{2+}$ by $[Cr(OH_2)_6]^{2+}$). The following factors explain the higher rate of reduction of M_{ox}—SCN in the ISET path.

  (i) **M_{ox}—SCN:** Both the *S* (adjacent attack) and *N* (remote attack) sites are available for coordination with the reducing agents (*e.g.* V^{2+}, Cr^{2+}, Fe^{2+}, etc.) but the remote attack on N gives the more favourable path for the given **hard reducing metal centres**.

  (ii) **M_{ox}—NCS:** Only the S (*i.e.* remote attack) site is available for coordination with the reducing agent.

(iii) **M_{ox}—SCN—M_{red}** > **M_{ox}—NCS—M_{red}** (in terms of stability, *cf.* HSAB principle).

Note: The above conclusions are true for the given hard reducing metal centres but not true for the reducing agent $[Co(CN)_5]^{3-}$ where the Co-centre is softened by the CN^- ligands.

- **(f) Chelation and coordinating power of the bridging ligand to determine the efficiency of the ISET path:** Stability of the binuclear precursor complex largely depends on the efficiency of coordination by the bridging ligand with the reducing centre. The steric effect and possibility of chelation are the important factors to determine the stability of the binuclear precursor complex. It is illustrated in the following examples:

(i) **Effect of steric crowding on the bridging ligand in the reduction of** $[(H_3N)_5Co^{III}—O—C(=O)R]^{2+}$ **by $[Cr(aq)]^{2+}$:**

R =	H		CH_3		CMe_3
k (dm^3 mol^{-1} s^{-1}) =	7.2	≫	0.35	≫	9.7×10^{-3}

Obviously with the increase of bulkiness of R, the increased steric crowding disfavours the coordination by the carbonyl oxygen (*i.e.* remote attack) with the reducing metal centre.

M_{ox}—O⋯C(R)⋯O*—M_{red}

(Remote attack) established by isotope labeling experiment

Note: For R = H (*i.e.* formate), the bridging ligand is itself reducible and it can lead to the **chemical mechanism of ISET.** This also partially contributes for the observed rate enhancement.

(ii) **Effect of chelation by the bridging ligand in the reduction of** $[(H_3N)_5Co^{III}—O—C(=O)R]^{2+}$ **by $[Cr(aq)]^{2+}$:**

R =	Ph		CH_2—OH
k (dm^3 mol^{-1} s^{-1}):	0.15	<	3.0

Co^{III}—O⋯C(Ph)⋯O—Cr^{II} < Co^{III}—O⋯C(CH₂—O(H)→)⋯O—Cr^{II} **(chelation)**

Stability sequence of the precursor complex

(**Note:** For the bridging ligand $^-OCOCH_2OH$ (*i.e.* glycolate), another factor comes to play. This reducible ligand can lead to the **chemical mechanism of electron transfer** to enhance the rate.)

(iii) For the **more powerful chelating bridging ligands,** the rates are dramatically increased. It is illustrated below for the reduction of $[(H_3N)_5Co^{III}—X]$ by $[Cr(aq)]^{2+}$.

X =	**glycolate**	**malonate**	**pyridine-2-carboxylate**	**salicylate**
k (dm^3 mol^{-1} s^{-1}) =	3.0	2.5×10^3	2×10^5	2×10^8

(iv) **Effect of chelation to stabilize the binuclear precursor complex** is illustrated in the efficiency of reduction of $[Co^{III}(NH_3)_5X]^{2+}$ by $[Fe(aq)]^{2+}$.

	$[CoCl(NH_3)_5]^{2+}$ + $[Fe(aq)]^{2+}$	$[Co(C_2O_4H)(NH_3)_5]^{2+}$ + $[Fe(aq)]^{2+}$	$[Co(NH_3)_5(ntaH_2)]^{2+}$ + $[Fe(aq)]^{2+}$
k ($dm^3\ mol^{-1}\ s^{-1}$):	1.3×10^{-3}	0.43	1×10^5

Chloride ligand cannot cause any chelation; coordinated bioxalate ($C_2O_4H^-$) can provide two bridging O-sites to make a chelation; unidentatedly coordinated $ntaH_2^-$ can provide three bridging sites (1N + 2-acetate–O) forming **two 5-membered chelate rings** with the Fe(II) centre to stabilize the 'binuclear precursor complex'. In fact, the binuclear precursor complex for $X^- = ntaH_2^-$ has been characterized and the formation constant ($= K = \frac{k_1}{k_{-1}} \approx 10^6\ mol^{-1}\ dm^3$) is very high. Effect of the stability of the binuclear precursor complex on the rate constant ($dm^3\ mol^{-1}\ s^{-1}$) is illustrated below (*Ref. Inorg. Chem.*, **9**, 500, 1970).

Oxidant	$[Cr(aq)]^{2+}$	$[V(aq)]^{2+}$	$[Fe(aq)]^{2+}$
$[Co(NH_3)_5(C_2O_4H)]^{2+}$	1×10^2	12.5	$\leq 5 \times 10^{-4}$
$[Co(NH_3)_5(C_2O_4)]^+$	4.6×10^4	45	0.43

The increased reactivity toward $[Cr(aq)]^{2+}$ and $[Fe(aq)]^{2+}$ for $[Co(NH_3)_5(C_2O_4)]^+$ as compared to $[Co(NH_3)_5(C_2O_4H)]^{2+}$ is due to more basicity of the bridging $C_2O_4^{2-}$ (oxalate) ligand than that of the bridging ligand $C_2O_4H^-$ (bioxalate) (*i.e.* **$C_2O_4^{2-}$ as a bridging ligand stabilises the precursor binuclear complex better compared to the less basic bridging ligand $C_2O_4H^-$**). For the reduction by $[V(aq)]^{2+}$, it passes through the **substitution controlled ISET path** and the rate of the overall reaction is not so much sensitive to the nature (*i.e.* basicity) of the bridging ligand.

(v) **Effect of the stability of the binuclear precursor complex** is illustrated in the reduction of $[Co(NH_3)_5X]^{2+}$ by $[Eu(aq)]^{2+}$ (see Sec. 6.4.5 for **halide effect**).

Stability: $EuF^+ > EuCl^+ > EuBr^+ > EuI^+$; **Rate for X:** F > Cl > Br > I (see Table 6.4.5.4)

- **(g) Identification of the nature of chelation by the bridging ligand from the product analysis:** Chelation by the bridging ligand is also known in many other cases. *The nature of chelation is very often identified from the product analysis.* Reduction of $[Co(C_2O_4)(NH_3)_4]^+$ by $[Cr(aq)]^{2+}$ produces $[Cr(C_2O_4)(OH_2)_4]^+$. It suggests the following precursor complex. It explains the presence of bidentate oxalate in the product Cr(III)-complex, *i.e.* $[Cr(C_2O_4)(OH_2)_4]^+$.

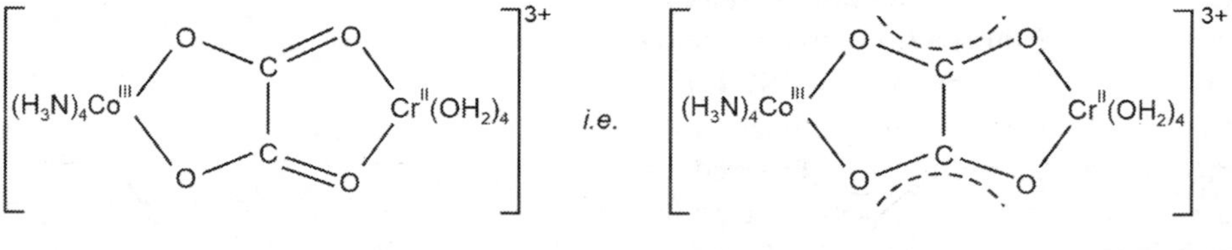

(**Note:** Reducible character of the bridging oxalate ligand may lead to the **chemical mechanism of ISET** to cause the rate enhancement.)

Reduction of $[Co(NH_3)_5(OCOCH_2CO_2H)]^{2+}$ by $[Cr(aq)]^{2+}$ gives the product $[Cr(malonate)(OH_2)_4]^+$ indicating formation of the following precursor complex.

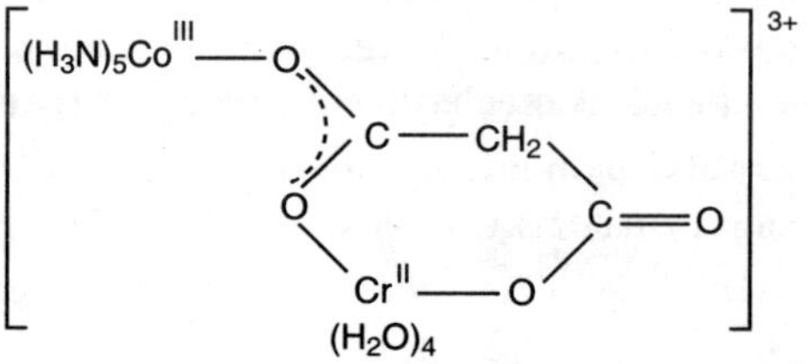

- **(h) Switch-over from the OSET to ISET path depending on the nature of the bridging ligand in the reduction of** $[(H_3N)_5M^{III}—N(C_5H_4R)]^{3+}$ **by $[Cr(aq)]^{2+}$ (M = Co, Ru):** Rate sequence is:

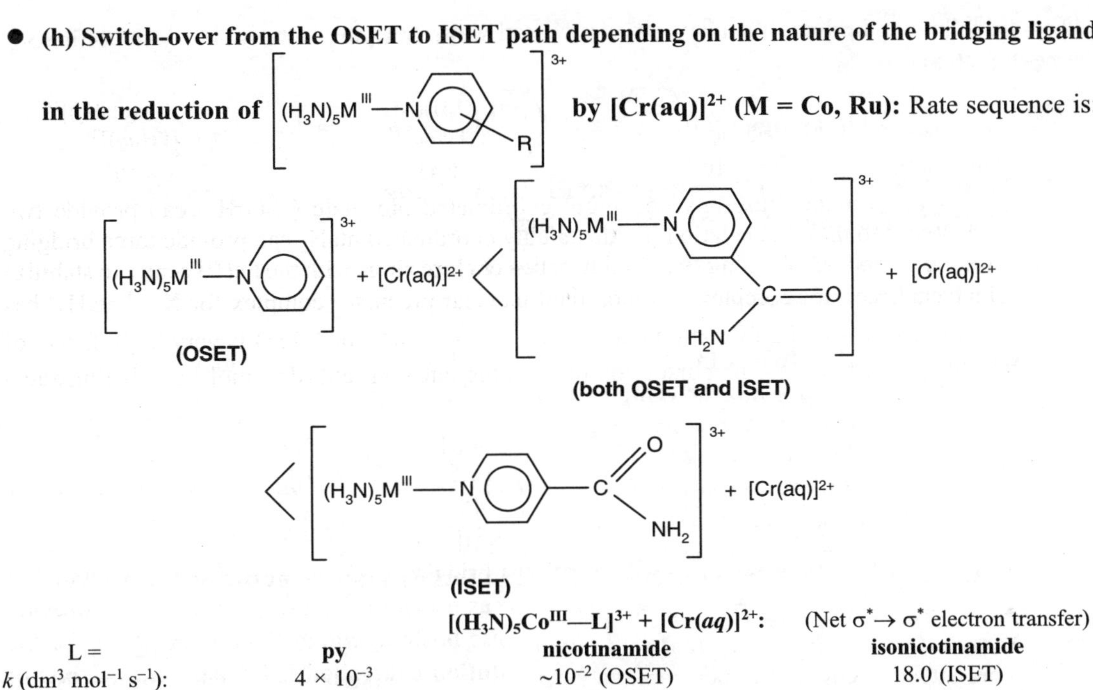

		$[(H_3N)_5Co^{III}—L]^{3+} + [Cr(aq)]^{2+}$:	(Net $\sigma^* \to \sigma^*$ electron transfer)
L =	**py**	**nicotinamide**	**isonicotinamide**
k (dm^3 mol^{-1} s^{-1}):	4×10^{-3} (OSET)	~10^{-2} (OSET) ~10^{-2} (ISET)	18.0 (ISET)
		$[(H_3N)_5Ru^{III}—L]^{3+} + [Cr(aq)]^{2+}$:	(Net $\sigma^* \to \pi^*$ electron transfer)
k (dm^3 mol^{-1} s^{-1}):	3.5×10^{3} (OSET)	~10^{4} (OSET) ~10^{4} (ISET)	4×10^{5} (ISET)

Pyridine cannot act as a bridging ligand. This is why, the complex $[(H_3N)_5M^{III}(py)]$ experiences the reduction through the **OSET process. Isonicotinamide** (amide group at the *p*-position in the pyridine ring) as a bridging ligand can experience a **remote attack on the carbonyl oxygen** by the reducing agent and it allows the *ISET path.* It may be noted that here the **remote attack on the amide-N site cannot occur** because the N-lone pair is engaged in resonance with the carbonyl group and it leads also to the failure of the ISET **resonance mechanism** when bridged through the amide-N site (*cf.* Secs. 6.4.9, 6.4.11). Nicotinamide (*i.e.* amide group at the *m*-position in the pyridine ring) as a bridging ligand allows **both the OSET and ISET paths simultaneously.** In fact, the bridging group (*i.e.* $CONH_2$) at the *meta* position experiences two obstacles in the ISET path: **steric crowding in the binuclear precursor complex and hindered conjugation.** This is why, both the OSET and ISET paths operate and the rate constants in both the paths are more or less comparable.

(i) Conjugated bridging organic ligand *vs.* nonconjugated organic bridging ligand: Electron transfer through the conjugated bridging ligand is favoured more than through the nonconjugated bridging ligand in the ISET path. Complete conjugation in the reductant-bridging ligand-oxidant segment, allows the highly favoured **resonance mechanism of electron transfer** in the ISET path. **For the long nonconjugated bridging ligand, the mechanism can switch over from the ISET path to the OSET path (discussed in the next Section).**

For the conjugated bridging ligands like oxalate, salicylate, the reaction through the ISET path goes on faster than for the nonconjugated succinate bridging ligand (for which even the OSET path may operate).

$CH_2—CO_2^-$ | $CH_2—CO_2^-$

(Oxalate) **(Salicylate)** **(Succinate)**

For the following bridging ligand, the favoured pathway of electron transfer has been discussed in the next Section.

Bridging ligand: N⟨py⟩–X–⟨py⟩N , X = $(CH_2)_n$ (n=1,2,3), C_2H_2, CO

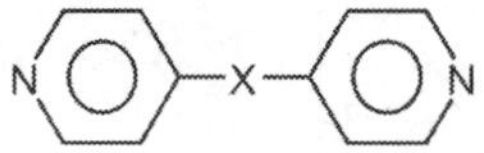

6.4.11 Role of the Bridging Ligand in the Electron Transfer Step – Tunneling Transfer, Resonance Transfer and Chemical Mechanism

- **Tunneling transfer:** Montoatomic bridging ligands like Cl^-, F^-, etc. simply help the **tunneling of electron** from the reductant to the oxidant through the three-centred σ-type delocalized MO (see Fig. 6.4.3.1).
- **Resonance transfer:** When there is a **conjugation** in the segment, oxidant–bridging ligand–reductant generated **through the remote attack** on the bridging ligand and the bridging ligand is **not reducible, tunneling of electron** is enhanced. This tunneling mechanism is described as the **resonance transfer.**

  It happens so when the ***unsaturated nonreducible bridging ligands*** (*e.g.* N_3^-, NCS^-, RCO_2^-, $RCONH_2$, etc.) experience the **remote attack** by the electrophilic reducing agent.

Outer Sphere Electron Transfer (OSET) in the Bridged Precursor Complex

This happening illustrates the fact that formation of the dinuclear precursor complex through the remote attack *does not always ensure the operation of an ISET mechanism.* An ISET path involving the remote attack needs the proper conjugation to allow the **resonance transfer mechanism** to operate. If this requirement is not fulfilled, then within the binuclear complex, the OSET path will operate instead of the ISET path. The following example of long-lived precursor complex illustrates the phenomenon.

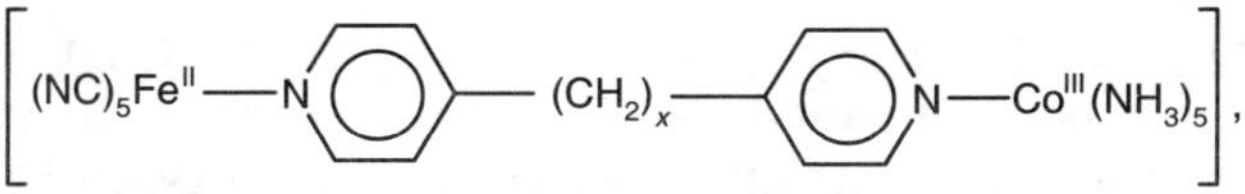

$[(NC)_5Fe^{II}—N⟨py⟩—(CH_2)_x—⟨py⟩N—Co^{III}(NH_3)_5]$, $x = 1 - 3$, **(OSET path)**

The $(CH_2)_x$ moiety is saturated and the *presence of it prevents the electron tunneling through the resonance transfer.* Thus the ISET path is prevented. However, **flexibility of the sufficiently long $(CH_2)_x$ moiety** can allow the oxidizing and reducing centres to come in proximity and this favours the simple OSET path.

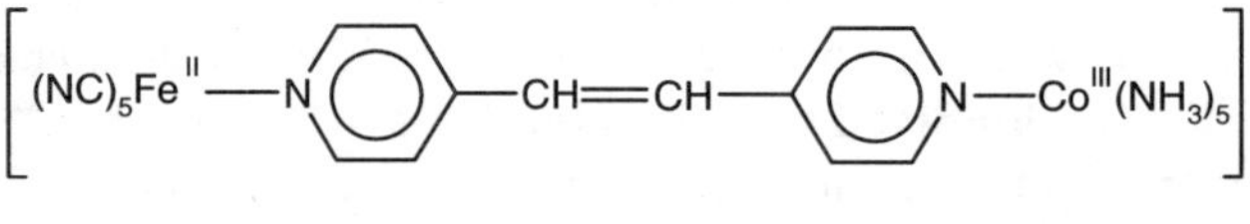

$[(NC)_5Fe^{II}—N⟨py⟩—CH{=}CH—⟨py⟩N—Co^{III}(NH_3)_5]$, **(ISET path)**

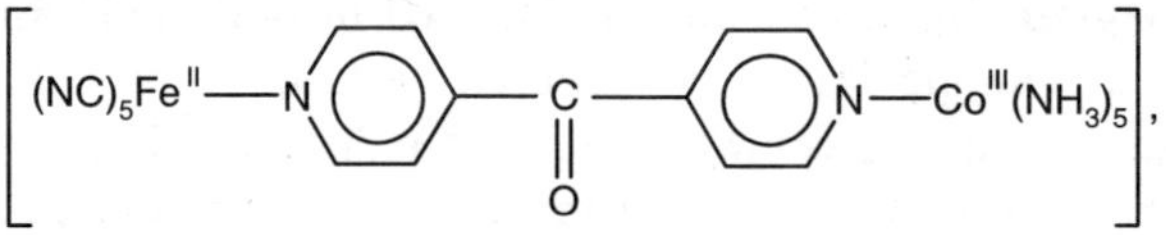

$[(NC)_5Fe^{II}—N⟨py⟩—C(=O)—⟨py⟩N—Co^{III}(NH_3)_5]$, **(OSET path)**

Here the presence of the >C=O group cannot extend the conjugation in the complete segment, **reductant-bridged ligand–oxidant.** Because of this incomplete conjugation in the segment, electron transfer through the resonance mechanism cannot occur. Interestingly, in contrast to the above example, the examples having the bridging ligand $X = N_3^-$, CN^-, NCS^-, RCO_2^-, etc. can allow the ISET path through the resonance transfer. The electron transfer mechanism for the bridging ligands **oxalate, salicylate** and **succinate** has been discussed and compared in Sec. 6.4.10.

- **Chemical mechanism:** When the bridging organic ligands (*e.g.* nicotinamide, isonicotinamide, formate, etc.) are reducible, then at the first step (*probably the rds*), the electron is transferred to the bridging ligand which is temporarily reduced to a *radical ion.* At the next step (faster step), the electron from the reduced bridging ligand (*i.e.* radical ion) is transferred to the oxidant. This stepwise **pathway** is described as the **chemical or radical ion mechanism.** It is illustrated below for the reduction of M^{III}—X by Cr^{II} (M = Co, Cr, Ru).

$$M^{III}\text{—}X + Cr^{II} \xrightarrow[\text{step-I}]{} M^{III}\text{—}X\text{—}Cr^{II} \xrightarrow[\text{step-II}]{(rds)} M^{III}\text{—}X^{\overline{\cdot}}\text{—}Cr^{III}$$

$$\downarrow \text{step-III}$$

$$M^{II} + Cr^{III}\text{—}X \xleftarrow[\text{step-IV}]{} M^{II}\text{—}X\text{—}Cr^{III}$$

Step-I: Formation of the binuclear precursor complexes.

Step-II (rate determining step): *e*-transfer to the bridging ligand from the reducing agent.

Step-III: *e*-transfer from the reduced ligand to the oxidant.

Step-IV: Fission of the binuclear successor complex.

(Note: If the life period of the **temporarily reduced bridging ligand** (*i.e.* $X^{\overline{\cdot}}$, **radical ion**) is too short, then the **radical ion mechanism** becomes equivalent to the **resonance tunneling** (*i.e.* resonance transfer) mechanism. It happens so (*i.e.* step-III, IV very fast) when the electron carrier orbital of the bridging ligand does not experience any **symmetry mismatch barrier** to transfer the electron to the receptor orbital of the oxidant. Such a barrier does not exist in the case of reduction of Ru^{III}—X by Cr(II) where the bridging ligand X is reducible. This aspect will be illustrated in detail later.)

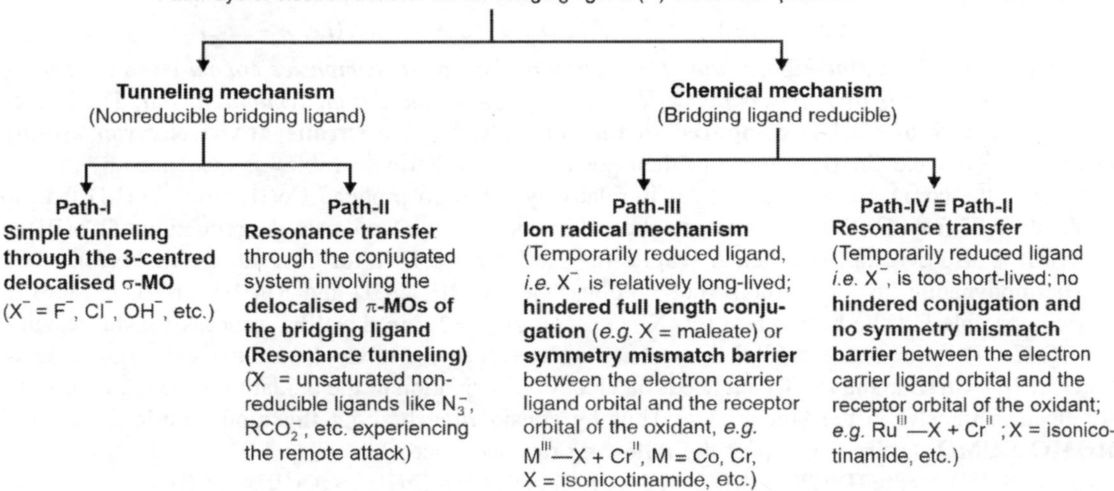

Scheme 6.4.11.1 Classification of the ISET reaction depending on the nature of mechanistic pathway of electron transfer through the bridging ligand.

(Note: Sometimes, the bridging ligand is itself reduced and stabilised. The reduced ligand fails in turn to reduce the oxidant of the binuclear complex. It may happen for the organic bridging ligand having a nitro group that receives the remote attack by the reducing agent. Subsequently, the nitro-compound is reduced to the stable amine compound (*i.e.* $-NO_2$ to $-NH_2$).

(a) **Reduction mechanism of the Co(III), Cr(III) and Ru(III) complexes by $[Cr(aq)]^{2+}$ through the ISET path controlled by the nature of the bridging ligand:**

Conclusion: $\begin{cases} k_{Co^{III}}/k_{Cr^{III}} \sim 10^6; k_{Co^{III}}/k_{Ru^{III}} \sim 10^1 \Rightarrow \textbf{tunneling mechanism} \\ k_{Co^{III}}/k_{Cr^{III}} \sim 10^1; k_{Co^{III}}/k_{Ru^{III}} \sim 10^{-5} \Rightarrow \textbf{chemical (ion radical) mechanism} \end{cases}$

$[M^{III}(NH_3)_5X]$—$[Cr(aq)]^{2+}$: $k_{Co^{III}}/k_{Cr^{III}} \sim 10^6$ (X = Cl, F, OH, $OCOCH_3$, etc.)

$k_{Co^{III}}/k_{Cr^{III}} \sim 10^1$ (X = isonicotinamide)

Thermodynamically, the Co(III) and Ru(III) complexes are more or less comparable (*cf.* E° values for their ammine complexes) as the oxidizing agents but the Cr(III) complexes are relatively the poorer oxidizing agents (thermodynamic sense). It is evident from their E^0 (standard reduction potentials) values of the aqua-complexes.

$$E^0 = 1.8 \text{ V for } \left[Co(aq)\right]^{3+}/\left[Co(aq)\right]^{2+}$$

$$= 0.06 \text{ V for } \left[Co(NH_3)_6\right]^{3+}/\left[Co(NH_3)_6\right]^{2+}$$

$$E^0 = -0.41\text{V for } \left[Cr(aq)\right]^{3+}/\left[Cr(aq)\right]^{2+}$$

$$E^0 = 0.25 \text{ V for } \left[Ru(aq)\right]^{3+}/\left[Ru(aq)\right]^{2+}$$

$$= 0.10 \text{ V for } \left[Ru(NH_3)_6\right]^{3+}/\left[Ru(NH_3)_6\right]^{2+}$$

The net electron transfer in the following redox reactions are:

	Co(III)–Cr(II)	**Cr(III)–Cr(II)**	**Ru(III)–Cr(II)**
Net transfer	$\sigma^* \rightarrow \sigma^*$	$\sigma^* \rightarrow \sigma^*$	$\sigma^* \rightarrow \pi^*$
	(*i.e.* $e_g \rightarrow e_g$)	(*i.e.* $e_g \rightarrow e_g$)	(*i.e.* $e_g \rightarrow t_{2g}$)

In the ISET process, the higher rate of reduction of the Co(III) complexes compared to the rate of reduction of the Cr(III)-complexes by Cr(II) is mainly due to the thermodynamic favour. The **higher electron affinity of Co(III) compared to that of Cr(III)** (*cf.* **electronegativity/electron affinity order** of Co(III) and Cr(III) in terms of their positions in the Periodic Table) is evidenced by the fact that under the comparable conditions, it is relatively easier to produce Co(II) from Co(III) than to produce Cr(II) from Cr(III). This gives the better thermodynamic favour for the reduction of Co(III) to Co(II). It is also evident from their E^0-values (*cf.* 1.8 V *vs.* –0.41 V for their aqua complexes).

Thermodynamic favour for both the reactions, *i.e.* Co(III)–Cr(II) and Ru(III)–Cr(II), is more or less comparable but the higher rate (*cf.* X = Cl, $k_{Co}/k_{Ru} \approx 20$) in the ISET process for the reaction Co(III)–Cr(II) is due to the **orbital favour** (*i.e.* **no requirement of electronic activation**). The reaction, Ru(III)–Cr(II) in the ISET path needs an electronic activation of the oxidant for the required $\sigma^* \rightarrow \sigma^*$ electron transfer in the kinetic step. Thus by considering the both **thermodynamic factor and HOMO-LUMO nature**, we expect the following rate sequence:

rate (ISET path): Co(III)–Cr(II) > Ru(III)–Cr(II) ≫ Cr(III)–Cr(II)

(b) **Tunneling mechanism for the nonreducible bridging ligand (*e.g.* OH^-, F^-, Cl^-, $CH_3CO_2^-$):**

M^{III}—Cl—Cr^{II}, M^{III}—O⋯C(CH₃)⋯O—Cr^{II}

For the chlorido or fluorido bridged binuclear complexes, the simple electron tunneling occurs through the bridge (*i.e.* **via the 3-centred delocalized σ-MO**). In such cases, the chemical mechanism, leading to F^{2-} or Cl^{2-} is not possible.

Table 6.4.11.1 Comparison of the rate constants for the reduction of Co(III), Cr(III) and Ru(III) complexes by $[Cr(aq)]^{2+}$ to illustrate the effect of the nature of the bridging ligand. (*Refs.* H. Taube, *Inorg. Chem.*, **8**, 2281, 1969; *ibid*, **9**, 2627, 1970; *J. Am. Chem. Soc.*, **90**, 1162, 1968; E. S. Gould, *J. Am. Chem. Soc.*, **96**, 2373, 1974).

Oxidant	***k* (dm³ mol⁻¹ s⁻¹)**	***Remark***
$[CoF(NH_3)_5]^{2+}$	2.5×10^5	**Simple tunneling** through the delocalised 3-centred σ-MO (*i.e.* bridging ligand acts as a mediator of electron flow)
$[CrF(OH_2)_5]^{2+}$	2.6×10^{-2}	
$[CoCl(NH_3)_5]^{2+}$	6×10^5	
$[CrCl(OH_2)_5]^{2+}$	90.0	
$[RuCl(NH_3)_5]^{2+}$	3.5×10^4	
$[Co(NH_3)_5(OCOCH_3)]^{2+}$	0.30	**Simple resonance tunneling ISET path** through the conjugated system involving the delocalised π-MOs of the bridging ligand (a remote attack leads to the resonance transfer).
$[Cr(OCOCH_3)(OH_2)_5]^{2+}$	$\sim 10^{-4}$	
$[Co(NH_3)_5(OCOCHO)]^{2+}$	$\sim 10^4$	
$[(H_3N)_5Co^{III}—N(C_5H_4)—C(=O)NH_2]^{3+}$	17.4	**Radical-ion mechanism (multistep process)** (remote attack)
$[(H_2O)_5Cr^{III}—N(C_5H_4)—C(=O)NH_2]^{3+}$	1.8	**Radical-ion mechanism (multistep process)**
$[(H_3N)_5Ru^{III}—N(C_5H_4)—C(=O)NH_2]^{3+}$	4×10^5	**Resonance transfer** (*i.e.* **direct tunneling process**)

For $CH_3CO_2^-$ as a bridging ligand, the remote attack can establish a conjugated system through which electron tunneling can occur by the **resonance transfer**. For such nonreducible bridging ligands, electron tunneling leads to an electron transfer. The observed sequence of the rate of reaction in such ISET processes is in conformity with the expected sequence from the consideration of thermodynamic favour and HOMO-LUMO nature. ***The Co(III)-complexes are reduced faster than the Cr(III) complexes by Cr(II) mainly due to the thermodynamic favour. Faster reduction of Co^{III}—X compared to that of Ru^{III}—X is mainly due to the kinetic favour (σ*→σ* vs. σ*→π*).***

(c) **Chemical mechanism for the reducible bridging ligands** (*e.g.* nicotinamide): In such cases, the chemical or ion radical mechanism may operate and the *rate determining step* is the transfer of an electron from Cr(II) to the bridging ligand and this involves the π*→ π* electron transfer. The reducing agent **$[Cr(aq)]^{2+}$ ($t_{2g}^3e_g^1$) needs an electronic activation** for the transfer of an electron into the vacant π*-MO of the ligand like isonicotinamide. At the next faster step, the *e*-transfer occurs from the π*-MO of the reduced bridging ligand (L^-) to the σ*–MO of the oxidant (*i.e.* Co^{III}, Cr^{III}). This step needs an electronic activation of M(III) (*cf.* π*→σ* electron transfer forbidden) so that the electron transfer from $L^{\overline{\cdot}}$ to M(III) can occur in the $\pi^*(L^{\overline{\cdot}}) \rightarrow \pi^*(M^{III})$ path. **This activation is not required for Ru(III)**. The whole process may be simply represented as follows:

$$Cr^{II}\left(t_{2g}^3 e_g^1\right) \xrightarrow{\text{activation}} Cr^{II}\left(t_{2g}^4 e_g^0\right) \xrightarrow[(\pi^* \rightarrow \pi^*)]{+L} Cr^{III}\left(t_{2g}^3\right) + L^{\overline{\cdot}}$$

$$\text{i.e. } (\pi^*)^3(\sigma^*)^1 \qquad \text{i.e. } (\pi^*)^4(\sigma^*)^0$$

$$M(III) \xrightarrow{\text{activation}} M(III)^*;\ L^{\overline{\cdot}} + M(III)^* \xrightarrow[(\pi^* \rightarrow \pi^*)]{\text{fast}} L + M(II),\ M = Co, Cr$$

Thus the reduction of both the Co^{III}– and Cr^{III}– complexes experiences the common rate determining step, i.e. electronic activation of Cr(II) and then e-transfer from the π–MO of Cr(II) to the vacant π*–MO of the bridging ligand.* In such cases, assuming the activation energy for Co(III) and Cr(III) (to provide the vacant π* orbital) more or less comparable, it is expected that both the Co(III) and Cr(III) complexes bearing the same bridging ligand will be reduced at the comparable rates in spite of the higher thermodynamic favour for the Co(III)-complexes. Experimental findings also support this view (*cf.* Table 6.4.11.1)

The above chemical mechanism of electron transfer is experimentally verified for the reduction of Co(III) and Cr(III) complexes bearing the same reducible ligands like isonicotinamide. *If the same mechanism is expected to operate for the reduction of the corresponding Ru(III)-complexes then its rate constant should be comparable with those for the reductions of the Co(III) and Cr(III)-complexes.*

In spite of the similar thermodynamic favour for the Co^{III}– and Ru^{III}–complexes, the Ru(III) complexes are reduced remarkably faster than the Co(III) and Cr(III) complexes through the chemical mechanism of ISET path. Interestingly, for the nonreducible inorganic bridging ligands like Cl^-, both the Ru(III) and Co(III) complexes are reduced by Cr(II) at comparable rates (within a factor of 10–20) where the **simple tunneling mechanism operates.** In the simple tunneling mechanism, the Co^{III}-complexes are reduced faster by a factor of 10– 20 compared to the Ru^{III}-complexes but the rates differ dramatically for the chemical mechanism. It demands an explanation.

For the reduction of the Co(III) or Cr(III) complex, the receptor orbital is σ^* (*i.e.* e_g) while for the reduction of the Ru(III) complex, the receptor orbital is π^* (*i.e.* t_{2g})

$$\underset{(t_{2g}^6 e_g^0,\ l.s.)}{Co(III)} + \underset{(t_{2g}^3 e_g^1)}{Cr(II)} \longrightarrow \underset{(t_{2g}^5 e_g^2,\ h.s.)}{Co(II)} + \underset{(t_{2g}^3 e_g^0)}{Cr(III)}\ ;\ (\sigma^* \rightarrow \sigma^* \text{ electron transfer})$$

$$\underset{(t_{2g}^3 e_g^0)}{Cr(III)} + \underset{(t_{2g}^3 e_g^1)}{Cr(II)} \longrightarrow \underset{(t_{2g}^3 e_g^1)}{Cr(II)} + \underset{(t_{2g}^3)}{Cr(III)}\ ;\ (\sigma^* \rightarrow \sigma^* \text{ electron transfer})$$

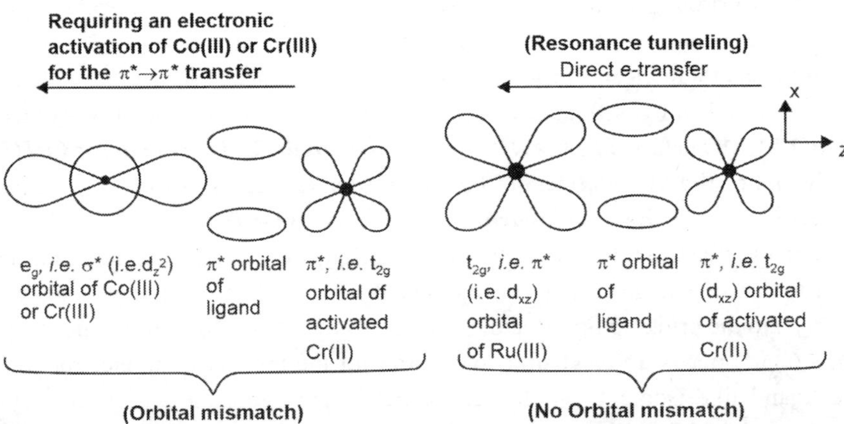

Fig. 6.4.11.1 Symmetries of the receptor orbitals of the oxidants (*i.e.* Co^{III}, Cr^{III}, Ru^{III}) and electron carrier orbital of the temporarily reduced bridging ligand

$$\underset{\left(t_{2g}^5 e_g^0,\ l.s.\right)}{\text{Ru(III)}} + \underset{\left(t_{2g}^3 e_g^1\right)}{\text{Cr(II)}} \longrightarrow \underset{\left(t_{2g}^6 e_g^0,\ l.s.\right)}{\text{Ru(II)}} + \underset{\left(t_{2g}^3 e_g^0\right)}{\text{Cr(III)}}\ ;\ \left(\sigma^* \rightarrow \pi^* \text{ electron transfer}\right)$$

In the chemical mechanism, *activated Cr(II)* transfers an electron from its π*–MO to the vacant π*–MO of the reducible ligand and thence to the receptor orbital of the oxidant. In the case of the Co(III) and Cr(III) oxidants, the receptor orbital is of σ*-symmetry while in the case of the Ru(III) oxidant, the receptor orbital is of π*-symmetry. Thus to transfer the electron from the ligand π*-orbital to the receptor orbital of oxidant, ***it needs some sort of activation in the case of the Co(III) or Cr(III) complexes.*** But no such activation is needed for the Ru(III) complex that provides the receptor orbital of π*-symmetry. ***This is why, the reduction of some Ru(III) complexes is faster than the reduction of the Co(III) or Cr(III) complexes through the chemical mechanism.***

- **(d) Chemical mechanism of the reduction of Co(III), Cr(III) *vs.* Ru(III) by $[Cr(aq)]^{2+}$ – nature of the receptor orbital:** By changing the bridging ligand Cl^- (a representative nonreducible ligand) to isonicotinamide (a representative reducible ligand), the rate constants are dramatically affected for the reduction of the Co(III) and Cr(III) complexes (*cf.* Table 6.4.11.1). But there is no such change for the reduction of the Ru(III)-complexes. In the cases of the Co(III) and Cr(III) complexes, there is a change of mechanistic path of electron transfer, *i.e.* **simple tunneling** (for $X = Cl^-$) **to chemical mechanism** (for X = isonicotinamide). *But, for the chemical mechanism to operate for the reduction of Co(III) and Cr(III) complexes, there is a mismatch in symmetry between the receptor orbital of the oxidant and electron carrier orbital of the reduced ligand.* ***Thus this chemical mechanism needs the stepwise passage of the electron for the reduction of the Co(III) and Cr(III) complexes.*** For the reduction of the Ru(III) complexes, no such barrier from the symmetry mismatch appears and the electron transfer can directly occur from the electron

Chemical mechanism of ISET: Radical-ion mechanism *vs.* resonance tunneling mechanism (*cf.* Fig. 6.4.11.1)

$$\left[Co^{III}(NH_3)_5X\right]^{3+} \xrightarrow{\left[Cr(aq)\right]^{2+}}$$

$$\left[Cr^{III}(OH_2)_5X\right]^{3+} \xrightarrow{\left[Cr(aq)\right]^{2+}}$$

—X = —N(C₅H₄)—C(=O)NH₂

(pyridine carboxamide)

$$\left[Ru(NH_3)_5X\right]^{3+} \xrightarrow{\left[Cr(aq)\right]^{2+}}$$

- **Radical ion mechanism** (*e*-transfer from the t_{2g}, *i.e.* π* orbital of **activated** Cr^{2+} to the e_g, *i.e.* σ* orbital of the oxidant (*i.e.* Co^{III}, Cr^{III}) through the π* orbital of the temporarily reduced bridging ligand. It leads to: **stepwise process** due to the **mismatch in symmetry** between the receptor orbital and electron carrier orbital of the temporarily reduced ligand. Here the oxidant is to be activated to provide the π*-orbital as the receptor orbital. The rate constant (0.85 $dm^3\ mol^{-1}\ s^{-1}$) for the reduction of $[Co(NH_3)_5X]^{3+}$ (X = isonicotinamide) by $[Eu(aq)]^{2+}$ indicates that the radical ion mechanism also works here as in the case of reduction by $[Cr(aq)]^{2+}$.
- **Direct tunneling through the resonance mechanism** (*e*-transfer from the π* orbital of activated Cr^{2+} to the π* orbital of Ru^{III} through the π* orbital of the temporarily reduced ligand: direct *electron transfer* experiencing ***no barrier from the symmetry mismatch of the orbitals.***
- In the **radical ion mechanism,** electron transfer occurs in a **multistep process** involving the participation of the bridging ligand **chemically** in the redox process.
- The chemical mechanism becomes equivalent to the **resonance tunneling mechanism** when the temporarily reduced bridging ligand is too short-lived. In this **resonance mechanism,** electron transfer occurs **directly** where the bridge simply acts as a mediator of electron flow. When the life time of the temporarily reduced bridging ligand is neither too short-lived nor long-lived (*i.e.* intermediate character as in the partially hindered conjugation), it becomes difficult to distinguish between the two types of chemical mechanism.

carrier ligand orbital to the receptor orbital of Ru(III). ***This favoured direct electron passage is very often described as the resonance transfer or resonance tunneling mechanism.***

- (e) **Distinction between the two possible routes of chemical mechanism of ISET process – electron transfer through resonance mechanism *vs.* radical ion mechanism:** From, the above discussion, it appears that the difference between the radical-ion mechanism (*i.e.* chemical mechanism) and resonance transfer mechanism lies in the ***life-time of the reduced radical ion, i.e.*** **temporarily reduced ligand**. In the case of **direct electron transfer**, *i.e.* **resonance electron transfer** (experiencing no barrier from the orbital symmetry mismatch), definitely the life time of the radical ion is too short to be detected.

  For the reduction of the Co(III) and Cr(III) complexes by $[Cr(aq)]^{2+}$, the radical ion (*e.g.* temporarily reduced isonicotinamide, $H_2NCOC_5H_4N$) is **relatively long-lived** and there are experimental evidences to support the existence of the radical ion, *i.e. radical ion chemical mechanism*. In the reduction of the Co(III) complexes by Eu(II), there are also evidences in favour of the proposed radical intermediate. But, for the reduction of analogous Ru(III) compounds by $[Cr(aq)]^{2+}$, the temporarily reduced bridging ligand is **too short-lived** to be detected. It occurs through the resonance transfer (*i.e.* **direct electron transfer**) where there is **no symmetry mismatch** between the electron carrier ligand orbital and the receptor orbital of Ru(III).

- (f) **Role of the 'R' group in dictating the mechanism of reduction of $[M(NH_3)_5(OCOR)]^{2+}$ by $[Cr(aq)]^{2+}$:**

$$\left[(H_3N)_5M^{III}\!-\!O\!-\!C(={}^{*}O)R\right]^{2+} + [Cr(aq)]^{2+} \longrightarrow [M(aq)]^{2+} + 5NH_3 + \left[(H_2O)_5Cr^{III}\!-\!O^{*}\!-\!C(=O)R\right]^{2+}$$

The important observations are:

(i) If the R-group is **saturated** and the $-O-C(=O)R$ moiety is **not reducible**, the rate constants are comparable for different R-groups. In such cases, $[Cr(aq)]^{2+}$ makes a remote attack on the carbonyl oxygen and then the electron transfer occurs through the simple resonance tunneling. Examples of such bridging groups (OCOR) are: *acetate, butyrate, succinate, etc.*

$$\left[(H_3N)_5M^{III}\!-\!O\cdots C(R)\cdots O\!-\!Cr^{II}(OH_2)_5\right]^{4+}$$

R = CH_3 (acetate)
= C_3H_7 (butyrate)
= $(CH_2)_2CO_2H$ (acid succinate)

(Binuclear precursor complex)

(ii) If the R-group is **unsaturated (specially to maintain the conjugation)** and the OCOR group is **reducible** then, the rate constants are remarkably affected depending on the nature of M(III) (*cf.* Table 6.4.11.1). It happens so for oxalate, maleate, fumarate, etc. In such cases, the free carboxylate group (which is a better coordinating group than the carbonyl group) coordinates with Cr(II) and the established conjugation (due to unsaturation) facilitates the *resonance transfer* if there is ***no symmetry mismatch barrier*** between the electron carrier bridging ligand orbital and the receptor orbital of M(III). If this symmetry mismatch barrier exists, then the ***radical ion mechanism*** — a multistep process is perferred (*cf.* reduction of Co^{III}, Cr^{III} *vs.* Ru^{III} by Cr^{II}, earlier discussion). However, the positively charged metal centres are kept well separated in such binuclear complexes and it reduces both the steric and electrostatic repulsion between the M(III) and Cr(II) centres. The bridge for the fumaric acid is shown below.

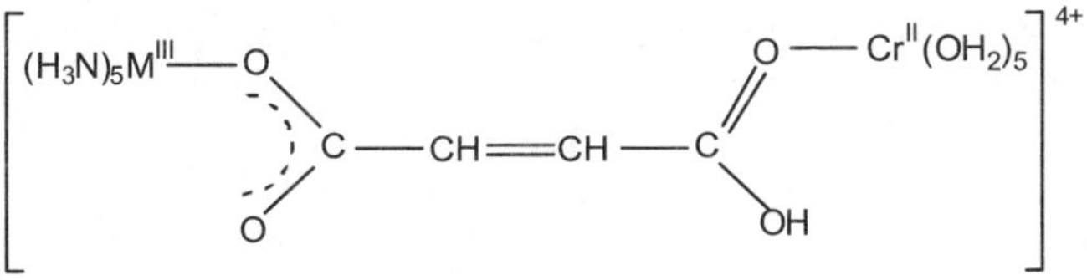

$$\left[(H_3N)_5M^{III}—O\!\!\!\!\!\begin{array}{c}\\ \end{array}\overset{\cdot\cdot}{\underset{O}{C}}—CH{=}CH—C\!\!\begin{array}{l}{=}O—Cr^{II}(OH_2)_5\\ —OH\end{array}\right]^{4+}$$

$(HO_2C)CH{=}CH(CO_2H)$
trans-form: fumaric acid
cis-form: maleic acid

Thus the increased rate (compared to those of the cases having the bridging ligand like acetate) arises due to the factors: **better bridging** by using the free carboxylate group, **unsaturation in the bridging ligand** allows the better electron flow (*i.e. resonance transfer through the conjugated system*) and avoiding the electrostatic and steric repulsion between the M(III) and Cr(II) centres which are kept separated by the bridging ligand.

It has been suggested by some workers that for oxalic acid, formic acid and maleic acid, in the bridge, M^{III}—ligand—Cr^{II}, ***full-length conjugation is not attained.*** For maleic acid, it is due to the steric factor. In such cases, the enhanced rate is believed to be due to their reducible character and the reaction is believed to pass through the ***chemical mechanism*** (for $M^{III} = Co^{III}$, Cr^{III}, Ru^{III}, *i.e.* irrespective of the symmetry mismatch barrier) where the bridging ligand is temporarily reduced as an intermediate one.

ISET: Chemical mechanism *vs.* Tunneling mechanism (*cf.* Fig. 6.4.11.1)

- Sometimes, in both the cases, the **vacant low lying orbital of the bridging ligand** participates and very often, it is very difficult to distinguish between these two mechanisms of electron transfer.

$$[M^{III}(NH_3)_5X] \xrightarrow{[Cr(aq)]^{2+}} M(II) + Cr^{III}X + 5NH_3$$

Rate of the above redox reaction depends on the reducibility of X. For X = –$OCOCH_3$ **(nonreducible bridging ligand)**, the rate is slow; for X = –OCOCHO (glyoxylate) **(reducible bridging ligand)**, the rate is fast. For X = –OCOCHO, the presence of a vacant low lying orbital in this bridging ligand facilities the *e*-transfer process through the chemical mechanism.

6.4.12 Multiple Bridging in the Activated Complex in the ISET Process

There are examples in which multiple bridges remain present in the activated complex in the ISET process. Some examples are given below. *The nature of bridging is very often identified from the product analysis.*

(i) **Reduction of *cis*-$[Cr(OH_2)_4X_2]^+$ by $[Cr(aq)]^{2+}$ (X = N_3^-, RCO_2^-)**

$$cis\text{-}[Cr(OH_2)_4X_2]^+ + [^*Cr(aq)]^{2+} \longrightarrow [Cr(aq)]^{2+} + cis\text{-}[^*Cr(OH_2)_4(X)_2]^+$$

Transfer of two *cis*-ligands (X) indicates the nature of bridging in the **activated complex** as follows:

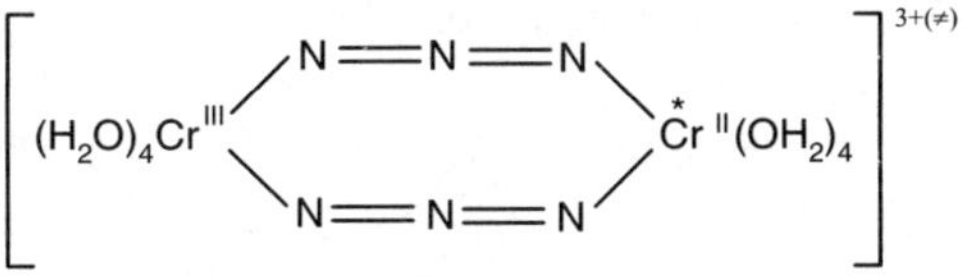

$$\left[(H_2O)_4Cr^{III}\begin{array}{c}\diagup N{=}N{=}N \diagdown\\ \diagdown N{=}N{=}N \diagup\end{array}\overset{*}{C}r^{II}(OH_2)_4\right]^{3+(\neq)}$$

Though there is a double bridge, but one electron transfer occurs. Probably, the double bridging stabilizes the binuclear precursor complex more. It has been found that for $X = N_3^-$, the rate constant increases by 30 times compared to the single bridge.

$$\frac{k_{(\text{double bridge})}}{k_{(\text{single bridge})}} = 30.$$

(ii) **Reduction of *cis*-$[Co(NH_3)_4(N_3)_2]^+$ or *cis*-$[Co(en)_2(N_3)_2]^+$ by $[Cr(aq)]^{2+}$** also passes through the double bridge made by the azides in *cis*-position.
Note: Reduction of ***cis*-$[CrF_2(OH_2)_4]^+$ by $[Cr(aq)]^{2+}$** occurs through the single bridge though the two *cis*-fluorides are in a suitable position to bridge.

$$cis\text{-}\left[CrF_2(OH_2)_4\right]^+ + \left[{}^*Cr(aq)\right]^{2+} \longrightarrow \left[Cr(aq)\right]^{2+} + F^- + \left[{}^*CrF(OH_2)_5\right]^{2+}$$

i.e. only one F^- is transferred.

Probably, for the monoatomic bridging ligands like F^-, the double bridge will produce a strain to destabilize the binuclear system.

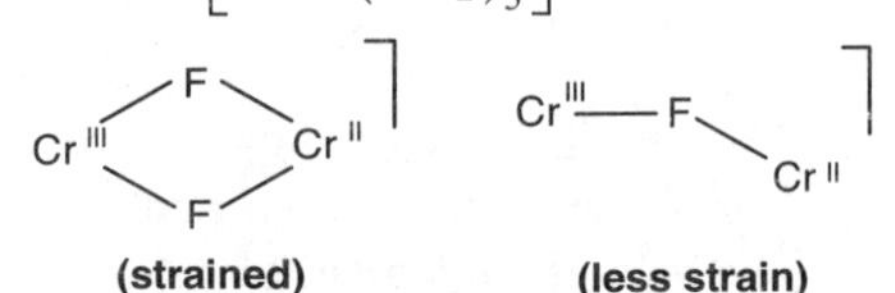

(iii) **Reduction of $[Co(edta)]^-$ by $[Cr(aq)]^{2+}$** gives the product $[Cr(edta)(OH_2)_3]^-$ (where edta acts as a tridentate ligand). It indicates that in the binuclear precursor complex, $edta^{4-}$ provides three binding sites to coordinate with the reducing agent.

(iv) The **reduction of $[Co(NH_3)_5(ntaH)]^+$ by $[Fe(aq)]^{2+}$** indicates that nta^{3-} provides three binding sites to coordinate with Fe(II) in the precursor complex. This aspect has been discussed earlier. The structure of the binuclear precursor complex is as follows:

(v) **Reduction of $[Co(C_2O_4)(NH_3)_4]^+$ by $[Cr(aq)]^{2+}$** produces $[Cr(C_2O_4)(OH_2)_4]^+$ suggesting the following binuclear precursor complex.

(vi) **Reduction of $[Co(NH_3)_5(OCOCH_2CO_2H)]^{2+}$ by $[Cr(aq)]^{2+}$** gives $[Cr(malonate)(OH_2)_4]^+$. It suggests the following binuclear precursor complex.

(vii) In the **reduction of $[Co^{III}(acac)_2(en)]^+$ by $[Cr(aq)]^{2+}$**, different products of Cr(III)-complexes are found. These are: $[Cr(aq)]^{3+}$, $[Cr(acac)(aq)]^{2+}$, $[Cr(acac)_2(aq)]^+$. It can be reasonably concluded that $[Cr(aq)]^{3+}$ is produced through the OSET path; $[Cr(acac)(aq)]^{2+}$ is produced through the single acetylacetonate bridged ISET path and $[Cr(acac)_2(aq)]^+$ is produced through the double acetylacetonate bridged ISET path. The probable structures of the binuclear bridged precursor complexes are given below.

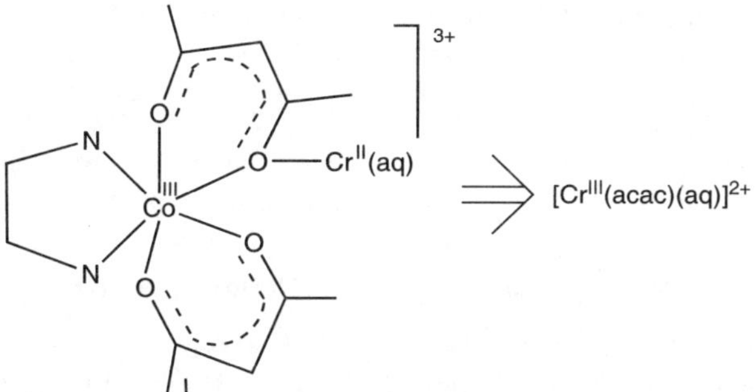

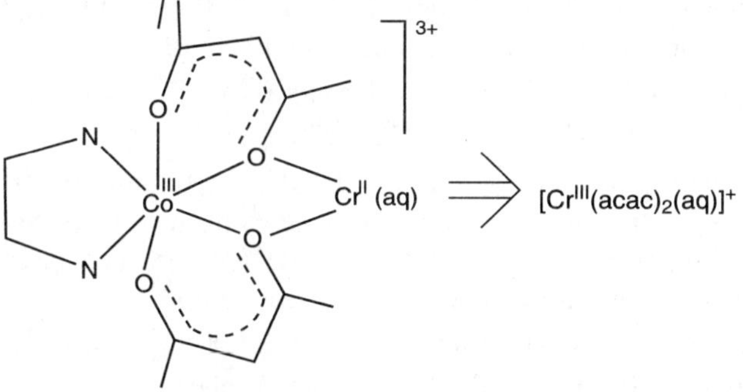

$\Rightarrow [Cr^{III}(acac)(aq)]^{2+}$

$\Rightarrow [Cr^{III}(acac)_2(aq)]^+$

Note: In all the above given examples, though multibridges are formed in the binulcear precursor complexes, but there is only one electron transfer. It may be pointed out that in the Pt(II)–Pt(IV) exchange path, two electron transfer can occur through a single bridge like Pt^{II}—Cl—Pt^{IV} (*cf.* Sec. 6.4.15).

6.4.13 Bridging Ligand Provided by the Reducing Agent in the ISET Process

In the examples discussed in all earlier sections, it was noted that the bridging ligand is provided by the oxidant (which is also substitutionally inert). However, in the reduction of the chromic acid ($HCrO_4^-$) by $[Fe(CN)_6]^{4-}$, it was noted that the *inert reductant provides the bridging ligand.* It is supported by the characterization of the following **uncharged product species** (*JACS*, **91**, 3189, 1969).

$$[(H_2O)_5Cr^{III}—N{\equiv}C—Fe^{III}(CN)_5]^0$$

Substitutionally inert

This is the **thermodynamically and kinetically stable product** produced at the 3rd step of reduction of Cr(VI) **through the ISET path**. This binuclear complex is actually the successor complex in the reduction of Cr(IV) by $[Fe(CN)_6]^{4-}$.

Cr(VI) experiences 3*e*-reduction to Cr(III) at the successive 1*e*-transfer steps giving rise to Cr(V) and Cr(IV) intermediates. All these higher valent species, *i.e.* Cr(VI), Cr(V) and Cr(IV), are labile while

Cr(III) is inert. The above mentioned product (binuclear) is produced through the reduction of Cr(IV) by $[Fe(CN)_6]^{4-}$ in the ISET path.

In fact, it is a rare example where the **reducing metal centre provides the bridging ligand for the ISET path**. The stepwise reduction of Cr(VI) by the 1*e*-reductant like $[Fe(CN)_6]^{4-}$ which is **substitutionally inert** should give the following stoichiometry.

$$Cr(VI) \xrightarrow{+1e} Cr(V) \xrightarrow{+1e} Cr(IV) \xrightarrow{+1e} Cr(III)$$

$$HCrO_4^- + 3[Fe(CN)_6]^{4-} + 7H^+ \longrightarrow Cr^{3+} + 3[Fe(CN)_6]^{3-} + 4H_2O$$

But, in the present case, Cr^{3+} does not exist as $[Cr(OH_2)_6]^{3+}$ but it exists as $[(H_2O)_5Cr^{III}(NC)Fe^{III}(CN)_5]$ that establishes the 3rd step of the reaction to go through the ISET path. Thus the stoichiometry of the reaction is:

$$HCrO_4^- + 3[Fe(CN)_6]^{4-} + 7H^+ \longrightarrow [CrFe(CN)_6] + 2[Fe(CN)_6]^{3-} + 4H_2O$$

$$[CrFe(CN)_6] \text{ denotes } [(H_2O)_5Cr^{III}(NC)Fe^{III}(CN)_5]$$

Similarly, the reduction of $HCrO_4^-$ by $[Fe(bpy)(CN)_4]^{2-}$ or $[Fe(bpy)_2(CN)_2]$, the **ISET path of the 3rd step** produces a CN^- bridged binuclear complex of Cr^{III} and Fe^{III}.

$$HCrO_4^- + 3[Fe(bpy)(CN)_4]^{2-} + 7H^+ \longrightarrow [CrFe(bpy)(CN)_4]^{2+} + 2[Fe(bpy)(CN)_4]^- + 4H_2O$$

$$[CrFe(bpy)(CN)_4]^{2+} \text{ denotes } [(H_2O)_5Cr^{III}(NC)Fe^{III}(bpy)(CN)_3]^{2+}$$

Here it is worth mentioning that the reduction of $HCrO_4^-$ by the the **substitutionally inert Fe(II)-complexes** like $[Fe(bpy)_3]^{2+}$, $[Fe(phen)_3]^{2+}$, the process goes through the **OSET path** because bpy or phen cannot act as the bridging ligand to mediate the ISET path. In such cases, no binuclear complex of Fe^{III} and Cr^{III} is produced. The stoichiometry of the reaction is:

$$HCrO_4^- + 3[Fe(L\text{—}L)_3]^{2+} + 7H^+ \longrightarrow Cr^{3+} + 3[Fe(L\text{—}L)_3]^{3+} + 4H_2O$$

$$(Cr^{3+} \text{ exists as } [Cr(OH_2)_6]^{3+};\ L\text{—}L = \text{bpy or phen})$$

Note: For the reduction of $HCrO_4^-$ by the **substitutionally inert** Fe^{II}-complexes like $[Fe(CN)_6]^{4-}$, $[Fe(bpy)(CN)_4]^{2-}$ or $[Fe(bpy)_2(CN)_2]$, the other steps of reduction (*i.e.* $Cr^{VI} \rightarrow Cr^{V}$, $Cr^{V} \rightarrow Cr^{IV}$) may also go through the **ISET path** by using the bridging CN^- ligand coordinated to Fe^{II}, but the binuclear successor complexes involving the **labile** Cr^V or Cr^{IV} centres undergo rapid aquation. For the reduction of $HCrO_4^-$ by the **labile reductant** like $[Fe(OH_2)_6]^{2+}$, the stepwise 1*e* reductions may pass through the **ISET path** involving the **Fe—O—Cr bridge** (where the bridging O-center is provided by the oxidant) but the successor binuclear complexes undergo rapid aquation (*cf.* lability of high spin Fe^{III}, Cr^V and Cr^{IV}). The first step of reduction of $HCrO_4^-$ by $[Fe(aq)]^{2+}$ through the ISET path probably involves the following binuclear precursor complex.

$$\underset{\text{(precursor complex)}}{[(HO)Cr^{VI}(O)_2\text{—}O\text{—}Fe^{II}(OH_2)_5]^+} \xrightarrow{e\text{-transfer}} \underset{\text{(successor complex)}}{[(HO)Cr^{V}(O)_2\text{—}O\text{—}Fe^{III}(OH_2)_5]^+}$$

In the above binuclear successor complex, both the metal centres are labile. It may be noted that the preferred coordination number is 4 (tetrahedral) for Cr^{VI} and Cr^V while it is 6 (octahedral) for $Cr^{IV}(d^2)$ and $Cr^{III}(d^3)$.

6.4.14 Effect of the Non-bridging Ligand on the Rate Process in the ISET Path: Ligand Field Strength Controlling Energy of the Receptor Orbital of the Oxidant and Franck-Condon Barrier

Energy of the receptor orbital (σ*) of the oxidant: It has been noted that the rate of reduction of *trans*-$[M^{III}(NH_3)_4(L)X]$ (M = Co and Cr) by $[Cr(aq)]^{2+}$ or $[Fe(aq)]^{2+}$ in the ISET process drastically depends on the ligand field strength of L (*trans* to the bridging ligand X). It is quite reasonable by considering the nature of the receptor orbital of the oxidant. For the oxidants Co(III) (t_{2g}^6) and Cr(III) (t_{2g}^3), the receptor orbital is σ* (*i.e.* e_g orbital which is d_{z^2} by considering the bridge along the

z-direction). Energy of this receptor orbital is very much sensitive to the nature of the *trans*-ligand L (when X is kept the same for comparison) compared to the *cis*-ligands. *With the increase of the ligand field strength of L, energy of the e_g-orbital (i.e. d_{z^2}) will increase, i.e.* ***the receptor orbital will be destabilized more by the stronger field ligand L and it will be stabilized more for the weaker field ligand L.***

To receive the electron in the vacant e_g (*i.e.* σ^*) orbital of the oxidant, energy of this receptor orbital must be lowered. This can be done by moving the *trans*-ligands L and X outwards (*i.e.* L and X move in opposite directions). If the ligand field strength of L is varied (for a particular bridging ligand X), then for the stronger field ligand, more distance is to be traveled outwards by L to lower the energy of the receptor orbital. **Consequently, more activation energy will be needed for the stronger field ligand L.** The ligand field strength is: en~NH_3 ≫ H_2O. Thus, *trans*-$[CoCl(en)_2(NH_3)]^{2+}$ should react slower than *trans*-$[CoCl(en)_2(OH_2)]^{2+}$. This expectation is experimentally verified.

$$trans\text{-}\left[CrCl(NH_3)_4(^*OH_2)\right]^{2+} + \left[Cr(aq)\right]^{2+} \longrightarrow \text{products (ISET path)}.$$

It has been noted that *trans*–$[CrCl(NH_3)_4(^{18}OH_2)]^{2+}$ *reacts slower than trans*–$[CrCl(NH_3)_4(^{16}OH_2)]^{2+}$. This observation supports the requirement of bond length elongation in the *trans*-axial directions to lower the energy of the receptor e_g-orbital. *Obviously, the movement of heavier isotope needs more energy.*

Table 6.4.14.1 Effect of the nonbridging ligands on rate constants for the ISET reactions (*cf.* A. Haim, *J. Am. Chem. Soc.*, **86**, 2352, 1964; R. D. Cannon *et al. J. Am. Chem. Soc.*, **87**, 5264, 1965)

Oxidant	***Reducing agent***	***k* (dm³ mol⁻¹ s⁻¹) 25°C**
trans-$[Co(en)_2(OH)(OH_2)]^{2+}$	$[Cr(aq)]^{2+}$	2.6×10^6
trans-$[Co(en)_2(NH_3)(OH)]^{2+}$	$[Cr(aq)]^{2+}$	2.2×10^5
trans-$[CoCl(en)_2(OH_2)]^{2+}$	$[Fe(aq)]^{2+}$	0.25
trans-$[CoCl(en)_2(NH_3)]^{2+}$	$[Fe(aq)]^{2+}$	6.5×10^{-5}
$[CrCl(NH_3)_5]^{2+}$	$[Cr(aq)]^{2+}$	8.8×10^{-2}
trans-$[CrCl(NH_3)_4(OH_2)]^{2+}$	$[Cr(aq)]^{2+}$	1.3
cis-$[Co(en)_2(OH)(OH_2)]^{2+}$	$[Cr(aq)]^{2+}$	7.9×10^5
cis-$[Co(en)_2(NH_3)(OH)]^{2+}$	$[Cr(aq)]^{2+}$	2.0×10^5
cis-$[CoCl(en)_2(OH_2)]^{2+}$	$[Fe(aq)]^{2+}$	4.6×10^{-4}
cis-$[CoCl(en)_2(NH_3)]^{2+}$	$[Fe(aq)]^{2+}$	1.8×10^{-5}

Here it is important to mention that for the corresponding ***cis*-oxidant complexes**, the rate constant is **relatively less sensitive** to the ligand field strength of the *cis*-ligand (L) to the bridging ligand (X). It happens so because in the *cis*-complex, the **chelated N-atom** *trans* to the **bridging ligand restrains its motion**.

It may be noted that the rate of reduction by $[Fe(aq)]^{2+}$ is much slower than the rate of reduction by $[Cr(aq)]^{2+}$. It is because of the more positive reduction potential of the $[Fe(aq)]^{3+}/[Fe(aq)]^{2+}$ couple and consequently, $\mathbf{[Fe(aq)]^{2+}}$ **is a poorer reducing agent** than $[Cr(aq)]^{2+}$ (*cf.* Marcus equation; E^0 value of the two couples differs by about 1.2 V). Besides this, reduction of Co^{III} or Cr^{III} complexes by $[Fe(aq)]^{2+}$ involes the net $\pi^* \rightarrow \sigma^*$ electron transfer while for the reduction by $[Cr(aq)]^{2+}$, it involves the $\sigma^* \rightarrow \sigma^*$ electron transfer. It favours the reduction by $[Cr(aq)]^{2+}$ in the ISET path.

Ligand field strength effect and Franck-Condon barrier: To overcome the Franck-Condon barrier, bond length is to be elongated in the oxidant and bond length is to be shortened in the reductant to attain the transition state. Thus in the reductant, presence of the stronger field ligand will require less shortening of the bond length. This will reduce the activation energy. In fact, it has been experimentally verified that the Co(III)-complexes are reduced by the Cr(II)-complexes in the following order of rate constants.

$$[Cr(en)_2(OH_2)_2]^{2+} > [Cr(en)(OH_2)_4]^{2+} > [Cr(OH_2)_6]^{2+}$$

$\xleftarrow{\hspace{3em}}$ Increasing ligand field strength

$\xleftarrow{\hspace{3em}}$ Reducing power (kinetically)

Two electron exchange reaction through OSET *vs.* ISET

Simultaneous transfer of more than one electron **through the OSET process is quite unlikely,** because the reorganisation energy to equalise the bond lengths **(to overcome the Franck Condon barrier)** at a metal centre differing by two units of oxidation states (*e.g.* Rh^{I}/Rh^{III}, Pt^{II}/Pt^{IV}, etc.) is too high (*i.e.* activation energy is too high). However, two electron transfer through the ISET path involving the Pt^{IV}—X—Pt^{II} bridge with the similar ligand environment around the Pt^{IV} and Pt^{II} centres in the binuclear complex is well known. Rh^{III}—Rh^{I} system can also react in the same way.

6.4.15 Two Electron Transfer in the ISET Process

Here we shall illustrate the case of electron exchange in the Pt(II)–Pt(IV) system where there is only one bridging ligand in the binuclear precursor complex but two electrons are transferred through the bridge followed by the transfer of the bridging ligand.

Pt(II) which is relatively more labile than Pt(IV) is known to **catalyze the exchange of free Cl^- with** the Cl^- bound to Pt(IV) in the following reaction.

$$trans\text{-}\left[PtCl_2(en)_2\right]^{2+} + {}^*Cl^- \xrightarrow[\text{(catalyst)}]{[Pt(en)_2]^{2+}} trans\text{-}\left[Pt^*ClCl(en)_2\right]^{2+} + Cl^-$$

The proposed mechanism involves the rapid coordination of free $*Cl^-$ to $[Pt(en)_2]^{2+}$ to give a five coordinate intermediate followed by the formation of the precursor bridged complex utilizing the vacant coordination site of $[PtCl^*(en)_2]^+$. Then electron transfer (transfer of two σ*-electrons) occurs within the activated precursor complex where the metal centres are in similar environments followed by the fission of the successor complex. This fission will transfer the bridging ligand.

$$[{}^*Pt^{II}(en)_2]^{2+} + {}^*Cl^- \rightleftharpoons [{}^*Pt({}^*Cl)(N)_4]^+ \text{ i.e., } [{}^*Pt({}^*Cl)(en)_2]^+$$

$$trans\text{-}[PtCl_2(en)_2]^{2+} + [{}^*PtCl^*(en)_2]^+ \rightleftharpoons [{}^*Cl\text{—}{}^*Pt^{II}(en)_2\text{—}Cl\text{—}Pt^{IV}(en)_2\text{—}Cl]^{3+}$$

(Relatively more labile centre)

2e-transfer

$$[{}^*Cl\text{—}{}^*Pt^{IV}(en)_2\text{—}Cl\text{—}Pt^{II}(en)_2\text{—}Cl]^{3+} \longrightarrow trans\text{-}[{}^*Pt^*ClCl(en)_2]^{2+} + [Pt(en)_2]^{2+} + Cl^-$$

By using the isotopically labelled Pt-centre, the above mechanism has been proved. The rate law is given by:

$$\text{rate} \propto [\text{Pt(II)}][\text{Pt(IV)}][\text{Cl}^-]$$

If the reaction is carried out by using a radiochloride ($^*Cl^-$) for a chiral Pt(IV)-complex, then the rate of exchange of $^*Cl^-$ is the same with the rate of the **change in optical rotation** in the reaction of *trans*-$[Pt(l\text{-}pn)_2Cl_2]^{2+}$ with $[Pt(en)_2]^{2+}$.

$$[\text{Pt(en)}_2]^{2+} + trans\text{-}[\text{Pt}(l\text{-pn})_2\text{Cl}_2]^{2+} + {}^*\text{Cl}^- \longrightarrow [\text{Pt}(l\text{-pn})_2]^{2+} + trans\text{-}[\text{Pt(en)}_2{}^*\text{ClCl}]^{2+} + \text{Cl}^-$$

The reaction passes through the involving $^*Cl^-$, four coordinate Pt(II) complex and six-coordinate Pt(IV)-complex as outlined above. *l*-pn is the *l*-form of $H_2N—CHMe—CH_2—NH_2$. Optical rotations of the reactant, *trans*-$[Pt(l\text{-}pn)_2Cl_2]^{2+}$ and product, $[Pt(l\text{-}pn)_2]^{2+}$ are different. Thus rate of change of optical rotation gives the measure of rate of the above reaction.

Here it is worth mentioning that the above reactions are prevented if the substituted ethylenediamine like $Me_2N—(CH_2)_2—NMe_2$ is used in place of *en* as the ligand. In fact, for such a bulky ligand, formation of the Cl^--bridged dinuclear complex of Pt(II) and Pt(IV) is not possible due to the steric reason.

In the same way, Rh(I) (d^8) being more labile can catalyse the substitution reaction at the inert Rh(III) (d^6) centre. In fact, by taking the advantage of this principle, Delepine synthesised *trans*-$[Rh(py)_4Cl_2]^+$ from $Na_3[RhCl_6]$.

$$\text{Na}_3[\text{RhCl}_6] \xrightarrow[\text{aq. soln}]{\text{py}} [\text{Rh(py)}_3\text{Cl}_3]\downarrow$$

$$\text{Na}_3[\text{RhCl}_6] \xrightarrow[\text{aq. soln + small amount of } C_2H_5OH \text{ as a reducing agent}]{\text{py}} trans\text{-}[\text{Rh(py)}_4\text{Cl}_2]^+$$

The reaction mechanism is **similar to that of Pt(IV)–Pt(II) electron exchange reaction.**

$$[\text{RhCl}_6]^{3-} \xrightarrow{\text{aquation}} [\text{RhCl}_5(\text{OH}_2)]^{2-} + \text{Cl}^-$$

$$\text{Rh(III)} + C_2H_5OH \longrightarrow \text{Rh(I)} + CH_3CHO, \text{ (reduction of Rh}^{III} \text{ to Rh}^{I})$$

$$\text{Rh(I) (labile)} + 4\text{py} \xrightarrow{\text{Fast}} [\text{Rh(py)}_4]^+ \text{ (square planar complex of Rh}^{I}, d^8 \text{ system)}$$

$$[\text{Rh(py)}_4]^+ + [\text{RhCl}_5(\text{OH}_2)]^{2-} + H_2O \xrightarrow{\text{Slow}} [H_2O—\text{Rh}^{I}(\text{py})_4—\text{Cl}—\text{Rh}^{III}(\text{Cl})_4—OH_2]^-$$

$$\downarrow \text{2e transfer, fast}$$

$$[H_2O—\text{Rh}^{III}(\text{py})_4—\text{Cl}—\text{Rh}^{I}(\text{Cl})_4—OH_2]^-$$

$$\xrightarrow{\text{Fast}} \text{Rh(I) (i.e. } [\text{Rh(aq)}]^+) + 4\text{Cl}^- + [\text{Rh}^{III}(\text{py})_4(\text{OH}_2)\text{Cl}]^{2+}$$

$$[\text{Rh}^{III}(\text{py})_4(\text{OH}_2)\text{Cl}]^{2+} \xrightarrow[\text{Slow}]{+\text{Cl}^-, -H_2O} [\text{Rh}^{III}(\text{py})_4\text{Cl}_2]^+$$

The following ligand substitution reaction actually represents a **Pt(II)-catalysed process**. Pt(II) may be generated in the reaction through the reduction of Pt(IV) by NO_2^- or it may be added externally.

$$trans\text{-}[PtCl_2(en)_2]^{2+} + NO_2^- \xrightarrow{[Pt(en)_2]^{2+}} [PtCl(en)_2(NO_2)]^{2+} + Cl^-$$

rate $\propto [Pt(II)][Pt(IV)][NO_2^-]$.

Here it may be mentioned that *between the Pt(II) and Pt(IV) centres, the Pt(II)-centre is relatively more labile.* In fact, it is the common observation that in a particular system, if the lower valent centre is more labile and the higher valent centre is inert then the lower valent centre catalyzes the ligand substitution process at the higher valent centre (*cf.* Sec. 6.6).

6.4.16 Electron Transfer in Mixed Valence Complexes where the Metal Centres Separated by the Bridging Ligands (*cf.* Secs. 7.22 of Vol. 5, 1.13 of Vol. 4)

In such bimetallic complexes, the oxidising and reducing metal centres with different oxidation states are separated by the bridging ligands and the electron transfer (ISET path) rate is largely controlled by the nature of the bridging ligand and the separation distance (*d*) (*i.e.* length of the bridging ligand) between the oxidising and reducing metal centres.

Different types of mixed valence complexes experiencing the **intervalence charge transfer (IVCT)** have been discussed in Sec. 1.13 (Vol. 4) and Sec. 7.22 (Vol. 5). For the **class-III type** mixed valence complexes $[Ru—X—Ru]^{5+}$ of Ru^{III} and Ru^{II}, the electron is fully delocalised and the separate +2 and +3 oxidation states are not sensed and they are best represented by the +2.5 oxidation state. In such cases, **no intervalence charge transfer (IVCT)** absorption band is possible.

In the following **class-II type** $Ru^{II}—X—Ru^{III}$ complex, having the relatively longer 4,4′-bpy as the bridging ligand, the intervalence charge transfer (IVCT) absorption band (1050 nm) arises in the near-infrared (NIR) region.

$$\left[(H_3N)_5Ru^{II}—N\langle\text{C}_5\text{H}_4\rangle—\langle\text{C}_5\text{H}_4\rangle N—Ru^{III}(NH_3)_5\right]^{5+}$$

$$\downarrow \text{IVCT } (\lambda = 1050 \text{ nm})$$

$$\left[(H_3N)_5R^{III}—N\langle\text{C}_5\text{H}_4\rangle—\langle\text{C}_5\text{H}_4\rangle N—Ru^{II}(NH_3)_5\right]^{5+}$$

The intervalence electron transfer rates for the following $Ru^{II}—X—Ru^{III}$ intervalence complex having the different bridging ligands are compared in Table 8.4.16.1 depending on the nature and length of the bridging ligand (X).

$$[(bpy)_2ClRu—L—L—RuCl(bpy)_2]^{3+}$$

It appears that for the conjugated bridging ligand, the intervalence electron transfer process is favoured and with the increase of separating distance (*d*) between the Ru^{II} and Ru^{III} centres, the rate decreases.

Table 6.4.16.1: Calculated electron transfer rate constants in $[Ru(bpy)_2Cl]_2^{3+}(\mu\text{–L—L})$ of the Ru^{III}–Ru^{II} bimetallic complex (Ref. C. Creutz, *Prog. Inorg. Chem.*, **30**, 1, 1983)

L—L	*d (pm)*	k_e (s^{-1})
N⟨ring⟩N (pyz)	680	3×10^9
$Ph_2P—CH_2—PPh_2$ (dppe)	710	1×10^8
N⟨ring⟩—⟨ring⟩N (4,4′-bpy)	1130	1×10^8
N⟨ring⟩—CH═CH—⟨ring⟩N	1380	2×10^7

Pyrazine (*pyz*) and diphosphine (*dppe*) are the bridging ligands of comparable length (*d*), but the electron transfer rate is about 30 times faster for the *pyz*-bridging ligand having the conjugation compared to *dppe* lacking in conjugation. In fact, for the bridging ligand *dppe* and 4,4′-*bpy*, the rate constants are the same though the bridging length for the conjugated 4,4′-*bpy* is much longer than that for nonconjugated *dppe*.

6.5 MECHANISTIC CRITERIA AND THE FACTORS FAVOURING THE OUTER AND INNER SPHERE ELECTRON TRANSFER PROCESSES

(A) Factors favouring the OSET process

- formation of the binuclear bridged precursor complex is denied;
- $\pi^* \rightarrow \pi^*$ net electron transfer;
- less inner sphere reorganization energy to overcome the Franck-Condon barrier;
- higher equilibrium constant for the overall reaction (*cf.* Marcus equation for cross-reaction).

Bridge formation leading to the dinuclear precursor complex is disfavoured. It may happen in many ways:

(i) The reductants like $\mathbf{[Cr(bpy)_3]^{2+}}$ (t_{2g}^4), $\mathbf{[Fe(bpy)_3]^{2+}}$ (t_{2g}^6), $\mathbf{[Ru(NH_3)_6]^{2+}}$ (t_{2g}^6), etc. are tremendously **substitutionally inert** and their associated ligands, *i.e.* bpy, NH_3, etc. cannot function as the bridging ligands. In such cases, the inert oxidants like $[Co^{III}(NH_3)_5X]$ can provide the bridging ligand (X) but *formation of the binuclear precursor complex is prevented because of the inertness of the reducing centre.*

$$\left.\begin{array}{l}[M^{III}(NH_3)_5X]^{2+} + [Cr(bpy)_3]^{2+} \longrightarrow \text{OSET} \\ [M^{III}(NH_3)_5X]^{2+} + [Cr(bpy)_3]^{2+} \longrightarrow \text{OSET}\end{array}\right\} \begin{array}{l} M^{III} = Co^{III}, Ru^{III} \\ X^- = Cl^-, Br^-, N_3^- \text{ etc.}\end{array}$$

(ii) The reductant may be substitutionally inert but **it may provide the bridging ligand.** For example, the inert reductant $[Fe(CN)_6]^{4-}$ can provide the bridging ligand, *i.e.* C-bound cyanido ligand but if the **oxidant is not labile** then formation of the dinuclear species will be prevented. Consequently, the reactions occur through the OSET process.

$$\left.\begin{array}{l}[IrCl_6]^{2-} + [Fe(CN)_6]^{4-} \longrightarrow \\ [Fe(phen)_3]^{3+} + [Fe(CN)_6]^{4-} \longrightarrow\end{array}\right\} \begin{array}{l}\text{OSET process.} \\ \text{(both the oxidant and reductant are inert)}\end{array}$$

(**Note:** The reaction of $HCrO_4^-$ with $[Fe(CN)_6]^{4-}$ goes through *the ISET process* where the —CN group bridges with the labile oxidant, see Sec. 6.4.13).

(iii) The oxidant is inert and it fails to provide any suitable bridging ligand. In such cases, formation of the dinuclear precursor complex is prevented even when the reductant is labile. Reduction of $[Co(phen)_3]^{3+}$, $[Fe(bpy)_3]^{3+}$, $[M(NH_3)_6]^{3+}$, $[M(NH_3)_5(py)]^{3+}$ (M = Co, Ru), by the even labile reductants like $[Cr(aq)]^{2+}$, $[Fe(aq)]^{2+}$ occurs through the OSET process.

$$\left.\begin{array}{r}[M(LL)_3]^{3+} + [M'(aq)]^{2+} \longrightarrow \\ [M(NH_3)_6]^{3+} + [M'(aq)]^{2+} \longrightarrow \\ [M(NH_3)_5(py)]^{3+} + [M'(aq)]^{2+} \longrightarrow\end{array}\right\} \begin{array}{l}\text{OSET process} \\ M = Co, Ru; LL = bpy, phen \\ M' = Cr, Fe\end{array}$$

cf. $[M(NH_3)_5X]^{2+} + [M'(aq)]^{2+} \rightarrow$ product; ISET; $X^- = Cl^-, Br^-, OH^-, SCN^-, N_3^-, RCO_2^-$, etc.

(B) Factors favouring the ISET process

- formation of the binuclear bridged complex is possible;
- bridging ligand is suitable to mediate the electron transfer process;

- $\sigma^* \rightarrow \sigma^*$ net electron transfer;
- less reorganization energy to overcome the Franck-Condon barrier;
- higher equilibrium constant for the overall reaction (*i.e.* **Marcus equation**);

Formation of the binuclear bridged precursor complex is favoured under the following conditions.

(i) One of the reactant is labile while the other reactant is inert but capable of providing a bridging ligand. *Generally, the inert oxidant provides the bridging ligand and the reductant is labile.*

Formation of this dinuclear bridged precursor complex (where the bridging ligand is suitable to mediate the electron transfer process) is the primary requirement for the operation of the ISET process. But fulfillment of this condition cannot ensure the ISET process and it happens so in the case of reduction of $[Ru(NH_3)_5X]^{2+}$ by $[V(aq)]^{2+}$ ($\pi^* \rightarrow \pi^*$ net electron transfer). This reaction occurs through the OSET path (*see* Sec. 6.4.5). The reaction experiences the net $\pi^* \rightarrow \pi^*$ electron transfer which is a favourable situation for the OSET path but **it is an unfavourable situation for the ISET path** because it needs the electronic activation of both the oxidant and reductant for the ISET path. Moreover, the inertness of the $[V(aq)]^{2+}$ centre produces also a barrier for the ISET path. These are the reasons for the OSET path.

If the bridging ligand is a long chain saturated moiety, then even within the binuclear precursor complex, the OSET path instead of the ISET path operates (*see* Sec. 6.4.11).

(C) Experimental Evidences in Favour of the ISET Process

(i) **Detection of the binuclear successor complex:** If it is possible to **detect experimentally the binuclear successor complex** under the experimental condition, then it can be concluded without any hesitation that the reaction is passing through the ISET process. Here it is worth mentioning that detection of the binuclear precursor complex cannot ensure always the ISET path (*cf.* bridging ligand with a long chain saturated moiety, see Sec. 6.4.11). Simple identification of a **short-lived binuclear complex** cannot be treated as an evidence of the ISET path (see Sec. 6.4.6).

(ii) **Ligand transfer** (*see* Sec. 6.4.8)**:** If the electron transfer is accompanied with the **ligand transfer** and the overall reaction rate is much faster than the substitution rate of the metal centre to which the ligand is transferred, *then it can be concluded without any hesitation that the reaction is passing through the ISET process.*

Here it may be mentioned that the ligand transfer is not a necessary condition for the ISET process. Thus, in absence of the ligand transfer, other arguments are needed to substantiate the occurrence of the ISET process.

(iii) **Rate dependence on the nature of the bridging ligand:** If nature of the bridging ligand affects the overall rate process, then it supports the occurrence of the ISET process. But, for the substitutionally controlled ISET process, the rate does not depend remarkably on the nature of the bridging ligand. It happens so for the reduction of Co^{III}-complexes by $[V(aq)]^{2+}$ (d^3, inert centre). It may be noted that the reduction of Co^{III}-complexes by $[Eu(aq)]^{2+}$ **(labile centre)** occurs through the ISET path, but the rate is not so much sensitive to the nature of the bridging ligand (X) (see Sec. 6.4.5).

- It has been already mentioned that the bridge, M_{ox}—N=N=N—M_{red} is more efficient than the bridge, M_{ox}—NCS—M_{red} for the ISET path when the reducing metal centre (M_{red}) is **a hard or border line one** (*e.g.* Cr^{2+}, Fe^{2+}, etc.). **Thus the criterion, $k_{N_3} \gg k_{NCS}$, is an argument in favour of the ISET process.** Similarly, for the hard and borderline reducing agents like $[Cr(aq)]^{2+}$, $[Fe(aq)]^{2+}$, the ***S*-bonded NCS⁻** (*i.e.* M_{ox}—SCN) **can act as a better bridging ligand than the *N*-bonded *NCS*⁻** (*i.e.* M_{ox}—NCS) and the observation, $k_{SCN} \gg k_{NCS}$, is an evidence in favour of the ISET process. The $k_{OH} > k_{H_2O}$ observation is also in favour of the ISET process. These aspects have been discussed in detail in Sec. 6.4.10.

$$[M^{III}(NH_3)_5X]+[M'(aq)]^{2+} \longrightarrow \text{Product}; \; M = Co^{III}, Ru^{III}, \textit{etc.}, \; M' = Fe^{II}, Cr^{II}.$$

$$X = N_3^-, NCS^-, SCN^-, OH^-, H_2O.$$

$$k_{N_3} \gg k_{NCS}; \; k_{SCN} \gg k_{NCS}; \; k_{OH} > k_{H_2O} \Rightarrow \textbf{ISET process.}$$

Note: For the **softened reductant** $[Co(CN)_5]^{3-}$, the criteria $k_{N_3} \gg k_{NCS}, k_{SCN} \gg k_{NCS}$ are not true for the ISET path.

- Sometimes, the overall reaction rate increases with the increase of ***coordinating power*** of the ***bridging ligand.*** *It supports the ISET process* (*cf.* Sec. 6.4.10).
- If the **remote electrophilic attack** on the bridging ligand by the reducing agent can establish a conjugated system (*i.e. if the bridging ligand bears the proper unsaturation*), then it favours *the* ***electron tunneling through the resonance mechanism.*** This observation supports the ISET process (*cf.* Secs. 6.4.9, 11).

  If the bridging ligand (*e.g.* oxalate, formate, glyoxylate, maleate, glycolate, isonicotinamide, etc.) is **reducible**, then the electron transfer process is favoured due to the ***chemical* or *ion-radical mechanism***. Such an observation is in conformity with the ISET process. (*cf.* Sec. 6.4.11).

  (**Note:** If the ISET process is ***substitution controlled,*** *i.e. rate determining step is the formation of the binuclear precursor complex,* and if the substitution process occurs through the ***dissociative path,*** then even in the ISET process, the overall rate is insensitive to the nature of the bridging ligands. If the substitution at the reducing centre occurs through the even **associative path** (I_a), the overall reaction rate is not so much sensitive to the nature of the bridging ligand. It happens so for the reduction by $[V(aq)]^{2+}$. Then the kinetic parameters of the overall reaction are found to be comparable with those of the substitution process (I_a path) at $[V(aq)]^{2+}$. This aspect has been discussed in detail in Sec. 6.4.5.

(iv) ***Volume of activation (ΔV^≠):*** It may be used to diagnose the reaction pathway. In the OSET path where formation of the precursor complex enhances the charge (*e.g.* electron exchange in the couple $[Fe(aq)]^{2+}$–$[Fe(aq)]^{3+}$), **$\Delta V^{\neq}$ is expected to be negative** because of the more electrorestriction imposed on the surrounding solvents by the activated complex than by the reactants. However, the charge destruction (*i.e.* ion pair formation) in the precursor complex formation may lead to the +ve value of $\Delta V^{\neq}$. In the ISET path, formation of the binuclear precursor complex releases solvent from the labile centre. ***It generally makes $\Delta V^{\neq}$ positive.*** These are illustrated in the following examples.

$[Fe(OH_2)_6]^{3+}$–$[Fe(OH_2)_6]^{2+}$ (electron exchange): $\Delta V^{\neq} = -12 \pm 2$ cm^3 mol^{-1}

(⇒ **OSET path**)

$[Fe(OH)(OH_2)_5]^{2+}$–$[Fe(OH_2)_6]^{2+}$ (electron exchange): $\Delta V^{\neq}$ is slightly positive.

(⇒ **ISET path**)

$[Co(NH_3)_5X]^{2+}$–$[Fe(OH_2)_6]^{2+}$ (X = F, Cl, Br, N_3): $\Delta V^{\neq} = 8 - 14$ cm^3 mol^{-1}

(⇒ **ISET path**)

6.6 LIGAND REPLACEMENT REACTIONS THROUGH ELECTRON TRANSFER

If for a particular metal ion, the higher valent centre is relatively more inert but the lower valent state is relatively more labile, then the lower valent state can catalyze the ligand substitution process at the higher valent inert centre. Such examples are found for:

Pt(II)–Pt(IV); Cr(III)–Cr(II); Co(III)–Co(II).

This aspect is illustrated in different examples.

(i) **Unusually high water exchange rate for $[Co(aq)]^{3+}$** (*see* Sec. 5.13): The water exchange rate at $[Co(OH_2)_6]^{3+}$ is unexpectedly high. The high reduction potential of the $[Co(aq)]^{3+}/[Co(aq)]^{2+}$

couple (E^0 = 1.8 V) allows the oxidation of H_2O (*cf.* E^0 for O_2/H_2O = 1.23 V) by $[Co(aq)]^{3+}$ but it occurs very slowly (kinetic reason, *i.e.* over-potential for gas evolution). Thus, $[Co(aq)]^{3+}$ solution is always contaminated by a trace amount of $[Co(aq)]^{2+}$ which is labile and its water exchange rate is very high.

Rapid water exchange: $\left[Co(OH_2)_6\right]^{2+} + 6H_2O^* \rightleftharpoons \left[Co(^*OH_2)_6\right]^{2+} + 6H_2O$

Fairly rapid electron exchange (OSET): $[Co(OH_2)_6]^{3+} + [Co(^*OH_2)_6]^{2+} \rightleftharpoons [Co(^*OH_2)_6]^{3+} + [Co(OH_2)_6]^{2+}$

The fairly rapid electron exchange between $[Co(OH_2)_6]^{3+}$ and $[Co(^*OH_2)_6]^{2+}$ leads to the apparently enhanced overall water exchange rate in $[Co(OH_2)_6]^{3+}$. The electron exchange rate in the $[Co(OH_2)_6]^{3+}/[Co(OH_2)_6]^{2+}$ couple is fairly fast even through the OSET process (*see* Sec. 6.3.6). The electron exchange through the **ISET path cannot be completely ruled out** as H_2O can act also as a poor bridging ligand (*cf.* Sec. 6.4.4, **ISET path**).

$$[Co(NH_3)_5(^*OH_2)]^{3+} + [Cr(aq)]^{2+} \longrightarrow [Co(aq)]^{2+} + 5NH_3 + [Cr(^*OH_2)(OH_2)_5]^{3+} \quad \textbf{(ISET path)}$$

$$[^*Co(OH_2)_6]^{3+} + [Co(OH_2)_6]^{2+} \rightleftharpoons [^*Co(OH_2)_6]^{2+} + [Co(OH_2)_6]^{3+} \quad \textbf{(OSET path)}$$

$$[(H_2O)_5][^*Co^{III}—OH_2]^{3+} + [Co(^*OH_2)_6]^{2+} \rightleftharpoons \left[(H_2O)_5Co^{III}—O(H)(H)—Co^{II}(^*OH_2)_5\right]^{5+} + H_2O^*$$

$$\xrightarrow[\text{ISET path}]{H_2O \;|\; \text{e-transfer}} [Co(OH_2)_6]^{2+} + [Co(OH_2)(^*OH_2)_5]^{3+}$$

Thus the **rapid water exchange** at $[Co(aq)]^{2+}$ (present in a catalytic amount) followed by the favourable electron exchange **(mainly by the OSET path, and partly by the ISET path)** in the couple $[Co(aq)]^{3+}$–$[Co(aq)]^{2+}$ explains the unusually high water exchange rate at $[Co(aq)]^{3+}$ (*cf.* Sec. 5.13).

Because of the critical 10 *Dq* value, $[Co(aq)]^{3+}$ remains in a low spin ⇌ high spin equilibrium and this is the another reason for the overall enhanced water exchange rate (*cf.* high spin $[Co(aq)]^{3+}$ is labile but low spin $[Co(aq)]^{3+}$ is inert). **This aspect has been discussed in Sec. 5.13.** Here it is worth mentioning that the said critical 10 Dq value leads to the said spin state isomerisation (see Vol. 4 for details) that also explains the rapid electron exchange in the couple through the OSET path (see Sec. 6.3.6).

(ii) **Rapid ligand exchange in $[Co(AA)_3]^{3+}$ (AA = en, bpy, phen) catalysed by Co(II):** The mechanism is the same as in the water exchange process of $[Co(aq)]^{3+}$ catalysed by $[Co(aq)]^{2+}$.

$[Co(OH_2)_6]^{2+} + 3{}^*AA \rightleftharpoons [Co(^*AA)_3]^{2+} + 6H_2O$ (fast), $[Co(aq)]^{2+}$ labile)

$[Co(AA)_3]^{3+} + [Co(^*AA)_3]^{2+} \rightleftharpoons [Co(AA)_3]^{2+} + [Co(^*AA)_3]^{3+}$ (electron exchange, OSET path)

(iii) **Catalysis by $[Cr(aq)]^{2+}$ in the water exchange process of $[Cr(aq)]^{3+}$:** The water exchange rate at $[Cr(aq)]^{3+}$ is exceedingly slow but in presence of a trace amount of $[Cr(aq)]^{2+}$, the water exchange process at $[Cr(aq)]^{3+}$ is tremendously accelerated. This route leading to this observed catalysis is the same as in the case of $[Co(aq)]^{3+}$ discussed above.

Note: In the case of $[Co(aq)]^{3+}$, the catalyst $[Co(aq)]^{2+}$ is spontaneously generated from $[Co(aq)]^{3+}$ but for $[Cr(aq)]^{3+}$, the catalyst $[Cr(aq)]^{2+}$ is to be externally added. This difference is reasonable in terms of the E^0-values of the M^{III}/M^{II} couples.

E^0 for $[Co(aq)]^{3+}/[Co(aq)]^{2+} = 1.8$ V; E^0 for $[Cr(aq)]^{3+}/[Cr(aq)]^{2+} = -0.41$ V; E^0 for $O_2/H_2O = 1.23$ V, *i.e.* $[Co(aq)]^{3+}$ is a much stronger oxidizing agent and it can slowly oxidize water and it is itself reduced to $[Co(aq)]^{2+}$.

(iv) **Solubilization of anhydrous $CrCl_3$:** Anhydrous $CrCl_3$ is insoluble in water but it can be solubilized in presence of a catalytic amount of $[Cr(aq)]^{2+}$. This can be explained in terms of the ISET path. $[Cr(aq)]^{2+}$ can be generated from $CrCl_3$ by using some suitable reducing agent.

$$-{}^{*}Cr^{III}-Cl + [Cr(aq)]^{2+} \rightleftharpoons -{}^{*}Cr^{III}-Cl-Cr^{II}(aq)$$

Polymeric network through the $-Cl-Cr^{III}-Cl-$ bridges

e-transfer followed by the bridging Cl^- ligand transfer

$$[{}^{*}Cr(aq)]^{2+} + [Cl-Cr^{III}(aq)]^{2+}$$

The polymeric network of solid $CrCl_3$ involving the $-Cl-Cr^{III}-Cl-$ segments (present in anhydrous $CrCl_3$) is insoluble because of the **inertness of Cr^{III}-centre**. However, the Cr^{III}-centre (shown as *Cr) of the polymer is converted into the labile Cr^{II}-centre through an ISET path in presence of $[Cr(aq)]^{2+}$. *Rapid substitution at this labile centre breaks down the polymer.*

(v) **Catalysis by Pt(II) in the ligand exchange or ligand replacement at Pt(IV):** This has been illustrated in Sec 6.4.15.

(vi) **Catalysis by $[Cr(aq)]^{2+}$ in the ligand substitution process of Cr(III):** This has been illustrated already on the fission of the inert successor complex, Ru^{II}–*X*–Cr^{III} (*cf.* Sec. 6.4.5c).

● The **aquation of NH_3 ligands in $[Cr(NH_3)_5X]^{2+}$** is very slow but the process is potentially catalyzed by $[Cr(aq)]^{2+}$.

$$\left[Cr^{III}(NH_3)_5X\right]^{2+} + 5H_3O^+ \longrightarrow \left[Cr(OH_2)_5X\right]^{2+} + 5NH_4^+;\ (X = Cl,\ Br,\ I)$$

The process is catalyzed by $[Cr(aq)]^{2+}$ as shown below.

$$\left[(H_3N)_5{}^{*}Cr^{III}-X\right]^{2+} + \left[Cr(aq)\right]^{2+} \rightleftharpoons \left[(H_3N)_5{}^{*}Cr^{III}-X-Cr^{II}(OH_2)_5\right]^{4+}$$

e-transfer (ISET path)

$$5NH_4^+ + \left[{}^{*}Cr(aq)\right]^{2+} + \left[Cr(OH_2)_5X\right]^{2+} \xleftarrow[(H_3O^+)]{\text{fast}} \left[(H_3N)_5{}^{*}Cr^{II}-X-Cr^{III}(OH_2)_5\right]^{4+}$$

● $[Cr(aq)]^{2+}$ can catalyze the **ligation or substitution at $[Cr(aq)]^{3+}$** (inert centre) as follows:

$$\left[Cr(aq)\right]^{2+} + L \rightleftharpoons \left[Cr^{II}(aq)(L)\right],\ \text{(rapid)}$$

Electron exchange:

$$\left[Cr^{II}(aq)(L)\right] + \left[{}^{*}Cr^{III}(aq)\right] \longrightarrow \left[Cr^{III}(aq)(L)\right] + \left[{}^{*}Cr^{II}(aq)\right];\ (\textbf{OSET path, Cr}^{\textbf{III}}\textbf{ - inert centre})$$

The rate is given by: rate = $k[Cr^{III}][Cr^{II}][L]$

Note: The electron transfer reaction:

$$\left[Cr^{II}(aq)L\right] + \left[Cr^{III}(aq)\right] \longrightarrow \left[Cr^{III}(aq)L\right] + \left[Cr^{II}(aq)\right],$$

can occur through the **OSET path or through the ISET path by using the weak bridging property of H_2O**. The process experiences a net $\sigma^* \to \sigma^*$ electron transfer, a favourable situation for the ISET path. In this ISET path, H_2O bound to Cr(III) makes a bridge with the labile Cr(II) centre.

• Cr(II) also catalyzes the reverse reaction, *i.e.* **aquation of L in $[Cr^{III}(L)(aq)]$ where L is a nonbridging ligand through the OSET path.**

Electron exchange:

$$[(aq)Cr^{III}\text{—}L]+[Cr^{II}(aq)] \longrightarrow [(aq)Cr^{II}\text{—}L]+[Cr^{III}(aq)], \textbf{(OSET path, L nonbridging ligand)}$$

$$\downarrow \text{fast}$$

$$[Cr^{II}(aq)] + L$$

The rate is given by: $\text{rate} = k[Cr^{III}(L)][Cr^{II}]$

Here 'L' is supposed to have ***no bridging property***. **If L can bridge**, then the above reaction will go through the ISET path and L will be present with the Cr^{III}-centre (*i.e.* **no effective aquation of L**). This is illustrated below:

$$[(aq)Cr^{III}\text{—}L]+[Cr^{II}(aq)] \rightleftharpoons [(aq)Cr^{III}\text{—}L\text{—}Cr^{II}(aq)]$$

$$\downarrow \textit{e}\text{-transfer and fission} \Big\} \textbf{ISET path}$$

$$[Cr(aq)]^{II}+[L\text{—}Cr^{III}(aq)]$$

(vi) **$[Co(CN)_5]^{3-}$ or Co(II) catalysed substitution reactions of $[Co^{III}(NH_3)_5X]$ by CN^-** (***Ref.*** *JACS*, **85**, 2517, 1963): There are two types of products depending on X.

$$[Co^{III}(NH_3)_5X] + 5CN^- \longrightarrow [Co^{III}(CN)_5X] + 5NH_3$$

$$\text{or } [Co^{III}(NH_3)_5X] + 6CN^- \longrightarrow [Co(CN)_6]^{3-} + 5NH_3 + X$$

These are shown below.

$$[Co^{III}(NH_3)_5X] + CN^-(\text{excess}) \xrightarrow[\text{(as a catalyst)}]{[Co(CN)_5]^{3-}}$$

or

$$[Co^{III}(NH_3)_5X] + CN^-(\text{excess}) \xrightarrow[\text{(as a catalyst)}]{Co(II)}$$

→ $[Co^{III}(CN)_5X]$; when X (having good bridging property) = Cl^-, Br^-, I^-, N_3^-, OH^-, $S_2O_3^{2-}$, etc.

→ $[Co^{III}(CN)_6]^{3-}$; when X (no bridging or poor bridging property) = NH_3, CO_3^{2-}, PO_4^{3-}, SO_4^{2-}, RCO_2^-, etc.

→ $[Co^{III}(CN)_5X]$ + $[Co(CN)_6]^{3-}$; when X (intermediate in bridging property) = F^-, NO_3^-, etc.

In the above reaction, $[Co(CN)_5]^{3-}$ or $[Co(CN)_6]^{4-}$ is the **effective catalyst**. They remain in an equilibrium mixture in presence of excess CN^-. It is illustrated for Co(II) added as the catalyst. Here it may be mentioned that a trace amount of Co(II) (**may be as an impurity**) present in the system may also act as the catalyst. In presence of CN^-, Co(II) gives an equilibrium mixture of $[Co(CN)_5]^{3-}$ and $[Co(CN)_6]^{4-}$, *i.e.*:

$$Co(II) + 5CN^- \rightleftharpoons [Co(CN)_5]^{3-}; (\text{fast and high } K_{eq})$$

$$[Co(CN)_5]^{3-} + CN^- \rightleftharpoons [Co(CN)_6]^{4-} (\text{very low } K_{eq})$$

$[Co(CN)_5]^{3-}$ as a reducing agent can experience the ISET path while $[Co(CN)_6]^{4-}$ as a reducing agent can experience only the OSET path. Participation of the catalyst $[Co(CN)_5]^{3-}$ or $[Co(CN)_6]^{4-}$ as the reducing agent depends on the bridging property of X in $[Co^{III}(NH_3)_5X]$.

In the reduction of $[Co^{III}(NH_3)_5X]$ by $[Co(CN)_5]^{3-}$ **in presence of free CN^-**, formation of an equilibrium mixture of $[Co(CN)_6]^{4-}$ and $[Co(CN)_5]^{3-}$ leads to both the OSET path involving $[Co(CN)_6]^{4-}$ (**X = poor bridging ligand**) and ISET path involving $[Co(CN)_5]^{3-}$ (X = good **bridging ligand**) depending on the bridging property of X (*Ref. JACS*, **85**, 2517, 1963).

- **ISET path:** $\left[(H_3N)_5Co^{III}—X\right] + \left[Co(CN)_5\right]^{3-} \rightleftharpoons \left[(H_3N)_5Co^{III}—\mathbf{X}—Co^{II}(CN)_5\right]$

$$\downarrow e\text{-transfer}$$

$$5NH_3 + \left[Co(aq)\right]^{2+} + \left[X—Co^{III}(CN)_5\right] \xleftarrow{\text{fast}} \left[(H_3N)_5Co^{II}—X—Co^{III}(CN)_5\right]$$

This ISET path is favoured, if X is a good bridging ligand. In fact, in this route $[Co^{III}(CN)_5(S_2O_3)]^{4-}$ was synthesized by **Prof. Ray (1927)**. For this process, the rate is given by:

$$\text{rate} = k_{IS}[(H_3N)_5Co^{III}X][Co(CN)_5^{3-}]$$

- **OSET path:** $\left[(H_3N)_5Co^{III}—X\right] + \left[Co^{II}(CN)_6\right]^{4-} \rightleftharpoons \underbrace{\left[(H_3N)_5Co^{III}—X\right]\cdot\left[Co^{II}(CN)_6\right]^{4-}}_{\text{O.S. precursor complex}}$

$$\downarrow e\text{-transfer}$$

$$\left[Co^{III}(CN)_6\right]^{3-} + 5NH_3 + X + \left[Co(aq)\right]^{2+} \xleftarrow{\text{(fast)}} \underbrace{\left[(H_3N)_5Co^{II}—X\right]\cdot\left[Co^{III}(CN)_6\right]^{3-}}_{\text{O.S. successor complex}}$$

The OSET path is favoured when X is having no bridging or very poor bridging property. For this process, the rate is given by:

$$\text{rate} = k_{OS}[(H_3N)_5Co^{III}X][Co(CN)_6^{4-}]$$

When the bridging property of X is of an intermediate nature, both the OSET and ISET paths can operate.

Note: Co(II) in $[Co(CN)_5]^{3-}$ is **softened** by the 5 CN^- ligands (**symbiotic effect**). This is why, in the ISET path, the coordinatively unsaturated reductant $[Co(CN)_5]^{3-}$ **does not prefer the hard donor site** of the bridging ligand (X). **In such cases, the mechanism switches over to the OSET path involving the reductant $[Co(CN)_6]^{4-}$.**

6.7 NON-COMPLEMENTARY ELECTRON TRANSFER REACTIONS

The reactions (*e.g.* Fe^{III}–Fe^{II}, Co^{III}–Fe^{II}, Co^{III}–Cr^{II}, Co^{III}–V^{II}, Ru^{III}–Cr^{II}, Pt^{IV}–Pt^{II}, etc.) already discussed involve the oxidants and reductants which undergo the same formal oxidation state change. These are called the **complementary electron transfer reactions**. These involve the 1*e*-oxidant–1*e*-reductant, 2*e*-oxidant–2*e*-reductant, etc. and the stoichiometries are 1:1. But there are reactions in which the changes of oxidation states in the oxidant and reductant are not the same. These involve the 1*e*-oxidant–2*e*-reductant, 2*e*-oxidant–1*e*-reductant, 3*e*-oxidant–1*e*-reductant, etc. Such reactions are referred to as the **noncomplementary electron transfer reactions.** Consequently, in these reactions, the stoichiometries are not 1:1. Some examples are:

$$\begin{aligned} 2Fe(III) + Sn(II) &\longrightarrow 2Fe(II) + Sn(IV) \\ Tl(III) + 2Fe(II) &\longrightarrow Tl(I) + 2Fe(III) \\ Pt(IV) + 2Cr(II) &\longrightarrow Pt(II) + 2Cr(III) \\ Cr(VI) + 3Fe(II) &\longrightarrow Cr(III) + 3Fe(III) \\ Mn(VII) + 5Fe(II) &\longrightarrow Mn(II) + 5Fe(III) \end{aligned}$$

In such cases, the electron transfer *generally occurs **in consecutive 1e-transfer steps***. In fact, 2*e* or multi-electron transfer cannot occur in a single step. Stepwise 1e-transfer steps are illustrated below.

$$\mathrm{Tl(III) + Fe(II)} \underset{k_{-1}}{\overset{k_1}{\rightleftharpoons}} \mathrm{Tl(II) + Fe(III)}$$

$$\mathrm{Tl(II) + Fe(II)} \xrightarrow{k_2} \mathrm{Tl(I) + Fe(III)}$$

By considering the steady state condition to Tl(II), we get the following rate law:

$$\frac{d[\mathrm{Tl(II)}]}{dt} = 0 = k_1[\mathrm{Tl(III)}][\mathrm{Fe(II)}] - k_{-1}[\mathrm{Tl(II)}][\mathrm{Fe(III)}] - k_2[\mathrm{Tl(II)}][\mathrm{Fe(II)}]$$

i.e.

$$[\mathrm{Tl(II)}] = \frac{k_1[\mathrm{Tl(III)}][\mathrm{Fe(II)}]}{k_{-1}[\mathrm{Fe(III)}] + k_2[\mathrm{Fe(II)}]}$$

$$\text{Rate} = -\frac{d[\mathrm{Tl(III)}]}{dt} = k_2[\mathrm{Tl(II)}][\mathrm{Fe(II)}] = \frac{k_1 k_2[\mathrm{Tl(III)}][\mathrm{Fe(II)}]^2}{k_{-1}[\mathrm{Fe(III)}] + k_2[\mathrm{Fe(II)}]}$$

Reduction of Cr(VI) (3*e*-oxidant) by the 1e-reductant Fe(II) occurs as follows:

$$\mathrm{Cr(VI) + Fe(II)} \underset{k_{-1}}{\overset{k_1}{\rightleftharpoons}} \mathrm{Cr(V) + Fe(III)}$$

$$\mathrm{Cr(V) + Fe(II)} \xrightarrow{k_2} \mathrm{Cr(IV) + Fe(III)}$$

$$\mathrm{Cr(IV) + Fe(II)} \xrightarrow{\text{fast}} \mathrm{Cr(III) + Fe(III)}$$

$$-\frac{d[\mathrm{Cr(VI)}]}{dt} = \frac{k_1 k_2[\mathrm{Cr(VI)}][\mathrm{Fe(II)}]^2}{k_{-1}[\mathrm{Fe(III)}] + k_2[\mathrm{Fe(II)}]};$$ applying the steady condition to Cr(V).

The mechanisms (*i.e.* OSET, ISET) of reduction of Cr(VI) by the different types of Fe(II)-compounds like $[Fe(aq)]^{2+}$ (labile), and substitutionally inert complexes like $[Fe(CN)_6]^{4-}$, $[Fe(bpy)_3]^{2+}$, $[Fe(phen)_3]^{2+}$, etc. have been discussed in Sec. 6.4.13.

6.8 REDUCING ACTIVITY OF THE SOLVATED ELECTRON

Hydrated electrons can be generated in many ways:

(a) **Radiolysis of water** by the ionizing radiations like X-rays and γ-rays.

$$H_2O \xrightarrow{h\nu} H^+_{aq} + OH + e^-_{aq}$$

(b) **Irradiation by UV light** on the solution containing the reducing agents (Red) like I^-, $[Fe(CN)_6]^{4-}$, M^{2+} (bivalent metal ions of the 1st transition series), etc.

$$\text{Red} \xrightarrow{h\nu} \text{Red}^+ + e^-_{aq}$$

The hydrated electron has the following **properties**:

(i) absorbtion in the red region (λ_{max} = 720 nm; ε_M = 15,800 mol^{-1} cm^{-1}); (ii) high reducing power indicated by the estimated reduction potential –2.7 V; (iii) half life < 10^{-3} s; (iv) reducing activity – almost diffusion controlled rate.

Because of the very short life-time, its characterization is a relatively difficult task. The following reactions are responsible for the rapid destruction of the hydrated electron.

$2H_2O + 2e^-_{aq} \longrightarrow H_2 + 2OH^-_{aq}$ ($k \approx 10^{10}$ dm^3 mol^{-1} s^{-1} at 25°C, pH~11)

$H_3O^+_{aq} + e^-_{aq} \longrightarrow H + H_2O$ ($k \approx 10^{10}$ dm^3 mol^{-1} s^{-1} at 25°C, pH = 2 – 4)

Thus in the acidic condition, both the reducing species H and e^-_{aq} coexist.

$H_2O + e^-_{aq} \longrightarrow H + OH^-$ ($k \approx 16$ dm^3 mol^{-1} s^{-1} at 25°C, pH = 8.4)

H_2 can produce H as follows:

$H_2 + OH \longrightarrow H_2O + H$

The reactivities of H and e^-_{aq} are different. N_2O reacts with both e^-_{aq} and H to yield N_2 but their rates differ by a factor 10^4.

$$N_2O + e^-_{aq} \longrightarrow N_2 + O^-, \; O^- + H_2O \longrightarrow OH + OH^-$$

Hydrated electrons generally react very fast close to the diffusion controlled rate. Because of the very powerful reducing activity (both thermodynamically and kinetically) of the hydrated electrons, these are used to generate the unstable lower valent cations. Thus the generated lower valent cations can be studied.

EXERCISE 6 (PART I)

A. General Type Questions

1. Illustrate the electron exchange and redox reactions.
2. Compare the mechanistic steps of OSET and ISET process.
3. What do you mean by Franck-Condon barrier in the electron transfer process? How does it contribute to the activation energy of the electron transfer process. Illustrate with examples.
4. Discuss the factors favouring the OSET process. Give the rate law.
5. What are the types of receptor and donor orbitals in the OSET process?
6. Discuss the Marcus equation in relation to Cross-reactions passing through the OSET process.
   - Discuss the relationship between the activation energy ($\Delta G^{\neq}$) and free energy change (ΔG^o) of a reaction in the **normal and inverted Marcus regions**.
7. What do you mean by ISET process? Discuss the steps of ISET process. Deduce the rate law.
8. Classify the ISET reaction depending on the position of rate determining step.
9. Discuss the factors favouring the ISET process.
10. Discuss the possibility of characterization of binuclear complexes involved in the ISET process.
11. Discuss the ISET reaction experiencing the concomitant ligand transfer.
12. Give the examples where the ISET reaction experiences no ligand transfer.
13. What do you mean by the substitution controlled ISET reaction?
14. Discuss the role of a bridging ligand in the ISET process.
15. How does the bridging ligand help in the tunneling of electron from the reductant to the oxidant?
16. How can you distinguish between the chemical and tunneling mechanism of electron transfer in the ISET process?
17. What are the conditions of electron transfer through the resonance mechanism in the ISET process?
18. Discuss the effect of nonbridging ligands in the ISET process.
19. Discuss the characteristic features of reduction of $[Co^{III}(NH_3)_5X]$ by $[V(aq)]^{2+}$, $[Eu(aq)]^{2+}$ and $[Co(CN)_5]^{3-}$.
20. Discuss the mechanisms of reductioon of $HCrO_4^-$ by $[Fe(aq)]^{2+}$, $[Fe(CN)_6]^{4-}$, $[Fe(bpy)(CN)_4]^{2-}$ and $[Fe(bpy)_3]^{2+}$.
21. Compare the rates of reduction of $[M(NH_3)_5(SCN)]^{2+}$ and $[M(NH_3)_5(NCS)]^{2+}$ by $[Cr(aq)]^{2+}$, $[V(aq)]^{2+}$ and $[Co(CN)_5]^{3-}$ (M = Co).
22. Discuss the mechanistic criteria of the OSET and ISET processes.
23. How can you distinguish between the OSET and ISET processes?
24. Discuss the ligand replacement reactions through the electron transfer process.
25. Illustrate the complementary and noncomplementary electron transfer reactions.
26. Discuss the properties of solvated electron.

B. Explain the following statements

1. Electron exchange and redox reactions are basically of the same type.
2. The rate law of an OSET reaction is : rate = kK_{os}[Ox][Red].
3. In the electron exchange process, energies of the participating donor and receptor orbitals should be the same.
4. Potential energy diagram of the electron transfer process can give us the measure of the activation energy.
5. Marcus equation is a representation of linear free energy relationship (LFER).
   - Marcus equation can theoretically calculate the electron exchange rate constant.
   - For the electron transfer cross reaction, the rate may decrease if the reaction is thermodynamically highly favoured (*i.e.* $\Delta G°$ too negative).
6. In the electron exchange process, Franck-Condon barrier gives the measure of activation energy.
7. In the electron carrier metalloproteins like ferredoxins, blue proteins, nature has placed the metal centres in the distorted geometries rather than in the regular geometries.
8. Electronic activation may be an important factor to determine the activation energy of an electron transfer process.
9. In the OSET process, $\pi^* \rightarrow \pi^*$ (*i.e.* $t_{2g} \rightarrow t_{2g}$) electron transfer is the most favourable pathway for the octahedral complexes.
10. The electron exchange rate in the $[Fe(aq)]^{3+}$—$[Fe(aq)]^{2+}$ couple is very slow though it involves the $\pi^* \rightarrow \pi^*$ electron transfer, a favourable situation for an OSET path.
11. The electron exchange rates in the couples $[Fe(CN)_6]^{3-}$—$[Fe(CN)_6]^{4-}$, $[Fe(phen)_3]^{3+}$—$[Fe(phen)_3]^{2+}$ are fairly high.
12. The electron exchange rate in the couple $[Co(aq)]^{3+}$—$[Co(aq)]^{2+}$ is fairly fast though the electron exchange rate in the couple $[Co(NH_3)_6]^{3+}$—$[Co(NH_3)_6]^{2+}$ is very slow.
13. Electron exchange rate in $[Ru(NH_3)_6]^{3+}$— $[Ru(NH_3)_6]^{2+}$ is much higher than that in $[Co(NH_3)_6]^{3+}$— $[Co(NH_3)_6]^{2+}$.
14. The electron exchange rate in $[Mn(phen)_3]^{3+}$—$[Mn(phen)_3]^{2+}$ is fairly high.
15. The electron exchange rate in the Co(III)—Co(II) couple depends on the ligand field strength of the involved ligand. *Crystal field strength*: H_2O ⟨ oxalate ⟨ edta ⟨ NH_3 ~ en ⟨ phen. The electron exchange rate is fast for both the weak (*e.g.* H_2O) and strong (*e.g.* phen) field ligands.
16. The electron exchange rates in the couples, $[Co(NH_3)_6]^{3+}$–$[Co(NH_3)_6]^{2+}$ and $[Co(phen)_3]^{3+}$ – $[Co(phen)_3]^{2+}$ differ widely.
17. The electron exchange rate in the $[Cr(aq)]^{3+}$–$[Cr(aq)]^{2+}$ couple is much slower than that in the $[V(aq)]^{3+}$–$[V(aq)]^{2+}$ couple.
18. In the OSET process, $[V(aq)]^{2+}$ is a better reducing agent than $[Cr(aq)]^{2+}$ kinetically but the reverse is true thermodynamically.
19. The rate of reduction of $[Ru(NH_3)_6]^{3+}$ by $[V(aq)]^{2+}$ is much faster than the rate of reduction of $[Co(NH_3)_6]^{3+}$ by the same reducing agent though the thermodynamic favour in both the reactions is more or less the same.
20. $[Ru(aq)]^{2+}$ and $[Ru(NH_3)_6]^{2+}$ are the good reducing agents in the OSET path.
21. The Ru(III)–Ru(II) reaction is faster than the Co(III)–Ru(II) reaction.
22. The electron exchange reactions in the couples like $[Fe(CN)_6]^{3-}$– $[Fe(CN)_6]^{4-}$, $[MnO_4]^-$– $[MnO_4]^{2-}$ are catalysed by the alkali metal ions. The catalyzing efficiency of the alkali metal ions runs as:
   $$Cs^+ > Rb^+ > K^+, NH_4^+ > Na^+ > Li^+$$
23. The electron exchange process in $[Co(NH_3)_6]^{3+}$–$[Co(NH_3)_6]^{2+}$ can be catalyzed by the anions.
24. The electron exchange rate in $[Co(NH_3)_6]^{3+}$– $[Co(NH_3)_6]^{2+}$ is remarkably increased in an alkaline solution.

25. Compared to $[Co(NH_3)_6]^{3+}$, the reduction of $[CoCl(NH_3)_5]^{2+}$ by $[Cr(aq)]^{2+}$ is much faster.
26. $[CoCl(NH_3)_5]^{2+} + [Cr(aq)]^{2+} \longrightarrow [Co(aq)]^{2+} + 5NH_3 + [Cr(Cl)(OH_2)_5]^{2+}$
The experimental findings of the above reaction cannot be explained by an OSET path but by an ISET path.
27. Depending on the position of rate determining step, the ISET reaction can be classified into three groups.
28. Rate law for the ISET reactions depends on the position of rate determining step.
29. In an ISET path, $\sigma^* \rightarrow \sigma^*$ (*i.e.* $e_g \rightarrow e_g$) electron transfer is the most favourable path for the M—X type oxidant (X = monoatomic bridging ligand).
30. There is no example of ISET reaction experiencing the net $\pi^* \rightarrow \pi^*$ electron transfer.
31. Reduction of $[Co(NH_3)_5(OH_2)]^{3+}$ by $[Cr(aq)]^{2+}$ occurs through the ISET path not through the OSET path. The rate of the reaction depends on pH.
    - Reduction of $[Co(NH_3)_5X]^{2+}$ (X = F, Cl, Br, I) by $[Cr(bpy)_3]^{2+}$ and $[Ru(NH_3)_6]^{2+}$ shows the reactivity order:
    X = I > Br > Cl > F **(halide effect)**
    - Reduction of Fe(III) complexes by $V(aq)^{2+}$ occurs through the ISET path.
32. $\left[Co(NH_3)_5X\right]^{2+} + \left[Cr(aq)\right]^{2+} \longrightarrow$ Product
    - Rate for $X = NH_3 < OH_2 < {}^-OH$; $X = N_3^- \approx SCN^- \gg NCS^-$
    - For X = —*N*CS, only one linkage isomer of Cr(III) is obtained; for X = –*S*CN, a mixture of linkage isomers of Cr(III) is obtained.
    - $[Co(NH_3)_5X]^{2+} + [Co(CN)_5]^{3-} \longrightarrow$ Products. For X = N_3^- and NCS^-, the rates do not differ remarkably. For X = NCS^- and SCN^-, the reactivities are different.
    - The kinetically controlled products in the reduction of $[Co(NH_3)_5(SCN)]^{2+}$ and $[Co(NH_3)_5(NO_2)]^{2+}$ by $[Co(CN)_5]^{3-}$ are not the thermodynamically controlled products.
33. $\left[Co(NH_3)_5X\right]^{2+} + \left[M(aq)\right]^{2+} \longrightarrow$ Product; M = V, Cr, Fe
    - For reduction by $[V(aq)^{2+}]$, the rate is more or less insensitive to the nature of X while for reduction by $[Cr(aq)]^{2+}$ or $[Fe(aq)]^{2+}$, the rate is sensitive to the nature of X (bridging ligand).
    - The reactivity order is: $[Cr(aq)]^{2+} \rangle [V(aq)]^{2+} \rangle [Fe(aq)]^{2+}$.
    - For M = V, Fe and Eu, the bridging ligand X^- sometimes gets released in the solution.
34. $\left[M(NH_3)_5X\right]^{2+} + \left[V(aq)\right]^{2+} \longrightarrow$ Products, M = Co, Ru.
    - If the rate is much faster than the ligand substitution rate of $[V(aq)]^{2+}$, the reaction may be argued to pass through the OSET path.
    - If the rate is comparable to the rate of nucleophilic substitution rate of $[V(aq)]^{2+}$, then the reaction may be argued to pass through the ISET path.
    - For M = Co(III), it is an ISET reaction while for M = Ru(III), it is an OSET reaction.
35. $\left[Co(NH_3)_5X\right]^{2+} + \left[Eu(aq)\right]^{2+} \longrightarrow$ Products
For X = Cl^-, N_3^- and NO_3^-, the rate is more or less the same and X^- is released to the solution.
36. $[RuCl_6]^{3-}$– $[Cr(aq)]^{2+}$; $[V(aq)]^{4+}$ + $[Cr(aq)]^{2+}$; $[IrCl_6]^{2-}$–$[Cr(aq)]^{2+}$.
The above reactions (ISET path) can generate the stable dinuclear complexes that can be decomposed easily in presence of $[Cr(aq)]^{2+}$ and ${}^-OH$.
    - If reduction of $[Co(NH_3)_5Cl]^{2+}$ by $Cr(aq)^{2+}$ is carried out in presence of $P_2O_7^{4-}$, both Cl^- and $P_2O_7^{4-}$ are found in the final Cr(III) product.

- Reduction of $[Co(NH_3)_5X]^{2+}$ (X = F, Cl, Br, I) by $V(aq)^{2+}$ occurs with the reactivity order (**halide effect, normal order**) X = I > Br > Cl > F. This reactivity order can be explained by considering both the OSET and ISET paths.
- For the reduction of $[Co(NH_3)_5X]^{2+}$ (X = F, Cl, Br) by $Eu(aq)^{2+}$ and $Fe(aq)^{2+}$, the reactivity order is:
  X = F > Cl > Br (**halide effect, inverse order**)
- Reduction of $[Co(NH_3)_5(X)]^{2+}$ (X = SCN^-, NCS^-) by $[Fe(aq)]^{2+}$ passes through the ISET path. The intermediate $Fe(NCS)^{2+}$ can be detected for the reduction of $[Co(NH_3)_5(SCN)]^{2+}$ but the intermediate $Fe(SCN)^{2+}$ cannot be detected for the reduction of $[Co(NH_3)_5(NCS)]^{2+}$.
- The reactivity order of reduction of $[Co(NH_3)_5Cl]^{2+}$ by different reducing agents follows the sequence:
  $[Co(CN)_5]^{3-} > [Cr(aq)]^{2+} > [Eu(aq)]^{2+} > [Fe(aq)]^{2+}$
- For reduction of $[Co(edta)Cl]^{2-}$ and $[Co(edta)(OH_2)]^-$ by $Ti(aq)^{3+}$ and $Cr(aq)^{2+}$ shows the following rate ratio:
  $k_{Cl}/k_{aq} = 31$ (for $Ti(aq)^{3+}$) and 2×10^3 (for $Cr(aq)^{2+}$)
- The intervalence electron transfer rates for the following binuclear complex depends on the nature of the bridging ligand (X).
  $[(bpy)_2ClRu—X—RuCl(bpy)_2]^{5+}$
  k_{el} (s^{-1}) = 3×10^9 (X = pyz), 1×10^8 (X = dppe), 1×10^8 (X = 4,4′-bpy)

37. $[IrCl_6]^{2-} + [Co(CN)_5]^{3-}$; $[Fe(CN_6]^{3-} + [Co(CN)_5]^{3-}$; $HCrO_4^- + [Fe(CN_6]^{4-}$
    The above reactions can produce the stable binuclear complexes.
38. There are many ISET reactions where there is no ligand transfer.
39. $[Co(NH_3)_5(ntaH_2)]^{2+}$ is reduced much faster than $[CoCl(NH_3)_5]^{2+}$ by $[Fe(aq)]^{2+}$.
40. There are some favourable situations for characterizing the binuclear precursor complexes in the ISET path.
    - Identification of a binuclear successor complex is a strong evidence in favour of the **ISET path**.
    - Detection of a binuclear precursor complex cannot always ensure the ISET path.
    - Detection of a binuclear complex (**short-lived**) is possible for both the OSET and ISET paths.
41. Electron transfer leading to the simultaneous ligand transfer is a convincing evidence in favour of an ISET path.
42. Ligand transfer is not an essential condition for an ISET path.
    - Sometimes, the bridging ligand of the ISET path is released to the solution.
    - $[Ru(NH_3)_5X]^{2+} + [Cr(aq)]^{2+} \longrightarrow$ Product. The ISET path for this reaction may lead to ligand (X) transfer or not depending on the nature of X.
43. Product analysis is helpful to predict the nature of the precursor complex in an ISET path.
44. Reduction of $[CoCl(NH_3)_5]^{2+}$, and $[Co(NCS)(NH_3)_5]^{2+}$ by $[Cr(aq)]^{2+}$ can occur through the adjacent attack and remote attack on the bridging ligand by the reducing agent respectively.
45. Reduction of $[Co(NH_3)_5(SCN)]^{2+}$ by $[Cr(aq)]^{2+}$ or $[Co(CN)_5]^{3-}$ can occur through the both remote and adjacent attacks.
46. Reduction of $\left[(H_3N)_5Co—N\langle\bigcirc\rangle—C(=O)NH_2\right]^{3+}$ by $[Cr(aq)]^{2+}$ experiences the remote attack on the carbonyl–O not on the NH_2 group.

47. $[(H_3N)_5Co^{III}—OCOR]^{2+}$ can experience the remote attack but $\left[(H_3N)_5Co^{III}—O{=}C\begin{smallmatrix}NH_2\\H\end{smallmatrix}\right]^{3+}$ cannot experience a remote attack in reduction through the ISET path.
48. Reduction of $[(H_3N)_5CoCN]^{2+}$ by $[Co(CN)_5]^{3-}$ in presence of excess CN^- gives a mixture of products $[Co(CN)_5(NC)]^{3-}$ and $[Co(CN)_6]^{3-}$ in the final solution. On standing, it gives only $[Co(CN)_6]^{3-}$.
49. $\left[Co(NH_3)_5(NO_2)\right]^{2+} + \left[Co(CN)_5\right]^{3-} \longrightarrow \left[Co(CN)_5(NO_2)\right]^{3-} + \left[Co(aq)\right]^{2+} + 5NH_3$

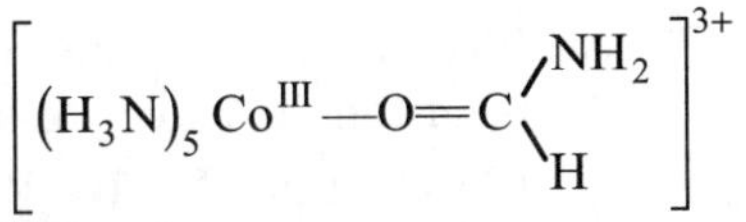

    The product can be explained in terms of remote attack not in terms of adjacent attack but apparently it indicates an adjacent attack.
    - $[Co(NH_3)_5(SCN)]^{2+} + [Co(CN)_5]^{3-} \longrightarrow [Co(aq)]^{2+} + 5NH_3 + [Co(CN)_5(SCN)]^{3-}$.
50. $\left[Co(NH_3)_5(OCOR)\right]^{2+} + \left[Cr(aq)\right]^{2+} \longrightarrow$

    The simple product analysis cannot distinguish between the remote and adjacent attacks.
51. The adjacent attack in ISET generally leads to electron transfer by a **tunneling mechanism** but the remote attack for the nonreducible bridging ligands leads to electron transfer through a **resonance transfer mechanism.**
52. The electron exchange rate of the $[Fe(aq)]^{3+}$–$[Fe(aq)]^{2+}$ couple increases with the increase of pH.
53. Reduction of $[CoCl(NH_3)_5]^{2+}$ by $[Cr(aq)]^{2+}$ in presence of N_3^- gives the product $[CrCl(N_3)(OH_2)_4]^+$.
54. ISET reduction of $[Co(NH_3)_5X]^{2+}$ by $[M(aq)]^{2+}$: (M = V, Cr, Eu, Fe)

    For $X = N_3^-$ and NCS^-: $k_{N_3}/k_{NCS} \gg 1 \Rightarrow$ ISET process.

    $k_{N_3}/k_{NCS} \approx 1 \Rightarrow$ OSET and/or ISET path.
    - ISET reduction of $[Co(NH_3)_5X]^{2+}$ by $[Co(CN)_5]^{3-}$ does not satisfy the condition $k_{N_3}/k_{NCS} \gg 1$ and $k_{SCN} \gg k_{NCS}$.
55. $\left[Co(NH_3)_5X\right]^{2+} + \left[M(aq)\right]^{2+} \longrightarrow$ Product, (M = Cr, Fe, V, etc.)
    - X = —*NCS*$^-$ or —*SCN*$^-$: $k_{N_3} \gg k_{NCS} \Rightarrow$ ISET process.

      $k_{N_3} \approx k_{NCS} \Rightarrow$ OSET and/or ISET process.
    - The rate constant varies as (M = Cr):

X =	$—OCOCH_3$ (acetate)	$—OCOCH_2OH$ (glycolate)	—OCOH (formate)	—OCOCHO (glyoxylate)	$—OCOCO_2H$ (acid oxalate)	$—OCOCO_2^-$ (oxalate)
k ($dm^3\ mol^{-1}\ s^{-1}$):	0.35	3.1	7.2	10^4	1×10^2	4.5×10^4

56. Reduction of $[Ru(NH_3)_5X]^{2+}$ by $[Ru(aq)]^{2+}$ occurs through the OSET path not by the ISET path.
57. $\left[(H_3N)_5M^{III}—X\right] + \left[Cr(aq)\right]^{2+} \longrightarrow$ Product, M = Co, Ru. Different mechanisms operate for:

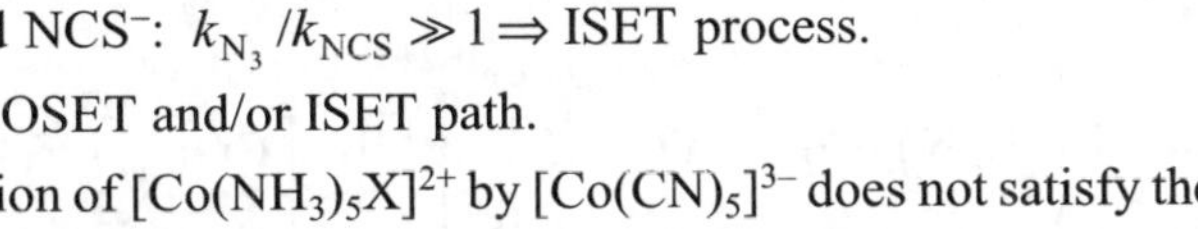

    X = N(pyridine), N(pyridine)—$CONH_2$, N(pyridine with $CONH_2$ at 3-position).
58. The following binuclear complexes experience the OSET path.

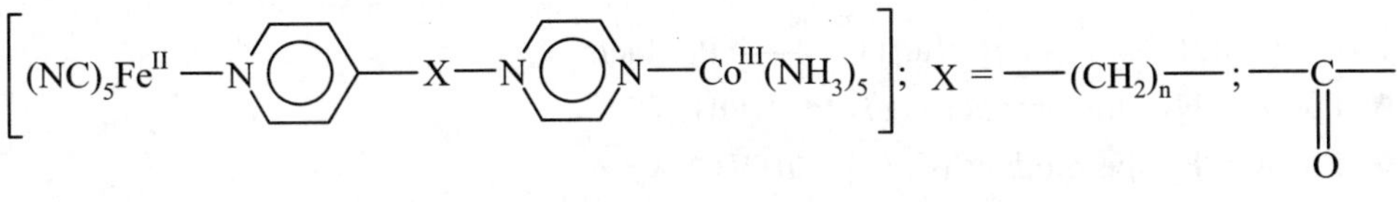

$$\left[(NC)_5Fe^{II}—N\bigcirc—X—\bigcirc N—Co^{III}(NH_3)_5\right];\ X = —(CH_2)_n—\ ;\ —\underset{\underset{O}{\|}}{C}—$$

- Electron transfer rate in the following mixed valence (Ru^{III}—Ru^{II}) bimetallic complex depends on the nature of bridging ligand (L—L).
  $[(bpy)_2ClRu—L—L—RuCl(bpy)_2]^{3+}$, $[(H_3N)_5Ru—L—L—Ru(NH_3)_5]^{5+}$

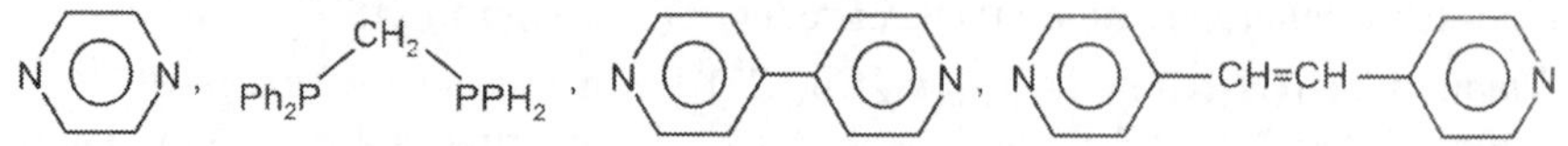

  N(pyrazine)N, $Ph_2P—CH_2—PPH_2$, N(4,4'-bipyridine)N, N(py)—CH=CH—(py)N

59. The chemical mechanism of electron transfer in the ISET path is a multistep process.
60. Sometimes, it may be difficult to distinguish between the resonance mechanism and chemical mechanism for electron transfer in the ISET process.
61. $\left[(H_3N)_5 Co^{III}—X\right]+\left[Cr(aq)\right]^{2+} \longrightarrow$ Product
    - The rate constants varies as:

X =	succinate	glycolate	malonate	pyridine-2-carboxylate	salicylate
k ($dm^3 mol^{-1} s^{-1}$) =	1.0	3.0	2.5×10^3	2×10^5	2×10^8

    - The rate of reduction of $[Co(NH_3)_5(C_2O_4)]^+$ by $[Cr(aq)]^{2+}$ and $[Fe(aq)]^{2+}$ is much higher than the reduction of $[Co(NH_3)_5(HC_2O_4)]^{2+}$.
    - The rates of reduction of $[Co(NH_3)_5(C_2O_4)]^+$ and $[Co(NH_3)_5(HC_2O_4)]^{2+}$ by $[V(aq)]^{2+}$ are not widely different.
    - For X = –O–COCHO (glyoxylate) the rate is faster than that for X = $–OCOCH_3$.
    - For X = benzoate, acetate and propionate, the rate is relatively slower than that for X = formate, oxalate, maleate, fumarate. ($HO_2CCH{=}CHCO_2H$, *cis*-form = fumaric acid; *trans*-form = maleic acid)
    - For the reduction of $\left[(H_3N)_5 Co^{III}—O—C(=O)R\right]^{2+}$ by $\left[Cr(aq)\right]^{2+}$, the rate varies as:

      R = CH_3 (acetate), C_3H_7 (butyrate), C_6H_5 (benzoate), $(CH_2)_2CO_2H$ (acid succinate) < CH_2OH (glycolate) < CHO (glyoxylate).
62. $\left[M^{III}(NH_3)_5 X\right]+\left[Cr(aq)\right]^{2+} \longrightarrow$ Products; M = Co, Cr, Ru.
    - For X = F, Cl, $OCOCH_3$ etc; $\frac{k_{Co(III)}}{k_{Cr(III)}} \geq 10^6$; $\frac{k_{Ru(III)}}{k_{Cr(III)}} \geq 10^5$; $\frac{k_{Co(III)}}{k_{Ru(III)}} \approx 10$
    - $\frac{k_{Co(III)}}{k_{Ru(III)}} \approx 10^1$ (for X = F, Cl) but $\frac{k_{Ru(III)}}{k_{Co(III)}} > 10^5$ (for X = isonicotinamide)
    - For X = N(py)—$CONH_2$, $\frac{k_{Co(III)}}{k_{Cr(III)}} \approx 10^1$; $\frac{k_{Ru(III)}}{k_{Cr(III)}} \geq 10^5$, $\frac{k_{Ru(III)}}{k_{Co(III)}} > 10^5$
    - For X = N(py)—$CONH_2$, Co(III) and Cr(III) experience the ion radical chemical mechanism of electron transfer while Ru(III) experiences the resonance tunneling mechanism for electron transfer.
63. *cis*-$\left[Cr(H_2O)_4 X_2\right]^+ + \left[{}^*Cr(aq)\right]^{2+} \longrightarrow$ Products
    - For X = N_3^-, the product is *cis*-$[{}^*Cr(OH_2)_4X_2]^+$
    - For X = F^-, the product is *cis*-$[{}^*Cr(OH_2)_5X]^{2+}$

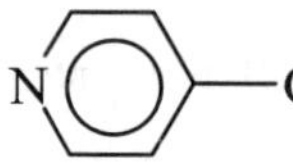

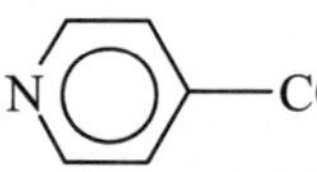

64. Reduction of $HCrO_4^-$ by $[Fe(CN)_6]^{4-}$ generates the product, $[(H_2O)_5Cr—N{\equiv}C—Fe(CN)_5]$ (*i.e.* $[CrFe(CN)_6)]$.
   - Stoichiometry of the reaction is:
     $HCrO_4^- + 3[Fe(CN)_6]^{4-} + 7H^+ \longrightarrow [CrFe(CN)_6] + 2[Fe(CN)_6]^{3-} + 4H_2O$
   - Reduction of $HCrO_4^-$ by $[Fe(bpy)_2(CN)_2]$ produces also a binuclear complex of Cr^{III} and Fe^{III}.
   - Reduction of $HCrO_4^-$ by $[Fe(aq)]^{2+}$ or $[Fe(bpy)_3]^{2+}$ produces no such binuclear complex.

65. $\left[Co^{III}(acac)_2(en)\right]^+ + \left[Cr(aq)\right]^{2+} \longrightarrow \left[Cr(aq)\right]^{3+} + \left[Cr(acac)(aq)\right]^{2+} + \left[Cr(acac)_2(aq)\right]^+$
   The product analysis indicates the different parallel paths of reduction of the Co(III) complex.
   - $\left[Co(edta)\right]^- + \left[Cr(aq)\right]^{2+} \longrightarrow \left[Co(aq)\right]^{2+} + \left[Cr(edta)(OH_2)_3\right]^-$
   The product can predict the structure of the activated complex.

66. Reduction of *trans*– $[Co^{III}(en)_2(L)(OH)]$ by $[Cr(aq)]^{2+}$.
   - Rate is faster for L = H_2O than for L = NH_3
   - $[Cr(aq)]^{2+}$ reduces *trans*-$[CrCl(NH_3)_4(OH_2)]^{2+}$ faster than $[CrCl(NH_3)_5]^{2+}$.
   - For the corresponding *cis*-complex of Co^{III}, the rate of reduction is relatively less sensitive to the nature of L.
   - $\dfrac{k(OH_2)}{k(NH_3)}(trans\text{-isomer}) > \dfrac{k(OH_2)}{k(NH_3)}(cis\text{-isomer}) > 1.$

67. Reduction rate of the Co(III)-complexes by Cr(II) follows the sequence:
   $[Cr(en)_2(OH_2)_2]^{2+} > [Cr(en)(OH_2)_4]^{2+} > [Cr(OH_2)_6]^{2+}$.

68. $[Pt(en)_2]^{2+}$ can catalyze the chloride exchange process in *trans*-$[PtCl_2(en)_2]^{2+}$.
   - 2*e*-transfer through the OSET path is highly improbable but it can occur through the ISET path in the $Pt^{IV}—X—Pt^{II}$, $Rh^{III}—X—Rh^{I}$ systems.
   - $Na_3[RhCl_6] \xrightarrow[\text{aq. soln}]{\text{py}} [Rh(py)_3Cl_3]\downarrow$
     $Na_3[RhCl_6] \xrightarrow[\text{aq. soln + small amount of } C_2H_5OH \text{ as a reducing agent}]{\text{py}}$ *trans*-$[Rh(py)_4Cl_2]^+$
   - Rates of change in optical rotation of $[Pt^{IV}(l\text{–}pn)_2Cl_2]^{2+}$ and radiochloride exchange in presence of a catalytic amount of $[Pt(en)_2]^{2+}$ are the same.

69. The reducing agents like $[Ru(aq)]^{2+}$, $[Ru(NH_3)_6]^{2+}$, $[Cr(bpy)_3]^{2+}$, $[Fe(CN)_6]^{4-}$, $[Fe(bpy)_3]^{2+}$, $[Co(CN)_6]^{4-}$ are selective for the OSET process when the oxidant metal centre is inert.

70. Reduction of $[M(phen)_3]^{3+}$, $[M(NH_3)_6]^{3+}$, $[M(NH_3)_5(py)]^{3+}$ (M = Co, Ru) cannot occur through the ISET path.

71. $\Delta V^{\neq}$ value may be helpful to distinguish between the OSET and ISET processes.

72. The water exchange rate at $[Co(aq)]^{3+}$ is unusually high.

73. The water exchange process at $[Cr(aq)]^{3+}$ can be catalyzed by $[Cr(aq)]^{2+}$.

74. Anhydrous $CrCl_3$ is insoluble in water but it goes into solution easily in the presence of a catalytic amount of $[Cr(aq)]^{2+}$.

75. Generally, the Co(III) complexes are more inert than the Cr(III) complexes but for the water exchange process, the reverse is true.

76. Aquation of NH_3 in $[Cr(NH_3)_5X]^{2+}$ (X with the bridging property) can be catalyzed by $[Cr(aq)]^{2+}$.
   - Aquation of L in $[Cr^{III}(aq)(L)]$ can be catalyzed by $[Cr(aq)]^{2+}$, if L does not have any bridging property.

77. Reduction of $[Co^{III}(NH_3)_5X]$ by $[Co(aq)]^{2+}$ in presence of excess CN^- leads to different products depending on the bridging property of X.

   - $[Co^{III}(NH_3)_5X] \xrightarrow[(+\text{excess } CN^-)]{+\text{trace amount of Co(II)}}$ different types of products depending on the nature of X.

78. Reduction of $[CoCl(edta)]^{2-}$ and $[Co(edta)(OH_2)]^-$ by different reducing agents gives the different rate ratios for the chlorido and aqua complexes (*Ref.* H. Ogino *et al, J. Chem. Soc. Dalton Trans.* 894, 1981)

Reducing agent	k_{Cl}/k_{aqua}
$[Cr(aq)]^{2+}$	2×10^3
$[Fe(aq)]^{2+}$	$>3 \times 10^2$
$[Fe(CN)_6]^{4-}$	33
$[Ti(aq)]^{3+}$	31

(**Hints:** ● ISET: The reductants $[Cr(aq)]^2$ ($\sigma^* \rightarrow \sigma^*$) and $[Fe(aq)]^{2+}$ ($\pi^* \rightarrow \sigma^*$) for the chlorido complex. ● For the aqua-complex, both ISET (Cl^-, a better bridging ligand than H_2O) and OSET for the reductants $[Cr(aq)]^{2+}$ and $[Fe(aq)]^{2+}$ ● OSET: both the chlorido- and aqua- complexes for $[Fe(CN)_6]^{4-}$ (substitutionally inert) ● $[Ti(aq)]^{3+}$ ($\pi^* \rightarrow \sigma^*$ net electron transfer, requiring an electronic activation of Ti(III) in the ISET path). This electronic activation energy for $[Fe(aq)]^{2+}$ is relatively less as 10Dq: Ti(III) > Fe(II).

79. Reduction of $[Co(NH_3)_5X]^{2+}$ (X = F, Cl, Br, I) by $[Eu(aq)]^{2+}$ gives the rate constant order as follows:
   k (dm^3 mol^{-1} s^{-1}) = 2.6×10^4 (F) > 4×10^2 (Cl) > 2.5×10^2 (Br) > 1.1×10^2 (I)
   (**Hints:** Thermodynamic stabilities: $EuF^+ > EuCl^+ > EuBr^+ > EuI^+$; ISET path; X^- is transferred to the solution; extreme lability of EuX^{2+} leads to its hydrolysis).

80. Reduction of $[Co^{III}(NH_3)_5X]$ (X = NH_3, H_2O, halide, nitrate, etc.) by $[Cr(bpy)_3]^{2+}$ and $[Ru(NH_3)_6]^{2+}$ occurs exclusively through the OSET path.

81. The products of reduction of $[Co^{III}(NH_3)_5X]$ by $[Co(CN)_5]^{3-}$ in presence of free CN^- may be $[Co^{III}(CN)_5X]$ or $[Co(CN)_6]^{3-}$ or a mixture of these two products depending on the bridging property of X.
   - The rate law also depends on the bridging property of X.
   - The rate constants may vary with X = Cl^-, Br^-, N_3^-, NCS^-, OH^-, etc. or may not vary with $X = NH_3, PO_4^{3-}, CO_3^{2-}, SO_4^{2-}$, etc.

82. The ligand exchange reaction in $[Co(AA)_3]^{3+}$ (AA = bpy, phen, en) is catalysed by free Co(II)-ion.

CHAPTER 6

PART II

Fundamental Aspects of Photochemistry: Photochemical Reactions of Inorganic Compounds and Metal Complexes

6.9 PHOTOEXCITATION AND FATE OF THE PHOTOCHEMICALLY EXCITED MOLECULAR SPECIES

6.9.1 Photoexcitation of Molecules

Each electronic energy state is associated with its vibrational and rotational energy levels characterised by the vibrational ($v = 0, 1, 2, 3, ...$) and rotational ($J = 0, 1, 2, 3, ...$) quantum numbers (Fig. 6.9.1.1). Electronic excitation is accompanied with the changes in the vibrational and rotational quantum numbers to produce the **fine structure in the absorption spectrum.** It gives the **molecular band spectra** (*cf.* atomic spectra: discrete lines, no question of vibrational level).

The electronic excitation occurs very fast in a time period of about 10^{-15} s while the vibration time period (*ca.* 10^{-13} s) is relatively higher due to the larger mass of the involved nuclei in the vibration compared with that of the electron involved in the electronic transition. This is why, **during the electronic excitation, the inter-nuclear separation remains almost constant** and the electron transitions can be represented by the vertical lines in the potential energy diagrams. This is the **Franck–Condon principle.**

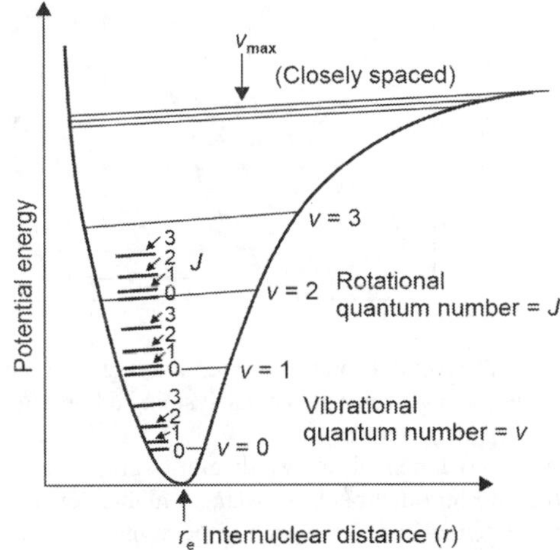

Fig. 6.9.1.1a: A particular electronic energy state with the rotational energy levels for the different vibrational levels

The primary activation through the absorption of a photon ($h\nu$) leads to the **electronic energy state change.** This electronic transition follows the **selection rules** (see Sec. 9.10.5, Vol. 2 and Vol. 7 for the selection rules of electronic transitions in the diatomic molecules). The primary activation in a molecule may lead to the different possibilities as discussed below.

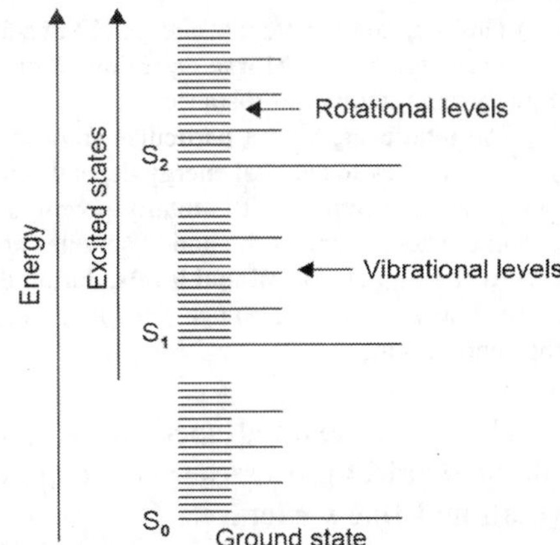

Fig. 6.9.1.1b: Representation of different electronic energy states (overlapping regions not shown)

(i) **Photoexcitation to the higher stable electronic state:** In the excited stable electronic energy state (Fig. 6.9.1.2), the molecule does not dissociate though the internuclear distance is increased over that of the ground electronic state.

$D_{el} = D_0 + \frac{1}{2}h\nu_o$ **(zero-point energy)**
D_{el} denoting bond energy
D_0 = Spectroscopic energy of dissociation of the molecule at $T = 0$ K
Zero-point energy is quite significant for the molecules consisting of light atoms (*e.g.* in H_2, it is about 26 kJ mol^{-1}) but it is insignificant for the heavier species (see Vol. 2, Sec. 10.1).

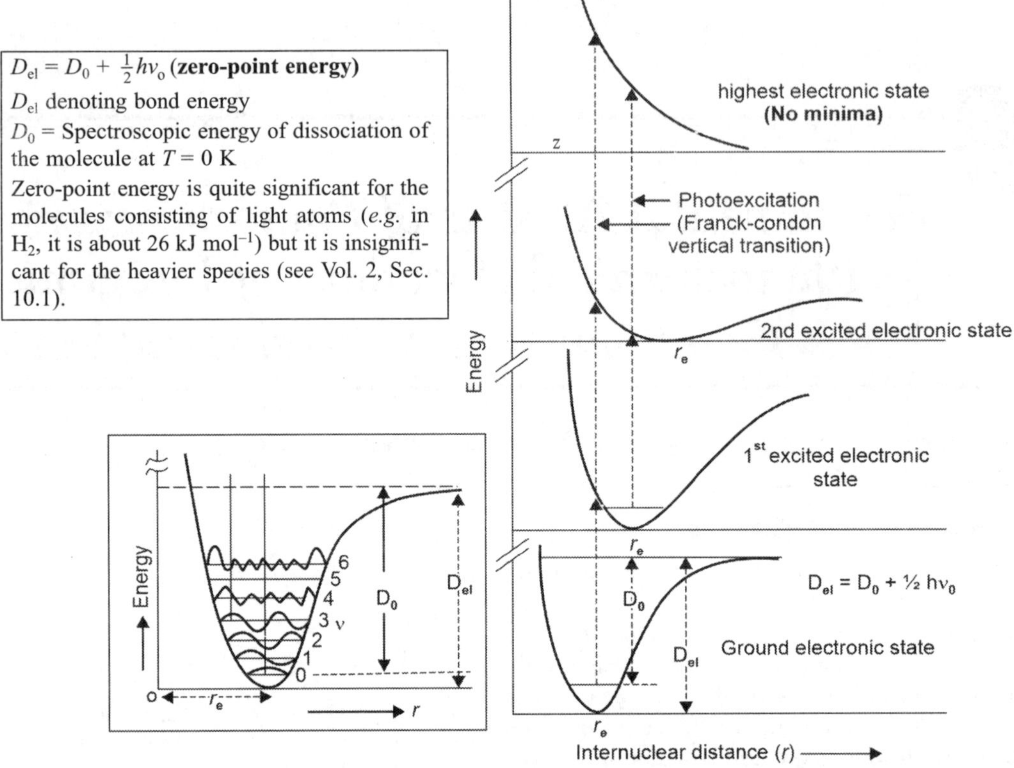

Fig. 6.9.1.1c: Schematic representation of the potential energy (PE) diagram of the ground and excited electronic energy states of a particular molecular system and photoexcitation (**Franck–Condon vertical transition**) to different electronic excited energy states.

Note: (i) Potential energy diagram shifts slightly towards the right for the higher electronic states and it makes r_e (equilibrium internuclear distance) higher for the higher electronic states.

(ii) Potential energy diagram at the higher electronic states becomes gradually **shallower** (*i.e.* depth gradually decreases) and consequently, **dissociation energy gradually decreases**.

(iii) At the highest electronic state, no alteractive zone of potential energy curve exists and only the repulsive part of the curve exists. The **curve does not exhibit minima** and it indicates that the **bond is not stable at this electronic energy state.**

(iv) The variation of potential energy (***V***) in a diatomic molecule with the displacement from the equilibrium position (r_e) is represented by the **Morse equation**. $V(\text{r}) = D[1 - \exp\{-a(r - r_e)^2]$, a = constant, $r - r_e$ = displacement from the equilibrium internuclear distance.

(v) The total energy for a particular vibrational level is constant but the relative contributions of kinetic energy and potential energy to the total energy depends on the displacement from the equilibrium position (r_e). The energy is totally kinetic at the centre but it is totally potential at the turning points. Thus the relative contributions of the kinetic and potential energy vary as the bond oscillates about the equilibrium position.

(vi) **At the higher vibrational levels**, during the oscillation about the equilibrium position, the molecule spends most of its time near the turning points. Thus its probability of its existence is very high near the turning points but very low at the centre points.

The stable excited species (A*) participates subsequently in different **photophysical** and **photochemical** processes. For such photoexcitation, the electronic spectra show the **vibrational and rotational fine structure**.

(ii) **Photoexcitation leading to the bond dissociation:** The molecule is excited to the **continuum level** in the higher electronic state where the molecule dissociates in its first oscillation (Fig. 6.9.1.3).

Consequently, the absorption spectrum does not show the discrete vibrational–rotational fine structure rather shows the continuum (*i.e.* diffuse spectrum). It is noted for HI.

$$HI \xrightarrow{h\nu} H + I$$
$$H + HI \longrightarrow H_2 + I$$
$$I + I = I_2$$

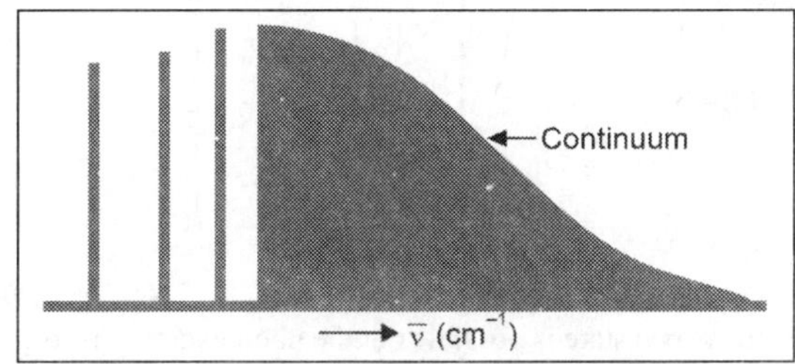

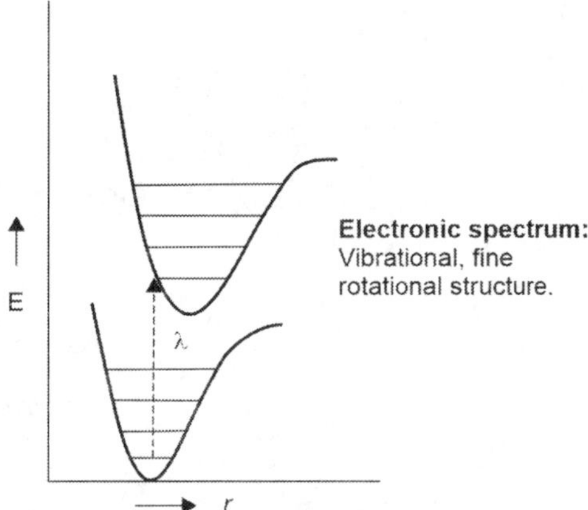

Fig. 6.9.1.2: Illustration of photoexcitation of a molecule to an upper stable electronic state

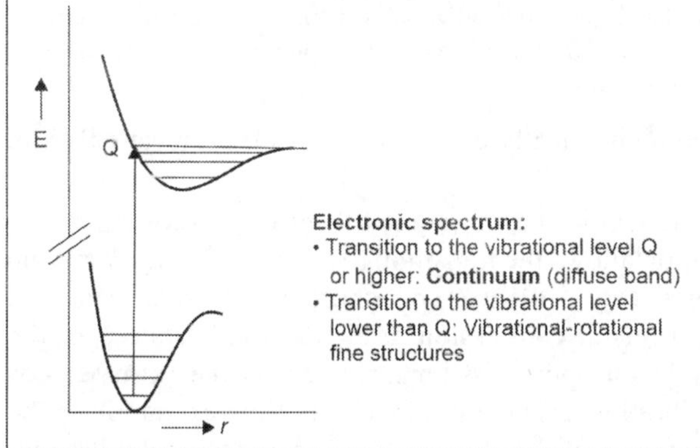

Continnum in atomic and molecular spectra: In atomic absorption spectra, each series (*e.g.* Lyman series in H atom) consisting of well separated lines ends with a continuum at higher frequencies **due to the ionisation** releasing free electrons which can absorb any energy to increase their kinetic energy (*i.e.* **kinetic energy not quantised)**. In molecular absorption band spectra, at higher frequency, the continuum arises **due to photodissociation** into the atoms that can absorb any energy to increase their kinetic energy.

Fig. 6.9.1.3: Illustration of photoexcitation to an upper electronic state where the bond undergoes dissociation

(iii) **Photoexcitation to the higher unstable electronic state:** In this situation, the molecule is excited to a higher electronic state which is unstable characterised by the absence of any minima in the potential energy curve (Fig. 6.9.1.4). The electronic spectrum consists of only continuum. In such cases, the excited molecule undergoes dissociation instantaneously. The high energy (kinetic energy) dissociated fragments can participate in secondary chemical processes (*e.g.* initiation of chain reaction as in the case of (ii)).

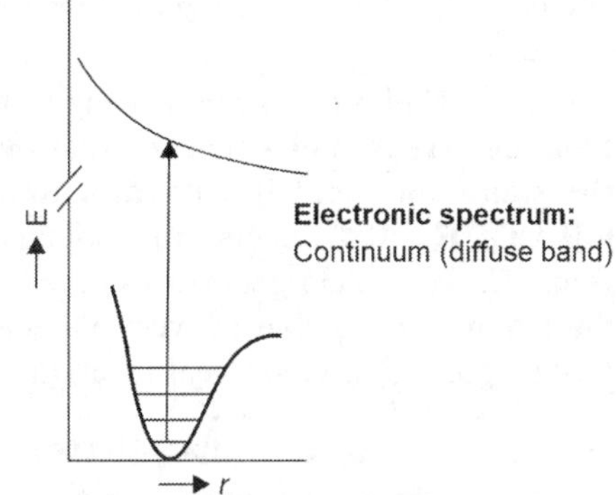

Fig. 6.9.1.4: Illustration of photoexcitation to the highest electronic state (**potential energy curve showing no minima**) where the excited species is unstable

(iv) **Predissociation:** Here in the excited state, potential energy diagram of a particular bond is quite stable but it intersects with the potential energy diagram of another electronically excited bond having the lower bond dissociation energy. Thus excitation energy of a particular bond can be transferred to the other bond of the same polyatomic molecule leading to the dissociation of the

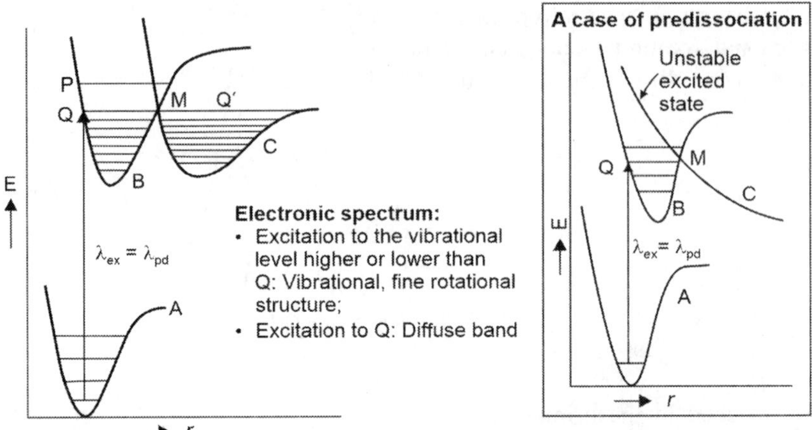

Fig. 6.9.1.5: Illustration of excitation for predissociation.

Note: (i) Photoexcitation from the A-electronic state to the B-electronic state is possible but the photoexcitation from the A-electronic state to the C-electronic state is forbidden (**Franck–Condon principle**).
(ii) Excitation to the vibrational level Q of the B-electronic state leads to the radiationless transition to another electronic state C along QMQ′. It causes the migration of excitational energy from the B-electronic state to the C-electronic state of another bond where the bond is unstable at the Q′-vibrational level

bond of lower dissociation energy (*cf.* the bond photochemically excited is not itself dissociated) (Fig. 6.9.1.5).

For the excitation from $v = 0$ of the ground state (A) to the vibrational level Q (say for $\lambda_{ex} = \lambda_{pd}$; Fig. 6.9.1.5) of the electronic state B, the first vibration allows the transition to the intersecting electronic state C where it undergoes dissociation characterised by a **diffuse spectrum** (*i.e.* no rotational fine structure). This transition probability through the **internal conversion** to the electronic state C of the same spin multiplicity is very high because the molecule spends its longer time near the extreme ends of its vibrational motion, specially for the larger vibrational quantum number. For the excitation to the higher vibrational level (say P for $\lambda_{ex} < \lambda_{pd}$; Fig. 6.9.1.5) of the B-state, the excited molecule does not dissociate. It moves rather rapidly to the lower vibrational levels through the **vibrational relaxation**. In fact, the probability of its crossing over to the unstable state C through the intersecting point of two energy states (*i.e.* B and C) is reduced. Thus for the photoexcitation (*i.e.* $\lambda_{ex} < \lambda_{pd}$), normal electronic spectrum (*i.e.* coarse vibrational associated with rotational fine structure) is observed. Similarly, for the photoexcitation ($\lambda_{ex} > \lambda_{pd}$) to the lower vibrational level (< Q) of the B-state, the excited state does not dissociate.

In fact, **if photoexcitation leads to the excited vibrational level close to the intersecting point, then the excited molecule can cross-over through the internal conversion to the excited state C of the same spin multiplicity and it leads to the predissociation illustrated for CH_3CHO or CH_3COCH_3.** At the intersecting point (which is **isoenergetic and kinetic energy ≈ 0**) of the two states B and C, the nuclear geometries of the two states are the same. This is why, at this intersecting point, the internal conversion between the states of **same spin multiplicity** occurs most readily as at the turning point of a vibrational level, the molecule spends larger time.

Characteristic features of a predissociation spectrum

There is a diffuse region (*i.e.* continuum) corresponding to the photoexcitation ($\lambda_{ex} = \lambda_{pd}$) leading to the cross-over of the excited molecule to the unstable electronic state allowing the dissociation of the excited molecule. But for the photoexcitation at the shorter or longer wavelength (*i.e.* $\lambda_{pd} < \lambda_{ex} < \lambda_{pd}$), the spectrum shows the normal vibrational and rotational fine structure.

In the case of CH_3CHO or CH_3COCH_3, photoexcitation of the C═O bond occurs first and then the excitation energy is transferred to the adjacent C—C bond which dissociates.

$$H_3C(H)C{=}O \xrightarrow{h\nu} H_3C(H)C\overset{*}{=}O \longrightarrow H_3C^{*}(H)C{=}O \longrightarrow {}^{\bullet}CH_3 + {}^{\bullet}CHO$$

$$\text{i.e. } H_3CCHO \xrightarrow{h\nu} {}^{\bullet}CH_3 + {}^{\bullet}CHO$$

$$H_3C^{*}(H_3C)C{=}O \xrightarrow{h\nu} {}^{\bullet}CH_3 + CH_3{}^{\bullet}CO$$

The radicals produced in the primary process can initiate the chain reaction in the subsequent **thermal secondary processes.**

Photochemically excited species may undergo deactivation through different **photophysical pathways** and may also participate in **photochemical reactions** discussed in the next sections.

6.9.2 Photodissociation of the Molecules and Thermodynamic Bond Dissociation Energy (BDE) (*cf.* A.K. Das *et al, Chemistry Teacher International* (IUPAC), 5, 91–105, 2023)

Generally, energy of the absorbed light/radiation to cause photodissociation of a molecule is significantly larger than the thermodynamic bond dissociation energy (BDE). The reason is that in the photodissociation, at least one of the dissociated fragments remains in an **electronically excited state.** It results from the **selection rules** that govern the photoexcitation process.

For the direct photodissociation of a molecule, the **wavelength of excitation must lie in the absorption continuum.**

(i) **Photodissociation of I_2:** Let us illustrate for I_2 having the ground state electronic configuration: $\cdots\pi_u^4(5p_{y,z})\pi_g^{*4}(5p_{y,z})$ (assuming x-axis as the molecular axis) that gives the electronic state $^1\Sigma_g^+$ (see Sec. 9.10.4, Vol. 2). The HOMO is $\pi_g^{*4}(5p_{y,z})$ while the LUMO is $\sigma_u^*(5p_x)$. The potential energy diagram (Fig. 6.9.2.1) of I_2 indicates that the photoexcitation, $^1A\rightarrow{}^3B$ (first excited stable electronic state) produces an **absorption continuum** that begins at 499.0 nm. For **direct photodissociation of a molecule, the wavelength of the radiation to be absorbed must lie in the absorption continuum region of the excited electronic state.** For I_2, the continuum at the ^3B excited state starts at 499.0 nm and it dissociates for $\lambda \leq 499.0$ nm. To produce two I atoms, one of these two I atoms remains in the excited state $^2P_{1/2}$ (see Vol. 1 for the **term symbols**). The transition, singlet to triplet (*i.e.* $^1A\rightarrow{}^3B$) is **spin forbidden** but it is partially relaxed for this heavy diatomic molecule consisting of a larger number of electrons.

In fact, for the molecules containing the **heavy atoms** like iodine having the **high spin-orbit coupling constant**, the spin-orbit coupling allows the mixing of the ground singlet state of I_2 with the excited triplet states. **This spin-orbit coupling relaxes the spin selection rule** (see Vol. 2). Among the halogens (X_2), the energy gap between the HOMO (π_g^*) and LUMO (σ_u^*) is minimum for I_2. This **smaller energy dsifference** between the ground and excited states further favours the spin-orbit coupling to **relax the spin selection rule better** (see Vol. 2).

The thermodynamic BDE of I_2 is 148.7 kJ mol^{-1} and it considers the production of two I-atoms at the ground state.

$$I_2 \rightarrow I(5p^5, {}^2P_{3/2}) + I(5p^5, {}^2P_{3/2}),\ \Delta H = BDE = +\ 148.7\ \text{kJ mol}^{-1}$$

The two **spin-orbit states** (differing in J-values) of the I atom are: I($5p^5$, $^2P_{3/2}$, ground state) and I($5p^5$, $^2P_{1/2}$, excited state).

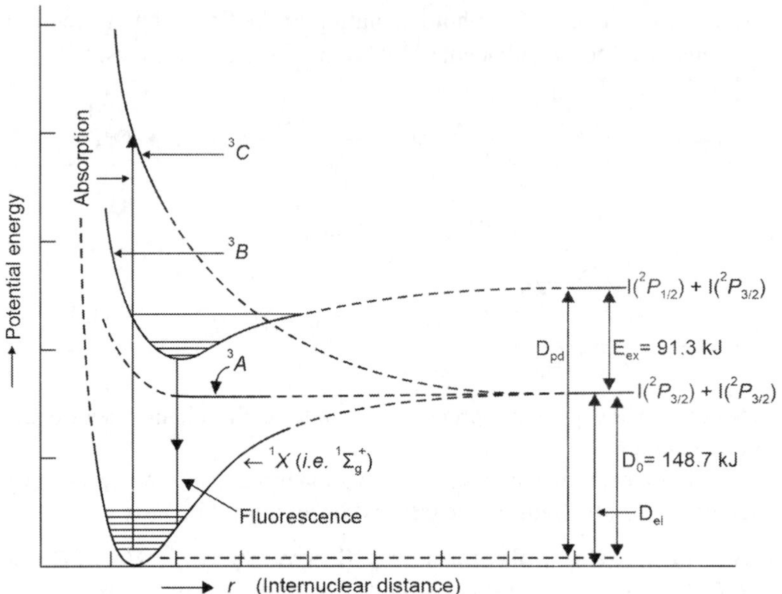

Fig. 6.9.2.1: Qualitative representation of the potential energy diagram of some electronic states of I_2 (g) molecule to illustrate the relationship between the ground state thermodynamic BDE (D_0) and photodissociation BDE (D_{pd})

Note: 1: The excited triplet states are **simply denoted** by ^3A, ^3B and ^3C. The photoexcitation to the dissociative (*i.e.* repulsive) ^3A and ^3C states from the ground state are not important. **This is why, photodissociation occurs from the excitation to the ^3B state.**

Note 2: The singlet → triplet photoexcitation is **spin forbidden**. But it is significantly relaxed through the **spin-orbit coupling** discussed in the text (see Sec. 6.10.8 for details).

The BDE value corresponds to λ =

$$\lambda_{\text{threshold}} = \frac{N_0 hc}{BDE} = \frac{(6.022\times10^{23}\text{ mol}^{-1})(6.626\times10^{-34}\text{ J s})(3\times10^{8}\text{ m s}^{-1})}{148.7\times10^{3}\text{ J mol}^{-1}}$$

$$= \frac{119.6\times10^{-6}\text{ m}}{148.7} = \frac{119.6\times10^{-6}\times10^{9}}{148.7}\text{ nm} \approx 804\text{ nm}$$

$$(cf.\ \ E = \frac{N_0 hc}{\lambda} \approx \frac{1.20\times10^{5}}{\lambda}\text{ kJ mol}^{-1}, \lambda\text{ in nm}$$

$$i.e.\ \ \lambda(\text{nm}) = \frac{1.20\times10^{5}}{\text{BDE}}, \text{BDE in kJ mol}^{-1})$$

The $\lambda_{\text{threshold}}$ (= 499.0 nm) for the continuum in the ^3B state corresponds to:

$$\text{BDE} = \frac{1.20\times10^{5}}{499.0}\text{ kJ mol}^{-1} \approx 240\text{ kJ mol}^{-1}$$

This difference of 91.3 kJ mol^{-1} (= 240.0 – 148.7) is the **excitation energy of the I-atom** from the $^2P_{3/2}$ to the $^2P_{1/2}$ state. From the potential energy diagram (Fig. 6.9.2.1), it is evident that direct photodissociation of I_2 produces two I-atoms, one at the ground state and the other at the excited state. This is why, it does not photodissociate at λ = 804 nm (corresponding to the thermodynamic BDE).

$$I_2 + h\nu \longrightarrow I(^2P_{3/2}) + I(^2P_{1/2}), \lambda \le 499\text{ nm}$$

Photodissociation BDE (D_{pd}) = thermodynamic BDE (D_0) + excitation energy of an I-atom = (148.7 + 91.3) kJ mol^{-1} = 240 kJ mol^{-1}.

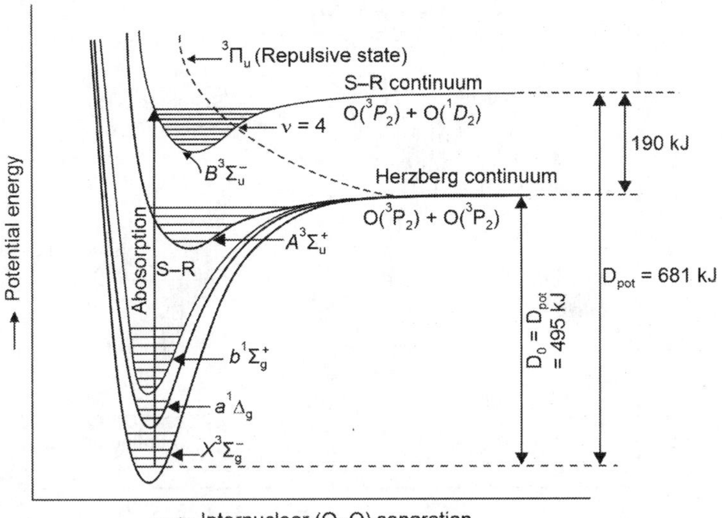

Fig. 6.9.2.2: Qualitative representation of the potential energy diagram of the electronic states of O_2 to illustrate the relationship between the ground state thermodynamic BDE (D_0) and photodissociation BDE (D_{pd})

(ii) **Photodissociation of O_2:** From the potential energy diagram of O_2 (Fig. 6.9.2.2), it is evident that its photodissociation occurs through the **Herzberg transition** to the weakly bound state $O_2(A^3\Sigma_u^+)$ that produces two triplet ($^3P_2$ ground state) O-atoms. The Herzberg transition merges to the **continuum** to cause the photodissociation of O_2 as follows:

$$O_2(X^3\sum\nolimits_g^-)\xrightarrow[(\lambda\approx 200\text{ nm})]{h\nu} O_2(A^3\sum\nolimits_u^+)\xrightarrow{\text{Dissociation}} 2O(^3P_2)$$

i.e. $$O_2(X^3\Sigma_g^-)+h\nu\longrightarrow O(^3P_2)+O(^3P_2),\ \Delta H = 495\text{ kJ mol}^{-1}$$

The ground state bond dissociation energy is the thermodynamic BDE and it corresponds to 242 nm.

$$\lambda=\frac{1.2\times10^5}{495}\approx 242\text{ nm}$$

The above **symmetry forbidden Herzberg transition** ($-\rightarrow+$ transition) leading to the above photodissociation gets partially relaxed through the spin-orbit coupling between the ground and excited states. In fact, photodissociation in the **Herzberg continuum** (200–242 nm) is the major source of O-atom for O_3 (ozone) formation in the stratosphere.

$$\text{O}(^3P_2)+\text{O}_2(X^3\Sigma_g^-)+\text{M}\xrightarrow{\text{three body reaction}}\text{O}_3+\text{M}^*\ (\text{M}=\text{N}_2,\text{O}_2)$$

The following **spin and symmetry allowed** photodissociation of O_2 involving the $B^3\Sigma_u^-$ state occurs at $\lambda \le 176$ nm (threshold value of the **Schumann–Runge continuum**) (see Fig. 6.9.2.3). **This process is more efficient for photodissociation of O_2.**

$$\text{O}_2(X^3\Sigma_g^-)\xrightarrow[(\lambda\le 176\text{ nm})]{h\nu}\text{O}_2(B^3\Sigma_u^-)\xrightarrow{\text{Dissociation}}\text{O}(^3P_2)+\text{O}(^1D_2)$$

i.e. $$\text{O}_2(^3\Sigma_g^-)+h\nu\longrightarrow\text{O}(^3P_2)+\text{O}(^1D_2)\ \text{(excited state)}$$

$$\text{(triplet}\longrightarrow\text{triplet + singlet)}$$

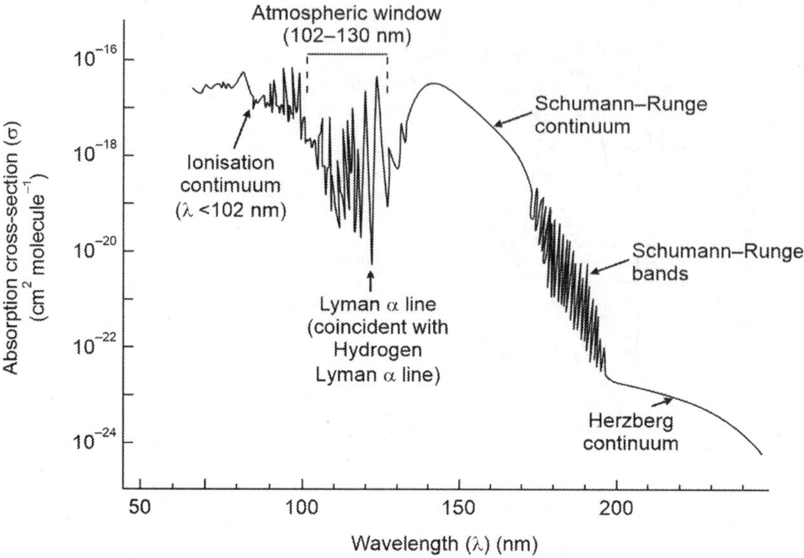

Fig. 6.9.2.3: Absorption spectrum of O_2 in the UV-range

Note: O_2 absorbs in three different wavelength regions: (a) **Herzberg continuum** (~ 242 – 200 nm), $(\pi_g)^4 (\pi_g)^2 \rightarrow (\pi_u)^3 (\pi_g)^3$, *i.e.* $X^3\Sigma_g^- \longrightarrow A^3\Sigma_u^+$ (weakly bound state); (ii) **Schumann–Runge (S–R) bands** (~ 200 – 176 nm), $O_2(X^3\Sigma_g^-) \longrightarrow O_2(B^3\Sigma_u^-)$ **(dissociative state)**, the S–R band converges to **S–R continuum** at λ = 176 nm; (iii) **S–R continuum** (~ 176 – 130 nm). **Atmospheric window** (~ 130 – 102 nm), discrete and diffuse bands instead of continuum

It produces one O-atom at the ground state and one O-atom at the electronically excited state (^1D). The threshold value (λ = 176 nm) for the S–R continuum leading to the above photodissociation of O_2 corresponds to 681 kJ mol^{-1}.

Thermodynamic BDE (producing two O-atoms at the ground triplet states) + excitation energy of an O-atom (*i.e.* $^3P_2 \rightarrow {}^1D_2$) = (495 + 190) kJ mol^{-1} = 685 kJ mol^{-1}. Thus the calculated value nicely agrees with the experimental value (= 681 kJ mol^{-1}).

(iii) **Photodissociation of O_3:** Theoretically, O_3 can photodissociate through the following two **spin conserved pathways**.

(a) $O_3(\text{singlet}) + h\nu \longrightarrow O_2(X^3\Sigma_g^-, \text{triplet}) + O(^3P_2, \text{triplet}), \Delta H = +106 \text{ kJ mol}^{-1}$

(singlet $\longrightarrow$ triplet + triplet)

The ground state thermodynamic BDE of the above process = 106 kJ mol^{-1} that corresponds to 1123 nm. This process of photodissociation is **very much inefficient (very low absorption cross-section)**.

(b) O_3 (singlet) + $h\nu \rightarrow O_2(^1\Delta_g$-singlet) + O($^1D_2$-singlet), ΔH = + 386 kJ mol^{-1}

(singlet → singlet + singlet)

The photodissociation pathway involves:

ΔH = ground state thermodynamic BDE + excitation energies of the products from their ground triplet states to their first excited singlet states = (106 + 90 + 190) kJ mol^{-1} = 386 kJ mol^{-1}.

Excitation energy for O($^3P_2$) to O($^1D_2$) = 190 kJ mol^{-1}

Excitation energy for $O_2(^3\Sigma_g^-)$ to $O_2(^1\Delta_g) = 90$ kJ mol^{-1}

In fact, in **stratosphere**, O_3 absorbs in the UV region between 200–315 nm (**Hartley continuum**) (See Fig. 6.9.2.4) and undergoes photodissociation to run the **Chapman ozone cycle**.

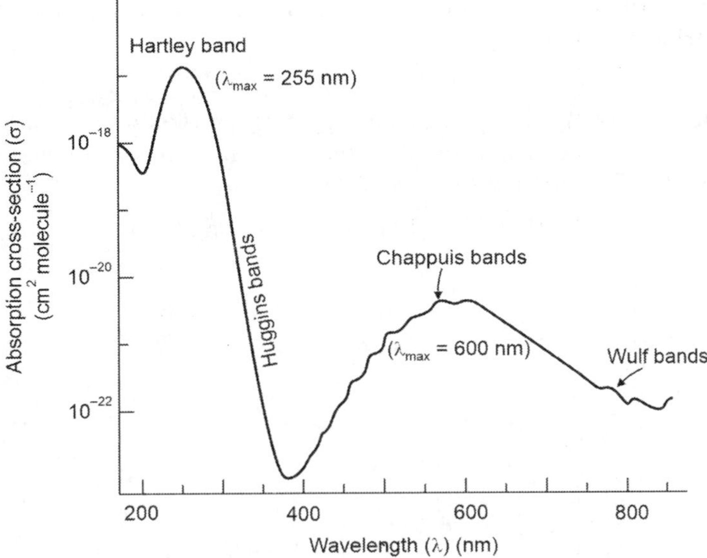

Fig. 6.9.2.4: Absorption spectrum of O_3. There are four major absorption bands: **Hartley band** (*ca.* 214–300 nm), **Huggins band** (*ca.* 300–354 nm), **Chappuis bands** (*ca.* 420–700 nm), **Wulf bands** (*ca.* 700–1048 nm, in the near infrared region). Hartley band (strongest one), $O_3(X^1A_1) \rightarrow O_3(^1B_2)$, λ_{max} = 255 nm, mainly responsible to **block the UV–B radiation**. Chappuis and Huggins bands (weaker), excitation to the other singlet states; Wulf bands – forbidden excitation to the low lying triplet states.

This is the main pathway of photodissociation of ozone. Ozone and O_2 combindely block the UV radiation to protect us (Fig. 6.9.2.5).

Note: Ozone is stable in the visible range of radiation and it dissociates in the UV-region (Hartley band) and the absorption continuum begins at about 290 nm. But in presence of a small amount of Cl_2, ozone dissociates at $\lambda \leq 478.5$ nm at which absorption continuum of Cl_2 begins.

$$Cl_2 + hv\ (\lambda \leq 478.5\ \text{nm}) \rightarrow 2Cl$$

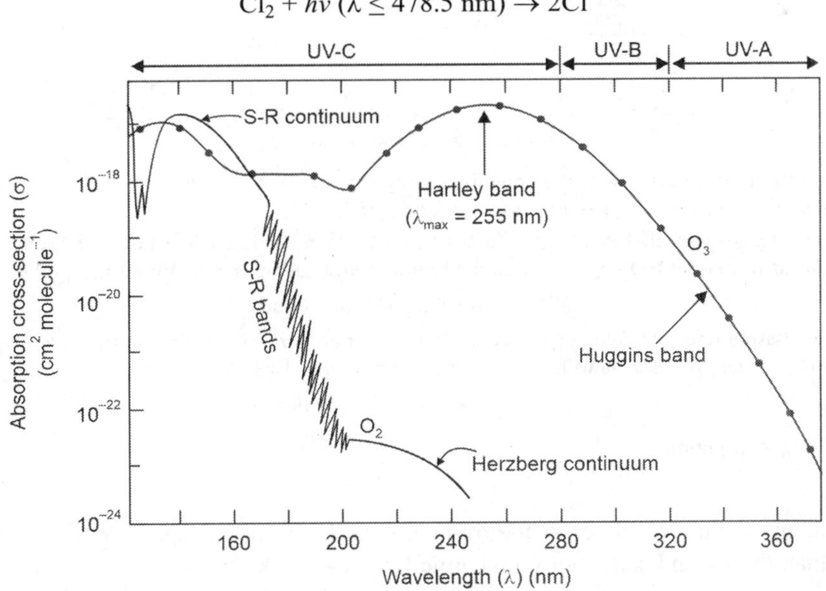

Fig. 6.9.2.5: Absorption of UV-radiation by atmospheric O_2 and O_3 (produced from O_2 in stratosphere) jointly block the highest energy UV–C radiation and O_3 mainly blocks the UV–B radiation

The Cl atom reacts with ozone to dissociate it in a complex **chain reaction** (*cf.* **ozone depletion** and **chlorofluorocarbons**, CFCs):

$$Cl + O_3 \rightarrow ClO + O_2, \ldots.$$

(iv) **Photodissociation of H_2:** The potential energy diagram (Fig. 6.9.2.6) of H_2 indicates that its photodissociation occurs when it is excited to the continuum region of the excited states and photodissociation products are not in the ground state.

$$H_2 + h\nu \rightarrow H(1s^1) + H(2p^1) \text{ or } H(2s^1)$$

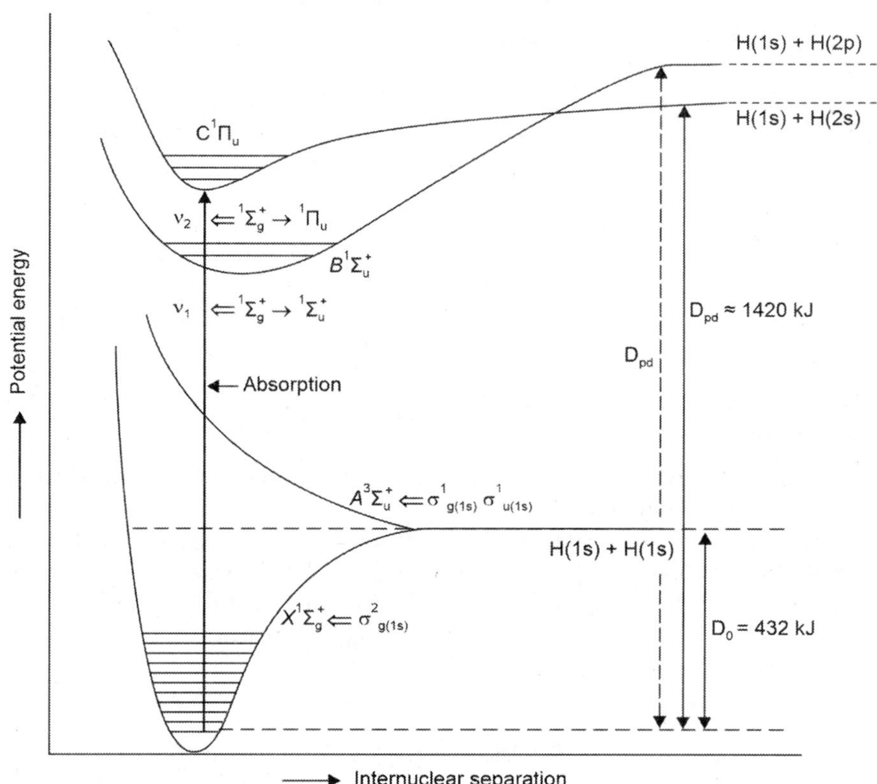

Fig. 6.9.2.6: Qualitative representation of the potential energy diagram of H_2 to illustrate the relationship between the ground state thermodynamic BDE (D_0) and photodissociation BDE (D_{pd})

Note: Absorption of photon (**parallel excitation**, electric field parallel to the internuclear axis) to excite an electron from the σ-BMO **to the lowest σ-ABMO** (σ_u) produces the **Lyman band** having the minimum energy 11.2 eV.

$$X^1\Sigma_g^+ \xrightarrow{h\nu} B^1\Sigma_u^+ \textbf{ (Lyman band)}$$

Perpendicular excitation (electric field perpendicular to the internuclear axis) of an electron from the σ-BMO to the **lowest Π_u orbital** produces **Werner band** having the minimum energy 12.3 eV.

$$X^1\Sigma_g^+ \xrightarrow{h\nu} C^1\Pi_u \textbf{ (Werner band)}$$

Emission in hydrogen arc lamp:

$$^3\Sigma_g^+(i.e.\ \sigma_{g(1s)}^1\sigma_{g(2s)}^1 \longrightarrow {}^3\Sigma_u^+(i.e.\ \sigma_{g(1s)}^1\sigma_{u(1s)}^1)$$

The continuum leading to the dissociation begins at 84.5 nm that corresponds to 1420 kJ mol^{-1} much higher than the ground state thermodynamic BDE (= 432 kJ mol^{-1})

$$H_2(^1\Sigma_g^+) \longrightarrow 2H(1s^1, 2S_{1/2}), \Delta H = +432 \text{ kJ mol}^{-1}$$

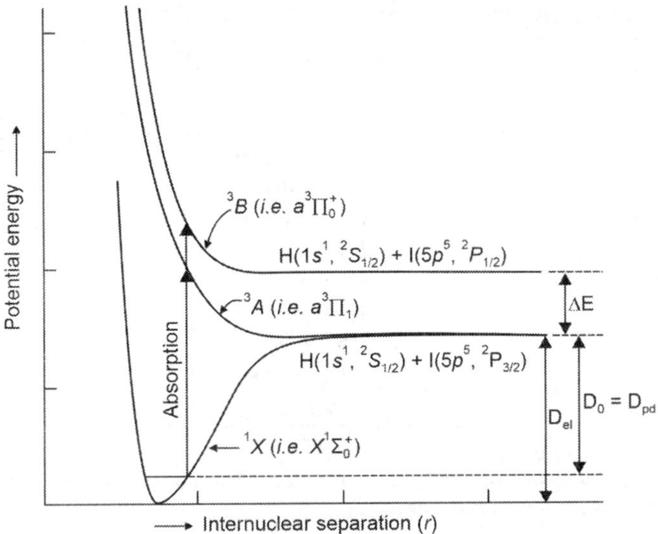

Fig. 6.9.2.7: Qualitative representation of the potential energy diagram of some electronic states of HI(g) to illustrate the relationship between the ground state thermodynamic BDE (D_0) and photodissociation BDE (D_{pd})

Note 1: Excited triplet electronic states are simply denoted by ^3A and ^3B. Here only two excited electronic states are shown.

Note 2: Here the shown photoexcitations (singlet → triplet) are **spin-forbidden**. But these are relaxed for the such many **electron diatomic molecules having the heavy atom** like iodine for which the spin-orbit coupling constant is very high. Spin-orbit coupling allowing the mixing of ground singlet state and excited triplet states relaxes the spin selection rule (see Sec. 6.10.8 for details). It happens so for the I_2 molecule discussed earlier.

(v) **Photodissociation of HI:** The case of photodissociation of HI is different from the above mentioned cases. Here **all the excited electronic states are dissociative/repulsive (*i.e.* no minima)** (Fig. 6.9.2.7). Its photodissociation products are:

HI + $h\nu_1 \rightarrow$ H(1s^1, $^2S_{1/2}$) + I(5p^5, $^2P_{3/2}$, ground state) **(perpendicular excitation)**

HI + $h\nu_2 \rightarrow$ H(1s^1, $^2S_{1/2}$) + I(5p^5, $^2P_{1/2}$), excited state **(parallel excitation)** ($\nu_2 > \nu_1$).

In both the cases, the products possess a large amount of kinetic energy. The lighter fragment (*i.e.* H atom) bears the larger fraction of the excess energy as the kinetic energy and it is described as the **hot atom**. The excess energy ΔE (Fig. 6.9.2.7) is required to dissociate HI into the products with the excited I atom with respect to the dissociation energy leading to the products at the ground electronic states.

Ground state I($^2P_{3/2}$) is produced from the **perpendicular excitations** (ΔΩ = 1, electric field perpendicular to the internuclear axis): $X^1\Sigma_o^+ \rightarrow A^1\Pi_1$ (not shown in Fig. 6.9.2.7) and $X^1\Sigma_o^+ \rightarrow a^3\Pi_1$. The excited I($^2P_{1/2}$) atom is produced from the **parallel excitation** (ΔΩ = 0, electric field parallel to the internuclear axis): $X^1\Sigma_0^+ \rightarrow a^3\Pi_0^+$

6.9.3 Direct Photodissociation vs. Photosensitizer Mediated Photodissociation (see Sec. 6.12)

In most of the cases (see Table 6.9.3.1), the photodissociation products are in the **excited electronic states**. It occurs so as the result of **selection rules** that govern the photoexcitation. For H_2, its direct photodissociation occurs at $\lambda \leq 84.5$ nm at which the **absorption continuum begins**. But in presence of Hg-vapour as the photosensitizer that strongly absorbs at 253.7 nm (≈ 472 kJ mol^{-1} > D_0 = 432 kJ mol^{-1}), the **quenching reaction** dissociates H_2. The excess energy (40 kJ mol^{-1}) imparts the kinetic energy to the H-atoms (see Sec. 6.12 for details):

Table 6.9.3.1: Required wavelength of light/radiation for photodissociation of some simple molecules

Photodissociation	*λ (nm) ≡ BDE (kJ mol⁻¹) = D_{pd}*	*Ground state thermodynamic BDE = D_0 (in kJ mol⁻¹)*
$H_2 + h\nu \rightarrow H(1s^1) + H^*(2s^1 \text{ or } 2p^1)$	< 84.5 ≡ 1420.0	432.0
$O_2 + h\nu \rightarrow O(^3P) + O(^3P)$	242.0 ≡ 495.0	495.0
$O_2 + h\nu \rightarrow O(^3P) + O^*(^1D)$	< 176.0 ≡ 681.0	495.0
$Cl_2 + h\nu \rightarrow Cl + Cl^*$	< 478.5 ≡ 250.0	239.0
$Br_2 + h\nu \rightarrow Br + Br^*$	< 511.0 ≡ 234.8	190.0
$I_2 + h\nu \rightarrow I(^2P_{3/2}) + I^*(^2P_{1/2})$	< 499.0 ≡ 240.0	148.7
$O_3 + h\nu \rightarrow O_2^*(^1\Delta_g) + O^*(^1D)$	< 309.0 ≡ 388.0	106.0
$HI + h\nu \rightarrow H + I$	< 327.0 ≡ 366	295.0

$$E\,(\text{kJ mol}^{-1}) = \frac{N_0hc}{\lambda\,(\text{nm})} = \frac{1.20\times10^5}{\lambda}$$

Note: In the absorption continuum, whatever may be the electronic states of the dissociated fragments, the dissociated fragments or atoms possess always some kinetic energy.

$Hg + h\nu\,(253.7\text{ nm}) \rightarrow Hg^*,\ Hg^* + H_2 \rightarrow Hg + H(1s) + H\,(1s)$, ($^2S_{1/2}$ ground electronic state of H-atom)

The energy transfer process through a collision as in the quenching reaction is not limited by any selection rule. Thus if the sensitizer possesses the sufficient energy to dissociate the quenching molecule then the dissociation of the quencher will occur. The wavelength of light used for photosensitisation cannot directly dissociate the quenching molecule. This is the unique advantage of photosensitization.

Through collision with the quencher (H_2), the excitation energy of Hg* is released as the **heat energy** that dissociates H_2 and the excess energy imparts the kinetic energy to the dissociated H-atoms (see Sec. 6.12).

6.10 PHOTOPHYSICAL PATHWAYS OF DEACTIVATION OF ELECTRONICALLY EXCITED MOLECULES, PRIMARY AND SECONDARY PROCESSES, QUANTUM YIELD OF PHOTOCHEMICAL REACTIONS

If the electronically excited molecule does not get involved in a photochemical reaction, then it must return to the ground state from which it is photochemically excited. Besides the photochemical reactions, the other possible pathways for dissipation of excitational energy are described as the **photophysical processes** which are illustrated assuming S_0 (singlet) as the ground state.

Primary absorption: $S_0 \xrightarrow{h\nu} S^*$ which may be S_1, S_2, S_3, etc. (*i.e.* excited singlet states)

Some of the photophysical processes to deactivate S* are **unimolecular** while some of the processes are **bimolecular**. The unimolecular processes are the **intrinsic properties** of the molecules while the bimolecular processes involving the collisions depend on the external conditions.

Unimolecular photophysical processes:

- **Internal conversion (IC):** $S_3 \rightarrow S_2 \rightarrow S_1 \rightarrow S_0$ (nonradiative)
- **Intersystem crossing (ISC):** $S^* \rightarrow T^*$ (nonradiative)
- **Fluorescence emission:** $S_1 \rightarrow S_0 + h\nu_f$ (radiative) ($\nu > \nu_f$)
- **Phosphorescence emission:** $T_1 \rightarrow S_0 + h\nu_p$ (radiative) ($\nu > \nu_f > \nu_p$)

(T* denotes the excited **triplet states** like T_2, T_1 in the present system where S_0 (singlet) is the ground state).

Energy transfer: Electronic transfer (Sec. 6.10.13), collisional transfer as the heat energy (Sec. 6.9.3).

Bimolecular photophysical processes

- $S^* + M \longrightarrow S_0 + M + h\nu_q$ (collision induced emission)
- $S^* + M \longrightarrow S_0 + M + \text{heat}$ and $T^* + M \longrightarrow S_0 + M + \text{heat}$ (collisional deactivation, *i.e.* collisional quenching, the released heat energy may initiate the chemical reactions, Sec. 6.9.3)
- $S^* + S_0 \longrightarrow S_0 + S^*$ (singlet-singlet) and $T^* + T_1 \longrightarrow T_1 + T^*$ (triplet-triplet) (electronic energy transfer, ET, Sec.6.10.13)

Mechanisms of bimolecular quenching of an excited state: Collisional deactivation, resonance energy transfer and electron transfer.

Collisional deactivation may dissociate the quencher. The transferred heat energy dissociates the quencher (see Sec. 6.9.3).

Nonradiative decay through collision with M is called **collisional quenching** or **external conversion (EC)**.

Different photophysical processes of deactivation of the photoexcited species are shown in **Jablonski diagram** (Fig. 6.10a,b).

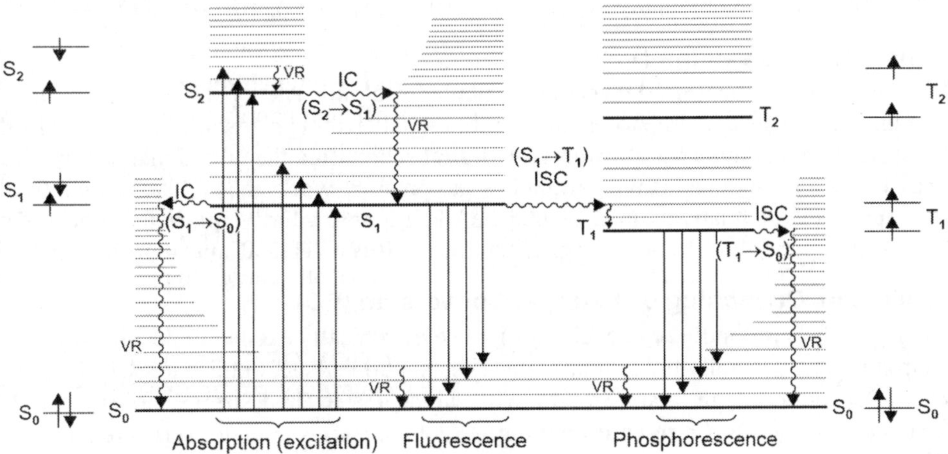

Fig. 6.10a: Jablonski diagram showing the excitation and photophysical processes of deactivation of the excited states in a typical organic molecule whose ground state is singlet (*i.e.* paired electrons)

- **Vibrational relaxation (VR)** represented by vertical wavy line
- **Internal conversion (IC)** and **intersystem crossing** (ISC) represented by horizontal wavy lines

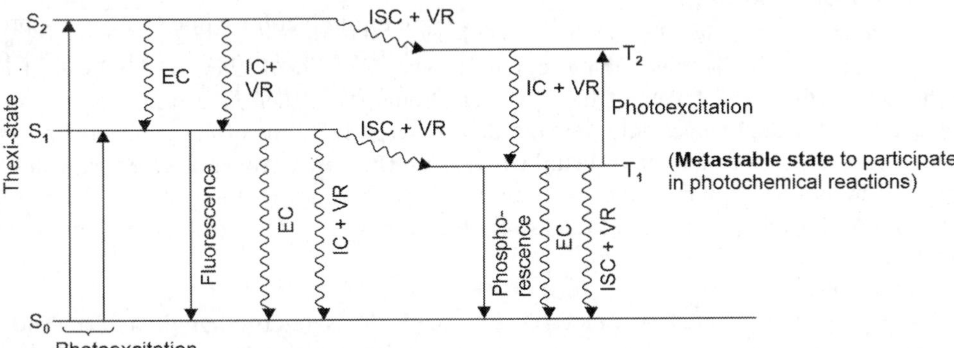

Fig. 6.10b: Simplified **Jablonski diagram** (without showing the vibrational levels) showing the photophysical processes

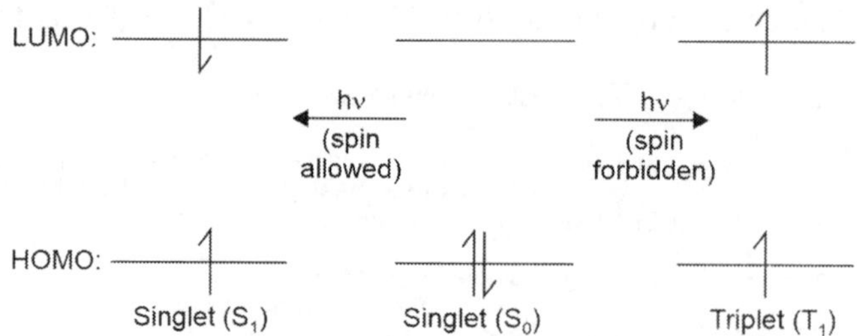

Fig. 6.10.2: Spin allowed and spin forbidden photoexcitations. In terms of exchange energy, T_1 is more stable than S_1 **spin-orbit coupling relaxes the spin selection rule** (see Sec. 6.10.8 for details and Vol. 2, Fig. 9.9.2.3)

Some special photophysical processes are:

$S^* + S_0 \longrightarrow S_0S^*$ **(Excimer formation)**, $S_0S^* \longrightarrow 2S_0 + h\nu_E$ **(excimer emission)**

$S^* + h\nu \longrightarrow S_0 + 2h\nu$ **(Stimulated emission as in LASER)**

Time scale

Light absorption (excitation, $S_0 \rightarrow S_n$): ~ 10^{-15} s; IC ($S_n \rightarrow S_1$) = 10^{-11}–10^{-14} s; VR ($S_n^* \rightarrow S_n$) = 10^{-10}–10^{-12} s; ISC ($S_1 \rightarrow T_1$) = 10^{-6}–10^{-11} s; Fluorescence ($S_1 \rightarrow S_0$) = 10^{-6}–10^{-9} s; Phosphorescence ($T_1 \rightarrow S_0$) = 10^{-3}–10^{2} s, Nonradiative decay: 10^{-5}–10^{-7} s ($S_1 \rightarrow S_0$), 10^{-2}–10^{3} s ($T_1 \rightarrow S_0$).

6.10.1 Vibrational Relaxation (VR)

Molecules in an **excited vibrational level** of a particular electronic energy state undergo **collisions** with the surrounding molecules and drop to the lower vibrational levels (*i.e.* towards $v = 0$) **within the same electronic state**. Through this **spin allowed vibrational relaxation**, the vibrational energy is lost as the **heat energy** to the surroundings. Obviously, at low gas pressure where the collisions are infrequent, the VR process appears to be relatively inefficient. The VR is a **nonradiative path** and it occurs **very fast** (10^{-12}–10^{-13} s time period). VR is often represented by a **wavy vertical arrow** in Jablonski diagram.

6.10.2 Collisonal Quenching or External Conversion (EC)

On collision with the surrounding molecules (M), the electronically excited molecule may return to the lower electronic state, generally the ground state. The excitational electronic energy is dissipated as the heat energy. Depending on the nature of M with which the excited species undergo collision, the process may be described as the **solvent quenching** (collision with solvent), **self-quenching** (collision among themselves) or **impurity quenching** (collision with an impurity molecule).

EC: $S^* + M \rightarrow S_0 + M$ + heat **(nonradiative process)**

Efficiency of this process depends on the collision frequency. This is why, this process is more efficient in the solid and liquid phase (10^{-12}–10^{-14} s) than in the gas phase (*ca.* 10^{-8} s).

6.10.3 Internal Conversion (IC)

Vibrational relaxation (VR) rapidly brings the molecule at $v = 0$ of the same electronic state. When two electronic states intersect, in the overlapping region of lower vibrational levels of the upper electronic state and the upper vibrational levels of the lower electronic state, there is a transition to the lower electronic state from the upper electronic state (at the **isoenergetic point**) of the same spin multiplicity (*i.e.* $\Delta S = 0$). This process is called the **internal conversion** (IC) where **there is no energy change**. In this process, generally transition from the upper electronic state occurs at $v = 0$ or very close to zero to the higher vibrational levels (*i.e.* $v > 0$) of the lower electronic state at the isoenergetic point present in the overlapping region.

Thus IC represents:

transition or cross-over to the lower electronic state (at the isoenergetic point) of the same spin multiplicities ($\Delta S = 0$, *e.g.* $S_2 \rightarrow S_1 \rightarrow S_0$; $T_2 \rightarrow T_1$, etc.) **through the spin allowed vibrational coupling**.

The process is extremely fast except for $S_1 \rightarrow S_0$ and $T_1 \rightarrow T_0$. For $S_3 \rightarrow S_2$, $S_2 \rightarrow S_1$, $k_{IC} \approx 10^{12}$–10^{13} s^{-1}; $S_1 \rightarrow S_0$, $k_{IC} \approx 10^8$ s^{-1}.

The IC process involves no energy change and consequently, the process is radiationless. Through the efficient VR, the excited molecules return rapidly to $v = 0$ (ground vibrational level) or v = close to zero and then it passes to the higher vibrational levels of the intersecting lower electronic state (of the same spin multiplicity). After transfer to the higher vibrational level of the lower electronic state, through the VR process, it comes rapidly to its ground vibrational level ($v = 0$). It is illustrated below:

$$S_2^{v>0} \xrightarrow{VR} S_2^{v=0} \xrightarrow{IC} S_1^{v>0} \xrightarrow{VR} S_1^{v=0}$$

(Isoenergetic: $S_2^{v=0}$ and $S_1^{v>0}$)

Here it is important to note that the IC (**radiationless**) between the states of same spin multiplicity is a **very rapid process** (~ 10^{12} s^{-1}) because it does not involve any radiation and consequently, **its rate is not limited by the quantum-mechanical factors that limit the radiative processes**.

Mechanism of the radiationless IC: In the overlapping region of the two intersecting electronic states, the horizontal transition to the lower electronic state occurs **at the isoenergetic point**. For a particular horizontal transition through IC, $S_2^{v=0} \rightarrow S_1^{v=4}$ (say), the efficiency of the process depends on the **overlap integral** (*i.e.* **Franck–Condon integral** called **Franck–Condon factor**) between $S_2^{v=0}$ and $S_1^{v=4}$ states and it is given by:

$$I_{2\rightarrow 1} = \int \Psi_2^{v=0} \Psi_1^{v=4} d\tau_V \quad (cf. \text{ Fig. } 6.10.3.1)$$

Qualitative understanding of the magnitude of the overlap integral can explain why the transition, $S_1^{v=0} \rightarrow S_0^{v>0}$ is slower than the transition, $S_2^{v=0} \rightarrow S_1^{v>0}$. The energy difference between the S_1 and S_2 states is relatively smaller than the energy difference between the S_1 and S_0 states. Consequently, the zero-vibrational level of S_1 can only overlap with the high vibrational levels of the ground state (S_0), say $S_0^{v=12}$ and the transition probability is determined by the magnitude of the following overlap integral.

$$I_{1\rightarrow 0} = \int \Psi_1^{v=0} \Psi_0^{v=12} d\tau_V \quad (cf. \text{ Fig. } 6.10.3.1)$$

Because of the larger energy difference between the S_1 and S_0 energy states, their potential energy (PE) surfaces do not intersect. In fact, the two curves run almost parallel and they overlap near the central position denoting the equilibrium nuclear geometry (*i.e.* $r = r_e$) **where the kinetic energy is large** (*cf.* kinetic energy is almost zero at the turning points of a vibrational level). In other words, $S_1^{v=0}$ can only overlap with the almost topmost vibrational levels (say $v = 12$) near the equilibrium position (*i.e.* $r = r_e$) of S_0. On the other hand, because of the smaller energy difference between the S_1 and S_2 states, the PE surfaces of the S_2 and S_1 states can intersect and let the isoenergetic point be $v = 0$ for S_2 and $v = 4$ for S_1. In other words, $S_2^{v=0}$ can overlap with the lower vibrational level (say, $v = 4$) of S_1 at the **extreme point where the probability of the said overlap is higher** compared to that at the central position of the vibrational level (*cf.* harmonic oscillator) (Figs 6.10.3.1 and 6.9.1.1c).

Here it is worth mentioning that transfer from the higher electronic energy state to the lower electronic energy state through the spin allowed IC-process may also occur from the higher vibrational level ($v > 0$) of the upper electronic state at the isoenergetic point (*i.e.* cross-over point) if the corresponding overlap integral is sufficiently high. But the efficient VR-process brings the upper electronic state at $v = 0$ very rapidly. This is why, the transition occurs generally from $v = 0$ or close to zero of the higher electronic energy state.

Pictorial representation indicates: $I_{2\rightarrow 1} \gg I_{1\rightarrow 0}$ (Fig. 6.10.3.1)

The lower value of $I_{1\rightarrow 0}$ arises mainly due to the overlap at the central position (where the kinetic energy is very high) of very high vibrational level (say $v = 12$, close to the topmost level) of the S_0-state. It may be noted that for such a high vibrational level, probability distribution at the central position is very small (*cf.* harmonic oscillator, Fig. 6.9.1.1c). This is why, the IC process

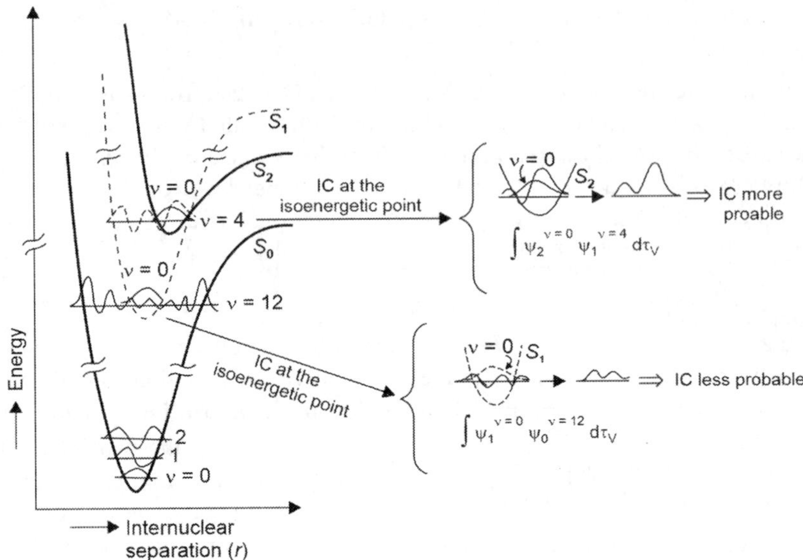

Fig. 6.10.3.1: Schematic representation of the potential energy diagrams of S_0, S_1 and S_2 electronic states and Franck–Condon overlap integral between $S_2^{v=0}$ and $S_1^{v=4}$ (for the horizontal energy transfer from $S_2 \rightarrow S_1$ through internal conversion); and between $S_0^{v=12}$ and $S_1^{v=4}$ (internal conversion, $S_1 \rightarrow S_0$). (*Ref.* Fundamentals of Photochemistry by K.K. Rohatgi–Mukherjee, New Age International Publishers, 3rd edition).

Note: For the higher vibrational level (say $v = 12$), during the oscillation about the equilibrium position, the molecule classically spends most of its time near the turning points. Thus, its probability of existence is very high at the turning point but very low at the central positions.

$S_1^{v=0} \rightarrow S_0^{v \gg 0}$ is relatively less efficient. In general, the larger energy difference between the states disfavours the IC-process.

The efficient VR and spin allowed IC-process combinedly bring the excited species (S_x, $x > 1$) rapidly to the zero-vibrational level of S_1 (assuming S_0 as the ground state) from which **IC to S_0 is not much favoured**. This is why, $S_1^{v=0}$ is relatively a stable state (*i.e.* **relatively long lifetime)** and described as the **thexi-state** (describing the *the*rmally equilibrated *exci*ted state). This is why, in general, **Kasha's rule** states as:

fluorescence emission occurs from the zero vibrational level of the first excited state (*e.g.* $S_1^{v=0}$ for the singlet ground state and $T_1^{v=0}$ for the triplet ground state) whatever may be the excitation.

6.10.4 Intersystem Crossing (ISC)

It is similar to IC ($\Delta S = 0$) but here $\Delta S \neq 0$ (*e.g.* $S_1 \rightarrow T_1$, $S_2 \rightarrow T_2$, etc.) (Singlet)($\uparrow\downarrow$) $\xrightarrow{\text{ISC}}$ ($\uparrow\uparrow$) (Triplet). Thus in this process, there is a **spin-forbidden transition** between the electronic states of different spin multiplicities. Because of the spin-forbiddenness, the ISC process is much slower than the spin allowed IC process. As in the IC process, in the ISC process, the $S_1 \rightarrow T_1$ transition is faster than the $T_1 \rightarrow S_0$ transition (called **reverse intersystem crossing**) because of the higher energy difference between the T_1 and S_0 states compared to that between the S_1 and T_1 states.

$$S_1 \xrightarrow{\text{ISC}} T_1 \ (k_{ISC} \approx 10^6 - 10^{11} \text{ s}^{-1}),\ T_1 \xrightarrow{\text{ISC}} S_0 \ (k_{ISC} \approx 10^{-2} - 10^4 \text{ s}^{-1})$$

The mechanism of ISC is similar to that of IC. But, because of the spin-forbiddenness, ISC is about 10^6 times slower than the spin allowed VR and IC. **Spin-orbit (SO) coupling** (important in transition metal ions and heavier atoms) **can partially relax the spin selection rule in ISC** (see Sec. 6.10.8 for

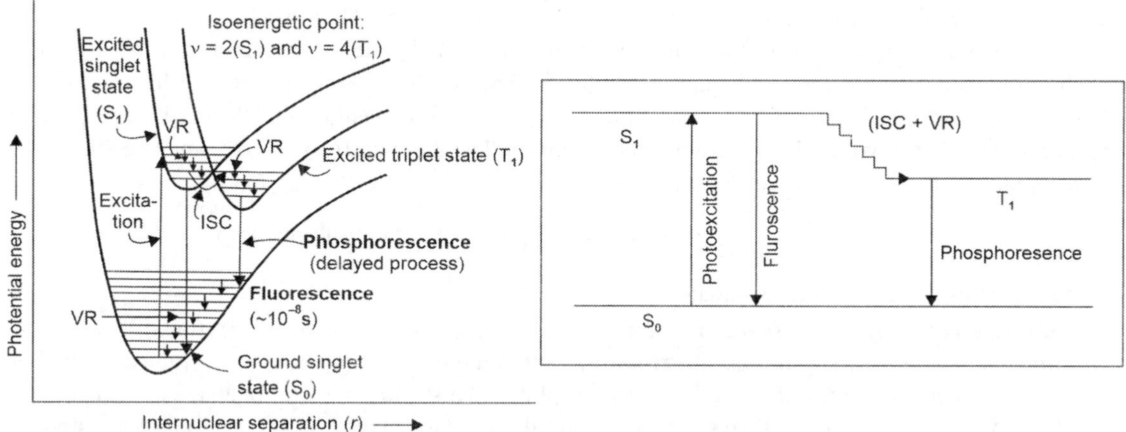

Fig. 6.10.4.1: Schematic illustration of **intersystem crossing** (ISC) from the excited singlet (S_1) state to the excited triplet state (T_1). Simple way of representation of ISC (Box); VR denotes the radiationless vibrational relaxation

details) Spin-orbit coupling leads to the vectorial addition of the spin and orbital angular momenta and thus in such systems, the spin angular momentum is not well defined and it partially removes the spin-forbiddenness of the $S_1 \rightarrow T_1$ transition. Thus efficiency of ISC depends upon the success of spin-orbit coupling interaction which can allow the spin pairing or unpairing. Thus, at the crossing point (*i.e.* **isoenergetic point**) of the PE surfaces of the excited singlet (say S_1) and excited triplet state (say T_1), both the singlet and triplet states have the **common nuclear geometry** and the spin-orbit coupling allows the spin conversion and the transfer, $S_1^{\nu=0} \rightarrow T_1^{\nu>0}$ occurs and then the rapid VR within the T_1-state produces the $T_1^{\nu=0}$ (*i.e.* zero-vibrational level of the T_1 state).

$$S_1^{\nu>0} \xrightarrow{\text{VR}} S_1^{\nu=0} \xrightarrow{\text{ISC}} T_1^{\nu>0} \xrightarrow{\text{VR}} T_1^{\nu=0}$$

(Isoenergetic: $S_1^{\nu=0}$ and $T_1^{\nu>0}$)

It may be noted that as in the spin-allowed IC-process, the transfer from the S_1-state to the T_1-state may also occur from the higher vibrational levels of S_1, if the isoenergetic condition at the cross-over point is attained (Fig. 6.10.4.1). But, very often the efficient VR process rapidly produces the $S_1^{\nu=0}$ state. This is why, ISC generally occurs from $\nu = 0$ or close to 0 of the upper electronic state as in IC.

- **Radiative photophysical processes:** Fluorescence (spin-allowed transition) and phosphorescence (spin-forbidden transition).
- **Nonradiative photophysical processes:** Spin allowed vibrational relaxation (VR), spin allowed internal conversion (IC), spin forbidden intersystem crossing (ISC), and collisional quenching or external conversion (EC).

6.10.5 Fluorescence Emission

Let us illustrate this radiative pathway of deexcitation by taking the singlet ground state (S_0) which is generally found in the spin-paired organic molecules. The initial excitation (*i.e.* primary absorption) occurs in the time period, *ca.* 10^{-15} s following the Franck–Condon principle (*i.e.* vertical transition) to produce the excited states of the same spin multiplicity (*i.e.* S_1, S_2, S_3, ...). Through the nonradiative **vibrational relaxation** (VR) and **internal conversion** (IC), the excited singlet states are brought to the S_1 state (1st excited singlet state) at the zero-vibrational level, *i.e.* $S_1^{\nu=0}$. In the VR-process through collision with the surrounding molecules, the excess vibrational energy over the zero-point vibrational energy ($v = 0$) in a particular excited electronic state is lost as heat. Then the IC converts an excited electronic state (generally with $v = 0$) into another lower excited state (of the same spin multiplicity) which immediately attains the $v = 0$ state through VR. The IC, $S_1^{v=0} \rightarrow S_0$ is not so efficient because of the larger energy difference between the S_1 and S_0 states. In conclusion, the VR and IC combinedly produce the $S_1^{v=0}$ state within 10^{-12}–10^{-13} s after excitation. The **thexi-state**, $S_1^{v=0}$ (*ther*mally equilibrated *exci*ted state) undergoes fluorescence emission.

$$S_1^{v=0} \xrightarrow[\text{(time scale} \sim 10^{-8}\text{ s)}]{\text{fluorescence}} S_0 + h\nu_f, (\nu_f < \nu = \text{excitation frequency})$$

The above observations are formulated as two rules:

(i) **Kasha's rule:** The fluorescent emission generally occurs from the zero-vibrational level of the first excited electronic state (*i.e.* **thexi-state**) though the photo-excitation may produce the higher excited electronic states of the same spin multiplicity. This rule can be rationalised by considering the radiationless efficient VR and IC processes of deexcitation. This rule indicates that the **radiative emission occurs from the lowest (*i.e.* first) excited electronic state**. It makes ν_f independent of the frequency of exciting radiation.

(ii) **Vavilov's rule:** Fluorescence quantum efficiency (ϕ_f) is independent of the excitation wavelength. In other words, efficiency of fluorescence will be the same, no matter how the excited states are produced. ϕ_f depends on the relative efficiency of other competitive pathways.

The important features of fluorescent emission are pointed out here.

● **Red-shifted fluorescent emission (Kasha's rule) and Stokes' shift:** Due to the efficient VR and IC processes, fluorescent emission always occurs from $v = 0$ of the first excited electronic state (*i.e.* **thexi-state**), irrespective of the nature of excitation. It makes ν_f less than the ν of the exciting radiation. The energy difference between the absorption maximum and the fluorescence emission maximum is called the **Stokes' shift** (Figs 6.10.5.1, 2).

Exceptions: At high temperatures, if the higher vibrational levels ($v > 0$) of the ground electronic state are thermally populated to allow the excitation from these higher vibrational levels, then **anti-Stokes effect** (*i.e.* $\nu_f > \nu$) may be noted. It is illustrated by taking the following transitions:

$$S_0^{v=1} \xrightarrow[\text{excitation}]{h\nu} S_1^{v=0} \xrightarrow[\text{fluorescence}]{h\nu_f} S_0^{v=0}, \nu_f > \nu$$

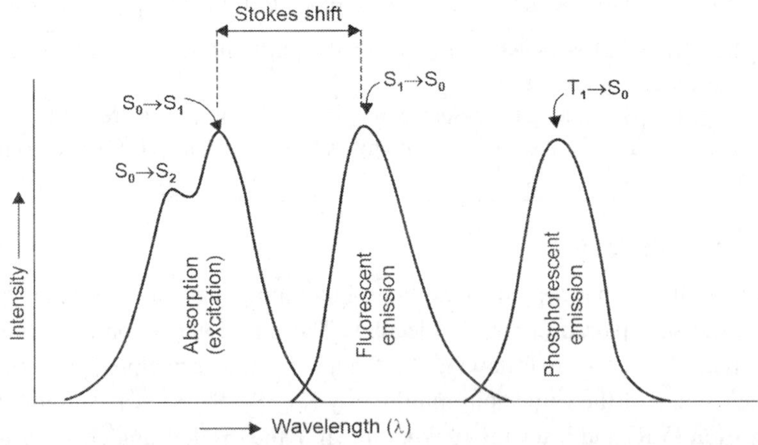

Fig. 6.10.5.1: Qualitative illustration of the absorption spectrum and **red-shifted** fluorescent and phosphorescent spectra (assuming S_0 as the ground electronic state)

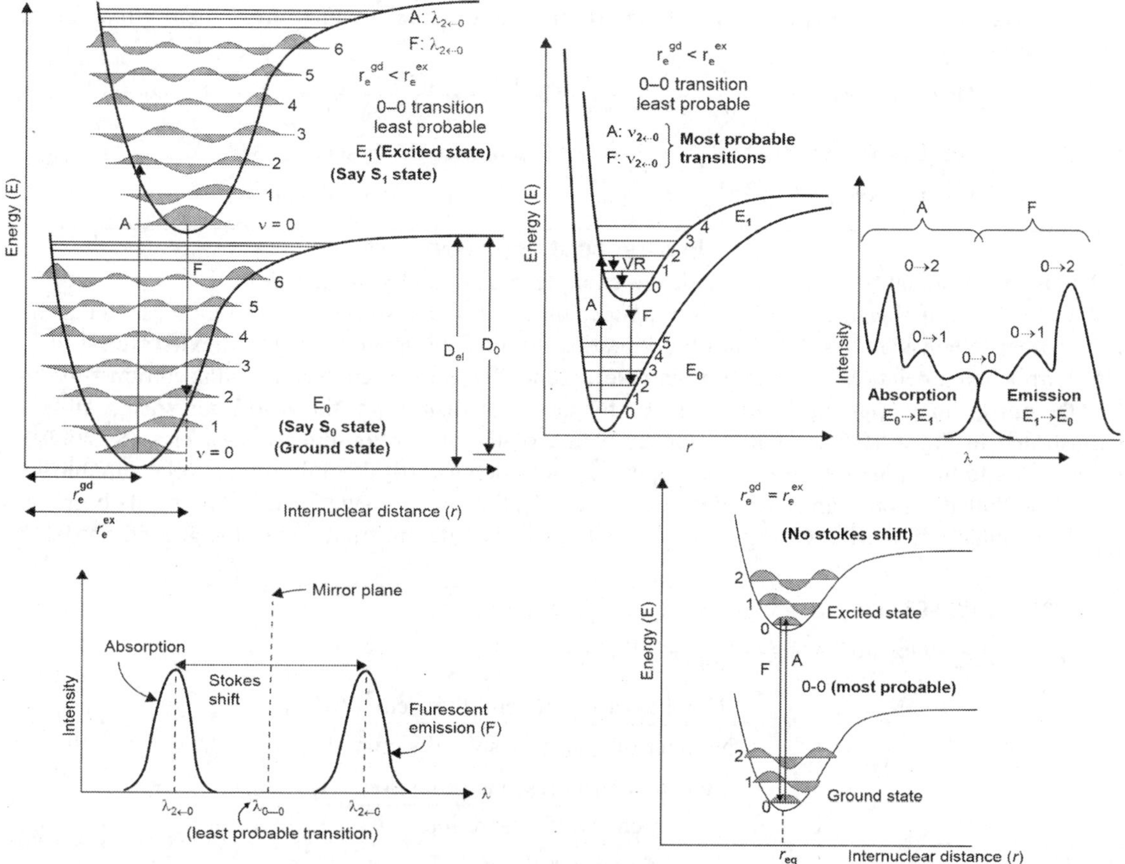

Fig. 6.10.5.2: Illustration of photoexcitation (A) and fluorescent emission (F) for the geometric relations: $r_e^{gd} = r_e^{ex}$ and $r_e^{ex} > r_e^{gd}$

- **Stokes shift:** Energy difference between the maximum of the absorption (*i.e.* excitation) band and that of the **red-shifted** fluorescent emission band; larger the difference between the internuclear equilibrium distances of the ground state and excited state geometries (*i.e.* $\Delta r_e = r_e^{ex} - r_e^{gd}$), larger the Stokes shift.
- Fluorescent emission spectra, generally independent of the excitation frequency (*cf.* **Kasha'a rule**).
- **Mirror image relationship (Levschin's rule)** between the fluorescent emission spectrum ($S_1 \rightarrow S_0$) and absorption spectrum for the absorption, $S_0 \rightarrow S_1$ and assuming the similarities among the vibrational energy levels of the S_0 and S_1 states.

This is why, very often, there is a **overlap region between the absorption and fluorescence emission spectra**. The following examples illustrate the **special cases of fluorescence** emission.

(i) In **azulene**, the IC, $S_2 \rightarrow S_1$ is relatively inefficient because of the higher energy difference between the S_2 and S_1 states. Consequently, **fluorescent emission occurs from both the S_2 and S_1 excited states** (*exception of Kasha's rule*).

(ii) In the **gaseous phase at low pressure**, VR through the collision becomes relatively inefficient because the probability of collision decreases under the conditions. When the interval between the collisions is attained about 10^{-8}–10^{-9} s, the collisional deactivation rate of excess vibrational energy becomes comparable with the fluorescence emission rate. Under the situation, **fluorescence emission may occur from the higher vibrational levels of the excited states** (*exception of Kasha's rule*).

(iii) If the wavelength of fluorescence emission is exactly the same as that absorbed (*i.e.* $\nu_f = \nu$, Stokes' shift = 0), then the observed fluorescence is called the **resonance fluorescence**.

Resonance fluorescence is noticed in **monoatomic vapours** at low pressures where vibrations are not possible.

$$\text{Na (gas phase)}: 3s^1 \xrightarrow[(\lambda = 589,\ 589.5\ \text{nm})]{h\nu} 3p^1 \xrightarrow[(\lambda = 589,\ 589.5\ \text{nm})]{h\nu_f} 3s^1$$

$$\left.\begin{array}{l}\text{Na–D}_1\ (589.5\ \text{nm}): 3\,^2P_{1/2} \rightarrow 3\,^2S_{1/2} \\ \text{Na–D}_2\ (589\ \text{nm}): 3\,^2P_{3/2} \rightarrow 3\,^2S_{1/2}\end{array}\right\} \begin{array}{l}\textbf{Resonance fluorescence}: 3p^1 \rightarrow 3s^1 \\ \text{Life-time of the } 3p^1 \text{ state} \sim 10^{-8}\ s\end{array}$$

Fluorescent emission

- ν_f is independent of the frequency of the exciting radiation (**Kasha'a rule**).
- Wavelength of fluorescence emission is independent of the absorption wavelength assuming the fluorescence emission to occur from the vibrationally relaxed upper state (*i.e.* **thexi-state**).
- Mirror image relationship (**Levschin's rule):** Generally, fluorescence emission spectrum appears as the mirror image (taking the mirror plane for the ν_{0-0} transition) of the absorption spectrum. It is illustrated in Fig. 6.10.5.2. This symmetric nature of the absorption-emission spectra ($S_0 \rightleftarrows S_1$) arises due to the same transitions (one upward, other downward) assuming the similar vibrational levels in both the ground and excited states. Obviously, the mirror image relationship is not observed, if the geometry of the fluorescing excited state (say S_1) is different from that of the ground state (*i.e.* S_0).

- **Quantum efficiency** of fluorescence (ϕ_f): It is defined as:

$$S_0 \xrightarrow[\text{absorption}]{I_a} \text{excited states} \xrightarrow[\text{(fast)}]{\text{VR + IC}} S_1 \xrightarrow[\text{(slow)}]{k_f} S_0$$

$$\phi_f = \frac{\text{Number of quanta emitted/s/cm}^2}{\text{Number of quanta absorbed/s/cm}^2}$$

$$= \frac{\text{Intensity of fluorescence emission}}{\text{Intensity of absorption}} = \frac{I_f}{I_a}$$

$$= \frac{\text{Rate of emission}}{\text{Rate of absorption}} = \frac{k_f[S_1]}{I_a}$$

The intensity of radiation absorbed (I_a) at any particular wavelength is given by:

$$I_a = I_0 - I = I_0 - I_0\exp(-2.303\varepsilon Cl) = I_0[1 - \exp(-2.303\varepsilon Cl)]$$

I_0 = incident light intensity, I = transmitted light intensity, ε = molar extinction coefficient, C = molar concentration, l = optical path length (*cf.* A = absorbabce = $\log(I_0/I) = \varepsilon Cl$, **Beer–Lambert's law).**

$$I_f = I_a\phi_f = \phi_f I_0\,[1 - \exp(-2.30\varepsilon Cl)]$$

$$\approx \phi_f I_0(1 - 1 + 2.303\varepsilon Cl) = 2.303\ I_0\phi_f\varepsilon Cl\ (\textit{cf.}\ \exp(-x) \approx 1 - x, \text{ when } x \text{ is small})$$

The above relation indicates:

(a) ϕ_f is a property of the fluorescing excited state.

(b) I_f at any wavelength depends on the molar concentration and molar absorptivity (= εCl) of the absorbing species in the ground state. It explains why the fluorescence spectrum reproduces the absorption spectrum.

(c) $I_f \propto C$ is the basis of the quantitative fluorometry.

6.10.6 Factors Favouring the Fluorescence Emission

For the organic fluorescent molecules, the ground state is generally singlet (S_0). The photoexcitation leads to the excited singlet states S_1, S_2, ... Then through the internal conversion and vibrational relaxation, the excited molecule rapidly comes to $v = 0$ of S_1 state (*i.e.* ground vibrational level of the first excited singlet state described as the **thexi-state**) from which it comes back to the ground singlet state

(S_0) through the fluorescence emission **provided the nonradiative deexcitation of S_1 to S_0 is efficiently prevented**. The following factors favour the fluorescence emission.

i. **Nature of the electronic transition ($\pi \rightarrow \pi^*$ *vs.* $n \rightarrow \pi^*$, *n* denoting NBMO) in excitation:** Organic molecules experiencing the extended delocalisation as in the **aromatic compounds**, aliphatic and alicyclic carbonyl compounds bearing the conjugated double bonds undergo $\pi \rightarrow \pi^*$ transitions **(photoexcitation)** with the high molar absorptivities (*i.e.* high molar extinction coefficients). The **heterocycles** like pyridine, furan, thiophene, etc. undergo the $n \rightarrow \pi^*$ transition **(photoexcitation)** with the lower molar absorptivities because the nonbonding MOs (*i.e. n*) do not overlap with the π^*–MOs as much as the bonding π–MOs do with the π^*–MOs (see Vol. 7). Thus excitation through the $\pi \rightarrow \pi^*$ transition is more efficient than through the $n \rightarrow \pi^*$ transition. Because of the same ground, the $\pi^* \rightarrow \pi$ fluorescence emission is more efficient than the $\pi^* \rightarrow n$ fluorescence emission. Moreover, in the **heterocycles**, the excited singlet state produced through the $n \rightarrow \pi^*$ transition rapidly **converts to the excited triplet state through ISC** because of the less energy difference between the excited singlet and triplet states (*cf.* ISC is less efficient for the $\pi \rightarrow \pi^*$ excitation). Because of these two factors, fluorescence efficiency follows the order: $\pi^* \rightarrow \pi \gg \pi^* \rightarrow n$.

Fluorescence is found in the compounds experiencing the intense and low energy $\pi \rightarrow \pi^*$ transition. Simple heterocycles experiencing the lowest energy and inefficient $n \rightarrow \pi^*$ transitions are nonfluorescent. **However, fusion of a benzene ring to such a heterocycle makes it fluorescent.**

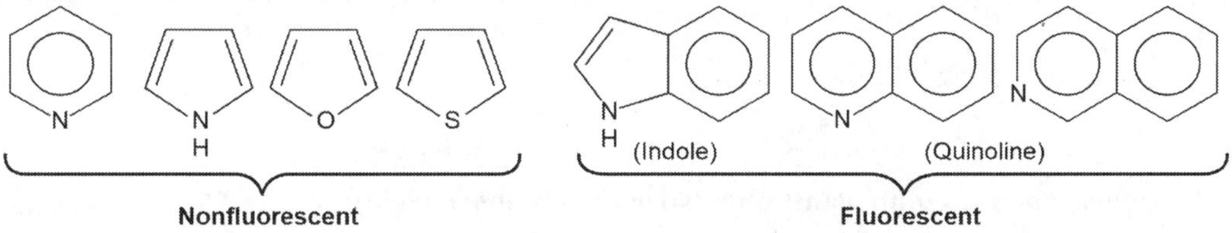

(Indole) (Quinoline)

Nonfluorescent **Fluorescent**

Among the **essential amino acids**, phenylalanine, tyrosine and tryptophan are fluorescent (*cf.* **effect of the presence of aromatic ring**).

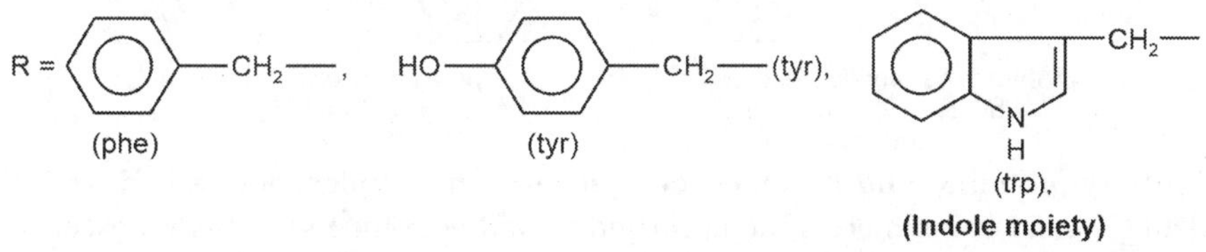

R = C_6H_5–CH_2–, HO–C_6H_4–CH_2–(tyr), (indolyl)–CH_2–

(phe) (tyr) (trp), **(Indole moiety)**

ii. **Nature of the excited electronic states:** The energy difference between the ground state (S_0) and the first excited state (S_1) should be sufficiently large **to prevent the nonradiative relaxation** (*i.e.* deexcitation) **to the ground state (S_0)** through the radiationless processes including the **internal conversion** (*cf.* efficiency of internal conversion depends on the energy difference between the concerned electronic states, see Sec. 6.10.5). The energy difference between the excited singlet and triplet states should be sufficiently large **to prevent the intersystem crossing (ISC)**

Extended conjugation and intensity of fluorescence emission ($\pi^* \rightarrow \pi$)

- **$\pi \rightarrow \pi^*$ excitation transition:** Extended conjugation reduces the energy gap between the HOMO (*i.e.* π-BMO) and LUMO (*i.e.* π-ABMO) and causes the **red shift** in the $\pi \rightarrow \pi^*$ excitation (see the **model, particle in a 1-D box**, Vol. 1, $\Delta E \propto 1/d^2$, d = length of the 1D-box). Because of this smaller energy gap (ΔE), the **spin selection rule** is relaxed better through the **spin-orbit coupling mechanism** leading to the mixing of the excited triplet and ground singlet states (see Fig. 6.10.2 and Vol. 7). Thus the red shifted $\pi \rightarrow \pi^*$ excitation favours the $\pi^* \rightarrow \pi$ fluorescence emission in the same way.
- **Structural rigidity:** Extended delocalisation introduces the **partial double bond character** in the C—C and C—N bonds of the molecular skeleton. Thus it reduces the flexibility of the molecule and **increases the structural rigidity** to favour the fluorescence.

which may favour the phosphorescence.

iii. **Structural rigidity to prevent the deexcitation to the ground state through the loss of rotational and vibrational energy:** If the excited molecule relaxes to the ground state through the loss of vibrational and rotational energies (*i.e.* nonradiative decay to the ground state through the conversion of excitational energy into the heat energy), the fluorescence (*i.e.* emission of UV-visible radiation) becomes an inefficient process. Any substituent favouring the molecular vibration or rotation quenches the fluorescence. **Fused ring system gives the structural rigidity** to prevent the deexcitation to the ground state through the loss of vibrational and rotational energy. This aspect is illustrated in the following examples:

- Fusion of a large number of benzene rings providing the molecular rigidity (effect of introduction of partial double bond character in the C—C bonds) and extended delocalisation favours the fluorescence ($\pi^* \rightarrow \pi$ transition) as in **naphthalene, anthracene, phenanthrene**, etc.
- Simple **heterocycles** (*e.g.* pyrrole, pyridine, furan, etc.) do not exhibit the fluorescence but fusion of benzene rings to a heterocyclic nucleus makes them fluorescent (*cf.* fluorescence of **quinoline, indole**, etc. *vs.* simple nitrogen heretocycles.
- The much better fluorescence of **fluorene** compared to that of **biphenyl** can be explained by considering their molecular rigidity (*cf.* CH_2 in **fluorene** imparts the molecular rigidity).

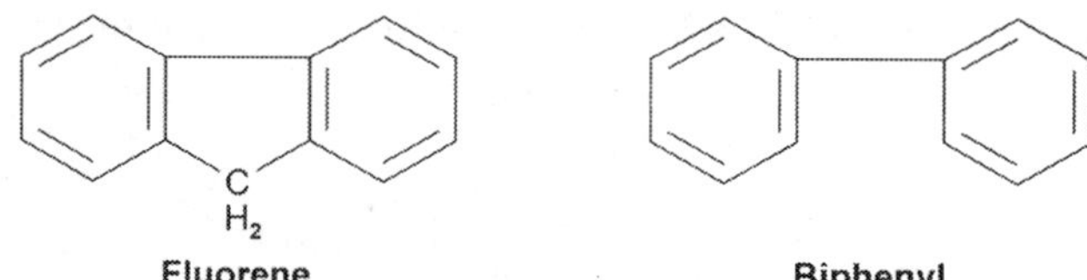

Fluorene **Biphenyl**

- **Azobenzene *vs.* azophenanthrene (structurally more rigid):**

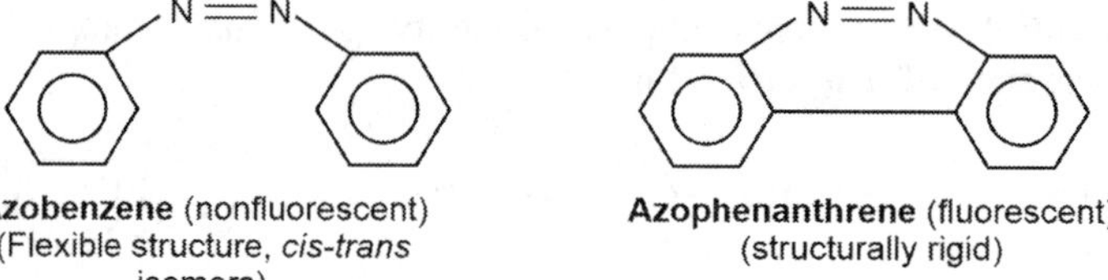

Azobenzene (nonfluorescent) (Flexible structure, *cis-trans* isomers)

Azophenanthrene (fluorescent) (structurally rigid)

- **8-Hydroxyquinoline and its metal complexes:** On complexation, 8-hydroxyquinoline as a chelating ligand gains more structural rigidity and consequently, its fluorescence property is increased.

N HO *vs.* N O Zn O N **(tetrahedral complex)**

(more fluorescent)

- **Phenolphthalein *vs.* fluorescein** (see Fig. 6.10.6.1): Both are coloured in alkaline condition but phenolphthalein is nonfluorescent while fluorescein is fluorescent (535 nm) under the identical condition. Both these compounds are structurally comparable and can act as the acid-base indicator.

In fluorescein, the **oxo-tricyclic ring imparts the structural rigidity** which prevents the rotation and loss of absorbed excitational energy via this route. On the other hand, under the identical

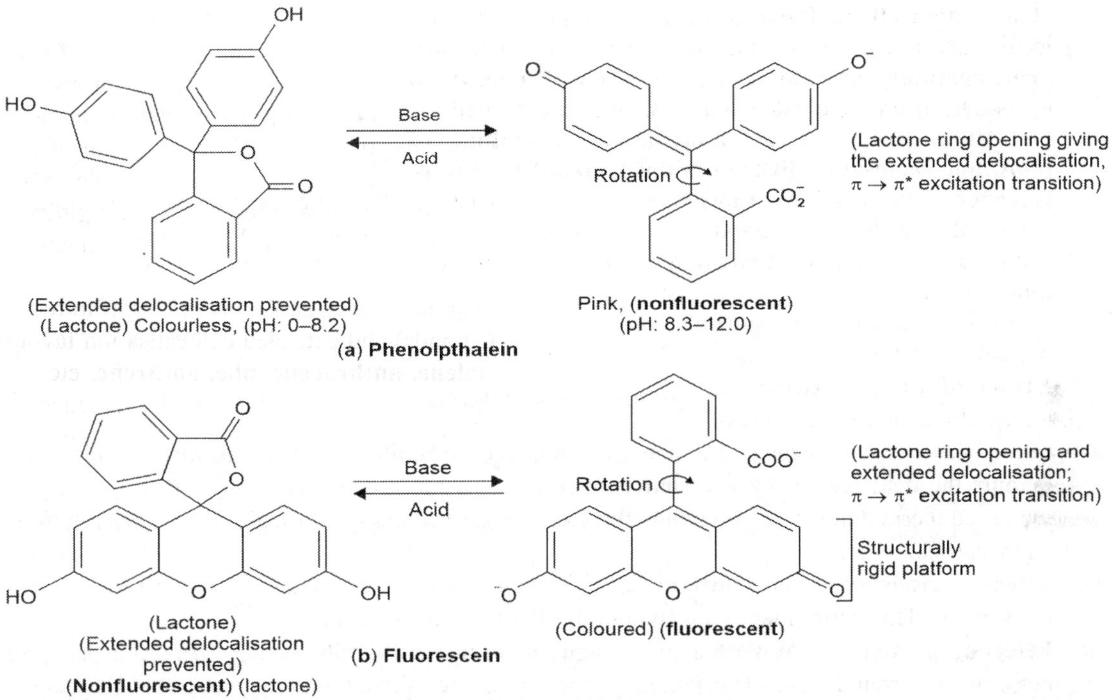

Fig. 6.10.6.1: Structural features of phenolphthalein and fluorescein (Fluorescence emission: $\pi^* \rightarrow \pi$ transition)

Note: Organic molecules having the **extended conjugation** experience the **structural rigidity and low energy $\pi \rightarrow \pi^*$ excitation** and show the fluorescence emission ($\pi^* \rightarrow \pi$). Any substituent which sterically prevents the π-electron delocalisation reduces the fluorescence emission.

condition, phenolphthalein lacking in such structural rigidity can lose the absorbed energy by this nonradiative pathway. This is why, **fluorescein is fluorescent while phenolphthalein is almost nonfluorescent.**

iv. **Electron donating groups enhancing the fluorescence while electron withdrawing groups inhibiting the fluorescence:** The electron donating group causes a **red shift** in the $\pi \rightarrow \pi^*$ transition while the electron withdrawing group causes a **blue shift** in the $\pi \rightarrow \pi^*$ transition (*cf.* electron withdrawing groups make the aromatic system electron deficient) (see Vol. 7).

The red-shifted $\pi \rightarrow \pi^*$ excitation for the electron donating groups makes the fluorescence emission ($\pi^* \rightarrow \pi$) more intense. Besides this, **it reduces the efficiency of $S_1 \rightarrow T_1$ ISC** as the energy difference between the S_1 and T_1 states is increased. On the other hand, for the electron withdrawing substituents, the reverse effects reduce the efficiency of fluorescence emission.

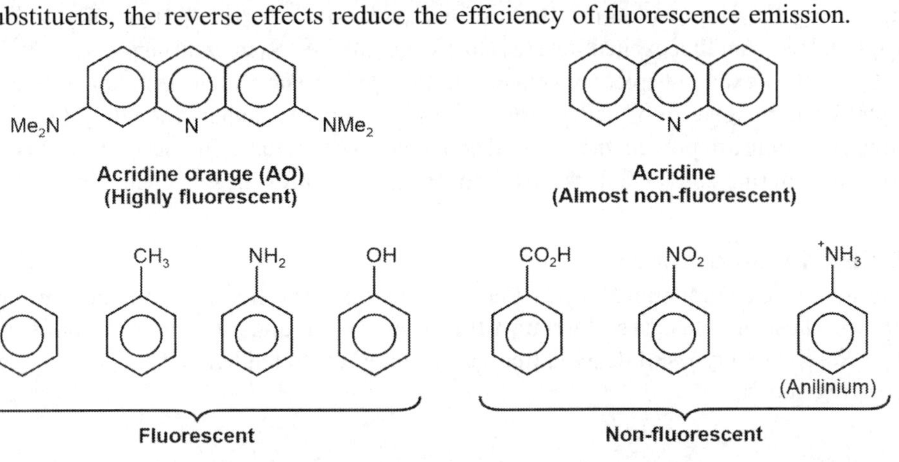

v. **Heavy atom effect:** Presence of heavy atoms in the molecular structure disfavours the fluorescence. The **high spin-orbit coupling constant** favours the excited molecule to cross-over from the excited singlet state to the excited triplet state via the ISC process. It **favours the probability of phosphorescence but disfavours the probability of fluorescence**. It has been experimentally verified for 1-halonaphthalene: the ratio of quantum efficiency of phosphorescence and fluorescence (ϕ_p/ϕ_f), goes to increase for X = F to I, *i.e.* the ratio becomes highest for X = I and minimum for X = F. This phenomenon is described as the **heavy atom effect**.

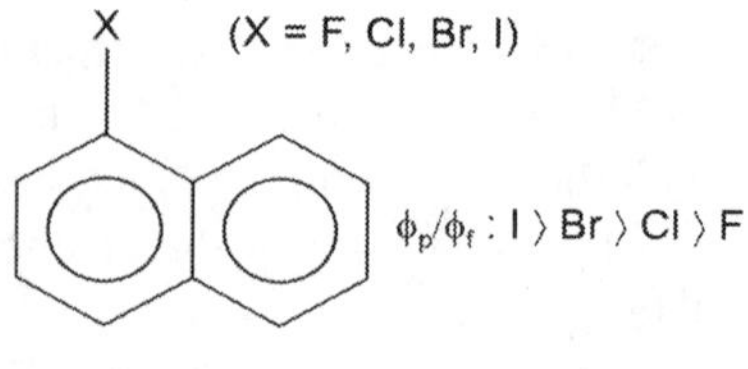

(X = F, Cl, Br, I)

ϕ_p/ϕ_f : I > Br > Cl > F

It also happens so for the simple halobenzenes.

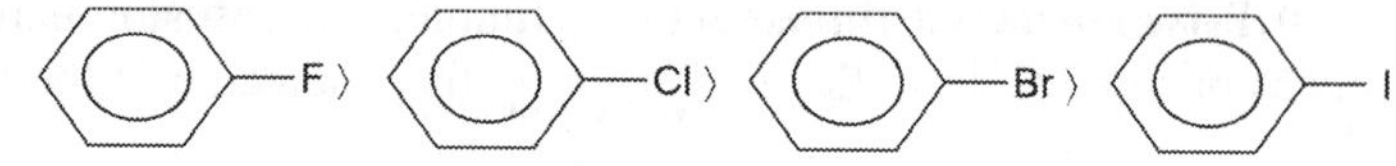

C₆H₅–F > C₆H₅–Cl > C₆H₅–Br > C₆H₅–I

Relative intensity of fluorescence

vi. **Effect of temperature:** At higher temperature, the fluorescence becomes less because the increased frequency of collisions at the elevated temperatures favours the deactivation by **external conversion** (*i.e.* collisional quenching). Besides this, the increased thermal motion also favours the loss of absorbed energy through the loss of vibrational and rotational energy.

vii. **Effect of viscosity of the solution:** Low viscosity favours the quenching through the external conversion. This consequently disfavours the fluorescence emission.

viii. **Effect of dissolved O_2 and other paramagnetic bodies in the solution:** O_2 (triplet) is paramagnetic at the ground state. The paramagnetic substances favour the conversion of the excited singlet state to the triplet state through **ISC** to disfavour the fluorescence. Dissolved O_2 may also photochemically destroy the fluorescing species.

ix. **Effect of pH:** Depending on the pH, the fluorescing species may have different structures having the different degrees of delocalisation. Thus the pH can control the efficiency of fluorescence (*cf.* Fig. 6.10.6.1). Aniline on being protonated in an acidic condition becomes nonfluorescent because in the anilinium cation, the degree of delocalisation of π-electron is reduced.

x. **Effect of concentration:** At higher concentration, **self-quenching** through the mutual collision can deactivate the molecules to disfavour the fluorescence. In some cases, the fluorescent emission may excite the other molecules remaining at the ground state (*cf.* energy transfer process).

Fluorescence: Extended life time of the 1st excited state (*i.e.* **thexi-state**) and prevention of other competitive pathways are the basic requirements.

Fluorescence favoured for: ● Molecular rigidity to prevent the vibrational and rotational energy loss. ● Rigid planar structure allowing the extended conjugation and experiencing the allowed **red-shifted** $\pi \rightarrow \pi^*$ excitation transition (*cf.* fluorescence emission: $\pi^* \rightarrow \pi$). ● High energy difference between the ground (S_0) and first excited state (S_1) to prevent the $S_1 \rightarrow S_0$ internal conversion. ● Disfavouring the excited singlet → excited triplet conversion via ISC. ● Presence of electron donating groups on the aromatic platform. ● Absence of heavy atoms and paramagnetic bodies (including O_2) which favour ISC. ● Low temperature to prevent the external conversion (*i.e.* collisional quenching). ● High viscocity to disfavour the external conversion. ● Low concentration to prevent self-quenching and self-absorbance.

6.10.7 Delayed Fluorescence

Usually, fluorescence occurs immediately after the excitation. The fluorescence rate constant is as high as 10^8–10^9 s^{-1}. These are the cases of **prompt fluorescence** but under certain conditions, fluorescence can be delayed. The two types of delayed fluorescence are illustrated here (assuming S_0 = ground state, S_1 = fluorescing state).

● **E-type delayed fluorescence:** ISC can convert S_1 into T_1 and if the life-time of T_1 state is sufficiently high, then the T_1 state may be reconverted into the S_1 state through the absorption of heat from the surrounding. In fact, **reverse VR** in T_1 with the help of thermal energy occurs and then ISC (*i.e.* $T_1 \rightarrow S_1$) occurs at the isoenergetic point. Obviously, conversion of the T_1 state to the fluorescing state S_1 is favoured at higher temperature.

$$S_1 \xrightarrow{\text{VR + ISC}} T_1 \xrightarrow[\text{(+ heat)}]{\text{Reverse VR + ISC}} S_1 \xrightarrow{h\nu_f\,(\text{ED})} S_0$$

It is named (E-type) so because it was first noted in **eosin**. Obviously, the reverse transition ($T_1 \rightarrow S_1$) occurs when the phosphorescence ($T_1 \rightarrow S_0 + h\nu_p$) emission is disfavoured.

● **P-type delayed fluorescence:** Bimolecular collision between the two molecules of T-state produces one molecule in the S_1-state and other molecule in the S_0-state.

$$S_1 \xrightarrow{\text{VR + ISC}} T_1,\ 2T_1 \xrightarrow{\text{Bimolecular collision}} S_1 + S_0$$

$$S_1 \longrightarrow S_0 + h\nu_f\ (\text{PD})$$

The process is named (P-type) so because it was first noted in **pyrene** and **phenanthrene**.

Note: Phosphorescence ($T_1 \rightarrow S_0 + h\nu_p$) is also a delayed process with respect to the prompt fluorescence. Thus the delayed emission may be either due to the delayed fluorescence or phosphorescence (difference: $\nu_{ex} > \nu_f > \nu_p$).

6.10.8 Phosphorescence Emission

It is illustrated for the systems having the S_0 as the ground state. Through the VR and IC processes, the excited singlet states (*i.e.* S_1, S_2, S_3, ...) attain rapidly (within 10^{-12}–10^{-13} s) the **thexi-state**, $S_1^{\nu=0}$ which may be transferred into into the T_1-state through ISC, if the S_1 and T_1 states do not differ much in energy. It has been already mentioned that for the heavy atoms and transition metal ions for which the **spin-orbit coupling is highly efficient**, the ISC process ($S_1 \rightarrow T_1$ when S_0 is the ground state or $T_1 \rightarrow S_1$ when T_0 is the ground state) is very much meaningful. Once, the lower energy state (say T_1) is attained from the S_1 state through the ISC process at the isoenergetic point of S_1 and T_1, it undergoes vibrational relaxation (VR) to give the $T_1^{\nu=0}$ state. From this $T_1^{\nu=0}$ state, the **spin-forbidden transition** to the S_0 state gives the phosphorescent emission ($h\nu_p$) (Fig. 6.10.1a, b).

$$S_1 \xrightarrow{\text{VR + ISC}} T_1 \xrightarrow{\text{VR}} T_1^{\nu=0} \xrightarrow[\text{Phosphorescence}]{-h\nu_p} S_0$$

$$S_1 \xrightarrow{\text{VR}} S_1^{\nu=0} \xrightarrow[\text{Fluorescence}]{-h\nu_f} S_0$$

($h\nu_f > h\nu_p$ as the energy of the S_1 state is higher than that of the T_1-state, *cf.* Fig. 6.10.1).

● **Relaxation of the spin selection rule:** In phosphorescence, triplet to singlet ($T_1 \rightarrow S_0$) transition is spin forbidden. But due to the spin-orbit (SO) coupling interaction, **the excited triplet state undergoes mixing with the ground singlet state**. Consequently, both the excited and ground states are the linear combination of the singlet and triplet states (see Fig. 6.10.2).

$$\psi_{\text{excited}} = a\psi_T + b\psi_s, (a > b)$$

$$\psi_{\text{ground}} = a'\psi_T + b'\psi_s, (b' > a')$$

Thus in the transition $\psi_{\text{excited}} \rightarrow \psi_{\text{ground}}$, the triplet→triplet and singlet→singlet transition components are spin allowed while the singlet→triplet and triplet→singlet transition components are spin-forbidden.

Thus the spin selection rule is relaxed through the spin-orbit coupling interaction. In the spin-orbit coupling interaction, the states of different spin multiplicities possessing the different spin and orbital angular momenta mix because **they possess the same total angular momenta**. It has been already mentioned that the spin-orbit coupling interaction becomes efficient when the energy difference between the states is not too large and the spin-orbit coupling constant is high as in the case of transition metal ions and heavy atoms.

The important features of phosphorescence are given here.

● **Red shifted phosphorescence:** The order is: $\nu_p < \nu_f < \nu$ (frequency of absorption). It can be realised by considering the position of energy states (Fig. 6.10.1a,b).

● **Time scale and low quantum yield (ϕ_p):** Compared to the spin-allowed transition in fluorescence, the spin-forbidden transition in phosphorescence is slower.

$$S_1 \xrightarrow{k_f} S_0 + h\nu_f \text{ (fluorescence)}, k_f \approx 10^8 \text{ s}^{-1}$$

$$T_1 \xrightarrow{k_p} S_0 + h\nu_p \text{ (phosphorescence)}, k_p \approx 10^{-2}–10^4 \text{ s}^{-1}$$

The time interval between absorption (*i.e.* excitation) and fluorescent emission is very small while the time interval in the case of phosphorescence is extended. In fact, a phosphorescent material may show the emission seconds, minutes or even hours after the initial absorption. **Because of this delayed emission, the process is termed phosphorescence.**

Because of the longer life-time of the phosphorescent state (say T_1 in $T_1 \rightarrow S_0$ transition), this **excited metastable state** (T_1) may be thermally re-excited to the S_1 state to show the **delayed fluorescence** (Sec. 6.10.7). The said metastable excited state (T_1) having the sufficiently longer life-time may undergo collision and get deactivated. Because of all these factors (*i.e.* collisional deactivation, re-excitation to the fluorescent state) at room temperature, the phosphorescence quantum yield (ϕ_p) is generally small (*i.e.* $\phi_p < \phi_f$). In the highly viscous glassy media and at low temperatures, the deactivating collisions can be significantly reduced to increase the ϕ_p-value.

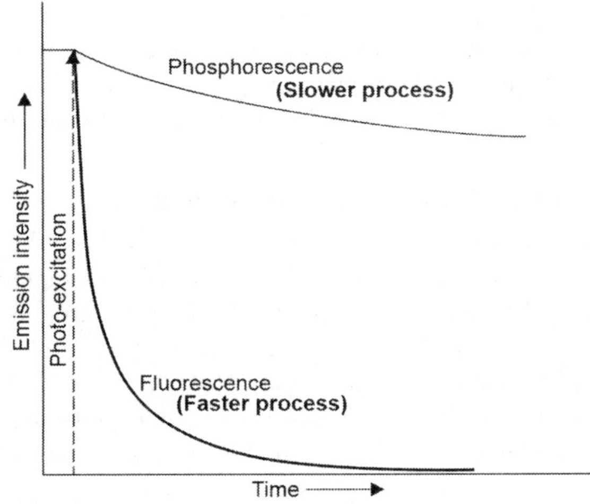

Fig. 6.10.8.1: Comparison of the time scale of fluorescence and phosphorescence processes

The long-lived phosphorescent state (T_1) may be quenched by the paramagnetic species or may undergo photochemical reactions to reduce the phosphorescence quantum yield (ϕ_p) further.

● **New absorption bands:** From the long-lived phosphorescent state (say T_1 where S_0 is the ground state), the new absorption bands for the $T_1 \rightarrow T_2$ and $T_1 \rightarrow T_3$ transitions (which are forbidden from the ground S_0-state) can be induced (Fig. 6.10.1b).

6.10.9 Factors Favouring the Phosphorescence Emission

The important factors favouring the phosphorescence are (assuming S_0 as the ground state):

(i) **ISC:** The conversion, excited singlet state→excited triplet state via ISC should be efficient. For this purpose, the energy difference between the **first excited singlet state (S_1) and first excited triplet state (T_1)** should be small and the spin-orbit coupling constant (*cf.* heavy atom effect) should also be high to favour the ISC process. The favoured $S_1 \rightarrow T_1$ ISC process obviously disfavours the fluorescence emission.

(ii) **Prevention of radiationless transition (*e.g.* EC) from T_1 to S_0:** For this purpose, the energy difference between the T_1 and S_0 states should be high. It will minimise the possibilities of different pathways of radiationless $T_1 \rightarrow S_0$ transitions (see Fig. 6.10.1b).

(iii) **Other factors** like initial excitation ($\pi \rightarrow \pi^*$), structural rigidity, temperature, solvent viscosity, pH, dissolved paramagnetic bodies including O_2 play the same roles as in the case of fluorescence (discussed earlier).

6.10.10 Luminescence: Phosphorescence *vs.* Fluorescence

● **Position of energy states (assuming S_0 as the ground state):** The high energy difference between the S_1 and T_1 states preventing the $S_1 \rightarrow T_1$ transition via ISC and high energy difference between the S_1 and S_0 states preventing the radiationless $S_1 \rightarrow S_0$ conversion **favour the fluorescence**. The small energy difference between the S_1 and T_1 states favouring the $S_1 \rightarrow T_1$ transition via ISC and high energy difference between the T_1 and S_0 states preventing the radiationless $T_1 \rightarrow S_0$ conversion **favour the phosphorescence**.

For example, in carbonyl compounds, the energy difference between the S_1 and T_1 states is only ~29 kJ mol^{-1} (*i.e.* ISC leading to $S_1 \rightarrow T_1$ is highly efficient) and consequently, it shows the **weak fluorescence**. In benzene, the said energy difference is about 170 kJ mol^{-1} (*i.e.* $S_1 \rightarrow T_1$ via ISC is less efficient) and it shows both fluorescence and weak phosphorescence.

● **Time scale:** Fluorescence transition being spin-allowed (say, $S_1^{v=0} \rightarrow S_0$) is much faster than the spin-forbidden phosphorescent transition (say, $T_1^{v=0} \rightarrow S_0$). Fluorescent emission generally occurs immediately after the absorption (exception: delayed fluorescence) while phosphorescence occurs slowly from the long-lived phosphorescent excited state.

● **Quantum yield:** $\phi_f > \phi_p$.

● **Red-shift:** Compared to the exciting radiation, both v_f and v_p are red shifted but v_p is more red-shifted (*i.e.* $v_{ex} > v_f > v_p$).

Note: In **chemiluminescene,** the required excited state is generated in a chemical reaction not by photon absorption. This aspect has been discussed in Sec. 7.9.5.

6.10.11 Primary Quantum Yield (ϕ)

Deactivation of the excited species can occur through the radiative, nonradiative and chemical processes. Primary quantum yield (ϕ) of a particular process is defined as the ratio of the number of events occurred in that process and the number of photons absorbed in the same time interval.

$$\phi = \frac{\text{Number of events}}{\text{Number of photons absorbed}}$$

$$= \frac{\text{Rate } (r) \text{ of the particular process}}{\text{Intensity of light absorbed}} = \frac{r}{I_{abs}} = \frac{r}{\sum r_i}$$

Thus ϕ is defined with respect to a particular process which may be radiative, nonradiative or photochemical. Obviously, sum of the ϕ_i-values of all the possible photophysical and photochemical processes is equal to 1.

$$\sum \phi_i = \sum \frac{r_i}{I_{abs}} = 1$$

It gives: $\phi_{pc} + \phi_{pp} = 1$, where ϕ_{pp} = sum of the ϕ-values of all photophysical processes and ϕ_{pc} = sum of the ϕ-values of all photochemical reactions. The quantum yield of fluorescence (ϕ_f) has already been defined.

6.10.12 Photostationary State and Stern-Volmer Equation

In absence of any photochemical reaction, deactivation of the excited fluorescence state (S_1) and phosphorescent state (T_1) (assuming S_0 as the ground state) occurs through the only photophysical processes (Figs 6.10.12.1, 6.10.1).

Excitation (*i.e.* light absorption): $S_0 + h\nu \rightarrow S_1$, Rate = I_{abs}

Fluorescence: $S_1 \xrightarrow{k_f} S_0 + h\nu_f$, Rate = $k_f[S_1]$

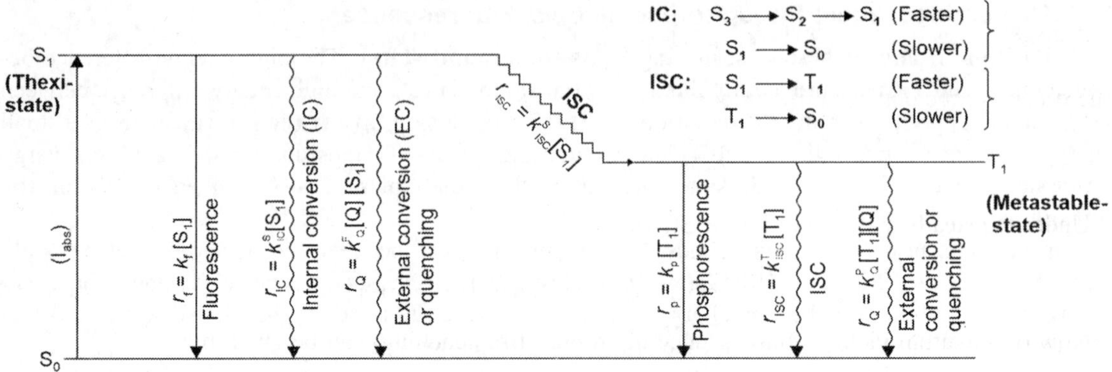

Fig. 6.10.12.1: Generation and deactivation (through the photophysical processes only) of the excited fluorescent (S_1) and phosphorescent (T_1) states

Internal conversion (IC) (nonradiative): $S_1 \xrightarrow{k_{IC}^{S}} S_0$, Rate $= k_{IC}^{s}[S_1]$

Fluorescence quenching (nonradiative): $S_1 + Q \xrightarrow{k_Q^F} S_0 + Q +$ kinetic energy, Rate $= k_Q^F[Q][S_1]$

Intersystem crossing (ISC) (nonradiative): $S_1 \xrightarrow{k_{ISC}^{S}} T_1$, Rate $= k_{ISC}^{s}[S_1]$

Phosphorescence: $T_1 \xrightarrow{k_p} S_0 + h\nu_p$, Rate $= k_p[T_1]$

Phosphorescence quenching (nonradiative):

$T_1 + Q \xrightarrow{k_q^P} S_0 + Q +$ kinetic energy, Rate $= k_Q^P[T_1][Q]$

Intersystem crossing (ISC) (nonradiative): $T_1 \xrightarrow{k_{ISC}^{T}} S_0$, Rate $= k_{ISC}^{T}[T_1]$

(I) **Analysis of the photostationary state of S_1 (*i.e.* fluorescent state called the thexi-state, see Kasha's rule):** Let us consider the rate of formation and decay (under the condition of $\phi_{pc} = 0$, *i.e.* no photochemical reaction of the S_1-state) of the fluorescent state, S_1.

Rate of formation of $S_1 = I_{abs}$

Rate of decay of $S_1 = (k_f + k_{IC}^S + k_{ISC}^S + k_Q^F[Q])[S_1]$. Under the condition of **steady illumination**, $[S_1]$ does not vary with time and it leads to:

$$\frac{d[S_1]}{dt} = 0 = I_{abs} - (k_f + k_{IC}^S + k_{ISC}^S + k_Q^S[Q])[S_1]$$

The fluorescence life-time (τ_F) of the excited state (S_1) is defined as:

$$\tau_F = \frac{1}{k_f + k_{IC}^S + k_{ISC}^S + k_Q^S[Q]}$$

It gives:

$$\frac{d[S_1]}{dt} = 0 = I_{abs} - \frac{[S_1]}{\tau_F}; \text{ or } [S_1] = \tau_F I_{abs}$$

The **quantum yield (ϕ_f) of fluorescence** is given by:

$$\phi_f = \frac{r_f}{I_{abs}} = \frac{r_f}{\sum r_i} = \frac{k_f[S_1]}{(k_f + k_{IC}^S + k_{ISC}^S + k_Q^F[Q])[S_1]} = k_f\tau_F$$

● **Under the condition of no quenching:** In absence of external conversion (*i.e.* no quenching), we have:

$$-\frac{d[S_1]}{dt} = (k_f + k_{IC}^S + k_{ISC}^S)[S_1] = k_d^S[S_1], \textbf{ first order path}$$

It gives: $$[S_1]_t = [S_1]_0 \exp(-k_d^S t) = [S_1]_0 \exp\left(-\frac{t}{\tau_0}\right)$$

where τ_0 gives the observed life time of the excited state S_1 in absence of quenching and it is given by:

$$\tau_0 = \frac{1}{k_d^S} = \frac{1}{k_f + k_{IC}^S + k_{ISC}^S}$$

Under the steady state condition of S_1, we can write:

$$\frac{d[S_1]}{dt} = 0 = I_{abs} - k_d^S[S_1]; \text{ or } I_{abs} = k_d^S[S_1]$$

Now ϕ_f (quantum yield of fluorescence in absence of quenching) can be given by:

$$\phi_{f,0} = \frac{r_f}{I_{abs}} = \frac{k_f[S_1]}{k_d^S[S_1]} = \frac{k_f}{k_f + k_{IC}^S + k_{ISC}^S} = k_f\tau_0$$

● **Stern-Volmer equation for the fluorescent state (S_1):** In absence of a quencher, *i.e.* [Q] = 0, we have the quantum yield (denoted by $\phi_{f,0}$):

$$\phi_{f,0} = \frac{k_f}{k_f + k_{IC}^S + k_{ISC}^S}$$

It gives: $$\frac{\phi_{f,0}}{\phi_f} = \frac{\text{Quantum yield in absence of a quencher}}{\text{Quantum yield in presence of the quencher}}$$

$$= \frac{k_f}{(k_f + k_{IC}^S + k_{ISC}^S)} \times \frac{(k_f + k_{IC}^S + k_{ISC}^S + k_Q^F[Q])}{k_f}$$

$$= \frac{k_s + k_{IC}^S + k_{ISC}^S + k_Q^F[Q]}{k_f + k_{IC}^S + k_{ISC}^S}$$

$$= 1 + \frac{k_Q^F[Q]}{k_f + k_{IC}^S + k_{ISC}^S} = 1 + \tau_0 k_Q^F[Q] = 1 + K_{SV}[Q]$$

The above equation is called the **Stern–Volmer equation** and it has been experimentally verified from the linear plot of $\phi_{f,0}/\phi_f$ *vs.* [Q]. K_{SV} is called the **Stern–Volmer constant** (*cf.* Fig. 6.10.12.2).

Upon substitution of τ_0/τ_F for $\phi_{f,0}/\phi$, the Stern–Volmer equation reduces as follows.

$$\frac{\tau_0}{\tau_F} = 1 + \tau_0 k_Q^F[Q] \text{ or } \frac{1}{\tau_F} = \frac{1}{\tau_0} + k_Q^F[Q]$$

Note: The mechanisms of **bimolecular quenching** of an excited state are (see Sec. 6.10): collisional deactivation, resonance energy transfer and electron transfer.

(II) **Analysis of the photostationary state of T_1 (*i.e.* phosphorescent state):** Under the steady state condition applied for T_1, we can write (assuming no photochemical reaction of T_1)

$$\frac{d[T_1]}{dt} = 0 = k_{ISC}^S[S_1] - k_p[T_1] - k_{ISC}^T[T_1] - k_Q^P[T_1][Q]$$

$$\tau_P \textbf{ (phosphorescence life-time)} = \frac{1}{k_p + k_{ISC}^T + k_Q^P[Q]}$$

It gives: $$\frac{d[T_1]}{dt} = 0 = k_{ISC}^S[S_1] - [T_1]/\tau_P, \; i.e. \; [T_1] = \tau_P k_{ISC}^S[S_1]$$

Using the relation, $[S_1] = \tau_F I_{abs}$, we have:

$$[T_1] = \tau_P k_{ISC}^S \tau_F I_{abs} = \tau_P \tau_F k_{ISC}^S I_{abs}$$

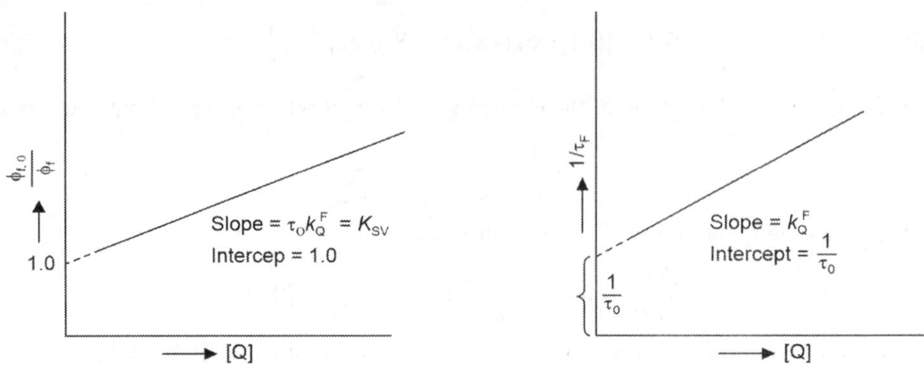

Fig. 6.10.12.2: Illustration of **Stern–Volmer plots,** $\phi_{f,o}/\phi_f$ *vs*. [Q] or $1/\tau_F$ *vs*. [Q] where [Q] denotes the quencher concentration.

Note: $[Ru(bpy)_3]^{2+}$ shows an intense metal-to-ligand charge transfer (**MLCT**) absorption band at 450 nm and gives the fluorescence emission at 600 nm. But quenching of the excited state of $*[Ru(bpy)_3]^{2+}$ by $[Fe(aq)]^{3+}$ in an acidic solution occurs **through the electron transfer.**

$$[Ru(bpy)_3]^{2+} \xrightarrow[(450\text{ nm})]{h\nu} *[Ru(bpy)_3]^{2+} \xrightarrow{[Fe(aq)]^{3+}} [Ru(bpy)_3]^{3+} + [Fe(aq)]^{2+}$$

Quenching rate constant (k_Q^F) can be ontained from the Stern–Volmer plot by measuring the emission lifetimes at 600 nm

The **quantum yield of phosphorescence** (ϕ_P) may be expressed as:

$$\phi_p = \frac{k_p[T_1]}{I_{abs}} = \frac{k_p\tau_F\tau_P k_{ISC}^S I_{abs}}{I_{abs}} = \frac{k_p\tau_P k_{ISC}^S[S_1]}{I_{abs}} \quad (cf.\ [S_1] = \tau_F I_{abs})$$

$$= \frac{\tau_P k_{ISC}^S[S_1]k_p}{[S_1](k_f + k_{IC}^S + k_{ISC}^S + k_Q^F[Q])}$$

$$= \frac{1}{(k_P + k_{ISC}^T + k_Q^P[Q])} \times \frac{k_{ISC}^S k_P}{(k_f + k_{IC}^S + k_{ISC}^S + k_Q^F[Q])}$$

In absence of a quencher, *i.e.* [Q] = 0, it gives:

$$\phi_{p,0} = \frac{k_p}{(k_p + k_{ISC}^T)} \times \frac{k_{ISC}^S}{(k_f + k_{IC}^S + k_{ISC}^S)}$$

The ratio of two quantum yields (**in absence of a quencher**) is given by:

$$\frac{\phi_{p,0}}{\phi_{f,0}} = \frac{k_p[T_1]}{k_f[S_1]}$$

$$= \frac{k_p}{(k_p + k_{ISC}^T)} \times \frac{k_{ISC}^S}{(k_f + k_{IC}^S + k_{ISC}^S)} \times \frac{(k_f + k_{IC}^S + k_{ISC}^S)}{k_f}$$

$$= \frac{k_p}{k_f} \times \frac{k_{ISC}^S}{(k_p + k_{ISC}^T)}$$

6.10.13 Resonance Energy Transfer (RET)

Bimolecular quenching of an excited state (singlet or triplet) occurs in **three possible ways**. It is illustrated by taking S_1 (*i.e.* first excited state of singlet ground state) as the excited state.

(i) $S_1 + Q \rightarrow S_0 + Q$ + kinetic energy (collisional deactivation)

(ii) $S_1 + Q \rightarrow S_0^+ + Q^-$ or $Q^+ + S_0^-$ (electron transfer) (*cf.* quenching of $*[Ru(bpy)_3]^{2+}$ by $[Fe(aq)]^{3+}$ where $[Fe(aq)]^{3+}$ is reduced to $[Fe(aq)]^{2+}$.

(iii) $S_1 + Q \rightarrow S_0 + Q^*$ (**resonance energy transfer**).

The relative efficiencies of the above mentioned pathways depend on the nature of Q for a particular excited state (*i.e.* S_1 in the present case). Collisional deactivation is favoured when Q is a **heavy atom** species (*e.g.* I^-). The favourable situation for the electron transfer can be rationalised in terms of the **Marcus theory** (Sec. 6.3.4).

The resonance energy transfer is one of the most important processes in photochemistry. **It can lead to the photosensitized physical and chemical processes in the energy acceptor molecule. Photosynthesis** is the classical example of such photosensitized reactions. The resonance energy transfer from the electronically excited donor molecule (say D) to the acceptor molecule (say A) can be simply represented as follows:

$D + h\nu \rightarrow D^*$ (Excitation of the donor by absorption of light)

$D^* + A \rightarrow D + A^*$ (**Nonradiative energy transfer** from the electronically excited donor to the acceptor).

Quantum mechanical condition of absorption light by D is: $\Delta E = E(D^*) - E(D)$ = Energy separation between the ground and excited electronic states of the donor $= h\nu$.

The electrical component of the incident electromagnetic radiation induces an **oscillating electric dipole moment** in D. If the frequency of oscillation of the induced electric dipole moment matches with the frequency (ν) of the incident radiation, then the radiation is absorbed. This is the **resonance condition of radiation absorption.**

The induced oscillating electric dipole moment in D also induces an oscillating dipole moment in the nearby quencher molecule, *i.e.* acceptor molecule (A). If the frequency (ν) of oscillation of the electric dipole moment in D matches the condition, $h\nu = \Delta E(A)$ = energy separation between the ground and excited electronic states of the acceptor molecule, then the nonradiative energy transfer from D* to A will occur.

In other words, for the effective resonance energy transfer (RET), **the fluorescence emission spectrum of the donor must significantly overlap with the absorption spectrum of the acceptor** (Fig. 6.10.13.1). Thus in RET, the excitation energy of the donor (D) is not simply lost through fluorescence. Rather, the excitation energy is transferred to the quencher, *i.e.* acceptor (A), under the above mentioned condition.

Radiative Energy Transfer *vs.* Nonradiative Energy Transfer (RET)

Radiative energy transfer: $D^* \rightarrow D + h\nu$; $A + h\nu \rightarrow A^*$

Nonradiative energy transfer (RET): $D^* + A \rightarrow A^*$

Radiative energy transfer represents a **trivial process** which does not require any energetic interaction between the donor and acceptor. The process simply represents the reabsorption of fluorescence emission and efficiency of the process shows the r^{-2} dependence where r denotes the donor-acceptor separation distance. On the other hand, the nonradiative energy transfer (RET) from the donor to the acceptor occurs through the intermolecular long-range **dipole-dipole coupling**. Efficiency of the energy transfer in this process shows the r^{-6} **dependence**.

Energy transfer through collision as the heat energy: In this collisional energy transfer, there is no quantum mechanical restriction (see Sec. 6.9.3).

In RET, the excitation energy (called **exciton**) is transferred nonradiatively through the **intermolecular long-range dipole–dipole coupling** from the donor to the acceptor only when the donor and acceptor molecules are **very closely positioned** (within the distance of 1–10 nm). This nonradiative

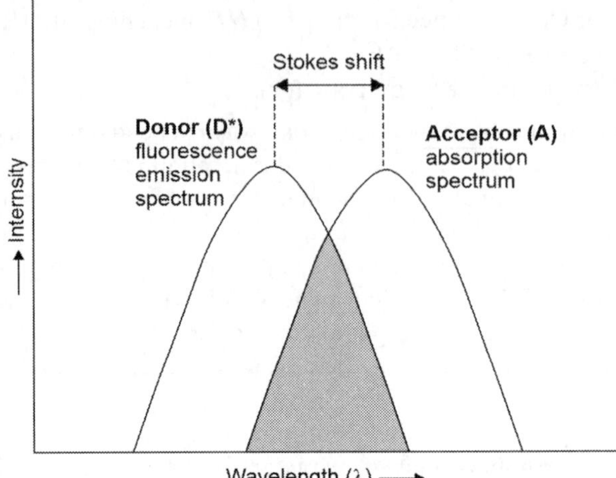

Fig. 6.10.13.1: Illustration of the condition of FRET

Note: Rate of energy transfer (through the FRET process) from D* to A is optimised at the wavelength region at which the fluorescence emission spectrum of D* overlaps with the absorption spectrum of A.

RET is also known as **Forster resonance energy transfer (FRET)**. To increase the efficiency of the exciton transfer in this process, the following **two conditions** must be maintained.

(i) The distance (r) between the energy donor and energy acceptor must be very small; efficiency of this process shows the r^{-6} dependence. More correctly, the rate constant for the energy transfer process (k_{RET}) shows the r^{-6} dependence, *i.e.* $k_{RET} \propto r^{-6}$. The efficiency denoted by η_{RET} is given by ($r \approx$ 1–10 nm).

$$\eta_{RET} = \frac{r_0^6}{r_0^6 + r^6},$$ r_0 = a characteristic distance parameter for a **particular donor-acceptor pair**; it is different for different pairs; it lies in the range 2–5 nm

$$= \frac{1}{1+\left(\dfrac{r}{r_0}\right)^6}$$

$$\approx \frac{r_0^6}{r^6} \text{ (if } r_0 \ll r)$$

(ii) The λ_{max} of the fluorescence emission spectrum of the donor must be close to the λ_{max} of the absorption spectrum of the acceptor, *i.e.* fluorescence emission spectrum of the donor and the absorption spectrum of the acceptor must overlap significantly (Fig. 6.10.13.1), *i.e.* **Stokes shift** should be very small. In fact, the HOMO-LUMO energy gap of the energy acceptor molecule (A) must be **equal or slightly less than that in the energy donor (D) molecule**. It may be noted that the excitation energy of the donor molecule may be partially lost through vibration (*cf.* **vibrational relaxation and Kasha's rule**).

Radiationless Froster resonance energy transfer (**FRET**) (*i.e.* exciton transfer) is illustrated in Fig. 6.10.3.2.

In **photosynthesis**, from the light harvesting pigments (**antenna chlorophylls**), the excitation energy (*i.e.* exciton) transfer occurs through FRET. In this process, the quencher, *i.e.* acceptor (A), is **photosensitized** and A* may participate in many photophysical processes and photochemical reactions. The species (*i.e.* D) absorbing and transferring the radiation energy is called the **photosensitizer**. The light energy absorbed by chlorophylls is transported by the RET mechanism through the antenna

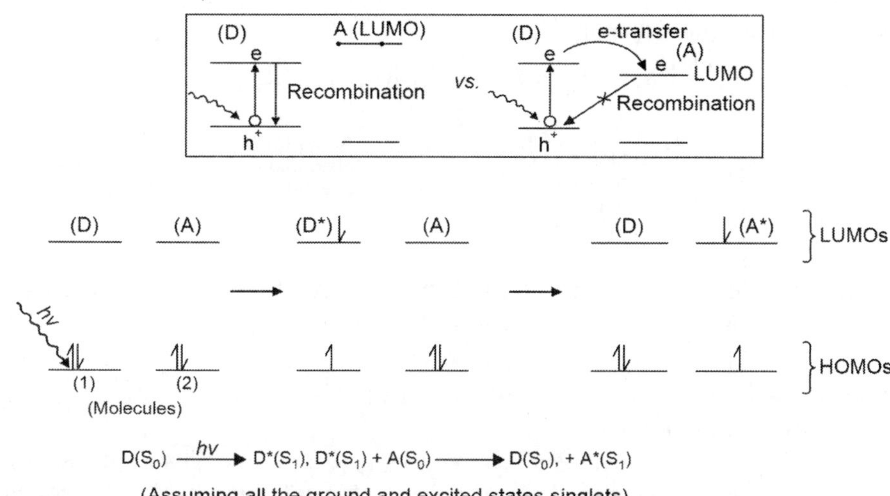

Fig. 6.10.13.2: Schematic representation of FRET mechanism. The excitation energy of the donor molecule (D, 1) is utilised to excite the acceptor molecule (A, 2). Inset (Box): **Charge transfer** *vs.* **charge recombination.** **Charge Recombination:** E(LUMO, D) < E(LUMO, A). **Charge Transfer:** E(LUMO, D) ≥ E(LUMO, A) (as in FRET).

chlorophylls arranged in an ordered way and ultimately, the light energy is transferred to the **reaction centre chlorophylls** (a dimer of Chl-a, called a **special pair** of Chl-a) that use the light energy in the chemical reactions (light phase reactions) of photosynthesis (see the authors' book, **Bioinorganic Chemistry**).

● **Spin selection rule for RET through the dipole–dipole interaction:** It states that the total spin multiplicity of D* and D must be the same and similarly, spin multiplicity of A and A* must be the same. Thus the **two commion spin allowed processes** of RET are:

Singlet-singlet RET: D* (singlet) + A (singlet) → D (singlet) + A* (singlet)

Singlet-triplet RET: D* (singlet) + A (triplet) → D (singlet) + A* (triplet)

It may be noted that the **spin forbidden RET** occurs also effectively because the low efficiency of the process is partially compensated by the longer life-time of D*.

$$D^* \text{ (triplet)} + A \text{ (singlet)} \rightarrow D \text{ (singlet)} + A^* \text{ (singlet)}$$

6.10.14 Primary and Secondary Processes in Photochemical Reactions

Photochemical reactions generally involve the **primary process of photoexcitation** followed by the **secondary process of thermal reaction**. The primary process of photoexcitation (*i.e.* Franck–Condon transition to the excited electronic states) may lead to the different possibilities including the predissociation discussed in Sec. 6.9.1.

The products of primary process may participate in the **subsequent thermal reactions** called **secondary processes**. The secondary process can occur in a single step or multi-steps. Sometimes, the secondary process may lead to the **chain reactions**.

The photochemical decomposition of HI occurs through the following primary and secondary processes.

Primary process: $HI \xrightarrow{h\nu} H + I$, rate = I_{abs} (*cf.* Fig. 6.9.2.7).

Secondary process: $H + HI \xrightarrow{k_2} H_2 + I, I + I \xrightarrow{k_3} I_2$

i.e. $2HI \xrightarrow{h\nu} H_2 + I_2, \phi \text{ (quantum yield)} = 2.0$

$$-\frac{d[HI]}{dt} = I_{abs} + k_2[H][HI]$$

Under the steady state approximation,

$$\frac{d[\mathrm{H}]}{dt} = 0 = \mathrm{I_{abs}} - k_2[\mathrm{H}][\mathrm{HI}], \text{ or } k_2[\mathrm{H}][\mathrm{HI}] = \mathrm{I_{abs}}$$

It gives: $-\dfrac{d[\mathrm{HI}]}{dt} = \mathrm{I_{abs}} + k_2[\mathrm{H}][\mathrm{HI}] = 2\mathrm{I_{abs}}$

Quantum yield (ϕ) $-\dfrac{d[\mathrm{HI}]/dt}{\mathrm{I_{abs}}} = \dfrac{2\mathrm{I_{abs}}}{\mathrm{I_{abs}}} = 2$

6.10.15 Laws of Photochemistry and Quantum Yield or Quantum Efficiency (ϕ) of Photochemical Reactions

The following two laws govern the photochemical reactions.

(i) **Grotthus–Draper law:** It states that **only those radiations which are absorbed can be effective to cause the chemical reactions.**

It indicates that for a photochemical reaction to occur, it requires the photoexcitation by a suitable incident radiation of appropriate wavelength. **The reverse statement is not true** because the photoexcited molecule may also participate in different **photophysical processes** (*e.g.* fluorescence, phosphorescence, etc.) including the collisional deexcitation to increase the thermal energy.

(ii) **Einstein's law of photochemical equivalence:** It states that **each quantum of radiation absorbed by a molecule activates the molecule in the primary process of a photochemical reaction (*i.e.* one photon excites only one molecule in the primary process)**. Quantum efficiency of this primary process is one according to this law of photochemical equivalence.

● **Quantum yield (ϕ):** The efficiency of a photochemical reaction is expressed in terms of **quantum yield** or **quantum efficiency** (ϕ) which is defined as:

$$\phi = \frac{\text{Number of molecules reacted}}{\text{Number of photons absorbed}}$$

According to the law of photochemical equivalence, quantum efficiency of the photoexcitation process is always one but the quantum efficiency of a photochemical reaction may vary in a wide range (*ca.* 10^{-3} to 10^{+6}) depending on the nature of the subsequent secondary thermal reactions.

● **Reasons for low quantum yield (ϕ):** The reasons for the low quantum efficiency (ϕ) of photochemical reactions are: deactivation of the photochemically excited species through the different phytophysical processes, **secondary thermal steps** requiring the **high activation energies**, etc.

In the photochemical reaction, $H_2 + Br_2 \rightarrow 2HBr$, at the primary process, Br_2 absorbs photon to produce the excited Br_2 molecule that dissociates to the Br atoms. The next **thermal secondary endothermic reaction**, $Br + H_2 \rightarrow HBr + H$, requires a **high activation energy** (*ca.* 70 kJ mol^{-1}) and consequently, most of the Br atoms produced at the primary process are deactivated and the quantum yield (ϕ) becomes very low (*ca.* 10^{-2}).

$$\mathrm{Br_2} \xrightarrow{h\nu} \mathrm{2Br},\ \mathrm{H_2 + Br} \longrightarrow \mathrm{HBr + H}\ (\textbf{high activation energy})$$

$$\mathrm{H + Br_2} \longrightarrow \mathrm{HBr + Br},\ \mathrm{H + HBr} \longrightarrow \mathrm{H_2 + Br},\ \mathrm{Br + Br} \longrightarrow \mathrm{Br_2}$$

For the secondary thermal reactions requiring **the low activation energy** and involving only a few steps, ϕ values appear in the range 1–3.

$$\mathrm{2HX} \xrightarrow{h\nu} \mathrm{H_2 + X_2}\ (\mathrm{X = Br, I}),\ \phi = 2$$

$$\mathrm{CH_3COCH_3} \xrightarrow{h\nu} \mathrm{CO + C_2H_6},\ \phi = 1$$

$$(\mathrm{CH_3COCH_3} \xrightarrow{h\nu} {}^{\bullet}\mathrm{CH_3} + {}^{\bullet}\mathrm{COCH_3},$$

$${}^{\bullet}\mathrm{COCH_3} \longrightarrow \mathrm{CO} + {}^{\bullet}\mathrm{CH_3},\ {}^{\bullet}\mathrm{CH_3} + {}^{\bullet}\mathrm{CH_3} \longrightarrow \mathrm{C_2H_6})$$

● **Reasons for high quantum yield (ϕ):** Secondary processes involve the **chain reactions** initiated by the atoms, or free radicals generated in the primary process. Obviously, such secondary thermal reactions must require the low activation energies. For example, in the photochemical reaction, $H_2 + Cl_2 \rightarrow 2HCl$, at the primary process, Cl_2 molecule dissociates into the two Cl atoms that initiate the chain reactions by coliding with the H_2 molecule. The quantum yield of the process is very high (*ca.* 10^6).

$Cl_2 \xrightarrow{hv} 2Cl$, $Cl + H_2 \rightarrow HCl + H$, $H + Cl_2 \rightarrow HCl + Cl$, and so on.

● $$\phi(\text{quantum yield}) = \frac{\text{No. of molecules reacting}}{\text{No. of photons absorbed}} = \frac{\text{No. of moles reacting}}{\text{No. of einsteins of energy absorbed}}$$

According to the law of photochemical equivalence, energy absorbed per mole of the reactant is:

$$E = N_o hv = \frac{N_o hc}{\lambda} = 1 \text{ einstein}$$

$$= \frac{(6.023 \times 10^{23} \text{ mol}^{-1})(6.626 \times 10^{-34} \text{ J s})(3 \times 10^8 \text{ m s}^{-1})}{\lambda \times 10^{-9} \text{ m}} (\lambda \text{ in nm})$$

$$= \frac{1.20 \times 10^5}{\lambda} \text{ kJ mol}^{-1} (\lambda = \text{in nm})$$

$$= \frac{1.20 \times 10^{16}}{\lambda} \text{ erg mol}^{-1} (\lambda \text{ in Å}, 1\text{Å} = 10^{-8} \text{ cm})$$

The energy of Avogadro number (N_0) of photons is called **one einstein.**

Illustration: In an uranyl oxalate actinometer, the cell containing 10 mL of 0.0495 M oxalic acid solution and 0.01 M uranyl sulfate is irradiated with the light (λ = 2540 Å). After absorption of 8.81×10^8 erg radiation, the concentration of oxalic acid is reduced to 0.0383 M. Find quantum yield of photochemical decomposition of oxalic acid.

$$HO_2C—CO_2H \xrightarrow[hv]{UO_2^{2+} \text{ (photosensitiser)}} H_2C_2O_4^* \longrightarrow CO + CO_2 + H_2O$$

$$1 \text{ einstein} = \text{energy of 1 mole photon}$$

$$= \frac{N_0 hc}{\lambda}$$

$$= \frac{6.023 \times 10^{23} \text{ mol}^{-1} \times 6.626 \times 10^{-27} \text{ erg s} \times 3 \times 10^{10} \text{ cm s}^{-1}}{2540 \times 10^{-8} \text{ cm}}$$

$$= 4.71 \times 10^{12} \text{ erg mol}^{-1}$$

$$8.81 \times 10^8 \text{ erg radiation} \equiv \frac{8.81 \times 10^8}{4.71 \times 10^{12}} \text{ einstein}$$

$$= 1.87 \times 10^{-4} \text{ einstein}$$

$$= 1.87 \times 10^{-4} \text{ mole photon}$$

No. of moles of oxalic acid decomposed

$$= 10(0.0495 - 0.0383) \times 10^{-3} \text{ mole}$$

$$= 0.0112 \times 10^{-2} \text{ mole}$$

$$\phi = \frac{0.0112 \times 10^{-2}}{1.87 \times 10^{-4}} \approx 0.60$$

Illustration: Irradiation of propionaldehyde produces CO.

$$\lambda = 3020 \text{ Å, intensity of light} = 15000 \text{ erg s}^{-1}$$

$$1 \text{ einstein} = \frac{N_0hc}{\lambda} = \frac{1.2\times10^{16}}{3020} \text{ erg mol}^{-1}$$

$$\text{Intensity of irradiation} = 15000 \text{ erg s}^{-1}$$

$$\equiv \frac{15000 \text{ erg s}^{-1} \times 3020}{1.2\times10^{16} \text{ erg mol}^{-1}}$$

$$= 3.78\times10^{-9} \text{ mol s}^{-1} = 3.78\times10^{-9} \text{ einstein per sec}$$

$$\phi = \frac{\text{No. of moles of CO produced per second}}{\text{No. of moles of photons absorbed per second}}$$

$$\text{or } 0.54 = \frac{\text{No. of moles of CO produced per second}}{3.78\times10^{-9}}$$

i.e. rate of formation of CO $= 0.54\times3.78\times10^{-9} \text{ mol s}^{-1} = 2.04\times10^{-9} \text{ mol s}^{-1}$

Illustration: $H_2 + Cl_2 \xrightarrow{hv} 2HCl, \phi = 1\times10^6, \lambda = 480$ nm. Find the amount of HCl produced per calories of the radiation absorbed.

$$1 \text{ einstein} = \frac{N_0hc}{\lambda}$$

$$= \frac{1.2\times10^5}{480} \text{ kJ mol}^{-1} = 2.5\ 10^5 \text{ J mol}^{-1}$$

$$1 \text{ cal} \approx 4.2 \text{ J}$$

$$\equiv \frac{4.2 \text{ J}}{2.5\times10^5 \text{ J mol}^{-1}} \approx 1.68\times10^{-5} \text{ mol photon}$$

$$\phi = \frac{\text{No. of moles reacted}}{\text{No. of moles of photon absorbed}}$$

$$\text{or } 1\times10^6 = \frac{\text{No. of moles of } Cl_2 \text{ reacted}}{1.68\times10^{-5} \text{ mol}}$$

i.e. No. of moles Cl_2 reacted $= 1\times10^6 \times 1.68\times10^{-5}$ moles
$= 16.8$ moles

1 mole of Cl_2 on reacting with 1 mole of H_2 produces 2 moles of HCl.
i.e. 16.8 moles of Cl_2 react with 16.8 moles of H_2 to produce 2 × 16.8 = 33.6 moles of HCl.

6.11 DIFFERENT TYPES OF PHOTOCHEMICAL REACTIONS TO DEACTIVATE THE EXCITED SPECIES

There are different types of photochemical reactions. These are illustrated here.

(i) **Ionisation:** $X \xrightarrow{hv} X^* \longrightarrow X^+ + e$

(*e.g.* $NO \xrightarrow[(134\text{ nm})]{hv} NO^* \longrightarrow NO^+ + e$)

(ii) **Electron transfer:** $X \xrightarrow{hv} X^* \xrightarrow{+Y} X^+ + Y^-$ or $X^- + Y^+$ (Y = quencher)

(*e.g.* $[Ru(bpy)_3]^{2+} \xrightarrow[(452\text{ nm})]{hv} {}^*[Ru(bpy)_3]^{2+} \xrightarrow{+[Fe(aq)]^{3+}} [Ru(bpy)_3]^{3+} + [Fe(aq)]^{2+}$

Quenching of $^*[Ru(bpy)_3]^{2+}$ (excited state giving the fluorescence emission at 600 nm) by $[Fe(aq)]^{3+}$ through the electron transfer).

(iii) **Dissociation:** $X \xrightarrow{hv} X^* \longrightarrow Y + Z$

(*e.g.* $O_2 \xrightarrow[(<242\text{ nm})]{\text{UV-C}(hv)} O_2^* \longrightarrow O + O; O_3 \xrightarrow[(240\text{–}320\text{ nm})]{\text{UV-B}(hv)} O_3^* \longrightarrow O_2 + O$)

$X \xrightarrow{hv} X^*, X^* + B\text{—}C \longrightarrow X + B + C$

(*e.g.* $Hg \xrightarrow{hv} Hg^*, Hg^* + CH_4 \longrightarrow Hg + H + CH_3, Hg^* + C_2H_4 \longrightarrow Hg + H + C_2H_3$

(iv) **Abstraction:** $X \xrightarrow{h\nu} X^* \xrightarrow{+B—C} X—B + C$ (*e.g.* $Hg^* + H_2 \longrightarrow HgH + H$)

(v) **Isomerisation:** $X \xrightarrow{h\nu} X^* \longrightarrow X'$ (isomer)

Ph–N=N–Ph $\underset{h\nu \text{ (450 nm) or heat}}{\overset{h\nu \text{ (365 nm)}}{\rightleftharpoons}}$ Ph–N=N–Ph (*cis-trans* isomerisation of azobenzene)

$O_2N–NO$ $\underset{380 \text{ nm}}{\overset{720 \text{ nm}}{\rightleftharpoons}}$ O=N–O–N=O

(N_2O_3, asymmetric) (N_2O_3, symmetric)

(vi) **Dimerisation:** $X \xrightarrow{h\nu} X^* \xrightarrow{+X} X_2$

$h\nu$

Photochemical dimerisation (*i.e.* $[2\pi + 2\pi]$ cycloaddition) of ethylene to cyclobutane (see Appendix 9A, Vol. 2, for details).

(vii) **Photoisomerisation—the key step behind the visual sensing** (Ref. Das *et al*, *ChemTexts*, 10:3, 2024; *Asian J. Org. Med. Chem.*, **8**(14), 309–320, 2022): The visual **rhodospin pigment** is the retinal-opsin Schiff base where the retinal moiety acts as the chromophore. Retinal is vitamin A_1. At the ground state, the retinal moiety of rhodospin exists as 11-*cis*-retinal (Z) which undergoes photoisomerisation to **all-*trans*-retinal** (E) through the rotation around the $C_{11}–C_{12}$ double bond leading to the movement of the Schiff base N-atom by about 500 pm (Fig. 6.11.1).

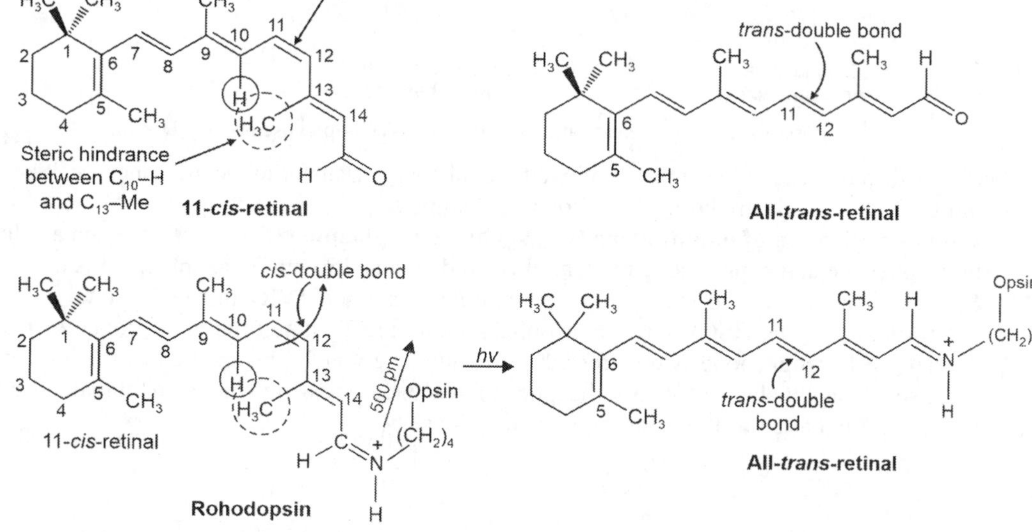

Fig. 6.11.1: Structure of the visual pigment, rhodopsin (present in the rod cell membrane) and its light induced isomerisation, 11-*cis*-retinal → all-*trans*-retinal moiety through the rotation around the $C_{11}{=}C_{12}$ double bond leading to the movement of Schiff base N-atom by about 500 pm

In fact, the $\pi \rightarrow \pi^*$ transition enhances the electron density in the ABMO (*i.e.* π^*–MO) and reduces the electron density in the π–BMO. It weakens the C_{11}–C_{12} π-bond. Consequently, the activation energy barrier for the *cis* (Z)→*trans* (E) isomerisation is significantly reduced. The skeleton of *cis*-retinal moiety is **bent** (producing the steric hinderance between the C_{13}–CH_3 and C_{10}–H groups) while the skeleton of all-*trans*-retinal moiety is straight and this *trans*-form requires a different type of space for its accommodation in the protein structure. It causes the **conformation change in the tertiary structure of opsin protein**. It generates a **nerve signal** to produce the sense of visualisation. This aspect has been discussed in detail in the authors' book, **Biophysical, Bioorganic and Bioinorganic Chemistry.**

(viii) **Photochemical reactions of metal complexes:** In the subsequent sections, we shall mainly discuss the different types of photochemical reactions of metal complexes including the metal carbonyls.

6.12 PHOTOSENSITIZED REACTIONS

An excited species (D*) can transer its excitation energy (*i.e.* exciton) nonradiatively to another species (A) which becomes activated as A*. This activated species (A*) may participate in different photophysical processes and photochemical reactions. **Thus the reactant (A) does not absorb radiation at the primary step**. The species (*i.e.* D) responsible to generate the active reactant (A*) is called the **photosensitizer** (Fig. 6.12.1) and the reactant (A) is called the **quencher**.

The step may be represented as follows:

$$D \xrightarrow{h\nu} D^*;\ D^* + A \longrightarrow D + A^* \text{ (energy transfer)}$$

$$A^* \longrightarrow \text{reaction products}$$

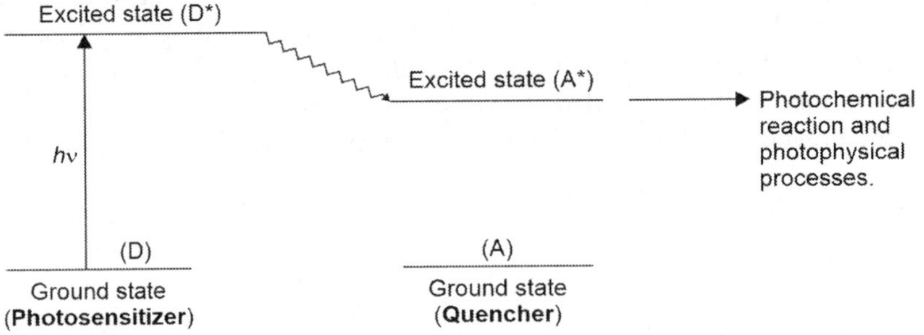

Fig. 6.12.1: Photosensitization of quencher (A) by photosensitizer (D)

There are different possible pathways of excitational energy transfer to the quencher (see Secs. 6.9.3, 6.10 and 6.10.3). It is illustrated in the following examples:

(i) **Photosensitization of naphthalene ($C_{10}H_8$) by benzophenone ($Ph_2C{=}O$):** When a solution of benzophenone and naphthalene in ethanol is irradiated at 350 nm, benzophenone is excited to its S_1-state from its ground state. Then vibrational relaxation (VR) in the S_1-state followed by intersystem crossing (ISC) brings benzoophenone in the T_1 state. The excitational energy in the T_1-state of benzophenone is transferred to naphthalene which gets excited to the T_1-state. This T_1-state of naphthalene can give the phosphorescence emission though naphthalene does not absorb any radiation at the primary activation process.

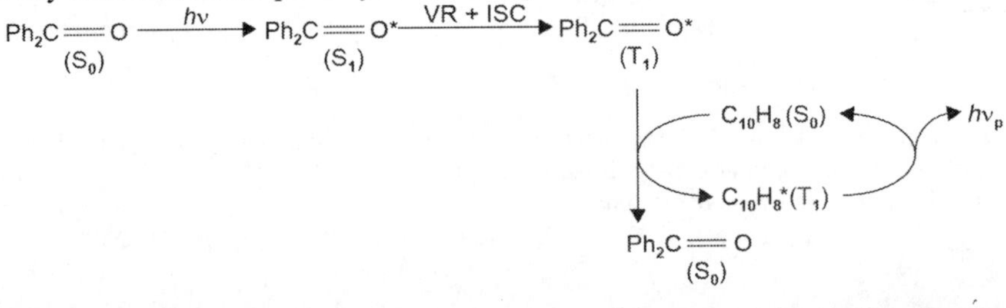

$$\underset{(S_0)}{Ph_2C{=}O} \xrightarrow{h\nu} \underset{(S_1)}{Ph_2C{=}O^*} \xrightarrow{VR + ISC} \underset{(T_1)}{Ph_2C{=}O^*}$$

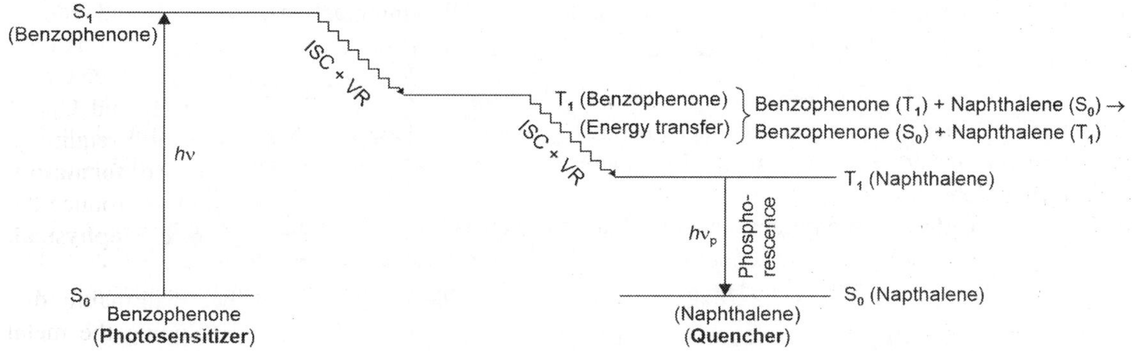

Fig. 6.12.2: Schematic illustration (not in energy scale) of photosensitization of naphthalene by benzophenone

The energy diagram of photosensitization of naphthalene by benzophenone is shown in Fig. 6.12.2.

(ii) **Hg-vapour as a photosensitizer:** When a mixture of H_2, O_2 and Hg-vapour is irradiated at 253.7 nm, it produces H_2O but without the Hg-vapor, the reaction cannot occur under the identical conditions. Here, Hg-vapour as a photosensitiser strongly absorbs at 253.7 nm to produce the excited Hg*-atom ($^3P_1$) and its excitational energy is transferred to H_2 which eventually undergoes dissociation.

$$\text{Hg}\,(6^1S_0) \xrightarrow[(253.7\ \text{nm})]{h\nu} \text{Hg}^*\,(6^3P_1);\ \text{Hg}^*\,(6^3P_1) + H_2\ (\text{quencher}) \longrightarrow \text{Hg}\,(6^1S_0) + \text{H} + \text{H}$$

$$\text{H} + O_2 \longrightarrow \text{chain reaction producing } H_2O$$

BDE of H_2 corresponds to 277.6 nm but H_2 gas does not absorb the radiation of wavelength of 277.6 nm. It requires the radiation of much lower wavelength to produce H-atoms (see Fig. 6.9.2.6). Hg as a photosensitiser (excitational energy = 47.5 kJ mol^{-1}) can dissociate H_2 at λ = 253.7 nm.

Now let us consider the energetics of the process.

$$\lambda = 253.7\ \text{nm} \equiv 471.5\ \text{kJ mol}^{-1};\ \text{BDE of } H_2 = 432\ \text{kJ mol}^{-1} \equiv 277.6\ \text{nm}$$

BDE of O_2 = 490.2 kJ mol^{-1}; BDE = bond dissociation energy.

Thus the excitational energy in Hg* is released as the **heat energy through collision with the quancher** and the released energy is sufficient to dissociate H_2 but insufficient to dissociate O_2. H_2 needs 432 kJ mol^{-1} for its dissociation from Hg* and the remaining energy (39.5 kJ mol^{-1}) of Hg* is used to impart the translational energy of the two dissociated H-atoms.

Here it is interesting to note that the energy associated with 253.7 nm radiation is sufficient to dissociate H_2 but the **radiation cannot do the task directly** (see Sec. 6.9.3), *i.e.*

$$H_2 \xrightarrow[(253.7\ \text{nm})]{h\nu} \!\!\times\!\!\rightarrow 2\text{H}\ \ (\text{see Fig. 6.9.2.6})$$

Direct Photodissociation *vs.* Photosensitiser Mediated Dissociation (see Sec. 6.9.3): Selection rules govern the photoexcitation leading to photodissociation but there is no such restriction in the energy transfer to the quencher from the excited photosensitizer **through collision**.

For the absorption of the incident radiation to directly dissociate H_2, the wavelength of radiation to be absorbed must lie in the absorption **continuum region of H_2** which starts at 84.9 nm. Thus H_2 remains insensitive with respect to the radiation of 253.7 nm. For incident radiation ($\leq$ 84.9 nm), obviously H_2 undergoes directly dissociated and at least one atom remains in excited state.

$$H_2 \xrightarrow[(\leq 84.9\ \text{nm})]{h\nu} \text{H}^* + \text{H}\ (\text{excess energy may impart kinetic energy to the H-atoms})$$

In the same way, Hg* can excite C_2H_4 to induce the following reactions,

$$Hg^* + C_2H_4 \rightarrow Hg + C_2H_4^*,\ Hg^* + C_2H_4 \rightarrow Hg + H + C_2H_3$$

$$C_2H_4^* \rightarrow C_2H_2 + H_2, H_2 + Hg^* \rightarrow 2H + Hg, H + C_2H_4 \rightarrow C_2H_5, 2C_2H_5 \rightarrow C_4H_{10}, \text{etc.}$$

Note: The above example illustrates that Hg as a photosensitizer can excite C_2H_4 through energy transfer to produce C_2H_4* (that subsequently dissociates) and can also directly dissociate C_2H_4 through collision.

(iii) **Uranyl oxalate actinometer:** A **chemical actinometer** is used to determine the number of photons absorbed. In uranyl oxalate actinometer, UO_2^{2+} acts as a **photosensitizer** by absorbing light in the range 250–440 nm. The excited uranyl ion transfers its energy to oxalic acid which then decomposes.

$$UO_2^{2+} \xrightarrow{h\nu} (UO_2^{2+})^*$$

$$(UO_2^{2+})^* + \underset{CO_2H}{\overset{CO_2H}{|}} \longrightarrow CO + CO_2 + H_2O + UO_2^{2+}$$

The amount of oxalic acid remaining undecomposed can be determined by titrating with $KMnO_4$ solution. The quantum yield of the reaction is 0.57. By estimating the amount of oxalic acid decomposed, the number of photons absorbed can be determined.

(iv) **Photosensitized oxidation by singlet oxygen:** It is well known that O_2 can destroy many organic substances including the microorganisms in presence of light and photosensitizing dyes. Sensitizing dyes through photoexcitation produce the excited singlet states which are eventually converted into the **long-lived excited triplet states** through the **intersystem crossing** (*cf.* Fig. 6.12.2 for the sensitizer benzophenone). These excited triplet sensitizers react with triplet O_2 (*i.e.* $^3\Sigma_g^-$) to produce the singlet O_2 ($^1\Delta_g$ and $^3\Sigma_g^+$) which is highly active. $^1\Delta_g$–O_2 (94 kJ mol^{-1}) is relatively more stable but less reactive than $^1\Sigma_g^+$–O_2 (157 kJ mol^{-1}).

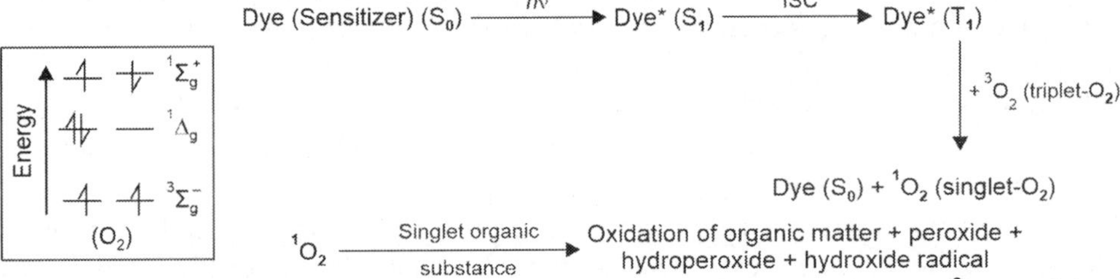

Thus the produced **reactive oxygen species (ROS)** like $^\bullet OH$, O_2^-, ^-OOH, O_2^{2-}, participate in the secondary reactions. ROS being the powerful oxidants can destroy the organic matter including the microorganisms.

Triplet dye* can directly react with triplet-O_2 to produce O_2^- (superoxide) through electron transfer.

$$^3Dye^* + {}^3O_2 \longrightarrow Dye^+ + O_2^- \text{ (superoxide)}$$

Thus the superoxide (O_2^-) produced can also destroy the organic matter. In fact, O_2^- subsequently generates different ROS (discussed later) to decompose the organic matter.

(v) **Protective role of the carotenoids in photosynthesis:** In photosynthesis, chlorophylls act as the **photosensitizers**. A photosensitizer absorbs the light energy and gets excited and the excitation energy (*i.e.* exciton) is transferred to a **quencher** (see Fig. 6.10.3.2 for energy transfer to the reaction centre chlorophyll dimer). If the quencher is nonabsorbing, then the interaction between the excited sensitizer and quencher may lead to the **photochemical electron transfer**.

In the photochemical process of photosynthesis, through the primary photon absorption, the special pair of chlorophyll (simply denoted by Chl) present in the reaction centre is excited to the singlet state (S_1 denoted by ^{1}Chl*). **If this excitation energy is not completely used in photo-**

synthesis, $^1Chl^*$ is gradually transferred into the triplet state (T_1) and produces $^3Chl^*$ through the **intersystem crossing** (ISC). $^3Chl^*$ is not itself destructive but it reacts with the triplet oxygen ($^3O_2$ denoted by $^3\Sigma_g^-$, ground state) to produce the **singlet oxygen** ($^1O_2$ characterised by $^1\Delta_g$ and $^1\Sigma_g^2$) and many other **reactive oxygen species (ROS).** $^1\Delta_g$-singlet oxygen of relatively low energy (about 94 kJ mol^{-1}) can be populated significantly by the energy transfer from the triplet chlorophyll. $^1\Sigma_g^+$-singlet oxygen of higher energy (about 157 kJ mol^{-1}) is less populated but it is more reactive.

- **Generation of the singlet oxygen ($^1O_2$) through energy transfer:**

$$^1\text{Chl } (S_0, \text{ singlet ground state}) \xrightarrow{h\nu} {}^1\text{Chl}^* \ (S_1 \text{ excited singlet state})$$

$$^1\text{Chl}^* \ (S_1) \xrightarrow{\text{ISC}} {}^3\text{Chl}^* \ (T_1 \text{ excited triplet state})$$

$$^3\text{Chl}^* \ (T_1) + {}^3O_2 \ (\textbf{quencher}) \longrightarrow {}^1\text{Chl } (S_0) + {}^1O_2 \ (\text{energy transfer})$$

- **Generation of ROS through electron transfer from $^3Chl^*$ and subsequent reactions:**

$$^3O_2 \xrightarrow[e\text{-transfer}]{^3\text{Chl}^*} O_2^- \xrightarrow{H^+} HOO^\bullet \xrightarrow{\text{Disproportionation}} {}^1O_2 + H_2O_2$$

$$O_2^- + H_2O_2 \longrightarrow {}^1O_2 + OH^- + OH^\bullet, \ O_2^- + HO_2^\bullet \longrightarrow HO_2^- + O_2$$

The **hydroxyl radical ($^\bullet OH$) and hydroperoxide radical ($HO_2^\bullet$) are notoriously toxic**. In presence of the Fe^{III}/Fe^{II} system (biologically relevant), from H_2O_2, the highly toxic $^\bullet OH$ radical may also be generated (*cf.* **Fenton's reagent**, $Fe^{2+} + H_2O_2$).

$$Fe^{3+} + H_2O_2 \longrightarrow Fe(OOH)^{2+} + H^+; \ Fe(OOH)^{2+} \longrightarrow Fe^{2+} + HO_2^\bullet$$

$$Fe^{2+} + H_2O_2 \longrightarrow Fe^{3+} + OH^- + {}^\bullet OH; \ Fe^{2+} + O_2^- \longrightarrow Fe^{3+} + O_2^{2-}$$

$$Fe^{2+} + H_2O_2 \longrightarrow FeO^{2+} \ (\textit{i.e. } Fe^{IV}O^{2-}, \textbf{ ferryl species}) + H_2O$$

$$FeO^{2+} + H_2O_2 \longrightarrow Fe^{2+} + O_2 \ ({}^1\Delta_g) + H_2O$$

The above reactions indicate that formation of O_2^- can pave the way for the generation of the ROS like $^\bullet OH$, $HO_2^\bullet$, etc. Obviously, the electron transfer from the excited chlorophyll to triplet oxygen to generate O_2^- is favoured when the photochemically excited electrons of chlorophyll are not fully utilised in photosynthesis to produce NADPH which subsequently reduces CO_2 to CH_2O in dark phase. **Thus O_2^- and many other ROS are mainly generated when the photosynthesis occurs under very intensive sunlight.**

Thus the produced singlet O_2 and many other ROS can cause the **oxidative damage** to the biological system. But fortunately, the carotenoids (Car) which always coexist with the photosynthetic pigment chlorophyll can prevent the said oxidative damage. In fact, **β-carotene is an efficient quencher of singlet O_2 and triplet chlorophyll** and the involved quenching mechanism occurs via the efficient resonance energy transfer process ($k_{RET} \approx 10^{8-10}$ M^{-1} s^{-1}).

$$^3\text{Chl}^* + {}^1\text{Car} \longrightarrow {}^1\text{Chl} + {}^3\text{Car}^*, \ {}^1O_2 + {}^1\text{Car} \longrightarrow {}^3O_2 + {}^3\text{Car}^*$$

$$^3\text{Car}^* \xrightarrow[\text{energy transfer}]{\text{Radiationless}} {}^1\text{Car (generation of the quencher)}$$

Thus carotene can act as a **catalyst and it can be recycled**. It is evident that carotenoids remove not only the singlet O_2 and triplet chlorophyll but also dissipate the unutilised excess energy through the **radiationless energy transfer process**. In fact, **the plants lacking in a sufficient amount of carotenoids die quickly on exposure to bright sunlight and oxygen**.

Note: The mechanism of **antioxidant activity of tocopherol** (vit. E) is different from that of β-carotene. The tocopherols present in the membrane remove ROS (*e.g.* $^1O_2$, O_2^-, $^\bullet OH$, $HO_2^\bullet$, etc.)

by reducing the ROS. Thus the tocopherols as the reducing agents act here as the **sacrifical donors** and the oxidised forms of tocopherols are ultimately destroyed. This is why, the tocopherols cannot show the **catalytic activity** in the antioxidant activity (*cf.* carotenoids act as the catalysts in showing the antioxidant activity).

Carotenoids: This yellow pigment absorbs light in the range 400–500 nm (**named after the orange coloured carrots**) and it also acts as a **light harvesting pigment** (*i.e.* molecular antenna) in photosynthesis. In fact, its role in photosynthesis becomes very much important at the later age of the leaves when the chlorophyll is almost degraded (*i.e.* **leaves become yellow coloured**). Thus carotenoids have two functions: as a **light harvesting pigment** and **protection from ROS.**

6.13 PHOTOEXCITATION AND LIFE-TIME OF THE EXCITED SPECIES EXPERIENCING THE COMPETITIVE PHOTOPHYSICAL PROCESSES AND CHEMICAL REACTIONS

In photochemical reactions, the following steps are involved.

(i) **Photochemical excitation to provide the activation energy:** Absorption of photon from the UV-visible regions (*i.e.* 200–600 nm) is equivalent *ca.* 600–200 kJ mol^{-1}. Thus absorption of such a photon by a metal complex can raise the energy by a few hundred kilojules per mole. **These energies are quite larger than the typical activation energies** ($\leq$ 30 kJ mol^{-1}) **required for most of the reactions.** The photochemically excited species requires only a negligible amount of activation energy ($\approx$ 8–17 kJ mol^{-1}) to give the reaction product (*cf.* Fig. 6.14.1). In fact, absorption of photon may lead to both nonredox (*e.g.* substitution, isomerisation, racemisation) and redox reactions. But very often, **the backward thermal reaction may prevent the net progress of the reaction**. This is why, it is required to carry out the process in such a condition that can avoid the backward reaction.

Cage-effect

The species generated in the primary photochemical process may remain confined in a solvent cage at the moment of their generation. If the species cannot escape from the solvent case to the bulk, they will recombine (as in the case of recombination of photodissociation products) to give back the starting species to reduce the quantum yields. This phenomenon is called the **cage-effect**.

(ii) **Competition between the photophysical processes and photochemical reactions:** The photoexcited species (generated through the photon absorption) may undergo deactivation either through the photophysical processes (Sec. 6.10) or through the photochemical reactions (see Sec. 6.11). Thus the time required for a photoexcited species to participate in a reaction must be of the order of the life time of the particular excited state or less so that the reaction can compete with the possible photophysical processes.

The competitive photophysical processes are: vibrational relaxation (VR), internal conversion (IC), intersystem crossing (ISC) and luminescence (*i.e.* fluorescence and phosphorescence). The life-time of the initially photoexcited species is generally very small and it produces the more stable excited states through the internal conversion (coupled with VR) and intersystem crossing (coupled with VR). For example, if singlet state (S_0) is the ground state, the primary photoexcitation may produce the excited singlet states like S_1, S_2, S_3, ..., which are soon converted into the $S_1^{v=0}$ state (**thexi-state**) through the IC + VR process. The $S_1^{v=0}$ thexi-state may be converted into the $T_1^{v=0}$ state (**metastable state**, relatively more long-lived) through the ISC + VR steps (see Fig. 6.10.1). **These S_1 and T_1 states having the relatively longer life times generally** participate in chemical reactions. Let us consider the photoexcitation of $A(S_0)$.

$$A(S_0) \xrightarrow{h\nu} [^1A^*] \xrightarrow{IC + VR} {}^1A^*(S_1) \xrightarrow{ISC + VR} {}^3A^*(T_1)$$

(Excited singlet state)

$^1A^*(S_1)$ and $^3A^*(T_1)$ are expected to participate in chemical reactions which may be **unimolecular** or **bimolecular** (*i.e.* reacting with the other species present in the system).

$$\begin{cases} ^1A^*(S_1) \longrightarrow C + D \text{ (unimolecular)} \\ ^3A^*(T_1) \longrightarrow C' + D' \text{ (unimolecular)} \end{cases}$$

$$\begin{cases} ^1A^*(S_1) + B \longrightarrow \text{Product (bimolecular)} \\ ^3A^*(T_1) + B \longrightarrow \text{Product (bimolecular)} \end{cases}$$

High quantum yield arises due to the **secondary thermal reactions** (see Sec. 6.10.15). The high quantum yield also arises when the initially excited molecule can start a **chain reaction** and giving rise to a larger number of product molecules before the chain is finally terminated (see Sec. 6.10.15). For the **nonchain reactions**, the quantum yields for various competitive photophysical (ϕ_{pp}) and photochemical reactions (ϕ_{pc} or ϕ_R) arising from a particular excited state must add up to unity, *i.e.*

$$\phi_{pp} + \phi_{pc} = 1$$

$$\left.\begin{array}{l} \text{For the } ^1A^*(S_1) \text{ state, it is:} \\ \phi_R^S + \phi_f^S + \phi_{IC}^S + \phi_{ISC}^S = 1.0 \\ \text{For the } ^3A^*(T_1) \text{ state, it is:} \\ \phi_R^T + \phi_p^T + \phi_{ISC}^T = 1.0 \end{array}\right\} \text{Photophysical deactivation by \textbf{external conversion} (EC) has been ignored}$$

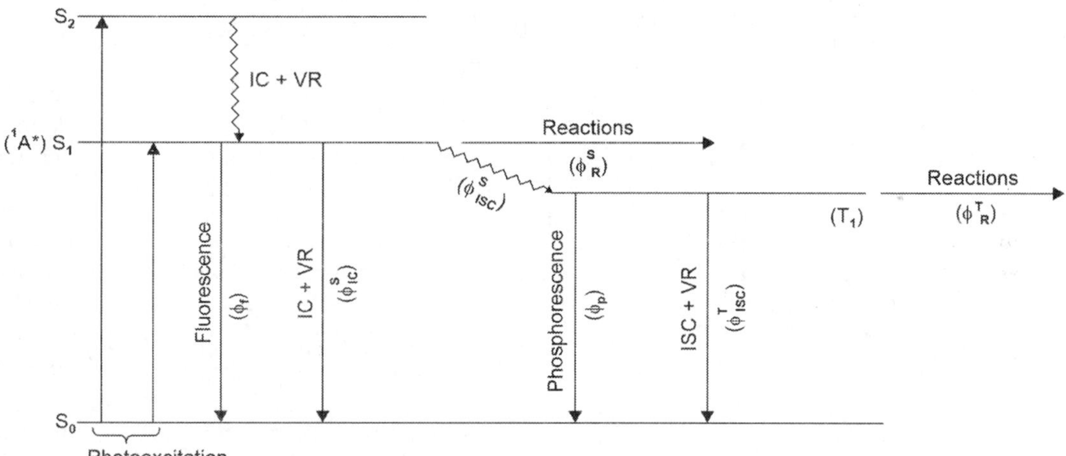

Fig. 6.13.1: Excited states (generated through the photon absorption) participating in competitive photophysical and photochemical reactions

Note: If the primary process of photoexcitation generates the **radicals**, then they may lead to the secondary processes **in the dark**. Thus the radicals produced in photoexcitation initiate the **chain reaction** and the resultant quantum yields (ϕ_R) becomes very high.

6.14 PHOTOCHEMICAL REACTIONS *VERSUS* THERMAL REACTIONS

Reactions starting from the **ground electronic states** of the reactants through the formation of activated species or transition state are called the **thermal reactions** as the reaction rate depends on the temperature (*cf.* Arrhenius equation). Such reactions are also described as the **dark reactions** (to distinguish from the photochemical reactions) as they can go on in absence of light. On the other hand, in photochemical reactions, the reactant is electronically excited (by the amount, 150–600 kJ mol^{-1}) through the absorption of light. Such electronically excited species require a little amount of activation energy (*ca.* 8^{-17} kJ mol^{-1}) to give the product(s) (*cf.* activation energy in thermal reactions is generally much higher; and it may be as high as 30 kJ mol^{-1}).

● **Electronic state of the reactant:** Thermal reactions start from the ground electronic states of the reactants while the photochemical reactions start from the electronically excited states of the reactants.

The photochemically excited species may react to give the different products **through the different transition states**. Alternatively, the photochemically excited species **through a common transition state** may produce first the **initial product (IP) intermediate**. Then, the **initial product (IP) intermediate** is ultimately converted into the more stable product(s) (Fig. 6.14.1) which may be the same as in the thermal reactions. The IP intermediate can produce different products with respect to those of the thermal reactions and some of the products may be thermodynamically quite inaccessible.

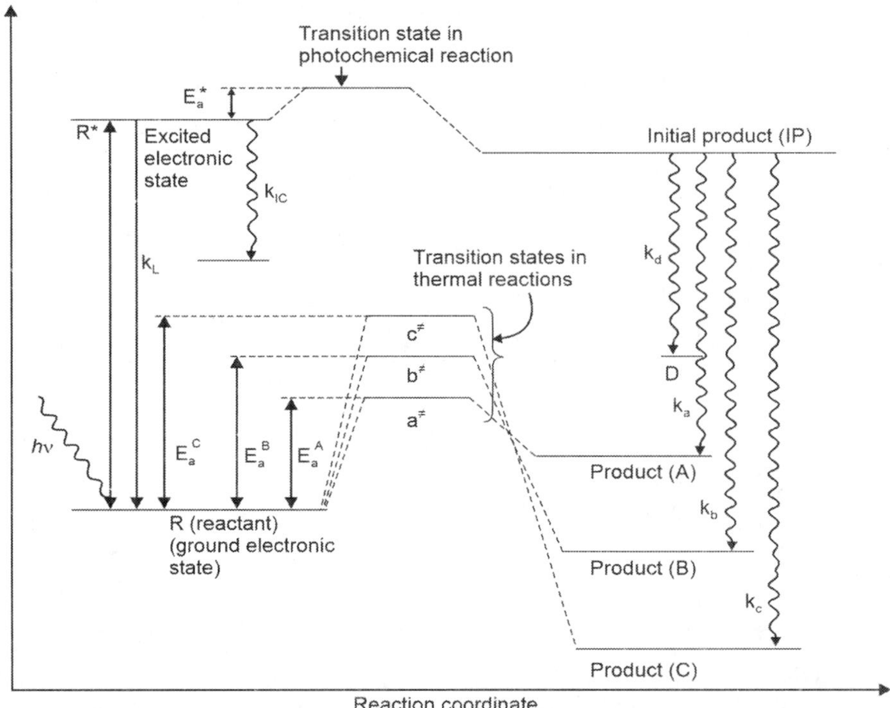

Fig. 6.14.1: Comparison of the energy profile diagrams of the photochemical and thermal reactions (*cf.* Figs 6.14.2, 3) (activated species $a^{\neq}$, $b^{\neq}$ and $c^{\neq}$ for the products A, B and C respectively in thermal reactions; E_a and E_a^* denoting the activation energies for the ground and excited electronic states of the reactant; $E_a \gg E_a^*$ *i.e.* **E_a (thermal reactions) $\gg E_a^*$ (photochemical reactions)**. k_L = luminesce rate constant; k_{IC} = internal conversion rate constant; k_a, k_b, k_c and k_d denoting the photochemical reaction rate constants for the products A, B, C and D respectively.

Note: Photochemically all the products A, B, C and D are accessible though A and D being thermodynamically unstable are inaccessible in thermal reactions. Product C having very high activation energy is inaccessible in thermal reactions.

Thermal reactions to go on must satisfy the **thermodynamic condition** (*i.e.* $\Delta G_{P,T} < 0$; simply products must be more stable than the reactants) and **kinetic condition** (*i.e.* thermal energy must be sufficiently high to provide the activation energy) but no such thermodynamic and kinetic conditions are required for the photochemical reactions which can produce the **thermodynamically and kinetically inaccessible products** (in terms of the energy profile diagram of the thermal reactions, Fig. 6.14.1).

The characteristic difference between the thermal and photochemical reactions are discussed here.

(i) **Electronic state of the reactant:** Thermal reactions start from the ground electronic state of the reactant while photochemical reactions start from the excited electronic state of the reactant.

(ii) **Activation energy:** In ordinary thermal reactions (called **dark reactions** as they can occur in absence of light), the required activation energy ($\leq$ 30 kJ mol^{-1}) is attained through collisions where the kinetic energy of the colliding species is used up to give the thermal activation (*i.e.* heat energy is consumed to provide the thermal activation energy). In photochemical reactions, the excited species is generated through the light absorption (*i.e.* photoexcitation) and the required activation energy for such photochemically excited species is negligibly small (*ca.* 8–17 kJ mol^{-1}).

(iii) **Temperature dependence:** In a thermal reaction, the activation energy is obtained from the heating. At a particular temperature, only a certain fraction (determined by Boltzmann distribution law) of the reactant molecules is possessing the minimum activation energy required (Fig. 6.14.2). With the increase of temperature, this fraction increases rapidly and the rate of the reaction increases (*cf.* rate $\propto$ exp ($-E_a/RT$), E_a = activation energy, **Arrhenius equation**). In photochemical reactions, the molecules experiencing the photoexcitation can only participate in photochemical reactions (which remain in competition with the photophysical processes). Thus, the photochemical reactions are **temperature independent** (assuming the competitive photophysical processes and the **secondary thermal steps** to be temperature independent). The primary process of photoexcitation is **temperature independent**. If the photochemically excited species requires a very small or no thermal activation, then the process becomes temperature independent. But, if the **secondary process** requires the thermal activation, then the overall photochemical reaction becomes temperature dependent.

Note: Here it may be pointed out that in some cases, after the photoexcitation, **thermal activation is required**. It occurs so **when the photochemically excited intermediate is long-lived**. In such cases, **quantum yield becomes temperature dependent**. It happens so in the photochemical reactions of Cr(III) complexes (see Sec. 6.15.6).

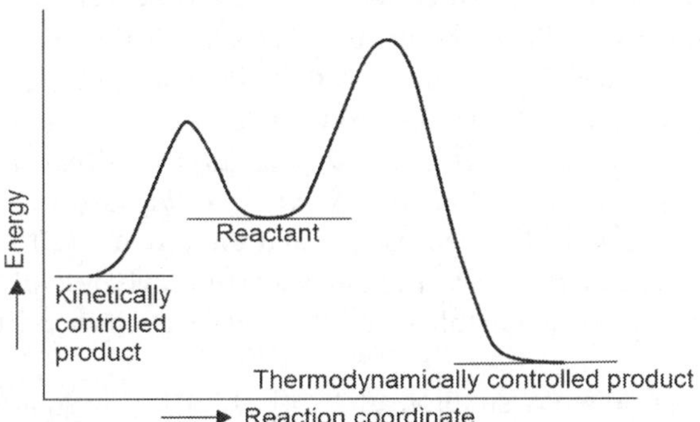

Fig. 6.14.2: Illustration of kinetically and thermodynamically controlled products. At higher temperature, thermodynamically controlled product (higher activation energy) predominates

Note: It may be noted that the thermodynamically controlled product may have the lower activation energy and then **kinetically controlled product** is the thermodynamically controlled product.

(iv) **Importance of thermodynamics and $\Delta G_{P,T}$:** The reactions with only $\Delta G_{P,T} < 0$ can occur thermally. The reactions with $\Delta G_{P,T} > 0$ can only occur, if such thermodynamically unfavoured reactions are coupled with the thermodynamically favoured reactions (*i.e.* $\Delta G_{P,T} < 0$) so that for the overall coupled reactions, $\Delta G_{P,T}$ becomes negative. On the other hand, the reactions with $\Delta G_{P,T} > 0$ can occur photochemically (Fig. 6.14.1) where the photoabsorption produces the activated species and finally the **light energy is stored as the free energy into the product. In other words, light energy is required to drive such thermodynamically unfavoured reactions.** As the product has higher free energy than the reactant, in absence of light, the product may convert into the reactants releasing heat (*i.e.* the backward thermal reaction is exothermic). Photosynthesis produces glucose from CO_2 and H_2O ($\Delta G_{P,T}$ = +ve) while the backward thermal reaction (*i.e.* respiration) gives back the reactants ($\Delta G_{P,T}$ = –ve).

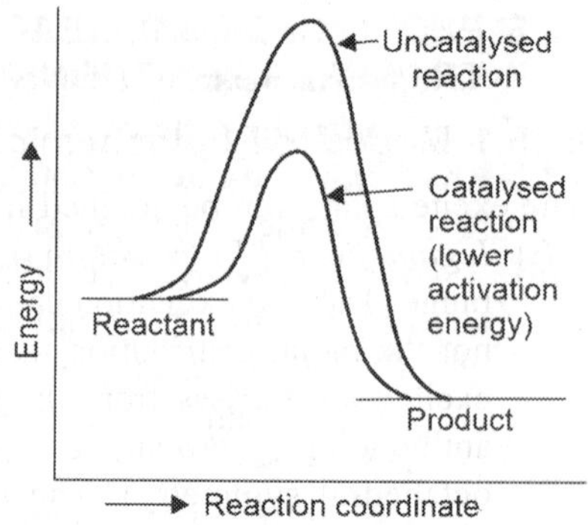

Fig. 6.14.3: Comparison of catalysed and uncatalysed thermal reactions

(v) **Product nature:** The same reaction when carried out photochemically and thermally, different products may be obtained due to the different mechanistic pathways of reactions. In thermal reactions, thermodynamically unstable products are inaccessible and the products requiring very high activation energy are also kinetically inaccessible. In photochemical reactions, all such products (which are either thermodynamically or kinetically inaccessible in thermal reactions) are accessible (*cf.* Fig. 6.14.1). It may be noted that in thermal reactions, distribution of the kinetically controlled and thermodynamically controlled products depends on the working temperature and nature of the energy profile diagram (*i.e.* energy *vs.* reaction coordinate).

(vi) **Catalyst *vs.* photosensitizer:** Thermal reactions can be kinetically accelerated by the catalyst but a catalyst works if only the reaction is thermodynamically allowed. It indicates that **in absence of the catalyst, the reaction occurs but slowly** (Fig. 6.14.3). In photochemical reactions, the photosensitizer can initiate the reaction and it is eventually regenerated. **In absence of the photosensitizer, under the identical conditions, the reaction cannot at all occur** (Sec. 6.12). *Thus a photosensitizer can both initiate and accelerate a reaction while a catalyst can only accelerate the rate of the reaction but cannot initiate the reaction.*

Photochemical *vs.* Thermal reactions (*i.e.* Dark reactions)

● **Product:** Figs. 6.14.1, 6.15.1.1 illustrate that the photochemically excited state can participate in the reactions to give the products which may be thermodynamically less stable compared to the starting reactants. Sometimes, it may give the products for which the activation energy (in thermal reactions) may be very high. *In fact, many thermodynamically and kinetically unfavourable thermal reactions can occur through the photochemical assistance.* In Figs 6.14.1, 6.15.1.1, in thermal path, the product (A) is thermodynamically inaccessible and the product (C) is kinetically inaccessible (*i.e.* very high activation energy). **But these products are photochemically accessible.** Here it must be mentioned that the paths of photochemical and thermal reactions are different. It is illustrated in the following example:

● **$[Co(NH_3)_6]^{3+}$:** When an aqueous solution of $[Co(NH_3)_6]^{3+}$ is irradiated by the light of lower wavelength (λ < 330 nm), it readily aquates to $[Co(NH_3)_5(OH_2)]^{3+}$. But, an acidic solution of $[Co(NH_3)_6]^{3+}$ can be boiled for hours without any change because the process is kinetically disfavoured in spite of the thermodynamic favour.

● **$[CoBr(NH_3)_5]^{2+}$:** Thermal reaction of $[CrBr(NH_3)_5]^{2+}$ leads to slow aquation of Br^- but photochemical reaction of the same complex leads to aquation of NH_3 (*cf.* **Adamson's rule**).

● **Franck-Condon principle:** In the photochemical electron transfer process, there is no requirement of bond length adjustment and other geometric reorganisation before the electron transfer, *i.e.* **Franck-Condon principle does not operate.** But in the thermal electron transfer process, Franck-Condon principle is to be obeyed.

6.15 PHOTOCHEMICAL REACTIONS (BOTH SUBSTITUTION AND ELECTRON TRANSFER REACTIONS) IN COORDINATION COMPLEXES

6.15.1 Modes of Photochemical Excitation in the Transition Metal Complexes

The excited state can be attained in different possible routes.

(i) *Ligand field (LF) transition (i.e. d–d transition):* The *d–d* transitions generally occur in the visible range. The *d–d* excitation causes the change in the **angular distribution of the excited electron** not the radial distribution that arises from the charge transfer (CT) excitation. The ligand field excited state arises from the promotion of an electron from the nonbonding t_{2g} orbitals to the antibonding e_g^* orbital, *i.e.* $t_{2g} \rightarrow e_g$ in terms of CFT $\equiv \pi_d \rightarrow \sigma^*$ in terms of MOT (considering the octahedral symmetry). The increased electron density in the antibonding orbital causes the M—L bond weakening. This bond weakening is more severe for the stronger field ligands for which the e_g orbital possesses the more antibonding character. *This bond weakening process can lead to various types of reactions, e.g.* ligand replacement, isomerization and racemization. The *d–d* excitation mainly causes the photosubstitution while the CT excitation causes the redox reaction. Thus to promote the photosubstitution, the frequency of CT band should be avoided.

Photodissociation: $[ML_n] \xrightarrow{h\nu} [ML_{n-1}] + L$

Photosubstitution: $[ML_n] + L' \xrightarrow{h\nu} [ML_{n-1}\,L'] + L$

Photoisomerisation: $cis\text{-}[ML_4\,L'_2] \xrightarrow{h\nu} trans\text{-}[ML_4L'_2]$

Photorecimisation: $\Delta\text{-}[M(L\text{—}L)_3] \xrightarrow{h\nu} \Lambda\text{-}[M(L\text{—}L)_3]$

Examples of these types of photochemical reactions have been discussed in the next Sections.

> **Note (orbital designation in the O_h-complexes)** (see Sec. 6.3.2)**:** In terms of MOT, the e_g-orbitals (CFT) always appear as the σ-ABMOs (*i.e.* σ*-MOs) but the t_{2g} orbitals (CFT) appear as the **NBMOs** (for the pure σ-bonding ligands as in $[Co(NH_3)_6]^{3+}$) or π-BMOs (π-t_{2g} or π*-t_{2g}) (depending on the π-bonding nature of the ligands). Thus the t_{2g} orbitals of **π-symmetry** are of different characters in terms of MOT depending on the nature of the ligands (see Fig. 3.17.3.4, Vol. 4).

(ii) *Charge transfer (CT) transition*: Sometimes, absorption of radition can lead to the charge transfer transition (*i.e.* metal → ligand, *i.e.* MLCT or ligand → metal, *i.e.* LMCT) to generate the *charge transfer excited state.* Such excited states very often lead to the photo-redox reactions. These charge transfer (*i.e.* MLCT and LMCT) excitations are the consequences of **inner-sphere charge transfer** (ISCT) transitions which **change the radial charge distribution**. These may result in redox reactions involving the metal centre and the ligands. The change in radial charge distribution very often makes the excited complex susceptible towards protonation, isomerisation, redox processes or nucleophile/electrophile depending on the condition. The **radial charge redistribution** in the excited states generated from ISCT may lead to the **inner-sphere redox processes** as illustrated (in general) below.

Photooxidation of the metal centre and photoreduction of the ligand

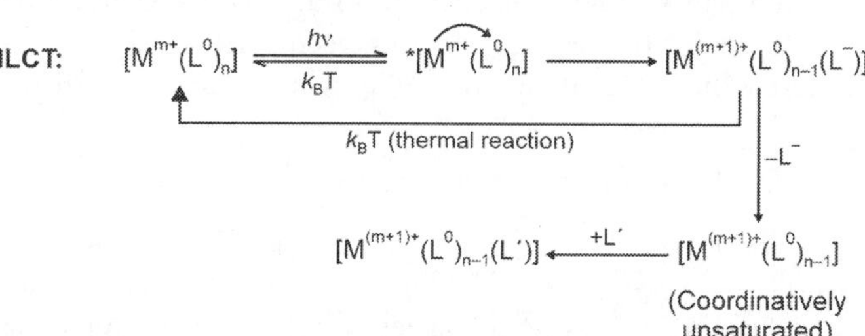

MLCT excitation generally occurs in the complexes of metals at their **low oxidation states** (*e.g.* Cu^I, Au^I, Fe^{II}, Ru^{II}, Mo^{IV}, etc.) and complexing with the **good π-acceptor ligands** (*e.g.* bpy, phen,

NO^+, CN^-, etc.). Some representative examples of such reactions have been discussed in the next sections.

Photooxidation of the ligand and photoreduction of the metal

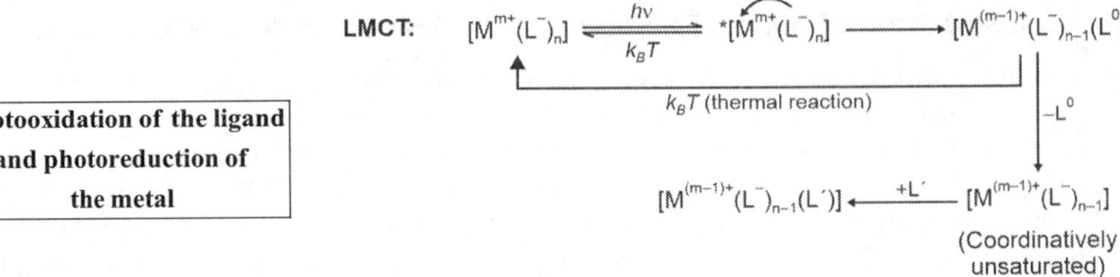

LMCT excitations generally occur in the metal complexes of metals at higher oxidation states (*e.g.* Cu^{II}, Fe^{III}, Ru^{III}, Co^{III}, Ir^{IV}, Pt^{IV}, etc.) complexing with the good donor ligands (*e.g.* S^{2-}, N_3^-, O_2^{2-}, RS^-, edta, nta, heavier halides like Br^-, I^-, etc.). Some representative examples of such reactions have been discussed in the subsequent sections.

(iii) **Ligand centred (LC) charge transfer-intraligand and interligand charge transfer transitions:** The intraligand CT transition is mainly localised on the ligand to generate the intraligand excited state that may initiate the ligand rearrangement or intraligand bond rupture or reaction of the ligand with the other substrates depending on the conditions. The most common intra-ligand charge transfer transition produces the ligand-centres (LC) excited state in a polyatomic ligand (say XY) to **weaken the intraligand bond**. In fact, this ligand centred **intraligand CT transition** leads to the **excitation of electron into the ligand antibonding orbitals** and it weakens the intraligand bond to assist the intraligand bond cleavage. It is illustrated below.

$$[L_nM(XY)] \xrightarrow[(LC)]{h\nu} [L_nM(XY)^*] \longrightarrow [L_nMX] + Y$$

$$[(H_3N)_5M(N_3)]^{2+} \xrightarrow[(LC)]{h\nu,\, H^+} [(H_3N)_5M(NH)]^{3+} + N_2, (M = Rh^{III} \text{ or } Ir^{III})$$

($:\dot{N}H$-**nitrene** ligand generated through the splitting of azide with the concomitant release of N_2). Similarly, intraligand CT transitions lead to the cleavage of the C—C bond of the oxalato ligand present in the Co^{III}-complexes to release CO_2.

In the case of **interligand CT transitions**, *i.e.* **ligand to ligand charge transfer (LLCT) transitions**, charge can be transferred from one ligand to another ligand to produce the excited state. It generally happens so in the complexes bearing both the π-donor ligands (*e.g.* aromatic thiols) and good π-acceptor ligands (*e.g.* bpy, phen). The LLCT transition leads to the charge transfer from the π-donor ligand to the π-acceptor ligand. LLCT also belongs to the class of ISCT-transitions. In the Ti^{IV}, Mo^{VI} or Re^{VII} complexes with the O_2^{2-} ligands, the LLCT transition from one O_2^{2-}-ligand (acting as the CT donor or reductant) to another O_2^{2-} ligand (**acting as the CT acceptor or oxidant**) leads to the **photodismutation of** O_2^{2-} to $2O^{2-}$ and O_2.

(iv) **Outer-sphere charge transfer (OSCT) transition:** Metal to solvent charge transfer (MSCT) and ion-pair charge transter (IPCT) are the most important CT transitions in the category of OSCT. In MSCT, charge transfer occurs from the metal centre to the surrounding solvent to generate the solvated electron, e^-_{solv} which can lead to different types of redox reactions (see Sec. 6.15.9).

In general, $[M^{m+}(L)_n] \xrightarrow[(MSCT)]{h\nu} [M^{(m+1)+}(L)_n] + e^-_{solv}$

e.g. $\left[Fe(OH_2)_6\right]^{2+} \xrightarrow[(MSCT)]{h\nu} \left[Fe(OH_2)_6\right]^{3+} + e^-_{aq}$

$[Fe(CN)_6]^{4-} \xrightarrow[(MSCT)]{h\nu} [Fe(CN)_6]^{3-} + e^-_{aq}$ (in aqueous solvent)

In fact, many other cyano complexes like $[Ru(CN)_6]^{4-}$, $[M(CN)_8]^{4-}$ ($M = Mo^{IV}$, W^{IV}) can also undergo the MSCT excitations to produce the hydrated electrons. In acidic condition, the hydrated electron may combine with the proton (*i.e.* $e_{aq}^- + H^+ \rightarrow H, 2H \rightarrow H_2$) **to produce the molecular hydrogen**. This route may be exploited to produce H_2 (**a clean fuel**) in a commercial scale from the acidified water.

ICPT occurs within the ion-pair (an **outer-sphere association**). Thus IPCT belongs to the group of OSCT. IPCT excitation may lead to the redox reactions when the ion-pair is constituted by a relatively strong reducing and an oxidising agent. It is illustrated in the following example.

$$\underset{\textit{Ion-pair}}{\{Co(NH_3)_6^{3+}\cdot I^-\}} \xrightarrow[\text{IPCT}]{h\nu} \{[Co(NH_3)_6]^{3+}\cdot I^-\} \longrightarrow [Co(NH_3)_6^{2+}\cdot I] \longrightarrow [Co(aq)]^{2+} + 6NH_3 + \tfrac{1}{2} I_2$$

$\left[Co(NH_3)_6\right]^{2+}$ ($t_{2g}^5 e_g^2$) is **labile** and it undergoes aquation rapidly. .

(v) **Intervalence transfer (IVT or IT):** MMCT (metal to metal charge transfer) or IVCT (intervalence charge transfer) (see Vol. 4, and Secs 6.4.16, 7.22, Vol. 5) transition is also an ISET process. Absorption of irradiation may lead to electron transfer from one metal centre to another metal centre in the dinuclear and polynuclear complexes. It can initiate the redox reaction. It is illustrated in the following examples.

$$\left[(H_3N)_5Co^{III}—NC—Ru^{II}(CN)_5\right]^- \xrightarrow[\text{aq. solution}]{h\nu} [Co(aq)]^{2+} + 5NH_3 + \left[Ru(CN)_6\right]^{3-}$$

CT excitation *vs.* LF excitation

- In octahedral complexes, **ligand field excitation:** $t_{2g} \rightarrow e_g$ in terms of **CFT** $\equiv \pi_d \rightarrow \sigma^*$, simply $\pi \rightarrow \sigma^*$ in terms of MOT.
- **Ligand field (LF) excitation** (*d* → *d* excitation) changes the **angular distribution** of the electron but not the **radial distribution.** In the **charge-transfer excitation** (M → L or L → M), the radial distribution of the electron density is changed. LF excitation mainly promotes the photosubstitution in the complexes.
- When the ligand field band and charge transfer band overlap, the LF excitation may also cause the **photoredox reaction** as in the Co(III)-complexes.
- **Ligand field (LF) excitation** generally leads to photo-substitution, photo-isomerization and photo-racemization reactions.
- **Charge-transfer (CT) excitation** generally initiates the photo-redox reactions. However, a CT excitation in $[CoCl(NH_3)_5]^{2+}$ leads to aquation by an *indirect path.*

$$\left[(H_3N)_5Co^{III}—Cl\right]^{2+} + H_2O \xrightarrow[(\lambda < 350\ nm)]{h\nu} \left[(H_3N)_5Co^{III}—OH_2\right]^{3+} + Cl^-$$

In fact, the primary step of photo-excitation leads to the homolytic fission of the Co^{III}—Cl bond causing the reduction of Co^{III} to Co^{II} which is labile.

The mixed valence compounds like $\left[Cl(bpy)_2Ru^{II}—N\bigcirc N—Ru^{III}(bpy)_2Cl\right]^{3+}$ experience the intervalence transfer through the **bridging ligand pyrazine** (see Sec. 1.13 of Vol. 4; Secs 6.4.16, 7.22, Vol. 5).

$$\left[(H_3N)_5Ru^{II}—N\bigcirc—\bigcirc N—Ru^{III}(NH_3)_5\right]^{5+} \xrightarrow[\lambda = 1050\ nm]{h\nu} \left[(H_3N)_5R^{III}—N\bigcirc—\bigcirc N—Ru^{II}(NH_3)_5\right]^{5+}$$

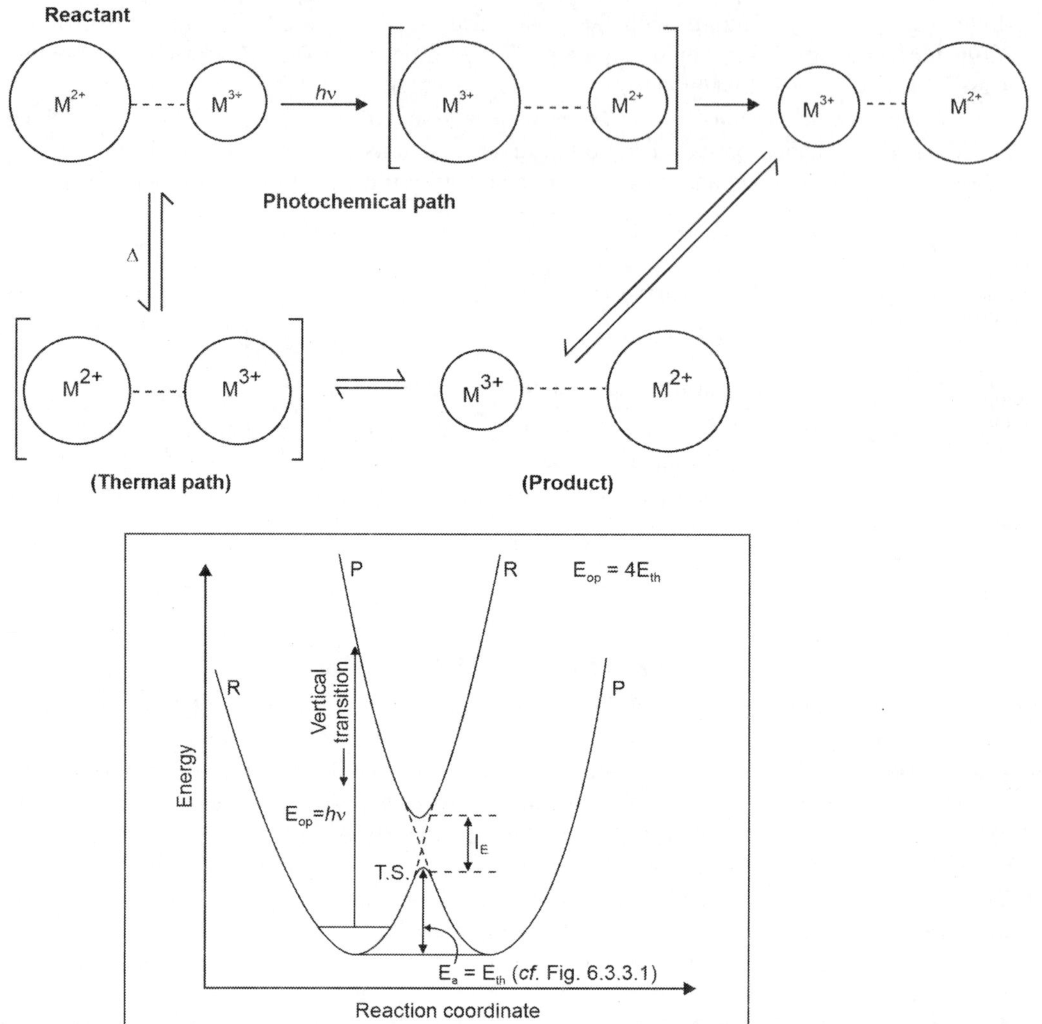

Fig. 6.15.1.1 Comparison of the photochemical and thermal homonuclear electron transfer in a homonuclear mixed valence systems, *e.g.* $M^{2+} \cdots M^{3+}$ (photochemical path experiencing **no Franck-Condon barrier** while thermal path experiencing the Franck–Condon barrier) (*Ref.* Creutz *et al. Prog. Inorg. Chem.*, **30**, 1–73, 1983)

This IT occurs in the near infrared region through the **bridging ligand 4, 4′-bipyridine.**

Photochemical and thermal IT in a homonuclear mixed valence system, say $M^{2+} \cdots M^{3+}$ can be compared. In the photochemical process, electron transfer from M^{2+} to M^{3+} (*i.e.* electron transfer from the potential energy surface of $M^{2+} \cdots M^{3+}$ to that of $M^{3+} \cdots M^{2+}$) can occur **without any preorganization (*i.e.* bond length adjustment).** But in the thermal process, before the electron transfer, the bond length must be adjusted (as demanded by **Franck-Condon principle**, see Fig. 6.3.2.1). These aspects are illustrated in Fig. 6.15.1.1. For a symmetrical mixed valence complex, E_{op} (for **vertical transition** in photochemical IT) = $4E_{th}$ (for thermal IT). This difference is due to the Franck-Condon principle.

(vi) **Photodissociation of the metal-metal bond in the dinuclear complexes having the M—M multiple bond:** Irradiation by light may lead to $\delta \rightarrow \delta^*$ transition to cause photodissociation of the metal-metal bond. Irradiation by light of 500 nm on $[(PhO)_2PO_2]Mo \equiv Mo\left[O_2P(OPh)_2\right]$ having the $Mo \equiv Mo$ quadruple bonding can cause fission of the metal-metal bond under the suitable conditions to favour the oxidation of the Mo centre. M—M single bond in the dinuclear carbonyl complexes may undergo photoinduced cleavage through the $\sigma \rightarrow \sigma^*$ transition. This aspect has been discussed in detail in Sec. 6.15.4.

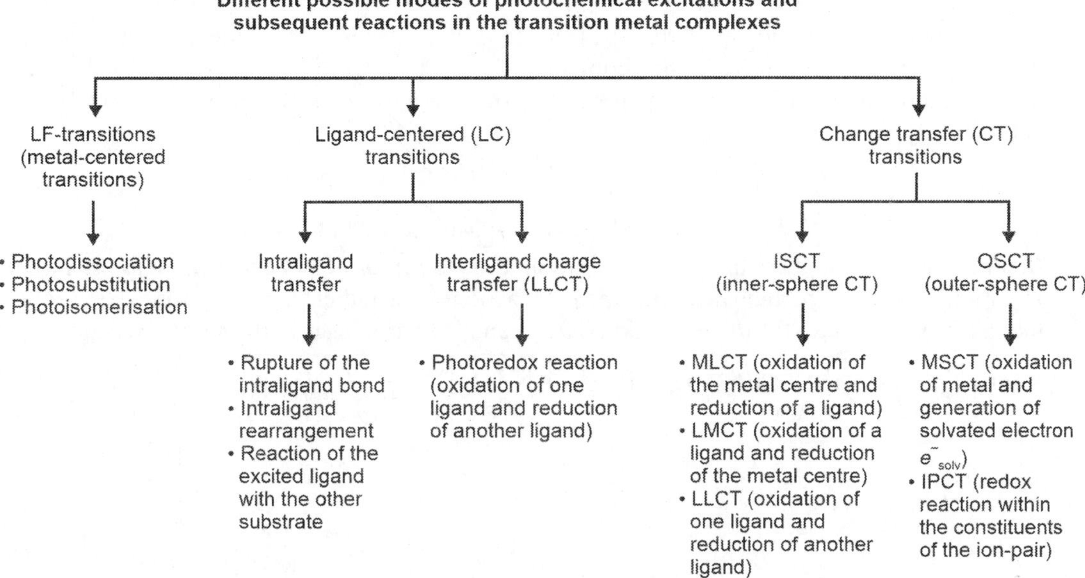

6.15.2 Controlled Reactions — Application of the Photochemical Reactions

Selectivity of the photochemical reactions may be used to control the reactions. It is illustrated in the following examples.

(i) LF excitation of $[Co(NH_3)_5N_3]^{2+}$ causes the aquation of NH_3 while CT excitation of the same complex causes the redox decomposition of the complex.

$$\left[Co(NH_3)_5(N_3)\right]^{2+} \xrightarrow[\text{(green light)}]{\text{LF excitation}} \left[Co(NH_3)_4(N_3)(OH_2)\right]^{2+} + NH_3,$$ (*cf.* **Adamson's rule**).

$$\left[Co(NH_3)_5(N_3)\right]^{2+} \xrightarrow[\text{(UV light)}]{\text{CT excitation}} Co(II) + NH_3 + N_2$$

Thus selection of wavelength of the irradiation can give the different products.

(ii) In thermal reaction (*i.e.* dark reaction), $[CrBr(NH_3)_5]^{2+}$ aquates Br^- while in photochemical reaction, it aquates NH_3 (*see* **Adamson's rule**).

$$\left[CrBr(NH_3)_5\right]^{2+} \xrightarrow[\text{(thermal reaction)}]{k_BT\text{(slow)}} \left[Cr(NH_3)_5(OH_2)\right]^{3+} + Br^-$$

$$\left[CrBr(NH_3)_5\right]^{2+} \xrightarrow[\text{(rapid)}]{h\nu} \left[CrBr(NH_3)_4(OH_2)\right]^{2+} + NH_3$$

The details of the mechanistic aspects of the above reactions will be discussed later.

6.15.3 Prompt and Delayed Photochemical Reactions

In some cases, the photochemically excited states dissociate promptly and in some cases, the excited states have the relatively long lifetimes.

- (a) **Prompt reactions** (*occurring very fast and in less than 10 ps*): Some representative examples are discussed below.

(i) **Substitution reactions in metal carbonyls:**

$$\underset{(18e)}{[Cr(CO)_6]} \xrightarrow{h\nu\ (UV)} \underset{(16e)}{[Cr(CO)_5]} + CO,$$

$$[Cr(CO)_5] + L \longrightarrow [Cr(CO)_5 L]\ (18e)\ (L = \text{THF solvent})$$

Absorption of UV radiation by $[Cr(CO)_6]$ leads to $\pi \rightarrow \sigma^*$ electronic transition and this excitation populates the electron density in the antibonding orbital to weaken the Cr—CO bond. This leads to dissociation of the Cr—CO bond to produce a **16*e* species** $[Cr(CO)_5]$ (square pyramidal) which is kinetically very much active.

$$[Mn_2(CO)_{10}] \xrightarrow[\text{(fast)}]{h\nu} 2[{}^{\bullet}Mn(CO)_5]\ (17e\ \text{species})$$

$$[{}^{\bullet}Mn(CO)_5] + CCl_4 \xrightarrow{\text{slow}} [Mn(CO)_5Cl] + {}^{\bullet}CCl_3$$

In this binuclear complex, the $\sigma \rightarrow \sigma^*$ transition leads to metal-metal bond rupture to produce the free radicals that can participate in free radical reactions (free radical mechanism) with the solvent like CCl_4, $CHCl_3$, etc. $[Fe_2(CO)_9]$, $[Co_2(CO)_8]$ can also participate in the similar reactions.

(ii) **Photo-aquation of $[CoCl(NH_3)_5]^{2+}$:** $[Co^{III}Cl(NH_3)_5]^{2+} \xrightarrow[(\lambda < 350\ nm)]{h\nu} \underset{(cage)}{\{[Co^{II}(NH_3)_5]^{2+} + {}^{\bullet}Cl\}}$

$$\xrightarrow{H_2O} [Co(NH_3)_5(OH_2)]^{3+} + Cl^-$$

Scission (*i.e.* **homolytic fission**) of the Co^{III}—Cl bond leads to reduction of Co^{III} to Co^{II} which is a labile centre and experiences a rapid aquation followed by reoxidation by Cl. This type of ***photoredox process*** leading to substitution is common with the Co(III) and Ru(II) complexes. This actually involves the *charge transfer* (CT) leading to M^-—L^+ or M^+—L^-.

The ***quantum yield*** (measuring the amount of reaction per mole of photons absorbed) of the above reaction increases with the increase of wavelength towards the **limiting/threshold value** (λ_{lim} required just to dissociate the bond). *In fact, the excess energy (= hν – bond energy) remaining after the bond fission is available to the dissociated fragments to increase their kinetic energy and it increases the probability of escaping away of the dissociated fragments from each other in solution* (*cf.* **cage effect**).

(iii) ***Photosubstitution in $[Cr^{III}(NH_3)_xX_y]^{n+}$ (x + y = 6):*** Very often, photosubstitution in these cases leads to different products compared to those obtained in the thermal substitutions. In the thermal substitution reactions, halides (X^-) are substituted while in photosubstitutions, NH_3 may also be replaced. The photoexcitation may lead to the following excited states:

${}^4T_{2g}$, ${}^4T_{1g}$ (*i.e.* $t_{2g}^2e_g^1$) and ${}^2E_g$ (**low spin** t_{2g}^3, *i.e.* (↑↓)(↑)).

$$Cr^{3+}(t_{2g}^3, {}^4A_{2g}) \xrightarrow{h\nu} Cr^{3+}(t_{2g}^2e_g^1, {}^4T_{1g})$$

$$\xrightarrow{h\nu}$$

$({}^4A_{2g}$, *i.e.* high spin $t_{2g}^3)$ $\qquad$ $({}^2E_g$, *i.e.* low spin $t_{2g}^3)$

In the ${}^4T_{2g}$ or ${}^4T_{1g}$ state (see Fig. 6.15.6.1), accumulation of electron density in the antibonding orbital weakens the metal-ligand bond to favour the **dissociative process of substitution** while in the ${}^2E_g$ state, a vacancy created in the t_{2g} level can favour the **associative process of substitution**. Here it is worth mentioning that the Cr(III) complexes in ground state (*i.e.* $^4A_{2g}$ state, t_{2g}^3) experience the **associative activation** in thermal or dark reaction for ligand substitution with a high crystal field activation energy (CFAE).

The following photosubstitution reaction is very rapid and it occurs within less than 5 ps.

$$\left[Cr(NH_3)_6\right]^{3+} + H_2O \xrightarrow{h\nu} \left[Cr(NH_3)_5(OH_2)\right]^{3+} + NH_3$$

- (b) **Delayed Reactions:** Sometimes, the photochemically excited state is long-lived and it can participate in delayed reactions.

$$\left[Ru(bpy)_3\right]^{2+} \xrightarrow[\lambda=590\text{ nm}]{h\nu} \left[Ru(bpy)_3^*\right]^{2+}$$

$$\left[Cr(bpy)_3\right]^{3+} \xrightarrow{h\nu} \left[Cr(bpy)_3^*\right]^{3+}$$

$[Ru(bpy)_3^*]^{2+}$ is relatively long-lived to participate in different types of redox reactions to be discussed later (Sec. 6.15.13). Photochemically, the excited $[Cr(bpy)_3^*]^{3+}$ state possesses a long-lived ^2E state and it participates in several delayed redox reactions. In fact, $[Cr(bpy)_3^*]^{3+}$ is a good oxidizing agent (E^0 = +1.3 V). **These are discussed in detail in Secs. 16.16.1–2, Vol. 3B.**

6.15.4 Photochemical Reactions of Metal Carbonyls (*Ref.* G.L. Geoffroy, *J. Chem. Edu.*, **60**, 861, 1983)

Different types of photochemical reactions in metal carbonyls are known. We shall illustrate these examples and rationalise the reactions in terms of MOT/CFT.

Different Types of Photochemical Reactions of Metal Carbonyls

(i) **Photosubstitution in mononuclear carbonyls:**

$$\underset{(18e)}{[M(CO)_6]} \xrightarrow[(\pi\to\sigma^*)]{h\nu} [M(CO)_6^*] \text{ (excited state)} \rightarrow \underset{(16e)}{[M(CO)_5]} + CO, \ (M = Cr, Mo, W)$$

$[M(CO)_5] + L \rightarrow [M(CO)_5L]$, ($L$ = *CO for ligand exchange, PPh_3, PR_3, THF as the coordinating solvent)

The photoexcitation, $\pi\to\sigma^*$ (in terms of MOT), *i.e.* $t_{2g}\to e_g$ (in terms of CFT, t_{2g}^6 to $t_{2g}^5e_g^1$) depletes the bonding electron density but increases the electron density in the strongly antibonding orbital (σ^*) to weaken the M—CO bond to favour the **dissociative path** (Fig. 6.15.4.1).

At the ground state (t_{2g}^6, $^1A_{1g}$) of $[M(CO)_6]$, the thermal substitution reaction occurs also through the **dissociative path** with the rate constant in the range 10^{-6} s^{-1}. The life-time of the photoexcited state, $[M(CO)_6^*]$ is expected to be $<10^{-10}$ s. This excited state dissociates to the reactive intermediate pentacarbonyl, and the corresponding dissociation rate constant is likely to be $\sim10^{10}$ s^{-1}. **Thus the photosubstitution rate constant is higher than the corresponding thermal rate constant by a factor about 10^{16}.**

Thermal reaction: $[M(CO)_6](t_{2g}^6) \xrightarrow[(k\approx10^{-6}\text{ s}^{-1})]{rds} [M(CO)_5] + CO$

$$[M(CO)_5] + L \xrightarrow{\text{fast}} [M(CO)_5L]$$

Photochemical reaction: $[M(CO)_6](t_{2g}^6) \xrightarrow{h\nu} [M(CO)_6^*] \xrightarrow[k\approx10^{10}\text{ s}^{-1}]{rds} [M(CO)_5] + CO$

$$[M(CO)_5] + L \xrightarrow{\text{fast}} [M(CO)_5(L)]$$

The rate constant for the photosubstitution is **about 10^{16} times faster** than that for the thermal substitution in $[M(CO)_6]$.

Here it may be noted that working in noble gas matrices at about 15 K, spectroscopic evidences support the existence of $[M(CO)_5]$ intermediate in the photolysis of $[M(CO)_6]$. The intermediate $[M(CO)_5]$ (16*e*) is highly reactive and it may recombine readily with the dissociated ligand (CO) to give back the reactant. In solution, $[M(CO)_5]$ and the dissociated CO ligand are held in a **solvent cage**. To avoid their recombination, they are to be separated.
Other examples of photosubstitutions in **mononuclear carbonyls** are:

$$[Fe(CO)_5] + \text{(butadine)} \xrightarrow{h\nu} [\text{(butadiene)}\text{-----}Fe(CO)_3] + 2CO$$

$$[Ni(CO)_4] + PR_3 \xrightarrow{h\nu} [Ni(CO)_3(PR_3)] + CO$$
$$[W(CO)_6] + PPh_3 \xrightarrow{h\nu} [W(CO)_5(PPh_3)] + CO$$
$$[Co(CO)_3(NO)] + PPh_3 \xrightarrow{h\nu} [Co(CO)_2(NO)\,(PPh_3)] + CO$$
$$[Mn(\eta^5\text{–}Cp)(CO)_3] + py \xrightarrow{h\nu} [Mn(\eta^5\text{–}Cp)(CO)_2(py)] + CO$$
$$[V(\eta^5\text{–}Cp)\,(CO)_4] + PhC{\equiv}CPh \xrightarrow{h\nu} [V(\eta^5\text{–}Cp)\,(CO)_3\,(PhC{\equiv}CPh)] + CO$$
$$[Mo(\eta^6\text{–mesitylene})(CO)_3] + H_2C{=}CH_2 \xrightarrow{h\nu} [Mo(\eta^6\text{–mesitylene})(CO)_2(C_2H_4)] + CO$$
$$[ReCl(CO)_5] + py \xrightarrow{h\nu} [ReCl(CO)_4(py)] + CO$$

(ii) **Extent of photosubstitution depending on the nature of the entering ligand (L):** If L binds weakly than CO, then after substitution of one CO group, the **M—L bond formed being weaker becomes more labile** (in the **dissociative path**) than the M—CO bond. Consequently, further substitution of the CO group does not occur and it happens so in the thermal reactions also (see Sec. 5.31). In fact, if L is a **weaker π-acceptor ligand** than CO, then after substitution of one or two CO groups, the remaining M—CO bonds become stronger because of **less competitive π-back bonding** with the metal *d*-orbitals ($d_M \rightarrow {}^*\pi_{CO}$) among the remaining CO groups. However, if the M—L and M—CO bonds are of comparable bond strength (*i.e.* L is a **good π-acceptor ligand** like PF_3, $P(OMe)_3$), then further CO group substitution can occur.
It is illustrated in the following examples.

$$[Mo(CO)_6] + P(OMe)_3\ (\text{excess}) \xrightarrow{h\nu} [Mo(P(OMe)_3)_6] + 6CO$$
$$[Co(CO)_3(NO)] + PF_3\ (\text{excess}) \xrightarrow{h\nu} [Co(CO)_{3-x}(NO)(PF_3)_x] + xCO\ (x = 1, 2, 3)$$
$$[Mn(CO)_5H] + PF_3\ (\text{excess}) \xrightarrow{h\nu} [Mn(CO)_{5-x}(PF_3)_xH] + xCO,\ (x = 1-4)$$
$$[Fe(CO)_2(PF_3)_3] + PF_3\ (\text{excess}) \xrightarrow{h\nu} [Fe(PF_3)_5] + 2CO$$

Note: If the entering ligand (non π-bonding ligand like L = py, THF, PR_3, etc.) binds more weakly than CO, then the M—L bond formed undergoes photosubstitution.

$$[M(CO)_6] + L \xrightarrow{h\nu} [M(CO)_5L] + CO$$
$$[W(CO)_5L] + \text{1-pentene} \xrightarrow{h\nu} [W(CO)_5(\text{1-pentene})] + L\ (= py, NH_3)$$
$$[M(CO)_5(L)] \xrightarrow{h\nu} [M(CO)_5] + L;\ [M(CO)_5L] \overset{h\nu}{\not\rightarrow} [M(CO)_4L] + CO$$

This principle is utilised in the synthesis of mixed ligand carbonyl complexes illustrated below.

(iii) **Synthesis of metal carbonyl mixed ligand complexes through photosubstitution:** The photosubstitution is carried out first in a weakly coordinating solvent like THF (tetrahydrofuran), acetonitrile, pyridine etc. to produce the solvent substituted derivative which is subsequently subjected to the thermal substitution reaction to replace the solvent ligand by the desired ligand. In competition, the weaker M—solvent bond is easily substituted in the **dissociative path** It is illustrated below.

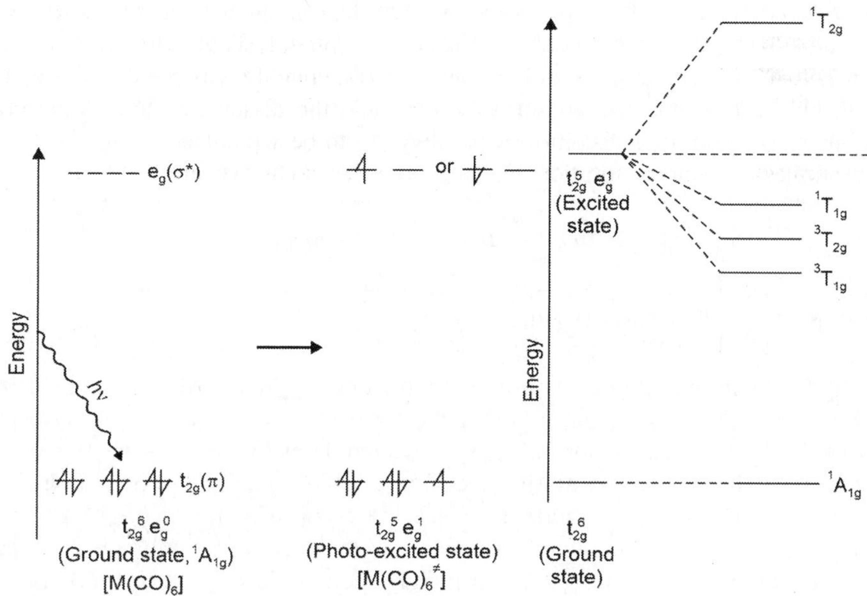

Fig. 6.15.4.1: Schematic illustration of orbital and energy state diagrams (for both the ground state and photoexcited state of $[M(CO)_6]$ (M = Cr, Mo, W) and $[V(CO)_6]^-$ (all 18*e* species) of d^6-configuration (**low-spin** octahedral complex) (see Figs 7.15.1.1, 2).

Note 1: In the photoexcited state ($t_{2g}^5 e_g^1$), the increased electron density in the σ*-orbitals (*i.e.* e_g-orbitals) and decreased electron density in the bonding π-orbital weakens the M—CO bond to favour its dissociation. The photoexcitation $t_{2g}^6 \rightarrow t_{2g}^5 e_g^1$, *i.e.* ${}^1A_{1g} \rightarrow {}^{1,3}T_{1g}, {}^{1,3}T_{2g}$ produces the excited species $[M(CO)_6^*]$ (lifetime < 10^{-10} s). **From the ${}^3T_{1g}$ excited state (lowest energy), the dissociation of a CO group occurs.**

Note 2: In olefins, the photoexcitation, π→π* weakens the C=C double bond and consequently, the **rotation energy barrier around the C=C bond is lowered down**. It causes the *cis-trans*-photoisomerisation.

Note 3: In $[Cr(CO)_6]$ and $[Mo(CO)_6]$, the **MLCT bands** (~35000 cm^{-1}, ~ 44,000 cm^{-1}) appear for the transition from the filled t_{2g}^6 level of metal to the vacant π*-MO of CO. The weak *d*→*d* transition bands appear as the shoulders on the strong CT bands (see Sec. 7.21.2, Vol. 5). **For the photosubstitution in $[M(CO)_6]$, the ligand field excitation** (*i.e. d*→*d* transition) **is important.**

$$[Mn(\eta^5\text{-Cp})(CO)_3] + \text{THF (most commonly used solvent)} \xrightarrow{h\nu} [Mn(\eta^5\text{-Cp})(CO)_2(THF)] + CO$$

$$[Mn(\eta^5\text{-Cp})\,(CO)_2(THF)] + L \xrightarrow[\text{reaction}]{\text{Thermal}} [Mn(\eta^5\text{-Cp})(CO)_2(L)] + THF$$

(iv) **Photosubstitution in $[M(CO)_5L]$ (18*e*):** If L is a weaker field ligand than CO, then between the two *trans*-axial bonds (L—M—CO, say along the *z*-axis), the M—L bond is relatively longer. It will split both the t_{2g} and e_g levels of O_h-crystal field in the fashion of *z*-out distortion (Fig. 6.15.4.2). Obviously, in this C_{4v} symmetry of $[M(CO)_5L]$, the splitting in the e_g-level is more important than in the t_{2g}-level because the lobes of e_g-orbitals are directly projected towards the ligands. The photoexcitation, $d_\pi \rightarrow d_{z^2}$ (**antibonding character**, σ_z^*) will weaken/labilise the *trans*-axial bonds L—M—CO along the *z*-axis.

$$[M(CO)_5L] \xrightarrow{h\nu} [M(CO)_5L^*] \begin{cases} \nearrow [M(CO)_5] + L & \dots(a) \\ \searrow [M(CO)_4L] + CO & \dots(b) \end{cases}$$

The relative efficiencies of the paths (a) and (b) depend on the relative inherent M—L and M—CO bond strengths mainly controlled by their relative π-acidic characters (*cf.* competition between L and CO for the metal (*d*)→ligand π-back bonding, stronger π-acid ligands are the stronger field ligands). For the weaker field ligands like THF, py etc. the path (a) predominates. This is why, it produces $[M(CO)_5L]$ exclusively and $[M(CO)_4L_2]$ becomes an insignificant product because it requires the path (b). On the other hand, for the stronger field ligand (*i.e.* stronger π-acidic ligand) like PF_3, the path (b) becomes important to produce the species $[M(CO)_{6-x}(PF_3)_x]$ (x = 1–6) (already illustrated).

$$\textbf{a-path}\begin{cases}[Mo(CO)_5(THF)] + PPh_3 \xrightarrow{hv} [Mo(CO)_5(PPh_3)] + THF \\ [W(CO)_5(L)] + \text{1-pentene} \xrightarrow{hv} [W(CO)_5(\text{1-pentene})] + L(=py, NH_3)\end{cases}$$

- **Wavelength dependent mode of photosubstitution in $[M(CO)_5L]$:** Here it is important to mention that the relative importance of ligand dissociation paths (a and b) from the photoexcited states of $[M(CO)_5L]$ depends not only on their bond strengths but **also on the frequency (*i.e.* wavelength) of the incident radiation** (see Table 6.15.4.1). Higher energy radiation promotes the M—CO bond dissociation **(path b) from the equatorial plane of $[M(CO)_4]$ moiety** of $[M(CO)_5L]$. In fact, higher energy radiation leads to the photoexcition to the $d_{x^2-y^2}(\sigma^*_{xy})$ orbital while the lower energy radiation leads to mainly the photoexcitation to the $d_{z^2}(\sigma^*_z)$ (see Fig. 6.15.4.3).

$$d_\pi \xrightarrow{hv_1} d_{z^2}(\sigma^*_z) \text{ and } d_\pi \xrightarrow{hv_2} d_{x^2-y^2}(\sigma^*_{xy}) \quad (hv_2 > hv_1)$$

Table 6.15.4.1: Wavelength dependence of quantum yield (ϕ) of the reaction $[M(CO)_5(py)] + py \rightarrow cis\text{-}[M(CO)_4(py)_2] + CO_{equatorial}$ (*i.e.* path b) in THF

M	$\phi_{366\ nm}$	$\phi_{436\ nm}$
Cr	0.28	0.17
Mo	0.35	0.22
W	0.06	0.02

Note: $\phi_{366\ nm} > \phi_{436\ nm}$. It indicates that for the higher energy radiation, the probability of photo-substitution of the stronger M—CO (equatorial) bond increases

The photoexcitation to the $d_{z^2}(\sigma^*_z)$ orbital weakens the *trans*-axial bonds (*i.e.* L—M—CO) along the *z*-axis and the **weaker M—L bond is preferably labilised in the dissociative path.** On the other hand, the photoexcitation to the $d_{x^2-y^2}(\sigma^*_{xy})$ orbital weakens the M—CO bonds at the equatorial plane (*xy*-plane) and it facilitates the M—CO bond dissociation from the equatorial plane. However, the **internal conversion** (IC) of the higher excited state to the lower excited state can lead to the M—L bond dissociation (Fig. 6.15.4.3). Thus the higher energy radiation can lead to both the M—CO (equatorial) and M—L bond substitution while the lower energy radiation mainly leads to the substitution of M—L bond.

(v) **Photodissociation leading to ligand rearrangement:** After photodissociation of a CO group, the reactive intermediate (coordinatively unsaturated) may lead to ligand rearrangement through the **heptacity change** (*e.g.* η^1-allyl to η^3-allyl, η^3-Cp to η^5-Cp).

$$[(\eta^1\text{-allyl})Mn(CO)_5] \xrightarrow{hv} [(\eta^3\text{-allyl})Mn(CO)_4] + CO, \ (\eta^1\text{-allyl} \rightarrow \eta^3\text{-allyl})$$

(vi) **Photochemical dimerisation or trimerisation depending on the condition:** The reactive intermediate (coordinatively unsaturated) produced through the photodissociation of a CO group may

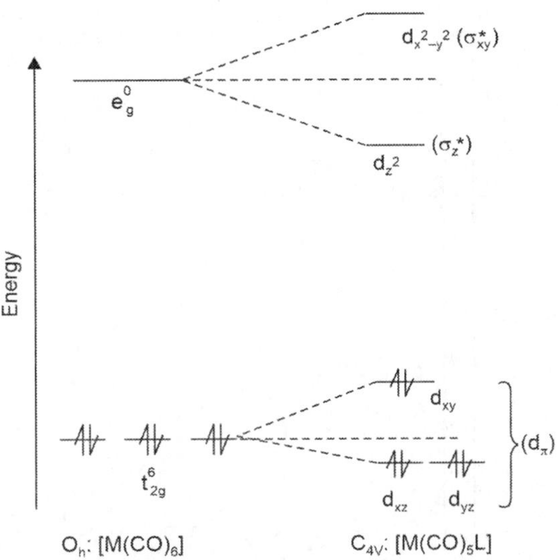

Fig. 6.15.4.2: Orbital energy diagrams of $[M(CO)_6]$ (O_h) and $[M(CO)_5L]$ (C_{4v})

undergo polymerisation to attain the 18*e* configuration. It is illustrated for dimerisation in the case of $[Fe(CO)_5]$.

$$[Fe(CO)_5] \xrightarrow{hv} [Fe(CO)_5^*] \longrightarrow [Fe(CO)_4] + CO$$

$$[Fe(CO)_4] + [Fe(CO)_5] \longrightarrow [Fe_2(CO)_9], \text{ i.e. } 2[Fe(CO)_5] \xrightarrow{hv} [Fe_2(CO)_9] + CO$$

Similarly, formation of $[Os_2(CO)_9]$ and $[Os_3(CO)_{12}]$ through the photolysis of $[Os(CO)_5]$ can be explained. In this case, the reactive intermediate is $[Os(CO)_4]$.

(vii) **Decomposition of $[Ni(CO)_4]$ by flash photolysis:** Flash photolysis of $[Ni(CO)_4]$ by using the high intensity UV-radiation gives the following result.

$$[Ni(CO)_4] \xrightarrow{hv} [Ni(CO)_3] + CO; [Ni(CO)_3] \xrightarrow{hv} [Ni(CO)_2] + CO$$

$$[Ni(CO)_2] \longrightarrow \text{solid product}$$

Decomposition of $[Ni(CO)_4]$ can be prevented by adding CO because the intermediates $[Ni(CO)_3]$ and $[Ni(CO)_2]$ can recombine with CO to give back $[Ni(CO)_4]$.

(viii) **Photoinduced homolytic cleavage of the metal-metal bonds in dinuclear carbonyls:** This is illustrated in the following examples.

$$[Mn_2(CO)_{10}] \xrightarrow[(\sigma\to\sigma^*)]{hv} [Mn_2(CO)_{10}^*] \longrightarrow 2[^{\bullet}Mn(CO)_5]\ (17e)$$

$$([Mn_2(CO)_{10}], \text{ i.e. } [(OC)_5Mn{-}Mn(CO)_5])$$

The $\sigma \to \sigma^*$ *e*-transfer weakens the metal-metal sigma bond to favour the metal-metal bond cleavage (see Fig. 6.15.4.4).

$$[^{\bullet}Mn(CO)_5] + Cl{-}CCl_3(\text{solvent}) \longrightarrow [MnCl(CO)_5]\ (18e) + {}^{\bullet}CCl_3$$

$$(\text{i.e. } [Mn_2(CO)_{10}] \xrightarrow[CCl_4]{hv} 2[MnCl(CO)_5])$$

$$[Re_2(CO)_{10}] \xrightarrow[I_2]{hv} 2[Re(CO)_5I]$$

The metal fragments (radicals) like $[^{\bullet}Mn(CO)_5]$ and $[^{\bullet}Re(CO)_5]$ (17*e* species) (**active radicals**) in **inert solvents** recombine in absence of any suitable free radical scavanger at diffusion controlled

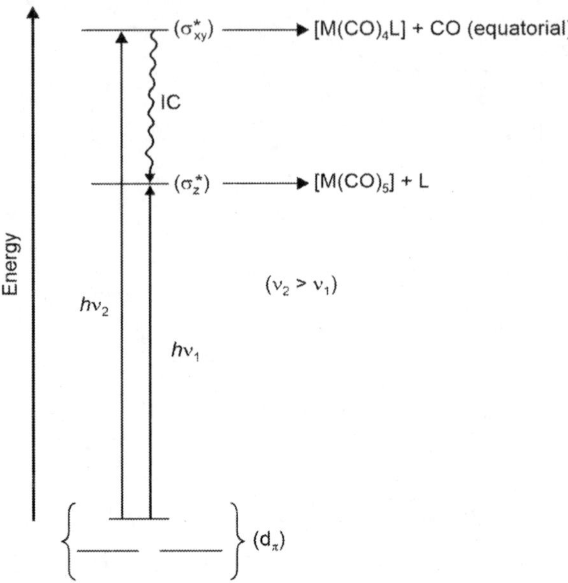

Fig. 6.15.4.3: Photoexcited states of [M(CO)$_5$L] and ligand dissociation

rates. They can abstract Cl-radical from different **halocarbons** like CCl_4, $CHCl_3$, CH_2Cl_2 and $PhCH_2Cl$.

$$[(\eta^5\text{-Cp})(CO)_3W\text{—}W(CO)_3(\eta^5\text{-Cp})] \xrightarrow{h\nu} 2[^{\bullet}W(\eta^5\text{-Cp})(CO)_3]$$

$$-2^{\bullet}R \downarrow R\text{—}Cl, R = Ph_3C, PhCH_2$$

$$2[Cl\text{—}W(\eta^5\text{-Cp})(CO)_3]$$

$$2\,^{\bullet}R \rightarrow R\text{—}R$$

Photoinduced homolytic single metal—metal bond cleavage can lead to metal exchange reactions as illustrated below.

$$[Mn_2(CO)_{10}] + [Re_2(CO)_{10}] \underset{h\nu}{\overset{h\nu}{\rightleftharpoons}} 2[(MnRe(CO)_{10}],\ i.e.\ 2[(CO)_5Mn\text{ — }Re(CO)_5]$$

$$[M_2(\eta^5\text{-Cp})_2(CO)_6] + [M'_2(CO)_{10}] \underset{h\nu}{\overset{h\nu}{\rightleftharpoons}} 2[(\eta^5\text{-Cp})(CO)_3M\text{—}M'(CO)_5]$$

$$M = Mo, W \qquad M' = Mn, Re$$

(ix) **Photoinduced metal-metal bond homolytic cleavage leading to photosubstitution of CO:** It is illustrated in the following example.

$$[Mn_2(CO)_{10}] + PPh_3 \xrightarrow{h\nu} [Mn_2(CO)_8(PPh_3)_2]\ \textbf{(major product)}$$

$$+ [Mn_2(CO)_9(PPh_3)]\ \textbf{(minor product)}$$

The above photosubstitution reaction occurs through the following route.

$$[Mn_2(CO)_{10}] \xrightarrow{h\nu} 2[^{\bullet}Mn(CO)_5] \xrightarrow[-2CO]{2PPh_3} 2[^{\bullet}Mn(CO)_4(PPh_3)]$$

$$\downarrow \text{Recombination}$$

$$[Mn_2(CO)_8(PPh_3)_2]$$

$$i.e.\ [(OC)_4(Ph_3P)Mn\text{—}Mn(PPh_3)(CO)_4$$

(Major product)

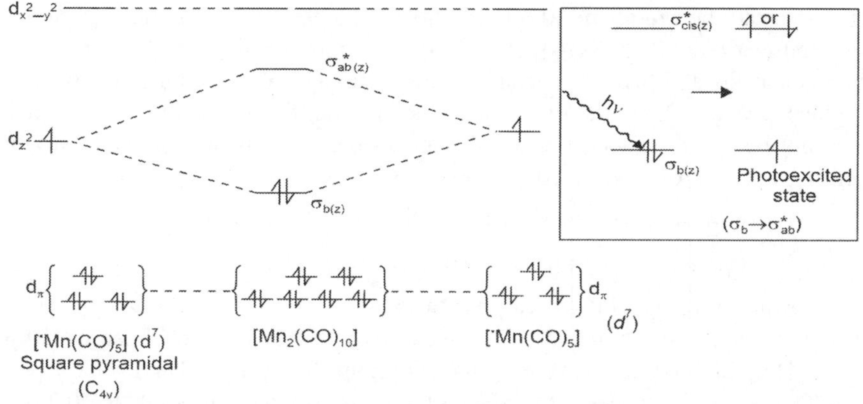

Fig. 6.15.4.4: Qualitative representation of orbital energy diagram of $[Mn_2(CO)_{10}]$

Note 1: Combination of two $[^{\bullet}Mn(CO)_5]$ square pyramidal (C_{4v}) fragments (d^7, 17*e* system) produces the dimeric species, $[Mn_2(CO)_{10}]$. In forming the dimeric species, the $d^1_{z^2}$ orbital of each fragment overlaps to produce the Mn—Mn bonding (σ_z) and antibonding (σ^*_z) orbitals with the electron distribution as $\sigma_z^2\sigma_z^{*0}$ (*i.e.* Mn—Mn bond order one, the bond along the *z*-axis in the diamagnetic binuclear cluster). The d_π-orbitals (*i.e.* d_{xy}, d_{yz} and d_{xz}) remain localised on each metal fragment in the binuclear cluster.

The orbital energy diagram is applicable for both the **homodinuclear** and **heterodinuclear clusters** constituted by the 17*e* fragments like $[Mn(CO)_5]$, $[Re(CO)_5]$, $[Co(CO)_4]$, $[Fe(\eta^5\text{-}Cp)(CO)_2]$, $[Mo(\eta^5\text{-}Cp)\,(CO)_3]$, etc.

Note 2: The transition, $d_\pi \rightarrow \sigma^*_z$ is **symmetry forbidden**. The allowed lowest energy transition is: $\sigma_{b(z)} \rightarrow \sigma^*_{ab(z)}$ (an **intense band** at 336 nm). This σ→σ* photoexcitation **weakens the Mn—Mn bond and it undergoes photodissociation**. In fact, this photoexcitation causes the metal—metal bond fission in different homodinuclear and heterodinuclear diamagnetic clusters constituted by the coupling of two 17*e* fragments.

$$[^{\bullet}Mn(CO)_5] + [^{\bullet}Mn(CO)_4(PPh_3)] \longrightarrow [Mn_2(CO)_9(PPh_3)]$$
$$i.e.\ [(OC)_4(Ph_3P)Mn\text{—}Mn(CO)_5]$$

(Minor product)

The photo-generated 17*e* metal fragment $[^{\bullet}Mn(CO)_5]$ (a **labile** species) undergoes substitution by L for CO followed by recombination of the radical metal fragments leading to the binuclear product.

Here it is worth mentioning that if the direct photosubstitution mechanism were operating, then it would lead to the $[Mn_2(CO)_9(PPh_3)]$ as the major product but it is not observed

$$\left.\begin{aligned}&[Mn_2(CO)_{10}] \xrightarrow{h\nu} [Mn_2(CO)^*_{10}] \longrightarrow [Mn_2(CO)_9] + CO\\&[Mn_2(CO)_9] + L\ (i.e.\ PPh_3) \longrightarrow [Mn_2(CO)_9(L)]\end{aligned}\right\}\ \textbf{Not in agreement with the observed result}$$

(x) **Photoinduced heterolytic metal-metal bond cleavge in dinuclear carbonyls:** The σ→σ* photoexcitation generally causes the homolytic metal-metal bond fission as discussed above but in some limited cases, the heterolytic cleavage has been noted. Probably, excitation at higher frequencies may produce a singlet excited state leading to the heterolytic cleavage. It happens so in the following reaction.

$$[(\eta^5\text{-}Cp)(OC)_3Mo\text{—}Mo(CO)_3(\eta^5\text{-}Cp)] + 2PR_3 \xrightarrow{\lambda < 290\text{ nm}}$$
$$[(\eta^5\text{-}Cp)Mo(CO)_2(PR_3)_2]^+ + [(\eta^5\text{-}Cp)Mo(CO)_3]^- + CO$$

(xi) **Photochemical reactions of polynuclear carbonyls:** In the **mononuclear carbonyls**, photodissociation of CO groups leads to the **photosubstitution** while in the dinuclear carbonyls, photolysis leads to the **metal-metal bond cleavage** (generally homolytic cleavage). In the polynuclear carbonyls like $[Ru_3(CO)_{12}]$, $[Os_3(CO)_{12}]$, etc. depending on the condition both metal-metal bond cleavage and photodissociation may occur. Absorption of a single photon is expected not to induce cleavage more than one metal-metal bond. Consequently, the fragmentation reactions may not be noted, specially when the metal-metal bonds are relatively stronger as in the polynuclear carbonyls of heavier congeners.

$$[M_3(CO)_{12}] \xrightarrow{PPh_3,\, h\nu} \text{Photo-fragmentation } (M = Fe, Ru)$$

$$[Os_3(CO)_{12}] + PPh_3 \xrightarrow{h\nu} [Os_3(CO)_{11}(PPh_3)] + CO$$

i.e. photosubstitution, **resistant to fragmentation.**

Here it may be noted that structure of $[Fe_3(CO)_{12}]$ is different from those of $[Ru_3(CO)_{12}]$ and $[Os_3(CO)_{12}]$ $[Fe_3(CO)_{12}]$ bears two bridging CO groups, *i.e.* $[Fe_3(CO)_{10}(\mu_2\text{–}CO)_2]$ but there is no bridging CO group in $[M_3(CO)_{12}]$ (M = Ru, Os). Photo-fragmentation of $[Ru_3(CO)_{12}]$ may occur in the following route.

$$\text{cyclo-}[(OC)_4Ru\text{–}Ru(CO)_4\text{–}Ru(CO)_4] \underset{}{\overset{h\nu}{\rightleftharpoons}} (OC)_4Ru^{\bullet}\text{–}Ru(CO)_4\text{–}{}^{\bullet}Ru(CO)_4 \xrightarrow{+3L} 3[Ru(CO)_4L]$$

(Diradical)

It explains the photo-fragmentation of $[Ru_3(CO)_{12}]$ in presence of CO to produce $[Ru(CO)_5]$

$$[Ru_3(CO)_{12}] + 3CO \xrightarrow{h\nu} 3[Ru(CO)_5]$$

6.15.5 Photochemical Reactions in the Co(III) Complexes

In the Co(III)-complexes, light of about 550 nm causes the ***ligand field excitation*** while light of about 370 nm causes the ***charge transfer excitation,*** *i.e.*

$[Co(NH_3)_6]^{3+}$, $[Co(NH_3)_5X]^{2+}$, $[Co(NH_3)_5(SO_4)]^{+}$, (X = Cl, Br, I, N_3, NO_2, etc.)

$h\nu$ (λ = 370 nm) → **Charge transfer (CT) excitation**

$h\nu$ (λ = 550 nm) → $t_{2g} \rightarrow e_g$ (CFT), *i.e.* $\pi \rightarrow \sigma^*$(MOT)
Ligand field (LF) excitation
(increasing electron density in e_g^*)

In the O_h complexes, in terms of MOT, the t_{2g} orbitals (CFT) are of π-symmetry. When only σ-bonding is considered, the t_{2g} orbitals remain as the nonbonding orbitals and e_g orbitals form the antibonding σ-orbital (*i.e.* σ^* or e_g^*) in terms of MOT.

(**Note:** For the Co(III)-complexes, the CT band very often overlaps with the LF band. Hence, even for the LF excitation, it is very difficult to prevent the photoredox reaction that generates the reactive Co(II)-species. Though, in general, the CT excitation leads to the photoredox reaction.)

Photochemical LF excitation ($t_{2g} \rightarrow e_g$ transition) leads to promotion of an electron from the nonbonding t_{2g} orbital (σ-bonding MO diagram, see Vol. 4) to the antibonding e_g^* orbital (MOT). It weakens the metal-ligand bond. **Thus the LF excitation generally favours the substitution and isomerisation reaction through the dissociative process.** For the stronger field ligands, the e_g-level possesses the more antibonding character and the LF excitation weakens the metal-ligand bond more. In fact, in general, **quantum yield increases with the ligand field strength.**

For $[Co(CN)_6]^{3-}$ and $[Co(CN)_5(X)]^{3-}$, the ligand field strength is sufficiently high and the light of 370 nm causes the *ligand field excitation rather than the charge transfer excitation.* Ligand field excitation weakens the metal-ligand bond in the excited state because of the enhanced electron density in the antibonding orbital.

Generally, CT excitation favours the redox process. However, it depends on the nature of X (*i.e.* leaving group). For example, CT excitation in $[Co(NH_3)_5X]^{2+}$ causes aquation of X^- for X = Cl and redox decomposition of the complex for X = I.

The photochemical reactions (say photoaquation) of Co(III)-complexes are believed to proceed through the following steps. It is illustrated for the aquation reaction.

Step-I: $$\underset{\left(\substack{X^- = \text{halide or}\\ \text{pseudohalide}}\right)}{\left[L_5Co^{III}{-}X\right]} \underset{}{\overset{h\nu}{\rightleftharpoons}} \underset{\left(\substack{\text{outer sphere}\\ \text{complex}}\right)}{\left\{\left[L_5Co^{II}\right]{}^{\bullet}X\right\}},$$ ***(i.e. CT excitation* ($\lambda \le 350$ *nm*) *causing the homolytic bond cleavage*)**

The first-step (*i.e.* homolytic bond cleavage) is basically a ***photoredox process*.** Where Co(III) is reduced to Co(II) and X^- is oxidized to X. The free radicals ($^{\bullet}X$) can be detected by Flash-photolysis.

Similar reactions are also known for the Rh(III) and Ir(IV)-complexes.

$$\left[Rh(NCS)(NH_3)_5\right]^{2+} \xrightarrow{h\nu} \left[Rh(NH_3)_4\right]^{2+} + NH_3 + {}^{\bullet}NCS;$$

$$\left[IrCl_6\right]^{2-} + H_2O \xrightarrow{h\nu} \left[IrCl_5(OH_2)\right]^{2-} + {}^{\bullet}Cl$$

In all these cases, one electron transfer occurs to give the changes:

$$[Co^{III} \rightarrow Co^{II},\ Rh^{III} \rightarrow Rh^{II},\ Ir^{IV} \rightarrow Ir^{III}]$$

Step-II: $$\left[L_5Co^{II}\right]{}^{\bullet}X + H_2O \longrightarrow \underset{(\text{outer sphere complex})}{\left\{\left[L_5Co^{II}(OH_2)\right]{}^{\bullet}X\right\}},$$ (high-spin Co^{II} is labile)

Step-III:

$\{[L_5Co^{II}(OH_2){}^{\bullet}X\}$
Outer sphere complex

$[L_5Co^{II}(OH_2)] + {}^{\bullet}X$ (if $^{\bullet}X$ is not a good electron acceptor, *i.e.* oxidation of Co^{II} to Co^{III} is not possible).

$[L_5Co^{III}(OH_2)] + X^-$ (if $^{\bullet}X$ is a good electron acceptor)

If $^{\bullet}X$ cannot oxidize Co(II) to Co(III), then the free radical $^{\bullet}X$ (=1, N_3, etc.) comes out from the outer-sphere complex to participate in different reactions. $^{\bullet}X$ can combine to generate X_2. It happens so for X = I as in the case of $[Co(NH_3)_5I]^{2+}$. For X = Br, released Br_2 will oxidise the available oxidisable substances.

$$\left[Co(NH_3)_5I\right]^{2+} \xrightarrow[\lambda = 370,550\ \text{nm}]{h\nu} Co(II) + I_2 + 5NH_3$$

For $[Co(NH_3)_5I]^{2+}$, at both $\lambda = 370$, 550 nm, the redox reaction is predominant while for $[CoCl(NH_3)_5]^{2+}$ the aquation is the predominant reaction, *i.e.*

$$\left[CoCl(NH_3)_5\right]^{2+} \xrightarrow[\lambda = 370,550\ \text{nm}]{h\nu} \left[Co(NH_3)_5(OH_2)\right]^{3+} + Cl^-$$

It is reasonable as Cl is a much better electron acceptor than I. It is worth mentioning that aquation of $[Co(CN)_5X]^{3-}$ gives the following results.

$$\left[Co(CN)_5X\right]^{3-} \xrightarrow[\lambda = 370\ \text{nm}]{h\nu} \left[Co(CN)_5(OH_2)\right]^{2-} + X^-$$

Quantum yield (ϕ): $I^- > Br^- > Cl^-$; *Co^{III}—X bond strength:* $Cl^- \rangle Br^- \rangle I^-$

It indicates that the first step leading to the homolytic cleavage of the Co^{III}—X bond is the important step.

$$\left[(NC)_5\,Co^{III}—X\right] \xrightarrow{h\nu} \left\{\left[(NC)_5\,Co^{II}\right]{}^{\bullet}X\right\},\ (i.e.\ \text{reduction of } Co^{III})$$

Here the electron transfer from X^- to Co^{III} is most favoured for the least electronegative X. It follows the sequence: $I^- > Br^- > Cl^-$. It is also the **sequence of quantum yield.**

Table 6.15.5.1 Photochemical reactions of Co(III) complexes (*cf.* A. W. Adamoson, *Disc. Faraday Soc.*, **29**, 163, 1960).

Complex	*Irradiating radiation* (nm)	*Product*	*Quantum Yield* ϕ (mol einstein^{-1})
$[Co(NH_3)_6]^{3+}$	370	No reaction	–
$[Co(NH_3)_5X]^{2+}$			
X = Cl	370	$[Co(NH_3)_5(OH_2)]^{3+}$	0.011
	550	–do–	0.001
X = I	370	Co(II), NH_3, I_2	0.66
	550	–do–	0.10
X = N_3^*	370	Co(II)	0.44
	550	–do–	0.011
$[Co(CN)_5X]^{3-}$			
X = CN	370	$[Co(CN)_5(OH_2)]^{2-}$	0.9
X = Cl	370	–do–	0.3
X = Br	370	–do–	0.7
X = I	370	–do–	0.95
$[Co(C_2O_4)_3]^{3-}$	370	Co(II), CO_2	1.0

* Azide may be oxidised to N_2.

● **LF excitation *vs* CT excitation:** Generally, the CT excitation causes the photoredox reaction while the LF excitation causes the photosubstitution reaction. In the Co(III)-complexes, very often, the LF band and CT band overlap to complicate the observation. However, the following examples illustrate the effect of wavelength of irradiation.

$$\left[Co(NH_3)_5(N_3)\right]^{2+} \xrightarrow[\text{(green light)}]{\text{(LF excitation)}} \left[Co(NH_3)_4(N_3)(OH_2)\right]^{2+} + NH_3,$$

(aquation of NH_3) (*cf.* **Adamson's rule)**

$$\left[Co(NH_3)_5(N_3)\right]^{2+} \xrightarrow[\text{UV-light}]{\text{(CT excitation)}} \left[Co(NH_3)_5\right]^{2+} + {}^{\bullet}N_3 \longrightarrow Co(II) + NH_3 + N_2$$

(Redox decomposition).

In the following example, the CT (LMCT) excitation causes the **photoisomerization.**

$$\left[Co(NH_3)_5(NO_2)\right]^{2+} \xrightarrow{h\nu(CT)} \left\{\left[Co(NH_3)_5\right]^{2+}{}^{\bullet}NO_2\right\} \xrightarrow{\text{Isomerization}} \left[Co(NH_3)_5(ONO)\right]^{2+};$$

(Linkage isomer)

6.15.6 Photochemical Reactions of the Cr(III) Complexes

For the Cr(III) complexes, analogous to those of the Co(III) complexes, **there is no redox reaction under the comparable conditions.** It is reasonable. For the Co(III)-complexes, Co(II) can be generated as an intermediate, *i.e.* reduction of Co(III) to Co(II) is feasible. On the other hand, for the Cr(III) complexes, it is unlikely to generate the Cr(II) intermediate state. This can be rationalised in terms of their E^0-values, *e.g.* 1.8 V for $[Co(aq)]^{3+}/[Co(aq)]^{2+}$ and –0.42 V for $[Cr(aq)]^{3+}/[Cr(aq)]^{2+}$ couple, *i.e.* reduction of Co(III) to Co(II) is thermodynamically easier. Other characteristic features of the photochemical reactions of the Cr(III) complexes are discussed below.

Differences in photochemical reactions of the analogous Co(III) and Cr(III) complexes

- For the Co(III)-complexes, there may be both substitution and redox reactions depending on the conditions, but for the Cr(III) complexes, redox reaction does not occur.
- For the Co(III)-complexes, photoexcitation can generate **a reactive intermediate Co(II)** (which is labile and reducing) while for the Cr(III)-complexes no such reactive intermediate like Cr(II) is generated. In fact, photoexcitation of the Cr(III)-complexes, probably gives a **long-lived** ^2E_g state that subsequently participates in the reactions ***requiring*** *a the thermal activation energy.*
- For the Co(III)-complexes, after the primary photoexcitation, *there is no need of thermal activation energy for the subsequent reactions.* ***The Cr(III) complexes need a thermal activation energy for the subsequent reactions.***
- Reactions of the Cr(III)-complexes are slower.
- For the Cr(III)-complexes, **quantum yield is more or less insensitive to the wavelength in the range 300-700 nm.** For the Co(III)-complexes, light of 370 nm favours the CT excitation for the redox process while light of 550 nm leads to ligand field excitation to favour the substitution reaction.
- For the Cr(III)-complexes, two possible **reactive intermediates,** *i.e.* ^2E_g and $^4T_{2g}$ are suggested. The ^2E_g state (low-spin t_{2g}^3) having a vacant orbital in the t_{2g}-level experiences an **associative activation** while the $^4T_{2g}$ **state** ($t_{2g}^2\ e_g^1$) having an enhanced electron density in the antibonding orbital experiences the **dissociative activation** in the subtitution process. LF excitation in the Co(III)-complexes favours the **dissociative activation** due to the weakening of metal ligand bond.

The quantum yield increases with the increase of temperature. It indicates the ***requirement of a thermal activation energy***. It is suggested that after the *primary photochemical process,* it requires the thermal activation energy for the subsequent reactions. Consequently, the overall process is slow. **Here it may be mentioned that for the analogous Co(III)-complexes, there is no need of such thermal activation energy**. Because, for the Co(III)-complexes, after the primary photochemical process, a very reactive centre which is Co(II) is generated. In the case of Cr(III) complexes, there is no such possibility of generation of Cr(II) species. In fact, even after the photoexcitation, the Cr(III)-state (an inert centre) is still maintained. It is suggested that photoexcitation of the Cr(III) complexes produces **a relatively long-lived doublet state** ^2E_g **which subsequently participates in the substitution reactions.** This long-lived ^2E_g state requires the thermal activation energy for its subsequent substitution reactions.

(A) **High activation energy (*ca.* 60 kJ mol^{-1}) and effect of temperature:** The quantum yield (ϕ) is *almost insensitive to the wavelength of irradiation in the range 300-700 nm* but *the quantum yield increases with the rise of temperature.* In fact, this high activation energy (*ca.* 60 kJ mol^{-1}) is unusual for a photochemical reaction. This can be explained by considering the requirement of a thermal activation after the photoexcitation. This step of thermal activation is only possible, if the photochemically produced intermediate is sufficiently long-lived (life-time ≥ 1 ms). Thus to explain the effect of temperature on ϕ, we must consider an intermediate (produced photochemically) that will survive at least 1 ms. The probable intermediate ^2E_g and/or $^4T_{2g}$ (to be discussed later) is still Cr(III) (d^3) which is inert and it needs a thermal activation for substitution or racemisation.

Table 6.15.6.1 Photochemical reactions of the Cr(III) complexes (*cf.* A. W. Adamson *et al*, *J. Am. Chem. Soc.*, **80**, 3865, 1958; R. A. Plane *et al*, *ibid*, **79**, 3343, 1957; *Inorg. Chem.*, **3**, 231, 1964).

Complex/System	***Wavelength of irradiation λ (nm)***	***Reaction product***	***Quantum Yield* ϕ (mol einstein^{-1})**
$[Cr(NH_3)_6]^{3+}$	320–600	$[Cr(NH_3)_5(OH_2)]^{3+}$	0.32
$[Cr(NH_3)_5(OH_2)]^{3+}$	320–600	$[Cr(NH_3)_4(OH_2)_2]^{3+}$	0.25
$[Cr(NH_3)_4(OH_2)_2]^{3+}$	320–700	$[Cr(NH_3)_3(OH_2)_3]^{3+}$	0.16
$[Cr(NH_3)_3(OH_2)_3]^{3+}$	320–700	$[Cr(NH_3)_2(OH_2)_4]^{3+}$	0.014
$[Cr(NH_3)_2(OH_2)_4]^{3+}$	320–700	$[Cr(NH_3)(OH_2)_5]^{3+}$	0.0018
$[Cr(NH_3)_5(NCS)]^{2+}$	560	$[Cr(NH_3)_5(OH_2)]^{5+}$	0.013
$[Cr(OH_2)_6]^{3+} + Cl^-$	370	$[CrCl(OH_2)_5]^{2+}$	0.006
$[Cr(OH_2)_6]^{3+} + NCS^-$	400	$[CrNCS(OH_2)_5]^{2+}$	0.0024
$(+)[Cr(C_2O_4)_3]^{3-}$	370	Racemization (no decomposition)	0.045

(B) Nature of the intermediate state produced through the primary photoexcitation process in the case of Cr(III)-complexes: To understand this aspect, it is required to consider the energy level diagrams of the Cr(III)-complex (Fig. 6.15.6.1).

The *spin allowed transitions* in the Cr(III) complexes are:

$$^4A_{2g}(F) \xrightarrow{\nu_1} {^4T_{2g}}(F),\ ^4A_{2g}(F) \xrightarrow{\nu_2} {^4T_{1g}}(F),\ ^4A_{2g}(F) \xrightarrow{\nu_3} {^4T_{1g}}(P).$$

The high energy third transition (*i.e.* ν_3) very often merges with the strong charge transfer band. The spin forbidden transition is:

$$^4A_{2g}(F) \longrightarrow {^2E_g}(G)$$

The ^2E_g state arises from the low spin configuration of t_{2g}^3, *i.e.* (↑↓)(↑)( ) where one level is remaining vacant.

The spin-forbidden transition from the ^2E_g state to the ground state gives the longer lifetime (about 10^{-3}–10^{-4} s) of the ^2E_g state (*i.e.* **metastable state**). The upper two excited quartet states (*i.e.* $^4T_{1g}$) rapidly drop to the lowest excited quartet state ($^4T_{2g}$) in 10^{-10}–10^{-11} s in the nonradiative paths IC and EC (**collisional deactivation**). The $^4T_{2g}$ state reverts to the ground state mainly **through the nonradiative (EC) path** because no fluorescence is observed. The radiationless transition (ISC) from the $^4T_{2g}$ state to the ^2E_g state occurs within 10^{-7}–10^{-8} s.

(a) **Doublet state, *i.e.* ^2E_g, as the reactive intermediate:** It is suggested that the excited quartet states [*i.e.* $^4T_{1g}(P)$, $^4T_{1g}(F)$ and $^4T_{2g}(F)$] will rapidly revert to the ground state but in this deexcitation process, population density of the doublet state (*i.e.* ^2E_g) will increase to some extent through the **ISC path** and *this doublet state is relatively long-lived because the transition from the ^2E_g state to the ground $^4A_{2g}$ state is spin forbidden.* ***This is why, some workers believe that the ^2E_g state is the intermediate state generated from the photoexcitation process.***

The ^2E_g excited state (low spin t_{2g}^3) generates a vacant orbital at the t_{2g} level. Availability of this vacancy at the t_{2g} level favours the **associative activation in the photosubstitution process.** It may be noted that in the thermal substitution reactions, Cr(III) also adopts the associative path with a high activation energy.

The quantum yield (ϕ) is not sensitive to the wavelength of irradiaion in the region 300-700 nm. It indicates that the reactive intermediate is the **lowest excited state** which is not produced directly through the photochemical excitation. **It supports the ^2E_g state as the reactive intermediate state.**

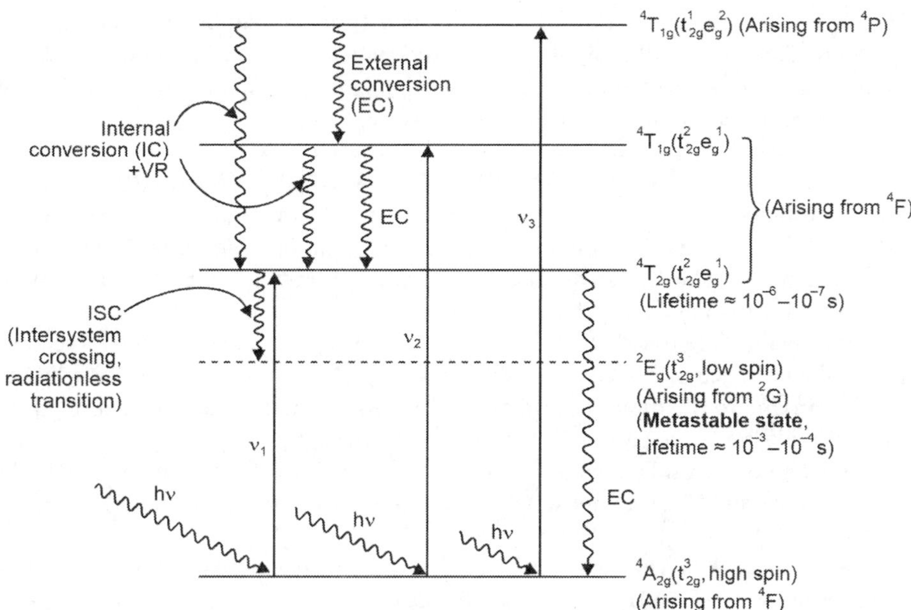

Fig. 6.15.6.1 Energy level diagram of the octahedral complexes of Cr(III) and electronic transitions in terms of **Jablonski diagram.** The ${}^2E_g$ state actually represents two states ${}^2E_g$ and ${}^2T_{1g}$ arising from ${}^2G$, *i.e.* ${}^2T_{1g}(G)$, ${}^2E_g(G) = {}^2E_g(G)$ (*cf.* Tanabe–Sugano diagram, Fig. 7.14.5.6).

This observation as a rule is analogous to **Kasha's rule** for the radiative emission from the lowest excited state (**thexi-state**). Thus, if thc ${}^2E_g$ state is considered to be responsible as the reactive intermediate then the photochemical yield should depend on the life period of this doublet state. **It has been noted that the *quantum yield for the photoaquation of* $[Cr(NH_3)_x(OH_2)_{6-x}]^{3+}$ *decreases gradually from* $[Cr(NH_3)_6]^{3+}$ *to* $[Cr(OH_2)_6]^{3+}$.** This observation can be explained by considering the energy difference between the states ${}^2E_g$ and ${}^4T_{2g}$. This energy difference is maximum for $[Cr(NH_3)_6]^{3+}$ while it is minimum (*ca.* 30 kJ mol^{-1}) for $[Cr(OH_2)_6]^{3+}$ (*cf.* NH_3 is a stronger field ligand than H_2O). **In the case of $[Cr(OH_2)_6]^{3+}$, by using the thermal energy, population density of the ${}^2E_g$ state may decrease due to the promotion to the ${}^4T_{2g}$ state and thence to the ${}^4A_{2g}$ state** (ground state).

$$ {}^2E_g \xrightarrow{\text{thermally}} {}^4T_{2g} \xrightarrow{\text{(allowed)}} {}^4A_{2g}\,\text{(ground state)} $$

This ***indirect path*** (described as the **leakage path**) reduces the life-time or population density of the ${}^2E_g$ state which is supposed to be the reactive intermediate. This indirect path to reduce the life time of the ${}^2E_g$ state is most efficient for the $[Cr(OH_2)_6]^{3+}$ complex in the series $[Cr(NH_3)_x(OH_2)_{6-x}]^{3+}$. *The efficiency of this leakage mechanism increases with the decrease of x in the series.* It explains the increasing trend of quantum yield for the photoaquation of $[Cr(NH_3)_x(OH_2)_{6-x}]^{3+}$ with the increase of x.

(b) **Quartet state as the reactive intermediate: *Some workers believe that the ${}^4T_{2g}$ state, i.e. quartet state, is the reactive intermediate*** having the electronic configuration $t_{2g}^2 e_g^1$. This state experiences the **Jahn-Teller distortion.** Because of this J.T. distortion, the geometry of the excited quartet state is different from that of the nondistorted ground state. *Because of this difference in symmetry between the distorted excited state and nondistorted ground state, the life-time of the excited quartet state increases.* Because of the J.T. distortion, the excited quartet state is also labile.

In the $^4T_{2g}$ state ($t_{2g}^2e_g^1$), accumulation of electron density in the antibonding orbital weakens the metal-ligand bond. This favours the **dissociative activation** in photosubstitution. The J.T. distortion also favours the **dissociative activation.** Here it is worth mentioning that the ^2E_g excited state and $^4A_{2g}$ ground state (*i.e.* thermal reaction) favour the associative activation.

Nature of the reactive intermediate: ^2E_g (G) *vs.* $^4T_{2g}$ (F)

- Enhanced life-time of the ^2E_g-state arises due to the **spin-forbidden transition** to the $^4A_{2g}$ ground state. However, its life-time may decrease through the ***indirect leakage path.***
- Enhanced life-time of the $^4T_{2g}$-state arises due to the **symmetry forbidden transition** to the ground state which is nondistorted while the excited $^4T_{2g}$ state is distorted (J.T. distortion).
- In the photosubstitution process, the ^2E_g state (low spin t_{2g}^3) favours the **associative activation** because of the presence of a vacant t_{2g}-orbital while the $^4T_{2g}$ state ($t_{2g}^2e_g^1$) favours the **dissociative activation due to the metal-ligand bond weakening.** In the thermal reaction, the ground state adopts the associative path with a high activation energy.
- In fact, in the photochemical reactions of the Cr(III) complexes, there are evidences for both the ^2E_g (doublet state) and $^4T_{2g}$ (quartet state) states to represent the reactive intermediate.

(C) **Adamson's rules:** Adamson had formulated some **empirical rules** to explain the results of photochemical reactions of $[Cr(NH_3)_5X]^{2+}$ (X^- = halide).

Rule 1: ***In $[Cr(NH_3)_5X]^{2+}$, the ligands are at the ends of three mutually perpendicular axes. On irradiation, the ligands will be labilized most along the axis having the lowest average ligand field strength.***

Rule 2: ***If there are two different ligands along the labilized axis, the ligand having the higher ligand field strength will be preferably labilized.***

Thus for $[Cr(NH_3)_5X]^{2+}$, the *trans*-ligands NH_3 and X will be labilized most and between these two ligands, NH_3 (having the higher ligand field strength) will be preferably labilized (**antithermal behaviour**).

$$[CrBr(NH_3)_5]^{2+} + H_2O \xrightarrow{h\nu} [CrBr(NH_3)_4(OH_2)]^{2+} + NH_3$$

However, there are exceptions to these rule.

The $\Delta V^{\neq}$ values for photoaquation of $[Cr(NH_3)_5X]^{2+}$ and $[Cr(NH_3)_6]^{3+}$ are negative for aquation of both NH_3 and X^-. **These can be interpreted in terms of the I_d-mechanism where the dissociated fragments in the T.S. put an enhanced electrorestriction over the solvent.** This is reasonable as the excited state ($^4T_{2g}$) accumulates more electron density in the antibonding e_g^* orbital from the nonbonding t_{2g}-orbital and this will weaken the metal-ligand bond to favour the dissociative activation. It may be noted that in the thermal reaction, the Cr(III)-complexes **adopt the associative activation** for aquation that also leads to the negative value of $\Delta V^{\neq}$.

(D) **Interpretation of Adamson's Rule in photosubstitution reaction in terms of crystal field theory** (Ref. A.W. Adamson, *Coord. Chem. Rev.*, **3**, 169-188, 1968): The rules are applicable for both the Co(III) (d^6, low-spin, *i.e.* t_{2g}^6) and Cr(III) (d^3, *i.e.* t_{2g}^3) complexes for which the ligand field transitions are quite similar (*cf.* Figs. 7.14.5.5, 7.15.2.4). But, for the Co(III) complexes, the LF band and CT band very often overlap and the Co(III)-complexes very often experience the *photo-redox reactions* because of the ease of reduction of Co(III) to Co(II). Here we shall illustrate the rule taking the Cr(III)-complexes as the representative ones.

It is suggested that the *lowest quartet excited state* ($^4T_{2g}$ for $t_{2g}^2e_g^1$) of the Cr(III)-complexes is the **reactive intermediate** for the photosubstitution process. The excited quartet state ($t_{2g}^2e_g^1$) experiences the *Jahn-Teller distortion* generating the elongated two *trans*-L—M—L bonds (*cf.* *z*-out distortion).

Because of this significant difference in the geometries of the undistorted ground state ($^4A_{2g}$, t_{2g}^3) and excited state ($^4T_{2g}$, $t_{2g}^2e_g^1$), *lifetime of the excited quartet state is increased as the transition,* $^4T_{2g} \rightarrow {}^4A_{2g}$ *experiences the* ***Franck-Condon barrier.*** For the difference in the structures of $^4A_{2g}$ and $^4T_{2g}$ states, the Franck-Condon vertical transition ($^4A_{2g} \rightarrow {}^4T_{2g}$) produces the excited state (*i.e.* $^4T_{2g}$ state) at a higher vibrational level (*cf.* Fig. 7.4.1.3).

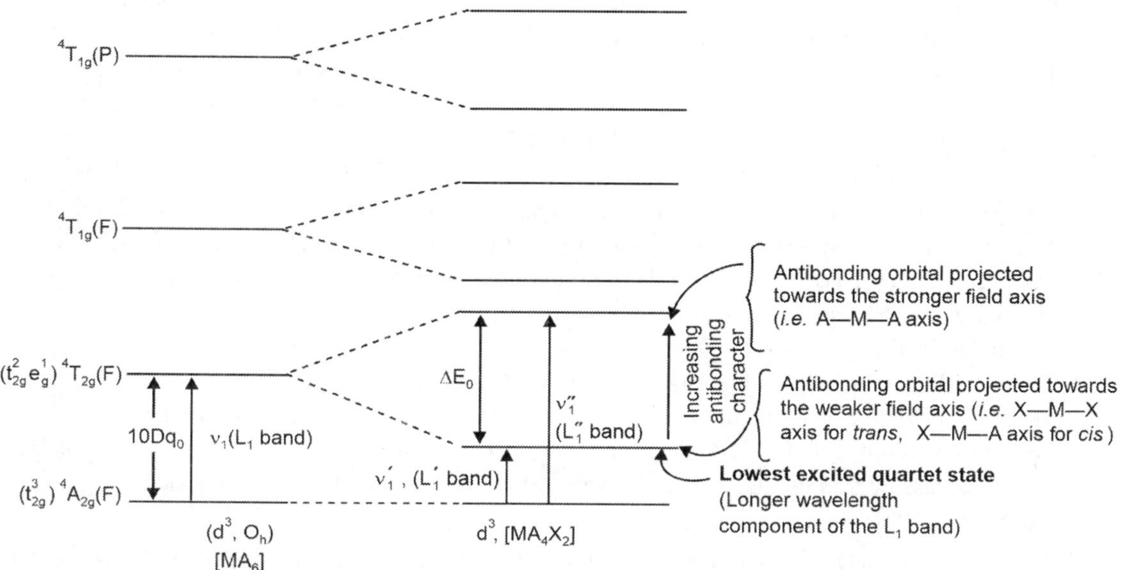

Fig. 6.15.6.2: Energy diagram for the d^3 (*e.g.* Cr^{III}) octabedral (O_h) and less symmetric (D_{4h}, C_{2v}) complexes (*cf.* Figs. 7.14.5.5, 7.15.2.1).

Note: Even for the perfectly octahedral complexes like [Cr^{III}L$_6$], the $^4T_{2g}$ (*i.e.* $t_{2g}^2e_g^1$) state experiences a distortion (J.T. distortion) and in fact, this excited state is structurally different from the ground state ($^4A_{2g}$, t_{2g}^3). Thus, the vertical Franck-condon transition, $^4A_{2g} \rightarrow {}^4T_{2g}$ produces the excited state at a higher vibrational level. The excited state may thermally equilibrate before emission to have the lower vibrational level at the excited state. Then emission (due to the return to the $^4A_{2g}$ state) will occur to a higher vibrational level of the ground state (*cf.* Fig. 7.4.1.3).

In the *unsymmetrical octahedral complexes,* the average ligand field strength along the *three mutually perpendicular axes* will not be the same and the $^4T_{1g}$ and $^4T_{2g}$ states will split. This splitting pattern for the *cis-/trans-*[Cr^{III}A$_4$X$_2$] complex is shown in Fig. 6.15.6.2. For the [Cr^{III}A$_4$X$_2$] complex, orientations of the ligands along the mutually perpendicular axes are as follows:

(i) ***cis*-complex** (two types of axes): A—A, A—X, A—X; **Average ligand field strength:** A—A axis > A—X axis (assuming the ligand field strength: A > X, *e.g.* NH_3 > Cl^-).

(ii) ***trans*-complex** (two types of axes): A—A, A—A, X—X; **Average ligand field strength:** A—A axis > X—X axis.

(*cf.* positions of A (=NH_3, say) and X^- (halide, pseudohalide, say) in the spectrochemical series; NH_3 is a stronger field ligand than X^-).

The magnitude of splitting (ΔE_0) of the quartet state ($^4T_{2g}$) depends on the difference in the average ligand field strength between the two types of axes. The average ligand field strength difference between the A—A and X—X axes in the *trans*-complex is twice the difference between the A—A and A—X axes in the *cis*-complex. This is why, in the *trans*-complex, splitting of the quartet state is more. Because of this splitting, the lowest ligand field band (say L$_1$, $\nu_1 = {}^4A_{2g} \rightarrow {}^4T_{2g}$) will undergo splitting into two components, say $L_1'(\nu_1')$ and $L_1''(\nu_1'')$ where $\nu_1'' > \nu_1'$. The ν_1' transition (longer wavelength component of the L$_1$ band) gives the **lowest quartet excited state** (in terms of the spin allowed ligand field excitation) which accumulates the electron density in the antibonding orbital to weaken the metal-ligand bond. The L_1' **band (*i.e.* ν_1' transition) leads to the accumulation of electron density in the antibonding orbital projected towards the weaker ligand field axes X—M—X (for the *trans*-complex) and A—M—X (for the *cis*-complex)** (*cf.* **Rule 1**). Less splitting of the $^4T_{2g}$ state (*i.e.* lower average ligand field strength along the said axis) corresponds to the higher value of ν_1'. Obviously, the higher value of ν_1' indicates the higher antibonding electron density in the lowest excited quartet state, *i.e.* higher metal-ligand bond weakening in the lowest excited quartet state (*cf.* **Rule 1**).

- **Rule 1:** Thus for the unsymmetrical octahedral complexes, we may say that, in general, *the longer wavelength component of the L_1 band corresponds to the accumulation of electron density in the antibonding orbital directed along the weaker field axis. Thus the axis having the lower average ligand field strength will be labilised more in terms of the dissociative activation.* **It leads to the Rule 1,** if we assume that **all the photochemically excited states will rapidly convert into the lowest quartet excited state** through the different possible ways as shown in Jablonski diagram (*cf.* Fig. 6.9.5.1). *Thus it is reasonable to expect that regardless of the wavelength of irradiation, the complex will attain the same lowest excited quartet state (i.e.* ***same quantum yield****)* according to **Kasha's rule**.
- **Rule 2:** Higher ligand field strength (*i.e.* more Δ_o) makes, in general, the e_g orbitals of more antibonding character. Thus the ligand field excitation (*i.e.* $t_{2g} \rightarrow e_g$) weakens the metal-ligand bond more where Δ_o is more. This is why, in general, the quantum yield for the photochemical reactions passing through the dissociative activation increases with the ligand field strength. This explains the Rule 2.

  Note: The rules specify the ligands to be replaced but cannot specify the stereochemistry of the final product (see limitations, discussed later).

Illustration: (i) Anion of **Reinecke's salt,** *i.e.* ***trans*-[Cr(NCS)$_4$(NH$_3$)$_2$]$^-$** possess two different types of axes, *i.e.* H_3N—Cr—NH_3 (**stronger field axis**) and SCN—Cr—NCS (**weaker field axis** which experiences the better labilisation). According to Rule 1, it should have the **photoaquation of thiocyanate** not NH_3. In fact, it happens so.

$$trans\text{-}\left[Cr(NCS)_4(NH_3)_2\right]^- \xrightarrow[(+H_2O)]{h\nu} \left[Cr(NCS)_3(NH_3)_2(OH_2)\right] + NCS^-$$

(ii) In **[Cr(NCS)(NH$_3$)$_5$]$^{2+}$**, the two axes are: H_3N—Cr—NH_3 (**stronger field axis**) and H_3N—Cr—NCS (**weaker field axis** which experiences the better labilisation). In fact, it experiences the **photo-aquation of NH$_3$** (an **antithermal behaviour**) which is a stronger field ligand than NCS^- (Rule 1, 2).

$$\left[Cr(NCS)(NH_3)_5\right]^{2+} \xrightarrow[(+H_2O)]{h\nu} \left[Cr(NCS)(NH_3)_4(OH_2)\right]^{2+} + NH_3$$

(*cf.* Dark or thermal reaction leads to aquation of NCS^-).

(iii) In **[Cr(NH$_3$)$_5$(N$_3$)]$^{2+}$**, the two axes are: H_3N—Cr—NH_3 (**stronger field axis**) and H_3N—Cr—N_3 (**weaker field axis** which experiences the better labilisation), and NH_3 is a stronger field ligand than N_3^-. It leads to the **photoaquation of NH$_3$** (Rule 1 and 2).

$$\left[Co(NH_3)_5(N_3)\right]^{2+} \xrightarrow[\left(\substack{\text{green light}\\ \text{irradiation}}\right)]{LF \text{ excitation}} \left[Co(NH_3)_4(N_3)(OH_2)\right]^{2+} + NH_3$$

(iv) **Photoracemisation of [Cr(ox)$_3$]** occurs through the **one ended dissociation of the oxalate chelate ring** (see Sec. 5.29).

(v) Photochemical behaviour of the *cis-trans* isomers of **$[Cr(en)_2(OH)_2]^+$:**

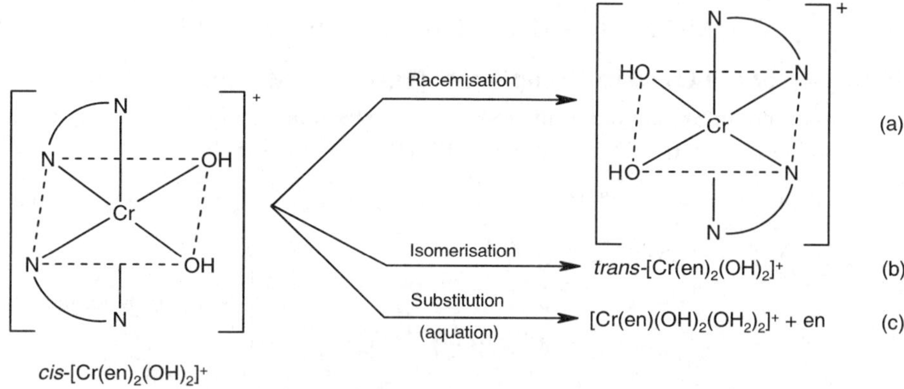

Racemisation (a) and isomerisation (b) of *cis*-$[Cr(en)_2(OH)_2]^+$ occur through the **one-ended dissociation** of the ethylenediamine (en) chelate ring (producing an intermediate of lower coordination number, see Secs. 5.28.2 and 5.29) followed by the reattachment by the dissociated end of en to the metal centre. *Photochemically the en-aquation and isomerisation reactions (b,c) are important while the predominant thermal reaction is only isomerisation.* **Thermal reaction cannot aquate en.**

In *cis*-$[Cr(en)_2(OH)_2]^+$ the two ligand field axes are:

(en)N—Cr—N(en) **(stronger field axis)** and (en)N—Cr—OH **(weaker field axis)**. According to Rule 1, the more labilised axis is (en)N—Cr—OH and according to Rule 2, the (en)N—Cr bond is labilised more. Thus **photoaquation of en** (reaction c) and **photo-isomerisation** (reaction b) can be explained through the rupture of (en)N—Cr bond. In the corresponding *trans*-complex, the weaker field axis is HO—Cr—OH. It predicts (Rule 1) the **O-exchange** and isomerisation through the formation of a stereomobile intermediate of lower coordination number. For the *trans*-compound, **en-aquation is not expected** and it has been experimentally verified.

- *cis*-$[Cr(en)_2(OH)_2]^+$ **(weaker field axis:** (en)N—Cr—OH): **en-aquation** and **isomerisation but no O-exchange.**
- *trans*-$[Cr(en)_2(OH)_2]^+$ **(weaker field axis**: HO—Cr—OH): O-exchange and isomerisation but no en-aquation.

These predictions have been experimentally verified.

Limitations of the Adamson's empirical rules: There are many experimental findings to contradict the Adamson's rules. For example, in photoaquation of $[Cr(NH_3)_5X]^{2+}$ (X^- = halide or pseudo-halide), photoaquation occurs along the *trans*-H_3N—Cr—X axis (Rule 1) and it should give the *trans*-$[Cr(NH_3)_4(OH_2)(X)]^{2+}$ but this expectation is not always satisfied. Formation of the *cis*-product (*i.e.* **antithermal behaviour**) has been explained by considering the *stereomobile character* of the intermediate of lower coordination number (CN = 5) produced through the ligand dissociation from the **photochemically excited states** of Cr(III). Here it should be pointed out that the ground state participating in **thermal reactions is stereorigid**.

6.15.7 Photolysis of $[M(CN)_8]^{3-}$ and $[M(CN)_8]^{4-}$ (M = Mo, W) in Aqueous Solution

The quantum yield (ϕ) for photolysis of $[M(CN)_8]^{3-}$ is greater than 1. To explain this higher value of ϕ, it has been argued for the generation of OH radical (in the primary photochemical process) which subsequently participates in the reaction. The suggested reaction paths are:

$$[M(CN)_8]^{3-} + H_2O \xrightarrow{h\nu} [M(CN)_8]^{4-} + H^+ + OH$$

$$[M(CN)_8]^{3-} + OH \longrightarrow [M(CN)_8]^{4-} + H^+ + \tfrac{1}{2}O_2$$
$$[M(CN)_8]^{3-} + H_2O + OH \longrightarrow [M(CN)_8]^{4-} + H^+ + 2OH$$

6.15.8 Photolysis of the Oxalato-Complexes, $[M(ox)_3]^{3-}$ (M = Fe, Mn, Co)

The said *tris*-oxalato-complexes undergo the redox decomposition during the photolysis. The proposed mechanism involving the intermediates M(IV) and M(II) is outlined below.

$$[M^{III}(C_2O_4)_3]^{3-} \xrightarrow{h\nu} [M^{III}(C_2O_4)_3^*]^{3-} \longrightarrow [M^{II}(C_2O_4)_2]^{2-} + C_2O_4^{\overline{\cdot}} \text{ (oxalate radical)}$$

$$\left.\begin{array}{l} [M^{III}(C_2O_4)_3]^{3-} + C_2O_4^{\overline{\cdot}} \longrightarrow [M^{IV}(C_2O_4)_3]^{2-} + C_2O_4^{2-} \\ \qquad\qquad\qquad\qquad\qquad \downarrow \\ \qquad\qquad\qquad\qquad [M^{II}(C_2O_4)_2]^{2-} + 2CO_2 \end{array}\right\} \text{Nonphotochemical path}$$

The overall process of photolysis of $[M(C_2O_4)_3]^{3-}$ may be shown as follows:

$$2[M(C_2O_4)_3]^{3-} \xrightarrow{h\nu} 2[M(C_2O_4)_2]^{2-} + 2CO_2 + C_2O_4^{2-}$$

(**Net reaction:** Oxidation of $C_2O_4^{2-}$ by M(III)).

The quantum yield (ϕ) for such photochemical redox decomposition processes depends on the wavelength of irradiation but it remains more or less constant in the charge-transfer region and it decreases in the *d–d* band region. For example, for $[Fe(C_2O_4)_3]^{3-}$, the ϕ value (0.4–0.6) is fairly high in the CT region while the ϕ value is very small in the region of spin-forbidden *d–d* band region (*ca.* 〉 650 nm). **The higher quantum yield (ϕ) for the light of shorter wavelength arises due to the fact:** allowed transition in the CT band region, excess energy allows the degraded species $[M(C_2O_4)_2]^{2-}$ and $C_2O_4^{\overline{\cdot}}$ to escape away from each other and preventing their recombination. At higher wavelength (*i.e.* low-energy light), recombination of the oxalate radical with $[M(C_2O_4)_2]^{2-}$ is more probable (**cage effect**) to reduce the ϕ value.

Blueprinting by using $K_3[Fe(C_2O_4)_3]$ (see Sec. 1.16.18, Vol. 4): A white paper soaked with a solution of the complex salt is **air dried in dark.** Now place an opaque substance (say a coin) on the paper and place it to the bright sunlight or artificial light for a few minutes. Now dip the paper into a $K_3[Fe(CN)_6]$ solution. Now take out the paper and wash off the ferricyanide solution and see the **blueprint of the area exposed to light**. The area exposed to light experiences the photochemical decomposition of $[Fe(C_2O_4)_3]^{3-}$ to produce Fe(II) that reacts with $K_3[Fe(CN)_6]$ to produce the blue stain of **Prussian blue**.

Note: ● Some of the oxalato-complexes find uses in ***actiometer***. $[Fe(ox)_3]^{3-}$ can be used in this purpose. UO_2^{2+} – oxalate system can also be used in acitometer. It may be noted that UO_2^{2+} can cause photochemical oxidation of different types of organic and inorganic substrates.

● The complexes of Co(III), Mn(III) and Fe(III), *i.e.* **$[M(C_2O_4)_3]^{3-}$ (M = Co, Mn, Fe) experience the redox decomposition photochemically** but the corresponding Cr(III)-complex, *i.e.* $[Cr(C_2O_4)_3]^{3-}$ does not experience any such photochemical redox decomposition. The proposed mechanistic paths involve the participation of the intermediates M(IV) and M(II). Thus formation of these oxidation states should be energetically feasible. In fact, **photochemical stability of $[Cr(C_2O_4)_3]^{3-}$** arises due to the fact that the generation of Cr(II) is not likely. However, the Cr(III)-complex experience the *photorecimization* (as discussed in Sec. 6.15.6).

6.15.9 Photochemical Reactions of the Fe(II) and Fe(III) Complexes

● Photolysis of an acidic aqueous solution of Fe(II) can lead to oxidation of Fe(II) to Fe(III).

$$[Fe(OH_2)_6]^{2+} \underset{}{\overset{h\nu}{\rightleftharpoons}} [Fe(OH_2)_6]^{3+} + e_{aq}^- \text{ (charge transfer to solvent, **MSCT**)}$$

After this photochemical primary process, different sets of reactions depending on the pH and concentration of the solution are set up.

- Photochemical excitation of $[Fe(OH_2)_6]^{3+}$ or $[Fe(OH_2)_5X]^{2+}$ (X = OH, Cl, Br, N_3, etc.) in the CT region (ligand to metal) will lead to the oxidation of the coordinated ligands including the solvent H_2O.

$$\left[Fe(OH_2)_5X\right]^{2+} \xrightarrow[(CT\ \text{band})]{h\nu} \left[Fe(OH_2)_5\right]^{2+} + X$$

X may subsequently oxidize other reducing agents including water.

- Photolysis of $[Fe(C_2O_4)_3]^{3-}$ has been discussed in Sec. 6.15.8.
- Photolysis of $[Fe(CN)_6]^{4-}$ in aqueous solution involves the following processes.

$$\left[Fe(CN)_6\right]^{4-} \underset{}{\overset{h\nu}{\rightleftharpoons}} \left[Fe(CN)_6\right]^{3-} + e^-_{aq}, \textbf{ charge transfer to solvent } \text{(MSCT), (mainly CT excitation)}$$

$$\left[Fe(CN)_6\right]^{4-} \xrightarrow[H_2O]{h\nu} \left[Fe(CN)_5(OH_2)\right]^{3-} + CN^-, \text{ (mainly LF excitation)}$$

The released electron, *i.e.* e^-_{aq} can be trapped by using **N_2O as a scavanger** to generate the OH radical which can subsequently oxidize $[Fe(CN)_6]^{4-}$.

$$N_2O + e^-_{aq} \longrightarrow N_2 + O^-,\ O^- + H^+ \longrightarrow OH;\ \left[Fe(CN)_6\right]^{4-} + OH \longrightarrow \left[Fe(CN)_6\right]^{3-} + OH^-$$

- Photochemical reaction of $[Fe(CN)_6]^{3-}$ in aqueous solution is more complicated. The events are shown below.

$$\left[Fe(CN)_5(OH_2)\right]^{3-} + HCN + OH \xleftarrow[(H_2O)]{h\nu} \left[Fe(CN)_6\right]^{3-} \xrightarrow[(H_2O)]{h\nu} \left[Fe(CN)_5(OH_2)\right]^{2-}$$

(**Redox aquation**) (**Aquation**)

$$\left[Fe(CN)_5(OH_2)\right]^{3-} \downarrow \cdots \downarrow \left[Fe(OH_2)_6\right]^{2+} \xrightarrow{[Fe(CN)_6]^{3-}} \textbf{Prussian blue}$$

$$\left[Fe(CN)_5(OH_2)\right]^{2-} \downarrow \cdots \downarrow \left[Fe(OH_2)_6\right]^{3+} \xrightarrow{OH^-} Fe(OH)_3 \downarrow$$

$$H_2O + OH + \left[Fe(CN)_6\right]^{3-} \longrightarrow \left[Fe(CN)_6\right]^{4-} + H^+ + 2OH$$

Thus the produced $[Fe(CN)_6]^{4-}$ may also experience the photochemical reactions as stated above. The above photochemical reactions explain why the **$[Fe(CN)_6]^{4-}$ and $[Fe(CN)_6]^{3-}$ solutions on standing give the blue precipitate of Prussian blue.**

6.15.10 Photochemical Redox Process in $[M(aq)]^{n+}$

It can lead to oxidation or reduction of the metal centre depending on the conditions.

$$\textbf{Reducing metal centre: } \left[M(OH_2)_x\right]^{n+} \xrightarrow[(CT:\ M\rightarrow L)]{h\nu} \left[M(OH)(OH_2)_{x-1}\right]^{n+} + H$$

It happens so for $[Fe(OH_2)_6]^{2+}$. Subsequently, the H-radical can show the reducing activity.

$$\textbf{Oxidising metal centre: } \left[M(OH_2)_x\right]^{n+} \xrightarrow[(CT:\ L\rightarrow M)]{h\nu} \left[M(OH_2)_5\right]^{(n-1)+} + H^+ + OH$$

It happens so for $Tl(aq)^{3+}$, $Ce(aq)^{4+}$. Then the OH radical can oxidize the available reducing agents.

$$OH + e \longrightarrow OH^-$$

6.15.11 Photochemical Substitution Process in Pt(IV)

The ligand exchange and substitution reactions are photochemically carried out at the Pt(IV) centre which is expected to be inert in thermal activation process.

$$[PtCl_6]^{2-} + {}^*Cl^- \xrightarrow{h\nu} [PtCl_5^*Cl]^{2-} + Cl^- \; ; \; [PtCl_6]^{2-} + X^- \xrightarrow{-h\nu} [PtCl_5X]^{2-} + Cl^-$$

$$[PtCl_6]^{2-} + H_2O \xrightarrow{h\nu} [PtCl_5(OH)]^{2-} + H^+ + Cl^-$$

It is believed that in the primary photochemical step, homolytic bond cleavage generates the Pt(III) centre (d^7) which is labile.

$$[PtCl_6]^{2-} \xrightarrow{h\nu} [Pt^{III}Cl_5]^{2-} + Cl, \; [Pt^{III}Cl_5]^{2-} + H_2O \rightarrow [PtCl_4(OH)]^{2-} + H^+ + Cl^-,$$

$$[Pt^{III}Cl_5]^{2-} + X^- \longrightarrow [PtCl_4X]^{2-} + Cl^-$$

$$[PtCl_4X]^{2-} + [PtCl_6]^{2-} \xrightarrow[\text{(inner sphere)}]{e\text{-transfer}} [PtCl_5X]^{2-} + [PtCl_5]^{2-}$$

The process proceeds as follows (**ISET** with the ligand transfer):

$$[PtCl_4X]^{2-} + [PtCl_6]^{2-} \longrightarrow \left[\underset{\substack{(d^6)\\ \text{(inert)}}}{Cl_5Pt^{IV}} - Cl - \underset{\substack{(d^7)\\ \text{(labile)}}}{Pt^{III}Cl_4X}\right]^{4-}$$

$$\xrightarrow{1e\text{-transfer}} [Cl_5Pt^{III} - Cl - Pt^{IV}Cl_4X]^{4-} \longrightarrow [PtCl_5]^{2-} + [PtCl_5X]^{4-}$$

($X^- = {}^*Cl^-$ or other nucleophile)

Thus the generated $[PtCl_5]^{2-}$ species can participate in chain reactions and it leads to the high quantum yields.

The ligand exchange reaction in $[PtBr_6]^{2-}$ passes through the similar path generating Pt(II) at the primary photochemical step. The Pt(II) centre (distorted structure due to the removal of a *cis–* pair of Br-atoms through the **photochemical reductive elimination** from $[PtBr_6]^{2-}$) is labile to experience the rapid ligand exchange.

$$[PtBr_6]^{2-} \xrightarrow{h\nu} [PtBr_4]^{2-} + Br_2 \; ; \; [PtBr_4]^{2-} \xrightarrow[\text{(ligand exchange)}]{{}^*Br^-} [Pt^*Br_4]^{2-}$$

$${}^*Br^- + [Pt^*Br_4]^{2-} + [PtBr_6]^{2-} \longrightarrow [{}^*Br - {}^*Br_4Pt^{II} - Br - Pt^{IV}Br_5]^{5-}$$

$$\xrightarrow[\text{(ISET)}]{2e\text{-transfer}} [Pt^*Br_5Br]^{2-} + Br^- + [PtBr_4]^{2-}$$

The mechanistic aspects of the above reaction leading to the bromide exchange have been discussed in Sec. 6.4.15.

Thus, $[PtBr_4]^{2-}$ is regenerated to carry out the **chain reaction**. It explains the high quantum yield.

6.15.12 Photoisomerisation of $[Pt(gly)_2]$

It passes through the photoexcited tetrahedral geometry (**ring twisting mechanism**, see Sec. 5.30.11).

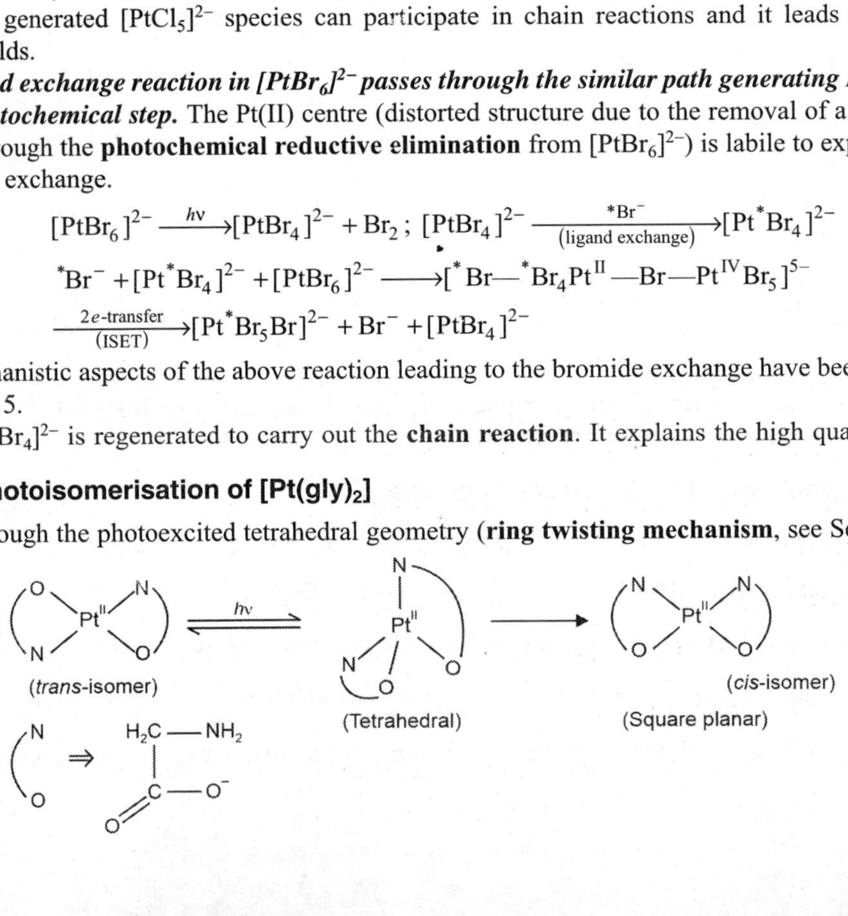

(*trans*-isomer) $\xrightarrow{h\nu}$ (Tetrahedral) $\longrightarrow$ (*cis*-isomer) (Square planar)

N⌒O ⇒ $H_2C—NH_2$, $C(=O)—O^-$

The *cis*-isomer is **less stable** because of the lone pair-lone pair repulsions between the two *cis*-donor O-sites (see Sec. 1.16.26, Vol. 4).

6.15.13 Photochemical Splitting of Water (*i.e.* Artificial Photosynthesis) and Photochemistry of $[Ru(bpy)_3]^{2+}$

In the present section, we shall restrict our attention to the complexes which can catalyse the splitting of water to generate the **fuel gas** ($2H_2 + O_2$) which can be utilised in fuel cells to produce electricity. **Artificial photosynthesis** - solar splitting of H_2O to produce the fuel (*i.e.* H_2) has attracted the attention of chemists. In fact, such reactions are practised in running the spacecrafts.

- **Photochemistry of** $[Ru(bpy)_3]^{2+}$ and related complexes like $[Ru(\text{phen})_3]^{2+}$ are important.

$$[Ru(bpy)_3]^{2+} \xrightarrow[(\lambda = 452\ nm)]{h\nu} [Ru(bpy)_3^{**}]^{2+} \xrightarrow{\text{relaxation}} [Ru(bpy)_3^{*}]^{2+}$$

$[Ru(bpy)_3]^{2+}$ absorbs photon (λ = 452 nm) and is excited to $[Ru(bpy)_3^{**}]^{2+}$ which relaxes to the **relatively long-lived $[Ru(bpy)_3^{*}]^{2+}$ species**. In the photoexcitation process of $Ru(II)$ (d^6), one electron from the t_{2g}^6 orbital (*i.e.* HOMO) is excited to the empty π^* orbital (*i.e.* LUMO). Here, the HOMO is predominantly enriched with the *d*-orbital (t_{2g}) character of metal while the LUMO (*i.e.* π_L^*) is mainly constituted by the π^*-orbitals of *bpy*. Thus $[Ru(bpy)_3]^{2+}$ undergoes photoexcitation easily through the ***MLCT* (metal to ligand charge transfer)** transition to promote the excited electron to the π^*–*MO* of the ligand (*i.e.* π_L^*). The excited singlet state undergoes radiationless transition (called **intersystem crossing, ISC**) to produce **the long lived phosphorescent excited triplet state**. The photochemical electronic transition and deexcitation of the excited state are illustrated in Fig. 6.15.13.1.

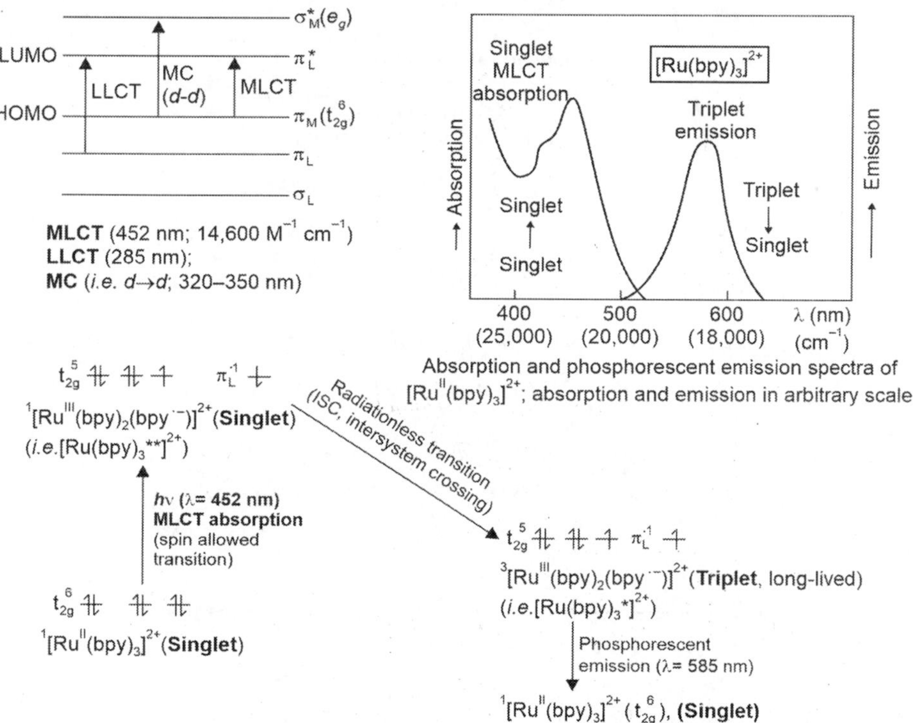

Fig. 6.15.13.1. Illustration of electronic transitions, absorption and phosphorescent emission spectra, photoexcitation and deexcitation steps of $[Ru^{II}(bpy)_3]^{2+}$ (see the authors' book. **"An Introduction to Supramolecular Chemistry".)**

The photochemically **excited triplet,** $[Ru(bpy)_3^*]^{2+}$ is long-lived and can participate in *delayed reactions*. It may show *phosphorescence* (~ 585 nm) and the wavelength of phosphorescent emission is longer than the wavelength of the absorption causing the photoexcitation.

• **Redox properties of $[Ru(bpy)_3^*]^{2+}$:** In generating the photochemically excited species, $[Ru(bpy)_3^*]^{2+}$ (which is triplet), the charge transfer (CT) process leads to Ru^{3+} and an electron gets localised on the *bpy* ligand leading to a ***radical anion of the ligand (i.e. bpy•−). The excited species can act as a better oxidising agent as well as a better reducing agent.*** These are evident from the reduction potentials of the coupels.

$$[Ru(bpy)_3]^{3+} + e \rightleftharpoons [Ru(bpy)_3]^{2+}, E^\circ = +1.26\text{ V}$$
$$[Ru(bpy)_3]^{3+} + e \rightleftharpoons [Ru(bpy)_3^*]^{2+}, E^\circ = -0.86\text{ V}$$
$$[Ru(bpy)_3^*]^{2+} + e \rightleftharpoons [Ru(bpy)_3]^{+}, E^\circ = +0.84\text{ V}$$
$$[Ru(bpy)_3]^{2+} + e \rightleftharpoons [Ru(bpy)_3]^{+}, E^\circ = -1.28\text{ V}$$

It is evident from the E° values (standard reduction potentials), that the photoexcitation producing $[Ru(bpy)_3^*]^{2+}$, lowers the $E^\circ_{3^+/2^{*+}}$ value (= –0.86 V) by 2.12 V (*i.e.* +1.26 V to –0.86 V) and raises the $E^\circ_{2^{*+}/1^+}$ value (*i.e.* +0.84 V) by 2.12 V (*i.e.* –1.28 V to +0.84 V). It makes the excited state a better oxidising agent (= +0.84 V) and a better reducing agent (E° = –0.86 V) than the ground state. The potentials are shown in the modified Latimer diagram (Fig. 16.15.13.2).

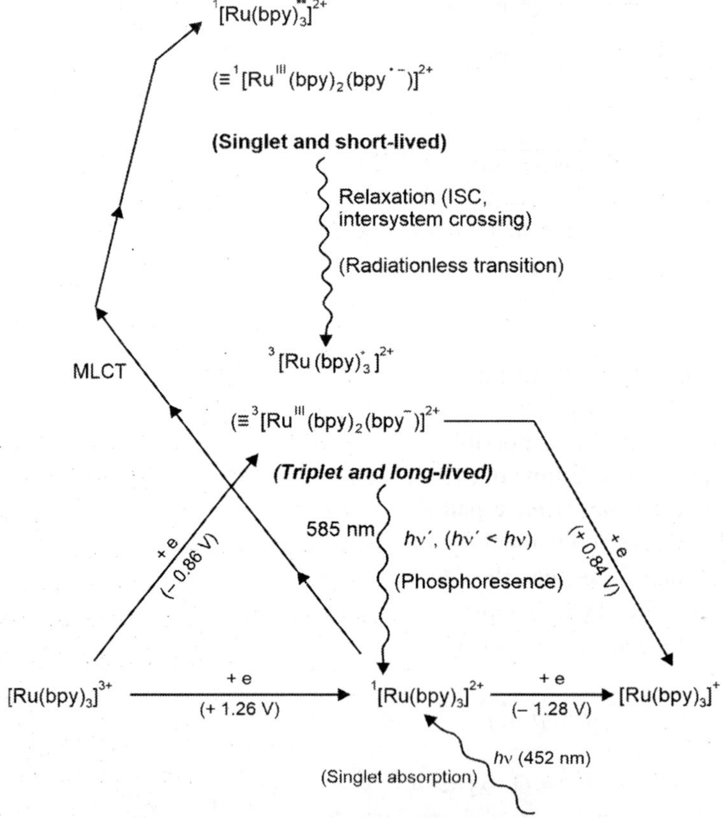

Fig. 16.15.13.2: Schematic representation of photoexcitation of $[Ru(bpy)_3]^{2+}$ and deexcitation steps and modified Latimer diagram of ruthemium-bpy complex (*cf. Coord. Chem. Rev.*, **84**, 85, 1986).

In the long-lived **triplet** intermediate $^3[Ru(bpy)_3^*]^{2+}$, a *d*-electron from the *Ru*-centre is transferred into the vacant antibonding orbital (*i.e.* LUMO) of *bpy*, *i.e.* $^3[Ru^{III}(bpy)_2(bpy^{\bullet -})^*]^{2+}$. The electron present in the vacant π^*-MO of *bpy* can be easily donated to an electron acceptor (*i.e.* oxidant) and *it makes triplet* $^3[Ru^{III}(bpy)_2(bpy^{\bullet -})^*]^{2+}$ ***a powerful reducing agent.*** On the other hand, the vacancy (*i.e.* positive hole) created at the t_{2g} level (*i.e.* HOMO) of *Ru*-centre makes it *electron greedy* and this is why, *the excited complex can also act* ***as a better oxidising agent.*** *A similar effect in photosynthesis is known. Photoexcitation of chlorophyll makes it a better oxidising agent as well as a better reducing agent.*

The photoexcitation process of $[Ru(bpy)_3]^{2+}$ and modified Latimer diagram (*cf.* A. Juris, *et al.*, *Coord. Chem. Rev.*, **84,** 85, 1986) are shown in Fig. 6.15.13.2.

From the modified Latimer diagram, it is evident that $[Ru(bpy)_3]^{2+}$ in ground state is thermodynamically stable with respect to its disproportionation but the excited triplet state, *i.e.* $^3[Ru(bpy)_3^*]^{2+}$ can spontaneously disproportionate to produce the strong oxidant, $[Ru(bpy)_3]^{3+}$ ($E° = +1.26$ V) and the strong reductant, $[Ru(bpy)_3]^+$ ($E° = -1.28$ V). However, generation of these powerful oxidant (*i.e.* $[Ru(bpy)_3]^{3+}$) and reductant (*i.e.* $[Ru(bpy)_3]^+$) from $^3[Ru(bpy)_3^*]^{2+}$ occurs through a reaction with the other electron acceptor and donor molecules in solution.

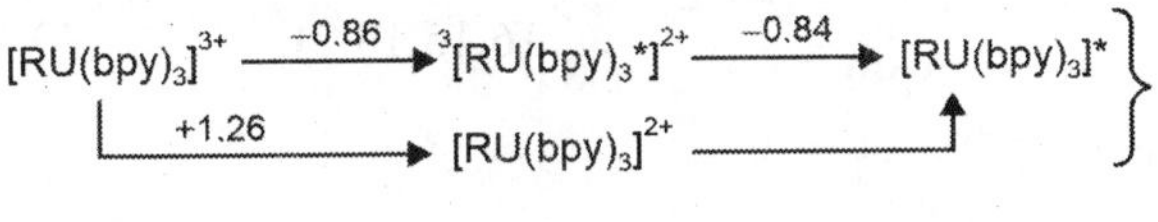

$$[RU(bpy)_3]^{3+} \xrightarrow{-0.86} {}^3[RU(bpy)_3{}^*]^{2+} \xrightarrow{-0.84} [RU(bpy)_3]^+$$

$$[RU(bpy)_3]^{3+} \xrightarrow{+1.26} [RU(bpy)_3]^{2+} \longrightarrow [RU(bpy)_3]^+$$

Modified Latimer diagram due to photoexcitation

$$2\,{}^3[RU(bpy)_3{}^*]^{2+} \xrightarrow{\text{Disproportionation}} [RU(bpy)_3]^{3+} + [RU(bpy)_3]^+$$
$$E°_{cell} = (0.84 + 0.86)V = 1.70\ V$$

$$2[RU(bpy)_3]^{2+} \xrightarrow[\text{(Not possible)}]{\text{Disproportionation}} [RU(bpy)_3]^{3+} + [RU(bpy)_3]^+,$$
$$E°_{cell} = (-1.28 - 1.26)V = -2.54\ V$$

Chemical synthesis of $[Ru(bpy)_3^*]^{2+}$ showing the chemiluminescence

$$[Ru(bpy)_3]^{3+}(t_{2g}^5) \xrightarrow[N_2H_4 \text{ or hydrated electron}]{\text{Reduction by:}} {}^3[Ru(bpy)_3^*]^{2+}\ \text{(triplet)} \xrightarrow[\text{Chemiluminescence (phosphorescence) of orange light}]{-h\nu} {}^1[Ru(bpy)_3]^{2+}(t_{2g}^6)\ \text{(singlet)}$$

• **Life-time of $^3[Ru(bpy)_3^*]^{2+}$ and its possibility to participate in chemical reactions:** The long-lived excited complex $^3[Ru(bpy)_3^*]^{2+}$ may return to the ground state through radiation (*i.e. phosphorescence*) or participate in different possible redox reactions **depending on the presence of suitable quenching agent (*Q*).** The life-time of $^3[Ru(bpy)_3^*]^{2+}$ in aqueous solution is 0.6 μs at 25°C (in terms of luminescence or phosphorescence pathway) and the life-time can be enhanced by lowering the temperature to minimise the collisional deactivation (*i.e.* EC). The electron transfer reactions involving $^3[Ru(bpy)_3^*]^{2+}$ is much faster than the decay rate through luminescence (*e.g.* electron exchange rate constant ~ 10^8 dm^3 mol^{-1} s^{-1} in the complexes $[Ru(bpy)_3]^{3+}/[Ru(bpy)_3^*]^{2+}$, $[Ru(bpy)_3^*]^{2+}/[Ru(bpy)_3]^+$). **It indicates that $^3[Ru(bpy)_3^*]^{2+}$ is sufficiently stable to allow its participation in several types of redox reactions.**

Energy transfer from $^3[Ru(bpy)_3^*]^{2+}$ may occur in the presence of a quenching agent (Q) in the following ways:

$$^3[Ru(bpy)_3^*]^{2+} + Q \rightarrow [Ru(bpy)_3]^{2+} + Q^*, \text{(simple energy transfer)}$$

$$^3[Ru(bpy)_3^*]^{2+} + Q \rightarrow [Ru(bpy)_3]^{+} + Q^+, \text{(oxidation of } Q)$$

$$^3[Ru(bpy)_3^*]^{2+} + Q \rightarrow [Ru(bpy)_3]^{3+} + Q^-, \text{(reduction of } Q)$$

Formation of $[Ru(bpy)_3]^{3+}$ and $[Ru(bpy)_3]^{+}$ from $^3[Ru(bpy)_3{}^*]^{2+}$ are illustrated in the following quenching reactions.

$$^3[Ru(bpy)_3^*]^{2+} + [Ru(NH_3)_6]^{3+} \longrightarrow [Ru(bpy)_3]^{3+} + [Ru(NH_3)_6]^{2+}$$

$$^3[Ru(bpy)_3^*]^{2+} + [Ru(NH_3)_6]^{2+} \longrightarrow [Ru(bpy)_3]^{+} + [Ru(NH_3)_6]^{3+}$$

$$(E^\circ \text{ for } [Ru(NH_3)_6]^{3+}/[Ru(NH_3)_6]^{2+} = 0.1\text{ V})$$

But the efficient back-reactions in the above quenching reactions dissipate the stored energy as the heat energy to the solution. Thus it requires the efficient **quencher to produce the stable products $[Ru(bpy)_3]^{3+}$ and $[Ru(bpy)_3]^{+}$ to store the light enery.**

Thus the produced *$[Ru(bpy)_3]^{+}(d^7)$ is an excellent reducing agent* and *$[Ru(bpy)_3]^{3+}(d^5)$* is *a good oxidising agent.* ***Thus by using the suitable quenching agents, it is possible to covert the light energy into chemical energy.***

- **Artificial photosynthesis by using the photoexcitation of $[Ru(bpy)_3]^{2+}$: Artificial photosynthesis** describes the photolytic production of H_2 (to be used as a **green fuel**) and O_2 from water. It may be noted that the **natural photosynthesis** produces O_2 and $C_6H_{12}O_6$, *i.e.* $(CH_2O)_6$ (a bio-fuel) from H_2O and CO_2 by using the solar radiaton.

Water does not itself absorb any visible light to get photoexcited but the electrode potential values indicate that $[Ru(bpy)_3^*]^{2+}$ can catalyse to split water by using the visible radiation (Creutz and Sutin). It might help trap the solar energy to produce the fuel (*i.e.* $H_2 + O_2$). At $pH \sim 7.0$, and 10^5 Pa pressure, the electrode potentials are :

$$2H_2O + 2e \rightarrow H_2 + 2OH^-, E_N^\circ = -0.42\text{ V}$$

$$4H^+ + O_2 + 4e \rightarrow 2H_2O, E_N^\circ = +0.82\text{ V}$$

The **thermodynamic requirements** for carrying out the artificial photosynthesis (*i.e.* solar splitting of H_2O) are: the photochemically excited species itself or the species generated from the photochemically excited species is capable of reducing H_2O to H_2 and capable of oxidising water to O_2. At $pH = 7$, to reduce H_2O, the reductant must have the E° value less than –0.42 V ($= E^\circ_{H_2O/H_2}$ at $pH = 7$) and to oxidise H_2O, the oxidant must have the E° value greater +0.82 V ($= E^\circ_{O_2/H_2O}$ at $pH = 7$). The photochemically excited species, $^3[Ru(bpy)_3^*]^{2+}$ can satisfy the said thermodynamic conditions. The possible routes are discused below:

Possible pathways I: The following sequence of reactions involving the disproportionation of $^3[Ru(bpy)_3^*]^{2+}$ to produce the required oxidant and reductant to oxidise water and reduce water respectively can directly split water.

$$4\,^1[Ru(bpy)_3]^{2+} \xrightarrow{4h\nu} 4\,^3[Ru(bpy)_3^*]^{2+}, \text{ (photoexcitation followed by } ISC\text{, see Fig. 6.5.13.2)}$$

$$4\,^3[Ru(bpy)_3^*]^{2+} \xrightarrow{\text{Disproportionation}} 2[Ru(bpy)_3]^{3+}\ (E^\circ = +1.26\text{ V}) + 2[Ru(bpy)_3]^{+}\ (E^\circ = -1.28\text{ V})$$

$$2[Ru(bpy)_3]^{3+} + H_2O \longrightarrow 2[Ru(bpy)_3]^{2+} + \tfrac{1}{2}O_2 + 2H^+, E^\circ_{\text{cell}} = (1.26 - 0.82)\text{ V} = +0.44\text{ V}$$

$$2[Ru(bpy)_3]^{+} + 2H_2O \longrightarrow 2[Ru(bpy)_3]^{2+} + H_2 + 2OH^-, E^\circ_{\text{cell}} = [-0.42 - (-1.28)]\text{ V} = +0.86\text{ V}$$

Net reaction: $$H_2O \xrightarrow[{[Ru(bpy)_3]^{2+}}]{4h\nu} H_2 + \frac{1}{2}O_2$$

This route faces two problems. (i) Without the separate production of H_2 and O_2, they will simply recombine releasing the heat energy in solution. This is of no practical significance. (ii) There may be a kinetic barrier. The deexcitation rate, disproportionation rate, H_2-gas evolution rate, etc. are to be synchronised properly. This aspect has been discussed later.

Possible pathways II: The alternative thermodynamically possible pathways to split water are outlined below :

$$^{1}[Ru(bpy)_3]^{2+} \xrightarrow{h\nu} {}^{3}[Ru(bpy)_3^*]^{2+}$$

$$[Ru(bpy)_3]^{3+} + e \rightarrow {}^{3}[Ru(bpy)_3^*]^{2+}, E^\circ = -0.86 \text{ V}$$

Thus the following cell reactions are quite thermodynamically feasible.

$$2\ {}^{3}[Ru(bpy)_3^*]^{2+} + 2H_2O \rightarrow 2[Ru(bpy)_3]^{3+} + 2OH^- + H_2$$

$$E^\circ_{cell} = -0.42 - (-0.86) = 0.44 \text{ V}$$

Thus the produced $[Ru(bpy)_3]^{3+}$ $(E^\circ = 1.26 \text{ V})$ is a potential oxidising agent and it can oxidise H_2O to O_2.

$$2[Ru(bpy)_3]^{3+} + H_2O \rightarrow 2[Ru(bpy)_3]^{2+} + \tfrac{1}{2}O_2 + 2H^+$$

$$E^\circ_{cell} = 1.26 - 0.82 = 0.44 \text{ V}$$

Thus splitting of water by $^{3}[Ru(bpy)_3^*]^{2+}$ *is thermodynamically quite feasible at normal conditions.* But this route also cannot prevent the **recombination of the products** in solution. There are also ***kinetic barriers*** which prevent the direct use of $[Ru(bpy)_3]^{2+}$ to split water photochemically. These are discussed below.

In fact, we need a **suitable quencher** in solution so that oxidation and reduction of H_2O can occur at different locations. This avoids the recombination of products. At the same time, the suitable quencher must be able to remove the kinetic barriers. These are discussed here.

The oxidation of $^{3}[Ru(bipy)_3^*]^{2+}$ *in the reaction with* H_2O *to release* H_2 *is much slower than the spontaneous decay rate of* $^{3}[Ru(bpy)_3^*]^{2+}$ *through luminescence.* The *Ru*-complex can act as 1*e* transfer agents (*i.e.* Ru^{3+}/Ru^{2+}, Ru^{2+}/Ru^{+} couples) while H_2O requires 4*e* and 2*e* transfer for oxidation and reduction steps respectively. Overall, for production of one molecule of oxygen, it needs 4*e* transfer and for production of one molecule of hydrogen, it needs 2*e* transfer. Thus, during photolysis of water, different intermediate states are to be produced. To stabilise these intermediates and to provide sufficient electrons, suitable quenching agents are required. These are illustrated in pathway III.

Possible pathways III (by using a quencher): For photochemical splitting of water by $[Ru(bpy)_3]^{2+}$, in the presence of sunlight, the suggested quenching agent is **paraquat**, known as **methyl viologen** (MV^{2+}). The *Ru*-complex harvests the solar radiation through photochemical excitation and then the quencher (MV^{2+}) sets up a redox reaction leading to decomposition of water. The redox reactions are catalysed by finely divided *Pt* and RuO_2 particles supported on TiO_2. The involved reactions are :

$$[Ru(bpy)_3]^{2+} \xrightarrow{h\nu} {}^{3}[Ru(bpy)_3^*]^{2+} \xrightarrow{MV^{2+}} [Ru(bpy)_3]^{3+} + MV^+$$

(MV^{2+} = Me—$\overset{+}{N}$(C₅H₄)—(C₅H₄)$\overset{+}{N}$—Me, methyl viologen known as *paraquat*)

$$MV^+ + H_2O \xrightarrow{Pt \text{ (catalyst)}} MV^{2+} + OH^- + \frac{1}{2}H_2$$

(regeneration of the quencher)

Oxidation of H_2O by $[Ru(bpy)_3]^{3+}$ occurs as follows :

$$2[Ru(bpy)_3]^{3+} + H_2O \xrightarrow{RuO_2 \text{ (catalyst)}} 2[Ru(bpy)_3]^{2+} + 2H^+ + \frac{1}{2}O_2$$

(regeneration of the light harvesting agent)

Here it is worth mentioning that in the case of reduction of H_2O by MV^+, to compromise with the kinetic problem, ***Pt*-metal is used as the catalyst** (*i.e.* heterogenous catalysis). Similarly, in the oxidation of water by $[Ru(bpy)_3]^{3+}$, a kinetic problem exists and $\boldsymbol{RuO_2}$ **is used as the required heterogenous catalyst**. But, the use of RuO_2 reduces the efficiency of the process as it itself partially oxidises $[Ru(bpy)_3]^{2+}$ and the catalyst becomes inactive.

The reactions can be carried out in an electrochemical cell to generate electricity with the generation of $\boldsymbol{H_2}$ **and** $\boldsymbol{O_2}$ **separately** at the cathode and anode respectively.

To regenerate $[Ru(bpy)_3]^{2+}$ from $[Ru(bpy)_3]^{3+}$, **triethanolamine** [TEA, $N(CH_2CH_2OH)_3$] may be added and then **only** $\boldsymbol{H_2}$ is produced as follows (Fig. 6.15.13.3).

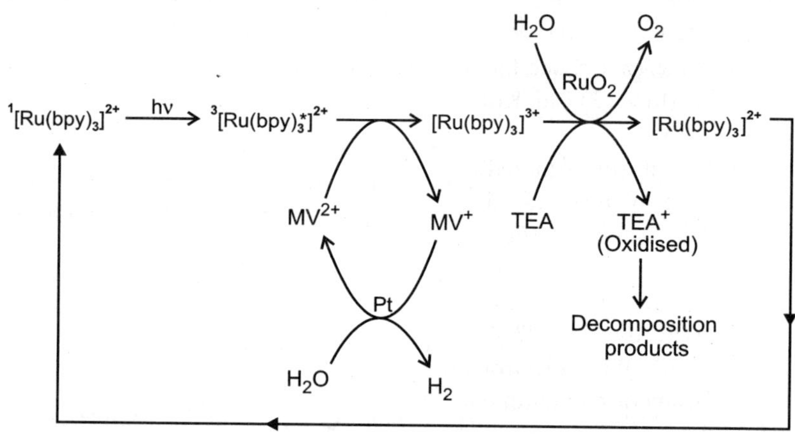

Fig. 6.15.13.3: Schematic representation of photochemical generation of H_2 and O_2 from water by using $[Ru(bpy)_3]^{2+}$ as the light harvesting agent (*cf.* C.J. Kutal, *J. Chem. Educ.*, **60**, 882, 1983).

Other compounds used in photochemical splitting of water are: $[Rh_2(\mu_2\text{–}NCCH_2CH_2CN)_4]^{2+}$ on irradiation can release H_2 from H^+; $SrTiO_{3-x}$ (mixed oxide with stoichiometric defect), lead incorporated potassium niobate. $SrTiO_{3-x}$ on being irradiated with the radiation of wavelength 350-360 nm can produce $2H_2 + O_2$ from a dilute aqueous $NaOH$ solution (*ca.* 5%). Lead incorporated potassium niobate can produce H_2 from an aqueous solution of methanol in the presence of solar radiation.

• **Photochemically oxidation of water by $\boldsymbol{[Ru(bpy)_3]^{2+}}$ in presence of a sacrifical oxidant like persulfate ($\boldsymbol{S_2O_8^{2-}}$):** The photochemically excited species, $^3[Ru(bpy)_3^*]^{2+}$ being a powerful reducing agent ($E° = -0.86$ V) can transfer its excited electron from π_L^* (*i.e.* π*-MO of *bpy*) to $S_2O_8^{2-}$ and it produces the strong oxidant $[Ru(bpy)_3]^{3+}$ ($E° = +1.26$ V) which can oxidise water to O_2.

$$2[Ru(bpy)_3]^{2+} \xrightarrow{2hv} 2\,^3[Ru(bpy)_3^*]^{2+} \text{ (photoexcitation)}$$

$$\left.\begin{aligned} 2\,^3[Ru(bpy)_3^*]^{2+} + 2S_2O_8^{2-} &\longrightarrow 2[Ru(bpy)_3]^{3+} + 2SO_4^{2-} + 2SO_4^{\bullet -} \\ 2[Ru(bpy)_3]^{2+} + 2SO_4^{\bullet -} &\longrightarrow 2[Ru(bpy)_3]^{3+} + 2SO_4^{2-} \end{aligned}\right\} \textbf{Oxidant generation}$$

$$4[Ru(bpy)_3]^{3+} + 2H_2O \longrightarrow 4[Ru(bpy)_3]^{2+} + 4H^+ + O_2\uparrow \text{ (water oxidation)}$$

Net reaction: $2H_2O + 2S_2O_8^{2-} \xrightarrow{[Ru(bpy)_3]^{2+},\,hv} 4H^+ + O_2 + 4SO_4^{2-}$

This reaction (*i.e.* photochemical oxidation of water) may be compared with the reaction occurring in **PS-II of natural photosynthesis.**

• **Possibility of the use of $\boldsymbol{[Ru(bpy)_3]^{2+}}$ in photogalvanic cells (*PGCs*) to harness solar energy:** Photogalvanic cell is a **photoelectrochemical cell** (*PEC*) where light energy is used **to drive a**

nonspontaneous redox reaction to generate the electrical energy. The reverse spontaneous reaction is allowed to take place in dark. Thus by coupling the light phase and dark phase reactions, it is possible to convert the solar energy into the electrical energy in a *PGC*.

It has been stated that the photochemically excited $^3[Ru(bpy)_3^*]^{2+}$ complex is a better reducing agent than $[Ru(bpy)_3]^{2+}$ (*cf.* $E°$-values: –0.86 V *vs.* 1.26 V). In fact, $[Ru(bpy)_3]^{2+}$ can reduce Fe^{3+} in presence of light but the reverse reaction occurs in dark.

$$[Ru(bpy)_3]^{2+} + Fe^{3+} \underset{\text{Dark}}{\overset{\text{Light}}{\rightleftharpoons}} [Ru(bpy)_3]^{3+} + Fe^{2+}$$

$$E^\circ_{cell} \text{ (in dark)} = (0.77 - 1.26) \text{ V} = -0.49 \text{ V } (\textit{i.e.} \text{ reverse reaction spontaneous})$$

$$E^\circ_{cell} \text{ (in light)} = (0.77 + 0.86) \text{ V} = +1.36 \text{ V } (\textit{i.e.} \text{ spontaneous reaction}).$$

Note: For details of the *PGCs*, see the authors' book, **"Environmental Chemistry with Green Chemistry"**, Books and Allied (P) Ltd, Kolkata. One typical *PGC* using the photobleaching of the **dye thionine** is discussed below.

The **thionine dye** (T^+, purple coloured) is reduced by Fe^{II} in presence of light while the reverse reaction occurs in absence of light (*i.e.* in dark). To construct the *PG* cell, the anode compartment (containing both the T^+/TH_2^+ and Fe^{3+}/Fe^{2+} systems) is kept separated by a semipermeable membrane (permeable to the Fe^{3+}/Fe^{2+} system but impermeable to the T^+/TH_2^+ system) from the cathode compartment (containing only the Fe^{3+}/Fe^{2+} system). When the anode compartment is irradiated by visible light, Fe^{2+} reduces the dye T^+ to TH_2^+ which undergoes oxidation at the anode to release electrons. Through the external circuit, these electrons reach the cathode (kept in dark) where Fe^{3+} reduces to Fe^{2+} which diffuses to the anode compartment to complete the redox cycle. The *PG* cell is schematically shown in Fig. 6.15.13.4.

$$T^+ + 2Fe^{2+} + 2H^+ \underset{\text{Dark}}{\overset{\text{Light}}{\rightleftharpoons}} TH_2^+ + 2Fe^{3+}$$

(Purple) $+ 2H^+ + 2e \rightleftharpoons$ (Colourless)

T^+ (Thionine) TH_2^+ (Leucothionine)

(R = H for thionine; R = Me for methylene blue)

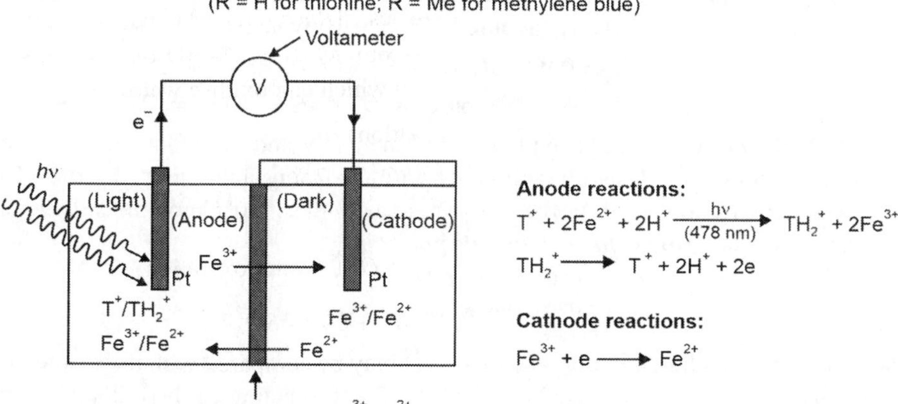

Membrane (permeable to Fe^{3+}/Fe^{2+})
but impermeable to T^+/TH_2^+

Fig. 6.15.13.4: Schematic representation of a photogalvanic cell (PGC) using the thionine dye

6.15.14 TiO_2, an Important Photocatalyst: Oxidation of Organic Matter and Photoelectrolysis of Water

The oxides (*e.g.* TiO, VO) of bivalent metal ions of early members of the 1st transition series show the metal-like conductivity due to the overlap of t_{2g} orbitals of the adjacent cations. **Thus the produced t_{2g}-band remains partially filled in to act as a conduction band (CB) and it is responsible for metallic conductivity.** The late members of the 3*d*-series and the members with higher positive charge (where the *d*-orbitals are sufficiently contracted to deny the overlapping interaction) show the **semiconductivity** in their oxides (*cf.* Chapter 13, Vol. 3A). ***TiO*** (*cf.* Ti(II), a d^2-system) **shows the metallic conduction while *TiO*$_2$ shows the semiconductivity** (*cf.* Ti(IV), a d^0-system).

Anatase *TiO*$_2$ acts as an **indirect band gap semiconductor** (*cf.* **rutile *TiO*$_2$ acts a direct band gap semiconductor**). In an indirect band gap semiconductor, direct transitions (*i.e.* return) of the photoexcited electrons from the CB (conduction band) to the VB (valence band) accommodating the holes are not possible (*cf.* Sec. 13.9.4, Vol. 3A). This is why, in anatase TiO_2 semiconductor, the **electron-hole pairs generated through photoexcitation possess the longer life time to participate in surface reactions.**

TiO_2, a *n*-type semiconductor (see Sec. 12.15.2 and Fig. 13.10.2, Vol. 3A for the band structure of TiO_2), possesses the band gap (*i.e.* energy difference between the valence band, VB and conduction band, CB) of about 3.23 eV and it can act as a **promising photocatalyst**. This band gap gives the energy difference between the HOMO (≡ VB) and LUMO (≡ CB). In photoexcitation, an electron is promoted from the HOMO housing the oxide (O^{2-}) lone pairs to the delocalised LUMO based on the *Ti* 3*d*-orbitals. Actually, in this photoexcitation, an electron from the oxide lone pairs is excited to the delocalised LUMO constituted mainly by the *Ti* 3*d*-orbitals. This excited electron from the LUMO is easily transferred to the molecular oxygen (O_2) of atmosphere to produce the superoxide (O_2^-) ion. This LUMO of TiO_2 and π*-MOs (*i.e.* SOMOs) of O_2 energetically match to favour the said electron transfer process.

$$(Ti^{4+})(O^{2-})(O^{2-}) \xrightarrow{h\nu} (Ti^{4+})(e, \text{LUMO})(O^{2-})(O^-)$$

$$\text{LUMO}(TiO_2) \xrightarrow{e-\text{transfer}} \pi^*\text{-}MO(O_2), \textit{i.e. } e\,(\text{LUMO, TiO}_2) + O_2 \rightarrow O_2^- \textbf{ (powerful oxidising agent)}$$

O^- (7 valence electron) ≡ h^+ (positive hole in *VB*) acting as the electron acceptor generates the **powerful oxidising agent $^\bullet$*OH* radical** from water (*i.e.* moisture).

Inertness of O_2 (triplet ground state) as an oxidant (4*e* transfer in steps) **arises from the thermodynamically unfavourable step,** $O_2 + e \rightleftharpoons O_2^-$ ($E^o \approx -0.35$ V). **In presence of *TiO*$_2$ as a photocatalyst, this thermodynamically hindered step is overcome.** In fact, subsequent electron transfer reactions are thermodynamically favoured (E^o values given at $pH = 7.0$).

$$O_2^- + 2H^+ + e \rightleftharpoons H_2O_2\ (E^o \approx 0.9\text{ V});\ H_2O_2 + H^+ + e \rightleftharpoons H_2O + {}^\bullet OH\ (E^o \approx 0.38\text{ V});$$

$${}^\bullet OH + H^+ + e \rightleftharpoons H_2O\ (E^o \approx 2.30\text{ V});\ \text{Net process: } O_2 + 4H^+ + 4e \rightleftharpoons 2H_2O\ (E^o \approx 0.80\text{ V}).$$

Thus the produced O_2^- (a potential oxidant both kinetically and thermodynamically) can oxidise the organic hydrocarbons present as air pollutants without facing any energy barrier. The electron deficient O^- (7*e* system) (≡ hole, h^+ in VB) can also react with the atmospheric moisture to ***produce the $^\bullet$OH radical that can also destroy the organic air pollutants.***

$$O^- + H_2O\ (\text{moisture}) \rightarrow {}^-OH\ (\text{surface bound}) + {}^\bullet OH,$$
$$O^- + {}^-OH\ (\text{surface bound}) \rightarrow O^{2-} + {}^\bullet OH$$

$$\text{Organic matter (including microorganisms)} \xrightarrow{{}^\bullet OH} \text{Oxidised products of organic matter}$$

Thus, electrons are captured by these holes (h^+ in $VB \equiv O^-$) acting as the oxidants from OH^- and H_2O (moisture) to generate the **powerful oxidant $^\bullet$*OH* radical** which is highly efficient to extract *H*-atom from the *C*—*H*, *N*—*H* bonds. Thus the $^\bullet OH$ radical abstracts the *H*-atom to generate the radical which can initiate the **oxidative degradation** of organic matter including the microorganisms. In fact, the **two potent oxidants** O_2^- (generated by the excited electron in *LUMO* or *CB*) and **$^{\bullet}OH$ radical** (generated by the holes in *HOMO* or *VB*) can oxidise almost anything (*e.g.* NO_x, smokes, organics, microorganisms, etc.). This is why, TiO_2 semiconductor has earned the attention of environmental chemists.

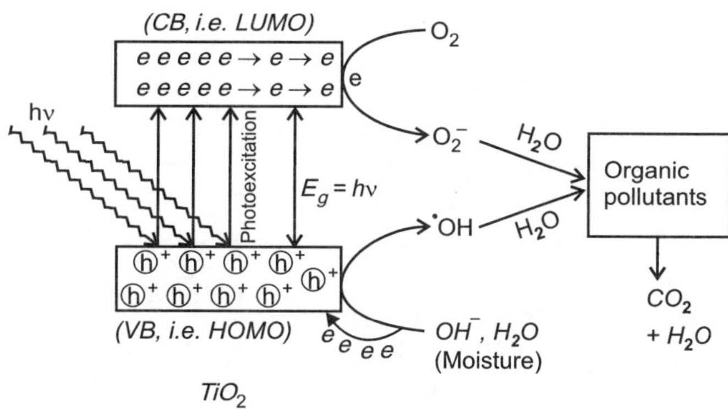

Fig. 6.15.14.1: Schematic representation of photocatalytic activity of TiO_2.

TiO_2 as a photocatalyst may be of tremendous importance to the environmental chemists to remove the air pollutants. Photocatalysis by TiO_2 is schematically shown in Fig. 6.15.14.1.

• **Photoelectrolysis of H_2O by using TiO_2 semiconductor as the anode:** Here, solar energy is converted into electrical energy which is used for electrolysis of water. This is an important method of harnessing solar energy. The electrolytic decomposition of H_2O under the standard condition ($[H^+] = 1$ M) requires the decomposition potential (E_d) of 1.23 V (*cf.* $E^o_{H^+/H_2} = 0$ V and $E^o_{O_2/H_2O} = 1.23$ V; *i.e.* $E^o_{cell} = 1.23$ V for the cell reaction, $2H_2 + O_2 \rightarrow 2H_2O$). $E^o_{cell} = 1.23$ V for the galvanic cell constituted by the electrolysis products, ignoring the **hydrogen overvoltage** and **oxygen overvoltage** (see Vol. 3B for details). E_d value for water electrolysis is greater than 1.23 V due to overvoltage. For photoelectrolysis of water, the photoanode used is made of TiO_2 semiconductor while the cathode is platinized-*Pt*. At the anode, O_2 is evolved while at the cathode, H_2 is evolved.

The photoanode (made of TiO_2 semiconductor) on being exposed to *UV*-radiation (solar light) experiences the **electron-hole separation**. In other words, photoexcitation promotes electrons from the *VB* to *CB* producing positive holes (h^+) in the *VB* (*cf.* Fig. 6.15.14.2a). The holes generated in the *VB* oxidise water to O_2 while the excited electrons in the *CB* pass to the cathode through the external circuit. These electrons are utilised at the cathode to reduce H^+ to H_2. The electrode reactions are:

At the photoanode: $TiO_2 + 2h\nu \longrightarrow 2e$ (in the *CB*) $+ 2h^+$ (in the *VB*)

$$2h^+ + H_2O \longrightarrow 2H^+ + \tfrac{1}{2}O_2$$

At the cathode (*Pt*): $2H^+ + 2e \longrightarrow H_2$

Net reaction: $H_2O \xrightarrow[(TiO_2)]{2h\nu} H_2 + \tfrac{1}{2}O_2$

A schematic representation of the photoelectrolytic cell for splitting water is shown in Fig. 6.15.14.2.

Note: (1) The electrolytic decomposition potential (E_d) of acidified water is 1.23 V **ignoring the overpotential factor**. In reality, due to the overpotential factor, E_d value is greater than 1.23 V (see Vol. 3B). Thus $E_g \geq 1.23$ eV should be able to split water pholytically. But, in reality, the overpotential factor must be considered along with the thermodynamic potential. Besides this, complete utilisation of solar energy is not possible. Considering all these factors, for an effective photoelectrolysis of water, E_g should be at least greater than 2.5 eV.

2. TiO_2, a ***n*-type semiconductor** (see Sec. 12.15.2, Vol. 3A) shows the band gap about 3.0 eV (3.0 eV for rutile and 3.2 eV for anatase). Anatase **TiO_2 shows the better photocatalytic activities than rutile TiO_2**. For **anatase TiO_2 acting as an indirect band gap semiconductor** (see Sec. 13.9.4, Vol. 3A), the electron-hole pair possesses the longer lifetime as their recombination possibility is less favoured.

It is due to the fact that in an indirect band gap semiconductor, direct transitions (*i.e.* returns) of the photogenerated electrons of *CB* (conduction band) to the valence band (*VB*) accommodating the holes (h^+) are not possible (see Fig. 13.9.4.1, Vol. 3A). While, for **rutile acting as a direct band gap semiconductor**, the lifetime of the separated electron-hole pair is relatively less. This is why, the charge carriers of anatase are more likely to participate, in surface reactions. This is one of the reasons for the **better photocatalytic activity of anatase TiO_2** (*Refs*. T. Luttrell *et al*, *Sci., Rep.*, **4**, 4043, 2014; J. Zhang *et al*, *Phys. Chem. Chem. Phys.*, **16**, 20382, 2014).

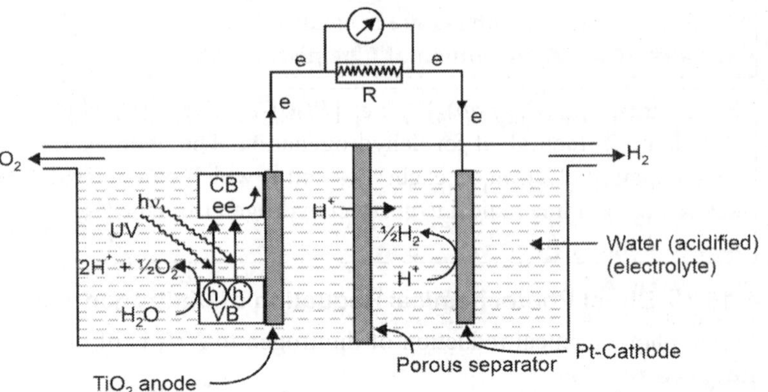

Fig. 6.15.14.2a. Schematic representation of photoelectrolytic cell for splitting water by using TiO_2 semiconductor as the anode.

3. ***ZnO*** (*n*-type) semiconductor shows $E_g \approx 3.3$ eV but it easily corrodes when excited in presence of aqueous environment. But it works well as a photocatalyst in vapour phase.

Prevention of electron-hole recombination in TiO_2 through bias induced band bending: For an effective photocatalytic activity, the lifetime of separated electron-hole pair must be sufficiently large so that the charge carriers can participate in surface reactions. But, the electron-hole recombination reduces the lifetime of the pair. To minimise the possibility of electron-hole recombination, band bending plays the crucial role at the *p-n* junction (see Fig. 13.9.8.2, Vol. 3A). In a **photoelectrochemical cell (*PEC*)**, band bending at the semiconductor-electrolyte interface can be induced by applying a potential between the working electrode and reference electrode. For the *n*-type semiconductor like TiO_2, band bending (*i.e.* upward bending) can be induced by applying a positive bias to the semiconductor. This reduces the possibility of electron-hole pair recombination (see Fig. 6.15.14.2b). This band bending at the junction of semiconductor-electrolyte interface (called **Schottky junction**) determines the electron-hole separation kinetics (*Refs*. C. Jiang *et al*, *Chem. Soc. Rev.*, **46**, 4645, 2017; R. van de Krol *et al*. *J. Mater Chem.*, **18**, 2311, 2008).

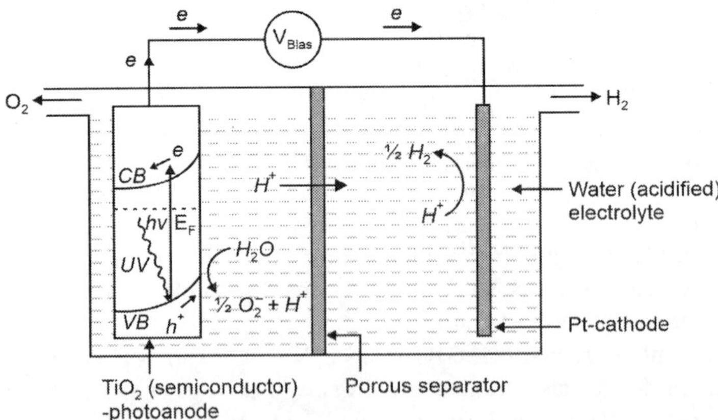

Fig. 6.15.14.2b: Schematic illustration of bias induced band bending in TiO_2 semiconductor used as the photoanode in the photoelectrolytic cell for splitting water

TiO_2 – A green photocatalyst

Use of TiO_2 as a promising green catalyst in different purposes has been discussed in detail in the authors' book, **Environmental Chemistry with Green Chemistry,** Books & Allied, Kolkata.

Platinum-pop; $[Pt_2(pop)_4]^{4-}$, *i.e.* $[Pt_2(\mu\text{-}P_2O_5H_2)_4]^{4-}$ (where $H_2P_2O_5^{2-}$ = pyrophosphite) – an excellent **photocatalyst** for dehydrogenation. The photoexcitation produces a **triplet state** of $^*Pt_2$ (one unpaired electron on each *Pt-centre) which **can activate the C—H bond**. In fact, each *Pt-centre acts as a free radical to abstract hydrogen from the C—H bond. This aspect has been discussed in Sec. 13.8.1 (Vol. 3A).

6.15.15 Direct Photochemical Reduction of Dinitrogen

Direct photochemical reduction of dinitrogen to NH_3 is of much significance from the standpoint of **nitrogen fixation.**

$$N_2 + 3H_2O \xrightarrow{h\nu} 2NH_3 + \tfrac{3}{2}O_2$$

The process is catalysed by $[Ru(Hedta)(H_2O)]^-$ (where $edtaH_4$ = ethylenediaminetetraacetic acid) under normal condition (25°C, 10^5Pa) in the presence of $CdS/Pt/RuO_2$ mixture where CdS acts as a ***direct bandgap semiconductor*** in photoexcitation. It is suggested that photoexcitation by the visible light (λ = 505 nm) leads to the promotion of electrons from the valency band (VB) of CdS to its conduction band (CB). The promoted electrons into the CB are ultimately transferred to the Ru-complex via platinum metal while the electrons released in the photo-oxidation of H_2O catalysed by RuO_2 are trapped in the positive holes of VB of the CdS semiconductor. $[Ru(Hedta)(H_2O)]^-$ is highly labile to allow the formation of a dinitrogen complex $[Ru(Hedta)(N_2)]^-$ and this bound dinitrogen is actually reduced to NH_3. Thus, photo-oxidation of H_2O catalysed by RuO_2 is the source of electron and dinitrogen bound in the Ru-complex is the sink of the electron; and electron travels through the CdS semiconductor. The probable route is schematically shown in Fig. 6.15.15.1.

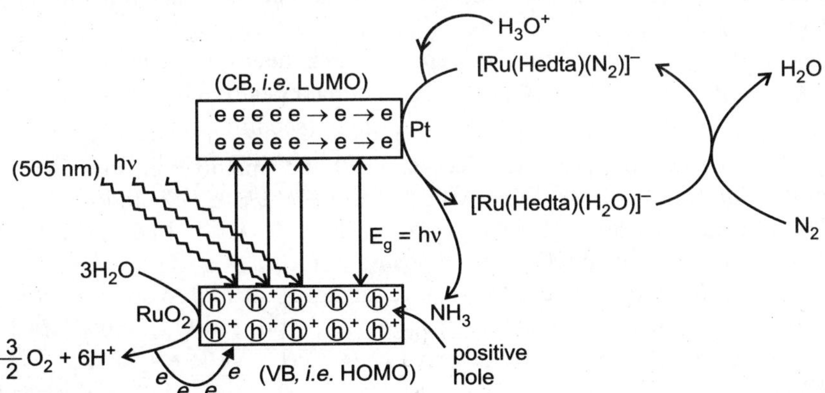

Fig. 6.15.15.1: Schematic representation of direct photochemical reduction of N_2 (*cf.* M.M.T. Khan, *et al., Angew. Chem. Int. Ed. Engl.*, **27**, 923, 1988).

6.15.16. Charge Transfer Band and Redox Stability

The ligand to metal charge transfer (LMCT) transitions can sometimes determine the redox stability of the compounds. The colour due to this charge transfer band (where the electron is transferred from one atom to another atom, more correctly HOMO to LUMO where the HOMO is enriched with the ligand orbital character and the LUMO is enriched with the metal orbital character) is generally more intense than that due to the *d–d* or *f–f* transitions (where the electron transition occurs between the same type of orbitals on the same atom, *cf.* **Laporte selection rule**). The *d–d* or *f–f* transitions are the Laporte forbidden transitions but this orbital selection rule is not applicable for the CT transitions. It makes the CT bands more intense (see Sec. 7.21).

The **relationships among the parameters like** $E°$, ν_{CT} **(*LMCT*) and η (hardness)** of the oxyanions (XO_4^{n-}) have been discussed in Sec. 7.21.1 (Vol. 5).

Compounds showing the visible charge transfer spectra are highly coloured and very often they possess only marginal redox stability. **This is illustrated in the following cases.**

- ***Fe*(III)*-halides*** : These are:

FeF_3 (**white**), $FeCl_3$ and $FeBr_3$ (**intensely red brown**), FeI_3 (**intensely black**, if isolated).

The colour of *Fe*(III)-halides originates from the electron transition from the halide to the metal. It is the most favoured process in FeI_3 (I^- is the most polarisable one and the most reducing centre) and it is the least favoured process in FeF_3 (F^- is the least polarisable one and the least reducing centre). In FeI_3, the electron transfer is so prominant that it causes the following redox reaction.

$$2FeI_3 \rightarrow 2FeI_2 + I_2 \text{ (i.e. } Fe^{3+} \text{ is reduced and } I^- \text{ is oxidised)}$$

- ***Nitrogen trihalides*** **(NX_3)** : NF_3 is colourless, nonexplosive and stable and NI_3 is intensely coloured, explosive and unstable. **The colour intensity increases from NF_3 to NI_3 and *the stability decreases from* NF_3 to NI_3.** In fact, NCl_3 and NBr_3 are of intermediate stability and of intermediate colour intensity. The colour appears due to the charge transfer from X^- to *N(III)* and it is the least favoured process in NF_3 and it is the most favoured process in NI_3.

- ***Silver bromide*** **(*AgBr*)** : The charge transfer band (Br^- to Ag^+) makes the compound *photosensitive* and it is *photochemically decomposed*. This gives the basic process in photography and **photochromatic lenses.**

$$AgBr \xrightarrow{h\nu} Ag + Br$$

(***Note***: The redox stability of a compound can also be explained by considering the electrode potential values of the involved redoox couples, *cf.* Sec. 16.21, Vol. 3B).

If the compound does not experience the redox decomposition, the transferred electron subsequently comes back to the original atom.

- ***Chemistry of photography*** : At the **1st step**, a latent image is produced on the film through the deposition of silver atom produced in the photochemical reaction.

$$AgBr \xrightarrow{h\nu} Ag + Br, 2Br \rightarrow Br_2 \text{ (absorbed by gelatine)}$$

The energy difference (E_g) between the filled **valence band** (mainly constituted by the *Br*-orbitals) and **vacant conduction band** (mainly lying with *Ag*) allows the photoexcitation of an electron from the *VB* (*i.e.* HOMO band) to *CB* (*i.e.* LUMO band) (*cf.* Figs. 9.11.4.1, 12.5.5.3; Sec. 10.6.4, Vol. 2, 3A). This makes the photochemical redox decomposition of *AgBr*.

At the **2nd step, *a developing agent*** (a weak reducing agent like hydroquinone) intensifies the latent image through the reduction of more *AgBr* present only on the light exposed area.

$$2AgBr \text{ (present on the light exposed area)} + C_6H_4(OH)_2 \rightarrow 2Ag + 2HBr + C_6H_4O_2$$

At the **3rd step**, unchanged *AgBr* (present in the area not exposed to light) is washed away by the treatment with sodium thiosulfate (hypo).

$$2AgBr + 2Na_2S_2O_3 \rightarrow Na_3[Ag(S_2O_3)_2] + NaX$$

Thus a negative is produced.

Then printing is done from the negative on a photosensitive paper by repeating all the above mentioned steps.

6.15.17 Light Emiting Diode (LED) (*cf.* Fig. 6.15.17.1)

When an *n–p* junction is *forward biased*, the potential energy barrier is decreased. It favours the uphill migration of electrons from the *n*-side to the *p*-side in the CB and the uphill migration of the holes from

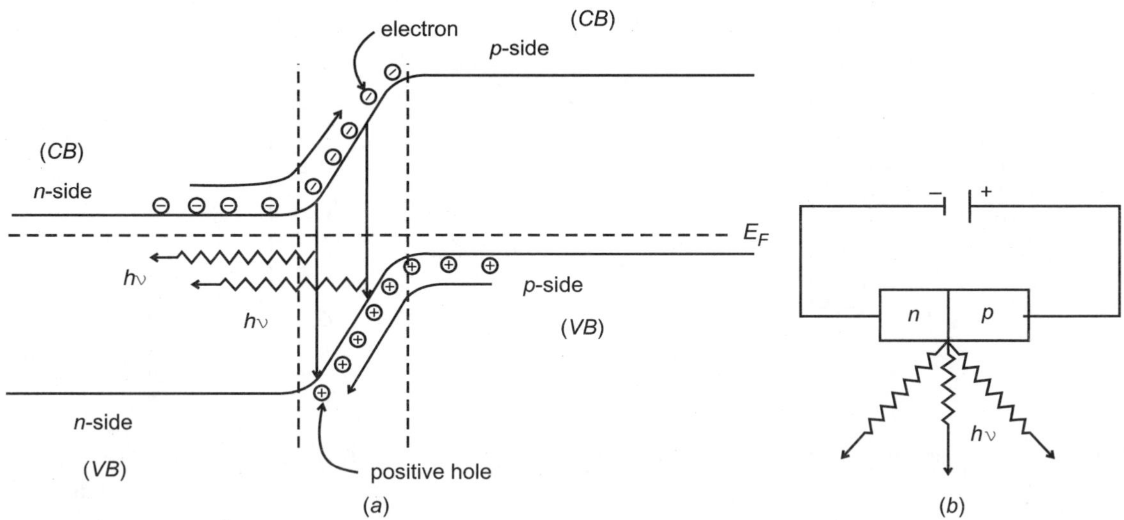

Fig. 6.15.17.1: Electron-hole recombination in a forward biased p–n junction diode (LED). (b) Construction of a LED.

the p-side to the n-side in the VB. At the *junction*, they recombine through the downward transition of the electrons from the CB to the VB to release the electromagnetic radiation. *The electron-hole recombination to release the electromagnetic radiation of wavelength* (λ) *gives the following relationship.*

$$E_g = h\nu;\ \text{or},\ \lambda = ch/E_g = \frac{1.24}{E_g}\ \mu\text{m},\ (E_g \text{ in eV})$$

$$(c = \text{velocity of light} \approx 3 \times 10^8 \text{ m s}^{-1},\ 1\ \mu\text{m} = 10^{-6} \text{ m})$$

The optical energy in LED is obtained from the biasing electrical energy. ***The downward transition of an electron from the CB to VB is favoured for the direct gap materials.*** In this regard, *GaAs* is an useful *direct gap semiconductor*. It releases the infrared radiation (λ = 890 nm) but the **alloy semiconductor** $GaAs_{1-x}P_x$ can release the radiation in the visible range. LEDs made of $GaAs_{1-x}P_x$ are commercially used. For, $x = 0.4$, E_g becomes 1.9 eV and the corresponding radiation is red (λ = 650 nm). *By varying x, different lights like orange, yellow and green can be obtained.* ***Such LEDs are used in signals and displays.***

Other important semiconducting materials used in LED are : *GaP* (2.3 eV), *GaSb* (0.7 eV), $Al_xGa_{1-x}As$ $(0 \leq x \leq 1)$ with E_g = 1.4 eV (x = 0) to 2.1 eV (x = 1), etc. For the production of n–p junction, proper substitution is required. It is illustrated for *GaAs* (13-15 semiconductor), ***partial substitution of Ga by Zn leads to p-type; partial substitution of As by Se leads to n-type.***

Phosphors: Light-emitting solids

These are used in television screen and fluorescent tubes. The phosphors absorb energy and re-emit it as light. In TV, phosphors are bombarded by cathode rays. In fluorescent tubes, UV-radiation (254 nm) produced by electric discharge through *Hg*-vapour at low pressure is absorbed by phosphors (*e.g.* $3Ca_3(PO_4)_2.CaF_2$, white powder) coated inside the tube. This re-emits the visible light. (Application of Semiconductors: See Vol. 3A).

6.15.18 Solar Cell

A solar cell is basically a p–n junction which converts the sunlight into electricity. We have already mentioned that the forward biasing can reduce the height of the potential barrier. *By illuminating a*

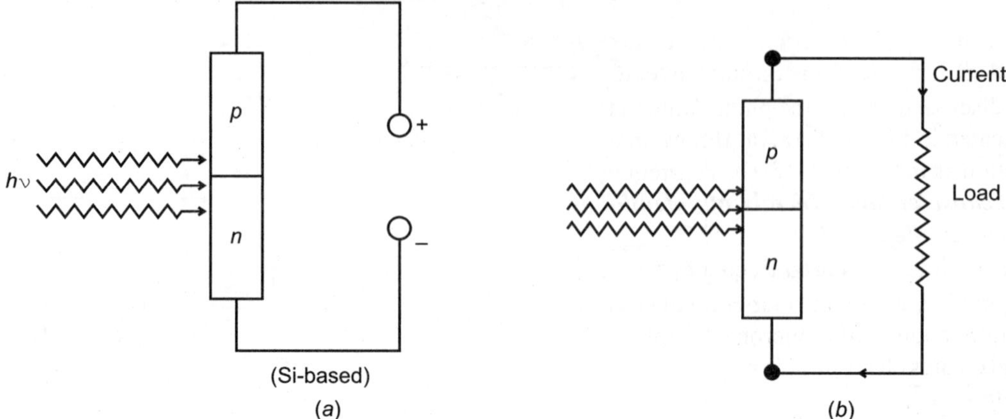

Fig. 6.15.18.1: Construction of a solar cell, (a) Photovoltaic effect in Si-solar cell. (b) Direct conversion of solar energy into electricity.

junction with light of proper frequency, it is also possible to reduce the potential barrier. This phenomenon is called ***photovoltaic effect*** which is used in developing a solar cell.

If a *p–n* junction is illuminated (while disconnected from any external source of voltage), as in Fig. 6.15.18.1a), at the vicinity of the junction, the electron-hole pairs are separated. The hole slides down (*i.e.* downhill migration) to the *p*-side in the VB and the promoted electron slides down (*i.e.* downhill migration) to the *n*-side in the CB (*cf.* Fig. 6.15.18.2a). ***Thus the p-region becomes more positive with respect to the n-region.* In fact, separation of electrons and holes as mentioned above produces a *forward voltage* across the junction** and this forward voltage opposes the inherent built-

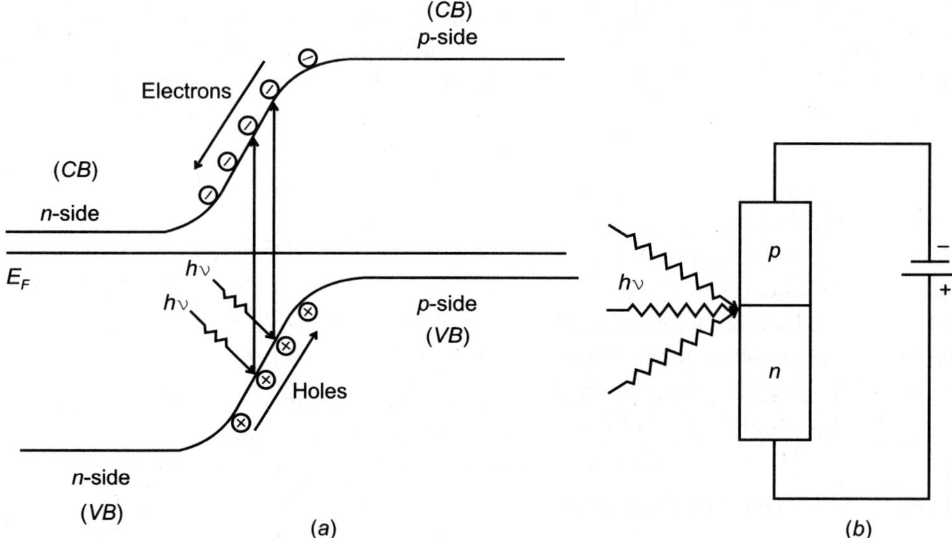

Fig. 6.15.18.2: (a) Electron-hole separation in a photodiode by radiation in a reversed biased *p–n* junction. (b) Construction of a semiconductor photodiode.

Note: When a *n–p* junction is **reverse biased**, the potential energy barrier at the junction is increased and the small reverse current arises from the downhill migration of the minority carriers. Incident radiation on the junction generates the electron-hole pairs which are separated by their **downhill migrations**. Thus, the incident light enhances the reverse current due to the increase of minority carriers in photodiodes (see Vol. 3A for details) (see photodiode *vs*. LED).

in field of the p–n junction. In fact, the photovoltaic effect leads to charge separation that can be utilised to drive electrons through an external circuit for doing an electrical work.

As discussed above, the photovoltaic effect produces an **electromotive force (emf)** and the polarity is indicated in Fig. 6.15.18.1a. If now an external load (*e.g.* resistor) is connected to the illuminated p–n junction as in Fig. 6.15.18.1b, a current will flow from the p-side to the n-side. ***Thus we can conclude that because of the reduction of potential barrier by photovoltaic effect, $I_{forward}$* (forward current) *increases in the circuit.***

The minimum required frequency (ν) of the light depends on the E_g value as : $h\nu = E_g$.

In practice, a silicon solar cell consists of a *wafer* of the p-type silicon with a n-type silicon layer (very thin, about half a micron). Visible light falls on the n-layer and light penetrates into the junction site and creates electron-hole pairs and these are separated as discussed above. To facilitate the absorption of photons, an *anti-reflecting coating* on the n-layer is used. To increase the efficiency, the surface area of the junction should be made very large. However, manufacturing cost of the *Si*-solar cell is very high.

6.15.19 Dye Sensitised Mesoporous Titania Solar Cells (Gratzel Cell) (see the authors' book, 'An Introduction to Nanomaterials and Nanoscience' for details)

In the conventional solar cells, the p–n junction being illuminated produces the electrical energy. But in terms of its maintenance cost and durability, this type of solar cells is not economically viable. **Dye sensitised solar cells** (DSSCs, developed by O'Regan and Grätzel, Gratzel awarded 2010 Millenium Technology Prize) are less efficient in terms of conversion of solar energy into electrical energy but the cell construction cost and maintenance cost are relatively small. DSSCs are environment-friendly and cost-effective to convert solar energy into electrical energy. It represents the **third generation solar cells.** The anode of a DSSC (Fig. 6.15.9.1) consists of a glass plate coated with a **transparent conductive oxide (TCO)** like indium-tin-oxide (ITO). A thin layer of porous nanocrystalline titania (anatase) is applied on the conducting glass plate. The semiconductor TiO_2 (E_g = 3.2 eV ≡ 387 nm of light), film having the mesoporous structure possesses a high surface area. Then the anode is soaked with a dye solution and the adsorbed dye on the TiO_2-film acts as a **photosensitizer**.

The dyes are generally organic pigments including the natural pigments like chlorophyll ($\lambda_{max} \approx$ 400–450 nm and 650–700 nm), beta-carotene (($\lambda_{max} \approx$ 400–500 nm) and anthocyanin ($\lambda_{max} \approx$ 460–550 nm, pH dependent). The Ru^{II}-complexes (photochemically active) can substitute the organic dyes. The **cathode** is a glass plate coated with a thin *Pt*-film that also acts as a catalyst in the electrode reaction. The **electrolyte** uses an iodide/triiodide (I^-/I_3^-) solution. The following steps explain the photo-electrochemical process of DSSCs.

Step I: When the anode is illuminated, the dye is excited. The energy of the HOMO of the dye is higher than that of the non-conducting VB (valence band) of the TiO_2 semiconductor while energy of the LUMO of the dye is higher than that of the CB (conduction band) of the TiO_2 semiconductor (*i.e.* $E_{LUMO\,(dye)} > E_{CB\,(TiO_2)}$; E_{CB} of $TiO_2 \approx$ 4.2 eV and $E_{HOMO\,(dye)} > E_{VB\,(TiO_2)}$, **Type II band alignment**).

$$\text{Excitation of the dye: } S \xrightarrow{h\nu} S^*$$

Step II: The excited electron is then injected into the vacant CB of TiO_2 from the LUMO of the dye.

$$S^* \longrightarrow S^+ + e_{CB}$$

Then the electrons (e_{CB}) are transported from the anode to the cathode through an external circuit for doing an electrical work. Thus the dye acts like a photosensitizer **chlorophyll in photosynthesis.**

Step III: The dye (S) is regenerated from S^+ through the electron capture from I^- present in the electrolyte.

$$S^+ + \frac{3}{2}I^- \longrightarrow S + \frac{1}{2}I_3^-$$

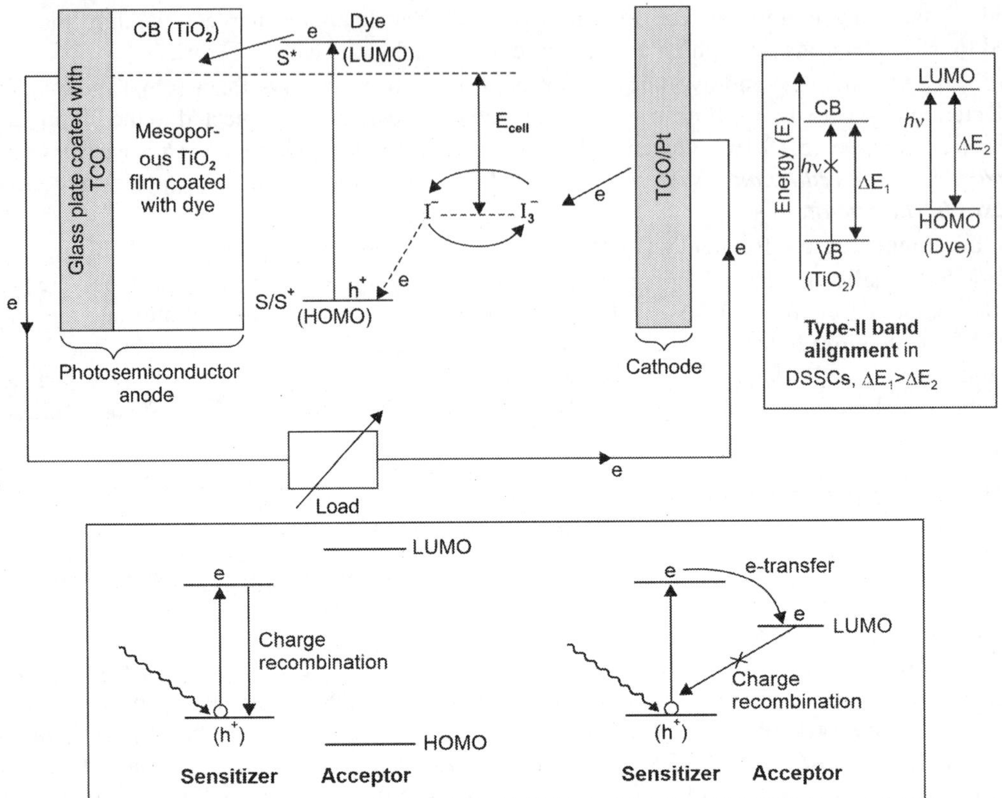

Fig. 6.15.19.1: Schematic representation of a dye-sensitised mesoporous titania photovoltaic cell (solar cell). Inset indicates the condition of **charge transfer** and **charge recombination** (*Ref.* M. Gartzel, *Nature*, **414**, 338, 2001)

Reduction potential: $E_{S^+/S} > E_{I_3^-/I^-}$

Step IV: Through the reduction of I_3^- at the cathode surface, I^- is regenerated.

$$\tfrac{1}{2}I_3^- + e_{CB} \longrightarrow \tfrac{3}{2}I^-$$

Step III and IV combinedly involve the hole transfer to the electrolyte and electron capture by the I_3^-/I^- redox couple at the cathode-electrolyte interface.

The voltage output (E_{cell}) (*i.e.* **highest open circuit voltage**) for doing the external work is approximately given by the energy difference between the energy of CB (E_{CB} expressed in terms of E *vs*. NHE in volt) of TiO_2 semiconductor and redox potential of the redox electrolyte ($E_{cell} \approx 0.7$ V).

Fundamental requirements in DSSCs: (i) E_{LUMO} (S) > E_{CB} (TiO_2) and E_{HOMO}(S) > E_{VB} (TiO_2), *i.e.* **Type II band alignment** leading to the photoexcitation of the dye preferably (*i.e.* not photoexcitation of TiO_2). (ii) $E_{(S^+/S)} > E_{(I_3^-/I^-)}$ (E denoting the reduction potential of the couple).

Advantages of DSSCs (third generation solar cells)**:** The large band gap (3.2 eV) TiO_2 semiconductor (**with a higher probability of charge recombination**) absorbs only the ultraviolet part of the solar radiation and consequently, it has low conversion efficiencies while in a DSSC, the dye (**with a lower HOMO–LUMO gap**) absorbed on mesoporous TiO_2 having a large surface area absorbs far more of the incident solar light (**mainly in the visible range**) to increase the conversion efficiency. Besides this, cost-effective (free from expensive silicon), light weight, easy fabrication, environment-friendly, fairly high solar energy conversion efficiency.

Traditional Si-based solar cell *vs.* DSSC: In the Si-based solar cells, Si-semiconductor (an **expensive material**) has **two functions**: as the source of photoelectrons and to provide the electric field to separate the charges to generate the current. On the other hand, the semiconductor used in DSSC acts as the **charge transporter** and photoelectrons generated in the dyes are injected into the CB of the semiconductor. The movement of these injected photoelectrons produces the current. The dyes get the electrons from the nearby electrolyte rapidly to return back to the ground state. Thus DSSC mimicks the **photosynthetic activity**.

In the traditional Si-based solar cells, the photoexcited electrons recombine with its holes giving off light energy (*i.e.* **electron-hole recombination to produce photons**). Thus all the photoexcited electrons cannot be used to generate the electricity. In DSSCs, injection of electrons into the CB of TiO_2 semiconductor from the photoexcited dye does not introduce any hole in the VB of TiO_2. **Thus there is no possibility of electron-hole recombination within the TiO_2 semiconductor**. The rate of return back of the injected electron to the dye is much slower than the rate at which the dye regains the electron from the surrounding electrolyte. Thus the possibility of loss of the injected photoelectrons into the TiO_2 semiconductor is significantly reduced and thus **these injected photoelectrons can be more efficiently used to generate the electricity**. This is why, DSSCs can work even under the cloudy skies and non-direct sunlight. On the other hand, the traditional Si-based solar cells require the bright sunlight because in poor illumination, the charge carrier mobility to generate the electricity is low and the electron-hole recombination becomes more important.

DSSCs use the thin layers of **conductive plastic** on the front layer and **it favours to radiate away the heat energy more easily** compared to the Si-based cells that use glass boxes for encasing. Besides this, the production/fabrication cost of DSSCs is relatively more economic.

The **major disadvantage** in DSSCs is the use of **liquid electrolyte** whose stability depends on temperature. It may also freeze at low temperatures. The electrolyte solvent may vaporise. The presence of volatile organic compounds (VOCs) in the liquid electrolyte may create the environmental pollution. The replacement of the liquid electrolyte with the solid electrolyte can solve these problems. Active research is going on in this direction.

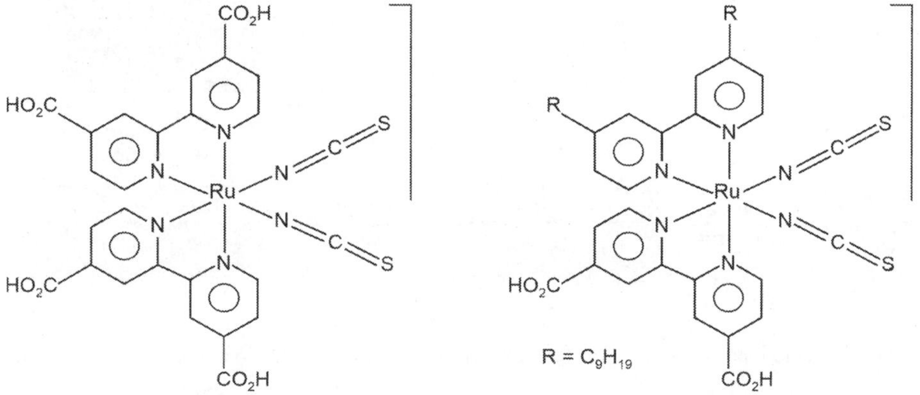

cis-Bis(isothiocynato) *bis*(2,2′-bipyridyl–4,4′–dicarboxylic acid)ruthenium(II)

cis-Bis(isothiocynato) (2,2′-bipyridyl–4,4′–dicarboxylic acid) (–4,4′–dinonyl –2,2′-bipyridyl)ruthenium(II)

Fig. 6.15.19.2: Some representative examples of Ru(II)-based dyes used in DSSCs. They show **MLCT** (*i.e.* $4d \rightarrow \pi^*$) at the relatively lower energies and **LCCT** (ligand centred charge transfer, $\pi \rightarrow \pi^*$) at higher energies (see Sec. 16.15.13).

Photoexcitation of the Ru(II) complex produces the excited state of Ru-complex (see Sec. 6.15.13) and this photoexcited electron is transferred to the CB of TiO_2 semiconductor. This leads to **charge**

separation: negative charge on the titania phase and positive charge on the surface absorbed Ru^{3+} which is subsequently reduced rapidly by I^- present in the surrounding electrolyte.

6.15.20 Chemistry of Photocopying Process

Amorphous selenium (*Se*) (E_g = 1.8 eV) is a photoconductor and for this purpose, visible light can be used. In a photocopying machine, a thin film of **amorphous selenium** is deposited by using the **vapour deposition** technique on a metal drum (generally made of aluminium). On the *Se*-surface, an electrostatic **positive charge** is generated by a **corona discharge.** In dark, selenium is nonconductor and the electrostatic charge does not dissipate. The page to be photocopied is then imaged on the *Se*-surface and the portions on which light falls becomes photo-conductive. The incident light creates the **electron-hole pairs** which are separated as follows:

The electrons are attracted towards the electrostatic positive charges residing on the *Se*-surface and the positive holes move inwards towards the metallic drum which is earthed.

Thus, the portions on which light falls become neutralised while the dark portions of the *Se*-surface bears the electrostatic positive charges. Now the *Se*-surface is dusted with the **negatively charged powdered carbon black (called toner)** which gets adhered on the positively charged placed on the *Se*-surface. Thus the adhered ink is then transferred to a clean paper with the help of a second corona discharge. Then the paper is heated and the image is fixed. The steps are illustrated in Fig. 16.15.20.1.

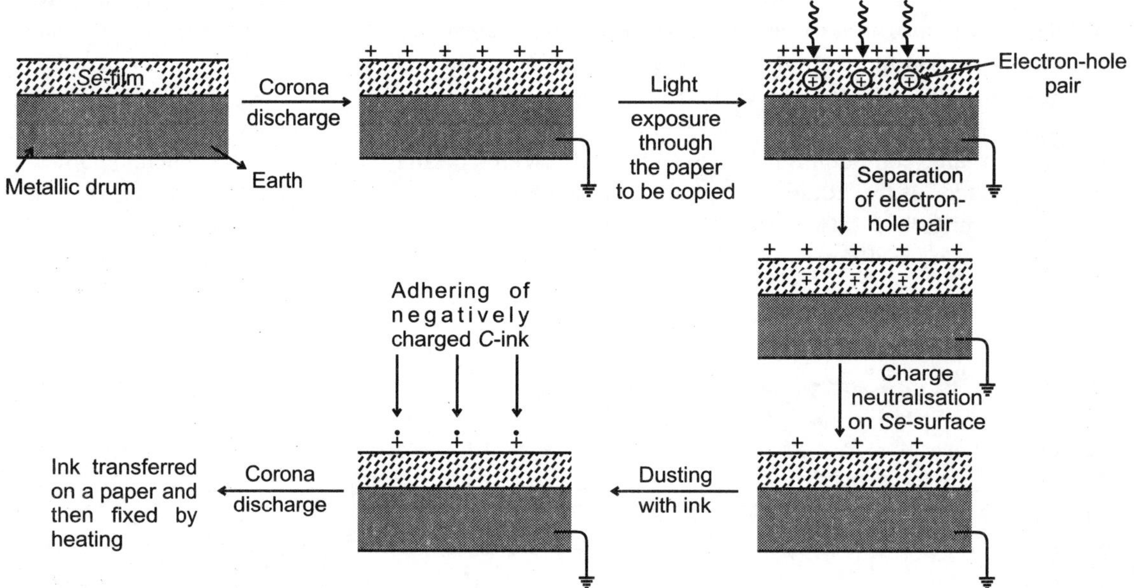

Fig. 6.15.20.1: Schematic representation of different steps of photocopying (**xeroxing**) process. The metallic drum is cylindrical.

EXERCISE 6 (PART II)

A. General Type Questions

1. What do you mean by photoexcitation? Discuss the fate of the photochemically excited species.
2. Discuss the characteristic features of predissociation.
3. What are the different photophysical pathways of deactivation of the electronically excited molecules. Discuss their characteristic features.

4. Describe the **Jablonski diagram.**
5. Discuss the mechanism of internal conversion (IC) and intersystem crossing (ISC).
6. What are the characteristic features of fluorescence emission? Discuss the factors favouring the fluorescence emission.
7. What do you mean by delayed fluorescence? How can you distinguish between the **delayed fluorescence** and **phosphorescence**?
8. What are the characteristic features of fluorescence and phosphorescence emission.
9. Compare the characteristic features of fluorescence and phosphorescence emission.
10. What do you mean by photostationary state? Write down the **Stern-Volmer equation** and its importance.
11. What do you mean by primary quantum yield, fluorescence quantum yield and phosphorescence quantum yield?
12. Discuss the characteristic features of **resonance energy transfer** (RET). How does it work in photosynthesis.
13. Give the different types of photochemical reactions to deactivate the photochemically excited species.
    - What are the primary and secondry processes of photochemical reactions.
    - Comment on the **temperature dependence of a photochemical reaction**.
14. Discuss the characteristic features of **photosensitized reactions.** Illustrate with examples with special reference to photosynthesis.
15. Discuss the chemistry and uses of $[Ru(bpy)_3]^{2+}$ and TiO_2 semiconductor as the photocatalysts.
16. What do you mean by **artificial photosynthesis**? How can you use $[Ru(bpy)_3]^{2+}$ in artificial photosynthesis?
17. How can you use TiO_2 semiconductor in photoelectrolysis of water and oxidation of organic matter?
    - What do you mean by **Pt-pop**? How does it photochemically **activate the C—H bonds**?
18. What do you mean by **photogalvanic cell** (PGC)? Discuss the electrochemistry of thionine dye based PG cell. How can you construct a PG cell by using $[Ru(bpy)_3]^{2+}$?
19. Compare the photocatalysis by rutile TiO_2 and anatase TiO_2.
20. Discuss the possibility of direct photochemical reduction of dinitrogen (N_2).
21. Discuss the charge transfer band and redox stability of some compounds.
22. Discuss the chemistry of photography.
23. Discuss the working principle of light emitting diode (LED).
24. Discuss the mechanism of **photocopying (xeroxing).**
25. Discuss the working principle of Si-based solar cells.
26. Discuss the working principle of **dye sensitised solar cells** (DSSCs).
27. Compare the Si-based solar cells and DSSCs in terms of working principle, advantages and disadvantages.
28. Discuss the mechanism of photosubstitution in metal carbonyls like $[M(CO)_6]$ (M = Cr, Mo, W) in terms of CFT.
29. Explain the following two possibilities:
    (i) $[M(CO)_5L] + Nu \xrightarrow{h\nu} [M(CO)_5(Nu)] + L$
    (ii) $[M(CO)_5L] + Nu \xrightarrow{h\nu} [M(CO)_4(Nu)L] + CO$
30. Explain the mechanism of photochemical reaction of polynuclear carbonyls like $[M_3(CO)_{12}]$ (M = Fe, Ru, Os).

31. Discuss the possible mechanisms of photochemical reaction of polynuclear carbonyls like $[M_3(CO)_{12}]$ (M = Fe, Ru, Os).
32. Discuss the chemistry behind **blue printing** by using $K_3[Fe(C_2O_4)_3]$.
33. Discuss the protective role of carotenoids in photosynthesis.
34. Discuss the photoexcited species experiencing the competitive photophysical and photochemical deactivations.
35. Compare the characteristic features of photochemical and thermal reactions.
    - What are the primary and secondary processes in photochemical reactions.
    - Define the photochemical yield of a photochemical reaction. Explain the reasons of **low and high values of quantum yield**.
36. Discuss and compare the different modes of photoexcitations and photochemical reactions in the transition metal complexes.
37. Write notes on:
    (i) Franck–Condon principle, (ii) predissociation, (iii) internal conversion (IC), (iv) intersystem crossing (ISC), (v) vibrational relaxation (VR), (vi) external conversion (EC), (vii) Kasha's rule, (viii) Vavilov's rule, (ix) Levschin's rule, (x) Heavy atom effect in fluorescence and phosphorescence, (xii) Adamson's rule.
38. What do you mean by the photochemical reaction? How does it differ from the thermal reaction?
39. What are the different types of photochemical reaction? What do you mean by the delayed and prompt photochemical reaction?
40. What are the ways of photochemical excitation in the coordination compounds?
41. Discuss the photochemical substitution reaction in the metal carbonyls.
42. Compare the photochemical reactions of the Co(III) and Cr(III) complexes.
43. Discuss the mechanism of photolysis of $[M(ox)_3]^{3-}$ (M = Co, Fe).
44. Discuss the photochemical reactions of the Pt(II), Pt(IV), Fe(II) and Fe(III) complexes.
45. Discuss the photochemistry of $[Ru(bpy)_3]^{2+}$. Discuss the photolysis of water by using the photocatalyst $[Ru(bpy)_3]^{2+}$.
46. Discuss the use of TiO_2 as a green photocatalyst. Why is it considered as a green catalyst in removing the air pollution?
    - Discuss the principle of photoelectrolysis of water by using the TiO_2 semiconductor.
47. State the Adamson's rules in connection with the photosubstitution processes in the Cr(III)- and Co(III)- complexes. Explain the Rules in terms of CFT and illustrate with the suitable examples.
48. Comment on the photochemical stability of $K_3[Fe(CN)_6]$ and $K_4[Fe(CN)_6]$ aqueous solutions.
49. Compare the photochemical stability of $[M(ox)_3]^{3-}$ for M = Fe, Co and Cr.

B. Explain the following statements

1. Generally the *ligand field* (LF) excitation in the coordination compounds leads to the nonredox processes, *e.g.* substitution, isomerization, racemization while the *charge transfer* (CT) excitation leads to the redox reactions.
    - Photochemically, it is possible to have the products which are thermodynamically and kinetically inaccessible in thermal reactions.
    - In the photochemical reactions, the Co(III) complexes require no thermal activation but the analogous Cr(III) complexes require a thermal activation.
    - In the photochemical reactions of the Cr(III)- complexes, the quantum yield increases with the increase of temperature.
2. Charge transfer excitation in $[CoCl(NH_3)_5]^{2+}$ leads to aquation while charge transfer excitation in $[Co(C_2O_4)_3]^{3-}$ leads to redox decomposition.

3. $\left[Co^{III}(NH_3)_5X\right]^{2+} \xrightarrow[(H_2O)]{h\nu}$ Product.

   The nature of product depends largely on the nature of X.
   - For X = Cl, it leads to aquation while for X = I, it leads to redox decomposition.
   - The quantum yield for the photochemical aquation of $[CoCl(NH_3)_5]^{2+}$ increases with the energy of the irradiating radiation.
4. The nature of photosubstitution in $[Cr^{III}(NH_3)_x(Y)_y]^{n+}$ ($x + y = 6$) is different from the thermal substitution process.
   - Adamson's rule can predict the nature of product in the photosubstitution of the Cr(III) and Co(III) complexes.
5. In the photochemical reaction of $[Cr^{III}(NH_3)_xX_y]^{n+}$ ($x + y = 6$), the reactive intermediate state may be characterized by $^4T_{2g}$ and/or ^2E_g state.
6. Photochemical aquation of $[Co(CN)_5X]^{3-}$:

   $\left[Co(CN)_5(X)\right]^{3-} \xrightarrow[(\lambda=370\text{ nm})]{h\nu} \left[Co(CN)_5(OH_2)\right]^{2-} + X^-$; Quantum Yield: X = I > Br > Cl.

- $\left[Co(NH_3)_5(N_3)\right]^{2+} \xrightarrow{h\nu(LF)} \left[Co(NH_3)_4(N_3)(OH_2)\right]^{2+} + NH_3$; **(Adamson's rule)**
- $\left[Co(NH_3)_5(N_3)\right]^{2+} \xrightarrow{h\nu(CT)} Co(II) + NH_3 + N_2$
- $\left[CrBr(NH_3)_5\right]^{2+} \xrightarrow[\text{(dark reaction)}]{k_BT} \left[Cr(NH_3)_5(OH_2)\right]^{3+} + Br^-$
- $\left[CrBr(NH_3)_5\right]^{2+} \xrightarrow[\text{(photochemical reaction)}]{h\nu} \left[CrBr(NH_3)_4(OH_2)\right]^{2+} + NH_3$; **(Adamson's rule)**.

7. Photochemical reactions of the Co(III) complexes are generally faster than those of the Cr(III) complexes though the reverse is true for the thermal reactions.
   - In the photochemical reactions of the Cr(III)-complexes, the quantum yield depends on the temperature, but such a situation does not prevail for the Co(III)-complexes.
   - Primary photochemical process in the Cr(III)-complexes generates a relatively long-lived reactive intermediate which requires a thermal activation.
8. Photochemical reaction of $[Cr(NH_3)_x(OH_2)_{6-x}]^{3+}$.

   $$\left[Cr(NH_3)_x(OH_2)_{6-x}\right]^{3+} \xrightarrow[(H_2O)]{h\nu} \left[Cr(NH_3)_{x-1}(OH_2)_{6-x+1}\right]^{3+}$$

   The quantum yield increases with the increase of x
   - The quantum yield is maximum for $[Cr(NH_3)_6]^{3+}$ (*i.e.* $x = 6$) and it is minimum for $[Cr(H_2O)_6]^{3+}$ (*i.e.* $x = 0$).
   - For the Cr(III)-complexes, the excited states ^2E_g and $^4T_{2g}$ favour different pathways of photosubstitution.
9. Photolysis of $[M(C_2O_4)_3]^{3-}$ (M = Co, Mn, Fe):
   - The quantum yield depends on the λ of the irradiation.
   - The quantum yield is maximum for the CT excitation and it is minimum for the LF excitation.
   - The said complexes experience the photo-redox decomposition while the corresponding Cr(III) complex experiences the photoracemization.
10. On standing, $[Fe(CN)_6]^{3-}$ produces a blue stain.
11. The photochemical redox process in $[M(OH_2)_x]^{n+}$ depends on the nature of the charge transfer excitation, *i.e.* M→ L or L → M.
12. Photolysis of $[M(CN)_8]^{3-}$ (M = Mo, W) in aqueous solution shows the quantum yield > 1.
13. $[PtCl_6]^{2-}$ and $[PtBr_6]^{2-}$ can experience the photochemical ligand exchange and substitution reactions where quantum yield is very often greater than unity.

14. Fluorene shows the better fluorescence emission than biphenyl.
15. **Phenolphthalein** is nonfluorescent while **fluorescein** is fluorescent in an alkaline condition.
16. Carotenoids play a protective role in photosynthesis.
17. **Artificial photosynthesis** can be carried out by using $[Ru(bpy)_3]^{2+}$.
18. Photochemical oxidation of water by $[Ru(bpy)_3]^{2+}$ is possible in presence of a sacrifical oxidant.
19. $[Ru(bpy)_3]^{2+}$ can be used to design a photogalvanic cell (PGC) to harness the solar energy.
20. TiO_2 as a photocatalyst can be used for electrolysis of water.
21. Photochemical reduction of N_2 is possible in presence of a suitable photocatalyst.
22. Photochemistry of AgBr controls the chemistry of photography and photochromatic sunglasses.
23. $[M(CO)_6] + L \longrightarrow [M(CO)_5L] + CO$, M = Cr, Mo, W. Both the thermal and photosubstitution reactions go on through the **dissociative path** but the photosubstitution is about 10^{16} times faster.
24. $[M(CO)_6] + L\ (\text{excess}) \xrightarrow{h\nu}$ products (photosubstitution). The extent of substitution depends on the relative strength of the M—CO and M—L bonds.
25. $[Mn_2(CO)_{10}] \xrightarrow{h\nu,\ RCl} 2[MnCl(CO)_5]\ (R = CCl_3, Ph_3C, PhCH_2)$

$$[M_2(\eta^5\text{-Cp})_2(CO)_6] \xrightarrow{h\nu,\ RCl} 2[Cl\text{—}M(\eta^5\text{-Cp})(CO)_3]$$

$$[Mn_2(CO)_{10}] + [Re_2(CO)_{10}] \xrightarrow{h\nu} 2[Mn\,Re(CO)_{10}]$$

$$\underset{(M = Mo,\ W)}{[M_2(\eta^5\text{–Cp})_2(CO)_6]} + \underset{(M' = Mn,\ Re)}{[M'_2(CO)_{10}]} \underset{h\nu}{\overset{h\nu}{\rightleftharpoons}} 2[(\eta^5\text{–Cp})(CO)_3M\text{—}M'(CO)_5]$$

All the above reactions occur through the photoinduced metal–metal homolytic bond cleavage.
26. $[Mn_2(CO)_{10}] + PPh_3 \xrightarrow{h\nu} [Mn_2(CO)_8(PPh_3)_2]$ (major product) + $[Mn_2(CO)_9(PPh_3)]$ (minor product)

The above photosubstitution reaction occurs through the photoinduced metal–metal bond cleavage.
27 $[M(CO)_6] + py \xrightarrow{h\nu} [M(CO)_5(py)] + CO$ (at the relatively longer wavelength)
$[M(CO)_6] + py \xrightarrow[-CO]{h\nu} [M(CO)_5(py)] + cis\text{-}[M(CO)_4(py)_2]$ (at the shorter wavelength)

Quantum yield (ϕ) of the product, *cis*-$[M(CO)_4(py)_2]$, is wavelength dependent (*i.e.* **higher ϕ value for the shorter wavelength)** (M = Mo, W).
28. In the dye sensitised solar cells (DSSCs), there is no possibility of hole-electron recombination but this recombination is an inherent drawback of the Si-based solar cells.
29. DSSCs can work even under the cloudy sky but the traditional Si-based solar cells require the bright sunlight.
30. TiO_2 is a **green photocatalyst** and it can be used for self-cleaning.
31. TiO_2 as a photocatalyst can be used for the photoelectrolysis of water.
32. The mechanism of DSSCs is comparable to that of photosynthesis.
33. Quantum yield of the photochemical reactions may vary in a wide range.
34. The internal conversion (IC), $S_1 \rightarrow S_0$ is less probable than $S_2 \rightarrow S_1$.
35. The activation energy of the photochemical reactions may be very low as well as very high.
    - Temperature dependence of a photochemical reaction rate depends on the nature of **thermal secondary reactions.**
36. There may be exceptions to **Kasha'a rule** for the fluorescence emission.

37. Photodissociation bond dissociation energy very often is much larger than the thermodynamic bond dissociation energy.
38. Direct photodissociation of H_2 requires the radiation of wavelength $\leq$ 84.5 nm but in presence of Hg-vapour photosensitizer, H_2 dissociates at higher wavelength of radiation.
39. Introduction of an electron pushing group into the aromatic platform favours the fluorescence emission.
    - C_6H_5–NH_2 is fluorescent but C_6H_5–NH_3^+ is nonfluorescent.
40. Photoisomerisation of the **retinal chromophore** is the key step of our vision process.
41. Fluorescent efficiency order runs as: $\pi^* \rightarrow \pi > \pi^* \rightarrow n$.
42. Fusion of a benzene ring makes the nonfluorescent heterocycles (*e.g.* pyridine, pyrrole, etc.) fluorescent.
43. Fusion of a large number of benzene rings as in anthracene, phenanthrene, etc. favours the fluorescence.
44. Compared to the free chelating ligand 8-hydroxyquinoline, its complexes are more fluorescent.

CHAPTER 7

Electronic Spectra of Metal Complexes

7.1 INTRODUCTION TO ELECTROMAGNETIC RADIATION AND ABSORPTION SPECTRUM

When an electromagnetic radiation is allowed to pass through a matter, the radiation may be partly absorbed and the remaining portion of the radiation will be transmitted. The portion absorbed can be identified through the analysis of the radiation (obtained after interaction with the matter) by a prism. This analysis shows a spectrum with gaps at certain wavelengths where the interacting matter has absorbed the incident electromagnetic radiation. Such an analysed spectrum is called **absorption spectrum** where absorption versus wavelength or frequency is plotted (Fig. 7.1.1).

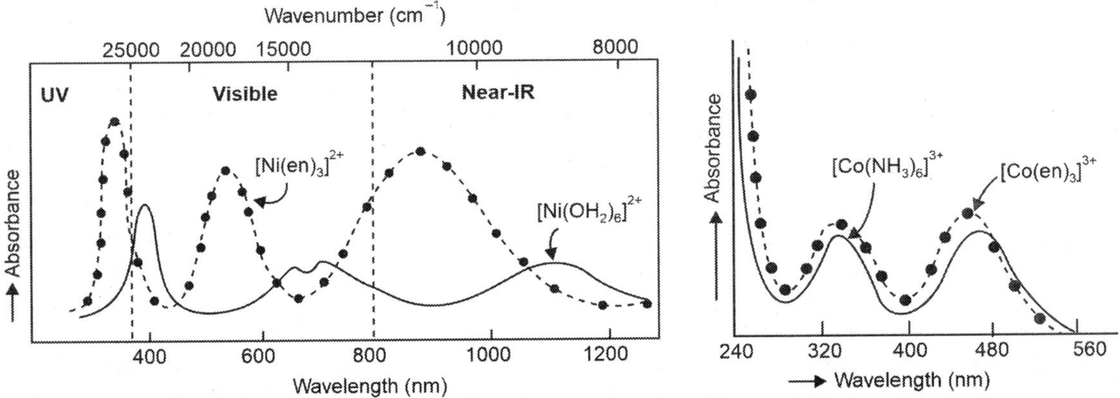

Fig. 7.1.1: Representation of electronic absorption spectra of some representative complexes like $[Ni(OH_2)_6]^{2+}$, $[Ni(en)_3]^{2+}$, $[Co(NH_3)_6]^{3+}$ and $[Co(en)_3]^{3+}$ in aqueous solution. (*See* Fig. 7.14.4.2 for the Ni(II)-complexes; Fig. 7.15.1.3 for the Co(III) complexes).

Depending on the nature of interaction of the electromagnetic radiation with matter, the absorption peaks may be sharp or broad.

During the process of absorption, the molecules or atoms of the matter are excited from a lower energy state to a higher energy state. If the molecules or atoms return to the lower energy state by radiating the excess energy, then an emission spectrum is generated.

The different regions of electromagnetic radiation and the process of interaction with matter are shown in Table 7.1.1.

Table 7.1.1: Range of electromagnetic radiation and process of interaction with matter.

Spectral region	*Wavelength* (nm)	*Wavenumber* (cm^{-1})	*Frequency* ($\times 10^{-12}$ Hz)	*Process of interaction*
(i) γ–radiation	< 0.1	$> 10^8$	$> 3 \times 10^6$	Nuclear reactions
(ii) X–rays	0.1–10	10^6–10^8	3×10^4–3×10^6	Transition of inner electrons in atoms or molecules
(iii) Ultraviolet	10–360	2.5×10^4–1×10^6	8×10^2–3×10^4	Transition of outer electrons
(iv) Visible	360–800	12.5×10^3–2.5×10^4	4×10^2–8×10^2	Transition of outer electrons
(v) Near infrared	800–2500	3×10^3–12.5×10^3	100–400	Bond vibration
(vi) Infrared	3×10^3–3×10^4	330–3300	10–100	Bond vibration
(vii) Far-infrared	3×10^4–3×10^5	10–330	1–10	Molecular rotation
(viii) Microwave	3×10^5–1×10^9	0.01–10	3×10^{-4} – 1.0	Molecular rotation
(ix) Radio-frequency	1×10^9–1×10^{13}	1×10^{-6}–1×10^{-2}	1×10^{-8} – 3×10^{-4}	Change of spin (NMR, ESR)

$c = \nu\lambda = 3 \times 10^{10}$ cm s^{-1}; $\overline{\nu}$ (wavenumber) = $1/\lambda$

A. Energy associated with the radiation

- $E = h\nu = hc/\lambda = hc\overline{\nu}$ *i.e.* $\overline{\nu} \Rightarrow E = hc\overline{\nu}$
- $\overline{\nu} = \dfrac{1}{\lambda}$, 1 kilokaiser (kK) = 1000 cm^{-1}
- 1 kK per molecule $= 10^3$ cm^{-1} molecule^{-1} $\equiv hc \times 1000$ cm^{-1} molecule^{-1}

$$= 6.626 \times 10^{-34}\ \text{J s} \times 3 \times 10^8\ \text{m s}^{-1} \times 10^3 \times 10^2\ \text{m}^{-1}\ \text{molecule}^{-1}$$
$$= 1.987 \times 10^{-20}\ \text{J molecule}^{-1}$$
$$= 1.987 \times 10^{-20} \times N_0\ \text{J mol}^{-1} \approx 11.96\ \text{kJ mol}^{-1}$$

$$\boxed{i.e.\ 1\ \text{kK} = 1000\ \text{cm}^{-1} = 11.96\ \text{kJ mol}^{-1}}$$

- 1 eV = 1.602×10^{-19} J; 1 eV molecule^{-1} = 1.602×10^{-19} J $\times N_0$ mol^{-1}

$$= 96.484\ \text{kJ mol}^{-1}$$

Similarly, 1 eV molecule^{-1} = 23.06 kcal mol^{-1}

- Say, λ = 200 nm = 200×10^{-9} m

$$\nu = \frac{c}{\lambda} = \frac{3 \times 10^8\ \text{m s}^{-1}}{200 \times 10^{-9}\ \text{m}} = 1.5 \times 10^{15}\ \text{Hz}$$

$$E = hc\overline{\nu} = h\nu = 6.626 \times 10^{-34}\ \text{J s} \times 1.5 \times 10^{15}\ \text{Hz}$$
$$= 9.94 \times 10^{-19}\ \text{J} = 9.94 \times 10^{-19} \times N_0\ \text{J mol}^{-1}$$
$$= 9.94 \times 10^{-19} \times 6.023 \times 10^{23}\ \text{J mol}^{-1}$$
$$= \frac{60.2 \times 10^4}{10^3}\ \text{kJ mol}^{-1} = 602\ \text{kJ mol}^{-1}$$

i.e. electromagnetic radiation of 200 nm corresponds to 602 kJ mol^{-1}

$$E = \frac{N_o hc}{\lambda} = \frac{(6.022 \times 10^{23}\ \text{mol}^{-1})(6.626 \times 10^{-34}\ \text{J s})(3 \times 10^8\ \text{m s}^{-1})}{\lambda\ \text{m}}$$

$$\approx \frac{1.2 \times 10^5}{\lambda}\ \text{kJ mol}^{-1},\ \lambda \text{ in nm},\ 1\ \text{nm} = 10^{-9}\ \text{m}$$

B. Colour observed

When the visible light (380–750 nm) is absorbed by a matter, the observed colour of the matter is given by the transmitted light which is not absorbed. ***Thus we see the complementary colour***. This is illustrated below.

λ (nm)	*$\bar{\nu}$ (kK)*	*Colour of the absorbed light*	*Colour of the transmitted light (i.e. complementary colour)*
380–450	26.3–22.2	Violet	Yellow-Green
450–495	22.2–20.0	Blue	Yellow
495–570	20.2–17.5	Green	Purple
570–590	17.5–16.9	Yellow	Blue
590–620	16.9–16.1	Orange	Green-blue
620–750	16.1–12.5	Red	Blue-green

Thus the yellow colour of K_2CrO_4 solution is due to the absorption of blue light and the blue colour of $CuSO_4$ solution arises due to the absorption of yellow colour.

The relationship between the absorbed colour and the observed colour (*i.e.* complimentary colour) is represented in **artist's wheel** (Fig. 7.1.2). According to the additive **RGB model** of colour, appropriate combination of the **three different primary colours**, red (R), green (G) and blue (B) can produce all colours. For example, B + R = M (magenta), R + G = Y (yellow), B + G = C (cyan means 'brilliant greenish-blue', *i.e.* turquoise), R + G + B = white.

By considering the absorption spectrum and the **artist's wheel,** we can predict the colour of a substance. The absorption spectra of $CuSO_4$ in presence of excess NH_3 solution and aqueous solution of $NiSO_4$ are given in Fig. 7.1.3.

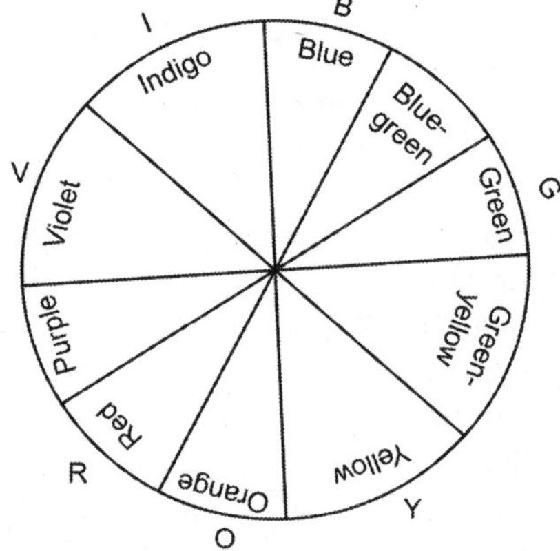

Fig. 7.1.2 Representation of the **artist's wheel** indicating the colours and their corresponding complementary colour

● **Cu^{2+} and excess NH_3 solution:** The strong absorption occurs above 450 nm and the absorption maximum lies in the range of 600 nm (*i.e.* yellow light). Thus it looks blue. In fact, the solution absorbs minimum in the range 380 – 500 nm (*i.e.* blue light).

● **Aqueous solution of $NiSO_4$:** Spectrum of $NiSO_4$ in aqueous solution represents the electronic transitions in $[Ni(OH_2)_6]^{3+}$. The transitions (*cf.* Figs. 7.1.1, 3) are:

strongest absorption band at 26,000 cm^{-1}; the band at 14,000 cm^{-1} is actually a doublet one (15,200 cm^{-1}; 13,800 cm^{-1}) and the third band at 8,500 cm^{-1} (near infrared region).

Thus the absorption occurs in three regions: (i) strongest band at around 26,000 cm^{-1} (*i.e.* violet region) giving the **complementary colour: yellow-green;** (ii) second absorption band at around 14,000 cm^{-1} (*i.e.* red region) giving the **complementary colour: blue-green;** (iii) third absorption band at around near infrared region (at about 8,000 cm^{-1}) contributes nothing to the observed colour.

Thus, the resultant colour is determined by the absorption bands at 26,000 cm^{-1} and 14,000 cm^{-1} and the **solution looks green (yellowish).** The solution absorbs minimum in the region of blue-green (*i.e.* around 20,000 cm^{-1}).

Spectra in solids — polarisation studies, diffuse reflectance spectrum: Generally, molecules of lower symmetry (compared with that of cubic symmetry) possess the **dichroic properties,** *i.e.* **anisotropic absorption spectra.** It arises so when the electronic transition is allowed in certain directions. A beam

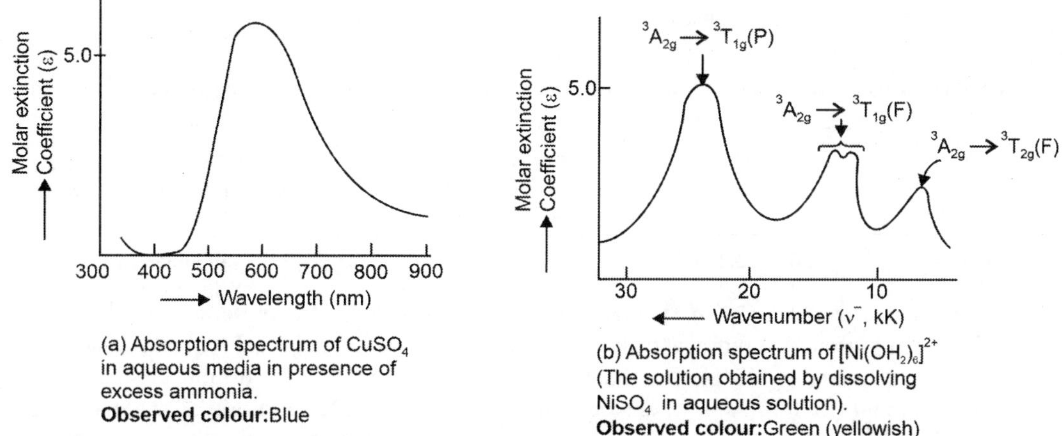

(a) Absorption spectrum of $CuSO_4$ in aqueous media in presence of excess ammonia.
Observed colour: Blue

(b) Absorption spectrum of $[Ni(OH_2)_6]^{2+}$ (The solution obtained by dissolving $NiSO_4$ in aqueous solution).
Observed colour: Green (yellowish)

Fig. 7.1.3: Understanding of the observed colour of (a) aqueous slution of $CuSO_4$ in presence of excess NH_3; (b) aqueous solution of $NiSO_4$ in terms of their absorption spectra (*cf.* Fig. 7.1.1).

of polarised light is used to record the absorption spectra in different crystal orientations. By knowing the orientation of the molecules with respect to the crystal axes, polarisation properties of the electronic spectra are determined. **Thus the single crystal may be studied with the polarised light in different directions to identify the allowed electronic transitions in different directions.** *Obviously, the question of polarisation does not arise if the studies are carried out in a solution phase because in the solution phase, the molecules are randomly oriented with respect to the incident light beam.*

For recording the **diffuse reflectance spectrum,** the sample is ground finely and is generally mixed with a suitable white material like $MgCO_3$ powder and the light reflected from the mixture is examined to identify the absorption bands. It is not possible to obtain the molar extinction coefficient in this method of study. This method is generally practised when the sample is insoluble or unstable in solution.

Note: (Water colour due to vibrational transition, *J. Chem. Edu.*, **70**, 612–614, 1993): Vibrational transitions generally occur in the IR range, but **in the hydrogen bonded condensed phase** like liquid state and solid state of H_2O, some vibrational transitions occur in the red end of visible range (*i.e.* bands at 760 nm and two weak bands at 660 nm and 605 nm) and we see the **light** blue colour, **(blue colour of oceans)** which is the complementary colour of red (very small molar extinction coefficient, **water vibrationally coloured**—unique example, see Vol. 7 for details).

7.2 LAWS OF LIGHT ABSORPTION: LAMBERT-BEER LAW

When a beam of light is incident on a homogeneous solution, a part of the light is absorbed by the solution and the rest is transmitted (ignoring the reflection by the solution). The absorption of the incident light of a particular wavelength is governed by Lambert-Beer law which may be expressed in the following form.

$$\log\left(\frac{I}{I_0}\right) = \log T = -\varepsilon cl, \text{ or } A = \log\left(\frac{1}{T}\right) = \varepsilon cl$$

where I_0 = intensity of the incident radiation of a particular wavelength. I = intensity of the transmitted light of the same wavelenth. l = path-length through which the incident radiation passes through the absorbing solution.

$T = \frac{I}{I_0}$ = transmittance (dimensionless), $A = \log(1/T) = \log\left(\frac{I_0}{I}\right)$ = absorbance or optical density (OD) **(dimensionless)**.

ε = molar absorptivity or molar absorption coefficient or molar extinction coefficient.
dimension: $(\text{concentration})^{-1}\ (\text{length})^{-1} = (\text{mol dm}^{-3})^{-1}(\text{cm})^{-1} = \text{dm}^3\ \text{mol}^{-1}\ \text{cm}^{-1}$.

$$= 10^3\ \text{cm}^3\ \text{mol}^{-1}\ \text{cm}^{-1} = \text{cm}^2 \times \frac{1}{10^{-3}\ \text{mol}} = \text{cm}^2\ \text{mmol}^{-1}$$

The higher value of ε indicates the higher absorbance.

7.3 TYPES OF ELECTRONIC SPECTRA OF THE TRANSITION METAL COMPLEXES

(i) **Ligand Field Spectra,** *i.e.* ***d–d*** **spectra for the transition metal complexes:** For the transition metal ions, in presence of the ligands, the degeneracy of the *d*–orbitals is lifted and these split into different sets of energy levels. In the light of CFT, these different orbital sets are the pure *d*-orbitals though LFT or ACFT considers the incorporation of ligand orbital character into the metal orbital sets. However, these orbitals are predominantly enriched in the character of the *d*–orbitals of metals. Thus both the CFT and LFT describe roughly the same nature of these orbitals. The electron transition occurs among these energetically nondegenerate orbital sets. In terms of the crystal field terminology, these transitions at the metal centre are generally described as the *d–d* transitions.

These *d–d* transitions may cover the range infrared, visible and ultraviolet regions but the bands in the infrared region are not experimentally accessible and the transitions in the ultraviolet region are very often overshadowed by the charge transfer (CT) bands. In the octahedral geometry, the transition may be simply represented as follows:

$$d^n\text{–system: } t_{2g}^{n-m}\ e_g^{m} \rightarrow t_{2g}^{n-(m+1)}\ e_g^{m+1}.$$

Number of absorption peaks for the $t_{2g} \rightarrow e_g$ electron transition: This single transition should involve a fixed amount of energy change and a **single absorption peak corresponding to ν as given by the following relation is expected:**

$$\Delta E = h\nu = \Delta_0,\ i.e.\ \nu = \Delta_0/h$$

But for a complex having **more than one electron**, the $t_{2g} \rightarrow e_g$ single electron transition shows **more than one ligand field absorption peak**. This apparently confronts. The single transition splits into the multiple absorption peaks due to the **electron-electron repulsion**. For example, for the $t_{2g}^2 \rightarrow t_{2g}^1 e_g^1$ transition, the **excited state $t_{2g}^1 e_g^1$ gives two energy states where the electron-electron repulsion energy differs**. This is why, the transition gives rise to two absorption peaks. This aspect has been discussed in detail in Sec. 7.9.4.

(ii) **Charge Transfer (CT) Spectra:** In such transitions, an electron transition occurs between the two different moieties, *i.e.* metal and ligand in the complex. These are of two types:
metal to ligand ($M \rightarrow L$), *i.e.* **MLCT**; ligand to metal ($L \rightarrow M$), *i.e.* **LMCT**

In the $M \rightarrow L$ transitions, the metal electrons suffer a transition into the vacant ligand orbital while in the $L \rightarrow M$ transitions, the ligand electrons move into the vacant metal orbitals. This type of electron movement is equivalent to the redox process (*i.e.* in the $M \rightarrow L$ transitions, the metal centre is oxidised and the ligand is reduced while in the $L \rightarrow M$ transitions, the ligand is oxidised and the metal centre is reduced). This is why, the charge transfer (CT) bands are also described as the **redox bands.** Sometimes, the CT band arises in the ion-pairs consisting of an oxidisable ion and a reducible ion as in the ion pair $[Co(NH_3)_6]^{3+}\cdot I^-$ and such bands are called the **IPCT bands** (Sec. 7.21.3). Metal to metal charge transfer **(MMCT)** bands are important in the mixed-valence compounds (*e.g.* **Prussian blue**) (see Vol. 4).

In terms of the CFT model, origin of a CT band in a coordination compound cannot be explained but the MOT can explain this. In the terms of MOT, in the $M \rightarrow L$ transitions, the electrons move from the MOs having predominantly the metal orbital character into the vacant MOs primarily constructed by the ligand orbitals. Similarly, in the $L \rightarrow M$CT band, the electrons move from the MOs primarily centred on the ligands to the MOs primarily centred on the metal.

(iii) **Ligand Spectra (*i.e.* Ligand centred transitions, intraligand CT transitions):** Such spectra arise due to the characteristic absorptions by the ligands themselves, *e.g.* $n \rightarrow \sigma^*$ transition as for water as a coordinated ligand; $\pi \rightarrow \pi^*$ and $n \rightarrow \pi^*$ transition as for the unsaturated ligands. However, these transitions may be partly modified in the complex. The $\pi \rightarrow \pi^*$ ligand centred transitions in the **porphyrin complexes like chlorophyll** (Mg-complex), **hemoglobin and myoglobin** (Fe-complexes) give their characteristic colours.
Besides the intraligand CT transitions, in some cases possessing both the oxidisable and reducible ligands, the **interligand CT transitions** may appear. These aspects are discussed in Sec. 6.15.1.

(iv) **Counter-ion spectra:** Sometimes, the characteristic absorption by the counter-ions residing outside the coordination sphere gives the counter-ion spectra (different from the **IPCT bands**). For example, the yellow coloured chromate ion (CrO_4^{2-}) forming an outer-sphere (OS) complex gives the colour of the OS complex.

7.4 SELECTION RULES FOR THE ELECTRONIC TRANSITIONS

7.4.1 Selection Rules and Illustrations

● **Laporte or orbital selection rule in electric dipole transitions :** The electronic transitions having $\Delta l = \pm 1$ are only allowed. The $s \rightarrow s$, $p \rightarrow p$, $d \rightarrow d$, $f \rightarrow f$ transitions having $\Delta l = 0$ are not allowed. In terms of symmetry, the orbitals can be classified as *g* (*gerade* for the presence of centre of inversion) and *u* (*ungerade* for the absence of centre of inversion). The *d* and *s* orbitals are of the *g*–symmetry while the *p*–orbitals are of the *u*–symmetry. In a centrosymmetric environment, electron transition can only occur between the orbitals having the different symmetry properties, *i.e.* ***g → u and u → g transitions are allowed but the g → g and u → u transitions are not allowed.***

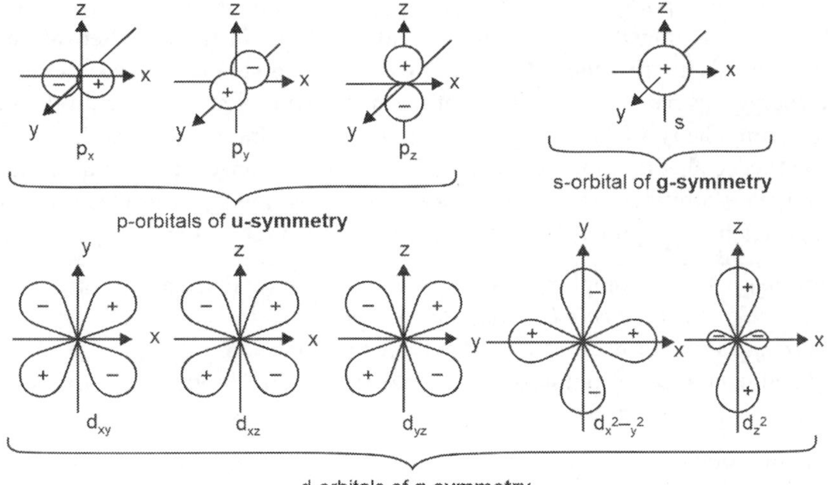

Fig. 7.4.1.1: Symmetry properties of the *s, p-* and *d*-orbitals.

The orbital selection rule is also referred to as the **parity selection rule** which **states that electron transition can only occur between the states of opposite parity.**

Parity is defined in terms of the behaviour of the orbital with respect to inversion of the space coordinates through the origin, *i.e.* x, y, z are replaced by $-x, -y, -z$. If $\psi(x, y, z) = \psi(-x, -y, -z)$ as in $\cos(x) = \cos(-x)$, then the wave function is described to have the **even parity.** On the other hand, if $\psi(x, y, z) = -\psi(-x, -y, -z)$, then the orbital is described to have the **odd parity**. Obviously, the even parity indicates the presence of centre of inversion (i) and the odd parity indicates the absence of centre of inversion (i). Thus the even parity means the g symmetry and the odd parity means the u symmetry.

Electric dipole (ED) transition *vs.* magnetic dipole (MD) transition: In the ED transition model, interaction of the electron of the absorbing species with the electric vector of the incident electromagnetic radiation causes a **change in electric dipole moment**. In MD transition, interaction of the electron of the absorbing species with the magnetic part of incident radiation causes a **change in magnetic dipole moment**. Allowed MD transitions are **weaker** (by a factor $\geq 10^5$) than the allowed ED transitions. In MD transitions, it occurs between the **states of same parity** (*i.e.*. $g \leftrightarrow g$, $u \leftrightarrow u$ allowed transitions, *cf.* difference from the ED transitions). For the **rare earths**, the weak MD transitions contribute.

● **Laporte selection rule for the tetrahedral and octahedral complexes:** The Laporte selection rule is **not strictly applicable for the noncentrosymmetric tetrahedral or trigonal bipyramidal complexes**. In fact, for such noncentrosymmetric complexes, the $d \to d$ transition gives the more intense absorption bands ($\varepsilon \approx 100$–200) compared to those in the centrosymmetric octahedral complexes ($\varepsilon \leq 80$) but they are still much weaker than the Laporte allowed CT-bands ($\varepsilon \approx 10^3$) (see Table 7.6.1).

In the noncentrosymmetric tetrahedral complexes, the orbitals should not have the g or u designation. However, the parity character of the atomic orbitals (from which the MOs are generated) is retained in the MOs. In terms of MOT, in the tetrahedral complexes, the electron transition occurs from the **nonbonding *e*-orbitals** which are purely the metal d-orbitals to the t_2 MOs which are formed from the metal d (g character) and p (u character) orbitals and ligand p (u-character) orbitals. Thus in the t_2-MOs, both the g- and u-character remain present due to the d–p mixing. Thus the $e \to t_2$ electron transition consists of the Laporte allowed $d \to p$ (*i.e.* $g \to u$) component. Obviously, the $d(g) \to d(g)$ component of the $e \to t_2$ transition is Laporte forbidden. Thus, in the **tetrahedral complexes, the Laporte selection rule is partly allowed (*i.e.* Laporte selection rule is relaxed)**. It may be noted that because of the energy mismatch, participation of the metal np orbitals along with the metal $(n-1)d$ orbitals in constructing the t_2–MOs is insignificant in the tetrahedral complexes. On the other hand, in the centrosymmetric octahedral complexes, the electron transition occurs from the nonbonding t_{2g} orbitals (pure atomic d-orbitals) to the d-based e_g–MOs. Thus, the $t_{2g} \to e_g$ electron transition in the octahedral complexes being a pure d (*i.e.* g) $\to$ d (*i.e.* g) transition is Laporte forbidden.

O_h *vs.* T_d complexes: If in a complex, there is no centre of inversion, the Laporte selection rule is not strictly applicable. *Thus the rule is applicable for an octahedral complex but not applicable for a tetrahedral complex.* More correctly, the Laporte selection rule is relaxed better in the noncentrosymmetric tetrahedral complexes compared to that in the centrosymmetric octahedral complexes (see Sec. 7.5 for more details).

● **Spin selection rule:** The electron transition can only occur between the states having the same spin multiplicity, *i.e.* $\Delta S = 0$.

This selection rule is illustrated in Fig. 7.4.1.2.

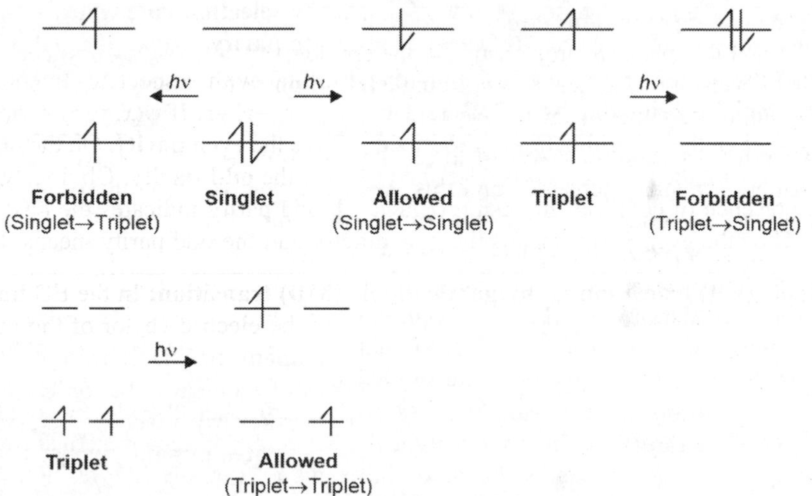

Fig. 7.4.1.2: Illustration of the spin allowed and spin forbidden transitions.

● **Number of electron participating in transition:** One electron transfer at a single step is allowed but more than one electron transfer at a single step is forbidden.

$t_{2g}^{n}\, e_g^{m} \rightarrow t_{2g}^{(n-1)}\, e_g^{(m+1)}$, *i.e.* 1*e* transition **(allowed)**

$t_{2g}^{n}\, e_g^{m} \rightarrow t_{2g}^{(n-2)}\, e_g^{(m+2)}$, *i.e.* 2*e* transition **(forbidden)**.

● **Examples of allowed and forbidden transitions:**

The following transition is forbidden.

$t_{2g}^{3}\, e_g^{2}$ (*e.g.* $[Mn(OH_2)_6]^{2+}$) $\rightarrow t_{2g}^{2}\, e_g^{3}$ **(spin forbidden and Laporte forbidden)**

($S = 5/2$, *i.e.* $2S + 1 = 6$) ($S = 3/2$, *i.e.* $2S + 1 = 4$)

The following transition is spin allowed.

$t_{2g}^{1}\, e_g^{0}$ (*e.g.* $[Ti(OH_2)_6]^{3+}$) $\rightarrow t_{2g}^{0}\, e_g^{1}$ **(spin allowed but Laporte forbidden)**

However, both the ***d-d transitions are the Laporte forbidden transitions.***

In the case of $[V(OH_2)_6]^{3+}$, between the following two transitions, the transition leading to the simultaneous transition of two electrons is forbidden.

$t_{2g}^{2}\, e_g^{0}$ (*e.g.* $[V(OH_2)_6]^{3+}$) $\rightarrow t_{2g}^{0}\, e_g^{2}$ **(forbidden)**

$t_{2g}^{2}\, e_g^{0} \rightarrow t_{2g}^{1}\, e_g^{1}$ **(spin allowed but Laporte forbidden).**

● **Reason behind the orbital selection rule:** In terms of **electric dipole (ED) model**, absorption of electromagnetic radiation leads to a change in the electric dipole moment of the molecules as it happens so in the molecular vibration excited in the **infrared red spectra** (see Vol. 7). In the electronic spectra, an electron interacts with the electric field of the incident electromagnetic radiation and moves in the direction of the electric field (*i.e.* **spatial redistribution of the electron being excited**). During the electronic excitation, **the algebric signs of the involved orbitals must change in a dipole like fashion** in moving from the ground state orbital to the excited state orbital. For example, the +/– sign arrangement of the orbital lobes must change to the –/+ sign arrangement during the electron transition.

CT transition *vs.* *d–d* transition: It may be noted that in the **charge-transfer (CT) transitions,** the electron redistribution makes a large change in the dipole moment and they produce the very intense bands. The *d–d* transition changes the **angular distribution of the electron** while the CT transition changes the **radial distribution** of the electron.

If an electron moves from an *s*–orbital to another *s*–orbital (say $1s \rightarrow 2s$), then spherical charge distribution pattern remains unchanged and this migration cannot create any orbital sign change in a dipole like fashion. Consequently, the ***s → s transition is not allowed*** (Fig. 7.4.1.3).

Similarly, in the ***p → p transition*** (say, $2p \rightarrow 3p$), the charge distribution pattern remains unchanged and consequently, this transition cannot make any orbital sign change (Fig. 7.4.1.3). This is why, this transition is also forbidden. The *d*-orbitals are **symmetric** in terms of sign of the lobes and consequently, a $d \rightarrow d$ transition cannot make the required orbital sign change and it is a forbidden transition. In general, the transitions possessing $\Delta l = 0$ are forbidden.

Now let us consider the ***s → p transition*** (say, $1s \rightarrow 2p_z$ when the electric field of the incident radition lies along the *z*–direction). In this transition, the charge distribution pattern changes (*i.e.* spherical to dumbell-shaped) to cause a change in the orbital signs in a dipole like fashion. Thus the ***s → p transition (i.e. Δl = –1) is allowed*** (Fig. 7.4.1.3).

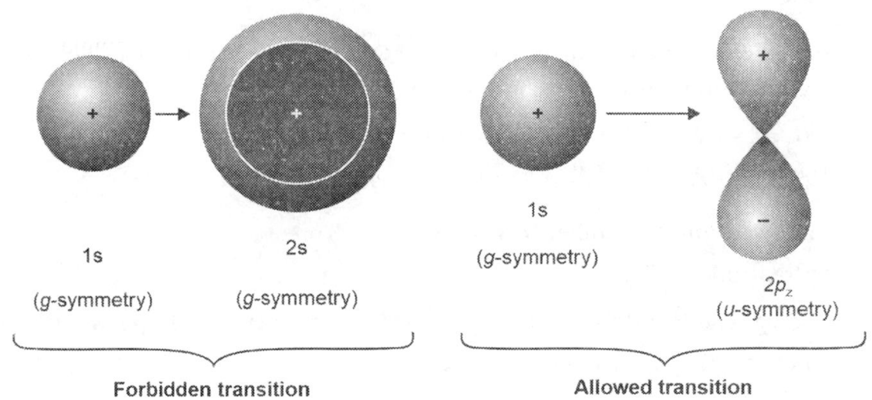

Fig. 7.4.1.3: Illustration of Laporte-selection rule.

● **Franck-Condon transition and internuclear separation:** If the **internuclear separation** at equilibrium (r_e) remains the same for both the ground and excited electronic states then ($v'' = 0 \rightarrow v' = 0$) transition is the most probable one as in the case of *f–f* transition in the lanthanides for which the deeply seated 4*f* levels hardly participate in metal-ligand bonding. But, in most of the cases (*e.g.* $t_{2g} \rightarrow e_g$ in transition metal O_h complexes), the internuclear separation in the excited state is generally longer than that in the ground state. **In such cases, the Franck-Condon vertical transition leads to the higher vibrational levels in the excited state (*cf.* Fig. 7.4.1.4).** During the Franck-Condon vertical transition, the quantum mechanical restriction (if any) for $\upsilon'' \rightarrow \upsilon'$ must be taken into consideration, *i.e.* for a particular value of *r*, all the possible vertical transitions between the vibrational levels of the two electronic states may not be quantum mechanically allowed (*cf.* condition of vibronic coupling). It may be noted that in the vertical Franck-Condon electronic transition, there is **no restriction on the change of vibrational quantum number** (*i.e.* $\Delta v = 0, \pm 1, \pm 2, \pm 3, ...$) provided the electronic transition is allowed by its own right.

The time taken for an electronic transition is much smaller than the time taken to execute a vibration to change the internuclear separation. In other words, during the electronic transition (*i.e.* transition from one electronic state to another), the nuclei can be reasonably assumed to remain fixed because the

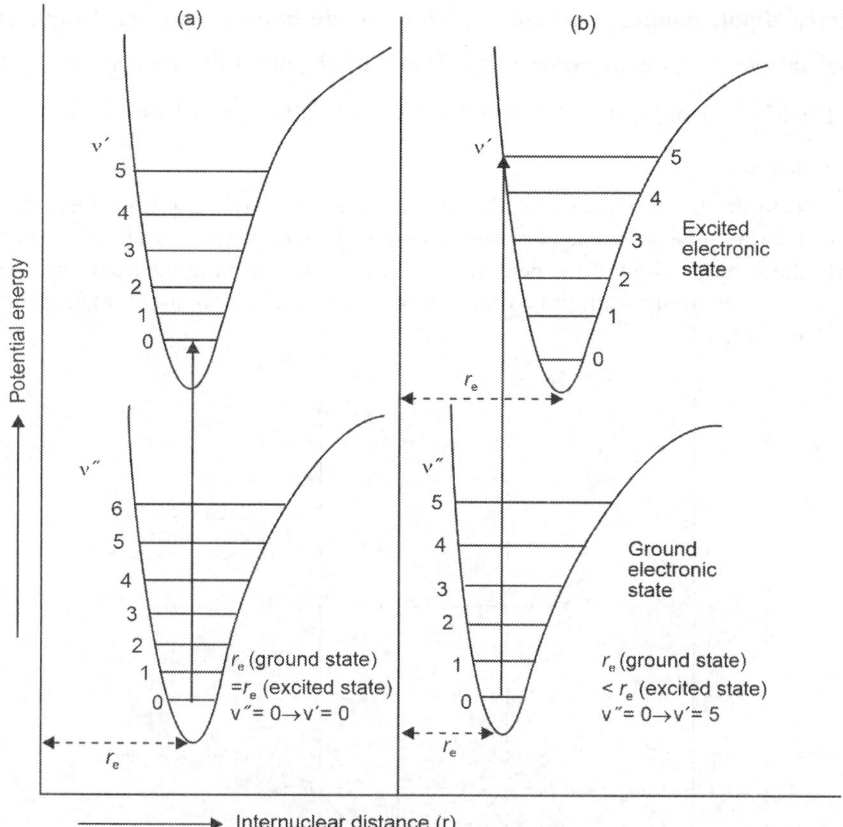

Fig. 7.4.1.4: Illustration of Franck-Condon **vertical transition** by considering the **Morse potential energy curve** (*i.e.* plot of potential energy *versus* internuclear separation). (a) r_e (ground state) = r_e (excited state), (b) r_e (ground state) < r_e (excited state) where r_e denotes the equilibrium internuclear separation (see Figs 6.10.5.2 and 7.7.2.1).

electronic transition is much faster than the molecular vibration. This is the basis for **Franck-Condon principle** and **it demands the vertical transition** as shown in Fig. 7.4.1.4.

Magnetic dipole (MD) model of absorption of electromagnetic radiation: In this model, when a molecule absorbs the incident radiation, the magnetic dipole of the absorbing molecule oscillates. However, this mechanism of absorption of electromagnetic radiation is very much inefficient (by a factor about 10^{5-8}) compared to the **electric dipole (ED) model**. This is why, this mode of absorption of light by the matter is not discussed in spectroscopy. Interestingly, in the magnetic dipole model, the orbital selection rule is opposite, *i.e.* $p \to p$, $d \to d$, $f \to f$, etc. are allowed while $p \to d$ is forbidden. The *f*-block elements probably partly absorb light in this route.

7.4.2 Quantum Mechanical Explanation of the Orbital Selection Rule

Any electronic transition must be associated with a change in the dipole moment in the system. Mathematically, the **transition dipole moment integral** or in short, **transition moment integral** (TMI, I) for one-electron transition from the orbital ψ_m to the orbital ψ_n must be nonzero, *i.e.*

$$\text{TMI} = I = \int_{-\infty}^{+\infty} \psi_m \mu \psi_n d\tau \neq 0 \text{ (For the electric dipole (ED) model)}$$

where μ = **electric dipole moment operator which is an odd function.** μ is an **electric dipole vector** that can be resolved into the *x*, *y* and *z*-components. Thus the integral has the components $\int_{-\infty}^{+\infty} \psi_m \mu_x \psi_n d\tau$, $\int_{-\infty}^{+\infty} \psi_m \mu_y \psi_n d\tau$ and $\int_{-\infty}^{+\infty} \psi_m \mu_z \psi_n d\tau$. To make the transition allowed, at least one of these three integrals must be nonzero.

The **dipole strength** (*D*), the square of the integral (*i.e.* $D = [I]^2$) is proportional to the absolute intensity of the absorption peak for the correspondiong electronic transition. The *D* measures the **peak area** rather than the simply height of the peak (*i.e.* ε). However, assuming the peak width to be roughly same for the peaks under comparison, it is quite reasonable to consider a direct relationship between *D* and ε ($\varepsilon \propto D$) (see Sec. 7.6).

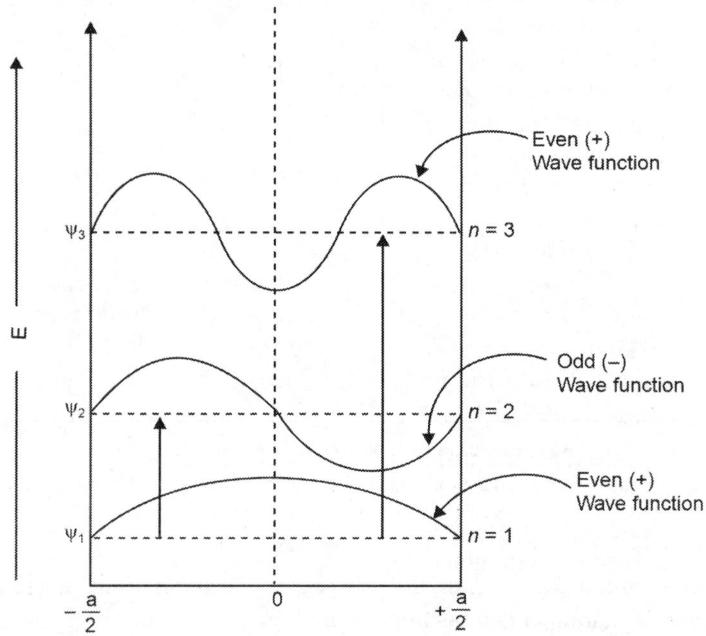

Fig. 7.4.2.1 Even and odd wave functions and allowed electronic transitions (*cf.* particle in a one dimensional box).

Odd and even parity function in terms of centre of inversion

Even parity function: $\psi(x, y, z) = \psi(-x, -y, -z)$; *e.g.* $\cos x = \cos(-x)$

Odd parity function: $\psi(x, y, z) = -\psi(-x, -y, -z)$; *e.g.* $\sin x = -\sin(-x)$

Now let us consider the electron (to experience the electronic transition) to reside as a particle in a one-dimensional box where the nucleus is placed at the centre of the box (Fig. 7.4.2.1). Here the dipole

Properties of the odd (*i.e. u*) and even (*i.e. g*) functions

- The total integration of an **odd function** is zero because the result of integration over the one-half of the function will produce the same result with an opposite sign obtained from the integration over the other half of the function.
- The total integration of an even function is nonzero.

moment given by '*ex*' where e = charge of the electron and x = distance between the electron and nucleus ***changes its sign at the mid point of the box.*** **Thus, the dipole moment operator (μ) is an odd function.**

The result of integration over an odd (*i.e.* anti-symmetric) function is zero while the integration over an even (*i.e.* symmetric) function is nonzero. ***Thus to make the transition moment integral (I) nonzero, the integrand must be overall an even function.***

● **Physical significance of the transition moment integral (TMI):** When an electron interacts with the electric field of the incident light wave, the electron will move in the direction of the electric field. This movement of an electronic charge will cause a change in the dipole moment. Thus the *transition dipole moment integral (I)* gives a measure of the impulse created by an electron transition to the electromagnetic field of light. A large impulse indicates an intense transition and a zero impulse indicates a forbidden transition.

As the dipole moment operator (μ) is an odd function, the product of ψ_m and ψ_n must be an odd one so that (odd) × (odd) becomes even. This is why, if ψ_m is a symmetric wave function (*i.e. g*) then ψ_n must be an antisymmetric wave function (*i.e. u*) because (even) × (odd) becomes odd. ***Thus in an electronic transition, between the orbitals involved, if one orbital is a symmetric one, then the other orbital must be an antisymmetric one.*** This is why, the electronic transitions between the wave functions having the same *l* values are fobidden, *i.e. s–s, p–p, d–d, f–f* transitions are forbidden but the *s–p, p–d* and *d–f* transitions are allowed. Here it may be noted that the *p*–orbitals are the odd wave functions (*i.e. u*–symmetry) while the *d*–orbitals are the even wave functions (*i.e. g*–symmetry).

Note: For the **magnetic dipole** (MD) transition, the corresponding TMI is given by:

$$I = \int_{-\infty}^{+\infty} \psi_m \mu_m \psi_n d\tau \neq 0, \mu_{MD} = \text{magnetic dipole moment operator}$$

Intensity of MD transition peak is very weak compared to that of **electric dipole (ED) transition** (stronger by a factor about 10^5–10^8). But in MD transition, *s–s, p–p, d–d, f–f* transitions are allowed. For the **lanthanides and actinides**, the MD transitions (*i.e. f–f* transitions) probably contribute.

7.4.3 Quantum Mechanical Explanation for the Forbidden Many-electron Transition

For many-electron transition, the transition moment integral $\int \psi_m \mu \psi_n \partial\tau$ becomes zero when ψ_m and ψ_n are expressed as the products of one-electron wave functions. **It happens so because the one-electron wave functions are mutually orthogonal.** This is why, in general, many-electron transitions are forbidden.

Note: Here it may be mentioned that the many-electron wave functions cannot be strictly expressed as the products of one-electron wave functions. In fact, they are the linear combinations of the one-electron wave functions. Thus, the integral *I* is nonvanishing but vanishingly small. This is why, the two-electron transitions are noted with the weak intensities in some cases as in $[V(aq)]^{2+}$.

7.4.4 Quantum Mechanical Explanation of the Spin Selection Rule

To understand this selection rule, the total wave function is to be expressed in terms of the orbital wave function (ψ^o) and spin wave function (ψ^s).

i.e. $\psi_{total} = \psi_{orbital} \times \psi_{spin} = \psi^o \times \psi^s$.

It leads to:

$$I = \int \psi_m \mu \psi_n \partial\tau$$
$$= \int \psi_m^o \psi_m^s \mu \psi_n^o \psi_n^s \partial\tau$$

The **dipole moment operator (μ)** is an **electric dipole vector** and it cannot change the spin orientation of the electron that experiences only the **spatial redistribution** in the electronic transition. This is why, the dipole moment oiperator has nothing to do with the spin wave function and we can express the above integral as follows:

$$I = \int \psi^o_m \mu \psi^o_n \partial\tau \int \psi^s_m \psi^s_n \partial\tau$$

Here the second integral involving the spin wave functions (*i.e.* $\int \psi^s_m \psi^s_n \partial\tau$) is zero when the spin states differ. Thus for $\Delta S = 0$, I becomes only nonzero, *i.e.* electron transition can occur only between the states having the same spin multiplicity.

7.5 RELAXATION OF THE SELECTION RULES

7.5.1 Relaxation of the Laporte Selection Rule: Vibronic Coupling, *d–p* Mixing and Symmetry Lowering

In the transition metal complexes, the electronic transitions like $t_{2g} \rightarrow e_g$ (in the O_h complexes), $e \rightarrow t_2$ (in the T_d complexes) basically involve the $d \rightarrow d$ transitions. These are the Laporte forbidden transitions and they should be colourless. But in reality, many octahedral and tetrahedral complexes like $[Ti(OH_2)_6]^{2+}$, $[Ni(NH_3)_6]^{2+}$, $[CoCl_4]^{2-}$, etc. are sufficiently coloured to be characterised even by our naked eyes. Actually, the Laporte selection rule may be relaxed in different ways. These are discussed below.

Electronic Energy State and Vibronic Transition

In an electronic state, there are many vibrational levels (*see* Fig. 7.4.1.3). Distribution of the molecules in the different vibrational energy levels is governed by the Maxwell-Boltzmann distribution law:

$$N_{ex} = N_0 \exp(-\Delta E/k_B T)$$

N_{ex} and N_0 denote the number of molecules in the excited state and total number of molecules respectively, T is the temperature in Kelvin scale, k_B is the Boltzmann constant, ΔE denotes the energy difference between the ground and excited states. At room temperature, the thermal energy measured by $k_B T$ is about 200 cm^{-1}. ***Thus at room temperature, upper or excited electronic states are seldom populated thermally and almost all the molecules are in the lowest vibrational state of the ground electronic state.***

The vibrational transitions require much lower energy than that required for an electronic transition. ***Thus the vibrational transitions are expected to be mixed up with the electronic transition. This gives the vibronic transition.***

(A) Vibronic coupling (Van-Vleck) in the centrosymmetric systems: In a centrosymmetric system, *e.g.* $[Ni(NH_3)_6]^{2+}$, $[Co(NH_3)_6]^{3+}$, etc. all the *d*-orbitals are considered to have the *g*-symmetry (*i.e.* gerade symmetry) and the $d \rightarrow d$ (*i.e.* $g \rightarrow g$) transitions are forbidden. However, this restriction can be relaxed through the coupling of electronic and vibrational wave functions. This coupling is described as **vibronic coupling** ('*vib*' from the term *vibrational* and '*ronic*' from the term *electronic*). The metal-ligand bonds are always vibrating. These vibrations may be both *symmetric* (*i.e.* *g*–symmetry or even function) and *antisymmetric* (*i.e.* *u*–symmetry or odd function) with respect to the centre of inversion (*i*). Each electronic energy level is associated with the several vibrational energy levels (Fig. 7.4.1.3). During **the vibronic transition, the transition probability** between the lower electronic state (say ground state) and upper electronic state (say, excited state) can be determined by considering the vibronic wave functions ($\psi_{vibronic}$) of the two states.

$$\psi_{vibronic} = \psi_{el} \times \psi_{vib}$$

[**Note:** The total wave function (ψ) is the product of an electronic wave function (ψ_{el}), a vibrational wave function (ψ_{vib}) and a rotational wave function (ψ_{rot}) (*cf.* **Born-Oppenheimer approximation** which states that none of these three components of the complete wave function are interdependent).

$$\psi = \psi_{el}\psi_{vib}\psi_{rot}$$

The wave functions ψ_{el} and ψ_{vib} are not totally independent but ψ_{rot} can be treated as independent of the other two wave functions, *i.e.* ψ_{el} and ψ_{vib}. Thus, reasonably, we can write the vibronic wave function as follows:

$$\psi_{vibronic} = \psi_{el}\psi_{vib}$$

Thus the vibronic coupling (*i.e.* an electronic transition coupled with the vibrational transition) indicates the **breakdown of the Born-Oppenheimer approximation because the vibronic coupling leads to the conclusion that an electronic transition is not quite independent of a vibrational transition.**

Assuming the vibronic transition to be spin-allowed (*i.e.* $\int \psi_s^{gd}\psi_s^{ex} d\tau$ is nonzero) and omitting the integral involving the spin wave functions, the following **transition moment integral** (I) can be considered to understand the **transition probability** in a vibronic transition.

$I = \int \psi_{el}^{gd}\psi_{vib}^{gd}\mu\psi_{el}^{ex}\psi_{vib}^{ex} d\tau$, ('gd' for ground state, 'ex' for excited state).

To make the transition allowed, the above integral must be nonzero, *i.e.* the product $\psi_{el}^{gd}\psi_{vib}^{gd}\mu\psi_{el}^{ex}\psi_{vib}^{ex}$ must be a symmetric function (*g*-symmetry).

In a centrosymmetric system (*e.g.* O_h-complex), ψ_{el} (*i.e.* electronic wave function) is symmetric, *i.e.* both ψ_{el}^{gd} and ψ_{el}^{ex} are of *g*-symmetry.

Thus the product of electronic wave functions of the ground state and excited state is of *g*-symmetry, *i.e.* $\psi_{el}^{gd} \times \psi_{el}^{ex} \Rightarrow g \times g \Rightarrow g$–symmetry (*i.e.* even function).

The dipole moment operator (μ) is itself of *u*–symmetry (*i.e.* odd function). It leads to:

$$\psi_{el}^{gd} \times \psi_{el}^{ex} \times \mu \Rightarrow g \times g \times u \Rightarrow u\text{–symmetry (i.e. odd function)}$$

Thus to make the integral nonzero, the product of vibrational wave functions of the two electronic states must be an odd function, *i.e.*

$$\psi_{vib}^{gd} \times \psi_{vib}^{ex} \Rightarrow u\text{–symmetry (i.e. odd function)}$$

Thus if one vibrational wave function is of g–type then the vibrational wave function of the other electronic state must be of u–type.

Then we can have (assuming ψ_{vib}^{gd} of *g*-symmetry):

$$\underbrace{\psi_{el}^{gd} \times \psi_{vib}^{gd} \times \mu}\times\underbrace{\psi_{el}^{ex} \times \psi_{vib}^{ex}} \Rightarrow (g\times g\times u)\times(g\times u) \Rightarrow u\times u \Rightarrow g\ (\text{i.e. even function})$$

Thus, the transition is possible only when the g-type vibration of the ground electronic state is coupled with the u–type vibration of the excited electronic state and vice-versa.

At the lowest vibrational state (*i.e.* $v = 0$) *of an electronic state,* ψ_{vib} *is totally symmetric* (*i.e. g-type symmetry*, see Figs 6.10.5.2 and 7.7.2.1). Thus, if we consider the transition from the lowest vibrational state (*i.e.* ψ_{vib}^{gd} is of *g*-type symmetry) in an octahedral complex (a centrosymmetric system), then for the nonzero value of the transition moment integral, ψ_{vib}^{ex} must be of *u*-type symmetry because ψ_{vib}^{gd} is of *g*-type symmetry. Thus the following product leads to a symmetric or even function.

$$\underbrace{\psi_{el}^{gd} \times \psi_{vib}^{gd} \times \mu \times \psi_{el}^{ex} \times \psi_{vib}^{ex}}_{\text{(Allowed vibronic transitions)}} \Rightarrow (g\times g\times u)\times(g\times u) \Rightarrow g\ (\text{i.e. even function})$$

In an octahedral complex, the T_{1u} and T_{2u} normal modes of vibration can destroy the centre of symmetry (i), *i.e.* these vibrations are of *u*-symmetry. This is why, in the octahedral (O_h) complexes, the electronic transitions are allowed through the coupling of a T_{1u} or T_{2u} vibration (*i.e.* simultaneous excitation to a T_{1u} or T_{2u} vibration).

ψ_{vib} is always totally symmetric (*i.e.* *g*-symmetry) for the lowest vibration level (*i.e.* $v = 0$) in an electronic state (*e.g.* $\psi_{\text{vib}}^{\text{gd}}$ is of A_{1g} symmetry for O_h). Thus we can conclude that the electronic transition coupled with the **$v^{gd} = 0 \rightarrow v^{ex} = 0$ vibrational transition is not allowed** because both $\psi_{\text{vib}}^{\text{gd}}$ and $\psi_{\text{vib}}^{\text{ex}}$ are of *g*-symmetry, *i.e.* the product $\psi_{\text{vib}}^{\text{gd}} \times \psi_{\text{vib}}^{\text{ex}}$ is of *g*-symmetry. **Thus the 0 → 0 vibrational transition makes the transition moment integral zero** when the electronic transition is itself forbidden.

$\psi_{\text{vib}}^{\text{gd}} \times \psi_{\text{vib}}^{\text{ex}} \Rightarrow g \times g \Rightarrow g$-symmetry; $\psi_{\text{el}}^{\text{gd}} \mu \psi_{\text{el}}^{\text{ex}} \Rightarrow g \times u \times g \Rightarrow u$-symmetry;

$\psi_{\text{el}}^{\text{gd}} \mu \psi_{\text{el}}^{\text{ex}} \times \psi_{\text{vib}}^{\text{gd}} \psi_{\text{vib}}^{\text{ex}} \Rightarrow g \times u \times g \times g \times g \Rightarrow u$-symmetry (for the $0 \rightarrow 0$ transition)

It may be noted that for the **tetrahedral complexes,** the $0 \rightarrow 0$ transition may be allowed (**provided** $r_e^{ex} \approx r_e^{gd}$, *cf.* Fig. 7.7.2.1) as the electronic transition is not always forbidden.

Note: The vibrational wave functions $\psi_{\text{vib}}^{\text{gd}}$ and $\psi_{\text{vib}}^{\text{ex}}$ of different electronic states are not orthogonal. The vibrational overlap integral, $\int \psi_{\text{vib}}^{\text{gd}} \psi_{\text{vib}}^{\text{ex}} \, d\tau$ for the vibrational wave functions of the two electronic states is called **Franck-Condon overlap** and square of the vibrational overlap integral is described as the **Franck-Condon factor (FCF) to be discussed later.**

(B) Physical picture of vibronic coupling: Even in a centrosymmetric system, some vibrations can destroy the centre of symmetry (*i*) temporarily. In an octahedral complex, the T_{1u} and T_{2u} vibrations can distort the octahedral complex in such a way that the centre of symmetry is removed temporarily. *When the vibration removes the centre of symmetry, g-character of the electronic states is lost and at that condition, the orbital selection rule, i.e. parity selection rule, is not applicable and the electronic transition becomes allowed.*

Here it should be mentioned that the loss of centre of symmetry **(temporarily by an odd vibration and permanently by a ligand environment)** *does not ensure all the electronic transitions allowed* (*e.g.* in a tetrahedral complex, the $A_2 \rightarrow T_1$ transition is itself allowed but the $A_2 \rightarrow T_2$ transition is forbidden; *see* Solved Problem 15). However, loss of centre of symmetry and lowering of symmetry by any mechanism makes the electronic transition more allowed **(in terms of probability),** in general. But, if incorporation of the odd vibrational representation in the transition moment integral produces a **totally symmetric represention A_1 or A_{1g},** then the transition is definitely allowed. This aspect will be illustrated later.

For a nonlinear *n*-atomic molecule, there will be $3n - 6$ **normal modes of vibration** and these can be characterised by means of their symmetry properties. For an octahedral complex, the **15 normal modes of vibration** are distributed as (*cf.* Table 7.23.3 and Appendix 2B, Vol. 7).

$$A_{1g} + E_g + 2T_{1u} + T_{2u} + T_{2g}$$

The odd vibrational modes destroying the centre of sysmmetry of the octahedral complex are T_{1u} and T_{2u} (Fig. 7.5.1.1).

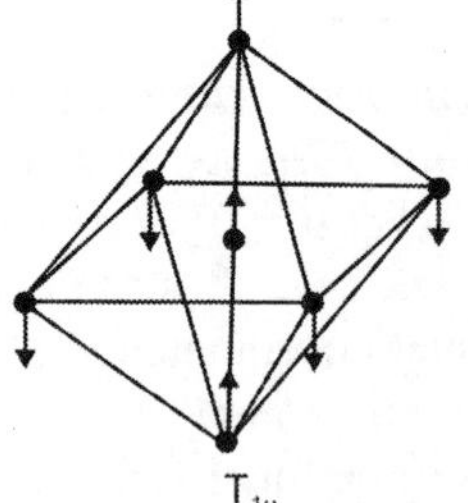

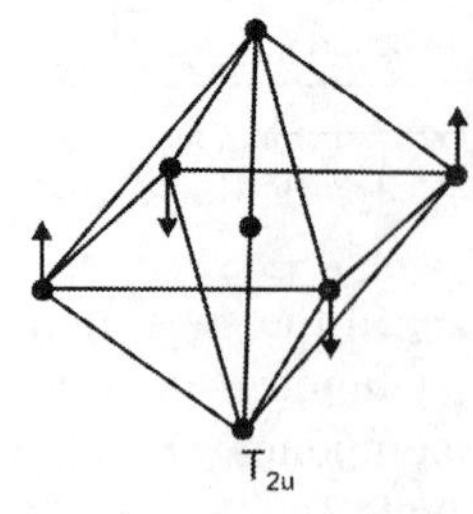

Fig. 7.5.1.1 T_{1u} and T_{2u} normal modes of vibration (destroying the centre of symmetry through the displacement of atoms) in an octahedral complex ML_6 belonging to the O_h point group. **Note:** There is another type of T_{1u} vibration (in O_h) removing the centre of symmetry (*see* Appendix 2B, Vol. 7).

(C) Understanding of vibronic coupling in terms of group theory: The condition for an allowed vibronic transition can be realised in terms of group theory. However, it is easier to understand this in terms of the symmetry arguments. For an allowed vibronic transition, the following **transition moment integral (TMI)** must be nonzero.

$$I = \int(\psi_{el}^{gd}\ \psi_{vib}^{gd})\mu(\psi_{el}^{ex}\ \psi_{vib}^{ex})d\tau$$

For the nonzero value of I, the product $(\psi_{el}^{gd}\psi_{vib}^{gd})\mu(\psi_{el}^{ex}\psi_{vib}^{ex})$ must be totally symmetric or the product will contain at least one **totally symmetric representation term of the point group** (*e.g.* A_{1g} term for the O_h system). Assuming the vibrational transition from the lowest vibrational state (*i.e.* $v = 0$, at the ground electronic state), **we can conclude ψ_{vib}^{gd} to be totally symmetric** (*e.g.* ψ_{vib}^{gd} is A_{1g} for O_h). The direct product of $(\psi_{el}^{gd}\mu\psi_{el}^{ex} \equiv g \times u \times g \Rightarrow u)$ does not contain any term of totally symmetric representation but to make the integral I nonzero, the direct product of $\psi_{el}^{gd}\mu\psi_{el}^{ex}\psi_{vib}^{ex}$ must contain at least one term of totally symmetric representation of the same point group (*e.g.* A_{1g} for O_h). *It is possible, if there is any normal mode of vibration of ψ_{vib}^{ex} whose representation belongs to at least one of the representations resulted from the direct product $\psi_{el}^{gd}\mu\psi_{el}^{ex}$.* In other words, as the direct product of $\psi_{el}^{gd}\mu\psi_{el}^{ex}$ is of u-symmetry (*cf.* $g \times u \times g \Rightarrow u$), then ψ_{vib}^{ex} must be of **identical *u*-symmetry** to make I nonzero.

Condition for allowedness of electronic transitions: In terms of group theory treatment, the function present in TMI (transition moment integral) **must be totally symmetric**, *i.e.* it must possess the **totally symmetric representation** (*e.g.* A_{1g} for the O_h complexes, A_1 for the T_d complexes). For the O_h complexes, the **totally symmetric irreducible representation A_{1g} is a product of only two identical representations.** This condition of TMI is applicable for **both the centrosymmetric and noncentrosymmetric systems** while Laporte rule is applicable only for the centrosymmetric systems. In the noncentrosymmetric systems, all the possible electronic transitions may not be allowed. For example, in the T_d complexes, the $A_2 \rightarrow T_1$ transition is **orbitally allowed** (TMI contains A_1) but the $A_2 \rightarrow T_2$ transition is **orbitally forbidden** (TMI lacks in A_1).

● **Vibronic coupling in the centrosymmetric O_h-system:** Let us illustrate for the transition, **$^3T_{1g}(F) \rightarrow {}^3T_{2g}(F)$ in the d^2–case.** To get the result of direct product, we can consider the Table 7.5.1.1. The derivation of the Table is beyond the scope of the book. For this purpose, the readers may consult any book on Group Theory.

Table 7.5.1.1 Direct products applicable for the tetrahedral (T_d) and octahedral (O_h) complexes.

	A_1	A_2	E	T_1	T_2
A_1	A_1	A_2	E	T_1	T_2
A_2	A_2	A_1	E	T_2	T_1
E	E	E	$A_1 + A_2 + E$	$T_1 + T_2$	$T_1 + T_2$
T_1	T_1	T_2	$T_1 + T_2$	$A_1 + E + T_1 + T_2$	$A_2 + E + T_1 + T_2$
T_2	T_2	T_1	$T_1 + T_2$	$A_2 + E + T_1 + T_2$	$A_1 + E + T_1 + T_2$

By using the above Table, we get the results of direct product of T_{1g} and T_{2g}.

$$T_{1g} \times T_{2g} = A_{2g} + E_g + T_{1g} + T_{2g}$$

In the O_h system, the Cartesian coordinate axes have the same symmetry property resembling the symmetry property of the p–orbitals. It is T_{1u} which is the symmetry property of the dipole moment component, *i.e.* $\mu_x = \mu_y = \mu_z = \mu$ transforms into the T_{1u} species (*see* Character Table, Appendix 2A of Vol. 7).

$$\begin{aligned} T_{1g} \times \mu \times T_{2g} &= T_{1g} \times T_{1u} \times T_{2g} = T_{1g} \times T_{2g} \times T_{1u} = (A_{2g} + E_g + T_{1g} + T_{2g}) \times T_{1u} \\ &= A_{2g} \times T_{1u} + E_g \times T_{1u} + T_{1g} \times T_{1u} + T_{2g} \times T_{1u} \\ &= (T_{2u}) + (T_{1u} + T_{2u}) + (A_{1u} + E_u + T_{1u} + T_{2u}) + (A_{2u} + E_u + T_{1u} + T_{2u}) \\ &= A_{1u} + A_{2u} + 2E_u + 3T_{1u} + 4T_{2u} \end{aligned}$$

Thus for the transition, ${}^3T_{1g} \rightarrow {}^3T_{2g}$ in the octahedral d^2–case, all the terms of the direct product $T_{1g} \times T_{1u} \times T_{2g} (\equiv \psi_{el}^{gd} \times \mu \times \psi_{el}^{ex})$ are of u-symmetry, as expected. If any one of the terms present in the direct product $T_{1g} \times T_{1u} \times T_{2g}$ matches in terms of symmetry with a normal mode of vibration of an O_h-system, then the direct product of $\psi_{el}^{gd}\mu\psi_{el}^{ex}\psi_{vib}^{ex}$ (assuming ψ_{vib}^{gd} to be totally symmetric, *i.e.* ${}^1A_{1g}$ in O_h) will contain the term which is totally symmetric (*i.e.* A_{1g} for O_h) and I becomes nonzero. The normal modes of vibration of an octahedral (O_h) complex are of the following symmetries (*cf.* Table 7.23.3).

$$A_{1g}, E_g, 2T_{1u}, T_{2g}, T_{2u}$$

Thus the T_{1u} and T_{2u} vibrations match with the terms obtained from the direct product $T_{1g} \times T_{1u} \times T_{2g}$ $(\equiv \psi_{el}^{gd}\mu\psi_{el}^{ex})$ for the $T_{1g} \rightarrow T_{2g}$ transition. ***This is why, the electronic transition is vibronically allowed through the possible coupling of the T_{1u} and T_{2u} normal mode of vibration*** (*i.e.* ψ_{vib}^{ex} must be of T_{1u} and/or T_{2u} symmetry) and then the direct $\psi_{el}^{gd}\mu\psi_{el}^{ex}\psi_{vib}^{ex}$ will contain the totally symmetric A_{1g} term (*irreducible representation*) (*cf.* both the direct products $T_{1u} \times T_{1u}$ and $T_{2u} \times T_{2u}$ contain the A_{1g} term; *see* Table 7.5.1.1).

In other words, we may conclude that the direct product $\psi_{el}^{gd}\mu\psi_{el}^{ex}$ for an electronic transition in an octahedral system will produce the terms of u-symmetry (*cf.* $g \times u \times g \Rightarrow u$) and at least there must be one term of T_{1u} or T_{2u} symmetry in the direct product and then coupling of the T_{1u} or T_{2u} symmetry vibration (ψ_{vib}^{ex}) will make the electronic transition vibronically allowed. **If there is no T_{1u} or T_{2u} term in the direct $\psi_{el}^{gd}\mu\psi_{el}^{ex}$ (electronic transition in an O_h system), then the transition is not vibronically allowed.** Here it is worth mentioning that in making the above conclusion, it has been assumed that the electronic transition occurs from $v = 0$ (*i.e.* lowest vibrational level at the ground electronic state) for which ψ_{vib}^{gd} is totally symmetric (*i.e.* ${}^1A_{1g}$ for O_h). Here we shall illustrate this principle by taking some examples.

- **Vibronic coupling for the $A_2 \rightarrow T_2$ Transition in the noncentrosymmetric T_d system (d^7)**

We have already mentioned that even in a **noncentrosymmetic system**, all the possible electronic transitions may not be allowed (*e.g.* $A_2 \rightarrow T_2$ transition in the noncentrosymmetric d^7 T_d-system is

Table 7.5.1.2 Illustration of the allowdness of electronic transitions in the octahedral complexes assuming the transition from $v = 0$ (lowest vibrational level of the ground electronic state, *i.e.* ψ_{vib}^{gd} is totally symmetric)

	Transition ($\psi_{el}^{gd} \rightarrow \psi_{el}^{ex}$)	***Direct product*** $\Gamma(\psi_{el}^{gd})\Gamma\mu\Gamma(\psi_{el}^{ex})$	***Conclusion***
Electronic transitions in the low-spin octahedral Co(III)-complex, *e.g.* $[Co(NH_3)_6]^{3+}$	${}^1A_{1g} \rightarrow {}^1T_{1g}$	$A_{1g} \times T_{1u} \times T_{1g} = A_{1u} + E_u + T_{1u} + T_{2u}$	The direct of $\psi_{el}^{gd}\mu\psi_{el}^{ex}$ contains the term T_{1u} and/or T_{2u}; the electronic transitions are vibronically allowed through the coupling of T_{1u} and/or T_{2u} vibration, ψ_{vib}^{ex}. ***($T_{1u} \times T_{1u}$ or $T_{2u} \times T_{2u}$ always contains the A_{1g} term in the direct product result).***
	${}^1A_{1g} \rightarrow {}^1T_{2g}$	$A_{1g} \times T_{1u} \times T_{2g} = A_{2u} + E_u + T_{1u} + T_{2u}$	
Electronic transition in the octahedral d^1-system, *e.g.* $[Ti(OH_2)_6]^{3+}$	${}^2T_{2g} \rightarrow {}^2E_g$	$T_{2g} \times T_{1u} \times E_g = A_{1u} + A_{2u} + 2E_u + 2T_{1u} + 2T_{2u}$	
Electronic transition in $[Cr(NH_3)_6]^{3+}$(d^3, O_h)	${}^4A_{2g} \rightarrow {}^4T_{2g}$	$A_{2g} \times T_{1u} \times T_{2g} = A_{1u} + E_u + T_{1u} + T_{2u}$	

forbidden) and the forbidden transitions may be made allowed through the vibronic coupling. The condition is the same, *i.e.*

$\int \psi_{el}^{gd} \mu \psi_{el}^{ex} \psi_{vib}^{ex} d\tau \neq 0$, assuming ψ_{vib}^{gd} to be totally symmetric (*e.g.* A_1 for T_d)

i.e. the above **matrix element** must contain the totally symmetric representation of the group.

In a T_d system, the dipole moment operator (μ) transforms into the T_2 species. The following direct product result is to be analysed for the **$A_2 \rightarrow T_2$ transition.**

$A_2 \times \mu \times T_2 = A_2 \times T_2 \times T_2 = T_1 \times T_2 = A_2 + E + T_1 + T_2$, (by using the Table 7.5.1.1).

The direct product result does not bear the **totally symmetric irreducible representation A_1** and consequently, the transition is forbidden by its own right. But, coupling of any vibration of A_2, E, T_1 and T_2 symmetry will make the transition allowed. For the T_d point group, normal modes of vibration (*cf.* Table 7.23.3) belong to $A_1 + E + 2T_2$, *i.e.* E or T_2 vibration can be coupled to make the transition vibronically allowed.

Condition for Vibronic Transition in a Centrosymmetric System

- A suitable combination of the metal-ligand vibrations of the lower state (*i.e.* ground state) and upper state (*i.e.* excited state) can allow the vibronic transition partially allowed even in a centrosymmetric system. Thus in a centrosymmetric octahedral complex, the $d \rightarrow d$ transition (*i.e.* $g \rightarrow g$ transition) becomes partially allowed, if the ground state electronic wave function is combined with a g–type (*i.e.* symmetric) vibrational wave function and the excited state (*i.e.* upper state) electronic wave function is combined with a u–type vibrational wave function and vice-versa. In such cases, the so called $d \rightarrow d$ transition (*i.e.* $g \rightarrow g$) will have the components like $g \rightarrow u$ or $u \rightarrow g$ which make the transition partially allowed. It is reasonable to assume that the electronic transition occurs from the lowest vibrational state ($\upsilon = 0$ at the ground electronic state) that makes ψ_{vib}^{gd} totally symmetric (*i.e.* A_{1g}-symmetry in O_h). ***Under this condition, the electronic transition becomes vibronically allowed through the coupling of an u-type vibration of the excited electronic state (i.e. ψ_{vib}^{ex} is of u-type symmetry) — so that the matrix element $\langle \psi_{el}^{gd} \times \mu \times \psi_{el}^{ex} \times \psi_{vib}^{ex} \rangle$ contains the totally symmetric representation term of the point group.***
- Vibronic couping is applicable to both the centrosymmetric and noncentrosymmetric systems.

It may be noted that in a T_d-system, the **$A_2 \rightarrow T_1$ transition is allowed by its own right** and it is evident from the following triple product result.

$A_2 \times \mu \times T_1 = A_2 \times T_2 \times T_1 = A_2 \times (A_2 + E + T_1 + T_2) = A_1 + E + T_2 + T_1$ (*cf.* Table 7.5.1.1).

The result contains the **totally symmetric representation term A_1** of the T_d point group and it makes the transition moment integral nonzero.

(D) Understanding of peak broadening through vibronic coupling: Here we shall illustrate this aspect by considering the transition, ${}^3T_{1g}(F) \rightarrow {}^3T_{2g}(F)$ (in a d^2-octahedral complex). The direct product $T_{1g} \times T_{1u} \times T_{2g} (\equiv \psi_{el}^{gd} \times \mu \times \psi_{el}^{ex})$ $(= A_{1u} + A_{2u} + 2E_u + 3T_{1u} + 4T_{2u})$ contains the T_{1u} and T_{2u} terms.

Thus the **vibrations of T_{1u} symmetry and T_{2u} symmetry** when coupled with the electronic transition, ${}^3T_{1g} \rightarrow {}^3T_{2g}$, the transition becomes vibronically allowed. The final product contains three T_{1u} terms and four T_{2u} terms. Thus, coupling of a set of T_{2u} metal-ligand vibrations splits the peak into 4 components (*i.e.* subpeaks). Similarly, coupling of a set of T_{1u} metal-ligand vibration splits the peak into three components. *In an octahedral complex, there are two distinct sets of T_{1u} vibration and one set of T_{2u} vibration.* Thus coupling of 2 sets of T_{1u} vibration and one set of T_{2u} vibration produces **10 components or sub-peaks (= 3 + 3 + 4) of the main peak.** These subpeaks are separated by only *ca.* 200 cm^{-1} which represents the energy of metal-ligand bond vibrations. ***Thus the sub-peaks simply broaden the peak. This is the mechanism of peak broadening by vibronic coupling.***

(E) Allowedness of electronic transition in the tetrahedral complexes (noncentrosymmetric systems) in terms of group theory and *d-p* mixing: We have seen that in the octahedral complexes (centrosymmetric systems), the pure electronic transitions are not allowed because the direct product $\psi_{el}^{gd}\mu\psi_{el}^{ex}$ does not contain the totally symmetric term A_{1g}. However, the vibronic coupling (*i.e.* electronic transition coupled with the T_{1u} and /or T_{2u} vibration of the excited state) can make the transitions allowed. Let us consider, the noncentrosymmetric tetrahedral complexes, *e.g.* $[CoX_4]^{2-}$, $X^- = Cl^-$, Br^-, SCN^- where the spin-allowed electronic transitions are:

$$^4A_2 \rightarrow {}^4T_1 \text{ and } {}^4A_2 \rightarrow {}^4T_2$$

In a T_d complex, the coordinates *x*, *y* and *z* generate a basis for the T_2 representation (*see* Character Table, Appendix 2A, Vol. 7). **Thus μ transforms as T_2.** To understand the allowedness of the above electronic transitions, we are to consider the results of direct products.

$A_2 \times T_2 \times T_1$ (for $A_2 \rightarrow T_1$ transition) and $A_2 \times T_2 \times T_2$ (for $A_2 \rightarrow T_2$ transition).

$^4A_2 \rightarrow {}^4T_1$ transition: $\psi_{el}^{gd} \times \mu \times \psi_{el}^{ex} \equiv A_2 \times T_2 \times T_1 = A_1 + E + T_1 + T_2$ (*cf.* Table 7.5.1.1).

The **totally symmetric** A_1 irreducible representation is present in the direct product and consequently, the **electronic transition is orbitally allowed.**

Now let us consider the **$^4A_2 \rightarrow {}^4T_2$ transition** for which the direct product $A_2 \times T_2 \times T_2$ reduces as follows:

$\psi_{el}^{gd} \times \mu \times \psi_{el}^{ex} = A_2 \times T_2 \times T_2 = A_2 + E + T_1 + T_2$, (*cf.* Table 7.5.1.1.)

Since there is no A_1 **(totally symmetric irreducible representation)** term in the direct product, **the transition is forbidden.** However, it gains some intensity through the vibronic coupling (*i.e.* coupling of *E* and/or T_2 normal mode of vibration of the T_d point group). ***This is why, intensity of the $A_2 \rightarrow T_1$ transition is about 10 - 100 times higher than that of the $A_2 \rightarrow T_2$ transition.***

In fact, in a T_d complex, the *p*-orbitals, d_{xy}, d_{yz} and d_{xz} orbitals belong to the T_2 representation and it allows the ***d-p* mixing**. This is why, the electronic transition consists of different allowed components: $d \rightarrow p$ or $p \rightarrow d$. These aspects will be discussed later in detail (*cf.* Sec. 7.5.1G-K).

In the **spectrum of $[CoCl_4]^{2-}$** (in $Cs_2[CoCl_4]$), the $0 \rightarrow 0$ line is the most intense one and it indicates that ***r_e is more or less the same for both the ground ($^4A_2$) state and excited ($^4T_1$) state*** (*cf.* $^4A_2 \rightarrow {}^4T_1$ transition is orbitally allowed) (*cf.* Fig. 7.7.2.1).

Let us illustrate and compare the electronic transitions, $T_{2g} \rightarrow E_g$ **(d^1, d^6; O_h complexes)** and $E \rightarrow T_2$ **(d^1, d^6; T_d complexes).**

- **$T_{2g} \rightarrow E_g(O_h)$:** $\psi_{el}^{gd} \times \mu \times \psi_{el}^{ex} = T_{2g} \times T_{1u} \times E_g = A_{1u} + A_{2u} + 2T_{1u} + 2T_{2u} + 2E_u$

Absence of the totally symmetric representation (A_{1g}) in the direct product makes the transition orbitally forbidden but vibronic coupling (coupling of T_{1u} and/or T_{2u} vibration, *i.e.* ψ_{vib}^{ex} is of T_{1u} or T_{2u} symmetry) makes the transition allowed.

- **$E \rightarrow T_2(T_d)$:** $\psi_{el}^{gd} \times \mu \times \psi_{el}^{ex} = E \times T_2 \times T_2 = A_1 + A_2 + 2E + 2T_1 + 2T_2$ (*cf.* Table 7.5.1.1). Presence of the totally symmetric representation (A_1) in the direct product makes the transition orbitally allowed.

Note: In an electronic transition, there is ***no restriction on the change of vibrational quantum number*** (*i.e.* $\Delta\upsilon = 0, \pm1, \pm2, ...$) for the ***vertical Franck-Condon transition provided the electronic transition (e.g. $A_2 \rightarrow T_1$ in a tetrahedral complex) is itself allowed by its own right.*** Thus, the most probable vibrational transition $\upsilon = 0 \rightarrow \upsilon = 0$ is possible under the conditions:

(i) Internuclear distances in the ground and excited electronic states are more or less the same, *i.e.* r_e(ground) ≡ r_e(excited).

If r_e(excited) > r_e(ground), then the $0 \rightarrow 0$ transition is no longer the most probable one in terms of **Franck-Condon vertical transition** (*cf.* Fig. 7.4.1.4). In such cases, vibrational transition occurs to the higher vibrational states of the excited electronic state.

(ii) The electronic transition is itself allowed (*i.e.* vibronic coupling is not required to make the transition allowed).

(F) Noncentrosymmetric condition imposed in the octahedral complexes by crystal packing and intrinsic asymmetry in the structure of polyatomic ligands: It has been already mentioned that the asymmetrical vibrations (*i.e.* T_{1u} and T_{2u}) can destroy the centre of symmetry in the O_h-systems and then the Laporte selection rule is not applicable. Sometimes, crystal packing in solid state and intrinsic asymmetry in the structure of polyatomic ligands may lead the O_h complexes to depart from the perfect centrosymmetric condition. It allows the electronic transition.

(G) *d–p* mixing (*i.e.* *g – u* mixing) in the noncentrosymmetric systems like the T_d complexes: Let us illustrate the case with the tetrahedral system which lacks in centre of symmetry. This is why, the orbital groups t_2 (= d_{xy}, d_{yz}, d_{xz}) and $e (= d_{x^2-y^2}, d_{z^2})$ do not possess the *g* or *u* properties (more correctly, the MOs of a tetrahedral complex do not possess the *g* or *u* property). However, the parity properties of *d*-orbitals (*i.e.* *g*-symmetry) are retained in the MOs constituted by the *d*-orbitals.

In a tetrahedral system, in terms of MOT, the *e*-set is basically nonbonding and these are purely metal *d*-orbitals (*i.e.*, $d_{x^2-y^2}$ and d_{z^2}). ***On the other hand, the t_2–MOs are constituted by the overlap of metal d–orbitals (i.e. d_{xy}, d_{yz} and d_{xz}) and metal p-orbitals with the ligand p–orbitals (cf. Fig. 3.17.7.1 of Vol. 4).*** However, the extent of participation of the metal *p*-orbitals is relatively less important because of the large energy difference between the $(n-1)d$ and *np* orbitals of metal. Thus, the transition, $e \rightarrow t_2$ bears the following components:

$$d(g) \rightarrow d(g)\ (i.e.\ \Delta l = 0),\ d(g) \rightarrow p(u)\ (i.e.\ \Delta l \neq 0),$$

Thus, though the d → d transition component is forbidden but the d → p transition component is allowed. This is why, the intensity of spectral transitions in the tetrahedral complexes is about 10^2 times higher than that of the centrosymmetric octahedral complexes.

Note: In a tetrahedral complex, in terms of MOT, the transition is: $e \rightarrow t_2^*$ and it is actually **MLCT** (metal → ligand CT), *i.e.* $d(\text{metal}) \rightarrow p(\text{ligand})$.

(H) Characteristics of *d–p* mixing: In a noncentrosymmetric systems, the *d–p* mixing is possible because in such systems, it is not realistic to consider the *d*–orbitals as *g*–type and *p*–orbitals as *u*–type. In fact, in a T_d-complex, the metal d_{xy}, d_{yz} and d_{xz} orbitals and *p*-orbitals of both the metal and ligand belong to the T_2-representation. But, in a centrosymmetric system, the *d–p* mixing is not possible because of their different symmetry properties, *i.e.* *d* as *g*–type and *p*–as *u*–type (*cf.* O_h *vs.* T_d).

(I) Possibility of *d–p* mixing in the octahedral complexes: An asymmetric metal-ligand bond vibration can destroy the centrosymmetric property of the complex at that particular moment. At that moment, the metal $(n-1)d$ orbital can mix with the metal *np* orbital and ligand *p*-orbital. This **instantaneous *d–p* mixing** in the metal centre can relax the Laporte selection rule. But efficiency of this *d–p* mixing can be reviewed in terms of their energy difference.

_d–p_ Mixing: Energetics of mixing of the metal and ligand _p_-orbitals with the metal _d_-orbitals

In the noncentrosymmetric systems, the *d-p* mixing is possible. In the transition metal complexes, the *np* orbitals are of much higher energy than the $(n-1)d$ orbitals. **Thus the intermixing of the metal *np* and $(n-1)d$ orbitals is energetically inefficient.** In fact, for the 3*d*- and 4*p* orbitals, it is about only 10%. **Relativistic effect** (Das *et al.* *Coord. Chem. Rev.*, 2023; *CTI* (IUPAC), 2023) may favour degree of this mixing for the heavier congeners. However, **in terms of MOT, the ligand *p*-orbitals can mix efficiently with the metal *d*-orbitals** (see Fig 3.17.7.1, Vol. 4).

The $(n - 1)d$ orbitals and np orbitals of the metal centre differ significantly in energy and their mixing is not very much efficient. **In fact, for the first transition series, intermixing of the 3*d* and 4*p* orbitals is not very much efficient.** It may be mentioned that because of the **relativistic effect** (*i.e.* relativistic destabilisation of the *d*-orbitals and relativistic stabilisation of the *p*-orbitals), the energy difference between the $(n - 1)d$ orbitals and np orbitals decreases for the heavier congeners (see Vol. 1 for details). Thus energetically, the mixing of metal $(n - 1)d$ orbitals and np orbitals becomes more efficient for the heavier congeners (*i..e.* 4*d*- and 5*d*-members).

(J) Possibility of *d–p* mixing in terms of MOT in the tetrahedral complexes: The *d–p* mixing, *i.e.* **mixing of the metal *d*–orbitals with the ligand *p*–orbitals** leading to the MOs is quite efficient in the noncentrosymmetric systems like the tetrahedral complexes. In a tetrahedral complex, the metal t_2-orbitals (*i.e.* d_{xy}, d_{yz} and d_{xz}) can mix with the ligand *p*–orbitals to produce the t_2-MOs.

In fact, the t_2-MOs of a tetrahedral complex are constructed by combining both the metal $(n–1)d$ and np orbitals with the ligand *p*-orbitals (*cf.* Fig. 3.17.7.1, Vol. 4). But, due to the energy mismatch (*i.e.* metal np orbitals are of much higher energy than the $(n–1)d$ orbitals), participation of the metal *p*–orbitals is relatively less important in constructing the t_2-MOs. This is why, for the sake of simplicity, here we are considering formation of the t_2-MOs through the combination of only metal *d*-orbitals (*i.e.* d_{xy}, d_{yz} and d_{xz}) with the ligand *p*-orbitals.

$$t_2\text{–MO: } \psi_{MO} = x\psi_d^{t_2} + \sqrt{(1-x^2)}\sum\psi_{LGO},$$

x gives the measure of mixing, *i.e.* covalence and $\Sigma\psi_{LGO}$ represents the appropriate combination of the ligand *p*-orbitals in the formation of MO (molecular orbital).

The *e* level is constituted by the pure metal *d*–orbitals (*i.e.* $d_{x^2-y^2}, d_{z^2}$). Thus the $e \rightarrow t_2$ transition leads to the following **transition moment integral** (*I*).

$$I = \int \psi_d^e \,\mu\psi_{MO} d\tau, \ \mu = \text{dipole moment operator}$$

$$= x\int \psi_d^e \mu \psi_d^{t_2} d\tau + \sqrt{(1-x^2)}\int \psi_d^e \mu \sum \psi_{LGO} d\tau$$

The first integral consisting of pure $d \rightarrow d$ transition is zero but the second integral is nonzero because of the *d* (metal) $\rightarrow$ *p* (ligand) transition which basically represents a **MLCT band**.

- **Laporte selection rule in the noncentrosymmetric tetrahedral complexes:** Laporte selection rule is applicable only for the centrosymmetric systems. **Thus, strictly, the rule is not applicable for the noncentrosymmetric tetrahedral and trigonal bipyramidal complexes.** The above analysis indicates that the rule is **partly relaxed** in the tetrahedral complexes. In fact, intensity of the ligand field absorption peaks in the tetrahedral complexes is higher than that of the centrosymmetric octahedral complexes but is much less than that of the fully Laporte allowed charge transfer (CT) peaks (*cf.* Table 7.6.1).

(K) Relaxation of the Laporte selection rule through the *d–p* mixing in the octahedral *vs* tetrahedral complexes:

- ***The ligand field transition ($t_{2g} \rightarrow e_g$) is forbidden in the octahedral complexes but the ligand field transition ($e \rightarrow t_2$) in the tetrahedral complexes is partly allowed because of the efficient d–p mixing (mixing of metal d–orbital and ligand p–orbital) in the noncentrosymmetric tetrahedral complexes.*** For the tetrahedral complexes, the electronic transition should be interpreted in terms of MOT not in terms of CFT. In the noncentrosymmetric T_d-complexes, the *e*-set ($d_{x^2-y^2}$, d_{z^2}) is basically representing the **nonbonding metal *d*-orbitals** while the t_2-set is constituted by the extensive overlap between the metal *d*-orbitals (*i.e.* d_{xy}, d_{yz}, d_{xz}) and ligand orbitals (*i.e.* *p* or hybrid orbitals). Thus the ***e* $\rightarrow$ t_2 *transition should not be considered as the pure d–d transition.***

- Here it may be mentioned that in both the octahedral (*i.e.* centrosymmetric) and tetrahedral (*i.e.* noncentrosymmetric) systems, vibronic coupling to relax the Laporte selection rule is applicable. Mixing of the $(n - 1)d$ orbitals with the *np* orbitals of metals is not efficient because of an energetic ground but the metal *d*–orbitals can mix with the ligand *p* orbitals efficiently as in the noncentrosymmetric tetrahedral complexes. ***This d–p mixing mechanism is more efficient than the vibronic coupling to relax the Laporte selection rule. This d-p mixing in the centrosymmetric octahedral complexes is possible only at the moments when the asymmetric metal-ligand bond vibration destroys the centre of symmetry. This is why, the ligand field transitions in the tetrahedral systems are more intense compared to those of the octahedral systems.***
- **$[Co(NH_3)_6]^{3+}$** is centrosymmetric while **$[Co(NH_3)_5X]^{2+}$** is noncentrosymmetric (C_{4v} symmetry). Thus the *d–p* mixing (*i.e.* mixing of the metal *d*-orbitals with the metal *p*-orbitals and ligand *p*-orbitals) is possible in $[Co(NH_3)_5X]^{2+}$ while it is not possible in $[Co(NH_3)_6]^{3+}$. This is why, the ligand field transitions in $[Co(NH_3)_5X]^{2+}$ are more intense. However, the mixing of 3*d* and 4*p* orbitals is not efficient because of an energetic ground. Thus, this only can weakly relax the selection rule.
- *Cis*-$[Co(NH_3)_4X_2]^+$ (noncentrosymmetric) gives the **more intense ligand field bands** than *trans*-$[Co(NH_3)_4X_2]^+$ (centrosymmetric).
- In the same way, we can explain the relatively higher intensity of the ligand field bands in **$[M(en)_3]^{3+}$** (M = Cr, Co) having the D_3 symmetry compared to those of $[M(NH_3)_6]^{3+}$ of O_h symmetry. **In fact, lowering of symmetry can allow the *d–d* transition partially allowed.**

In the D_3 point group, the orbitals p_x, p_y, d_{xz}, d_{yz}, d_{z^2} and $d_{x^2-y^2}$ belong to the *E* species. This allows the *d–p* mixing and the electronic transition becomes partly allowed.

> **Relativistic effect and *d–p* mixing:** Relativistic effect is more important for the heavier congeners (*cf.* importance: 3*d*-series < 4*d*-series < 5*d*-series) (Das *et al., Coord. Chem. Rev.*, 2023; *CTI* (IUPAC), 2023). In fact, relativistic expansion/destabilisation of the $(n - 1)d$ orbitals and relativistic contraction/stabilisation of the *np* orbitals favour the mixing of the $(n - 1)d$ and *np* metal orbitals energetically for the heavier congeners. **Thus, the Laporte selection rule is relaxed better for the heavier congeners in terms of *d–p* mixing.**

(L) Effect of temperature on vibronic coupling: In vibronic coupling, the metal-ligand bond vibration is important and thermal energy is required for the metal-ligand bond vibration. According to Franck-Condon principle, only **vertical transitions** (Fig. 7.4.1.3) between the two electronic states can occur. If the temperature is lowered (*i.e.* low thermal energy), only fewer vibrational levels of the electronic states are populated. ***To allow the vertical Franck-Condon transition between the two electronic states, inhabitated portions of the two potential energy curves must lie vertically above or below each other.*** When fewer vibrational levels are inhabited at the ground electronic state (*i.e.* ψ_{el}^{gd}), *the number of ways of vibrational transitions from the ground electronic state to the upper electronic state is decreased, i.e.* the number of ways of combination of ψ_{vib}^{gd} and ψ_{vib}^{ex} decreases.

For an ***allowed vibronic transition,*** the product of $\psi_{el}^{gd}\psi_{vib}^{gd}\mu\psi_{el}^{ex}\psi_{vib}^{ex}$ must be of *g*-symmetry. For an octahedral complex, ψ_{el}^{gd} and ψ_{el}^{ex} are of *g*-symmetry and μ is of *u*-symmetry. Hence for an allowed vibronic transition, the product of $\psi_{vib}^{ex} \times \psi_{vib}^{gd}$ must be of *u*-symmetry. It is possible, if ψ_{vib}^{gd} is of *g*-symmetry and ψ_{vib}^{ex} is of *u*-symmetry or ψ_{vib}^{gd} is of *u*-symmetry and ψ_{vib}^{ex} is of *g*-symmetry (*cf.* $g \times u \Rightarrow u$). *If different vibrational levels in the ground electronic state are occupied then the number of possible ways to satisfy the required condition of vibronic coupling through the combination of ψ_{vib}^{gd} and ψ_{vib}^{ex} increases, i.e.* chance of the allowed vibronic transition through the coupling of suitable vibrations of the ground and excited electronic states increases. At a lower temperature, fewer vibrational levels at the ground electronic state are occupied and it reduces the possibilities of the appropriate combinations of ψ_{vib}^{gd} and ψ_{vib}^{ex} to make the product $\psi_{vib}^{gd} \times \psi_{vib}^{ex}$ of *u*-symmetry (*cf.* $v = 0$, ψ_{vib} is of *g*-symmetry; with the lowering of temperature, population density at $\upsilon = 0$ increases **but $\upsilon = 0 \rightarrow \upsilon' = 0$ is a forbidden transition**) (*cf.* $\psi_{el}^{ex}\mu\psi_{el}^{gd}$ is of *u* symmetry and $\psi_{vib}^{ex}\psi_{vib}^{gd}$ is of *g* symmetry for $\upsilon = 0 \rightarrow \upsilon' = 0$ transition). Besides this, the $\upsilon = 0 \rightarrow \upsilon' = 0$ Franck–Condon transition is forbidden for $r_e^{ex} > r_e^{gd}$ (see Figs 6.10.5.2, 7.4.1.4 and 7.7.2.1). Thus, relaxation of the Laporte selection rule through the **vibronic coupling becomes less efficient at a lower temperature** (*cf.* Sec. 7.5.1R). Intensity (measured by **oscillator strength** *f*) of a vibronic transition depends on temperature (*i.e.* **intensity increases with the increase of temperature**) as follows:

$$f\,(\text{measuring intensity}) \propto \coth\,(h\Delta\nu/2k_BT), \quad \coth x = \frac{e^x + e^{-x}}{e^x - e^{-x}}$$

where $h\Delta\nu$ = difference in energy between the levels corresponding to the two wave functions.

Thus the relaxation of Laporte selection rule through the vibronic coupling becomes less efficient at a lower temperature. In other words, the metal-ligand bond vibration is the key player for relaxation in vibronic transition. Lowering of temperature restricts the metal-ligand bond vibration to disfavour the relaxation through vibronic coupling. This is why, in such cases, the intensity of the band decreases with the lowering of temperature. **Thus, intensity of such bands depends on temperature** and this is why, such bands are called the **hot bands.**

Here it may be noted that the *d–p* mixing (mixing of the metal *d*–orbitals with the *p*–orbitals of ligands through the molecular orbital formation) to relax the Laporte selection rule does not depend on the temperature.

Note: Electronic transition is much faster than the time required to execute a molecular vibration. Thus during the electronic transition, the nuclei may be considered to remain fixed. Thus it is considered that the electronic transition occurs so rapidly that during the electronic transition, the metal-ligand bond distance remains unchanged. This is the basis of **Franck-Condon vertical transition (*cf.* Fig. 7.4.1.4).**

(M) Intensity stealing — a special case of vibronic coupling: If there are two excited electronic states (*i.e.* upper states) and to one of which a transtion from a particular electronic level is allowed while the transition to the other excited state is forbidden, then the forbidden transition becomes partially allowed, if the energy of the two excited states is comparable. When the forbidden and allowed excited states lie close, ***some suitable mode of vibration can allow the mixing of the electronic wave functions of the two excited states.*** In this vibronic coupling process, transition to the forbidden excited state becomes partially allowed and intensity of the forbidden band is **increased** but intensity of the allowed band is relatively **decreased**. This phenomenon is called **intensity stealing.** This mixing between the forbidden and allowed excited states becomes less efficient if the energy difference between the states is high.

Generally when the ligand field band (d–d transition) lies close to a charge-transfer band (which is allowed), the intensity of the ligand field band is increased through the intensity stealing phenomenon. For this reason, the ligand field bands moving towards the higher energy region of wavelength gain more intensity because in this region, the high energy charge transfer bands generally prevail.

(N) Lowering of symmetry to relax the Laporte selection rule in terms of group theory: It has been already mentioned that in the noncentrosymmetric complexes like $[M(NH_3)_5X]^{2+}$ (M = Co, Cr) (C_{4v} symmetry) and $[Co(en)_3]^{3+}$ (D_3 symmetry), the ligand field bands are more intense compared to those in $[M(NH_3)_6]^{3+}$ of O_h symmetry. This has been explained by considering the possibility of *d-p* mixing. This can also be understood in the following manner.

● **$O_h \rightarrow D_3$:** In $[Cr(NH_3)_6]^{3+}(O_h)$, the spin-allowed transition, $^4A_{2g} \rightarrow {}^4T_{2g}$ may be considered. In this centrosymmetric O_h complex, both the excited and ground electronic states are of *g*–symmetry and the dipole moment operator is of *u*–symmetry (*i.e.* T_{1u}). It indicates: $g \times u \times g = u$ and the corresponding

Intensity of ligand field bands: O_h *vs.* T_d *Complexes*

Vibronic coupling (O_h) *vs. d–p* mixing (T_d): In the centrosymmetric octahedral complex, intensity of the ligand field band is earned through the **vibronic coupling** by relaxing the Laporte Selection Rule. In the noncentrosymmetric tetrahedral complex, this vibronic coupling is also applicable. ***But, in the tetrahedral system, the more efficient mechanism is the d–p mixing (mixing of the metal t_2 d-orbitals with the ligand p-orbitals) to relax the Laporte Selection Rule***. This *d–p* mixing is not efficient in the centrosymmetric octahedral system.

Thus, an octahedral system enjoys only vibronic coupling to relax the Laporte selection Rule while a tetrahedral system enjoys both the vibronic coupling and *d–p* mixing to relax the Laporte Selection Rule. But the *d–p* mixing route is more efficient than the vibronic coupling to relax the Laporte Selection Rule. ***This is why, the tetrahedral complexes are more intense in colour.*** Here it is worth mentioning that an asymmetric metal-ligand bond vibration can instantaneously destroy the centrosymmetric character and at that moment, metal $(n-1)d$ orbitals can mix with the metal *np*–orbital and ligand *p*-orbitals. This instantaneous *d-p* mixing can partly relax the Laporte selection rule. It may be noted that the large energy difference between the metal $(n-1)d$ orbitals and metal *np* orbitals does not allow this *d–p* mixing at the metal centre significantly.

Intensities of the spectral transitions in the T_d complexes **are not affected much by temperature** because efficiency of the *d–p* mixing which relaxes the Laporte selection rule does not depend on temperature. But intensities of the spectral transitions in the O_h complexes decrease with temperature as the efficiency of the vibronic coupling to relax the selection rule decreases with the decrease of temperature.

Thus mixing of the metal d-orbitals with the ligand p-orbitals is favoured both mechanistically and energetically in the noncentrosymmetric T_d complexes while the instantaneous mixing of the metal *d*-orbitals with the metal and ligand *p*-orbitals is definitely less efficient in the centrosymmetric O_h complexes. It may be noted that mixing of the metal $(n-1)d$ orbitals and *np* orbitals is inefficient in the centrosymmetric O_h complexes both mechanistically and energetically.

transition moment integral is zero. Thus it represents a **forbidden electronic transition.** In moving from the O_h to D_3 point group of D_3 symmetry (*e.g.* $[Cr(en)_3]^{3+}$), the T_{2g} term of O_h system splits into the A_1 and E states and the A_{2g} term transforms into A_2 in D_3 (*cf.* Fig. 7.5.1.2). Thus, the relevant transitions in the D_3-point group are:

$$^4A_2 \rightarrow {^4A_1};\ ^4A_2 \rightarrow {^4E}.$$

Now let us consider the transition moment integral for the transition, $^4A_2 \rightarrow {^4A_1}$ where $\psi_{el}^{gd} \times \psi_{el}^{ex} = A_2 \times A_1 = A_2$. The dipole moment operator components μ_x and μ_y are of E symmetry while the μ_z component is of A_2 symmetry (*see* **Character Table, Appendix 2A, Vol. 7**). Thus we can write:

$\psi_{el}^{gd}\ \psi_{el}^{ex}\ \mu = A_2 \times A_1 \times E = E$ (for the μ_x and μ_y component); (A_1 term absent in the product)

$= A_2 \times A_1 \times A_2 = \mathbf{A_1}$ (for the μ_z component)

It indicates that the transition is allowed for the μ_z component of the dipole moment operator because in the direct product result, the totally symmetric irreducible representation (A_1) is present. But the transition for the μ_x and μ_y components is forbidden.

In moving from **$[Cr(NH_3)_6]^{3+}$ (O_h) to $[Cr(en)_3]^{3+}$ (D_3)**, the symmetry is reduced and we can compare and analyse the spectral transitions in these two point groups (*cf.* Fig. 7.5.1.2).

The results of **all the electronic transitions in the D_3 point group** are given below (*cf.* Table 7.23.1 for the direct product result) to understand the electronic polarisation.

Transitions	$\Gamma\psi_{el}^{gd}\Gamma\mu_{x,y}\Gamma\psi_{el}^{ex}$	$\Gamma\psi_{el}^{gd}\Gamma\mu_z\Gamma\psi_{el}^{ex}$
$A_2 \rightarrow A_1$	$A_2 \times E \times A_1 = E$	$A_2 \times A_2 \times A_1 = \mathbf{A_1}$
$A_2 \rightarrow A_2$	$A_2 \times E \times A_2 = E$	$A_2 \times A_2 \times A_2 = A_2$
$A_2 \rightarrow E$	$A_2 \times E \times E = \mathbf{A_1} + A_2 + E$	$A_2 \times A_2 \times E = E$

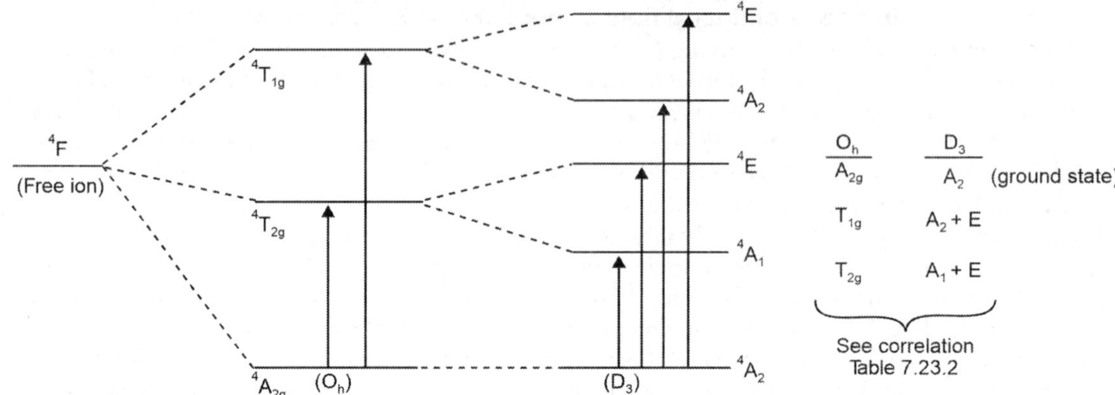

Fig. 7.5.1.2 Spectral transitions in the O_h and D_3 point groups for Cr^{3+} (d^3). The 3rd ligand field transition, ${}^4A_{2g} \rightarrow {}^4T_{1g}$ (P) not shown here and this transition gets very often obscured due to the CT band. In the O_h-system, all the electronic transitions are orbitally forbidden.

The direct product contains the totally symmetric irreducible representation A_1 in two transitions and these are allowed and remainings are forbidden. These are:

	$A_2 \rightarrow A_1$	$A_2 \rightarrow A_2$	$A_2 \rightarrow E$
For μ_z (II to C_3 axis):	**Allowed**	Forbidden	Forbidden
For $\mu_{x,y}$ ($\perp$ to C_3 axis):	Forbidden	Forbidden	**Allowed**

It indicates the **electronic polarisation** of the incident radiation. These predictions have been experimentally verified.

Note: The details of direct product (*see* Table 7.23.1) and character table (*see* Appendix 2A, Vol. 7) are not discussed here.

Possibility of *d–p* mixing in D_3-symmetry: Here it may be pointed out that in the D_3 point group, the orbitals (p_x, p_y), (d_{xz}, d_{yz}) and $\left(d_{x^2-y^2}, d_{z^2}\right)$ transform as the *E*-species. This allows the $(n-1)d - np$ mixing to some extent depending on the energy difference between the metal $(n-1)d$ and *np* orbitals. This route can partially relax the orbital selection rule but it is absent in the O_h-complex.

$[Cr(NH_3)_6]^{3+}$ *vs.* $[Cr(en)_3]^{3+}$, *i.e.* O_h *vs.* D_3: Now let us compare the **spectral characteristics of $[Cr(NH_3)_6]^{3+}$ (O_h) and $[Cr(en)_3]^{3+}$ (D_3).**

	${}^4A_{2g} \rightarrow {}^4T_{2g}$	${}^4A_{2g} \rightarrow {}^4T_{1g}$
$[Cr(NH_3)_6]^{3+}$	21,550 cm^{-1}; (30)	28,500 cm^{-1}, (27)
$[Cr(en_3)]^{3+}$	21,850 cm^{-1}; (76)	28,450 cm^{-1}, (60)

It is evident that in moving from the O_h to D_3 symmetry, the ***spectral bands are not split***, *i.e.* both $[Cr(NH_3)_6]^{3+}$ and $[Cr(en)_3]^{3+}$ are showing the same spectral pattern (*see* Fig. 7.14.5.3). In fact, in the D_3 symmetry of $[Cr(en)_3]^{3+}$, the splitting of the energy states is so small that the separate absorption peaks are not noticed, *i.e.* they overlap giving rise to two peaks as in the O_h-system. Thus, ***approximately, both the complexes are of octahedral symmetry,*** *i.e.* $[M(N)_6]^{3+}(O_h)$. The intensities (measured by ε, given in parenthesis, *cf.* Table 7.14.5.1) of the en-complex of D_3 symmetry are approximately twice as large as those of the ammine complex of O_h symmetry. *This **slightly enhanced intensity** for the en-complex arises from the **possibility of d–p mixing in D_3-symmetry***. Thus, basically in both the ammine and en complexes, the electronic transitions are the **vibronic transitions.**

● $\boldsymbol{O_h \rightarrow D_3}$ ***vs.*** $\boldsymbol{O_h \rightarrow C_{2v}, D_{4h}}$**:** Though in moving from $[M(NH_3)_6]^{3+}$ (O_h, M = Co, Cr) to $[M(en)_3]^{3+}$ (D_3), the spectral characteristics are not significantly changed (*i.e.* ***chelation by en does not effectively change the*** $\boldsymbol{O_h}$ ***symmetry***), in moving from the O_h to D_{4h} symmetry (as in *trans*-$[MCl_2(NH_3)_4]^+$, *trans*-$[MCl_2(en)_2]^+$) and C_{2v} symmetry (as in *cis*-$[MCl_2(NH_3)_4]^+$, *cis*-$[MCl_2(en)_2]^+$), the spectral features are remarkably changed (*cf.* Secs. 7.14.4, 7.14.5, 7.15.2, 7.15.3). The *trans*-complexes of $[MCl_2(NH_3)_4]^+$ or $[MCl_2(en)_2]^+$ are still centrosymmetric and these are more symmetric than the corresponding *cis*-complexes of C_{2v} symmetry. ***In fact, the cis-complexes show the more intense bands.*** In general, we can conclude:

the d–d transitions are more allowed for the complexes of lower symmetry.

Vibronic Coupling in the Transitions in $[Co(NH_3)_6]^{3+}$ (*cf.* Table 7.5.1.2)

${}^1A_{1g} \rightarrow {}^1T_{1g}$ and ${}^1A_{1g} \rightarrow {}^1T_{2g}$ transitions; normal modes of vibration = $3n - 6 = 15$ for the MN_6 moiety and these are: A_{1g}, $2E_g$, $2T_{1u}$, T_{2u}, T_{2g} where the odd *M—N* vibrations are T_{1u} and T_{2u}; $A_{1g} \times T_{1g} = T_{1g}$; $A_{1g} \times T_{2g} = T_{2g}$; dipole moment operator = T_{1u}. It leads to:

$$A_{1g} \times T_{1g} \times \mu = T_{1g} \times T_{1u} = A_{1u} + E_u + T_{1u} + T_{2u};$$

$$A_{1g} \times T_{2g} \times \mu = T_{2g} \times T_{1u} = A_{2u} + E_u + T_{1u} + T_{2u}$$

Combination of the odd vibration leads to: $T_{1u} \times T_{1u} \Rightarrow A_{1g}$ (one term) and $T_{2u} \times T_{2u} \Rightarrow A_{1g}$ (one term). Thus the transitions are **vibronically allowed** (*see* Table 7.5.1.1 for the direct product).

● $\boldsymbol{O_h \rightarrow D_{4h}}$**:** Now let us consider the lowering of symmetry leading to O_h to D_{4h}. The *z*–out or *z*–in tetragonal distortion converts the O_h point group into the D_{4h} point group. Let us consider the transition, ${}^3T_{1g} \rightarrow {}^3A_{2g}$ in a d^2**–octahedral complex**. The dipole moment operator component μ ($= \mu_x = \mu_y = \mu_z$) is a T_{1u} species. The direct product is:

$$T_{1g} \times \mu \times A_{2g} = T_{1g} \times T_{1u} \times A_{2g} = T_{1u} + T_{2u} + E_u + A_{2u} \text{ (cf. Table 7.5.1.1)}$$

The direct product does not contain the totally symmetric A_{1g} term and consequently, the **electronic transition is forbidden.** However, coupling of the odd T_{1u} and T_{2u} vibrations with the electronic transition will allow the transition.

Now let us consider the fate of the transition in the D_{4h} point group which still maintains the centre of symmetry (*i.e.* centrosymmetric).

$T_{1g}(O_h)$ is converted into $E_g + A_{2g}$ (in D_{4h}); $A_{2g}(O_h)$ is converted into B_{1g} (in D_{4h}); the dipole moment operator components in D_{4h} symmetry are: μ_x, μ_y of E_u symmetry and μ_z of A_{2u} symmetry.

Thus the ${}^3T_{1g} \rightarrow {}^3A_{2g}$ transition in O_h represents the transitions, E_g, $A_{2g} \rightarrow B_{1g}$ in D_{4h}. The direct product for the transitions in the D_{4h} point group becomes:

$$\left.\begin{aligned} &(E_g + A_{2g})E_u \times B_{1g} + (E_g + A_{2g}) \times A_{2u} \times B_{1g} \\ &= \underbrace{(A_{1u} + A_{2u} + B_{1u} + B_{2u} + E_u)}_{(\text{For } \mu_x, \text{ and } \mu_y)} + \underbrace{(E_u + B_{1u})}_{(\text{For } \mu_z)} \end{aligned}\right\} \text{(cf. Table 7.23.1)}$$

For moving from the O_h to D_{4h} symmetry, though the symmetry is lowered, the **above transitions are still forbidden** because the direct products of $\psi_{el}^{gd} \times \mu \times \psi_{el}^{ex}$ do not contain the totally symmetric irreducible representation A_{1g} (*cf.* D_{4h} still maintains the centre of inversion). However, the **vibronic coupling can allow the transitions.** The normal modes of vibration of D_{4h} symmetry are.

$2A_{1g}$, B_{1g}, B_{2g}, E_g, $2A_{2u}$, B_{1u}, $3E_u$ (*cf.* Table 7.23.3 and Appendix 2B, Vol. 7).

Coupling of the odd vibrations, *i.e.* A_{2u}, B_{1u} and E_u, can remove the forbiddeness of the transitions.

(O) Vibronic polarisation and dichroism: Let us illustrate it for the electronic transitions in *trans*-$[CoCl_2(en)_2]^+$ (D_{4h} symmetry) where the spin allowed transitions are (*cf.* Fig. 7.15.2.4c, d^6 system):

$$\left.\begin{array}{l} {}^1A_{1g}\rightarrow{}^1E_g \text{ and } {}^1A_{1g}\rightarrow{}^1A_{2g} \text{ (Corresponding transition in } O_h\text{: } {}^1A_{1g}\rightarrow{}^1T_{1g}) \\ {}^1A_{1g}\rightarrow{}^1B_{2g} \text{ and } {}^1A_{1g}\rightarrow{}^1E_g \text{ (Corresponding transition in } O_h\text{: } {}^1A_{1g}\rightarrow{}^1T_{2g}) \end{array}\right\} (cf. \text{ Fig. 7.15.2.1})$$

In D_{4h}, μ_z (along the C_4 axis) transforms as A_{2u} while μ_x and μ_y perpendicular to z (C_4) axis transform as E_u (*see* Character Table, Appendix 2A, Vol. 7). The normal modes of vibration are:

$$2A_{1g}, B_{1g}, B_{2g}, E_g, 2A_{2u}, B_{1u}, 3E_u \text{ (cf. Table 7.23.3).}$$

Transitions	$\Gamma\psi_{el}^{gd}\Gamma\mu_{x,y}\Gamma\psi_{el}^{ex}$	$\Gamma\psi_{el}^{gd}\Gamma\mu_z\Gamma\psi_{el}^{ex}$	
${}^1A_{1g}\rightarrow{}^1A_{2g}$	$A_{1g}\times E_u\times A_{2g}=E_u$	$A_{1g}\times A_{2u}\times A_{2g}=A_{1u}$	
${}^1A_{1g}\rightarrow{}^1B_{2g}$	$A_{1g}\times E_u\times B_{2g}=E_u$	$A_{1g}\times A_{2u}\times B_{2g}=B_{1u}$	(*cf.* Table 7.23.1)
${}^1A_{1g}\rightarrow{}^1E_g$	$A_{1g}\times E_u\times E_g=A_{1u}+A_{2u}+B_{1u}+B_{2u}$	$A_{1g}\times A_{2u}\times E_g=E_u$	

The **odd vibrations** are: A_{2u}, B_{1u} and E_u. Hence, if in the direct product $\psi_{el}^{gd}\mu\psi_{el}^{ex}$, there is any term matching with the symmetrie of these odd vibrations, then the transition is vibronically allowed. *Thus the transition $A_{1g}\rightarrow A_{2g}$ is vibronically forbidden for z-polarisation while all other transitions are vibronically allowed.*

The band at ~16000 cm^{-1} is due to ${}^1A_{1g}\rightarrow{}^1E_g$ and the band at ~22,000 cm^{-1} is due to ${}^1A_{1g}\rightarrow{}^1A_{2g}$ (*cf.* Fig. 7.15.2.4c). These E_g and A_{2g} components are derived from the T_{1g} state of O_h symmetry. The ${}^1A_{1g}\rightarrow{}^1E_g$ transition is vibronically allowed in all directions while the ${}^1A_{1g}\rightarrow{}^1A_{2g}$ transition is **vibronically forbidden for z-polarisation.** *In fact, the band at ~22,000 cm^{-1} is strongly polarised and this band remains absent for the light polarised parallel to z (C_4) axis but it exists for the polarised light perpendicular to z-axis. The band at ~16,000 cm^{-1} shows no such polarisation effect.* The broad bad at ~27,000 cm^{-1} is due to the ${}^1A_{1g}\rightarrow E_g$, B_{2g} (states arising from the T_{2g} state of O_h symmetry). This dichroism of *trans*-$[CoCl_2(en)_2]^+$ has been experimentally verified (*cf.* Fig. 7.15.2.4c).

Note: For the results of direct product in D_{4h} point group, the readers are to consult any standard book on group theory (*see* Table 7.23.1).

(P) Different factors (summary) responsible for the origin of colour in the octahedral complexes: These are summarised here.

(i) **Loss of centre of symmetry:** Electronic transitions become more allowed when the centre of symmetry is lost. It may be mentioned that in absence of the centre of symmetry, *probability of the allowedness of the electronic transition is increased* though all the possible electronic transitions may not be allowed (*cf.* in the noncentrosymmetric T_d system, the $A_2\rightarrow T_2$ transition is not itself allowed by its own right, *see* solvd problem 15).

Loss of the centrosymmetric character allows the ***d–p* mixing** which in turn makes the electronic transitions more allowed due to the $d\rightarrow p$ and/or $p\rightarrow d$ component in the electronic transition.

Loss of the centre of symmetry can occur in different ways:

- **asymmetric metal-ligand bond vibration** (temporary loss of centre of symmetry); ● **crystal packing** in solid state may lead to distortion to remove the centre of symmetry; ● **intrinsic asymmetry in the polyatomic ligands** (generally organic ligands); ● symmetry reduction or **symmetry lowering** due to the ligand environment (*e.g.* $O_h\rightarrow C_{4v}$, C_{2v}, D_3, etc.)

(ii) **Vibronic coupling:** Coupling of a suitable odd vibration may lead to the electronic transition allowed under the condition: transition moment integral with the incorporation of the odd vibrational representation leads to the totally symmetric representation of the group, *i.e.* the matrix element $\langle\psi_{el}^{gd}\times\mu\times\psi_{el}^{ex}\times\psi_{vib}^{ex}\rangle$ must contain the totally symmetric representation of the group (*e.g.* ${}^1A_{1g}$ in O_h) assuming ψ_{vib}^{gd} to be totally symmetric.

(iii) ***cis- vs. trans*-complex:** The *trans*-complexes being more symmetric give the less intense bands than the *cis*-complexes, *e.g.* for $[CoCl_2(en)_2]^+$, the *trans*-isomer (D_{4h}, centrosymmetric) gives the less intense bands than the *cis*-isomer (C_{2v}).

(iv) **Greater covalency:** It will establish the overlap between the ligand orbital and metal orbital. It will generate the more intense bands. For example, in a **tetrahedral complex**, the $e \rightarrow t_2$ transition does not really describe the $d \rightarrow d$ transition. The *e*-level is basically constituted by the nonbonding $d_{x^2-y^2}$ and d_{z^2} metal orbitals but the t_2-level is constituted by the overlap of metal *d*-orbitals (d_{xy}, d_{yz}, d_{xz}) and ligand *p*-orbitals or *s–p* hybrid orbitals. If the covalency can be introduced into the octahedral complexes, then the intensity of the bands will increase. In fact, the **higher oxidation state** will favour the better covalency (*cf.* Fajan's rule) to give the more intense bands.

(v) **Intensity stealing effect:** If the strong CT band appears close to the ligand field bands, then the so called weak ligand field bands become more intense due to the intensity stealing effect.

(Q) Vibronic spectra and selection rule for the vibrational transition: Vibronic spectra represent the electronic spectra associated with the vibrational transitions from the ground electronic state to the excited electronic state (*cf.* Figs. 7.4.1.4, 7.7.2.1). Vibronic spectra appear for all the allowed electronic transitions. At normal temperature, the lowest vibrational level (*i.e.* $v = 0$) is most populated and the vibrational transition will predominantly occur from $v = 0$. For the vibrational transition, $v = 0 \rightarrow v'$ (*cf.* Fig. 7.7.2.1), the **condition of vertical Franck-Condon transition** (*i.e.* r_e^{gd} for $v_0 \approx r^{ex}$ for v') **must be maintained.** If the electronic transition is itself allowed, then there is no quantum restriction for the vibrational transition but if the electronic transition is to be allowed through the vibronic coupling then there will be a ***quantum restriction for the vibrational transition.***

Case I (Electronic transition allowed itself): Some of the electronic transitions become allowed by their own right in the noncentrosymmetric systems. In such cases, there is no selection rule for Δv (= $v \sim v'$, v denotes the vibrational level at the ground electronic state and v' denotes the vibrational level at the excited electronic state) provided the condition of Franck-Condon vertical transition is satisfied. In such cases, Δv may be 0, ±1, ±2, ... However, the **Franck-Condon factor** will determine the relative intensities (*cf.* Sec. 7.7.2) of the transitions.

Case II (Electronic transition allowed through the vibronic coupling): In such cases, besides the condition of Franck-Condon vertical transition, the matrix element $\langle \psi_{el}^{gd}\ \psi_{vib}^{gd}\ \mu\ \psi_{el}^{ex}\ \psi_{vib}^{ex} \rangle$ must contain the **totally symmetric representation term** of the group. In an octahedral system (which is centrosymmetric), for the vibronically allowed electronic transition, the product $\psi_{vib}^{gd} \times \psi_{vib}^{ex}$ must be of *u*-symmetry (*i.e.* T_{1u} or T_{2u}). In such cases, the transitions with $\Delta v = 0$ (*e.g.* $\boldsymbol{v} = 0 \rightarrow \boldsymbol{v}' = 0$, $v = 1 \rightarrow \boldsymbol{v}'$ = 1, etc.) are the Laporte forbidden transitions but the vibrational transitions like $\boldsymbol{v} = 0 \rightarrow \boldsymbol{v}' = 1$, $\boldsymbol{v} = 1 \rightarrow \boldsymbol{v}' = 2$, $\boldsymbol{v} = 1 \rightarrow \boldsymbol{v}' = 0$, etc. may relax the Laporte selection rule if $\psi_{vib}^{gd} \times \psi_{vib}^{ex}$ is of *u*-symmetry.

(R) Peak broadening due to the vibrational transitions and effect of temperature (*cf.* Secs. 7.7.2, 3): If the internuclear distance (*i.e.* metal-ligand bond distance) does not change during the electronic transition (*i.e.* $r_e^{gd} = r_e^{ex}$), then **the peak becomes sharp** while the **broader peaks** will appear under the condition, $r_e^{gd} < r_e^{ex}$, because of the more possible vibrational transitions.

For the **allowed electronic transition and $\boldsymbol{r_e^{gd} = r_e^{ex}}$,** at normal temperature, $\boldsymbol{v} = 0 \rightarrow \boldsymbol{v}' = 0$ (*i.e.* $0 \rightarrow 0$) transition is the most intense one (giving a sharp peak) but at the higher temperature, $\boldsymbol{v} = 1 \rightarrow \boldsymbol{v}' = 0, 1, 2, ...$, $\boldsymbol{v} = 2 \rightarrow \boldsymbol{v}' = 0, 1, 2, ...$, etc. transitions become important (*cf.* $\Delta \boldsymbol{v}$ = any value) when the higher vibrational levels at the ground electronic state become more populated. It should be remembered that for the **vibronically allowed electronic transitions, even for $\boldsymbol{r_e^{gd} = r_e^{ex}}$, the $\boldsymbol{0 \rightarrow 0}$ transition is forbidden,** and the $0 \rightarrow 1$ transition may become important (*cf.* $\Delta \boldsymbol{v} = 0$ forbidden). At higher temperature, 1 → 0, 1 → 2, etc. transitions may become important provided $\psi_{vib}^{gd} \times \psi_{vib}^{ex}$ is of *u*-symmetry.

At the low temperature, the lowest vibrational level (*i.e.* v = 0) is the most populated one and the vibrational transition will occur from the v = 0 level. But at higher temperatures, the population density at the higher vibrational levels (*i.e.* v = 1, 2,...) increases according to the Boltzmann distribution law and the vibrational transitions from these higher levels (*i.e.* v = 1, 2,...) will **broaden the peak. Thus at the lower temperature, the peak becomes relatively sharper.**

Case I (electronically allowed transition, Δv = any value **and $r_e^{gd} \approx r_e^{ex}$):** At the low temperature, the v = 0 level (at the ground electronic state) is the most populated one and the $v = 0 \rightarrow v' = 0$ transition (having the highest Franck-Condon factor, *cf.* Sec. 7.7.2) is the most intense one. **At the higher temperatures**, population density at the higher vibrational levels (*i.e.* v = 1, 2, ...) will increase according to Boltzmann's distribution law and the transitions like, 1 → 0 (*i.e.* v = 1, v' = 0); 2 → 1; 3 → 2; 1 → 1; 2 → 2; etc. will occur at the lower frequencies compared to that of the 0 → 0 transition (*cf.* ΔE_{ex} = energy difference between the successive vibrational levels at the excited electronic state < ΔE_{gd} = energy difference between the successive vibrational levels at the ground state). Thus at higher temperature, then intensity of the 0 → 0 transition will decrease because the population density at v = 0 of the ground electronic state will decrease and the **band will be broadened towards the longer wavelength** but *the total oscillator strength (f) will remain unchanged.* Thus in such cases, we can conclude:

with the rise of temperature, the peak will be broadened towards the longer wavelength, and the molar extinction coefficient will decrease for the 0 → 0 transition but the total oscillator strength f (determined by the area of the peak) will remain unchanged.

Case II (vibronically allowed but electronically forbidden transition with $r_e^{gd} \approx r_e^{ex}$): In such cases, the vibrational transition must satisfy the condition: $\psi_{vib}^{ex} \times \psi_{vib}^{gd}$ is of *u*-symmetry. Here the 0 → 0 transition (most probable one in terms of Franck-Condon Factor) is forbidden but the 0 → 1 transition is allowed. With the lowering of temperature, population density at v = 0 will increase and it will make the 0 → 1 transition more intense. At the higher temperature, population density at the higher vibrational levels (*i.e.* v = 1, 2,...) of the ground electronic state will increase and it will allow the transitions like 1 → 0; 1 → 2; 2 → 1; 2 → 3′; etc. that will occur at the longer wavelength compared to that of the 0 → 1 transition. With the lowering of temperature, the population density at v = 0 increases but the $v = 0 \rightarrow v' = 0$ transition having the highest Franck-Condon factor is forbidden. *It will definitely enhance the intensity of the v = 0 → v′ = 1 transition but it will not be able to make up the loss of intensity earned from the transitions occurring from the higher levels, i.e. v = 1, 2, 3,... This is why, with the lowering of temperature the peak becomes sharper but the oscillator strength (f) decreases for the vibronically allowed electronic transitions.* (*cf.* ***f*** **is independent of temperature for the allowed electronic transitions**). In such cases, we can conclude:

with the rise of temperature, the peak broadens towards the longer wavelength and intensity measured by the total oscillator strength (f) will also increase (i.e. vibronic coupling becomes more efficient at the high temperature).

Note 1: If $\Delta E_{gd} = \Delta E_{ex}$, *i.e.* energy separation between the adjacent vibrational levels in the ground and excited electronic states is the same, then all the transition 0 → 0; 1 → 1; 2 → 2; ... (*i.e.* Δv = 0) will coincide, *i.e.* no peak broadening effect. If $\Delta E_{gd} > \Delta E_{ex}$, then 1 → 1; 2 → 2; 1 → 0; etc. transitions will broaden the peak towards the longer wavelength **compared to that of the 0 → 0 transition.**

Similarly, for the vibronically allowed transition, under the condition, $\Delta E_{ex} < \Delta E_{gd}$, the transitions $1 \rightarrow 2$; $2 \rightarrow 3$; $1 \rightarrow 0$; $2 \rightarrow 1$; $3 \rightarrow 2$ will occur at the longer wavelength compared to that of the **$0 \rightarrow 1$ transition.**

Note 2: Spin allowed $t_{2g} \rightarrow e_g$ transition makes $r_e^{ex} > r_e^{gd}$ and it causes **peak broadening**. **Spin forbidden transition** arising from the rearrangement of electron in the t_{2g} level as in the Cr^{III}-complexes (d^3-system) makes $r_e^{ex} \approx r_e^{gd}$ and the **peak becomes sharper** due to the transition from the lower vibrational energy levels of the ground state to the excited state (see Secs 7.7.2, 3 for details).

7.5.2 Relaxation of the Spin Selection Rule

According to this rule, the singlet → triplet or triplet → singlet transition is not possible. In other words, the electronic transition can occur only between the states having the same spin multiplicity (see Fig. 7.4.1.2). When the ground state and excited state differ in spin multiplicity, the **spin-orbit (SO) coupling phenomenon** can allow the mixing of two states. In such cases, the transiton becomes partially allowed. It may be noted that the ***spin-orbit (SO) coupling is efficient when the energy difference between the ground state and excited state is small.***

The states of different spin multiplicity possess the different spin and orbital angular momenta but they can mix through the SO coupling because they possess the **same total angular momenta**. This mixing interaction can be represented as follows (see Sec. 6.10.8 for details):

$$\psi = a\psi_S + b\psi_T$$

where ψ_S and ψ_T describe the pure singlet and triplet states respectively, a and b are the mixing coefficients measuring the relative contributions of the pure states to the resultant state ψ. For $a \gg b$, the ground state is essentially a singlet one with a small amount of triplet character while the excited state is essentially a triplet one with a small amount of singlet character. Because of this SO coupling, both the ground and excited states possess the singlet and triplet characters in different ratios. **It makes the singlet-triplet transition partially allowed** (see Sec. 6.10.8 for details).

If we consider that the ground state is singlet and the excited state is triplet, then spin-orbit coupling can mix some triplet character of the excited state into the ground state and similarly some character of the singlet ground state can be introduced into the excited state. Consequently, the so called, singlet → triplet transition will have the components singlet → singlet and tripet → triplet which are allowed. Thus, the so called spin forbidden transition becomes partially allowed. For example, in $[Mn(OH_2)_6]^{2+}$ the ground state bears $S = 5/2$ while the excited state bears $S = 3/2$. The mixing through spin-orbit coupling can introduce some character of $S = 3/2$ into the ground state and some character of $S = 5/2$ into the excited state. Thus the transition becomes partially allowed. Thus, the spin forbidden transition becomes partially allowed.

Here it is important to mention that the strength of spin-orbit coupling (*i.e.* coupling of spin and orbital angular momenta) is measured by the spin-orbit coupling constant **which is higher for the heavier atoms** (see Sec. 8.13, Vol. 6). This is why, **intensity of the spin forbidden transition is higher for the heavier metal centres (heavy atom effect)**. For the 3*d*-series, the spin-orbit coupling constant is relatively smaller and their spin forbidden peaks are weaker while the spin forbidden transition peaks are stronger the the 4*d* and 5*d* series.

Note: Russel–Saunders (R–S) coupling considers the well defined values of L and S but the SO coupling indicates that the assumption of definite value of S is not correct. Thus interpretation of the electronic spectra by using the R–S coupling for the free ion states becomes uncertain and less reliable specially, for the heavy atoms having the high SO coupling constants.

7.6 BAND INTENSITY AND THE PARAMETERS MEASURING THE BAND INTENSITY

Intensities of the different types of transitions: These are compared in Table 7.6.1.

Parameters measuring the band intensity: Intensities of different bands are very often compared by their ε_{max} (molar absorptivity or molar extinction coefficient at λ_{max} of absorption) values. This procedure of comparison of band intensities is true if the bands under comparison are *sharp, symmetric* and of *the same bandwidth.* But very often, the bands are *unsymmetric* and they are of *different bandwidths.* Bandwidth may change with temperature. Thus, to measure the band intensity, ε_{max} is not an accurate parameter and for this purpose, **we should measure the area under the band** and this can be obtained by integrating as follows:

$$I = \int_{\bar{\nu}_1}^{\bar{\nu}_2} \varepsilon_{\bar{\nu}} d\bar{\nu}$$ (*cf.* $I = D = (\text{TMI})^2$, transition moment integral = **TMI**, see Se. 7.4.2)

I = Integrated band intensity; $\bar{\nu}_1$ and $\bar{\nu}_2$ are the starting and ending wave numbers of the band; $\varepsilon_{\bar{\nu}}$ (in litre mol^{-1} cm^{-1}) = molar absorptivity at $\bar{\nu}$. *I represents the area under the absorption curve.* If $\bar{\nu}$ is measured in cm^{-1}, I has the unit litre mol^{-1} cm^{-2} or 10^3 cm mol^{-1}. For a symmetrical band, we can reasonably write:

$$I = \int_{\bar{\nu}_1}^{\bar{\nu}_2} \varepsilon_{\bar{\nu}} d\bar{\nu} = \varepsilon_{max} \Delta\bar{\nu}_{1/2},$$

($\Delta\bar{\nu}_{1/2}$ called **half intensity bandwidth,** *i.e.* width of the band at $\frac{1}{2}\varepsilon_{max}$ and it is measured in cm^{-1}, *cf.* Sec. 7.7).

I **(integrated band intensity)** can be expressed in terms of **oscillator strength** (*f* number) as follows:

$$f \propto I = D = [\text{TMI}]^2 \text{ or, } f \text{ (dimensionless number)} = 4.315 \times 10^{-9} \text{ (mol cm}^{-1}) \int_{\bar{\nu}_1}^{\bar{\nu}_2} \varepsilon_{\bar{\nu}} d\bar{\nu}$$

$$= 4.315 \times 10^{-9} \varepsilon_{max} \Delta\bar{\nu}_{1/2}, \text{ (for a symmetric band).}$$

The quantity *f* is determined graphically by plotting ε *vs* $\bar{\nu}$ (in cm^{-1}) and calculating area of the peak.

Note: Oscillator strength (*f*) for a particular transition is defined as the **ratio of the actual and ideal transition probabilities,** *i.e.* $f = \frac{\text{Actual transition probability}}{\text{Ideal transition probability}}$. It indicates that for a perfectly allowed

Table 7.6.1 Intensities in different types of electronic transitions

Type of transition	*Example*	**ε (mol^{-1} dm^3 cm^{-1}) (or mol^{-1} litre cm^{-1})**
(i) Both Laporte forbidden* and spin forbidden	$[Mn(OH_2)_6]^{2+}$, $[Fe(OH_2)_6]^{3+}$	10^{-1} to 10^{-2}
(ii) Spin allowed but Laporte forbidden*	$[Ti(OH_2)_6]^{3+}$ $[Ni(OH_2)_6]^{2+}$	~10^1
(iii) Spin allowed and Laporte partly allowed (through *d–p* mixing in the noncentrosymmetric systems as in the T_d complexes)	$[CoCl_4]^{2-}$	10^2 to 10^3
(iv) Spin allowed, Laporte forbidden (but **intensity stealing** phenomenon)	Octahedral complexes with the ligands like acetylacetonate, P or As the donor ligands	~10^3
(v) Charge transfer band (allowed transition)	$[TiCl_6]^{2-}$, $[MnO_4]^-$ $[Fe(OH_2)_5(SCN)]^{2+}$, etc.	10^3 to 10^5

* Partly allowed due to vibronic coupling.

transition, $f = 1$. The values of f in the range 0.1 to 1.0 correspond to ε (molar absorptivity) in the range 10^3 to 10^4 depending on the bandwidth.

7.7 BANDWIDTH: SYMMERTRIC AND ASYMMETRIC BANDS

Half bandwidth $(\Delta\bar{v}_{1/2})$ or simply called bandwidth of a symmetric absorption peak is defined as the width of the absorption peak in wave number at $\varepsilon = \frac{1}{2}\varepsilon_{max}$. Thus width of the peak at half-intensity is described as **full width at half maximum intensity** (FWHM). This definition works well for the symmetric bands (Fig. 7.7.1).

Very often, bandwidth of the electronic transitions in the complexes is of the order of 1000 cm^{-1}. But the **Franck-Condon vertical transition** between the two states of energy E_2 and E_1 should lead to a sharp peak of a definite frequency $(\nu)\left(=\frac{E_2 \sim E_1}{h}\right)$. Many factors are responsible for widening the bands. These are discussed here.

Note: For the free ion (*i.e.* gaseous ion), the width of the spectral lines is less than 1 cm^{-1} while for a spin allowed ligand field transition, the band is quite broad and the bandwidth is in the order of 1000 cm^{-1}.

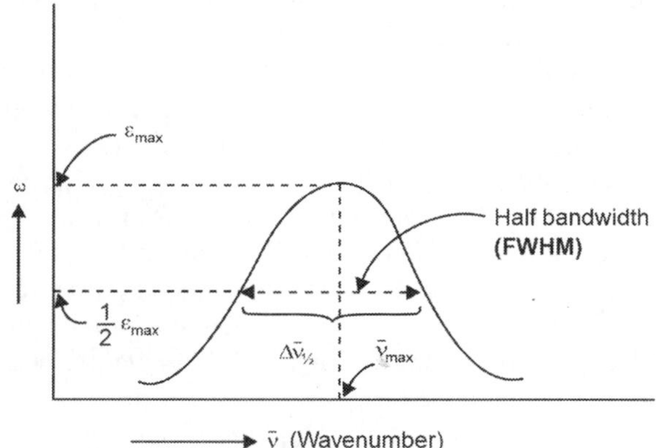

Fig. 7.7.1: Representative illustration of bandwidth for a symmetric peak

Here we shall point out some phenomena causing peak broadening in general. These are not important in the electronic spectra of metal complexes.

- **Peak broadening due to the uncertainty principle** (*i.e.* natural **peak broadening**) (see Vol. 1 for details): In the electronic transition between the ground and excited states, the uncertainty in the energy difference between these two states is determined by the uncertainty in the energy of the excited state for which the lifetime is relatively short. Here it is assumed that the lifetime in the ground state is very high and consequently, the **uncertainty in the energy of the ground state is almost zero** ($\Delta t \rightarrow \infty$ and $\Delta E \rightarrow 0$). Uncertainty principle gives:

$$\Delta E \times \Delta t \approx \frac{h}{4\pi} \approx 10^{-34} \text{ J s}$$

Δt = lifetime of the excited state, ΔE = uncertainty of energy of the excited state. Lifetime of an **excited electronic state** is about 10^{-8} s which gives the value of $\Delta E \approx 10^{-34}/10^{-8} = 10^{-26}$ J. The uncertainty ($\Delta\nu$) in the associated radiation emitted or absorbed is given by:

$$\Delta\nu = \frac{\Delta E}{h} \approx \frac{h}{4\pi h \Delta t} = \frac{1}{4\pi\Delta t} \approx 10^8 \text{ Hz } (\text{s}^{-1})$$

This uncertainty ($\approx 10^8$ Hz) is **apparently very large but it is negligible with respect to the radiation frequency** ($\approx 10^{14}$–10^{16} Hz) of the electronic transition (*cf.* $v/\Delta v \approx 10^6$). Thus **natural peak broadening is not important in the electronic transitions**.

To convert the above uncertainty in wavelength ($\Delta\lambda$), we can write:

$$v = \frac{c}{\lambda}, \text{ i.e. } dv = -\frac{c}{\lambda^2} d\lambda, \text{ i.e. } |\Delta\lambda| = \frac{\Delta v \lambda^2}{c}$$

$$\text{say} \quad \lambda = 500 \text{ nm}, |\Delta\lambda| = \frac{(500 \times 10^{-9} \text{ m})^2 (10^8 \text{ s}^{-1})}{3 \times 10^8 \text{ m s}^{-1}} \approx 10^{-14} \text{ m} = 10^{-5} \text{ nm}$$

Natural line width $\approx 10^{-5}$ nm **is negligible in the electronic transitions**.

However, this **natural peak broadening is important for the NMR** (10^6–10^8 Hz) and **ESR** (~ 10^8–10^{10} Hz) spectroscopy for which $\Delta t \approx 10^{-7}$ s. It leads to $\Delta v = 10^7$ Hz which is not negligible with respect to the usual frequency of transition in NMR and ESR (see Vol. 7).

● **Collisional (pressure) peak broadening:** At high pressure (*i.e.* high concentration) and high temperature, due to the mutual collision in liquid and vapour phase, the excited state undergoes deactivation, *i.e.* Δt (lifetime of the excited state) decreases to increase Δv according to the **uncertainty principle**. Besides this, through collision, energy transfer may occur to **change the energy of the excited state**. It causes the variation in the frequency of emitted or absorbed radiation.

● **Doppler peak broadening:** The observed frequency of emitted photon depends on the velocity (u) of the emitting atom/molecule. If the emitting source is moving towards the detector, the observed frequency becomes higher (*i.e.* **blue shift**) than the frequency of radiation emitted by a stationary source. The reverse (*i.e.* **red shift**) is true, if the emitting source is moving away from the detector. The wavelength (λ) of emitted radiation appears $[1 - (u/c)]\lambda$ (blue shift) and $[1 + (u/c)]\lambda$ (red shift) (c = velocity of light). The radiating atoms/molecules of mass (m) in gas phase travel randomly at very high velocity (temperature and mass dependent) and follow the **distribution of velocities given by Maxwell distribution**. It makes the Doppler broadening of a **peak bell-shaped (Gaussian curve)** and the width ($\Delta\lambda$ at FWHM) of the peak can be expressed as:

$$\boldsymbol{\Delta\lambda = 2(\lambda/c)(2k_BT/m)^{1/2}}$$

In fact, high temperature of the radiating source like the sun can be determined by measuring the Doppler broadening ($\Delta\lambda$) of a particular peak (*e.g.* $\lambda = 677.4$ nm of the ionised ^{57}Fe atom from the sun's surface, $\Delta\lambda = 0.053$ nm, T = 4.7×10^5 K).

Doppler effect is very much important in **Mossbauer spectroscopy** (see Vol. 7).

$$\frac{Dv}{v_0} = \frac{u}{c}$$

Δv = frequency shift (Doppler shift), v_0 = frequency when the source is not moving, u = velocity of the source, c = velocity of light

7.7.1 Bandwidth and Difference (Δx) between the Internuclear Distances of the Ground State and Excited States

The bandwidth is a function of $\Delta x\,(= r_e^{gd} \sim r_e^{ex})$, difference between the internuclear distances (*i.e.* metal-ligand bond distances) of the ground and excited states. Δx depends largely on the electronic configurations in the e_g levels of the ground state and excited state. In the transition, $t_{2g} \rightarrow e_g$, the electron density in the e_g level of the excited state inccreases and it elongates the metal-ligand bond distance. Consequently, in such transitions, the metal-ligand bond distance in the excited state is longer than that in the ground state. Here it is worth mentioning that during the electronic transition, according to Franck-Condon principle, there will be no change in the internuclear distance. ***This is why, the electronically excited molecules are in the vibrationally excited levels having the same bond distance of the ground state electronic configuration to execute the Franck-Condon vertical transition*** (*cf.* Fig. 7.4.1.4).

For $[Cr(OH_2)_6]^{3+}$, the transition, ${}^4A_{2g}$ ($t_{2g}^3e_g^0$) → ${}^4T_{2g}$ ($t_{2g}^2 e_g^1$) at 17,400 cm^{-1} shows the **bandwidth of about 1600 cm^{-1}.** On the other hand, the spin forbidden transition, ${}^4A_{2g}$ ($t_{2g}^3e_g^0$) → ${}^2E_g$ (low-spin t_{2g}^3) shows a **sharp peak.** For the spin-allowed transition, $t_{2g}^3 e_g^0 \rightarrow t_{2g}^2 e_g^1$, the bond length in the excited state is longer, *i.e.* Δx is high while for the spin-forbidden transition, t_{2g}^3 (high-spin, *i.e.* three unpaired electron) → t_{2g}^3 (low-spin, *i.e.* two paired electrons and one unpaired electron), there is no change in the electron density in the e_g level, *i.e.* bond length in both the excited state and ground state is more or less the same and Δx is almost zero. In fact, in such cases, the potential energy diagrams of the excited state and ground state are **similar in shape** and sharp peaks arise from the transitions **to a low energy vibrational level of the excited state** (*cf.* for $r_e^{ex} > r_e^{gd}$, the transition occurs to a higher energy vibrational level of the excited state according to the Franck–Condon vertical transition principle, see Fig. 7.4.1.4).

In fact, the higher value of Δx indicates the higher bandwidth. **In gereral, the spin-allowed transitions (causing a change in the electron density in the e_g–level) produce the broad peaks while the spin-forbidden transitions (generally causing no change in the electron density in the e_g-level) produce the sharp peaks.** For example, the spin-forbidden transitions in $[Mn(OH_2)_6]^{2+}$ (maintaining the same $t_{2g}^3 e_g^2$ electronic configurations both in the ground and excited states but with the different spin multiplicities, *i.e.* ${}^6A_{1g} \rightarrow {}^4A_{1g}$, ${}^4E_g$) produce the weak but sharp peaks (see Fig. 7.14.9.2). These aspects of band broadening have been discussed in Sec. 7.7.2.

Very often, structural parameters of the excited state in the $t_{2g} \rightarrow e_g$ transition differ significantly from those of the ground state and energy levels of the excited electronic state may undergo splitting (*cf.* for d^3, the excited state $t_{2g}^2e_g^1$ experiences the J.T. distortion to split the energy levels). ***Splitting of the energy levels causes the peak broadening and sometimes peak splitting.***

7.7.2 Peak Broadening Vibrational Transitions in Terms of Franck-Condon Vertical Transitions and Franck-Condon Factor

At normal temperature, the ground vibrational level (*i.e.* $\upsilon = 0$) will be most populated at any electronic state and for $\upsilon = 0$, the highest **vibrational probability density** (determined $\psi_{vib}\psi_{vib}^*$ or ψ_{vib}^2, Fig. 7.7.2.1) is **localised in a relatively smaller region.** On the other hand, for the higher values of υ, the probability density is distributed in different regions with some nodal points. ***For the higher vibrational levels, higher probability density regions prevail towards the walls of the potential energy well. The intense transition is possible when the high vibrational probability regions for the ground and excited electronic states occur at the same internuclear distance.*** This is why, when the minima of the Morse potential curves lie for the same internuclear distance (*i.e.* $r_e^{gd} = r_e^{ex}$), the 0 → 0 transition (*i.e.* $\upsilon = 0 \rightarrow \upsilon'_{min} = 0$) gives the most intense transition for the allowed electronic transition. In such cases, value of the **Franck-Condon overlap integral** ($= \int \psi_{vib}^{gd}\psi_{vib}^{ex} d\tau$) becomes high and consequently, the **Franck-Condon Factor (FCF)** which is the square of the vibrational overlap integral becomes high. Here it is worth mentioning that even for an allowed electronic transition, if the overlap between the initial and final vibrational wave functions is negligibly small (*i.e.* FCF is small), the transition becomes practically forbidden. This is understandable in terms of the following **transition moment integral** (R) for an allowed electronic transition.

$$R = \int \psi_{vib}^{gd}\psi_{vib}^{ex}\left[\int \psi_{el}^{gd}\mu_{el}\psi_{el}^{ex}d\tau\right]d\tau = R_e\int \psi_{vib}^{gd}\psi_{vib}^{ex}d\tau$$

$$R_e = \int \psi_{el}^{gd}\mu_{el}\psi_{el}^{ex}d\tau$$

FCF for $r_e^{gd} \approx r_e^{ex}$: In such cases, the shapes of potential energy diagrams of both the excited and ground states are the same and their minima lie for the same r value (Fig. 7.7.2.1). In such cases, the FCF value is very high and the **absorption peak becomes sharp and intense.**

● **Variation of FCF for the most probable vibrational transition:** The most probable (*i.e.* most intense) vibrational transition is: $\upsilon = 0 \rightarrow \upsilon'_{min}$, where υ'_{min} gives the minimum value of vibrational level (at the excited electronic state) for which r_e^{ex} becomes equal to r_e^{gd}, *i.e.* $r_e^{gd} \approx r_e^{ex}$. The FCF value varies as:

FCF: $\upsilon = 0 \rightarrow \upsilon'_{min} = 0 \gg \upsilon = 0 \rightarrow \upsilon'_{min} = 1 \gg \upsilon = 0 \rightarrow \upsilon'_{min} = 2 \gg ...$

(FCF becomes maximum for the $\upsilon \rightarrow \upsilon'$ transition when υ and υ' become minimum).

With the increase of the vibrational number, the number of probability regions increases. This is why, FCF decreases with the increase of υ' and then the peak broadening transitions with the less intensity become important to maintain the constancy of total oscillator strength (f) for an **allowed electronic transition.**

● **Peak broadening vibrational transitions:** Besides the most probable vibrational transition, *i.e.* $\upsilon = 0 \rightarrow \upsilon'_{min}$ (the minimum value of υ' for which $r_e^{ex} \approx r_e^{gd}$), other vibrational transitions of the lower intensity will cause the peak broadening. If $0 \rightarrow 2$ transition is the most probable (*i.e.* most intense) transition, then the less intense transitions like $0 \rightarrow 3$, $0 \rightarrow 4$, $0 \rightarrow 5$, ... **will broaden the peak towards the lower wavelength** (*i.e.* higher frequency) and the less intense transitions like $0 \rightarrow 1$, will broaden the peak towards the longer wavelength (*i.e.* lower frequency) region. In such cases, the peak is **almost symmetrically** broadened (Fig. 7.7.2.1b).

Case I ($r_e^{gd} \approx r_e^{ex}$): In such cases (*cf.* Fig. 7.7.2.1a) of the allowed electronic transition, FCF for the most probable transition, *i.e.* $0 \rightarrow 0$ transition (at the ordinary condition when the higher vibrational levels, *i.e.* $\upsilon = 1, 2, ...$ at the ground electronic state are not appreciably populated) is almost unity and consequently, the other possible peak broadening transitions (towards the higher frequency) are not important. It **makes the peak sharp** (*i.e.* high ε-value) to maintain the constancy in the total oscillator strength f for the **allowed electronic transition.** It happens so for the **$f \rightarrow f$ transitions** in the lanthanides, **spin forbidden $d \rightarrow d$ transitions** in the d^3 system (O_h complexes) and high spin d^5 system where the condition $r_e^{gd} \approx r_e^{ex}$ is maintained (*i.e.* **electron density in the e_g level remains unchanged**, as in the **spin forbidden transitions** in the high spin d^5 system and high spin $t_{2g}^n \rightarrow$ low spin t_{2g}^n transitions for the d^2 and d^3 systems). It may be noted that in such cases, the less intense transitions like $0 \rightarrow 1$, $0 \rightarrow 2$, $0 \rightarrow 3$, $0 \rightarrow 4$, ... are only possible to broaden the peak towards the lower wavelength (*i.e.* higher frequency). **Thus, the peak is asymmetrically broadened (*i.e.* skewed)** (Fig. 7.7.2.1b).

Here it may be mentioned that for an electronic transition allowed through the **vibronic coupling** (*i.e.* **$\psi_{vib}^{gd} \times \psi_{vib}^{ex}$ must be of u-symmetry**), even for the $r_e^{gd} = r_e^{ex}$ condition, the $\upsilon = 0 \rightarrow \upsilon' = 0$ transition is forbidden and in such cases the most probable (*i.e.* most intense) transition is $\upsilon = 0 \rightarrow \upsilon' = 1$ that makes $\psi_{vib}^{gd} \times \psi_{vib}^{ex}$ of u symmetry.

Note: There is no quantum restriction for $\Delta\upsilon$, if, the electronic transitions are assumed to be allowed by their own right. If vibronic coupling allows the electronic transition, then besides the condition of $r_e^{gd} = r_e^{ex}$ (for υ'), there will be a quantum restriction for υ'. In an octahedral system, $\psi_{vib}^{gd} \times \psi_{vib}^{ex}$ must be of u-symmetry. For $\upsilon = 0$, in an octahedral system, ψ_{vib}^{gd} is of g-symmetry (more correctly A_{1g}) and υ' must make ψ_{vib}^{ex} of u-symmetry (*i.e.* T_{1u} or T_{2u} symmetry).

Thus in an octahedral system, $\upsilon = 0 \rightarrow \upsilon' = 0$, $\upsilon = 1 \rightarrow \upsilon' = 1$, etc. transitions are Laporte forbidden even when the vertical transition is possible. However, $\upsilon = 0 \rightarrow \upsilon' = 1$, $\upsilon = 1 \rightarrow \upsilon' = 2$, $\upsilon = 1 \rightarrow \upsilon' = 0$ etc. transitions may not be Laporte forbidden.

Case II ($r_e^{ex} > r_e^{gd}$): In such cases (*cf.* Fig. 7.7.2.1b), FCF for the most probable transition, $\upsilon = 0 \rightarrow \upsilon'_{min}(> 0)$, is relatively smaller (*i.e.* away from unity). Consequently, the peak broadening vibrational

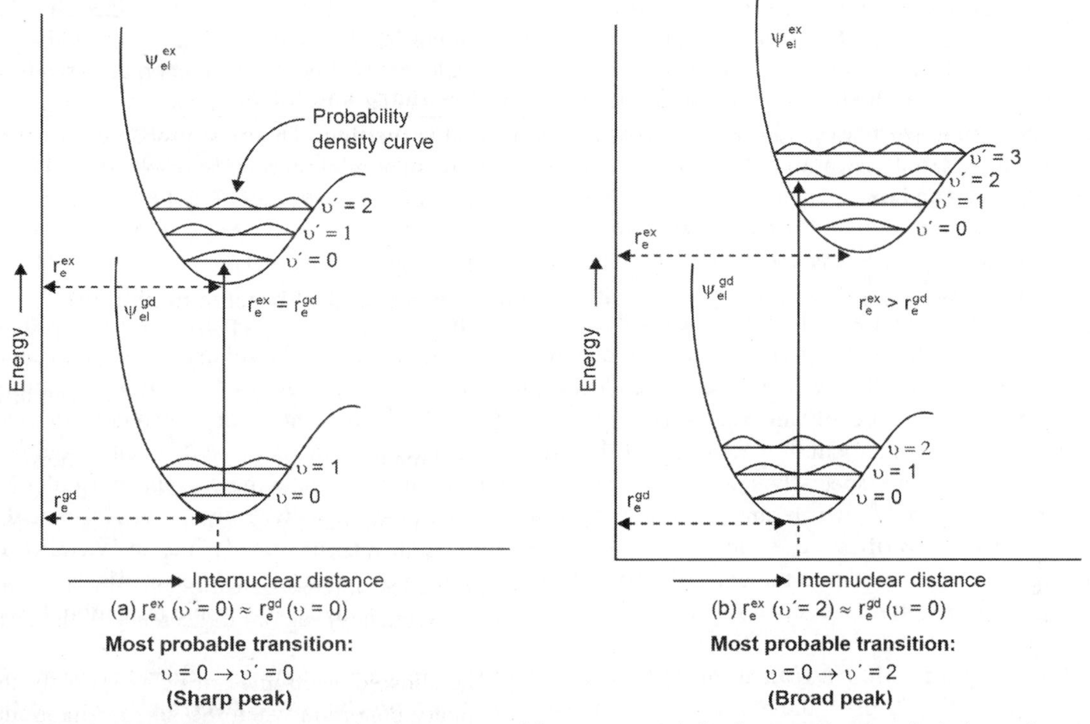

Fig. 7.7.2.1a: Potential energy curves (*i.e.* Morse potential energy curres) for the ground and excited electronic states illustrating the Franck-Condon vertical transition. (a) No change in internuclear distance (*i.e.* metal-ligand bond distance) in the electronic transition. (b) Intenuclear distance (r_e) is longer for the excited electronic state.

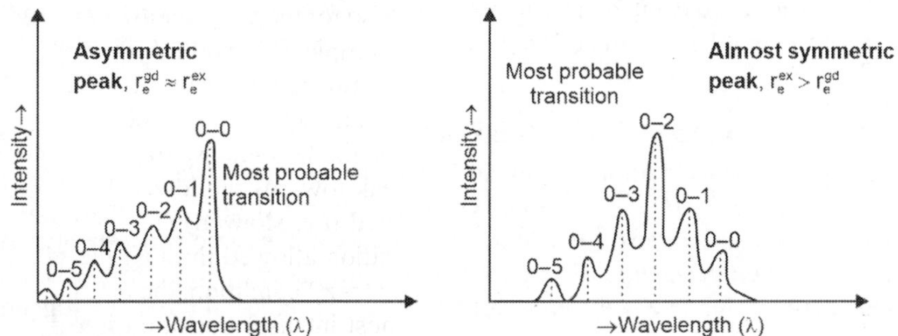

Fig. 7.7.2.1b: Illustration of peak broadening due to vibrational transitions coupled with the electronic transition

transitions (*i.e.* $0 \rightarrow \upsilon' > \upsilon'_{min}$ causing peak broadening towards the higher frequency; $0 \rightarrow \upsilon' < \upsilon_{min}$ causing peak broadening towards the lower frequency) become important to maintain the constancy in the total oscillator strength f for an allowed electronic transition. **It makes the peak broad.**

For the **octahedral complexes**, t_{2g} (≡ NBMO) → e_g (≡ ABMO) transitions make the metal-ligand bond distance longer in the excited state, *i.e.* $r_e^{ex} > r_e^{gd}$ ***and the peaks become broader.***

Δυ (υ → υ′): No quantum restriction for an allowed electronic transition; if the electronic transition is allowed through the **vibronic coupling**, then υ and υ′ levels must be of opposite symmetries to make $\psi_{vib}^{gd} \times \psi_{vib}^{ex}$ of *u*-symmetry.

Effect of temperature on the bandwidth (*cf.* Sec. 7.5.1R): It has been already explained that due to the increase of population density at the higher vibrational levels (*i.e.* $\upsilon = 1, 2, ...$) at the higher temperature, the peak is broadened towards the longer wavelength. At the lower temperature, the peak becomes relatively sharper. The total oscillator strength (f) remains constant for the allowed electronic transitions but f-decreases with the lowering of temperatures for the electronic transitions allowed through the vibronic coupling (*cf.* higher temperature favours the metal-ligand bond vibration to favour the vibronic coupling).

● **Peak broadening in terms of solvation of the excited state:** For the $t_{2g} \rightarrow e_g$ electronic transition in the octahedral complexes, $r_e^{ex} > r_e^{gd}$ (*cf.* bond length increases with the increase of electron density in the e_g level), the transition occurs to the higher energy vibrational level (υ') of the excited state (see Fig. 7.7.2.1b). Because of the difference between the r_e^{gd} and r_e^{ex} values, for a particular excited state, different molecules will experience different degrees of solvation because the solvent molecules cannot rearrange to solvate the excited state in the electron transition time (*cf.* **electron transition is a much faster process than the solvent reorganisation process through the translational movement**). Because of the difference in structural parameters of the ground and excited states, the excited state requires a different orientation (specially solvent-excited state interaction distance) of the solvent molecules for solvation compared to that required for solvation of the ground state. **In fact, the solvent molecules remain at various distances away from the excited state when it is produced**. Thus the variation of solvation energies produces the **excited states in a variable energy range**. In other words, solvation effect produces different excited states of different energies for a particular electronic transition. It will broaden the peak.

For the **spin forbidden transitions** (*e.g.* in the high spin d^5 system and the transitions like high spin $t_{2g}^n \rightarrow$ low spin t_{2g}^n as in the d^2 and d^3 systems) $r_e^{ex} \approx r_e^{gd}$ (*cf.* **no change in electron density in the e_g level**), the **solvent molecules require no rearrangement for the solvation of the excited state** which is almost identical in geometry to that of the ground state. Thus solvation effect does not produce the different excited states. In such cases, sharp spectral peaks appear for the transition to the low energy vibrational level of the excited state whose Morse potential energy curve is similar in shape to that of the ground state and the minima of their potential energy curves lie at the same r value (*cf.* Fig. 7.7.2.1a).

7.7.3 Molecular Vibration to Broaden the Peak

Due to the metal-ligand bond vibration, the metal-ligand distance (r) varies. The $10Dq$ value is highly sensitive to this change (*cf.* $10Dq \propto \frac{1}{r^6}$; approximate relation). If energies of the ground state and excited state between which the transition occurs, do not vary in the same way (quantitatively) with the change of $10Dq$ value, then the energy difference between the states is different for the different values of $10Dq$ (*i.e.* the energy difference is different for the different values of the metal–ligand distance). In such cases, the transition between the states will cover a wide range of wavenumber. But if the concerned states run parallel with the $10Dq$ value (*i.e.* ligand field strength), then the energy difference between the states remains fixed during the change of metal-ligand distance due to the metal-ligand bond vibration. In such cases, the metal-ligand bond vibration cannot widen the peak. These are qualtitatively illustrated in Fig. 7.7.3.1.

The effect of metal-ligand bond vibration on the bandwidth can be better understood by considering the **Tanabe-Sugano diagram**. This diagram will be discussed later.

It has been already mentioned that the **vibronic coupling** is responsible for the relaxation of Laporte selection rule in the centrosymmetric octahedral complexes. Thus intensity of the band/peak depends on the success of vibronic coupling in such cases but the vibronic coupling will broaden the peak at the same time. *With the lowering of temperature, the vibronic coupling becomes less efficient* (*cf.* Sec. 7.5.1) *and the intensity of the peak decreases but the peak becomes sharper.*

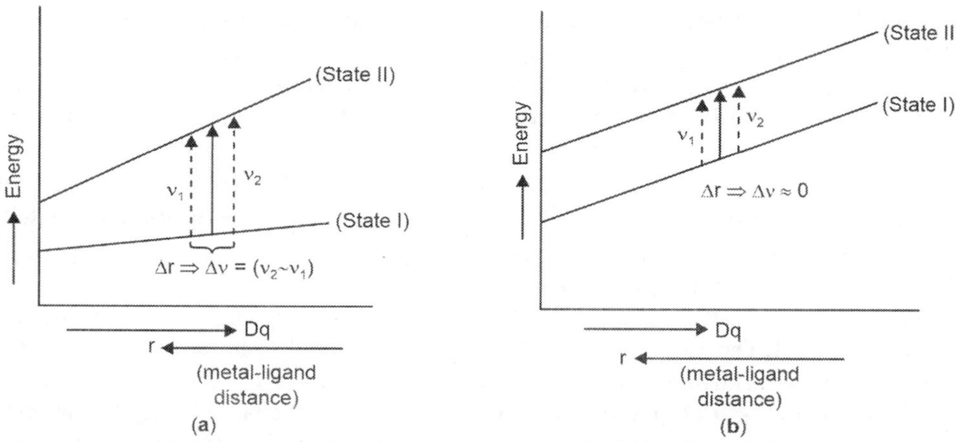

Fig. 7.7.3.1a: Qualitative representation of the variation of energy of different states with *Dq* which is a function of the metal-ligand distance (*r*). (a) If the metal-ligand symmetric vibration causes the change by Δr (= 0)in the *r* value, it would lead to $\Delta\nu$ band-width. (b) The change of Δr ($\neq$ 0)causes no change in the ν value of transition.

● **Interpretation of band broadening due to molecular vibration in terms of Tanabe-Sugano diagram:** The T.S. diagram records the relative change of energy of the different states with the change of crystal field strength (10*Dq*). Here we shall illustrate the effect in some representative cases.

(a) d^2–case: The **spin-allowed transitions** are:

$$^3T_{1g}(F) \rightarrow {}^3T_{1g}(P);\ {}^3T_{1g}(F) \rightarrow {}^3T_{2g}(F);\ {}^3T_{1g}(F) \rightarrow {}^3A_{2g}(F)$$

From the T.S. diagram (Fig. 7.7.3.2), it is evident that the energies of the states $^3T_{1g}(P)$ and $^3T_{2g}(F)$ vary almost in the same way with the change of 10*Dq* (*i.e.* the $^3T_{1g}(P)$ and $^3T_{2g}(F)$ states run parallel). On the other hand, the $^3A_{2g}(F)$ state is steeper. **This is why, the peak due to the transition to $^3A_{2g}(F)$ is expected to be broader than the peaks due to the transitions to $^3T_{1g}(P)$ and $^3T_{2g}(F)$.** In fact, the peaks due to the transitions to $^3T_{1g}(P)$ and $^3T_{2g}(F)$ are of almost comparable bandwidth—as expected from the T.S. diagram.

In the Tanabe-Sugano diagram, the steeper energy state $^3A_{2g}$ (representation of the $t_{2g}^0\, e_g^2$ configuration) compared to the $^3T_{1g}(P)$ and $^3T_{2g}(F)$ states (representing the $t_{2g}^1\, e_g^1$ configuration) can be rationalised in terms of the electronic configuration affecting the metal-ligand bond length. Compared to the ground state $^3T_{1g}$ (*i.e.* $t_{2g}^2\, e_g^0$), the bond length in the excited states $^3A_{2g}$ (*i.e.* $t_{2g}^0\, e_g^2$) and $^3T_{2g}$, $^3T_{1g}(P)$ (*i.e.* $t_{2g}^1\, e_g^1$) is longer. It is due to the fact that with the increase of electron density in the e_g level, the approaching ligands experience the more repulsion. It makes the metal-ligand bond distance longer. The bond length sequence is:

$$t_{2g}^0\, e_g^2 > t_{2g}^1\, e_g^1 > t_{2g}^2\, e_g^0$$

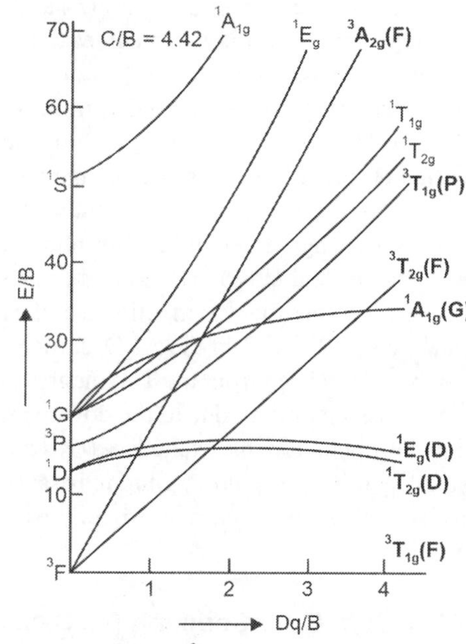

Fig. 7.7.3.2: Tanabe-Sugano energy level diagram of d^2–system, in an octahedral field

Thus the bond length difference (Δx) between the ground state and excited state is maximum for the $t_{2g}^2 \rightarrow t_{2g}^0 e_g^2$ transition while it is relatively less for the $t_{2g}^2 \rightarrow t_{2g}^1 e_g^1$ transition. It explains why the bandwidth in the ${}^3T_{1g}(F) \rightarrow {}^3A_{2g}$ transition is much higher than that in the ${}^3T_{1g} \rightarrow {}^3T_{2g}$ transition.

The **spin-forbidden transitions to the states** ${}^1A_{1g}$(G) and ${}^1E_g(D)$, ${}^1T_{2g}(D)$ produce the very **sharp peaks.**

$${}^3T_{1g}(F) \rightarrow {}^1A_{1g}(G);\ {}^3T_{1g}(F) \rightarrow {}^1E_g(D);\ {}^3T_{1g}(F) \rightarrow {}^1T_{2g}(D)$$

In fact, in the T.S. diagram, the states **${}^1A_{1g}(G)$, ${}^1E_g(D)$ and ${}^1T_{2g}(D)$ run almost parallel to the ground ${}^3T_{1g}(F)$ state.** This is why, the metal-ligand bond vibration cannot broaden the peaks of these spin-forbidden transitions and **the peaks are sharp.**

These four states, *i.e.* ${}^3T_{1g}(F)$, ${}^1A_{1g}(G)$, ${}^1E_g(D)$ and ${}^1T_{2g}(D)$ arise from the **rearrangement of the two electrons in the t_{2g} level.** The ground state ${}^3T_{1g}(F)$ represents the high-spin electronic configuration t_{2g}^2, *i.e.* the two unpaired electrons are distributed in the three orbitals of t_{2g} level, while the said singlet states arise from the **low-spin electronic configuration of t_{2g}^2,** *i.e.* the paired electrons are distributed in the t_{2g}–orbitals. *It is obvious that the electronic transition from the ${}^3T_{1g}(F)$ ground state to the said singlet states merely involves the rearrangement of the electrons within the t_{2g} orbitals and it does not depend on Δ which gives the measure of energy separation between the t_{2g} and e_g levels.*

(b) d^3–system: The following **spin-forbidden transitions are very sharp.**

$${}^4A_{2g}(F) \rightarrow {}^2E_g(G);\ {}^4A_{2g}(F) \rightarrow {}^2T_{1g}(G);\ {}^4A_{2g}(F) \rightarrow {}^2T_{2g}(G).$$

The sharpness of the peaks of the above spin-forbidden transitions arises because the excited states ${}^2E_g(G)$, ${}^2T_{1g}(G)$ and ${}^2T_{2g}(G)$ run parallel to the ground state ${}^4A_{2g}(F)$ in the T.S. diagram. The reason behind this observation can be explained by considering the electronic configurations generating the ${}^4A_{2g}(F)$ and ${}^2E_g(G)$ states. The high spin t_{2g}^3 configuration gives the ${}^4A_{2g}$ state while the low spin t_{2g}^3 configuration (*i.e.* one level in t_{2g} remains vacant) gives the **excited doublet states** (see Sec. 7.14.5). Thus the ${}^4A_{2g} \rightarrow {}^2E_g$ transition **(intra-configurational transition)** does not change the electron density in the e_g-level and consequently, the bond length does not change, *i.e.* bond length remains the same in both the ground (${}^4A_{2g}$) and excited (${}^2E_g$) states.

Here it is important to mention that the role of Cr(III) (a d^3–system) doped in the Al_2O_3 lattice in **ruby laser** can be rationalised from the T.S. diagram of d^3–system. The emission of light due to the downward spin forbidden transition, from the excited ${}^2E_g(G)$ state to the ground state ${}^4A_{2g}(F)$ gives almost **a monochromatic laser beam** because these two states run parallel in the T.S. diagram.

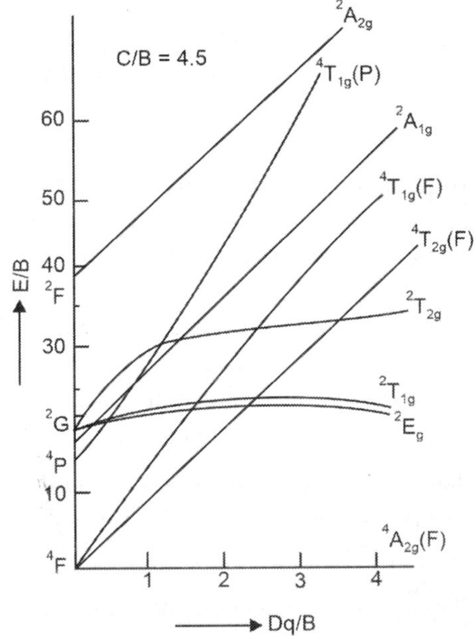

Fig. 7.7.3.3 Tanabe-Sugano energy level diagram of the d^3–system in an octahedral field

7.7.4 Spin-Orbit Coupling (*cf.* Sec. 8.13, Vol. 6) to Broaden the Peak

Spin-orbit coupling is more important for the heavier transition metals (*i.e.* 4*d* and 5*d* metals) and λ (spin-orbit coupling constant) is only in the range of 100–1000 cm^{-1} for the 3*d* metal ions. Spin-orbit coupling can broaden the peak in two ways:

Spin-orbit coupling constant (*cf.* Sec. 8.13, Vol. 6)

It is generally denoted by ζ (zeta) which is the proportionality constant to express the spin-orbit coupling interaction energy, *i.e.*

$$E_{\text{spin-orbit}} \propto \vec{l}\,\vec{s}\cos(\vec{l},\vec{s})$$
$$= \zeta\vec{l}\,\vec{s}\cos(\vec{l},\vec{s})$$

The $\vec{l}$ and $\vec{s}$ vectors couple to generate the $\vec{j}$ vector. This proportionality constant ζ is used to express the spin-orbit interaction in a single electron. For a multielectronic system, it is replaced by λ which is related with ζ as follows:

$$\lambda = \pm\,\zeta/(2S) = \pm\,\zeta/n, \quad S = \frac{n}{2}$$

The + sign is applicable for the half-filled and less than half-filled shell (*e.g.* d^{1-4}) and – sign is applicable for more than half-filled shell (*e.g.* $d^{6\text{-}9}$).

J ranges from $|L - S|$ to $|L + S|$ and energy separation between the two successive levels specified by J and $J + 1$ is given by **Lande Interval Rule.**

$$\Delta E_{J,\,J+1} = \lambda(J + 1).$$

Energy (with respect to that of the unsplit Russel-saunders term of a particular level) of a level characterised by the J value is given by:

$$E_J = \tfrac{1}{2}\lambda\left[J(J+1) - L(L+1) - S(S+1)\right]$$

Thus for the $^3F_2$ term (ground level, d^2-system), its energy is given by:

$$\tfrac{1}{2}\lambda\left[2(2+1) - 3(3+1) - 1(1+1)\right] = -4\lambda.$$

(i) **Allowing the spin-forbidden transitions:** The spin selection rule is strictly applicable for the systems obeying the Russell-Saunders coupling. **The spin-orbit coupling can allow the mixing of different spin states and in such cases S is not a valid quantum number.** In fact, spin-orbit coupling makes the spin-forbidden transitions more intense and the intensity increases as the efficiency of spin-orbit coupling increases. Very often, the spin forbiddent and spin allowed allowed transitions give the overlapping bands (which may not be resloved) to broaden the peak.

(ii) **Splitting of *T*-state:** Spin-orbit coupling can split the T-state to give the fine structure band or broadened band. This aspect is discussed in this section at a greater length.

In an atom, the terms like ^3P, ^3F can split through the spin-orbit coupling.

$$\underset{(L=1,\,S=1)}{^3P} \longrightarrow {}^3P_2,\ {}^3P_1,\ {}^3P_0; \qquad \underset{(L=3,\,S=1)}{^3F} \longrightarrow {}^3F_4,\ {}^3F_3,\ {}^3F_2$$

The energy difference between these successive energy levels depends on the values of J and it is governed by **Lande-interval rule,** $\Delta E = E_J - E_{J-1} \propto J$ or, $\Delta E = \lambda J$.

Energy of an state generated through the spin-orbit coupling depends on the value of spin-orbit coupling constant λ. The energy is given by: $E_J = \tfrac{1}{2}\lambda\left[J(J+1) - L(L+1) - S(S+1)\right]$, with respect to the energy of the unsplit Russet-Saunders term.

Spin-orbit coupling can also split the crystal field ***T*-states into different energy levels.** When a *T*-state splits, energy difference between the lowest and highest states may lie in the range 100–1000 cm^{-1}. The splitting of *T*-states through spin-orbit coupling is shown in Figs. 7.7.4.1 and 2.

If in the electronic transition, one state is of T-type, then the particular transition will cover the range of some hundred cm^{-1}. Thus it should create a **fine structure** components of the band. But for the first transition series (specially in solution), it is overshadowed by the **vibrational effect which produces the continuous bands** (*cf.* Fig. 7.7.3.1). This is why, the fine structure components of the bands due to spin-orbit coupling cannot be seen.

For the spin-forbidden transitions where peak broadening by vibration is not important, the fine structure of the peak can be detected. For the heavier transition metal ions of the 2nd and 3rd transition series, where the spin-orbit coupling constant is very high, peak broadening due to spin-orbit coupling is of much importance.

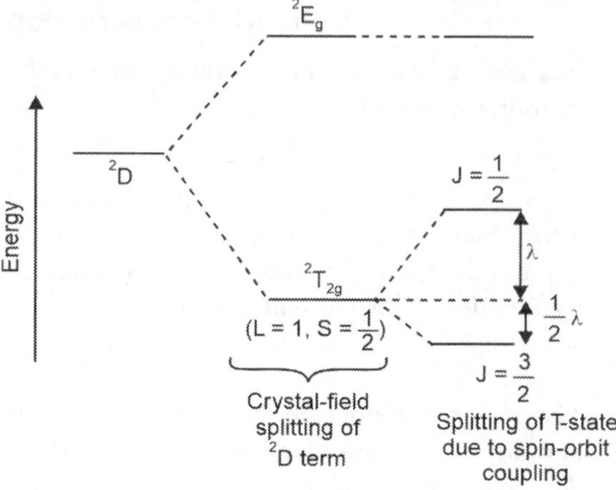

Fig. 7.7.4.1 Spin-orbit splitting of the $^2T_{2g}$ state of ^2D term of d^1–configuration in the octahedral field (*cf.* Fig. 8.20.3.1 of Vol. 6 for details).

Here it may be mentioned that **the splitting of A and E states due to spin-orbit coupling does not occur.**

Thus we can conclude as follows:

$$\textbf{Crystal field state:} \quad \underbrace{A_{1g}, A_{2g}, E_g}$$
$$\textbf{Spin-orbit coupling components:} \quad 1\,(i.e.\text{ no splitting})$$

Crystal field state:	$^2T_{2g}$	$^3T_{1g}$	$^4T_{1g}$	$^5T_{2g}$	$^3T_{2g}$	$^4T_{2g}$	$^5T_{2g}$
Spin-orbit coupling components:	2	3	3	3	3	3	3

Splitting of T–states in the tetrahedral fields also follows the same pattern.

(**Note:** ● It may be noted that in presence of a magnetic field, each level of a particular value of J can further split into the levels characterised by the M_J value which can range from $+J$ to $-J$.

● The spin-orbit coupling constant λ is positive for the half-filled and less than half-filled level (*i.e.* d^1 to d^5) and it is negative for more than half-filled (*i.e.* d^6 to d^9) level.

● For the states having the different J values, their energy order is determined by **Hund's 3rd rule** (see Vol. 1) which states that for the half-filled or less than half-filled level, the state having the lowest J value is the most stable one and for the more than half-filled level, the state having the highest J value is the most stable one.)

Now let us consider the following transitions in the d^2 and d^8–octahedral systems.

$$^3A_{2g} \rightarrow {}^3T_{2g}, {}^3T_{1g}(F), {}^3T_{1g}(P) \qquad (d^8\text{–system})$$
$$^3T_{1g} \rightarrow {}^3T_{2g}, {}^3T_{1g}(P), {}^3A_{2g}(P) \qquad (d^2\text{–system}).$$

Because of the splitting of T–states due to spin-orbit coupling, the peak in each case of the above mentioned systems should contain the fine structure components. This will broaden the peak.

7.7.5 Peak Broadening due to Jahn-Teller Distortion (*i.e.* Departure from Cubic Symmetry) and due to Reduction in Symmetry by the Ligands

(A) Jahn-Teller distortion effect: It has been mentioned that in an octahedral system, if the t_{2g} or e_g level is unsymmetrically filled in then degeneracy of that level is lifted. If the transition,

$$t_{2g}^m\, e_g^n \rightarrow t_{2g}^{m-1}\, e_g^{n+1}$$

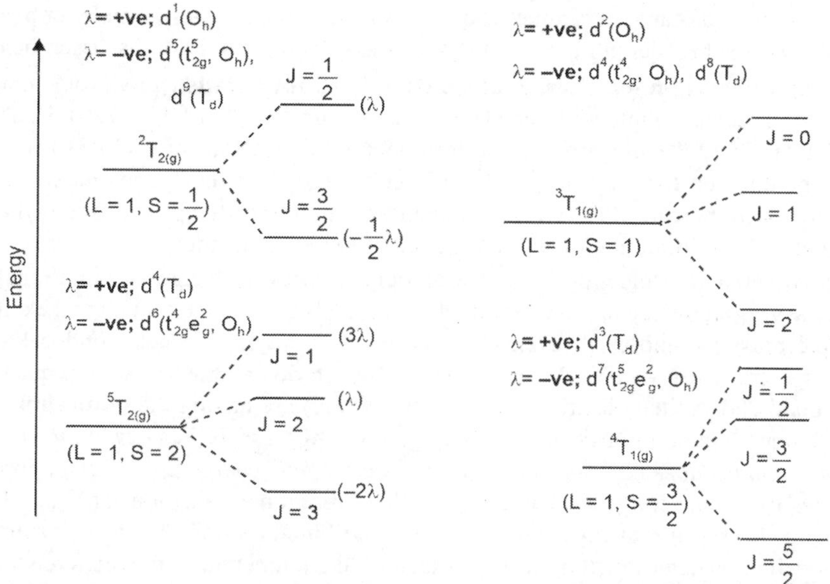

Fig. 7.7.4.2: Splitting of different the T-states due to spin-orbit coupling. ($J = |L + S|$ to $|L - S|$) (*cf.* **Sec. 8.20.3,** Vol. 6 for determination of L and S values of a T-state). **Note:** Energy level order is shown for λ = +ve. For λ = –ve, the energy level order will be inverted.

experiences the J.T. distortion in the ground state and/or excited state, then splitting of the energy levels will allow more than one electronic transition. These multiple transitions will broaden the peak. In some cases, the peak may split into different components. These aspects have been illustrated in Sec. 3.11.3, Vol. 4. The J.T. distortion is also possible in the tetrahedral systems and the J.T. distortion can also broaden the peak in a tetrahedral system. However, this effect is relatively less important in the tetrahedral systems.

Here it should be mentioned that in the case of **dynamic J.T. distortion,** the crystallographic data indicate no distortion but the spectral property can reveal the distortion. **Thus effects of both the dynamic and static J.T. distortion are reflected on the spectra.** This aspect has been discussed in Secs. 7.11, 7.14, and 3.11.3, 4.

(B) Reduction in Symmetry: When the symmetry of an octahedral ($O_h \rightarrow C_{4v}, C_{2v}, D_{4h}, D_3$, etc.) or a tetrahedral ($T_d \rightarrow D_{2d}$, etc.) complex is lowered by the ligand environment, additional bands (which may not be always resolved) appear in their spectra and very often, they broaden the peak. It occurs due to the splitting of energy levels. For example, the bands of $[CoCl(NH_3)_5]^{2+}$ (C_{4v}) (purple coloured) are broader than those of $[Co(NH_3)_6]^{3+}(O_h)$ (orange yellow coloured) though their band positions are almost similar indicating the effective O_h symmetry.

7.7.6 Band Shapes: Factors Controlling the Band Shape

Very often, the bands are symmetrical of the **Gaussian shape** and in such cases, the distribution of ε for the different values of frequency (ν) is given by:

$$\varepsilon = \varepsilon_0 \exp[-\alpha(\nu - \nu_0)^2]$$

where α is a constant, ν_0 is the frequency at the centre of the band having the extinction coefficient ε_0. The area of a such symmetric peak is given by $\varepsilon_0 \Delta\nu_{1/2}$ (which is related with the oscillator strength) where $\Delta\nu_{1/2}$ is the half bandwidth (see Secs 7.6, 7.7).

But sometimes, the peaks are not symmetric and the factors which are responsible for band broadening are also responsible for broadening the asymmetric peaks. The factors are discussed below.

(i) **Vibrational interaction (*cf.* Secs. 7.5.1 Q, R; 7.7.1–3 for details):** In vibronic coupling, vibrational quantun levels couple with the electronic quantum levels (*cf.* Fig. 7.4.1.3). The transitions like (0, 0), (1, 1), (2, 2), etc. are Laporte forbidden while the transitions like (0, 1), (1, 0), (2, 1), etc. may be allowed for the centrosymmetric complexes like the O_h complexes. In the bracket ( ), the first term indicates the vibrational quantum number of the ground state while the second term indicates the vibrational quantum number of the excited state.

Vibronic spectra (*i.e.* electronic spectra associated with a change in vibrational energy) are generally **asymmetric** due to the vibrational transitions between the ground and excited electronic states. The **most probable vibrational transition** (*i.e.* highest Franck-Condon Factor, FCF) in vibronic spectra is: $\upsilon^{gd} \rightarrow \upsilon^{ex}$ where υ^{gd} and υ^{ex} (which denote the vibrational quantum numbers at the ground and excited electronic states respectively) are having the **minimum values** for the allowed Franck-Condon transition. This will give the **highest FCF**. *Thus in the vibronic spectra, the band maximum does not represent exactly the electronic energy difference between the ground and excited electronic states.* Probabilities of other vibrational transitions of higher frequency and lower frequency are not always symmetrically distributed around the most probable vibrational transition which is determined by different factors like temperature, the relative values of r_e^{gd} and r_e^{ex}, the possible values of υ^{gd} and υ^{ex} for the vibronic coupling to relax the Laporte selection rule (*cf.* Fig. 7.7.2.1). These aspects are discussed in detail in Secs. 7.5.1, Q, R; 7.7.1–3.

With the increase of temperature, excited vibrational levels of the ground electronic state are populated according to Boltzmann's distribution law. **Transition from the higher vibrational levels of the ground electronic state will occur at a lower frequency.** It will produce a tail towards the lower frequency to introduce the asymmetric shape of the peak. This aspect has been illustrated in detail in Sec. 7.5.1Q, R.

(ii) **Spin-orbit coupling:** It has been already mentioned that due to spin-orbit coupling, the *T*–state is split and the splitting very often occurs asymmetrically. Thus the peak becomes asymmetric.

(iii) **Low-symmetry introduced by the ligands and Jahn-Teller effect:** It splits the *E* and *T*–states. The *T*–state splits into two components: one component maintains the two fold orbital degeneracy of a particular energy; another component is nondegenerate of a different energy. **Thus the T-state splits asymmetrically. The *E*-state splits itself symmetrically.** But the resultant transitions due to splitting of the *E* and *T*-states give the asymmetric shape of the peak. In fact, very often, the resultant peak is the sum of different peaks (which are themselves symmteric individually and differing slightly in energy). When the individual peaks are of large difference in energy, separate symmetrical peaks will arise. Effect of J.T. distortion on spectra has been separately discussed in Sec. 3.11.

Conclusion: The above discussion indicates that there are many reasons for an asymmetric shape of the peaks. But, spectral bands of transition metal complexes in solution are very often symmetrical. It is believed that the resultant effects of the above mentioned causes make the peaks not very much asymmetric.

7.8 CHARACTERISTIC FEATURES OF THE ISOSBESTIC POINT

If two species remain in a dynamic equilibrium and their absorption spectra overlap at a point, the point is called **isosbestic point** that does not change with the position of the equilibrium. **Concentrations of the species remaining in an equilibrium may change depending on the condition but absorption**

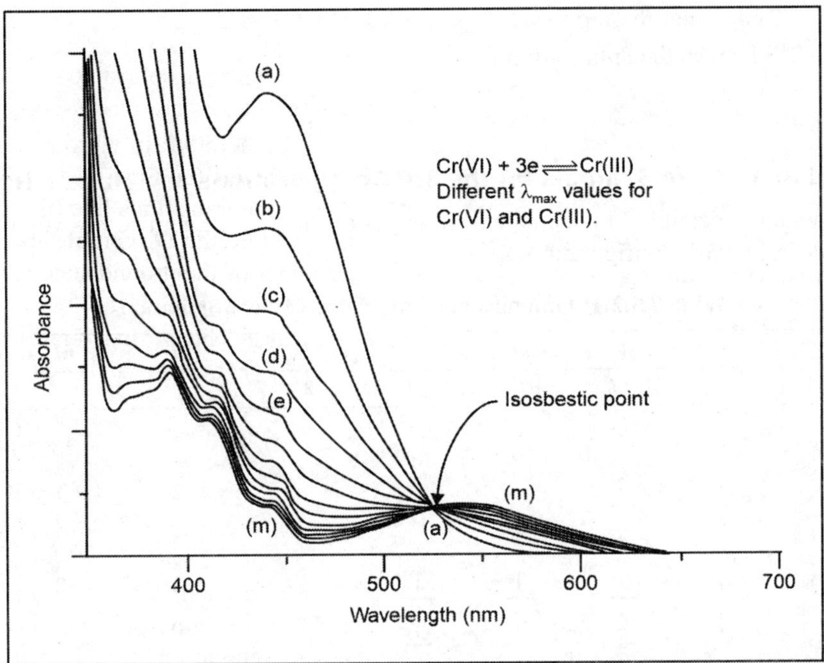

Fig. 7.8.1: A representative example of scanned spectrum in the oxidation of hexitols by chromic acid in presence of 2, 2′–bipyridine as a catalyst. (a) at the beginning; (m) at about after 50 min. Representing the gradual conversion of Cr(VI) into Cr(III)–species (Recorded in the author's laboratory)

at the isosbestic point will not change. ***In fact, at the isosbestic point, the molar extinction coefficients of the species are the same.***

If in a particular reaction, the reactant and product show an isosbestic point, the dynamic spectra nicely explain the characteristic property of the isosbestic point. As the reaction proceeds, spectra at both the sides of the isosbestic point will change, *i.e.* absorbance in one side (*i.e.* product) will increase while absorbance in the other side (*i.e.* reactant) will decrease but the absorbance at the isosbestic point will not change.

In general, the acid-base indicators show the different colours (*i.e.* spectra) for the acid (InH^+) and base (In) forms remaining in a pH dependent equilibrium.

$$In + H^+ \rightleftharpoons InH^+$$

In such cases, if the spectra of an indicator are followed at different pH values, then the isosbestic point will be illustrated.

7.9 CRYSTAL FIELD TERMINOLOGY IN ELECTRONIC SPECTRA: ORIGIN OF THE ELECTRONIC SPECTRA AND LUMINESCENCE

7.9.1 Russel-Saunders Coupling (L–S Coupling) and Term Symbols

The details of Russel-Saunders coupling scheme have been discussed in "Fundamental Concepts of Inorganic Chemistry"– Volume 1. This scheme is applicable for the lighter elements ($Z \leq 30$). Thus it works well for the 1st transition series while for the heavier elements having the higher spin-orbit coupling constants, *j–j* coupling scheme is required.

In terms of L–S coupling, the term symbol is expressed as:
$^{2s+1}L_J$ where $2S$+1 gives the spin multiplicity,

	L =	0	1	2	3	4	5	6 ....
Term	**Symbol** =	S	P	D	F	G	H	I ....

7.9.2 Ground State Term Symbols for the 3d^n Configurations and Hund's Rules

These are discussed in detail in "Fundamental Concepts of Inorganic Chemistry"– Volume 1. Here we give the results for the 3d^n configurations.

Table 7.9.2.1: Ground state terms for the d^n configurations.

3d^n	*m_l*					*M_L*	*M_S*	*Term symbol*
	+2	*+1*	*0*	*−1*	*−2*			
3d^1(e.g. Ti^{3+})	↿	–	–	–	–	2	1/2	$^2D_{3/2}$
3d^2(e.g. V^{3+})	↿	↿				3	1	$^3F_2$
3d^3(e.g. Cr^{3+})	↿	↿	↿			3	3/2	$^4F_{3/2}$
3d^4(e.g. Cr^{2+})	↿	↿	↿	↿		2	2	$^5D_0$
3d^5(e.g. Mn^{2+})	↿	↿	↿	↿	↿	0	5/2	$^6S_{5/2}$
3d^6(e.g. Fe^{2+})	⥮	↿	↿	↿	↿	2	2	$^5D_4$
3d^7(e.g. Co^{2+})	⥮	⥮	↿	↿	↿	3	3/2	$^4F_{9/2}$
3d^8(e.g. Ni^{2+})	⥮	⥮	⥮	↿	↿	3	1	$^3F_4$
3d^9(e.g. Cu^{2+})	⥮	⥮	⥮	⥮	↿	2	1/2	$^2D_{5/2}$

- **Terms for the d^n-configurations in complexes:** We have learnt how to determine the free ion terms for the d^n-configurations. For the **high spin complexes** where there is no additional pairing of electrons, the terms are the same as for the free ion terms. But the terms are different in the **low spin complexes** where there is more electron pairing with respect to that of the free d^n-configurations. It is illustrated for the d^3 and d^6 configurations in the octahedral complexes.

$d^3\,(t_{2g}^3 e_g^0)$ system: (e_g) — — ; (t_{2g}) ↿ ↿ ↿

$$M_{L(\max)} = \sum m_l = +2+1+0 \Rightarrow F \text{ term } (L=3)$$

$$S = \tfrac{3}{2}, \text{ Ground state term} = {}^4F$$

High spin $d^6\,(t_{2g}^4 e_g^2)$ system: (e_g) ↿ ↿ ; (t_{2g}) ⥮ ↿ ↿

$$M_{L(\max)} = \sum m_l = +2+2+1+0-1-2$$

$$= 2 \Rightarrow D \textit{ term } (L=2)$$

$$S = 5\times\frac{1}{2} - \frac{1}{2} = 2, \text{ Ground state term} = {}^5D$$

Low spin $d^6\,(t_{2g}^6 e_g^0)$ system: (e_g) — — ; (t_{2g}) ⥮ ⥮ ⥮

$$M_{L(\max)} = \sum m_l = +2+2+1+1+0+0$$

$$= 6 \Rightarrow I \textit{ term } (L=6)$$

$$S = \frac{3}{2} - \frac{3}{2} = 0, \text{ Ground state term} = {}^1I$$

Similarly, it can be shown that the d^4 system generates the ^5D and ^3H terms in the high spin ($t_{2g}^3e_g^1$) and low spin (t_{2g}^4) complexes respectively.

Note: In the octahedral complexes, for the three t_{2g}-orbitals of lower energy, the m_l values are taken as +2, +1, and 0 while for the two e_g-orbitals of higher energy, the m_l values are taken as –1 and –2. In the tetrahedral complexes, for the two *e*-orbitals of lower energy, the m_l values are taken as +2 and +1 while for the three t_2-orbitals of higher energy, the m_l-values are taken as 0, –1, –2. It is illustrated in Fig. 7.9.2.1 (See Sec. 1.10.10, Vol. 1). Thus for the nd^6 system, the high spin state (*i.e.* $t_{2g}^4e_g^2$ in the octahedral crystal field) corresponds to the ground state ^5D (*i.e.* $L_{max} = 2$ and $S = 2$) while the low spin state (*i.e.* $t_{2g}^6e_g^0$) represents the ground state term ^1I (*i.e.* $L_{max} = 6$ and $S = 0$).

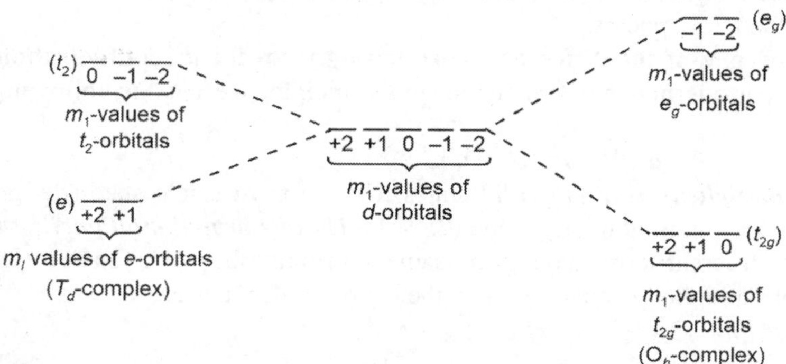

Fig. 7.9.2.1: m_l-values of different *d*-orbitals in the octahedral and tetrahedral complexes

7.9.3 Terms Arising from the d^n–Configuration

The term symbol of the ground state can be easily obtained with the help of Hund's rules. Besides the ground state, there are many possible excited states. To obtain the term symbols for such excited states, we are to consider the M_L and M_S values for the corresponding excited states. A particular electronic arrangement with the specific m_l and m_s values represents a **microstate.** Thus a microstate represents the allowed combination of the m_l and m_s values of a particular electronic configuration.

The procedure for determination of the microstates for different d^n configurations has been discussed in "Fundamental Concepts of Inorganic Chemistry"—Volume 1. Here it may be mentioned that according to the **hole formalism rule,** d^n and d^{10-n} configurations give the same terms. The possible terms arising from the different d^n–configurations are given in Table 7.9.3.1.

(a) Hole formalism or equivalency: Russel-Saunders terms for the d^n and d^{10-n} configurations are the same. Let us illustrate for the d^2 and d^8 systems. The d^8 system may result by adding two positrons in the filled d^{10} electronic configuration. A filled shell does not contribute anything towards the resultant angular momenta, *i.e.* $L = 0$, $S = 0$. Thus, the addition of two positrons to the d^{10} electronic configuration will give the values of *L* and *S* determined by the two positrons, *i.e. the Russell-Saunders terms due to the* d^{10}*-electron* ($L = 0$, $S = 0$) + d^2*-positron will arise solely from the* d^2*-positronic configuration.* In terms of the interelectronic repulsion, the positrons are equivalent to the electrons. Thus the Russel-Saunders terms for the d^2 and d^8 systems will be the same. In other words, in terms of *L–S* coupling, **the d^8–electronic configuration is equivalent to the d^2 positronic system which is equivalent to the d^2 electronic configuration.** Because of the same ground, the p^n and p^{6-n} configurations will have the same Russel-Saunders terms. But **an electron is most stable where a positron is least stable.**

Table 7.9.3.1 Terms for different d^n–configurations

Configuration	*Terms (after considering Pauli exclusion principle)*
d^1, d^9	$\mathbf{^2D}$
d^2, d^8	$\mathbf{^3F}$, ^3P, ^1G, ^1D, ^1S
d^3, d^7	$\mathbf{^4F}$, ^4P, ^2H, ^2G, ^2F, $^2D(2)$, ^2P
d^4, d^6	$\mathbf{^5D}$, ^3H, ^3G, $^3F(2)$, ^3D, $^3P(2)$, ^1I, $^1G(2)$, ^1F, $^1D(2)$, $^1S(2)$
d^5	$\mathbf{^6S}$, ^4G, ^4F, ^4D, ^4P, ^2I, ^2H, $^2G(2)$, $^2F(2)$, $^2D(3)$, ^2P, ^2S

The ground state terms are shown in bold letters. The number 2, 3, etc. in parentheses denotes the number of times the term appears.

(b) Relative energies of the different terms arising from the d^2–configuration (an illustrative example): After considering the Pauli exclusion principle, we get the following terms for a d^2 configuration.

$$d^2: {}^3F, {}^3P, {}^1G, {}^1D, {}^1S$$

Hund's spin multiplicity rule (1st rule) states that the most stable state (*i.e.* ground state term) possesses the highest spin multiplicity value $(2S + 1)$. ***Hund's orbital multiplicity rule (i.e. 2nd Rule)*** states that for the different terms having the same spin multiplicity, the stability increases with the increase of orbital multiplicity $(2L + 1)$. Thus the energy order runs as:

$$^3F < {}^3P < {}^1G < {}^1D < {}^1S$$

The energy order may be $^3F < {}^3P < {}^1D < {}^1G < {}^1S$ under some conditions (to be discussed later in this section). Under some conditions, $^3P > {}^1D$ energy order **(contradicting the Hund's rule)** may also appear.

Each term can split depending upon the J-values. The energy order for the states differing only in the J values (*i.e.* same L and S values) can be determined by using the ***Hund's 3rd rule*** (*i.e.* rule of J) which states that for a half-filled or less than half-filled level (*e.g.* p^1 to p^3, d^1 to d^5), the **lowest J value** will indicate the most stable state while for a more than half-filled level (*e.g.* p^4 to p^6, d^6 to d^{10}), the **highest J value** indicates the most stable state. Thus the term ^3F generates the states $^3F_4$, $^3F_3$ and $^3F_2$ through the spin-orbit coupling interaction and their energy order is:

$$\mathbf{^3F}: {}^3F_2 < {}^3F_3 < {}^3F_4 \ (d^2 \text{ — less than half-filled})$$

Similarly, we can write for ^3P:

$$\mathbf{^3P}: {}^3P_0 < {}^3P_1 < {}^3P_2$$

Considering the **further splitting of the J–values** in presence of a magnetic field into the M_J values which can range from $+J$ to $-J$ (*i.e.* splitting into $2J + 1$ levels, *i.e.* each state is $(2J + 1)$ fold degenerate), the energy states arising from the d^2–configuration are shown in Fig. 7.9.3.1.

Terms, States and Microstates (see Sec. 1.10.11, Vol. 1)

Electronic configuration (nd^2, say) $\xrightarrow{\text{Interelectronic repulsion}}$ Terms (^{3}F, ^{3}P, ^{1}G, ^{1}D, ^{1}S) $\xrightarrow{\text{Spin-orbit coupling}}$ States ($^3F_{2,3,4}$, $^3P_{0,1,2}$, $^1G_4$, $^1D_2$, $^1S_0$) $\xrightarrow{\text{Magnetic field}}$ Microstates

Note: Very often, in description, the words 'terms' and 'states' are used considering no difference between them. But, it is not correct.

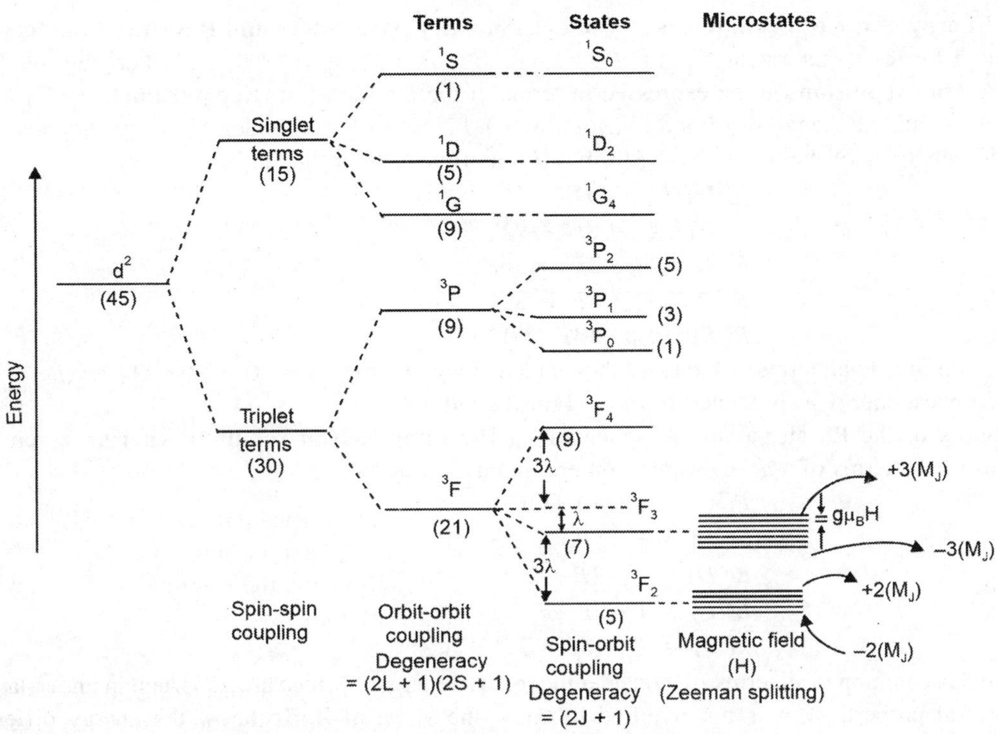

Fig. 7.9.3.1: Schematic representation of the splitting of energy states for the d^2-system (Total degeneracy of each level is indicated in the parenthesis). Energy order according to Hund's rule but the actual energy order of the different states depend on the *B/C* ratio to be discussed later (see Sec. 1.10.11, Vol. 1).

The energy separation between the states differing by only *J* values is governed by **Lande Interval Rule** (*cf.* Sec. 8.13, Vol. 6). It states that the energy separation between the successive levels having the *J* values as *J* and *J* + 1 is proportional to the higher value of *J*, *i.e.* *J* + 1, *i.e.*

$$\Delta E_{J,\ J+1} \propto (J+1), \text{ or } \Delta E_{J,\ J+1} = \lambda(J+1)$$

The proportionality constant λ is called the **spin-orbit coupling constant.** The spin-orbit coupling leads to the energy levels characterised by the *J* values and energy of a such level is determined by the following relation.

$$E_J = \frac{1}{2}\lambda[J(J+1) - L(L+1) - S(S+1)], \text{ (with respect to that of the unsplit term)}$$

Thus for $^3F_2$ (ground level, d^2 configuration), its energy is given by:

$$\frac{1}{2}\lambda[2(2+1) - 3(3+1) - 1(1+1)] = -4\lambda.$$

Splitting of a level having a particular *J* value leads to 2*J* + 1 sublevels which are equally spaced and separation energy is $g\mu_B H$ where *g* is called the **Lande splitting factor,** μ_B stands for the **Bohr magneton** and *H* gives the magnetic field strength. The Lande splitting factor *g* is given by the **Lande formula.**

$$g = 1 + \frac{J(J+1) + S(S+1) - L(L+1)}{2J(J+1)}$$

(c) Energy of the terms in terms of Condon-Shortley parameters and Racah parameters: It is illustrated for the terms arising from a d^2 configuration. The energies of the terms arising due to the **interelectron repulsion** can be expressed in terms of the **Condon-Shortley parameters** (F_0, F_2 and F_4) or the empirical **Racah parameters** (A, B and C). In terms of the Condon-Shortley parameters, the repulsion energies of the terms arising from a d^2–configuration are given below:

$$E(^1G) = F_0 + 4F_2 + F_4$$
$$E(^3F) = F_0 - 8F_2 - 9F_4$$
$$E(^1D) = F_0 - 3F_2 + 36F_4$$
$$E(^3P) = F_0 + 7F_2 - 84F_4$$
$$E(^1S) = F_0 + 14F_2 + 126F_4$$

F_0 is common to all terms. It indicates the repulsion energy order: $^1S > {}^1G > {}^3P > {}^1D > {}^3F$ (*i.e.* ^3F term has the lowest energy as predicted from the **Hund's rule**).

In terms of the Racah parameters **(measuring the interelectron repulsion energies),** repulsion energies of the terms of a d^2–configuration are given as follows:

$$E(^1G) = A + 4B + 2C$$
$$E(^3F) = A - 8B$$
$$E(^1D) = A - 3B + 2C$$
$$E(^3P) = A + 7B$$
$$E(^1S) = A + 14B + 7C$$

Here A is common to all terms. To consider the relative energies (which are important in understanding the spectral properties), it is not required to know the value of A. To know the energy difference (= $15B$) between the ^3P and ^3F terms, we need only the knowledge of the B parameter. Similarly, we need the knowledge of only the B parameter for the energy difference between the ^1D and ^1G terms. To know the energy difference between the ^3F and ^1D terms (differing in spin multiplicities), we require the knowledge of both B and C.

Under the condition, $C > 5B$, energy order of the terms is :

$^3F < {}^3P < {}^1D < {}^1G < {}^1S$ (*cf.* Fig. 7.9.3.1 and energy order of ^1D and ^1G)

(*cf.* expected order from Hund's rule: $^3F < {}^3P < {}^1G < {}^1D < {}^1S$)

This order is **approximately** the same as obtained by considering the Hund's rules. However, under the condition, $C < 5B$, ^3P lies above ^1D (*cf.* **contradicting the Hund's rules**) and it happens so for Ti^{2+}. Thus for Ti^{2+}, the observed order is: $^3F < {}^1D < {}^3P < {}^1G < {}^1S$ (*cf.* the prediction by considering the Condon-Shortley parameters). ***In fact, the orbital multiplicity rule (i.e. Hund's second rule) is not reliable for predicting the order of excited states but it can predict the ground state (i.e. highest L).*** It has been noted that for most of the metal ions of 1st transition series, the condition $C \approx 4B$ is maintained.

The Condon-Shortley parameters are related with the Racah parameters as follows:

$$A = F_0 - 49F_4,\ B = F_2 - 5F_4,\ C = 35F_4,\ F_2 = B + \frac{C}{7}$$

The Racah parameters can be obtained from a linear combination of the exchange and Coulomb integrals. For the first transition metal ions, the ratio C/B is essentially constant at a value of 4 and B is in the order of 1000 cm^{-1}. The interelectron repulsion parameters B and C are decreased in the complexes compared to their values in the free ions. The parameter B is decreased to about 70% during the complexation.

The energy difference between the two terms for a d^2–configuration can be given as follows:

$$E(^3P) - E(^3F) = 15(F_2 - 5F_4) = 15B$$

$$E(^1D) - E(^3F) = 5(F_2 + 9F_4) = 5B + 2C.$$

Similarly, it can be shown that for a d^3–configuration (*e.g.* V^{2+}), the energy difference between the ^4F and ^4P terms is $15B$.

Table 7.9.3.1 Energy differences for some of the terms for different d^n configurations

		Energy difference	
Configuration	***Term interval***	***F parameter***	***B and C Parameters***
d^2, d^8	$^1S - ^3F$	$22F_2 + 135F_4$	$22B + 7C$
	$^1G - ^3F$	$12F_2 + 10F_4$	$12B + 2C$
	$^3P - ^3F$	$15F_2 - 75F_4$	$15B$
	$^1D - ^3F$	$5F_2 + 45F_4$	$5B + 2C$
	$^1S - ^1D$	$17F_2 + 90F_4$	$17B + 5C$
	$^3P - ^1D$	$10F_2 - 120F_4$	$10B - 2C$
d^3, d^7	$^2P - ^4F$	$9F_2 + 60F_4$	$9B + 3C$
	$^2G - ^4F$	$4F_2 + 85F_4$	$4B + 3C$
	$^2H - ^4F$	$9F_2 + 60F_4$	$9B + 3C$
	$^4P - ^4F$	$15F_2 - 75F_4$	$15B$
	$^2G - ^2H$	$-5F_2 + 25F_4$	$-5B$
d^4, d^6	$^3D - ^5D$	$16F_2 + 60F_4$	$16B + 4C$
	$^3G - ^5D$	$9F_2 + 95F_4$	$9B + 4C$
	$^3H - ^5D$	$4F_2 + 120F_4$	$4B + 4C$
	$^3D - ^3H$	$12F_2 - 60F_4$	$12B$
	$^3G - ^3H$	$5F_2 - 25F_4$	$5B$
d^5	$^4G - ^6S$	$10F_2 + 125F_4$	$10B + 5C$
	$^4F - ^6S$	$22F_2 + 135F_4$	$22B + 7C$
	$^4D - ^6S$	$17F_2 + 90F_4$	$17B + 5C$
	$^4P - ^6S$	$7F_2 + 210F_4$	$7B + 7C$
	$^2I - ^6S$	$11F_2 + 225F_4$	$11B + 8C$
	$^2H - ^6S$	$13F_2 + 285F_4$	$13B + 10C$

Thus it is evident that the energy difference between the terms having the same spin multiplicity is a function of the single Racah parameter B while the energy difference between the terms of different spin multiplicities is a function of two Racah parameters B and C (cf. Table 7.9.3.1).

In Tanabe-Sugano diagram, energy (E) and ligand field strength ($10Dq$) are to be expressed in terms of the parameter B and the plot is E/B *vs.* $10Dq/B$.

7.9.4 Energy States Arising in Presence of Ligand Fields and Electronic Transitions Developing the Ligand Field Bands: Crystal Field Splitting of Free Ion Terms

The terms of free ions undergo splitting in the presence of ligands. Splitting of terms of sulphur free ions in **cubic fields**—cubic, octahedral and tetrahedral is shown in Table 7.9.4.1.

Table 7.9.4.1: Free ion terms and their splitting in cubic crystal fields

Free ion term	*Orbital multiplicity (2L + 1)*	*Terms arising in crystal field (Spectroscopic States)*
S	1	A_1
P	3	T_1
D	5	$E + T_2$
F	7	$A_2 + T_1 + T_2$
G	9	$A_1 + E + T_1 + T_2$
H	11	$E + T_1 + T_2 + T_1$
I	13	$A_1 + A_2 + E + T_1 + T_2 + T_2$

Note:

(i) Splitting of the terms in the presence of a crystal field, may be compared with the splitting pattern of the orbitals in a **cubic crystal field.**
Single s–orbital (spherically symmetrical): no splitting
Three p–orbitals: no splitting
Five d–orbitals: Splitting into two groups t_2 and e (one triply degenerate and another one doubly degenerate)
Seven f–orbitals: Splitting into three groups: t_1 (triply degenerate), t_2 (triply degenerate) and a_2 (nondegenerate)

(ii) The capital letter *T, E, A, B* denote the **orbital degeneracy.** *T* stands for 3–fold degeneracy, *E* stands for 2–fold degeneracy; *A* or *B* stands for no degeneracy.

(iii) The *numerical supercripts* (e.g. 2, 3, 4) represent the spin multiplicity.

(iv) The numerical subscripts in T_1, T_2, A_2, etc. are originated from group theory. These are beyond the scope of this book.

(v) The subscripts *g* and *u* denote *gerade* (*i.e.* even) and *ungerade* (*u*) respectively. The label *g* is applicable for the systems having an inversion centre, *e.g.* regular octahedron and the label *u* appears for the systems lacking in centre of symmetry as in a tetrahedron.

(vi) During the splitting of free ion term, **orbital multiplicity (2*L* + 1) is retained in the sum of the energy states** (*i.e.* spectroscopic states). It is illustrated below:

$$F\ (L = 3,\ i.e.\ 2L + 1 = 7) \rightarrow A_2\ (1) + T_1\ (3) + T_2\ (3)$$

i.e. total orbital multiplicity = 1 + 3 + 3 = 7.

(b) Splitting pattern of the free ion terms (*D, F*) in a cubic crystal field: Orbital splitting of the *D* and *F* terms are shown in Fig. 7.9.4.1. (*See* Solved Problem 14 for **energy calculation**).

Let us illustrate for the d^1 and d^2 configurations which give the ground state terms *D* and *F* respectively. For the d^1 system, t_{2g}^1 corresponds to $^2T_{2g}$ (crystal field energy = $-4Dq$) and $t_{2g}^0\, e_g^1$ (excited state) corresponds to ^2E_g (crystal field energy = $+\,6Dq$). These energy values are given with respect to that of the unsplit ^2D state.

For the d^2 system, the ^3F term splits into the $^3T_{1g}$(F), $^3T_{2g}$(F) and $^3A_{2g}$(F) states and there is a $^3T_{1g}$(P) state arising from the excited ^3P term. **Thus there are two $^3T_{1g}$ states and they tend to mix.** This mixing is insignificant in a strong crystal field where the states are widely apart but the mixing is possible to some extent in a weak crystal field. No such problem exists for the $^3T_{2g}$(F) ($\equiv t_{2g}^1 e_g^1$, *i.e.* first excited state) and $^3A_{2g}$(F) ($\equiv t_{2g}^0\, e_g^2$, *i.e.* second excited state) states. Thus for the strong field ligand, the ground state $^3T_{1g}$(F) corresponds to $t_{2g}^2\,e_g^0$ but in a weak field, **considering the mixing between the $^3T_{1g}$(F) and $^3T_{1g}$(P) states**, the ground state may be written as $t_{2g}^x\,e_g^y$ where $x + y = 2$ and $x < 2$. From Fig. 7.9.4.1, we can write:

Ground state $^3T_{1g}$(F) ≡ $t_{2g}^x e_g^y$, crystal field energy = $(-4x + 6y)Dq = pDq$ (say).

1st excited state $^3T_{2g}$(F) ≡ $t_{2g}^1\,e_g^1$, crystal field energy = $+2Dq$.

2nd excited state $^3A_{2g}$(F) ≡ $t_{2g}^0\,e_g^2$, crystal field energy = $+12Dq$.

i.e. $3p + 3(2Dq) + 1(12Dq) = 0$,
or $3p = -18Dq$; or, $p = -6Dq$.
(**Considering the orbital degeneracy** and energy conservation principle).

It leads to: $(-4x + 6y)Dq = p = -6Dq$; and $x + y = 2$

Solving these two equations we get: $x = 1.8$, $y = 0.2$. **Thus, in a weak crystal field, the ground state $^3T_{1g}$(F) represents $t_{2g}^{1.8}\,e_g^{0.2}$ not t_{2g}^2** (*cf.* **Solved Problem 14**). In fact, in a weak crystal field, the inter-electronic repulsion shifts some electron density from the t_{2g} level to the e_g level.

In a strong crystal field, the ground state (*i.e.* t_{2g}^2) is so stabilised that no electron cloud moves from the t_{2g} level to the e_g level.

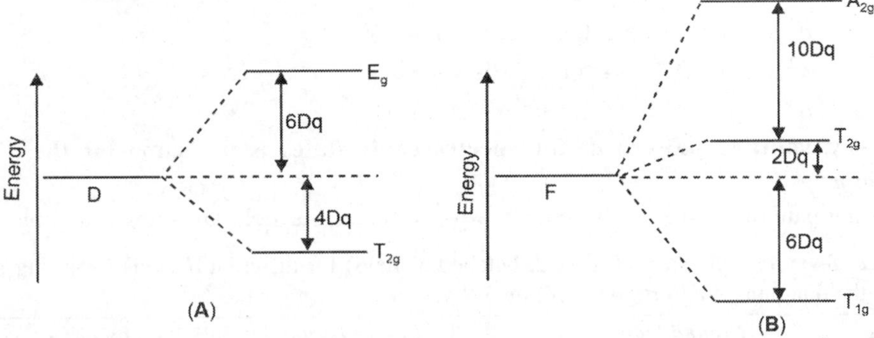

Fig. 7.9.4.1: Splitting of $D(d^1)$ and $F(d^2)$ terms in **a weak octahedral field** (*cf. see* Solved Problem 14 for the energy values)

(**Note:** Energy values are given in Fig. 7.9.4.1 without considering the symmetry interaction energy in the case of splitting of the *F*–term).

- **Strong field ligand for the d^2-configuration (assuming $\Delta_o \gg B'$):**

$$E(t_{2g}^2, T_{1g}) = -8\,Dq,\ E(t_{2g}^1 e_g^1, T_{2g} \text{ and } T_{1g}(P)) = +2\,Dq$$

$$E(e_g^2, A_{2g}) = +12\,Dq$$

(see Sec. 7.14.3 for detailed mathematical treatment)

Under this condition:

$$^3T_{1g}(F) \xrightarrow{\nu_1} {}^3T_{2g}(F),\ {}^3T_{1g}(P),\ \nu_1 = 10\,Dq$$

$$^3T_{1g}(F) \xrightarrow{\nu_2} {}^3A_{2g}(F),\ \nu_2 = 20\,Dq$$

Here it is interesting to note that under the assumption $10\,Dq \gg$ B′, **the basis for resolution of the $t_{2g}^1 e_g^1$ microstates into the $^3T_{1g}$ (P) and $^3T_{2g}$ (F) is ignored!**

(c) d^n and d^{10-n} systems; d^n and d^{n+5} systems: According to the rule of **hole formalism, the d^n and d^{10-n} systems** should have the same Russel-Saunders terms and it arises due to the equivalence of the **d^n (electronic) configuration** with the **d^n (positronic configuration)**. It is illustrated for the d^2 and d^8 *i.e.* d^{10-2} electronic configurations. A d^8 electronic configuration is obtained by adding 2 positrons in a filled d^{10} electron subshell for which L = 0, S = 0. Thus, the Russel-saunders terms for d^{10} electron + d^2 positron will be determined by the d^2 positronic configuration, *i.e.* we can write: Russel-Saunders terms of the d^8 electronic configuration and d^{10-8}, *i.e.* d^2 positronic configuration are the same. **Russel-Saunders terms for the d^n electronic and d^n positronic systems are the same.** *But an electron is most stable where a positron is least stable.*

Thus both d^8 and d^2 electronic configurations will give the same Russel-Saunders terms. It leads to: d^n electronic configuration ≡ d^n positronic configuration ≡ d^{10-n} positronic configuration ≡ d^{10-n} electronic configuration. It leads to: in terms of Russel-Saunders terms, d^n electronic configuration ≡ (10 – n) holes in the d^{10} electronic configuration, *i.e.* **d^n configuration is hole equivalent of d^{10-n} configuration. But an electron is most stable where a positron is least stable.** This is why, the pattern of splitting of the spectroscopic states arising from the d^n and d^{10-n} electronic configurations is of opposite nature, *i.e.*

d^n–diagram = inverse of d^{10-n} diagram

Russel-Saunders terms (*i.e.* L value) for **d^n and d^{n+5}** are the same in an **weak field condition** because the half-filled d^5 system generates a zero orbital angular momentum (L = 0). However, they differ in spin multiplicities only. It is illustrated below:

d^2 and d^7: ${}^xF(x = 3$ for d^2, $x = 4$ for $d^7)$;

d^3 and d^8: ${}^xF(x = 4$ for d^3, $x = 3$ for $d^8)$;

d^4 and d^9: ${}^xD(x = 5$ for d^4, $x = 2$ for $d^9)$,

This is why, **splitting pattern of the spectroscopic states is the same for the d^n and d^{n+5} configurations.**

The splitting patterns of the d^n–systems in a weak octahedral field are shown in Table 7.9.4.2.

Table 7.9.4.2 Spectroscopic states (in weak octahedral field) for different d^n–ions (Splitting pattern with reference to the diagrams A and B given in Fig. 7.9.4.1).

d^n	*Ground State Term (free ion)*	*Energy level diagrams (Spetroscopic terms)*	*Important Excited State Term (free ion)*
d^1	^{2}D	A	–
d^2	^{3}F	B	^{3}P
d^3	^{4}F	Inverted B	^{4}P
d^4	^{5}D	Inverted A	–
d^5	^{6}S	No splitting (*i.e.* $^6A_{1g}$)	–
d^6	^{5}D	A	–
d^7	^{4}F	B	^{4}P
d^8	^{3}F	Inverted B	^{3}P
d^9	^{2}D	Inverted A	–
d^{10}	^{1}S	No splitting	–

(d) Electronic configurations of the spectroscopic states developed in a cubic crystal field for the d^2– and d^3– configurations: Now let us illustrate the possible spectral transitions for a **d^2–configuration** for which the important terms in free ion are ^3F and ^3P. The possible electronic configurations are:

e_g:		
t_{2g}:		
($t_{2g}^2 e_g^0$, ground state, $^3T_{1g}(F)$, *i.e.* orbital degeneracy = 3)	($t_{2g}^1 e_g^1$, excited state; $^3T_{1g}$(P) + $^3T_{2g}$(F) *i.e.* total orbital degeneracy = 6)	($t_{2g}^0 e_g^2$, excited state; $^3A_{2g}$ *i.e.* orbital degeneracy = 1)

Now let us consider the mode of distribution of electrons in different states.

● In the **ground state t_{2g}^2**, the two electrons will be singly occupying any two of the three t_{2g} orbitals keeping their spins parallel (*i.e.* spin multiplicity = 3). These two electrons can be distributed in three ways (*i.e.* orbital degeneracy = 3).

$$(d_{xy})^1(d_{xz})^1;\ (d_{xz})^1(d_{yz})^1;\ (d_{xy})^1(d_{yz})^1$$

This state (orbital degeneracy = 3 denoted by T, spin multiplicity = 3) is represented by $^3T_{1g}(F)$

● For the ***d*³-system** (S = 3/2), the important free ion terms are ^4F and ^4P. The ground state and two excited states are as follows:

Ground state: $t_{2g}^3 e_g^0$, $^4A_{2g}(F)$ (orbitally singlet state); **First excited state:** $t_{2g}^2 e_g^1$, $^4T_{1g}(F) + {}^4T_{2g}(F)$, *i.e.* orbitally 6-fold degenerate; **Second excited state:** $t_{2g}^1 e_g^2$, $^4T_{1g}(P)$ (*i.e.* orbitally 3-fold degenerate).

● For the ***d*²-system, in the excited state $t_{2g}^1 e_g^1$,** two different energy states differing in the degree of interelectron repulsion develop. **If the electron in the t_{2g} level resides in the d_{xy} orbital, then the electron in the $d_{x^2-y^2}$ orbital of the e_g-set will experience the more inter-electron repulsion compared to the electron placed in the d_{z^2} orbital of the e_g-set. Thus the electronic configuration, $(d_{xy})^1(d_{x^2-y^2})^1$, is less stable than the electronic configuration $(d_{xy})^1(d_{z^2})^1$.**

Similarly, $(d_{xz})^1\left(d_{z^2}\right)^1$ is less stable than $(d_{xz})^1\left(d_{x^2-y^2}\right)^1$ and $(d_{yz})^1\left(d_{z^2}\right)^1$ is less stable than $(d_{yz})^1\left(d_{x^2-y^2}\right)^1$. Thus, we can have two different sets of electron distribution for the excited state $t_{2g}^1 e_g^1$.

$$\left.\begin{array}{l}\textbf{Lower energy set:}\ \underbrace{\left(d_{xy}\right)^1\left(d_{z^2}\right)^1;\ \left(d_{xz}\right)^1\left(d_{x^2-y^2}\right)^1;\ \left(d_{yz}\right)^1\left(d_{x^2-y^2}\right)^1}_{\text{Orbital degeneracy=3. Spin multiplicity=3; symbol= }^3\mathbf{T_{2g}(F)}.}\end{array}\right\}\ t_{2g}^1 e_g^1$$

$$\left.\begin{array}{l}\textbf{Higher energy set:}\ \underbrace{\left(d_{xy}\right)^1\left(d_{x^2-y^2}\right)^1;\ \left(d_{xz}\right)^1\left(d_{z^2}\right)^1;\ \left(d_{yz}\right)^1\left(d_{z^2}\right)^1}_{\text{Orbital degeneracy=3. Spin multiplicity=3; symbol=}^3\mathbf{T_{1g}(P)}.}\end{array}\right\}\ t_{2g}^1 e_g^1$$

The higher interelectron repulsion in the second set represented by $^3T_{1g}(P)$ makes the energy state less stable compared to the $^3T_{2g}(F)$ state.

● ***d*², *d*³ systems: The $t_{2g} \rightarrow e_g$ electron transition gives two peaks—why?** The energy difference between the t_{2g} level and e_g level is Δ_o. Thus for the $t_{2g}^x e_g^0 \rightarrow t_{2g}^{x-1} e_g^1$ (where x = 2, 3) transition (*e.g.* initial state: $t_{2g}^2 e_g^0$; final state; $t_{2g}^1 e_g^1$), we can expect one peak ($h\nu = \Delta_o$). But the electron-electron repulsion in the excited state $t_{2g}^1 e_g^1$ gives two different energy states, *i.e.* $^3T_{1g}$(P) and $^3T_{2g}$(F). It explains the origin of two peaks for the $t_{2g} \rightarrow e_g$ electron promotion. Similarly, the $t_{2g}^3 e_g^0 \rightarrow t_{2g}^2 e_g^1$ transition gives two peaks because the $t_{2g}^2 e_g^1$ excited state gives two energy states $^4T_{1g}(F)$ and $^4T_{2g}(F)$ which differ in electron-electron repulsion.

● For the d^2-system, in the **excited state e_g^2,** the unpaired two electrons can be distributed in a single way, *i.e.* $\left(d_{x^2-y^2}\right)^1\left(d_{z^2}\right)^1$. It leads to orbital degeneracy = 1 and spin multiplicity = 3. The state is respresented by $^3A_{2g}$.

The total orbital degeneracy of the $^3F(L = 3)$ and $^3P(L = 1)$ terms is: $(2 \times 3 + 1) + (2 \times 1 + 1) = 10$.

Table 7.9.4.3 Spectroscopic transitions for the d^n–configuration in a weak octahedral field

		Complex (O_h)				***Transitions (Spin allowed)***
Free ion		***Ground state***		***Excited state***		
Elect. config.	***Term***	***Elect. config.***	***State***	***Elect. config.***	***State***	
d^1	^2D	$t_{2g}^1\ e_g^0$	$^2T_{2g}$	$t_{2g}^0\ e_g^1$	^2E_g	$^2T_{2g}(F) \rightarrow {}^2E_g$
d^2	^3F, ^3P*	$t_{2g}^2\ e_g^0$	$^3T_{1g}(F)$	$t_{2g}^1\ e_g^1$ $t_{2g}^0\ e_g^2$	$^3T_{2g}(F) + {}^3T_{1g}(P)$ $^3A_{2g}$ (F)	$^3T_{1g}(F) \rightarrow {}^3T_{2g}(F)$ $^3T_{1g}(F) \rightarrow {}^3T_{1g}(P)$ $^3T_{1g}(F) \rightarrow {}^3A_{2g}$ **(Forbidden transition)****
d^3	^4F, ^4P*	$t_{2g}^3\ e_g^0$	$^4A_{2g}(F)$	$t_{2g}^1\ e_g^2$ $t_{2g}^2\ e_g^1$	$^4T_{1g}(P)$ $^4T_{1g}(F) + {}^4T_{2g}(F)$	$^4A_{2g} \rightarrow {}^4T_{2g}(F)$ $^4A_{2g} \rightarrow {}^4T_{1g}(F)$ $^4A_{2g} \rightarrow {}^4T_{1g}(P)$
d^4	^5D	$t_{2g}^3\ e_g^1$	^5E_g	$t_{2g}^2\ e_g^2$	$^5T_{2g}$	$^5E_g \rightarrow {}^5T_{2g}$
d^5	^6S	$t_{2g}^3\ e_g^2$	$^6A_{1g}$	$t_{2g}^2\ e_g^3$	$^4T_{1g} + {}^4T_{2g}$	**Spin-forbidden transitions**
d^6	^5D	$t_{2g}^4\ e_g^2$	$^5T_{2g}$	$t_{2g}^3\ e_g^3$	^5E_g	$^5T_{2g} \rightarrow {}^5E_g$
d^7	^4F, ^4P*	$t_{2g}^5\ e_g^2$	$^4T_{1g}(F)$	$t_{2g}^3\ e_g^4$ $t_{2g}^4\ e_g^3$	$^4A_{2g}$ $^4T_{1g}(P) + {}^4T_{2g}(F)$	$^4T_{1g}(F) \rightarrow {}^4T_{2g}(F)$ $^4T_{1g}(F) \rightarrow {}^4A_{2g}(F)$ $^4T_{1g}(F) \rightarrow {}^4T_{1g}(P)$
d^8	^3F, ^3P* $^3T_{1g}(F)$	$t_{2g}^6\ e_g^2$	$^3A_{2g}(F)$	$t_{2g}^4\ e_g^4$ $t_{2g}^5\ e_g^3$	$^3T_{1g}(P)$ $^3T_{1g}(F) + {}^3T_{2g}(F)$	$^3A_{2g} \rightarrow {}^3T_{2g}(F)$ $^3A_{2g} \rightarrow {}^3T_{1g}(F)$ $^3A_{2g} \rightarrow {}^3T_{1g}(P)$
d^9	^2D	$t_{2g}^6\ e_g^3$	^2E_g	$t_{2g}^5\ e_g^4$	$^2T_{2g}$	$^2E_g \rightarrow {}^2T_{2g}$.

* Excited state with the same spin multiplicity of the ground state. ** 2*e* transition not allowed

Note: In representing the transition, the **standard convention** is: excited state ← ground state, *i.e.* excited state is to be written first. ***Thus the transition in the d^9–system should be written as: $^2T_{2g} \leftarrow {}^2E_g$.*** However, in this book, this rigid convention is not followed.

The total orbital degeneracy arising from the $^3T_{1g}(F)$, $^3T_{1g}(P)$, $^3T_{2g}(F)$ and $^3A_{2g}(F)$ states is also 10. This is illustrated afterwards.

The development of energy states for the ^3F and ^3P terms of the free ion is as follows:

^3F (orbital degeneracy = 2 × 3 + 1 = 7)

(i) $(d_{xy})^1 (d_{xz})^1$; $(d_{xz})^1 (d_{yz})^1$; $(d_{xy})^1 (d_{yz})^1$; $\Rightarrow {}^3T_{1g}(F)$

(ii) $(d_{xy})^1 (d_{z^2})^1$; $(d_{xz})^1 (d_{x^2-y^2})^1$; $(d_{yz})^1 (d_{x^2-y^2})^1 \Rightarrow {}^3T_{2g}(F)$

(iii) $(d_{x^2-y^2})^1 (d_{z^2})^1 \Rightarrow {}^3A_{2g}(F)$

(Total orbital degeneracy = 3 + 3 + 1)

^3P (orbital degeneracy = 2 × 1 + 1 = 3)

(i) $(d_{xy})^1 (d_{x^2-y^2})^1$; $(d_{xz})^1 (d_{z^2})^1$; $(d_{yz})^1 (d_{z^2})^1 \Rightarrow {}^3T_{1g}(P)$

(Total orbital degeneracy = 3).

(e) Group theoretical treatment to determine the spectroscopic states from the electronic configuration: Spectroscopic states arising from the electronic configurations can be determined theoretically from the direct product rule (*cf.* Table 7.23.1). It is illustrated below. For details, the readers are advised to consult the book on group theory.

$t_{2g}^2 \to t_{2g}^1 \times t_{2g}^1 \to {}^3T_{1g} + {}^1T_{2g} + {}^1E_g + {}^1A_{1g}$; Total degeneracy = 3 × 3 + 1 × 3 + 1 × 2 + 1 × 1 = 15

$t_{2g}^1 e_g^1 \to t_{2g}^1 \times e_g^1 \to {}^1T_{1g} + {}^1T_{2g} + {}^3T_{1g} + {}^3T_{2g}$; Total degeneracy = 1 × 3 + 1 × 3 + 3 × 3 + 3 × 3 = 24

$e_g^2 \to e_g^1 \times e_g^1 \to {}^1E_g + {}^3A_{2g} + {}^1A_{1g}$; Total degeneracy = 1 × 2 + 3 × 1 + 1 × 1 = 6

$t_{2g}^3 \to t_{2g}^1 \times t_{2g}^2 \to T_{2g} \times \left({}^3T_{1g} + {}^1T_{2g} + {}^1E_g + {}^1A_{1g}\right) \to {}^4A_{2g} + {}^2T_{1g} + {}^2T_{2g} + {}^2E_g$;

$t_{2g}^2 e_g^1 \to E_g \times \left({}^3T_{1g} + {}^1T_{2g} + {}^1E_g + {}^1A_{1g}\right) \to {}^4T_{1g} + {}^4T_{2g} + 2{}^2T_{1g} + 2{}^2T_{2g} + 2{}^2E_g + {}^2A_{1g} + {}^2A_{2g}$

$t_{2g}^1 e_g^2 \to T_{2g} \times \left({}^3A_{2g} + {}^1E_g + {}^1A_{1g}\right) \to {}^4T_{1g} + 2{}^2T_{1g} + 2{}^2T_{2g}$.

The results are summarised as follows:

$t_{2g}^6 \Rightarrow {}^1A_{1g}$; t_{2g}^5 and $t_{2g}^1 \Rightarrow {}^2T_{2g}$; t_{2g}^2 and $t_{2g}^4 \Rightarrow {}^3T_{1g} + {}^1T_{2g} + {}^1E_g + {}^1A_{1g}$;

$t_{2g}^3 \Rightarrow {}^4A_{2g} + {}^2E_g + {}^2T_{2g} + {}^2T_{1g}$; $e_g^4 \Rightarrow {}^1A_{1g}$; e_g^3 and $e_g^1 \Rightarrow {}^2E_g$; $e_g^2 \Rightarrow {}^3A_{2g} + {}^1E_g + {}^1A_{1g}$

In a tetrahedral field, the *g*-suffixes are dropped.

7.9.5 Origin of Luminescence: Phosphorescence, Fluorescence, Jablonski Diagram (see Part II of Chapter 6 for details)

When an electron is excited to a higher energy level (*i.e.* spectroscopic state), the excitation energy may be released in two ways:

- **radiative decay** through photon emission.
- **nonradiative path through the thermal degradation of energy as heat to the surrounding; the lattice vibration or collisional quenching** (*i.e.* **external conversion) is the main cause of nonradiative decay;** this nonradiative path leading to the ground state (generally) is very fast (within 10^{-8} s in gas phase, within 10^{-12}–10^{-14} s in liquid phase).

When the nonradiative decay path is relatively inefficient (*i.e.* lifetime for the nonradiative path is elongated), the radiative path and nonradiative path may compete mutually and the radiative decay path may predominate also. In such cases, the system can show the **luminescence.** For the transition metal complexes, generally, the nonradiative decay path is quite efficient even at low temperature. **This is why, the transition metal complexes are not generally luminescent.** For the lanthanides, the 4*f*–orbitals are deeply seated in the inner core and the 4*f*–orbitals are hardly influenced by the environment. In such cases, the nonradiative path (*i.e.* through lattice vibration) becomes comparatively less efficient. **In fact, many lanthanide complexes can show the luminescene.**

Chemiluminescence: If the excited state required for luminescence is generated by a chemical reaction not by light absorption, then the term chemiluminescence is used. It is illustrated below through the production of singlet O_2 which is very much reactive towards the organic substrates. Singlet oxygen ($^1\Delta_g$) can be prepared in the following reactions.

$$Cl_2 + H_2O_2 \rightarrow 2Cl^- + 2H^+ + O_2(^1\Delta_g)$$

$$ClO^- + H_2O_2 \rightarrow Cl^- + H_2O + O_2(^1\Delta_g)$$

The mechanisms of the given reactions do not allow the production of O_2 with the unpaired electrons, *i.e.* the given reactions do not allow the formation of triplet O_2 ($^3\Sigma_g^-$). Thus the produced singlet oxygen produces **red chemiluminescent glow** when it returns to its ground state, *i.e.* triplet oxygen.

Luminescence leading to fluorescence and phosphorescence (see Sec. 6.10)**:** Thus **luminescence (*i.e.* radiative decay)** occurs only when the **nonradiative path is quite inefficient.** Depending on the life time of the excited state, luminescence can be of two types:

Luminescence (Inefficient non-radiative path) }
- **Fluorescence** – rapid radiative decay
- **Phosphorescence** – delayed radiative decay.

Thus, fluorescence stops if the exiciting radiation is removed but phosphorescence exists even after the removal of the exciting radiation. **For phosphorescence, lifetime of the excited state must be relatively higher.** In fact, for fluorescence, both the excited and ground state are having the same spin-multiplicity and the return to the ground state from the excited state is a **spin-allowed transition** occurring in the scale of nanoseconds. On the other hand, the radiative decay of phosphorescence occurs through the **spin-forbidden transition** (*i.e.* excited state and ground state differ in spin multiplicity). This is why, the life-time of the excited state is relatively longer for the phosphorescence.

Intersystem crossing (ISC) and internal conversion (IC)—two important nonradiative paths of deexcitation (Secs 6.10.3, 4)**:** Generally, photoexcitation from the ground state into the excited state occurs through a spin-allowed transition. This initial excited state may be converted into another excited state of different multiplicity through **intersystem crossing (ISC) (a nonradiative path).** This intersystem crossing occurs when the potential energy curves of the two excited states cross. At the point of crossing of the two curves, this conversion occurs through a **radiationless transition. Spin-orbit coupling allows the intersystem crossing**

Sometimes, theremay be two exited states of the same spin multiplicity. Then the higher excited state may be converted into the lower excited state of the same spin-multiplicity through the **internal conversion (IC)** or internal quenching. In internal conversion, the electronic energy is converted into the vibrational energy of the molecule itself. For example, in the case of Cr(III)-complexes, the higer excited state $^4T_{1g}(F)$ may be converted into the lower excited state $^4T_{2g}(F)$ through internal conversion. This excited $^4T_{2g}(F)$ state is converted into the ^2E_g state through **intersystem crossing (ISC)** (a radiationless transition) and the ^2E_g state acts as the energy reservoir. It may be noted that in general, **IC is faster than ISC**. ISC is favoured in the heavier atoms for which the spin-orbit coupling is more efficient (**heavy atom effect**).

The **spin-orbit coupling** allows the intersystem crossing (*e.g.* triplet to singlet, quartet to doublet, etc.). The second excited state obtained through the intersystem crossing survives relatively longer because its transition to the ground state is spin-forbidden. **This spin forbidden transition** (*i.e.* delayed transition) **to the ground state gives the phosphorescence.**

In the case of **fluorescence** (a radiative decay), return from the excited state to the ground state occurs through the spin-allowed transition. It is very rapid and it completes within $10^{-8} - 10^{-9}$ s.

Jablonski diagram describing the fate of excited electronic states: *Jablonski diagram* (Fig. 7.9.5.1–2) can explain the phenomena like external and internal conversion, phosphorescence and fluorescence.

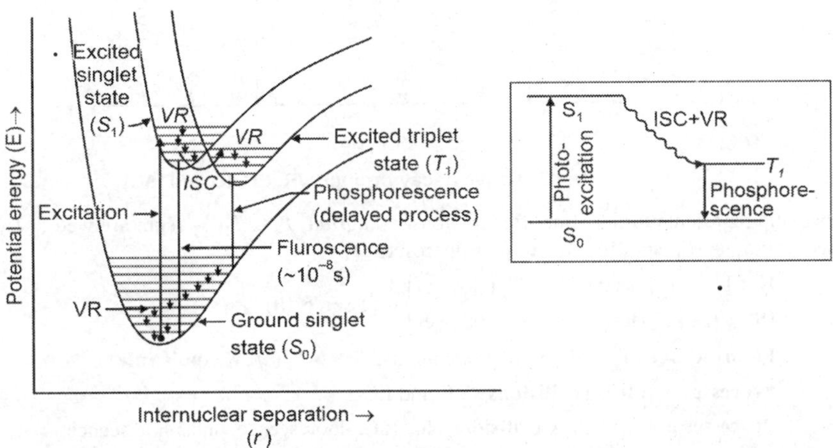

Fig. 7.9.5.1: Schematic illustration of **intersystem crossing** (ISC) from the excited singlet state (S_1) to the excited triplet state (T_1) at the isoenergetic point, simple way of representation of ISC (Box)—a simplified version of **Jaslonski diagram** (see Sec. 6.10.4)

Photochemistry of the Cr(III) complexes (*cf.* Sec. 6.15.6) and phosphorescence of the Cr(III) complexes: To understand the **photochemistry of Cr(III) complexes, and the ruby laser action**, the excited states $^4T_{1g}(F)$, $^4T_{2g}(F)$, $^2E_g(G)$ and the ground state $^4A_{2g}$ (Fig. 7.7.3.3) are to be considered (*cf.* Sec. 6.15.6). The excitations, $^4A_{2g}(F) \rightarrow {}^4T_{1g}(F)$, $^4T_{2g}(F)$ followed by **intersystem crossing** (ISC) populates the ^2E_g state. Then, the transition $^2E_g \rightarrow {}^4A_{2g}(F)$ gives the phosphorescence. These aspects have been discussed in Sec 7.14.5. **Photochemistry of $[Ru(bpy)_3]^{2+}$** can also be rationalised in terms of its phosphoresence behaviour (*cf.* Sec. 6.15.13).

In the case of Cr(III), energy difference between the ground state $^4A_{2g}$ ***(high spin*** t_{2g}^3***)*** and the **phosphorescent state** ^2E_g ***(low spin*** t_{2g}^3***)*** does not depend on the crystal field strength (see Fig. 7.7.3.3). In presence of very weak field ligands, the $^4T_{1g}$ and $^4T_{2g}$ states may reside below the ^2E_g state and *under this condition, Cr(III) complexes will show the* ***fluorescence not phosphorescence.***

Conclusion

- Luminescence (*i.e.* radiative decay of the excited state) leading to fluorescence and phosphorescence is important only when the nonradiative decay of the excited state is relatively inefficient.
- Intersystem crossing (ISC) through spin-orbit coupling enhances the population density of an excited state differing in spin-multiplicity with the ground state. Return from this excited state to the ground state through the **spin-forbidden transition gives the phosphorescence.**
- Return from the excited state to the ground state through the **spin-allowed transition gives fluorescence.**
- Fluorescence is an **instantaneous emission** while phosphorescence is a **delayed emission.**

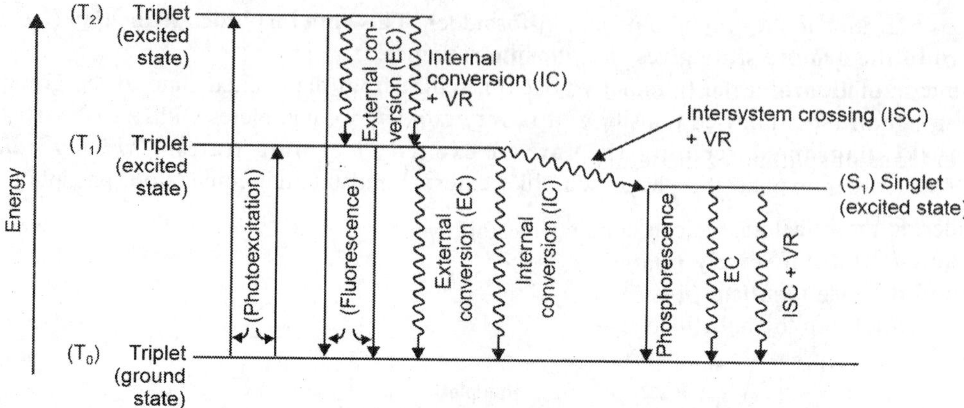

Fig. 7.9.5.2 Schematic representation of simplified *Jablonski diagram.* $T_0 \rightarrow T_1$, T_2 (spin allowed); $T_1, T_2 \rightarrow T_0$ (spin allowed, fluorescence); $S_1 \rightarrow T_0$ (spin forbidden, phosphorescence).

$$\left.\begin{array}{l}\textbf{IC}: T_2 \rightarrow T_1 \text{ (faster)}, T_1 \rightarrow T_0 \text{ (slower)} \\ \textbf{ISC}: T_1 \rightarrow S_1 \text{ (faster)}, S_1 \rightarrow T_0 \text{ (slower}\end{array}\right\} \text{(see Sec. 6.10)}$$

Fluorescence : $T_1 \rightarrow T_0$ more probable than $T_2 \rightarrow T_0$ (*cf*. **Kasha's rule**)

Processes involving collisions : VR and EC

Processes not involving collisions : IC, ISC, fluorescence, phosphorescence

- **External Conversion (EC) or collisional quenching** (see Sec. 6.10.2): Loss of excitation energy occurs as the **translational energy** or **heat energy** through the lattice vibration or collision with the surroundings; *It is a nonoradiative decay.* Time scale: ~10^{-8} s (gas phase), ~10^{-12}–10^{-14} s (solid or liquid phase).
- **Internal conversion (IC) or internal quenching** (see Sec. 6.10.3): It occurs between the states of **same spin multiplicity** (*i.e.* $\Delta S = 0$). This self quenching does not involve collision and electronic energy is converted into the **vibrational energy** of the molecule iteself. Time scale: ~10^{-11}–10^{-14} s.
- **Intersystem crossing (ISC) (radiationless transition)** (see Sec. 6.10.4): Spin-orbit coupling allows the conversion of one excited state to another excited state differing in spin-multiplicity. It occurs when the potential energy curves of the two excited states of different spin multiplicities cross. Time scale: ~10^{-6}–10^{-10} s.
- **VR (vibrational relaxation)** (see Sec. 6.10.1): Transition from the higher vibrational levels to the lower vibrational levels in a particular electronic state (*i.e.* $\Delta S = 0$) and the vibrational energy is lost as the heat energy. It occurs through collision with the surroundings. Time scale: ~10^{-10}–10^{-12} s.

7.10 POSSIBLE SPIN-ALLOWED ELECTRONIC TRANSITIONS IN THE WEAK FIELD OCTAHEDRAL AND TETRAHEDRAL COMPLEXES AND ORGEL DIAGRAM

***O_h vs. T_d* system:** The crystal field splitting patterns of the *d*–orbitals in the octahedral and tetrahedral complexes are of opposite nature for a particular d^n configuration (*cf.* Figs. 7.10.1, 2)

d^n *vs.* d^{10-n} systems: The splitting patterns of the terms for d^n and d^{10-n} configurations in an octahedral or tetrahedral field are of opposite nature. This is interpreted in terms of *the hole formalism rule* which states: d^n electronic system ≡ d^n positronic system ≡ d^{10-n} positronic system ≡ d^{10-n} electronic system. It leads to: d^n electronic configuration ≡ $(10 - n)$ holes in the d^{10} electronic system, *i.e.* d^n configuration is **hole equivalent** of d^{10-n} configuration. **But an electron is most stable where a positron is least stable.**

d^n and d^{n+5} systems: In a weak field, d^5–system has a zero orbital angular momentum (*i.e.* $L = 0$). Hence, the orbital angular momentum for the d^{5+n} and d^n systems will be identical. Thus for the d^n and

d^{n+5} systems, L will be the same. Consequently, the ground state terms for the d^n and d^{5+n} systems will be the same but they will differ in S and J values. This is illustrated below:

$$d^1(^2D_{3/2}),\ d^6(^5D_4);\ d^2(^3F_2),\ d^7(^4F_{9/2});\ d^3(^4F_{3/2}),\ d^8(^3F_4);\ d^4(^5D_0),\ d^9(^2D_{5/2})$$

Thus in terms of splitting of the terms in an octahedral or a tetrahedral field, it is follows:

$$d^1 \equiv d^6;\ d^2 \equiv d^7;\ d^3 \equiv d^8;\ d^4 \equiv d^9$$

Considering these factors, we can conclude as follows:

(i) **d^n and d^{10-n} are inversely related.**

(ii) **d^n and d^{n+5} are identical ($n < 5$)**

(iii) **d^{5-n}** ($\equiv n$ holes in the half-filled shell, *i.e.* d^5) and **d^{10-n}** ($\equiv n$ holes in the full filled shell, *i.e.* d^{10}) **are identical** and **reverse of those for d^n.**

(iv) **d^n (O_h) and d^{10-n} (T_d) are identical.**

(v) **d^n (O_h) and d^n (T_d) are inversely related.**

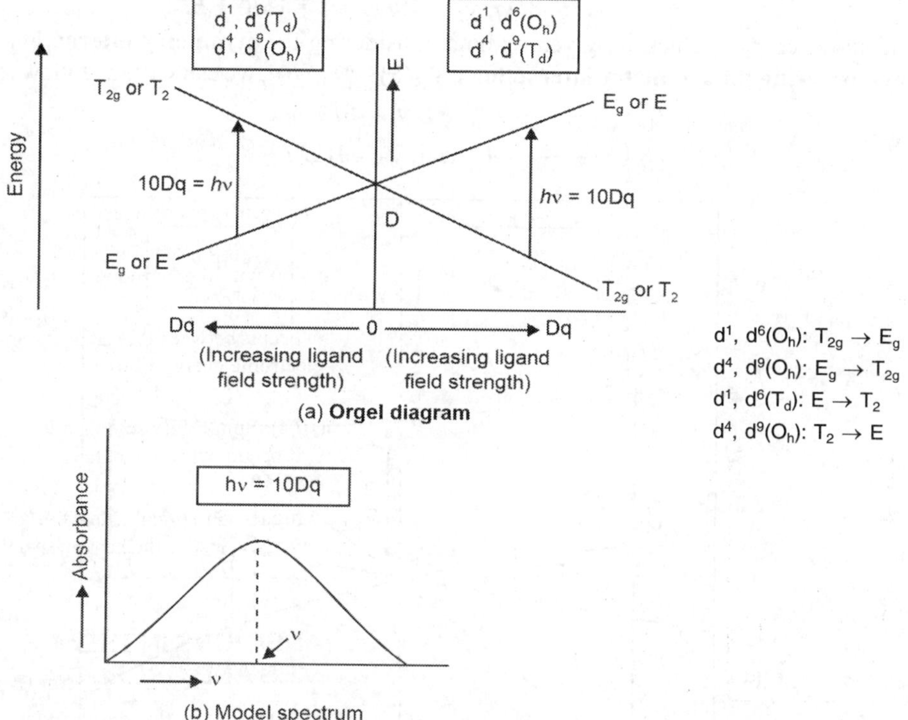

Fig. 7.10.1: Orgel diagram for the d^1, d^4, d^6 and d^9 systems (weak field ligands, high spin complexes) and spectral transition in the octahedral (O_h) and tetrahedral (T_d) systems. (**Note:** Here J. Teller effect has been ignored).

The O_h–systems (having the centre of symmetry) will bear the 'g'-subscript while for the T_d–systems (lacking in centre of symmetry), the g–subscript is dropped.

These are illustrated below.

$$\left.\begin{aligned} d^2(O_h) &\equiv \text{inverse of } d^8(O_h) \equiv \text{inverse of } d^2(T_d) \\ &\equiv d^8(T_d) \equiv d^7(O_h) \equiv \text{inverse of } d^3(O_h) \\ &\equiv \text{inverse of } d^7(T_d) \\ &\equiv d^3(T_d). \end{aligned}\right\} \quad (cf.\ \text{Fig. 7.10.2})$$

Similarly, we can write:

$$\left.\begin{aligned} d^1(O_h) &\equiv d^6(O_h) \equiv \text{inverse of } d^4(O_h) \equiv \text{inverse of } d^1(T_d) \\ &\equiv \text{inverse of } d^9(O_h) \\ &\equiv d^4(T_d) \equiv d^9(T_d) \end{aligned}\right\} \quad (cf.\ \text{Fig. 7.10.1})$$

Possible spectral transitions for the d^n–configurations are shown in Figs. 7.10.1 and 7.10.2.

(A) Spectral transitions in d^2, d^7, d^3 and d^8 systems: These can be formulated as follows from the Orgel diagram.

$$\mathbf{d^2, d^7\,(O_h)}:\ {}^nT_{1g}(F) \xrightarrow{\nu_1} {}^nT_{2g}(F) \qquad \mathbf{d^3, d^8\,(T_d)}:\ {}^nT_1(F) \xrightarrow{\nu_1} {}^nT_2(F)$$

$${}^nT_{1g}(F) \xrightarrow{\nu_2} {}^nA_{2g}(F) \qquad {}^nT_1(F) \xrightarrow{\nu_2} {}^nA_2(F)$$

$${}^nT_{1g}(F) \xrightarrow{\nu_3} {}^nT_{1g}(P) \qquad {}^nT_1(F) \xrightarrow{\nu_3} {}^nT_1(P)$$

- **Without considering the symmetry interaction energy**, we can write:

$$\nu_1 = 8Dq,\ \nu_2 = 18Dq,\ \nu_3 = 15B' + 6Dq$$

The above energy values are given without considering the **symmetry interaction energy**.

- **By considering the symmetry interaction** [${}^nT_{1g}(F) - {}^nT_{1g}(P)$], we can write as follows (*cf.* Fig. 7.10.2).

$$T_{1g}(F) \xrightarrow{\nu_1} T_{2g}(F),\ \nu_1 = 8Dq + x'$$

$$T_{1g}(F) \xrightarrow{\nu_2} A_{2g}(F),\ \nu_2 = 18Dq + x'$$

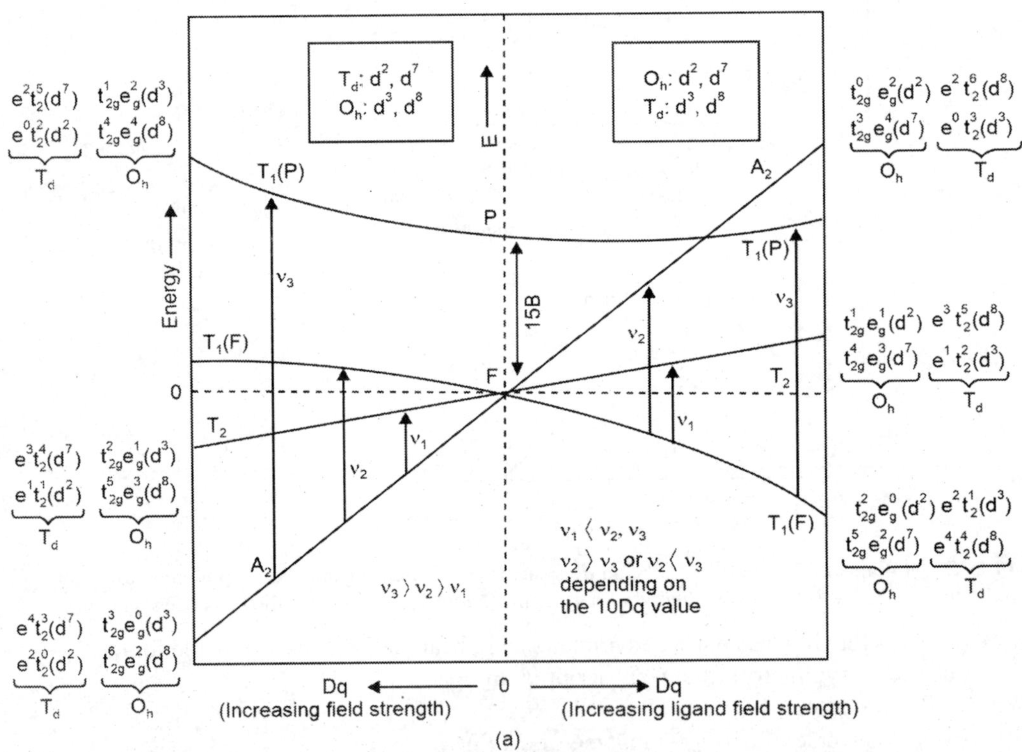

Fig. 7.10.2: Orgel diagram for the d^2, d^7, d^3 and d^8 systems in high spin octahedral and tetrahedral crystal fields (considering configurational interaction between the $T_1(P)$ and $T_1(F)$ states).

Note: In a strong field limit, $T_1(F)$ and $T_2(F)$ states for the $d^{2,7}(T_d)$, $d^{3,8}(O_h)$ systems merge; similarly, the $T_2(F)$ and $T_1(P)$ states merge for the $d^{3,8}(T_d)$ and $d^{2,7}(O_h)$ systems (*cf.* Fig. 7.10.2c).

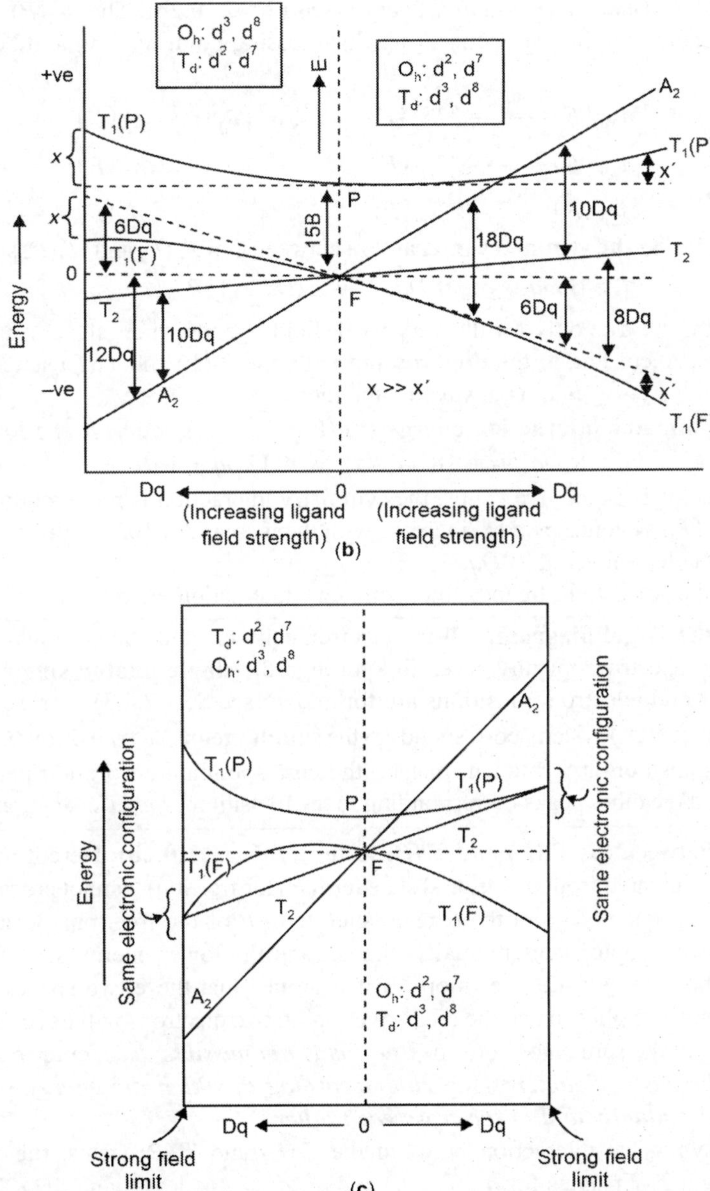

Fig. 7.10.2 (Contd.): Orgel diagram for the d^2, d^7, d^3 and d^8 systems in high spin octahedral and tetrahedral crystal fields (considering configurational interaction between the $T_1(P)$ and $T_1(F)$ states).

Note: In a **strong field limit**, $T_1(F)$ and $T_2(F)$ states for the $d^{2,7}(T_d)$, $d^{3,8}(O_h)$ systems merge; similarly, the $T_2(F)$ and $T_1(P)$ states merge for the $d^{3,8}(T_d)$ and $d^{2,7}(O_h)$ systems.

$$T_{1g}(F) \xrightarrow{\nu_3} T_{1g}(P),\ \nu_3 = 15B' + 6Dq + 2x'$$

$$\nu_2 - \nu_1 = 10Dq;\ \nu_2 + \nu_3 - 3\nu_1 = 15B'$$

Here x' denotes the *symmetry interaction energy* (see Fig. 7.10.2b). The $^3T_1(P)$ and $^3A_2$ states can mutually cross depending on the Dq–value. After this crossing point, $\nu_3 < \nu_2$ and before this crossing point $\nu_2 < \nu_3$.

$$d^3, d^8 (O_h): \quad {}^nA_{2g}(F) \xrightarrow{\nu_1} {}^nT_{2g}(F) \qquad d^2, d^7 (T_d): \quad {}^nA_2(F) \xrightarrow{\nu_1} {}^nT_2(F)$$

$$\quad {}^nA_{2g}(F) \xrightarrow{\nu_2} {}^nT_{1g}(F) \qquad \quad {}^nA_2(F) \xrightarrow{\nu_2} {}^nT_1(F)$$

$$\quad {}^nA_{2g}(F) \xrightarrow{\nu_3} {}^nT_{1g}(P) \qquad \quad {}^nA_2(F) \xrightarrow{\nu_3} {}^nT_1(P)$$

● **Without considering the symmetry interaction** between the $^nT_1$ (F) and $^nT_1(P)$ states, we can write:

$$\nu_1 = 10Dq, \ \nu_2 = 18Dq, \ \nu_3 = 12Dq + 15B'$$

These relationships are true only for the very weak field ligands where the configurational interaction is almost absent. According to the **noncrossing rule**, the $T_1(P)$ and $T_1(F)$ states can never cross. This is why, the trend, $\nu_3 > \nu_2 > \nu_1$ is always maintained.

● **In terms of symmetry interaction energy (*x*)** (Fig. 7.10.2b), we have the following relations:

$$\nu_1 = 10Dq, \ \nu_2 = 18Dq - x, \ \ \nu_3 = 12Dq + 15B' + x$$

For a certain value of $10Dq$, $\boldsymbol{x \gg x'}$. Thus, the symmetry interaction is relatively more important for the $d^{3,8}$ (O_h) and $d^{2,7}$ (T_d) systems. x and x' increase with the increase of $10Dq$ values. x and x' are almost negligible for the smaller values of $10Dq$.

This is why, for the weak field ligands, the symmetry interaction energy x is not so important.

● **Limitations of the Orgel diagram:** All the electronically excited states shown in the Orgel diagram are not always spectroscopically accessible through the **single photon single electron excitation**. In fact, two or multielectron transitions are forbidden (see Sec. 7.4.3). For example, the transition, $^3T_{1g} \rightarrow {}^3A_{2g}$ for the $d^2(O_h)$ system, corresponds to the **simultaneous 2*e* transition**, *i.e.* $t_{2g}^2 e_g^0 \rightarrow t_{2g}^0 e_g^2$, and the electronic transition is forbidden. In fact, the corresponding absorption peak is of very low intensity compared to the other peaks corresponding to the transitions, *i.e.* $^3T_{1g} \rightarrow {}^3T_{2g}$ and $^3T_{1g} \rightarrow {}^3T_{1g}$ (P).

(B) Interaction between the $T_1(F)$ and $T_1(P)$ states (*cf.* Fig. 7.10.2b): Here it may be pointed out that the $T_{1g}(F)$ state originated from the ground state term F (for free ion), cannot cross (*i.e.* **noncrossing rule**) the $T_{1g}(P)$ state originated from the excited state term P of the free ion. Because of their same symmetry, these spectroscopic states mutually interact and the lower energy state is further lowered down while the higher energy state rises more. If it is argued that these two spectroscopic states will mutually intersect then it implies that at the cross-over point, two electrons will have the same symmetry and the same energy in the same atom or ion. ***But this is not possible according to Pauli's exclusion principle. This is the basis of noncrossing rule according to which the energy states of the same symmetry will repell mutually and never cross each other.***

Because of the symmetry interaction between the $T_1(F)$ and $T_1(P)$ states, the energy difference between the $A_2(F)$ and $T_1(F)$ states for d^3, d^8 (O_h) and d^2, d^7 (T_d) is less than $18Dq$. But for d^3, d^8 (T_d) and d^2, $d^7(O_h)$, energy difference between the $A_2(F)$ and $T_1(F)$ states is slightly greater than $18Dq$. But, for the weak field ligands where the symmetry interaction energy is very small, the energy difference is almost equal to $18Dq$.

(C) Interelectronic repulsion parameter (*B*, *B'*) (see Sec. 7.9.3): In general, the excited ^{2s+1}P state lies above the ground state ^{2s+1}F and the energy difference between these two states of free ion is given by $15B$ where B (called Racah parameter) gives the measure of interelectronic repulsion. Generally, in the complexes, the degree of interelectronic repulsion reduces and consequently, the value of B is reduced by about 20%. In the complexes, the reduced B is denoted by B'.

(D) Spectral transitions in d^1, d^4, d^6 and d^9 systems: Octahedral and tetrahedral complexes of these electronic configurations are characterised by a single absorption band and the corresponding energy change is equal to $10Dq$. In fact, it happens so due to the splitting of the *D*–term (*cf.* Figs. 7.10.1 and 3). However, the J.T. distortion can split the band.

(E) Spectral transitions in d^2, d^3, d^7 and d^8 systems: Octahedral and tetrahedral complexes of these electronic configurations are characterised by three absorption bands.

The lowest band (ν_1) [$A_2 \rightarrow T_2(F)$] for the d^3, d^8 (O_h) and d^2, $d^7(T_d)$ complexes correspond to $10Dq$ (*cf.* Figs. 7.10.2–3). For the d^3, d^8 (T_d) and d^2, d^7 (O_h) systems, the energy difference between the lowest band (ν_1) [$T_1 \rightarrow T_2(F)$] and the band (ν_2) [$T_1 \rightarrow A_2$ (F)] corresponds to $10Dq$ (*cf.* Figs. 7.10.2 – 7.10.3). The ν_2–band is the highest energy band if the Dq value allows the crossing of the $A_2(F)$ and $T_1(P)$ states.

Summary of the ligand field transitions in terms of the Orgel Diagram

- **d^1, d^4, d^6, d^9 (O_h, T_d):** One transition (Δ_o or Δ_t) occurs (*cf.* Fig. 7.10.1):

$$d^1, d^6\left(O_h\right): T_{2g} \xrightarrow{10Dq_o} E_g \quad \Big| \quad d^4, d^9\left(O_h\right): E_g \xrightarrow{10Dq_o} T_{2g}$$

$$d^4, d^9\left(T_d\right): T_2 \xrightarrow{10Dq_t} E \quad \Big| \quad d^1, d^6\left(T_d\right): E \xrightarrow{10Dq_t} T_2$$

- **d^2, d^3, d^7, d^8 (O_h and T_d):** Three transitions are expected (*g* suffix omitted for T_d) (*cf.* Fig. 7.10.2).

$\mathbf{d^2, d^7\left(O_h\right)}$ **and** $\mathbf{d^3, d^8\left(T_d\right)}$:
$\nu_1 < \nu_2, \nu_3; \nu_2 - \nu_1 = \Delta_o$ or Δ_t
Relative values of ν_2 and ν_3 depend on $10Dq$ value.
$\nu_2 + \nu_3 - 3\nu_1 = 15B'$

$$\begin{cases} T_{1(g)}(F) \xrightarrow{\nu_1} T_{2(g)}(F), \ \nu_1 \approx 8Dq \\ T_{1(g)}(F) \xrightarrow{\nu_2} A_{2(g)}(F), \ \nu_2 \approx 18Dq \\ T_{1(g)}(F) \xrightarrow{\nu_3} T_{1(g)}(P), \ \nu_3 \approx 6Dq + 15B' \end{cases}$$

$\mathbf{d^3, d^8\left(O_h\right)}$ **and** $\mathbf{d^2, d^7\left(T_d\right)}$:
$\nu_1 = \Delta_o$ or $\Delta_t < \nu_2 < \nu_3$

$$\begin{cases} A_{2(g)}(F) \xrightarrow{\nu_1} T_{2(g)}(F), \ \nu_1 \approx 10Dq \\ A_{2(g)}(F) \xrightarrow{\nu_2} T_{1(g)}(F), \ \nu_2 \approx 18Dq \\ A_{2(g)}(F) \xrightarrow{\nu_3} T_{1(g)}(P), \ \nu_3 \approx 12Dq + 15B' \end{cases}$$

(F) Techniques to analyse the electronic spectra of the octahedral and tetrahedral complexes d^n (n = 2, 3, 7, 8) having the ground spectroscopic state $A_2(F)$ or $T_1(F)$: The spectral analysis, involves: band assignments; evaluation of $10Dq$, B' and $\beta\left(=\frac{B'}{B_0}\right)$. To determine these spectral parameters, different techniques are known. If all the three ligand field spectral transitions are identified, then the spectral analysis becomes straight-forward. When this is not possible, the **transition energy ratios,** *i.e.* ν_3/ν_1, ν_3/ν_2, ν_2/ν_1 are calculated and these are fitted with the standard curves (*cf.* Fig. 7.10.4a, b) of the transition energy ratios *vs.* Dq/B derived from the Tanabe-Sugano diagram. These theoretical plots (*i.e.* standard curves) of ν_2/ν_1, ν_3/ν_2 *vs.* Dq/B (for the A_2 and T_1 ground states) are constructed from the Tanabe-Sugano diagram. Standard Tables for the theoretical transition energy ratios, ν_1/B, ν_2/B, ν_3/B and Dq/B values are also available (*cf.* A.B.P. Lever, *J. Chem. Edu.,* **45,** 711, 1968; A.B.P. Lever, Inorganic Electronic Spectroscopy, 2nd Edn., 1984; Elsevier).

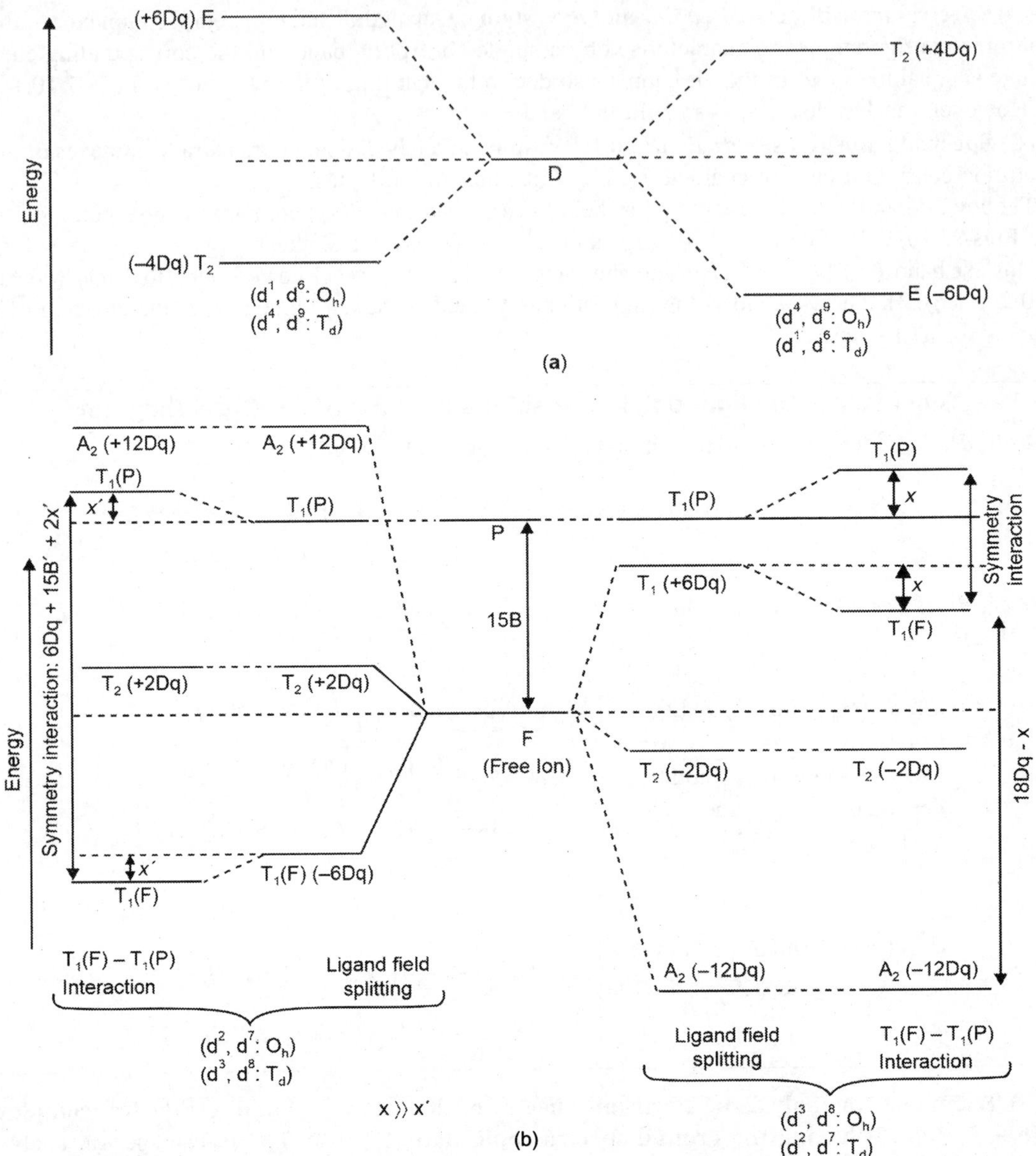

Fig. 7.10.3 Splitting of the *D* and *F* terms in the octahedral (O_h) and tetrahedral (T_d) fields for the weak field ligands. *x* or *x*′ denotes the symmetry interaction energy for the repulsive interaction between the $T_1(F)$ and $T_1(P)$ states (*see* Fig. 7.9.4.1 for the energy values of different states).

Note: ● *g*–subscript to be used for the octahedral systems. ● For the d^2, d^7 (O_h) and d^3, d^8 (T_d) systems, here the crystal field strength is considered to be sufficiently high so that **the $A_2(F)$ state can cross the $T_1(P)$ state.**

7.11 SPLITTING OF THE ENERGY LEVELS DUE TO JAHN TELLER DISTORTION (*cf.* Sec. 3.11.3–4, Vol. 4 and Sec. 7.14 for details)

Here we shall illustrate the case of octahedral complexes where the J.T. distortion is more important than in the case of tetrahedral complexes. In the octahedral stereochemistry, if any level (t_{2g} or e_g)

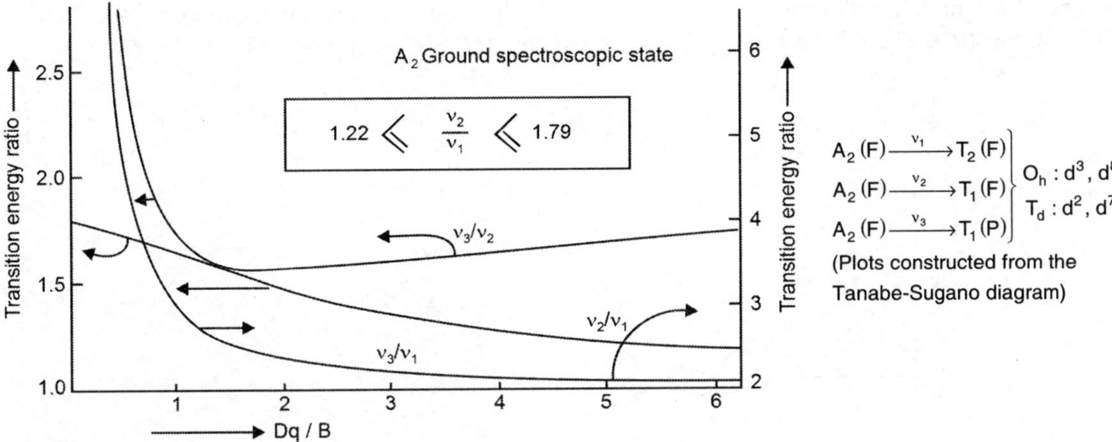

Fig. 7.10.4: (a) Plots of transition energy ratio *vs.* Dq/B for the A_2 ground spectroscopic state.

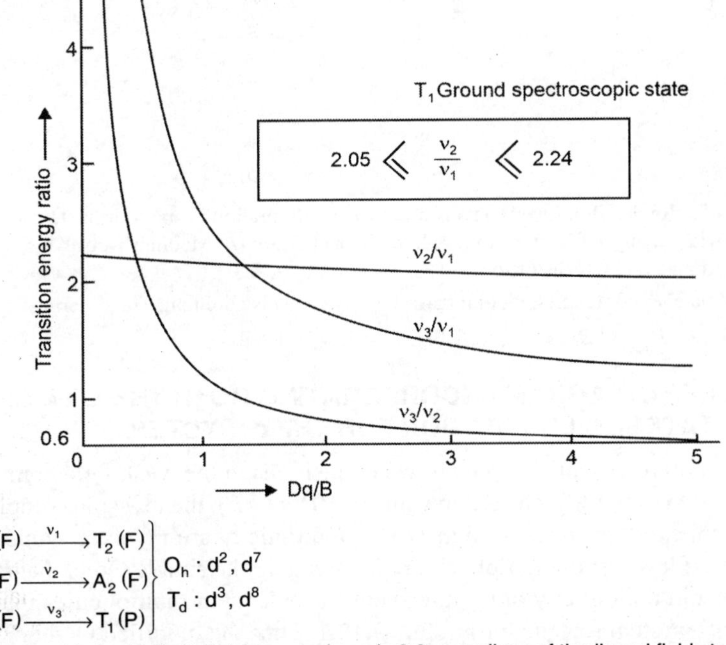

T$_1$ (F) —ν$_1$→ T$_2$ (F)
T$_1$ (F) —ν$_2$→ A$_2$ (F)
T$_1$ (F) —ν$_3$→ T$_1$ (P)
O$_h$: d^2, d^7
T$_d$: d^3, d^8

Note: ν_2/ν_1 ($\approx$ 18 Dq/8 Dq) is approximately 2.0 regardless of the ligand field strength.
(Plots constructed from the Tanabe-Sugano diagram)

Fig. 7.10.4: (b) Plots of transition energy ratio *vs.* Dq/B for the T_1 ground spectroscopic state.

remains unsymmetrically filled in before or after electron transition, then the levels undergo splitting. Such a splitting for the d^9 system in an octahedral field is shown in Fig. 7.11.1.

The splitting of energy levels due to the J.T. distortion may broaden the band or sometimes generate the close spaced bands. *Here it is important to note that there is some freedom about the choice of B_1 and B_2 and they may get interchanged* (*i.e.* the state denoted by B_1 by one author may be represented by B_2 by another author).

The effect of J.T. distortion is well noticed in the high-spin octahedral complexes of Cr^{2+} (d^4), Ti^{3+}(d^1), Fe^{2+}(d^6). These have been discussed separately (*cf.* Sec. 3.11.3–4, Vol. 4).

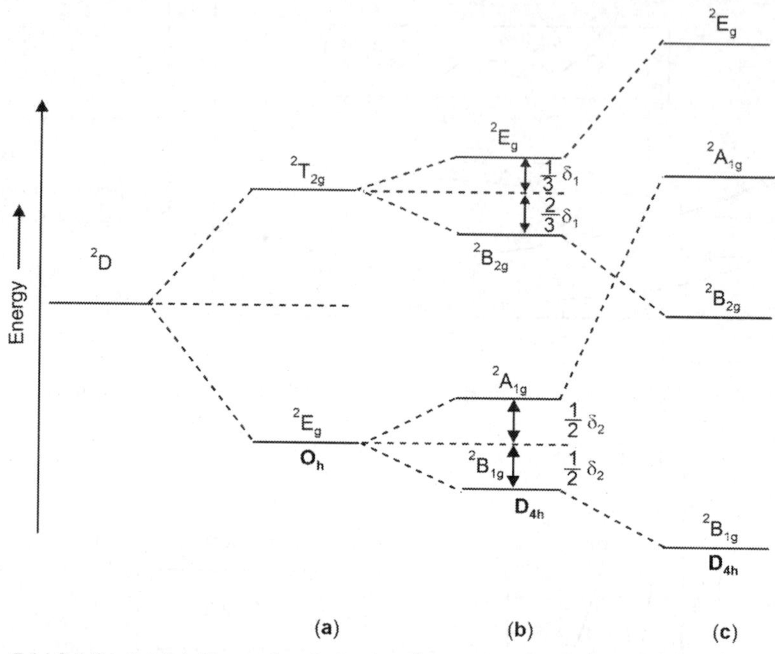

Fig. 7.11.1 (*cf.* Fig. 7.14.2.1 for details): Energy levels for the d^9 configuration in six-coordinate complexes. (a) Perfectly octahedral field (O_h)—without any J.T. distortion (*i.e.* hypothetical case); (b) Medium *z*-out J.T. distortion (*i.e.* medium axial elongation); (c) Strong *z*-out J.T. distortion.

Note: For the six-coordinate d^1–complexes, medium *z*-in J.T. distortion is common. The ordering of energy levels is:

$$^2B_{2g} < {}^2E_g << {}^2B_{1g} < {}^2A_{1g} \text{ (cf. Fig. 7.14.1.2 for details)}$$

7.12 MODIFIED ORGEL DIAGRAM (INCORPORATING BOTH THE LOW SPIN AND HIGH SPIN STATES): ILLUSTRATION FOR THE d⁶ SYSTEM

Orgel diagrams generally represent the spin allowed transitions in the weak field complexes. However, for the d^4, d^5, d^6 and d^7 systems, after a certain value of $10Dq$ (= P), the low spin complexes are formed. In the high-spin complexes, the states of highest spin multiplicity are involved while for the low-spin complexes, the states of lower spin multiplicity are involved. In the simple Orgel diagram (dealing with the states of highest spin multiplicity only), the states of lower spin multiplicities may be included. It is illustrated for the d^6–system (octahedral) in Fig. 7.12.1. Thus the modified Orgel diagram represents both the spin-allowed and spin-forbidden transitions.

The ground state term for the free the ion of d^6 configuration (*e.g.* Co^{3+}) is ^5D which splits into $^5T_{2g}$ and ^5E_g in presence of a weak octahedral field. After the critical $10Dq$ value (= P), it attains the low-spin state ($t_{2g}^6\, e_g^0$). The **first excited state** of the low-spin complex is $t_{2g}^5\, e_g^1$. Here it is assumed that in the excited state, the two unpaired electrons in the t_{2g} and e_g levels remain in an antiparallel condition to give the spin multiplicity 1. It will allow the spin allowed transitions (*cf.* Sec. 7.15.1 for details). The distribution of 5 electrons (≡ one hole) in the d_{xy}, d_{yz} and d_{xz} orbitals of the t_{2g} level can be made in three possible ways. The single electron in the $d_{x^2-y^2}$ and d_{z^2} orbitals of the e_g level can be distributed in two ways. **Thus the excited state $t_{2g}^5\, e_g^1$ can have 6 (= 3 × 2) probable ways of electron distribution.**

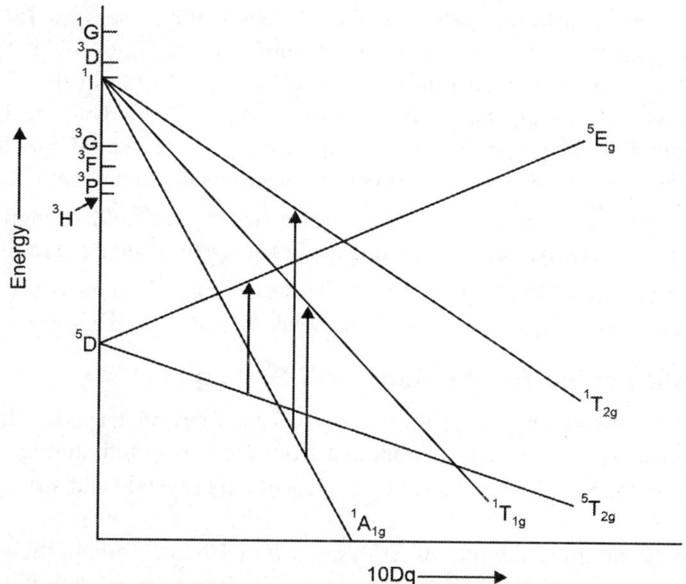

Fig. 7.12.1: Modified Orgel diagram for the d^6 configuration in the octahedral complexes showing the transitions in both the high-spin (*i.e.* lower $10Dq$ value) and low-spin (*i.e.* higher $10Dq$ value) states.

These 6 probable ways (*i.e.* orbital multiplicity = 6) of electronic configuration will give the $^1T_{1g}$ and $^1T_{2g}$ states and each of them is having the orbital degeneracy 3. Thus the t_{2g}^6 configuration gives the $^1A_{1g}$ level (ground state) while the excited state $t_{2g}^5\,e_g^1$ leads to the $^1T_{1g}$ and $^1T_{2g}$ levels. These can be correlated with the singlet term ^1I of the free ion.

$$^1I\,(\text{free ion term}) \xrightarrow[\text{(strong field)}]{O_h-\text{field}} \underbrace{^1A_{1g}}_{\text{ground state}}\ ,\ \underbrace{^1T_{1g},\ ^1T_{2g}}_{\text{excited states}} \quad (\textit{cf.}\ \text{Sec. 7.15.1})$$

Now in the simple Orgel diagram of d^6 configuration (involving only the ^5D free ion term), the singlet free ion term may be included (Fig. 7.12.1). The diagram shows that up to a certain value of $10Dq$ ($< P$), the ground state is $^5T_{2g}$ while after the critical $10Dq$ value, the ground state is represented by the $^1A_{1g}$ level. By considering the $^5T_{2g}$ level as the ground state, the spin allowed transition is: $^5T_{2g} \rightarrow {}^5E_g$. On the other hand, by considering the $^1A_{1g}$ level as the ground state, there are two possible spin allowed transitions:

$^1A_{1g} \rightarrow {}^1T_{1g}$ and $^1A_{1g} \rightarrow {}^1T_{2g}$. These predictions are experimentally verified.

7.13 LIMITATIONS OF THE ORGEL DIAGRAM: TANABE—SUGANO DIAGRAMS

7.13.1 Limitations of the Orgel Diagram

The simple Orgel diagram deals with the highest spin multiplicity states (*i.e.* high spin complexes) and the spin allowed transitions only. For the high-spin d^5 system, there is no spin allowed transition. For the d^1, d^9, d^4 (h.s.) and d^6 (h.s.) systems, **only one ligand field transition** is expected while for the d^2, d^3, d^7 (h.s.) and d^8 systems, there are only **three spin allowed transitions.** But, very often, the actual spectra cannot be quantitatively interpreted in terms of the only spin-allowed transitions predicted from the simple Orgel diagram because besides the spin allowed transitions, there may be some spin-forbidden transitions of weaker intensity.

Some specific cases are illustrated here. For the d^8-system (O_h), the **spin forbidden transition,** $^3A_{2g}(F) \rightarrow {}^1E_g(D)$ may gain the sufficient intensity (*cf.* intensity stealing effect, Sec. 7.5.1M), if the transition lies close to the spin allowed transition, $^3A_{2g}(F) \rightarrow {}^3T_{1g}(F)$ (as in the case of $[Ni(OH_2)_6]^{2+}$, Fig. 7.14.4.2). For Cr(III) (d^3, O_h), the weak transition, $^4A_{2g} \rightarrow {}^2E_g$ finds some importance. For the heavier metals, the spin-forbidden transitions are quite important (see Sec. 7.5.2, **heavy atom effect**).

- Simple Orgel diagram fails to predict the spectral transitions in the low-spin octahedral complexes.
- Orgel diagram plots the energies of different states (*vs.* crystal field strength) in absolute units. Consequently, quantitatively, **each diagram applies only to a particular metal ion system.**

These are the main limitations of the Orgel diagram. These are majorly removed in the Tanabe-Sugano diagram (*cf.* Y. Tanabe and S. Sugano, *J. Phys. Soc. Jpn.*, **9**, 753–779, 1954).

7.13.2 Characteristics of the Tanabe-Sugano (T-S) Diagram

(i) The ordinate (*i.e. y*-axis) represents the energy of the free ion terms in the unit of B (Racah parameter). This energy scale may be obtained from the concerned atomic spectra.

(ii) The abscissa or horizontal axis (*i.e. x*–axis) represents the crystal field strength ($10Dq$ or Dq) in the unit of B.

(iii) Thus, the *T–S* diagram gives the plot of E/B *vs.* Dq/B or $10Dq/B$ for a particular d^n configuration. The *T–S* diagram shows the variation of energy of different states with the variation of crystal field strength. Thus it includes the *energy states of both the high spin and low-spin* (if possible, as in the cases of d^4, d^5, d^6, d^7 systems) complexes.

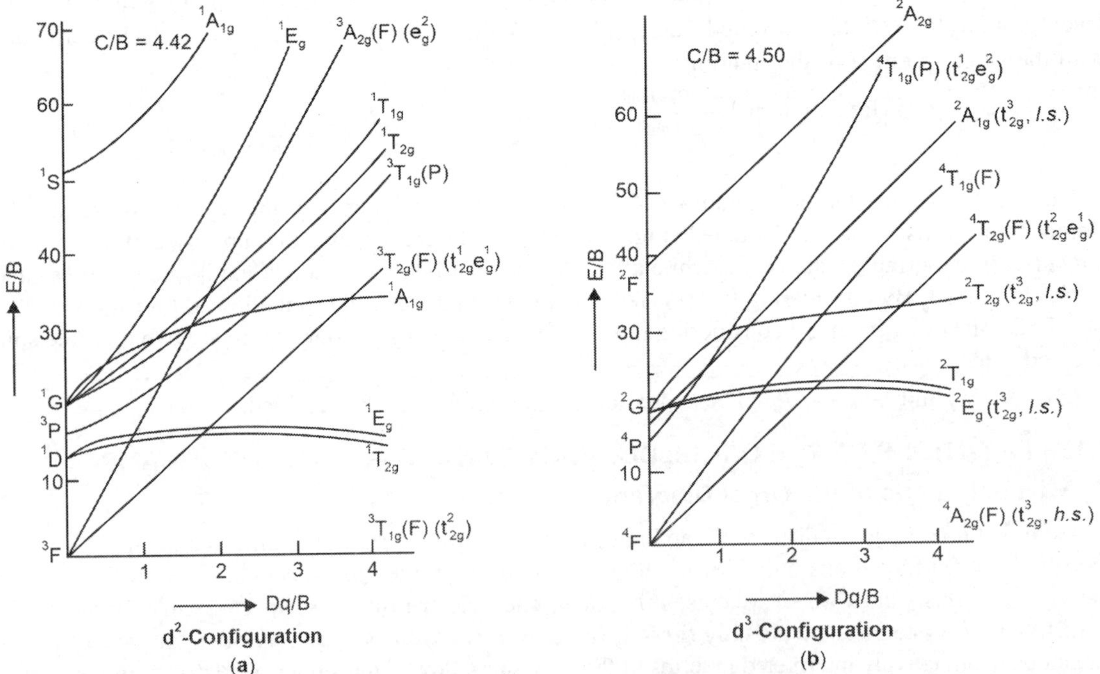

Fig. 7.13.2.1: (a) T.S. diagram of the d^2–configuration in an octahedral field. (b) T.S. diagram of the d^3–configuration in an octahedral field.

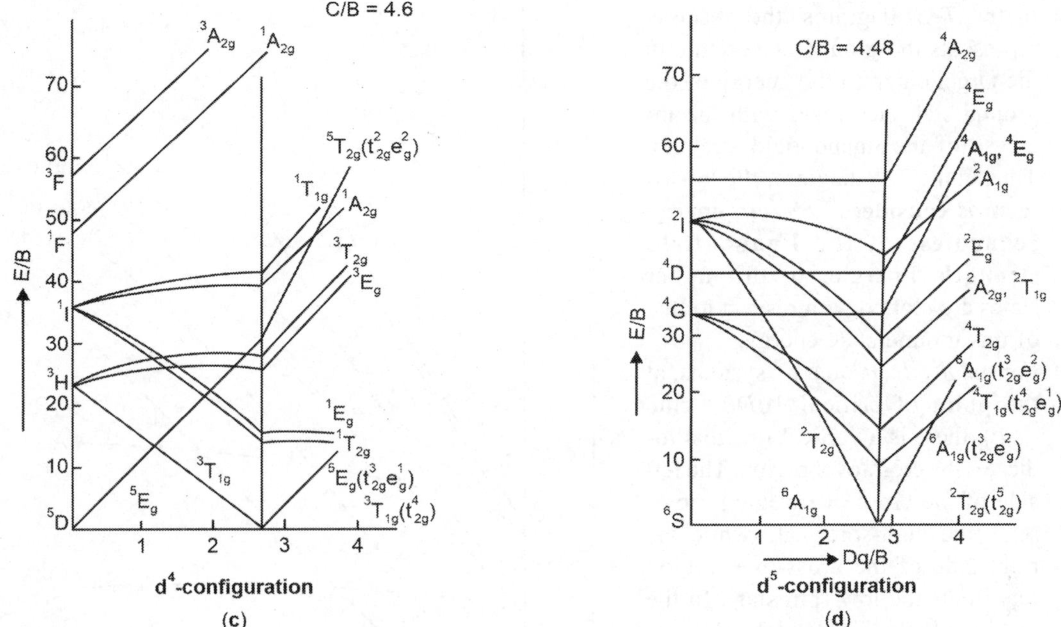

(c)

(d)

Fig. 7.13.2.1: (c) T.S. diagram of the d^4–configuration in an octahedral field; (d) T.S. diagram of the d^5–configuration in an octahedral field.

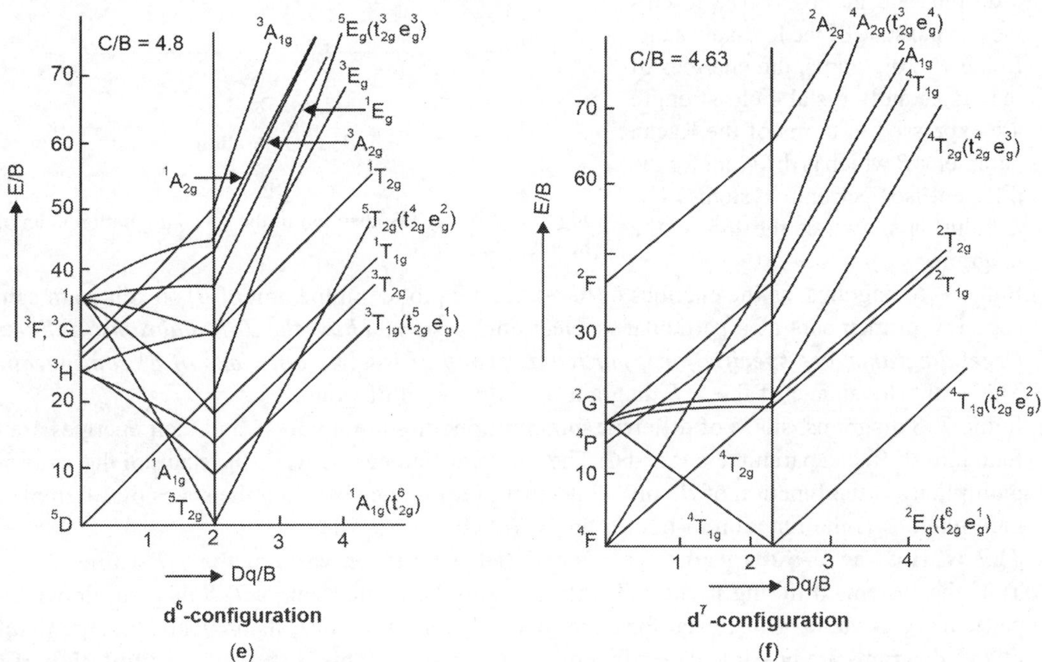

(e)

(f)

Fig. 7.13.2.1 (e) T.S. diagram of the d^6–configuration in the octahedral field; (f) T.S. diagram of the d^7–configuration in an octahedral field.

(iv) In the *T–S* diagrams, the abscissa represents the ground state while in the Orgel diagram the energy of the ground state decreases with the increase of the ligand field strength. Thus in the *T–S* diagram, **the lowest term is considered of zero energy regardless of the ligand field strength.** Energies of the higher states are plotted with respect to that of this ground state energy.

(v) For the d^4, d^5, d^6 and d^7 systems, at the point of critical $10Dq$ value (= *P*), there is a vertical line that indicates the cross-over region. The left side of the cross-over region represents the high-spin state while the right side of the cross-over region represents the low-spin state. In the region of high-spin state, the abscissa represents the ground state of the high-spin state while in the region of low-spin state, the abscissa represents the ground state of the low-spin state.

(vi) In the *T–S* diagrams, the energies of the states and crystal field strength are expressed in terms of the Racah parameter *B* which is different for the different isoelectronic systems. The *B* value is shown at the top of the diagram.

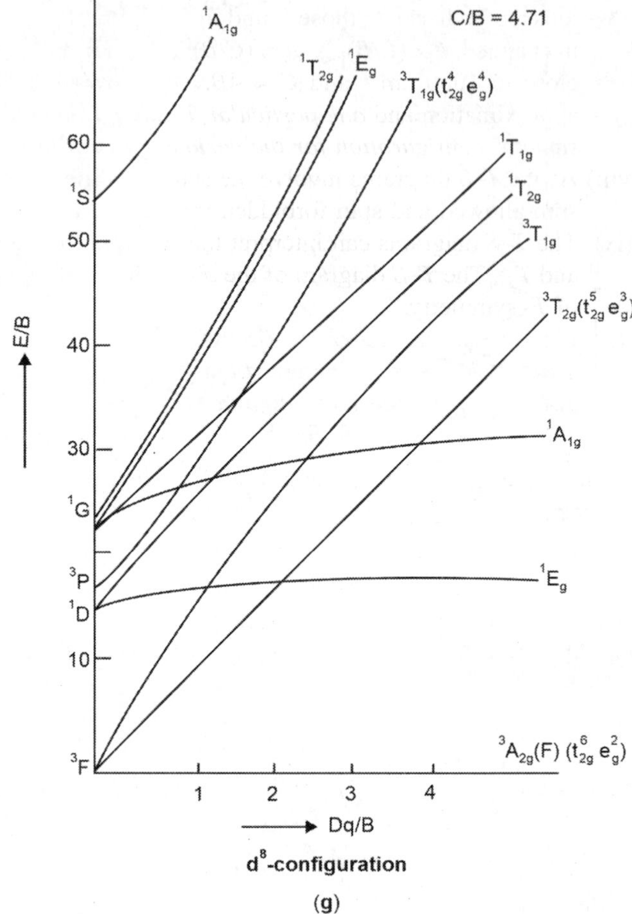

Fig. 7.13.2.1 (g) T.S. diagram of the d^8–configuration in an octahedral field.

In the *T–S* diagrams, as the energies of the states are plotted in the unit of *B*, one diagram can be used for all members of a particular isoelectronic system. ***Thus, the T–S diagrams, unlike the Orgel diagrams, are specific for a particular group of isoelectronic ions of d^n configuration.*** This is the advantage of the *T–S* diagram over the Orgel diagram.

(vii) In the *T–S* diagrams, states of different spin multiplicities are involved and their energies are the functions of Racah parameters *B* and *C*. The energy difference between the states of the same spin multiplicity is the function of *B* only while energy difference between the states of different spin multiplicities is the function of both *B* and *C* (*cf.* Table 7.9.3.1).

This is why, the T–S diagrams are constructed for a given ratio of the *C*/*B* value.

Thus the systems differing in the *C*/*B* values cannot be treated with a *T–S* diagram drawn for a particular *C*/*B* value. Thus, even for a particular d^n configurations, theoretically, a large number of *T–S* diagrams are possible depending on the *C*/*B* values. **This is the serious limitation of the *T–S* diagram.** However, in reality, it has been found that for most of the transition metal ions, the *C*/*B* ratio has been found close to 4.0. This aspect is discussed below.

The B and C values are decreased during the complexation, *i.e.* the values of B and C for the free ions are higher than those found in their complexes. However, the ratio C/B remains more or less unchanged, *i.e.* $(C/B)_{\text{Free ion}} \approx (C/B)_{\text{Complex}}$. For most of the transition metal ions, the value of B is close to 1000 cm^{-1} and $C \approx 4B$, *i.e.* $C/B \approx 4.0$. The *T–S* diagrams are drawn based on this approximation and *one particular T–S diagram can be used for the different metal ions with the same d^n configuration for the various types of ligands.*

(viii) As the *T–S* diagrams involve the states of different spin multiplicities, these can predict both the spin allowed and spin forbidden transitions.

(ix) The *T–S* diagrams can interpret the spectra of the metal complexes of cubic symmetry (both O_h and T_d). The *T–S* diagram of the d^n–complex of O_h symmetry is applicable for the d^{10-n} complex of T_d symmetry.

(x) If the electronic transition occurs between the states which run parallel (*i.e.* having the same slopes) in the *T–S* diagram, ***then the peak becomes sharp because of the absence of peak broadening by the metal-ligand bond vibration*** (*cf.* Secs. 7.7.1–2). On the other hand, if the slopes of two states differ then the peak will be broadened by the metal-ligand bond vibration.

The *T–S* diagrams are given in Fig. 7.13.2.1. Application of the *T–S* diagram in the spectral analysis has been discussed in Secs 7.7.3, 7.10 (see Fig. 7.10.4) and will be discussed later.

7.14 LIGAND FIELD SPECTRA OF THE HIGH-SPIN OCTAHEDRAL COMPLEXES OF FIRST TRANSITION SERIES

The aqua metal ions, *i.e.* $[M(OH_2)_6]^{n+}$ of 1st transition series are octahedral and their colour originates from their ligand field bands. Their spectra and shown in Fig. 7.14.1.

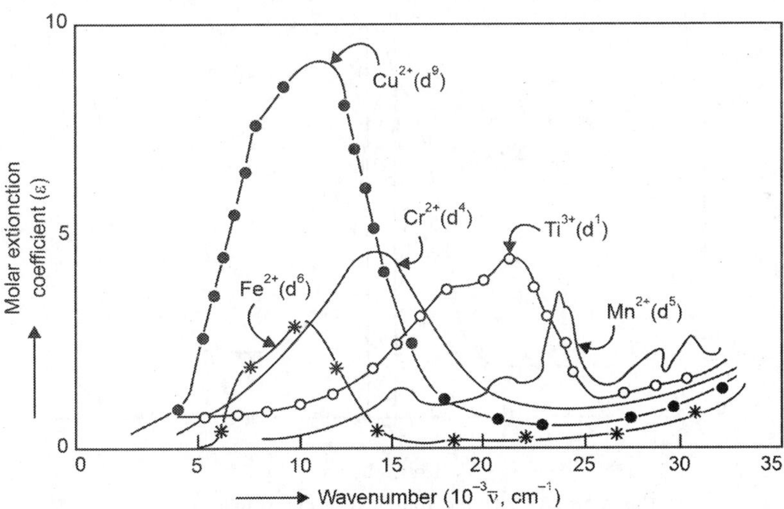

Fig. 7.14.1 (a) Electronic absorption spectra of the **aqua complexes (high-spin)** of some bivalent and trivalent metal ions of 1st transition series.

7.14.1 Octahedral Complexes of 3d^1 Metal Ions

The representative examples are the complexes of Ti(III) (3d^1), V(IV) (3d^1), VO^{2+} (3d^1).

(A) Octahedral complexes of Ti(III) (*cf.* Sec. 3.11.4, Vol. 4)**:** The aqueous solution of titanous salt, *i.e.* $[Ti(OH_2)_6]^{3+}$, looks purple or violet coloured having the absorption maximum at ~20,000 cm^{-1} (*i.e.*

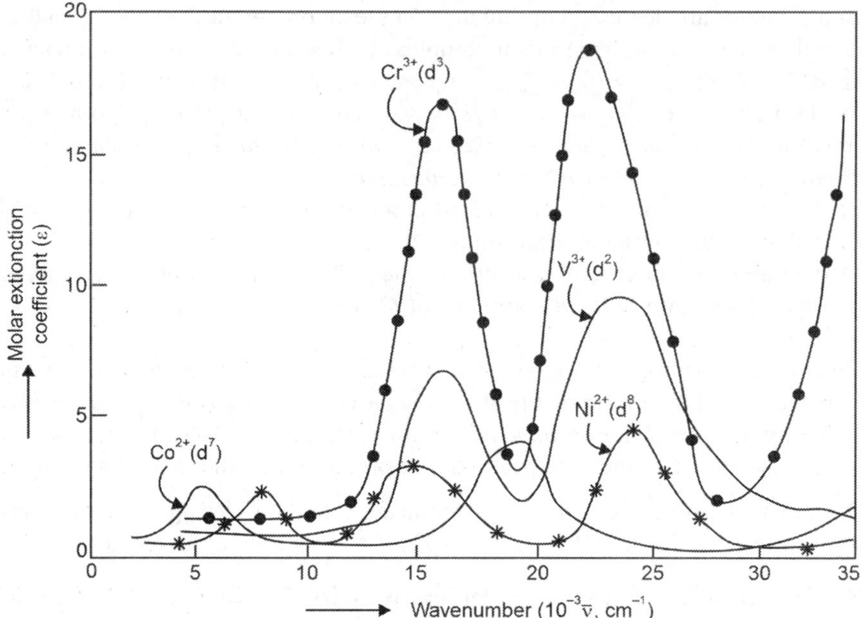

Fig. 7.14.1 (b) Electronic absorption spectra of the **aqua complexes (high-spin)** of some bivalent and trivalent metal ions of 1st transition series.

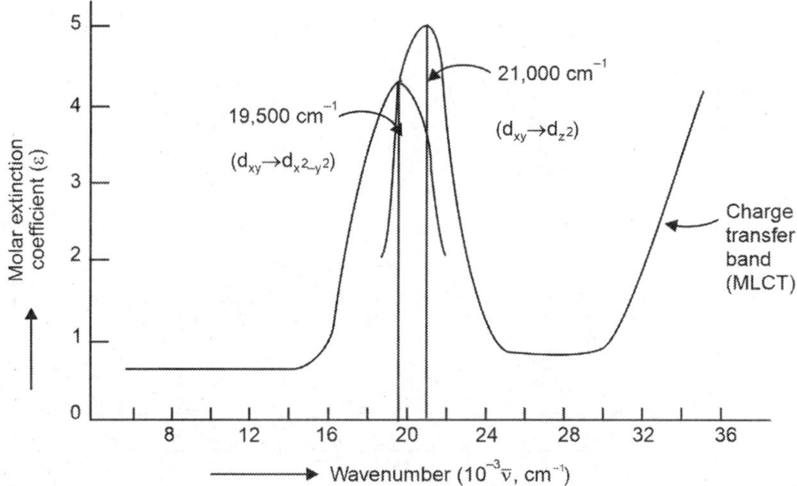

Fig. 7.14.1.1: The absorption spectrum of $[Ti(OH_2)_6]^{3+}$

~500 nm). This purple coloured solution is obtained by dissolving $TiCl_3$ in dilute HCl. The purple colour of $[Ti(OH_2)_6]^{3+}$ indicates that its broad absorption peak occupies in such a way that most of the red and violet light of the visible spectrum remain unabsorbed, *i.e.* blue and green light are absorbed. *Thus the unabsorbed red-violet colour gives the purple colour of the solution.*

The **characteristic features** of the absorption spectrum of $[Ti(OH_2)_6]^{3+}$ are: **weak, broad, and unsymmetrical.** Thus we are to explain: ***position, poor intensity (ε ≈ 5), bandwidth and shape of the absorption peak.***

(i) **Position:** Analysis of the spectrum of $[Ti(OH_2)_6]^{3+}$ indicates that the absorption band appears at 21,000 cm^{-1} **with a shoulder** at 19,500 cm^{-1}. The **average maximum** lies at about 20,300 cm^{-1} (*i.e.* the band is centred at about 20,300 cm^{-1} $\approx \frac{1}{2}(21{,}000 + 19{,}500)\text{cm}^{-1}$).
From the Orgel diagram, the electron transition responsible for the electronic spectrum of $[Ti(OH_2)_6]^{3+}$ is:

$$^2T_{2g} \rightarrow {}^2E_g,\ i.e.\ t_{2g}^1\, e_g^0 \rightarrow t_{2g}^0\, e_g^1$$

The involved energy change is $10Dq$. Thus we can write:

$$10Dq = 20{,}300 \text{ cm}^{-1} = 493 \text{ nm} \approx 243 \text{ kJ mol}^{-1}$$

$$\left(1000 \text{ cm}^{-1} = 12.0 \text{ kJ mol}^{-1},\ E \approx \frac{1.2\times10^5}{\lambda(\text{in nm})} \text{kJ mol}^{-1}, \text{Sec. 6.10.14}\right).$$

(ii) **Poor intensity:** The poor intensity of the peak is characterised by $\varepsilon \approx 4$–5. It is due to the Laporte forbidden *d–d* transition. It is just partially allowed due to the vibronic coupling.

(iii) **Bandwidth and asymmetric shape:** It gives a wide and unsymmetric band. Analysis of the spectrum indicates that the **broad peak actually consists of two overlapping symmetric peaks.** It takes place due to the splitting of the ^2E_g state, a consequence of Jahn-Teller effect.

According to Jahn Teller theorem, the t_{2g}^1(**ground state**) level will undergo splitting and it will definitely suffer the **z–in distortion** rather than the *z*-out distortion to gain the additional cfse $\frac{2}{3}\delta_2$ compared to the undistorted octahedron. The upper e_g state will undergo splitting but it will have no energetic contribution towards the ground state because it remains unoccupied in the ground state ($t_{2g}^1 e_g^0$). But the **excited state** (*i.e.* $t_{2g}^0\, e_g^1$) will suffer the distortion and will have the energetic effect of a larger magnitude compared to that of the ground state. The possible electron transitions are shown in Fig. 7.14.1.2. These are:

$$^2B_{2g} \xrightarrow{\nu_3} {}^2E_g;\ {}^2B_{2g} \xrightarrow{\nu_1} {}^2B_{1g};\ {}^2B_{2g} \xrightarrow{\nu_2} {}^2A_{1g}, \textbf{(in terms of energy state)}$$

$$d_{xy} \xrightarrow{\nu_3} (d_{xz}, d_{yz});\ d_{xy} \xrightarrow{\nu_1} d_{x^2-y^2};\ d_{xy} \xrightarrow{\nu_2} d_{z^2}, \textbf{(in terms of orbitals)}$$

$$b_{2g} \xrightarrow{\nu_3} e_g;\ b_{2g} \xrightarrow{\nu_1} b_{1g};\ b_{2g} \xrightarrow{\nu_2} a_{1g}, \textbf{(in terms of orbitals)}$$

The peaks due to ν_1 and ν_2 (cf. Fig. 7.14.1.2) are closely spaced and they are overlapping to produce a resultant broad peak with a shoulder. The ν_3 transition lies only in the infra-red region and the corresponding energy change (δ_2) is very small (*ca.* 400 – 800 cm^{-1}). In fact, splitting of the E_g level is more significant (*ca.* δ_1 is of the order of 1000 – 3000 cm^{-1}).

The main peak at about 21,000 cm^{-1} ($d_{xy} \xrightarrow{\nu_2} d_{z^2}$, *i.e.* $^2B_{2g} \longrightarrow {}^2A_{1g}$) and the other peak at about 19,500 cm^{-1} ($d_{xy} \xrightarrow{\nu_1} d_{x^2-y^2}$, *i.e.* $^2B_{2g} \longrightarrow {}^2B_{1g}$) showing a shoulder overlap to produce a resultant broad peak having the average maximum at about 20,300 cm^{-1}. *The ν_1 transition actually involves the 10Dq value of the corresponding hypothetical octahedral complex.* An ideal octahedral complex (which is a hypothetical case for Ti(III)) of a d^1–system should have a single peak. By considering this fact, average peak position of ν_1 ($\approx$ 19,500 cm^{-1}) and ν_2 ($\approx$ 21,000 cm^{-1}) has been *conventionally regarded* as the peak position of the **hypothetical octahedral complex,** $[Ti(OH_2)_6]^{3+}$ and this peak position has been considered to estimate the 10Dq value. However, this is an **approximation** but it does not affect the analysis seriously because the ν_1 and ν_2 bands are very close to each other. For $[Ti(OH_2)_6]^{3+}$, the average peak position appears at about 20,300 cm^{-1} which is considered as the 10Dq value of the corresponding **hypothetical octahedral complex.** With the increase of the crystal field strength, the 10Dq value increases.

	$[TiCl_6]^{3-}$	$[TiF_6]^{3-}$	$[Ti(OH_2)_6]^{3+}$	$[Ti(CN)_6]^{3-}$	$[Ti(bpy)_3]^{3+}$
10Dq (cm^{-1}):	13,000	18,900	20,300	22,300	25,000

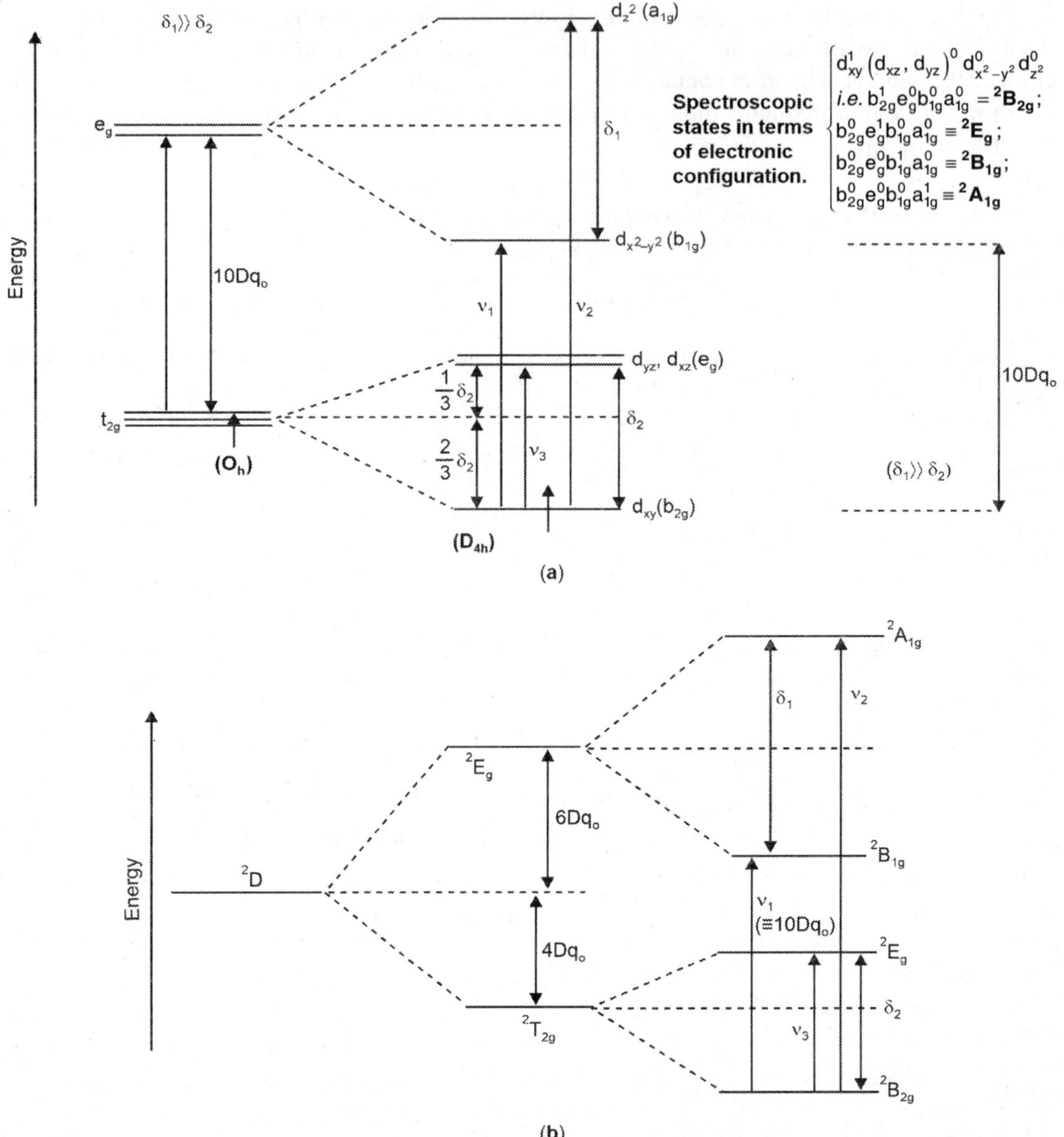

Fig. 7.14.1.2: Possible ways of electron transition in the octahedral d^1 system due to **z-in Jahn-Teller distortion (Note:** b_{1g}, b_{2g}, e_g and a_{1g} are used **to describe the orbital** while B_{1g}, B_{2g}, E_g and A_{1g} denote the **spectroscopic states**).

[Note: Some authors analyse the peaks of $[Ti(OH_2)_6]^{3+}$ as follows:

absorption band at 20,300 cm^{-1} and shoulder at 18,400 cm^{-1}.

Without the J.T. distortion, a single peak ($^2T_{2g} \rightarrow {}^2E_g$) will appear at $\frac{1}{2}$(20,300 + 18,400) cm^{-1}, *i.e.* 19350 cm^{-1}, which is considered as the $10Dq$ value of the hypothetical octahedral $[Ti(OH_2)_6]^{3+}$ complex].

Here it is worth mentioning that if the peaks due to ν_1 and ν_2 are widely separated (*i.e.* δ_1 is relatively high) then the two peaks may not overlap and **two distinct peaks will appear.** It happens so in the high-spin $[Co^{III}F_6]^{3-}$ complex which involves the transition:

$$t_{2g}^4\, e_g^2 \rightarrow t_{2g}^3\, e_g^3$$

This aspect will be discussed separately for the d^6–system.

The spectral features of other octahedral Ti(III)–complexes like $[TiCl_6]^{3-}$, $[TiF_6]^{3-}$, $[Ti(urea)_6]^{3+}$, Ti(III) in Al_2O_3 matrix or in $CoTi(SO_4)_3.12H_2O$ are similar to those of $[Ti(OH_2)_6]^{3+}$.

(B) Possibility of spin-orbit coupling to split the energy levels to explain the broad peak of Ti(III)–complexes: Splitting of the $^2T_{2g}$ state (originated from ^2D) due to spin-orbit coupling has been shown in Fig. 7.7.4.1. The ^2E_g state does not split due to spin-orbit coupling but the $^2T_{2g}$ state splits into two components characterised by $J = \frac{3}{2}$ and $J = \frac{1}{2}$. In the case of Ti(III), the spin-orbit coupling constant is quite small (*cf.* $\lambda \approx +155$ cm^{-1}) and consequently, the energy difference between the states ($J = \frac{3}{2}$ and $J = \frac{1}{2}$. for ground state $J = \frac{3}{2}$) is $3\lambda/2$ which is approximately 232 cm^{-1}. This value is negligibly small compared to the crystal field splitting parameter $10Dq$ ($\approx$ 20,000 cm^{-1}). **Thus the spin-orbit coupling leading to the splitting of the $^2T_{2g}$ state can hardly explain the peak broadening phenomenon in the case of Ti(III).** This can be explained better by J.T. distortion as discussed above.

Note: Spin-orbit coupling constants of the 3*d*-metal ions are quite negligible compared to the $10Dq_o$ value of the octahedral complexes. Thus the peak broadening by this mechanism is not significant. **But, for the tetrahedral complexes where $10Dq_t$ is relatively small, spin-orbit coupling phenomenon can broaden the peak appreciably.** In fact, splitting of the spectrum of the tetrahedral $[CoCl_4]^{2-}$ complex by spin-orbit coupling is well documented. This aspect will be discussed later.

(C) Charge transfer band in $[Ti(OH_2)_6]^{3+}$: The metal to ligand charge transfer (MLCT) band due to the electron transfer from Ti(III) (which is reducing in character) to oxygen (O) appears at the ultraviolet range (beyond 30,000 cm^{-1}). In that region, there is a strong peak for this MLCT band.

(D) Blue sapphire gemstone (blue corundum): Corundum (Al_2O_3, E_g = 8.3 eV) is itself transparent but presence of only 0.01% of titanium and iron in it gives the intense blue colour of **blue sapphire** (*cf.* **deep red ruby colour of corundum**, due to the $d \rightarrow d$ transition of Cr(III) impurity, at least 1% required to see the red colour). The $d \rightarrow d$ transition cannot explain this intense blue colour at such low concentration levels. The **heteronuclear MMCT transitions like** Fe(II)→Ti(IV) or Ti(III)→Fe(III) can explain this intense blue colour (see Sec. 7.14.5 for details) (Das *et al.*, *Resonance*, 2025).

(E) Six–coordinate complexes of V(IV) and six-coordinate complexes of oxovanadium, *i.e.* vanadyl cation, VO^{2+}: Vanadium(IV) or oxovanadium(IV) represents the $3d^1$ system. The electronic spectra of the six-coordinate V(IV) complexes like $[VCl_6]^{2-}$ in $K_2[VCl_6]$ or $[VF_6]^{2-}$ in $(NH_4)_2[VF_6]$ can be analysed as in the case of six-coordinate Ti(III) complexes. For $[VCl_6]^{2-}$ and $[VF_6]^{2-}$, their absorption bands are centred at around 15,400 cm^{-1} and 20,100 cm^{-1} respectively.

The spectra of the octahedral (O_h) and distorted octahedral (D_{4h}, *z*-in distortion) complexes of V(IV), *e.g.* $[VCl_6]^{2-}$ etc. can be explained as in the cases of Ti(III) complexes. The six coordinate complexes of oxovanadium(IV) are of C_{4v} symmetry and their spectra can be explained by using the Fig. 7.14.1.3 (which is similar to Fig. 7.14.1.2a). In the oxovanadium(IV) complexes like $[VOL_5]$, the V=O unit remains projected to an axial direction (say *z*-direction). In $[VO(OH_2)_5]^{2+}$, the V—O (for VO^{2+} unit) bond length is 167 pm while the V—$O(OH_2)$ bond length is in the range of 203–230 pm. This supports the C_{4v} symmetry **in $[VOL_5]$**.

The six-coordinate complexes of oxovanadium (*i.e.* **vanadyl cation**, VO^{2+}) are not octahedral and they experience the compression in one axial bond (say *z*-direction) (*cf.* in D_{4h} symmetry of *z*-in tetragonal distortion, both the *z*-axial bonds are compressed) and the complexes like $[VO(OH_2)_5]^{2+}$,

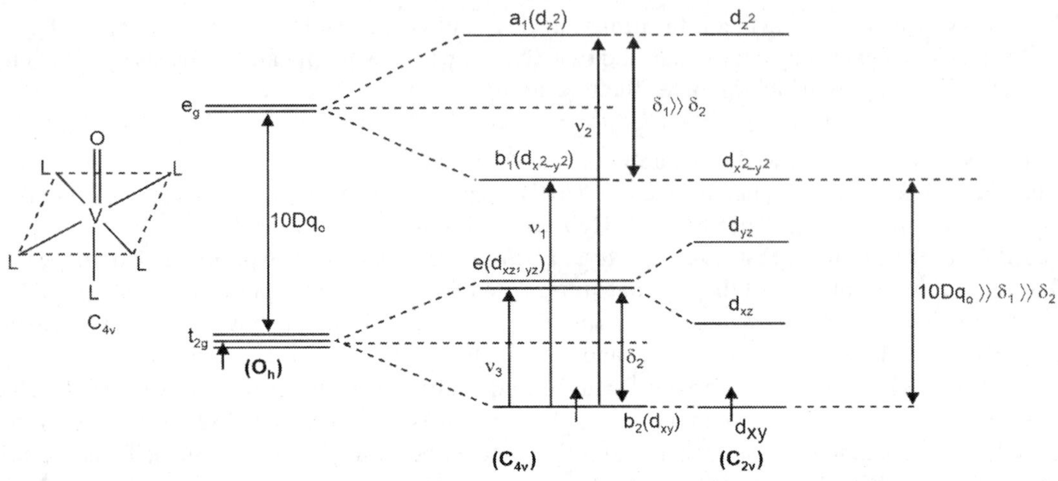

Fig. 7.14.1.3: Splitting of *d*–orbitals in six-coordinate oxovanadium(IV) complexes, *i.e.* $[VOL_5]$ and possible electron transitions. **Note:** The splitting pattern is the same as in z-in distortion but the 'g' symmetry is dropped as the C_{4v} symmetry lacks in the centre of inversion.

Table 7.14.1.1: Ligand field spectra of oxovanadium(IV) complexes

	$^2B_2 \xrightarrow{\nu_3} {}^2E$; *i.e.*, $d_{xy} \longrightarrow d_{xz}, d_{yz}$ ν_3 (cm^{-1})	$^2B_2 \xrightarrow[(\equiv 10Dq)]{\nu_1} {}^2B_1$ $d_{xy} \longrightarrow d_{x^2-y^2}$ ν_1 (cm^{-1})	$^2B_2 \xrightarrow{\nu_2} {}^2A_1$ $d_{xy} \longrightarrow d_{z^2}$ ν_2 (cm^{-1})
$[VO(OH_2)_5]^{2+}$	13,100	16,000	–
$[VO(NCS)_5]^{3-}$	13,500	17,200	23,900
$[VO(ox)_2]^{2-}$	12,600	16,500	29,400
$[VO(acac)_2]$, (in methanol)	13,000	17,450	25,630

Here it may be noted that the $^2B_2 \rightarrow {}^2E$ transition is **allowed** but the $^2B_2 \rightarrow {}^2B_1$ and $^2B_2 \rightarrow {}^2A_1$ transitions are **forbidden.** However, these can be allowed by vibronic coupling. These two transitions give the **relatively weaker bands.**
Note: The charge transfer bands of oxovanadium(IV) complexes are similar in both solid state and solution. It indicates the similar fate of the vanadyl $(V{=}O)^{2+}$ group both in solution and solid state. In fact, **basicity of the oxygen site in the VO^{2+} group is drastically reduced due to π–bonding with V(IV)** and consequently, it is not protonated easily.

$[VO(NCS)_5]^{3-}$ etc. belong to the C_{4v} point group. The orbital splitting in the six-coordinate complexes of VO^{2+} is shown in Fig. 7.14.1.3 which is similar to the Fig. 7.14.1.2a for the D_{4h} symmetry (*z*-in distortion, *i.e.* tetragonal compression where both the *z*-axial bonds are compressed).

The possible electronic transitions are:

$$^2B_2 \xrightarrow{\nu_3} {}^2E, \text{ i.e. } b_2 \xrightarrow{\nu_3} e; {}^2B_2 \xrightarrow[(\equiv 10Dq)]{\nu_1} {}^2B_1, \text{ i.e. } b_2 \xrightarrow{\nu_1} b_1; {}^2B_2 \xrightarrow{\nu_2} {}^2A_1, \text{ i.e. } b_2 \xrightarrow{\nu_2} a_1$$

The electronic transitions in terms of the energy states and d^n-configurations are correlated for the C_{4v} oxovanadium(IV) complexes as follows:

$$d^1_{xy}(d_{xz}, d_{yz})^0(d_{x^2-y^2})^0 d^0_{z^2}, \text{ i.e. } b^1_2 e^0 b^0_1 a^0_1 \equiv {}^2B_2 \Big\} \textbf{ ground state}$$

$$\left.\begin{aligned} &d^0_{xy}(d_{xz}, d_{yz})^1(d_{x^2-y^2})^0 d^0_{z^2}, \text{ i.e. } b^0_2 e^1 b^0_1 a^0_1 \equiv {}^2E \\ &d^0_{xy}(d_{xz}, d_{yz})^0(d_{x^2-y^2})^1 d^0_{z^2}, \text{ i.e. } b^0_2 e^0 b^1_1 a^0_1 \equiv {}^2B_1 \\ &d^0_{xy}(d_{xz}, d_{yz})^0(d_{x^2-y^2})^0 d^1_{z^2}, \text{ i.e. } b^0_2 e^0 b^0_1 a^1_1 \equiv {}^2A_1 \end{aligned}\right\} \textbf{ excited states}$$

● The oxovanadium complexes like $[VO(NCS)_5]^{3-}$, $[VO(acac)_2]$, etc. should be written as $[V(NCS)_5O]^{3-}$, $[V(acac)_2O]$, ...respectively (*See* Chapter 1, IUPAC recommendation).

(F) Five-coordinate complexes of oxovanadium(IV): The spectra of five-coordinate complexes like $[VO(ox)_2]^{2-}$, $[VO(acac)_2]$ of C_{4v} symmetry can also be explained in terms of Fig. 7.14.1.3.

(G) Spectral properties of the oxovanadium(IV) complexes of C_{2v} symmetry: Sometimes, the oxovanadium complexes, *e.g.* vanadyl tartarate, show **four bands instead of three bands** (expected from the C_{4v} symmetry). Origin of the four spectral transitions has been explained by considering the **C_{2v} symmetry**. In moving from the C_{4v} symmetry to the C_{2v} symmetry, degeneracy of the *e*-level (*i.e.* d_{xz} and d_{yz} orbitals) is lifted. Then the possible transitions are:

$$\mathbf{C_{2v}}: d_{xy} \xrightarrow{11\text{ kK}} d_{xz};\ d_{xy} \xrightarrow{17\text{ kK}} d_{yz};\ d_{xy} \xrightarrow{18.8\text{ kK}} d_{x^2-y^2};\ d_{xy} \xrightarrow{25.3\text{ kK}} d_{z^2}$$

(For vanadyl tartarate).

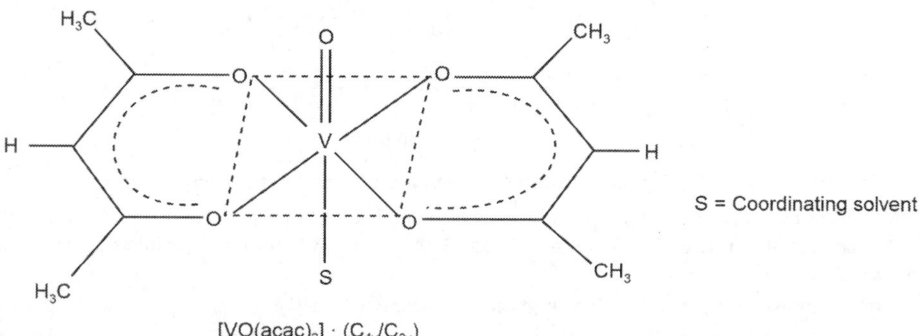

$[VO(acac)_2] \cdot (C_{4v}/C_{2v})$

Blue-green $[VO(acac)_2]$: When dissolved in various organic solvents varying in coordinating power, it shows the spectral bands of different energies. It happens so due to the coordination by the solvent molecule (S) at the *trans*-axial site to the V=O bond to give the 6-coordinate species. Thus for different solvents, the average $10Dq$ value changes and the spectral transitions occur at different energies. In fact, the extents of splitting of the *d*-orbitals in C_{4v} symmetry depend on the average crystal field strength (*cf.* spectral behaviour of $[Cu(acac)_2]$ in different solvents, Fig. 7.14.2.5).

Detailed studies (both is solid state and mixed solvents) on **$[VO(acac)_2]$** indicate that it also adopts the C_{2v} symmery and the transition $d_{xy} \rightarrow d_{xz}, d_{yz}$ (*i.e.* $b_2 \rightarrow e$ in C_{4v}) gives two spectral components, *i.e.* two separate transitions as expected for the C_{2v} symmetry. However, some authors have attempted to explain the four peaks (*i.e.* components) in the electronic absorption spectrum of $[VO(acac)_2]$ in terms of vibrational structure.

O
~160 pm ($\nu_{V=O} \approx 1000\ cm^{-1}$) (cf. $\nu_{V=O} \approx 950\ cm^{-1}$ in $[V(acac)_2O(py)]$)
H3C V CH3
(cf. equatorial V—O ≈ 210 pm)
H3C 210 pm CH3

C_{4v} symmetry (in **solid state and noncoordinating solvents**) (square pyramidal, archetypal geometry; **V sitting above the square plane**).

In coordinating solvents or in presence of any suitable ligand, six coordinate distorted octahedral geometry is attained through the coordination by the sixth ligand, *e.g.* [V(acac)$_2$O(py)] at the *trans*-axial position. Electron donation by this **sixth ligand** reduces the π-back bonding ($O_{2p} \rightarrow V_{3d}$) by the vanadyl oxygen and the $\nu_{V=O}$ stretching frequency reduces by about 50 cm^{-1}.

Vanadyl cation (VO^{2+}) of V(IV) acts as a **class-a cation** (*i.e.* **hard acid**) and forms complexes with the hard donor ligands like F, O, N, Cl, etc.

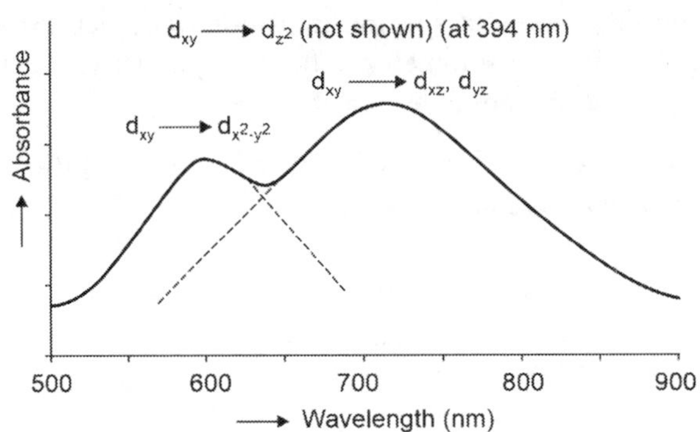

Fig. 7.14.1.4: Absorption spectrum of [VO(acac)$_2$] in CH_3CN (acetonitrile)

Note 1: [VO(acac)$_2$] in the coordinating CH_3CN solvent, shows the *d–d* transitions at 14200 cm^{-1} (704 nm), 16950 cm^{-1} (590 nm), 25400 cm^{-1} (394 nm) and LMCT ($O^{2-} \rightarrow V^{IV}$) at 33400 cm^{-1} (≈ 300 nm) (tailing into the visible range) (*Ref. Inorg. Chem.*, **35**, 3885, 1996)

Note 2: Vanadyl complexes are generally **blue to green** in colour. Besides the *d–d* transitions (d^1-system), the lowest energy LMCT ($O^{2-} \rightarrow V^{IV}$, more correctly, V—O π-electron to the metal *d*-orbital enriched MO) transition occurs at about 33000 cm^{-1} (303 nm) tailing into the visible range.

7.14.2 Six-Coordinate Complexes of 3d^9–System

(A) Possible spectral transitions in the distorted octahedral complexes of Cu(II): Cu(II) complexes are the representative examples of the d^9–system. Energy level diagram for the spectral transitions in a d^9–system is the inverted energy level diagram of the d^1–system. It is expected so from the **hole formalism rule.** Thus by using the Orgel diagram, for the octahedral complexes of Cu(II), the spectral transition is:

$$^2E_g \rightarrow {}^2T_{2g}, \text{ (cf. for the } d^1\text{–system: } {}^2T_{2g} \rightarrow {}^2E_g)$$

In terms of the electronic configuration, the transition is:

$$t_{2g}^6 e_g^3 \rightarrow t_{2g}^5 e_g^4$$

The ground state is **orbitally doubly degenerate.**

$$t_{2g}^6 e_g^3 \left({}^2E_g\right): t_{2g}^6 \left(d_{x^2-y^2}\right)^2 \left(d_{z^2}\right)^1, t_{2g}^6 \left(d_{x^2-y^2}\right)^1 \left(d_{z^2}\right)^2$$

The excited state is **orbitally triply degenerate.**

$$t_{2g}^5 e_g^4 \left({}^2T_{2g}\right): (d_{xy})^2 (d_{yz})^2 (d_{xz})^1 e_g^4; (d_{xy})^2 (d_{yz})^1 (d_{xz})^2 e_g^4; (d_{xy})^1 (d_{yz})^2 (d_{xz})^2 e_g^4$$

The ground state electronic configuration, $t_{2g}^6 e_g^3$, indicates that the unsymmetrical filling in the e_g level causes the J.T. distortion (*i.e.* tetragonal distortion) at the ground state. Thus it can never have the perfectly octahedral complexes. The six-coordinate complexes of Cu(II) are the distorted octahedrals. The tetragonal distortion can produce both the *z*-out (*i.e.* tetragonal elongation) and *z*–in (*i.e.* tetragonal compression) distortions with an equal probability because both the possibilities lead to the same additional cfse $\frac{1}{2}\delta_1$ (δ_1 = splitting energy of the e_g level) for a d^9–system. But in reality, most of the Cu(II)–complexes are experiencing the *z*–out distortion. The tetragonal compression is known only in some rare cases. One such example is mentioned here. In solid state, **$K_2[CuF_4]$ shows the *z*–in distortion** (*i.e.* compressed octahedral) around Cu(II).

The *d*–orbital splitting in the **elongated octahedral complexes of Cu(II)** is shown in Fig. 7.14.2.1. The possible transitions are:

$$^2B_{1g} \xrightarrow{\nu_3} {}^2A_{1g},\ i.e.\ d_{z^2} \xrightarrow{\nu_3} d_{x^2-y^2}\ \text{or},\ a_{1g} \xrightarrow{\nu_3} b_{1g},\ (12{,}000 - 17{,}000\ \text{cm}^{-1})$$

$$^2B_{1g} \xrightarrow{\nu_2} {}^2B_{2g},\ i.e.\ d_{xy} \xrightarrow[(\equiv 10Dq)]{\nu_2} d_{x^2-y^2}\ \text{or},\ b_{2g} \xrightarrow{\nu_2} b_{1g},\ (15{,}500 - 18{,}000\ \text{cm}^{-1})$$

$$^2B_{1g} \xrightarrow{\nu_1} {}^2E_g,\ i.e.\ d_{xz}, d_{yz} \xrightarrow{\nu_1} d_{x^2-y^2}\ \text{or},\ e_g \xrightarrow{\nu_1} b_{1g},\ (17{,}000 - 20{,}000\ \text{cm}^{-1})$$

The ν_3 transition actually gives the measure of splitting of the e_g level denoted by δ_1.

The extents of splitting of the t_{2g} and e_g levels measured by δ_2 and δ_1 respectively (*cf.* Fig. 7.14.2.1.) due to the J.T. distortion depend on the ligand field strength of the ligands. In most of the cases, the splittings are not too large. Consequently, the peaks due to ν_1, ν_2 and ν_3 very often overlap to give ***a broad resultant peak of unsymmetrical shape***. In some cases, where the splittings are quite large, separate peaks may appear.

(B) Peak broadening by spin-orbit coupling in Cu(II) complexes: The absorption spectra of the Cu(II)–complexes are broadened due to J.T. distortion discussed above. Besides this, the peaks may be further broadened due to the spin-orbit coupling. The T_{2g} state is split (*cf.* Secs. 7.7.4, 8.20.3) due to the spin-orbit coupling to broaden the peak. For the 3*d*-metal ions, the spin-orbit coupling constant is quite small and quite negligible compared to the crystal field splitting energy. Thus peak broadening/splitting by spin-orbit coupling is relatively less important compared to that by J.T. distortion. However, the spin-orbit coupling effect is relatively more important in the tetrahedral complexes having the smaller crystal field splitting. Thus in the tetrahedral complexes, the spin-orbit coupling interaction energies are more or less comparable to the ligand field energies.

(C) Colour of the Cu(II)–complexes: Most of the Cu(II) complexes are **blue or green** in colour. Some complexes look **red or brown**. Such red or brown complexes experience a strong absorption in the near ultraviolet region due to the charge transfer (CT) bands. The tails of the CT bands extend into the blue end of the visible spectrum. It makes the red-brown colour. The possible electronic transitions in the distorted octahedral complexes of Cu(II) are given below.

$$e_g^4 b_{2g}^2 a_{1g}^2 b_{1g}^1 \left(i.e.\ {}^2B_{1g}\right) \xrightarrow{\nu_1} e_g^3 b_{2g}^2 a_{1g}^2 b_{1g}^2 \left(i.e.\ {}^2E_g\right)\ i.e.\ {}^2B_{1g} \longrightarrow {}^2E_g$$

$$e_g^4 b_{2g}^2 a_{1g}^2 b_{1g}^1 \left(i.e.\ {}^2B_{1g}\right) \xrightarrow{\nu_2} e_g^4 b_{2g}^1 a_{1g}^2 b_{1g}^2 \left(i.e.\ {}^2B_{2g}\right)\ i.e.\ {}^2B_{1g} \longrightarrow {}^2B_{2g}$$

$$e_g^4 b_{2g}^2 a_{1g}^2 b_{1g}^1 \left(i.e.\ {}^2B_{1g}\right) \xrightarrow{\nu_3} e_g^4 b_{2g}^2 a_{1g}^1 b_{1g}^2 \left(i.e.\ {}^2A_{1g}\right)\ i.e.\ {}^2B_{1g} \longrightarrow {}^2A_{1g}$$

(D) Spectral properties of $[Cu(OH_2)_6]^{2+}$: It possesses the tetragonally elongated octahedral structure and splitting of the energy levels occurs as shown in Fig. 7.14.2.1. H_2O is a weak field ligand and consequently, splitting of the t_{2g} and e_g levels is not very high and, the three possible spectral transitions give three peaks (ν_1, ν_2 and ν_3) and they overlap to produce a **broad and unsymmetric peak** having the average maximum at 12,600 cm^{-1}, *i.e.* ~800 nm ($\varepsilon \approx 11$). In $(NH_4)_2[Cu(OH_2)_6](SO_4)_2$ (solid state), three transitions have been identified,

$$d_{z^2} \longrightarrow d_{x^2-y^2}\ (6{,}400\ \text{cm}^{-1});\ d_{xy} \longrightarrow d_{x^2-y^2}\ (10{,}650\ \text{cm}^{-1});\ d_{xz}, d_{yz} \longrightarrow d_{x^2-y^2}\ (11{,}550\ \text{cm}^{-1}).$$

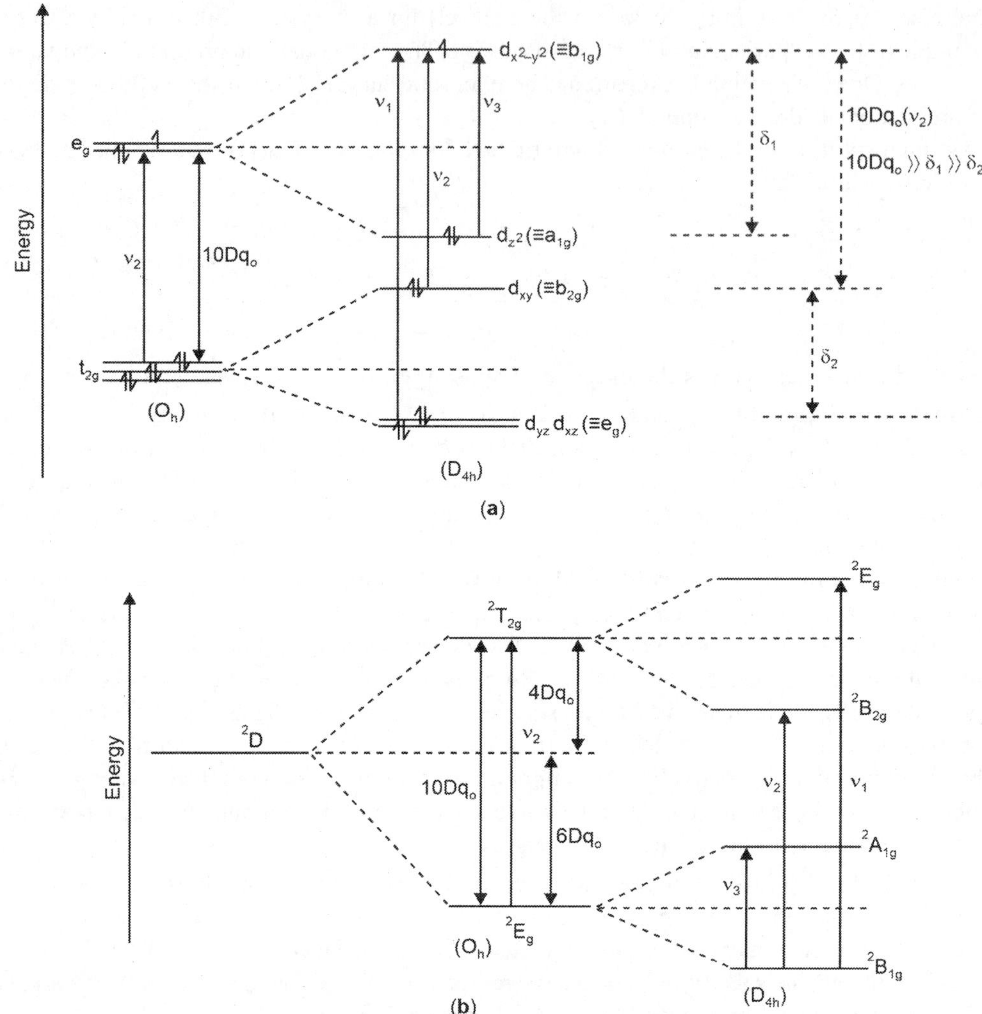

Fig. 7.14.2.1: Splitting of the energy levels due to the medium *z*-out tetragonal distortion in the six-coordinate complexes of Cu(II) (d^9-system). (a) Splitting of the *d*-orbitals; (b) splitting of the spectroscopic states.
Note: The ^2D term splits in the same way both in the C_{4v} and D_{4h} symmetry.

The crystal structure of $CuSO_4 \cdot 5H_2O$ is quite interesting (*cf.* Fig. 11.1.7.3, Vol. 2). Here each Cu(II)–centre is coordinated by four water molecules in a square planar geometry and two *trans*-axial positions are occupied by two sulfate O–atoms. These two *trans*–axial bonds are relatively longer as expected from the *z*-out distortion. The fifth water molecule (called **anion water**) is H-bonded with the SO_4–moiety and the water molecule coordinated with the metal centre. Analysis of the polarised crystal spectrum of $CuSO_4 \cdot 5H_2O$ shows three absorptions peaks as expected (Fig. 7.14.2.2).

$$10{,}500\ \text{cm}^{-1}\ (\nu_3),\ 13{,}000\ \text{cm}^{-1}\ (\nu_2)\ \text{and}\ 14{,}500\ \text{cm}^{-1}\ (\nu_1)$$

Note: Hydrated copper sulfate ($CuSO_4 \cdot 5H_2O$ called **blue vitriol**) becomes white/colourless on heating (see Fig. 11.1.7.3, Chapter 11, Vol. 2).

$$CuSO_4 \cdot 5H_2O\ (\textbf{blue}) \xrightarrow{\sim 100°C} CuSO_4 \cdot 3H_2O \xrightarrow{\sim 150°C} CuSO_4 \cdot H_2O \xrightarrow{\sim 250°C} CuSO_4\ (\textbf{white})$$

The X-ray crystal structures of the different forms of copper sulfate indicate that in all cases, the **tetragonally (*z*-out) distorted octahedral geometry is maintained.**

$CuSO_4{\cdot}5H_2O$	$CuSO_4{\cdot}3H_2O$	$CuSO_4$
Cu: $2O_s + 4O_w$	Cu: $3O_s + 3O_w$	Cu: $6O_s$
13,200 cm^{-1} (757 nm)	13,000 cm^{-1} (709 nm)	12,500 cm^{-1} (800 nm)

O_w → Water–O
O_s → Sulfate–O
(Ref: K. Nagase *et al*, *Thermochimica Acta*, **23,** 283, 1978)

The above spectral data for the $d \rightarrow d$ transitions are recorded from their **reflectence spectra** (solid state electronic spectra) (Fig. 7.14.2.2). In terms of the spectrochemical series, SO_4^{2-} **is a weaker field ligand than water.** In fact, for the replacement of one water–O (O_w) by a sulfate–O (O_s), there is a **red shift** by about 150–200 cm^{-1} for the λ_{max} of *d–d* transition. Another observation is: the **red shifted peak becomes narrower for the anhydrous form.** Besides the *z*-out tetragonal J.T. distortion prevailing in all cases, in the hydrated forms, the structures are less symmetric because of the different types of O-donor. For example, in $CuSO_4{\cdot}5H_2O$ (see Fig. 11.1.7.3, Vol. 2), the coordinated water molecules are of two types (*i.e.* three types of O-donor: $2O_s + 2O_w + 2O_w$) while in the anhydrous form $CuSO_4$, all the O-donor sites are sulfate-O. **For the more asymmetric crystal field splitting, the resultant peak (consisting of several *d–d* transitions) becomes broader (already discussed).**

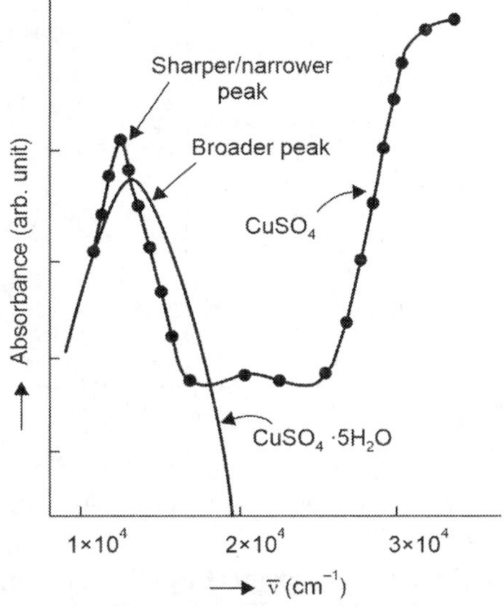

Fig. 7.14.2.2: Reflectance spectra of $CuSO_4{\cdot}5H_2O$ and anhydrous $CuSO_4$

The **blue** colour disappears in the dehydrated form **due to the red shifting towards the IR-region and narrowing of the peak of *d–d* transition.**

Interestingly, there is an intense **red shifted LMCT** band (at about 3000 cm^{-1} for the anhydrous form (*cf.* sulfate-O is more polarisable than H_2O, anion *vs.* neutral). In fact, in the whole UV-visible range, the anhydrous copper sulfate absorbs more light than the hydrated form. **Thus if our eyes could record all the radiations in the UV-visible range, the anhydrous form would appear more coloured.**

(E) Spectral properties of $[Cu(NH_3)_x(OH_2)_y]^{2+}$ (x + y = 6): With the replacement of weak field ligands by the relatively stronger field ligands, the magnitude of crystal field splitting increases and consequently, the absorption peaks move towards the lower wavelength, *i.e.* higher frequency. NH_3 is a stronger field ligand than H_2O. Thus, the gradual replacement of H_2O by NH_3 (upto $x = 4$), the absorption maximum shifts towards the higher frequency (*i.e.* blue shift towards UV).

$[Cu(OH_2)_6]^{2+}$: $\lambda_{max} \approx 800$ nm, $\varepsilon_{max} \approx 11.0$ **$[Cu(NH_3)_4(OH_2)_2]^{2+}$:** $\lambda_{max} \approx 590$ nm, $\varepsilon_{max} \approx 50.0$

In the mixed ligand complex $[Cu(NH_3)_4(OH_2)_2]^{2+}$, the average ligand field strength is the average of the fields due to water and ammonia.

$$\bar{\nu}_{[MA_xB_{6-x}]} = \frac{x}{6}\bar{\nu}_{[MA_6]} + \frac{(6-x)}{6}\bar{\nu}_{[MB_6]} \quad \text{or,} \quad 10Dq_{[MA_xB_{6-x}]} = \frac{x}{6}10Dq_{[MA_6]} + \frac{6-x}{6}10Dq_{[MB_6]}$$

Here it is interesting to note that **entry of the fifth NH_3 molecule** in $[Cu(NH_3)_4(OH_2)_2]^{2+}$ shifts the peak **towards the longer wavelength** (*i.e.* red shift). No evidence for the formation of $[Cu(NH_3)_6]^{2+}$ in aqueous media is reported. On entering the stronger field ligand, the peak is expected to shift towards the lower wavelength but on placing the fifth NH_3 molecule, there is a **reversal of spectral shift.** The detailed explanation of the *reversal of spectral shift* will be discussed later in connection with the spectral shift of $[Cu(acac)_2]$ in different solvents.

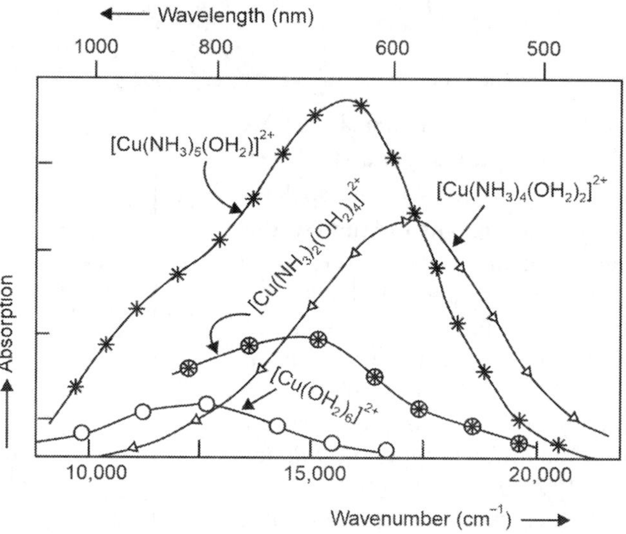

Fig. 7.14.2.3: Comparison of the electronic spectra of $[Cu(OH_2)_6]^{2+}$ and its different ammine complexes in aqueous solution.

(F) Spectral properties of $[Cu(acac)_2]$ in different solvents: Belford and cowerkers (*cf. J. Chem. Phys.*, **26,** 1165, 1957) studied the spectral properties of *bis*-acetylacetonatocopper(II), *i.e.* $[Cu(acac)_2]$, in different solvents varying in coordinating power to investigate **the effect of the degree of tetragonal distortion on the spectral transitions.** $[Cu(acac)_2]$ adopts the square planar geometry (a limiting situation of *z*–out distortion). When $[Cu(acac)_2]$ is dissolved in different solvents, the solvent molecules (S) may coordinate along the *trans*– axial directions to give a tetragonally distorted octahedral structure in solution (Fig. 7.14.2.4).

Ligation by the solvent molecules in the *trans*–axial positions will convert the **square planar $[Cu(acac)_2]$ complex** into a **distorted octahedral $[Cu(acac)_2(S)_2]$ complex**. If the coordinating power of solvent is poor, then the complex mainly remains as a square planar one. Limiting situation of the splitting of the t_{2g} and e_g levels due to the *z*–out distortion will give the square planar geometry. **In a poor coordinating solvent (*e.g.* $CHCl_3$), $[Cu(acac)_2]$ remains practically as a square planar complex,** *i.e.* splitting of the t_{2g} and e_g levels will be the maximum. On the other hand, in a good coordinating solvent (*e.g.* piperidine), coordination along the *trans*–axial directions by S will give the distorted octahedral geometry where the splittings of the t_{2g} and e_g levels are relatively less compared to those in the poor coordinating solvents.

- δ_1, δ_2 (*cf.* Fig. 7.14.2.1): $CHCl_3$ > Dioxane > Pentanol > Pyridine > Piperidine
- Coordinating power and basicity of the solvent (S): $CHCl_3$ < Dioxane < Pentanol < Pyridine < Piperidine
- Tendency to adopt the octahedral geometry from the square planar geometry (in different solvents): $CHCl_3$ < Dioxane < Pentanol < Pyridine < Piperidine

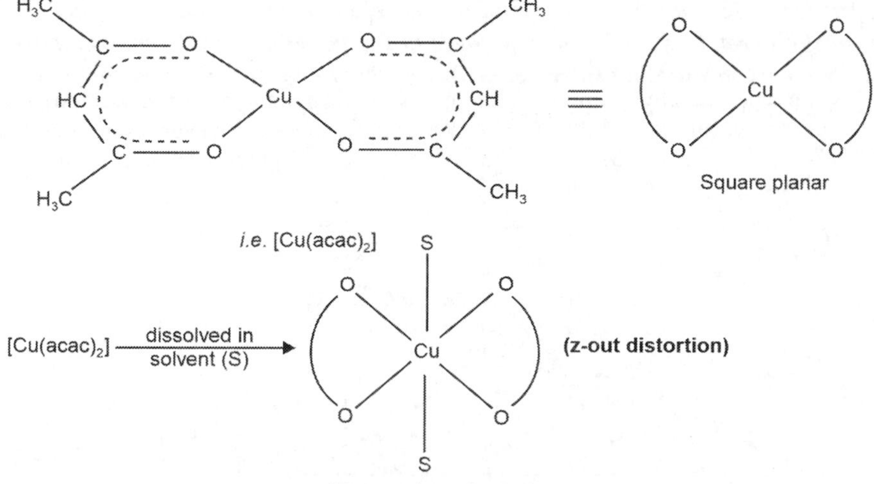

Fig. 7.14.2.4: Structural features of $[Cu(acac)_2]$ in the coordinating solvents.

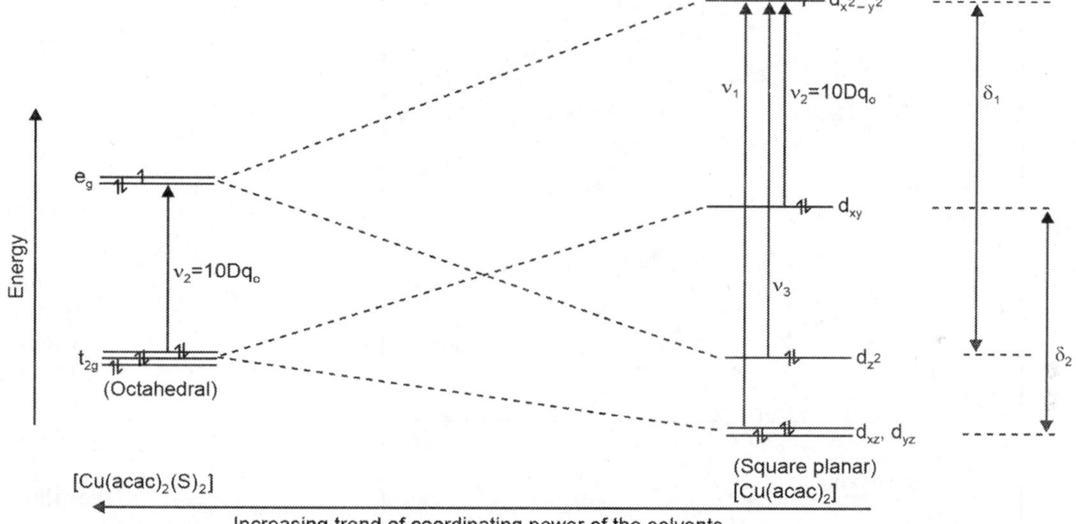

Fig. 7.14.2.5: Gradual transition of the square planar geometry of $[Cu(acac)_2]$ into the octahedral complex $[Cu(acac)_2(S)_2]$ with the increase of coordinating power (*i.e.* basicity) of the solvents (S).

The splittings measured by δ_1 and δ_2 (*cf.* Fig. 7.14.2.1a) are minimum in the most octahedral geometry as in the piperidine solvent. With the decrease of δ_1 and δ_2, ν_3 (frequency) decreases towards the lower frequency while ν_1 (frequency) decreases towards ν_2 that remains unchanged. For the ν_2 transition, the involved orbitals possess no z–component (*cf.* effect of J.T. distortion along the $\pm z$–directions).

$$\left.\begin{array}{ll} \nu_3\left(\equiv \delta_1\right): & d_{z^2} \longrightarrow d_{x^2-y^2} \\ \nu_2\left(\equiv 10Dq_o\right): & d_{xy} \longrightarrow d_{x^2-y^2} \\ \nu_1\left(\equiv 10Dq_o+\delta_2\right): & d_{yz},\, d_{xz} \longrightarrow d_{x^2-y^2} \\ \delta_2,\, \delta_1 \longrightarrow 0,\ \nu_3 \longrightarrow 0;\ \nu_1 \longrightarrow \nu_2 & \end{array}\right\} cf.\ \text{Fig. 7.14.2.1a}$$

Thus we can conclude: ***with the increase of the basicity (i.e. coordinating power) of the solvents, v_3 decreases to the lower frequency and v_1 moves to merge with v_2 that remains more or less unchanged; in the poor coordinating solvents, the peaks due to v_1, v_2 and v_3 become more manifested.***

The above predictions have been experimentally verified (*cf.* Fig. 7.14.2.6 and Table 7.14.2.1).

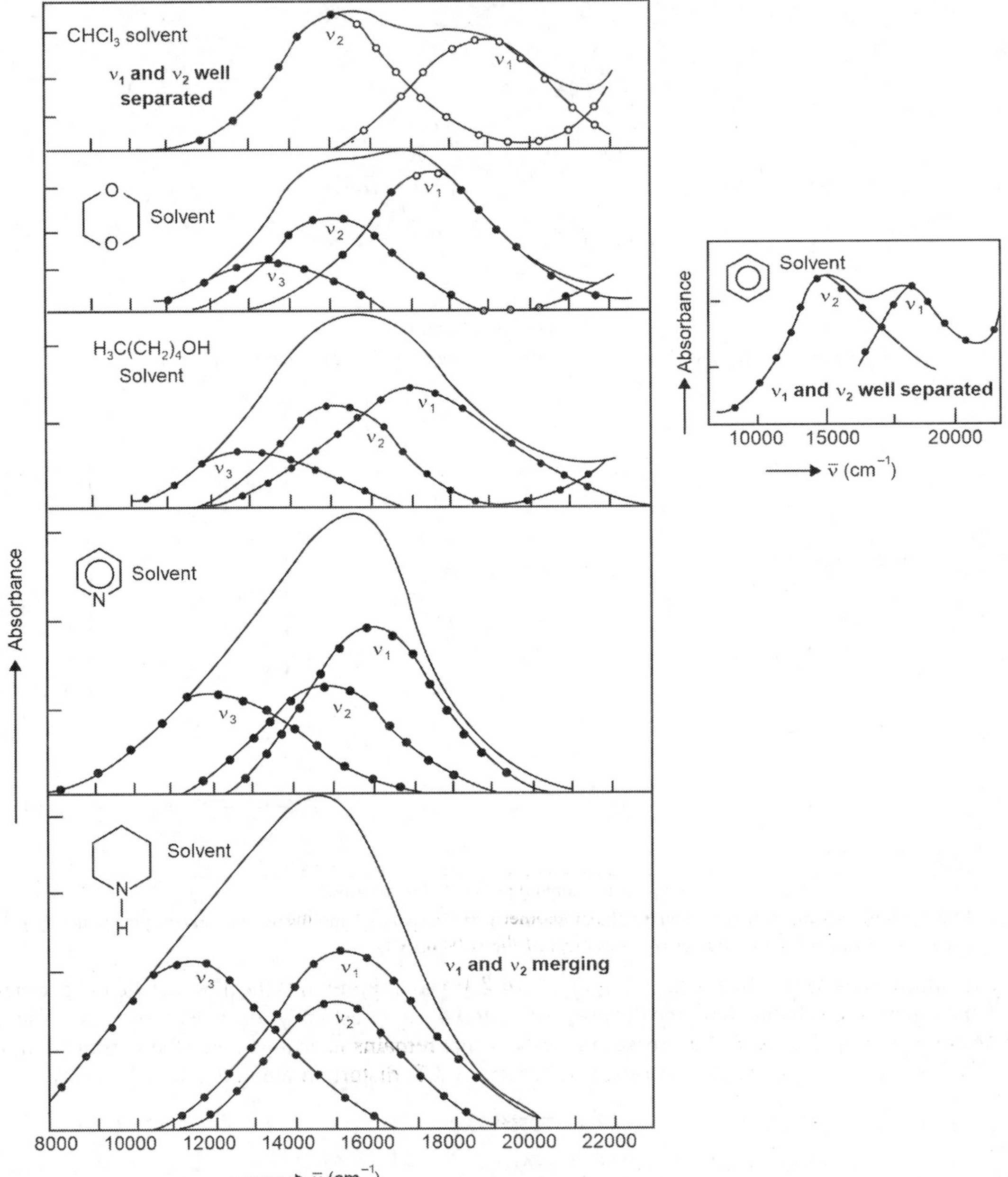

Fig. 7.14.2.6: Spectral behaviour of [Cu(acac)$_2$] in different solvents differing in basicities (*i.e.* coordinating power as a ligand).

Table 7.14.2.1: Effect of coordinating power of the solvents on the spectra of $[Cu(acac)_2]$.

Solvent	$d_{xz}, d_{yz} \xrightarrow[(cm^{-1})]{\nu_1} d_{x^2-y^2}$	$d_{xy} \xrightarrow[(cm^{-1})]{\nu_2} d_{x^2-y^2}$	$d_{z^2} \xrightarrow[(cm^{-1})]{\nu_3} d_{x^2-y^2}$
Chloroform	18,810	15,190	–
1,4–Dioxane	17,500	15,100	13,500
n–Pentanol	17,100	15,200	13,000
Pyridine	15,900	14,800	12,100
Piperidine	15,100	14,800	11,300
Increasing coordinating power ↓	Decreasing frequency ↓	More or less unchanged ↓	Decreasing frequency ↓

Chloroform: $CHCl_3$; 1, 4–Dioxane: O(CH$_2$CH$_2$)$_2$O ring; *n*–Pentanol: $H_3C(CH_2)_4OH$; Pyridine: N

Piperidine: 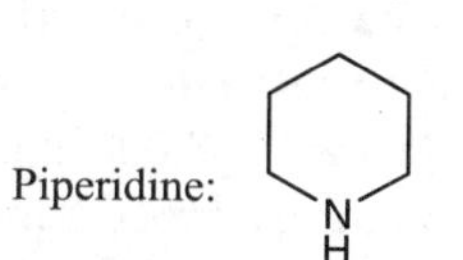

(G) Spectral shift (*i.e.* red shift) in moving from $[Cu(NH_3)_4(OH_2)_2]^{2+}$ to $[Cu(NH_3)_5(OH_2)]^{2+}$: Now let us consider the effect of entry of the fifth NH_3 molecule in $[Cu(NH_3)_4(OH_2)_2]^{2+}$. *Trans*–$[Cu(NH_3)_4(OH_2)_2]^{2+}$ is more or less square planar in geometry (where two H_2O molecules are weakly held at the *trans*-axial directions) while in $[Cu(NH_3)_5(OH_2)]^{2+}$, one axial position is occupied by NH_3 giving rise to a distorted octahedral geometry. It is reasonable because NH_3 is a better base than H_2O (*cf.* $[Cu(acac)_2]$ in the solvent, piperidine *vs.* dioxane). Thus $[Cu(NH_3)_5(OH_2)]^{2+}$ is going to have approximately the splitting of *d*–orbitals as in the octahedral geometry while *trans*- $[Cu(NH_3)_4(OH_2)_2]^{2+}$ is going to have the splitting pattern as in the square planar geometry. The extent of splitting of *d*-orbitals in the square planar geometry is more than that in the octahedral geometry. This is why, when $[Cu(NH_3)_4(OH_2)_2]^{2+}$ is converted into $[Cu(NH_3)_5(OH_2)]^{2+}$, **the maximum of absorption spectrum is shifted towards the longer wavelength (*i.e.* red shift).**

(H) Spectral properties of the square pyramidal (C_{4v}) complexes of Cu(II): Splitting pattern of the *d*–orbitals in the square pyramidal geometries is the same as in the case of *z*–out distortion. In fact, the ^2D term splits in the same way both in C_{4v} (lacking in centre of symmetry) and D_{4h} (*z*-out tetragonal distortion having the centre of inversion). Thus the spectral transitions are the same as given in Fig. 7.14.2.1 However, the subscript '*g*' is dropped because the square pyramidal complexes lack in the centre of symmetry. Thus in the **square pyramidal complexes** like $[Cu(NH_3)_5]^{2+}$, $[Cu(HN_3)_4(OH_2)]^{2+}$, etc. the transitions are:

$$\left.\begin{aligned}
d_{z^2} &\xrightarrow{\nu_3} d_{x^2-y^2},\ i.e.\ {}^2B_1 \xrightarrow{\nu_3} {}^2A_1 \\
d_{xy} &\xrightarrow{\nu_2} d_{x^2-y^2},\ i.e.\ {}^2B_1 \xrightarrow{\nu_2} {}^2B_2 \\
d_{xz}, d_{yz} &\xrightarrow{\nu_1} d_{x^2-y^2},\ i.e.\ {}^2B_1 \xrightarrow{\nu_1} {}^2E
\end{aligned}\right\}
\begin{aligned}
&(9{,}000 - 10{,}000\ cm^{-1}),\\
&(11{,}500 - 16{,}000\ cm^{-1}),\\
&(15{,}000 - 19{,}000\ cm^{-1}),
\end{aligned}$$

Very often, the three bands (ν_1, ν_2 and ν_3) are not detected separately. They can overlap to produce a **broad band.** For $[Cu(NH_3)_5]^{2+}$ in $NH_4[Cu(NH_3)_5]^{3+}(PF_6^-)_3$, two bands at 15000 cm^{-1} and 11400 cm^{-1} are noted.

(I) Spectral properties of the square planar complexes (D_{4h}) of copper(II): The possible electronic transitions are:

$$\left.\begin{array}{l} {}^2B_{1g} \xrightarrow{\nu_3} {}^2A_{1g},\ i.e.\ d_{z^2} \xrightarrow{\nu_3} d_{x^2-y^2};\ {}^2B_{1g} \xrightarrow{\nu_2} {}^2B_{2g},\ i.e.\ d_{xy} \xrightarrow{\nu_2} d_{x^2-y^2} \\ \text{and } {}^2B_{1g} \xrightarrow{\nu_1} {}^2E_g,\ i.e.\ d_{xz}, d_{yz} \xrightarrow{\nu_1} d_{x^2-y^2} \end{array}\right\} cf.\ \text{Fig. 7.14.2.1}$$

In the square planar complexes (where there is a limiting situation of z-out distortion), the common energy order is:

$$ {}^2B_{1g} < {}^2B_{2g} < {}^2E_g < {}^2A_{1g},\ i.e.\ d_{z^2} < d_{xz}, d_{yz} < d_{xy} < d_{x^2-y^2}$$

Note: It is worth mentioning that in a square planar geometry, there is an uncertainty in the energy order of the orbitals d_{z^2} and d_{xz}, d_{yz} (see Vol. 4 for details). It may happen as: $d_{xz}, d_{yz} < d_{z^2}$ or $d_{xz}, d_{yz} \approx d_{z^2}$ depending on the conditions.

The spectral properties of [Cu(acac)$_2$] have been already discussed. The rose red *bis*(biguanide)copper(II) chloride dihydrate, *i.e.* **[Cu(bigH)$_2$]Cl$_2$·2H$_2$O, shows a broad spectral band** (Fig. 7.14.2.7) at about 19,000 cm^{-1} (which consists of three components: 20,200 cm^{-1}, 18,000 cm^{-1} and 17,200 cm^{-1}). In fact, the expected three bands (*i.e.* ν_1, ν_2 and ν_3) overlap to produce a broad band and it happens so for many square planar [CuN$_4$] (D_{4h}) structures. In a lower symmetry like D_{2h} (*e.g.* CuN$_2$O$_2$ chromophore), the E_g level further splits into the A_g and B_g components.

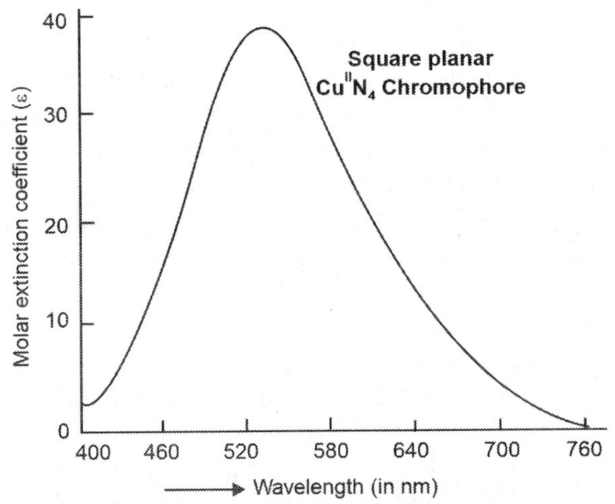

Fig. 7.14.2.7: Absorption sectrum of [Cu(bigH)$_2$]$^{2+}$ in solution.

(J) Spectral properties of *tris*-chelates, *e.g.* [Cu(bpy)$_3$]$^{2+}$, [Cu(phen)$_3$]$^{2+}$ of the D_3 symmetry: In moving from the O_h to D_3 symmetry, the ${}^2T_{2g}$ level splits into the ${}^2E$ and ${}^2A_1$ levels while the ${}^2E_g$ state remains as ${}^2E$ (without any spliting). Thus the possible transitions are:

$$\boldsymbol{O_h}: {}^2E_g \rightarrow {}^2T_{2g};\ \boldsymbol{D_3}: {}^2E \rightarrow {}^2A_1 \text{ and } {}^2E \rightarrow {}^2E$$

In fact, [Cu(bpy)$_3$]$^{2+}$ and [Cu(phen)$_3$]$^{2+}$ show two well defined peaks near 6000 cm^{-1} and 15,000 cm^{-1} (Fig. 7.14.2.8). The higher energy band corresponds to the transition ${}^2E \rightarrow {}^2E$. This higher energy band shows a doublet structure which may be due to the vibrational transition and/or further splitting of the E-term giving rise to the C_2 symmetry.

(K) Spectral properties of the tetrahedral (T_d) and distorted (D_{2d}) tetrahedral complexes of Cu(II): These are discussed in Sec. 7.16.

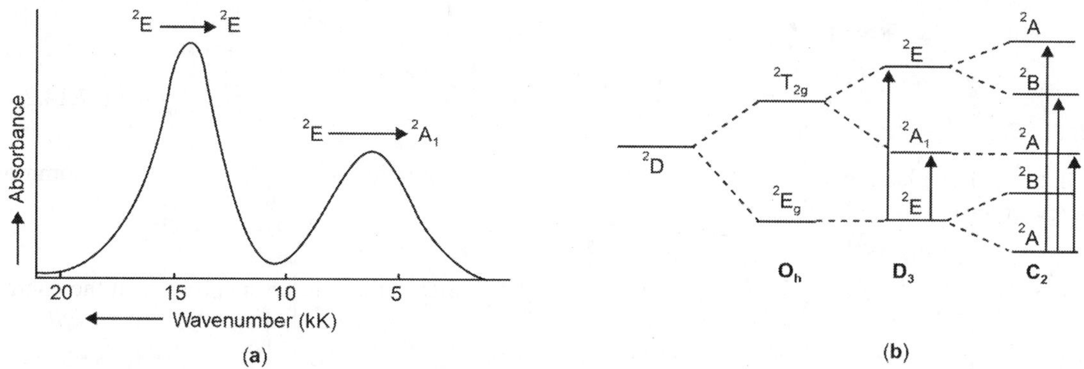

Fig. 7.14.2.8: (a) Absorption spectrum of trigonally distorted $[Cu(bpy)_3]^{2+}$ (D_3 symmetry); (b) Energy level diagram for $O_h \to D_3 \to C_2$ in a d^9 system.

(L) Spectral transitions in the dimeric form of Cu(II)-acetate, *i.e.* $[Cu(CH_3CO_2)_2(OH_2)]_2$: These are discussed in Chapter 8 (*cf.* Fig. 8.31.1.2, Vol. 6) in terms of the square pyramidal symmetry around Cu(II).

7.14.3 Six–Coordinate Complexes of $3d^2$–System

$[V(OH_2)_6]^{3+}$ ($3d^2$) is the representative example of this group. In terms of Orgel diagram (Fig. 7.10.2), the spin allowed transitions are:

$$\left.\begin{aligned} &\nu_1: {}^3T_{1g}(F) \to {}^3T_{2g}(F);\ \nu_1 = 8Dq + x' \approx 8Dq \\ &\nu_2: {}^3T_{1g}(F) \to {}^3A_{2g}(F);\ \nu_2 = 18Dq + x' \approx 18Dq \\ &\nu_3: {}^3T_{1g}(F) \to {}^3T_{1g}(P);\ \nu_3 = 15B' + 6Dq + 2x' \approx 15B' + 6Dq \end{aligned}\right\} \begin{aligned} &\nu_3 > \nu_2 > \nu_1 \text{ or } \nu_2 > \nu_3 > \nu_1 \\ &\text{depending on the conditions.} \end{aligned}$$

The ν_2-transition involves the ***2e transition*** (*cf.* ${}^3T_{1g}(F) \Rightarrow t_{2g}^2$; ${}^3A_{2g}(F) \Rightarrow e_g^2$) and thus, it is a **forbidden transition.**

x' denotes the symmetry interaction energy (*cf.* Figs. 7.10.2; 7.14.3.1) and it can be neglected **for the relatively weak field ligands for the sake of simplicity.** B' denotes the Racah parmeter in the complex and it gives the measure of interelectron repulsion.

The above relations lead to:

$$10Dq = \nu_2 - \nu_1;\ 15B' = (\nu_2 + \nu_3 - 3\nu_1);\ \nu_3 - 2\nu_1 = 15B' - 10Dq$$

Here it is important to mention that the spectroscopic state ${}^3A_{2g}$ arises from the electronic configuration e_g^2. Thus the ν_2–peak due to the transition, ${}^3T_{1g} \to {}^3A_{2g}$, involves **a two-electron excitation** (*i.e.* simultaneous transition of two electrons) and this is a **forbidden transition**. Generally, it occurs in the UV–region where the charge transfer band prevails. Because of the strong charge transfer band, the forbidden peak (ν_2) of poor intensity very often remains uncharacterised.

A. Spectral behaviour of $[V(OH_2)_6]^{3+}$: It looks green with two strong absorption peaks at 17,200 cm^{-1} ($\varepsilon \approx 6$) and 25, 600 cm^{-1} ($\varepsilon \approx 8$) (*cf.* Fig. 7.14.3.2).

Now let us try to assign the peaks with the help of Tanabe-Sugano diagram (*cf.* Figs. 7.13.2.1, 7.14.3.3). Let us assume arbitrarily that the peak at 17,200 cm^{-1} is due to the following transition.

$${}^3T_{1g}(F) \xrightarrow[(17{,}200\ \text{cm}^{-1})]{\nu_1} {}^3T_{2g}(F)\ (\textbf{lowest energy band});\ \nu_1 \approx 8Dq = 17{,}200\ \text{cm}^{-1}$$

Then, the following transition should occur at about 38,500 cm^{-1}.

$${}^3T_{1g}(F) \xrightarrow{\nu_2} {}^3A_{2g}(F);\ \nu_2\,(\text{expected}) \approx 18Dq = 38{,}500\ \text{cm}^{-1}$$

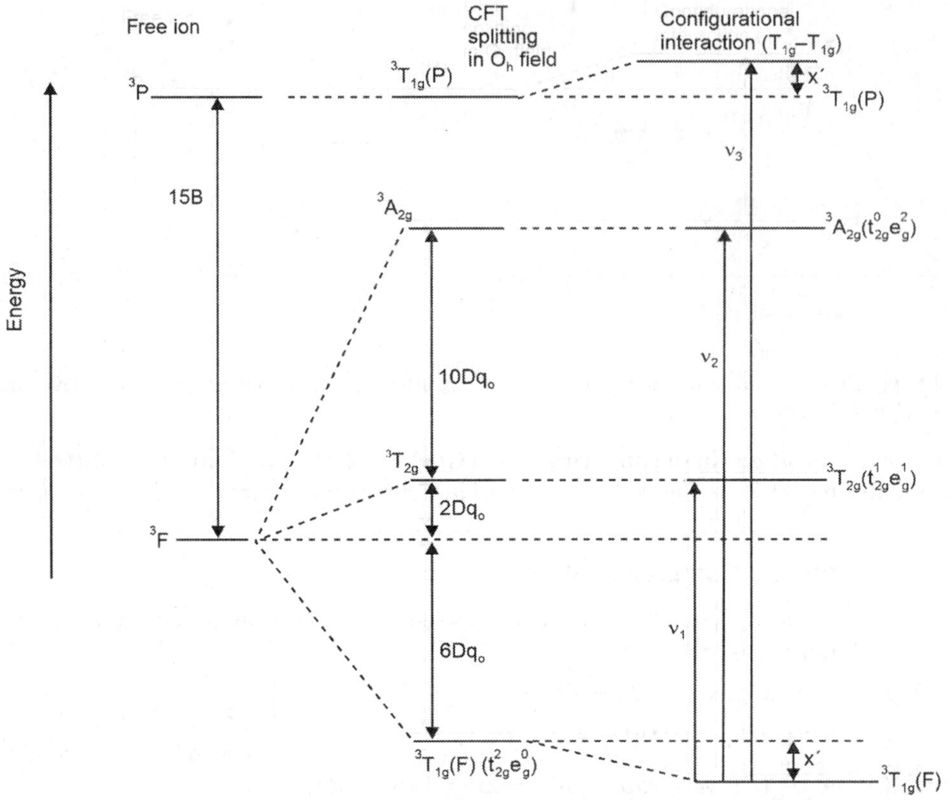

Fig. 7.14.3.1: Energy level diagram for a d^2–octahedral system. ($^3T_{1g}(F) \Rightarrow t_{2g}^{1.8}e_g^{0.2}$ of energy $-6Dq$)

Note: The $^3T_{1g}(P)$ and $^3A_{2g}(F)$ states may intersect depending on the $10Dq$ value. Here it is assumed that the $^3T_{1g}(P)$ state lies above the $^3A_{2g}(F)$ state; x' denotes the configurational interaction energy for the $^3T_{1g}(P)$ – $^3T_{1g}(F)$ interaction. (*See* Fig. 7.9.4.1 for energy values of different states and **$-6Dq$ energy of the $^3T_{1g}(F)$ state corresponds to $t_{2g}^{1.8}e_g^{0.2}$**).

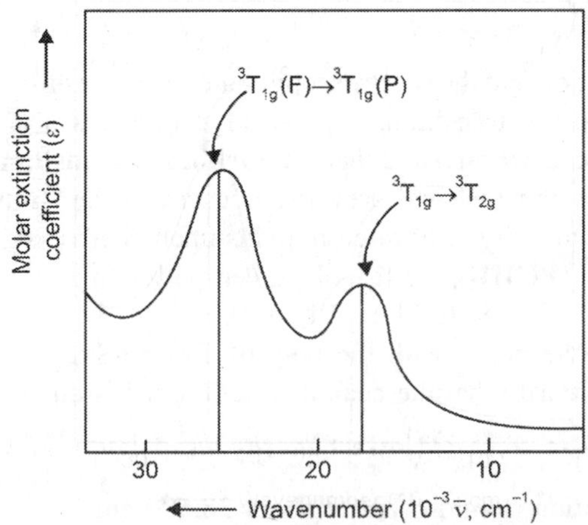

Fig. 7.14.3.2: The electronic absorption spectrum of $[V(OH_2)_6]^{3+}$.

Thus the peak at 25,600 cm^{-1} is not due to the $^3T_{1g} \rightarrow {}^3A_{2g}$ transition, if we consider that the peak due to $^3T_{1g} \rightarrow {}^3T_{2g}$ occurs at 17,200 cm^{-1}. Under the situation, we may assume that the peak at 25,600 cm^{-1} is due to the following transition.

$$^3T_{1g}(F) \xrightarrow[(25{,}600\ \text{cm}^{-1})]{\nu_3} {}^3T_{1g}(P);\ \nu_3 \approx 6Dq + 15B' = 25{,}600\ \text{cm}^{-1}$$

Note: From Fig. 7.10.4b, it is evident that for the T_1(F) ground state, the transition energy ratio, **ν_2/ν_1 is close to 2 regardless of the ligand field strength.** If ν_2 is taken as 25,600 cm^{-1}, then **the ratio ν_2/ν_1 becomes 1.48 < 2.** It indicates that ν_2 does not correspond to 25,600 cm^{-1}. Thus it is reasonable to assume ν_3 = 25,600 cm^{-1} (*see* Solved Problem 12).

Now let us verify whether these assignments are correct or not. The above assignments lead to:

$$\frac{^3T_{1g} \longrightarrow {}^3T_{1g}(\text{P})}{^3T_{1g} \longrightarrow {}^3T_{2g}} = \frac{\nu_3}{\nu_1} = \frac{25{,}000}{17{,}200} \approx 1.50$$

This ratio of ν_3 and ν_1 can be fitted in the enlarged Tanabe-Sugano diagram (*cf.* Fig. 7.14.3.3) and this ratio fits for the value of $10Dq/B'$ as 28 (*i.e.* ***Dq/B'* = 2.8**). This ν_3/ν_1 ratio can also be fitted in the standard curve, ν_3/ν_1 *vs.* Dq/B' to find Dq/B' (*cf.* Fig. 7.10.4). For this $10Dq/B'$ or Dq/B' value, we find:

$$\frac{\Delta E\left[{}^3T_{1g}(F) \rightarrow {}^3T_{2g}(F)\right]}{B'} = 25.9$$

$$\text{i.e.}\quad \frac{\nu_1}{B'} \approx \frac{8Dq}{B'} = 25.9$$

$$\text{or,}\quad \frac{17{,}200}{B'} = 25.9$$

$$\text{or,}\quad B' = 664.1\ \text{cm}^{-1}$$

$$\frac{\Delta E\left[{}^3T_{1g}(F) \rightarrow {}^3T_{1g}(P)\right]}{B'} = 38.6$$

$$\text{i.e.}\quad \frac{\nu_3}{B'} \approx \frac{15B' + 6Dq}{B'} = 38.6$$

$$\text{or,}\quad \frac{25{,}600}{B'} = 38.6$$

$$\text{or,}\quad B' \approx 663.2$$

Thus the evaluated B' becomes about 664 cm^{-1}(*cf.* B_0 = 860 cm^{-1}). This reduced value of B' compared to that of B_0 (for free ion) is quite reasonable in terms of the ***nephelauxetic effect***. Thus the nephelauxetic ratio ($\beta = B'/B_0$) becomes about 0.80, *i.e.* B is reduced by about 20% in $[V(OH_2)_6]^{3+}$. The $10Dq$ value can be obtained by using the following relationship.

$$\frac{10Dq}{B'} = 28 \text{ or, } 10Dq \approx 28 \times 664\ \text{cm}^{-1} = 18{,}592\ \text{cm}^{-1}$$

For $\frac{Dq}{B'} = 2.8$, theoretically predicted energy ratios (*cf.* Fig. 7.10.4) are:

$$\frac{\nu_2}{\nu_1} = 2.08 \text{ and } \frac{\nu_3}{\nu_2} = 0.72,\ \text{i.e. } \nu_2 = \nu_1 \times 2.08 = 17{,}200 \times 2.08 = 35{,}776\ \text{cm}^{-1}$$

$$\text{or,}\quad \nu_2 = \frac{\nu_3}{0.72} = \frac{25{,}600}{0.72} = 35{,}555\ \text{cm}^{-1}$$

Thus the predicted ν_2 should appear at about 35,666 cm^{-1}($\approx \frac{1}{2}$(35,776 + 35,555) cm^{-1}).

Now let us consider the peak due to the transition, $^3T_{1g}(\text{F}) \xrightarrow{\nu_2} {}^3A_{2g}(\text{F})$ and it should occur at about 35,666 cm^{-1} from the above assignments. *We have already mentioned that the transition is forbidden because it involves the simultaneous transition of two electrons.* Such a weak ligand field band in the near ultraviolet region is generally obscured by the intense charge transfer bands. Thus, by taking the spectrum of $[V(OH_2)_6]^{3+}$ in solution, it is not possible to verify the prediction of the existence of ν_2 ($\approx 18Dq + x'$) at about 35,666 cm^{-1}. However, in solid state when V(III) is present in the matrix crystal of Al_2O_3 where the charge transfer band appears at still higher energy, a weak band occurs at

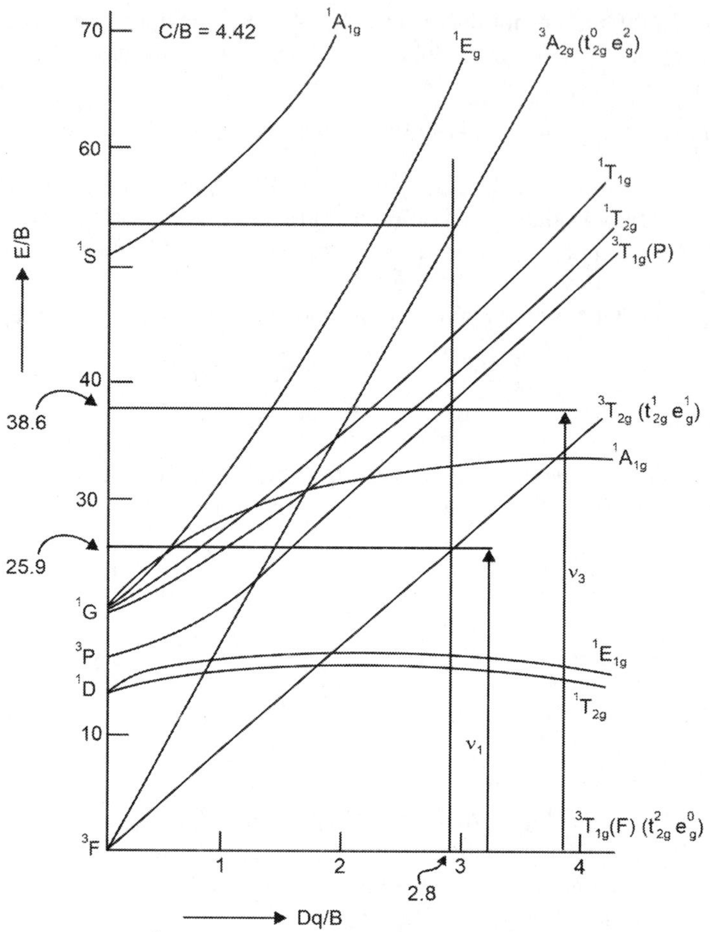

Fig. 7.14.3.3: T.S. diagram for a d^2–octahedral system and analysis of the absorption spectrum of $[V(OH_2)_6]^{3+}$.

about 34,500 cm^{-1}. In the solid matrix of Al_2O_3, V(III) centre is octahedrally surrounded by the O–atoms. Thus the $10Dq$ of V(III) in the solid matrix of Al_2O_3 is comparable to that of $[V(OH_2)_6]^{3+}$. The observed peak at 34,500 cm^{-1} can reasonably be considered due to ${}^3T_{1g}(F) \rightarrow {}^3A_{2g}(F)$.

- Thus the assignments, *i.e.* ${}^3T_{1g}(F) \xrightarrow{17{,}200\ cm^{-1}} {}^3T_{2g}(F)$ and ${}^3T_{1g}(F) \xrightarrow{25{,}600\ cm^{-1}} {}^3T_{1g}(P)$ are correct. These are supported by the fact: estimated B value (= 664 cm^{-1}) is less than that of the free ion ($B' \approx 80\%$ of B_0); the predicted ν_2 value for ${}^3T_{1g}(F) \rightarrow {}^3A_{2g}(F)$ is experimentally verified.
- If the assignments, ${}^3T_{1g}(F) \xrightarrow{17{,}200\ cm^{-1}} {}^3T_{1g}(P)$ and ${}^3T_{1g}(F) \xrightarrow{25{,}600\ cm^{-1}} {}^3A_{2g}(F)$ are done, **then it will lead to B' value which is higher than that of free ion (*i.e.* $B' \rangle B_0$) and this is not acceptable.** Besides this, it predicts the band for the ${}^3T_{1g}(F) \rightarrow {}^3T_{2g}(F)$ transition in the infra-red region and this prediction is not experimentally verified.

Note: If ν_1 (17,200 cm^{-1}) is considered to be $8Dq$, then it leads to $10Dq$ to be 21,500 cm^{-1}. But this value is quite different from that (18, 592 cm^{-1}) obtained from fitting in the Tanabe-Sugano diagram. It indicates the *importance of symmetry interaction energy x'*.

$$\nu_1 = 17{,}200\ cm^{-1} = 8Dq + x';\ \nu_3 = 25{,}600\ cm^{-1} = 15B' + 6Dq + 2x'$$

$$\nu_2 \approx 35{,}666\ cm^{-1}\ \text{(predicted value)}$$

$$= 18Dq + x'$$

$$\nu_2 - \nu_1 = 10Dq = 35{,}666 - 17{,}200 = 18{,}466\ \text{cm}^{-1}$$

i.e. $Dq \approx 1846\ \text{cm}^{-1}$

$$\nu_1 = 17{,}200\ \text{cm}^{-1} = 8Dq + x' = 8 \times 1846 + x'$$

i.e. $x' = 2432\ \text{cm}^{-1}$

$$\nu_3 = 25,600\ \text{cm}^{-1} = 15B' + 6Dq + 2x' = 15B' + 6 \times 1846 + 2 \times 2432$$

$$= 15B' + 15{,}940$$

or, $15B' = 9660$ or, $B' \approx 644\ \text{cm}^{-1}$ **(acceptable value)**

Spectral properties of some other V(III)–complexes are given in Table 7.14.3.1.

Table 7.14.3.1: Characteristic features of the ligand field bands of some complexes of V(III)

Complex	$^3T_{1g}(F) \rightarrow {}^3T_{2g}(F)$ ν_1 (cm^{-1})	$^3T_{1g} \rightarrow {}^3A_{2g}(F)$ ν_2 (cm^{-1})	$^3T_{1g}(F) \rightarrow {}^3T_{1g}(P)$ ν_3 (cm^{-1})	B' (cm^{-1})	$10Dq$ (cm^{-1})
$[V(OH_2)_6]^{3+}$	17,200	(36000)*	25,200	665	18,600
$[VF_6]^{3-}$	14,800	(31,000)*	23,000	650	16,200
$[V(urea)_6]^{3+}$	16,200	(33,500)*	24,000	610	17,500

* Predicted approximate values. $B_0 = 862\ \text{cm}^{-1}$.

B. Energy of the spectroscopic states arising from the ^3F–term (for d^2):

(i) *Without considering the symmetry interaction between $^3T_{1g}(F)$ and $^3T_{1g}(P)$*

$$E[^3T_{1g}(F)] = -6Dq; \quad E[^3T_{1g}(P)] = 15B'$$

$$E[^3T_{2g}(F)] = 2Dq; \quad E[^3A_{2g}(F)] = 12Dq$$

(ii) *After considering the symmetry interaction between $^3T_{1g}(F)$ and $^3T_{1g}(P)$* (*cf.* Fig. 7.14.3.1)

$$E[^3T_{1g}(F)] = -6Dq - x'; \quad E[^3T_{1g}(P)] = 15B' + x'$$

$$E[^3T_{2g}] = 2Dq; \quad E[^3A_{2g}] = 12Dq$$

where x' is the energy due to the symmetry interaction.

In a very weak field, the symmetry interaction energy (x') is negligibly small and the $^3T_{1g}(F)$ state is stabilised by $6Dq$ compared to the ^3F term of free ion. In the absence of crystal field, the $^3T_{1g}(P)$ state is destabilised by $15B_0$ compared to the ^3F term. In absence of the crystal field, the $^3T_{1g}(F)$ state has the same energy as that of unsplit ^3F term. In absence of symmetry interaction, energy of the $^3T_{1g}(P)$ state is the same as that of the ^3P term of free ion, *i.e.* energy of the $^3T_{1g}(P)$ state is zero (0) with respect to that of the ^3P term. Thus we can write for $H = H_0 + V_{oct}$:

$$\left.\begin{aligned}\int {}^3T_{1g}(F)|H_0|{}^3T_{1g}(F)\partial\tau &= 0\\ \int {}^3T_{1g}(P)|H_0|{}^3T_{1g}(P)\partial\tau &= 15B\end{aligned}\right\}(\textbf{No crystal field}) \qquad \left.\begin{aligned}\int {}^3T_{1g}(F)|V_{oct}|{}^3T_{1g}(F)\partial\tau &= -6Dq\\ \int {}^3T_{1g}(P)|V_{oct}|{}^3T_{1g}(P)\partial\tau &= 0\end{aligned}\right\}(\textbf{Crystal field})$$

The $^3T_{1g}(F)$ and $^3T_{1g}(P)$ states having the same group theoretical representation will experience a symmetry interaction or **configurational interaction** (*i.e.* **noncrossing rule** which states that the same symmetry states cannot cross one another). If the interaction energy is x', then we can write;

$$\int {}^3T_{1g}(F)|V_{oct}|{}^3T_{1g}(P)\partial\tau = x'$$

The relevant secular determinant with $H = H_0 + V_{oct}$ is given by:

$$\begin{array}{c|cc|} & {}^3T_{1g}(F) & {}^3T_{1g}(P) \\ {}^3T_{1g}(F) & -6Dq - E & x' \\ {}^3T_{1g}(P) & x' & 15B' - E \end{array} = 0$$

i.e. $\qquad (-6Dq - E)\,(15B' - E) = x'^2 \qquad$ ...(7.14.3.1)

when the limiting strong field ligands are considered, the interelectron repulsion parameter B' becomes negligible (*i.e.* when $10Dq$ or Δ is very large, B' becomes negligible). Under this condition, *i.e.* $E \gg 15B'$, we can write:

$$E^2 + 6DqE - x'^2 = 0 \qquad \text{...(7.14.3.2)}$$

For $x' = 4Dq$, the above quadratic equation gives: $E = -8Dq,\ 2Dq$.

$E = -8Dq$ corresponds to the crystal field energy of t_{2g}^2 configuration and $E = +2Dq$ corresponds to the crystal field energy of $t_{2g}^1e_g^1$ configuration.

Putting $x' = 4Dq$ in Eqn. 7.14.3.1, we can write:

$$(-6Dq - E)(15B' - E) - (4Dq)^2 = 0$$

or, $\qquad E^2 + E(6Dq - 15B') - 90DqB' - 16(Dq)^2 = 0$

Solution of the above quadratic equation gives two roots of the state ${}^3T_{1g}$. These are:

$$E = \frac{1}{2}\left[15B' - 6Dq \pm \left\{(6Dq - 15B)^2 + 4\left(90DqB' + 16(Dq)^2\right)\right\}^{1/2}\right]$$

$$= \frac{1}{2}\left[15B' - 6Dq \pm \left\{225B'^2 + 100(Dq)^2 + 180DqB'\right\}^{1/2}\right]$$

$$= \frac{1}{2}\left[15B' - \frac{3}{5}\Delta \pm \left\{225B'^2 + \Delta^2 + 18B'\Delta\right\}^{1/2}\right]$$

$$\left(cf.\ ax^2 + bx + c = 0,\quad x = \frac{-b \pm \sqrt{b^2 - 4ac}}{2a}\right)$$

Energies of the two ${}^3T_{1g}$ states are given as follows:

$$E\left[{}^3T_{1g}(F)\right] = \frac{1}{2}\left[15B' - \frac{3}{5}\Delta - \left(225B'^2 + 18B'\Delta + \Delta^2\right)^{1/2}\right], \text{ (corresponding to } t_{2g}^2e_g^0\text{)}$$

$$E\left({}^3T_{1g}(P)\right) = \frac{1}{2}\left[15B' - \frac{3}{5}\Delta + \left(225B'^2 + 18B'\Delta + \Delta^2\right)^{1/2}\right], \text{ (corresponding to } t_{2g}^1e_g^1\text{)}$$

Energies of the other two states do not involve x' (*i.e.* symmetry interaction energy term). These are:

$$E\left[{}^3T_{2g}(F)\right] = \frac{1}{5}\Delta,\quad E\left[{}^3A_{2g}(F)\right] = \frac{6}{5}\Delta$$

- **In the case of limiting weak field ligand** (*i.e.* $\Delta \rightarrow 0$), energies of the ${}^3T_{1g}(F)$, ${}^3T_{2g}(F)$ and ${}^3A_{2g}(F)$ states become zero with respect to that of the ${}^3F$ term, *i.e.* the ${}^3T_{1g}(F)$, ${}^3T_{2g}(F)$ and ${}^3A_{2g}(F)$ states merge with the ${}^3F$ term. However, energy of the ${}^3T_{1g}(P)$ state becomes $15B$ with respect to that of the ${}^3F$ term.

● **In the case of limiting strong field ligand,** B' can be ignored and we can write:

$$E\left[{}^3T_{1g}(F)\right] = -\frac{1}{2}\times\frac{8}{5}\Delta = -\frac{4}{5}\Delta = -8Dq = \text{energy of } t_{2g}^2$$

$$E\left[{}^3T_{2g}(F)\right] = \frac{1}{5}\Delta = 2Dq = \text{energy of } t_{2g}^1e_g^1$$

$$E\left[{}^3T_{1g}(P)\right] = \frac{1}{2}\times\frac{2}{5}\Delta = \frac{1}{5}\Delta = 2Dq = \text{energy of } t_{2g}^1e_g^1,\ E\left[{}^3A_{2g}(F)\right] = \frac{6}{5}\Delta = 12Dq = \text{energy of } e_g^2$$

Thus the transition energies can be expressed as follows:

$${}^3T_{1g}(F)\xrightarrow{\nu_1}{}^3T_{2g}(F),\ \nu_1 = \frac{1}{2}\left[\Delta - 15B' + \left(225B'^2 + 18B'\Delta + \Delta^2\right)^{1/2}\right]$$

i.e.
$$\frac{E(\nu_1)}{B'} = \frac{1}{2}\left[\frac{\Delta}{B'} - 15 + \left(225 + 18\frac{\Delta}{B'} + \frac{\Delta^2}{B'^2}\right)^{1/2}\right]$$

$${}^3T_{1g}(F)\xrightarrow{\nu_3}{}^3T_{1g}(P);\ \nu_3 = \left(225B'^2 + 18B'^2\Delta + \Delta^2\right)^{1/2}$$

i.e.
$$\frac{E(\nu_3)}{B'} = \left(225 + 18\frac{\Delta}{B'} + \frac{\Delta^2}{B'^2}\right)^{1/2}$$

Energy change in the spin-allowed ligand field bands for the high-spin octahedral d^2, d^7 octahedral and d^3, d^8 tetrahedral systems

$$T_1(F)\xrightarrow{\nu_1}T_2(F):\ \nu_1 = \frac{1}{2}\left[\Delta - 15B' + \left(225B'^2 + 18B'\Delta + \Delta^2\right)^{1/2}\right]$$

$$T_1(F)\xrightarrow{\nu_2}A_2(F):\ \nu_2 = \frac{1}{2}\left[3\Delta - 15B' + \left(225B'^2 + 18B'^2\Delta + \Delta^2\right)^{1/2}\right]$$

$$T_1(F)\xrightarrow{\nu_3}T_1(P):\ \nu_3 = \left(225B'^2 + 18B'\Delta + \Delta^2\right)^{1/2};\ \nu_3 - 2\nu_1 = 15B' - 10Dq$$

It is evident from the expressions of $E(\nu_1)/B'$ and $E(\nu_3)/B'$ that these are the functions of Δ/B'. The transition energies can be graphically represented by plotting E/B' against Δ/B'. This practice was introduced by Y. Tanabe and S. Sugano and consequently, the diagrams (*i.e.* plot of E/B *vs.* Δ/B) are named as **Tanabe-Sugano diagrams** (*cf.* Sec. 7.13) to honour the scientists.

7.14.4 Six-Coordinate Complexes of 3d^8–System

Octahedral complexes of Ni(II) are the representative examples of this group. In terms of Orgel diagram (Fig. 7.10.2) and energy level diagram shown in Fig. 7.14.4.1, the spin-allowed ligand field transitions are:

$$\left.\begin{aligned}
&{}^3A_{2g}(F)\longrightarrow{}^3T_{2g}(F),\quad \nu_1 = 10Dq\\
&{}^3A_{2g}(F)\longrightarrow{}^3T_{1g}(F),\quad \nu_2 = 18Dq - x \approx 18Dq\\
&{}^3A_{2g}(F)\longrightarrow{}^3T_{1g}(P),\quad \nu_3 = 12Dq + 15B' + x \approx 12Dq + 15B'
\end{aligned}\right\}\nu_3 > \nu_2 > \nu_1$$

The ν_3-transition involves the **2*e*-transition** (*cf.* ${}^3A_{2g}$(F) ⇒ $t_{2g}^6e_g^2$; ${}^3T_{1g}$(P) ⇒ $t_{2g}^4e_g^4$) and thus it is a **forbidden transition.**

(**Note:** Sometimes two spin-forbidden transitions, ${}^3A_{2g}$(F) → ${}^1E_g$(D) and ${}^3A_{2g}$(F) → ${}^1T_{2g}$(D) are observed (*cf.* Fig. 7.13.2.1)).

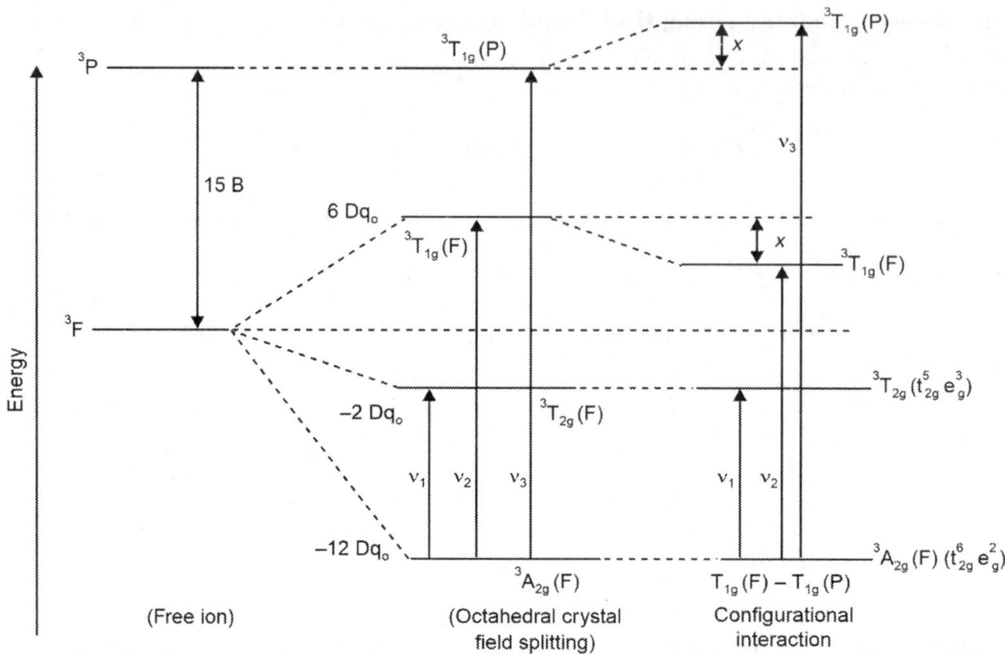

Fig. 7.14.4.1: Splitting of the ^3F and ^3P terms of $3d^8$ system in an octahedral field and electronic transitions.

Here x denotes the **configurational interaction energy.** Because of the configurational interaction (*i.e.* **symmetry interaction**, noncrossing rule) between the states $^3T_{1g}(F)$ and $^3T_{1g}(P)$, energy of the $^3T_{1g}(F)$ state is lowered by x and energy of the $^3T_{1g}(P)$ state goes up by x.

The configurational interaction between the $^3T_{1g}(F)$ and $^3T_{1g}(P)$ states becomes accountable when the ligands are fairly strong (*see* Fig. 7.9.4.1 for the different energy values).

For the octahedral Ni(II)–complexes, the three spin-allowed ligand field bands appear in the range:

$$\nu_1(8{,}000 - 13{,}000\ \text{cm}^{-1}),\ \nu_2(15{,}000 - 19{,}000\ \text{cm}^{-1}),\ \nu_3(25{,}000 - 30{,}000\ \text{cm}^{-1})$$

Table 7.14.4.1 Positions of ligand field bands in some octahedral Ni(II) complexes.

Complex	$^3A_{2g} \xrightarrow{\nu_1} {}^3T_{2g}(F)$	$^3A_{2g} \xrightarrow{\nu_2} {}^3T_{1g}(F)$	$^3A_{2g} \xrightarrow{\nu_3} {}^3T_{1g}(P)$	*10Dq*	*B′*
	(cm^{-1})	**(cm^{-1})**	**(cm^{-1})**	**(cm^{-1})**	**(cm^{-1})**
$[Ni(OH_2)_6]^{2+}$	8,700	13,800	25,300	8,700	905
$[Ni(NH_3)_6]^{2+}$	10,750	17,500	28,200	10,750	895
$[Ni(en)_3]^{2+}$	11,200	18,350	29,000	11,200	920
$[Ni(bpy)_3]^{2+}$	12,650	19,200	Obscured	12,650	–
$[Ni(DMSO)_6]^{2+}$	7,730	12,970	24,040	7,730	920
$[Ni(DMF)]_6]^{2+}$	8,500	13,600	25,000	8,500	875

DMSO = dimethyl sulfoxide, *DMF* = *N, N*–dimethyl formamide, B_0 (for free ion) = 1030 cm^{-1}

The energy change in ν_1 is independent of the configurational interaction. It gives the measure of 10*Dq*. If all the three ligand field bands (ν_1, ν_2 and ν_3) are characterised then B' can be estimated as follows:

$$15B' = (\nu_2 + \nu_3 - 3\nu_1)$$

In many cases, the ν_3 band gets obscured due to overlapping with the strong charge transfer band. In such cases, B' value can be estimated with the help of Tanabe–Sugano diagram.

Energy of the spectroscopic states are:

$$E\left({}^3T_{2g}\right) = -2Dq;$$

$$E\left({}^3A_{2g}\right) = -12Dq;\ E\left({}^3T_{1g}(F)\right) = 7.5B' + 3Dq - \frac{1}{2}\left(225B'^2 + 100Dq^2 - 180DqB'\right)^{1/2};$$

$$E\left({}^3T_{1g}(P)\right) = 7.5B' + 3Dq + \frac{1}{2}\left(225B'^2 + 100Dq^2 - 180DqB'\right)^{1/2}$$

$${}^3A_{2g} \Rightarrow t_{2g}^6 e_g^2;\ {}^3T_{2g} \Rightarrow t_{2g}^5 e_g^3;\ {}^3T_{1g}(F) \Rightarrow t_{2g}^5 e_g^3;\ {}^3T_{1g}(P) \Rightarrow t_{2g}^4 e_g^4 \text{ (see Fig. 7.10.2)}$$

For the said spin-allowed ligand field transitions, the energy changes are as follows:

$$\nu_1 = E\left({}^3T_{2g}\right) - E\left({}^3A_{2g}\right) = -2Dq + 12Dq = 10Dq$$

$$\nu_2 = E\left[{}^3T_{1g}(F)\right] - E\left({}^3A_{2g}\right) = 15Dq + 7.5B' - \frac{1}{2}\left[100Dq^2 + 225B'^2 - 180B'Dq\right]^{1/2}$$

$$\nu_3 = E\left[{}^3T_{1g}(P)\right] - E\left({}^3A_{2g}\right) = 15Dq + 7.5B' + \frac{1}{2}\left[100Dq^2 + 225B'^2 - 180B'Dq\right]^{1/2}$$

$$Q\,(\text{say}) = \left[100Dq^2 + 225B'^2 - 180B'Dq\right]^{1/2} = \nu_3 - \nu_2$$

$$10Dq = \nu_1 = \frac{1}{3}\left[\nu_2 + \nu_3 - 15B'\right]$$

These relations are valid for the octahedral d^3 and d^8 systems and for the tetrahedral d^2, d^7 systems. From the above relations, it is evident that when ν_3 is not characterised due to overlapping with the charge transfer band, Dq and B' can be estimated from ν_1 and ν_2 and position of the ν_3–band can be predicted.

● **Transition, ${}^3A_{2g}$(F) → ${}^3T_{1g}$(P):** The ground state ${}^3A_{2g}$ represents the electronic configuration $t_{2g}^6 e_g^2$ and the excited state ${}^3T_{1g}$(P) arises from the $t_{2g}^4 e_g^4$ configuration. Thus the said transition involves the **promotion of two electrons** from the t_{2g} level. ***This is a forbidden transition*** (*cf.* ${}^3T_{1g} \rightarrow {}^3A_{2g}$ in d^2 configuration). Thus it is reasonable to expect that intensity of the band for the ${}^3A_{2g} \rightarrow {}^3T_{1g}$(P) transition should be poor. ***However intensity of this band is comparable to that of other allowed transitions.*** It is believed that mixing with the allowed transitions gives the intensity to this forbidden transition. This phenomenon is referred to as ***intensity borrowing or intensity stealing.***

● **Prediction of the coordinating behaviour of NO_3^- in $[Ni(NO_3)_4]^{2-}$ from the spectral characteristics:** Electronic absorption spectrum of $[Ni(NO_3)_4]^{2-}$ represents the characteristic features of the ligand field spectra of the octahedral complexes of Ni(II). It indicates that some of the NO_3^- groups act as the didentate ligands in $[Ni(NO_3)_4]^{2-}$ to provide the octahedral coordination geometry of Ni(II).

For most of the octahedral Ni(II) complexes, the ratio ν_2/ν_1 lies in the range 1.5 to 1.7 (*cf.* Fig. 7.10.4) and ***it is the characteristic feature of the octahedral complexes of Ni(II).***

(A) Spitting of the ν_2–band [*i.e.* ${}^3A_{2g}$ → ${}^3T_{1g}$(F)] in Ni(II)-octahedral complexes: In some cases as in $[Ni(DMSO)_6]^{2+}$ and $[Ni(OH_2)_6]^{2+}$, the ν_2-band undergoes splitting, *i.e.* the ν_2–band appears as a doublet. Analysis of the bands indicates the overlapping of two peaks. It is believed that there is **a spin-forbidden transition,** ${}^3A_{2g}(F) \rightarrow {}^1E_g(D)$, in the region of the spin-allowed transition, ${}^3A_{2g}(F) \rightarrow {}^3T_{1g}(F)$, in $[Ni(OH_2)_6]^{2+}$ and $[Ni(DMSO)_6]^{2+}$. In fact, spin-orbit coupling allows the mixing of the singlet state (${}^1E_g$) and the triplet state (${}^3T_{1g}$(F) to some extent when these two states are of comparable energy. These singlet and triplet states are of comparable energy when the crystal field strength leads to $\frac{Dq}{B'} \approx 1$ (*cf.*

Figs. 7.13.2.1g, 7.14.4.2d). *This condition (i.e.* $\frac{Dq}{B'} \approx 1$) *of spin-orbit coupling mixing of the* $^1E_g(D)$ *and* $^3T_{1g}(F)$ *states is attained for the ligands like* H_2O *and* DMSO. For the stronger field ligands like en, the two energy states, *i.e.* $^1E_g(D)$ and $^3T_{1g}$(F), are so widely apart that no significant mixing between the states occurs. This is why, in $[Ni(OH_2)_6]^{2+}$ and $[Ni(DMSO)_6]^{2+}$, the spin forbidden transition gains some intensity **but in** $\mathbf{[Ni(en)_3]^{2+}}$**, the spin forbidden transition is not at all important**. The said spin-forbidden transition gains some intensity through the interaction with the $^3T_{1g}(F)$ state in the complexes like $[Ni(OH_2)_6]^{2+}$, $[Ni(DMSO)_6]^{2+}$ where $\frac{Dq}{B} \approx 1$. In fact, at low temperature, these two overlapping peaks have been characterised in $[Ni(OH_2)_6]^{2+}$.

$$^3A_{2g}(F) \xrightarrow{14,000\ \text{cm}^{-1}} {}^1E_g(D);\ {}^3A_{2g}(F) \xrightarrow{15,200\ \text{cm}^{-1}} {}^3T_{1g}(F).$$

In $[Ni(en)_3]^{2+}$ (where $\frac{Dq}{B'} \gg 1$), there is no splitting in the ν_2–band and it is also expected so. Here it may be pointed out that some authors have attempted to explain the splitting of the ν_2-band of $[Ni(OH_2)_6]^{2+}$ through the **spin-orbit coupling splitting of the** $\mathbf{^3T_{1g}(F)}$ state (*see* Sec. 7.7.4, 8.20.3). If this pathway were correct, then for all the octahedral complexes of Ni(II), the splitting of the ν_2-band would be observed irrespective of the Dq/B' ratio. But, in reality, splitting of the ν_2-band is noticed for the condition only, $Dq/B' \approx 1$. It supports the spin-orbit coupling mixing of the $^3T_{1g}$(F) and ^1E_g(D) states is the reason for the splitting of the ν_2-band.

(B) Assignment of the ligand field bands in $\mathbf{[Ni(OH_2)_6]^{2+}}$**:** Aqueous solution of Ni(II) looks light green in colour (*i.e.* absorption occurs in the red and blue portion of the visible spectrum). The observed strong bands are:

$$8,700\ \text{cm}^{-1}\ (\varepsilon = 1.6),\ 13,800\ \text{cm}^{-1}\ (\varepsilon = 2.0),\ 25,300\ \text{cm}^{-1}\ (\varepsilon = 4.6)$$

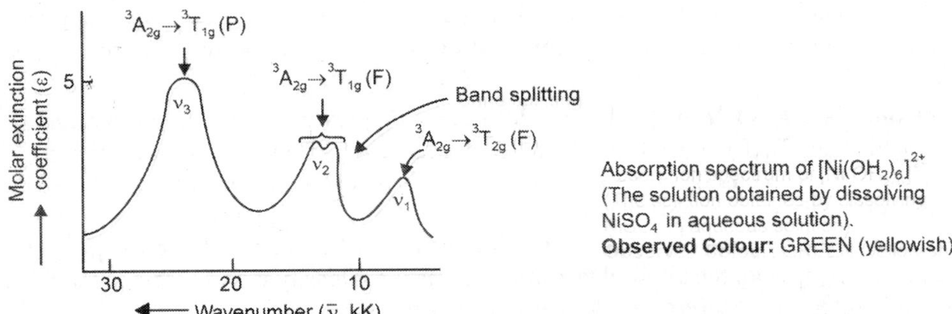

Fig. 7.14.4.2: (a) Electronic absorption spectrum of $[Ni(OH_2)_6]^{2+}$ (*cf.* Fig. 7.1.1).

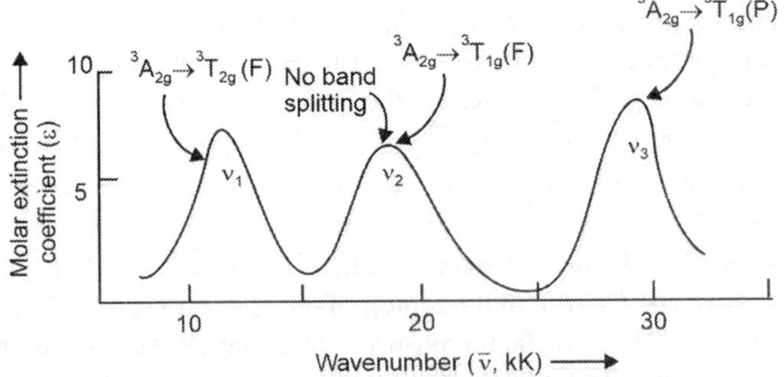

Fig. 7.14.4.2: (b) Electronic absorption spectrum of $[Ni(en)_3]^{2+}$ (*cf.* Fig. 7.1.1).

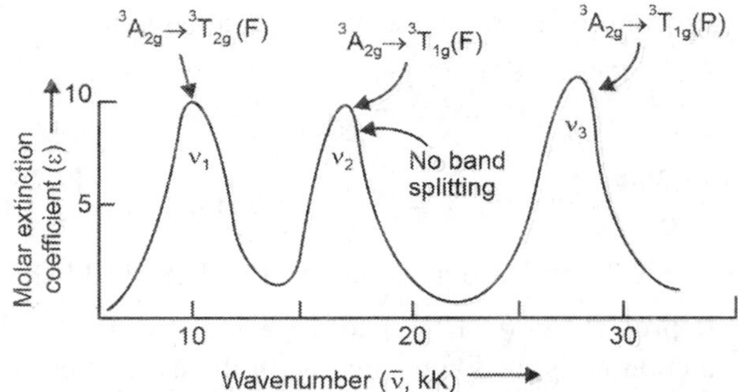

Fig. 7.14.4.2 (c) Electronic absorption spectrum of $[Ni(gly)_3]^-$, gly = glycinate.

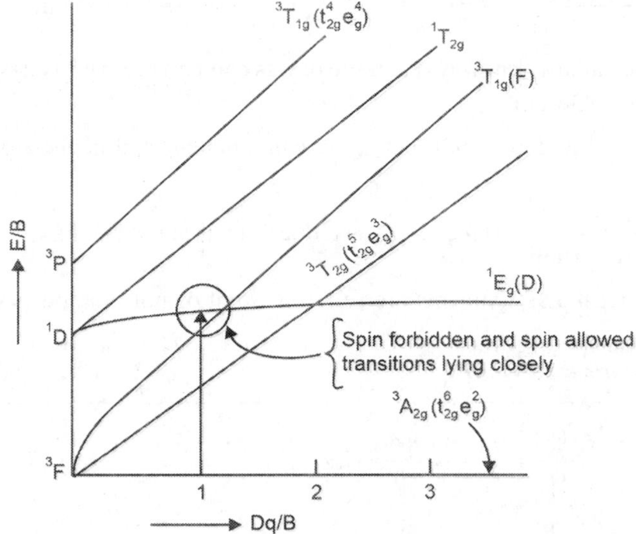

Fig. 7.14.4.2 (d) For d^8 (O_h); condition (*i.e.* $Dq/B' \approx 1.0$) for gaining the intensity in the spin forbidden transition to ^1E_g through the spin-orbit coupling **mixing of the $^3T_{1g}$ (F) and ^1E_g (D) states.**

The bands assignments are:

$$^3A_{2g}(F) \rightarrow {}^3T_{2g}(F),\ \nu_1 = 8{,}700\ \text{cm}^{-1},\ (= 10Dq)$$
$$^3A_{2g}(F) \rightarrow {}^3T_{1g}(F),\ \nu_2 = 13{,}800\ \text{cm}^{-1}$$
$$^3A_{2g}(F) \rightarrow {}^3T_{1g}(P),\ \nu_3 = 25{,}300\ \text{cm}^{-1}$$

Now let us verify the above assignments by considering the Tanabe-Sugano diagram.

$$\frac{^3A_{2g}(F) \longrightarrow {}^3T_{1g}(P)}{^3A_{2g}(F) \longrightarrow {}^3T_{1g}(F)} = \frac{\nu_3}{\nu_2} = \frac{25{,}300}{13{,}800} = 1.83$$

The above ratio is fitted into the Tanabe-Sugano diagram of the d^8–system. The above ν_3/ν_2 ratio corresponds to $Dq/B' = 0.88$ (*cf.* Fig. 7.10.4).

The ν_1 band, *i.e.* $^3A_{2g}(F) \rightarrow {}^3T_{2g}(F)$, gives the $10Dq$ value (= 8,700 cm^{-1}).

$$10Dq = 8{,}700\ \text{cm}^{-1},\ i.e.\ Dq = 870\ \text{cm}^{-1}$$

$$\frac{Dq}{B'} = 0.88 \quad \text{or,}\ B' = \frac{870}{0.88} \approx 989\ \text{cm}^{-1}$$

For $Dq/B' = 0.88$, we get from the relevant Tanabe-Sugano diagram:

$$\frac{^3A_{2g}(F) \longrightarrow {}^3T_{1g}(P)}{B'} = \frac{\nu_3}{B'} = 26.6$$

i.e.
$$\frac{25,300}{B'} = 26.6, \text{ or } B' = \frac{25,300}{26.6} = 951 \text{ cm}^{-1}.$$

Taking $\frac{Dq}{B'} = 0.88$ and $B' = 951$ cm^{-1}, we get: $Dq = 951 \times 0.88 \approx 837$ cm^{-1}

i.e. $10Dq = 8370$ cm^{-1}. It predicts ν_1 (= $10Dq$) to appear at 8,370 cm^{-1} (*cf.* experimental value 8,700 cm^{-1}). Thus the prediction is in good agreement with the experimental finding.

By taking all the three ligand field bands, B' can be also calculated.

$$B' = \frac{\nu_2 + \nu_3 - 3\nu_1}{15} = \frac{13,800 + 25,300 - 3 \times 8,700}{15} = 867 \text{ cm}^{-1}$$

The B' values obtained in different ways are more or less comparable. The average value is: B'(average) $= \frac{1}{3}(867 + 989 + 951) \approx 936$ cm^{-1}.

Taking $B_0 = 1030$ cm^{-1} and $B' = 936$ cm^{-1}, we can calculate the nephelauxetic parameter (β) as follows:

$$\beta = \frac{B'}{B_0} = \frac{936}{1030} \approx 0.91, \textit{ i.e. } B' \text{ is reduced to about 90\% of } B_0.$$

We can verify whether the assignments are self consistent or not in another way.

Transition energy ratio (experimental value)	*Theoretically calculated value for $\frac{Dq}{B'} = 0.88$ from Fig. 7.10.4*
$\frac{^3A_{2g}(F) \longrightarrow {}^3T_{1g}(F)}{^3A_{2g}(F) \longrightarrow {}^3T_{2g}(F)} = \frac{\nu_2}{\nu_1} = \frac{13,800}{8,700} \approx 1.59$	1.67
$\frac{^3A_{2g}(F) \longrightarrow {}^3T_{1g}(P)}{^3A_{2g}(F) \longrightarrow {}^3T_{2g}(F)} = \frac{\nu_3}{\nu_1} = \frac{25,300}{8,700} \approx 2.9$	3.0

Thus the calculated energy ratios are in good agreement with the experimental values and the B' value determined ($< B_0$) is also quite reasonable. All these support that the band assignments are correct.

(**Note:** The ν_2 band of $[Ni(OH_2)_6]^{2+}$ splits into a doublet and it has been already explained by considering the spin-forbidden transition to the $^1E_g(D)$ state.)

The $10Dq$ value can be determined from the ν_1 band without any uncertainty. It leads to: $10Dq$ = 8,700 cm^{-1}. If the energy change (without considering the configurational interaction) in the ν_2 band is considered to be $18Dq$ then we get:

$$\nu_2 = 18Dq = 13,800 \text{ cm}^{-1}, \textit{ i.e. } 10Dq = 7,667 \text{ cm}^{-1}.$$

This value is different from the value of $10Dq$ obtained from the ν_1–band. ***It indicates that in $[Ni(OH_2)_6]^{2+}$, the configurational interaction between the $^3T_{1g}(F)$ and $^3T_{1g}(P)$ states is quite important.*** Configurational interaction energy (x) can be calculated as follows:

$\nu_2 = 18Dq - x = 18 \times 870 - x = 13,800$

i.e. $x = 18 \times 870 - 13,800 = 15,660 - 13,800 = 1860$ cm^{-1}

(C) More rigorous analysis of the spectral parameters of Ni(II)–octahedral complexes: It has been shown that in $[Ni(OH_2)_6]^{2+}$, the $10Dq$ value calculated from the ν_2 (taking as $18Dq$) band is significantly different from the value obtained from the ν_1–band. It indicates the importance of configurational interaction. It leads to:

$$\nu_2 = 18Dq - x$$

Energies of the $^3A_{2g}(F)$ and $^3T_{2g}(F)$ states do not depend on the configurational interaction.

$$E[^3A_{2g}(F)] = -12Dq;\ (t_{2g}^6e_g^2\text{: cfse} = -12Dq)$$
$$E[^3T_{2g}(F)] = -2Dq;\ (t_{2g}^5e_g^3\text{: cfse} = -2Dq)$$

The excited states $^3T_{1g}(F)$ and $^3T_{1g}(P)$ experience the configurational interaction. Their energies are given by the following *second order secular determinant.*

$$\begin{array}{c} \\ ^3T_{1g}(F) \\ ^3T_{1g}(P) \end{array}\begin{array}{cc} ^3T_{1g}(F) & ^3T_{1g}(P) \\ \left|\begin{array}{cc} 6Dq-E & 4Dq \\ 4Dq & p-E \end{array}\right| \end{array} = 0,\ p = 15B' = \text{separation energy between the } ^3F \text{ and } ^3P \text{ terms in the complex.}$$

i.e. $(6Dq)p - 16(Dq)^2 - (6Dq)E - pE + E^2 = 0$

Roots of this equation give the energy values of $^3T_{1g}(P)$ and $^3T_{1g}(F)$ states.

If energy of the ^3F term (free ion) is considered to be zero, then energy of the ^3P term is p.

Energy of the $^3T_{1g}(P)$ state is given by:

$$\nu_3 = E[^3T_{1g}(P)] - E[^3A_{2g}(F)]$$
$$= E\,[^3T_{1g}(P)] - (-12Dq)$$

or, $E\,[^3T_{1g}(P)] = \nu_3 - 12Dq$

For $\mathbf{[Ni(\mathit{DMSO})_6]^{2+}}$: $\nu_1 = 7{,}730\ cm^{-1} = 10Dq$; *i.e.* $Dq = 773\ cm^{-1}$

$$\nu_2 = 12{,}970\ cm^{-1}$$
$$\nu_3 = 24{,}040\ cm^{-1}$$

We get: $E[^3T_{1g}(P)] = \nu_3 - 12Dq = 24{,}040 - 12 \times 773$
$= 14,\ 764\ cm^{-1}$

Now let us use this energy value ($=14{,}764\ cm^{-1}$) and Dq ($= 773\ cm^{-1}$) in the following equation.

$$(6Dq)p - 16(Dq)^2 - (6Dq)E - pE + E^2 = 0.$$

or $\quad 6 \times 773p - 16 \times (773)^2 - 6 \times 773 \times 14{,}764 - p \times 14{,}764 + (14{,}764)^2 = 0.$

It leads to: $\quad p = 13{,}818 = 15B'$

i.e. $\quad B' \approx 921\ cm^{-1}$

By using the values of Dq and p in the following equation, we get the two values of E.

$$(6Dq)p - 16(Dq)^2 - 6Dq(E) - pE + E^2 = 0$$
$$p = 13{,}818\ cm^{-1},\ Dq = 773\ cm^{-1}$$

The solutions are: $E_1 = 3,\ 694\ cm^{-1}$, $E_2 = 14{,}762\ cm^{-1}$

i.e. $\quad E[^3T_{1g}(P)] = 14{,}762\ cm^{-1}$,
$E[^3T_{1g}(F)] = 3{,}694\ cm^{-1}$.

Now let us calculate the ν_2 and ν_3 values.

$$\nu_2 = E[^3T_{1g}(F)] - E[^3A_{2g}(F)] = 3{,}694\ cm^{-1} - (-12 \times Dq)\ cm^{-1}$$
$$= 3{,}694\ cm^{-1} + 12 \times 773\ cm^{-1} = 12{,}970\ cm^{-1}$$
$$\nu_3 = E[^3T_{1g}(P)] - E[^3A_{2g}(F)] = 14{,}762 - (-12 \times 773) = 24{,}038\ cm^{-1}$$

These calculated values of ν_2 and ν_3 are in good agreement with the experimental values. Thus, the assignments of the spectral transitions are self-consistent.

(D) Spectral analysis of $[Ni(bpy)_3]^{2+}$ where the third band, *i.e.* ν_3-band is obscured: The two well defined peaks are:

$$\nu_1 = 12{,}650 \text{ cm}^{-1} \ (= 10Dq)$$
$$\nu_2 = 19{,}200 \text{ cm}^{-1}$$

There is a strong band at 26,000 cm^{-1} having a very high extinction coefficient. Thus, the band at 26,000 cm^{-1} is reasonably expected to be a charge transfer band into which the third ligand field band of weak intensity may be merged. However, by considering the positions of ν_1 and ν_2, the ligand field bands can be characterised with the help of Tanabe-Sugano diagram.

$$\frac{^3A_{2g}(F) \longrightarrow {}^3T_{1g}(F)}{^3A_{2g}(F) \longrightarrow {}^3T_{2g}(F)} = \frac{\nu_2}{\nu_1} = \frac{19{,}200}{12{,}650} \approx 1.52.$$

This ratio fits into the Tanabe-Sugano diagram for $Dq/B' \approx 1.7$ (*cf.* Fig. 7.10.4)

The ν_1 (= 10*Dq*) band gives $Dq = 1265$ cm^{-1}.

$$\frac{Dq}{B'} = 1.7 = \frac{1265}{B'}; \quad \text{or} \quad B' = \frac{1265}{1.7} = 744 \text{ cm}^{-1}.$$

i.e.
$$\beta = \frac{B'}{B_0} = \frac{744}{1030} = 0.72.$$

Note: $[Ni(bpy)_3]^{2+}$ belongs to the D_3 point group but in terms of electronic transitions, it effectively adopts the O_h symmetry of the chromophore $[Ni(N)_6]$. A similar situation arises for $[Cr(ox)_3]^{3-}$, $[Cr(en)_3]^{3+}$, etc. (*cf.* Secs. 7.14.5 and 7.15.4 for details).

(E) Spectral properties of the tetragonally distorted high spin Ni(II)-complexes like *trans*-$[Ni^{II}A_4B_2]$ of D_{4h} symmetry: In moving from the O_h to D_{4h} symmetry, the following changes occur.

$^3A_{2g} \rightarrow {}^3B_{1g}$; $^3T_{2g} \rightarrow {}^3E_g + {}^3B_{2g}$; $^3T_{1g} \rightarrow {}^3E_g + {}^3A_{2g}$ (*cf.* **Correlation Table** 7.23.2).

Let us consider the *z*-out distortion to give the D_{4h} symmetry. In the limiting situation, it will produce the diamagnetic square planar complex (which is discussed separately in Sec. 7.17). The six coordinate tetragonally distorted Ni(II) cmplexes are paramagnetic and in such cases, generally the energy order of the *d*-orbitals is as follows:

$e_g^4 < b_{2g}^2 < a_{1g}^1 < b_{1g}^1$ or $e_g^4 < a_{1g}^2 < b_{2g}^1 < b_{1g}^1$ (*i.e.* the relative positions of b_{2g} and a_{1g} depend on the degree of *z*-out distortion)

The different spectroscopic states are related with the electronic configurations as follows:

$$e_g^4b_{2g}^2a_{1g}^1b_{1g}^1 \Rightarrow {}^3B_{1g};\ e_g^4b_{2g}^1a_{1g}^2b_{1g}^1 \Rightarrow {}^3A_{2g}\left({}^3T_{1g}(F)\right);\ e_g^3b_{2g}^2a_{1g}^2b_{1g}^1 \Rightarrow {}^3E_g\left({}^3T_{2g}\right).$$

$$e_g^4b_{2g}^1a_{1g}^1b_{1g}^2 \Rightarrow {}^3B_{2g};\ e_g^3b_{2g}^2a_{1g}^1b_{1g}^2 \Rightarrow {}^3E_g\left({}^3T_{1g}(F)\right);\ e_g^3b_{2g}^3a_{1g}^2b_{1g}^2 \Rightarrow {}^3A_{2g}\left({}^3T_{1g}(P)\right);$$

$$e_g^2b_{2g}^2a_{1g}^2b_{1g}^2 \Rightarrow {}^3A_{2g}\left({}^3T_{1g}(P)\right).$$

The possible spin allowed **electronic transitions** in the 6-coordinate Ni(II) **paramagnetic complexes** of D_{4h} symmetry are as follows:

$$^3B_{1g} \xrightarrow{\nu_1} {}^3E_g^{(a)};\ {}^3B_{1g} \xrightarrow{\nu_2} {}^3B_{2g};\ {}^3B_{1g} \xrightarrow{\nu_3} {}^3A_{2g}(F);\ {}^3B_{1g} \xrightarrow{\nu_4} {}^3E_g^{(b)};$$

$$^3B_{1g} \xrightarrow{\nu_5} {}^3E_g(P);\ {}^3B_{1g} \xrightarrow{\nu_6} {}^3A_{2g}(P)$$

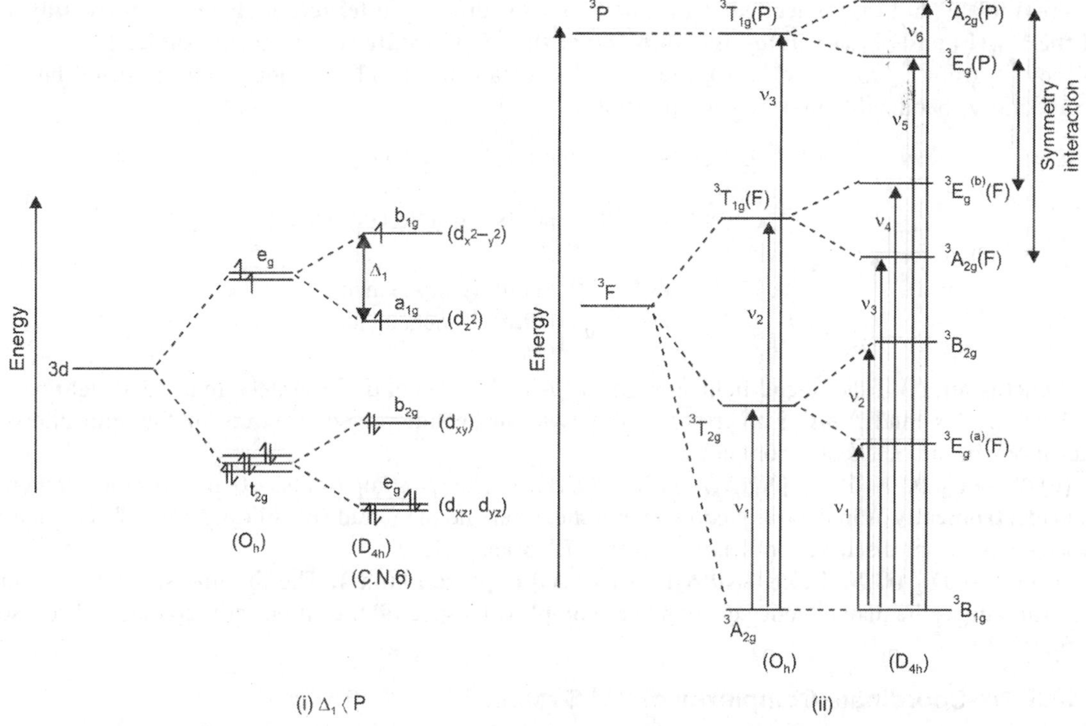

Fig. 7.14.4.3 Effect of tetragonal elongation on (i) *d*-orbital splitting, (ii) energy levels for the 6 coordinate Ni(II) complexes. **Note:** Energy order: $^3A_{2g}$ (P) $\rangle$ ^3E_g (P) but the reverse order for ^3E_g and $^3A_{2g}$ generated from the splitting of the $^3T_{1g}$(F) level. It occurs due to the **configurational interaction** between the $^3A_{2g}$(P) and $^3A_{2g}$(F) levels. This interaction minimises the splitting of both the $^3T_{1g}$(P) and $^3T_{1g}$(F) levels. Thus, in moving from $O_h \rightarrow D_{4h}$, **splitting of the $^3T_{2g}$(F) level is only appreciable.** ***This is why, generally in moving from O_h to D_{4h} (C.N. 6), the octahedral ν_2 and ν_3 bands are not affected much but the octahedral ν_1 band spits into two components.***

The splitting of the $^3T_{1g}(P)$ state is very small (*cf.* $^3E_g(P) \sim {}^3A_{2g}(P) \sim 500$ cm^{-1}) and generally, **the ν_5 and ν_6 transitions merge to produce a relatively broader peak.** The magnitude of tetragonal splitting in *trans*-[Ni^{II}A$_4$B$_2$](D_{4h}) depends on the difference of the ligand field strength of A and B. Peak positions of some representative 6-coordinate Ni(II) complexes of D_{4h} symmetry are given below:

[NiBr$_2$(py)$_4$]: ν_1(8.42 kK); ν_2(11.55 kK); ν_3(14.10 kK); ν_4(16.37 kK); $\nu_{5,6}$(25.9 kK);

[Ni(NH$_3$)$_4$(NO$_2$)$_2$]: ν_1(11.2 kK); ν_2(12.0 kK); ν_3(19.9 kK); ν_4(20.35 kK);

Note: (i) ***$O_h \rightarrow D_{4h}$* (C.N. = 6),** *i.e.* **[Ni^{II}A$_6$] → *trans*-[Ni^{II}A$_4$B$_2$]:** In moving from ***O_h* to *D_{4h}* (C.N. 6),** splitting of the energy levels can generate ***more than 3 ligand field bands*** (*cf.* Fig. 17.4.4.3), but in reality, in most of the cases, the ligand field bands can be interpreted in terms of the ν_1, ν_2 and ν_3 transitions of O_h symmetry, *i.e.* splitting of the energy levels is not sufficiently large to give all the separate peaks as expected in D_{4h} (C.N. 6). This occurs due to two reasons: splitting of the T-term is not large specially when the ligand field strengths of A and B in *trans*-[Ni^{II} A_4B_2] do not differ widely; both the $^3T_{1g}(F)$ and $^3T_{1g}(P)$ states yield $^3A_{2g}$ + ^3E_g in D_{4h} and symmetry interaction (*i.e.* **configurational interaction**) between the $^3A_{2g}$(F) and $^3A_{2g}$(P) states; and ^3E_g(F) and ^3E_g(P) states reduce the splitting. This configurational interaction leads to the energy order $^3E_g > {}^3A_{2g}$ for $^3T_{1g}$(F) while the order is inverted for ^3E_g and $^3A_{2g}$ generated from $^3T_{1g}$(P). **This symmetry interaction also minimises splitting of the $^3T_{1g}$(P) and $^3T_{1g}$(F) states and splitting of the $^3T_{2g}$(F) state is only accountable.** This is why, in moving from O_h to D_{4h} (C.N. 6), the ν_2 and ν_3 octahedral bands are not greatly affected but the octahedral ν_1 band splits into two components.

$$\left[\mathrm{Ni^{II}}A_6\right](O_h) \longrightarrow trans\text{-}\left[\mathrm{Ni^{II}}A_4B_2\right](D_{4h})$$

$$^3A_{2g} \longrightarrow {}^3T_{2g}(F)\text{: } \nu_1 \text{ splits into two components}$$

$$\left.\begin{array}{l} ^3A_{2g} \longrightarrow {}^3T_{1g}(F){:}\nu_2 \\ ^3A_{2g} \longrightarrow {}^3T_{1g}(P){:}\nu_3 \end{array}\right\} \begin{array}{l}\text{(slightly broadened)} \\ \text{almost unchanged.}\end{array}$$

Conclusion: (i) If the ligand field strengths of A and B do not differ largely, the ligand field bands of both the *cis*- and *trans*- isomers of [$Ni^{II}A_4B_2$] can be interpreted in terms of the spin allowed transitions in the octahedral complexes.

(ii) $\boldsymbol{O_h \rightarrow C_{2v}}$ **(C.N. 6),** *i.e.* **[$Ni^{II}A_6$] → *cis*-[$Ni^{II}A_4B_2$]:** The *cis*-complex also experiences the tetragonal field **electronically** as in the *trans*-complex but the order and magnitude of splitting are different. These aspects have been discussed in detail in Secs. 17.4.5 and 7.15.2.

(iii) $\boldsymbol{O_h \rightarrow D_{4h}}$ **(C.N. 4),** *i.e.* **[$Ni^{II}A_6$] → [$Ni^{II}A_4$] (square planar):** The limiting situation of *z*-out distortion gives the diamagnetic square planar complex. Its spectral transitions are separately discussed in Sec. 7.17.

7.14.5 Six-Coordinate Complexes of 3*d*³-System

Complexes of Cr(III) and V(II) are the representative examples of this group. From the Orgel diagram, the spin-allowed transitions are:

$$^4A_{2g}(F) \xrightarrow{\nu_1} {}^4T_{2g}(F)\,;\ \nu_1 = 10Dq\ \text{(independent of configurational interaction)}$$

$$^4A_{2g}(F) \xrightarrow{\nu_2} {}^4T_{1g}(F);\ \nu_2 = 18Dq - x \approx 18Dq$$

$$^4A_{2g}(F) \xrightarrow{\nu_3} {}^4T_{1g}(P);\ \nu_3 = 15B' + 12Dq + x$$

$$\approx 15B' + 12Dq$$

ν_3 involves **2*e* transition** $\left(i.e.\ t_{2g}^3 \longrightarrow t_{2g}^1 e_g^2\right)$ and it is a **forbidden transition**.

x is the symmetry interaction energy for the configurational interaction between the $^4T_{1g}(F)$ and $^4T_{1g}(P)$ states. However, in a weak field ligand, *x* becomes negligibly small (*cf.* Figs. 7.10.2; 7.14.5.2);

The *B*′ and 10*Dq* value may be calculated as follows:

$$\nu_1 = 10Dq,\ 15B' = (\nu_2 + \nu_3 - 3\nu_1).$$

Calculation of *B*′ by this method is applicable in the cases where the three ligand field bands (ν_1, ν_2 and ν_3) are well characterised. But in most of the cases, the third ligand field band (ν_3) is obscured by the strong charge transfer band. Moreover, the band due to ν_3 is **forbidden** because it involves a **two-electron transfer,** *i.e.* $^4A_{2g}(t_{2g}^3 e_g^0) \rightarrow {}^4T_{1g}$(P) $(t_{2g}^1 e_g^2)$. Thus it is a weak band. In the cases like [CrF_6]$^{3-}$ where the charge transfer band lies at still higher energy, all the three ligand field bands are well characterised and *B*′ can be calculated as follows:

[CrF_6]$^{3-}$: ν_1 = 14,900 cm^{-1}, ν_2 = 22,700 cm^{-1}, ν_3 = 34,400 cm^{-1}

It leads to: $10Dq = \nu_1 = 14{,}900 \text{ cm}^{-1}$

$$B' = \frac{34{,}400 + 22{,}700 - 3\times 14{,}900}{15} = 827 \text{ cm}^{-1}.$$

The energy change in the spin allowed ligand field bands for the high-spin octahedral d^3 and d^8 and tetrahedral d^2 and d^7 systems

$$^4A_2 \Rightarrow t_2^3 \Rightarrow E = -12Dq;\ ^4T_2(\text{F}) \Rightarrow t_2^2e^1 \Rightarrow E = -2Dq,$$

$$^4T_1(F) \Rightarrow t_2^2e^1 \Rightarrow E = 7.5B' + 3Dq - X;\ ^4T_1(P) \Rightarrow t_2^1e^2 \Rightarrow E = 7.5B' + 3Dq + X$$

$$X = \tfrac{1}{2}\left(225B'^2 + 100Dq^2 - 180DqB'\right)^{1/2}$$

$$A_2(F) \longrightarrow T_2(F),\ \nu_1 = \Delta,$$

$$A_2(F) \longrightarrow T_1(F),\ \nu_2 = 7.5B' + 1.5\Delta - \tfrac{1}{2}\left[225B'^2 + \Delta^2 - 18B'\Delta\right]^{1/2}$$

$$A_2(F) \longrightarrow T_1(P),\ \nu_3 = 7.5B' + 1.5\Delta + \tfrac{1}{2}\left[225B'^2 + \Delta^2 - 18B'\Delta\right]^{1/2}$$

$$B' = \frac{\nu_2 + \nu_3 - 3\nu_1}{15} = \frac{2\nu_1^2 + \nu_2^2 - 3\nu_1\nu_2}{(15\nu_2 - 27\nu_1)} = \frac{2\nu_1^2 + \nu_3^2 - 3\nu_1\nu_3}{15\nu_3 - 27\nu_1}$$

(A) Analysis of the spectrum of $[Cr(OH_2)_6]^{3+}$: Aqueous solution of $[Cr(OH_2)_6]^{3+}$ looks blue violet due to the absorption of yellow and blue parts of the visible spectrum. There are two intense bands at 17,000 cm^{-1} ($\varepsilon \approx 14$) and 24,000 cm^{-1} ($\varepsilon \approx 15$). A weak band at 37,000 cm^{-1} ($\varepsilon \approx 4$) appears as a shoulder of a strong charge transfer band. Besides these, there are two weak bands at about 15,000 cm^{-1} and 22,000 cm^{-1} and these are the **forbidden transitions.** These are not shown in Fig. 7.14.5.1.

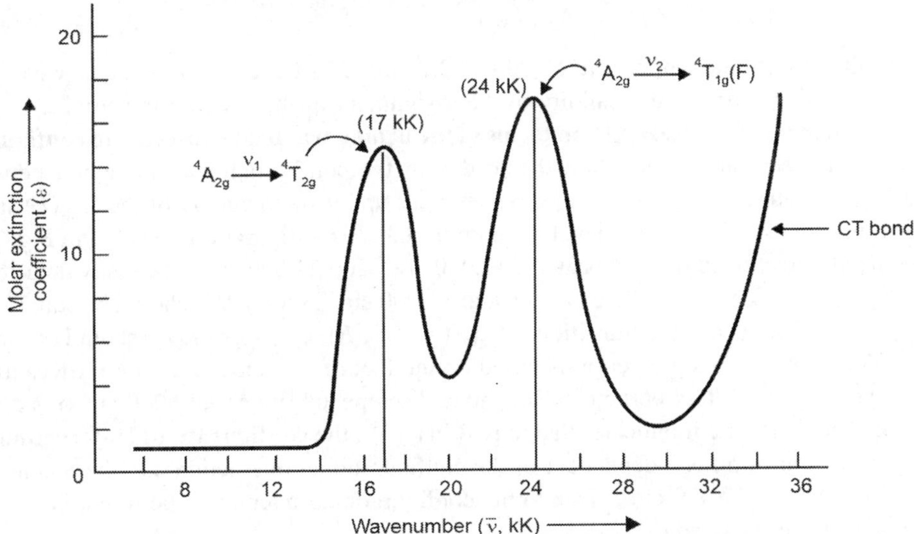

Fig. 7.14.5.1 The electronic absorption spectrum of $[Cr(OH_2)_6]^{3+}$.

Note: The **green colour** of Cr(III) noted during the titration of Fe(II) by $K_2Cr_2O_7$ is not due to $[Cr(OH_2)_6]^{3+}$ that looks **blue violet**. The green colour is due to the sulfato Cr(III)-complex produced in the reaction medium during the stepwise reduction of Cr(VI) to Cr(III). This aspect has been discussed in Sec. 1.12H, Vol. 4.

To assign the peaks, we can start with the intense peaks at 17,000 cm^{-1} and 24,000 cm^{-1}. Let us assign the transitions as follows:

$^4A_{2g}(F) \rightarrow {}^4T_{2g}(F)$; ν_1 = 17,000 cm^{-1} (= $10Dq$)

$^4A_{2g}(F) \rightarrow {}^4T_{1g}(F)$; ν_2 = 24,000 cm^{-1} (= $18Dq - x$).

We can verify whether these assignments are correct or not by comparing with the Tanabe-Sugano diagram. The ratio, $\nu_2/\nu_1 \approx 1.41$ fits with the Tanabe-Sugano diagram for the value $Dq/B' = 2.45$ (*cf.* Fig. 7.10.4). This Dq/B' value corresponds to $E(\nu_1)/B' = 24.5$ for the ν_1 transition, *i.e.*

$$\frac{{}^4A_{2g}(F) \longrightarrow {}^4T_{2g}(F)}{B'} = \frac{E(\nu_1)}{B'} = \frac{17{,}000}{B'} = 24.5$$

i.e. $$B' = \frac{17{,}000}{24.5} \approx 694 \text{ cm}^{-1}$$

and $$10Dq = \nu_1 = 17{,}000 \text{ cm}^{-1}$$

The B' value may compared with the B_0 value (of free ion).

$$B_0 = 1030 \text{ cm}^{-1}, \text{ i.e. } \beta \text{ (nephelauxetic ratio)} = \frac{B'}{B_0} = \frac{694}{1030} = 0.67$$

Thus in $[Cr(OH_2)_6]^{3+}$, the B value is reduced by about 33%.

Fitting of the ν_1 and ν_3 bands at Dq/B' = 2.45 in the Tanabe-Sugano diagram gives the following result.

$$\frac{{}^4A_{2g}(F) \longrightarrow {}^4T_{1g}(P)}{{}^4A_{2g}(F) \longrightarrow {}^4T_{2g}(F)} = \frac{E(\nu_3)}{E(\nu_1)} = 2.2. \text{ (cf. Fig. 7.10.4)}$$

It predicts that ν_3 should appear at 17,000 × 2.2, *i.e.* 37,400 cm^{-1}. In fact, a weak band at 37,000 cm^{-1} appears. It supports the validity of the assignment of the spectral transitions.

(B) Spectral analysis of $[Cr(OH_2)_6]^{3+}$ in terms of the octahedral field subjected to configurational interaction: It has been already mentioned that due to the configurational interaction between the $^4T_{1g}(F)$ and $^4T_{1g}(P)$ states, energy of the $^4T_{1g}(P)$ state goes up by x and energy of the $^4T_{1g}(F)$ state goes down by the same amount. The energy level diagram for the d^3–configuration is shown in Fig. 7.14.5.2.

We have already mentioned that the peaks at 17,000 cm^{-1} and 24,000 cm^{-1} are easily detected. If the $^4A_{2g}(F) \rightarrow {}^4T_{2g}(F)$ transition is considered to occur at 17,000 cm^{-1}, then $10Dq$ should be equal to 17,000 cm^{-1}, *i.e.* Dq = 1700 cm^{-1}. Thus the transition, $^4A_{2g}(F) \rightarrow {}^4T_{1g}(F)$, *i.e.* ν_2 (= $18Dq$) should occur at 1700 × 18 cm^{-1} = 30,600 cm^{-1}. This prediction is based on the fact of **no configurational interaction**. But in reality, the peak appears at 24,000 cm^{-1}. This huge discrepency (by about 6000 cm^{-1}) is not due to the experimental error. Rather it indicates that in $[Cr(OH_2)_6]^{3+}$, the **configurational interaction** is quite significant. The ν_1–transition is independent of the configurational interaction, *i.e.* there is no error in the relation, $10Dq = \nu_1$ = 17,000 cm^{-1}. Due to the configurational interaction between the $^4T_{1g}(F)$ and $^4T_{1g}(P)$ terms, we have the condition:

$$^4A_{2g}(F) \rightarrow {}^4T_{1g}(F),\ \nu_2 = 24{,}000 \text{ cm}^{-1} = 18Dq - x$$

$$^4A_{2g}(F) \rightarrow {}^4T_{1g}(P),\ \nu_3 = 37{,}000 \text{ cm}^{-1} = 15B' + 12Dq + x$$

It leads to:

$$24{,}000 = 18Dq - x = 18 \times 1700 - x$$

i.e. $$x = 6{,}600 \text{ cm}^{-1}$$

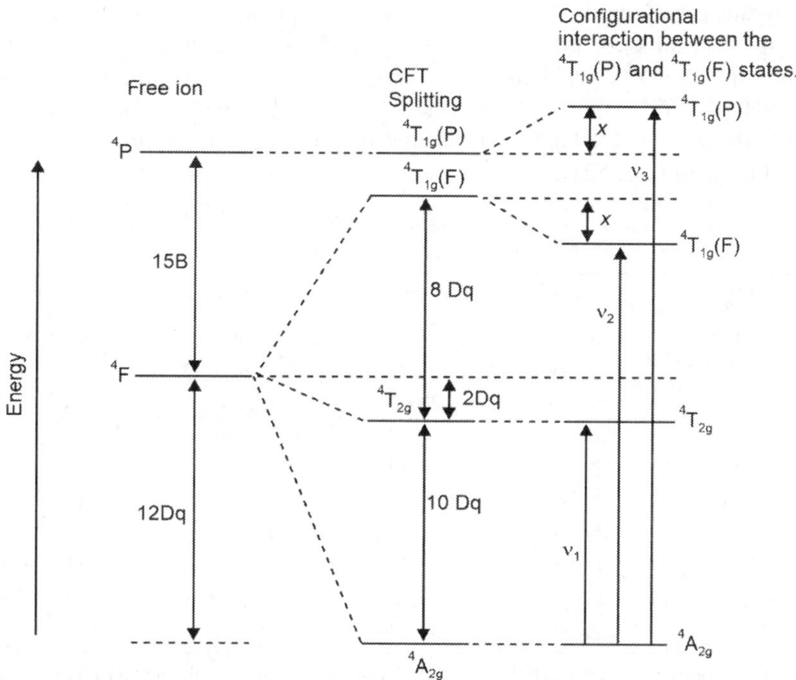

Fig. 7.14.5.2 Energy level diagram and spectral transitions (spin-allowed) in the d^3–octahedral system (x denotes the configurational interaction energy for the $^4T_{1g}(P) - {}^4T_{1g}(F)$ interaction) (*See* Fig. 7.9.4.1 and Solved Problem 14 for the energy values of different states).

Note: The electronic configurations and crystal field energies of the $^4T_{2g}$ and $^4T_{2g}$ states are: $^4A_{2g}(t_{2g}^3, -12Dq)$, $^4T_{2g}(t_{2g}^2e_g^1, -2Dq)$. Due to the mixing between the $^4T_{1g}(P)$ and $^4T_{1g}(F)$ states, the $^4T_{1g}(F)$ corresponds to $t_{2g}^x e_g^y$ (pDq, say; $x + y = 3$). Considering the energy conservation and orbital degeneray: $1 \times 12\ Dq + 3 \times 2Dq = 3pDq$, *i.e.* $p = 6$; $(-4x + 6y)Dq = pDq = 6Dq$ and $x + y = 3$ lead to $x = 1.2$ and $y = 1.8$. Thus in a weak field, $^4T_{1g}(F)$ corresponds to $t_{2g}^{1.2}\ e_g^{1.8}$ (not $t_{2g}^1 e_g^2$).

Assuming $\nu_3 = 37{,}000\ \text{cm}^{-1}$, we can find out the value of B' as follows:

$$37{,}000 = 15B' + 12Dq + x = 15B' + 12 \times 1700 + 6{,}600$$

or, $$30{,}400 - 20{,}400 = 15B'$$

or, $$B' \approx 667\ \text{cm}^{-1}\ (cf.\ B_0 = 1030\ \text{cm}^{-1})$$

i.e. $$\beta = \frac{B'}{B_0} = \frac{667}{1030} = 0.64$$

i.e. B is reduced to about 64% of the free ion value.

(C) O_h *vs.* C_{4v}, *e.g.* $[Cr(NH_3)_6]^{3+}$ (O_h) *vs.* $[Cr(NCS)(NH_3)_5]^{2+}$, $[CrCl(NH_3)_5]^{2+}$ (C_{4v}): Reduction in symmetry in moving from O_h to C_{4v} splits the energy levels as follows (*See* Correlation Table, Table 7.23.2) :

$A_{2g} \rightarrow B_1$; $T_{1g} \rightarrow A_2 + E$; $T_{2g} \rightarrow B_2 + E\}$ $\Rightarrow$ **splitting of the octahedral peaks into the doublets**

If we compare the spectral transitions in $[Cr(NH_3)_6]^{3+}$ and $[Cr(NH_3)_5X]^{2+}$, the degree of splitting of the T_1 and T_2 states depends on the **difference of the ligand field strength of NH_3 and X^-.** The splitting of the T_{2g} state is, in general, higher than that of the T_{1g} state (*cf.* Sec. 7.14.4). In general, the spectral transitions of $[Cr(NH_3)_5X]^{2+}$ can be explained in terms of the octahedral transitions as splitting of the *T*-states in the C_{4v} symmetry is not sufficiently high to develop the separate peaks. For $[Cr(NCS)(NH_3)_5]^{2+}$, the peaks do not split. For $[CrCl(NH_3)_5]^{2+}$, the peak due to the $^4A_{2g} \rightarrow {}^4T_{1g}(F)$ transition does not split but the peak due to the $^4A_{2g} \rightarrow {}^4T_{2g}(F)$ transition splits into a doublet as expected. In $[Cr(NH_3)_5X]^{2+}$(X = Cl, Br, NCS), the *LMCT bands are shifted towards the longer wavelength* compared to those of $[Cr(NH_3)_6]^{3+}$. The ligand field bands shift *towards the lower frequence,* if X^- is a weaker field ligand than NH_3.

Table 7.14.5.1a Spectral characteristics of the spin-allowed ligand field bands of the d^3-systems in an octahedral field.

Complex	ν_1 (cm^{-1})	ν_2 (cm^{-1})	ν_3 (cm^{-1})	*10Dq* (cm^{-1})	*B′* (cm^{-1})	β
$[Cr(OH_2)_6]^{3+}$	17,000	24,000	37,000	17,000	694	0.67
$[Cr(NH_3)_6]^{3+}$	21,550	28,500	–	21,550	–	–
$[CrCl_6]^{3-}$	13,200	18,700	28,600	13,200	513	0.56
$[Cr(CN)_6]^{3-}$	26,700	32,200	–	26,700	–	–
$[CrF_6]^{3-}$	14,900	22,700	34,400	14,900	827	0.80
$[Cr(ox)_3]^{3-}$	17,400	23,600	(38,100)*	17,400	600	0.58
$[Cr(en)_3]^{3+}$	21,800	28,500	(46,500)*	21,850	–	–
$[V(OH_2)_6]^{2+}$	12,300	18,500	27,900	12,300	633	0.84
$[VCl_6]^{4-}$	7,200	12,000	19,000	7,200	627	0.83
$[V(en)_3]^{2+}$	15,600	21,500	31,700	15,600	420	0.57

* The values are predicted but not experimentally verified. These are masked by the strong charge transfer bands.

Table 7.14.5.1b Comparison of the spectral properties of $[Cr^{III}(NH_3)_5X]^{2+}$ and $[Cr(NH_3)_6]^{3+}$ ($\bar{\nu}$ in cm^{-1}) (*cf.* Figs. 7.14.5.3a, 7.5.1.2).

	$^4A_{2g} \rightarrow {}^4T_{2g}(F)$		$^4A_{2g} \rightarrow {}^4T_{1g}(F)$	LMCT
	$^4B_1 \rightarrow {}^4E$	$^4B_1 \rightarrow {}^4B_2$	$^4B_1 \rightarrow {}^4A_2, {}^4E$	
$[CrCl(NH_3)_5]^{2+}$	19,400	22,100 (shoulder)	26,600 (no peak splitting)	42,000
$[Cr(NCS)(NH_3)_5]^{2+}$	20,370 (no peak splitting)		27,000 (no peak splitting)	
$[Cr(NH_3)_6]^{3+}$	21,500		28,500	50,000 (tail)

$^4A_{2g} \rightarrow {}^4T_{1g}(P)$ transition is not characterised.

Spectral characteristics of $[Co(NH_3)_6]^{3+}$ (O_h) and $[CoCl(NH_3)_5]^{2+}$ have been compared and discussed in Sec. 7.15.3 (*cf.* Figs. 7.15.3.1, 2).

(D) Spectral behaviour of $[Cr(ox)_3]^{3-}$ and $[Cr(en)_3]^{3+}$: For green $[Cr(ox)_3]^{3-}$, the two spin allowed transitions are at 17,400 cm^{-1} (ν_1) and 23,600 cm^{-1} (ν_2). The 3rd ligand field band (ν_3) is predicted to appear at about 38,100 cm^{-1} but this is not experimentally verified because in that region, there is a strong charge transfer band to mask the ligand field band (ν_3).

For yellow $[Cr(en)_3]^{3+}$, the spin allowed transitions are: 21,850 cm^{-1} (ν_1) and 28,500 cm^{-1} (ν_2). The predicted third ligand field band (ν_3) at about 46,500 cm^{-1} is obscured by the strong charge transfer band. In addition to the spin-allowed transitions, there are **two spin-forbidden transitions** at 15,100 cm^{-1} and 15,600 cm^{-1}. These will be discussed later on.

Here it is worth mentioning that the said chelated complexes like $[Cr(en)_3]^{3+}$, $[Cr(ox)_3]^{3-}$, etc. are belonging to the **lower symmetry point group D_3.** They are not perfectly octahedral. However, their

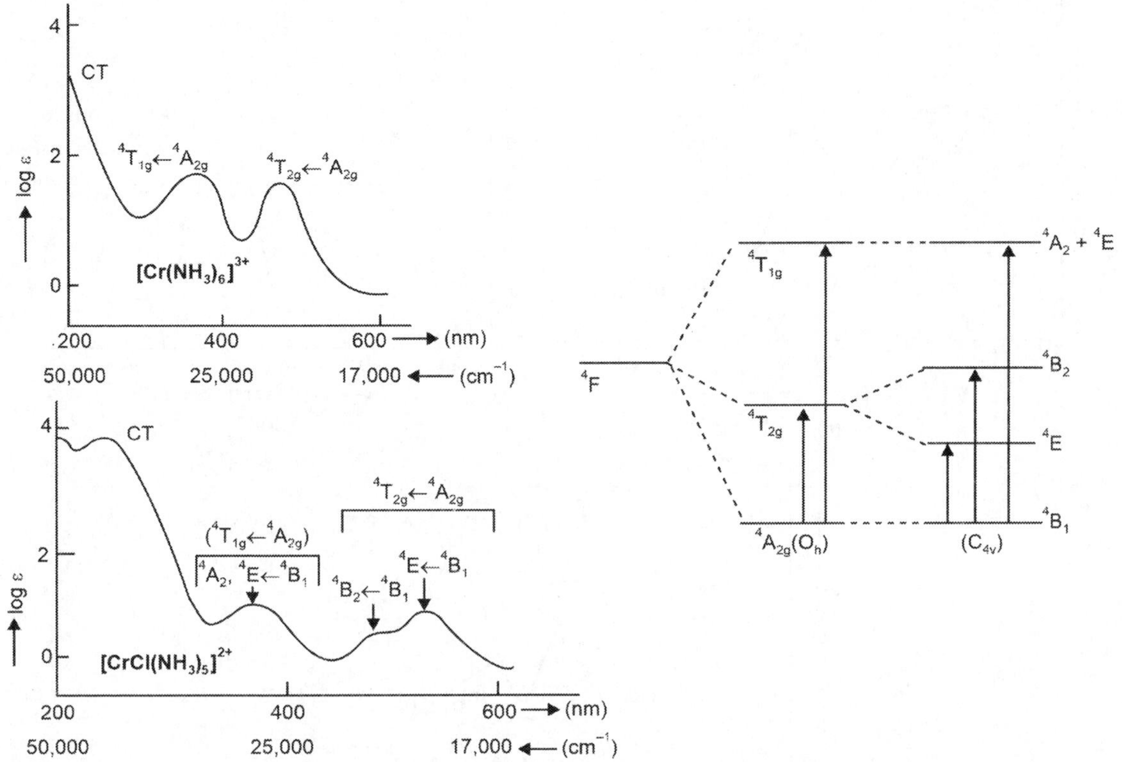

Fig. 7.14.5.3: (a) Spectral transitions of $[Cr(NH_3)_6]^{3+}$ (O_h) and $[CrCl(NH_3)_5]^{2+}$ (C_{4v}). The ligand field transition, $^4A_{2g} \rightarrow {^4T_{1g}}(P)$ is masked by the CT band. Splitting of the $^4T_{1g}(F)$ state in C_{4v} is not observed.

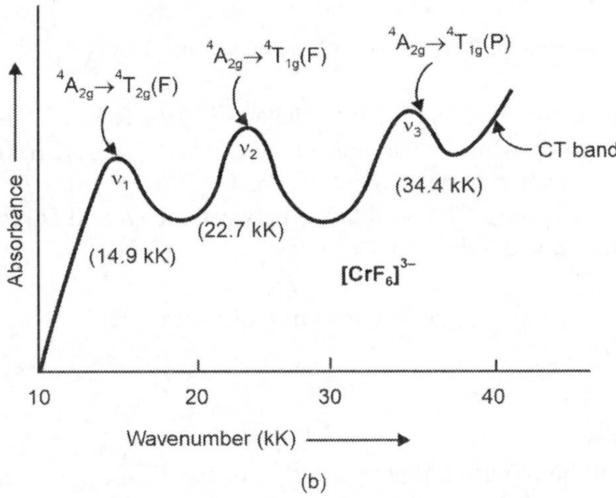

(b)

Fig. 7.14.5.3: (b) Absorption spectram of $[CrF_6]^{3-}$.

ligand field bands can be successfully explained by considering the octahedral field as in $[CrL_6]^{3+}$. It indicates that the perturbation is insignificant in these chelated complexes to destroy the ideal octahedral symmetry.

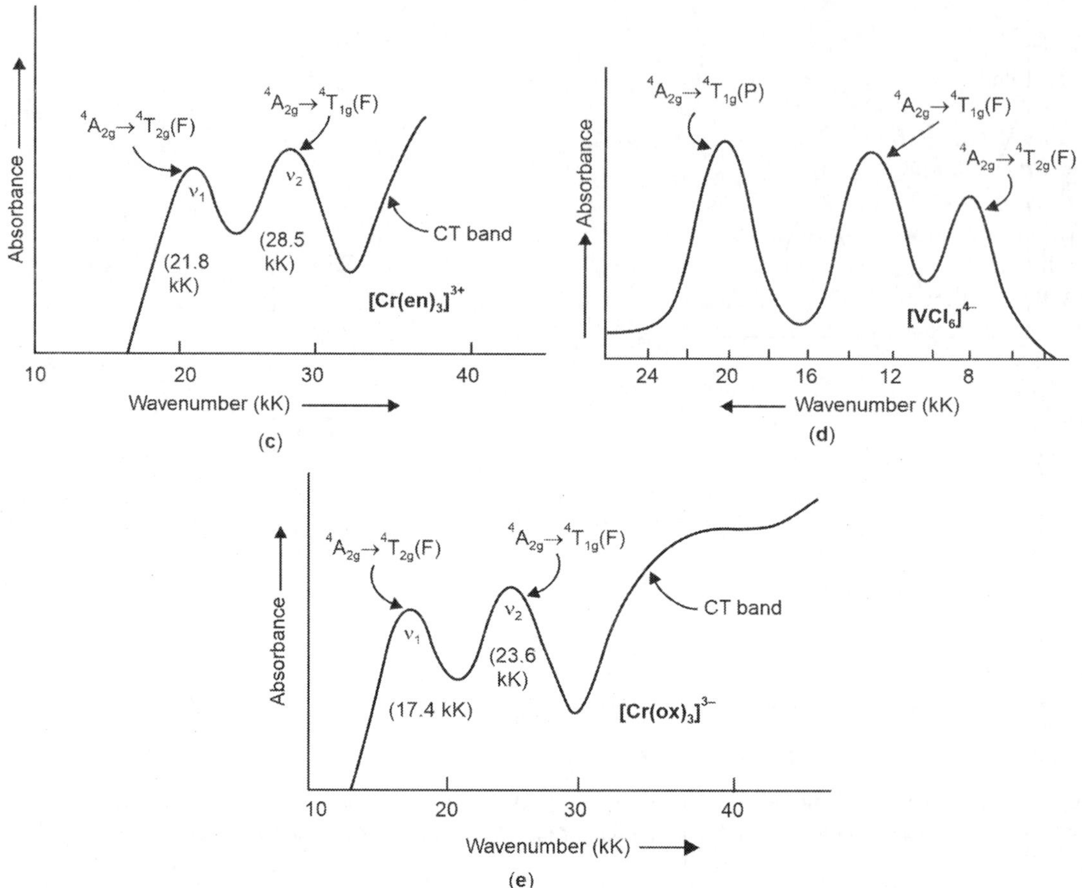

Fig. 7.14.5.3: (c) Absorption spectrum of $[Cr(en)_3]^{3+}$. (d) Absorption spectrum of $[VCl_6]^{4-}$. (e) Absorption spectrum of $[Cr(ox)_3]^{3-}$.

Note: Here it should be mentioned that the *tris*–chelates $[M(A\!-\!A)_3]$, *e.g.* $[Cr(en)_3]^{3+}$, $[Cr(ox)_3]^{3-}$ are belonging to the D_3 point group while the complexes like $[M(A)_6]$, *e.g.* $[Cr(NH_3)_6]^{3+}$ are belonging to the O_h point group. D_3 is a lower symmetry point group. The energy states T_{1g} and T_{2g} of the O_h point group are split in the D_3 point group. The correlation between the O_h and D_3 point groups is as follows (*cf.* **Correlation Table 7.23.2** and Table 7.15.4.2):

O_h	D_3 (lacking in centre of symmetry)	
$^4A_{2g}$	$^4A_2$	(see Fig. 7.5.1.2)
$^4T_{1g}$	$^4A_2 + {}^4E$	
$^4T_{2g}$	$^4A_1 + {}^4E$	

Thus the transitions will be modified due to splitting of the T_1 and T_2 states (see Fig. 7.5.1.2).

	O_h	D_3	
(*See* Fig. 7.5.1.2)	$^4A_{2g} \xrightarrow{\nu_1} {}^4T_{2g}$	$^4A_2 \to {}^4A_1, {}^4E$	(doublet)
	$^4A_{2g} \xrightarrow{\nu_2} {}^4T_{1g}(F)$	$^4A_2 \to {}^4A_2, {}^4E(F)$	(doublet)
	$^4A_{2g} \xrightarrow{\nu_3} {}^4T_{1g}(P)$	$^4A_2 \to {}^4A_2, {}^4E(P)$	(doublet)

Thus, all the three bands (ν_1, ν_2, ν_3) of the O_h ligand field should split into the doublets in the D_3 point group. But, in the ligand field spectra of $[Cr(en)_3]^{3+}$ and $[Cr(ox)_3]^{3-}$, no such splitting is noticed. (*cf.* Fig. 7.14.5.3c, e). In other words, the degree of splitting is too small to be reflected in the spectral transitions. ***This indicates that chelation has not effectively changed the effective octahedral symmetry in understanding the ligand field spectra (cf.* Sec. 7.15.4).**

D_3 symmetry (lacking in the centre of inversion) can allow the ***d–p* mixing** and the electronic transitions are **Laporte allowed**. But the molar extinction coefficients of $[Cr(ox)_3]^{3-}$ and $[Cr(en)_3]^{3+}$ are not remarkably increased compared with those of the O_h-complexes (*cf.* Table 7.15.4.1). Thus effectively they maintain the O_h-symmetry, *i.e.* 'CrO_6' in $[Cr(ox)_3]^{3-}$ and 'CrN_6' in $[Cr(en)_3]^{3+}$.

● **Group theory treatment of the D_3 point group and spectral transitions in $[Cr^{III}(AA)_3]$ (*cf.* Sec. 7.5.1N):** The dipole moment operator μ_z (along the C_3-axis) transforms as A_2 and μ_x, μ_y ($\perp$ to the C_3-axis) transform as E in a D_3 point group (*cf.* Character Table, Appendix 2A, Vol. 7).

The electronic transitions in a noncentrosymmetric system are allowed when the direct product $\psi_{el}^{gd}\mu\psi_{el}^{ex}$ contains the **totally symmetric A_1 irreducible representation**. The following electronic transitions are examined.

$A_2 \to A_1$ transition: $A_2 \times \mu_z \times A_1 = A_2 \times A_2 \times A_1 = A_1 \times A_1 = A_1$; *i.e.* **allowed transition** (||)
$A_2 \to A_2$ transition: $A_2 \times \mu_z \times A_2 = A_2 \times A_2 \times A_2 = A_2 \times A_1 = A_2$; *i.e.* forbidden transition (||)
$A_2 \to E$ transition: $A_2 \times \mu_z \times E = A_2 \times A_2 \times E = E \times A_1 = E$; *i.e.* forbidden transition (||)
$A_2 \to A_1$ transition: $A_2 \times \mu_{x,y} \times A_1 = A_2 \times E \times A_1 = E \times A_2 = E$; *i.e.* forbidden transition ($\perp$)
$A_2 \to A_2$ transition: $A_2 \times \mu_{x,y} \times A_2 = A_2 \times E \times A_2 = A_1 \times E = E$; *i.e.* forbidden transition ($\perp$)
$A_2 \to E$ transition: $A_2 \times \mu_{x,y} \times E = E \times E \times A_2 = E \times E = A_1 + A_2 + E$; *i.e.* **allowed transition** ($\perp$)

It indicates the **polarisation** (*i.e.* transitions are allowed only in certain directions) of the electronically allowed transitions in $[Cr^{III}(AA)_3]$ of D_3 symmetry (*see* 7.5.1, N). It may be noted that in *trans*-$[CoCl_2(en)_2]^+$, **vibronic polarisation** (dichroism) occurs. This aspect has been discussed earlier (Sec. 7.5.1, O).

(E) Spectral transitions in the tetragonally distorted Cr(III)–complexes (*cf.* Sec. 7.15.2 for details): Spectral transitions in the perfectly octahedral and tetrahedral complexes can be explained by considering the Orgel or Tanabe-Sugano diagrams. When the complexes are deviated from the cubic symmetry, many energy levels are split to complicate the interpretation. ***This is why, in the complexes of lower symmetry, more spectral transitions become possible.***

It has been already mentioned that the chelated complexes like $[Cr^{III}(LL)_3]$ possess the D_3–symmetry instead of the O_h symmetry. But spectra of such chelated complexes can be explained by considering the energy level diagrams of $[Cr^{III}(L)_6]$ complexes of O_h symmetry. However, the energy level diagrams of *cis/trans*–$[Cr^{III}A_4B_2]$ are quite different from that of the ideal octahedral complex of Cr(III). In fact, *cis*–$[Cr^{III}A_4B_2]$ belongs to the C_{2v} point group and *trans*–$[Cr^{III}A_4B_2]$ belongs to the D_{4h} point group. In moving from the O_h symmetry to the D_{4h} symmetry, many degenerate energy levels are split (*cf.* **Correlation Table** 7.15.4.2). It makes a larger number of spectral transitions possible in the D_{4h} molecules (*cf.* Fig. 7.14.5.5).

● **Splitting of energy levels in *cis*- and *trans*- isomers of $[MA_4B_2]$** (*cf.* Sec. 7.15.2): The *cis*- and *trans*- isomers of $[MA_4B_2]$ belong to the C_{2v} and D_{4h} symmetry point groups respectively but from the view point of electronic environment, **both the isomers, in practice, belong to the tetragonal one (D_{4h})** (*cf.* F. Teixidor and J. Casabo, *J. Chem. Edu.*, **64**, 461-62, 1987). **In reality, both the isomers show the tetragonal splitting of the O_h-symmetry but in the *cis*-isomers, the tetragonal splitting is only half of the splitting generated in the *trans*-isomer.** The order of splitting in the energy levels is of opposite nature for the isomers. Thus for a d^3-system, ***the three band spectrum of an octahedral complex becomes a six band spectrum in D_{4h} symmetry (i.e. tetragonal symmetry).*** However, the expected six bands are rarely seen. Moreover, the band splitting is noted for the *trans*-complexes not for the *cis*-complexes in which the tetragonal splitting is too small to generate the separate peaks.

In ***trans*–[Cr(en)$_2$F$_2$]$^+$**, the possible transitions are:

$$\left.\begin{aligned} &{}^4B_{1g} \xrightarrow{\nu_1} {}^4E_g(F)\,;\ {}^4B_{1g} \xrightarrow{\nu_2} {}^4B_{2g};\ {}^4B_{1g} \xrightarrow{\nu_3} {}^4E_g(F);\\ &{}^4B_{1g} \xrightarrow{\nu_4} {}^4A_{2g}(F)\,;\ {}^4B_{1g} \xrightarrow{\nu_5} {}^4A_{2g}(P);\ {}^4B_{1g} \xrightarrow{\nu_6} {}^4E_g(P) \end{aligned}\right\}\ (cf.\ \text{Fig. 7.14.5.5})$$

$\nu_1 = 18{,}500\ \text{cm}^{-1}$, $\nu_2 = 21{,}700\ \text{cm}^{-1}$, $\nu_3 = 25{,}300\ \text{cm}^{-1}$, $\nu_4 = 29{,}300\ \text{cm}^{-1}$, ν_5 and ν_6 are obscured by the strong charge transfer band in the far UV region. The expected values of ν_5 and ν_6 are:

$$\nu_5 = 41{,}000\ \text{cm}^{-1},\ \nu_6 = 43{,}750\ \text{cm}^{-1}$$

- **Comparison of the ligand field spectra of [Cr^{III}*A*$_4$*B*$_2$] (*cis*- and *trans*–isomers):** The geometrical isomers *cis*–[Cr^{III}A_4B_2] (C_{2v}) and *trans*–[Cr^{III}A_4B_2] (D_{4h}) can be distinguished by comparing their ligand field spectra. For the chelating ligand, the point group of *cis*-[Cr^{III}(en)$_2$Cl$_2$]$^+$ is C_2 and that of *trans*-isomer is D_{2h}. However, in terms of **local symmetry**, they can be considered as C_{2v} and D_{4h} respectively for explaining the spectral transitions. In such cases, splitting (measured by δ) of the *T* state is important. The splitting energy is determined by the average ligand field strength difference between the A—Cr—A and B—Cr—B axes (for the *trans*-complex) and between the A—Cr—A and A—Cr—B axes (for the *cis*- complex). It gives:

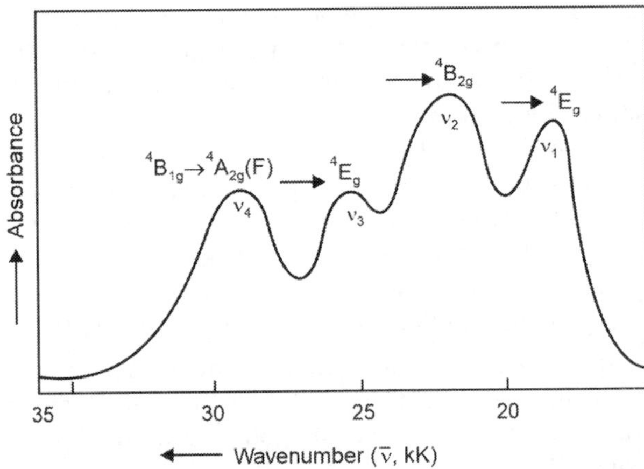

Fig. 7.14.5.4 Absorption spectrum of *trans*–[Cr(en)$_2$F$_2$]$^+$, (*cf.* Fig. 7.14.5.5 for the assignment of ν_1, ν_2, ν_3 and ν_4).

$$\delta(trans) : \delta(cis) = 2:1$$

In other words, we can say that this splitting energy (δ) depends on the relative positions of the ligands *A* and *B* in the spectrochemical series.

$$\boldsymbol{\delta(trans) = 2C(\Delta_A \sim \Delta_B),\ \delta(cis) = -C(\Delta_A \sim \Delta_B)}$$

$\Delta_A \sim \Delta_B$ gives the measure of the difference of the crystal field splitting powers of the ligands *A* and *B*. For the *trans*–isomer, the splitting is twice the splitting occurring in the corresponding *cis*–isomer and the minus sign indicates that the splitting pattern is inverted in the *cis*–isomer compared to that in the *trans*–isomer. **We can expect that for the *trans*–isomer, the peaks due to the transition to the $^4T_{1g}$ and $^4T_{2g}$ states (in the octahedral complex) will show the doublet character or two separate peaks but in the corresponding *cis*–isomer, this peak splitting is minimised because the *T* state does not split much in the *cis*–complex.** Thus, we can distinguish the *cis*– and *trans*– isomers by comparing their ligand field spectra. *It may be mentioned that the* ***peaks are more intense in the cis–complex*** *lacking in the centre of symmetry.* In the **noncentrosymmetric cis-isomers,** the electronic transitions are more allowed. *It gives also a better d–p mixing to relax the* ***Laporte selection rule*** *in the cis–complex compared to that in the trans–isomer having the centre of symmetry.*

If the ligands *A* and *B* occupy the positions close to each other in the spectrochemical series, *i.e.* $\Delta_A \sim \Delta_B$ is quite small, then analysis of the ligand field spectra cannot differentiate the *cis*- and

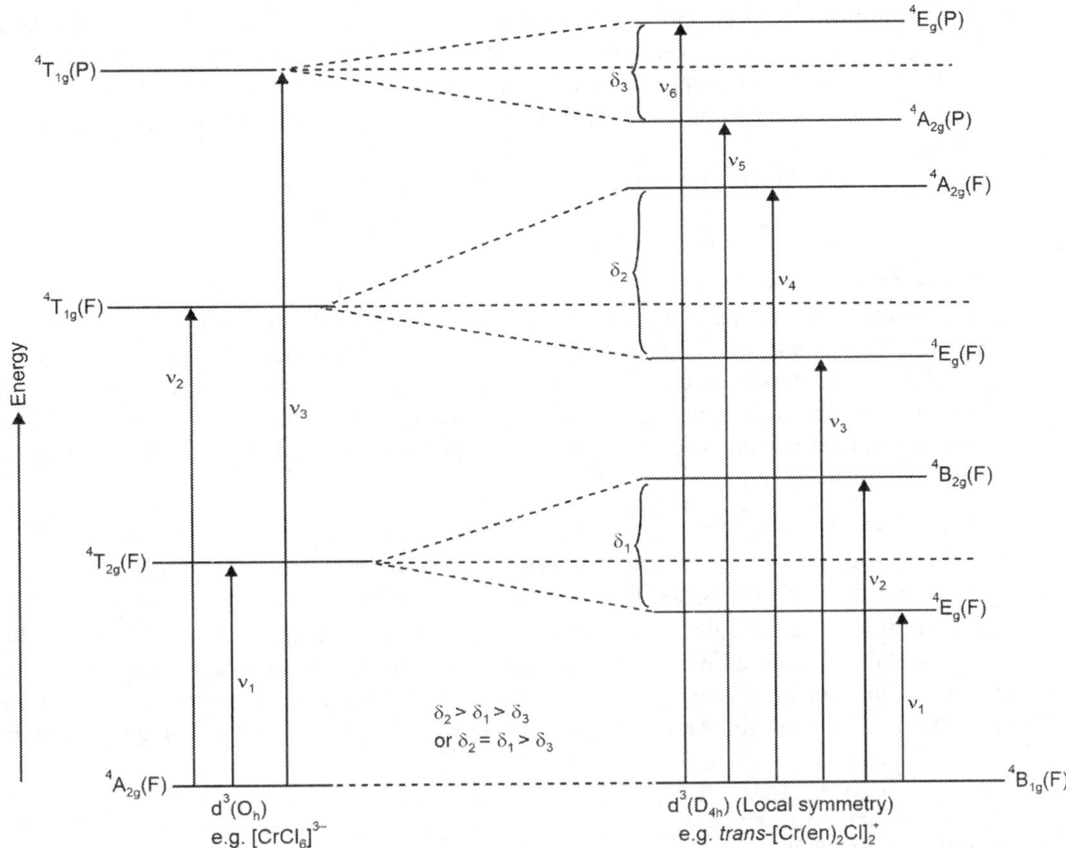

Fig. 7.14.5.5 Comparison of the spin-allowed transitions in the octahedral (O_h) and tetragonal (D_{4h}) complexes of d^3 system. (**Note:** There is a difference in opinion regarding the extent of splitting of the T_{1g} states and the order of A_{2g} and E_g states generated from the two T_{1g} states. These are discussed in Sec. 7.14.4 and Fig. 7.14.4.3).

trans-isomers. In fact, the geometrical isomers of **$[Cr(en)_2F_2]^+$** can be distinguished in this way but the isomers of **$[Cr(NCS)_2(OH_2)_4]^+$** cannot be distinguished in this way. This aspect has been discussed in Sec. 7.15.2a.

Table 7.14.5.2 Comparison of the spectral characteristics of *cis*- and *trans*- isomers of **$[Cr(en)_2F_2]^+$** (*cf.* F. Teixidor et al. *J. Chem. Edu.*, **64**, 461, 1987).

	Transitions in terms of the O_h symmetry		
Complex	$^4A_{2g} \xrightarrow[\text{(nm)}]{\nu_1} {}^4T_{2g}(F)$	$^4A_{2g} \xrightarrow[\text{(nm)}]{\nu_2} {}^4T_{1g}(F)$	$^4A_{2g} \xrightarrow[\text{(nm)}]{\nu_3} {}^4T_{1g}(P)$
cis-$[Cr(en)_2F_2]^+$	517	378	235
trans-$[Cr(en)_2F_2]^+$	530^a, 465^b	397^c, 351^d	248^e, 229^f

$\underbrace{a \Rightarrow {}^4B_{1g} \longrightarrow {}^4E_g,\ b \Rightarrow {}^4B_{1g} \longrightarrow {}^4B_{2g}}_{\text{Components of } \nu_1 \text{ in terms of the } D_{4h} \text{ symmetry}}$; $\underbrace{c \Rightarrow {}^4B_{1g} \longrightarrow {}^4E_g,\ d \Rightarrow {}^4B_{1g} \longrightarrow {}^4A_{2g}}_{\text{Components of } \nu_2 \text{ in terms of the } D_{4h} \text{ symmetry}}$;

$\underbrace{e \Rightarrow {}^4B_{1g} \longrightarrow {}^4A_{2g},\ f \Rightarrow {}^4B_{1g} \longrightarrow {}^4E_g}_{\text{Components of } \nu_3 \text{ in terms of the } D_{4h} \text{ symmetry}}$

Thus it is evident that the peaks undergo splitting for the *trans*-isomer but the corresponding peaks of the *cis*-compound undergo no splitting, *i.e.* O_h symmetry is practically maintained for the *cis*-isomer. The ν_1 transition corresponds to the $10Dq$ value (*i.e.* 517 nm ≡ 19,342 cm^{-1}) for the *cis*-isomer). By applying the **rule of average environment,** $10Dq$ value for *cis*-$[Cr(en)_2F_2]^+$ can be calculated as follows:

$$\nu_1 = 14{,}900 \text{ cm}^{-1} \text{ for } [CrF_6]^{3-};\ \nu_1 = 21{,}800 \text{ cm}^{-1} \text{ for } [Cr(en)_3]^{3+}$$

$$\nu_1 \equiv 10Dq \text{ of } cis\text{-}[Cr(en)_2F_2]^+ = \frac{2\times 14{,}900 + 4\times 21{,}800}{6} = 19{,}500 \text{ cm}^{-1}$$

Thus the calculated ν_1 value is in good conformity with the experimental value.

(F) Spin-forbidden transitions in d^3 (*e.g.* Cr^{III})–system: The spin-forbidden transitions (quartet to doublet) in the Cr(III)–complexes are quite important to understand their behaviour in **photochemistry** (*cf.* Chapter 6, Part II) and in ***ruby laser*.** Obviously, these spin-forbidden transitions are of low intensity and these are not important in interpreting the colour and absorption spectra of the Cr(III)–complexes.

In this connection, we are to consider the crystal field splitting of the ^2G term (of free ion) along with the ^4F and ^4P terms. The ^2G term splits as follows:

$$^2G(\text{free ion}, d^3) \xrightarrow{O_h \text{ field}} {}^2E_g, {}^2T_{1g}, {}^2T_{2g}, {}^2A_{1g}$$

These doublet terms obviously arise from the *spin-pairing* in the t_{2g} level. The $^2E_g(G)$ and $^2T_{1g}(G)$ states are of almost identical energy and thse two states are generally jointly represented by $^2E_g(G)$.

The ^2G term of free ion arises from the **low-spin configuration** of the t_{2g} level (*i.e.*, ⥮ ↿ –) where one of the t_{2g} orbitals is occupied by a pair of electrons and the 3rd electron remains in an unpaired condition in one of the remaining two t_{2g} orbitals. In the octahedral field, the ^2G term splits into 4-states: $^2E_g(G)$ and $^2T_{1g}(G)$ of almost the same energy, and $^2T_{2g}(G)$, $^2A_{1g}(G)$ of higher energy. The energy order is:

$$\underbrace{^2E_g(G),\ ^2T_{1g}(G)}_{(\equiv\, ^2E_g(G))} < {}^2T_{2g}(G) < {}^2A_{1g}(G).$$

The $^2E_g(G)$ and $^2T_{1g}(G)$ states are of almost the same energy and these are jointly represented by $^2E_g(G)$. The energy level diagram for both the quartet and doublet states of a d^3 (octahedral) system is shown in Fig. 7.14.5.6.

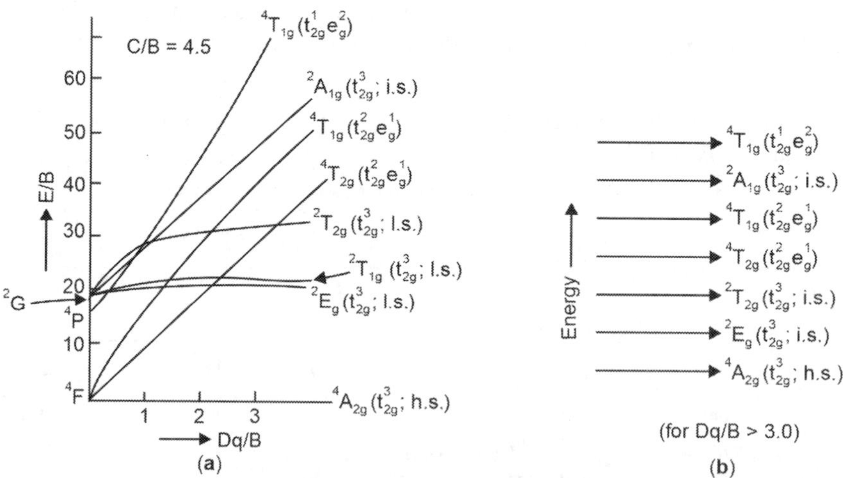

Fig. 7.14.5.6 (a) Tanabe-Sugano diagram of a d^3–octahedral system. (b) Relative energy order of different energy states of a d^3–octahedral system ($^2T_{1g}$, $^2E_g \equiv {}^2E_g$).

The characteristic features of the ^2E_g, $^2T_{1g}$ and $^2T_{2g}$ states are that **they run parallel to the ground state** ($^4A_{2g}$). These doublet states arise due to the rearrangement of electrons in the t_{2g} level. This is why, transition from the $^4A_{2g}$ state to these doublet states requires simply the electron rearrangement in the t_{2g} level and the transition energy does not depend on Δ which gives the measure of $t_{2g} - e_g$ separation. ***The metal-ligand bond distance remains more or less the same for both the ground $^4A_{2g}$ and excited ^2E_g, $^2T_{1g}$ and $^2T_{2g}$ states as the electronic configuration in the e_g level remains unchanged.*** Consequently, peak broadening due to the metal-ligand bond vibration does not occur in such spectral transitions (*cf.* Fig. 7.7.3.1 and Secs. 7.7.2–3). Δ value depends on the metal-ligand distance and a symmetric vibration (called **breathing mode**) (*i.e.* periodic contraction and elongation of the *M*—*L* bond) causes a continuous change of Δ within a limit. This is why, when energy of the two states (between which the spectral transition occurs) depends on Δ in different ways, the peak broadening will occur due to the *M*—*L* bond vibration. The mechanism of peak broadening by the *M*—*L* bond vibration has been discussed in detail in Secs. 7.7.2–3.

The following **spin-forbidden transitions** do not experience any peak broadening phenomenon due to the metal-ligand bond vibration. **These are very sharp peaks.**

$$^4A_{2g} \rightarrow {}^2E_g;\ {}^4A_{2g} \rightarrow {}^2T_{1g};\ {}^4A_{2g} \rightarrow {}^2T_{2g}$$

Photochemistry of the Cr(III) Complexes: Importance of the Excited ^2E_g State

The excited ^2E_g state plays a crucial role in controlling the photochemistry of the Cr(III) complexes. Photoexcitation gives a significant amount of population density in the ^2E_g state whose life time is relatively higher (*ca.* 10^{-3}–10^{-4} s) because the transition $^2E_g \rightarrow {}^4A_{2g}$ is spin-forbidden. Photochemical excitation of the ground $^4A_{2g}$ state to the quartet states will cause some population to the ^2E_g state (**ISC** intersystem crossing, quartet state to doublet state). Nonradiative deactivation of the higher quartet state will give the population density in the ^2E_g state. This ^2E_g state (low-spin t_{2g}^3 configuration) **bears a vacant t_{2g} orbital capable of receiving a nucleophilic attack and it is the cause of reactivity.** This aspect has been discussed in Chapter 6 (Part-II) in detail.

(G) **Role of Cr(III) in ruby laser: Role of the excited ^2E_g state (*cf.* Sec. 7.9.5):** In a ruby laser, a small amount of Cr(III) is embedded in α–Al_2O_3 lattice (corundum) in which Cr(III) experiences an octahedral field of 6 oxide O–sites. When a ruby crystal is irradiated to the radiation of an appropriate wavelength to excite Cr(III) from the $^4A_{2g}$ ground state to the $^4T_{2g}(F)$ state and partly to the $^4T_{1g}(F)$ state, the $^4T_{2g}(F)$ state deexcites through an **intersystem crossing** (ISC, a radiationless pathway of deexcitation) and comes to the ^2E_g state which possesses a relatively longer half-life time. The ISC process involves the spin-orbit coupling mechanism (Sec. 7.9.5). Thus the paths,

$$^4A_{2g}(F) \xrightarrow{h\nu} {}^4T_{2g}(F) \xrightarrow[\text{(lattice vibration)}]{\text{dexcitation } (ISC)} {}^2E_g(G)$$

$$^4A_{2g}(F) \xrightarrow{h\nu} {}^4T_{1g}(F) \xrightarrow[\text{conversion}]{\text{internal}} {}^4T_{2g}(F) \xrightarrow[\text{(intersystem crossing)}]{\text{Dexcitation}} {}^2E_g(G).$$

increase the population density (*i.e.* **population inversion**) in the excited $^2E_g(G)$ state. The $^4T_{2g}$(F) state corresponds to the $t_{2g}^2e_g^1$ configuration; the $^4T_{1g}(F)$ state also corresponds to the $t_{2g}^2e_g^1$ configuration; and the ground state term corresponds to the t_{2g}^3 configuration. It may be noted that the same electronic configuration is attained for both the $^4T_{2g}(F)$ and $^4T_{1g}(F)$ states for the relatively stronger crystal field (Fig. 7.10.2) and it happens so when Cr^{III} is doped in the corundum crystal; this aspect is discussed later. The transition to $^4T_{1g}(F)$ needs the absorption of violet light while the transition to $^4T_{2g}(F)$ needs the absorption of green light. **The absorption of violet and green light makes the ruby red.** The excited $^4T_{1g}(F)$ state is converted into the lower excited state $^4T_{2g}(F)$ through an **internal conversion.** The $^4T_{2g}(F)$ state is close to the ^2E_g state but they are of different spin multiplicities.

The excited ${}^4T_{2g}(F)$ state is converted through the lattice vibration to another excited state ${}^2E_g$ of different spin multiplicity. This conversion is described as the **nonradiative intersystem crossing (ISC)**. In the case of ruby, **the energy difference between the excited states ${}^4T_{2g}(F)$ and ${}^2E_g(G)$ matches appropriately the energy of a particular lattice vibrational mode.** This allows the ISC efficiently. Thus, the population density in the ${}^2E_g$ state is increased and this excited state acts as the **reservoir of energy.** This excited state can experience a radiative decay to the ground state through a spin-forbidden transition (*i.e.* **phosphorescence, *cf.* Sec. 7.9.5**). The return from the ${}^2E_g$ state to the ground state ${}^4A_{2g}$ is spin-forbidden. This enhances the life time of the ${}^2E_g$ state and at the cost of the exciting radiation, population density in the ${}^2E_g$ state may be more than that in the ground state. ***It leads to a population inversion.*** When the ${}^2E_g(G)$ state comes back to the ground state ${}^4A_{2g}(F)$, it gives the **monochromatic radiation** (16,000 cm^{-1}, 627 nm) because both the ${}^2E_g(G)$ and ${}^4A_{2g}(F)$ states run parallel in the Tanabe-Sugano diagram.

The spin-forbidden transition, ${}^2E_g \rightarrow {}^4A_{2g}$, gives the **red emission.** Both the ${}^2E_g$ (low-spin t_{2g}^3) and ${}^4A_{2g}$ (high-spin t_{2g}^3) states correspond to the t_{2g}^3configuration but spin orientations of the electrons in the two states are different. As the transition leads to the rearrangement of electrons in the t_{2g} level, the transition energy is independent of crystal field strength measuring the energy difference between the t_{2g} and e_g levels. **In fact, the spin-forbidden transition, ${}^2E_g \rightarrow {}^4A_{2g}$, in many octahedral Cr(III) complexes gives the phosphorescence of red light emission as in ruby emission.** If the ligands are rigid then life-time of the ${}^2E_g$ state may be several microseconds in solution. It happens so in the case of $[Cr(bpy)_3]^{3+}$.

In fact, return to the ground ${}^4A_{2g}(F)$ state from the ${}^2E_g(G)$ state is accompanied by *a stimulated emission of monochromatic radiation in phase with the stimulating radiation.* This gives the **laser beam.**

7.14.6 Six-Coordinate High-spin Complexes of 3d^7–Systems

Co(II) complexes are the main examples of this group. In fact, Co(II) can adopt both the high-spin complexes ($t_{2g}^5\,e_g^2$, ${}^4T_{1g}$ state) and low-spin complexes ($t_{2g}^6\,e_g^1$, ${}^2E_g$ state) depending on the relative values of 10Dq and pairing energy. In fact, only for the very strong field ligands (10Dq 〉 15000 cm^{-1}), the low-spin complexes of Co(II) may be formed. However, most of the six-coordinate Co(II) complexes are the high-spin complexes.

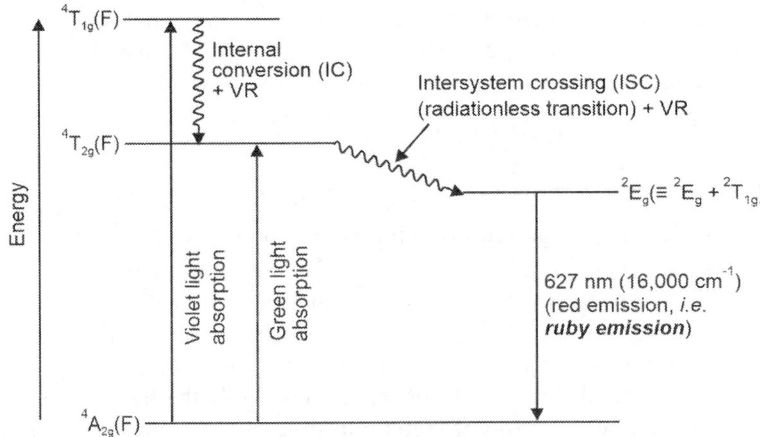

Fig. 7.14.5.7: Role of the ${}^2E_g$ state in **ruby emission** for the octahedral Cr(III) complexes in terms of Jablonski diagram. ${}^2E_g$, ${}^2T_{1g} = {}^2E_g$ (*cf.* Fig. 7.14.5.6). VR for vibrational relaxation (see Chapter 6, Part II)

Special colours through the isomorphous substitution of Al^{3+} by Cr^{3+} in colourless corundum (Al_2O_3) and spinel ($MgAl_2O_4$ and $BeAl_2O_4$) and beryl (*Ref.* Das *et al.*, *Resonance*, 2025)

The trivalent cations like Cr^{3+}, Ti^{3+} can lead to the isomorphous substitution of Al^{3+}, in Al_2O_3. But their sizes slightly differ: $r(Al^{3+}) = 67$ pm; $r(Cr^{3+}) = 75$ pm; $r(Ti^{+3}) = 81$ pm. Thus, Cr^{3+} and Ti^{3+} are larger than Al^{3+} and consequently, when Cr^{3+} and Ti^{3+} are placed in the positions of Al^{3+} in corundum, the Cr^{III}—O and Ti^{III}—O distances become shorter than the normal Cr^{III}—O and Ti^{III}—O distances. ***Because of this shorter M—O distance, Cr^{3+} and Ti^{3+} experience a larger crystal field splitting than the normal crystal field splitting in presence of the O^{2-} ligands.*** This is why, Cr^{3+} in corundum gives the unusual *red colour of ruby* (**blue shift,** *cf.* Cr_2O_3 green). This explains the ***red colour of ruby***. The **blue shift** in the ligand field transitions of Cr(III) doped into the Al_2O_3 structure increases under compression and this responsiveness can be utilised to measure pressure in high pressure experiments. In ruby, the **red phosphoresence emission** ($^2E_g \rightarrow {}^4A_{2g}$) intensifies the red colour of **ruby (red corundum)**. However, if any iron is present as an impurity, it will usually absorb the red phosphoresence emission. Hence, the finest rubies should be free from iron impurity in their crystal structure.

When Cr_2O_3 (about 1%) is added to spinel $MgAl_2O_4$, it produces **ruby red spinel** due to the **blue shifted** ligand field transitions in the $Cr^{III}O_6$ moieties present as the impurities in the host lattice structure of $MgAl_2O_4$ (*i.e.* partial replacement of Al(III) by Cr(III) at the octahedral sites of spinel structure). Cr(III) as an impurity in **alexandrite gemstone (*chrysoberyl mineral*,** spinel $BeAl_2O_4$), in the octahedral field, is responsible for its common greenish-yellow colour (*cf.* green colour of Cr_2O_3).

When Cr^{3+} is placed into the **beryl,** $Al_2Be_3[Si_6O_{18}]$ in the positions of Al^{3+}, the Cr—O distance becomes longer than the Cr—O distance in ruby. Thus Cr(III) in beryl experiences a weaker crystal field splitting. It explains the ***green colour of gemstone emerald*** (*cf.* green colour of Cr_2O_3).

For the high-spin d^7–system, Orgel diagram is similar to that of the d^2 system and the ground state is $^4T_{1g}$.

The spin-allowed transitions are:

$$^4T_{1g}(F) \xrightarrow{\nu_1} {}^4T_{2g}(F),\ \nu_1 = 8Dq + x' \approx 8Dq$$

$$^4T_{1g}(F) \xrightarrow{\nu_2} {}^4A_{2g}(F),\ \nu_2 = 18Dq + x' \approx 18Dq$$

$$^4T_{1g}(F) \xrightarrow{\nu_3} {}^4T_{1g}(P),\ \nu_3 = 6Dq + 15B' + 2x' \approx 6Dq + 15B'$$

$\nu_2, \nu_3 > \nu_1$; $\nu_2 >$ or $< \nu_3$; depending on the $10Dq$ values

(The symmetry interaction energy x' is quite negligible for the weak field ligands)

$$\nu_2 - \nu_1 = 10Dq,\ 15B' = \nu_3 + \nu_2 - 3\nu_1$$

Here it is worth mentioning that the transition (ν_2), $^4T_{1g}(F) \rightarrow {}^4A_{2g}(F)$, denotes the transition, $t_{2g}^5\, e_g^2 \rightarrow t_{2g}^3\, e_g^4$, *i.e.* a **two-electron transition. This is a forbidden transition.** This is why, it is a weak band.

(A) **Analysis of the ligand field spectra of $[Co(OH_2)_6]^{2+}$:** The absorption spectrum of light pink coloured $[Co(OH_2)_6]^{2+}$ gives a well defined weak peak at 8,100 cm^{-1} and a broad band at about 20,000 cm^{-1}. **Analysis of the broad band indicates the presence of three overlapping peaks at** 16,000 cm^{-1}, 19,400 cm^{-1} and 21,600 cm^{-1}. The peak assignments are as follows:

$$^4T_{1g}(F) \rightarrow {}^4T_{2g}(F),\ \nu_1 = 8{,}100 \text{ cm}^{-1},\ (\varepsilon = 1.3)$$

$$^4T_{1g}(F) \rightarrow {}^4A_{2g}(F),\ \nu_2 = 16{,}100 \text{ cm}^{-1},$$

$$^4T_{1g}(F) \rightarrow {}^4T_{1g}(P),\ \nu_3 = 19{,}400 \text{ cm}^{-1},\ (\varepsilon = 4.8)$$

The peak at 21,600 cm^{-1} is also believed to be originated either due to the spin-orbit coupling effect (*i.e.* splitting of the T–state, *i.e.* excited $^4T_{1g}$(P) state) or the spin-forbidden transition to a doublet state and spinel ($MgAl_2O_4$, $BeAl_2O_4$) and beryl (*Ref.* Das *et al.*, *Resonance*, 2025).

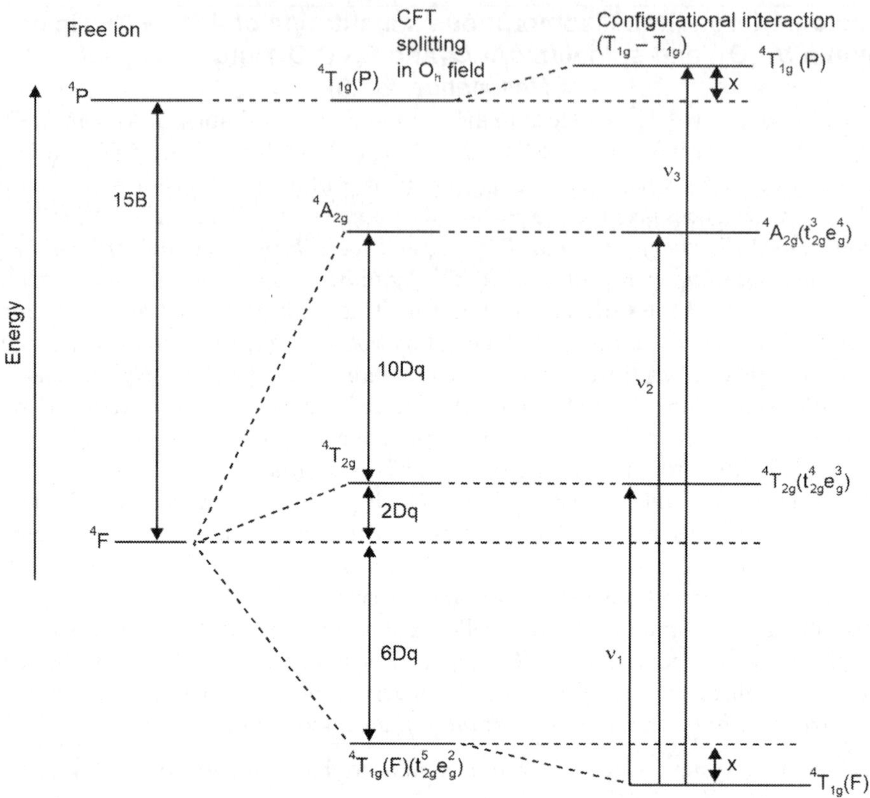

Fig. 7.14.6.1: Energy level diagram of the high-spin octahedral d^7–systems. **Note:** The $^4T_{1g}(P)$ and $^4A_{2g}(F)$ states may intersect depending on the Dq values. **Note:** $^4T_{1g}(P)$ and $^4A_{2g}(F)$ states may cross each other at a particular $10Dq$ value. When the $10Dq$ value crosses this particular value, $^4A_{2g}$ becomes of higher energy compared to $^4T_{1g}(P)$. x' indicates the configurational interaction energy for the symmetry interaction between the $^4T_{1g}(P)$ and $^4T_{1g}(F)$ states.

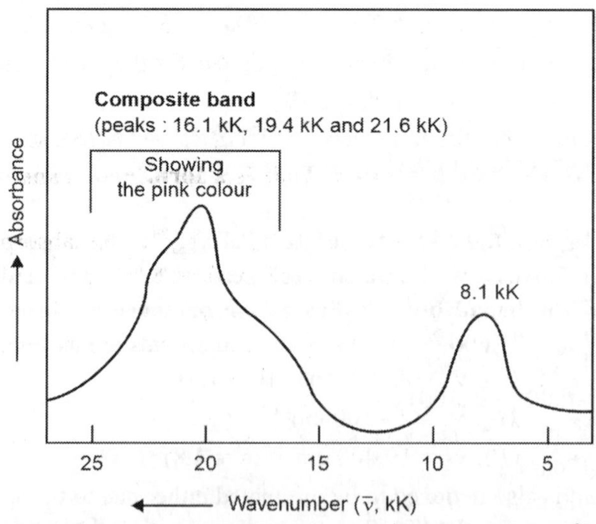

Fig. 7.14.6.2: Electronic absorption spectrum of $[Co(OH_2)_6]^{2+}$.

(**Note:** Some authors believe that the broad band at about 20,000 cm^{-1} consists of two peaks not three peaks. They suggest the following two transitions:

$$^4T_{1g}(F) \rightarrow {}^4A_{2g}(F),\ 19{,}600\ cm^{-1}$$

$$^4T_{1g}(F) \rightarrow {}^4T_{1g}(P),\ 21{,}600\ cm^{-1})$$

The said ν_2 (16,100 cm^{-1}) and ν_3 (19,400 cm^{-1}) peaks are closely spaced and they overlap. The ν_2-peak (2*e* transition, **a forbidden transition**) earns the intensity due to mixing with the ν_3 spin allowed transition (*cf.* **intensity stealing effect**). Close positions of the ν_2 and ν_3 peaks indicate that in $[Co(OH_2)_6]^{2+}$, the two states, *i.e.* $^4A_{2g}(F)$ and $^4T_{1g}(P)$, are in the **cross-over region.**

Now let us try to verify the assignments in terms of the Tanabe-Sugano diagram of the d^7-system.

$$\frac{\nu_3}{\nu_1} = \frac{^4T_{1g}(F) \longrightarrow {}^4T_{1g}(P)}{^4T_{1g}(F) - {}^4T_{2g}(F)} = \frac{19{,}400}{8{,}100} \approx 2.40$$

The above ratio fits in the corresponding Tanabe-Sugano diagram for $Dq/B' = 1.13$ (Fig. 7.10.4). This gives the following result.

$$\frac{^4T_{1g}(F) \longrightarrow {}^4T_{1g}(P)}{B'} = \frac{\nu_3}{B'} = 23.54 \quad \text{(obtained from the Tanabe-Sugano diagram for } Dq/B' = 1.13\text{).}$$

$$\frac{\nu_3}{B'} = \frac{19{,}400}{B'} = 23.54,\ i.e.\ B' = \frac{19{,}400}{23.54} = 824\ cm^{-1}\left(cf.\ B_0 = 971\ cm^{-1}\right)$$

$$\beta = \frac{B'}{B_0} = \frac{824}{971} = 0.85,\ i.e.\ B' \text{ is reduced to about 85\% of the free ion value } \left(B_0\right).$$

$$\frac{Dq}{B'} = 1.13 \text{ or, } Dq = B' \times 1.13 = 824 \times 1.13 = 931\ cm^{-1}$$

i.e. $10Dq = 9{,}310\ cm^{-1}$

From the experimental data, we get: $\frac{\nu_2}{\nu_1} = \frac{16{,}100}{8{,}100} = 1.99;\ \frac{\nu_3}{\nu_2} = \frac{19{,}400}{16{,}100} = 1.20$

For $\frac{Dq}{B'} = 1.13$, the calculated ratios (from the Tanabe-Sugano curve; *cf.* Fig. 7.10.4) are: $\frac{\nu_2}{\nu_1} = 2.14,\ \frac{\nu_3}{\nu_2} = 1.13$. These are in good agreement with the band assignment data.

We can calculate $10Dq$ as follows:

$$10Dq = \nu_2 - \nu_1 = 16{,}000 - 8{,}100 = 7{,}900\ cm^{-1}$$

This value is approximately close to 9,310 cm^{-1}. *But the agreement is not very good.* However, the given spectral assignments provide a reasonable account of the spectrum.

From the above peak assignments, we can calculate the configurational interaction energy (x') and other parameters as follows:

$$\nu_1 = 8{,}100\ cm^{-1} = 8Dq + x'$$

$$\nu_2 = 16{,}000\ cm^{-1} = 18Dq + x'$$

$$\nu_3 = 19{,}400\ cm^{-1} = 6Dq + 15B' + 2x'$$

$$10Dq = \nu_2 - \nu_1 = (16{,}000 - 8{,}100)\ cm^{-1} = 7{,}900\ cm^{-1};\ i.e.\ Dq = 790\ cm^{-1}$$

$$\nu_1 = 8{,}100\ cm^{-1} = 8Dq + x' = 8 \times 790 + x'$$

i.e. $x' = 1780\ cm^{-1}$

$$\nu_3 = 19{,}400\ cm^{-1} = 6 \times 790 + 15B' + 2 \times 1780 = 15B' + 8{,}300$$

or, $15B' = 11{,}100\ cm^{-1}$, *i.e.* $B' = 740\ cm^{-1}$

B. **Analysis of the ligand field spectra of $[CoF_6]^{4-}$ in the crystals of $KCoF_3$:** The spectral transitions are:

$$^4T_{1g}(F) \xrightarrow{\nu_1} {}^4T_{2g}(F),\ \nu_1 = 8Dq + x' = 7{,}150\ \text{cm}^{-1}$$
$$^4T_{1g}(F) \xrightarrow{\nu_2} {}^4A_{2g}(F),\ \nu_2 = 18Dq + x' = 15{,}200\ \text{cm}^{-1}$$
$$^4T_{1g}(F) \xrightarrow{\nu_3} {}^4T_{1g}(P),\ \nu_3 = 6Dq + 15B' + 2x' = 19{,}200\ \text{cm}^{-1}$$

Besides, the above three spin-allowed transitions, there are some **spin-forbidden transitions** (quartet to doublet). Analysis of the above results yields:

$$\nu_2 - \nu_1 = 10Dq = (15{,}200 - 7{,}150)\ \text{cm}^{-1} = 8{,}050\ \text{cm}^{-1},\ i.e.\ Dq = 805\ \text{cm}^{-1}$$

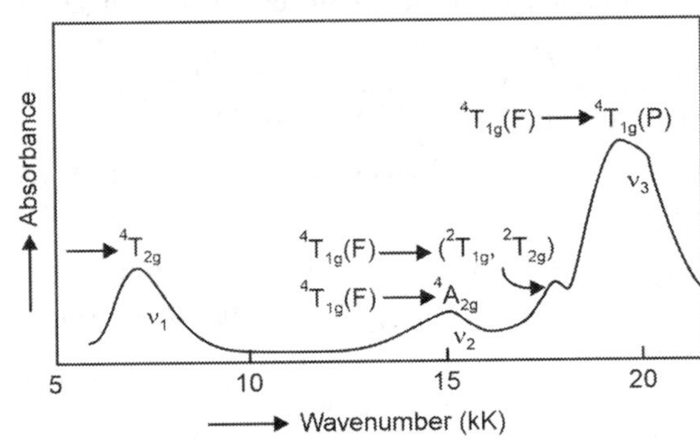

Fig. 7.14.6.3: Absorption spectrum of $[CoF_6]^{4-}$ in solid crystals of $KCoF_3$.

Calculation of $10Dq$ in this way is not affected by the configurational interaction.

$\nu_1 = 8Dq + x'$, or, $x' = \nu_1 - 8Dq = (7{,}150 - 8 \times 805)\ \text{cm}^{-1} = 710\ \text{cm}^{-1}$

$$\nu_3 = 15B' + 6Dq + 2x'.\ \text{or}\ B' = \frac{\nu_3 - 6Dq - 2x'}{15} = \frac{(19{,}200 - 6\times 805 - 2\times 710)\ \text{cm}^{-1}}{15} = 863\ \text{cm}^{-1}$$

$$\beta = \frac{B'}{B_0} = \frac{863}{970} = 0.88,\ i.e.\ B \text{ is reduced by } 12\%..$$

Here it is worth mentioning that Co(II) (h.s.) is highly **labile** and the species like $[CoF_6]^{4-}$ **cannot be studied in solution.**

C. **More rigorous analysis of the spectrum of high-spin octahedral d^7–systems:** Energies of the $^4T_{2g}(F)$ and $^4A_{2g}(F)$ states do not depend on the configurational interaction.

$$^4T_{2g}(F) \Rightarrow t_{2g}^4 e_g^3;\ {}^4A_{2g}(F) \Rightarrow t_{2g}^3 e_g^4;\ {}^4T_{1g}(P) \Rightarrow t_{2g}^4 e_g^3;\ {}^4T_{1g}(F) \Rightarrow t_{2g}^5 e_g^2$$

$$E\,[^4T_{2g}(F)] = +2Dq,\ E\,[^4A_{2g}(F)] = +12Dq.$$

The states $^4T_{1g}(F)$ and $^4T_{1g}(P)$ experience the configurational interaction and energy matrix for the two $^4T_{1g}$ states is given by:

$$\begin{vmatrix} -6Dq - E & 4Dq \\ 40Dq & 15B' - E \end{vmatrix} = 0$$

$$E = \tfrac{1}{2}[15B' - 6Dq \pm \{225B'^2 + 100(Dq)^2 + 180DqB'\}^{1/2}]$$
$$\nu_1[^4T_{1g}(F) \longrightarrow {}^4T_{2g}(F)] = \tfrac{1}{2}[\Delta - 15B' + (225B'^2 + 18B'\Delta + \Delta^2)^{1/2}]$$
$$\nu_2[^4T_{1g}(F) \longrightarrow {}^4A_{2g}(F)] = \tfrac{1}{2}[3\Delta - 15B' + (225B'^2 + 18B'\Delta + \Delta^2)^{1/2}]$$
$$\nu_3[^4T_{1g}(F) \longrightarrow {}^4T_{1g}(P)] = [(225B'^2 + 18B'\Delta + \Delta^2)^{1/2}]$$
$$\nu_3 - 2\nu_1 = 15B' - \Delta = 15B' - 10Dq$$

The ν_2–band involves a **two electron transition,** *i.e.* $t_{2g}^5 e_g^2 \rightarrow t_{2g}^3 e_g^4$ and this **forbidden transition** is of poor intensity. **This is why, for the spectral characterisation of the high-spin d^7 system, the ν_1 and ν_3 bands are quite important.**

Let us illustrate the above equations to analyse the spectrum of the high-spin octahedral complex, [Co(salicylaldehydebenzothiazolecarbohydrazide)$(OH_2)_2$] having $\nu_1 = 9100$ cm^{-1} and $\nu_3 = 18{,}500$ cm^{-1}.

$$\nu_3 - 2\nu_1 = 15B' - 10Dq$$

or, $18{,}500 - 2 \times 9{,}100 = 15B' - 10Dq$;

or, $18{,}500 - 18{,}200 = 15B' - 10Dq$ or, $15B' = 10Dq + 300$; or $B' = \dfrac{10Dq + 300}{15}$

$2\nu_1 = 10Dq - 15B' + [225B'^2 + 100(Dq)^2 + 180DqB']^{1/2}$.

Putting the values of ν_1 (= 9,100) and B' we get:

$$32(Dq)^2 + 960Dq - 34{,}21{,}600 = 0$$

or, $\quad Dq \approx 1019$ cm^{-1} (taking the positive value of Dq).

$$B' = \frac{10Dq + 300}{15} = \frac{10190 + 300}{15} = \frac{10{,}490}{15} \approx 699 \text{ cm}^{-1}$$

i.e. $\quad \beta = \dfrac{B'}{B_0} = \dfrac{699}{971} = 0.72.$

7.14.7 Six-Coordinate High-spin Complexes of 3d^4-System (*cf.* Sec. 3.11.4)

Complexes of Cr(II) and Mn(III) (with the weak field ligands) are the representative examples of this group. The ground state electronic configuration $t_{2g}^3 e_g^1$ indicates that it experiences the *z*-out Jahn-Teller distortion. The free ion term is ^5D which splits into the ^5E_g (lower) and $^5T_{2g}$ (higher) states in a weak octahedral field. If no distortion is considered, then we have the spectral transition:

$$^5E_g \rightarrow {}^5T_{2g}.$$

However, Jahn-Teller splitting of the states will occur as in the case of Cu^{2+}(d^9-system). The energy level diagram for a d^4 (high-spin) system is given in Fig. 7.14.7.1.

The possible electronic transitions are shown in Fig. 7.14.7.1. These transitions are:

$$^5B_{1g} \rightarrow {}^5A_{1g}\ (\nu_1);\ {}^5B_{1g} \rightarrow {}^5B_{2g}\ (\nu_2);\ {}^5B_{1g} \rightarrow {}^5E_g\ (\nu_3).$$

Table 7.14.7.1 Spectral features of some d^4-(high-spin) complexes.

Complex	$d_{z^2} \xrightarrow{\nu_1} d_{x^2-y^2}$ $^5B_{1g} \xrightarrow{\nu_1} {}^5A_{1g}$ **(cm^{-1})**	$d_{xy} \xrightarrow{\nu_2} d_{x^2-y^2}$ $^5B_{1g} \xrightarrow[(=10Dq)]{\nu_2} {}^5B_{2g}$ **(cm^{-1})**	$d_{xz}, d_{yz} \xrightarrow{\nu_3} d_{x^2-y^2}$ $^5B_{1g} \xrightarrow{\nu_3} {}^5E_g$ **(cm^{-1})**
$[Cr(OH_2)_6]^{2+}$	9,500	14,000	–
CrF_2 (solid crystal)	11,500	14,700	–
$[MnF_6]^{3-}$	9,000	17,400	19,600
*$[Mn(DMSO)_6]^{3+}$ (O-donor)	14,500	17,000	20,000
$[Mn(OH_2)_6]^{3+}$	–	–	21,000

* There are two types ν_{S-O} ir-stretching frequencies. This aspect has been discussed in Secs. 3.11.3–4, Vol. 4 in detail in terms of Jahn-Teller effect.

The blue coloured aqua complex $[Cr(OH_2)_6]^{2+}$ (chromus solution **unstable with respect to oxidation**) shows a peak at about 14,000 cm^{-1} which is generally considered as:

$$^5E_g \xrightarrow{\nu} {}^5T_{2g} \ (\nu = 14{,}000 \text{ cm}^{-1} = 10Dq).$$

But the finer analysis shows another band at 9,500 cm^{-1}. The peak assignments are:

$$\textbf{[Cr(OH}_2\textbf{)}_6\textbf{]}^{2+}\textbf{:}\ {}^5B_{1g} \xrightarrow{\nu_1} {}^5A_{1g} \ (9{,}500 \text{ cm}^{-1}),$$

$$^5B_{1g} \xrightarrow{\nu_2} {}^5B_{2g} \ (14{,}000 \text{ cm}^{-1}).$$

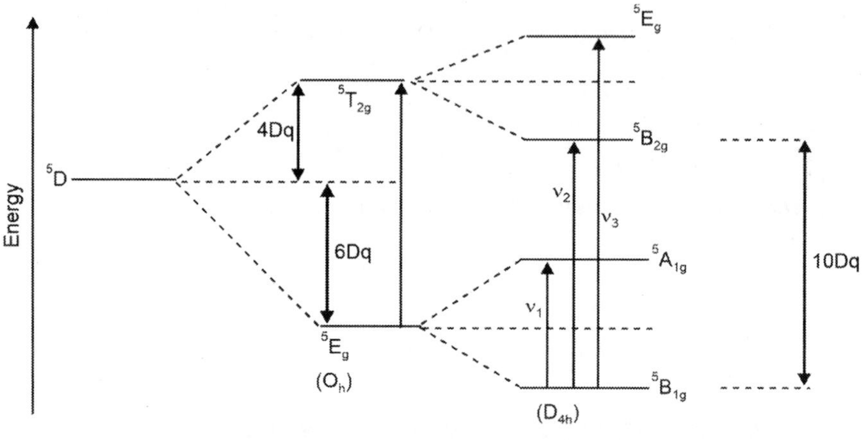

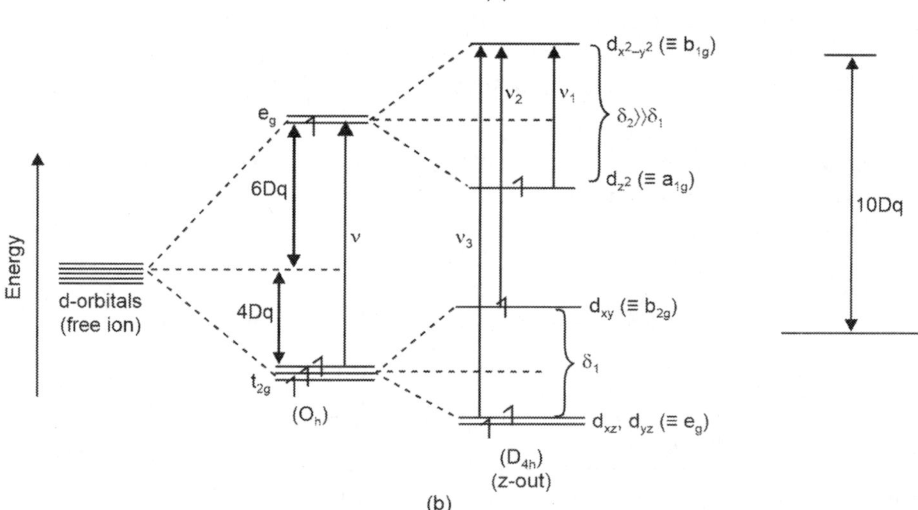

Relationship between the spectroscopic states and electronic configurations

$e_g^2b_{2g}^1a_{1g}^1b_{1g}^0 \equiv {}^5B_{1g}$; $e_g^2b_{2g}^1a_{1g}^0b_{1g}^1 \equiv {}^5A_{1g}$; $e_g^2b_{2g}^0a_{1g}^1b_{1g}^1 \equiv {}^5B_{2g}$; $e_g^1b_{2g}^1a_{1g}^1b_{1g}^1 \equiv {}^5E_g$

Fig. 7.14.7.1: Energy level diagram in the six-coordinate high-spin complexes of d^4-system. (a) Splitting of the spectroscopic states, (b) Splitting of the *d*-orbitals *Further splitting of ^5E_g of D_{4h} into ^5A_g and ^5B_g gives the C_{2h} (rhombic) symmetry.

Probably, the ν_2 ($d_{xy} \rightarrow d_{x^2-y^2}$) and ν_3 ($d_{xz}, d_{yz} \rightarrow d_{x^2-y^2}$) peaks are close in positions because splitting of the t_{2g} level is less important, *i.e.* $\delta_2 \rangle\rangle \delta_1$ (*cf.* splitting of the e_g level is more significant). **This**

is why, it is difficult to characterize the ν_2 and ν_3 bands separately. Spectral characteristics of some d^4-systems (high-spin) are given in Table 7.14.7.1.

A regular octahedral geometry for the Cr(II) complexes should show a **single spin allowed transition** (*i.e.* $^5E_g \rightarrow {}^5T_{2g}$) but very often, the Cr(II) complex (high spin) experiences the distortion (giving rise to splitting of the energy levels) to show the multiple peaks which may overlap. However, the spectrum of $[CrCl_6]^{4-}$ ($CrCl_2$ in molten $AlCl_3$ at 230°C) shows a single absorption (as expected for the O_h symmetry) at 10.2 kK without any apparent splitting (Fig. 7.14.7.2).

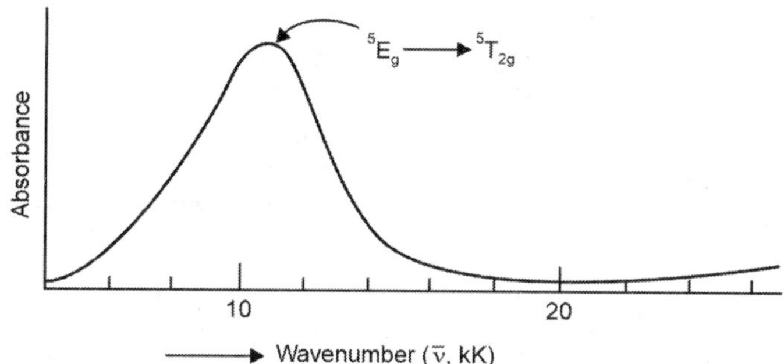

Fig. 7.14.7.2: Absorption spectrum of $[CrCl_6]^{4-}$ ($CrCl_2$ in molten $AlCl_3$ at 230°C and 5.5 atm pressure).

Here it may be mentioned that Cr(II) (d^4 system, h.s.) is highly **labile.** Consequently, the species like $[CrCl_6]^{4-}$ cannot be studied and chacterised in aqueous solution.

7.14.8 Six-Coordinate High-Spin Complexes of 3d^6-System (*cf.* Sec. 3.11.4, Vol. 4)

Fe(II) – complexes are the important examples in this group. **Two high-spin complexes of Co(III), *i.e.* $[CoF_6]^{3-}$ and $[CoF_3(OH_2)_3]$** are also known to belong to this group.

Both the ground state ($t_{2g}^4e_g^2$) and excited state ($t_{2g}^3e_g^3$) are prone to experience the J.T. distortion. Obviously, the excited state is more sensitive to the J.T. distortion because in the excited state, origin of the distortion lies at the e_g-level while in the ground state, origin of the distortion lies at the t_{2g} level.

If the high-spin complexes like $[Fe(OH_2)_6]^{2+}$, $[CoF_6]^{3-}$, etc. are supposed to have the perfectly octahedral geometry, we can predict the following spectral transition.

$$^5T_{2g}\left(t_{2g}^4e_g^2\right) \xrightarrow[(=10Dq)]{\nu} {}^5E_2\left(t_{2g}^3e_g^3\right)$$

However, in terms of Jahn-Teller distortion, there are **three possible spin allowed spectral transitions (Fig. 7.14.8.1)**:

the ν_1 (due to splitting of the t_{2g} level) lies in the ir-range, ν_2 and ν_3 lie in the visible range.

Very often, ν_2 and ν_3 overlap to produce a **broad peak with a shoulder.** But in some cases, the resultant broad peak may show a clear doublet structure as in $[CoF_6]^{3-}$.

$[Fe(OH_2)_6]^{2+}$ (pale green) shows a doublet peak (10,400 cm^{-1} and a shoulder at 8,300 cm^{-1}). It indicates the peak due to the transition, $^5T_{2g} \rightarrow {}^5E_g$ (in the hypothetical octahedral complex), is split into a doublet by about 2000 cm^{-1}. **$[CoF_3(OH_2)_3]$** shows a broad peak at about 13,000 cm^{-1}. For **$[CoF_6]^{3-}$**, the two well defined peaks appear at 11,800 cm^{-1} and 14,400 cm^{-1}. ***In such cases, conventionally, 10Dq is determined from the average position of the two peaks.*** Thus, for **$[CoF_6]^{3-}$**, the 10Dq value is taken as $\frac{1}{2}(14,400+11,800)$ cm^{-1} = 13,100 cm^{-1}. It is assumed that the **single ideal octahedral peak** of $[CoF_6]^{3-}$ arises at 13,100 cm^{-1}.

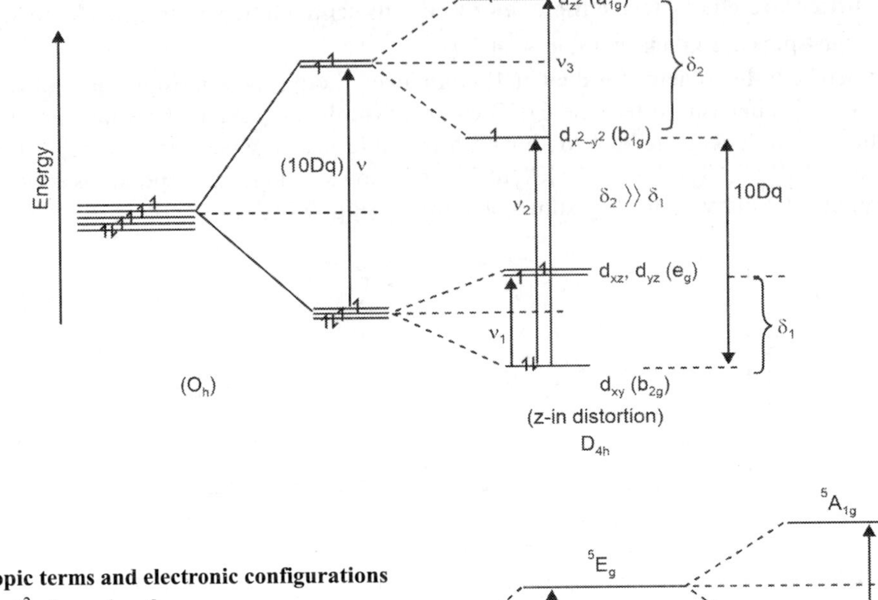

Spectroscopic terms and electronic configurations

$d_{xy}^2\left(d_{xz}, d_{yz}\right)^2 d_{x^2-y^2}^1 d_{z^2}^1 \equiv {}^5B_{2g}$;

$d_{xy}^1\left(d_{xz}, d_{yz}\right)^3 d_{x^2-y^2}^1 d_{z^2}^1 \equiv {}^5E_{g}$;

$d_{xy}^1\left(d_{xz}, d_{yz}\right)^2 d_{x^2-y^2}^2 d_{z^2}^1 \equiv {}^5B_{1g}$;

$d_{xy}^1\left(d_{xz}, d_{yz}\right)^2 d_{x^2-y^2}^1 d_{z^2}^2 \equiv {}^5A_{1g}$

Energy ${}^5D$ ${}^5E_g$ 6Dq 4Dq ${}^5T_{2g}$ ${}^5A_{1g}$ ${}^5B_{1g}$ ${}^5E_g$ ${}^5B_{2g}$ $ν_1$ $ν_2$ $ν_3$ 10Dq

Fig. 7.14.8.1a: Possible spectral transitions in high-spin six-coordinate complexes of d^6-system.

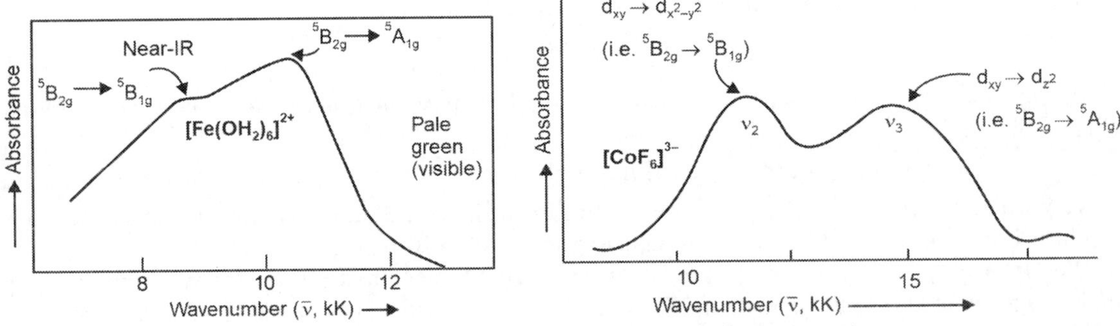

Fig. 7.14.8.1b: Absorption spectra of $[Fe(OH_2)_6]^{2+}$ and $[CoF_6]^{3-}$.

Pale green $\mathbf{[Fe(OH_2)_6]^{2+}}$ shows a broad and weak absorption band at about 10,000 cm^{-1} which actually consists of two peaks at 10,400 cm^{-1} and 8,300 cm^{-1} (appearing as a shoulder) (J.T. effect). It leads to the 10Dq value as $\frac{1}{2}$(10,400 + 8,300) ≈ 9,400 cm^{-1}. The case may be compared with the d^1-systems, *e.g.* $[Ti(OH_2)_6]^{3+}$.

However, it may be pointed out that from the splitting pattern of the *d*-orbitals, the ν_2 peak (*i.e.* lower energy peak, *e.g.* 8,300 cm^{-1} for $[Fe(OH_2)_6]^{2+}$, 11,800 cm^{-1} for $[CoF_6]^{3-}$) should represent the 10*Dq* value.

Spectral features of some representative high-spin d^6-complexes are given in Table 7.14.8.1.

7.14.8.1 Spectral features of some high-spin Fe(II) and Co(III) complexes.

Complex	$^5T_{2g} \rightarrow {}^5E_{2g}$ (cm^{-1}) (with a doublet structure)	
	$d_{xy} \xrightarrow{\nu_2} d_{x^2-y^2}$ $^5B_{2g} \longrightarrow {}^5B_{1g}$	$d_{xy} \xrightarrow{\nu_3} d_{z^2}$ $^5B_{2g} \longrightarrow {}^5A_{1g}$
$[Fe(OH_2)_6]^{2+}$	8,300	10,400
$[FeCl_2, 4H_2O]$ *i.e.* $[FeCl_2(OH_2)_4]$	6,800	11,400
$[FeCl_2(phen)_2]$	8,470	10,510
$[FeCl_2(py)_4]$	8,550	10,300
$[FeBr_2(py)_4]$	7,750	10,650
$[CoF_6]^{3-}$	11,800	14,400

Note: The octahedral complexes of Fe(II) with the unsaturated amines like bpy, phen, etc. are highly coloured due to the **metal to ligand charge transfer band**. These transitions are: $t_{2g} \rightarrow \pi^*$ (of the unsaturated ligand.)

- **Presence of Fe(II) as an impurity** in beryl ($Be_3Al_2[Si_6O_{18}]$), gives **blue-green aquamarine** while presence of Mn(II) produces **pink morganite**. The colour of blue-green aquamarine and pink morganite can be explained from the octahedral ligand field transitions of Fe(II) (h.s. d^6) and Mn(II) (h.s. d^5) respectively. Fe(II) as an impurity replacing 6-coordinate Mg(II) in $Mg(SiO_4)$ is responsible for the **yellow-green colour of peridot**. In **red garnet**, Fe(II) as an impurity replacing 8-coordinate Mg(II) in $Mg_3Al_2(SiO_4)_3$ is responsible for the colour (*Ref.* Das *et al.*, *Resonance*, 2025).

7.14.9 Six-Coordinate High-Spin Complexes of $3d^5$-System

The Mn(II) and Fe(III) complexes are the major examples of this group. The free ion ground state term ^6S generates the $^6A_{1g}$ ground state level. There is no excited state level having the spin multiplicity 6. Thus, ***there is no spin-allowed transition.*** The important excited states bear the spin multiplicity 4 (*i.e.* quartet). Before to discuss the spin-forbidden transitions, we are to consider the characteristic features of the energy level diagram of the high-spin complexes of d^5 system.

- **Free ion-terms:** The important free ion terms are ^6S (ground state), ^4G, ^4D, ^4P and ^4F. Their energy order is:

  $$\underset{(L=0)}{^6S} < \underset{(L=4)}{^4G} < \underset{(L=1)}{^4P} < \underset{(L=2)}{^4D} < \underset{(L=3)}{^4F}$$

  The first excited state (^4G) and the ground state (^6S) are in the expectation of Hund's rule. *But the higher excited states do not follow the energy order as predicted from the Hund's rules* (*see* Sec. 7.9.3).
- **Splitting of the free ion terms in an octahedral field:**
  $^6S \rightarrow {}^6A_{1g}$; $^4G \rightarrow {}^4A_{1g}$, ^4E_g, $^4T_{1g}$, $^4T_{2g}$; $^4P \rightarrow {}^4T_{1g}$; $^4F \rightarrow {}^4A_{2g}$, $^4T_{1g}$, $^4T_{2g}$; $^4D \rightarrow {}^4E_g$, $^4T_{2g}$.
- **Configurational interaction** (*i.e.* **symmetry interaction**):
  (i) $^4T_{1g}(G)$, $^4T_{1g}(P)$, $^4T_{1g}(F)$; (ii) $^4T_{2g}(G)$, $^4T_{2g}(D)$, $^4T_{2g}(F)$; (iii) $^4E_g(G)$, $^4E_g(D)$

  The $^4T_{1g}$ *states experience a higher configurational interaction than the* $^4T_{2g}$ *states* because of the higher symmetry of $^4T_{1g}$ states. The $^4T_{1g}(P)$ *state is sandwiched between the upper* $^4T_{1g}(F)$ *and lower* $^4T_{1g}(G)$ *states.* The interaction between the $^4T_{1g}(P)$ and $^4T_{1g}(G)$ states is relatively more important than the interaction between the $^4T_{1g}(P)$ and $^4T_{1g}(F)$ states because the $^4T_{1g}(P)$ and $^4T_{1g}(G)$ states are closer in energy. The repulsive interaction between the $^4T_{1g}(P)$ and $^4T_{1g}(G)$

states raises the $^4T_{1g}(P)$ state with the increase of Dq value. Then the raised $^4T_{1g}(P)$ state begins to experience a repulsive interaction (*i.e.* lowering of its energy) from the $^4T_{1g}(F)$ state. Thus, at the higher Dq values, the $^4T_{1g}(P)$ state experiences the repulsive interactions of opposite nature from the higher energy $^4T_{1g}(F)$ state and the lower energy $^4T_{1g}(G)$ state. ***This is why, after a certain Dq value, the $^4T_{1g}(P)$ state runs parallel to the ground $^6A_{1g}$ state. This is why, energy curve of the sandwiched $^4T_{1g}(P)$ state is of sigmoidal nature.***

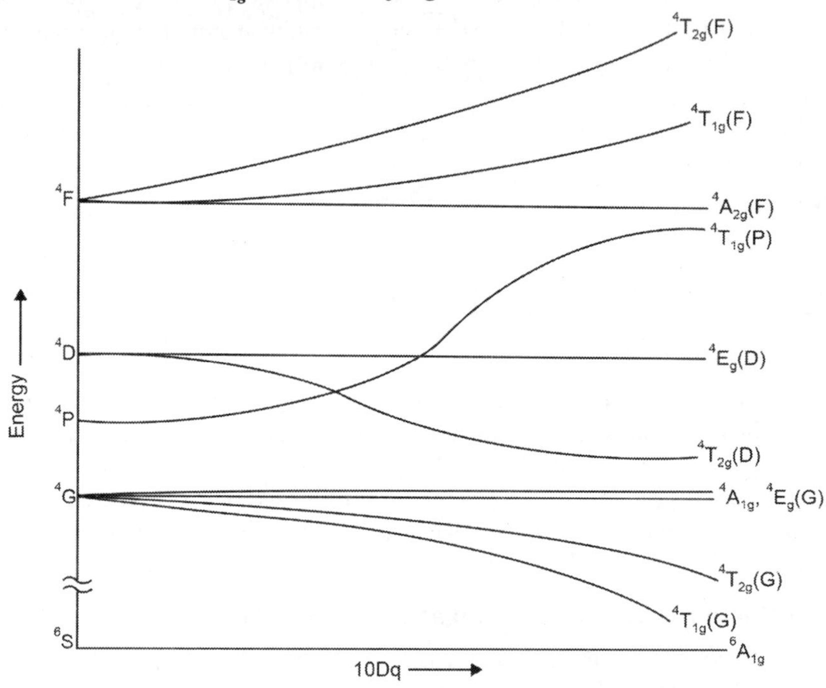

Fig. 7.14.9.1 Energy diagram for the d^5-ion in high-spin octahedral complexes.

Because of the same ground (discussed above), the **sandwiched** $^4T_{2g}(D)$ state between the $^4T_{2g}(F)$ (higher energy) and $^4T_{2g}(G)$ (lower energy) states experiences the **sigmoidal variation.** Initially, the repulsive interaction between the $^4T_{2g}(F)$ and $^4T_{2g}(D)$ states is more important. The configurational interaction between the $^4E_g(G)$ and $^4E_g(D)$ states is not so important. *Here it may be pointed out that the configurational interaction increases with the orbital degeneracy and it follows the order: T > E > A.* ***In fact, the two E_g states run parallel to the ground state, $^6A_{1g}$.***

- **Degenerate character of the $^4A_{1g}(G)$ and $^4E_g(G)$ states:** These two excited states are accidentally degenerate. Consequently, these two transitions produce a single peak.

$$^6A_{1g} \rightarrow {}^4A_{1g}(G),\ {}^4E_g(G)$$

- **Bandwidth:** Band broadening due to the vibronic coupling depends on the relative slopes (in the plot, energy *vs.* Dq) of the two states between which the transition occurs (*cf.* Fig. 7.7.3.1). The four excited energy states, *i.e.* $^4E_g(G)$, $^4A_{1g}(G)$, $^4E_g(D)$, $^4A_{2g}(F)$ run parallel to the ground state $^6A_{1g}(S)$. In such cases, the relative slopes of the states between which the transition occurs is zero and **the peaks are sharp** (see Fig. 7.7.3.1).

$$^6A_{1g} \rightarrow {}^4E_g(G),\ {}^4A_{1g}(G);\ {}^6A_{1g} \rightarrow {}^4E_g(D);\ {}^6A_{1g} \rightarrow {}^4A_{2g}(F).$$

In other words, the above transition energies are independent of the magnitude of Dq. Here it is worth mentioning that among the above mentioned three transitions, the last two transitions produce the bands which are not as sharp as the band produced in the first transition.

On the other hand, the excited states $^4T_{1g}(G)$ and $^4T_{2g}(G)$ do not run parallel to the ground state $^6A_{1g}$. Consequently, transitions to the $^4T_{1g}(G)$ and $^4T_{2g}(G)$ states produce the **relatively broader peaks.**

In fact, the excited states $^4A_{1g}$ and ^4E_g, and the ground state $^6A_{1g}$ originate from the **same electronic configuration** $t_{2g}^3e_g^2$ but different spin multiplicities. Consequently, equilibrium bond distance remains more or less the same for both the ground and excited states (*i.e.* peak broadening by metal-ligand bond vibration does not occur). But the excited states $^4T_{1g}$ and $^4T_{2g}$ originate from the electronic configuration $t_{2g}^4e_g^1$ which is different from the ground state electronic configuration $t_{2g}^3e_g^2$. Thus, the transitions $^6A_{1g} \rightarrow {}^4T_{1g}$; $^6A_{1g} \rightarrow {}^4T_{2g}$ involve a change in the equilibrium metal-ligand bond distance. It causes **peak broadening** through the metal-ligand bond vibration (*cf.* Secs. 7.7.1-3).

In fact, for high spin Mn(II), some of the transitions involve only the **electronic rearrangement** (in terms of spin orientation) without any change of electron density in the t_{2g} and e_g levels. Different possible electronic arrangements of a d^5-system in an octahedral field (high spin) are shown in Fig. 7.14.9.2.

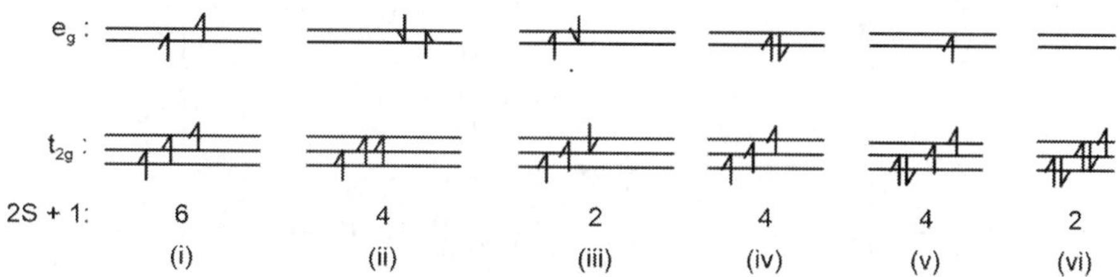

Fig. 7.14.9.2: Different possible electronic arrangements for a d^5-system in an octahedral field (high spin).

The spin forbidden transitions, (i) → (ii), (iii), (iv) do not require any electron excitation from the t_{2g} level to the e_g-level and all these states are having the same electronic configuration $t_{2g}^3e_g^2$ with different spin multiplicities. These transitions will produce the **sharp peaks** (*cf.* Secs. 7.7.2–3). The spin forbidden transitions, (i) → (v), (vi) will produce the **broader peaks** because electron density in the e_g-level is different in the states between which the electron transition occurs. In such cases, the metal-ligand bond length is different in the excited and ground electronic states.

The characteristics of ligand field spectra of $[Mn(OH_2)_6]^{2+}$, $[Fe(OH_2)_6]^{3+}$, $[FeF_6]^{3-}$ are given in Table 7.14.9.1.

- **Intensity:** In the high-spin d^5-complexes, the transitions are both Laporte forbidden and spin forbidden. Consequently, the peaks are of **very poor intensity.**
- **Spectral analysis of $[Mn(OH_2)_6]^{2+}$:** The peak positions given in Table 7.14.9.1 fit in the Tanabe-Sugano diagram for $Dq/B' = 1.1$. For the following transitions, we get the result:

$$\frac{^6A_{1g} \longrightarrow {}^4T_{1g}(G)}{B'} = \frac{18{,}900}{B'} = 24.0 \text{ (from the Tanabe-Sugano diagram, for } Dq/B' = 1.1)$$

It leads to: $B' = 787.5 \text{ cm}^{-1}$

and $Dq = 1.1 \times 787.5 \approx 866 \text{ cm}^{-1}$, *i.e.* 10Dq = 8660 cm^{-1}

$$\beta = \frac{B'}{B_0} = \frac{787.5}{960} \approx 0.82$$

i.e. B is reduced to about 80% of the free ion value.

Table 7.14.9.1: Characteristics of ligand field transitions of some high-spin d^5-complexes

Transitions **(cm^{-1})**	***[Mn(OH$_2$)$_6$]$^{2+}$*** **(cm^{-1})**	***[Fe(OH$_2$)$_6$]$^{3+}$*** **(cm^{-1})**	***[FeF$_6$]$^{3-}$*** **(cm^{-1})**
$^6A_{1g}\rightarrow{}^4T_{1g}(G)$	18,900 (0.013)	12,600	16,600
$^6A_{1g}\rightarrow{}^4T_{2g}(G)$	23,100 (0.009)	18,500	21,000
$^6A_{1g}\rightarrow{}^4E_g(G)$	24,970 (0.031)	24,300	26,500
$^6A_{1g}\rightarrow{}^4A_{1g}(G)$	25,300 (0.014)	24,600	–
$^6A_{1g}\rightarrow{}^4T_{2g}(D)$	28,000 (0.018)	28,800	29,800
$^6A_{1g}\rightarrow{}^4E_g(D)$	29,750 (0.013)	–	32,000
$^6A_{1g}\rightarrow{}^4T_{1g}(P)$	32,960 (0.02)	30,200	–
$^6A_{1g}\rightarrow{}^4A_{2g}(F)$	35,400	–	
$^6A_{1g}\rightarrow{}^4T_{1g}(F)$	36,900	–	
$^6A_{1g}\rightarrow{}^4T_{2g}(F)$	40,820	–	

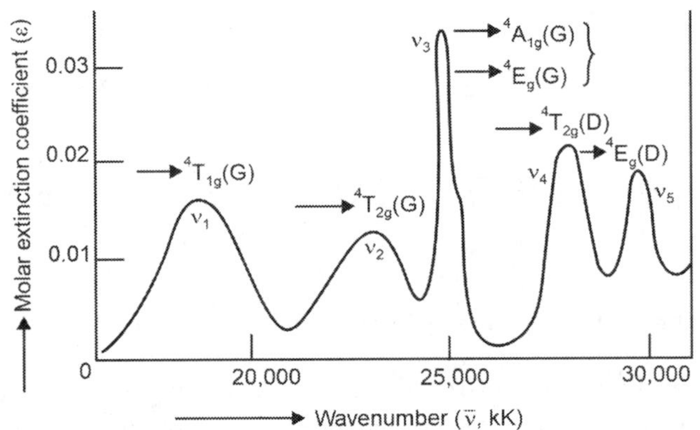

Fig. 7.14.9.2: Electronic absorption spectrum of $[Mn(H_2O)_6]^{2+}$.

Generally, for the high-spin Mn(II) complexes, the three lowest energy transitions are assigned and then the values of Dq, B and C are calculated therefrom to predict the positions of the higher energy transitions. The three lowest energy transitions are:

$$^6A_{1g}\xrightarrow{\nu_1}{}^4T_{1g}(G),\quad {}^6A_{1g}\xrightarrow{\nu_2}{}^4T_{2g}(G),\quad {}^6A_{1g}\xrightarrow{\nu_3}{}^4E_g,\ {}^4A_{1g}(G)$$

For $[Mn(OH_2)_6]^{2+}$, the ν_3-band (24,970 cm^{-1}) involves the energy change $10B' + 5C$, *i.e.* $10B' + 5C = 24{,}970$ cm^{-1}.

Taking $Dq = 866$ cm^{-1} and $B' = 787$ cm^{-1} (obtained by fitting the spectral data in the Tanabe-Sugano diagram), we can evaluate C.

$$10B' + 5C = 24{,}970;\ \text{or, } 10 \times 787 + 5C = 24{,}970;\ \text{or, } C = \frac{24{,}970 - 7{,}870}{5} = 3{,}420\ \text{cm}^{-1}.$$

For the transition, $^6A_{1g} \rightarrow {}^4E_g(D)$, the energy change is given by: $17B' + 5C = 17 \times 787 + 5 \times 3{,}420 = 30{,}479$ cm^{-1} (*cf.* experimental value, 29,750 cm^{-1}).

Note: Energy difference between the different states in terms of Racah parameters (A, B, C) or Condon-Shortley parameters (F_0, F_2, F_4) can be obtained from Table 7.9.3.1.

- **Spectral properties of the high-spin Fe(III)-complexes:** In the Fe(III) complexes, because of the ***ligand to metal charge transfer band,*** interpretation of the weak ligand field bands becomes very difficult. Ligand field bands of light violet $[Fe(OH_2)_6]^{3+}$ (in a fairly strong $HClO_4$ media) have been characterized. It is still complicated due to the strong propensity of $[Fe(OH_2)_6]^{3+}$ to hydrolyse to $[Fe(OH)(OH_2)_5]^{2+}$ which looks brown due to the charge transfer band (LMCT) (*cf.* ^{-}OH group is more polarizable than H_2O). The interpretation is further complicated due to the tendency of $[Fe(OH)(OH_2)_5]^{2+}$ to undergo dimerization.

The ligand field bands of **$[Fe(ox)_3]^{3-}$, $[Fe(acac)_3]$** can be explained by considering the spin-forbidden transitions of a d^5 (high-spin, O_h) system. But lowering of symmetry (compared to that of an O_h system) in such *tris*-chelates (D_3) splits the bands. The **strong CT bands** obscure some of the ligand field bands. Red $[Fe(acac)_3]$ in acetone solution shows four bands in the infrared and visible regions (molar extinction coefficient in parenthesis):

$$9{,}760\ cm^{-1}\ (0.45),\ 13{,}160\ cm^{-1}(0.70),\ 23{,}120\ cm^{-1}\ (3{,}250),\ 28{,}410\ cm^{-1}\ (3{,}420)$$

The strong last two energy peaks are due to the CT transitions (see Fig. 1.16.19.1, Vol. 4). $[Fe(ox)_3]^{3-}$ **(light green)** undergoes the **photochemical redox decomposition** due to LMCT transitions (Sec. 6.15.8).

7.15 LIGAND FIELD BANDS IN SPIN PAIRED SIX-COORDINATE COMPLEXES OF FIRST TRANSITION SERIES

Under the condition, $10Dq > P$, the d^4, d^5, d^6 and d^7 systems can produce the low-spin complexes. The possible spectral transitions can be assigned by analyzing the right hand portion of the Tanabe-Sugano diagram.

Very often, the π-bonding ligands (*i.e.* π-acid ligands like CN^-, NO, etc.) are the strong field ligands to induce the spin pairing. Such π-bonding ligands are highly polarizable. This is why, very often the charge transfer bands are extended in the visible range where the ligand field bands are overshadowed by the intense charge transfer bands. In such cases, the interpretation of the ligand field bands becomes complicated.

7.15.1 Ligand Field Spectra of the Low-spin Octahedral d^6 Systems

The Fe(II) and Co(III) complexes are the representative examples. For the d^6-configuration, the spin-pairing is relatively easier compared to many other d^n-configurations. Both the high cfse in the low spin complexes and relatively less loss of exchange energy in pairing the electrons favour the d^6 system to adopt the low-spin configuration.

The ground state electronic configuration t_{2g}^6 gives the $^1A_{1g}$ state. The excited state electronic configuration $t_{2g}^5 e_g^1$ generates the following states (*cf.* Figs. 7.15.1.1, 2 and Sec. 7.12).

$$t_{2g}^5\, e_g^1:\ {}^3T_{1g} + {}^3T_{2g} + {}^1T_{1g} + {}^1T_{2g}$$

The **energy order** is: ${}^3T_{1g} < {}^3T_{2g} < {}^1T_{1g} < {}^1T_{2g}$

The singlet states ${}^1T_{1g}$ and ${}^1T_{2g}$ are developed due to the antiparallel arrangement of the spins of the two unpaired electrons (one in the t_{2g} level and the other in the e_g level).

- t_{2g}^5, *i.e.* one hole in the d_{xy}, d_{yz} and d_{xz} orbitals: **3 possible ways of distribution.**
- e_g^1, *i.e.* one electron in the $d_{x^2-y^2}$ and d_{z^2} orbitals: **2 possible ways of distribution.**
- $t_{2g}^5 e_g^1 : 3 \times 2 = 6$ possible ways of electron distribution, *i.e.* **orbital multiplicity = 6.**
- ${}^1T_{1g} + {}^1T_{2g}$: total orbital multiplicity = 3 + 3 = 6

Because of the antiparallel arrangement of the spins, the **singlet states are of higher energy**.

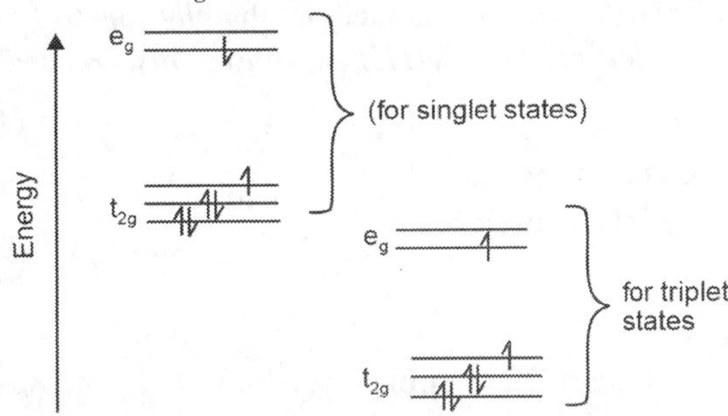

Fig. 7.15.1.1: Singlet and triplet states arising from the $t_{2g}^5 e_g^1$ configuration

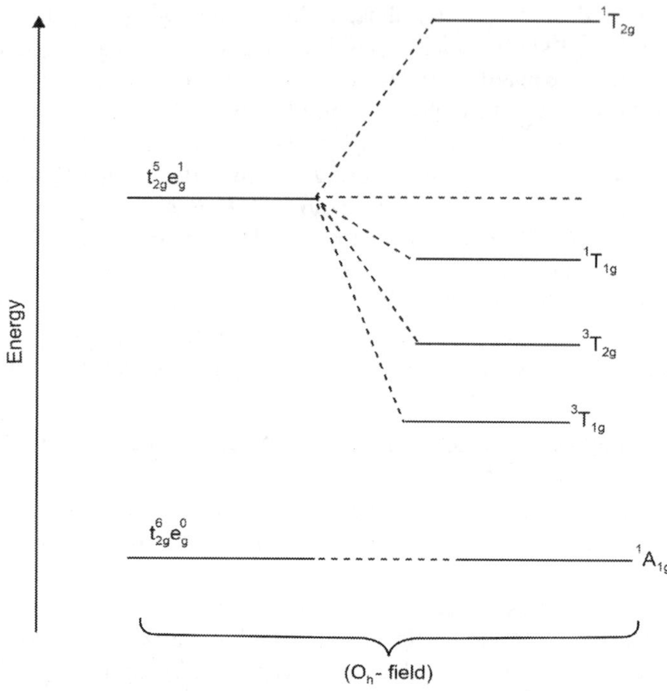

Fig. 7.15.1.2: (a) States arising from the low-spin d^6-configuration in an octahedral field.

From Fig. 7.15.1.2, the possible electronic transitions giving rise to the ligand field bands in the low-spin octahedral complexes of d^6-configuration are as follows:

$$^1A_{1g}\rightarrow{}^3T_{1g};\ {}^1A_{1g}\rightarrow{}^3T_{2g}\ \}\ \text{spin-forbidden transitions.}$$

$$^1A_{1g}\rightarrow{}^1T_{1g};\ {}^1A_{1g}\rightarrow{}^1T_{2g}\ \}\ \text{spin-allowed transitions.}$$

Obviously, transitions to the triplet states give the peaks of poor intensity and these spin-forbidden transitions occur at a lower energy. Energies of the $^3T_{2g}$ and $^1T_{1g}$ states are close to each other. This is why, the strong band arising from the spin-allowed $^1A_{1g}\rightarrow{}^1T_{1g}$ transition very often obscures the spin-forbidden transition $^1A_{1g}\rightarrow{}^3T_{2g}$. In fact, characterization and analysis of the said weak peaks due to the spin-forbidden transitions are not of much importance. On the other hand, analysis of the spin-allowed transitions is quite important. However, in some cases, all the four peaks have been characterized (*cf.* Table 7.15.1.1).

By considering the relatively stronger spin allowed transitions in the low-spin octahedral complexes of Co(III), it may be pointed out that ***the spectral behaviours of the Co(III) and Cr(III) complexes are quite similar. The Cr(III) complexes also record two strong ligand field bands.***

Octahedral Cr(III) complexes $\left\{ \begin{array}{c} ^4A_{2g} \xrightarrow{\nu_1} {}^4T_{2g}(F) \\ \text{and} \\ ^4A_{2g} \xrightarrow{\nu_2} {}^4T_{1g}(F) \end{array} \right.$ (The high energy 3rd ligand field band, $^4A_{2g} \xrightarrow{\nu_3} {}^4T_{1g}(P)$, is very often obscured by the strong charge transfer band).

Octahedral low spin Co(III) complexes $\left\{ \begin{array}{c} ^1A_{1g} \xrightarrow{\nu_1} {}^1T_{1g} \\ ^1A_{1g} \xrightarrow{\nu_2} {}^1T_{2g} \end{array} \right.$

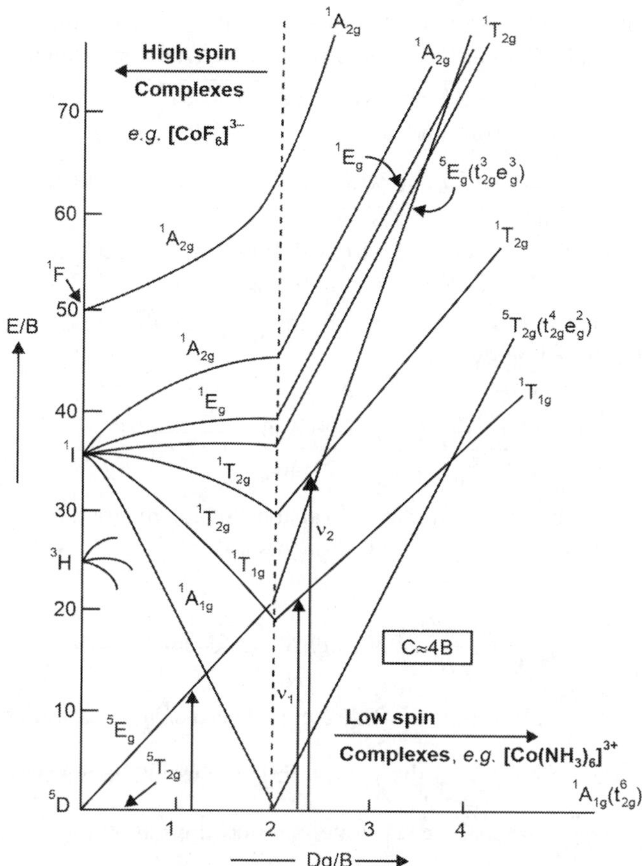

Fig. 7.15.1.2: (b) Tanabe-Sugano diagram for the d^6-octahedral complexes showing the spin-allowed transitions.

Table 7.15.1.1 Characteristic features of ligand field spectra of low-spin octahedral complexes of d^6-system (For $[M(CO)_6]$, the ligand field bands appear as the **shoulders of the strong MLCT bands**).

Complex	$^1A_{1g} \to {}^3T_{1g}$ (cm^{-1})	$^1A_{1g} \to {}^3T_{2g}$ (cm^{-1})	$^1A_{1g} \to {}^1T_{1g}$ (cm^{-1})	$^1A_{1g} \to {}^1T_{2g}$ (cm^{-1})
$[Co(en)_3]^{3+}$	13,700	17,500	21,400	29,400
$[Co(NH_3)_6]^{3+}$	13,000	–	21,100	29,500
$[Co(OH_2)_6]^{3+}$	8,000	12,500	16,500	24,600
$[Fe(CN)_6]^{4-}$	23,700	–	31,000	37,000
$[Fe(phen)_3]^{2+}$	8,950	9,250	12,600	14,700
$[RhF_6]^{3-}$	–	–	21,300	27,800
$[Cr(CO)_6]$	–	–	29,500	31,500
$[Mo(CO)_6]$	–	–	30,150	31,950

For the octahedral low spin Co(III)-complexes, the energy changes in the ν_1 and ν_2 transitions are given by:

$$\left.\begin{aligned} \nu_1 &= 10Dq - 4B' + 86B'^2/10Dq = \Delta - 4B' + 86B'^2/\Delta \\ \nu_2 &= 10Dq + 12B' + 2B'^2/10Dq = \Delta + 12B' + 2B'^2/\Delta \end{aligned}\right\} (\text{Taking, } C \approx 4B')$$

In the low spin complexes, the B'-parameter (measuring the interelectronic repulsion) is relatively less important and the terms containing B'^2 can be neglected. This approximation leads to:

$$\nu_1 \approx 10Dq - 4B' \text{ and } \nu_2 = 10Dq + 12B'$$
$$\text{i.e.} \quad \nu_2 - \nu_1 \approx 16B' \text{ and, } 10Dq \approx \nu_1 + 4B' \approx \nu_2 - 12B'$$

Let us analyze the spectral results of **$[Co(en)_3]^{3+}$** by using the approximate relations.

$$\nu_1 = 21{,}400 \text{ cm}^{-1},\ \nu_2 = 29{,}400 \text{ cm}^{-1}$$
$$16B' \approx \nu_2 - \nu_1 = 29{,}400 - 21{,}400 = 8{,}000, \text{ i.e. } B' \approx 500 \text{ cm}^{-1} = 0.5 \text{ kK}$$
$$10Dq \approx \nu_1 + 4B' = 21{,}400 + 4 \times 500 = 23{,}400 \text{ cm}^{-1} = 23.4 \text{ kK}$$

Here it may be mentioned that the said spectral analysis can also be done by using the Tanabe-Sugano diagram. It is illustrated below.

The ratio of ν_2 and ν_1 is given by:

$$\frac{{}^1A_{1g} \longrightarrow {}^1T_{2g}}{{}^1A_{1g} \longrightarrow {}^1T_{1g}} = \frac{\nu_2}{\nu_1} = \frac{29{,}400}{21{,}400} \approx 1.37$$

The above mentioned ratio can be fitted into the Tanabe-Sugano diagram by sliding a ruler along the abscissa. This fitting is achieved for $Dq/B = 4.0$. For this value of Dq/B, ν_1/B' becomes 38 (obtained from the diagram).

$$\frac{{}^1A_{1g} \longrightarrow {}^1T_{1g}}{B'} = \frac{\nu_1}{B'} \approx \frac{21{,}400}{B'} = 38, \text{ i.e. } B' \approx 563 \text{ cm}^{-1} = 0.563 \text{ kK}$$

$$\frac{Dq}{B'} = 4.0 \text{ or, } Dq = B' \times 4 = 2{,}252 \text{ cm}^{-1} \text{ (i.e. } 10Dq = 22{,}520 \text{ cm}^{-1} = 22.52 \text{ kK)}$$

These $10Dq$ and B' values are close to the values obtained from the approximate relations.

$$\beta = \frac{B'}{B_0} = \frac{563}{1100} \approx 0.51, \text{ i.e. } B \text{ value is reduced by about 50\%.}$$

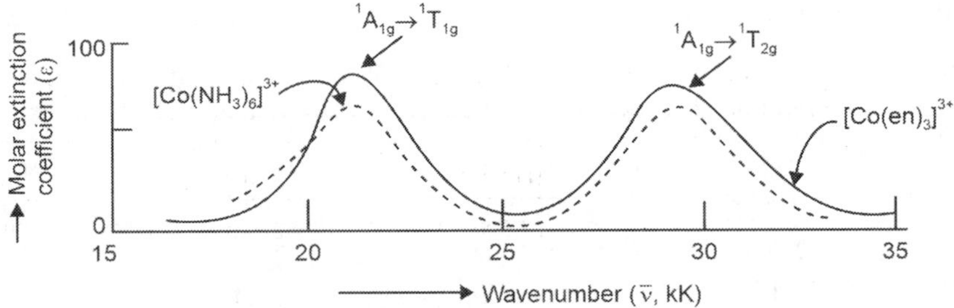

Fig. 7.15.1.3 Electronic absorption spectra of $[Co(NH_3)_6]^{3+}$ (dotted line) and $[Co(en)_3]^{3+}$ (solid line).

● **$[Co(en)_3]^{3+}$ *vs.* $[Co(ox)_3]^{3-}$:** The spectra of **yellow** $[Co(en)_3]^{3+}$ and **green** $[Co(ox)_3]^{3-}$ complexes are compared in Fig. 7.15.1.4. Both are of D_3 symmetry but their spectral behaviours indicate that they maintain effectively the O_h symmetry, *i.e.* $[M(N)_6]$ and $[M(O)_6]$. The two spin allowed transitions are: ${}^1A_{1g} \xrightarrow{\nu_1} {}^1T_{1g}$ and ${}^1A_{1g} \xrightarrow{\nu_2} {}^1T_{2g}$. For $[Co(ox)_3]^{3-}$, the spin allowed transition (ν_2) at the higher energy is close to the strong CT band. The **higher polarizable character of the anionic oxalate ligand** leads the LMCT band to appear in the lower energy (*cf.* ν_{LMCT} : $[Co(en)_3]^{3+} \gg [Co(ox)_3]^{3-}$). In other words, oxalate is a **more oxidisable** ligand than the ethylene diamine ligand and it produces a red shifted LMCT band in $[Co(ox)_3]^{3-}$ (*cf.* **photochemical redox decomposition** through the LMCT

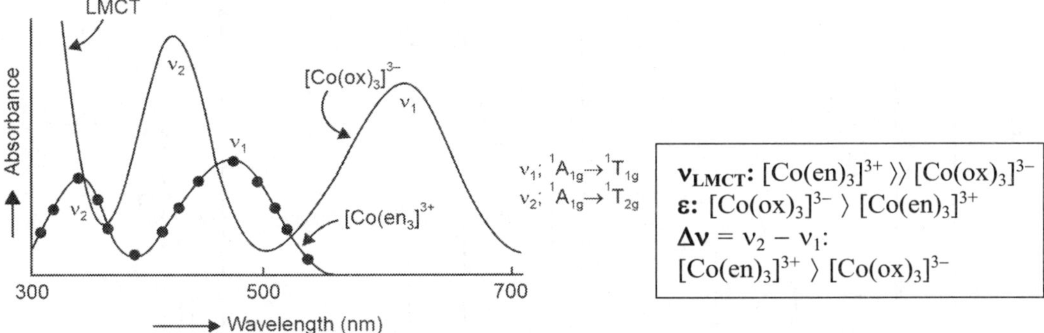

Fig. 7.15.1.4: Comparison of the absorption spectra of $[Co(ox)_3]^{3-}$ and $[Co(en)_3]^{3+}$.

Note: Higher intensity for the $[Co(ox)_3]^{3-}$ complex may arise from the allowed charge transfer band (*cf.* **intensity stealing effect**).

transitions, Sec. 6.15.8). Thus the Laporte selection rule is relaxed better in $[Co(ox)_3]^{3-}$ to give the *higher intensity* probably due to the **intensity stealing phenomenon** from the nearby allowed LMCT band. *The slope of the* $^1T_{2g}$ *state is higher than that of the* $^1T_{1g}$ state (*cf.* Fig. 7.15.1.2b). Thus separation between the two peaks (*i.e.* $\Delta\nu = \nu_2 - \nu_1$) will increase with the ligand field strength. *This is why, these two peaks are separated more in* $[Co(en)_3]^{3+}$ (*cf.* crystal field strength: en > ox^{2-}).

7.15.2 Ligand Field Spectra of Low-Symmetry Six-coordinate Complexes of Co(III). (*cf.* Sec. 7.14.5 E for Cr(III) Complexes and Sec. 7.14.4 for Ni(II) Complexes)

(A) Splitting of the octahedral spectroscopic states in tetragonal and rhombic symmetry: When the ligands are nonequivalent, this situation will arise. Among the six-coordinate complexes, the three types of field symmetries are quite important. These are:

cubic (*i.e.* perfectly octahedral), **tetragonal** and **rhombic.**

These are classified in terms of μ which is the sum of charges (q) in the directions of x, y and z axes along which the ligands are placed. If the μ–*s* along the respective axes are denoted by μ_x, μ_y, and μ_z, then we can write:

cubic and octahedral: $\mu_x = \mu_y = \mu_z$, *e.g.* $[Co(NH_3)_6]^{3+}$

tetragonal: $\mu_x = \mu_y \neq \mu_z$, *e.g.* $[CoCl_2(NH_3)_4]^+$

rhombic: $\mu_x \neq \mu_y \neq \mu_z$, *e.g.* *mer*-$[CoCl_3(NH_3)_3]$

Table 7.15.2.1: Symmetries of some common types of six-coordinate complexes.

Symmetry	*Complex*
Cubic	$[MA_6]$ (O_h); 1,2,3-$[MA_3B_3]$ (*fac* or *cis*)(C_{3v});
$(\mu_x = \mu_y = \mu_z)$	
Tetragonal	
$(\mu_x = \mu_y \neq \mu_z)$	
C_{2v}	*cis*-$[MA_4B_2]$
D_{4h}	*trans*-$[MA_4B_2]$
Rhombic	1,2,6-$[MA_3B_3]$ (*mer* or *trans*) (C_{2v})
$(\mu_x \neq \mu_y \neq \mu_z)$	

In moving from the octahedral to tetragonal (D_{4h}) field symmetry, the *T*-state is split into two states A_2 and E or B_2 and E., *i.e.* $T_{1g}(O_h) \rightarrow A_{2g}$, $E_g(D_{4h})$ and $T_{2g} \rightarrow B_{2g}$, $E_g(D_{4h})$ (*cf.* Correlation Table

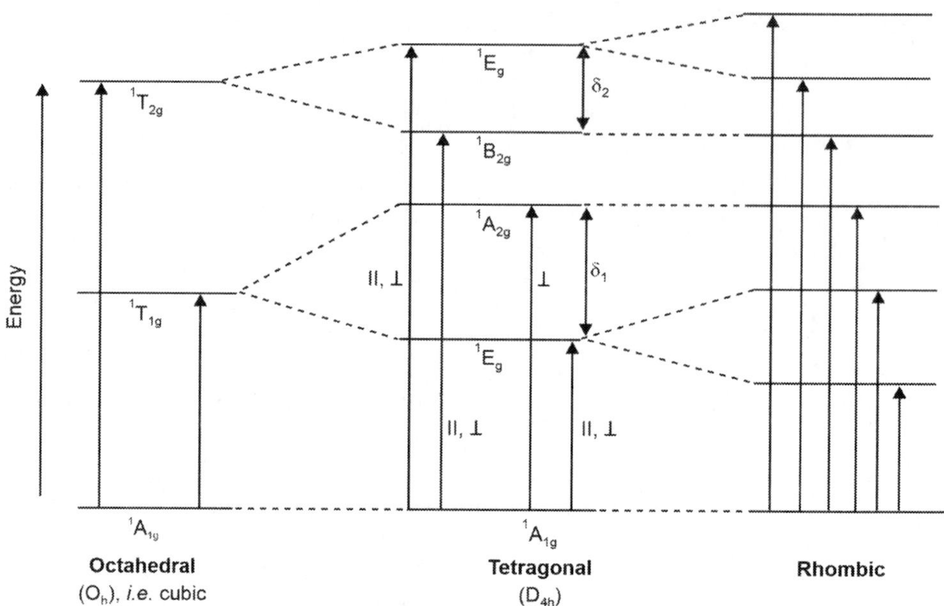

Fig. 7.15.2.1: Energy level diagrams for the spin-allowed spectral transitions in the octahedral, tetragonal and rhombic field symmetries of the low-spin Co(III) complexes.

7.15.4.2). Again the E-state is split into B_1 and B_2 in moving from the tetragonal to rhombic field symmetry. These are illustrated in Fig. 7.15.2.1. ***In terms of electronic environment, both the cis- and trans-isomers of [MA_4B_2] may be considered to have the tetragonal symmetry.*** This aspect has been discussed later.

Note: Some authors represent the energy order of the ^1E_g and $^1A_{2g}$ states generated from the splitting of $^1T_{1g}$ as: $^1E_g > {}^1A_{2g}$. However, polarisation (*i.e.* dichroism) studies on *trans*-[$CoCl_2(en)_2$]$^+$ (D_{4h}) indicate that $^1A_{2g}$ is the higher energy state (*cf.* S. Yamada *et al. Bull. Chem. Soc. Japan*, **25**, 127, 1952; **28**, 222, 1955; C.J. Ballhausen et. al. *J. Inorg. Nucl, Chem.*, **3,** 178, 1956). This aspect will be discussed later.

Note: ● $\delta_1 \gg \delta_2$ ● For the D_{4h} tetragonal symmetry (*e.g. trans*-[MA_4B_2]), the ordering is $^1E_g < {}^1A_{2g}$ while for the C_{2v} tetragonal symmetry (*e.g. cis*- [MA_4B_2]), the ordering is $^1A_2 < {}^1E$. Splitting of the $^1T_{1g}$ state in the *trans*-complex is almost double to that of the corresponding *cis*-complex (*cf.* Fig. 7.15.2.3). In an octahedral complex, there are **two possible electronic transitions;** in a tetragonal complex, there are **four possible electronic transitions** while in a rhombic system, there are **six possible electronic transitions**).

From the energy level diagram given in Fig. 7.15.2.1, it is evident that for the low-spin complexes of Co(III), we can expect the following electronic transitions (*if the splitting is sufficiently large to develop the separate peaks*).

Cubic: two peaks; **Tetragonal:** four peaks; **Rhombic:** six peaks.

However, in most of the cases, splitting of the $^1T_{2g}$ state is too small to be detected. Similarly, splitting of the ^1E_g state in a rhombic field is again found to be too small to be detected. Considering this fact, we can expect the following peaks:

Cubic: two peaks; **Tetragonal:** three peaks; **Rhombic:** three peaks.

(B) ***Characterization of the cis- and trans-isomers of low-spin [$Co^{III}A_4B_2$] from their ligand field spectra:*** Both the *cis- and trans*-isomers are expected to give the three ligand field bands. **Splitting of the $^1T_{2g}$ state for the both isomers is too small to be detected, but splitting of the $^1T_{1g}$ state is more**

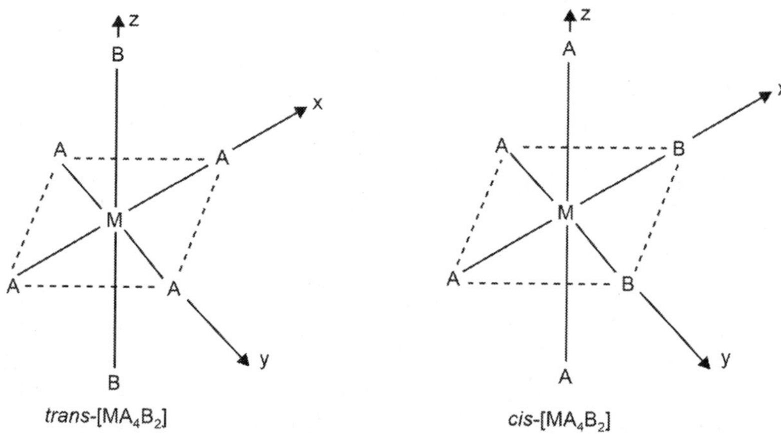

Fig. 7.15.2.2: Geometrical isomers of $[MA_4B_2]$

for the *trans*-isomer compared to that of the corresponding *cis*-isomer. These can be understood in terms of the μ-values. Here we shall consider the splitting (δ_1) of the T_{1g} state.

The **tetragonality parameter** $\mu_x - \mu_z$ gives the measure of splitting of the *T*-state. It is obtained as follows:

***Trans*-$[MA_4B_2]$**

$\mu_x = \mu_y = 2q_A$

$\mu_z = 2q_B$

i.e. $\mu_x - \mu_z = 2(q_A - q_B)$

and $\delta_1(trans) \propto 2(q_A - q_B)$

***Cis*-$[MA_4B_2]$**

$\mu_x = \mu_y = q_A + q_B$

$\mu_z = 2q_A$

i.e. $\mu_x - \mu_z = -(q_A - q_B)$

and $\delta_1(cis) \propto -(q_A - q_B)$

From the relative values of the **tetragonality parameter** $\mu_x - \mu_z$, *it is evident that for the trans-complex, the T_{1g} state undergoes splitting twice that of the cis-complex.* The minus sign indicates the ordering of the ^1E_g and $^1A_{2g}$ states originated due to splitting of the $^1T_{1g}$ state is inverted in the *cis*-complex compared to that of the *trans*- complex. In fact, the energy order of ^1E_g and $^1A_{2g}$ states follows the sequence:

$$trans\text{-}[MA_4B_2]:\ ^1E_g < {}^1A_{2g};\ cis\text{-}[MA_4B_2]:\ ^1E > {}^1A_2 \text{ and, } \delta_1\ (trans) : \delta_1\ (cis) = 2:1$$

The magnitude of splitting (δ_1) of the $^1T_{1g}$ state can be expressed in terms of positions of the ligands *A* and *B* in the spectrochemical series. The splitting energy is determined by the average ligand field strength difference between the A—Co—A and B—Co—B axes (for the *trans*-complex) and between the A—Co—A and A—Co—B axes (for the *cis*-complex). It gives:

$$\boldsymbol{\delta_1(trans) = 2C(\Delta_A \sim \Delta_B),\ \delta_1(cis) = -C(\Delta_A \sim \Delta_B).}$$

Here *C* is a proportionality constant and Δ_A and Δ_B denote the crystal field splitting power of the ligands *A* and *B* respectively. **Thus the $\Delta_A \sim \Delta_B$ parameter represents their difference in position in the spectrochemical series.**

Note: ● Though the *cis*- and *trans*- isomers belong to the C_{2v} and D_{4h} symmetry point groups respectively in terms of local symmetry, **from the view point of electronic environment around the metal centre, both the isomers experience the tetragonal splitting only differing in the magnitude of splitting and energy order of the splitted states** (*cf.* F. Teixidor *et al, J. Chem. Edu.*, **64,** 461, 1987).

● Some authors indicate the energy order as $^1E_g > {}^1A_{2g}$ (for the *trans*-isomer) but the **polarisation studies** on *trans*-$[CoCl_2(en)_2]^+$ indicate the energy order: $^1A_{2g} > {}^1E_g$ (for the *trans*-isomer) and this practice is followed in this book.

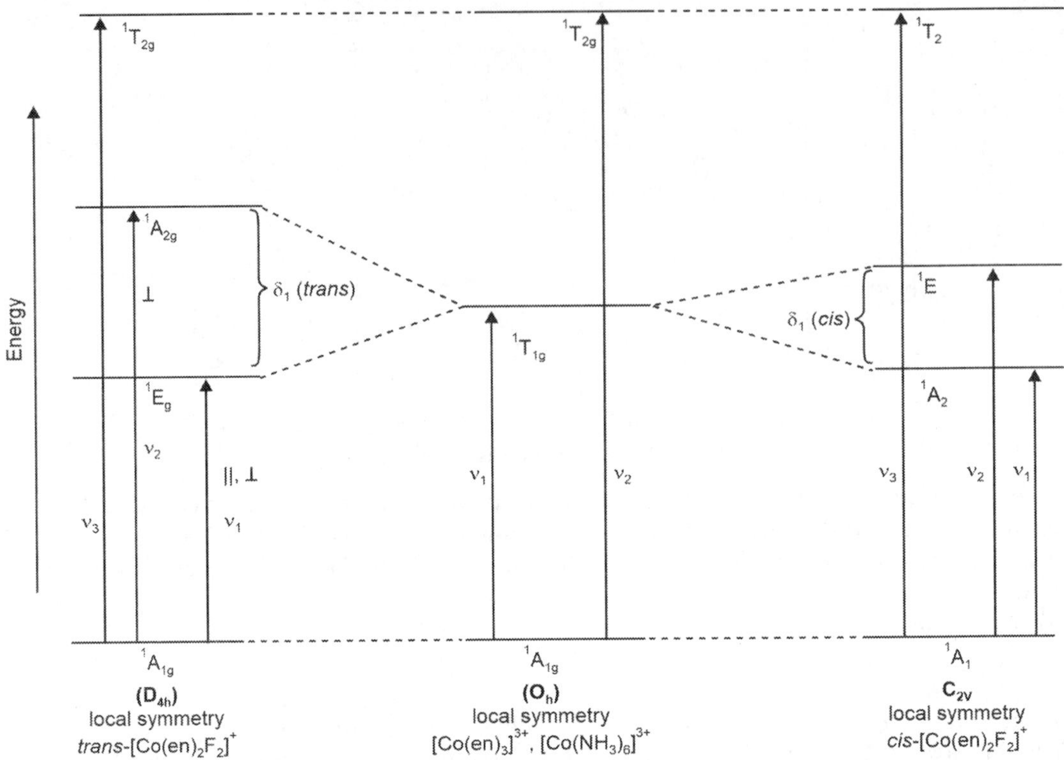

Fig. 7.15.2.3: The spin allowed transitions in the low spin $[Co^{III}A_6]$ (O_h) and *cis-trans-* isomers of $[Co^{III}A_4B_2]$ type complexes: $\delta_1(trans) \approx 2\delta_1(cis)$; **Splitting of the $^1T_{2g}$ is negligibly small and it is ignored.**

> **Note:** The point groups are (see Vol 2): $[Co(NH_3)_6]^{3+}(O_h)$, $[Co(en)_3]^{3+}(D_3)$, *cis*-$[Co(NH_3)_4F_2]^+$ (C_{2v}), *cis*-$[Co(en)_2F_2]^+$ (C_2), *trans*-$[Co(NH_3)_4F_2]^+$ (D_{4h}), *trans*-$[Co(en)_2F_2]$ (D_{2h}). The D_{4h} and C_{2v} symmetry for the *trans-* and *cis*-isomers of $[Co(en)_2F_2]^+$ have been considered respectively in Fig. 7.15.2.3 in terms of local symmetry. Similarly, for $[Co(en)_3]^{3+}$, O_h symmetry has been considered.

Now let us consider the relative positions of the peaks due to the following transitions.

$$(cf.\ \text{Fig. 7.15.2.3})\begin{cases} {}^1A_1 \xrightarrow{\nu_1} {}^1A_2;\ {}^1A_1 \xrightarrow{\nu_2} {}^1E,\ {}^1A_1 \xrightarrow{\nu_3} {}^1T_2\ \textbf{(for \textit{cis}-isomer)} \\ {}^1A_{1g} \xrightarrow{\nu_1} {}^1E_g;\ {}^1A_{1g} \xrightarrow{\nu_2} {}^1A_{2g},\ {}^1A_{1g} \xrightarrow{\nu_3} {}^1T_{2g}\ \textbf{(for \textit{trans}-isomer)} \end{cases}$$

The difference, $\nu_1 \sim \nu_2$, in the *trans*-isomer is twice that of the corresponding *cis*-complex, *i.e.*

$$\frac{(\nu_1 \sim \nu_2)_{trans}}{(\nu_1 \sim \nu_2)_{cis}} \approx 2.0.$$

In reality, in the *cis*-complexes, *e.g.* *cis*-$[Co(en)_2F_2]^+$, *cis*-$[Co(C_2H_5COO)_2(en)_2]^+$, the positions of ν_1 and ν_2 are so close that two separate peaks are not observed, *i.e.* they merge or overlap with each other to give a **relatively broad peak**. On the other hand, for their corresponding *trans-isomers*, the transitions due to ν_1 and ν_2 are well separated to be characterized as the separate peaks. This is why, the *trans-* isomer gives three peaks (*cf.* Fig. 7.15.2.4a).

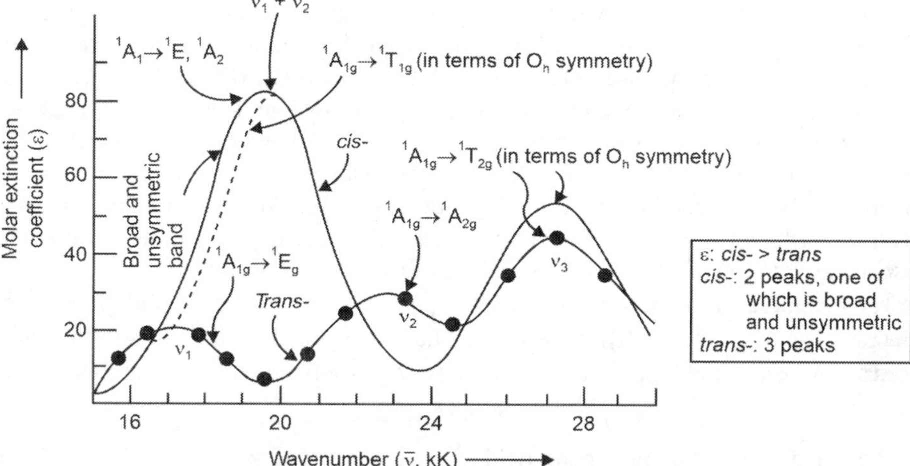

Fig. 7.15.2.4 (a) Electronic absorption spectra of *cis*- and *trans*- isomers of $[Co(en)_2F_2]^+$ (ν_1, ν_2, ν_3 in terms of tetragonal symmetry).

(**Note:** The dotted line in Fig. 7.15.2.4a for the *cis*-isomer indicates the position of the symmetrical band for the $^1A_{1g} \rightarrow {^1T_{1g}}$ transition without any splitting of the $^1T_{1g}$ state (assuming the O_h symmetry).

Positions of the ligand field bands in the *cis*- and *trans*- isomers of $[Co(en)_2F_2]^+$ are given below:
(***cis*-$[Co(en)_2F_2]^+$:** $\underbrace{\nu_1, \nu_2}$ (19,000 cm^{-1}, ε = 85), ν_3 (27,500 cm^{-1}, ε = 50)

Merged to give a broad and asymmetric peak.

***trans*-$[Co(en)_2F_2]^+$:** ν_1 (17,500 cm^{-1}, ε = 20), ν_2 (22,500 cm^{-1}, ε = 30), ν_3 (27,800 cm^{-1}, ε = 50).

Here it may be mentioned that the ligand field spectra can distinguish the geometrical isomers of $[Co^{III}A_4B_2]$ only when the positions of *A* and *B* in the spectrochemical series are widely apart. Because, the extent of splitting of the $^1T_{1g}$ state depends on the relative crystal field splitting powers of *A* and *B*.

Here it is interesting to note that **the intensity of the peaks of the *cis*-$[Co^{III}A_4B_2]$ complexes is much higher than those of the *trans*-$[Co^{III}A_4B_2]$ complexes.** The *cis*-complexes lack in centre of symmetry. Consequently, the *d-p* mixing is more efficient in relaxing the Laporte selection rule in the *cis*-complex compared to that in the *trans*- isomer which is possessing the centre of symmetry. Besides this, in the ***noncentrosymmetric cis-complex,*** some of the electronic transitions should be allowed by their own right because in such cases, the Laporte selection rule is not applicable. However, the molar extinction coefficient values (ε) do not differ widely for the *cis*- and *trans*- complexes and the ε-values are also more or less comparable with those of the octahedral complexes (*cf.* for $[Co(NH_3)_6]^{3+}$: ν_1(21.2 kK, ε = 56); ν_2(29.55 kK, ε = 46)). These observations suggest that in the given complexes, ***effective octahedral symmetry is maintained.***

In $[Co(en)_2F_2]^+$, the parameter $\Delta_{en} \sim \Delta_F$ giving the measure of difference in crystal field splitting power of the ligands en and F^- is significantly high. Consequently, their *cis*- and *trans*- isomers can be distinguished by comparing their ligand field spectra. For the *cis*- complex, the octahedral ν_1 transition does not experience any splitting but in the *trans*-complex, the corresponding octahedral ν_1 transition gives two separate peaks as expected for the D_{4h} symmetry (*cf.* Fig. 7.15.2.4a). ***In the cis-complex, the octahedral ν_1-band is only broadened to have an asymmetric shape. However, even in the trans-complex, the octahedral ν_2 band does not experience any splitting.***

F^- and H_2O are very close in their positions in the spectrochemical series (*i.e.* $\Delta_{en} - \Delta_F \approx \Delta_{en} - \Delta_{H_2O}$) and the geometrical isomers of $[Co(en)_2(OH_2)_2]^{3+}$ can also be distinguished by comparing their ligand field spectra as in the case of $[Co(en)_2F_2]^+$. *Trans*- $[Co(en)_2(OH_2)_2]^{3+}$ shows one absorption band with two components at 18,200 cm^{-1} and 22,500 cm^{-1} and another absorption band at 29,000 cm^{-1}. The positions of H_2O and NCS^- in the spectrochemical series are very close and in fact, the geometrical isomers of $[Cr(NCS)_2(OH_2)_4]^+$ cannot be distinguished in this way (*cf.* Sec. 7.14.5D). Here it is worth mentioning that the crystal field splitting power of Co(III) is greater than that of Cr(III) (*cf.* $10Dq = fg$, $g = 18.2$ for Co^{3+}; $g = 17.4$ for Cr^{3+}). It causes more splitting of the T-state in the Co(III) complexes compared to that in the Cr(III)-complexes for the same ligands.

(C) Characterization of the *fac-mer* isomers of $[MA_3B_3]$: It has been already mentioned that the *fac*-isomer (*i.e. cis*-or 1,2,3-$[MA_3B_3]$) experiences the octahedral ligand field splitting of *d*-orbitals ($\mu_x = \mu_y = \mu_z$). In an octahedral ligand field, *fac*-$[Co^{III}A_3B_3]$ experiences two ligand field transitions, *i.e.*

$$^1A_{1g} \xrightarrow{\nu_1} {}^1T_{1g} \text{ and } {}^1A_{1g} \xrightarrow{\nu_2} {}^1T_{2g}$$

On the other hand, in the rhombic ligand field ($\mu_x \neq \mu_y \neq \mu_z$) of *mer*-$[Co^{III}A_3B_3]$, both the octahedral ν_1 and ν_2 peaks are expected to split into triplet components because of the splitting of T_{1g} and T_{2g} states into three energy levels (*i.e.* each *T*-state splits into 3 levels, *cf.* Fig. 7.15.2.1). In reality, splitting of the T_2-state is negligibly small. Consequently, the ν_2-band remains more or less the same for both the isomers. Splitting of the T_{1g} state in a rhombic field is relatively higher, but the splitting energy may not be sufficiently high to generate the three different peaks through the splitting of the octahedral ν_1-band. **However very often, the ν_1-band may be broadened in a rhombic field** (compared to that of the octahedral field).

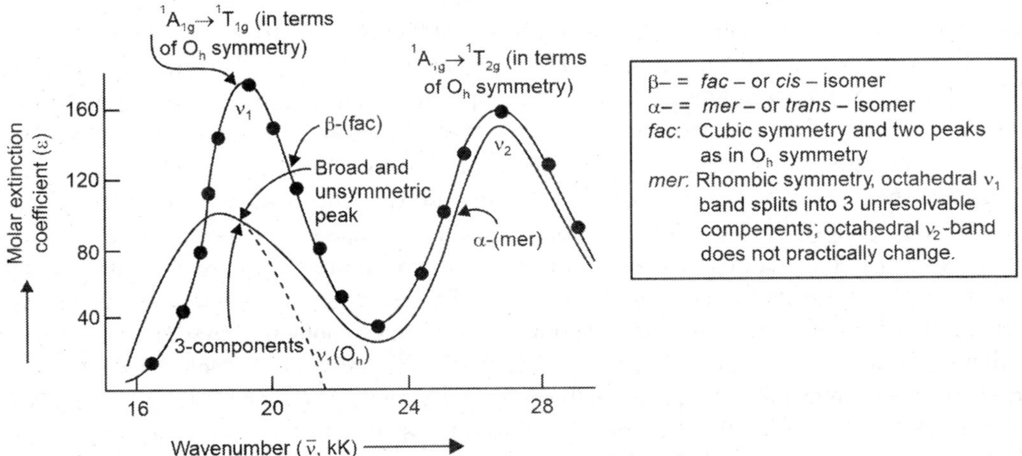

Fig. 7.15.2.4: (b) Electronic absorption spectra of α- (*mer*) and β (*fac*)-isomers of $[Co(gly)_3]$, *i.e.* $[Co(N,O)_3]$.

(**Note:** The broken line in Fig. 7.15.2.4b for the α-isomer indicates the position of the the symmetrical band for the $^1A_{1g} \rightarrow {}^1T_{1g}$ transition (assuming the O_h symmetry) without any splitting of the $^1T_{1g}$ state.)

Now let us consider to compare the ligand field spectra (Fig. 7.15.2.4b) of the α (*violet, dihydrate*) and β (*red, monohydrate*) isomers of $[Co(glycinate)_3]$, *i.e.* $[Co(N,O)_3]$. The α-isomer is actually *mer*- or 1,2,6- or, *trans*-isomer while the β-isomer is *fac*- or, 1,2,3- or *cis*-isomer. ***Thus the α-isomer experiences the rhombic ligand field splitting while the β-isomer experiences the octahedral ligand field***

***splitting*.** In fact, for the α-isomer, the lower energy band (*i.e.* ν_1-band) is broadened with a shoulder (*i.e.* broad and unsymmetric band) while the corresponding band for the β-isomer is symmetrical. However, the higher energy band (*i.e.* ν_2-band) is having more or less of the same shape for both the α- and β- isomers.

Here it is worth mentioning that characterization of the geometrical isomers by considering the band splitting property in their ligand field spectra is well documented for ***the inert complexes of Cr(III), Co(III), Rh(III), Ir(III), Ru(III),*** etc. But the technique is not applicable for the labile complexes which are expected to experience the rearrangement and solvolysis quickly.

(D) Vibronic polarisation in ***trans*-[CoCl$_2$(en)$_2$]$^+$ (D_{4h}) and characterisation of the energy states derived from the splitting of the $^1T_{1g}$ state (*cf.* Figs. 7.15.2.1, 3):** It has been already indicated in Sec. 7.5.1 that *trans*-[CoCl$_2$(en)$_2$]$^+$ shows the **dichroism** (*i.e.* vibronic polarisation). In this centrosymmetric system, the forbidden electronic transitions are partially allowed through the vibronic coupling but the transitions experience a **vibronic polarisation** (*see* Sec. 7.5.1 for details) as indicated below:

	$^1A_{1g} \longrightarrow {}^1A_{2g}$	$^1A_{1g} \longrightarrow {}^1E_g$	$^1A_{1g} \longrightarrow {}^1B_{2g}$ (arising from the splitting of $^1T_{2g}$ state).
For μ_z (II to C_4 axis):	Forbidden	Allowed	Allowed
For $\mu_{x,y}$ (⊥ to C_4 axis):	Allowed	Allowed	Allowed

When the spectra are recorded with the polarised light perpendicular (⊥) to the C_4 axis (*i.e.* Cl—Co—Cl), **three peaks are observed** but when the spectra are recorded with the polarised light parallel (||) to the C_4-axis, only **two peaks are noted,** *i.e.* one peak is missing. The group theory treatment predicts that the missing peak is due to the $^1A_{1g} \longrightarrow {}^1A_{2g}$ transition. The **three peaks (⊥ polarisation)** appear at ~16 kK, ~22 kK and ~27 kK (which merges with the LMCT band; *cf.* Cl$^-$, a polarisable ligand). For the polarised light (|| to the C_4 axis), the peak at about ~22 kK is mssing and it is definitely due to the $^1A_{1g} \longrightarrow {}^1A_{2g}$ transition. Thus the peaks are reasonably assigned as:

$$^1A_{1g} \longrightarrow {}^1E_g\ (\sim 16\ \text{kK});\ {}^1A_{1g} \longrightarrow {}^1A_{2g}\ (\sim 22\ \text{kK});\ {}^1A_{1g} \longrightarrow {}^1B_{2g}\ (\sim 27\ \text{kK}).$$

It indicates the energy order as $^1B_{2g} > {}^1A_{2g} > {}^1E_g$. As the splitting in the unsplit $^1T_{2g}$ state is negligibly small, energy of the $^1B_{2g}$ state corresponds to that of the unsplit $^1T_{2g}$ state (approximately). For details of the dichroism studies on *trans*-[CoCl$_2$(en)$_2$]$^+$ (*Ref.* S. Yamada et. al. *Bull. Chem. Soc. Japan*, **25**, 127, 1952; **28**, 222, 1955) and spectral analysis/interpretation (*Ref.* C.J. Ballhausen et. al. *J. Inorg. Nucl. Chem.*, **3**, 178, 1956), the readers may consult the original references. Dichroism of *trans*-[CoCl$_2$(en)$_2$]$^+$ is shown in Fig. 7.15.2.4c.

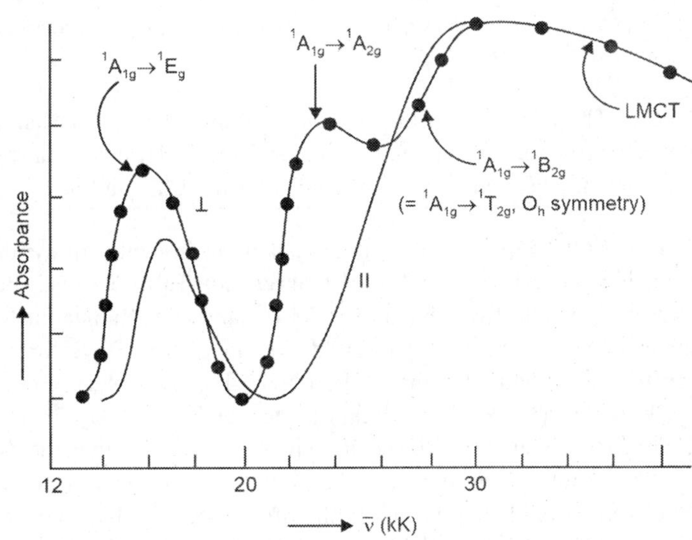

Fig. 7.15.2.4: (c) Dichroism (*i.e.* vibronic polarisation) of *trans*-[CoCl$_2$(en)$_2$]$^+$.

7.15.3 Spectral Transitions in Six-Coordinate Low-spin Complexes of Co(III) Belonging to the O_h, C_{4v} and D_{4h} Point Groups (*cf.* Sec. 7.14.4E for the Ni(II) complexes; Sec. 7.14.5E for the Cr(III) complexes)

We have already mentioned that lowering of symmetry (with respect to O_h) leads to splitting of the ${}^1T_{1g}$ and ${}^1T_{2g}$ states. But splitting of the ${}^1T_{2g}$ state is too small to be detected in the ligand field spectra. On the other hand, splitting of the T_{1g} state is quite significant to introduce its effect on the ligand field transitions. The representative examples are:

O_h: $[Co(NH_3)_6]^{3+}$; C_{4v}: $[CoCl(NH_3)_5]^{2+}$

D_{4h}: *trans*-$[CoCl_2(NH_3)_4]^+$;

We have already discussed and compared the ligand field spectra of *cis* (C_{2v})- and *trans* (D_{4h})-isomers of $[Co(en)_2F_2]^+$ both of which **electronically maintain the tetragonal symmetry** (*cf.* earlier discussion). In the same way, the ligand field spectra of the *cis*- and *trans*- isomers of $[CoCl_2(NH_3)_4]^+$ can be explained. The *cis*- isomer is violet while the *trans*- isomer is bright green. Spectral characteristics of $[Co^{III}(NH_3)_5X]^{2+}$ (X = F, Cl, Br, I) have been discussed in detail in Sec. 7.21.1 (*cf.* Fig. 7.21.1.1).

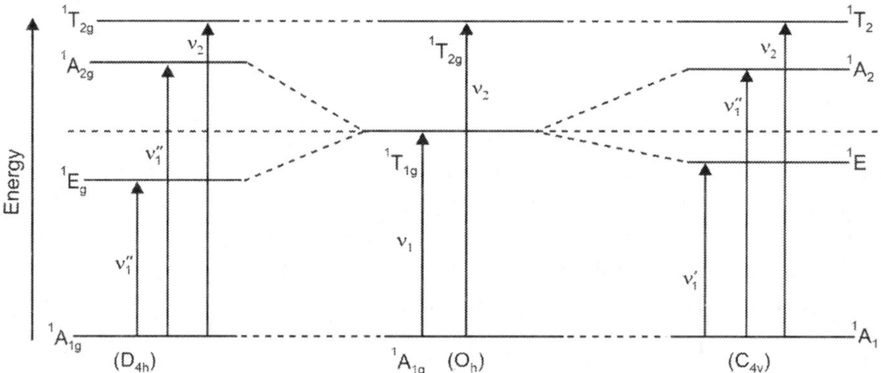

Fig. 7.15.3.1 Electronic transitions in O_h, D_{4h} and C_{4v} point group complexes of Co(III) (*cf.* Correlation Table 7.15.4.2) (Splitting of the T_{2g} state is ignored).

> **Note:** The peak due to the ${}^1A_{1g} \rightarrow {}^1T_{1g}$ transition is **broadened with an unsymmetrical shape** in moving from the O_h symmetry to C_{4v} symmetry. However, the peak shape for the ${}^1A_{1g} \rightarrow {}^1T_{2g}$ transition remains unchanged (if not influenced by the CT band in C_{4v} symmetry).

In $\mathbf{[CoCl(NH_3)_5]^{2+}}$ (C_{4v} symmetry lacking in centre of symmetry), the ${}^1T_{1g}$ state of the O_h system splits into the ${}^1E$ and ${}^1A_2$ states (*cf.* **Correlation Table** 7.15.4.2) but the splitting energy is not sufficiently large to generate two separate peaks (ν_1' and ν_1''). Rather the ν_1 band shows the **doublet character** indicated by the presence of a shoulder (*cf.* Fig. 7.15.3.2). Splitting of the energy levels in $[CoCl(NH_3)_5]^{2+}$ can be understood from that of $[Co(NH_3)_6]^{3+}$. By replacing one NH_3 ligand (stronger field ligand) by a Cl^- ligand (weaker field ligand), we get $[CoCl(NH_3)_5]^{2+}$ (C_{4v}) from $[Co(NH_3)_6]^{3+}$ (O_h). If Cl^- is assumed to occupy one axial position in the *z*-direction, then the *d*-orbitals will split as in ***z*-out distortion,** *i.e.* the orbitals bearing the *z*-component will be stabilised more. Splitting patterns of the energy levels are of similar type for both *trans*-$[CoCl_2(NH_3)_4]^+$ and $[CoCl(NH_3)_5]^{2+}$ (both experiencing the *z*-out distortion, ligand field strength: $NH_3 > Cl^-$). However, the T_{1g} state is splitted more in *trans*-$[CoCl_2(NH_3)_4]^+$ (*cf.* effect of two Cl^- *vs.* one Cl^-).

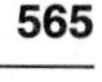

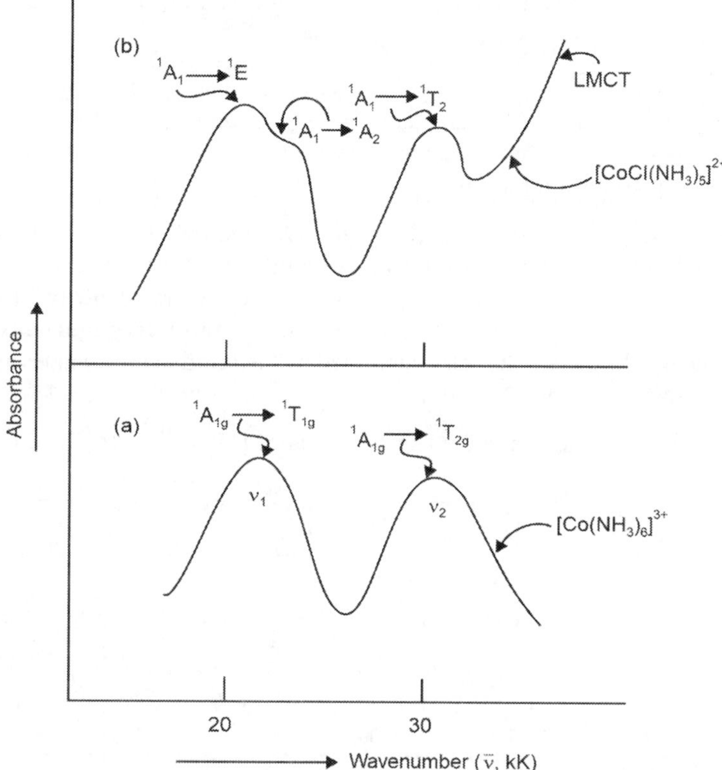

Fig. 7.15.3.2: Comparison of the electronic spectra of $[Co(NH_3)_6]^{3+}(O_h)$ and $[CoCl(NH_3)_5]^{2+}$ (C_{4v}).

In moving from **$[Co(NH_3)_6]^{3+}$ to $[Co(NH_3)_5X]^{2+}$ (X = F, Cl, Br, I),** the T_{1g} level splits into the components E and A_2 levels and the T_{2g} level does not practically split. The magnitude of splitting of the T_{1g} level depends on the difference of the average ligand field strength between the N—Co^{III}—N and N—Co^{III}—X axes. This difference is **maximum** for X = I and **minimum** for X = F. In fact, for X = F, the band does not practically split (*cf.* ligand field strength difference between NH_3 and X^- is **minimum** for X = F) and for X = Cl, Br, only a shoulder (towards the higher energy) appears. For, X = I, practically all the ligand field bands are masked by the LMCT bands. These aspects are discussed in detail in Sec. 7.21.1.

7.15.4 Comparison of the Ligand Field Spectra in the Co(III) and Cr(III) Complexes of O_h and D_3 symmetry (*cf.* Sec. 7.14.5 for group theoretical treatment)

Let us illustrate with the $[M(NH_3)_6]^{3+}(O_h)$ and $[M(en)_3]^{3+}$ (D_3) complexes for M = Co, Cr. In an O_h symmetry, we have the following ligand field transitions.

Cr(III): $^4A_{2g} \xrightarrow{\nu_1} {}^4T_{2g};\ {}^4A_{2g} \xrightarrow{\nu_2} {}^4T_{1g}(F);\ {}^4A_{2g} \xrightarrow{\nu_3} {}^4T_{1g}(P)$

(The high energy ν_3 band is very often obscured by the strong CT band).

Co(III): $^1A_{1g} \xrightarrow{\nu_1} {}^1T_{1g};\ {}^1A_{1g} \xrightarrow{\nu_2} {}^1T_{2g}$

The T_{1g} and T_{2g} states of an O_h symmetry are split in a D_3 symmetry (*cf.* Correlation Table 7.15.4.2):

$$T_{1g}(O_h) \rightarrow A_2 + E;\quad T_{2g}(O_h) \rightarrow A_1 + E$$
$$A_{1g}(O_h) \rightarrow A_1;\quad A_{2g}(O_h) \rightarrow A_2$$

Thus the ν_1-peak of an O_h system should split into two peaks in a D_3 symmetry.

For Cr(III): $\underbrace{{}^4A_{2g} \xrightarrow{\nu_1} {}^4T_{2g}}_{O_h}$; $\underbrace{{}^4A_2 \xrightarrow{\nu_1'} {}^4A_1;\ {}^4A_2 \xrightarrow{\nu_1''} {}^4E}_{D_3}$

For Co(III): $\underbrace{{}^1A_{1g} \xrightarrow{\nu_1} {}^1T_{1g}}_{O_h}$; $\underbrace{{}^1A_1 \xrightarrow{\nu_1'} {}^1A_2;\ {}^1A_1 \xrightarrow{\nu_1''} {}^1E}_{D_3}$

Similarly, the ν_2 band of an O_h-symmetry should also split into two components in a D_3 symmetry. Comparison of the ligand field spectra of $[M(NH_3)_6]^{3+}$ (O_h) and $[M(en)_3]^{3+}$(D_3) indicates that in both the cases, the two bands are symmetrical and **no splitting into the doublets is indicated in** $[M(en)_3]^{3+}$. Thus we can consider that in both types of complexes, the **effective octahedral symmetry, $[M(N)_6]^{3+}$ is maintained.** The chelation has not effectively changed the effective octahedral symmetry in understanding the ligand field spectra. It is also supported by the fact that the ε-values are not dramatically increased in moving from the O_h to D_3 symmetry.

Table 7.15.4.1: Comparison of the ligand field spectral data of $[M(NH_3)_6]^{3+}(O_h)$ and $[M(en)_3]^{3+}(D_3)$ complexes.

Complexes	${}^4A_{2g} \xrightarrow[(cm^{-1})]{\nu_1} {}^4T_{2g}$		${}^4A_{2g} \xrightarrow[(cm^{-1})]{\nu_2} {}^4T_{1g}(F)$	
$[Cr(NH_3)_6]^{3+}$	21,550	(ε = 30)	28,500	(ε = 30)
$[Cr(en)_3]^{3+}$	21,850	(ε = 75)	28,450	(ε = 60)
	${}^1A_{1g} \xrightarrow[(cm^{-1})]{\nu_1} {}^1T_{1g}$		${}^1A_{1g} \xrightarrow[(cm^{-1})]{\nu_2} {}^1T_{2g}$	
$[Co(NH_3)_6]^{3+}$	21,200	(ε = 55)	29,550	(ε = 45)
$[Co(en)_3]^{3+}$	21,500	(ε = 90)	29,600	(ε = 80)

Relative intensities of the peaks of $[M(NH_3)_6]^{3+}$ and $[M(en)_3]^{3+}$ complexes: It is evident that the ε-values of the peaks of $[M(en)_3]^{3+}$ complexes are relatively higher compared to those of the $[M(NH_3)_6]^{3+}$ complexes. **It is due to the fact that in a D_3 symmetry, relaxation of Laporte selection rule is more efficient than that in an O_h symmetry.** In the D_3 point group, there is no centre of symmetry. Thus for the complexes of D_3 symmetry, Laporte selection rule is not applicable and in fact, some of the electronic transitions are allowed by their own rights. However, some of these electronic transitions experience the polarisation (*i.e.* **dichroic properties**). These aspects have been illustrated in terms of group theory treatment (*cf.* Secs. 7.14.5, 7.5.1). In the D_3 point group, the orbitals p_x; p_y; d_{xz}, d_{yz}; $d_{x^2-y^2}$, d_{z^2} belong to the *E*-species. In other words, these *p* and *d* orbitals are of the same symmetry and they can undergo *d-p* mixing. On the other hand, in an O_h point group, the asymmetric vibration can only allow the *d-p* mixing momentarily. Thus, the vibronic coupling is the main pathway to relax the Laporte selection rule

Table 7.15.4.2 Correlation table for O_h group.

O_h	T_d	D_{4h}	C_{4v}	C_{2v}	D_3^*
A_{1g}	A_1	A_{1g}	A_1	A_1	A_1
A_{2g}	A_2	B_{1g}	B_1	A_2	A_2
E_g	E	$A_{1g} + B_{1g}$	$A_1 + B_1$	$A_1 + A_2$	E
T_{1g}	T_1	$A_{2g} + E_g$	$A_2 + E$	$A_2 + B_1 + B_2$	$A_2 + E$
T_{2g}	T_2	$B_{2g} + E_g$	$B_2 + E$	$A_1 + B_1 + B_2$	$A_1 + E$

(*For D_{3d} (centrosymmetric), same result with the *g*-subscripts.)

in the O_h complexes. **In fact, in general, the lowering of symmetry favours the relaxation of Laporte selection rule** (*cf.* T_d *vs.* O_h; D_3 *vs.* O_h; D_{4h} *vs.* C_{2v}). Table 7.15.4.1 indicates that in moving for the O_h symmetry to the D_3 symmetry, the molar extinction coefficients are increased to some extent but the increase is not significantly high (*cf.* ε values: *d–d* transition *vs.* CT transition). Thus it is reasonable to conclude that in both the cases, the effective O_h symmetry is maintained at least roughly.

7.15.5 Ligand Field Spectra of the Low-Spin Complexes of other d^n-Configurations

***d*⁴-system:** The low spin complex $[Mn(CN)_6]^{3-}$ (t_{2g}^4) shows the following two spin allowed ligand field transitions.

$$^3T_{1g} \longrightarrow {}^3E_g;\ {}^3T_{1g} \longrightarrow {}^3A_{1g}\ (\textit{see}\ \text{Tanabe-Sugano diagram})$$

Theoretically the possible transitions are: $^3T_{1g}\rightarrow{}^3E_g$; $^3T_{1g}\rightarrow{}^3T_{2g}$; $^3T_{1g}\rightarrow{}^3A_{1g}$; $^3T_{1g}\rightarrow{}^3A_{2g}$

$[Mn(CN)_6]^{3-}$ in solution shows a band at about 31,000 cm^{-1} with a shoulder. However, in solid state, the spectrum becomes a complex one due to the spin forbidden transitions along with the spin allowed transitions.

***d*⁵-system:** The low spin complex, **$[Fe(CN)_6]^{3-}$ (t_{2g}^5)** experiences the following spin-allowed transitions.

$$^2T_{2g}\rightarrow{}^2A_{2g},\ {}^2T_{1g};\ {}^2T_{2g}\rightarrow{}^2E_g;\ {}^2T_{2g}\rightarrow{}^2A_{1g}.\ (\textit{cf.}\ \text{Tanabe-Sugano diagram})$$

In reality, two bands are noted for $K_3[Fe(CN)_6]$.

$$^2T_{2g} \xrightarrow[30{,}250\ \text{cm}^{-1}]{} {}^2A_{2g},\ {}^2T_{1g};\ {}^2T_{2g} \xrightarrow[36{,}700\ \text{cm}^{-1}]{} {}^2E_g$$

***d*⁶-system:** The low-spin $[Fe(CN)_6]^{4-}$(t_{2g}^6) records two spin-allowed transitions:

$$^1A_{1g} \xrightarrow[31{,}000\ \text{cm}^{-1}]{} {}^1T_{1g};\ {}^1A_{1g} \xrightarrow[37{,}000\ \text{cm}^{-1}]{} {}^1T_{2g};\ (\textit{cf.}\ \text{Table 7.15.1.1})$$

***d*⁷ system:** The low spin ($t_{2g}^6e_g^1$) octahedral complex gives the spin allowed transition, $^2E_g \rightarrow {}^2T_{1g}$, $^2T_{2g}$ (*cf.* Tanabe-Sugano diagram). But the complexes (*e.g.* $[Co(NO_2)_6]^{4-}$ (yellow), $[NiF_6]^{3-}$ (violet), etc.) invariably experience the J.T. distortion (generally *z*-out distortion). Depending on the nature of distortion, the complex may adopt the D_{4h} or C_{4v} symmetry. The electron distribution in the ground state is:

$$\left(d_{xy}, d_{yz}\right)^4 < d_{xy}^2 < d_{z^2}^1 < d_{x^2-y^2}^0$$

Several electronic transitions are possible (*cf.* low-spin square planar complexes of Co(II)).

7.16 LIGAND FIELD SPECTRAL TRANSITIONS IN THE TETRAHEDRAL COMPLEXES OF FIRST TRANSITION SERIES

● **High intensity:** The tetrahedral complexes lacking in centre of symmetry allow some of **the electronic transitions by their own right.** Besides this, the noncentrosymmetric tetrahedral complex can also allow the ***d-p* mixing**. Thus, the Laporte selection rule can be more effectively broken down and **the ligand field bands gain more intensity** (*cf.* Sec. 7.5.1 for details). Sometimes, compared to the octahedral complexes, it may be difficult to differentiate the ligand field bands from the charge transfer bands, because the ligand field bands are also quite strong. In the tetrahedral complexes, the transitions should be interpreted **in terms of MOT.** In the T_d-complexes, the *e*-set (*i.e.* $d_{x^2-y^2}$ and d_{z^2}) is basically constituted by the **nonbonding metal *d*-orbitals** while the t_2-set is constituted by the overlap of the metal *d*-orbitals (*i.e.* d_{xy}, d_{yz}, d_{xz}) and ligand orbitals (*i.e.* *p* or hybrid orbitals), thus the **$e \rightarrow t_2$ transition is not a pure *d–d* transition.**

● **Spin-orbit coupling splitting: Peak splitting or peak broadening by spin-orbit coupling interaction is quite important in understanding the ligand field spectra of the tetrahedral complexes**

(*cf.* Secs. 7.7.4, 8.20.3): The spin-orbit coupling leads to splitting of the *T*-state and the splitting energy is comparable to that of the tetrahedral crystal field splitting energy (Δ_t). Effect of spin-orbit coupling is relatively less important in interpreting the ligand field spectra of the octahedral complexes of 1st transition series because the octahedral crystal field splitting energy (Δ_o) is much higher than the spin-orbit coupling splitting energy.

● **Orgel diagram:** The spin-allowed transitions can be easily predicted from the Orgel Diagram (Figs. 7.10.1,2). In reality, the spin-paired tetrahedral complexes are rarely known.

The octahedral energy diagram of d^{10-n} configuration is applicable for the tetrahedral energy diagram for d^n configuration. In fact, for the sake of simplicity, the Tanabe-Sugano diagrams of high-spin octahedral complexes of d^{10-n} configuration are used to interpret the ligand field transitions in the tetrahedral complexes of d^n-configuration.

(1) *d^1 and d^6 systems* (*D* term): The spin allowed transition is:

$$^2E \rightarrow {}^2T_2\ (d^1\text{-system});\ {}^5E \rightarrow {}^5T_2\ (d^6\text{-system}).$$

$[VCl_4]$ (d^1-configuration) shows the ligand field absorption band at 9,000 cm^{-1} ($\varepsilon \sim 100$) for the $^2E \rightarrow {}^2T_2$ transition. This absorption band can be resolved **into three components** (6,600; 7,850 and 9,250 cm^{-1}). These may arise in different possible ways: ***vibrational fine structure, splitting of the excited state by spin-orbit coupling or Jahn-Teller effect.*** However, theoretical calculations support the Jahn-Teller distortion. The complex also shows the strong LMCT band (Cl → V). Some tetrahedral complexes like $[V(NR_2)_4]$ (R = Me, Et) may adopt the distorted tetrahedral structure of D_{2d} symmetry. It may be noted that in moving from the T_d to D_{2d} symmetry, *E* splits into A_1 and B_1; and T_2 splits into B_2 and *E*. For the d^1 system, B_1 is the ground state for the D_{2d} symmetry (*cf.* Fig. 7.16.1). The transitions are:

D_{2d} (d^1 system): $^2B_1 \rightarrow {}^2E$; $^2B_1 \rightarrow {}^2B_2$ (orbitally forbidden; Solved Problem 15); **$[FeCl_4]^{2-}$, $[FeBr_4]^{2-}$**, etc. are the representative examples of tetrahedral complexes of d^6-system. The spin allowed transition is:

$$^5E \xrightarrow{\nu} {}^5T_2,\ \nu = 4{,}000\ \text{cm}^{-1}\ ([FeCl_4]^{2-})$$
$$= 3{,}000\ \text{cm}^{-1}\ ([FeBr_4]^{2-})$$
$$= 2{,}500\ \text{cm}^{-1}\ ([FeI_4]^{2-})$$

For **$[FeL_2X_2]$ (C_{2v}, rhombic symmetry)**, the low symmetry ligand field splits the *E* and T_2 states as follows:

$$E(T_d) \rightarrow A_1 + A_2\ (C_{2v})$$
$$T_2(T_d) \rightarrow A_1 + B_1 + B_2\ (C_{2v})$$

In practice, in such cases, only **two broad bands** are observed. It indicates that the splitting of the peaks simply broadens the main peaks.

(2) *d^4 and d^9 systems* (*D* term): The spin allowed transition is:

$$^5T_2 \rightarrow {}^5E\ (\text{for } d^4 \text{ system});\ {}^2T_2 \rightarrow {}^2E\ (\text{for } d^9 \text{ system}).$$

$[CrCl_4]^{2-}$, $[CrBr_4]^{2-}$ are the representative examples of the tetrahedral complexes of d^4-system. They show the absorption band at about 10,000 cm^{-1} for the $^5T_2 \rightarrow {}^5E$ transition.

Perfectly tetrahedral complexes of Cu(II) (d^9-system) should record **only one absorption band** for the transition, $^2T_2 \rightarrow {}^2E$. The absorption bands at ~8,000 cm^{-1} and ~9,000 cm^{-1} for **$[CuBr_4]^{2-}$ and $[CuCl_4]^{2-}$** respectively have been considered for the $^2T_2 \rightarrow {}^2E$ transitions (assuming the tetrahedral geometry).

Undistorted tetrahedral Cu(II) complex ($e^4t_2^5$) is unexpeced because it will experience the **Jahn-Teller distortion** (*cf.* t_2 level is unsymmetrically filled in). Cu(II) has a tendency to adopt the square planar geometry and in such cases, the tetrahedral geometry experiences a compression (*cf.* Table 3.11.3.3) along one of the S_4-axis (*i.e.* flattening of the tetrahedron along a C_2-axis leading towards a square planar geometry). This distortion will generate a **structure of D_{2d} symmetry** (which is an intermediate state of the two limiting geometries, *i.e.* tetrahedron and square planar). Splitting of the energy levels and *d*-orbitals in the T_d and D_{2d} symmetry are shown in Fig. 7.16.1 (*cf.* Fig. 3.11.2.1).

The possible electronic transitions are:

$$\mathbf{T_d}: {}^2T_2 \rightarrow {}^2E;\ i.e.\ (e \rightarrow t_2)$$

$\mathbf{D_{2d}}$ (*cf.* **Fig. 7.16.1**): ${}^2B_2 \xrightarrow{\nu_3} {}^2E; {}^2B_2 \xrightarrow{\nu_2} {}^2B_1$ (**orbitally forbidden,** *see* Solved Problem 15);

$${}^2B_2 \xrightarrow{\nu_1} {}^2A_1$$

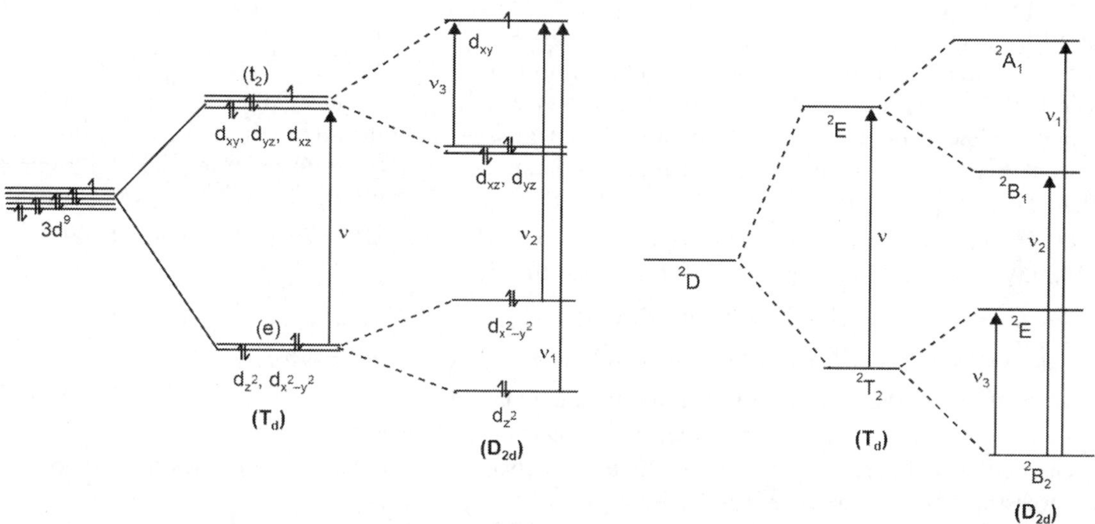

Fig. 7.16.1 Splitting of the *d*-orbitals and energy states in T_d and D_{2d} (compressed tetrahedron) symmetry for Cu(II) (d^9 system).

In terms of the *d*-orbitals, the possible transitions in the D_{2d} symmetry are:

$$d_{z^2}^2 d_{x^2-y^2}^2 \left(d_{xz}d_{yz}\right)^4 d_{xy}^1 \left(i.e.\ {}^2B_2\right) \xrightarrow[\left({}^2B_2 \longrightarrow {}^2A_1\right)]{\nu_1} d_{z^2}^1 d_{x^2-y^2}^2 \left(d_{xz}d_{yz}\right)^4 d_{xy}^2 \left(i.e.\ {}^2A_1\right)$$

$$d_{z^2}^2 d_{x^2-y^2}^2 \left(d_{xz}d_{yz}\right)^4 d_{xy}^1 \left(i.e.\ {}^2B_2\right) \xrightarrow[\left({}^2B_2 \longrightarrow {}^2B_1\right)]{\nu_2} d_{z^2}^2 d_{x^2-y^2}^1 \left(d_{xz}d_{yz}\right)^4 d_{xy}^2 \left(i.e.\ {}^2B_1\right)$$

$$d_{z^2}^2 d_{x^2-y^2}^2 \left(d_{xz}d_{yz}\right)^4 d_{xy}^1 \left(i.e.\ {}^2B_2\right) \xrightarrow[\left({}^2B_2 \longrightarrow {}^2E\right)]{\nu_3} d_{z^2}^2 d_{x^2-y^2}^2 \left(d_{xz}d_{yz}\right)^3 d_{xy}^2 \left(i.e.\ {}^2E\right)$$

It may be noted that the ν_1 transition is *z*-polarised and the ν_3 transition is *x*, *y*-polarised. The ν_2-transition is orbitally forbidden. In short, the transitions in D_{2d} symmetry can be represented as follows:

$$d_{z^2} \xrightarrow{\nu_1} d_{xy}; d_{x^2-y^2} \xrightarrow{\nu_2} d_{xy}; d_{xz}, d_{yz} \xrightarrow{\nu_3} d_{xy}$$

In **$Cs_2[CuCl_4]$,** transitions for $[CuCl_4]^{2-}$ (in D_{2d} symmetry) are:

$$\nu_1 = 9050\ cm^{-1};\ \nu_2 = 7900\ cm^{-1};\ \nu_3 \approx 5{,}000\ cm^{-1}$$

The finer resolution indicates that the ν_3-band consists of two components (5,500 cm^{-1} and 4,800 cm^{-1}). It is suggested that further distortion leads to D_{2d} to C_s symmetry through the splitting of the d_{xz} and d_{yz} orbitals. This splitting leads to the following transitions.

$$d_{xz} \rightarrow d_{xy}\ (5{,}500\ cm^{-1});\ d_{yz} \rightarrow d_{xy}\ (4{,}800\ cm^{-1}).$$

There is another view of **spin-orbit coupling effect** ($\lambda = -830\ cm^{-1}$) to split the $^2T_2$ state. This can explain the possible four transitions in $[CuCl_4]^{2-}$. The orbitally **forbidden transition, $^2B_2 \rightarrow {}^2B_1$,** gains the intensity through the mixing with one component of the ^2E state (which splits through the spin-orbit coupling). Thus the spectral features of $[CuCl_4]^{2-}$ may be explained by the spin-orbit coupling and/or further distortion of D_{2d} to C_s symmetry.

● **Amethyst** is a kind of α-quartz with Fe(III) replacing Si(IV) partly in SiO_2 tetrahedra and introducing alkali metals to balance the charge. Alternatively, replacement of Si(IV) by Fe(III) produces a hole (*i.e.* one missing electron) that can be excited by the ionizing radiation to produce Fe(IV) in the quartz matrix. Amethyst possesses a beautiful **violet colour due to the ligand field transitions of Fe(IV)** (d^4 system) in a tetrahedral coordination (*Ref.* Das *et al.*, *Resonance*, 2025).

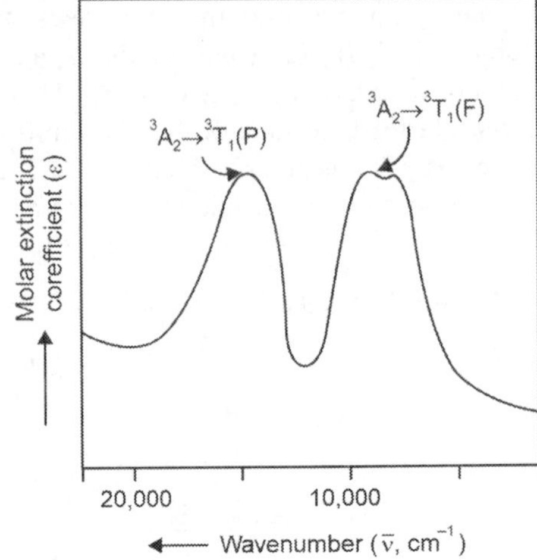

Fig. 7.16.2 Electronic absorption spectrum of $[VCl_4]^-$ (a d^2-tetrahedral complex).

(3) *d^2 and d^7 systems (ground state term F, 1st excited state term P)*: From the Orgel diagram (Fig. 7.10.2), we have the following three spin-allowed transitions.

$^xA_2(F) \xrightarrow{\nu_1} {}^xT_2(F)$ (**orbitally forbidden**);

$^xA_2(F) \xrightarrow{\nu_2} {}^xT_1(F)$ and $^xA_2(F) \xrightarrow{\nu_3} {}^xT_1(P)$. ($x = 3$ for d^2; $= 4$ for d^7)

Without considering any configurational interaction,

$$\nu_1 = 10Dq,\ \nu_2 = 18Dq,\ \nu_3 = 12Dq + 15B'$$

Due to the configurational repulsive interaction between the $T_1(P)$ and $T_1(F)$ states, they follow the **noncrossing rule** and the energy order is: $\nu_3 > \nu_2 > \nu_1$.

Tetrahedral complexes of the d^2-system (*e.g.* **$[VCl_4]^-$**) are rarely known. On the other hand, tetrahedral complexes of Co(II) (d^7-system), have been studied in more detail to assign their spectral transitions.

(A) $[VCl_4]^-$: Here, one band consists of three components (8,000 cm^{-1}, 9,500 cm^{-1} and 10,500 cm^{-1}) while the high energy band appears at 15,000 cm^{-1}. Splitting of the *T*-state through the spin-orbit coupling is to be considered to interpret the spectra.

(B) $[CoCl_4]^{2-}$: The spectrum of **$[CoCl_4]^{2-}$** in aqueous solution indicates two bands:

$$5{,}800\ cm^{-1}\ (\varepsilon = 64),\ 15{,}000\ cm^{-1}\ (\varepsilon = 520).$$

The bands have been assigned as follows:

$$^4A_2(F) \xrightarrow[(5{,}800\ cm^{-1})]{\nu_2} {}^4T_1(F);\ {}^4A_2(F) \xrightarrow[(\sim 15{,}000\ cm^{-1})]{\nu_3} {}^4T_1(P)$$

If the band at 5,800 cm^{-1} is considered for the transition, $^4A_2(F) \rightarrow {}^4T_2(F)$, *i.e.* ν_1, then the $10Dq$ value becomes 5,800 cm^{-1} ***which is too high for the tetrahedral complexes of Co(II)***. In fact, this value cannot be fitted in the energy diagram. This is illustrated below:

if $\nu_1 = 5{,}800\ cm^{-1}$, and $\nu_2 = 15{,}000\ cm^{-1}$, then $\frac{\nu_2}{\nu_1}$ becomes 2.58.

From Fig. 7.10.4, it is evident that for the A_2-ground spectroscopic state, the $\frac{\nu_2}{\nu_1}$ ratio lies within the range 1.8 – 1.22. Thus the band assignments, $\nu_1 = 5{,}800\ cm^{-1}$ and $\nu_2 = 15{,}000\ cm^{-1}$ are not theoretically acceptable.

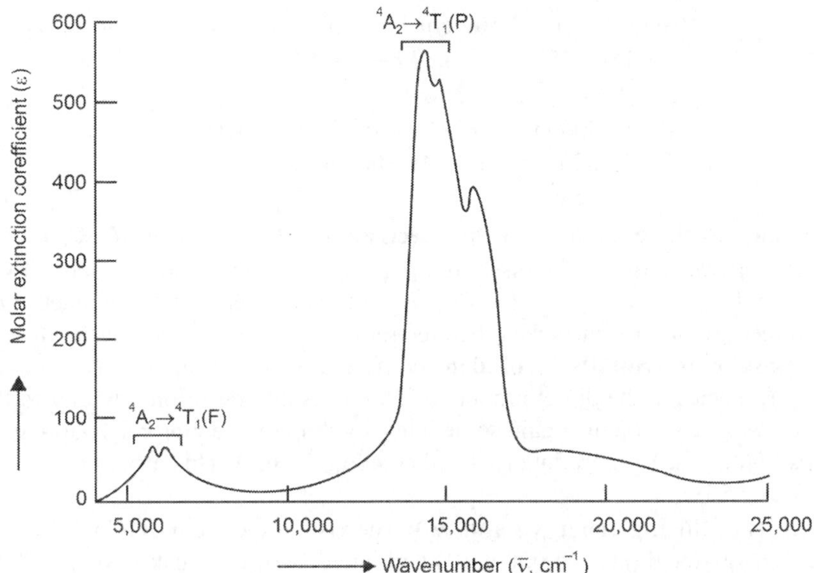

Fig. 7.16.3 Electronic absorption spectrum of $[CoCl_4]^{2-}$ (a d^7-tetrahedral complex) (see Fig. 1.10.1, Vol. 4)

In the energy diagram, the ν_3/ν_2 ratio $\left(\frac{15,000}{5,800} = 2.58\right)$ can be fitted for $Dq/B' = 0.45$ and $\nu_2/B' = 7.9$.

It leads to: $$\frac{5,800}{B'} = 7.9 \text{ or, } B' = 734 \text{ cm}^{-1}$$

and $$Dq = B' \times 0.45 \approx 330 \text{ cm}^{-1} \quad \text{or, } 10Dq = 3,300 \text{ cm}^{-1}$$

This B' value is about 70% of the free ion value (*cf.* $B_0 = 1033$ cm^{-1}). It indicates the presence of an extensive ***covalent interaction*** in the complex, $[CoCl_4]^{2-}$.

We can verify the band assignments from Fig. 7.10.4.

$$\left.\frac{\nu_3}{\nu_1} = \frac{15,000}{3,300} = 4.54; \frac{\nu_2}{\nu_1} = \frac{5,800}{3,300} = 1.75\right\}\text{(Experimental values)}$$

For $\frac{Dq}{B'} = 0.45$, Fig. 7.10.4 theoretically gives the values 4.59 and 1.74 for ν_3/ν_1 and ν_2/ν_1 respectively. These calculated values are in good agreement with the experimental values.

Thus the transition, $^4A_2(F) \rightarrow {}^4T_2(F)$, *i.e.* ν_1, is expected to appear at 3,300 cm^{-1}. In fact, studies in the infra-red region indicate the predicted band to appear at about 3,500 cm^{-1}.

The ν_1 (= $10Dq$) band at 3,500 cm^{-1} indicates: $10Dq = 3,500$ cm^{-1}.

Thus, the calculated positions of ν_2 becomes as follows:

$$\nu_2 = 18Dq = 18 \times 350 \text{ cm}^{-1} = 6,300 \text{ cm}^{-1} \text{ (without configurational interaction)}$$

This calculated value is significantly different from the experimental value (= 5,800 cm^{-1}). Similarly, if ν_3 (= $12Dq + 15B_0$) is calculated, it becomes:

$$\nu_3 = 12 \times 350 + 15,500 = 19,700 \text{ cm}^{-1}; \text{ taking } 15B_0 = 15,500 \text{ cm}^{-1}.$$

This calculated value differs significantly from the experimental value (= 15,000 cm^{-1}). Apparently, it is due to the reduction of B-value in the complex (*i.e.* $15B'$ value is dropped to about 10,800 cm^{-1} which is about 70% of $15B_0$). ***The above discrepancy can be explained by considering the configurational interaction between the $^4T_1(F)$ and $^4T_1(P)$ states and the reduced B'-value.*** The modified relations are:

$$\nu_1 = 10Dq, \ \nu_2 = 18Dq - x, \ \nu_3 = 12Dq + 15B' + x \ (x = \text{configurational interaction energy}).$$

Taking $10Dq\ (= \nu_1) = 3{,}500\ cm^{-1}$ (experimental value), ν_2 becomes:

$$\nu_2 = 18 \times 350 - x = 6{,}300 - x = 5{,}800\ cm^{-1}$$

i.e. $$x = 500\ cm^{-1}$$

$$\nu_3 = 15{,}000\ cm^{-1} = 12 \times 350 + 15B' + 500$$

or, $$15B' = 15{,}000 - 4{,}700 = 10{,}300\ cm^{-1}$$

or, $$B' \approx 687\ cm^{-1}$$

(C) Important characteristic features of the spectrum of blue coloured $[CoCl_4]^{2-}$

(i) **Allowedness of the spectral transitions:** The $A_2 \rightarrow T_1$ transition is **orbitally allowed** (*cf.* $A_2 \times \mu \times T_1 = A_2 \times T_2 \times T_1 = A_1 + E + T_1 + T_2$; presence of the totally symmetric representation A_1 in the direct product result makes the electronic transition allowed by its own right) but the $A_2 \rightarrow T_2$ transition is **orbitally forbidden** (*cf.* $A_2 \times \mu \times T_2 = A_2 \times T_2 \times T_2 = A_2 + E + T_1 + T_2$, absence of A_1 species in the direct product result makes the transition orbitally forbidden). However, the $A_2 \rightarrow T_2$ transition may gain some intensity through the vibronic coupling. In fact, the $A_2 \rightarrow T_1$ transition gives the more intense band (*see* Fig. 7.16.3). This aspect has been discussed in detail in Sec. 7.5.1.

(ii) **High intensity of the high energy band (ν_3):** The visible spectrum of $[CoCl_4]^{2-}$ is dominated by the highest energy band (at ~ $15{,}000\ cm^{-1}$) and intensity of this peak is very high ($\varepsilon \approx 500$). On the other hand, for $[Co(OH_2)_6]^{2+}$, intensity of the spectrum is very low and it looks pale pink. The **high intensity** of the absorption bands in the tetrahedral complexes is the common feature, because some of the electronic transitions are allowed by their own right in the tetrahedral complexes. Some forbidden transitions may be allowed through the vibronic coupling. Besides these, for the noncentrosymmetric tetrahedral complexes, ***d–p* mixing allows the relaxation** of Laporte Selection rule better. All these aspects have been discussed in detail in Sec. 7.5.1.

(iii) **Relatively lower intensity of the lower energy band ν_2:** Here it is worth mentioning that, the transition, $^4A_2(F) \rightarrow {}^4T_1(F)$, *i.e.* ν_2 band is relatively weak in intensity ($\varepsilon \approx 60$).

(iv) **ν_1-band in the infra-red region:** The $^4A_2(F) \rightarrow {}^4T_2(F)$ orbitally forbidden transition generally occurs in the infra-red region and this band (*i.e.* ν_1) is very often obscured by the vibrational transitions. ***Thus spectrum of the tetrahedral Co(II)-complexes is mainly characterized by the ν_2 and ν_3 ligand field transitions.***

(v) **Bandwidth and fine-structure components in the ν_2 and ν_3 bands:** For $[CoCl_4]^{2-}$, both the ν_2 *i.e.* $^4A_2(F) \rightarrow {}^4T_1(F)$, and ν_3, *i.e.* $^4A_2(F) \rightarrow {}^4T_1(P)$, bands appear with some **structure** (*i.e.* splitting). **Spin-orbit coupling can split the *T*-states** (*cf.* Secs. 7.7.4 and 8.20.3). The fine-structure in the bands are believed to be caused due to spin-orbit coupling. Splitting of the *T*-states may be also caused by **lowering of symmetry,** *e.g.* $T_d \rightarrow D_{2d} \rightarrow C_s$, (*cf.* Fig. 7.16.1). The spin-orbit coupling also partially allows the **spin-forbidden quartet → doublet transitions.** Some of these spin-forbidden transitions occur in the range of spin-allowed transitions. These factors like *vibronic coupling, symmetry lowering, spin-orbit coupling, spin-forbidden transitions are responsible to give* the **fine structure components** of the ν_2 and ν_3 bands.

(vi) **Complex envelopes of the spectral bands:** The visible transitions in the tetrahedral complexes, generally have the **complex envelopes** because of the presence of several spin-forbidden transitions (to the doublet states) occurring in the same region of spin-allowed transitions. These spin-forbidden transitions earn some intensity through the *spin-orbit coupling allowing the mixing of quartet and doublet states and intensity stealing from the nearby spin allowed transitions.* The spin-forbidden transitions in $[CoCl_4]^{2-}$ also produce the bands near $20{,}000\ cm^{-1}$. Intensities of these bands are almost comparable to those of the spin-allowed bands of the octahedral complexes.

In fact, bandwidths of the ν_2 and ν_3 bands are in the range of about 1500–3000 cm^{-1}. For the computational purpose, ν_2 and ν_3 are assumed to represent the **centre of gravity of the respective broad bands.**

Peak broadening through the spin-orbit coupling in T_d vs. O_h complexes: Peak splitting or peak broadening effects by spin-orbit coupling is not so important for the octahedral complex of 1st transition series because the energy separation caused by crystal field splitting in the O_h systems is much larger than the energy separation caused by spin-orbit coupling to split the *T*-state (*cf.* Sec. 7.7.4).

Positions of the ligand field bands of tetrahalidocobalt(II) complexes: Ligand field strength: $Cl^- > Br^- > I^-$

	ν_2 ($\approx 18Dq$) (cm^{-1})	ν_3 ($\approx 12Dq + 15B'$)
$[CoCl_4]^{2-}$	5,800	15,000
$[CoBr_4]^{2-}$	5,300	13,700
$[CoI_4]^{2-}$	5,000	12,500

(D) Characteristics of the ligand field bands of $[Co(NCS)_4]^{2-}$: $\nu_2 \approx 18Dq \approx 7{,}780\ cm^{-1}$, $\nu_3 \approx 12Dq + 15B' \approx 16{,}250\ cm^{-1}$. The ν_1 transition is not definitely 7,780 cm^{-1} because it will lead to $10Dq = 7{,}780\ cm^{-1}$ which is too large for the tetrahedral complex of Co(II). Thus it is quite reasonable to consider:

$$\nu_2 = 7{,}780\ cm^{-1} \text{ and } \nu_3 = 16{,}250\ cm^{-1}$$

Without considering the symmetry interaction, we have:

$$\nu_2 = 18Dq = 7{,}780\ cm^{-1},\ i.e.\ Dq \approx 432\ cm^{-1} \text{ or } 10Dq = 4320\ cm^{-1}$$

$$\nu_3 = 12Dq + 15B' = 16{,}250\ cm^{-1},\ i.e.\ B' = \frac{16{,}250 - 12 \times 432}{15}\ cm^{-1} = 738\ cm^{-1}$$

More refined values of B' and $10Dq$ can be obtained by using the Tanabe-Sugano diagram or Fig. 7.10.4

$$\frac{\nu_3}{\nu_2} = \frac{{}^4A_2(F) \to {}^4T_1(P)}{{}^4A_2(F) \to {}^4T_1(F)} = \frac{16{,}250}{7{,}780} = 2.1$$

This ratio leads to the following values (obtained from Figs. 7.10.4 and T.S. diagram):

$$\frac{Dq}{B'} = 0.65,\ \frac{\nu_3}{B'} = 23.38,\ \frac{\nu_2}{\nu_1} = 1.71 \text{ and } \frac{\nu_3}{\nu_1} = 3.6$$

$$\frac{\nu_3}{B'} = \frac{16{,}250}{B'} = 23.38,\ i.e.\ B' = \frac{16{,}250}{23.38} = 695\ cm^{-1}$$

$$\frac{Dq}{B'} = 0.65,\ i.e.\ Dq = 695 \times 0.65 \approx 452\ cm^{-1},\ i.e.\ \nu_1 = 10Dq = 4{,}520\ cm^{-1}$$

$$\frac{\nu_2}{\nu_1} = 1.71,\ i.e.\ \frac{7{,}780}{\nu_1} = 1.71, \text{ or } \nu_1 = 4{,}550\ cm^{-1}$$

$\nu_2 = 18Dq - x = 18 \times 452 - x = 7{,}780\ cm^{-1}$, *i.e.* $x = 18 \times 452 - 7{,}780 = 8{,}136 - 7{,}780 = 356\ cm^{-1}$

x (configurational interaction energy) = 356 cm^{-1}

(E) More rigorous energy expression for the transitions in the tetrahedral d^7-complexes: Here, it may be mentioned that the detailed calculations of ν_1, ν_2 and ν_3 bands can be done as follows:

$$\nu_1 = \Delta$$

$$\nu_2 = 1.5\Delta + 7.5B' - \tfrac{1}{2}\left[225B'^2 + \Delta^2 - 18B'\Delta\right]^{1/2} = 1.5\Delta + 7.5B' - Q$$

$$\nu_3 = 1.5\Delta + 7.5B' + \tfrac{1}{2}\left[225B'^2 + \Delta^2 - 18B'\Delta\right]^{1/2} = 1.5\Delta + 7.5B' + Q$$

$$\Delta = \frac{1}{3}[\nu_2 + \nu_3 - 15B']$$

$$Q = \frac{1}{2}[\nu_3 - \nu_2]$$

$$4Q^2 = \Delta^2 - 18\Delta B' + 225B'^2$$

Let us illustrate the application of the above relations for a particular tetrahedral complex of Co(II):

(F) Bis(kojato)cobalt(II): One band at 7,140 cm^{-1}; a doublet peak (18,180 cm^{-1} and 19,600 cm^{-1}). *For the Co(II)-tetrahedral complexes, the ν_1 band lies in the infra-red region.* Thus it is reasonable to assume: ν_2 = 7,140 cm^{-1}, ν_3 (split by spin-orbit coupling) as the mean of 18,180 cm^{-1} and 19,600 cm^{-1}, *i.e.* $\nu_3 = \frac{1}{2}(18{,}180 + 19{,}600) = 18{,}890\ cm^{-1}$

$$\Delta = \frac{1}{3}[\nu_2 + \nu_3 - 15B']$$

$$= \frac{1}{3}[7{,}140 + 18{,}890 - 15B'] = 8{,}676.7 - 5B'$$

$$Q = \frac{1}{2}[\nu_3 - \nu_2] = \frac{1}{2}[18{,}890 - 7{,}140] = 5{,}875\ cm^{-1}$$

$$4Q^2 = \Delta^2 - 18\Delta B' + 225B'^2$$

Co(II)-complex of **kojic acid**

Using, $\Delta = 8{,}676.7 - 5B'$ and $Q = 5{,}875\ cm^{-1}$, the acceptable solution of the above equation leads to $B' = 916\ cm^{-1}$ (*cf.* other root is negative which is not acceptable).

Thus, $\Delta = 8{,}676.7 - 5 \times 916 \approx 4{,}100\ cm^{-1}$.

(G) Blue spinel and **cobalt blue glass** (*Ref.* Das *et al.*, *Resonance*, 2025)**:** Spinel $MgAl_2O_4$ is colourless but replacing Mg(II) (a d^0 system), partly by Co(II) (a d^7 system), blue spinels can be prepared (blue colour due to the ligand field transitions in the $Co^{II}O_4$ tetrahedra moieties). **Cobalt blue glass** is prepared by adding a small amount of CoO or $CoCO_3$ which thermally decomposes to CoO and CO_2 in the molten glass during the manufacturing step. The tetrahedral CoO_4 units present in the tetrahedral silicate matrix are responsible to impart the blue colour due to the $d \to d$ transitions. The tetrahedral CoO_4 units show the **relatively intense colour** in the range **blue to green** depending on the Co(II)—O bond length while the octahedral CoO_6 units show the **weak pink colour** due to the $d \to d$ transition (*cf.* $10Dq_{Oh} \gg 10Dq_{Td}$).

(4) d^5-*system*: For the d^5 system ($\equiv d^{10-5}$), the same energy level diagram (Fig. 7.14.9.1) is applicable for the octahedral, tetrahedral or cubic fields. **$[MnCl_4]^{2-}$** and **$[MnBr_4]^{2-}$** are the representative examples in this group. The ground state is $^6A_1$. There is no excited state having the spin multiplicity 6. ***Thus there are only spin-forbidden transitions.***

These are: $^6A_1 \xrightarrow{\nu_1} {}^4T_1(G)$; $^6A_1 \xrightarrow{\nu_2} {}^4T_2(G)$;

$^6A_1 \xrightarrow{\nu_3} {}^4E(G)$; $^6A_1 \xrightarrow{\nu_4} {}^4A_1(G)$, etc.

The ν_3 and ν_4 transitions are merged, *i.e.* $^6A_1 \to {}^4A_1(G)$, $^4E(G)$. These transitions are independent of 10Dq and consequently, these are **sharp** (*cf.* Fig. 7.7.3.1). It may be noted that in the high-spin Mn(II)-octahedral complexes, the transitions are also *spin-forbidden*. But compared to the octahedral complexes, in general, the spectral bands of the tetrahedral complexes are about 200 times stronger in intensity (*cf.* the effect of *d–p* mixing and allowedness of the electronic transitions by their own rights).

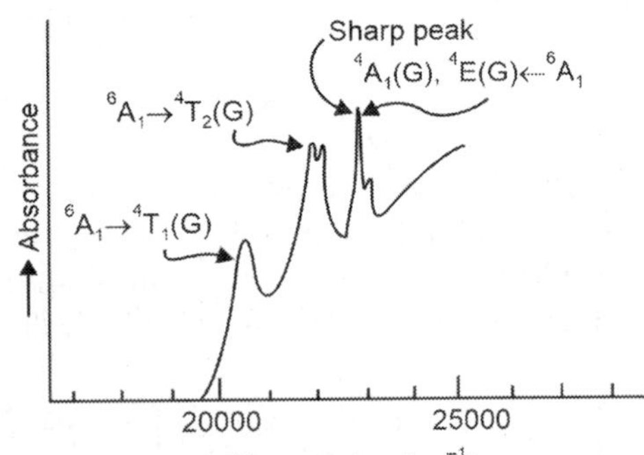

Fig. 7.16.5 Electronic absorption spectrum of $[MnCl_4]^{2-}$ in $Cs_2[MnCl_4]$.

$[FeCl_4]^-$ is also an example of tetrahedral complex of d^5-system. The strong charge transfer bands in $[FeCl_4]^-$ complicate the interpretation of the ligand field bands. Three bands have been characterized.

(5) *d^3 and d^8-system*: The possible transitions are:

$$\left.\begin{array}{l} {}^xT_1(F)\xrightarrow{\nu_1}{}^xT_2(F),\ \nu_1 = 8Dq + x' \approx 8Dq \\ {}^xT_1(F)\xrightarrow{\nu_2}{}^xA_2(F),\ \nu_2 = 18Dq + x' \approx 18Dq \\ {}^xT_1(F)\xrightarrow{\nu_3}{}^xT_1(P),\ \nu_3 = 15B' + 6Dq + 2x' \approx 15B' + 6Dq \end{array}\right\}\left(x = 4 \text{ for } d^3,\ x = 3 \text{ for } d^8\right)$$

$$i.e.\ \nu_2 - \nu_1 = 10Dq;\ \nu_3 - 2\nu_1 = 15B' - 10Dq \text{ and } 15B' = \nu_2 + \nu_3 - 3\nu_1$$

Here x' gives the measure of configurational energy due to the **symmetry interaction** between ${}^xT_1(F)$ and ${}^xT_1(P)$. More rigorously, we can write:

$$\nu_1[{}^xT_1(F) \rightarrow {}^xT_2] = \frac{1}{2}\left[10Dq - 15B' + \left\{225B'^2 + 100(Dq)^2 + 180DqB'\right\}^{1/2}\right]$$

$$\nu_2[{}^xT_1(F) \rightarrow {}^xA_2(\mathrm{F})] = \frac{1}{2}\left[30Dq - 15B' + \left\{225B'^2 + 100(Dq)^2 + 180DqB'\right\}^{1/2}\right]$$

$$\nu_3[{}^xT_1(F) \rightarrow {}^xT_1(\mathrm{P})] = \left[225B'^2 + 100(Dq)^2 + 180DqB'\right]^{1/2}$$

The ν_2 transition represents a **two-electron transition**. Thus it is a **forbidden transition** and consequently, its intensity is relatively less.

The energy levels ${}^xA_2(F)$ and ${}^xT_1(P)$ can cross each other. Thus, the relative positions of ν_2 and ν_3 depend on the crystal field strength. At the lower crystal field strength, $\nu_2 < \nu_3$ while at the higher crystal field strength, $\nu_3 < \nu_2$. In most of the tetrahedral complexes (where the crystal field splitting is relatively smaller), $\nu_2 < \nu_3$ condition is maintained.

Tetrahedral complexes of Ni(II) are the common examples of the d^8-system. The ν_1-band $[{}^3T_1(F) \rightarrow {}^3T_2(F)]$ generally lies in the *infrared region* (3,000–5,000 cm^{-1}). The ν_2 band $[{}^3T_1(F) \rightarrow {}^3A_2(F)]$ generally lies in the near infra-red region (6,500 to 9,000 cm^{-1}). This two-electron transition

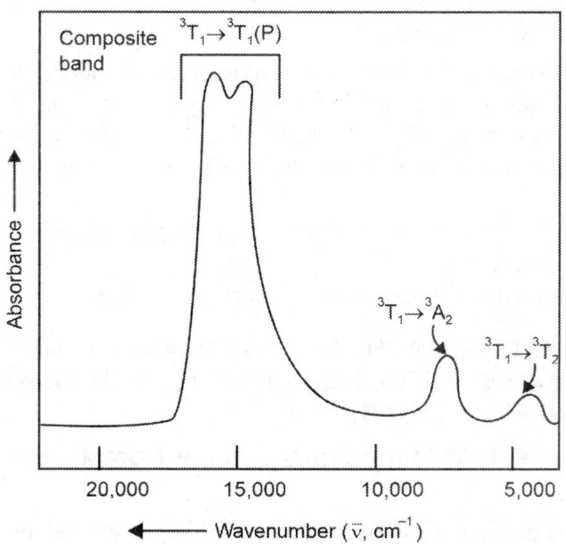

Fig. 7.16.6 Electronic absorption spectrum of $[NiCl_4]^{2-}$ (a d^8-tetrahedral complex).

band is a forbidden transition and it is weak. The ν_3-band [$^3T_1(F) \rightarrow {}^3T_1(P)$] lies in the *visible range* (12,000–17,000 cm^{-1}, $\varepsilon \approx$ 100–200). Some **spin forbidden transitions** to the components of ^1D and ^1G terms have been noted. These are very weak bands.

$[NiCl_4]^{2-}$ shows two bands at 7,800 cm^{-1} and at about 16,000 cm^{-1}. In fact, the band at 16,000 cm^{-1} represents a **multiplet band (*i.e.* splitting pattern)**. The band splitting may be due to **splitting of the *T*-state due to spin-orbit coupling and/or Jahn-Teller distortion.** The centre of gravity of this multiplet band lies at 16,000 cm^{-1}. The assigned transitions are:

$$^3T_1(F) \xrightarrow{\nu_2} {}^3A_2(F), \ \nu_2 = 7{,}800 \text{ cm}^{-1} \ (\varepsilon \approx 12)$$
$$^3T_1(F) \xrightarrow{\nu_3} {}^3T_1(P), \ \nu_3 = 16{,}000 \text{ cm}^{-1} \ (\varepsilon \approx 70)$$

The band (at 7,800 cm^{-1}) cannot be assigned for ν_1, *i.e.* $^3T_1(F) \rightarrow {}^3T_2(F)$. If it is considered for ν_1, then it leads to: $\nu_1 = 8Dq$, *i.e.* 7,800 cm^{-1} = $8Dq$, or $10Dq$ = 9,750 cm^{-1} which is too high for the tetrahedral $[NiCl_4]^{2-}$ complex (*cf.* $10Dq$ for octahedral $[Ni(OH_2)_6]^{2+}$ is 8,500 cm^{-1}). The ratio ν_3/ν_2 $= \frac{16{,}000}{7{,}800} = 2.05$ fits with the following value:

$Dq/B' = 0.5$ (obtained from T.S. diagram or Fig. 7.10.4)

and	$\dfrac{^3T_1(F) \longrightarrow {}^3A_2(F)}{B'} = \dfrac{\nu_2}{B'} = 9.2$	$\dfrac{^3T_1(F) \rightarrow {}^3T_1(P)}{B'} = \dfrac{\nu_3}{B'} = 18.44$
or,	$\dfrac{7{,}800}{B'} = 9.2$; or $B' \approx 848$ cm^{-1}	or, $\dfrac{16{,}000}{18.44} = B' \approx 868$ cm^{-1}
and	$Dq = B' \times 0.5 = 848 \times 0.5 = 424$ cm^{-1}	and $Dq = B' \times 0.5 = 868 \times 0.5 = 434$ cm^{-1}
i.e.	$10Dq = 4{,}240$ cm^{-1}	*i.e.* $10Dq = 4{,}340$ cm^{-1}

Taking, $\nu_2 \approx 18Dq$ (without considering the configurational interaction energy), we get:

$$7{,}800 \text{ cm}^{-1} = 18Dq, \text{ or } 10Dq \approx 4{,}330 \text{ cm}^{-1}$$

and $\quad \nu_3 \approx 6Dq + 15B'$; or, $16{,}000 = 6 \times 433 + 15B'$

or, $\quad 15B' = 13{,}402$

or, $\quad B' = 893.5$ cm^{-1}

The results (without considering the configurational interaction energy) are fairly in good agreement with the results obtained by the more refined method, *i.e.* analysis by using the Tanabe-Sugano diagram and transition energy ratio values (*i.e.* Fig. 7.10.4). It indicates that the symmetry interaction energy (x') is not very much significant and in fact, it happens so for the weaker crystal field splitting (*cf.* Fig. 7.10.2).

Taking, $Dq \approx 425$ cm^{-1}, the ν_1 (= $8Dq$) band is predicted to appear at about 3,400 cm^{-1} (*cf.* experimentally found at 4,080 cm^{-1}).

For **$[NiBr_4]^{2-}$**, the spectral bands appear at: ν_2 = 7,000 cm^{-1} (ε = 30), ν_3 = 15,200 cm^{-1} ($\varepsilon \approx 200$).

7.17 SPECTRAL TRANSITIONS IN THE SQUARE PLANAR COMPLEXES: SPECTRAL CHARACTERISTICS OF THE OCTAHEDRAL, TETRAHEDRAL AND SQUARE PLANAR COMPLEXES OF NICKEL(II)

7.17.1 Ligand Field Transitions in the Square Planar Complexes of Ni(II) and other d^8-Systems

We have already discussed the spin-allowed transitions in the octahedral and tetrahedral complexes of Ni(II). Now we shall discuss the spectral transitions in the square planar complexes of Ni(II).

In moving from the O_h to square planar geometry (D_{4h}), the following changes occur (see Table 7.23.2):

$$A_{1g} \to A_{1g};\ A_{2g} \to B_{1g};\ E_g \to A_{1g} + B_{1g}$$
$$T_{1g} \to A_{2g} + E_g;\ T_{2g} \to B_{2g} + E_g$$

The limiting situation of z-out distortion leads to the square planar geometry (D_{4h}). Positions of the different d-orbitals in the square planar geometry may depend on the condition of ligand field strength (see Vol. 4). However, the uppermost orbital is $d_{x^2-y^2}$ (b_{1g}). The possible ways of d-orbital splitting in a square planar geometry are shown in Fig. 7.17.1.1.

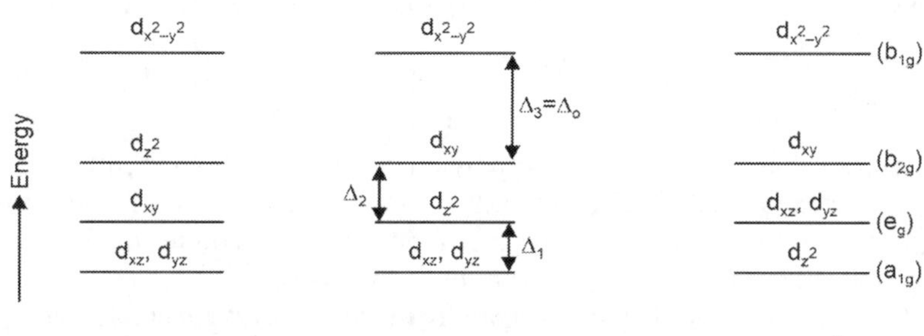

Fig. 7.17.1.1 Relative positions (not in scale) of the d-orbitals in the square planar complexes of d^8-system depending on the crystal field strength. **Note:** In all situations, for moving from the O_h to D_{4h} symmetry, energy separation between the d_{xy} and $d_{x^2-y^2}$ orbitals will remain the same (= Δ_o or $10Dq_o$).

From Fig. 7.17.1.1, it is evident that the $d_{x^2-y^2}$ orbital is of the highest energy while there is some uncertainty in the relative positions of the remaining four d-orbitals. For the **diamagnetic** Ni(II) (d^8 system) square planar complexes, the lower four d-orbitals are filled in and these four d-orbitals are so close in energy that it may be difficult to identify individually the three possible electronic transitions to the topmost vacant $d_{x^2-y^2}$ orbital. *The possible transitions may overlap to produce a single absorption band.*

Here it is reasonable to point out that in most of the square planar complexes of Ni(II), the energy difference (Δ_1) between a_{1g} (d_{z^2}) and e_g (d_{xz}, d_{yz}) is quite small compared to the energy difference (Δ_2) between $a_{1g}\left(d_{z^2}\right)$ and b_{2g} (d_{xy}); and the energy difference (Δ_3) between b_{2g} (d_{xy}) and $b_{1g}\left(d_{x^2-y^2}\right)$. Thus we can write:

$$\Delta_3\ (\approx \Delta_o \text{ for the hypothetical octahedral complex}) > \Delta_2 \gg \Delta_1$$

For most of the square planar complexes of Co(II), Ni(II), and Cu(II), the following relationship is maintained (*cf.* Secs. 3.5.6, 3.13, Vol. 4).

$$\Delta_3\ (\approx \Delta_o),\ \Delta_2 \approx \tfrac{2}{3}\Delta_3,\ \Delta_1 \approx \tfrac{1}{2}\Delta_3$$

But, for **[Ni(CN)$_4$]$^{2-}$**, the following relationship is found from the spectroscopic evidence.

$$\Delta_3\ (\approx \Delta_o),\ \Delta_2 \approx \frac{2}{5}\Delta_3,\ \Delta_1 \approx \frac{1}{38}\Delta_3$$

For the tetragonal (D_{4h}) Ni(II) (d^8) complexes, theoretically two possibilities may arise:

$e_g^4\, a_{1g}^2\, b_{2g}^2$ (*i.e.* **diamagnetic; $^1A_{1g}$ state**) when $\Delta_3 > P$; and $e_g^4\, a_{1g}^2\, b_{2g}^1\, b_{1g}^1$ (*i.e.* **paramagnetic, $^3A_{2g}$ state**) when $P > \Delta_3$. But in most of the cases, the ground state is characterised by the singlet (diamagnetic state) $^1A_{1g}$ sate for the d^8 system. For the **paramagnetic 6-coordinate Ni(II) complexes of D_{4h} symmetry,** the possible electronic transitions are shown in Fig. 7.14.4.3.

Now let us consider all the **possible electronic transitions in the diamagnetic square planar complexes of Ni(II)** (*cf.* Fig. 7.17.1.2).

$$e_g^4 a_{1g}^2 b_{2g}^2 b_{1g}^0, \text{ i.e. } {}^1A_{1g} \xrightarrow[({}^1\mathbf{A_{1g}} \to {}^1\mathbf{A_{2g}})]{\nu_1} e_g^4 a_{1g}^2 b_{2g}^1 b_{1g}^1, \text{ i.e. } {}^1A_{2g};\ \nu_1 \approx \Delta_3\ (\approx \Delta_o)$$

$$e_g^4 a_{1g}^2 b_{2g}^2 b_{1g}^0, \text{ i.e. } {}^1A_{1g} \xrightarrow[({}^1\mathbf{A_{1g}} \to {}^1\mathbf{B_{1g}})]{\nu_2} e_g^4 a_{1g}^1 b_{2g}^2 b_{1g}^1, \text{ i.e. } {}^1B_{1g};\ \nu_2 \approx \Delta_3 + \Delta_2$$

$$e_g^4 a_{1g}^2 b_{2g}^2 b_{1g}^0, \text{ i.e. } {}^1A_{1g} \xrightarrow[({}^1\mathbf{A_{1g}} \to {}^1\mathbf{E_g})]{\nu_3} e_g^3 a_{1g}^2 b_{2g}^2 b_{1g}^1, \text{ i.e. } {}^1E_g;\ \nu_3 \approx \Delta_3 + \Delta_2 + \Delta_1 \approx \Delta_3 + \Delta_2 = \nu_2$$

(Δ_1 is quite small)

The said electronic transitions **in terms of the orbitals** can be simply represented as follows:

$$e_g\left(i.e.\ d_{xz}, d_{yz}\right) \xrightarrow{\nu_3} b_{1g}\left(i.e.\ d_{x^2-y^2}\right);\ a_{1g}\left(i.e.\ d_{z^2}\right) \xrightarrow{\nu_2} b_{1g}\left(i.e.\ d_{x^2-y^2}\right);$$

$$b_{2g}\left(i.e.\ d_{xy}\right) \xrightarrow{\nu_1} b_{1g}\left(i.e.\ d_{x^2-y^2}\right);\ \nu_3 \approx \nu_2 > \nu_1 (= \Delta_o)$$

Thus it is difficult to characterize ν_2 and ν_3 individually. They are expected to overlap. The ν_2 and ν_3 transitions are expected to occur at a higher energy compared to ν_1. Thus, it is quite expected that ν_2 and ν_3 may be overlapped by the strong charge transfer band. If Δ_2 is quite small compared to Δ_3, then it is difficult to characterize the ν_1 and ν_2 transitions individually.

The possible electronic transitions in the **tetragonally distorted octahedral (D_{4h}) complexes of Cu(II)** (d^9-system) have been discussed earlier (*cf.* Sec. 7.14.2). The possible transitions are (*cf.* Fig. 7.14.2.1):

$$b_{2g}^2 b_{1g}^1 \xrightarrow[(=\Delta_o)]{\nu_1} b_{2g}^1 b_{1g}^2;\ a_{1g}^2 b_{1g}^1 \xrightarrow{\nu_2} a_{1g}^1 b_{1g}^2;\ e_g^4 b_{1g}^1 \xrightarrow{\nu_3} e_g^3 b_{1g}^2.$$

The relative positions of ν_2 and ν_3 depend on the degree of tetragonal distortion (*see* Fig. 7.14.2.1).

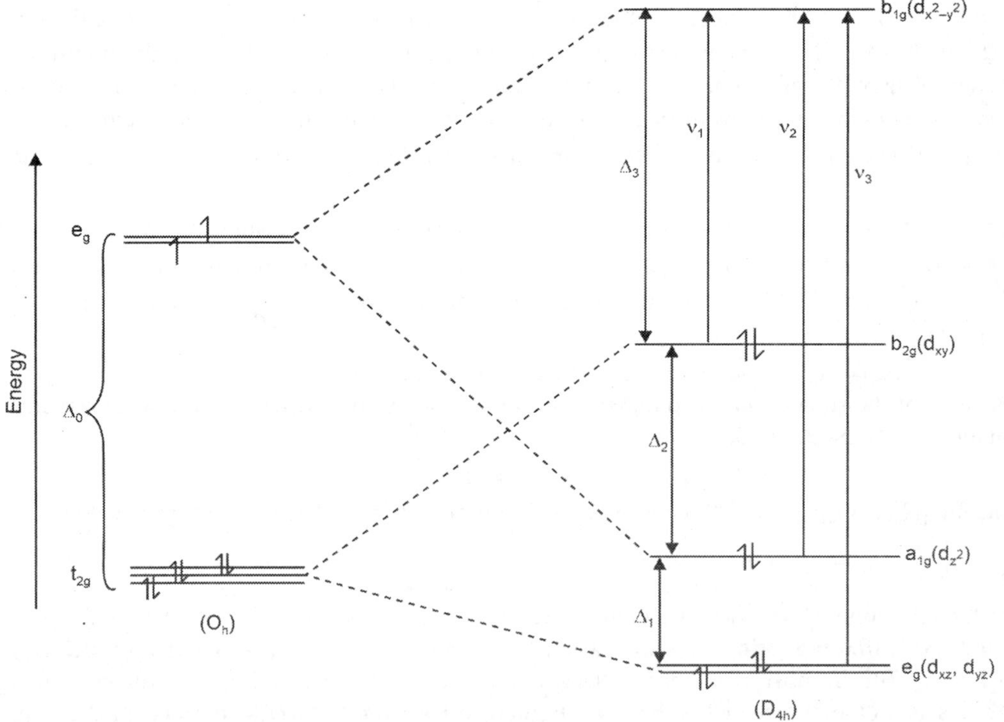

Fig. 7.17.1.2: Possible electron transitions (*i.e.* ligand field bands) in the square planar complexes of Ni(II).

Conventionally, for the square planar complexes of Ni(II), the band is assigned to the $b_{2g}(d_{xy}) \rightarrow b_{1g}(d_{x^2-y^2})$ transition. This transition energy is the Δ_o of the corresponding hypothetical octahedral geometry. **Most of the square planar complexes of Ni(II) are usually red or orange (sometimes green)** and the intense spectral band appears in the range 15,000 cm^{-1}–25,000 cm^{-1} ($\varepsilon \approx$ 50–500). This transition is generally assigned as:

$$b_{2g}(d_{xy}) \longrightarrow b_{1g}(d_{x^2-y^2}),\ i.e.\ {}^1A_{1g} \longrightarrow {}^1A_{2g}$$

There is also a **second more intense** band appearing in the range 25,000–35,000 cm^{-1}. This is a charge transfer band. For example, in **$[Ni(CN)_4]^{2-}$**, the *d–d* transitions making the shoulders occur at 22,500 cm^{-1} and 30,500 cm^{-1} and the MLCT bands have been assigned as follows:

$$d_{xy} \xrightarrow{32,000\ cm^{-1}} \pi^*(CN);\ d_{z^2} \xrightarrow{35,000\ cm^{-1}} \pi^*(CN);\ (d_{xz}, d_{yz}) \xrightarrow{37,600\ cm^{-1}} \pi^*(CN).$$

Some diamagnetic square planar complexes of Ni(II) are: **$[Ni(dmgH)_2]$**, *i.e.* bis(dimethylglyoximato)-nickel(II) (red; absorption at 21,000 cm^{-1}, 24,390 cm^{-1} and 27,000 cm^{-1}); **$[Ni(bigH)_2]Cl_2$**, *i.e. bis*(biguanide)nickel(II) chloride (orange); **$[Ni(CN)_4]^{2-}$**, *i.e.* tetracyanidonickelate(II) (yellow).

For the square planar biguanide complexes of Ni(II), Pd(II) and Pt(II), the absorption maxima are as follows:

$[Ni(bigH)_2]Cl_2$	**$[Pd(bigH)_2]Cl_2$**	**$[Pt(bigH)_2]Cl_2$**
22,472 cm^{-1} (445 nm)	33,400 cm^{-1} (300 nm)	45,500 cm^{-1} (220 nm)

Positions of the absorption maxima (*i.e.* ν: $Pt^{II} > Pd^{II} > Ni^{II}$) are in conformity with the crystal field splitting power of Pt^{II}, Pd^{II} and Ni^{II}.

Ni(II)-Complexes: Ligand field transitions

Octahedral	***Tetrahedral***	***Square planar***
${}^3A_{2g}(F) \xrightarrow[(\approx 10Dq_o)]{\nu_1} {}^3T_{2g}(F)$	${}^3T_1(F) \xrightarrow[(\approx 8Dq_t)]{\nu_1} {}^3T_2(F)$	${}^1A_{1g} \xrightarrow[(\sim 10Dq_o)]{\nu} {}^1A_{2g}$ *i.e.* $d_{xy} \longrightarrow d_{x^2-y^2}$
${}^3A_{2g}(F) \xrightarrow[(\approx 18Dq_o)]{\nu_2} {}^3T_{1g}(F)$	${}^3T_1(F) \xrightarrow[(\approx 18Dq_t)]{\nu_2} {}^3A_2(F)$	*$10Dq_o$ for the hypothetical O_h complex
${}^3A_{2g}(F) \xrightarrow[(\approx 15B'+12Dq_o)]{\nu_3} {}^3T_{1g}(P)$	${}^3T_1(F) \xrightarrow[(\approx 15B'+6Dq_t)]{\nu_3} {}^3T_1(P)$	
ν_1 (8,000–14,000 cm^{-1})	ν_1 (infra-red region)	ν (15,000–25,000 cm^{-1})
ν_2 (12,000–20,000 cm^{-1})	ν_2 (7,000–9,000 cm^{-1})	
ν_3 (20,000–30,000 cm^{-1})	ν_3 (14,000–16,000 cm^{-1})	$\varepsilon \approx$ 50–500
$\varepsilon \approx$ 1–5	$\varepsilon \approx$ 20 for ν_2,	**Colour:** red, orange
	$\varepsilon \approx$ 100–200 for ν_3	
Colour: blue-green	**Colour:** brown, green	

Intensity of the spectral bands of square planar complexes: Square planar geometry is centrosymmetric. According to Laporte Selection rule, the $g \rightarrow g$ transition is forbidden. Thus, the ligand field bands of the square planar complexes are expected to be of poor intensity. But in reality, these bands are quite strong. It arises due to the following facts:

(i) extensive mixing of the metal *d*-orbitals and ligand-orbitals, specially the π-bonding interaction.

(ii) the ligand field bands generally occur at high energy because of the high crystal field splitting in the square planar complexes; consequently, the ligand field bands very often lie close to the strong charge transfer bands and the ligand field bands gain the intensity through the **intensity stealing phenomenon.**

The characteristics of electronic spectra of some representative square planar complexes of Ni(II) are given below.

	$\bar{\nu}$ (in cm^{-1})		
	$^1A_{1g} \rightarrow {}^1A_{2g}$ i.e. $d_{xy} \rightarrow d_{x^2-y^2}$	$^1A_{1g} \rightarrow {}^1B_{1g}$ i.e. $d_{z^2} \rightarrow d_{x^2-y^2}$	$^1A_{1g} \rightarrow {}^1E_g$ i.e. $d_{xz}, d_{yz} \rightarrow d_{x^2-y^2}$
$[Ni(dmgH)_2]$:	21,000	24,390	27,000
$[Ni(dtc)_2]$:	17,000	21,000	23,910

dtc ⇒ dithiocarbamate.

$[Ni(dmgH)_2]$ in $CHCl_3$ solution shows the important bands at 253, 328, 373 and 420 nm along with the absorptions at longer wavelengths. Regarding the origin of different bands, different opinions are available. These are MLCT ($d \rightarrow \pi^*$), d–p ($3d_{z^2} \rightarrow 4p_z$), ligand centred $\pi \rightarrow \pi^*$, etc. (see *Inorganics*, **9**, 47, 2021 for details). It is important to mention that in solid state, H-bonded planar neutral complex $[Ni(dmgH)_2]$ **stacks in a one dimensional (1D) fashion** with the Ni····Ni separation distance 325 nm establishing a stabilising **pi-pi stacking interaction** (see Sec. 1.16.31, Vol. 4). This makes it so insoluble in water.

Conditions Favouring the Square Planar Complex Formation for a d^8-System

For Ni(II), only the strong field ligands favour the formation of square planar complex. But, sometimes, the weak field *S*-donor ligands can form the square planar complex. The *S*-donor sites can use the vacant *d*-orbitals to withdraw the metal *d*-electrons through π-bonding. Due to this **pronounced nephelauxetic effect,** the pairing energy (P) is reduced. Consequently, the splitting energy of the e_g orbitals due to the *z*-out distortion can exceed the pairing energy easily. It gives the spin-paired, diamagnetic square planar complexes of Ni(II). This aspect has been discussed in detail in Chapter 3, Vol. 4. For the heavier congeners Pd(II) and Pt(II), even for the relatively weak field ligands, the conditions for square planar complex formation are satisfied because of their higher crystal field splitting power.

In $[PdCl_4]^{2-}$, the ligand field band appears at about 23,000 cm^{-1} ($\varepsilon \approx 200$). Another high energy band at about 30,000 cm^{-1} ($\varepsilon \approx 400$) is overlapped by the shoulder of a charge transfer band at about 36,000 cm^{-1} ($\varepsilon \approx 3,000$). The ligand field bands in $[PdCl_4]^{2-}$ may be assigned as follows:

($a_{1g}^2 \rightarrow a_{1g}^1\, b_{1g}^1$, i.e. $^1A_{1g} \rightarrow {}^1B_{1g}$: 23,000 cm^{-1}) and ($e_g^4 a_{1g}^2 \rightarrow e_g^3 a_{1g}^2 b_{1g}^1$, i.e. $^1A_{1g} \rightarrow {}^1E_g$).

However, some authors have assigned the peak at 23,000 cm^{-1} due to the $^1A_{1g} \rightarrow {}^1A_{2g}$ (i.e. $d_{xy} \rightarrow d_{x^2-y^2}$) transition and the shoulder at 30,000 cm^{-1} due to the higher energy ligand field bands merging with the CT band.

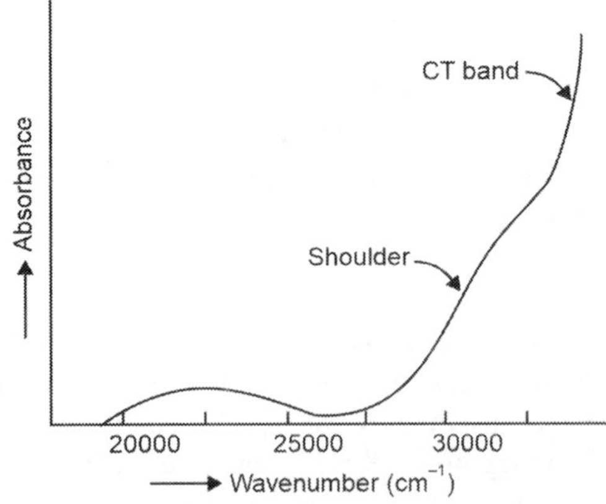

Fig. 7.17.1.3: Absorption spectrum of $[PdCl_4]^{2-}$ in dilute aqueous solution of HCl.

In general, for the **square planar complexes of d^8-systems,** *e.g.* Ni(II), Pd(II), Pt(II), Cu(III), Ag(III), the **spin allowed ligand field transitions** are:

$$\left.\begin{array}{l} d_{xy}\left(b_{2g}\right) \longrightarrow d_{x^2-y^2}\left(b_{1g}\right), \text{ i.e. } {}^1A_{1g} \xrightarrow{\nu_1} {}^1A_{2g},\ \nu_1 \approx \Delta_3 \approx \Delta_o \\ d_{z^2}\left(a_{1g}\right) \longrightarrow d_{x^2-y^2}\left(b_{1g}\right), \text{ i.e. } {}^1A_{1g} \xrightarrow{\nu_2} {}^1B_{1g},\ \nu_2 \approx \Delta_3 + \Delta_2 \\ d_{xz}, d_{yz}\left(e_g\right) \longrightarrow d_{x^2-y^2}\left(b_{1g}\right), \text{ i.e. } {}^1A_{1g} \xrightarrow{\nu_3} {}^1E_g,\ \nu_3 \approx \Delta_3 + \Delta_2 + \Delta_1 \end{array}\right\} (cf. \text{ Fig. } 7.17.1.2)$$

In addition to these, the weak **spin forbidden ligand field transitions** are:

$^1A_{1g} \rightarrow {}^3A_{2g}$; $^1A_{1g} \rightarrow {}^3B_{1g}$; $^1A_{1g} \rightarrow {}^3E_g$ } (lying in the near infra-red region).

The triplet energy states are as follows (where the unpaired two electrons are with the parallel spins):

$$^3A_{2g} \rightarrow e_g^4 a_{1g}^2\, b_{2g}^1 b_{1g}^1;\ {}^3B_{1g} \Rightarrow e_g^4 a_{1g}^1 b_{2g}^2 b_{1g}^1;\ {}^3E_g \Rightarrow e_g^3 a_{1g}^2 b_{2g}^2 b_{1g}^1$$

7.17.2 Coordination Equilibria in Ni(II)-Complexes and Thermochromism (*cf.* Secs. 8.26.1-4, Vol. 6)

The coordination equilibria are known in many Ni(II)-complexes.

$$\text{octahedral} \rightleftharpoons \text{planar} \rightleftharpoons \text{tetrahedral}$$

Existence of the equilibria is established by considering the magnetic properties of the complexes and ligand field spectra. These are illustrated in the following examples.

The temperature dependent coordination equilibria are the examples of **thermochromism** experiencing the colour change with the change of temperature. The octahedral $\rightleftharpoons$ tetrahedral $\rightleftharpoons$ square planar equilibria of the Ni(II) complexes are the classical examples of thermochromism. Some examples of thermochromism in Ni(II) complexes have been discussed here.

(A) N-substituted-salicylaldiminato complexes of Ni(II) have been studied in detail.

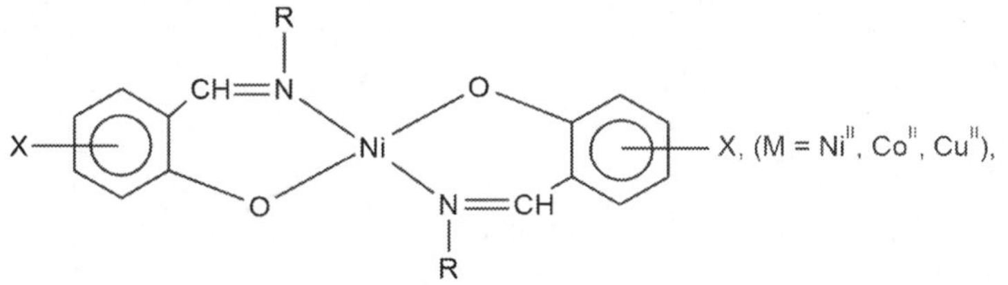

X, (M = Ni^{II}, Co^{II}, Cu^{II}),

$$\underset{\nu \approx 16000\ \text{cm}^{-1}}{\text{square planar (\textbf{olive green})}} \underset{\text{cooling}}{\overset{\text{heating}}{\rightleftharpoons}} \underset{\nu \approx 7200\ \text{cm}^{-1},\ 11200\ \text{cm}^{-1}}{\text{tetrahedral (\textbf{brown})}} \text{(thermochromism for Ni}^{\text{II}}\text{-complexes)}$$

For the above mentioned complex, existence of the temperature dependent equilibrium (*i.e.* **thermochromism**), square planar $\rightleftharpoons$ tetrahedral, has been established. In solution, the Ni(II) complex shows three absorption bands:

$$\sim 7{,}200\ \text{cm}^{-1},\ \sim 11{,}200\ \text{cm}^{-1},\ \sim 16{,}000\ \text{cm}^{-1}$$

With the increase of temperature, intensity of the bands at 7,200 cm^{-1} and 11,200 cm^{-1} increases while the intensity of the band at 16,000 cm^{-1} decreases. **The bands at 7,200 cm^{-1} and 11,200 cm^{-1} are the characteristic bands of the tetrahedral geometry for Ni(II)** and they correspond to the transitions, $^3T_1(F) \rightarrow {}^3A_2(F)$ and $^3T_1(F) \rightarrow {}^3T_1(P)$ respectively. On the other hand, the band at 16,000 cm^{-1} represents the characteristic spectral band of the square planar complexes and this represents the $^1A_{1g} \rightarrow {}^1A_{2g}$ transition, *i.e.* $d_{xy} \rightarrow d_{x^2-y^2}$. The change of intensity of different bands with the increase of temperature indicates that at higher temperature, the tetrahedral geometry is more preferred. *The tetrahedral complexes look brown while the square planar complexes look olive green.*

For the substituted N(sec-alkyl)-salicylaldiminato-Ni(II) complexes, the **bulky R-groups favour the tetrahedral geomtery** (where the steric repulsion at ~ 109° bond angle is relatively less). In fact, with the increase of the size of the R-group, the brown colour predominates and the magnetic moment also increases. It may be noted that the square planar complexes of Ni(II) are diamagnetic while the tetrahedral complexes are paramagnetic.

(B) The tetrahedral $\rightleftharpoons$ square planar equilibrium is well documented for **$[Ni(PR_3)_2X_2]$**.

Br, Ni, Br, $(Ph)_2EtP$, $PEt(Ph)_2$ $\underset{\text{Heating}}{\overset{\text{Cooling}}{\rightleftharpoons}}$ Br, $PEt(Ph)_2$, Ni, $(Ph)_2EtP$, Br , **(thermochromism)**

(**Green**, $\mu \approx 3.20$ BM)
$\lambda \approx 550$ nm, 880 nm
$\log \varepsilon \approx 2.3, 1.2)$

(**Red**, diamagnetic)
($\lambda = 400$ nm
$\log \varepsilon \approx 4.0)$

See Sec. 2.3 of Vol. 4 for the electronic spectra and PMR spectra of the con–formational isomers of the Ni(II) complex

In benzene solution, all the three absorption bands are found. On cooling, the red coloured square planar complex predominates while at a relatively higher temperature towards the room temperature, the green complex predominates. See Sec. 2.3 of Vol. 4 for more details of these conformational isomers.

(C) For the **Ni(II)-N-arylsalicylaldimine complexes,** the equilibrium, tetrahedral (monomeric) $\rightleftharpoons$ square planar (monomeric) $\rightleftharpoons$ octahedral (polymeric) has been established.

(D) The square planar $\rightleftharpoons$ octahedral equilibrium has been found for *cis*-**$[Ni(OH_2)_2(trien)]^{2+}$**, *cis*-**$[Ni(en)_2(OH_2)_2]^{2+}$** and *cis*-**$[Ni(OH_2)_2(pn)_2]^{2+}$**. The octahedral *cis*-**$[Ni(OH_2)_2(trien)]^{2+}$** is blue in colour while the planar **$[Ni(trien)]^{2+}$** is yellow in colour.

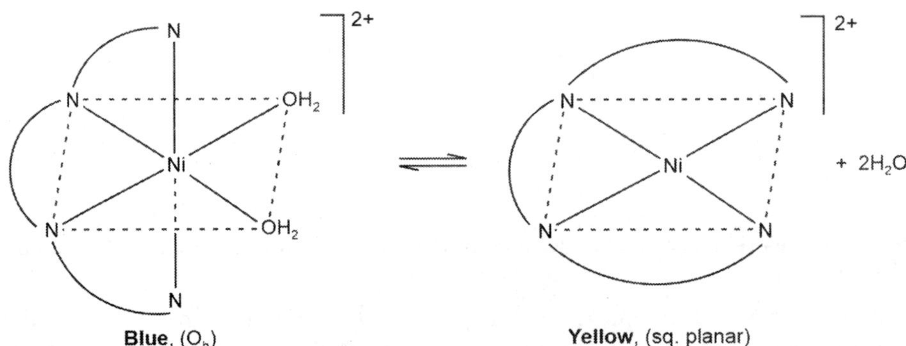

Blue, (O_h) **Yellow**, (sq. planar)

(E) In fact, dissolution of many square planar Ni(II)-complexes in the coordinating solvents like water, pyridine, etc. leads to the following **square planar $\rightleftharpoons$ octahedral equilibrium.**

$$\underset{\text{(square planar)}}{\left[NiL_4\right]} + 2S \rightleftharpoons \underset{\text{(octahedral)}}{\left[NiL_4(S)_2\right]},\ S = \text{coordinating solvent molecule.}$$

Thus, in solution both the absorption peaks of octahedral geometry and square planar geometry are noted. The equilibrium also leads to an **anomalous magnetic moment.**

(F) In **Lifschitz salts, $[Ni(N—N)_2]X_2$** (N—N = substituted en, X = different types of anions), the square planar $\rightleftharpoons$ octahedral equilibrium is well documented.

$$\underset{\substack{\textbf{Square planar, yellow}\\ \textbf{diamagnetic}}}{\left[Ni(N—N)_2\right]X_2} \rightleftharpoons \underset{\substack{\textbf{Octahedral, blue}\\ \textbf{paramagnetic}}}{\left[Ni(N—N)_2X_2\right] \text{ or } \left[Ni(N—N)_2S_2\right]X_2}$$

(G) $(NR_xH_{4-x})_2[NiCl_4]$ (x = 1,2,3) experiences the reversible thermocromism.

At a lower temperature, they have the yellow brown or green colour but with the increasing temperature, they become blue. This colour change occurs due to the structural change, octahedral (through the bridging Cl-ligands) to tetrahedral, *i.e.*

$$\underset{\substack{\text{(Yellow-brown/green)}\\ \text{(Lower temperature)}}}{\text{octahedral}} \underset{\text{cooling}}{\overset{\text{heating}}{\rightleftharpoons}} \underset{\substack{\text{(blue)}\\ \text{(Higher temperature)}}}{\text{tetrahedral,}} \quad (\textbf{thermochromism})$$

7.18 SPECTRAL TRANSITIONS IN THE SQUARE PLANAR COMPLEXES OF COBALT(II) (d^7)

The well known square planar complexes of Co(II) are:

[Co(salen)], [Co(acacen)], $[Co(bigH)_2]$, etc.

The d-orbital splitting pattern leading to the square planar complexes of Co(II) depends on the crystal field strength of the ligands. The $d_{x^2-y^2}$ orbital (b_{1g}) is of the highest energy while energy order of the remaining four d-orbitals depends on the ligand field strength. The ground state electronic configuration of Co(II) in the low-spin square planar complex is:

$e_g^4\, a_{1g}^2\, b_{2g}^1\, b_{1g}^0$ (low spin) (when tetragonality is very high)

$e_g^4\, b_{2g}^2\, a_{1g}^1\, b_{1g}^0$ (low spin) (when tetragonality is small)

If the tetragonality is extreme then d_{z^2} (a_{1g}) becomes the orbital of the lowest energy. High-spin square planar complexes of Co(II) with 3 unpaired electrons are theoretically possible. But no such complex is known. The possible electronic transitions are shown in Fig. 7.18.1.

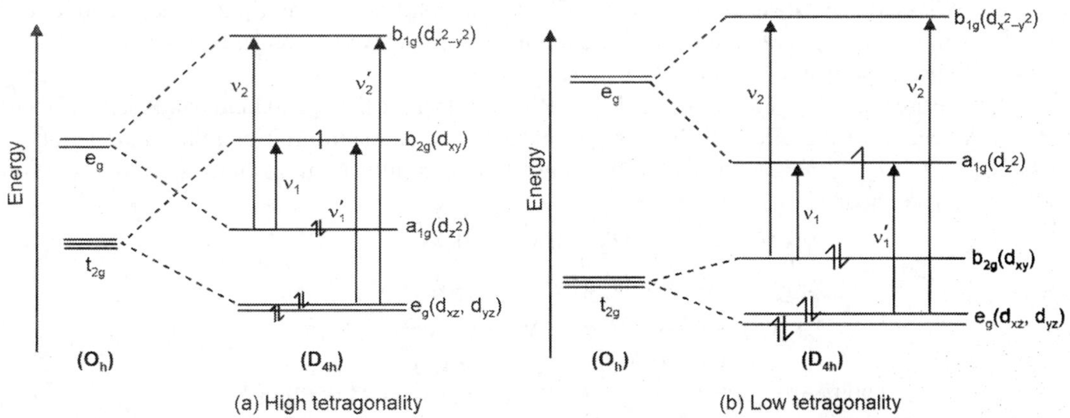

Fig. 7.18.1: Possible electronic transitions in the low-spin square planar complexes of Co(II).

In reality, the lowest energy levels, *i.e.* e_g and a_{1g} (for high tetragonality) or e_g and b_{2g} (for low tetragonality) are so close in energy that the ν_1 and ν_1' transitions cannot be characterized individually. Similarly, the ν_2 and ν_2' transitions cannot be characterized individually because of the same ground.

A low-spin square planar complex of Co(II) bearing the **chromophor CoO_2N_2 (1, 2)** [*e.g.* N,N′-ethylene*bis*(salicylideneiminato)cobalt(II), **1**] generally shows two absorption bands at around 8,500 cm^{-1} ($\varepsilon \approx 20$), and 20,000 cm^{-1} ($\varepsilon \approx 300$).

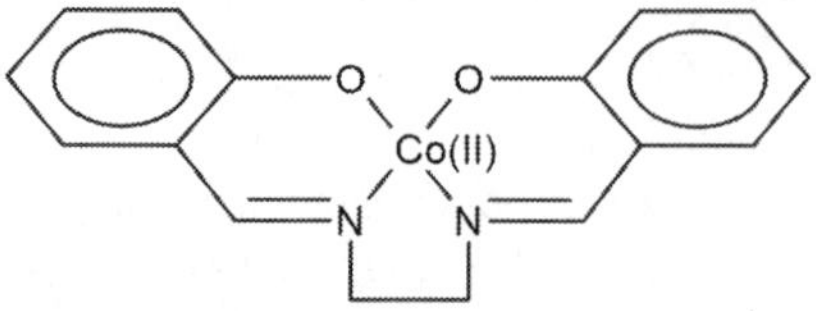

1 [Co(salen)]
N, N′-ethylene-*bis*(salicylideneiminato)cobalt(II)

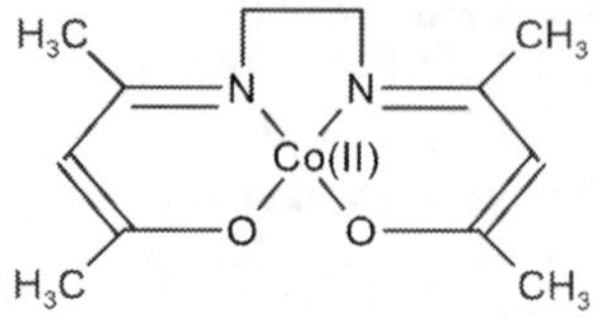

2 [Co(acacen)]
N, N′-ethylene-*bis*(acetylacetoneiminato)cobalt(II)

$$\left.\begin{array}{l} \nu_1, \nu_1' \ (i.e., a_{1g}, e_g \longrightarrow b_{2g}): \sim 8{,}500 \text{ cm}^{-1} \\ \nu_2, \nu_2' \ (i.e., a_{1g}, e_g \longrightarrow b_{1g}): \sim 20{,}000 \text{ cm}^{-1} \end{array}\right\} \begin{array}{l} \text{for the square planar } CoO_2N_2 \text{ chromophore} \\ \text{assuming \textbf{high tetragonality}, } cf. \text{ Fig. 7.18.1a)} \end{array}$$

7.19 LIGAND FIELD SPECTRA OF THE SECOND AND THIRD TRANSITION SERIES

The heavier congeners generally form the **low-spin complexes** because of their higher crystal field splitting power and reduced spin pairing energy. The heavier congeners also more readily can attain their **higher oxidation states** because the 4*d* and 5*d* electrons are relatively loosely bound compared to the 3*d* electron. It is partly due to the **relatavistic expansion of *d*-orbitals** (*cf.* A.K. Das *et al., Coord. Chem., Rev.*, 2023; *CTI* (IUPAC), 2023 and Ch. 8, Vol. 1).

Interpretation and characterization of the ligand field bands of the 2nd and 3rd transition series are complicated because of the following reasons.

(i) For the heavier congeners, the **charge transfer bands** appear at a relatively lower energy and these intense charge transfer bands overlap with the ligand field bands which appear at a relatively higher energy (*cf.* $10Dq$ value increases for the heavier congeners). It complicates the characterization of the ligand field bands.

(ii) The **spin-orbit coupling interaction** increases for the heavier congeners and it runs in the sequence: 5*d*-series > 4*d*-series > 3*d*-series. Thus peak splitting (*i.e* multiplet structure) due to the spin-orbit coupling interaction is quite predominant for the heavier congeners. It also complicates the interpretation of the ligand field bands.

In fact, there are only very few cases (*e.g.* Mo^{3+}, Rh^{3+}, Ir^{3+}) where the ligand field bands are sufficiently free from the complicating effect of the charge transfer band and spin-orbit coupling interaction.

$[MoCl_6]^{3-}$ (d^3, *i.e.* t_{2g}^3) in aqueous solution shows two spin-allowed transitions at 19,000 cm^{-1} and 24,000 cm^{-1}. The bands are assigned as:

$$^4A_{2g}(F) \xrightarrow{\nu_1} {}^4T_{2g}(F),\ \nu_1 = 10Dq$$

$$^4A_{2g}(F) \xrightarrow{\nu_2} {}^4T_{1g}(F),\ \nu_2 = 18Dq - x$$

$$^4A_{2g}(F) \xrightarrow{\nu_3} {}^4T_{1g}(P),\ \nu_3 = 12Dq + 15B' + x$$

(x = symmetry interaction energy)

$$\nu_1 = 19{,}000 \text{ cm}^{-1},\ i.e.\ 10Dq = 19{,}000 \text{ cm}^{-1} \text{ and } \nu_2 = 24{,}000 \text{ cm}^{-1}$$

The third band (*i.e.* ν_3) appears at a much higher energy and it is obscured by the overlapping charge transfer band.

The ligand field spectra of **Rh(III) low-spin (d^6**, *i.e.* t_{2g}^6) complexes like $[RhCl_6]^{3-}$, $[Rh(NH_3)_6]^{3+}$, etc. can be explained as in the case of low-spin Co(III)-complexes (Sec. 7.15.1). The Rh(III) complexes generally record two absorption peaks ($^1A_{1g} \rightarrow {}^1T_{1g}$ and $^1A_{1g} \rightarrow {}^1T_{2g}$) towards the blue end of the visible spectrum. This is very often accompanied with the charge transfer bands in the blue region. These make the colour of the Rh(III) complexes orange, red, yellow or brown. For $[Rh(NH_3)_6]^{3+}$, the assignments are:

$$^1A_{1g} \xrightarrow{\nu_1} {}^1T_{1g},\ \nu_1 = 32{,}700\ \text{cm}^{-1}$$

$$^1A_{1g} \xrightarrow{\nu_2} {}^1T_{2g},\ \nu_2 = 39{,}100\ \text{cm}^{-1}$$

The ligand field bands in $[RhF_6]^{3-}$ are:

$$^1A_{1g} \xrightarrow{\nu_1} {}^1T_{1g},\ \nu_1 = 21{,}300\ \text{cm}^{-1}$$

$$^1A_{1g} \xrightarrow{\nu_2} {}^1T_{2g},\ \nu_2 = 27{,}800\ \text{cm}^{-1}$$

Thus in terms of band assignments, Rh(III)-octahedral complexes are similar to the low-spin octahedral complexes of Co(III).

In the same way, spectra of the Ir^{III} (t_{2g}^6) complexes can be explained.

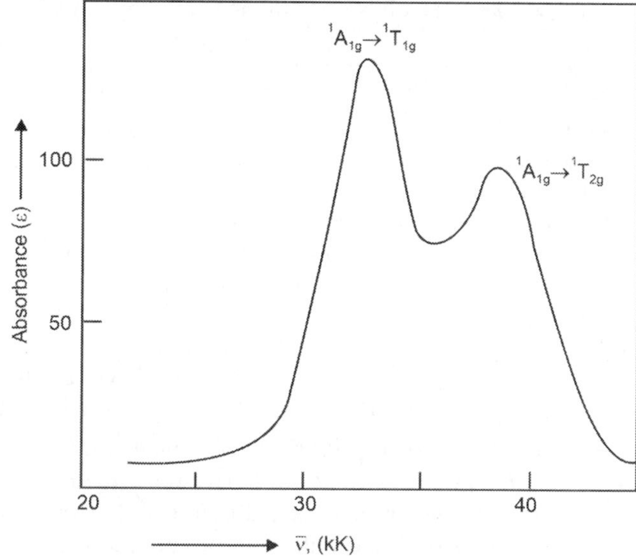

Fig. 7.19.1: Absorption spectrum of $[Rh(NH_3)_6]^{3+}$.

7.20 ELECTRONIC SPECTRA OF THE LANTHANIDE AND ACTINIDE COMPLEXES

(*Ref.* A.K. Das *et al.*, *Coord. Chem. Rev.*, 2023 and *CTI* (IUPAC), 2023)

7.20.1 Characteristics of the Electronic Spectra of the Lanthanides

In general, +3 is the **most common oxidation state** of the lanthanides. The occurance of +2 oxidation state (*e.g.* Eu^{2+}, f^7) and +4 oxidation state (*e.g.* Ce^{4+}, f^0) can be rationalized in terms of half-filled, full-filled and empty *f*-shell.

Except the lanthanides (Ln^{3+}) having the f^0, f^1, f^7, f^{13} and f^{14} configurations, all other lanthanides are having their characteristic colours.

It is evident that the ***lanthanides (Ln^{3+}) with the f^n and f^{14-n} configurations are having the same colour.***

Thus the colour sequence in the Lu–Gd series ($f^{14}-f^7$) is repeated in the La–Gd series (f^0-f^7).

f^n	La(III) f^0	Ce(III) f^1	Pr(III) f^2	Nd(III) f^3	Pm(III) f^4	Sm(III) f^5	Eu(III) f^6	Gd(III) f^7
	colourless	**colourless**	**green**	**blue-violet**	**pink**	**yellow**	**pale pink**	**colourless**
f^{14-n}	Lu(III) f^{14}	Yb(III) f^{13}	Tm(III) f^{12}	Er(III) f^{11}	Ho(III) f^{10}	Dy(III) f^9	Tb(III) f^8	Gd(III) f^7

(i) *f-f* transition in Ln^{3+} producing the sharp (*i.e.* Rydberg type spectra; *cf.* line spectra for atoms) and less intense peaks: In the lanthanides (Ln^{3+}), the *f-f* transition is forbidden in terms of Laporte selection rule (*cf. d-d* transition). For the transition elements, the *d-d* transition is made partially allowed through the vibronic coupling **which is a crystal field effect** (*i.e.* asymmetric stretching of the metal-ligand bond). For Ln^{3+}, the 4*f*-orbital is deeply seated and the 4*f*-electrons are screened well by the filled shells of higher principal quantum number. **Because of this fact, the (*n* – 2)*f* orbitals, in general, are hardly exposed to the ligands or crystal field, and consequently, the *f-f* transitions are hardly influenced by the ligands or crystal field effect.** The vibronic coupling arising from the metal-ligand bond vibration partially allows the electronic transition and at the same time, it broadens

Table 7.20.1.1: Colours, electronic configurations and ground state terms of trivalent lanthanides

M^{3+} ions (f^n)	***Ground state***	***Colours***
$La^{3+}(f^0)$, $Lu^{3+}(f^{14})$	$^1S_0\,(f^0)$, $^1S_0\,(f^{14})$	Colourless (no *f-f* transition)
$Ce^{3+}(f^1)$, $Yb^{3+}(f^{13})$	$^2F_{5/2}\,(f^1)$, $^2F_{7/2}\,(f^{13})$	Colourless
$Pr^{3+}(f^2)$, $Tm^{3+}(f^{12})$	$^3H_4\,(f^2)$, $^3H_6\,(f^{12})$	Green
$Nd^{3+}(f^3)$, $Er^{3+}(f^{11})$	$^4I_{9/2}\,(f^3)$, $^4I_{15/2}\,(f^{11})$	Lilac (blue-violet)
$Pm^{3+}(f^4)$, $Ho^{3+}(f^{10})$	$^5I_4\,(f^4)$, $^5I_8\,(f^{10})$	Pink, yellow
$Sm^{3+}(f^5)$, $Dy^{3+}(f^9)$	$^6H_{5/2}\,(f^5)$, $^6H_{15/2}\,(f^9)$	Yellow
$Eu^{3+}(f^6)$, $Tb^{3+}(f^8)$	$^7F_0\,(f^6)$, $^7F_6\,(f^8)$	Pale pink
$Gd^{3+}(f^7)$	$^8S_{7/2}\,(f^7)$	Colouless

the peak (see Secs. 7.5–7.7). **In absence of this effect, the peaks arising from the *f–f* transitions are sharp (*i.e.* line like Rydberg spectra) and of low intensity.**

***f–f* transitions in rare earths:** The *f*-orbitals are of *u* symmetry in a centrosymmetric point group and the **intra *f–f* transitions** are **parity forbidden** in terms of the **electric dipole (ED) transition**. However, the **weaker magnetic dipole (MD) transition** allows the *f–f* transitions and these MD transitions probably contribute in the electronic spectra of *f*-block elements.

Here it is to be mentioned that in the lanthanides, the **spin-orbit coupling splitting is more important than the crystal-field splitting.** In general, the crystal field effect on the deeply seated $(n-2)f$ orbitals is negligibly small and in fact, in the lanthanides, the crystal field splitting can split the various spectroscopic states by about 100 cm^{-1} only. **Thus the crystal field splitting cannot appreciably split or broaden the peak.** Absence of vibronic coupling (*i.e.* metal-ligand bond vibration is not important) and very weak crystal field effect are responsible in making the peaks (due to *f–f* transition) very sharp

Transition metal ions (*d–d* transitions): Spectroscopic states are generated by the **crystal field splitting** of the ground and excited state terms. Vibronic coupling makes the peaks stronger and broader. Peak broadening by spin-orbit coupling is important for the tetrahedral complexes but not for the octahedral complexes. ***Colour (i.e. electronic transition energies) of a particular metal ion depends on the nature of the ligands.***

Lanthanides (*f–f* transitions): Spectroscopic states are generated by the **spin-orbit (SO) coupling splitting** of the ground and excited state terms. Absence of vibronic coupling and poor crystal field effect (compared to spin orbit coupling splitting) make the peaks sharper. Due to the absence of vibronic coupling, it is difficult to relax the orbital selectron rule, *i.e.* Laporte selection rule, and **the peaks become weak.** In fact, besides the **electric dipole (ED) model** mechanism, other mechanisms like magnetic dipole, electric quarrupole transitions can only partially allow the *f–f* transition. It may be noted that in terms of the **magnetic dipole (MD) model of absorption of electromagnetic radiation**, the $p \rightarrow p$, $d \rightarrow d$ and $f \rightarrow f$ transitions are allowed but this mechanism is very much inefficient (by a factor about 10^5–10^8) compared to the electric dipole mechanism governed by the Laporte selection rule (see Sec. 7.4.2). Because of the no effect of ligand field on the *f–f* transitions, ***colour of a particular metal ion does not depend on the nature of the ligands.***

Lanthanides ($4f \rightarrow 5d$ transitions): These **orbitally allowed** stronger absorption bands generally appear in the UV region, sometimes tailing in the visible range. Ce(III) ($4f^1$, one electron more than the empty *f*-shell) and Tb(III) (f^8, one electron more than the half filled *f*-shell) allow the excitation of the **extra electron** in the $4f \rightarrow 5d$ transition more easily than for the other Ln^{3+} ions. It may be noted that for the ***Ln^{2+}* ions**, the $4f$ orbital is more expanded (*i.e.* more destabilised), and the $4f \rightarrow 5d$ transitions are **red shifted.**

in the lanthanides. In other words, **the metal-ligand bond distance is hardly affected due to the *f–f* transition** (*i.e.* metal-ligand bond distance remains more or less the same for the ground and excited electronic state) and consequently, the peaks become sharp (*cf.* Fig. 7.7.3.1; Secs. 7.7.2–3). **The narrow (*i.e.* sharp) absorption bands of the lanthanides are very often used to calibrate the wavelength in different instruments.**

Here it may be mentioned that though the $4f \to 4f$ transitions are forbidden, the $4f \to 5d$ transitions are allowed. **The peaks arising from the Laporte allowed $4f \to 5d$ transitions are more intense and relatively broader.** The *d*-orbitals are more exposed to the ligands and the metal-ligand bond vibration can broaden the peaks as usual. In such cases, the *f–f* transitions may earn some intensity from the *f–d* allowed transitions through the **intensity stealing mechanism.** It may be noted that the allowed $4f \to 5d$ transitions generally occur in the UV-region (*e.g.* $4f^{11} \to 4f^{10}5d^1$) in Er^{3+} occurs at about 150 nm).

(ii) **Minimum effect of crystal field on the *f–f* transition:** The crystal field effect has no significant effect on the *f-f* transitions. External fields can split the states arising from the f^n configurations to the magnitude of **100 cm^{-1}** (*cf.* spin-orbit coupling splitting energy is quite high; spin-orbit coupling constant is of the order of **1000 cm^{-1} for Ln^{3+}**). Thus the crystal field effect which depends on the nature of ligands is of minimum importance for the *f–f* transition. **Thus the colour and absorption spectrum of a particular Ln^{3+} ion are more or less the same for the various types of ligands. In other words, each Ln^{3+} ion has its own characteristic colour.** On the other hand, for the transition metals, the nature of the electronic transitions largely depends on the nature of the ligands and a particular transition metal ion shows different types of colours and absorption spectra for the different types of ligands.

(iii) **Spectroscopic states generated through the splitting of energy states by spin-orbit (SO) coupling interaction and a large number of transitions in the higher energy end:** For the lanthanides, the spin-orbit coupling constant **(λ) is fairly high** (~1000 cm^{-1}) and the $\vec{L}$ and $\vec{S}$ vectors couple to produce the $\vec{J}$ vector. In other words, *J* is a good quantum number for Ln^{3+}. Because of **the high spin-orbit (SO) coupling constant,** the energy difference among the different spectroscopic states generated through the splitting of the energy states by spin-orbit coupling is very high (*cf.* ***for the transition metal ions, the spectroscopic states are generated by the crystal field effect***). However, the less important **crystal field effect** can lift some of the $(2J + 1)$ degeneracy of the states of Ln^{3+} ions by the order of 100 cm^{-1}. It gives the **fine structure components** of some bands of the Ln^{3+} ions.

For the Ln^{3+} ions, generally, the energy difference between the ground state and first excited state (*J* values differ by one unit between these levels) is sufficiently higher than the thermal energy k_BT ($\approx$ 200 cm^{-1}). However, for Eu(III) and Sm(III), the spin-orbit coupling constants are fairly small (*cf.* $\lambda \approx 230$ cm^{-1} for Eu^{3+}, $\lambda \approx 240$ cm^{-1} for Sm^{3+}). Thus for Sm^{3+} and Eu^{3+}, the energy separation between the ground state and excited state is relatively small.

$$\textbf{\textit{Sm}}^{3+}\ (f^5)\text{: } E\,(^6H_{7/2}, \text{ first excited state}) - E\,(^6H_{5/2}, \text{ ground state})$$
$$= \tfrac{7}{2} \times 240 = 840 \text{ cm}^{-1}$$
$$\textbf{\textit{Eu}}^{3+}\ (f^6)\text{: } E(^7E_1) - E(^7E_0) = 1 \times 230 = 230 \text{ cm}^{-1}$$

Energy difference (ΔE) between the successive states differing in *J* values (*i.e.* *J* and $J + 1$) by one unit can be obtained by **Lande interval rule**: $\Delta E = (J + 1)\lambda$; $J = |L + S|$ to $|L - S|$

$$E_J = \tfrac{1}{2}\lambda\left[J(J+1) - L(L+1) - S(S+1)\right]$$

By considering the λ-values, it is expected that except for Sm^{3+} and Eu^{3+}, the excited states remain virtually unpopulated at ordinary temperature under the condition,

$$\Delta E_{J,J+1} \gg k_BT.$$

Note: For the lanthanides, the spectroscopic states are generated through the SO coupling splitting of the **ground state term and excited state terms** (*cf.* for the *d*-block elements, the spectrocopic states are generated through the **crystal field splitting** of the ground state and excited state terms). The energy separation among these component states generated from a particular term is about a few 1000 cm^{-1}. Thus for the ions having **no excited term of the same multiplicity of the ground state term**, the *f–f* transitions occur among the component states developed from the ground state term and the transitions occur in the *ir*-range (*i.e.* small transition energies) and consequently, **they look colourless** (see Table 7.20.1.1).

Selection Rules
For **electric dipole (ED) transition**, $\Delta S = 0$, $\Delta L \leq 6$, $\Delta J \leq 6$ and **opposite parity**; for **magnetic dipole (MD) transition**, $\Delta S = 0$, $\Delta L = 0$, $\Delta J = 0, \pm 1$ (except $J = 0 \rightarrow J = 0$), **same parity**.

(iv) **Energy states and selection rules for the *f–f* transitions:** For a particular Ln^{3+} ion of f^n configuration, the *f–f* transition occurs from one state of a particular *J* value to the excited state of another *J* value [between the *J*-states $\Delta J \leq 2l$, $l = 3$]. As the energy separation is fairly large, **the transitions generally occur in the high energy end of the visible spectrum.**

For an *f*-electron, $l = 3$, $m_l = 3, 2, 1, 0, -1, -2, -3$. **Thus a large number of transitions are possible and it gives a larger number of peaks.** The spectral transitions are governed by the following **selection rule:**

$$\Delta L \leq 2l = 6,\ \Delta S = 0,\ \Delta J \leq 2l\ (l = 3)$$

Application of the selection rules is illustrated for Dy^{3+}.

$$^6H_{15/2} \longrightarrow {}^6F_{1/2}\ \ \textbf{(forbidden transition;}\ \Delta J = 7 > 2 \times 3)$$

$$^6H_{15/2} \longrightarrow {}^6F_{3/2}\ \ \textbf{(allowed transition;}\ \Delta J = 6 = 2 \times 3)$$

If $J = 0$ for the initial state or final state, then the restriction for ΔJ is as follows:

$$\Delta J = 2, 4 \text{ or } 6$$

Thus for Eu^{3+}, the transitions, ${}^7F_0 \rightarrow {}^7F_3$ ($\Delta J = 3$), ${}^7F_0 \rightarrow {}^7F_5$ ($\Delta J = 5$) are **forbidden**.

Let us consider Pr^{3+} for its allowed (but Laporte forbidden) *f*→*f* transitions. For its $4f^2$-electronic configuration, the ground state term is ${}^3H_4$ (*i.e.* L = 5, S = 1, and J = 6, 5 and 4). Thus the lowest energy state is represented by ${}^3H_4$. The other excited state terms are: ${}^3F$, ${}^1G$, ${}^1D$ and ${}^3P$. Thus the energy order is:

$${}^3H < {}^3F < {}^1G < {}^1D < {}^3P \text{ (see Fig. 7.20.1.1)}$$

The **green-yellow solution of Pr^{3+}** shows the absorption in the range near IR to the UV-region (Fig. 7.20.1.1). It strongly absorbs between 450–500 nm (blue light) and 580 nm (yellow). Thus the residual colours that reach to the observer are mainly green and red. It gives the green-yellow colour of Pr(III)-compounds.

The **Laporte forbidden**, 4*f*→4*f* electronic transitions for the Pr(III)-compounds are:

$${}^3H_4 \rightarrow {}^3P_2\ (\Delta J = 2 < 2l = 6),\ {}^3H_4 \rightarrow {}^3P_1\ (\Delta J = 3 < 2l = 6)$$

$${}^3H_4 \rightarrow {}^3P_0\ (\Delta J = 4 < 6),\ {}^3H_4 \rightarrow {}^1D_2\ (\Delta J = 2 < 6)\ \textbf{(spin-forbidden)}$$

$${}^3H_4 \rightarrow {}^1G_4\ (\Delta J = 0 < 6)\ \textbf{(spin-forbidden)}$$

Nd^{3+} shows a strong absoroption peak at about 580 nm for the **hypersensitive electronic transition,** ${}^2G_{7/2} + {}^4G_{5/2} \leftarrow {}^4I_{9/2}$. This wavelength of absorption nicely matches with the **yellow emission** from the excited sodium atoms. This is why, **Nd^{3+} is incorporated into the goggles used by the glassblowers**. It reduces the yellow glare from the hot/molten sodium silicate glass and the glassblowers experience a reduced yellow glare.

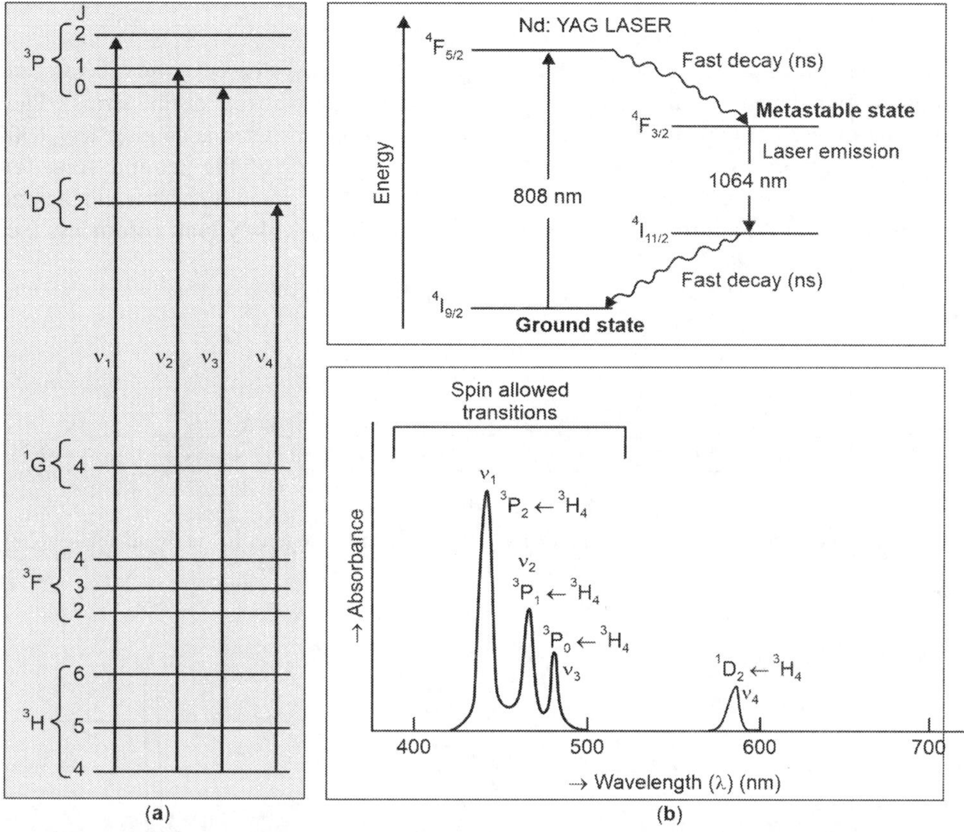

Fig. 7.20.1.1: (a) Energy level diagram of Pr(III). (b) Absorption spectrum of $[Pr(aq)]^{3+}$ in the visible range

The crystal field effect can lift some of the $(2J+1)$ fold degeneracy of the states of f^n configuration but the crystal field splitting can cause only fine structure in some of the bands of lanthanide complexes ***because the spin-orbit coupling effect is more important than the crystal field splitting effect.***

(v) **Spectral behaviour of Ln(II) ions (Ln^{2+}):** The spectral properties of Ln(II) ions stabilised in CaF_2 crystals have been studied. Because of the lower cationic charge in Ln^{2+}, the $4f$-orbitals are more expanded (*i.e.* destabilised) compared to those of Ln^{3+}). Thus for Ln(II), the **orbitally allowed 4*f*–5*d* transitions occur at the relatively lower energy** covering the near UV-visible range (*cf.* $4f \rightarrow 5d$ transitions occur generally in the UV region for Ln^{3+}). These low energy $4f \rightarrow 5d$ transitions of Ln(II) ions give the relatively stronger and broader peaks overlaid with the weaker and sharper peaks due to the orbitally forbidden *f*–*f* transitions.

(vi) **Luminescence (*cf.* Sec. 7.9.5 and Chapter 6, Part-II):** It has been already mentioned that the ligand field environment does not have any strong influence on the *f*–*f* transition. The excited *f*-electron can weakly interact with the ligand environment. ***Thus the nonradiative path (i.e. thermal degradation of the excitational energy as heat in the surroundings) is quite inefficient to bring back the excited electron to the ground state*. Thus the nonradiative lifetime of the excited state is quite long and luminescence is observed in many cases.**

In the lanthanides, there are many excited states of different multiplicities. These excited states are not normally populated but in presence of some exciting radiation, population density in such excited

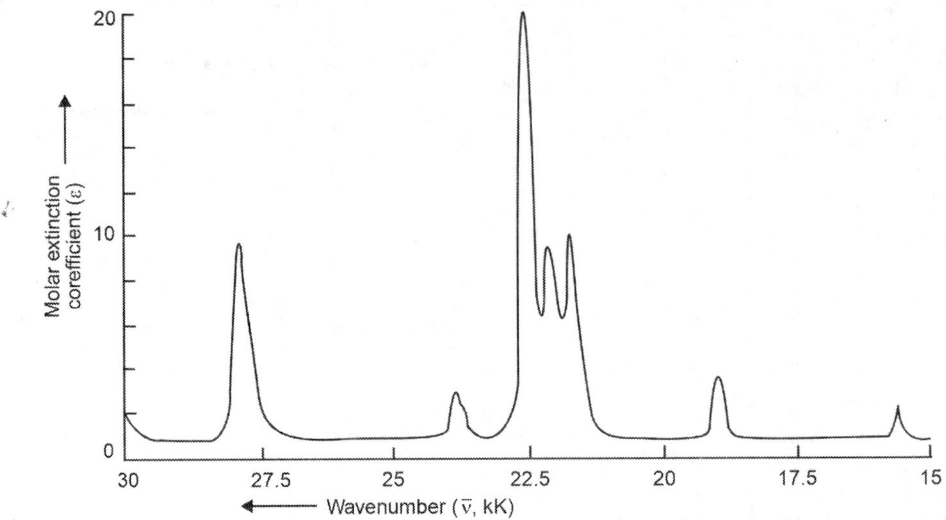

Fig. 7.20.1.2: Electronic absorption spectrum of $\mathbf{Ho^{3+}}$ ($4f^{10}$, ground state term $^5I_8$) in dilute acidic solution.

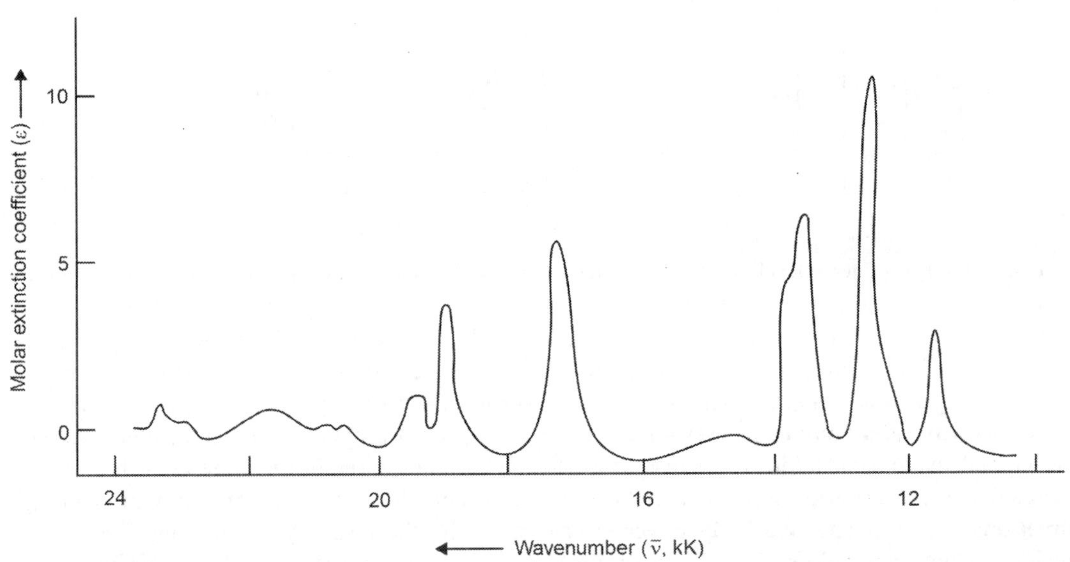

Fig. 7.20.1.3: Electronic absorption spectrum of $\mathbf{Nd^{3+}}$ ($4f^3$, ground state term $^4I_{9/2}$) in aqueous acidic medium.

states can be enhanced through ISC (intersystem crossing) in absence of any efficient nonradiative deexcitation. For the lanthanides having the relatively **higher value of spin-orbit coupling constants**, the ISC process is very much efficient to produce the **excited states of different multiplicities** from the ground state. It makes the **phosphorescence process highly favoured**.

Based on this property, the oxides of some lanthanides (*e.g.* Ho, Tb, Eu, etc.) are used in **oxide phosphors for colour TV screens and similar devices.** For displays, Eu^{3+} is used in **red phosphors**. The fluorescence emission spectrum of Eu^{3+} spans in the range 580 nm (orange) to 700 nm (red). Tb^{3+} is also used in the similar phosphors. Tb^{3+} gives emissions in the range 480 nm to 580 nm. Eu^{3+} doped

> **Hypersensative transitions in the trivalent lanthanides**
>
> It has been already mentioned that the absorption spectra of lanthanides (Ln^{3+}) are more or less insensitive to the ligand field environment. **Thus colour of the aqua-ions,** *i.e.* **$[Ln(aq)]^{3+}$ remains unchanged in presence of the complexing ligands.** However, there are some transitions obeying the selection rules ($|\Delta J| \leq 2$, $|\Delta L| \leq 2$), very much sensitive to the ligand environment. Such transitions are called the **hypersensitive transitions.** Representative examples of hypersensitive transitions are:
>
> Pr^{3+} ($^3H_4 \rightarrow {}^3F_2$), Nd^{3+} ($^4I_{9/2} \rightarrow {}^4G_{5/2}$ at 580 nm), Eu^{3+}($^5F_0 \rightarrow {}^5D_2$), Ho^{3+}($^5I_8 \rightarrow {}^3H_6$, $^5I_8 \rightarrow {}^5G_6$).
>
> The mechanism of hypersensitivity of these transitions will not be discussed here. This lies beyond the scope of this book.

YVO_4 is used for coating on the interior surface of **fluorescent light tubes for converting the UV radiation generated from the mercury discharge into the visible light**. The UV-radiation is used in the excitations for Eu^{3+} and this excitational energy is released as the fluorescent emission in the visible range.

In fact, because of the inefficient nonradiative decay of the excited states, **population inversion state** can also be easily achieved and stimulated emission can produce the **laser beams** in such cases.

● **Nd:YAG laser** (see INSET of Fig. 7.20.1.1)**:** In this connection, **Nd:YAG laser** is quite important. In Nd:YAG laser, Nd^{3+} is doped (*ca.* 1%) through substitution of Y^{3+} in the **garnet structure of yttrium aluminium oxide** ($Y_3Al_5O_{12}$) (YAG). In Nd:YAG laser, doped Nd^{3+} strongly absorbs in the wavelength region 730–760 nm and 790–820 nm from the high intensity light produced by krypton-containing flashlamps. The excitations occur from the ground state $^4I_{9/2}$ to a number of excited states from which the **relatively long-lived excited $^4F_{3/2}$ state (metastable state) is produced**. Radiative transition from this excited state to the $^4I_{11/2}$ state that lies just above the ground state $^4I_{9/2}$ is **stimulated** by a photon of the same frequency to generate the stimulated emission as a laser beam in Nd:YAG laser (*i.e.* **laser emission** of 1064 nm, $^4F_{3/2} \rightarrow {}^4I_{11/2}$). After this **laser emission**, the $^4I_{11/2}$ state rapidly decays to the $^4I_{9/2}$ ground state. The Nd:YAG laser can generate the laser emission at 532 nm (**frequency doubling principle**) and also higher harmonics at 355 and 266 nm.

> **Unique properties of lanthanides to favour luminescence:**
>
> (i) **Inefficient nonradiative decay of the excited states** through the vibrational modes (*i.e.* metal-ligand bond vibration) because the deeply seated 4*f*-orbitals overlap only inefficiently with the ligand orbitals. It gives a long **nonradiative lifetime** (milliseconds to nanoseconds) of the excited states (*i.e.* excited *f*-electrons).
>
> (ii) Because of the **large number of excited states**, several excited states are possible.
>
> (iii) **Efficient ISC** (because of the high spin-orbit coupling constant) allows the formation of **excited states of different multiplicites** from the ground state. It extends the lifetime of the excited states of different multiplicities from the ground state. It gives the ideal condition of phosphorescence.

By making the use of phosphorescence behaviour of the lanthanide ions, it has been possible to replace the ruby lasers and phosphors in commercial applications. **Lanthanide oxide phosphors are almost ubiquitously used in colour televisions.** Isomorphous substitution of Ca^{2+} ions in many **metalloproteins and metalloenzymes** by the lanthanide ions has been quite useful for characterization and detection by making the use of phosphorescence property of the lanthanide ions.

(vii) **Charge transfer (CT) bands:** The charge transfer bands are **relatively less important** in the lanthanides (Ln^{3+}) compared to those in the transition metal complexes. *This is due to the fact that the ligand orbitals can hardly interact with the 4f orbitals of the lanthanides.* LMCT (ligand to metal CT) bands are only important in the stable divalent lanthanides (*e.g.* Eu^{2+}, Sm^{2+}, Yb^{2+}, Tm^{3+}) and tetravalent cerium (*i.e.* Ce^{4+}). Thus the **yellow colour of Ce^{4+} and blood red colour of Sm^{2+}** are due to the charge transfer bands.

Pr^{3+}, Eu^{3+}, Yb^{3+} as the NMR shift reagents (*See* Chapter 12, Vol. 7)

Several highly paramagnetic lanthanides (Ln^{3+}) are used as the NMR shift reagents. Some organic molecules possess very complicated NMR spectra and the analysis is very difficult. When such organic molecules are coordinated to the highly paramagnetic lanthanides, the large magnetic moment of these ions can cause splitting and spreading of the NMR signals of the coordinated organic moieties. This technique helps analyze the NMR signals (*cf.* Sec. 12.2.13).

Spectral Properties of the Lanthanides in terms of the Special Properties of the 4*f*-Orbitals: A Summary

The 4*f* orbitals are more deeply seated compared to the 5*s*, 5*p* and valence 6*s* orbitals, *i.e.* $<r_{max}>_{4f} \ll <r_{max}>_{5s,\ 5p,\ 6s}$. This is why, the ligands can hardly interact with the 4*f* orbitals. It imparts the following properties:

- Spectral properties and colours are more or less **independent of the crystal field effects.**
- Electronically excited states (*i.e.* excited *f*-electrons) do not easily deexcite by the nonradiative paths, *i.e.* through lattice vibration; it imparts the properties of **luminescence.**
- Inefficient vibronic coupling (*i.e.* inefficient coupling with the metal-ligand bond vibration) makes the spectral **peaks sharp** (*i.e.* Rydberg type spectra, or line-like spectra of atoms) and of **poor intensity** (*i.e.* low ε value).
- In absence of crystal field splitting, the electronic states can be described by the **free-ion states.** The **spectroscopic states are generated by spin orbit coupling splitting of the energy states** (*cf.* crystal field splitting generates the spectrospcopic states for the transition metal ions).
- No interaction of the ligand orbitals with the deeply seated 4*f*-orbitals makes the **CT-bands less important.**
- Because of the **large number of excited states**, the number of absorption peaks is large (*cf.* fewer absorption peaks for the transition metal complexes).
- In some cases, the Laporte allowed **4*f* → 5*d* electronic transitions** give the relatively broader absorption peaks of higher intensity. These 4*f* → 5*d* transitions are red shifted for Ln(II) ions.

(viii) **Colourless $La^{3+}(4f^0)$ and $Lu^{3+}(4f^{14})$:** Their ground state term is ^1S. No *f*-*f* transition is possible and the CT bands are absent. These make them colourless.

(ix) **Very poor intensity in the absorption peaks of Gd(III) ($4f^7$):** The ground state term is $^8S_{7/2}$ ($L = 0$, $S = 7/2$, $J = 7/2$). There is no excited state having the same spin multiplicity. Thus, in Gd^{3+}, the *f*-*f* transitions are orbitally forbidden as well as spin forbidden (*cf.* Mn^{2+}). This is why, intensity of the absorption peaks of Gd^{3+} is very weak.

(x) **Colourless *vs.* coloured Ln^{3+} ions: Only one term** arises for the f^1 and f^{13} configurations and the *f*-*f* transitions within the component states generated from the **ground state term** occur in the *ir*-range, and consequently, **they look colourless**. Ce(III) shows the strong absorption bands in the UV-region for the 4*f* → 5*d* transitions.

Colourless Ln^{3+} ions like Ln(III), Gd(III), Lu(III), Ce(III), Yb(III): In such cases, the spin allowed *f*-*f* transitions occur among the component states generated through the SO coupling splitting of the ground state term. The corresponding energy separation is about only a few 1000 cm^{-1} and consequently, the corresponding *f*-*f* transitions occur in the *ir*-range (*i.e.* no absorption in the visible or near ultraviolet region). Consequently, **such Ln^{3+} ions look colourless**. In fact, for the other **coloured**

Ln^{3+} ions, the *f–f* transitions occur from the ground state component to the component states developed from the **higher energy terms** and these *f–f* transitions may occur in the UV-visible range. For the said colourless Ln^{3+} ions, the $4f \rightarrow 5d$ transitions occur in the UV-region.

7.20.2 Characteristics of Electronic Spectra of the Actinides

(i) **Two types of actinides:** The electronic spectra of the trivalent actinides can be broadly classified into two groups.

(a) **Heavier actinides (starting from Am^{3+})** show the electronic absorption spectra that resemble those of the lanthanides.

(b) **Lighter actinides (upto Pu^{3+})** show the electronic absorption spectra that approximately resemble those of the transition metals.

(ii) ***f–f* transition:** Just like the lanthanides (*i.e.* members of the 4*f*-series), the actinides (*i.e.* members of the 5*f* series) experience the $5f \rightarrow 5f$ transitions to produce the electronic absorption spectra.

Compared to the 4*f*-orbitals, the 5*f*-orbitals are more exposed towards the ligand environment (*i.e.* 4*f* orbitals are more deeply seated). **Thus the crystal field effect is more important to influence the *f–f* transitions for the actinides compared to that for the lanthanides.**

This crystal field effect leading to the splitting of the energy levels, distortion of the symmetry, vibronic coupling and metal-ligand bond vibration becomes relatively more important for the actinides and consequently, this effect will **broaden the peak and the peaks will get more intensity.**

Thus the peaks of the absorption spectra of the actinides are expected to be more intense and broader compared to those of the lanthanides.

For the lighter members of the actinides for which the effective nuclear charge on the metal ion is relatively less, the 5*f*-orbitals are relatively more exposed towards the environment compared to those of the heavier members of the series. **Thus the ligand field effects leading to peak broadening and peak intensifying are more important for the lighter members (*e.g.*** Th, U, Pu, etc.) and their spectral characteristics resemble to those of the transition metal ions (for which the *d*-orbitals experience the crystal field effect heavily) to some extent. In fact, in many other aspects like participation of the 5*f*-orbitals in bonding, attainment of higher oxidation states and variable oxidation states, etc. the lighter actinide members resemble to those of the transition metal ions. ***In fact, because of this ground, at the early days, the lighter members of the actinides were mistakenly considered as the members of the transition series.***

In conclusion, for the heavier members of the actinide series, the 5*f*-orbitals are more deeply seated compared to those of the earlier members. **Thus in this respect, the 5*f* orbitals of the heavier members of the series behave spectroscopically just like the 4*f* orbitals of the lanthanides.** It leads to:

- **Heavier actinides:** Peaks are sharp (*i.e.* Rydberg type line spectra) and of poor intensity for the heavier members of the actinides where the crystal field effect on the 5*f*-orbitals is insignificant; thus their spectra qualitatively resemble those of the lanthanides for which the **deeply seated 4*f* orbitals do not practically experience the ligand field effect;** however, these peaks are about 10 times more intense than those of the lanthanides; it indicates that even for the heavier actinides, the 5*f*-orbitals experience the ligand field effect to some extent.
- **Lighter actinides:** Peaks are broad and of higher intensity for the **lighter members** of the actinides where the crystal field effect on the 5*f*-orbitals is more important.

(iii) **$5f \rightarrow 6d$ transitions:** These electronic transitions are not orbitally forbidden. These peaks are relatively broader and of higher intensity as expected. These $5f \rightarrow 6d$ transitions occur at the relatively lower energies compared to those of the $4f \rightarrow 5d$ transitions of the lanthanides (*cf.* **relativistice expansion** of the *d* and *f* orbitals of the heavier congeners).

For U^{3+}, the $5f^3 \rightarrow 5f^2 6d^1$ transition occurs in the range 300–500 nm giving rise to the **deep-orange to red colour**. For Np^{3+} and Pu^{3+} having the higher effective nuclear charge, the energy separation between the 5*f* and 6*d* orbitals (*cf.* 5*f*-orbitals are more contracted) increases and consequently, the corresponding $5f \rightarrow 6d$ transitions occur in the UV region (*i.e.* **blue shift**). The **violet and violet-blue colour** of Np^{3+} and Pu^{3+} solutions arise respectively due to the $5f \rightarrow 5f$ electron transitions.

(iv) **Charge transfer bands:** Charge transfer bands are more important (specially for the lighter members of the actinides) in the actinides than in the lanthanides because the 5*f* orbitals are exposed better than the 4*f* orbitals towards the ligands. The **intense yellow colour** of UO_2^{2+} ion and its compounds is due to the **LMCT transitions**.

(v) **Fluorescence property of UO_2^{2+}:** UO_2^{2+} ion strongly fluoresces in the range 500–550 nm upon excitation by UV-radiation. In the past, this property was utilised for **colouring glass**. Addition of a small amount of uranyl salt (*ca.* 0.5–2%) to glass gives the **bright golden yellow colour**. In **sun light**, the glass looks **green-yellow** due to the fluorescence emission from UO_2^{2+}. When the glass is kept under **UV-radiation, it glows an intense green-yellow colour** due to this fluorescence. However, because of the **radioactivity of uranium**, this practice of colouring glass has been stopped to avoid the health hazard.

Lanthanides *vs.* actinides (*Ref.* Das *et al. Coord. Chem. Rev.*, **479**, 2023)**:** Relativistic expansion, 5*f*-orbital > 4*f*-orbital; crystal field effect more important for the actinides, crystal field factor *vs.* spin-orbit (SO) coupling factor for the lanthanides: 100 cm^{-1} *vs.* 2000 cm^{-1}; for the actinides: SO interaction (2000–4000 cm^{-1}) stronger and comparable to the crystal field effect and *J*, **not a good quantum number for the actinides.**

7.20.3 Comparison of the Absorption Spectra of the Transition Metal Ions and Inner-transition Metal Ions (*cf.* Sec. 8.23, Vol. 1)

(i) ***d-d vs. f–f* Transitions:** For the transition metal ions, the ligand field bands arises from the $d \rightarrow d$ transitions while for inner transition series members, the bands arise from the $f \rightarrow f$ transitions.

(ii) **Shape and intensity:** Both the *d-d* and *f-f* transitions are **orbitally forbidden** (*i.e.* Laporte forbidden transitions). Crystal field effect introducing the lower symmetry and metal-ligand bond vibration (*i.e.* vibronic coupling) are the important phenomena to relax the Laporte selection rule. These are important for the transition metals where the $(n-1)d$ orbitals are exposed towards the ligand environment but these are not important for the inner-transition series where the $(n-2)f$ orbitals are deeply seated and they are hardly affected by the ligands (Fig. 7.20.3.1). The peak broadening by the metal-ligand bond vibration and crystal field effect are important for the metal ions of transition series. ***Because of these reasons, the peaks due to the d-d transitions are relatively more intense and broader compared to those of the inner transition series.*** In fact, the weak spectra of the metal ions of the inner transition series, in general, resemble the **Rydberg type line spectra of the atoms.**

(iii) **Spin-orbit coupling splitting *vs.* crystal field splitting:** The deeply seated $(n-2)f$ orbitals are poorly influenced by the ligand field. Thus, the crystal field splitting of the energy states is less important for the inner transition series. On the other hand, crystal field splitting of the energy states arising from the d^n-configuration of the transition metal ions is highly significant.

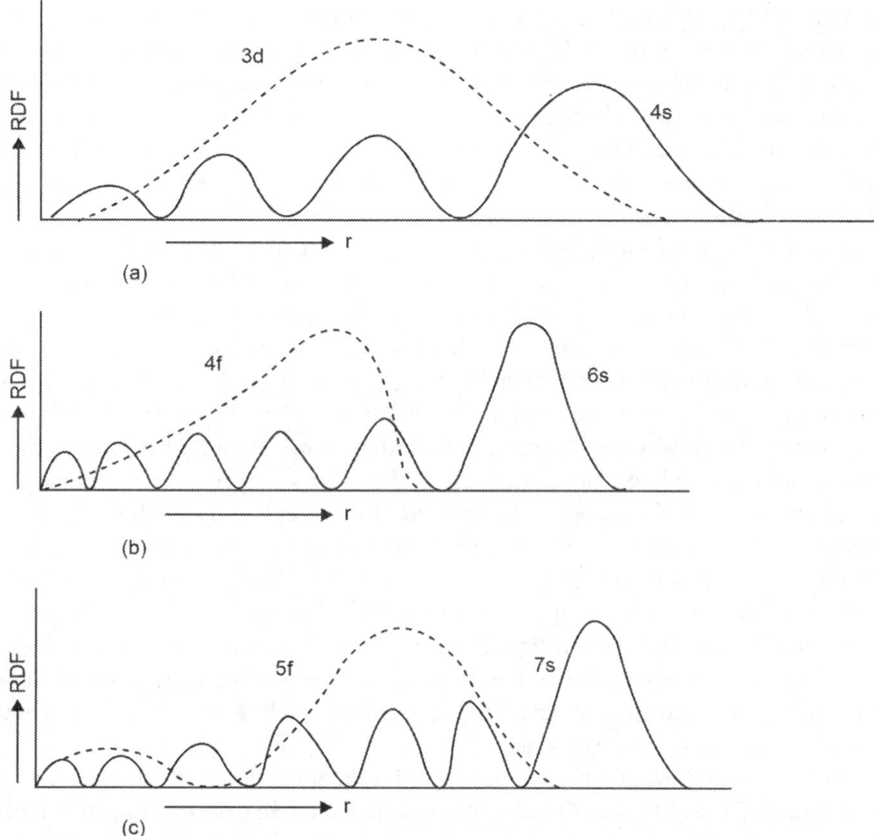

Fig. 7.20.3.1: Qualitative representation of the RDF curves for (a) 3d *vs.* 4s, (b) 4f *vs.* 6s; (c) 5f *vs.* 7s. RDF = $4\pi r^2[R_{n,l}(r)]^2$, r = distance from the nucleus.

Spin-orbit coupling constants for the inner transition series members are of much higher magnitude compared to those of the transition series members.

In conclusion, the spectroscopic states arising from the splitting of states by spin-orbit coupling interaction are important to explain the electronic absorption spectra of the inner transition series. On the other hand, the spectroscopic states arising from the crystal field effect are important to rationalize the electronic spectra of the transition series.

- Free ion term $\xrightarrow{\textbf{Crystal field splitting}}$ spectroscopic states **(transition metal ions)**.
- Free ion term $\xrightarrow{\textbf{Spin-orbit coupling splitting}}$ spectroscopic states **(inner transition metal ions)**.

(iv) **Energy and number of electronic transitions for the metal ions of the transition series and inner transition series:** For the inner transition series, the spin-orbit coupling interaction splits the energy states to generate the spectroscopic states, and the ***energy differences among the spectroscopic states differing in the J values*** **(condition for electronic transition between the spectroscopic states:** $\Delta J \leq 2l$, $l = 3$) ***are very high.*** In fact, the electronic transitions for the *f*-series generally occur in the high energy end of the visible spectrum. For the transition series, splitting of the energy states to generate the spectroscopic states occurs by the crystal field and ***the energy differences among the spectroscopic states are relatively smaller.*** In fact, for the transition metal ions, the electronic transition energy covers the whole visible spectrum.

For the transition metal ions, the number of spin allowed transitions among the ***spectroscopic states generated by the crystal field splitting*** is very few (generally not more than > 3). On the other hand, for the inner-transition metal ions, the ***spectroscopic states are generated by spin-orbit coupling*** and for the *f*-orbitals, $l = 3$, $m_l = +3, +2, +1, 0, -1, -2, -3$, the *f-f* transition can occur in many more possible ways under the conditions, $\Delta L \leq 2l$ ($l = 3$), $\Delta J \leq 2l$. In fact, in general, the number of spin allowed spectral transitions for the inner transition series is more than those found for the transition series.

(v) **Colour dependence on the nature of the ligands:** Crystal field splitting is of no much significance in the case of inner transition series. This is why, the energy of electron transition for a particular metal ion of *f*-series does not depend on the nature of the ligands. Thus colour of a particular metal ion of the inner transition series is **its own characteristic property** (*i.e.* ***colour does not depend on the nature of the ligands***). Crystal field splitting generating the spectroscopic states is very much important for the transition metal ions and the energy of electronic transition depends on the ligand field strength for a particular metal ion. **This is why, colour of a particular transition metal ion largely depends on the nature of the ligand** (*i.e.* crystal field splitting power of the ligand) and symmetry of the complex determined to some extent by the ligands.

(vi) **Luminescence:** Deexcitation of the excited states through the nonradiative path (*i.e.* thermal degradation of the excitational energy as the heat energy to the surroundings through the **metal-ligand bond vibration**) is less important for the inner transition series because the $(n-2)f$-orbitals are hardly affected by the ligand environment (*i.e.* practically no metal-ligand bond formation). Thus the lifetime of the excited state, *i.e.* excited *f*-electron, is relatively longer for the inner transition series and luminescence is expected for such members. Luminescence is only rarely found in the transition metal ions.

Note: The above comparison is strictly applicable for the transition metal ions *vs.* lanthanides and heavier actinides. The lighter actinides behave spectroscopically like the transition metals at least to some extent. The reason behind this has been already discussed in Sec. 7.20.2.

7.21 CHARGE TRANSFER (CT) TRANSITIONS

We have already discussed the *d-d* transitions in the transition metal complexes and *f-f* transitions in the lanthanides and actinides. In some complexes, the metal ion and the ligand may have the **complimentary redox properties** (*i.e.* metal centre is oxidizing and ligand moiety is oxidizable or vice-versa). In such cases, the electron transfer can occur from one component to another component of the same complex. These two components of a complex are: metal and ligand. If the electron transfer occurs from the metal to ligand, *i.e.* $M—L \xrightarrow{h\nu} M^+L^-$, then the absorption band developed is described as the **metal to ligand charge transfer (MLCT) band.** On the other hand, if the electron transfer occurs from the ligand to metal, *i.e.* $M—L \xrightarrow{h\nu} M^-L^+$, then it produces the **ligand to metal charge transfer (LMCT) band**. Obviously, in such charge transfer transitions, **a temporary redox reaction** takes place and the absorption bands developed in this way are also described as the **redox bands.** In MLCT, the metal centre is oxidized while the ligand moiety is reduced. **Thus, MLCT occurs when the metal centre is oxidizable and the ligand moiety is reducible.** In LMCT, the ligand is oxidized and the metal is reduced. **Thus LMCT occurs when the metal centre is oxidizing and the ligand moiety is oxidizable.**

Here it is important to note that the CT band may also appear in the d^{10} (*e.g.* HgI_2, HgS) and d^0 (*e.g.* MnO_4^-, CrO_4^{2-}) systems also. It may be noted that the d^0 and d^{10} systems cannot give any ligand field band.

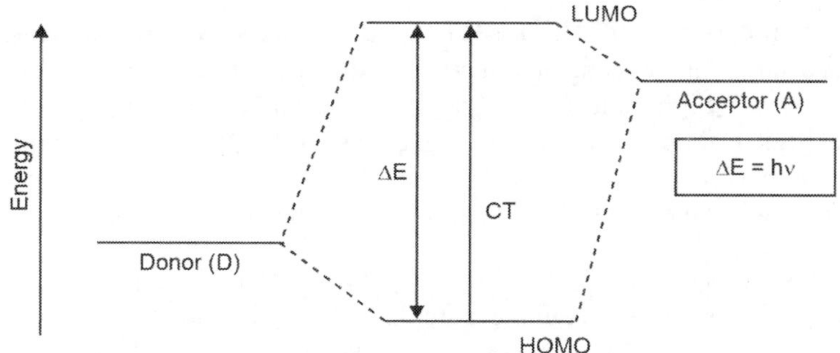

Fig. 7.21.1: Development of a CT band in terms of MOs.

Thus to develop a CT band or redox band, there must be an electron donor (D) and an electron acceptor (A). The energy of the CT band depends on the ionization energy (IE_D) of the donor and electron affinity (EA_A) of the acceptor.

$$\Delta E = h\nu = IE_D - EA_A - C, \ C \text{ is a constant}$$

The origin of the CT band can be simply explained by considering the MO orbitals formed in the DA (donor-acceptor) complex (Fig. 7.21.1).

The CT spectra can be best explained in terms of **molecular orbital theory** (MOT) not in terms of **crystal field theory** (CFT). In the transition metal complexes, the molecular orbitals are formed by the combination of the metal *d*-orbitals and ligand orbitals. Because of the energy difference between the metal and ligand orbitals, the metal and ligand orbital character are not uniformly distributed among the molecular orbitals. When an electron is transferred from a MO enriched in the ligand orbital character to a MO enriched with the metal orbital character, it is referred to as $L \rightarrow M$, *i.e.* LMCT transition.

Similarly, $M \rightarrow L$, *i.e.* MLCT transition, can be assigned for the electron transition from a MO enriched with the metal orbital character to a MO enriched with the ligand orbital character.

Here it is worth mentioning that compared to the ligand field bands arising from the *d–d* transition, **the charge transfer bands are more intense.** The MOs bear the character of both the *d*-orbital (coming from the metal centre) and *p*-orbital or *s-p* hybrid orbital (coming from the ligand). Thus, in the CT bands, the $p \rightarrow d$ and $d \rightarrow p$ transition components prevail along with the $p \rightarrow p$ and $d \rightarrow d$ transition components. The $p \rightarrow p$ and $d \rightarrow d$ transitions are not orbitally allowed but the $p \rightarrow d$ and $d \rightarrow p$ transition components are orbitally allowed. The CT bands are also the spin-allowed transitions. Thus the CT bands are both orbitally and spin allowed. **This is why, the CT bands are highly intense.** The intense colours developed due to the CT bands find many important applications in analytical chemistry.

7.21.1 LMCT Bands

The $L \rightarrow M$ charge transfer bands are very much common. In this transition, the ligand is oxidized and the metal centre is reduced. This CT transition is favoured when the metal centre is oxidizing and the ligand moiety is oxidizable. **Thus with the increase of the oxidizing power of the metal centre and oxidizability of the ligand, the charge transfer band will move towards the longer wavelength (*i.e.* red shift).** The LMCT spectra are common when the ligands have the filled π-orbitals (*e.g.* heavier halides and chalcogenides) and the metal centre with a relatively high ionization energy. The transition and post transition metals with the high oxidation states are the suitable metal centres for LMCT. In such cases, the electrons from the filled orbitals of relatively higher energy of the ligands will move towards the low energy orbitals of the metal.

Favourable conditions to shift the LMCT band towards the higher wavelength region
• Oxidizing metal centre; • lighter congener of more oxidising property (*e.g.* $Fe^{3+} > Ru^{3+} > Os^{+3}$; MnO_4^- > TcO_4^- > ReO_4^-; etc.); • high ionization energy of the metal centre; • high polarizing power of the metal centre; • high polarizablity of the ligand; • higher reducing power of the ligands; • filled π-orbitals of the ligand and vacant *d*-orbitals of the metals; • HOMO (enriched with the ligand orbital character) of high energy and LUMO (enriched with the metal orbital character) of low energy.

Examples of LMCT spectra: The halidocomplexes like $[MX_6]$, $[MX_4]$; Fe^{III}-SCN complex; Fe^{III}-phenolate complex; Fe^{III}-hydroxamate complex; V^V-hydroxamate complex; $[Co^{III}(NH_3)_5X]^{2+}$ (X = Cl, Br, I); $[CrO_4]^{2-}$, *i.e.* K_2CrO_4; $[MnO_4]^-$, *i.e.* $KMnO_4$, etc. show the LMCT bands.

The **ease** with which a **LMCT** in $[MX_6]$ (*i.e.* hexahalido complexes) occurs runs in the sequence:

$Rh^{4+} > Ru^{4+} > Cu^{2+} > Os^{4+} > Fe^{3+} > Ru^{3+} > Pd^{4+} > Pd^{2+} \sim Pt^{4+} \sim Rh^{3+} \rangle Pt^{2+} \rangle Ti^{4+} \sim Ir^{3+}$.

and, $I^- > Br^- > Cl^- > F^-$.

- $[Co^{III}(NH_3)_5X]^{2+}$**:** Charge transfer spectra of $[Co(NH_3)_5X]^{2+}$ (X = F, Cl, Br, I) are quite interesting. For X = F, the LMCT band lies in the far UV-region as in the case of $[Co(NH_3)_6]^{3+}$. For X = Cl, Br, I, the intense LMCT band overlaps with the higher energy weaker ligand field (*i.e. d-d* transition) band (*i.e.* ${}^1A_{1g} \rightarrow {}^1T_{2g}$) and the corresponding ligand field band gains the intensity from the allowed CT band through the **intensity stealing mechanism.**

Table 7.21.1.1 Spectral characteristics of $[Co^{III}(NH_3)_5X]^{2+}$ and $[Co(NH_3)_6]^{3+}$ (*cf.* Figs. 7.15.3.1–2).

Complex	O_h: ${}^1A_{1g} \rightarrow {}^1T_{1g}$		${}^1A_{1g} \rightarrow {}^1T_{2g}$	*LMCT bands*
	C_{4v}: ${}^1A_1 \rightarrow {}^1E$	${}^1A_1 \rightarrow {}^1A_2$	${}^1A_1 \rightarrow {}^1T_2$	
	$\bar{\nu}$ (cm^{-1})		$\bar{\nu}$ (cm^{-1})	$\bar{\nu}$ (cm^{-1})
$[Co(NH_3)_6]^{3+}$	21,200 (O_h)		29,550 (O_h)	–
$[CoF(NH_3)_5]^{2+}$	19,450	21,470	28,270	
$[CoCl(NH_3)_5]^{2+}$	18,720	21,350 (sh)	27,500	37,100, ~45,000
$[CoBr(NH_3)_5]^{2+}$	18,230	21,950 (sh)	Masked by CT band	32,000, ~39,000
$[CoI(NH_3)_5]^{2+}$	17,250	Masked by CT band	Masked by CT band	

$[Co(NH_3)_5X]^{2+}$ belongs to the C_{4v} point group while $[Co(NH_3)_6]^{3+}$ belongs to the O_h point group. However, the ligand field transitions in $[Co(NH_3)_5X]^{2+}$ can be roughly explained in terms of the octahedral transitions (*see* Sec. 7.15.3 for discussion). In moving from $[Co(NH_3)_6]^{3+}$ to $[Co(NH_3)_5X]^{2+}$, the T_{1g} state splits into the E and A_2 components but the T_{2g} state practically does not split (*cf.* Fig. 7.15.3.1). *The magnitude of splitting of the T_{1g} state depends on the difference of average ligand field strength between the N—Co^{III}—N and N—Co^{III}—X axes.* In terms of crystal field strength of the ligands (*cf.* $H_3N > F^- > Cl^- > Br^- > I^-$ ligand fied strength sequence), this difference is maximum for $[CoI(NH_3)_5]^{2+}$ and minimum for $[CoF(NH_3)_5]^{2+}$. In fact, the ligand field band related to the T_{1g} level practically does not split for X = F and only a shoulder appears on the higher energy side for X = Br and Cl; for *X* = I, these expected separate peaks are masked by the LMCT band. The peak related to the T_{2g} level does not split but it gets masked by the LMCT band for X = Br, I.

The LMCT band should shift towards the longer wavelength (*i.e.* **red shift**) in the sequence: X = I > Br > Cl > F and it happens so. For the **IPCT (ion-pair charge transfer)** band in $[Co(NH_3)_6]^{3+}\cdot X^-$, the similar spectral shift (*i.e.* **red shift** for the more polarisable anion) is also noted (*cf.* Fig. 7.21.3.1). For X = F and NH_3, the LMCT shoulder only appears at $\bar{\nu} \geq 40,000$ cm^{-1}. For X = I, practically all the

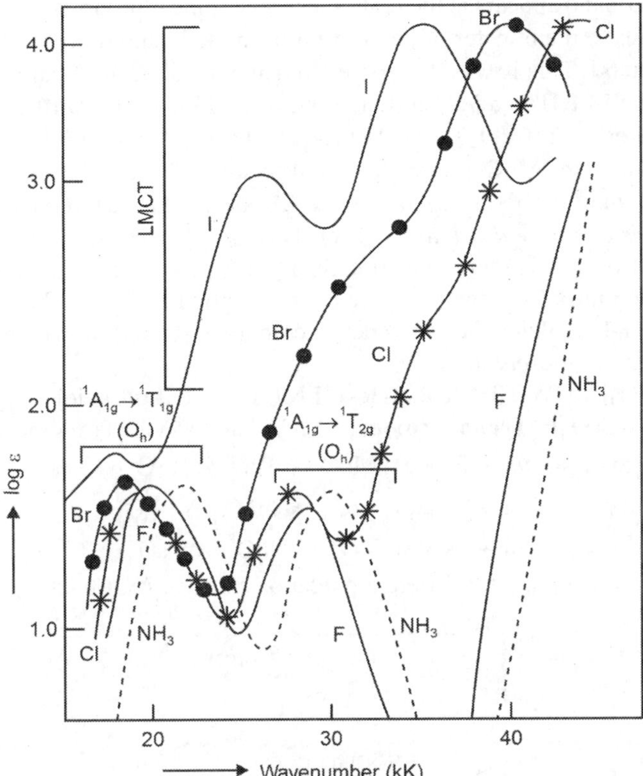

Fig. 7.21.1.1: Absorption spectra of $[Co^{III}(NH_3)_5(X)]^{n+}$ ($X = F^-, Cl^-, Br^-, I^-, NH_3$)

ligand field bands are masked by the strong LMCT bands; for X = F, Cl, the ligand field bands are not masked; for X = Br, the higher energy ligand field band (*i.e.* related to the $^1T_{2g}$ level) is only masked by the strong LMCT band. For X = Cl and Br, the lowest energy LMCT band appears as a shoulder at about 37,100 cm^{-1} and 32,000 cm^{-1} respectively. The next higher energy LMCT band appears distinctly with log $\varepsilon \approx 4.0$ for X = Cl and Br. For X = F, the lowest energy LMCT band shows a shoulder at 40,000 cm^{-1}.

In the same way, the **LMCT band in $[Cr^{III}(NH_3)_5X]^{2+}$** can be explained. For X = Cl, reduction in symmetry from O_h to C_{4v} does not much split the octahedral peaks but in general, the ligand field bands shift towards the longer wavelength (compared to those of $[Cr(NH_3)_6]^{3+}$) because Cl^- is a weaker field ligand than NH_3. The interesting observation is that for $[Cr(NH_3)_6]^{3+}$, a tail of the CT band appears at 50,000 cm^{-1} (*i.e.* 200 nm) while in $[CrCl(NH_3)_5]^{2+}$, the LMCT ($Cl^- \rightarrow Cr^{III}$) band appears at 42,000 cm^{-1} (*i.e.* 240 nm).

(B) Oxidizing power of the metal centre and position of the LMCT band: The colours of the following oxyanions with the d^0 metal centre are:

$[VO_4]^{3-}$ **(colourless)**, $[CrO_4]^{2-}$ **(yellow)**, $[MnO_4]^-$ **(pink)**
(V^{5+}) (Cr^{6+}) (Mn^{7+})

Among the said oxyanions, $[MnO_4]^-$ is the most oxidizing one (*i.e.* highest E^0 value, 1.52 V) and the LMCT band appears at the lowest energy while $[VO_4]^{3-}$ is the poorest oxidizing agent for which the LMCT band appears at the highest energy (in the ultraviolet region). In fact, the **LMCT redox band** leads to the transfer of an electron from the ligand to metal, **causing effectively the reduction of metal**. This is why, the more oxidising metal centre (that unergoes reduction more easily) shows the LMCT band at the longer wavelength (*i.e.* **red shift**).

(C) **$[MnO_4]^-$ *vs.* $[ReO_4]^-$; $[CrO_4]^{2-}$ *vs.* $[MoO_4]^{2-}$, $[WO_4]^{2-}$:** Now let us consider the positions of the LMCT bands in $[MnO_4]^-$ (**strongly pink coloured**) and $[ReO_4]^-$ (**colourless**). **We know that the higher oxidation state is more stabilized for the heavier congeners** (*cf. relativistic effect and expansion of the d-orbital*; Vol. 1). This is why, Re(VII) is more stable than Mn(VII). In other words, $[MnO_4]^-$ is more oxidizing than $[ReO_4]^-$. This is why, the LMCT band lies in the visible range (565 nm) for $[MnO_4]^-$ while the corresponding LMCT band of $[ReO_4]^-$ lies in the ultraviolet region. It explains the difference in colours of $[ReO_4]^-$ and $[MnO_4]^-$. Similarly, we can explain the following observation.

$[MoO_4]^{2-}$, $[WO_4]^{2-}$ (colourless, LMCT band at ultraviolet region),
$[CrO_4]^{2-}$ (yellow colour; LMCT band at visible region).

$$\nu_{LMCT} \text{ (for +7 state)}: MnO_4^- < TcO_4^- < ReO_4^-$$
$$\nu_{LMCT} \text{ (for +6 state)}: CrO_4^{2-} < MoO_4^{2-} < WO_4^{2-}$$
$$\nu_{LMCT} \text{ (for +5 state)}: VO_4^{3-} < NbO_4^{3-} < TaO_4^{3-}$$
$$(cf.\ \nu_{LMCT} \text{ for different oxidation states}: +7 < +6 < +5)$$

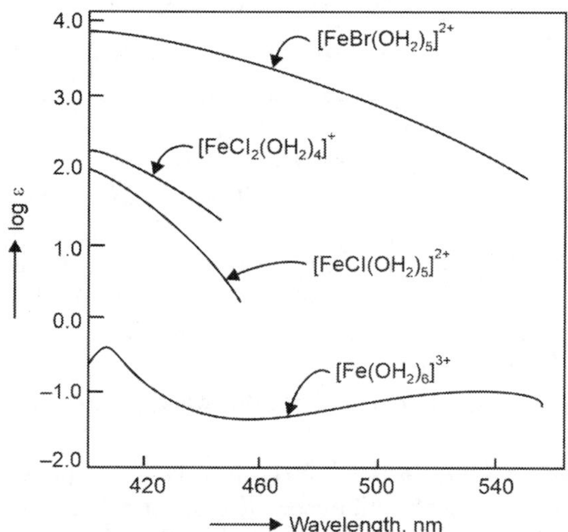

Fig. 7.21.1.2: Comparison of the electronic absorption spectra of different Fe(III) species in aqueous solution. $[Fe(OH_2)_6]^{3+}$ is almost colourless (forbidden *d–d* transition, both Laporte forbidden and spin forbidden). The halidoaqua-complexes are coloured due to the LMCT bands.

- **Fe(III)-aqua and halidoaqua-complexes:** The strong absorption band (*i.e.* LMCT band) of Fe(III) complexes moves towards the longer wavelength when the ligand becomes more polarizable and oxidizable. This is why, $[Fe(aq)]^{3+}$ is colourless, but $[Fe^{III}(OH)_x(OH_2)_{6-x}]$ (*i.e.* hydrolyzed species), $[Fe(aq)X]^{2+}$ (X = Cl, Br) are strongly coloured.
- **Fe^{III}–SCN^- complex:** From the filled π-orbitals of the ligand, electron transfer occurs to the t_{2g}-level of Fe(III). This gives the intense red colour and this colour formation reaction can be utilized for the qualitative and quantitative analysis of Fe(III).

(E) Colour of d^0 and d^{10} systems: The yellow colour of $[CrO_4]^{2-}$, pink colour of $[MnO_4]^-$, yellow colour of As_2S_3, orange colour of Sb_2S_3, the red colour of HgI_2, yellow colour of PbI_2, orange-red colour of BiI_3, etc. can be explained in terms of LMCT.

(F) Inorganic pigments: Colouration of some common inorganic pigments can be explained in terms of LMCT transfer (see Sec. 7.21.7 for the better explanation in terms of the **band theory of solids**).

CdS (bright yellow): $p_\pi(S) \rightarrow 5s(Cd)$

HgS (deep red, vermilion): $p_\pi(S) \rightarrow 6s(Hg)$

PbO (yellow): $p_\pi(O) \rightarrow 6s, 6p(Pb)$.

CrO_4^{2-} (yellow, as $PbCrO_4$): $p_\pi(O) \rightarrow 3d(Cr)$.

As_2S_3 (yellow): $p_\pi(S) \rightarrow 4p(As)$

Fe_2O_3 (brown): $p_\pi(O) \rightarrow 3d(Fe)$

(**Naples yellow**, as $Pb_3(SbO_4)_2$): $p_\pi(O) \rightarrow 5s, 5p(Sb)$.

Note: The $\pi^*(i.e.\ t_{2g}) \rightarrow \sigma^*(i.e.\ e_g)$ transitions, *i.e.* **metal centred (MC) transitions**, are the *d–d* or ligand field transitions. On the other hand, the **ligand centred (LC) transitions** are the $\sigma\,(i.e.\ \sigma_L) \rightarrow \pi^*(i.e.\ \pi_L^*)$ and $\pi\,(i.e.\ \pi_L) \rightarrow \pi^*(i.e.\ \pi_L^*)$ transitions appear in the much higher energy range (*i.e.* UV-range). The MC and LC transitions are not shown in the Fig. 7.21.1.3.

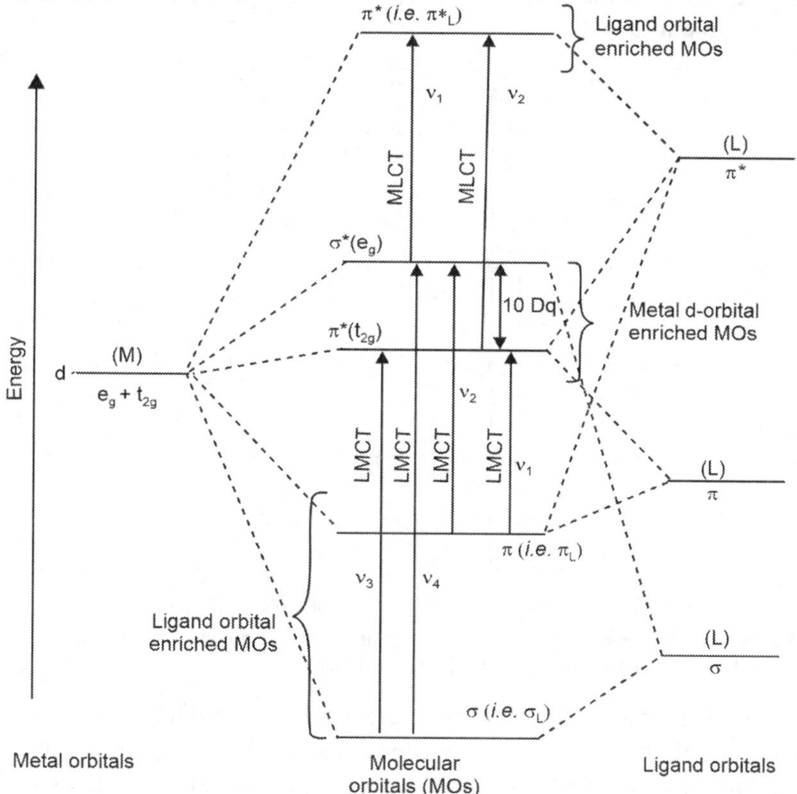

Fig. 7.21.1.3: (a) Simplified MO diagram for showing the LMCT and MLCT transitions in the octahedral transition metal complex

(G) LMCT inducing the instability of some compounds: If the LMCT is highly favoured (*i.e.* $\Delta E < 10{,}000\ \text{cm}^{-1}$, ΔE = energy difference between the HOMO predominantly enriched with the ligand orbital character and the LUMO enriched with the character of metal orbital), then the complete electron transfer from the ligand to the metal will establish a permanent redox decomposition of the complex. **It indicates that if the CT band arises at a sufficiently longer wavelength then a very small amount of energy will be necessary to allow the permanent redox decomposition of the complex.** If we consider the halido-complexes of Fe(III), the CT band will arise in the ultraviolet region for the fluorido-complex while for the other halido-complexes, the CT band will move to the visible range (*i.e.* towards the higher wavelength). For the iodo-complex, the electron transfer process is so favoured that the hypothetical $[FeI_4]^-$ complex undergoes a redox decomposition giving rise to Fe(II) and I_2. A similar situation arises for the halidocomplexes of Cu(II). In fact, CuI_2 or $[CuI_4]^{2-}$ undergoes a redox decomposition giving rise to CuI and I_2 while $[CuCl_4]^{2-}$ and $[CuBr_4]^{2-}$ are dark coloured and stable. The redox decomposition of $[Co(OH_2)_6]^{3+}$ can also be explained in this way.

$$Fe^{3+} + I^- \ (i.e.\ Fe^{III}\text{–}I) \rightarrow Fe^{2+} + I_2$$
$$Cu^{2+} + I^- \ (i.e.\ Cu^{II}\text{–}I) \rightarrow CuI + I_2$$
$$[Co(OH_2)_6]^{3+} \rightarrow [Co(OH_2)_6]^{2+} + O_2$$

Here it may be noted that the thermodynamically unstable $[Co(OH_2)_6]^{3+}$ can be characterized well because of its **kinetic stability.**

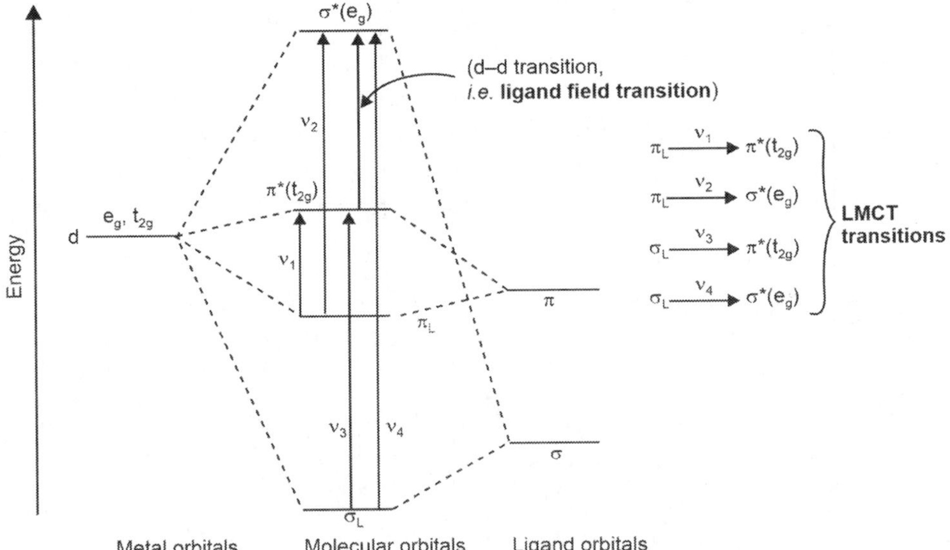

Fig. 7.21.1.3: (b) Simplified MO energy diagram describing the ligand to metal charge transfer (LMCT) transitions ν_1, ν_2, ν_3 and ν_4 in the octahedral complexes like $[MX_6]^{n-}$ (X^- = halide).

Note: By considering the *s*- and *p*- orbitals on the halides, the π-MOs transform as $t_{1g} + t_{2g} + t_{1u} + t_{2u}$ and the σ-MOs transform as $a_{1g} + e_g + t_{1u}$. Obviously, the $g \rightarrow g$ transitions, *e.g.* $t_{1g}(\pi)$, $t_{2g}(\pi) \rightarrow t_{2g}(\pi^*)$ are the **forbidden transitions** and $t_{1u}(\pi)$, $t_{2u}(\pi) \rightarrow t_{2g}(\pi^*)$; $t_{1u}(\sigma) \rightarrow e_g(\sigma^*)$, etc. transitions of the $u \rightarrow g$ type are the **allowed transitions**. It may be mentioned that the *p*(ligand) → *d*(metal) component of the LMCT band is **orbitally allowed**.

(H) Molecular orbital picture for LMCT in the octahedral complexes: A simplified MO energy diagram is shown in Fig. 7.21.1.3a, b for the octahedral complex [ML_6]. This simplified MO energy diagram has been extracted from the complete MO energy diagram of [ML_6] (*cf.* Fig. 3.17.2.5, Vol. 4).

The possible LMCT transitions are (*cf.* Fig. 7.21.1.3b):

$$\pi_L \xrightarrow{\nu_1} \pi^*(t_{2g});\ \pi_L \xrightarrow{\nu_2} \sigma^*(e_g);\ \sigma_L \xrightarrow{\nu_3} \pi^*(t_{2g}) \text{ and } \sigma_L \xrightarrow{\nu_4} \sigma^*(e_g)$$

The ν_1-transition gives the lowest energy band. In reality, the ν_1 and ν_2 LMCT bands are only important. The high energy ν_3 and ν_4 LMCT bands are rarely observed and these may appear in the far UV region. In fact, the broad and weak ν_3 band is not observed. Only in few cases, the ν_4 band is noticed. The important expectations are discussed below.

(i) **LMCT ($u \rightarrow g$):** The LMCTs (ν_1, ν_2, ν_3 and ν_4) arise for the transitions from the MOs predominantly enriched with the ligand orbital character of *u*-symmetry to the MOs (*i.e.* $t_{2g} = \pi^*$ and $e_g = \sigma^*$) enriched with the character of metal *d*-orbitals of *g*-symmetry. **If the t_{2g} level is completely filled in (*i.e.* t_{2g}^6), the lowest energy band (ν_1) remains absent and if both the t_{2g} and e_g levels are filled in (*i.e.* $t_{2g}^6 e_g^4$) then both ν_1 and ν_2 remain absent** and the high energy bands due to ν_3 and ν_4 are also not possible. In reality, for the d^{10} configuration, the tetrahedral complexes are preferred over the octahedral complexes and for such systems, the LMCTs can be explained by considering the MO energy diagram of the tetrahedral complexes.

(ii) **Narrowness of the ν_1 - band:** The π^*-MOs ($\equiv t_{2g}$ set of metal) are mainly nonbonding in character and consequently, the metal-ligand bond length does not remarkably change due to the excitation of electron to these MOs. In such cases, energy change for the electronic transition does not change remarkably due to the metal-ligand bond vibration (*cf.* Secs. 7.7.2–3). This is why, bandwidth broadening by the metal-ligand bond vibration is not of much importance for this lowest energy transition (*i.e.* ν_1 band). This band is quite sharp, *i.e.* narrow, and the bandwidth is within the range 400–1000 cm^{-1} (*cf.* bandwidth, 2000–4000 cm^{-1} for the ν_2-band). For the ν_2-band, the electron transition occurs to the σ^*-MOs. The σ^* (*i.e* e_g) MOs are antibonding in character and excitation of electron to these MOs leads to increase the bond length, *i.e.* $r_e^{ex} > r_e^{gd}$, and *this is why, the ν_2-band experiences peak broadening due to the metal-ligand bond vibration* (*cf.* Secs. 7.7.2–3).

(iii) **Absence of the ν_1 band ($\pi_L \rightarrow \pi^*$):** This lowest energy band is absent for the systems where π^* (*i.e.* t_{2g}) level is completely filled in as in the low-spin d^6-complexes and d^8-complexes. In such cases, the ν_2-band (*i.e.* $\pi_L \rightarrow \sigma^*$) is quite important. In fact, this is the lowest energy LMCT band in $[PtX_6]^{2-}$.

Let us consider the positions of the lowest energy LMCT bands in $[Fe(CN)_6]^{4-}$ ($t_{2g}^6 e_g^0$) and $[Fe(CN)_6]^{3-}$ ($t_{2g}^5 e_g^0$). The lowest energy band (*i.e.* ν_1) is possible for $[Fe(CN)_6]^{3-}$ (where the t_{2g}-level is remaining partially vacant) but ν_1 is absent in $[Fe(CN)_6]^{4-}$ for which ν_2 is the lowest energy band. **This is why, $[Fe(CN)_6]^{3-}$ shows the first LMCT band at a relatively longer wavelength compared to $[Fe(CN)_6]^{4-}$.** A similar explanation holds good for $[IrBr_6]^{2-}$ (t_{2g}^5) and $[IrBr_6]^{3-}$ (t_{2g}^6).

(iv) **Large bandwidth of the ν_2 band (*i.e.* $\pi_L \rightarrow \sigma^*$):** It has been mentioned that in the low-spin d^6 systems and d^8-systems, this is the lowest energy LMCT band. Though the π^*-MO (*i.e.* t_{2g}) is predominantly nonbonding but the σ^*–MO (*i.e.* e_g) is antibonding. Consequently, excitation of electron to the σ^*-MO elongates the metal-ligand bond (*i.e.* $r_e^{ex} > r_e^{gd}$). Thus the energy difference between these MOs (*i.e.* π_L-MO and σ^*-MO between which the electron transition occurs) depends on the metal-ligand bond vibration. It broadens the peak (*cf.* Secs. 7.7.2-3). In fact, the ν_2 band is broader than the ν_1-band (*cf.* band-width: for ν_2 band, 2000–4000 cm^{-1}; for ν_1 band 400–1000 cm^{-1}).

(v) **Positions of the LMCT bands:** The required energy for the LMCT transition largely depends on the oxidizing property of the metal centre and oxidizability of the ligand. The more oxidizing property of the metal indicates roughly its higher electron affinity, *i.e.* the corresponding metal orbitals are of lower energy. On the other hand, for the more oxidizable ligands, the ionization energy of the ligands is relatively less, *i.e.* the corresponding ligand orbital is of higher energy. Thus we can conclude:

- With the increase of the oxidizing power of the metal centre, the LUMO (predominant in the character of metal orbitals) comes down (energetically) and with the increase of the oxidizability of the ligand, the HOMO (predominant in the character of ligand orbitals) goes up (energetically). Consequently, the energy difference between the HOMO and LUMO decreases and the LMCT band moves towards the longer wavelength (*i.e.* **red shift**). This HOMO → LUMO transition gives the **lowest energy LMCT**. The higher energy LMCTs involve the transitions from the HOMO to the higher energy MOs enriched with the metal orbital character. These higher energy LMCTs also experience the red shift as for the lowest energy LMCT. These expectations are illustrated in the examples given in Table 7.21.1.2.

Table 7.21.1.2: Positions of the LMCT bands in some representative hexahalido complexes.

Complex		$\pi_L \xrightarrow{\nu_1} \pi^*(t_{2g})$ **(cm^{-1})**	$\pi_L \xrightarrow{\nu_2} \sigma^*(e_g)$ **(cm^{-1})**
$[TiCl_6]^{2-}$	d^0	31,800	42,500
$[TiBr_6]^{2-}$	d^0	25,200	36,500
$[ZrCl_6]^{2-}$	d^0	42,500	–
$[ZrBr_6]^{2-}$	d^0	39,000	–
$[VCl_6]^{2-}$	d^1	21,400	–
$[VBr_6]^{2-}$	d^1	(unstable)	

The shifting of both ν_1 and ν_2 towards the longer wavelength in moving from the chlorido complex to the bromido complex occurs due to the more oxidizability of Br^- compared to that of Cl^-.

$$\nu : [TiCl_6]^{2-} > [TiBr_6]^{2-}; [OsCl_6]^{2-} > [OsBr_6]^{2-} > [OsI_6]^{2-}.$$

It is noteworthy that though $[VCl_6]^{2-}$ is stable, $[VBr_6]^{2-}$ is unstable because of its spontaneous redox decomposition. In other words, the LMCT band of $[VBr_6]^{2-}$ is shifted to so longer wavelength that it cannot survive from its spontaneous redox decomposition.

The position of ν_1-band for the following chlorido complexes varies as follows:

ν_1 (cm^{-1}): $[ZrCl_6]^{2-} > [TiCl_6]^{2-} > [VCl_6]^{2-}$

This can be explained by considering the oxidizing power of the M(IV) centre.

Oxidizing power: V(IV) > Ti(IV) > Zr(IV)

Stability of the +4 oxidation state: Zr(IV) > Ti(IV) > V(IV) (*i.e.* higher oxidation state is stabilized better for the heavier congeners) due to the relativistic **expansion/destabilisation** of the *d*-orbitals of the heavier congeners, see Vol. 1.

(I) Molecular orbital picture of LMCT in the tetrahedral complexes: For this purpose, we need to consider the MO energy diagram of a tetrahedral complex with both σ- and π-bonding (Fig. 7.21.1.4).

The lowest 4-σ-bonding MOs ($a_1 + t_2$) can be predominantly enriched with the character of the ligand orbitals. They can accommodate 8-electrons. The next higher 8 π-MOs (*i.e.* $t_2 + e + t_1$; t_1 is basically a nonbonding one) can accommodate maximum 16 electrons. These 8 π-MOs are predominantly enriched with the character of ligand π orbitals. **The next higher MOs are $e^*(\pi^*)$ which are basically the metal $d_{x^2-y^2}$ and d_{z^2} orbitals.** These are predominantly enriched with the character of the metal *d*-orbitals and sometimes these are regarded as the **nonbonding orbitals.** The next higher t_2^* (σ^*, π^*) MOs are also predominantly enriched with the characters of metal d_{xy}, d_{yz} and d_{xz} orbitals. The energy difference between e^* and t_2^* gives the measure of Δ_t. The electronic configurations of some representative tetrahedral complexes are given below where each ligand or ligating site provides 6*p* electrons for the total electron count.

$[VCl_4](d^1)$: $[\sigma_b(L)]^8\ [\pi_{b,\,nb}\ (L)]^{16}\ [e^*(\pi^*)]^1\ [t_2^*(\sigma^*,\pi^*)]^0\ [a_1^*(\sigma^*)]^0\ [t_2^*(\sigma^*,\pi^*)]^0$

$[NiCl_4]^{2-}(d^8)$: $[\sigma_b(L)]^8\ [\pi_{b,\,nb}\ (L)]^{16}\ [e^*(\pi^*)]^4\ [t_2^*(\sigma^*,\pi^*)]^4\ [a_1^*(\sigma^*)]^0\ [t_2^*(\sigma^*,\pi^*)]^0$

$[CoCl_4]^{2-}(d^7)$: $[\sigma_b(L)]^8\ [\pi_{b,\,nb}\ (L)]^{16}\ [e^*(\pi^*)]^4\ [t_2^*(\sigma^*,\pi^*)]^3\ [a_1^*(\sigma^*)]^0\ [t_2^*(\sigma^*,\pi^*)]^0$

$[HgX_4]^{2-}(d^{10})$: $[\sigma_b(L)]^8\ [\pi_{b,\,nb}\ (L)]^{16}\ [e^*(\pi^*)]^4\ [t_2^*(\sigma^*,\pi^*)]^6\ [a_1^*(\sigma^*)]^0\ [t_2^*(\sigma^*,\pi^*)]^0$

$[MnO_4]^-$, $[CrO_4]^{2-}\ (d^0)$: $[\sigma_b(L)]^8\ [\pi_{b,\,nb}\ (L)]^{16}\ [e^*(\pi^*)]^0\ [t_2^*(\sigma^*,\pi^*)]^0\ [a_1^*(\sigma^*)]^0\ [t_2^*(\sigma^*,\pi^*)]^0$

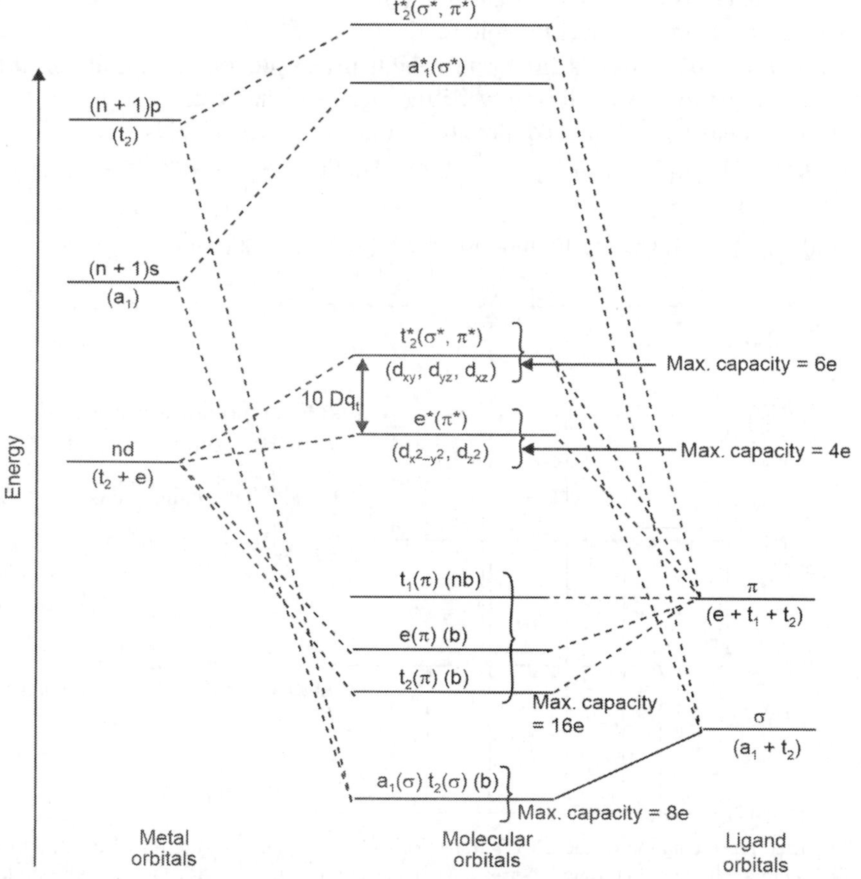

Fig. 7.21.1.4: Molecular orbital energy diagram of a representative tetrahedral complex.

Note: The MO energy diagram given in Fig. 7.21.1.4 is mainly applicable for the tetrahedral complexes of the lower oxidation states like +2, +3. There is a doubt whether the same diagram is applicable for the oxyanions like $[MnO_4]^-$, $[MnO_4]^{2-}$, $[CrO_4]^{2-}$, $[MoO_4]^{2-}$, etc. where the metal centre is in very high oxidation states.

The LMCT transitions occur from the $\pi_{b,\,nb}$ *(L) MOs* (which are predominantly enriched with the characters of ligand orbitals) to the higher vacant *MOs* (which are predominantly enriched with the character of metal orbitals). These are shown in Fig. 7.21.1.5.

Obviously, the ν_1 and ν_2 transitions (*cf.* Fig. 7.21.1.5) are not possible for the d^{10} systems; ν_1 transition is not possible for the d^7 and d^8 configurations; all the transitions are possible for the d^{0-6} configurations. It is understandable by considering the high-spin nature of the tetrahedral complexes (*i.e.* $P > 10Dq_t$). Thus for the d^0 to d^6 systems, vacancies at both the e^* and t_2^* levels remain present to allow all the possible LMCT transitions (*cf.* $d^5 \Rightarrow [e^*]^2[t_2^*]^3$; $d^6 \Rightarrow [e^*]^3[t_2^*]^3$; $d^7 \Rightarrow [e^*]^4[t_2^*]^3$, etc.). The diagram can also explain the origin of LMCT bands in d^0 (*e.g.* $[MnO_4]^-$) and d^{10} (*e.g.* $[HgX_4]^{2-}$) systems.

For $\mathbf{[NiX_4]^{2-}}$ and $\mathbf{[CoX_4]^{2-}}$ [X = Cl, Br, I), the lowest energy trasition is ν_2 (*cf.* Fig. 7.21.1.5). More correctly, this lowest energy transition represents the $t_1(\pi, nb) \rightarrow t_2(\sigma^*, \pi^*)$ transition (*cf.* Fig. 7.21.1.4). For CdS, HgS, etc. (*i.e.* d^{10} system), the ν_3 transition (*cf.* Fig. 7.21.1.5) is the lowest energy transition which is responsible for their characteristic colour.

(J) Positions of the LMCT bands in the tetrahedral complexes: It depends on the oxidizing property of the metal centre and oxidizability of the ligand in the same way as in the octahedral complexes. This aspect has been discussed already. The illustrative examples are:

ν(LMCT): $[ReO_4]^- > [TcO_4]^- > [MnO_4]^{2-} > [MnO_4]^-$; $[WO_4]^{2-} > [MoO_4]^{2-} > [CrO_4]^{2-}$; $[CrO_4]^{2-}$ (~ 27,000 cm^{-1}) > $[MnO_4]^-$ (~ 18,000 cm^{-1});

$[HgCl_4]^{2-}$ (43,700 cm^{-1}) > $[HgBr_4]^{2-}$ (40,000 cm^{-1}) > $[HgI_4]^{2-}$ (31,000 cm^{-1})

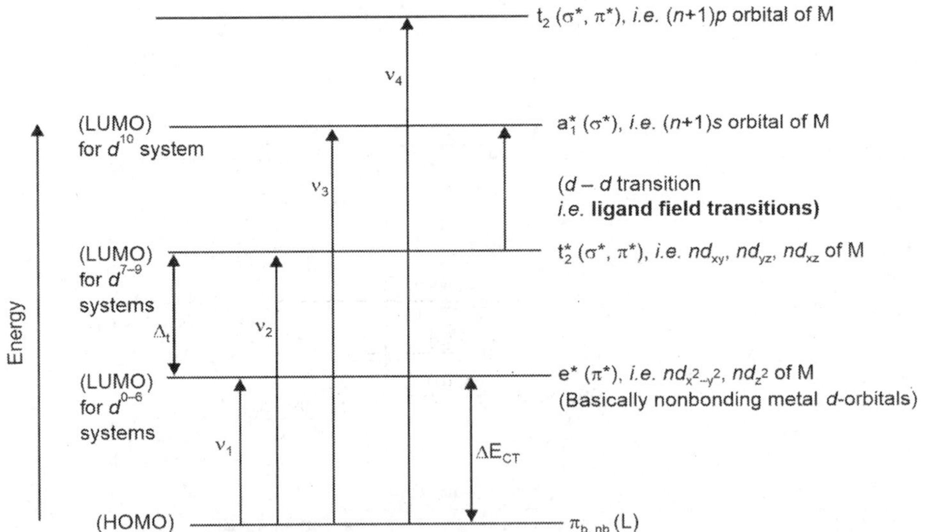

Fig. 7.21.1.5: Simplified representation of the LMCT transitions ν_1, ν_2, ν_3 and ν_4 (in terms of the *MO* energy diagram given in Fig. 7.21.1.4) in a tetrahedral complex. **Note:** By considering the π_b and π_{nb} MOs (of L) separately, ν_1 and ν_2 produce 4 transitions as illustrated in Fig. 7.21.1.6.

(K) LMCT bands in $[CrO_4]^{2-}$ and $[MnO_4]^-$: In $[CrO_4]^{2-}$, the intense LMCT bands appear at about 26,800 cm^{-1} (= 373 nm) and 36,500 cm^{-1} (= 274 nm) while in $[MnO_4]^-$, the corresponding bands appear at about 17,700 cm^{-1}(= 565 nm) and 29,500 cm^{-1} (= 339 nm). The strong purple colour of $[MnO_4]^-$ is due to ν_1 while the other transitions occur in the ultraviolet region. It is suggested that the lowest energy bands (at 17,700 cm^{-1} and 26,800 cm^{-1}) appear due to the excitation of non-bonding 2*p*-oxygen electron to the vacant *e*-orbitals (*i.e.* $d_{x^2-y^2}$ and d_{z^2}) of the $3d^0$ metal. There is a difference in opinion regarding the next higher energy band (ν_2). It may be due to $\pi_b(O) \rightarrow e$ or $\pi_{nb}(O) \rightarrow t_2^*$. The possible ways leading to the LMCT bands in $[MnO_4]^-$ and $[CrO_4]^{2-}$ are shown in Fig. 7.21.1.6.

(L) Simplified MOT for the isoelectronic $[MO_4]^{x-}$ species: E^0 value, η (hardness) value and λ_{max} for the LMCT band: For the tetrahedral isoelectronic species like $[VO_4]^{3-}$, $[CrO_4]^{2-}$ and $[MnO_4]^-$, energy of the orbitals of the central atom follows the order: $4p \rangle 4s \rangle 3d$. Thus, in terms of energy, compared to the 4*p* and 4*s* orbitals, the 3*d* orbitals of the central metal atom match better with the 2*p* orbitals of the O-atoms. But, the 4*s* and 4*p* orbitals of the central atom having a **greater spatial extent** can interact better with the 2*p* orbitals of oxygen. Thus, approximately, the central atom may be considered to use the 4*s* and 4*p* orbitals (or **$4sp^3$ hybrid orbitals**) to interact (*i.e.* σ-interaction) with the oxygen 2*p*-orbitals (σ-symmetry) and the contracted 3*d* orbitals of the central atom remain practically **as the nonbonding ones to act as the LUMOs.** It is reasonable to assume that for such high oxidation states, the $(n - 1)d$ orbitals are too contracted (compared to the *ns* and *np* orbitals) to participate in bonding through a good overlapping interaction. The filled in nonbonding 2*p* orbitals (π-symmetry) of oxygen atoms act as the **HOMOs.** Thus the lowest energy charge transfer band (LMCT) originates from the following transition.

HOMO (nonbonding $2p_\pi$-orbitals of oxygen) → LUMO (nonbonding 3*d* orbitals of M).

Thus HOMO is the same for all the isoelectronic species $[MO_4]^{x-}$ but energy of the LUMO varies as follows:

Energy (LUMO or 3*d*-orbital): V(V) > Cr(VI) > Mn(VII)

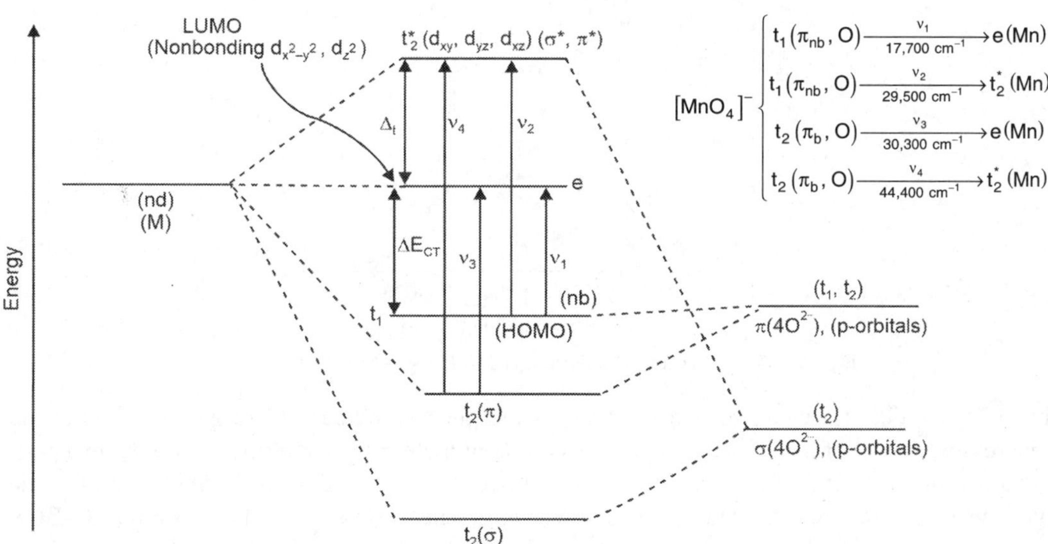

Fig. 7.21.1.6: A simplified partial MO energy diagram (through a partial modification in the diagram given in Fig. 7.21.1.4) of $[MnO_4]^-$ or $[CrO_4]^{2-}$ to show the LMCT transitions.

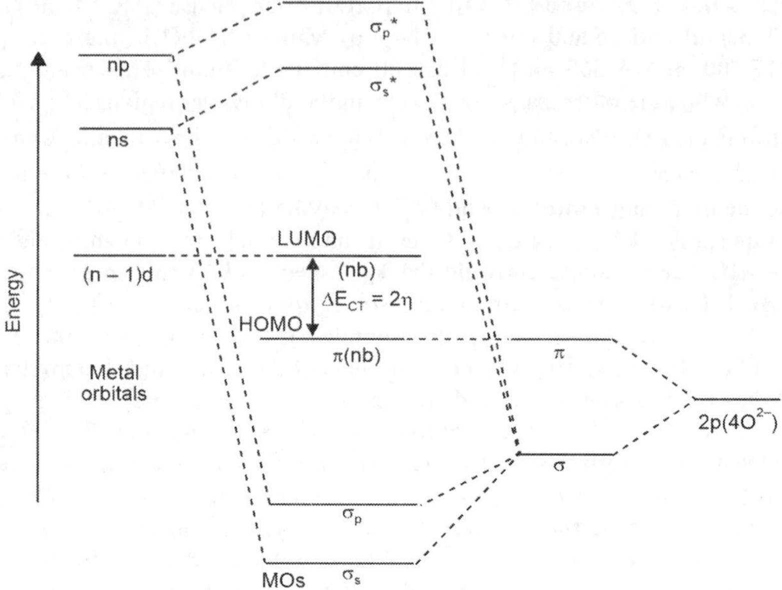

Fig. 7.21.1.7: Simplified MO energy diagram for the tetrahedral species $[MO_4]^{n-}$, *e.g.* $[VO_4]^{3-}$, $[CrO_4]^{2-}$, $[MnO_4]^{-}$, $[ReO_4]^{-}$, etc. assuming the approximate **sp^3 hybridisation of the central element. Note:** The 2*s*-orbitals of O^{2-} sites are so deeply seated (*i.e.* of very low energy) that they do not participate in bonding.

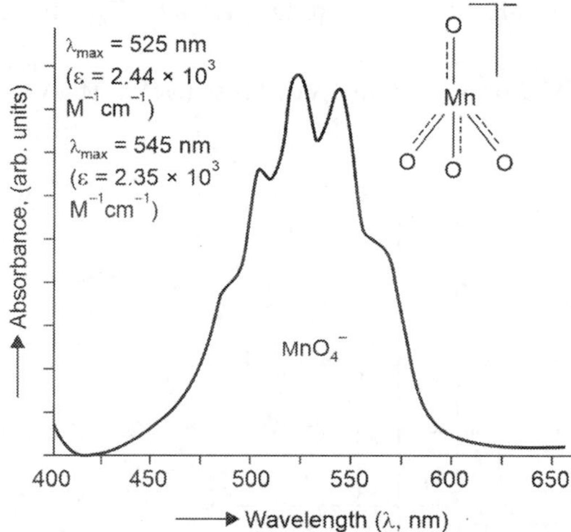

Fig. 7.21.1.8: Absorption spectrum (LMCT, $O_{2p} \rightarrow Mn_{3d}$) of MnO_4^-

This energy order is because of two reasons. As the **positive oxidation state** of the central metal atom increases, energy of the 3*d* orbital decreases (*cf.* **contraction of *d*-orbital**). Secondly, in a period, in moving from left to right, energy of the *d*-orbital decreases because with the **increase of nuclear charge**, the *d*-orbitals are attracted more to lower their energies. Thus, ΔE_{CT} (*i.e.* HOMO-LUMO energy gap) varies as:

$$\Delta E_{CT} \text{ or } \nu_{CT} : [VO_4]^{3-}(\text{colourless}) > [CrO_4]^{2-}(\text{yellow}) > [MnO_4]^{-} (\text{purple})$$

● **Hardness (η) and ν_{CT} (*i.e.* ΔE_{CT}):** Hardness of a molecular species is determined by the energy separation between the HOMO and LUMO. It is given by: $\eta = \frac{1}{2}(E_{LUMO} - E_{HOMO}) = \frac{1}{2}\Delta E_{CT}$, *i.e.* $\Delta E_{CT} = 2\eta$, (*cf.* Sec. 14.15. 3. Vol. 3A). Obviously, η or ν_{CT} runs as: $[MnO_4]^- < [CrO_4]^{2-} < [VO_4]^{3-}$.

● **E^0 and ν_{CT}:** The species $[MO_4]^{x-}$ to act as an oxidizing agent will have to use the LUMO to accpet the electron. Obviosuly, the LUMO of higher energy is more reluctant to accept the electron. It explains the order of oxidizing power.

E^0 or oxidizing power and λ_{CT}: $[VO_4]^{3-} < [CrO_4]^{2-} < [MnO_4]^-$

● **$[ReO_4]^-$ vs. $[MnO_4]^-$ (ν_{CT} and E^0):** In the same way, we can explain the higher oxidizing power of $[MnO_4]^-$ (purple) compared to that of $[ReO_4]^-$ (CT band appearing at the ultraviolet region). For $[MnO_4]^-$, the 3*d* orbital of Mn is the LUMO while for $[ReO_4]^-$, the 5*d* orbital of 'Re' is the LUMO, *i.e.* LUMO of $[ReO_4]^-$ is of relatively higher energy (*cf.* **relativistic expansion** of the *d*-orbitals: 5*d* > 4*d* >> 3*d*).

The E° values of the couples are:

$$MO_4^- + 8H^+ + 5e \rightleftharpoons M^{2+} + 4H_2O, M = Mn, Tc, and Re$$

$$E° = 1.50 V (MnO_4^- / Mn^{2+}) > 0.50 V (TcO_4^- / Tc^{2+}) and 0.42 V (ReO_4^- / Re^{2+})$$

Stability order : Re(VII) > Tc(VII) > Mn(VII)

(*cf.* Consequence of relativistic expansion of *d*-orbitals in the order, 5*d* > 4*d* ≫ 3*d*; Ch. 8, Vol. 1)

$$\nu_{LMCT} : MnO_4^- < TcO_4^- < ReO_4^-$$

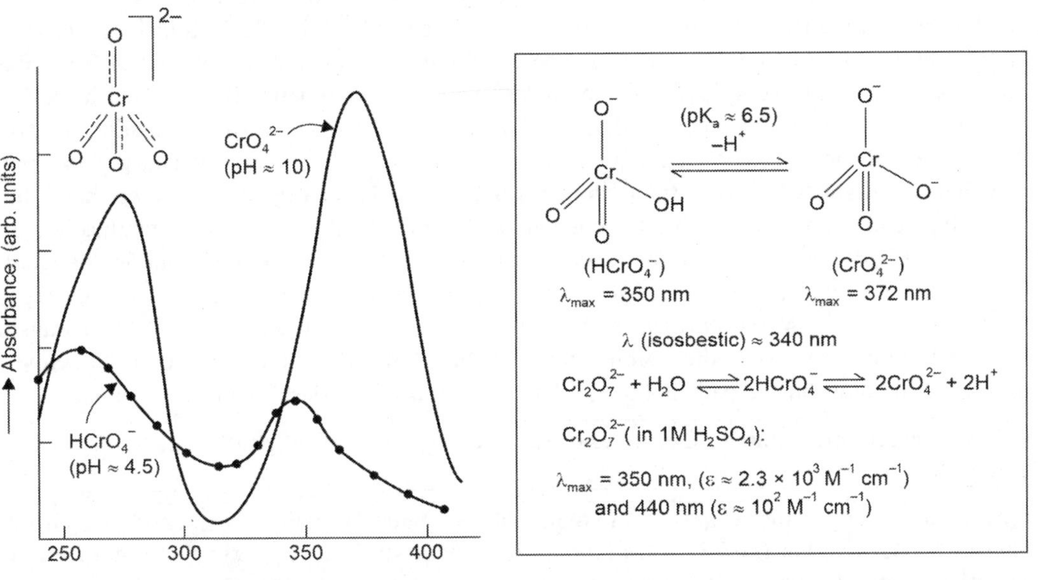

Fig. 7.21.1.9: Absorption spectra (LMCT, O_{2p}→Mn_{3d}) of $HCrO_4^-$ and CrO_4^{2-}

Note: Both $HCrO_4^-$ and CrO_4^{2-} experience the LMCT (O_{2p}→Cr_{3d}) transition. There is a **blue shift** of λ_{max} by about 22 nm (*i.e.* 372 nm to 350 nm) in going from CrO_4^{2-} to $HCrO_4^-$. In $HCrO_4^-$, the O-lone pair is protonated and thus the O-lone pairs are more tightly bound (*cf.* CrO_4^{2-} is more electron-rich and more polarisable than $HCrO_4^-$) and this is why, LMCT in $HCrO_4^-$ requires more energy. It **explains the blue shift**. The same thing happens for the other peak at about 250 nm. In the same way, the **blue shift** of λ_{max} (~350 nm) of $Cr_2O_7^{2-}$ compared to that of CrO_4^{2-} (λ_{max} ~372 nm) can be explained. In $Cr_2O_7^{2-}$ (*i.e.* $O_3Cr—O—CrO_3^{2-}$), the O-lone pairs are more tightly bound compared to those in the mononuclear species, CrO_4^{2-}.

(M) LMCT for the colourless anions like SiO_4^{2-}, PO_4^{3-}, SO_4^{2-} and ClO_4^- *vs.* coloured anions like MnO_4^- and CrO_4^{2-}: All these **tetrahedral oxyanions** (XO_4^{n-}) possess the similar bonding scheme involving the same number of valence electrons. For the oxyanions of the *p*-block elements, the central atom provides the orbitals 3*s*, 3*p* and 3*d* while for the oxyanions of the *d*-block elements, the central metal provides the 4*s*, 4*p* and 3*d* orbitals for bonding with the 2*p*-orbitals of terminal O-atoms. The energy order of the 3*d* orbitals is: 3*d* (X = Si, P, S, Cl) $\gg$ 3*d* (X = Cr, Mn) (effect of the **increased effetive nuclear charge** Z^* for the *d*-block elements). Consequently, the LMCT (*i.e.* $O_{2p} \rightarrow X_{3d}$) occurs in the higher energy (*i.e.* in the **UV region**) for X = Si, P, S, Cl (*i.e. p*-block elements) while it occurs in the lower energy **near UV-visible** region for X = Cr, Mn (*i.e. d*-block elements, 1st transition series elements). It makes SiO_4^{2-}, PO_4^{3-}, SO_4^{2-} and ClO_4^- **colourless** and MnO_4^- and CrO_4^{2-} **coloured.**

Note: Tendency of polymerization of oxyanions $[MO_4]^{x-}$: This can be explained in terms of the feasibility of the $p_\pi(O) \rightarrow d_\pi(M)$ bonding. This aspect has been discussed in Sec. 10.7.5.1 (Vol. 2).

(N) Relativistic effect and colours in the compounds of Pb(IV) *vs.* Sn(IV) in Gr. 14 and Bi(V) *vs.* Sb(V) in Gr. 15 (Refs. A.K. Das *et al, Coord. Chem. Rev.* 2023; *Chemistry Teacher International (IUPAC)*, 2023): The difference in colour is illustrated in the representative examples: **lead dioxide** (PbO_2) (dark brown marron colour) *vs.* **tin dioxide** (SnO_2) (white); $[PbCl_6]^{2-}$ (yellow) *vs.* $[SnCl_6]^{2-}$ (colourless); **sodium bismuthate** ($NaBiO_3$) (yellowish brown colour) *vs.* **sodium antimonate** ($NaSbO_3$) (white); etc. can be rationalised by considering the **LMCT** (ligand to metal charge transfer) transition and the relativistic stabilization of the lowest unoccupied molecular orbital (LUMO) enriched with the character of the heavy metal 6*s*-orbital (for the compounds of 6th period elements, *i.e.* Pb and Bi) or 5*s*-orbital (for the compounds of 5th period elements, *i.e.* Sn and Sb). The LMCT transition occurs from the HOMO enriched with the 2*p*-orbital of oxygen (for the oxides and oxyanions) or 3*p*-orbital of chlorine (for the chlorido-complexes) to the LUMO which is relativistically stabilised more for the heavier congeners (*cf.* relativistic contraction/stabilisation: 6*s* >> 5*s*). The LMCT transition is: 2*p*(O) or 3*p*(Cl) $\rightarrow$ 5*s*/6*s* of the central metal. Thus colour of Bi(V) and Pb(IV) compounds arises from the **relativistically red shifted LMCT transition towards the visible range**. On the other hand, for the corresponding compounds of the lighter congeners Sn and Sb, the relativistic stabilisation of the corresponding LUMO predominantly enriched with the 5*s*-orbital of the central metal is relatively less important. This is why, the HOMO-LUMO energy gap is relatively larger (> 3.2 eV that corresponds to the violet light with $\lambda \approx 390$ nm) for the Sn and Sb compounds. In reality, their LMCT transitions occur in the UV range but not in the visible range and **they look colourless or white**. In the same way, the difference in colour in the pair, **$BiPh_5$** (violet) *vs.* **$SbPh_5$** (colourless), can be explained.

Relativistic effect and colour of Cu, Ag and Au: The yellow colour of gold is due to the relativistic effect (see Vol. 1).

Shiny metallic lustre of metal surfaces: In metals, the VB and CB (both consisting of the **continuum of energy levels**) overlap (*i.e.* $E_g = 0$). Above the **Fermi surface** (E_F), there is a **continuum of unoccupied energy levels** and the metals can absorb the incident electromagnetic radiation in a wide range including the visible light. **As the entire visible range is absorbed, the metals should look black** as in the case of semiconductor solids with $E_g \leq 1.75$ eV (see Sec. 7.21.7). But in the case of metals, **all the visible lights absorbed are remitted back very efficiently without dissipation of energy** and it makes the metal surfaces shiny. In fact, collisions of the excited electrons (having the higher kinetic energy) with the lattice sites may lead to dissipate energy partly as heat but generally, it occurs for the incident photons of higher energy (above the UV region). The colour of some metals (*e.g.* **gold**-yellow, **copper**-reddish lustre) indicates that they absorb light of certain wavelengths only.

7.21.2 MLCT Bands

When the ligand possesses the low-lying vacant orbitals and the metal centres bear the filled orbitals of relatively higher energy, such a transition occurs. In this transition, the metal centre is oxidized while the ligand is reduced. Thus, for the more reducing metal centre and reducible ligands, the MLCT band shifts towards the longer wavelength (*i.e.* **red shift**).

This type of CT band predominates for the relatively lower valent metal centres and heteroaromatic N-bases as the ligands (*e.g.* py, bpy, phen, etc.) and other π-acid ligands like CN^-, CO, etc. In fact, the electron is transferred from the metal orbital to the vacant antibonding orbitals of the ligands like bpy, phen, etc. To observe the MLCT bands of such complexes having the said π-acid ligands, the MLCT bands must appear at lower energy than the **internal $\pi \rightarrow \pi^*$ transition** of the ligands themselves. Otherwises, these intense ligand bands will mask the MLCT bands.

The MLCT transitions in the octahedral complexes can be explained by considering the simplified MO energy diagram (Fig. 7.21.2.1) which is extracted from the complete MO-energy diagram (*cf.* 7.21.1.3) of $[ML_6]$ where L possesses the vacant π^*-MO.

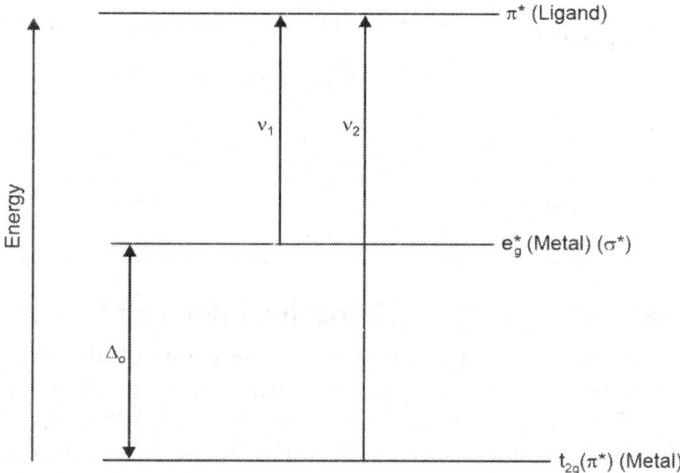

Fig. 7.21.2.1: Simple MO energy diagram to represent the origin of MLCT bands in the octahedral complexes. **Note:** $\nu_2 - \nu_1 = \Delta_o$.

Examples of MLCT Bands

- **$[M(CN)_4]^{2-}$ (M = Ni, Pd, Pt), $[Au(CN)_4]^-$:** In these square planar complexes, the MLCT bands appear due to the transitions to the vacant π*-MO of CN^-. The possible transitions are:

$$d_{xy} \longrightarrow \pi^*;\ d_{z^2} \longrightarrow \pi^*;\ d_{xz}, d_{yz} \longrightarrow \pi^* \text{ (see Sec. 7.17.1 and Fig. 7.17.1.2)}$$

  For $[Ni(CN)_4]^{2-}$, these MLCT bands appear at 32,300 cm^{-1}, 35,200 cm^{-1} and 37,600 cm^{-1}.
- **Fe(II)-bpy/phen complexes, *i.e.* $[Fe(L)_3]^{2+}$ (L = bpy, phen):** The red colour of these complexes arises for the ν_2 band (*i.e.* $t_{2g}^6 \rightarrow t_{2g}^5\pi^{*1}$) at about 19,600 cm^{-1}, *i.e.* 510 nm ($\varepsilon \approx 10^3$)
- **$[Ni(bpy)_3]^{2+}$ and $[Ni(phen)_3]^{2+}$:** For these complexes, the MLCT band appears at much higher energy (about 27,000 cm^{-1}) because Ni(II) is poorly oxidisable (*cf.* Fe(II) is more readily oxidisable). In terms of the energy level diagram, for Ni(II), the filled orbitals, *i.e.* $t_{2g}^6 e_g^2$, are

relatively more stabilised compared to the t_{2g}^6 orbitals of Fe(II) (*cf.* effect of increasing trend of Z^* along the period of 3*d*-series).

- **Fe(III)-bpy/phen system:** It is obvious that in the **Fe(III)-bpy/phen complex** (faint blue colour), the MLCT band should occur at higher energy (*i.e.* shorter wavelength) because oxidation of Fe(III) is more difficult than the oxidation of Fe(II). In terms of the energy level diagram, the filled t_{2g}^5 orbitals of Fe(III) are more stabilised compared to the t_{2g}^6 orbitals of Fe(II) (*cf.* effect of the increased positive oxidation state).
- **Cu(I)-phen system:** The intense colour of **Cu(I)-phen** complex is due to the MLCT transition.
- **Ir(III)-py complex:** The MLCT band appears in the region 30,000-34,000 cm^{-1} for the $t_{2g}^6 \rightarrow t_{2g}^5\pi^{*1}$ transition. (*i.e.* from the filled t_{2g} orbital of iridium to the vacant π^* of py).
- **[Cr(CO)$_6$] and [Mo(CO)$_6$]:** The MLCT bands (~35,000 cm^{-1}, ~44,000 cm^{-1}) appear for the electron transition from the filled t_{2g}^6 level of Cr or Mo to the vacant π*-MO of CO. The intense MLCT transitions are:

$$t_{2g}^6\left(i.e.\ ^1A_{1g}\right)\xrightarrow{\nu_1^{CT}} t_{2g}^5\pi^{*1}\left(t_{1u}\right);\ t_{2g}^6\xrightarrow{\nu_2^{CT}} t_{2g}^5\pi^{*1}\left(t_{2u}\right).$$

The weak ligand field bands ($d-d$ transitions) appear as **the shoulders on the strong CT bands.**

$$^1A_{1g}\xrightarrow{\nu_1^{d-d}}\ ^1T_{1g};\ ^1A_{1g}\xrightarrow{\nu_2^{d-d}}\ ^1T_{2g}\quad \text{(see Sec. 7.15.1)}$$

The peak positions are as follows:

	ν_1^{CT} (cm^{-1})	ν_2^{CT} (cm^{-1})	ν_1^{d-d} (cm^{-1})	ν_2^{d-d} (cm^{-1})
[Cr(CO)$_6$]	35,700	44,800	29,500	31,500
[Mo(CO)$_6$]	34,600	43,860	30,150	31,950

7.21.3 Charge Transfer Bands in the Ion-Paris: IPCT (Ion-Pair Charge Transfer) Bands

If the cationic complex is oxidizing and the counter anion residing outside the coordination sphere is reducing then a CT transition may occur from the reducing anion to the oxidizing cation (*cf.* Fig. 7.21.3.1 and Vol. 7) (*cf.* E.L. King *et al., J. Phys. Chem.,* **63**, 755, 1959; M.G. Evans, *et al., Trans. Faraday Soc.,* **49**, 363, 1953). It happens so in [Co(NH$_3$)$_6$]I$_3$. Such **intermolecular CT transitions** between the cation and anion are known in many suitable ion-pairs. Intensity and λ_{max} of such peaks depend on the outer sphere formation constant (K$_{OS}$) of the ion pairs.

From the measurement of the intensity of the **IPCT bands,** K$_{OS}$ can be estimated. The IPCT transition can be represented as follows:

$$\left[M\left(NH_3\right)_6\right]^{3+}\bullet X^-\,(\text{Ion-pair})\xrightarrow[\text{IPCT}]{h\nu}\left[M\left(NH_3\right)_6\right]^{2+}\bullet X;\ X = Cl, Br, I;\ M = Co, Ir$$

For the more oxidisable and polarisable anion (X$^-$), the intense IPCT band shifts towards the longer wavelength (*i.e.* **red shift**) *i.e.* ν: Cl$^-$ 〉 Br$^-$ 〉 I$^-$, (*cf.* Fig. 7.21.3.1). In fact, **[Ir(NH$_3$)$_6$]$^{3+}$ is itself colourless but it looks yellow coloured in presence of I$^-$.** Ion-pair formation does not affect the *d-d* transitions.

7.21.4 Metal to Metal Charge Transfer (MMCT) Bands or Intervalence Charge Transfer (IVCT) Bands

This aspect has been separately discussed in Sec. 7.22.1.

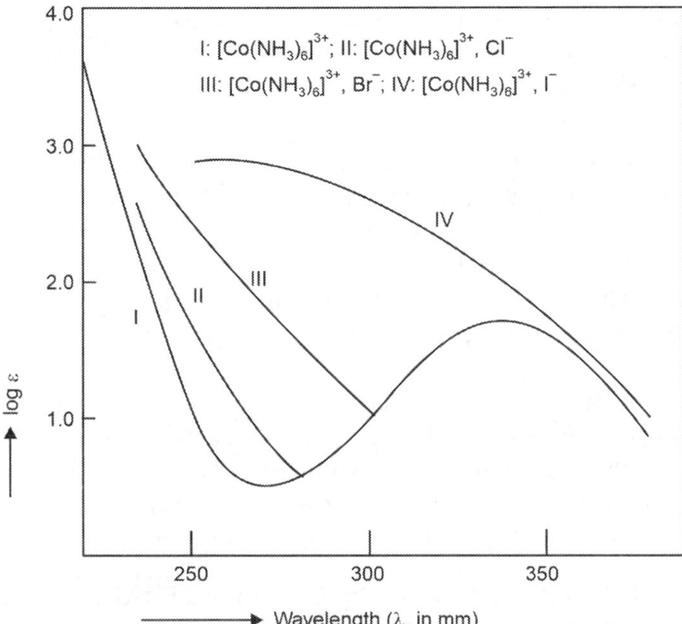

Fig. 7.21.3.1: Shifts of IPCT (*i.e.* ion-pair charge transfer) bands in $[Co(NH_3)_6]^{3+}$, X^- (X = Cl, Br, I).

7.21.5 Complication in Band Classification between the Ligand Field Bands (*i.e. d–d* transition) and Charge Transfer Bands

When the charge transfer band appears at the lower energy (specially in the visible range), it becomes difficult to distinguish between the CT and *d-d* bands. In fact, when a CT band overlaps with a *d-d* band, the *d-d* band also becomes intense through the **intensity stealing effect.** The problem is illustrated for the assignments of bands in $[Ir^{III}(NH_3)_5X]^{n+}$ (X = H_2O, NH_3, Cl, Br) (*cf.* Table 7.21.5.1).

Table 7.21.5.1 Assignments of spectral transitions in $[Ir^{III}(NH_3)_5X]^{n+}$ (see Sec. 7.15.1)

	X = NH₃	*H₂O*	*Cl⁻*	*Br⁻*
${}^1A_{1g} \xrightarrow{\nu_1} {}^1T_{1g}$ (in cm^{-1})	39,800 (ε = 90)	38,800 (ε = 85)	35,000 (ε = 72)	33,000 (ε = 100)
${}^1A_{1g} \xrightarrow{\nu_2} {}^1T_{2g}$ (in cm^{-1})	46,800 (ε = 160)	47,000 (ε = 130)	44,100 (ε =333)	43,500 (ε = 800)
$\nu_2 - \nu_1 \approx 16B'$ (in cm^{-1})	7,000	8,200	9,100	10,500
$\beta \left(= \frac{B'}{B_0}\right)$	0.71	0.84	0.95	1.12

From Table 7.21.5.1, it is evident that the band assignments in $[Ir(NH_3)_6]^{3+}$ and $[Ir(NH_3)_5(OH_2)]^{3+}$ do not face any complication but the assignments for ν_2 in the corresponding chlorido and bromido complexes are called in question because of the following ground.

(i) So high extinction coefficient is unlikely for a *d-d* transition.

(ii) The energy difference between the ${}^1T_{1g}$ and ${}^1T_{2g}$ states measured by $(\nu_2 - \nu_1)$ is increased in moving from the chlorido- to bromido-complexes (*i.e.* B': $Br^- > Cl^-$ but Br^- is more polarisable than Cl^-). **This is unusual.**

(iii) The β value ≥ 1 is unexpected.

In fact, the second band (ν_2) is affected by the low energy LMCT band. This complicates the interpretation. A similar situation arises for $[Co^{III}(NH_3)_5X]^{2+}$ (X = Br, I) (*cf.* Table 7.21.1.1 and Fig. 7.21.1.1).

In the complexes, if the ligand orbitals and metal orbitals are experiencing an extensive mixing to produce the **delocalized MOs (as in many tetrahedral complexes),** then it is not justified to assign the bands as the ligand field bands or charge transfer bands. The bands should be assigned in terms of the MOs involved in the transition.

In the tetrahedral complexes (which are noncentrosymmetric), the *e*-set (*i.e.* $d_{x^2-y^2}$ and d_{z^2}) is basically the nonbonding metal orbitals while the t_2-set is constituted by the overlap of metal *d*-orbitals (*i.e.* d_{xy}, d_{yz} and d_{xz}) with the ligand orbitals (*e.g.* *p*-orbitals or hybrid orbitals). ***Thus in the tetrahedral complexes, the e*** $\rightarrow$ ***t_2 transitions should not considered as the pure d–d transitions** (i.e. **ligand field bands). Such transitions should be described in terms of the MOs involved.***

7.21.6 Comparison Between the *d–d* Bands, *i.e.* Ligand Field Bands and Charge Transfer (CT) Bands

The basic differences are discussed below.

- **Extinction coefficient measuring the intensity of the peak:** The *d–d* transition may be spin-allowed but it is Laporte forbidden in the **centrosymmetric systems**. Consequently, its intensity is relatively less. On the other hand, the CT bands are the fully allowed transitions and the intensity is very high.
- **d^n configuration:** For the d^0 and d^{10} systems, there will be no *d-d* transition and it occurs only when the *d*-orbital is partially filled in. On the other hand, the CT bands may exist even for the d^0 (*e.g.* $[CrO_4]^{2-}$, $[MnO_4]^-$) and d^{10} (*e.g.* HgI_2) systems.
- **Relation with the redox reaction:** A CT band describes a temporary redox reaction. In a limiting situation (when the required energy for the CT band is sufficiently low), the CT process leads to a permanent redox reaction leading to the decomposition of the complexes.
  There is no relation between the redox reaction and *d-d* transition.
- **Involved orbitals:** In a *d-d* transition, the electron transition occurs from one *d*-orbital to another *d*-orbital of the same metal. On the other hand, in a CT transition, the electron transfer occurs either from the metal orbital to the ligand orbital or vice-versa.
- **Position of the bands:** In a CT band, position of the band depends on the oxidizing power of the electron acceptor component and oxidizability of the electron donor component. In a *d-d* transition, position of the band depends on the ligand field splitting of the spectroscopic terms. It depends on the crystal field strength of the ligand, stereochemistry of the complex and the inherent *d*-orbital splitting power of the metal ion under consideration.

In the MLCT transitions, the band shifts to the longer wavelength when the metal centre (*i.e.* donor) becomes more oxidizable and the ligand (*i.e.* acceptor) becomes more reducible. In the LMCT transition, the band shifts towards the longer wavelength when the metal centre (*i.e.* acceptor) becomes more reducible (*i.e.* more oxidizing) and the ligand (*i.e.* donor) becomes more oxidizable.

7.21.7 Colours of Semiconductor Type Solids like Solid Oxides and Sulfides of Metals of d^0, d^5 and d^{10} Systems and Band Theory of Solids: Variable Colours—Effect of Polymorphism and Particle Size (*cf.* Das *et al. Resonance*, 2025)

We have already mentioned the **LMCT transitions** to explain the colours of different oxides (*e.g.* PbO, HgO, Fe_2O_3, etc.) and sulfides (*e.g.* As_2S_3, HgS, etc.). For the better understanding of the origins of these colours, we should consider the **band theory of solids** (see Vol. 3A for band theory of solids). In terms of band theory of solids, the LMCT transitions are the excitations of electrons from the **HOMO band** (constituted mainly by the filled 2*p*-orbitals of oxygen for the oxides and filled 3*p*-orbitals of sulfur for the sulfides) to the **LUMO band** (constituted mainly by the vacant orbitals of metal). Each band consists of **continuum of energy levels** (*i.e.* closely spaced energy levels).

- **Colour depending on the value of band gap (E_g):**

(i) **White or colourless when band gap large (> 3.2 eV ≡ 390 nm ≈ 400 nm, say):** Visible range may be approximately considered to span the range 400 nm (≡ 3.2 eV) to 700 nm (≡ 1.75 eV). Thus, when the band gap is higher than 3.2 eV, no visible light will be absorbed for the HOMO band → LUMO band transition and such substances transmit all the incident light (*i.e.* all the incident visible light is reflected back to the observer) and **they appear colourless/white**. It explains the white colours of ZnO (E_g = 3.2 eV), pure ZnS (E_g = 3.6 eV), AgCl (E_g = 3.2 eV), TiO_2 (E_g = 3.3 eV), SnO_2 (rutile structure, E_g > 3.2 eV), Al_2O_3 (corundum, $E_g \gg$ 3.2 eV) etc. In fact, for such substances, higher energy UV light is absorbed for the HOMO band → LUMO band transition.

(ii) **Black colour when band gap small (≤ 1.75 eV ≡ 700 nm):** In such cases, all the incident visible light is absorbed for the HOMO band → LUMO band transition and **such substances look black** as in **black HgS** (metacinnabar, β-form of HgS of **Zn-blende structure**) with E_g = 0.5 eV. Other examples are: α-MnO_2 (E_g = 1.3 eV), PbS (E_g = 0.5 eV), PbO_2 (orthorhombic polymorph, $E_g \ll$ 1.75 eV), CdSe (E_g = 1.6 eV).

(iii) **Coloured when band gap in the range, 3.2 eV > E_g > 1.75 eV:** When the E_g value lies in the **intermediate range** between the highest limit (say 3.2 eV ≡ 400 nm of violet light) and the lowest limit (say, 1.75 eV ≡ 700 nm), a **part of the incident visible light ($h\nu \geq E_g$) is absorbed and we**

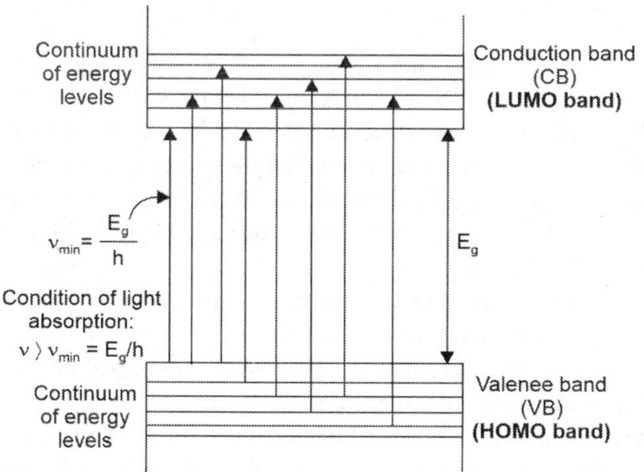

- Electron transition: $\nu \geq E_g/h$
- **Visible range (VIBGYOR):** Violet (400 nm) ≡ 3.2 eV Red (700 nm) ≡ 1.75 eV
- **White/colourless:** $E_g \geq$ 3.2 eV
- **Black:** $E_g \leq$ 1.75 eV
- **Coloured substance:** E_g in the range, 1.75–3.2 eV

Metals: VB and CB overlap, E_g = 0, but **shiny lustre (not black)**? All the visible lights absorbed re-emitted back without dissipation of any energy (discussed earlier)

Fig. 7.21.7.1: Schematic illustration of optical properties (electron transition from VB to CB) in terms of simple band theory of semiconductor type solids

Table 7.21.7.1: Colour and band gap (E_g) of some representative solids **(bulk phase)**

Substance	*Band gap (eV)	Colour	
C (diamond)	5.4	Colourless	**Large band** ($E_g \geq 3.2$ eV)
ZnS	3.6	White	
ZnO	3.2	White	
AgCl	3.2	White	
TiO_2	3.3	White	
AgI	2.8	Yellow	**Intermediate band gap**
CdS	2.6	Yellow	
$CdS_{1-x}Se_x$	2.3	Orange	
Cu_2O	2.1	Red	
HgS (cinnabar, α-polymorph)	2.0	Vermillion red	
CuO	1.7	Black	**Narrow band gap** ($E_g \leq 1.75$ eV)
HgS (metacinnabar, β-polymorph)	1.6	Black	
CdSe	1.6	Black	
α-MnO_2	1.3	Black	
Si	1.1	Black	
PbS	0.5	Black	

* E_g^{Bulk} for the bulk phase (*cf.* **Brus equation)**

$$E = h\nu = \frac{hc}{\lambda} = \frac{6.62 \times 10^{-34}\,\text{J s} \times 3.00 \times 10^{8}\,\text{m s}^{-1}}{\lambda\,(\text{in nm}) \times 10^{-9}\,\text{m}}\,\text{J}$$

$$= \frac{19.86 \times 10^{-17}}{\lambda}\,\text{J} = \frac{19.86 \times 10^{-17}}{1.6 \times 10^{-19}\lambda}\,\text{eV} \approx \frac{1240}{\lambda}\,\text{eV}$$

$$(1\,\text{eV} = 1.6 \times 10^{-19}\,\text{C} \times \text{V} = 1.6 \times 10^{-19}\,\text{J})$$

For $\lambda = 390\,\text{nm},\ E = \frac{1240}{390}\,\text{eV} \approx 3.2\,\text{eV}$

$\lambda = 700\,\text{nm},\ E = \frac{1240}{700}\,\text{eV} \approx 1.75\,\text{eV}$

see the complimentary colour. For CdS (E_g = 2.6 eV), only the light of energy higher than 2.6 eV is absorbed and in fact, it absorbs violet and some blue light. Consequently, **CdS looks yellow**. HgS (cinnabar, α-polymorph) with E_g = 2.0 eV absorbs almost all the visible light except red and it looks red (*cf.* VIBGYOR, **red is the lowest energy end of the visible spectrum)**. In the same way, AgI (E_g = 2.8 eV) looks yellow. $CdS_{1-x}Se_x$) E_g = 2.3 eV for the small values of x) absorbs violet, blue and green light and consequently, it looks **orange in colour** (*cf.* Se is less electronegative and more polarisable than S) (Table 7.21.7.1).

- **Different band gaps leading to different colours for the different polymorphs of a particular substance:** This aspect is illustrated by taking some representative examples.

(i) **HgS (red and black forms):** β-HgS (black, 4:4 coordination, **Zn-blende structure**) having the shorter Hg—S interaction distance shows the LMCT transition with the lower energy light (*i.e.* smaller band gap causing a **red-shift**) compared to that in α-HgS (red, 6:6 coordination, **rock salt structure**) having the relatively longer Hg—S interaction distance. Between these two polymorphs, the α-form is more stable and on heating, the black β-form is converted into the red form. E_g = 2.0 eV for red HgS (α-form) and E_g = 1.6 eV for black HgS (β-form).

(ii) **PbO (red and yellow forms):** PbO (**litharge**, red), tetragonal polymorph, stable at room temperature; PbO (**massicot**, yellow), orthorhombic polymorph, stable above 488°C. Obviously, **for the yellow form, the band gap is relatively larger than that of the red form**. In qualitative analysis, in the dry test tube heating, Pb-salts may undergo decomposition to produce the **yellow form of PbO** that remains unchanged on cooling.

(iii) **PbO_2 (maroon-coloured and black):** The maroon-coloured form has the tetragonal, rutile structure while the black form is the orthorhombic polymorph of relatively smaller E_g value (*cf.* black form exists under high pressure).

● **Variable band gaps of a particular substance depending on the particle size producing different colours:** E_g decreases with the increase of particle size (*i.e.* red shift of λ_{max} with the increase of particle size). This can be understood in terms of **Brus equation** (see the authors' book, **An Introduction to Nanomaterials and Nanoscience** for details).

$$E_g^{QD} \approx E_g^{Bulk} + \frac{h^2}{8R^2\mu}$$

μ = reduced mass, R = radius of the spherical quantum dot (QD) particle (*i.e.* nanoparticle)

It is evident from the above equation that the band gap of QD (*i.e.* E_g^{QD}) increases rapidly with the decrease of size (*i.e.* **quantum confinement**). It explains the red-shift of λ_{max} of optical absorption with the increase of particle size (*cf.* E_g values given in Table 7.21.7.1 refer to E_g^{bulk}). The effect is illustrated for the red and yellow forms of HgO.

HgO (LMCT: $O_{2p} \rightarrow Hg_{6s}$) can have the **red and yellow colours** depending on the particle size. Both forms have the same **zig-zag chain structure** with virtually linear O—Hg—O units in the —O—Hg—O—Hg— chains with Hg—O = 205 pm and Hg—$\hat{O}$—Hg = 107°.

CdX (X = O, S, Se) may have different colours depending on the particle size.

Note: In inorganic qualitative analysis, in the **reaction of Ag^+ with $S_2O_3^{2-}$**, white precipitate of $Ag_2S_2O_3$ slowly decomposes to Ag_2S (black) along with the gradual colour change: white→yellow→brown→black (Ag_2S).

This gradual colour change (mimicking the **sun-set colour change**) can be explained by considering the gradual increase of particle size of Ag_2S precipitate to bulk black precipitate.

A similar observation has been noted during the **precipitation of black HgS** by passing H_2S to an acidic solution of Hg^{2+} (see Chapter 17, Vol. 3B for details).

● **E_g value depending on the nature of orbitals to constitute the HOMO band (*i.e.* VB) and LUMO band (CB):** The HOMO band is constituted mainly by the filled *np*-orbitals of anions like O^{2-}, S^{2-}, Se^{2-}, halide, etc. while the LUMO band (*i.e.* CB) is mainly constituted by the vacant *ns*, *np* and (*n* – 1)*d* orbitals of the metal depending on the nature of metal (see Sec. 7.21.1F). For the more electronegative (and consequently, more nonpolarisable) anions, their *np* orbitals are of lower energy to increase the band gap leading to the **blue-shift** in the LMCT transition (*i.e.* HOMO band → LUMO band or VB → CB). It is illustrated for **CdX** (X = O, S, Se):

$$E_g : \text{CdO (brown)} > \text{CdS (yellow)} > \text{CdSe (black)} \quad \xrightarrow{\hspace{6em}} \text{Red-shift}$$

(*cf.* energy order: $O_{2p} < S_{3p} < Se_{4p}$; polarisability order: $Se^{2-} > S^{2-} > O^{2-}$).

Note: It is expected that E_g for ZnO will be higher than that of ZnS. But, E_g for ZnO is found to be slightly smaller than that of ZnS (*cf.* 3.2 eV vs. 3.6 eV). This is an exception. However, both are white as expected from their E_g values.

A similar situation arises for silver halides (**AgX**), lead halides (**PbX_2**),

$$E_g: \text{AgF (white)} > \text{AgCl (white)} > \text{AgBr (pale yellow)} > \text{AgI (yellow)} \xrightarrow{\hspace{2em}} \text{Red-shift}$$

$$E_g: PbF_2 \text{ (colourless)} > PbCl_2 \text{ (white)} > PbBr_2 \text{ (colourless)} > PbI_2 \text{ (yellow)} \xrightarrow{\hspace{2em}} \text{Red-shift}$$

$$E_g: SnF_2 \text{ (colourless)} > SnCl_2 \text{ (colourless)} > SnBr_2 \text{ (white)} > SnI_2 \text{ (red)} \xrightarrow{\hspace{2em}} \text{Red-shift}$$

(*cf.* energy order: $F_{2p} < Cl_{3p} < Br_{4p} < I_{5p}$; polarisability order: $I^- > Br^- > Cl^- > F^-$)

On the other hand, as the polarising power of the metal centre increases due to the increase of **effective nuclear charge** (Z*) and cationic charge (*i.e.* $M^{4+} > M^{3+} > M^{2+}$), energy of the LUMO band (*i.e.* CB) decreases to reduce the band gap (E_g) causing a red shift.

$$E_g: \text{ZnO (white)} > \text{CdO (brown)} > \text{HgO (red, yellow)} \xrightarrow{\hspace{2em}} \text{Red-shift; effect of Z*: Hg(II)} > \text{Cd(II)} > \text{Zn(II)}$$

$$E_g: Fe(OH)_2 \text{ (white)} > Fe(OH)_3 \text{ (brown)}; Cu_2O \text{ (red)} > CuO \text{ (black)} \xrightarrow{\hspace{2em}} \text{Red-shift, effect charge of the metal}$$

● Variable colours due to **crystal defects** like **F-centre** in ZnO, CdO, etc. (see Sec. 12.15, Vol. 3A; Chapter 19, Vol. 3B).

HgI_2 (solid, red) *vs.* $[HgI_4]^{2-}$ (colourless); BiI_3 (solid, black) *vs.* $[BiI_4]^-$ (orange):

$$Hg^{2+}\text{(soln.)} \xrightarrow{KI} HgI_2(\downarrow)\ \textbf{(red)} \xrightarrow{\text{excess KI}} [HgI_4]^{2-}\ \textbf{(colourless)}$$

$$Bi^{3+}\text{(soln.)} \xrightarrow{KI} BiI_3(\downarrow)\ \textbf{(black)} \xrightarrow{\text{excess KI}} [BiI_4]^-\ \textbf{(orange)}$$

For **solid** HgI_2 and BiI_3, their colours can be explained by considering their E_g values. For the corresponding **discrete complex tetrahedral anions** $[HgI_4]^{2-}$ (d^{10}) $[BiI_4]^-$ (d^0), the band theory of solids is not applicable and to understand their **LMCT transitions**, we are to consider their MO energy diagram (Fig. 7.21.1.4). For $[HgI_4]^{2-}$, the LMCT occurs in the **UV region** (31,000 cm^{-1}) while for $[BiI_4]^-$, the LMCT occurs in the visible range.

● **Doped semiconductors** (called **extrinsic semiconductors**, see Vol. 3A): In the *n*-type doped semiconductor, electron transition occurs from the **impurity donor levels** (close to the CB) to the CB (*e.g.* nitrogen doped **diamond looks yellow coloured**, *cf.* diamond with E_g = 5.5 eV looks transparent). In the *p*-type doped semiconductor, electron transition occurs from the VB to the **impurity acceptor levels** (close to the VB) (*e.g.* boron doped **diamond looks blue coloured** known as **hope diamond**) (Das *et al, Resonance*, 2025).

7.22 INTERVALENCE ELECTRON TRANSFER BANDS IN THE INTERVALENCE COMPOUNDS: METAL TO METAL CHARGE TRANSFER (MMCT) TRANSITIONS (see Secs. 1.13, 6.4.16)

In the inter-valence or mixed-valence compounds, the metal ions of different oxidation states are linked through some bridging ligands. Such compounds are generally highly coloured due to the electron transition from the lower valent metal centre to the higher valent metal centre. Some representative examples are discussed here:

(i) **Prussian blue** (*cf.* Sec. 1.12, Vol. 4): It is generally represented by $KFe^{III}[Fe^{II}(CN)_6]$ obtained by mixing a ferric salt with potassium ferrocyanide, $K_4[Fe(CN)_6]$. Prussian blue gives the deep blue colour but individually $K_4[Fe(CN)_6]$, simple ferric salt and $K_3[Fe(CN)_6]$ are devoid of any such deep colour. It is suggested that prussian blue is a polymeric compound in which each Fe(III)-centre is octahedrally surrounded by the N-ends of cyanides (CN^-) coordinated to Fe(II) through the C-ends. Thus both the Fe(III)- and Fe(II)-centres maintain the octahedral coordination. The bridging unit is:

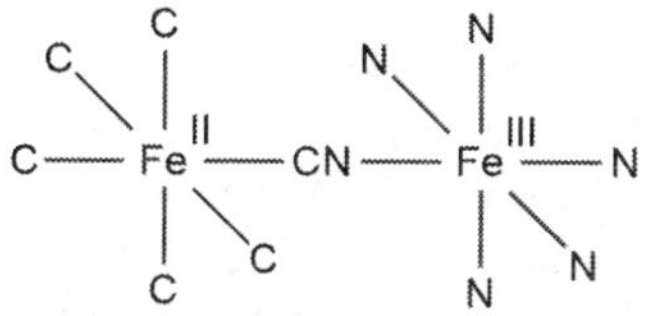

$$\left[Fe^{II}(CN)_6\right]: t_{2g}^6 e_g^0 \text{ (low-spin)}$$
$$\left[Fe^{III}(NC)_6\right]: t_{2g}^3 e_g^2 \text{ (high-spin)}$$

In Prussian blue, the absorption bands appear at 15,000 cm^{-1} and 25,000 cm^{-1}. These are assigned as:

from t_{2g} orbital of $Fe^{II}(t_{2g}^6)$ to t_{2g} orbital of $Fe(III)(t_{2g}^3 e_g^2)$: 15,000 cm^{-1}
from t_{2g} orbital of $Fe^{II}(t_{2g}^6)$ to e_g orbital of $Fe(III)(t_{2g}^3 e_g^2)$: 25,000 cm^{-1}

This electron transfer from Fe(II) to Fe(III) is mediated through the π-orbital of the bridging 'CN' ligand. It leads to temporary reduction of Fe(III) to Fe(II) and temporary oxidation of Fe(II) to Fe(III). Thus the electron transfer produces the charge transfer bands in the visible region and the compound is intensely coloured.

The colour of Prussian or Turnbull blue is due to the electron transition from Fe(II) to Fe(III) through the π-orbitals of the bridging ligand. It is supported by the fact that if in the polymeric compound, both the metal centres are in the same oxidation states (*e.g.*—Fe^{II}—C≡N—Fe^{II}— (**Everitt's white salt**) or —Fe^{III}—CN—Fe^{III}— (**Berlin green**) then no such strong colour is developed.

Turnbull blue is obtained by mixing a simple Fe(II) salt with $K_3[Fe(CN)_6]$. It is also a polymeric compound and its structure can also be realized in the same framework of Prussian blue. In Turnbull blue, the colour arises from the same IT. In fact, both the Prussian blue and Turnbull blue are having the **same structure** (evidenced by the same **Mossbauer spectra**, see Vols 4 and 7 for details).

(**Note:** For structural details of prussian and turnbull blue *see* Sec. 1.12, Vol 4).

MMCT (as in Prussian blue) is known to occur in many other CN-bridged polynuclear complexes like $M^ICu^{II}Fe^{III}(CN)_6$, $M^IRu^{II}Fe^{III}(CN)_6$ (**ruthenium purple**). In such compounds, M^{II} (like Cu^{II}, Ru^{II}) actually replaces Fe^{II} of Prussian blue.

(ii) **Creutz-Taube ion (*cf.* Sec. 1.13, Vol 4):** In this complex, Ru(II) and Ru(III) centres are linked through the bridging ligand pyrazine.

$$\left[(H_3N)_5Ru^{II}—N\langle\bigcirc\rangle N—Ru^{III}(NH_3)_5\right]^{5+}$$

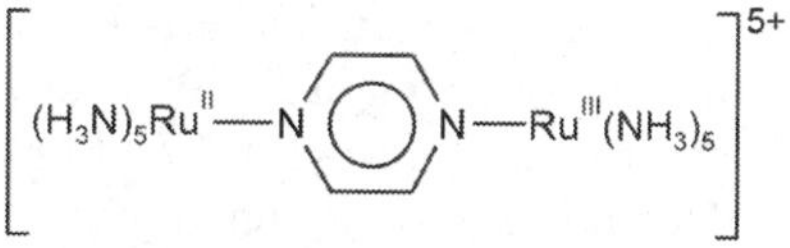

Here the electron transition from Ru(II) to Ru(III) occurs through the π-orbital of the bridging ligand pyrazine. This IT band appears at 1570 nm (*i.e.* in the near infrared region). This transition is schematically represented as:

$$[2, 3] \xrightarrow{h\nu} [3, 2],$$ 2 and 3 refer the oxidation states of ruthenium.

(iii) **Halide bridged mixed valence compounds:** Many compounds having the chain, —X—M^{II}—X—M^{IV}—X (X = Cl, Br; M = Pt, Pd) show the intense intervalence absorption of the polarized light along the chain. The examples are: $[Pt^{II}Br_2(NH_3)_2][Pt^{IV}Br_4(NH_3)_2]$, $[PdCl_2(NH_3)_2]$ $[PdCl_4(NH_3)_2]$; etc. **Wolfram's red salt** *trans*-$Pt(EtNH_2)_4Cl_3.2H_2O$ (Average oxidation state = +3) also belongs to this group. Here the octahedral *trans*-$[PtCl_2(EtNH_2)_4]^{2+}$ and planar $[Pt(EtNH_2)_4]^{2+}$ ions are linked in chains through the bridging chlorido ligands (see Vol. 4).

(iv) **Other examples of metal to metal charge transfer (MMCT) bands: Molybdenum blue** is a mixed oxide (MO_3, M_2O_5) having both Mo^V and Mo^{VI}. Its intense blue colour is due to the charge transfer between the metal centres of different oxidation states. Colour of **tungsten blue oxides** is probably due to such metal-metal charge transfer (MMCT) transition. Metal to metal charge transfer also occurs in **black gold**, $Cs_2Au^IAu^{III}Cl_6$ where the electron transfer occurs from the $AuCl_2^-$ moiety to the $Au^{III}Cl_4^-$ moiety. **Ruthenium red**, an oxido-bridged trinuclear mixed valence ($2Ru^{III}$, Ru^{IV}) complex, $[(H_3N)_5Ru^{III}—O—Ru^{IV}(NH_3)_4—O—Ru^{III}(NH_3)_5]^{6+}$ experiences the MMCT transition to give its colour. The dimeric Ru(II, III) species $[Ru_2Cl_{3+n}]^{(2-n)+}$ ($n = 0, 1, 2$) are called the **ruthenium blues.** In $[(H_3N)_3Ru(\mu—Cl)_3Ru(NH_3)_3]^{2+}$ (**ruthenium blue**), the metal centres are indistinguishable (*i.e.* average oxidation state, +2.5) (see Vol. 4 for details). The MMCT transition may occur also between the two different metals as in $[L_5Co^{III}—NC—Fe^{II}(CN)_5]^-$.

● **Blue sapphire gemstones (blue corundum)** (Das *et al. Resonance*, 2025)**:** Corundum ($E_g = 8.3$ eV) is itself transparent but presence of only 0.01% of titanium and iron in it gives the intense blue colour of **blue sapphire**. The $d \rightarrow d$ transition cannot explain this intense colour at such low concentration levels (*cf.* **deep red ruby colour of corundum**, due to the $d \rightarrow d$ transitions of Cr(III) impurity, at least 1% required to see the red colour) but the **heteronuclear MMCT transition** like Fe(II) → Ti(IV) or Ti(III) → Fe(III) can explain this intense blue colour. Alternatively, Fe^{3+}–Ti^{4+} **mixed acceptor states** residing close to the VB of corundum are responsible for the absorption bands (*i.e.* VB → acceptor levels as in the *p*-type semiconductor like **blue diamond**).

$\delta_b \rightarrow \delta^*$ Transition: Origin of colour in $[Re_2Cl_8]^{2-}$, $[Mo_2Cl_8]^{4-}$

In terms of MOT, the **quadruple bond** in $[Re_2Cl_8]^{2-}$ and $[Mo_2Cl_8]^{4-}$ is given by: $\sigma_b^2\pi_b^4\delta_b^2$. The energy difference between the δ_b and δ^*-MO is very small and the $\delta_b \rightarrow \delta^*$ transition gives the **intense red colour** in $[Mo_2Cl_8]^{4-}$ and the **blue colour** in $[Re_2Cl_8]^{4-}$. These are discussed in detail in Sec. 9.19, Vol. 2.

Colour of Magnus Green Salt $[Pt(NH_3)_4][PtCl_4]$

The constituent ions are having different colours: **$[Pt(NH_3)_4]^{2+}$ is colourless; $[PtCl_4]^{2-}$ is red.** The green colour of the salt $[Pt(NH_3)_4][PtCl_4]$ in **solid crystal** is due to the molecular interaction that leads to the **nonadditive colour.** In Magnus's green salt, the square planar cations $[Pt(NH_3)_4]^{2+}$ and square planar anions $[PtCl_4]^{2-}$ stack over one another to form a long –Pt–Pt– chain. Along the —Pt—Pt— chain, the Pt—Pt distance is 325 pm. This short Pt–Pt distance indicates the Pt—Pt interaction through the Pt–*d* orbitals. This **intermetallic interaction** is believed to shift the *d-d* absorption region of $[PtCl_4]^{2-}$ from green to red. Absorption of red light gives the green colour observed. This proposition is supported by its **dichoric behaviour:** intense absorption of the polarized light along the chain, and enhanced electrical conductivity along the chain. The said Pt—Pt interaction (*i.e.* stacking of square planar units) is prevented by the presence of bulky ligands. In fact, $[Pt(EtNH_2)_4][PtCl_4]$ shows a **pink colour which is sum of the colours of the constituent ions.** Here, the Pt—Pt interaction is absent. Here, it is worth mentioning that in the mixed valence compounds, the intervalence charge transfer (**IVCT**) bands also give the **nonadditive colour**. This aspect has been already discussed.

● Synthetic **ultramarine blue and natural lapis lazuli** (Das *et al. Resonance*, 2025): Both are the aluminosilicates (a form of zeolite) where the entrapped **polysulfide radical anions** like S_2^- and S_3^- are the chromophore centres (S_2^-: yellow and S_3^-: blue; **low energy electron transition within the anion radical**; these highly reactive radicals are stabilised on being entrapped in the zeolite cages).

$Na_8[Al_6Si_6O_{24}]Cl_2$ (**sodalite**) is colourless but partial substitution of Cl^- by S_3^- (19 valence electrons) makes it blue coloured (*cf.* colours of other inorganic radicals: brown $17e$-NO_2, yellow $19e$-ClO_2).

7.23 SOLVED PROBLEMS ON ELECTRONIC SPECTRA OF TRANSITION METAL COMPLEXES

Some Important Relations

d^2, d^7 (O_h) (High spin)
d^3, d^8 (T_d)

$^xT_1(F) \rightarrow {}^xT_2(F)$, $\nu_1 = 8Dq + x'$

$^xT_1(F) \rightarrow {}^xA_2(F)$, $\nu_2 = 18Dq + x'$

$^xT_1(F) \rightarrow {}^xT_1(P)$, $\nu_3 = 6Dq + 15B' + 2x'$

(x' = configurational inter zaction energy)

$10Dq = \nu_2 - \nu_1$; $10Dq = 2\nu_1 - \nu_3 + 15B'$; $15B' = \nu_2 + \nu_3 - 3\nu_1$

$$\nu_1 = \frac{1}{2}[10Dq - 15B' + Q]$$

$$\nu_2 = \frac{1}{2}[30Dq - 15B' + Q]$$

$\nu_3 = Q$

where $Q = [225B'^2 + 100Dq^2 + 180DqB']^{1/2}$

d^2, d^7 (T_d)
d^3, d^8 (O_h) (High spin)

$^xA_2(F) \rightarrow {}^xT_2(F)$, $\nu_1 = 10Dq$

$^xA_2(F) \rightarrow {}^xT_1(F)$, $\nu_2 = 18Dq - x$

$^xA_2(F) \rightarrow {}^xT_1(P)$, $\nu_3 = 12Dq + 15B' + x$

(x = configurational interaction energy)

$10Dq = \nu_1$, $15B' = \nu_2 + \nu_3 - 3\nu_1$

$$B' = \frac{2\nu_1^2 + \nu_2^2 - 3\nu_1\nu_2}{15\nu_2 - 27\nu_1}$$

$$= \frac{2\nu_1^2 + \nu_3^2 - 3\nu_1\nu_3}{15\nu_3 - 27\nu_1}$$

$\nu_1 = 10Dq$

$$\nu_2 = \frac{1}{2}[15B' + 30Dq - Q]$$

$$\nu_3 = \frac{1}{2}[15B' + 30Dq + Q]$$

$Q = [(15B' - 10Dq)^2 + 120DqB']^{1/2}$

$= [225B'^2 + 100Dq^2 - 180DqB']^{1/2}$

$= [\nu_3 - \nu_2]$

$$10Dq = \frac{1}{3}[\nu_2 + \nu_3 - 15B']$$

If ν_1 is not known, then the above expressions for ν_2 and ν_3 in terms of Q are sufficient to analyze the result.

d^6 (low spin) Octahedral complex

$$\begin{cases} {}^1A_{1g} \xrightarrow{\nu_1} {}^1T_{1g}, \nu_1 = 10Dq - 4B' + \dfrac{86B'^2}{10Dq} \\ \qquad \approx 10Dq - 4B' \\ {}^1A_{1g} \xrightarrow{\nu_2} {}^1T_{2g}, \nu_2 = 10Dq + 12B' + \dfrac{2B'^2}{10Dq} \\ \qquad \approx 10Dq + 12B' \end{cases} \quad \begin{Bmatrix} \nu_3 > \nu_2 > \nu_1 \\ \nu_1 = 10Dq \end{Bmatrix}$$

$$\begin{cases} 10Dq = f_{\text{ligand}} \times g_{\text{metal}} = fg; \\ \beta = \dfrac{B'}{B_0} = 1 - hk \end{cases}$$

Table 7.23.1 Rules of direct product.

For IRs				***For T × T**
$A \times A = A$	$B \times A = B$	$E \times A = E$	$T \times A = T$	$T_1 \times T_1 = T_2 \times T_2 = A_1 + E + T_1 + T_2$
$A \times B = B$	$B \times B = A$	$E \times B = E$	$T \times B = T$	$T_1 \times T_2 = T_2 \times T_1 = A_2 + E + T_1 + T_2$
$A \times E = E$	$B \times E = E$	$E \times E = *$	$T \times E = T_1 + T_2$	* For **E × E:** The product depends on the
$A \times T = T$	$B \times T = T$	$E \times T = T_1 + T_2$	$T \times T = *$	nature of point group

For subscripts and superscripts of IRs

	Primes	**Subscripts**		
$g \times g = g$	$' \times ' = '$	$1 \times 1 = 1$	For O, C_{3v} and T_d	$E_1 \times E_2 = E_2 \times E_1 = B_1 + B_2 + E_1$; $E_1 \times E_1 = E_2 \times E_2 = A_1 + A_2 + E_2$
$g \times g \times u = u$	$' \times '' = ''$	$1 \times 2 = 2$		
$u \times g = u$	$'' \times ' = ''$	$2 \times 1 = 2$	For C_{4v}, D_{2d} and D_4	$E \times E = A_1 + A_2 + B_1 + B_2$
$u \times u = g$	$'' \times '' = '$	$2 \times 2 = 1$		

Direct product Table for O point group (*e.g.* O_h, T_d)

	A_1	A_2	E	T_1	T_2
A_1	A_1	A_2	E	T_1	T_2
A_2	A_2	A_1	E	T_2	T_1
E	E	E	$A_1 + A_2 + E$	$T_1 + T_2$	$T_1 + T_2$
T_1	T_1	T_2	$T_1 + T_2$	$A_1 + E + T_1 + T_2$	$A_2 + E + T_1 + T_2$
T_2	T_2	T_1	$T_1 + T_2$	$A_2 + E + T_1 + T_2$	$A_1 + E + T_1 + T_2$

Examples

$\mathbf{D_{2d}}: E_g \times E_u = A_{1u} + A_{2u} + E_u$; $\mathbf{C_{3v}}: A_1 \times A_2 = A_2$; $E \times E = A_1 + A_2 + E$;
$\mathbf{T_d}: T_1 \times T_2 = A_2 + E + T_1 + T_2$; $\mathbf{D_{3h}}: E' \times E'' = A_1'' + A_2'' + E''$;
$\mathbf{D_{4h}}: E_g \times B_{2g} = E_g$; $E_g \times E_u = A_{1u} + A_{2u} + B_{1u} + B_{2u}$; $\mathbf{O_h}: E_g \times T_{1u} = T_{1u} + T_{2u}$;
$\mathbf{D_{3d}}: E_g \times E_g = A_{1g} + A_{2g} + E_g$; $\mathbf{T_d}: T_1 \times T_1 = A_1 + E + T_1 + T_2$

Table 7.23.2 Correlation Table for the O_h Point Group.

O_h	O	T_d	D_{4h}	D_{2d}	C_{4v}	D_3	D_{3d}	C_{2v}	C_{2h}
A_{1g}	A_1	A_1	A_{1g}	A_1	A_1	A_1	A_{1g}	A_1	A_g
A_{2g}	A_2	A_2	B_{1g}	B_1	B_1	A_2	A_{2g}	A_2	B_g
E_g	E	E	$A_{1g} + B_{1g}$	$A_1 + B_1$	$A_1 + B_1$	E	E_g	$A_1 + A_2$	$A_g + B_g$
T_{1g}	T_1	T_1	$A_{2g} + E_g$	$A_2 + E$	$A_2 + E$	$A_2 + E$	$A_{2g} + E_g$	$A_2 + B_1 + B_2$	$A_g + 2B_g$
T_{2g}	T_2	T_2	$B_{2g} + E_g$	$B_2 + E$	$B_2 + E$	$A_1 + E$	$A_{1g} + E_g$	$A_1 + B_1 + B_2$	$2A_g + B_g$
A_{1u}	A_1	A_2	A_{1u}	B_1	A_2	A_1	A_{1u}	A_2	A_u
A_{2u}	A_2	A_1	B_{1u}	A_1	B_2	A_2	A_{2u}	A_1	B_u
E_u	E	E	$A_{1u} + B_{1u}$	$A_1 + B_1$	$A_2 + B_2$	E	E_u	$A_1 + A_2$	$A_u + B_u$
T_{1u}	T_1	T_2	$A_{2u} + E_u$	$B_2 + E$	$A_1 + E$	$A_2 + E$	$A_{2u} + E_u$	$A_1 + B_1 + B_2$	$A_u + 2B_u$
T_{2u}	T_2	T_1	$B_{2u} + E_u$	$A_2 + E$	$B_1 + E$	$A_1 + E$	$A_{1u} + E_u$	$A_2 + B_1 + B_2$	$2A_u + B_u$

Table 7.23.3: Normal modes of vibration (*see* Appendix 12B).

O_h([ML_6]): $A_{1g} + E_g + T_{2g} + 2T_{1u} + T_{2u}$; **$T_d$([$ML_4$]):** $A_1 + E + 2T_2$
D_{4h}([ML_4]): $A_{1g} + B_{1g} + B_{2g} + A_{2u} + B_{2u} + 2E_u$; **$D_3$:** $3A_1 + 2A_2 + 5E$
D_{2d}([ML_4]): $2A_1 + B_1 + 2B_2 + 2E$; **D_{4h}([$ML_4L'_2$]):** $2A_{1g} + B_{1g} + B_{2g} + E_g + 2A_{2u} + B_{1u} + 3E_u$
C_{4v}([ML_5]): $3A_1 + 2B_1 + B_2 + 3E$; **C_{2v}([$ML_2L'_2$]):** $4A_1 + A_2 + 2B_1 + 2B_2$

SOLVED NUMERICAL PROBLEMS

Problem 1: Aqueous solution $[Ti(OH_2)_6]^{3+}$ shows a broad absorption peak having the centre at 20,300 cm^{-1}. Calculate the cfse.

Solution: Ti(III) is a d^1 system. The possible electronic transition (without considering the J.T. distortion) is:

$$t_{2g}^1 e_g^0 \xrightarrow[(=10Dq)]{h\nu} t_{2g} e_g^1, \ \bar{\nu} = 20{,}300 \text{ cm}^{-1}$$

$$= 20{,}300 \times 100 \text{ m}^{-1}$$

$$10Dq = h\nu = h\frac{c}{\lambda} = hc\bar{\nu}$$

$$= (6.62 \times 10^{-34} \text{ J s}) \times (2.99 \times 10^8 \text{ m s}^{-1}) \times 20{,}300 \times 100 \text{ m}^{-1}$$

$$= 4.03 \times 10^{-19} \text{ J / ion}$$

$$= \frac{4.03 \times 10^{-19} \times 6.022 \times 10^{23}}{1000} \text{ kJ/mole}$$

$$\approx 243 \text{ kJ/mole, } i.e. \ Dq \approx 24.3 \text{ kJ mol}^{-1}$$

cfse = $4Dq$ = 4 × 24.3 kJ mole = 97.2 kJ mol^{-1}

Problem 2: (a) Calculate the $10Dq$ value of $[TiL_6]^{3+}$ from the following data (L = CH_3COCH_3):
$[TiCl_3L_3]$: $10Dq$ = 15,400 cm^{-1}; $[TiCl_6]^{3-}$: $10Dq$ = 12,750 cm^{-1}

(b) Calculate $10Dq$ for $[Co(en)_2F_2]^+$

Given: $10Dq$ for $[Co(en)_3]^{3+}$ = 23,000 cm^{-1}

$10Dq$ for $[CoF_6]^{3-}$ = 13,000 cm^{-1}

Solution: (a) According to Jorgensen's rule of average environment, we have:

$$\bar{\nu}\left(\left[MA_xB_{6-x}\right]\right)=\frac{x}{6}\bar{\nu}\left(\left[MA_6\right]\right)+\frac{6-x}{6}\bar{\nu}\left(\left[MB_6\right]\right)$$

or,
$$10Dq\left(\left[MA_xB_{6-x}\right]\right)=\frac{x}{6}\times 10Dq\left(\left[MA_6\right]\right)+\frac{6-x}{6}10Dq\left(\left[MB_6\right]\right).$$

In the present case: $10Dq\left(\left[\mathrm{TiCl_3L_3}\right]\right)=\frac{3}{6}\times 10Dq\left(\left[\mathrm{TiCl_6}\right]^{3-}\right)+\frac{3}{6}\times 10Dq\left(\left[\mathrm{TiL_6}\right]^{3+}\right)$

or,
$$15{,}400=\frac{1}{2}\left[12{,}750+10Dq\left(\left[\mathrm{TiL_6}\right]^{3+}\right)\right]$$

or,
$$10Dq([\mathrm{TiL_6}]^{3+}) = 15{,}400 \times 2 - 12{,}750 = 18{,}000\ \mathrm{cm^{-1}}$$

(b) $10Dq\ [\mathrm{Co(en)_2F_2}]^+) = \frac{1}{6}[4\times 23{,}000+2\times 13{,}000]=20{,}000\ \mathrm{cm^{-1}}$

Problem 3: The absorption spectrum of $[\mathrm{Ni(OH_2)_6}]^{2+}$ shows peaks at 9,000 $\mathrm{cm^{-1}}$, 14,000 $\mathrm{cm^{-1}}$ and 25,000 $\mathrm{cm^{-1}}$. Empirical ligand (H_2O) parameter $h = 1$ and the empirical metal (Ni^{2+}) parameter $k = 0.12$. Comment on the band assignments and calculate the $10Dq$ value and configurational interaction energy (x). (B_0 = 1000 $\mathrm{cm^{-1}}$).

Solutions: The possible ligand field transitions are:

$$^3A_{2g}(F) \rightarrow {}^3T_{2g}(F),\ \nu_1 = 10Dq = 9{,}000\ \mathrm{cm^{-1}}$$
$$^3A_{2g}(F) \rightarrow {}^3T_{1g}(F),\ \nu_2 = 18Dq - x = 14{,}000\ \mathrm{cm^{-1}}$$
$$^3A_{2g}(F) \rightarrow {}^3T_1(P),\ \nu_3 = 12Dq + 15B' + x = 25{,}000\ \mathrm{cm^{-1}}$$

Thus, the lowest energy band gives the $10Dq$ value, *i.e.* $10Dq$ = 9,000 $\mathrm{cm^{-1}}$ and Dq = 900 $\mathrm{cm^{-1}}$

$$\nu_2 - \nu_1 = (18Dq - x) - (10Dq) = 8Dq - x$$

or, $14{,}000 - 9{,}000 = 8 \times 900 - x$

or, $5{,}000 = 7{,}200 - x$; or, $x = 2{,}200\ \mathrm{cm^{-1}}$

i.e. configurational interaction energy = 2,200 $\mathrm{cm^{-1}}$

$$15B' = \nu_2 + \nu_3 - 3\nu_1 = 14{,}000 + 25{,}000 - 3 \times 9{,}000$$
$$= 39{,}000 - 27{,}000$$
$$= 12{,}000\ \mathrm{cm^{-1}}$$

or,
$$B' = \frac{1}{15}\times 12{,}000 = 800\ \mathrm{cm^{-1}}$$

$$\text{Nephelauxetic parameter}\ (\beta)=\frac{B_{\text{complex}}}{B_{\text{free ion}}}=\frac{B'}{B_0}=1-hk$$

i.e.
$$\beta = 1 - 1 \times 0.12 = 0.88$$

$$\frac{B'}{B_0} = 0.88;\ \text{or},\ B' = B_0 \times 0.88 = 1000 \times 0.88 = 880\ \mathrm{cm^{-1}}$$

Thus the B' value calculated is in good agreement with B' value obtained from the spectral assignments.

Problem 4: The absorption spectrum of $[\mathrm{Co(NH_3)_6}]^{3+}$ shows two strong peaks at 21,000 $\mathrm{cm^{-1}}$ and 29,500 $\mathrm{cm^{-1}}$. Find the $10Dq$ value and B'. Compare with the $10Dq$ value obtained from $10Dq = fg$ (f = 1.25, g = 18,200 $\mathrm{cm^{-1}}$).

Solution: The spin allowed transitions in low-spin $[Co(NH_3)_6]^{3+}$ are:

$${}^1A_{1g} \xrightarrow{\nu_1} {}^1T_{1g};\ \nu_1 \approx 10Dq - 4B' = 21{,}000\ \text{cm}^{-1}$$

$${}^1A_{1g} \xrightarrow{\nu_2} {}^1T_{2g};\ \nu_2 \approx 10Dq + 12B' = 29{,}500\ \text{cm}^{-1}$$

$$\nu_2 - \nu_1 = (10Dq + 12B') - (10Dq - 4B') = 16B'$$

$$= 29{,}500 - 21{,}000$$

$$= 8{,}500\ \text{cm}^{-1}$$

i.e. $B' = 531\ \text{cm}^{-1}$

$$10Dq = \nu_1 + 4B' = 21{,}000 + 4 \times 531 = 23{,}124\ \text{cm}^{-1}$$

Note: For more correct results, the following relations are to be used:

$$\nu_1 = 10Dq - 4B' + \frac{86B'^2}{10Dq}$$

$$\nu_2 = 10Dq + 12B' + \frac{2B'^2}{10Dq}$$

- $10Dq = f_{\text{ligand}} \times g_{\text{metal}} = f_{NH_3} \times g_{Co^{3+}} = 1.25 \times 18{,}200$

$$= 22{,}750\ \text{cm}^{-1}$$

Thus, the calculated $10Dq$ value is approximately close to the value obtained from the spectral band assignment.

Problem 5: The spectral characteristics of high spin $[CoF_6]^{4-}$ (d^7 system) are:

$${}^4T_{1g} \xrightarrow{\nu_1} {}^4T_{2g}\ (\nu_1 = 7{,}150\ \text{cm}^{-1})$$

$${}^4T_{1g} \xrightarrow{\nu_2} {}^4A_{2g}\ (\nu_2 = 15{,}200\ \text{cm}^{-1})$$

$${}^4T_{1g} \xrightarrow{\nu_3} {}^4T_{1g}(P)\ (\nu_3 = 19{,}200\ \text{cm}^{-1})$$

Find $10Dq$, β and B' and configurational interaction energy (x'); Given: $B_0 = 970\ \text{cm}^{-1}$

Solution: $\nu_1 = 8Dq + x'$; $\nu_2 = 18Dq + x'$; $\nu_3 = 15B' + 6Dq + 2x'$

i.e. $\nu_2 - \nu_1 = 10Dq = 15{,}200 - 7{,}150 = 8{,}050\ \text{cm}^{-1}$

$$15B' = \nu_2 + \nu_3 - 3\nu_1 = 15{,}200 + 19{,}200 - 3 \times 7{,}150$$

$$= 34{,}400 - 21{,}450 = 12{,}950$$

or, $B' = 863\ \text{cm}^{-1}$

$$\nu_1 = 8Dq + x' = 8 \times 805 + x' = 6{,}440 + x'$$

$$x' = \nu_1 - 6{,}440 = 7{,}150 - 6{,}440$$

$$= 710\ \text{cm}^{-1}$$

$$\beta = \frac{B_{\text{complex}}}{B_{\text{free ion}}} = \frac{B'}{B_0} = \frac{863}{970} \approx 0.88.$$

Problem 6: The peaks of some Cr(III) complexes are as follows:

(i) $[Cr(OH_2)_6]^{3+}$: 17.4 kK, 24.6 kK, 37.8 kK and a very weak band at 15.0 kK

(ii) $[Cr(NCS)_6]^{3-}$: 16 kK, 17.7 kK, 23.8 kK, 32.4 kK.

Interpret the findings.

Solution: Cr(III) in O_h crystal field shows the following ligand field transitions.

$$\left.\begin{array}{ll} {}^4A_{2g}(F) \rightarrow {}^4T_{2g}(F), & \nu_1 = 17.4 \text{ kK} \\ {}^4A_{2g}(F) \rightarrow {}^4T_{1g}(F), & \nu_2 = 24.6 \text{ kK} \\ {}^4A_{2g}(F) \rightarrow {}^4T_{1g}(P), & \nu_3 = 37.8 \text{ kK} \end{array}\right\} \begin{array}{l} \nu_3 > \nu_2 > \nu_1 \\ \nu_1 = 10Dq \end{array}$$

In $[Cr(OH_2)_6]^{3+}$, the weak band at 15.0 kK is due to a spin-forbidden transition (*cf.* Tanabe-Sugano diagram):

$${}^4A_{2g} \rightarrow {}^2T_{2g}, {}^2E_g, \nu_4 = 15.0 \text{ kK}$$

These doublet states reside below ${}^4T_{2g}(F)$. This is why

$$\nu_4 < \nu_1$$

The poor intensity is due to the spin-forbidden transition.

In $[Cr(NCS)_6]^{3-}$, the ligand field peaks are:

$$\nu_1 = 16 \text{ kK}, \nu_2 = 17.7 \text{ kK and } \nu_3 = 23.8 \text{ kK}.$$

The high energy intense band at 32.4 kK is a CT band.

Problem 7: $[VO(acac)_2]$ shows three special bands at: 13 kK, 17.45 kK, 25.65 kK. Interpret the results.

Solution: VO^{2+} represents $3d^1$ where there is a strong covalent interaction with the oxo-ligand. It gives the tetragonally compressed $[VO(acac)_2]$ complex of C_{4v} symmetry. The orbital splitting (z-in J.T. distortion) is:

$$d_{xy}^1 < d_{xz}, d_{yz} < d_{x^2-y^2} < d_{z^2} \; (i.e. \; z\text{-in distortion})$$

The possible electronic transitions are (*cf.* Figs. 7.14.1.2, 3):

$$d_{xy} \xrightarrow{\nu_1} d_{xz}, d_{yz}; \; d_{xy} \xrightarrow{\nu_2} d_{x^2-y^2}; \; d_{xy} \xrightarrow{\nu_3} d_{z^2}$$

$$\nu_1 = 13 \text{ kK}; \nu_2 (= 10Dq) = 17.45 \text{ kK}; \nu_3 = 25.65 \text{ kK}.$$

Problem 8: A tetrahedral complex of Co(II) exhibits three ligand field bands. The lowest energy band in the visible range is 7,780 cm^{-1}. Can you predict the positions of other peaks? (Given: $\beta = 0.76$, $B_0 = 970$ cm^{-1})

Solution: The possible transitions are:

$$\left.\begin{array}{ll} {}^4A_2 \rightarrow {}^4T_2(F), & \nu_1 \approx 10Dq \\ {}^4A_2 \rightarrow {}^4T_1(F), & \nu_2 \approx 18Dq \\ {}^4A_2 \rightarrow {}^4T_1(P), & \nu_3 \approx 12Dq + 15B' \end{array}\right\} \begin{array}{l} \text{without considering} \\ \text{the symmetry} \\ \text{interaction energy} \end{array}$$

The absorption band at 7,780 cm^{-1} is not the ν_1 band. It would lead to $10Dq \approx 7.78$ kK which will be too high for a tetrahedral complex of Co(II). In fact, for the tetrahedral complexes of Co(II), the ν_1 band generally lies in the infrared region. Thus, the lowest energy band in the visible range is expected to be the ν_2 band, *i.e.*

$$\nu_2 \approx 18Dq = 7{,}780 \text{ cm}^{-1}, \; i.e. \; Dq \approx 432 \text{ cm}^{-1}$$

Thus, the ν_3 band should appear at:

$$\nu_3 = 12Dq + 15B'.$$

We have: $$\beta = \frac{B'}{B_0} = 0.76 = \frac{B'}{970}, \; i.e. \; B' \approx 737 \text{ cm}^{-1}$$

i.e. $$\nu_3 = 12 \times 432 + 15 \times 737 = 5{,}184 + 11{,}055 = 16{,}239 \text{ cm}^{-1}$$

Note: Taking $\beta \approx 1$, *i.e.* $B' \approx B_0 = 970$ cm^{-1}, $\nu_3 = 12Dq + 15B_0 = 12 \times 432 + 15 \times 970 = 19{,}734$ cm^{-1}

Problem 9. $[VCl_4]$ shows one absorption peak at 9,000 cm^{-1}. Predict the $10Dq$ for $[VCl_6]^{2-}$.

Solution: V(IV) is a $3d^1$ system. In the tetrahedral $[VCl_4]$ complex, the single peak arises from the following transition.

$$^2E \xrightarrow{\nu} {}^2T_2,\ \nu = \Delta_t = 10Dq = 9000\ cm^{-1}$$

$$\Delta_o = 10Dq_o \approx \tfrac{9}{4}10Dq_t = \tfrac{9}{4} \times 9000 = 20{,}250\ cm^{-1}$$

Note: The calculation done above has not considered the J.T distortion. The values of Δ_t and Δ_o appear too high for V(IV).

Problem 10: $[CoCl_4]^{2-}$ shows two spectral bands in the visible range at 5,800 cm^{-1} and 15,000 cm^{-1}. Comment on the $15B'$ parameter. (Given : $15B_0$ = 15,500 cm^{-1})

Solution: The ligand field transitions for the tetrahedral $[CoCl_4]^{2-}$ complex are:

$$^4A_2(F) \xrightarrow{\nu_1} {}^4T_2(F);\ {}^4A_2(F) \xrightarrow{\nu_2} {}^4T_1(F);\ {}^4A_2(F) \xrightarrow{\nu_3} {}^4T_1(P)$$

$$\nu_1 = 10Dq,\ \nu_2 \approx 18Dq,\ \nu_3 \approx 12Dq + 15B'$$

(without the symmetry interaction parameter)

If ν_1 = 5,800 cm^{-1} is considered, then $10Dq$ becomes 5,800 cm^{-1} which is too high for the tetrahedral $[CoCl_4]^{2-}$ complex. Thus it is reasonable to consider: ν_2 = 5,800 cm^{-1} ≈ $18Dq$;

i.e. $$Dq = \frac{5{,}800}{18} \approx 322\ cm^{-1}.$$

Thus, $\nu_3 \approx 12 \times 322 + 15B' = 3{,}864 + 15B' = 15{,}000\ cm^{-1}$

i.e. $15B' = 11{,}136\ cm^{-1}$

It indicates: $15B_0 = 15{,}500\ cm^{-1} > 15B' = 11{,}136\ cm^{-1}$, *i.e.* **nephelauxetic effect** is quite significant.

$$\beta = \frac{15B'}{15B_0} = \frac{11{,}136}{15{,}500} \approx 0.72,\ \textit{i.e. } B' \textit{ is } 72\% \text{ of } B_0 \text{ value.}$$

(**Note:** When only ν_2 and ν_3 bands are known, for more correct analysis, the procedure to be followed is illustrated in the following example).

Problem 11: A tetrahedral complex of Co(II) gives the two absorption bands in the visible range at 7,140 cm^{-1} and a doublet peak (18,180 cm^{-1}, 19,600 cm^{-1}). Determine its spectral parameters, *i.e.* $10Dq$, β, and assignment of all the possible ligand field bands. (Given: $15B_0$ = 15,500 cm^{-1})

Solution: The possible spin-allowed transitions in the tetrahedral Co(II) complexes are:

$$^4A_2(F) \xrightarrow{\nu_1} {}^4T_2(F);\ {}^4A_2(F) \xrightarrow{\nu_2} {}^4T_1(F);\ {}^4A_2(F) \xrightarrow{\nu_3} {}^4T_1(P)$$

$$\nu_1 = 10Dq,\ \nu_2 = 18Dq - x,\ \nu_3 = 12Dq + 15B' + x$$

[x = configurational interaction energy arising from the symmetry interaction between $^4T_1(F)$ and $^4T_1(P)$]

If the band at 7,140 cm^{-1} is considered to be due to ν_1 then the $10Dq$ (= 7,140 cm^{-1}) becomes too high for the tetrahedral Co(II) complexes. Thus it is reasonable to assign the bands as follows:

$$\nu_2 = 18Dq - x = 7{,}140\ cm^{-1};\ \nu_3 = 12Dq + 15B' + x = \tfrac{1}{2}(18{,}180 + 19{,}600) = 18{,}890\ cm^{-1}$$

The doublet character of the ν_3-band is due to the spin-orbit coupling splitting of the T_1 term. Thus, the centre of the doublets is taken as the ν_3 band, *i.e.*

$$\nu_3 = \tfrac{1}{2}(18{,}180 + 19{,}600) = 18{,}890\ cm^{-1}$$

If the values of all the three bands, *i.e.* ν_1 , ν_2 and ν_3 are known, the above relations are sufficient for the spectral analysis. But in absence of the ν_1-band value, we are to use the following relations:

$$\nu_2 = 1.5\Delta_t + 7.5B' - p,\ \nu_3 = 1.5\Delta_t + 7.5B' + p$$

$$p = \tfrac{1}{2}[225B'^2 + \Delta_t^2 - 18B'\Delta_t]^{1/2} = \tfrac{1}{2}[\nu_3 - \nu_2]$$

$$\nu_1 = \Delta_t = \tfrac{1}{3}[\nu_2 + \nu_3 - 15B'];$$

It leads to: $\Delta_t = \tfrac{1}{3}[\nu_2 + \nu_3 - 15B'] = \tfrac{1}{3}[7{,}140 + 18{,}890 - 15B']$

$$= 8676.7 - 5B'$$

$$p = \tfrac{1}{2}[\nu_3 - \nu_2] = \tfrac{1}{2}[18{,}890 - 7{,}140] = 5{,}875\ \text{cm}^{-1}$$

$$4p^2 = 225B'^2 - 18B'\Delta_t + \Delta_t^2$$

Using $p = 5{,}875$ cm^{-1} and $\Delta_t = 8676.7 - 5B'$, we get the acceptable solution: $B' = 916$ cm^{-1} and the the negative root for B' is not acceptable.

Thus, $\Delta_t = \nu_1 = 8676.7 - 5 \times 916 = 4100$ cm^{-1}

i.e. $10Dq = 4100$ cm^{-1}; $Dq = 410$ cm^{-1}

$\nu_2 = 18Dq - x = 18 \times 410 - x = 7{,}380 - x = 7{,}140$ cm^{-1}

i.e. $x = 240$ cm^{-1}

$15B_0 = 15{,}500$ cm^{-1}, *i.e.* $B_0 \approx 1033$ cm^{-1}

It leads to: $\beta = \dfrac{B_{\text{complex}}}{B_{\text{free ion}}} = \dfrac{B'}{B_0} = \dfrac{916}{1033} \approx 0.88.$

Note: If the configurational interaction energy is ignored, then we can write:

$\nu_2 \approx 18Dq = 7{,}140$ cm^{-1}, *i.e.* $Dq \approx 397$ cm^{-1}

$\nu_3 = 18{,}890$ cm^{-1} $= 12Dq + 15B' = 12 \times 397 + 15B' = 4764 + 15B'$

or, $15B' = 18{,}890 - 4764 = 14{,}126$ cm^{-1}, *i.e.* $B' \approx 942$ cm^{-1}

If leads to: $\beta = \dfrac{B'}{B_0} = \dfrac{942}{1033} \approx 0.9$

Problem 12: $[V(OH_2)_6]^{3+}$ gives two spectral bands at 17.2 kK and 25.6 kK. If these are supposed to be the first two ligand field bands. Calculate the position of the third ligand field band by using the Tanabe-Sugano diagram (given in Fig. 7.23.2).

Solution. The ligand field transitions for the d^2-system are:

$^3T_{1g}(F) \xrightarrow{\nu_1} {}^3T_{2g}(F)$, $\nu_1 = 17.2$ kK, **(lowest energy band)**

$^3T_{1g}(F) \xrightarrow{\nu_2} {}^3A_{2g}(F)$,

$^3T_{1g}(F) \xrightarrow{\nu_3} {}^3T_{1g}(P)$,

Between ν_2 and ν_3, one transition corresponds to 25.6 kK. For the $d^2(O_h)$ system, we have:

$\nu_1 = 8Dq + x' \approx 8Dq$; $\nu_2 = 18Dq + x' \approx 18Dq$; $\nu_3 = 6Dq + 15B' + 2x' \approx 6Dq + 15B'$

Taking, $\nu_1 \approx 8Dq = 17{,}200$ cm^{-1}, *i.e.* $Dq = 2{,}150$ cm^{-1}, ν_2 ($\approx 18Dq$) should appear at 38,700 cm^{-1}. Thus, the band at 25,600 cm^{-1} is not definitely due to the ν_2 transition (*i.e.* $^3T_{1g} \rightarrow {}^3A_{2g}$). Hence, it is reasonable to assume ν_3 to correspond to 25,600 cm^{-1}, *i.e.*

$$^3T_{1g}(F) \xrightarrow{\nu_3} {}^3T_{1g}(P),\ \nu_3 = 25.6\ \text{kK}.$$

Now let us verify our prediction from the Tanabe-Sugano diagram and the plot of ν_1/ν_3 *vs.* Dq/B' for the $T_{1g}(F)$ ground spectroscopic state (Figs. 7.23.1 and 7.10.4).

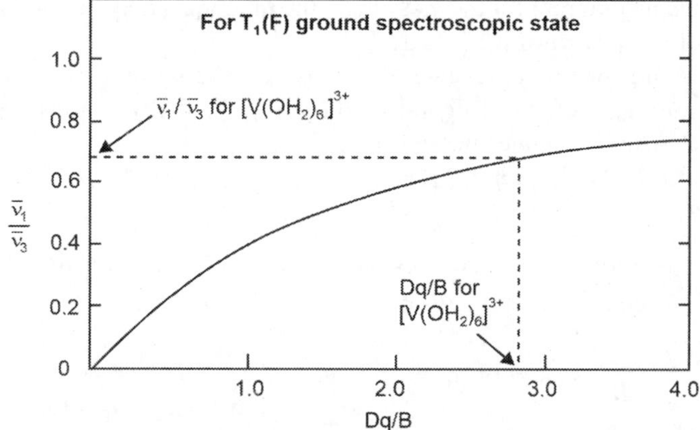

Fig. 7.23.1: Variation of the $\overline{\nu}_1/\overline{\nu}_3$ ratio with *Dq/B* for the d^2-octahedral complexes

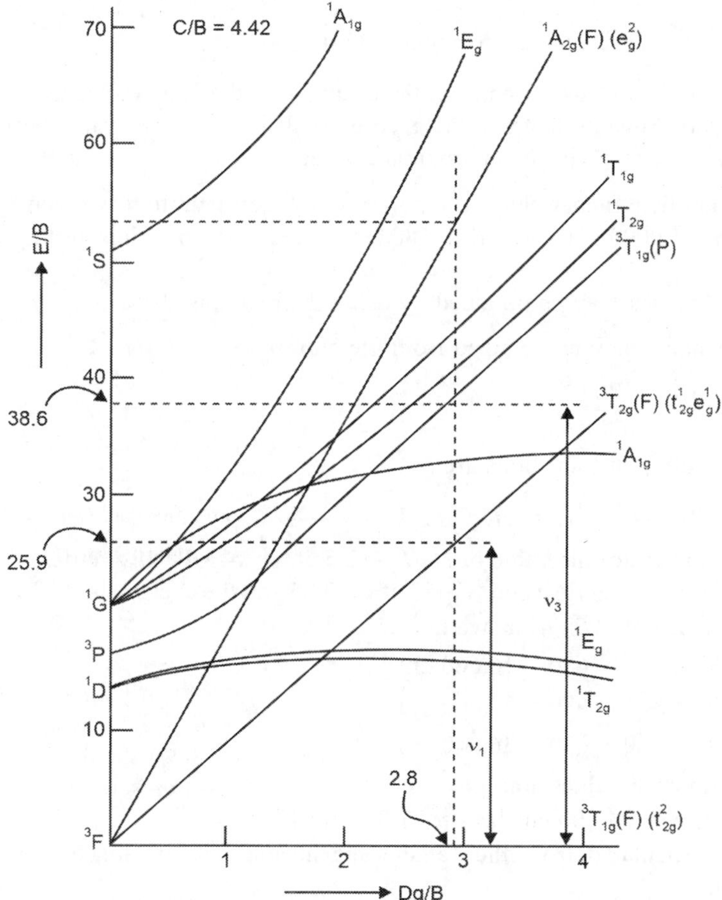

Fig. 7.23.2 T.S. diagram for a d^2–octahedral system and analysis of the absorption spectrum of $[V(OH_2)_6]^{3+}$ · ν_1: $^3T_{1g}(F) \rightarrow {}^3T_{2g}(F)$; ν_2: $^3T_{1g}(F) \rightarrow {}^3A_{2g}(F)$; ν_3: $^3T_{1g}(F) \rightarrow {}^3T_{1g}(P)$.

From the Tanabe-Sugano diagram for the d^2-system, the ratio ν_1/ν_3 may be calculated for the different Dq/B' values. This plot is shown in Fig. 7.23.1.

Note: For the T_1 ground spectrocopic state, the transition energy ratio, ν_2/ν_1 is close to 2 (but not less than 2) regardless of the strength of ligand field (*cf.* Fig. 7.10.4b). If ν_2 is taken as 25.6 kK, then ν_2/ν_1 becomes 1.48 $\langle$ 2. Thus ν_2 cannot be 25.6 kK.

$$\nu_1: {}^3T_{1g}(F) \rightarrow {}^3T_{2g}(F);$$
$$\nu_3: {}^3T_{1g}(F) \rightarrow {}^3T_{1g}(P);$$

In the present case, the ratio $\bar{\nu}_1/\bar{\nu}_3$ becomes 0.672 for which Dq/B' becomes 2.8. For this ratio of Dq/B', E/B' for ${}^3T_{2g}(F)$ becomes 26 and E/B' becomes 38 for ${}^3T_{1g}(P)$.

i.e. $$\frac{17{,}200}{B'} = \frac{\nu_1}{B'} = 26.0 \text{ or, } B' \approx 662 \text{ cm}^{-1}$$

and $$\frac{25{,}600}{B'} = \frac{\nu_3}{B'} = 38 \text{ or, } B' \approx 674 \text{ cm}^{-1}$$

average value of $B' = 668$ cm^{-1}

E/B' for ${}^3A_{2g}$(F) at $Dq/B' = 2.8$ becomes 52.5

i.e. $$\frac{\nu_2}{B'} = 52.5 \text{ or, } \nu_2 = 52.5 \times 668 = 35{,}070 \text{ cm}^{-1}$$

(**Note:** Accuracy of the results depends on the accuracy of the Tanabe-Sugano diagram used. Here an approximate Tanabe-Sugano diagram has been used. This is why, B' value becomes different for ${}^3T_{2g}(F)$ and ${}^3T_{1g}(P)$. Here ν_2 gives the approximate value.)

Problem 13: A simplified Tanabe-Sugano diagram for d^3 configuration is given in Fig. 7.23.3. For $[Cr(OH_2)_6]^{3+}$, $10Dq = 17{,}000$ cm^{-1}, and $B' = 680$ cm^{-1}. Predict the positions of the ligand field bands.

(**Note:** $10Dq$ and B' values may be obtained by using the relations. $10Dq = f \times g$, $\beta = \frac{B'}{B_0} = 1 - hk$, values of f, g, B_0, h and k may be obtained from literature.)

Solution. The ratio $\frac{Dq}{B'} = \frac{1700}{680} = 2.5$

The possible spin-allowed transitions are:

$${}^4A_{2g}(F) \xrightarrow{\nu_1} {}^4T_{2g}(F);\ {}^4A_{2g}(F) \xrightarrow{\nu_2} {}^4T_{1g}(F) \text{ and } {}^4A_{2g}(F) \xrightarrow{\nu_3} {}^4T_{1g}(P).$$

A vertical line is drawn for the value of $Dq/B' = 2.5$ at the x-axis. This vertical line cuts the energy state ${}^4T_{2g}(F)$ at $y = 24.5$, the energy state ${}^4T_{1g}(F)$ at $y = 34.5$ and the energy state ${}^4T_{1g}(P)$ at $y = 53.0$. The y-axis is plotted in terms of E/B'. Thus we have:

$$E(\nu_1) = 24.5 \times B' = 24.5 \times 680 = 16{,}660 \text{ cm}^{-1}$$
$$E(\nu_2) = 34.5 \times B' = 34.5 \times 680 = 23{,}460 \text{ cm}^{-1}$$
$$E(\nu_3) = 53.0 \times B' = 53.0 \times 680 = 36{,}040 \text{ cm}^{-1}$$

The actual experimental values are:

$\nu_1 = 17{,}000$ cm^{-1}, $\nu_2 = 24{,}000$ cm^{-1}, $\nu_3 = 37{,}000$ cm^{-1}

Thus the values calculated from the Tanabe-Sugano diagram are in good agreement with the experimental result.

(**Note:** In practice, the ratios ν_1/ν_2, ν_2/ν_3, ν_2/ν_3 are determined from the experimental peaks and these are fitted in the T.S. diagram or in Fig. 7.10.4 to determine the $10Dq$ and B' values.)

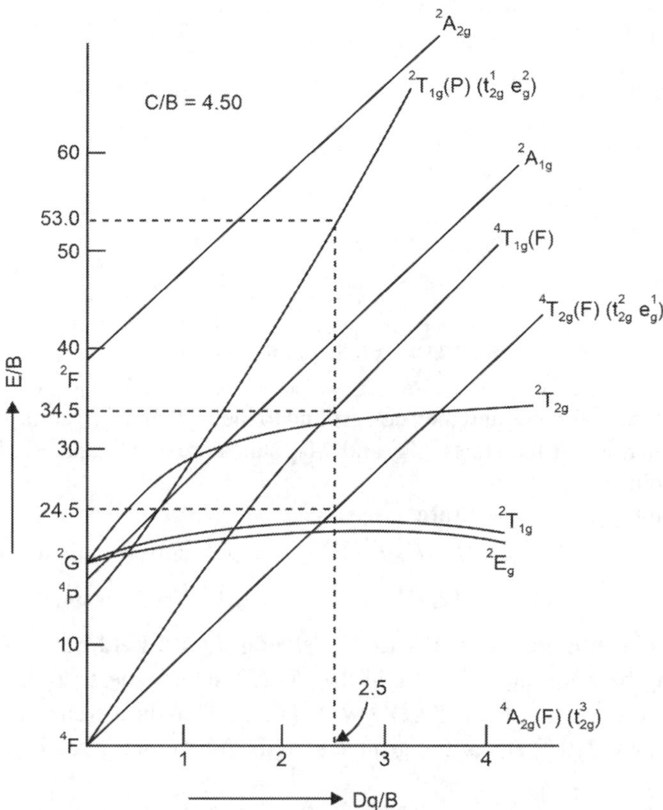

Fig. 7.23.3: Tanabe-Sugano diagram for the d^3-octahedral complex

For this purpose, the ratios $\nu_1/\nu_2, \nu_2/\nu_3$, ν_2/ν_3 are calculated for the different Dq/B' values from the *T-S* diagram and these ratios are plotted against the Dq/B' values (*cf.* Fig. 7.10.4). Then the experimental ratio is fitted in the curve to determine Dq/B'. This is illustrated in the Problem 12.)

Problem 14: Calculate energy of the states due to the crystal field splitting of the ^2D term (d^1 configuration), ^3F term (d^2-configuration) and ^4F term (d^3-configuration) in a weak crystal field. Comment on the ground state electronic configuration for the d^2 configuration.

Solution: (i) The ^2D term splits into the $^2T_{2g}$ and ^2E_g states. The $^2T_{2g}$ state corresponds to the t_{2g}^1 configuration (energy = – $4Dq$ with respect to that of the unsplit ^2D term) and ^2E_g corresponds to the e_g^1 configuration (energy = $+6Dq$ with respect to that of the unsplit ^2D term).

$$t_{2g}^1 : \underbrace{(d_{xy})^1(d_{yz})(d_{xz}),\ (d_{xy})(d_{yz})^1(d_{xz}),\ (d_{xy})(d_{yz})(d_{xz})^1}_{^2T_{2g}\text{(Orbital degeneracy = 3)}}$$

$$e_g^1 : \underbrace{(d_{x^2-y^2})^1(d_{z^2}),\ (d_{x^2-y^2})(d_{z^2})^1}_{^2E_g\text{(orbital degeneracy=2)}}$$

(ii) ^3F (for d^2) splits into $^3T_{1g}$, $^3T_{2g}$ and $^3A_{2g}$. From the Tanabe-Sugano diagram, it is seen that each of the $^3T_{2g}$ and $^3A_{2g}$ states appears only once but the $^3T_{1g}$ state appears twice, *i.e.* $^3T_{1g}(P)$ and $^3T_{1g}(F)$.

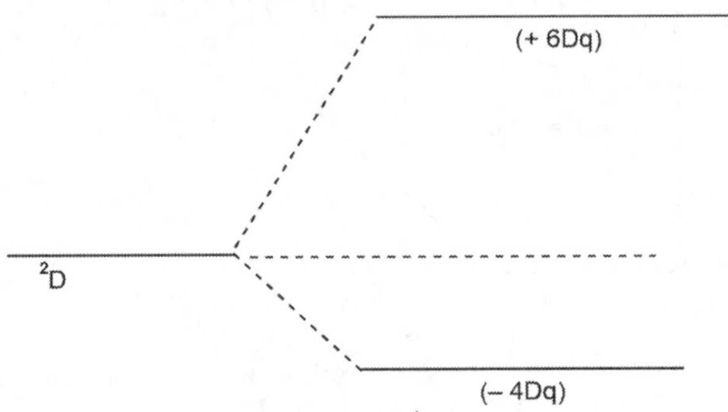

Fig. 7.23.4: Splitting of the ^2D term.

$^3T_{2g}$ and $^3A_{2g}$ are the excited states and they correspond to the $t_{2g}^1e_g^1$ and $t_{2g}^0e_g^2$ configurations respectively. The crystal field energies of the states $^3T_{2g}$ and $^3A_{2g}$ states are $+2Dq$ and $+12Dq$ respectively with respect to that of unsplit ^3F.

Configuration	**State**	
$t_{2g}^1e_g^1$	$^3T_{2g}(F)$	$^3T_{2g}$(F) corresponds to $(d_{xy})^1\ (d_{z^2})^1$, $(d_{xz})^1(d_{x^2-y^2})^1$, $(d_{yz})^1\ (d_{x^2-y^2})^1$
e_g^2	$^3A_{2g}(F)$	

The ground state t_{2g}^2 configuration works for the **strong ligand field** but in a **weak crystal field**, there will be a mixing between the $^3T_{1g}$(F) and the $^3T_{1g}(P)$ states where $^3T_{1g}(P)$ represents the $t_{2g}^1e_g^1$ configuration like $(d_{xy})^1\ (d_{x^2-y^2})^1$, $(d_{xz})^1\ (d_{z^2})^1$, $(d_{yz})^1\ (d_{z^2})^1$. Thus in a weak crystal field, due to the mixing of the $^3T_{1g}(P)$ and $^3T_{1g}(F)$ states, the ground state is to be represented by $t_{2g}^xe_g^y$ where $x + y = 2$ where x is less than 2).

Splitting of the ^3F term is shown in Fig. 7.23.5.

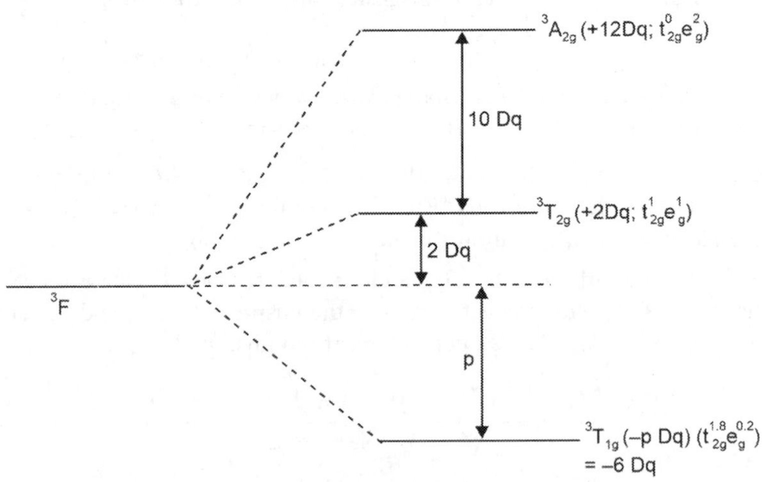

Fig. 7.23.5 Splitting of the ^3F term in a weak crystal field.

To maintain the Barrycentre, during splitting of the ^3F term, we have:

$-3p + 3 \times (2Dq) + 1 \times (12Dq) = 0$; orbital degeneracy for the T and A states are 3 and 1 respectively.

or $\qquad p = 6Dq$

Energy of the ground state $t_{2g}^x e_g^y = -4xDq + 6yDq = -6Dq$ and $x + y = 2$

Solution of these two equations gives: $x = 1.8$, $y = 0.2$, *i.e.* $t_{2g}^{1.8}e_g^{0.2}$

Note: In a strong field where the crystal field effect is more than the interelectron repulsion effect, the ground state is t_{2g}^2 (energy $= -8Dq$), but in a weak field, the ground state configuration is $t_{2g}^{1.8}e_g^{0.2}$ (energy $= -6Dq$). In a weak field, the interelectron repulsion is relatively more important. This is why, the interelectron repulsion forces some electron density to move into the e_g level.

***d*³ and ⁴*F* term:** By giving the same argument as in the d^2 system, we can write:

$^4A_{2g}(F)(t_{2g}^3, -12Dq)$, $^4T_{2g}(F)$ $(t_{2g}^2e_g^1, -2Dq)$ and $^4T_{1g}(F)$ $(t_{2g}^x e_g^y; pDq$ say, $x + y = 3)$. Here mixing of the $^4T_{1g}(F)$ and $^4T_{1g}(P)$ states in a weak crystal field has been considered. Applying the principle of energy conversion and orbital degeneracy, we can write (*cf.* Fig. 7.14.5.2):

$1 \times 12\,Dq + 3 \times 2\,Dq = 3\,pDq$, *i.e.* $p = 6$; $(-4x + 6y)Dq = pDq = 6Dq$ and $x + y = 3$ lead to $x = 1.2$, $y = 1.8$. Thus in a weak fied, $^4T_{1g}(F)$ corresponds to $t_{2g}^{1.2}e_g^{1.8}$ not $t_{2g}^1 e_g^2$.

Problem 15: In a noncentrosymmetric system, all the electronic transitions are not necessarily allowed. Illustrate for the complexes of T_d and D_{2d} symmetry.

Solution. (a) **$[CoCl_4]^{2-}$ (T_d):** Let us illustrate for the tetrahedral complex $[CoCl_4]^{2-}$ (d^7-system) where the spin allowed electronic transitions are:

$$^4A_2(F) \longrightarrow {}^4T_2(F);\ {}^4A_2(F) \longrightarrow {}^4T_1(F);\ {}^4A_2(F) \longrightarrow {}^4T_1(P).$$

The character table of T_d indicates that the coordinates x, y and z form the basis for the T_2 representation. It indicates that the dipole moment operator transforms as T_2.

- **$A_2 \rightarrow T_1$ transition:** $A_2 \times \mu \times T_1 = A_2 \times T_2 \times T_1 = A_2 \times (A_2 + E + T_1 + T_2) = A_1 + E + T_1 + T_2$

The triple product result contains the **totally symmetric** A_1 representation and consequently, the *transition is allowed.*

- **$A_2 \rightarrow T_2$ transition:** $A_2 \times \mu \times T_2 = A_2 \times T_2 \times T_2 = A_2 \times (A_1 + E + T_1 + T_2) = A_2 + E + T_1 + T_2$

There is no **totally symmetric** A_1 representation in the triple product result and thus the **transition is forbidden**. However, coupling of any one of the vibrations of A_2, E, T_1 or T_2 will make the transition allowed.

For the T_d point group, the normal modes of vibration are: $A_1 + E + 2T_2$. Thus coupling of E or T_2 vibration will make the $A_2 \rightarrow T_2$ transition allowed. (*cf.* A_1 is present in the direct products: $E \times E = A_1 + A_2 + E$; $T_2 \times T_2 = A_1 + E + T_1 + T_2$).

(b) **$[CuCl_4]^{2-}$(D_{2d}):** Compressed T_d leads to D_{2d} symmetry where the tetrahedral T_2 and E states split as follows:

T_d	D_{2d}
T_2	$B_2 + E$
E	$B_1 + A_1$

Thus the possible transitions in $[CuCl_4]^{2-}$ (D_{2d}) of d^9 system are:

$$^2B_2 \longrightarrow {}^2E;\ {}^2B_2 \longrightarrow {}^2B_1;\ {}^2B_2 \longrightarrow {}^2A_1$$

For the D_{2d} point group, the dipole moment operators transform as $B_2(\|)$ and E $(\perp)$.

Transition	*Triple product*	*Polarisation*
$B_2 \longrightarrow E$	$B_2 \times B_2 \times E = B_2 \times E = E$	Forbidden (‖)
	$B_2 \times E \times E = E \times E = \mathbf{A_1} + A_2 + B_1 + B_2$	**Allowed** (⊥)
$B_2 \longrightarrow B_1$	$B_2 \times B_2 \times B_1 = A_2 \times B_2 = B_1$	Forbidden (‖)
	$B_2 \times E \times B_1 = A_2 \times E = E$	Forbidden (⊥)
$B_2 \longrightarrow A_1$	$B_2 \times B_2 \times A_1 = B_2 \times B_2 = \mathbf{A_1}$	**Allowed** (‖)
	$B_2 \times E \times A_1 = B_2 \times E = E$	Forbidden (⊥)

Presence of the **totally symmetric A_1** representation in the triple product gives the **allowdness** while the absence of A_1 in the triple product indicates the **forbiddeness** in the respective electronic transitions. It indicates that the ${}^2B_2 \rightarrow {}^2B_1$ is orbitally forbidden and the other two transitions experience the electronic polarisation.

EXERCISE 7

A. General Type Questions

1. What is Lambert-Beer law? What are its limitations?
2. What are the different types of electronic spectra?
   - What do you mean by electric dipole (ED) transition and magnetic dipole (MD) transition in electronic spectra? Compare the selection rules for ED transition and MD transition.
3. What are the orbital and spin selection rules in electronic transitions? Give the quantum mechanical basis of these rules.
   - For the $t_{2g} \rightarrow e_g$ transition, sometimes, more than one peak is obtained. How can you explain this?
4. What are the possible mechanisms to relax the Laporte selection rule?
5. What are the effects of (i) lowering of symmetry and (ii) temperature on the intensity of electronic spectra.
6. What is the effect of spin-orbit coupling splitting of states on band intensity and bandwidth?
7. What are the important factors to determine the bandwidth of electronic spectra?
   - What do you mean by **Doppler effect** to broaden a peak? How can we use this effect to measure the temperature of the sun?
   - Comment on the importance of peak broadening (*i.e.* natural peak broadening) by **uncertainty principle** in electronic transitions.
   - Natural peak broadening is insignificant in electronic transitions but it is quite important in NMR and ESR spectroscopy. Explain.
8. How can you explain the asymmetric shape of the spectral bands?
9. What do you mean by isosbestic point? What is the significance of this point?
10. What do you mean by Condon-Shortley and Racah parameters?
11. Explain the origin of luminescence (with fluorescence and phosphorescence) in terms of Jablonski diagram.
12. Discuss the luminescence of Cr(III)-compounds.
13. Discuss the effect of J.T. distortion on electronic spectra.
14. Predict the spin-allowed spectral transitions in terms of Orgel diagram.

15. Discuss the effect of configurational interaction between the $^3T_1(P)$ and $^3T_1(F)$ states on the energies of spectral transitions of transition metal ions.
16. Calculate the energies of the states arising from the splitting of ^3F term of a d^2 configuration. Comment on the energies of the $^3T_{1g}$ state in a strong field and in a weak field. In a weak crystal field, the ground state electronic configuration t_{2g}^2 is not correct—comment. (**Hints:** *See* problem 14)
17. What are the limitations of Orgel diagram? Discuss the characteristic features of Tanabe-Sugano diagram.
18. How can you predict the spectral band-width from the Tanabe-Sugano diagram?
19. Generally, the spin-forbidden transitions give the sharp peaks—explain.
20. Give the energies of the spin-allowed transitions in the octahedral and tetrahedral geometries of $d^{1,2,3,8,9}$ and high-spin $d^{4,5,6,7}$ systems.
21. Discuss the structural and spectral spectral features of $[Ti(OH_2)_6]^{3+}$ with special reference to its intensity, band shape and band-width. How can you determine $10Dq_o$ for $[Ti(OH_2)_6]^{3+}$?
22. Discuss the structural and spectral characteristics of six-coordinate complexes of oxovanadium(IV). How do you determine the $10Dq_o$ value?
23. Compare the spectral behaviour of $[Cu(OH_2)_6]^{2+}$ and $[Cu(NH_3)_x(OH_2)_y]^{2+}$ $(x + y = 6)$.
24. Discuss the change of spectral properties of $[Cu(acac)_2]$ in different solvents differing in coordinating power.
25. Discuss the spectral properties of $[V(OH_2)_6]^{3+}$.
26. Discuss the origin of ruby laser action.
27. Discuss the spectral features of $[Cr(NH_3)_6]^{3+}$ and $[Cr(en)_3]^{3+}$. What is the effect of lowering of symmetry O_h to D_3 on the spectral transitions?
28. Discuss the spectra of $[Cr(OH_2)_6]^{2+}$, $[Mn(OH_2)_6]^{2+}$, $[Fe(OH_2)_6]^{2+}$, $[CoF_6]^{3-}$, $[Cr(ox)_3]^{3-}$, $[Co(OH_2)_6]^{2+}$, $[CoCl_4]^{2-}$, $[NiCl_4]^{2-}$, $[Ni(NH_3)_6]^{2+}$, $[Ni(CN)_4]^{2-}$, $[Ni(dmgH)_2]$, $[Co(NH_3)_6]^{3+}$, $[CoF_2(NH_3)_4]^+$ (both *cis*- and *trans* isomers), $[CoCl_3(NH_3)_3]$ (both *fac*- and *mer*-isomers), $[Co(gly)_3]$ (both *cis*- and *trans*- isomer), [Co(salen)], [Co(acacen)], $[CuCl_4]^{2-}$, $[FeCl_4]^-$, $[Ni(NH_3)_4(NO_2)_2]$, $[Cu(bpy)_3]^{2+}$.
29. Compare the spectral properties of (i) $[Cr(en)_3]^{3+}$ and $[Co(en)_3]^{3+}$; (ii) *cis*- and *trans*-isomers of $[Co(en)_2F_2]^+$; (iii) *cis*- and *trans*-isomers of $[Cr(en)_2F_2]^+$; (iv) $[Co(NH_3)_6]^{3+}$ and $[CoCl(NH_3)_5]^{2+}$.
30. Tetrahedral complexes of the 1st transition series metal ions are generally more intensely coloured than their octahedral complexes. Explain.
31. Compare the spectral properties of the octahedral, tetrahedral and square planar complexes of (i) Ni(II), (ii) Cu(II) and (iii) Co(II).
32. Illustrate the coordination equilibria and thermochromism in the Ni(II) complexes.
33. Comment on the ligand field spectra of the 2nd and 3rd transition series metal ions.
34. Discuss the general characteristic spectral features of the lanthanide and actinide ions.
    - Compare the electronic spectra of Ln^{3+} and Ln^{2+} ions.
    - Some Ln^{3+} ions are colourless. Explain.
35. Compare the electronic spectral properties (i) between the 4*f* and 5*f* series ions, (ii) *f*-series and *d*-series ions.
36. Explain the chemistry behind the use of lanthanides as the NMR shift reagents and as phosphors in colour TV.
37. Discuss the factors favouring the CT band towards the longer wavelength in LMCT and MLCT.
38. Discuss the CT-transitions in LMCT and MLCT in terms of molecular orbital diagram.

39. Illustrate the intravalance electron transfer and MMCT (metal to metal electron transfer).
40. Explain the origin of colour in **Prussian blue, Magnus green salt, Creutz-Taube ion, molebdenum blue, ruthenium red, ruthenium purple, ruthenium blue, $[Mo_2Cl_8]^{4-}$ and $[Re_2Cl_8]^{2+}$.**
41. Compare the characteristic features of the CT and ligand field bands.
42. Compare the spectral features of (i) $[Co(NH_3)_5X]^{2+}$ (X = F, Cl, Br, I), (ii) $[Ir^{III}(NH_3)_5X]$ (X = NH_3, H_2O, Cl, F).
43. $[Co(en)_3]^{3+}$ shows two peaks at 21.55 kK and 29.6 kK. Calculate the $10Dq$ and B' value.
44. $[CrF_6]^{3-}$ shows the peaks at 34.8 kK, 22.4 kK and 14.9 kK. Calculate $10Dq$, B' and configurational interaction energy.
45. $[V(OH_2)_6]^{3+}$ shows the ligand field bands at 17.2 kK, 25.7 kK and 36.0 kK. Calculate $10Dq$, B' and configurational interaction energy.
46. $[Co(NCS)_4]^{2-}$ gives peaks at 4.32 kK, 7.8 kK and 16.25 kK. Calculate $10Dq$, B' and configurational interaction energy.
    If the lowest energy band remains uncharacterized, then how do you calculate the values?
47. Discuss the change in spectral transitions due to the lowering of symmetry in the following cases.
    (i) Cu^{II} ($O_h \rightarrow D_{4h}$; C.N. 6);
    (ii) Ni^{II}($O_h \rightarrow D_{4h}$; C.N. 6) $\rightarrow D_{4h}$ (C.N. 4)).
    (iii) Co^{III}(l.s.) ($O_h \rightarrow D_{4h}$; $O_h \rightarrow C_{4v}$); ($O_h \rightarrow C_{2v}$);
    (iv) Cu^{II}($T_d \rightarrow D_{2d}$).
48. Explain the colours *due to impurities* in: red corundum, green coloured gemstone emerald, alexandrite gemstone (greenish-yellow colour), red garnet, yellow-green colour of peridot, blue-green aquamarine, pink morganite, blue spinels, ruby red spinel, cobalt blue glass, amethyst (violet colour), blue sapphire gemstone (blue corundum), ultramarine blue and natural lapis lazuli, blue and yellow diamond (*Ref.* Das *et al., Resonance,* 2025).

B. Justify the following statements

1. The electron transitions, $g \rightarrow g$, $u \rightarrow u$ are quantum mechanically forbidden while the $g \rightarrow u$ transitions are quantum mechanically allowed.
   - Presence of centre of symmetry generally disallows the electronic transitions orbitally but the absence of centre of symmetry (as in T_d point group) does not necessarily allow all the possible electronic transitions.
2. For an electron transition, $\Delta S = 0$.This is supported quantum mechanically.
   - The spin forbidden transitions give the relatively stronger absorption peaks for the 4*d* and 5*d* transition series.
   - In some cases, even for the 3*d* series, the spin forbidden transition peaks are of quite high intensity.
   - For $[Ni(OH_2)_6]^{2+}$ and $[Ni(dmso)_6]^{2+}$, the spin forbidden transitions give the quite intense absorption peaks but in $[Ni(NH_3)_6]^{2+}$ or $[Ni(en)_3]^{2+}$, the corresponding spin-forbidden transitions are not noted.
3. A 2*e*-transition is quantum mechanically forbidden.
   - 1*e* transition, $t_{2g} \rightarrow e_g$: d^1 system -one peak; d^2-system -two to three peaks; d^3-system -2 to 3 peaks. Apparently, $t_{2g} \rightarrow e_g$ transition should give one peak ($h\nu = \Delta_o$).

4. Franck-Condon vertical transition (0–0) is quite likely for the lanthanides experiencing the $f \rightarrow f$ transitions but for the transition metal complexes, 0–0 transition is the rarest possibility. (**Hints:** Difference in the internuclear distances between the ground and excited state)
   - For the transition metal complexes, vertical transition generally occurs to the higher vibrational levels of the excited electronic state.
   - Very often, the **spin-forbidden transitions** give the relatively **sharper peaks**.
5. Vibronic coupling is the main pathway for the relaxation of Laporte Selection rule for the centrosymmetric complexes while *d-p* mixing is the main pathway to to give the intensity of the ligand field bands in the noncentrosymmetric complexes.
6. The $T_{1g} \rightarrow T_{2g}$ transition is only allowed when this electronic transition is coupled with the T_{1u} and T_{2u} sets of metal-ligand vibration. (Use the direct product table)
   - For an O_h-system, all the electronic transitions are orbitally forbidden, but in a T_d-system, all the electronic transitions are not orbitally forbidden.
7. Vibronic coupling in the electronic transition, ${}^3T_{1g}(F) \rightarrow {}^3T_{2g}(F)$ (in octahedral system) should give a broad peak with 10 components or subpeaks. (**Hints:** Use the results of direct product)
8. In an octahedral system, the electronic transition is possible only when a *g*-type vibration of the ground state is coupled with the *u*-type vibration of the excited state and vice-versa.
9. Vibronic coupling is applicable for both the centrosymmetric and noncentrosymmetric complexes while *d-p* mixing is mainly applicable for the noncentrosymmetric systems.
10. Ligand field bands in the tetrahedral complexes are more intense than those in the octahedral complexes.
11. In the tetrahedral complexes, *d-p* mixing is the main pathway to relax the orbital selection rule.
    - In the tetrahedral complexes of 1st transition series, not the metal $(n-1)d$ and np orbital mixing but the metal $(n-1)d$ orbital and ligand *p*-orbital mixing is the more probable pathway.
12. Very often, the ligand field bands of the octahedral complexes become weaker with the lowering of temperature.
13. $[Co^{III}(NH_3)_5X]$ gives the more intense colour than $[Co^{III}(NH_3)_6]^{3+}$;
    - $[Co^{III}(en)_3]^{3+}$ gives the more intense colour than $[Co^{III}(NH_3)_6]^{3+}$;
    - *cis*-$[Co^{III}(NH_3)_4X_2]$ gives the more intense colour than *trans*-$[Co^{III}(NH_3)_4X_2]$.
14. Lowering of symmetry relaxes the orbital selection rule better.
    - Orbital selection rule is relaxed better in the C_{4v}, D_{4h} and D_3 point groups than in the O_h point group.
    - For the D_{4h}, D_3 and D_{2d} symmetry, electronic transitions are polarised.
    - Polarised electronic transitions cannot be studied in solution phase but in solid crystalline phase.
    - Tetrahedral (T_d) complexes are noncentrosymmetric but all the electronic transitions are not orbitally allowed.
15. A spin-forbidden transition lying close to a spin-allowed transition becomes relatively stronger.
16. A ligand field band lying close to the charge transfer band becomes relatively stronger.
17. Very often the spectral bands are symmetrical.

18. Metal-ligand bond vibration can broaden the peak.
    - The spin-allowed transition peaks are broad while the spin-forbidden transition peaks are sharp.
    - Metal-ligand bond vibration does not always broaden the peak.
19. In a d^2-system, the transition, ${}^3T_{1g}(F) \rightarrow {}^3A_{2g}(F)$ gives a very weak band.
20. In a d^2-system, the band due to the ${}^3T_{1g}(F) \rightarrow {}^3A_{2g}(F)$ transition is broader than the band due to the transition, ${}^3T_{1g}(F) \rightarrow {}^3T_{2g}(F)$.
    - In a d^2-system, transition from the triplet state to the singlet state gives the sharp peaks.
21. In the octahedral Cr(III)-system, the spin-forbidden transitions like, ${}^4A_{2g}(F) \rightarrow {}^2E_g(G)$, ${}^2T_{2g}(G)$ give the sharp peaks while the spin-allowed transitions, *i.e.* ${}^4A_{2g}(F) \rightarrow {}^4T_{2g}(F)$, ${}^4T_{1g}(F)$, ${}^4T_{1g}(P)$ give the broad peaks.
22. Spin-orbit coupling can broaden the ligand field bands. But this is more important in the tetrahedral complexes than in the octahedral complexes of 1st transition series.
23. J.T. distortion can broaden the ligand field bands but sometimes it may lead to new bands due to splitting of the bands.
24. Crystallographic data cannot distinguish between the dynamic and static J.T. distortion but spectral analysis can distinguish the both types of J.T. distortion.
25. At low temperature, the ligand field peaks are symmetrical but at higher temperature, the peaks become asymmetrical (*i.e.* tail towards the lower frequency).
    - If the condition of Franck-Condon vertical transition is satisfied, the $0 \rightarrow 0$ vibrational transition gives the most intense transition for an allowed electronic transition while the $0 \rightarrow 1$ transition gives the most intense transition for the electronic transition allowed through the vibronic coupling.
    - With the increase of temperature, the peaks generally become broader.
    - With the lowering of temperature, ε_{max} (at λ_{max}) of a peak increases for both the allowed electronic transitions and vibronically allowed electronic transitions, but the intensity measured by the total oscillator strength (f) remains unchanged for an allowed electronic transition while f- decreases for a vibronically allowed electronic transition with the lowering of temperature.
    - If the metal-ligand bond length remains unchanged (*e.g.* $f \rightarrow f$ transition, spin forbidden transition, etc.) during the electronic excitation, the peak becomes sharp but if the bond length in the excited state is longer then the peak becomes broader.
26. Existence of an isosbestic point indicates the presence of an equilibrium between the two absorbing species.
27. The d^n and d^{10-n} configurations are characterized by the same Russel-Saunders term.
    - The positions of energy states of d^n configuration are inverted in the case of d^{10-n} configuration.
28. The degree of interelectronic repulsion can be measured either by Condon-Shortley parameters (F_0, F_2, F_4) or Racah parameters (A, B, C).
    - For a particular d^n-configuration, the energy difference between the two states of the same spin multiplicity can be expressed by only one Racah parameter (B) while the energy difference between the states of different spin multiplicities can be expressed in terms of two Racah parameters (B, C).
29. During splitting of a free ion term in the crystal field, orbital multiplicity is retained.
30. For the d^2-configuration: high-spin t_{2g}^2 configuration is denoted by a T-state;
    - $t_{2g}^1e_g^1$ excited state leads two T-states of different energy.
    - $t_{2g}^0e_g^2$ excited state is represented by an A-state.

- ^3F term splits in a crystal field into $^3A_2$, $^3T_1$ and $^3T_2$ to maintain the orbital degeneracy 7.
- ^3P term converts in a crystal field into $^3T_1$ to maintain the orbital degeneracy 3.
- In a weak field, for a d^2 system, the ground state $^3T_{1g}(F)$ does not represent the $t_{2g}^2 e_g^0$ configuration. It represents the $t_{2g}^{1.8} e_g^{0.2}$ configuration. However, in a strong field, the ground state represents the t_{2g}^2 configuration.

31. Luminescence (*i.e.* radiative decay through fluorescence or phosphorescence) occurs when the nonradiative paths of decay are relatively inefficient.
    - In chemiluminescence, electronically excited state is produced at the cost of chemical energy.
    - Reaction of Cl_2 or ClO^- with H_2O_2 produces a red chemiluminescent glow.
32. Fluorescence involves a spin-allowed transition while phosphorescence involves a spin-forbidden transition.
    - Intersystem crossing (ISC) – radiationless transition between the two excited electronic states of different multiplicity is an essential condition for phosphorescence.
    - Jablonski diagram explains the luminescence.
33. Luminescence is well known for the lanthanide ions (Ln^{3+}) but it rarely occurs in transition metal ions.
34. Cr(III) can show the luminescence.
    - Existence of the $^2E_g(G)$ state close to the $^4T_{2g}(F)$ state is responsible for the luminescence property of Cr(III) in the octahedral geometry.
    - Depending on the crystal field strength, Cr(III) can show both the fluorescence and phosphorescence.
    - Participation of the $^2E_g(G)$ state – an excited state, controls the photochemical reactions of Cr(III).
    - In $[Cr^{III}(A)_6]$, the electronic transitions are orbitally forbidden but in $[Cr^{III}(A—A)_3]$, some of the electronic transitions are allowed and they show the electronic polarisation.
    - Very often, the spectral transitions in $[Cr^{III}(A—A)_3]$ can be explained in terms of the electronic transitions of Cr^{III}-octahedral complexes.
35. Photochemistry of $[Ru(bpy)_3]^{2+}$ can be rationalized in terms of its luminescence property.
36. Configurational interaction between the $^3T_1(P)$ and $^3T_1(F)$ states is relatively less important in the weak field ligands.
    - Configurational interaction energy is quite important to understand the spectral transition energies for the d^2, d^7 (T_d) and d^3, d^8 (O_h) systems while the configurational interaction energy is relatively less important for the d^2, d^7 (O_h) and d^3, d^8 (T_d) systems for a particular ligand.
    - For a d^3 (O_h) system, $10Dq$ values calculated from the two lowest energy bands as $\nu_1 = 10Dq$, $\nu_2 = 18Dq$ are not self consistent.
37. $^xA_2(F) \xrightarrow{\nu_1} {}^xT_2(F)$; $^xA_2(F) \xrightarrow{\nu_2} {}^xT_1(F)$ and $^xA_2(F) \xrightarrow{\nu_3} {}^xT_1(P)$.

    The energy order is: $\nu_3 > \nu_2 > \nu_1$, *i.e.* ν_3 is the highest energy band.
    - $^xT_1(F) \xrightarrow{\nu_1} {}^xT_2(F)$; $^xT_1(F) \xrightarrow{\nu_2} {}^xA_2(F)$; $^xT_1(F) \xrightarrow{\nu_3} {}^xT_1(P)$.

      ν_1 is always the lowest energy band but the highest energy band may be either ν_2 or ν_3 depending on the condition (*i.e.* ligand field strength).

38. Orgel diagram can only show the spin-allowed transitions in the high-spin complexes while Tanabe-Sugano diagram can show both the spin allowed and spin forbidden transitions in both the high-spin and low-spin complexes.
39. Orgel diagram cannot predict the relative broadness or sharpness of the peaks but Tanabe-Sugano diagram can predict this.
40. In constructing a Tanabe-Sugano diagram for a d^n-configuration, the ratio of C/B must be mentioned.
    - Different metal ions having the same d^n-configuration should have the different Tanabe-Sugano diagrams but in reality, the same Tanabe-Sugano diagram works for the different but isoelectronic metal ions.
41. $[Ti(OH_2)_6]^{3+}$ shows a weak but broad peak (at 21,000 cm^{-1}) with a shoulder at 19,500 cm^{-1}.
    - Δ_o for $[Ti(OH_2)_6]^{3+}$ is regarded as 20,300 cm^{-1}
    - Large band-width and asymmetric shape of the peak in $[Ti(OH_2)_6]^{3+}$ can be rationalized in terms of J.T. distortion but not in terms of spin-orbit coupling splitting of the T-state.
42. J.T. distortion cannot split the ligand field band into a doublet in $[TiF_6]^{3-}$ but in $[CoF_6]^{3-}$.
43. Six coordinate complexes of V(IV), *e.g.*$[VF_6]^{2-}$ and VO^{2+} (*e.g.* $[VO(NCS)_5]^{3-}$) are different in terms of the spectral features: complexes of V(IV) are characterized with a broad band while complexes of VO^{2+} show three different peaks.
    - 5 coordinate complexes of oxovanadium(IV) like $[VO(acac)_2]$ show 4 electronic transitions.
    - Spectral properties of both the 6- and 5-coordinate complexes of oxovanadium(IV) can be explained in the same way.
    - Sometimes, the oxovanadium(IV) complexes show 3 to 4 spectral bands. (**Hints:** C_{4v}, C_{2v} symmetry).
44. For a d^9 configuration, the ground state is orbitally doubly degenerate (*i.e.* ^2E_g) and the excited state is orbitally triply degenrate ($^2T_{2g}$).
45. Most of the Cu(II) complexes are blue or green but some of them may be red or brown.
46. Analysis of the polarized crystal spectrum of $CuSO_4.5H_2O$ shows three absorption peaks at 10,500 cm^{-1}, 13,000 cm^{-1} and 14,500 cm^{-1}.
47. λ_{max} for $[Cu(OH_2)_6]^{2+}$ is 800 nm while it is shifted to higher energy for $[Cu(NH_3)_4(OH_2)_2]^{2+}$ (λ_{max} = 590 nm). But in moving from $[Cu(NH_3)_4(OH_2)_2]^{2+}$ to $[Cu(NH_3)_5(OH_2)]^{2+}$, λ_{max} shifts towards the longer wavelength.
    - In spite of several possible spectral transitions in the square planar complexes of Cu(II), $[Cu(bigH)_2]^{2+}$ shows only a broad band.
    - $[Cu(bpy)_3]^{2+}$ or $[Cu(phen)_3]^{2+}$ shows two well defined peaks.
48. Ligand field bands of $[Cu(acac)_2]$ are of different types in different solvents having the different coordinating properties.
    - In the poor coordinating solvents, splitting of the spectral peaks of $[Cu(acac)_2]$ is prominent.
49. Square pyramidal (C_{4v}) complexes of Cu(II) are characterized by three ligand field bands.
50. $[V(OH_2)_6]^{3+}$ shows two spectral bands in the visible range. The 3rd peak due to the $^3T_{1g}(F) \rightarrow {}^3A_{2g}(F)$ transition is missing.
    - In the six-coordinate complexes of V(III), the intensity of the peak due to $^3T_{1g}(F) \rightarrow {}^3A_{2g}(F)$ is very weak.
51. In the Cr(III) octahedral complexes, the third ligand field band due to $^4A_{2g}(F) \rightarrow {}^4T_{1g}(P)$ is not always characterized.

- In the d^3 and d^8 octahedral complexes, $10Dq$ value is given by the lowest energy ligand field band (*i.e.* $\nu_1 = 10Dq$). But, $10Dq$ calculated from the next higher energy band assuming $\nu_2 = 18Dq$ is not correct.
- In the d^3, d^8 (O_h) systems, if all the three ligand field bands are characterized, then $10Dq$, B' and configurational interaction energy can be calculated easily.

52. $[Cr(OH_2)_6]^{3+}$ shows the following spectral features: two intense bands at 17 kK, and 24 kK; a weak band at 37 kK as a shoulder of a strong CT band; and two weak bands at 15 kK and 22 kK.
53. $[Cr(en)_3]^{3+}$ belongs to the D_3 point group while $[Cr(NH_3)_6]^{3+}$ belongs to the O_h point group. But their spectral characteristics are more or less the same.
    - In moving from the O_h to the D_3 symmetry of Cr(III) complexes, the three spectral ligand field bands should be three doublet peaks.
    - In the $[Cr(ox)_3]^{3-}$ and $[Cr(en)_3]^{3+}$, the third ligand field band cannot be easily characterized.
    - Colour of $[Cr(en)_3]^{3+}$ is more intense than that of $[Cr(NH_3)_6]^{3+}$.
    - $[Cr(ox)_3]^{3-}$ is more intensely coloured than $[Cr(en)_3]^{3+}$
54. The *trans*-$[Cr(en)_2F_2]^+$ complex should have the 6 ligand field bands.
    - Nature of the ligand field bands can distinguish the *cis*- and *trans*- isomers of $[Cr(en)_2F_2]^+$. But the ligand field bands cannot distinguish the geometrical isomers of $[Cr(NCS)_2(OH_2)_4]^+$.
    - *Cis*-$[Cr^{III}A_4B_2]$ complexes are more intense in colour than their *trans* isomers.
55. Spin forbidden transition of Cr(III) in the O_h field can explain the action of ruby laser.
    - Ruby (obtained through the isomorphous substitution of Al^{3+} by Cr^{3+} in corundum) looks red while Cr_2O_3 is green.
    - Gem-stone emerald (obtained through the isomorphous substitution of Al^{3+} by Cr^{3+} in beryl) looks green.
    - Ruby red phosphorescence emission does not depend on the nature of ligands.
    - Position of the $^2E_g(G)$ level with respect to that of the $^4T_{2g}(F)$ is crucial to explain the action of .. ruby laser.
    - Many octahedral complexes of Cr(III) give the phosphorescence of ruby emission (*i.e.* red light emission).
56. Heteronuclear MMCT transition is responsible for the **blue colour of sapphire gemstone (blue corundum).**
57. Most of the Ni(II)-octahedral complexes are characterized by the presence of three ligand field bands.
    - Spectral transitions in $[Ni^{II}A_4B_2]$ (D_{4h} symmetry) can be roughly explained in terms of the ligand field transitions of the octahedral complex of Ni(II).
    - In $[Ni^{II}A_4B_2]$ (D_{4h} symmetry), very often the lowest energy ligand field band shows a doublet character.
    - The octahedral, tetrahedral and square planar geometries of Ni(II) can be distinguished by considering their spectral and magnetic properties.
    - In the Ni(II) octahedral complexes, the peak due to the transition, $^3A_{2g}(F) \rightarrow {}^3T_{1g}(P)$ is expected to be weak because it is a **forbidden transition**. But intensity of the band is not so weak compared to that of the other spin-allowed ligand field bands.

      (**Hints:** The forbidden transition gains the intensity through the intensity stealing phenomenon.)

58. Nature of the ligand field bands in $[Ni(NO_3)_4]^{2-}$ indicates its octahedral structure (*i.e.* some of NO_3^- groups act as the didendate chelating ligands).
59. In $[Ni(OH_2)_6]^{2+}$, the band due to the ${}^3A_{2g}(F) \rightarrow {}^3T_{1g}(F)$ transition is a doublet. The same thing happens for $[Ni(dmso)_6]^{2+}$).
    - Always, the transition ${}^3A_{2g}(F) \rightarrow {}^3T_{1g}(F)$ does not give the doublet peak for the O_h Ni(II) complexes.
60. The 3rd ligand field band in $[Ni(bpy)_3]^{2+}$ is missing.
61. The absorption spectrum of $[Co(OH_2)_6]^{2+}$ shows the following features:
    One peak at 8.1 kK; a broad peak at about 20 kK constituted by three overlapping bands of 16 kK, 19.4 kK and 21.6 kK.
    - In $[Co(OH_2)_6]^{2+}$, the peak from the **forbidden transition**, ${}^4T_{1g}(F) \rightarrow {}^4A_{2g}(F)$, is not so weak.
    - In $[Co(OH_2)_6]^{2+}$, the two energy states ${}^4A_{2g}(F)$ and ${}^4T_{1g}(P)$ are in the cross-over region.
62. For the spectral analysis of the high-spin complexes of Co(II), instead of three ligand field bands, two ligand field bands are very often well characterized.
63. Absorption spectrum of $[Cr(OH_2)_6]^{3+}$ shows two bands at 9.5 kK and 14 kK.
64. $[Mn(DMSO)_6]^{3+}$ shows three ligand field transitions at 14.5 kK, 17 kK and 20 kK.
    - In $[Mn(DMSO)_6]^{3+}$, there are two types of ν_{S-O} stretching frequencies.
65. $[CrF_6]^{4-}$ (in CrF_2 crystal) shows two peaks while $[MnF_6]^{3-}$ shows three peaks but both Cr(II) and Mn(III) represent the d^4 system.
66. $[CoF_6]^{3-}$ shows a broad peak with a doublet splitting.
67. $[Fe(OH_2)_6]^{2+}$ shows a doublet peak.
68. The colours of $[Mn(OH_2)_6]^{2+}$ and $[Fe(OH_2)_6]^{3+}$ are very faint.
69. In the high-spin d^5-system, some of the ligand field bands are very much sharp.
70. In the d^5-system, the ${}^4T_{1g}(P)$ and ${}^4T_{2g}(D)$ states are of sigmoidal nature.
71. $[Fe(OH_2)_6]^{3+}$ in aq. $HClO_4$ media is almost colourless but it becomes coloured in aq. HCl media.
    - $[Fe(OH_2)_6]^{3+}$ is colourless but its hydrolyzed species is brown coloured.
72. Low-spin t_{2g}^6 gives the ${}^1A_{1g}$ state but the excited state $t_{2g}^5e_g^1$ gives the states, ${}^3T_{1g}$, ${}^3T_{2g}$, ${}^1T_{1g}$ and ${}^1T_{2g}$.
73. The low-spin Co(III) complexes show two spin-allowed ligand field bands and two spin-forbidden ligand field bands.
    - The spin-forbidden transitions are of lower energy while the spin-allowed transitions are of higher energy.
74. The low-spin Co(III) complexes and Cr(III) complexes show the similar spectral transitions.
75. For the low-spin Co(III) complexes, energy difference between the spin-allowed ligand field bands is approximately $16B'$.
76. Cubic (*i.e.* octahedral), tetragonal and rhombic symmetries of the low-spin Co(III) complexes can be distinguished by considering their ligand field spectra.
    - Cubic: two peaks; tetragonal: three peaks; rhombic: three peaks.
77. *cis*-$[CoCl_2(NH_3)_4]^+$ and *trans*-$[CoCl_2(NH_3)_4]^+$ can be distinguished by their ligand field spectra.
78. *fac-mer* isomers of $[CoCl_3(NH_3)_3]$ can be distinguished by their absorption spectra.
79. *trans*-$[Co(en)_2F_2]^+$ gives three spectral bands while *cis*-$[Co(en)_2F_2]^+$ gives two peaks.
    - Spectral features of the geometrical isomers of $[Co(en)_2F_2]^+$ are:
      cis-: 19 kK (ε = 85), 27.5 kK (ε = 50)
      trans-: 17.5 kK (ε = 20), 22.5 kK (ε = 30), 27.8 kK (ε = 50).

80. The *cis- trans* isomers of $[Co^{III}A_4B_2]$ (low-spin) may not be always distinguished by their absorption spectra.
   - The low spin *trans*-$[Co^{III}A_4B_2]$ complexes show the vibronic polarisation in terms of the spectral behaviour.
   - Polarisation studies (on spectral transitions) can be done by using the single crystal in solid state but not in solution.
   - They can be distinguished from their absorption spectra only when the ligands *A* and *B* are positioned widely apart in the spectrochemical series of the ligands.
81. The isomers of $[Co(gly)_3]$ can be distinguished by their absorption spectra.
82. The ligand field spectra of $[Co(NH_3)_6]^{3+}$, $[CoCl(NH_3)_5]^{2+}$ and $[CoCl_2(NH_3)_4]^+$ are of different types.
83. Though $[Co(NH_3)_6]^{3+}$ and $[Co(en)_3]^{3+}$ belong to the different point groups, their absorption spectra are of similar type. Only they differ in intensity slightly.
   - Electronic spectra of $[Co(ox)_3]^{3-}$ and $[Co(en)_3]^{3+}$ are very much comparable but their intensities and peak separations are different.
84. $K_4[Fe(CN)_6]$ gives two ligand field peaks at 31 kK and 37 kK.
85. In $[CuCl_4]^{2-}$, $[CuBr_4]^{2-}$, the ligand field band is broad.
   - Spectral behaviour of $[CuCl_4]^{2-}$ in $Cs_2[CuCl_4]$ can be best explained in terms of D_{2d} symmetry of $[CuCl_4]^{2-}$
   - In the tetrahedral complexes, the $e \rightarrow t_2$ transition does not really represent the *d–d* transition.
86. Tetrahedral complexes of Co(II) generally show two peaks in the visible range and one peak is more intense.
87. Pale pink colour of Co(II) aqueous solution becomes dark blue-green in presence of HCl.
88. $[CoCl_4]^{2-}$ gives a complex envelope of different spectral bands.
89. The peaks of $[CoCl_4]^{2-}$ in the visible range are broad and having the fine structure components.
90. Almost colourless aqueous solution of Mn(II) becomes slightly intensified on addition of HCl.
91. Both the spin-forbidden transitions and peak splitting by spin-orbit coupling are responsible to broaden the peaks to produce a complex envelope of peaks for the tetrahedral complexes of Co(II).
92. The octahedral, tetrahedral and square planar complexes of Co(II) can be distinguished by comparing their electronic spectra and magnetic moments.
93. Tetrahedral complexes of Ni(II) are generally characterized by one intense ligand field band in the visible range.
   - For the tetrahedral complexes of Ni(II), one ligand field band lies in the infra-red region, one weak band lies in the near-infrared region and one strong band lies in the visible range.
   - For $[NiCl_4]^{2-}$, the strong band at 15 kK shows the fine-structure components.
94. Most of the square planar complexes of Ni(II) are red or orange and the intense spectral band appears in the range 15 kK–25 kK and there is another more intense band that lies in the range 25 kK–35 kK.
95. Octahedral complexes of Ni(II) are generally blue-green, tetrahedral complexes of Ni(II) are brown, green while their square planar complexes are orange, red.
96. Spectral bands of the square planar complexes of Ni(II) are generally highly intense.
   - Square planar complexes of d^8 system are expected to give three ligand field bands.

97. Thermochromism is well documented among the Ni(II) complexes.
98. Ni(II) generally adopts the square planar geometry in presence of very strong field ligands but sometimes, the weak field S-donor ligands can favour the formation of square planar complexes of Ni(II).
99. In $[PdCl_4]^{2-}$, there are two ligand field bands at about 23 kK and 30 kK. The higher energy band is more intense.
100. N-substituted salicylaldiminato complexes of Ni(II) in solution give three absorption bands at: ~7.2 kK, ~ 11.2 kK and ~ 16 kK. With the increase of temperature of the solution, intensity of the band at 16 kK decreases while intensity of the other two bands increases.
101. At a lower temperature, $(NR_xH_{4-x})_2[NiCl_4]$ (x = 1,2,3) appears yellow-brown or green but with increasing the temperature, it becomes blue.
102. Lifschitz salts, *i.e.* $[Ni(N–N)_2]X_2$ (N–N = en or substituted en) are yellow in the noncoordinating solvents and blue in the coordinating solvents.
103. Solution of $[Ni(PR_3)_2X_2]$ shows the absorption peaks at 400 nm, 550 nm and 880 nm. On cooling, intensity of the peak at 400 nm increases while intensity of the other two peaks decreases.
104. In the noncoordinating solvents, $[Ni(trien)]^{2+}$ looks yellow while it becomes blue in the coordinating solvents.
105. [Co(salen)] and [Co(acacen)] complexes are characterized by the two spectral peaks at ~ 8,500 cm^{-1} and ~ 20,000 cm^{-1}.
106. Compared to the 1st transition series, it is more difficult to interpret the ligand field bands for the 2nd and 3rd transition series.
107. $[MoCl_6]^{3-}$ shows two absorptions peaks at 19,000 cm^{-1} and 24,000 cm^{-1} in the visible range. The 3rd ligand field band is probably obscured by the CT bands.
108. The orange-yellow-red complexes of Rh(III) are characterized by two absorption peaks.
109. The absorption peaks of the lanthanides are generally sharp (Rydberg type) and weak.
    - Except the lanthanides (M^{3+}) having f^0, f^1, f^7 and f^{13} and f^{14} configurations, all other lanthanides are coloured.
    - The colour of a particular lanthanide ion does not depend on the nature of the ligands.
    - For the lanthanide ions, a large number of absorption peaks are noticed in the higher energy end.
    - The $4f \rightarrow 5d$ transition peaks are red shifted for Ln(II) compared to those of Ln(III).
    - The magnetic dipole (MD) transitions may contribute to the $f \rightarrow f$ transitions in Ln^{3+}.
    - Luminescence property is common for the lanthanide ions while it is rarely observed for the transition metal ions.
    - Absorption peaks of Gd(III) are of very poor intensity.
    - For the lanthanide ions, spin-orbit coupling splitting of the energy states is more important than crystal field splitting of the energy states. For the 3*d*-metal ions, the reverse is true.
    - In the commercial uses, lanthanide compounds are replacing the ruby lasers and phosphors
    - Ce(IV) is a f^0-system, but its complexes are generally coloured.
    - CT bands are relatively less important for the lanthanide ions.
110. Compared to the lanthanides, the absorption peaks are more intense and broader for the actinides.
    - In terms of the spectroscopic properties, the actinides can be broadly classified into two groups.

- Early members of the actinides are very much similar to the transition metal ions rather than to the lanthanides in terms of chemical properties (*e.g.* variable oxidation states) and spectral properties.
- For the lanthanides, the $0 \rightarrow 0$ transition (*i.e.* $\upsilon = 0 \rightarrow \upsilon' = 0$) gives the most intense transition while this transition is forbidden (in general) for the transition metal complexes.
- The heavier actinides are similar to the lanthanides in terms of their spectral properties.
- The lanthanides (Ln^{3+}) have some unique properties to favour their luminescence properties.
- In **Nd:YAG laser**, the long-lived excited state ($^4F_{3/2}$) of Nd^{3+} plays the crucial rule.
- Ln^{3+} ions are used in phosphors used in displays.
- YVO_4 coating is used on the interior surface of fluorescent light tubes.
- Nd(III) salt is incorporated into the goggles of glass-blowers.
- The $5f \rightarrow 6d$ transitions in U^{3+}, Np^{3+} and Pu^{3+} occur in different ranges of electromagnetic radiation.
- UO_2^{2+} gives the intense yellow colour.
- UO_2^{2+} strongly fluoresces.
- Uranyl salt is utilised for colouring glass.

111. In terms of the spectral properties, Prussian blue and Turnbull blue are identical.
    - The product obtained through the mixing of Fe(II)-salts and $K_4[Fe(CN)_6]$ or mixing of Fe(III)-salts and $K_3[Fe(CN)_6]$ are colourless or light coloured.
    - Creutz-Taube ion shows an intense band in the near infrared region.
    - $[Pt(NH_3)_4]^{2+}$ is colourless and $[PtCl_4]^{2-}$ is red, but Magnus green salt constituted by these ions is green.
    - $[Pt(NH_3)_4][PtCl_4]$ (*i.e.* Magnus green salt) shows a colour which is different from the sum of the colours of the constituent ions. But $[Pt(EtNH_2)_4][PtCl_4]$ shows a colour which is the sum of the colours of the constituent ions.
    - Molybdenum blue or tungsten blue shows the colour due to the metal to metal charge transfer (MMCT)
    - Ruthenium blue, ruthenium red and ruthenium purple experience the MMCT/IVCT transitions.
    - $[Re_2Cl_8]^{2-}$ and $[Mo_2Cl_8]^{4-}$ are intensely coloured.
112. The charge transfer bands are more intense than the ligand field bands (*i.e. d-d* transitions).
113. Fe^{III}–SCN^-, Fe^{III}–phenolate, Fe^{III}–hydroxamate, V^V–hydroxamate complexes are highly coloured.
    - In $[Co^{III}(NH_3)_5X]^{2+}$ (X = F, Cl, Br, I), for X = F, the ligand field bands are not influenced by the CT band but for X = Br, I, the high energy ligand field band is overlapped by the CT band.
    - In $[Co^{III}(NH_3)_5X]^{2+}$ (X = F, Cl, Br, I), intensity of the ligand field bands is maximum for X = I.
    - In moving from $[Co(NH_3)_6]^{3+}$ (O_h) to $[Co(NH_3)_5X]^{2+}$ (C_{4v}), the octahedral ligand field bands undergo splitting and this peak splitting is maximum for X = I and minimum for X = F.
    - In $[Co(NH_3)_5X]^{2+}$, for X = I, all the ligand field bands are practically influenced by the LMCT band; for X = Cl, Br, only one ligand field band is affected.
    - $[VO_4]^{3-}$ is colourless, $[CrO_4]^{2-}$ is yellow while $[MnO_4]^-$ is pink.
    - $[ReO_4]^-$ is colourless but $[MnO_4]^-$ is coloured.

- $[CrO_4]^{2-}$ is coloured by $[MoO_4]^{2-}$, $[WO_4]^{2-}$ are colourless.
- The tetrahedral oxyanions like SiO_4^{2-}, PO_4^{3-}, SO_4^{2-} and ClO_4^- are colourless while CrO_4^{2-} and MnO_4^- are coloured.
- The oxidising power sequence, $[MnO_4]^- > [CrO_4]^{2-} > [VO_4]^{3-}$; $[MnO_4]^- > [ReO_4]^-$; $[CrO_4]^{2-} > [MoO_4]^{2-} > [WO_4]^{2-}$ can be explained in terms of the relative positions of their CT bands.
- $[CrO_4]^{2-}$ is harder than $[MnO_4]^-$ and the hardness is related with the ν_{CT} values.
- Colour of $[Fe(aq)]^{3+}$ depends on the pH of solution.
- Compounds of d^0 and d^{10} systems may be coloured.

114. Iodido complex of Cu(II) is unstable while the other halido complexes of Cu(II) are highly coloured and stable.
   - Iodido complex of Fe(III) is unstable while the other halido complexes of Fe(III) are dark coloured and stable (*cf.* fluorido complex of Fe(III) is stable but colourless).
   - Aqua-complex of Co(III) is unstable.
   - $[Fe(CN)_6]^{3-}$ shows the first LMCT band at a relatively longer wavelength compared to $[Fe(CN)_6]^{4-}$.
   - The CT band of $[IrBr_6]^{2-}$ is of lower energy compared to that of $[IrBr_6]^{3-}$.
   - $[VCl_6]^{2-}$ is stable but $[VBr_6]^{2-}$ is unstable.
   - The energy of the CT bands: $[OsCl_6]^{2-} > [OsBr_6]^{2-} > [OsI_6]^{2-}$; $[TiCl_6]^{2-} > [TiBr_6]^{2-}$
   - Energy of the LMCT band: $[HgCl_4]^{2-} > [HgBr_4]^{2-} > [HgI_4]^{2-}$.
   - $[TiX_6]^{2-}$ (X = Cl, Br) shows two CT bands. The lower energy band is relatively narrower and the band-width of the higher energy band is relatively larger.
   - Energy of the CT bands follows the sequence:
     $[ZrCl_6]^{2-} > [TiCl_6]^{2-} > [VCl_6]^{2-}$; $[HgCl_4]^{2-} > [HgBr_4]^{2-} > [HgI_4]^{2-}$; $[ReO_4]^- > [TcO_4]^- > [MnO_4]^-$

115. Fe(II)-bpy or phen complex is deep red coloured while the corresponding Fe(III)-complex shows a faint blue colour.
   - Cu(I)-bpy/phen complex shows an intense colour.
   - $[Cr(CO)_6]$ shows two MLCT bands.

116. For the complex, $[Ir^{III}(NH_3)_5X]$ ($X = NH_3$, H_2O, Cl^- and Br^-), there is no problem in assigning the ligand field bands for $X = NH_3$ and H_2O but the task is complicated for X = Cl and Br.
   - For X = Cl, Br, the ligand field peaks are more intense than for $X = NH_3$ and H_2O.

117. For the tetrahedral complexes, the ligand field transition represented by $e \rightarrow t_2$ is not a correct description, but for the octahedral complexes, the representation, $t_{2g} \rightarrow e_g$ is correct.
   - Colour of the complex cation $[M(NH_3)_6]^{3+}$ (M = Co, Ir) in solution may depend on the nature of anions present (**Hints:** IPCT band).
   - $[Ir(NH_3)_6]^{3+}$ is colourless but it looks yellow coloured in presence of I^-.

118. In the tetrahedral complexes, for the d^{0-6} system, 4 CT bands can arise; for the $d^{7,8}$ systems, 3 CT bands can arise while for the d^{10} system, only one CT band can arise.
   - In $[Cr(CO)_6]$ and $[Mo(CO)_6]$, the strong CT bands mask the expected two ligand field bands.
   - In $[Fe^{II}(L)_3]$, the MLCT band appears at about 19,500 cm^{-1} while in $[Ni^{II}(L)_3]$, the MLCT band appears at much higher energy (L = bpy, phen).
   - In $[M(CN)_4]^{2-}$ (M = Ni, Pd, Pt), there are three MLCT bands.
   - $[Fe(CN)_6]^{3-}$ shows two LMCT bands while $[Fe(CN)_6]^{4-}$ shows one LMCT band. The CT band in $[Fe(CN)_6]^{3-}$ appears at a relatively lower energy compared to that in $[Fe(CN)_6]^{4-}$.

119. The spectroscopic states in $[Co(en)_3]^{3+}$ and $[CoCl_2(en)_2]^+$ (both *cis*- and *trans*-isomers) are different.
    - Polarisation studies on the electronic transitions for *trans*-$[CoCl_2(en)_2]^+$ can identify the energy order of the ^1E_g and $^1A_{2g}$ levels (generated from the $^1T_{1g}$ level of O_h symmetry).
    - The *cis*- and *trans*- isomers of $[CoCl_2(en)_2]^+$ belong to the different point groups, *i.e.* C_{2v} and D_{4h} point groups (by considering the local symmetry) respectively but their spectral transitions can be explained by considering the tetragonal symmetry in both cases.
    - For $[Co^{III}A_4B_2]$ (low spin), the peak splitting is more, if the ligand field strengths of the ligands A and B differ widely.
    - For $[Co^{III}A_4B_2]$ (low spin), the peak splitting is more for the *trans*-complex compared to that in the *cis*-complex.
    - For $[Co^{III}A_5B]$ (low spin), the octahedral peak splitting becomes prominent, if the positions of A and B in the spectrochemical series are widely apart (*cf.* $[CoF(NH_3)_5]^{2+}$ *vs.* $[CoI(NH_3)_5]^{2+}$.
120. The band theory of solids can explain the colours of the **semiconductor type solids** (*e.g.* oxides, sulfides, selenides and halides) and **shiny metallic lustre** of the metal surfaces.
    - Most of the metals show the shiny metallic lustre but some metals like copper and gold show some specific colours.
    - ZnO, ZnS, AgCl, TiO_2, Al_2O_3, etc. ($E_g \geq 3.2$ eV) are colourless.
    - CuO, HgS (β-polymorph), PbS, CdSe, GaAs, MnO_2, etc. ($E_g \leq 1.75$ eV) are black.
    - AgI, CdS, HgS (α-polymorph) etc. (3.2 eV > E_g > 1.75 eV) are coloured
    - SnO_2 is colourless but PbO_2 is maroon coloured.
    - $NaBiO_3$ is light yellow coloured but $NaSbO_3$ is colourless.
    - $[PbCl_6]^{2-}$ is yellow coloured but $[SnCl_6]^{2-}$ is colourless. (**Hint: Relativistic red shift of the LMCT band**).
    - AgCl (white), AgBr (light yellow), AgI (deep yellow).
    - CdO (brown), CdS (yellow) but CdSe (black).
    - ZnO (white), CdO (brown), HgO (red).
    - PbF_2 (colourless), $PbCl_2$ (white), $PbBr_2$ (colourless), PbI_2 (yellow).
    - SnF_2 (colourless), $SnCl_2$ (colourless), $SnBr_2$ (white), SnI_2 (red).
    - PbO_2 has two forms: maroon coloured and black.
    - PbO has two forms: red (litharge) and yellow (massicot).
    - HgS has two forms: red and black forms.
    - CdX (X = O, S, Se) may have different colours depending on the particle size.
    - $HgCl_2$ (white) but HgI_2 (red).
    - HgI_2 (solid, red) in presence of excess KI solution forms the colourless $[HgI_4]^{2-}$ anion.
    - Hg^{2+} (solution) $\xrightarrow{\text{KI}}$ HgI_2 (↓) (red precipitate) $\xrightarrow{\text{Excess KI}}$ Colourless solution.
    - Bi^{3+} (solution) $\xrightarrow{\text{KI}}$ BiI_3(↓) (black precipitate) $\xrightarrow{\text{Excess KI}}$ $[BiI_4]^-$ (orange).
    - $BiPh_5$ (violet) while $SbPh_5$ (colourless) (*cf.* **relativistic effect** and red shift of LMCT).
    - CuO (black) but Cu_2O (red).
    - Both the semiconductor type solids like diamond, TiO_2, etc. with the higher band gap ($E_g \geq$ 3.2 eV) and the metals reflect back all the incident visible light but the **reasons are different**.

121. Transition metals as the impurities give the characteristic colour in: **red corundum, green coloured gemstone emerald, alexandrite gemstone (greenish-yellow colour), red garnet, yellow-green colour of peridot, blue-green aquamarine, pink morganite, blue spinels, ruby red spinel, cobalt blue glass, amethyst (violet colour), blue sapphire gemstone (blue corundum)** (*Ref.* Das *et al., Resonance*, 2025).
122. Colours of **blue and yellow diamond** are due to some impurity dopants (*Ref.* Das *et al., Resonance*, 2025).
123. Charge transfer transitions within the polysulfide anion radicals (S_n^-) give the colour of both synthetic **ultramarine blue and natural lapis lazuli** (*Ref.* Das *et al., Resonance*, 2025)
124. **Thenard blue** observed in cobalt nitrate test of Al-salts is due to the formation of $CoAl_2O_4$ **(blue spinel)** (*Ref.* Das *et al., Resonance*, 2025).
125. **Rinmann green** (cobalt green) observed in cobalt nitrate test of Zn-salt is due to the formation of the mixed oxide $Zn_{1-x}Co_xO$ (where $x < 0.3$). But for $x = 0.7$, it looks pink (*Ref.* Das *et al., Resonance*, 2025).
   Hints: For $x < 0.3$, it adopts the **wurtzite structure** (4:4) of the host ZnO in which Co(II) is doped in place of Zn(II) and it looks green due to the $d \rightarrow d$ transitions in the $Co^{II}O_4$ tetrahedra. For $x > 0.7$, the mixed oxide adopts the **rock salt structure** (6:6) of the host CoO in which Zn(II) is doped in place of Co(II) and it looks pink due to the $d \rightarrow d$ transitions in the $Co^{II}O_6$ octahedra (*Ref.* Das *et al., Resonance*, 2025).
126. Colour of **cobalt blue glass** is due to the tetrahedral 'CoO_4' units present in the tetrahedral silicate matrix.

APPENDICES

APPENDIX I

Units and Conversion Factors

Table 1: Basic physical quantities in SI units

Physical quantity	*Symbol for quantity*	*Name of unit*	*Symbol for unit*
Length	l	metre	m
Mass	m	kilogram	kg
Time	t	second	s
Electric current	I	ampere	A
Temperature	T	kelvin	K
Amount of substance	n	mole	mol
Luminous intensity	I_v	candela	cd

Table 2: Derived physical quantities in SI units

Physical quantity	*Name of SI unit*	*Symbol and definition of SI unit*	*Named after*
Force, weight	newton (N)	$N = kg\ m\ s^{-2} = J\ m^{-1}$	Isaac Newton (1642–1727)
Work, energy, quantity of heat	joule (J)	$J = N\ m$	James Prescott Joule (1818–1889)
Power	watt (W)	$W = J\ s^{-1}$	James Watt (1737–1819)
Pressure	pascal (Pa)	$Pa = N\ m^{-2}$	B. Pascal (1623–1662)
Electrical charge	coulomb (C)	$C = A\ s$	Charles Auguste de Coulomb (1736–1806)
Electrical potential	volt (V)	$V = kg\ m^2\ s^{-3}\ A^{-1}$ $= J\ A^{-1}\ s^{-1}$ $= J\ C^{-1} = W\ A^{-1}$	Allesandro Volta (1745–1827)
Electric capacitance	farad (F)	$F = C\ V^{-1}$	Michael Faraday (1791–1867)
Electric resistance	ohm (Ω)	$\Omega = V\ A^{-1}$	Georg Simon Ohm (1787–1854)
Electric conductance	siemens (S)	$S = \Omega^{-1}$	W. von Siemens (1816–1892)
Magnetic flux	weber (Wb)	$Wb = V\ s$	W. Weber (1804–1891)
Magnetic flux density	tesla (T)	$T = Wb\ m^{-2}$	Nikola Tesla (1856–1943)
Inductance	henry (H)	$H = V\ s\ A^{-1}$	Joseph Henry (1799–1878)
Frequency	hertz (Hz)	$Hz = s^{-1}$	Heinrich Hertz (1857–1894)
Radioactivity	becquerel (Bq)	$Bq = s^{-1}$	A. Henri Becquerel (1852–1908)
Absorbed dose of radiation	gray (Gy)	$Gy = m^2\ s^{-2}$ $= J\ kg^{-1}$	Louis H. Gray (1905–1965)
Area (A)	square metre	m^2	
Volume (V)	cubic metre	m^3	
Density (ρ)	kilogram per cubic metre	$kg\ m^{-3}$	
Velocity (u, v, w, c)	metre per second	$m\ s^{-1}$	
Acceleration	metre per square second	$m\ s^{-2}$	
*Concentration (c)	mole per cubic metre	$mol\ m^{-3}$	

*for concentration (c) in mol per litre, it is given by mol dm^{-3}

Table 3: SI prefixes

Fraction	*Prefix*	*Symbol*	*Multiple*	*Prefix*	*Symbol*
10^{-1}	deci	d	10	deka	da
10^{-2}	centi	c	10^{2}	hecto	h
10^{-3}	milli	m	10^{3}	kilo	k
10^{-6}	micro	μ	10^{6}	mega	M
10^{-9}	nano	n	10^{9}	giga	G
10^{-12}	pico	p	10^{12}	tera	T
10^{-15}	femto	f	10^{15}	peta	P

Table 4: CGS and SI units for some common physical quantities

Physical quantity	*CGS units*		*SI Units*	
	Name	*Symbol*	*Name*	*Symbol*
Length	centimetre	cm	metre	m
	Angstrom (10^{-8} cm)	Å		
Mass	gram	g	kilogram	kg
Time	second	sec	second	s
Temperature	celsius	°C	kelvin	K
	kelvin	°K		
Energy	calorie	cal	joule	J
	kilocalorie	kcal	kilojoule	kJ
	litre-atmosphere ergs	lit-atm erg		

* tonne (t) = 10^3 kg, lb = 453.59 g, Btu = 1.055×10^{10} erg = 252 cal

Table 5: Conversion of CGS units to SI units

Physical quantity	*Relation between the units*
Length	Angstrom (Å) = 10^{-10} m = 10^{-1} nm = 10^{2} pm
	micron (μ) = 10^{-6} m
Volume	llitre (L) = dm^3
Force	dyne = 10^{-5} N
Energy*	erg = 10^{-7} J
	cal = 4.185 J
	eV = 1.602×10^{-19} J; eV/molecule = 96.484 kJ mol^{-1}
Pressure	atm = 1.013×10^{6} dyne cm^{-2} = 101.33 kN m^{-2}
	atm = 760 torr
	mm Hg or torr = 133.32 N m^{-2}
	bar (10^6 dynes/cm^2) = 10^5 N m^{-2}
Viscosity	poise = 10^{-1} kg m^{-1} s^{-1}
Magnetic flux density	gauses = 10^{-4} T

* 1 cm^{-1} (wave number) = 2.86 cal mol^{-1}, reciprocal centimetre is known as **Kayser (K)** 1 kK (kilo Kayser) = 1000 cm^{-1}, 1 eV per molecule = 23.06 kcal mol^{-1}, 1 erg = 2.39×10^{-11} kcal. 1 eV = (charge of an electron) × 1 V = 1.6×10^{-19} C × V = 1.6×10^{-19} J,

$$1\ \text{eV/molecule} = \frac{1.6\times10^{-19}\times6.02\times10^{23}}{1000}\ \text{kJ/mole} = 96.488\ \text{kJ mol}^{-1}.$$

$$1\ \text{kK molecule}^{-1} = 10^3\ \text{cm}^{-1}\ \text{molecule}^{-1}$$

$$= hc \times 10^3\ \text{cm}^{-1}\ \text{molecule}^{-1} \left(cf.\ E = h\nu = \frac{hc}{\lambda}\right)$$

$$= 6.62 \times 10^{-34}\ \text{J s} \times 2.99 \times 10^{8}\ \text{m s}^{-1} \times 10^{5}\ \text{m}^{-1}\ \text{molecule}^{-1}$$

$$= 1.985 \times 10^{-20}\ \text{J molecule} = 11.96\ \text{kJ mol}^{-1}$$

Table 6: Atomic Units (*see* text, Vol. 1, Chapter 1, Sec. 1.8.2)

APPENDIX II
Some Physical and Chemical Constants

Table 1: Values of some physical and chemical constants

Constant	*CGS and other units*	*SI units*
Acceleration due to gravity (g)	980.665 cm sec^{-2}	9.80665 m s^{-2}
Avogadro's number (N_A)	6.02217×10^{23} molecules $mole^{-1}$	6.02217×10^{23} mol^{-1}
Bohr magneton (μ_B)	9.274×10^{-21} erg $gauss^{-1}$	9.274×10^{-24} J T^{-1}
Nuclear Magneton (μ_N)	5.04×10^{-24} erg G^{-1}	5.04×10^{-27} J T^{-1}
Bohr radius (a_o)	0.52918 Å	5.2918×10^{-11} m
Boltzmann constant (k_B or k)	1.3806×10^{-16} erg (degree K)$^{-1}$	1.3806×10^{-23} J K^{-1}
Debye	10^{-18} esu cm	3.3356×10^{-30} C m
Electronic charge (e)	4.80298×10^{-10} esu	1.60216×10^{-19} C
Electronic rest mass (m_e)	9.10953×10^{-28} g	9.10953×10^{-31} kg
Faraday (F)	96487 coulomb $equiv^{-1}$	9.6487×10^{4} C mol^{-1}
Gas constant (R)	8.3144×10^{7} erg (degree K)$^{-1}$ $mole^{-1}$	8.3144 J K^{-1} mol^{-1}
	8.31441 Joules (degree K)$^{-1}$ $mole^{-1}$	8.3144 N m K^{-1} mol^{-1}
	82.053×10^{-3} litre-atm (degree K)$^{-1}$ $mole^{-1}$	8.3144 Pa m^3 K^{-1} mol^{-1}
	1.987 cal (degree K)$^{-1}$ $mole^{-1}$	
Permitivity of vacuum (ε_o)		8.854188×10^{-12} C V^{-1} m^{-1} ($C V^{-1} m^{-1} = C^2 N^{-1} m^{-2}$)
Permeability of vacuum (μ_o)		$4\pi \times 10^{-7}$ H m^{-1}
Planck's constant (h)	6.62618×10^{-27} erg sec	6.62618×10^{-34} J s
Rest mass of proton (m_p)	1.672648×10^{-24} g	1.672648×10^{-27} kg
Rest mass of neutron (m_n)	1.674954×10^{-24} g	1.674954×10^{-27} kg
Atomic mass unit (u) ($1\ u = 10^{-3}$ kg mol^{-1}/N_A)	1.660565×10^{-24} g	1.660565×10^{-27} kg (≡ 931.48 MeV)
Vacuum speed of light (c)	2.99793×10^{10} cm sec^{-1}	2.99793×10^{8} m s^{-1}
Standard atmospheric pressure	76 cm Hg	101.325 kPa
	760 mm Hg (or torr)	1.01325 bar
	1.1032×10^{6} dyne cm^{-2}	
Hartree energy (E_h)		4.3597×10^{-18} J
Zero of the Celsius scale	0°C	273.15 K

APPENDIX III
Wavelength and Colours

Table 1: Spectral wavelength and colours

Range of the wavelength absorbed (nm)	*Colour of the radiation absorbed*	*Complementary colour observed*
400–450	Violet	Green-yellow
450–470	Indigo	Yellow
470–490	Blue	Orange
490–500	Blue-green	Red
500–560	Green	Purple
560–575	Green-yellow	Violet
575–590	Yellow	Indigo
590–640	Orange	Blue
640–730	Red	Blue-green

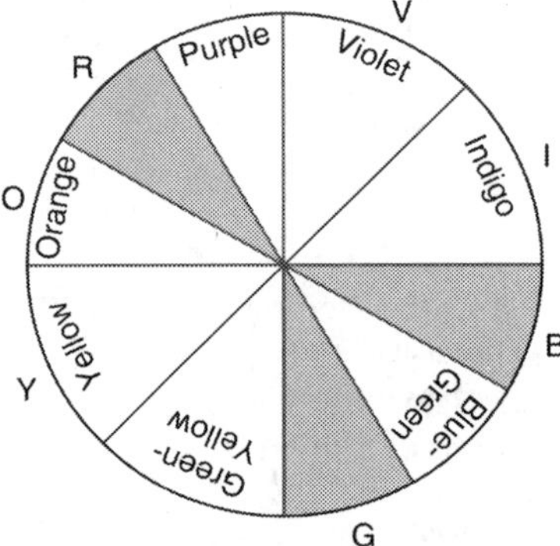

Artist's Colour Wheel: Colours and complementary colours.

Colours that stand opposite each other on the colour wheel are called as the **complimentary colours.** The amount of a colour in a particular image can be decreased by increasing its complimentary colour and vice-versa.

● **RGB colour model:** In this model, **red (R), green (G), and blue (B)** are considered as the **three primary colours** that can generate all the colours of the visible spectrum when mixed together in different proportions. Mixing of these three primary colours in equal parts gives **white** and complete absence of these three colours results in **black**. Computer monitor devices use these **three additive primaries** to produce the different colour shades. For example, B + R = M (**magenta**), R + G = Y (**yellow**), B + G = C (**cyan** means 'brilliant greenish-blue', *i.e.* turquoise), R + G + B = **white**.

APPENDIX IV

Names, symbols, atomic numbers and atomic weights* of the elements

Element	*Symbol*	*Atomic number*	*Atomic weight*	*Element*	*Symbol*	*Atomic number*	*Atomic weight*
Actinium	*Ac*	89	227.03	Lead	*Pb*	82	207.19
Aluminium	*Al*	13	26.98	Lithium	*Li*	3	6.94
Americium	*Am*	95	(243)	Lutetium	*Lu*	71	174.97
Antimony	*Sb*	51	121.75	Magnesium	*Mg*	12	24.31
Argon	*Ar*	18	39.95	Manganese	*Mn*	25	54.94
Arsenic	*As*	33	74.92	Mendelevium	*Md*	101	(258)
Astatine	*At*	85	(210)	Mercury	*Hg*	80	200.59
Barium	*Ba*	56	137.34	Molybdenum	*Mo*	42	95.94
Barkelium	*Bk*	97	(247)	Neodymium	*Nd*	60	144.24
Beryllium	*Be*	4	9.01	Neon	*Ne*	10	20.18
Bismuth	*Bi*	83	208.98	Neptunium	*Np*	93	237.05
Boron	*B*	5	10.81	Nickel	*Ni*	28	58.71
Bromine	*Br*	35	79.91	Niobium	*Nb*	41	92.91
Cadmium	*Cd*	48	112.40	Nitrogen	*N*	7	14.01
Calcium	*Ca*	20	40.08	Nobelium	*No*	102	(259)
Californium	*Cf*	98	(251)	Osmium	*Os*	76	190.2
Carbon	*C*	6	12.01	Oxygen	*O*	8	16.00
Cerium	*Ce*	58	140.12	Palladium	*Pd*	46	106.4
Cesium	*Cs*	55	132.91	Phosphorus	*P*	15	30.97
Chlorine	*Cl*	17	35.45	Platinum	*Pt*	78	195.09
Chromium	*Cr*	24	52.01	Plutonium	*Pu*	94	(244)
Cobalt	*Co*	27	58.93	Polonium	*Po*	84	(209)
Copper	*Cu*	29	63.54	Potassium	*K*	19	39.10
Curium	*Cm*	96	(247)	Praeseodymium	*Pr*	59	140.91
Dysprosium	*Dy*	66	162.50	Promethium	*Pm*	61	(145)
Einsteinium	*Es*	99	(252)	Protactinium	*Pa*	91	231.04
Erbium	*Er*	68	167.26	Radium	*Ra*	88	226.03
Europium	*Eu*	63	151.96	Radon	*Rn*	86	(222)
Fermium	*Fm*	100	(257)	Rhenium	*Re*	75	186.2
Fluorine	*F*	9	19.00	Rhodium	*Rh*	45	102.91
Francium	*Fr*	87	(223)	Rubidium	*Rb*	37	85.47
Gadolinium	*Gd*	64	157.25	Ruthenium	*Ru*	44	101.07
Gallium	*Ga*	31	69.72	Samarium	*Sm*	62	150.35
Germanium	*Ge*	32	72.59	Scandium	*Sc*	21	44.96
Gold	*Au*	79	196.97	Selenium	*Se*	34	78.96
Hafnium	*Hf*	72	178.49	Silicon	*Si*	14	28.09
Helium	*He*	2	4.00	Silver	*Ag*	47	107.87
Holmium	*Ho*	67	164.93	Sodium	*Na*	11	22.99
Hydrogen	*H*	1	1.008	Strontium	*Sr*	38	87.62
Indium	*In*	49	114.82	Sulfur	*S*	16	32.06
Iodine	*I*	53	126.90	Tantalum	*Ta*	73	180.95
Iridium	*Ir*	77	192.2	Technetium	*Tc*	43	(99)
Iron	*Fe*	26	55.85	Tellurium	*Te*	52	127.60
Krypton	*Kr*	36	83.80	Terbium	*Tb*	65	158.92
Lanthanum	*La*	57	138.91	Thallium	*Tl*	81	204.37
Lawrencium*	*Lr*	103	(260)	Thorium	*Th*	90	232.04

(Contd.)

* Previously it was represented by *Lw*. Now as per IUPAC recommendation, it is represented by *Lr*.

Element	*Symbol*	*Atomic number*	*Atomic weight*	*Element*	*Symbol*	*Atomic number*	*Atomic weight*
Thulium	*Tm*	69	168.93	Xenon	*Xe*	54	131.30
Tin	*Sn*	50	118.69	Ytterbium	*Yb*	70	173.04
Titanium	*Ti*	22	47.90	Yttrium	*Y*	39	88.91
Tungsten	*W*	74	183.85	Zinc	*Zn*	30	65.37
Uranium	*U*	92	238.03	Zirconium	*Zr*	40	91.22
Vanadium	*V*	23	50.94				

* Values are based on the presently accepted ***physical scale*** of $^{12}C = 12.00000$ as the standard. The ratio of these atomic weights to those on older ***chemical scale*** assuming the value 16.00000 for the atomic weight of natural oxygen (composition : ^{16}O, 99.757%; ^{17}O, 0.039%; ^{18}O, 0.204%) is 1.000279. The values given within the parentheses are the mass numbers of the most commonly known isotopes.

The true atomic weight of natural oxygen is given by : $16 \times 0.99757 + 17 \times 0.00039 + 18 \times 0.00204 = 16.00447$.

$$\text{It gives : } \frac{\text{Physical atomic weight}}{\text{Chemical atomic weight}} = \frac{16.00447}{16.00000} = 1.000279$$

Presently the term ***atomic weight*** is replaced by ***atomic mass*** or ***relative atomic mass*** (RAM). These are expressed in terms of *amu* or *u* or *Da* (Dalton). These aspects have been discussed in Sec. 5.2.3.

** For the ***translawrencium*** elements with $Z > 103$, the currently used names (cf. Chemistry of the Elements by N.N. Greenwood and A. Earnshaw and IUPAC, 1997) are : $_{104}Rf$ (rutherfordium), $_{105}Db$ (dubnium), $_{106}Sg$ (seaborgium), $_{107}Bh$ (bohrium), $_{108}Hs$ (hassium), $_{109}Mt$ (meitnerium). However, other names to these elements have also been suggested from time to time. These are : $Z = 104$ (kurchatovium, *Ku*; dubnium, *Db*), $Z = 105$ (hahnium, *Ha*, neilsbohrium, *Ns*, joliotium, *Jl*), $Z = 106$ (rutherfordium, *Rf*), $Z = 107$ (neilsbohrium, *Ns*), $Z = 108$ (hahnium, *Hn*). The IUPAC (1977) recommendation, i.e. a ***hybrid Latin-Greek numerical method*** (cf. Sec. 8.6), is free from such confusions. The elements with $Z = 110, 111, 112, 114, 116, 118$ have been characterised.

APPENDIX V

Some Useful Mathematical Relationship

(A) Differentials

(u and v are the functions of x; a and m are constants).

(i) $\frac{dx}{dx} = 1$ (ii) $\frac{d}{dx}(au) = a\frac{du}{dx}$ (iii) $\frac{d}{dx}(u+v) = \frac{du}{dx} + \frac{dv}{dx}$

(iv) $\frac{d}{dx}x^m = m\,x^{m-1}$ (v) $\frac{d}{dx}\ln x = \frac{1}{x}$ (vi) $\frac{d\ln u}{dx} = \frac{1}{u}\frac{du}{dx}$

(vii) $\frac{d}{dx}(e^x) = e^x$ (viii) $\frac{d}{dx}(e^u) = e^u\frac{du}{dx}$ (ix) $\frac{d}{dx}(uv) = u\frac{dv}{dx} + v\frac{du}{dx}$

(x) $\frac{d}{dx}\left(\frac{u}{v}\right) = \frac{v\left(\frac{du}{dx}\right) - u\left(\frac{dv}{dx}\right)}{v^2}$ (xi) $\frac{d}{dx}(\sin x) = \cos x$ (xii) $\frac{d}{dx}(\cos x) = -\sin x$

(xiii) $\frac{d}{dx}(\tan x) = \sec^2 x$ (xiv) $\frac{d}{dx}(\sin u) = \cos u\frac{du}{dx}$ (xv) $\frac{d}{dx}(\cos u) = -\sin u\frac{du}{dx}$

(B) Integrals

(u and v are the functions of x; a and m are constants; for the indefinite integrals, there should be an addition of *arbitrary constant of integration*).

(i) $\int dx = x$ (ii) $\int au\,dx = a\int u\,dx$ (iii) $\int (u+v)dx = \int u\,dx + \int v\,dx$

(iv) $\int x^m\,dx = \frac{x^{m+1}}{m+1}, (m \neq -1)$ (v) $\int \frac{dx}{x} = \ln x$ (vi) $\int e^x dx = e^x$

(vii) $\int e^{ax}\,dx = \frac{1}{a}e^{ax}$ (viii) $\int \ln ax\,dx = x\ln ax - x$ (ix) $\int \sin x\,dx = -\cos x$

(x) $\int \cos x\,dx = \sin x$ (xi) $\int uv dx = u\int v dx - \int\left[\frac{du}{dx}\int v dx\right]dx$, (by parts)

(xii) $\int u dv = uv - \int v du$ (xiii) $\int_0^\infty x^n e^{-ax} = \frac{n!}{a^{n+1}}$, ($n$ positive integer)

(C) Trigonometric relationship

(i) $\sin(90^\circ - \theta) = \cos\theta$ (ii) $\cos(90^\circ - \theta) = \sin\theta$

(iii) $\exp(\pm im\theta) = \cos(m\theta) \pm i\sin(m\theta)$, (***Demovier's theorem***)

$\cos(m\theta) = \frac{1}{2}[\exp(im\theta) + \exp(-im\theta)]$ $\sin(m\theta) = \frac{1}{2i}[\exp(im\theta) - \exp(-im\theta)]$

(iv) $\sin^2\theta + \cos^2\theta = 1$ (v) $\sin 2\theta = 2\sin\theta\cos\theta$

(vi) $\cos 2\theta = \cos^2\theta - \sin^2\theta = 2\cos^2\theta - 1 = 1 - 2\sin^2\theta$ (vii) $\sin(\alpha \pm \beta) = \sin\alpha\cos\beta \pm \cos\alpha\sin\beta$

(viii) $\cos(\alpha \pm \beta) = \cos\alpha\cos\beta \mp \sin\alpha\sin\beta$ (ix) $\sin\alpha \pm \sin\beta = 2\sin\frac{1}{2}(\alpha \pm \beta)\cos\frac{1}{2}(\alpha \mp \beta)$

(x) $\cos\alpha + \cos\beta = 2\cos\frac{1}{2}(\alpha + \beta)\cos\frac{1}{2}(\alpha - \beta)$ (xi) $\cos\alpha - \cos\beta = -2\sin\frac{1}{2}(\alpha + \beta)\sin\frac{1}{2}(\alpha - \beta)$

(xii) $\sin\alpha\sin\beta = \frac{1}{2}\cos(\alpha - \beta) - \frac{1}{2}\cos(\alpha + \beta)$ (xiii) $\cos\alpha\cos\beta = \frac{1}{2}\sin(\alpha - \beta) + \frac{1}{2}\cos(\alpha + \beta)$

(xiv) $\sin\alpha\cos\beta = \frac{1}{2}\sin(\alpha + \beta) + \frac{1}{2}\sin(\alpha - \beta)$

(D) Some Useful Expansions ($x^2 \leq 1$)

$(1 + x)^{-1} = 1 - x + x^2 - x^3 + \ldots..$ $(1 - x)^{-1} = 1 + x + x^2 + x^3 + \ldots.$ $(1 - x)^{-2} = 1 + 2x + 3x^2 + 4x^3 + \ldots.$

$$\ln(1 + x) = x - \frac{x^2}{2} + \frac{x^3}{3} - \frac{x^4}{4} + \ldots.$$

$$\ln(1 - x) = -x - \frac{x^2}{2} - \frac{x^3}{3} - \frac{x^4}{4} \ldots.$$

$$\exp(x) \text{ or } e^x = 1 + x + \frac{x^2}{2!} + \frac{x^3}{3!} + \frac{x^4}{4!} + \ldots., \text{ (for all } x)$$

$$\exp(-x) \text{ or } e^{-x} = 1 - x + \frac{x^2}{2!} - \frac{x^3}{3!} + \ldots.$$

$$\sin x = x - \frac{x^3}{3!} + \frac{x^5}{5!} - \frac{x^7}{7!} + \ldots., \text{ (for all } x)$$

$$\cos x = 1 - \frac{x^2}{2!} + \frac{x^4}{4!} - \frac{x^6}{6!} + \ldots., \text{ (for all } x)$$

$\ln x! = x\ln x - x$ or $x! = \frac{x^x}{e^x}$ (**Stirling approximation,** 1730)
(for $x > ca.\,10$)

(E) Progressions

(i) **Arithmatic progression :** $a, a + d, a + 2d, ...$

n-th term $(T_n) = a + (n - 1)d;$

Sum of first n terms $(S_n) = \frac{n}{2}(T_1 + T_n) = \frac{n}{2}\{2a + (n-1)d\}$

Sum of first n-natural numbers $= 1 + 2 + 3 + ... + n = \frac{n(1+n)}{2}$

(ii) **Geometric progression :** $a, ar, ar^2, ...$

n-th term $(T_n) = ar^{n-1}$; Sum of first n terms $(S_n) = \frac{a(1-r^n)}{(1-r)}$ when $r < 1$

$S_n = \frac{a(r^n - 1)}{(r-1)}$ when $r > 1$; sum to infinity $(S_\infty) = \frac{a}{1-r}$ when $|r| < 1$

(iii) **Useful results :**

$$\left.\begin{aligned} \sum_{x=1}^{n} x^2 &= \frac{n}{6}(n+1)(2n+1) \\ \sum_{x=1}^{n} x^3 &= \left[\frac{n}{2}(n+1)\right]^2 \end{aligned}\right\} \text{For first } n \text{ natural numbers, 1, 2, 3, ..., } n$$

(F) Permutations and combinations

$${}^nP_r = \frac{n!}{(n-r)!};\ {}^nP_n = \frac{n!}{0!} = \frac{n!}{1} = n!;\ {}^nC_r = \frac{n!}{r!(n-r)!};\ {}^nP_r = r!\,{}^nC_r;\ {}^nC_r = {}^nC_{n-r};\ {}^nC_n = {}^nC_0 = 1$$

$$n! = 1 \times 2 \times 3 \times ...... \times (n-2) \times (n-1) \times n;\ 0! = 1$$

(G) Logarithms

$\log(mn) = \log m + \log n$; $\log\left(\frac{m}{n}\right) = \log m - \log n$,
$\log(m^n) = n \log m$; $\log 1 = 0$; $\log_a a = 1$; $\ln(x) = 2.3 \log x$

Natural logarithm (base e); common logarithm (base 10).

(H) Mathematical Constants

$\pi = 3.14159265.....$, $e = 2.71828....$, $\ln x = 2.30258 ... \log_{10}^{x}$

APPENDIX VI
Latimer Diagrams (Standard Reduction Potentials) of Elements

ACIDIC SOLUTION (E^0_A)

$H^+ \xrightarrow{0} H_2 \xrightarrow{-2.25} H^-$

Group 1 (s-Block)

$Li^+ \xrightarrow{-3.04} Li$, $Na^+ \xrightarrow{-2.71} Na$, $K^+ \xrightarrow{-2.92} K$, $Rb^+ \xrightarrow{-2.92} Rb$, $Cs^+ \xrightarrow{-2.92} Cs$

Group 2 (s-Block)

$Be^{2+} \xrightarrow{-1.97} Be$, $.Mg^{2+} \xrightarrow{-2.36} Mg$, $Ca^{2+} \xrightarrow{-2.84} Ca$, $Sr^{2+} \xrightarrow{-2.89} Sr$, $Ba^{2+} \xrightarrow{-2.92} Ba$, $Ra^{2+} \xrightarrow{-2.92} Ra$

Group 3 (d-Block)

$Sc^{3+} \xrightarrow{-2.03} Sc$, $Y^{3+} \xrightarrow{-2.37} Y$, $La^{3+} \xrightarrow{-2.38} La$, $Ac^{3+} \xrightarrow{-2.13} Ac$

Lanthanides (f-Block)

$Ce^{4+} \xrightarrow{1.76} Ce^{3+} \xrightarrow{-2.34} Ce$, $Pr^{4+} \xrightarrow{3.20} Pr^{3+} \xrightarrow{-2.35} Pr$

$Sm^{3+} \xrightarrow{-1.55} Sm^{2+} \xrightarrow{-2.67} Sm$; $Sm^{3+} \xrightarrow{-2.30} Sm$

$Eu^{3+} \xrightarrow{-0.35} Eu^{2+} \xrightarrow{-2.80} Eu$; $Eu^{3+} \xrightarrow{-1.99} Eu$

$Yb^{3+} \xrightarrow{-1.05} Yb^{2+} \xrightarrow{-2.80} Yb$; $Yb^{3+} \xrightarrow{-2.22} Yb$

$Lu^{3+} \xrightarrow{-2.30} Lu$

$M^{3+} \xrightarrow{-2.30} M$ (M = La, Ce, Pr, Nd, Pm, Sm, Gd, Tb, Dy, Ho, Er, Tm, Yb, Lu), $Eu^{3+} \xrightarrow{-1.99} Eu$

Actinides (f-Block)

$Ac^{3+} \xrightarrow{-2.13} Ac$, $Th^{4+} \xrightarrow{-1.83} Th$, $Pa^{4+} \xrightarrow{-1.47} Pa$,

$UO_2^{2+} \xrightarrow{0.17} UO_2^{+} \xrightarrow{0.38} U^{4+} \xrightarrow{-0.52} U^{3+} \xrightarrow{-4.70} U^{2+} \xrightarrow{-0.10} U$

$UO_2^{2+} \xrightarrow{0.27} U^{4+}$; $U^{4+} \xrightarrow{-1.38} U$

$PuO_2^{2+} \xrightarrow{1.02} PuO_2^{+} \xrightarrow{1.04} Pu^{4+} \xrightarrow{1.01} Pu^{3+} \xrightarrow{-3.50} Pu^{2+} \xrightarrow{-1.20} Pu$

$PuO_2^{2+} \xrightarrow{1.03} Pu^{4+}$; $Pu^{4+} \xrightarrow{-1.25} Pu$; $Pu^{3+} \xrightarrow{-2.00} Pu$

Group 4 (d-Block)

$TiO^{2+} \xrightarrow{0.10} Ti^{3+} \xrightarrow{-0.37} Ti^{2+} \xrightarrow{-1.63} Ti$

$TiO^{2+} \xrightarrow{-0.86} Ti$; $Ti^{3+} \xrightarrow{-1.21} Ti$

$ZrO_2 \xrightarrow{-1.46} Zr$, $Zr^{4+} \xrightarrow{-1.55} Zr$, $HfO_2 \xrightarrow{-1.57} Hf$, $Hf^{4+} \xrightarrow{-1.70} Hf$

Group 5 (d-Block)

$VO_2^{+} / V(OH)_4^{+} \xrightarrow{1.00} VO^{2+} \xrightarrow{0.34} V^{3+} \xrightarrow{-0.26} V^{2+} \xrightarrow{-1.13} V$

$Nb_2O_5 \xrightarrow{-0.10} Nb^{3+} \xrightarrow{-1.10} Nb$

Group 6 (d-Block)

$Cr_2O_7^{2-} \xrightarrow{1.38} Cr^{3+} \xrightarrow{-0.42} Cr^{2+} \xrightarrow{-0.90} Cr$

$MoO_2^{2+} \xrightarrow{0.48} MoO_2^{+} \xrightarrow{0.09} Mo^{3+} \xrightarrow{-0.20} Mo$

$WO_3 \xrightarrow{-0.03} W_2O_5 \xrightarrow{-0.03} WO_2 \xrightarrow{-0.15} W^{3+} \xrightarrow{-0.11} W$

$WO_2 \xrightarrow{-0.12} W$

Group 7 (d-Block)

$MnO_4^{-} \xrightarrow{0.56} MnO_4^{2-} \xrightarrow{0.27} MnO_4^{3-} \xrightarrow{4.27} MnO_2 \xrightarrow{0.95} Mn^{3+} \xrightarrow{1.51} Mn^{2+} \xrightarrow{-1.18} Mn$

$MnO_4^{-} \xrightarrow{1.51} Mn^{2+}$

$MnO_4^{-} \xrightarrow{1.70} MnO_2$

$TcO_4^{-} \xrightarrow{0.70} TcO_3 \xrightarrow{0.76} TcO_2 \xrightarrow{0.14} Tc^{2+} \xrightarrow{-0.40} Tc \xrightarrow{-1.00} TcH_9^{2-}$

$ReO_4^{-} \xrightarrow{0.77} ReO_3 \xrightarrow{0.63} ReO_2 \xrightarrow{0.22} Re \xrightarrow{-0.10} ReH_9^{2-}$

Group 8 (d-Block)

$FeO_4^{2-} \xrightarrow{2.20} Fe^{3+} \xrightarrow{0.77} Fe^{2+} \xrightarrow{-0.44} Fe$, $[Fe(CN)_6]^{3-} \xrightarrow{0.36} [Fe(CN)_6]^{4-} \xrightarrow{-1.16} Fe$

$RuO_4 \xrightarrow{1.40} RuO_2 \cdot xH_2O \xrightarrow{1.99} Ru^{3+} \xrightarrow{0.25} Ru^{2+} \xrightarrow{0.45} Ru$

$RuO_2 \cdot xH_2O \xrightarrow{0.68} Ru$

$[Ru(NH_3)_6]^{3+} \xrightarrow{0.10} [Ru(NH_3)_6]^{2+}$, $[Ru(CN)_6]^{3-} \xrightarrow{0.85} [Ru(CN)_6]^{4-}$, $[Ru(bpy)_3]^{3+} \xrightarrow{1.53} [Ru(bpy)_3]^{2+}$

$OsO_4 \xrightarrow{1.00} OsO_2 \cdot 2H_2O \xrightarrow{0.68} Os$, $[Os(bpy)_3]^{3+} \xrightarrow{0.88} [Os(bpy)_3]^{2+}$

$OsO_4 \xrightarrow{0.85} Os$

Group 9 (d-Block)

$CoO_2 \xrightarrow{1.41} Co^{3+} \xrightarrow{1.92} Co^{2+} \xrightarrow{-0.27} Co$

$RhO_4^{2-} \xrightarrow{1.50} RhO^{2+} \xrightarrow{1.40} Rh^{3+} \xrightarrow{1.10} Rh^{2+} \xrightarrow{0.60} Rh^{+} \xrightarrow{0.60} Rh$

$IrO_4^{2-} \xrightarrow{2.05} IrO_2 \xrightarrow{0.22} Ir^{3+} \xrightarrow{1.73} IrO \xrightarrow{0.87} Ir$

$IrO_4^{2-} \xrightarrow{1.44} Ir^{3+}$

$Ir^{3+} \xrightarrow{1.15} Ir$

$IrO_2 \xrightarrow{0.92} Ir$

Group 10 (d-Block)

$NiO_4^{2-} \xrightarrow{>1.80} NiO_2 \xrightarrow{1.59} Ni^{2+} \xrightarrow{-0.25} Ni$

$NiO_4^{2-} \xrightarrow{>1.60} Ni$

$Pd^{4+} \xrightarrow{1.26} Pd^{2+} \xrightarrow{0.91} Pd$, $[PdCl_6]^{2-} \xrightarrow{1.47} [PdCl_4]^{2-} \xrightarrow{0.60} Pd$

$PtO_3 \xrightarrow{2.00} PtO_2 \xrightarrow{0.84} Pt^{2+} \xrightarrow{1.18} Pt$

$PtO_2 \xrightarrow{1.05} PtO \xrightarrow{0.98} Pt$

$[PtCl_6]^{2-} \xrightarrow{0.72} [PtCl_4]^{2-} \xrightarrow{0.76} Pt$

Group 11 (d-Block)

$CuO^+ \xrightarrow{1.80} Cu^{2+} \xrightarrow{0.16} Cu^+ \xrightarrow{0.52} Cu$, $Ag_2O_3(s) \xrightarrow{1.67} Ag^+$

$Cu^{2+} \rightarrow Cu$: 0.34

$AgO^+ \xrightarrow{2.10} Ag^{2+} \xrightarrow{1.98} Ag^+ \xrightarrow{0.80} Ag$, $[Ag_2O_3(s)] \xrightarrow{1.36} Ag^{2+} \xrightarrow{1.98} Ag^+ \xrightarrow{0.80} Ag$

$Au^{3+} \xrightarrow{1.36} Au^+ \xrightarrow{1.83} Au$, $[AuCl_4]^- \xrightarrow{0.93} [AuCl_2]^- \xrightarrow{1.15} Au$

$Au^{3+} \rightarrow Au$: 1.52; $[AuCl_4]^- \rightarrow Au$: 1.00

Group 12 (d-Block)

$Zn^{2+} \xrightarrow{-0.76} Zn$, $Cd^{2+} \xrightarrow{-0.40} Cd$

$Hg^{2+} \xrightarrow{0.91} Hg_2^{2-} \xrightarrow{0.79} Hg$, $Hg_2Cl_2 \xrightarrow{0.27} Hg$

$Hg^{2+} \rightarrow Hg$: 0.85

Group 13 (p-Block)

$H_3BO_3 \xrightarrow{-0.89} B \xrightarrow{0.08} B_{12}H_{12}^{2-} \xrightarrow{0.32} B_{10}H_{10}^{2-} \xrightarrow{-0.50} B_{10}H_{14} \xrightarrow{0.16} B_2H_6 \xrightarrow{0.36} BH_4^-$

$H_3BO_3 \rightarrow BH_4^-$: 0.47; $H_3BO_3 \rightarrow B_{10}H_{10}^{2-}$: 0.69; $H_3BO_3 \rightarrow B_{12}H_{12}^{2-}$: 0.67; $B_{12}H_{12}^{2-} \rightarrow BH_4^-$: 0.24

$Al^{3+} \xrightarrow{-1.68} Al$, $Ga^{3+} \xrightarrow{-0.65} Ga^{2+} \xrightarrow{-0.45} Ga$

$Ga^{3+} \rightarrow Ga$: –0.53

$In^{3+} \xrightarrow{-0.49} In^{2+} \xrightarrow{-0.40} In^+ \xrightarrow{-0.12} In$, $Tl^{3+} \xrightarrow{0.30} Tl^{2+} \xrightarrow{2.22} Tl^+ \xrightarrow{-0.34} Tl$

$In^{3+} \rightarrow In$: –0.34; $Tl^{3+} \rightarrow Tl^+$: 1.25

Group 14 (p-Block)

$CO_2 \xrightarrow{-0.20} HCOOH \xrightarrow{-0.03} HCHO \xrightarrow{0.24} CH_3OH \xrightarrow{0.59} CH_4$

$CO_2 \xrightarrow{0.10} CO \xrightarrow{0.51} C \xrightarrow{0.13} CH_4$, $SiO_2 \xrightarrow{-0.84} Si \xrightarrow{-0.14} SiH_4$

$GeO_2 \xrightarrow{-0.50} Ge^{2+} \xrightarrow{0.25} Ge \xrightarrow{-0.42} GeH_4$

$GeO_2 \rightarrow Ge$: –0.25

$Sn^{4+} \xrightarrow{0.15} Sn^{2+} \xrightarrow{-0.13} Sn$, $PbO_2 \xrightarrow{1.46} Pb^{2+} \xrightarrow{-0.12} Pb$, $PbO_2 \xrightarrow{1.70} PbSO_4 \xrightarrow{-0.36} Pb$

Group 15 (p-Block)

$NO_3^- \xrightarrow{0.80} N_2O_4 \xrightarrow{1.07} HNO_2 \xrightarrow{0.99} NO \xrightarrow{1.59} N_2O \xrightarrow{1.77} N_2 \xrightarrow{-1.87} NH_3OH^+ \xrightarrow{1.42} N_2H_5^+ \xrightarrow{1.27} NH_4^+$

$NO_3^- \xrightarrow{0.94} HNO_2$; $NO \xrightarrow{0.71} H_2N_2O_2 \xrightarrow{2.65} N_2$; $N_2 \xrightarrow{0.27} NH_4^+$

$H_3PO_4 \xrightarrow{-0.27} H_3PO_3 \xrightarrow{-0.50} H_3PO_2 \xrightarrow{-0.36} P_4 \xrightarrow{-0.10} P_2H_4 \xrightarrow{-0.006} PH_3$

$P_4 \xrightarrow{-0.06} PH_3$

$H_3AsO_4 \xrightarrow{0.56} H_3AsO_3 \xrightarrow{0.24} As \xrightarrow{-0.22} AsH_3$, $Sb_2O_5 \xrightarrow{0.60} SbO^+ \xrightarrow{0.20} Sb \xrightarrow{-0.51} SbH_3$

$Bi_2O_5 \xrightarrow{1.60} BiO^+ \xrightarrow{0.31} Bi \xrightarrow{-0.97} BiH_3$

Group 16 (p-Block)

$O_3 \xrightarrow{2.07} O_2(+H_2O)$ $O_2 \xrightarrow{-0.12} HO_2 \xrightarrow{1.51} H_2O_2 \xrightarrow{1.76} H_2O$

$O_2 \xrightarrow{0.69} H_2O_2$; $O_2 \xrightarrow{1.23} H_2O$

$SO_4^{2-} \xrightarrow{-0.25} S_2O_6^{2-} \xrightarrow{0.57} SO_2 \xrightarrow{0.51} S_4O_6^{2-} \xrightarrow{0.07} S_2O_3^{2-} \xrightarrow{0.60} S \xrightarrow{0.14} H_2S$, $S_2O_8^{2-} \xrightarrow{1.97} SO_4^{2-}$

$SO_4^{2-} \xrightarrow{0.16} SO_2$; $SO_2 \xrightarrow{0.45} S$

$SeO_4^{2-} \xrightarrow{1.10} H_2SeO_3 \xrightarrow{0.74} Se \xrightarrow{-0.11} H_2Se$, $H_2TeO_4 \xrightarrow{1.00} TeO_2(s) \xrightarrow{0.53} Te \xrightarrow{-0.64} H_2Te$

Group 17 (p-Block)

$F_2 \xrightarrow{2.87} F^-$

$ClO_4^- \xrightarrow{1.20} ClO_3^- \xrightarrow{1.18} HClO_2 \xrightarrow{1.70} HClO \xrightarrow{1.63} Cl_2 \xrightarrow{1.35} Cl^-$

$BrO_4^- \xrightarrow{1.85} BrO_3^- \xrightarrow{1.45} HBrO \xrightarrow{1.60} Br_2(l) \xrightarrow{1.06} Br^-$

$H_5IO_6 \xrightarrow{1.60} IO_3^- \xrightarrow{1.13} HIO \xrightarrow{1.44} I_2(I_3^-) \xrightarrow{0.54} I^-$

Group 18 (p-Block)

$H_4XeO_6(aq) \xrightarrow{2.40} XeO_3(aq) \xrightarrow{2.12} Xe(g)$

$H_4XeO_6(aq) \xrightarrow{2.18} Xe(g)$

Group 2 (s-Block)

$Be_2O_3^{2-} \xrightarrow{-2.62} Be$, $Sr(OH)_2 \xrightarrow{-2.88} Sr$, $Mg(OH)_2 \xrightarrow{-2.69} Mg$,

$Ba(OH)_2 \xrightarrow{-2.81} Ba$, $Ca(OH)_2 \xrightarrow{-3.03} Ca$

Group 3 (d-Block)

$Sc(OH)_3 \xrightarrow{-2.60} Sc$, $Y(OH)_3 \xrightarrow{-2.85} Y$, $La(OH)_3 \xrightarrow{-2.80} La$

Lanthanides (f-Block)

$M(OH)_3 \xrightarrow{\sim -2.80} M$, M = La, Ce, Pr, Nd, Pm, Sm, Eu, Gd, Tb, Dy, Ho, Er, Tm, Yb and Lu

Actinides (*f*-Block)

$$Ac(OH)_3 \xrightarrow{-2.50} Ac,\ ThO_2 \xrightarrow{-2.56} Th$$

$$UO_2(OH)_2 \xrightarrow{-0.30} UO_2 \xrightarrow{-2.60} U(OH)_3 \xrightarrow{-2.10} U$$

$$NpO_2(OH)_2 \xrightarrow{0.60} NpO_2(OH) \xrightarrow{0.30} Np(OH)_4 \xrightarrow{-2.10} Np(OH)_3 \xrightarrow{-2.23} Np$$

$$PuO_2(OH)_2 \xrightarrow{0.30} PuO_2(OH) \xrightarrow{0.90} Pu(OH)_4 \xrightarrow{-1.40} Pu(OH)_3 \xrightarrow{-2.46} Pu$$

$$AmO_2(OH)_2 \xrightarrow{0.90} AmO_2(OH) \xrightarrow{0.70} Am(OH)_4 \xrightarrow{0.50} Am(OH)_3 \xrightarrow{-2.53} Am$$

$$CmO_2 \xrightarrow{0.70} Cm(OH)_3 \xrightarrow{-2.53} Cm$$

Group 4 (d-Block)

$$TiO_2 \xrightarrow{-1.90} Ti,\ H_2ZrO_3 \xrightarrow{-2.36} Zr,\ HfO(OH)_2 \xrightarrow{-2.50} Hf$$

Group 5 (d-Block)

$$VO_4^{3-} \xrightarrow{0.12} V$$

Group 6 (d-Block)

$$CrO_4^{2-} \xrightarrow{-0.11} Cr(OH)_3 \xrightarrow{-1.10} Cr(OH)_2 \xrightarrow{-1.40} Cr$$

$$CrO_2^{-} \xrightarrow{-1.30} Cr$$

$$MoO_4^{2-} \xrightarrow{-0.78} MoO_2 \xrightarrow{-0.98} Mo,\ WO_4^{2-} \xrightarrow{-1.07} W$$

Group 7 (d-Block)

$$MnO_4^{-} \xrightarrow{0.56} MnO_4^{2-} \xrightarrow{0.27} MnO_4^{3-} \xrightarrow{0.96} MnO_2 \xrightarrow{0.15} Mn_2O_3 \xrightarrow{-0.25} Mn(OH)_2 \xrightarrow{-1.56} Mn$$

$$TcO_4^{-} \xrightarrow{-0.32} TcO_2 \xrightarrow{-0.55} Tc$$

$$ReO_4^{-} \xrightarrow{-0.70} ReO_4^{3-} \xrightarrow{-0.50} ReO_2 \xrightarrow{-0.53} Re(OH)_3 \xrightarrow{-0.60} Re \xrightarrow{-0.40} ReH_9^{2-}$$

Group 8 (d-Block)

$$FeO_4^{2-} \xrightarrow{0.72} Fe(OH)_3 \xrightarrow{-0.56} Fe(OH)_2 \xrightarrow{-0.88} Fe$$

$$RuO_4 \xrightarrow{1.00} RuO_4^{-} \xrightarrow{0.59} RuO_4^{2-} \xrightarrow{0.35} RuO_4 \cdot xH_2O \xrightarrow{-0.15} Ru$$

$$HOsO_5^{-} \xrightarrow{0.10} OsO_2 \cdot 2H_2O \xrightarrow{-0.12} Os$$

Group 9 (d-Block)

$$CoO_2 \xrightarrow{0.70} Co(OH)_3 \xrightarrow{0.17} Co(OH)_2 \xrightarrow{-0.73} Co$$

$$RhO_4^{2-} \xrightarrow{-0.10} Rh(OH)_3 \xrightarrow{0.00} Rh$$

$$IrO_4^{2-} \xrightarrow{0.40} IrO_2 \xrightarrow{0.10} Ir_2O_3 \xrightarrow{0.10} Ir$$

Group 10 (d-Block)

$$NiO_2^{2-} \xrightarrow{>0.40} NiO_2 \xrightarrow{0.49} Ni(OH)_2 \xrightarrow{-0.72} Ni,\ Pd(OH)_4 \xrightarrow{0.70} Pd(OH)_2 \xrightarrow{0.07} Pd$$

$$PtO_4^{2-} \xrightarrow{>0.40} Pt(OH)_6^{2-} \xrightarrow{-0.20} Pt(OH)_2 \xrightarrow{0.15} Pt$$

Group 11 (d-Block)]

$$CuO \xrightarrow{-0.29} Cu_2O \xrightarrow{-0.36} Cu,\ [Cu(NH_3)_4]^{2+} \xrightarrow{0.10} [Cu(NH_3)_2]^{+} \xrightarrow{-0.10} Cu$$

$$Ag_2O_3 \xrightarrow{0.74} AgO \xrightarrow{0.60} Ag_2O \xrightarrow{0.34} Ag,\ [Ag(NH_3)_2]^{+} \xrightarrow{0.37} Ag$$

$$HAuO_3^{2-} \xrightarrow{0.80} Au(OH)_2^{-} \xrightarrow{0.40} Au$$

Group 12 (d-Block)

$$ZnO_2^{2-} \xrightarrow{-1.28} Zn,\ Cd(OH)_2 \xrightarrow{-0.82} Cd,\ HgO \xrightarrow{0.10} Hg$$

Group 13 (p-Block)

$$B_4O_7^{2-} \xrightarrow{-1.76} B \xrightarrow{-1.04} B_{12}H_{12}^{2-} \xrightarrow{0.32} B_{10}H_{10}^{2-} \xrightarrow{1.15} B_{10}H_{14}(s) \xrightarrow{0.98} B_2H_6 \xrightarrow{0.78} BH_4^{-}$$

$$H_2AlO_3^{-} \xrightarrow{-2.31} Al,\ H_2GaO_3^{-} \xrightarrow{-1.22} Ga$$

$$Tl(OH)_3 \xrightarrow{-0.05} Tl(OH) \xrightarrow{-0.34} Tl$$

Group 14 (p-Block)

$$CO_3^{2-} \xrightarrow{-1.01} HCO_2^{-} \xrightarrow{-0.52} C \xrightarrow{-0.70} CH_4,\ SiO_3^{2-} \xrightarrow{-1.69} Si \xrightarrow{-0.93} SiH_4,$$

$$HGeO_3^{-} \xrightarrow{-0.89} Ge \xrightarrow{<-1.10} GeH_4,\ Sn(OH)_6^{2-} \xrightarrow{-0.90} HSnO_2^{-} \xrightarrow{-0.91} Sn,\ PbO_2 \xrightarrow{0.25} PbO \xrightarrow{-0.58} Pb$$

Group 15 (p-block)

$$NO_3^{-} \xrightarrow{-0.86} N_2O_4 \xrightarrow{0.86} NO_2^{-} \xrightarrow{-0.46} NO \xrightarrow{0.76} N_2O \xrightarrow{0.94} N_2 \xrightarrow{-3.04} NH_2OH \xrightarrow{0.73} N_2H_4 \xrightarrow{0.10} NH_3$$

$$PO_4^{3-} \xrightarrow{-1.12} HPO_3^{2-} \xrightarrow{-1.57} H_2PO_2^{-} \xrightarrow{-2.05} P_4 \xrightarrow{-0.89} PH_3$$

$$AsO_4^{3-} \xrightarrow{-0.67} H_2AsO_3^{-} \xrightarrow{-0.68} As \xrightarrow{-1.37} AsH_3$$

$$Sb(OH)_6^{-} \xrightarrow{-0.46} SbO_2^{-} \xrightarrow{-0.64} Sb \xrightarrow{-1.34} SbH_3$$

$$Bi_2O_5 \xrightarrow{0.78} Bi_2O_4 \xrightarrow{0.56} Bi_2O_3 \xrightarrow{-0.45} Bi \xrightarrow{<-1.60} BiH_3$$

Group 16 (p-Block)

$$O_3 \xrightarrow{1.24} O_2 + OH^{-}$$

$$O_2 \xrightarrow{-0.06} HO_2^{-} \xrightarrow{0.86} OH^{-}$$

$$O_2 \xrightarrow{0.40} OH^{-}$$

$$H_2O \xrightarrow{-0.83} H_2 + OH^{-}$$

$$SO_3^{2-} \xrightarrow{-0.66} S$$

$$SO_4^{2-} \xrightarrow{-0.94} SO_3^{2-} \xrightarrow{0.79} S_4O_6^{2-} \xrightarrow{0.08} S_2O_3^{2-} \xrightarrow{-0.74} S \xrightarrow{-0.45} S^{2-}$$

$$SeO_4^{2-} \xrightarrow{0.03} SeO_3^{2-} \xrightarrow{-0.36} Se \xrightarrow{-0.67} Se^{2-},\ TeO_2(OH)_4^{2-} \xrightarrow{0.07} TeO_3^{2-} \xrightarrow{-0.42} Te \xrightarrow{-1.14} Te^{2-}$$

Group 17 (p-Block)

$$F_2 \xrightarrow{2.87} F^{-}$$

$$ClO_4^{-} \xrightarrow{0.37} ClO_3^{-} \xrightarrow{0.29} ClO_2^{-} \xrightarrow{0.68} ClO^{-} \xrightarrow{0.42} Cl_2 \xrightarrow{1.35} Cl^{-}$$

$$BrO_4^{-} \xrightarrow{1.02} BrO_3^{-} \xrightarrow{0.49} BrO^{-} \xrightarrow{0.45} Br_2 \xrightarrow{1.06} Br^{-}$$

$$H_3IO_6^{2-} \xrightarrow{0.65} IO_3^{-} \xrightarrow{0.15} IO^{-} \xrightarrow{0.42} I_2 \xrightarrow{0.53} I^{-}$$

Group 18 (p-Block)

$$HXeO_6^{3-} \xrightarrow{0.94} HXeO_4^{-} \xrightarrow{1.24} Xe$$

References

1. B. Douglas, D. McDaniel and J. Alexander, Concepts and Models of Inorganic Chemistry, 3rd Edn, Wiley, India.
2. P. Alkins, T. Overton, J. Rourke, M. Weller and F. Armstrong. Shriver and Atkins' Inorganic Chemistry, 5th Edn., Oxford University Press.

APPENDIX VII
Books Consulted

Akhmetov, N.: General and Inorganic Chemistry
Akitt, J.W.: NMR and Chemistry
Anantharaman, R.: Fundamentals of Quantum Chemistry
Aruldhas, G.: Molecular Structure and Spectroscopy
Ballhausen, C.J.: Introduction to Ligand Field Theory
Banerjea, D.: Inorganic Chemistry—Elements and Compounds
Banerjea, D.: Coordination Chemistry
Banerjee, D.: Inorganic Chemistry (Principles)
Banerjea, D.: Fundamental Principles of Inorganic Chemistry
Banwell, C.N.: Fundamentals of Molecular Spectroscopy
Berrow, G.M.: Introduction to Molecular Spectroscopy
Basolo, F. and Pearson, R.G.: Mechanisms of Inorganic Reactions
Bertini, I., Gray, H.B., Lippard, S.J. and Valentine, J.S.: Bioinorganic Chemistry
Carlin, R.L.: Magnetochemistry
Chanda, M.: Atomic Structure and Chemical Bonding
Chatwal, G.R. and Anand, S.K.: Spectroscopy
Connelly, G., Damhus, T. (Editors): Nomenclature of Inorganic Chemistry: IUPAC Recommendations 2005.
Cotton, F.A. and Wilkinson, G.: Advanced Inorganic Chemistry
Cotton, F.A. and Wilkinson, G.: Basic Inorganic Chemistry
Cotton, F.A. and Wilkinson, G., Murillo, C.A. and Bochmann, M.: Advanced Inorganic Chemistry
Cotton, F.A.: Chemical Applications of Group Theory
Das, Asim K.: Bioinorganic Chemistry
Das, Asim K.: Inorganic Chemistry: Biological and Environmental Aspects
Das, Asim K.: Environmental Chemistry with Green Chemistry
Das, A.K.: A Textbook on Medicinal Aspects of Bioinorganic Chemistry
Day (Jr), D.A. and Underwood, A.L.: Quantitative Analysis
Day, M.C. and Selbin, J.: Theoretical Inorganic Chemistry
Dorain, P.B.: Symmetry in Inorganic Chemistry
Douglas, B.E. McDaniel, D.H. and Alexander, J.J.: Concepts and Models of Inorganic Chemistry
Drago, R.S,: Physical Methods in Chemistry
Drago, R.S.: Physical Methods for Chemists
Dyer, J.R.: Applications of Absorption Spectroscopy of Organic Compounds
Dutta, P.K.: General and Inorganic Chemistry
Dutta, R.L. and Syamal, A.: Elements of Magnetochemistry
Dutta, R.L.: Inorganic Chemistry (Part. I and II)
Earnshaw, A.: Introduction to Magnetochemistry
Ebsworth E.A.V., Rankin, D.W.H. and Cradock, S.: Structural Methods in Inorganic Chemistry:
Emeleus, H.J. and Anderson, J.S.: Modern Aspects of Inorganic Chemistry
Figgis, B.N.: Introduction to Ligand Fields
Glasstone, S.: Textbook of Physical Chemistry
Griffith J.S.: The Theory of Transition Metal Ions
Greenwood, N.N. and Earnshaw, A.: Chemistry of the Elements
Huheey, J.E.: Inorganic Chemistry—Principles of Structure and Reactivity
Huheey, J.E., Keiter, E.A., Keiter, R.L. and O.K. Mehdi: Inorganic Chemistry—Principles of Structure and Reactivity
Jolly, W.L.: Principles of Inorganic Chemistry
Jorgensen, C.K.: Absorption spectra and Chemical Bonding in Complexes
Kalshi, P.S.: Spectroscopy of Organic Compounds
Kapoor, K.L.: A Textbook of Physical Chemistry (Vol. 1-4)
Kettle, S.F.A.: Coordination Compounds
Langford, C.H. and Gray, H.B.: Ligand Substitution Processes
Laidler, K.J. and Meiser, J.H.: Physical Chemistry
Lee, J.D.: Concise Inorganic Chemistry
Lever, A.B.P.: Inorganic Electronic Spectroscopy
Lippard, S.J. and Berg, J.M.: Principles of Bioinorganic Chemistry
Mackay, K.M. and Mackay, R.A.: Introduction of Modern Inorganic Chemistry
Mahan, B.H.: University Chemistry
Miessler, G.L. and Tarr, D.A.: Inorganic Chemistry
Mohan, J.: Organic Spectroscopy: Principles and Applications.
Mukherjee, G.N. & Das, A.: Elements of Bioinorganic Chemistry
Nakamoto, K.: Infrared Spectra of Inorganic and Coordination Compounds
Orchin, M. and Jaffe, H.H.: The Importance of Antibonding Orbitals
Pavia, D.L., Lampman, G.M. and Kriz, G.S. : Introduction to Spectroscopy
Potterfied, W.W.: Inorganic Chemistry: A Unified Approach
Purcell, K.F. and Kotz, J.C.: Inorganic Chemistry
Rakshit, P.C.: Physical Chemistry
Rao, C.N.R.: Chemical Applications of Infrared Spectroscopy
Rao, C.N.R.: Ultraviolet and Visible Spectroscopy
Ray, R.K.: Electronic Spectra of Transition Metal Complexes
Reddy, K.V.: Symmetry and Spectroscopy of Molecules
Shriver, D.F., Atkins, P.W. and Langford, C.H.: Inorganic Chemistry
Sarkar, R.: General and Inorganic Chemistry (Part I and II)
Satyanaryana, D.N.: Electronic Absorption Spectroscopy and Related Techniques
Satyanarayana, D.N.: Vibrational Spectroscopy – Theory and Applications
Sharpe, A.G.: Inorganic Chemistry
Skoog, D.A., West, D.M., Holler, F.J. and Crouch, S.R.: Fundamental of Analytical Chemistry
Solomon, E.I. and Lever, A.B.P.: Inorganic Electronic Structure and Spectroscopy
Steed, J.W. and Atwood, J.L.: Supramolecular Chemistry
Topping, J.: Errors of Observation and Their Treatment
Vogel, A.I.: A Textbook of Quantitative Inorganic Analysis
Wilkins, R.G.: The study of Kinetics and Mechanism of Reactions of Transition Metal Complexes
Wulfsberg, G.: Inorganic Chemistry

Acknowledgement

The above listed references and sources have been freely consulted to borrow their views and ideas. The present authors are indebted to all these authors to an endless extent. The present author expresses his heartiest thanks and grateful acknowledgement to all of them.

Index

D

E

F

G

H

I

J

K

L

M

N

O

P

S

T

U

V

W

X

Y

Z